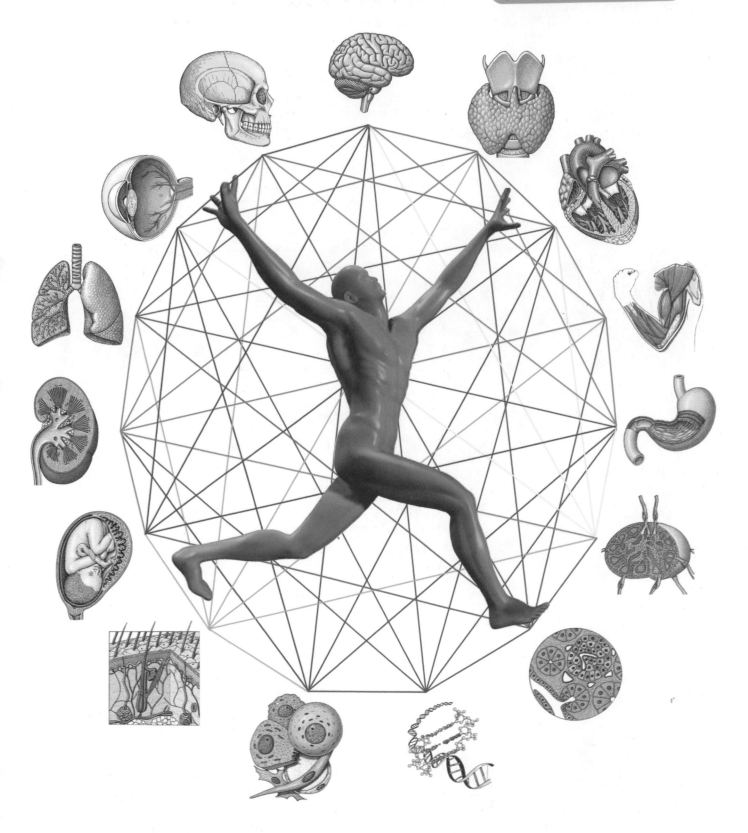

Human Physiology

An Integrated Approach

About the Author

Dee Unglaub Silverthorn

Dee Unglaub Silverthorn studied biology as an undergraduate at Tulane University and went on to earn a Ph.D. in marine science at the University of South Carolina. Her research interest is epithelial transport, and work in her laboratory currently focuses on transport properties of the chick allantoic membrane. She began her teaching career in the Physiology Department at the Medical University of South Carolina but over the years has taught a wide range of students, from medical and college students to those still preparing for higher education. At the University of Texas she lectures in physiology, coordinates student laboratories in physiology, and instructs graduate students in a course on developing teaching skills in the life sciences. She has substantial experience with active learning in the classroom, and has given workshops on this subject at regional, national, and international conferences. Dee is currently chair of the Teaching Section of the American Physiological Society and an associate editor of Advances in Physiology Education. She works with members of the International Union of Physiological Sciences to improve physiology education in developing countries. She is also a member of the Human Anatomy and Physiology Society, the Association for Biology Laboratory Education, the Society for College Science Teaching, the American Association for the Advancement of Science, and Sigma Xi. Her free time is spent creating multimedia fiber art and enjoying the Texas hill country with her husband, Andy, and their three dogs.

About the Illustrators

Dr. William C. Ober (art coordinator and illustrator)

Dr. William C. Ober studied as an undergraduate at the Washington and Lee University and obtained his M.S. from the University of Virginia in Charlottesville. In addition to his medical school education, he studied in the Department of Art as Applied to Medicine at Johns Hopkins University. After graduation, Dr. Ober completed training as a resident in family practice, He is now a Clinical Assistant Professor in the Department of Family Medicine, University of Virginia. During the summer, he teaches biological illustration at Shoals Marine Laboratory, where he is part of the core faculty. Dr. Ober's professional life now focuses on medical and scientific illustration.

Claire W. Garrison, R.N. (illustrator)

Claire Garrison chose to make medical illustration her career after almost two decades as a pediatric and obstetric nurse. After a five-year apprenticeship, she became Dr. Ober's associate in 1986. Ms. Garrison is also a core faculty member at Shoals. The texts illustrated by Dr. Ober and Ms. Garrison have won numerous awards from groups including the following: Association of Medical Illustrators (Award of Excellence); Chicago Book Clinic (Award for Art and Design); Printing Industries of America (Award of Excellence); and Bookbinders West. They have also received the Art Directors Award.

About the Clinical Consultant

Andrew C. Silverthorn

Andrew C. Silverthorn, M.D. is a physician in private practice in Austin, Texas. He is a graduate of the United States Military Academy (West Point), served in the infantry in Vietnam, and upon his return entered medical school at the Medical University of South Carolina in Charleston. He completed his family practice residency at the University of Texas Medical Branch, Galveston, where he was chief resident, and he is board-certified by the American Board of Family Practice. When Andy is not busy seeing patients he may be found on the golf course, playing golf or running with his chocolate lab, Lady Godiva.

About the Cover

Fluorescently tagged markers label a motor neuron (green), ACh receptors (red), and Schwann cells (blue) at a neuromuscular synapse. Two clusters of receptors at the top were experimentally denervated, but the remaining neuron has grown thin sprouts along a pathway provided by Schwann cells and will reinnervate them over time

Human Physiology
An Integrated Approach

Second Edition

Dee Unglaub Silverthorn, Ph.D.
University of Texas—Austin

with

William C. Ober, M.D.
Illustration Coordinator

Claire W. Garrison, R.N.
Illustrator

Andrew C. Silverthorn, M.D.
Clinical Consultant

Prentice Hall
Upper Saddle River, New Jersey 07458

Library of Congress Cataloging-in-Publication Data
SILVERTHORN, DEE UNGLAUB

Human physiology : an integrated approach / Dee Unglaub Silverthorn. [et al.].—2nd ed.

p. cm.

Includes index.
ISBN 0-13-017697-4
1. Human physiology. I. Silverthorn, Dee Unglaub
QP34.5.S55 2001   612—dc21   00-034685

Senior Acquisition Editor: Halee Dinsey
Development Editor: Anne A. Reid
Editor in Chief, Development: Carol Trueheart
Media Editor: Andrew Stull
Project Manager: Don O'Neal
Production Editor: Debra A. Wechsler
Editor in Chief: Sheri Snavely
Editorial Director: Paul F. Corey
Executive Managing Editor: Kathleen Schiaparelli
Assistant Vice President of Production and Manufacturing: David W. Riccardi
Marketing Manager: Martha McDonald
Director of Marketing, ESM: John Tweeddale
Manufacturing Buyer: Michael Bell
Manufacturing Manager: Trudy Pisciotti
Art Director: Heather Scott
Assistant to Art Director: John Christiana
Director of Creative Services: Paul Belfanti
Art Manager: Gus Vibal
Interior Designer: Joanne Del Ben
Cover Designer: Joseph Sengotta
Cover Image: Dr. Jane Lubischer
Photo Research: Kathleen Cameron
Photo Editor: Beth Boyd
Photo Coordinator: Nancy Seise
Illustrators: William C. Ober, M.D. and Claire W. Garrison, R.N.
Editorial Assistant: Susan Zeigler

© 2001, 1998 by Prentice-Hall, Inc.
Upper Saddle River, New Jersey 07458

Printed in the United States of America
10 9 8 7 6 5 4

ISBN 0-13-017697-4

Prentice-Hall International (UK) Limited, *London*
Prentice-Hall of Australia Pty. Limited, *Sydney*
Prentice-Hall Canada, Inc., *Toronto*
Prentice-Hall Hispanoamericana, S.A., *Mexico*
Prentice-Hall of India Private Limited, *New Delhi*
Prentice-Hall of Japan, Inc., *Tokyo*
Pearson Education Asia Pte. Ltd.
Editora Prentice-Hall do Brasil, Ltda., *Rio de Janeiro*

Brief Contents

v

OWNER'S MANUAL: How to use this Book

Welcome to Human Physiology!

As you begin your study of the human body, you should be prepared to make maximum use of the resources available to you, including your instructor, the library, the World Wide Web, and your textbook. One of my goals in this book is to provide you not only with information about how the human body functions, but also with tips for studying and learning to solve problems. Many of these study aids have been developed with the input of my students, so I think you may find them particularly helpful.

On the following pages, I have put together a brief tour of the special features of the book, especially those that you may not have encountered in textbooks previously. Please take a few minutes to read this section so that you can make optimum use of the text as you study. If you take advantage of features such as the Graph Questions, Concept Checks, and the Running Problems, you will find that you begin to think about physiology in a different way.

One of your tasks as you study will be to construct for yourself a global view of the body, its systems, and the many processes that keep the systems working. This "big picture" is what physiologists call the integration of systems, and it is a key theme in the book. In order to integrate information, however, you must do more than simply memorize it. You must truly understand it and be able to use it to solve problems that you have never encountered before. If you are headed for a career in the health professions, you will do this in the clinics. If you are headed for a career in biology, you must solve problems in the laboratory, field, or classroom. Learning to analyze, synthesize, and evaluate information are skills that you need to develop while you are in school, and I hope that the features of this book will help you with this task.

So find a comfortable chair and take a peek at what lies ahead!

Warmest regards,

Dr. Dee (as my students call me)

P.S. Join me on the book's World Wide Web site for additional information and activities that could help you be more successful in your physiology course. Who knows, it might even help you on the next test!

http://www.prenhall.com/silverthorn

CHAPTER OUTLINE

Each chapter begins with an outline listing the major headings, or topics, for that chapter. You will find a page reference next to each heading. Use this outline to quickly preview the chapter before you start reading. Do you see any connections between these topics and things you have learned in previous chapters or other courses?

The chapter you see here, "Integrative Physiology II: Fluid and Electrolyte Balance," also demonstrates another feature of this text—integrative chapters. Integrative chapters provide an opportunity to examine physiological processes as they actually occur in the body—as multiple systems interacting with each other to maintain homeostasis.

BACKGROUND BASICS

This list is a handy tool at the beginning of each chapter that shows you which important concepts you should have mastered prior to beginning the chapter. If any of these topics seem unfamiliar, use the page references to revisit the subjects you need to review.

---

19 Integrative Physiology II: Fluid and Electrolyte Balance

■ "At a 10% loss of body fluid, the patient will show signs of confusion, distress, and hallucinations and at 20%, death will occur." — *Poul Astrup, in Salt and Water in Culture and Medicine, 1993.* ■

The American businesswoman in Tokyo finished her workout and stopped at the snack bar of the fitness club to ask for a sports drink. The attendant handed her a bottle labeled "Pocari Sweat®." Although the thought of drinking sweat is not very appealing, the physiological basis for the name is sound.

During exercise, the body secretes sweat, a dilute solution of water and ions, particularly Na⁺, K⁺, and

---

WATER BALANCE AND THE REGULATION OF URINE CONCENTRATION 575

§ **DIABETES Diabetes and Osmotic Diuresis** The primary sign of diabetes mellitus is an elevated blood glucose concentration. In untreated diabetics, if blood glucose levels exceed the renal threshold for glucose reabsorption [⊂⊃ p. 00], glucose will be excreted in the urine. This may not seem like a big deal, but any additional non-reabsorbable solute that remains in the lumen will force additional water to be excreted, causing *osmotic diuresis*. For example, suppose the nephrons need to excrete 300 milliosmoles of NaCl. If the urine is maximally concentrated at 1200 mOsM, then the NaCl will be excreted in a volume of 0.25 L. If the NaCl is joined by 300 mosmoles of glucose that must be excreted, the volume of urine doubles, to 0.5 L. Osmotic diuresis in untreated diabetics (primarily type 1) will cause *polyuria* (excessive urination) and *polydipsia* [*dipsios*, thirst] due to dehydration and high plasma osmolarity.

six different types of aquaporins, including aquaporin-2 (AQP2), the water channel regulated by vasopressin.

AQP2 in collecting duct cells may be found in two locations: on the apical membrane facing the tubule lumen and in the membrane of cytoplasmic vesicles. (Two other types of aquaporin water channels are present in the basolateral membrane, but they are not regulated by vasopressin.)

When vasopressin levels and collecting duct water permeability are low, the tubule cell has few water pores in its apical membrane (Fig. 19-6 ■). It stores its AQP2 water pores in cytoplasmic vesicles.

When vasopressin arrives from the posterior pituitary, it binds to its receptor on the basolateral side of the collecting duct cell. Binding activates a G-protein/cAMP second messenger system [⊂⊃ p. 00]. Subsequent phosphorylation of intracellular proteins causes the AQP2 vesicles to move to the apical membrane and fuse with it. Exocytosis inserts the AQP2 water pores into the apical membrane. Now the cell is permeable to water. This process, in which parts of the cell membrane are alternately added by exocytosis and withdrawn by endocytosis, is known as **membrane recycling** [⊂⊃ Fig. 5-26, p. 00].

✓ Will the apical membrane of a collecting duct cell have more water channels when vasopressin is present or when it is absent?

✓ The collecting duct cells are surrounded by extremely high osmolarity on their basolateral sides, yet do not shrivel up. How can they maintain normal cell volume in the face of such high ECF osmolarity? (Hint: read the box on regulation of cell volume, p. 00.)

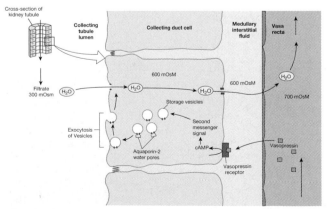

■ **Figure 19-6 The mechanism of action of vasopressin** In the collecting duct, vasopressin binds to a membrane receptor and activates a cAMP second messenger system. In response, the cell inserts AQP2 water pores into its apical membrane. In the absence of vasopressin, the pores are withdrawn and stored in cytosolic vesicles.

CONCEPT LINKS

These blue "chain-links" with page references are an extension of the Background Basics, linking you to concepts discussed earlier in the text. They help you find material that you might have forgotten or that might be helpful for understanding new concepts.

FIGURE REFERENCE LOCATORS

The red squares found next to every figure call out in the text function as place markers, making it easier for you to toggle between an illustration and the running narrative associated with that illustration.

CONCEPT CHECKS

The red check marks placed at intervals throughout the chapters act as stopping points where you can check your understanding of what you just read. Some of these questions are factual, while others ask you to think about or apply what you have learned. Answers to the Concept Check questions are in an appendix for instant feedback.

CHAPTER REVIEW

SUMMARY

1. The **nervous system** is a complex network of neurons that form the rapid control system of the body. (p. 215)

Organization of the Nervous System

2. The nervous system is divided into the **central nervous system (CNS),** composed of the **brain** and **spinal cord,** and the **peripheral nervous system.** (p. 215)

3. The peripheral nervous system has **afferent (sensory) neurons** that bring information into the CNS and **efferent neurons** that carry information away from the CNS back to various parts of the body. (p. 215)

4. The efferent neurons include **somatic motor neurons** that control skeletal muscles and **autonomic neurons** that control smooth and cardiac muscles, glands, and some adipose tissue. (p. 215)

5. Autonomic neurons are subdivided into **sympathetic** and **parasympathetic** neurons. (p. 215)

Cells of the Nervous System

6. Neurons have a **cell body** with a nucleus and organelles to direct cellular activity, **dendrites** to receive incoming signals, and an **axon** to transmit electrical signals from the cell

9. Material is transported between the cell body and axon terminal by **axonal transport.** (p. 218)

10. **Glial cells** provide physical support and direct the growth

CHAPTER SUMMARY

The end-of-chapter summary has enumerated points and page references, providing you with a quick summary of every major topic in the chapter.

END-OF-CHAPTER QUESTIONS

The end-of-chapter questions have been organized into four learning levels to allow you to develop your problem-solving skills through a logical progression of exercises—from factual to conceptual to real-world to quantitative.

LEVEL ONE questions review basic facts and terms. Successfully answering these questions gives you a foundation for more conceptual questions.

LEVEL TWO questions test your ability to understand the key concepts.

LEVEL THREE exercises are real-world scenarios designed to develop your problem-solving skills. You will be more successful dealing with Level Three questions if you have successfully worked through the first two levels. To check your answers or for additional help, you can go to the book's web site.

LEVEL FOUR questions are quantitative and require some mathematical manipulation. To successfully answer Level Four questions, you must first understand the facts and the concepts and then be able to manipulate relevant data to develop a quantitative picture of the physiology at hand.

QUESTIONS

LEVEL ONE   Reviewing Facts and Terms

1. List the four general functions of the cell membrane.

2. In 1972, Singer and Nicolson proposed a model of the cell

LEVEL TWO   Reviewing Concepts

20. Create a map using the following terms. You may add additional terms if you wish.

LEVEL THREE   Problem Solving

30. Sweat glands secrete a fluid into their lumen that is identical to interstitial fluid. As the fluid moves through the lumen on its way to the surface of the skin, the cells of the sweat gland's epithelium make the fluid hypotonic by removing Na^+ and leaving water behind. Design an epithe-

LEVEL FOUR   Quantitative Problems

34. The addition of dissolved solutes to water lowers the freezing point of water. A 1 OsM solution depresses the freezing point of water 1.86°. If a patient's plasma shows a freezing-point depression of 0.550° C, what is her plasma osmolarity? (Assume that 1 kg water = 1 L.)

COMBINED GLOSSARY AND INDEX

Students like the accessibility of definitions in a glossary, but sometimes you want additional information from the text itself. Usually this means flipping from the Glossary to the Index. Our combined Glossary-Index eliminates the need to flip back and forth between both.

Glossary/Index

Only the major terms and concepts are included. If a term is not here, consult the index.

ABC transporter   Membranes surround the contents of the cytoplasm and divide the interior of the cell into compartments such as the nucleus, mitochondria, endoplasmic reticulum, and Golgi complex,.486
abdominal aorta, 361
abdominal cavity, 12

Aerotolerant   The inner mitochondrial membrane, contains enzymes for oxidative phosphorylation,

Agglutination   A thin layer of lipids that acted as a between the aqueous interior and the watery env outside the cell, 224–25, 230.

Algae   The inner mitochondrial membrane, 224–25,

Alkaliphile   Properties of membranes found in diff types of cells.

Allergy 309–10

ANATOMY SUMMARIES

In order to understand physiology, you must have a good understanding of anatomy. To help you visulaize anatomy, this book includes several Anatomy Summary figures that show the anatomy of a physiological system, from a macro to micro perspective, in one large spread. "Zoom" arrows are used to enlarge the components of a structure, all the way down to the tissue level. These summaries allow you to see all of the essential features of each system in a single figure, whether as a review or learning it for the first time.

The Anatomy Summaries and other figures in the book were created by Bill Ober, M.D. and his associate Claire Garrison, R.N.

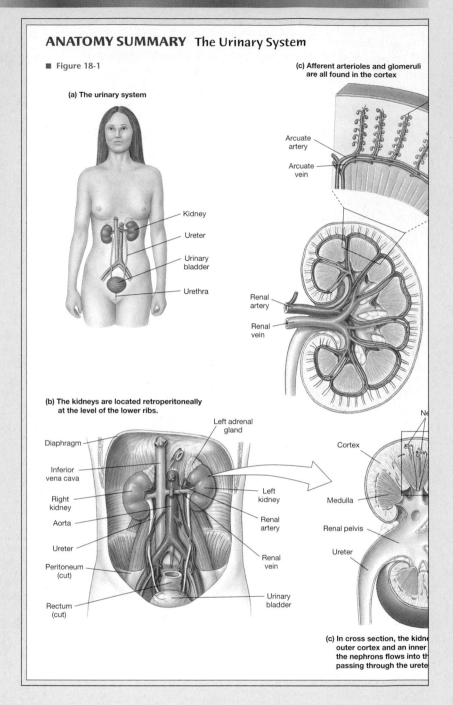

ANATOMY SUMMARY The Urinary System

■ Figure 18-1

(a) The urinary system

- Kidney
- Ureter
- Urinary bladder
- Urethra

(c) Afferent arterioles and glomeruli are all found in the cortex

- Arcuate artery
- Arcuate vein
- Renal artery
- Renal vein

(b) The kidneys are located retroperitoneally at the level of the lower ribs.

- Diaphragm
- Inferior vena cava
- Right kidney
- Aorta
- Ureter
- Peritoneum (cut)
- Rectum (cut)
- Left adrenal gland
- Left kidney
- Renal artery
- Renal vein
- Urinary bladder

- Cortex
- Medulla
- Renal pelvis
- Ureter

(c) In cross section, the kidn[...] outer cortex and an inner [...] the nephrons flows into th[...] passing through the urete[...]

FOCUS ON ORGAN BOXES

There are several large focus features highlighting the following organs: skin, liver, thymus, pineal gland, and spleen. Too often, the anatomy and physiology of these organs are overlooked in physiology texts.

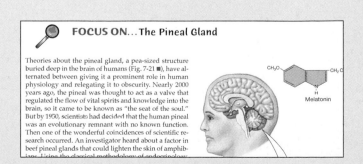

FOCUS ON...The Pineal Gland

Theories about the pineal gland, a pea-sized structure buried deep in the brain of humans (Fig. 7-21 ■), have alternated between giving it a prominent role in human physiology and relegating it to obscurity. Nearly 2000 years ago, the pineal was thought to act as a valve that regulated the flow of vital spirits and knowledge into the brain, so it came to be known as "the seat of the soul." But by 1950, scientists had decided that the human pineal was an evolutionary remnant with no known function. Then one of the wonderful coincidences of scientific research occurred. An investigator heard about a factor in beef pineal glands that could lighten the skin of amphib-[...]

Melatonin

x

REFLEX PATHWAY AND CONCEPT MAPS

The color-keyed reflex pathway maps and concept maps organize material into a logical, visual format that links both physiological themes and their supporting details into a unified reference. Like a schematic or a blueprint, these maps provide detailed visual accounts of systems and processes and guide you to a better understanding of coordinated physiological function. You will find it easier to remember complex pathways when you organize the information into maps. You will learn to create your own maps by completing the mapping questions found in the end-of-chapter exercises and following the mapping tips found inside the back cover.

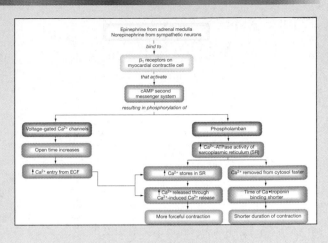

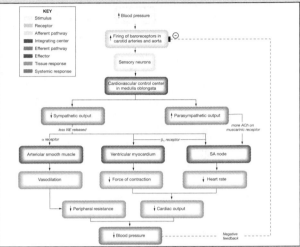

14. **Concept maps:** You may add any terms you like to the lists given.

   (a) Create a map showing the flow of blood through the heart and body. Label as many structures as you can.
   (b) Map the following terms to show how they influence cardiac output.

   β_1 receptor
   ACh
   adrenal medulla
   autorhythmic cells
   Ca^{2+}
   Ca^{2+}-induced Ca^{2+} release
   cardiac output
   contractile myocardium
   contractility
   force of contraction

   heart rate
   length-tension relationship
   muscarinic receptor
   norepinephrine
   parasympathetic neurons
   respiratory pump
   skeletal muscle pump
   stroke volume
   sympathetic neurons
   venous return

FIGURE AND GRAPH QUESTIONS

In an effort to help you interpret quantitative information and apply concepts, a number of figures and graphs throughout the book include questions that ask you to pause for a moment and consider what you really know about the information you just encountered. Answers to Graph and Figure Questions can be found in Appendix D.

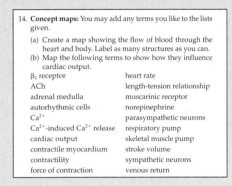

Figure question: Match the numbers on the figure to the boxes of the map.

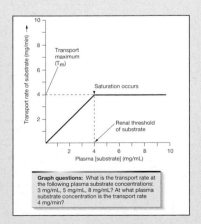

Graph questions: What is the transport rate at the following plasma substrate concentrations: 3 mg/mL, 5 mg/mL, 8 mg/mL? At what plasma substrate concentration is the transport rate 4 mg/min?

FOCUS BOXES

You will find four kinds of small boxes in this book. All of these boxes were developed with the goal of helping you understand the role of physiology in science and medicine today.

Separation of Proteins by Electrophoresis Many enzymes that have isozymes are complex structures with multiple protein chains. For example, lactate dehydrogenase has two kinds of subunits, named H and M, that are assembled into *tetramers*—groups of four. The isozymes include H_4, H_2M_2, and M_4. One way to determine which isozymes are present in a tissue sample is to use a technique known as electrophoresis. In this technique, a solution of proteins is placed at one end of a container filled with a polyacrylamide polymer gel. An electric

Sickle Cell Disease and Hydroxyurea One new treatment developed for sickle cell disease is the administration of hydroxyurea, a compound that inhibits DNA synthesis. In the bone marrow, hydroxyurea causes immature red blood cells to produce fetal hemoglobin (HbF), with alpha and gamma globin chains, instead of adult hemoglobin. The theory for hydroxyurea therapy is that fetal hemoglobin interferes with the crystallization of sickle cell hemoglobin (HbS), so the red blood cells no longer develop a sickle shape. However, some studies show

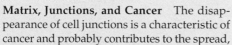

Matrix, Junctions, and Cancer The disappearance of cell junctions is a characteristic of cancer and probably contributes to the spread, or *metastasis*, of cancer throughout the body. Cancer cells lose the adherens junctions that tie them to adjacent cells because they have fewer of the cadherin molecules that make up the junctions. Once the cancer cell is released from its moorings, it secretes enzymes known as *proteases*. These enzymes, especially those called *matrix metalloproteinases* (MMPs), dissolve the extracellular matrix so that the cancer

DIABETES Diabetes and Amylin In 1987 researchers discovered that insulin is not the only hormone secreted by pancreatic beta cells. A second peptide named *amylin* is cosecreted with insulin. The functions of amylin are still being investigated, but at this time it appears that the hormone helps regulate glucose homeostasis following a meal. Amylin slows gastric emptying and gastric acid secretion, which delays digestion and absorption of carbohydrates. The combined actions of amylin and GIP thus set up a self-regulating cycle. Glucose in the intestine causes GIP release. GIP goes to the beta cells and initiates insulin and amylin secretion. Amylin then goes

BIOTECHNOLOGY BOXES
discuss physiology-related applications and laboratory techniques from the fast-moving world of biotechnology.

CLINICAL BOXES
focus on clinical applications and pathologies that clarify normal function. In addition to being interesting, these boxes will help you understand homeostasis and normal function.

EMERGING CONCEPTS
offer additional insight concerning physiological events at the cellular and molecular levels. Most physiological research today is being done at the cellular and molecular levels, so these boxes will help you integrate current research with the physiology you are learning.

DIABETES BOXES
illustrate the complexity of human physiology. A diabetes box might bear the icon of a Biotechnology box, a Clinical box, or an Emerging Concepts box, depending on the information presented. Because diabetes has such a widespread affect on the body, it makes a perfect example of integrated physiology. Some boxes explain the effects of diabetes and then show how the body must compensate to maintain homeostasis. Others describe biotechnological advances or breakthroughs in the treatment of diabetes.

RUNNING PROBLEMS

Each chapter includes a relevant, real-life problem that appears in segments throughout the chapter. Ch 14, p. 404 Problem opener. After the opening scenario is established, segments on subsequent pages present additional information and ask questions that prompt you to utilize information you have learned.

PROBLEM
Myocardial Infarction

At 9:06 A.M., the blockage that had been silently growing in Walter Parker's left coronary artery made its sinister presence known. The 53-year-old advertising executive had arrived at the Dallas Convention Center feeling fine, but suddenly a dull ac[...] [...] the [...] of his che[...] he became nauseate[...] of the convention b[...] it persisted, he ma[...] feeling very well," [...] digestion." The me[...] seeing his pale, sw[...] attack. "Let's get [...] checked out."

Continued from page 00

The blockage in Walter's coronary artery had restricted blood flow to his heart muscle, and its cells were beginning to die from lack of oxygen. When someone has a heart attack, medical intervention is critical to prevent additional damage and possibly save the patient's life. While waiting for the ambulance, the medic gave Walter oxygen, hooked him to a heart monitor, and started an intravenous (IV) injection of normal (isotonic) saline. With an intravenous injection line in place, other drugs could be given rapidly if Walter's condition should suddenly get worse.

Question 1: Why did the medic give Walter oxygen?

Question 2: What effect would the injection of isotonic saline have on Walter's extracellular fluid volume? On his intracellular fluid volume? On his total body osmolarity? [Hint: p. 00]

PROBLEM CONCLUSION

Walter's angiogram showed two blocked arteries, which were opened by balloon angioplasty. He returned home with instructions from his doctor for modifying his lifestyle to include a better diet, regular exercise, and no cigarette smoking.

In this running problem, you learned about some current techniques for diagnosing and treating heart attacks. You also learned that many of these treatments depend on speed to work effectively.

Further check your understanding of this running problem by comparing your answers to those in the summary table.

Question	Facts	Integration and Analysis
1 Why did the medic give Walter oxygen?	The medic suspects that Walter has had a heart attack. Blood flow and oxygen supply to the heart muscle may be blocked.	If the heart is not pumping effectively, the brain may not receive adequate oxygen. Administration of oxygen will raise the amount of oxygen that reaches both the heart and the brain.

Problem conclusion: At the end of the chapter, you can check your problem-solving strategy against the Problem Conclusion, which lists the questions from each segment, the relevant facts associated with each question, and the "integration and analysis" steps that were required to successfully analyze the running problem.

Question: To find out more about how to answer a question (or to check your own solution), examine the Facts and Integration and Analysis columns to the right of each question.

Facts: As you study the chapter, identify relevant facts from this and from earlier chapters. Use the facts listed here to check your assessment.

Integration and Analysis: As you work through the chapter, try to analyze the information you've learned to arrive at an answer. Using the explanation under Integration and Analysis, check to see if your reasoning is sound.

MEDIALABS

Found at the end of each chapter following the chapter exercises, MediaLabs are interactive questions that tame the vast jungle of information available on the Web. Each MediaLab session presents an Introduction that summarizes the themes of the chapter and then asks two Web Exploration questions relevant to the chapter material. To complete the Web Exploration questions, you will go to the page for that exercise at the free Silverthorn Companion Website (**www.prenhall.com/silverthorn**). There you will be guided through the Web to gather information about the problem at hand. Web Explorations also give you a key word to enter at the Website (for example, "GAP JUNCTIONS"). The key word will call up additional information (articles, electromicrographs, animations) about the topic.

E X P L **<MediaLab>**

Introduction

In this chapter you learned about the methods and mechanisms by which the body carries out internal communication, coordinates body functions, and maintains a consistent environment. Without these three functions—communication, integration, and homeostasis—large complicated organisms such as we are could not survive. The following Web Explorations will help you gain an appreciation of these coordinating functions. After reading the description below, visit the MediaLab for Chapter 6 in your Companion Website and select the appropriate keyword.

Web Exploration 1

Estimated time for completion = 5 minutes

Imagine yourself as a cell somewhere in your body. You are neighbors with a trillion other cells that must work together to maintain an internal environment (homeostasis) in the face of constant perturbation. As a cell, you neither have ears, eyes, or voice by which to send or receive messages. So how do you communicate? One of the most direct means of cell-to-cell communication is via "doorways" called gap junctions. Select the keyword **GAP JUNCTIONS** on the Website to see an electron micrograph (EM) of the intramembrane proteins of a gap junction. After you've viewed these micrographs and read the material, predict which molecules can freely pass between neighboring cells and which can not. As you begin to develop your model of the various ways cells communicate with

each other, it is important to factor in temporal (how fast does it work?) and spatial (over what distance does it work?) considerations as well as the nature of the substance being transported. What types of molecules can pass between cells? How rapid is this movement? What will be the consequence of the movement of this substance? As you learned in Chapter 5, molecules cross cell membranes only if there is a way across and if there is a driving force. What is the driving force for each molecule? To complete this exercise, visit MediaLab Web Exploration 1 in Chapter 6 of your Companion Website.

Web Exploration 2

Estimated time for completion = 5 minutes

The same chemical signal can cause very different responses throughout the body. How is this possible? One way is to use different receptors, while another is to use the same receptor but to couple that receptor to different second messenger cascades. A critical intermediate in many second messenger pathways is a G-protein. Select the keyword **G-PROTEIN** on your Website and read about the role of G-proteins in cellular signal transduction. As an interesting point of context, select the keyword **Rodbell** to read an example of how real scientific discoveries are made. The article describes how the actual discovery was worked out on a tablecloth in a bar. To complete this exercise, visit MediaLab Web Exploration 2 in Chapter 6 of your Companion Website.

Contents

4
Cellular Metabolism 73

5
Membrane Dynamics 109

8

The Nervous System 214

9

The Central Nervous System 252

Unit II   Homeostasis and Control

6

Communication, Integration, and Homeostasis   153

7

Introduction to the Endocrine System   185

13
Integrative Physiology I: Control of Body Movement   383

Unit III   Integration of Function

14
Cardiovascular Physiology   403

15

Blood Flow and the Control of Blood Pressure 443

16

Blood 474

19

Integrative Physiology II: Fluid and Electrolyte Balance 569

Unit IV   Metabolism, Growth, and Aging

20

Digestion 602

21
Energy Balance, Metabolism, and Growth 639

22
The Immune System 685

23
Integrative Physiology III: Exercise   716

24
Reproduction and Development   731

Preface

As we move into the 21st century, there has never been a more exciting time to study physiology. The human genome has almost been sequenced, but we are still a long way from fully understanding the physiology of the human body. We have the genetic sequence for many proteins, but we still don't know what many of them do in the living organism. Research with genetically engineered mice has shown us that we don't always understand the proteins we thought we knew. In more than one instance, researchers have created "knockout" mice that lack a certain protein, only to find that the protein was not as essential to the animal's function as they believed.

Scientists are turning back to whole animal and tissue studies in an effort to understand the integrated function of the body. Molecular biology and its technology are now merely tools to be used in the growing areas of physiology and molecular medicine. There is so much new and exciting information being reported daily, and one of the challenges of writing an undergraduate textbook is deciding what to include, and how to include it. However, as the alphabet soup of nicknamed proteins grows, it becomes more and more important for teachers and students to keep the basic themes of physiology in mind.

▶ KEY THEMES OF THIS BOOK

In the second edition we have kept the four themes that proved so popular in the first edition.

1. A Focus on Problem Solving.

One of the most valuable skills we can teach our students is the ability to think critically and use the information they learn to solve new problems. A number of features in the book are designed to help students practice these skills as they read, including Concept Checks, Running Problems, and Graph and Figure Questions. In this edition we added a fourth level with quantitative problems to the end-of-chapter questions.

2. An Emphasis on Integration.

The second edition of *Human Physiology* has retained the three chapters (Chapters 13, 19, and 23) that highlight the integrative nature of physiology and the cooperative function of multiple systems in the human body when homeostasis

is disturbed. In addition, I selected diabetes mellitus as a disease whose causes, complications, and treatments involve nearly every system of the body. You will find special diabetes focus boxes that run throughout the book to highlight the systemic effects of this common pathology.

3. Cellular and Molecular Physiology.

Most physiology research today is being done at the cellular and molecular level, and this edition continues to integrate physiology with cellular and molecular biology. There have been many exciting developments in molecular medicine and physiology since the first edition. For example, electrical signaling is not just for neurons and muscles any more, as you will learn in Chapter 5. And scientists have learned enough about membrane receptors to classify them into four major categories (Chapter 6). As a result, the first section of Chapter 6 has been reorganized, with seven new figures to explain signal transduction and second messenger pathways. The new information on membrane receptors is integrated with discussions of enzymes and membrane transporters to emphasize the commonalities of protein–protein binding: saturation, specificity, competition, and isoforms.

4. Physiology as a Dynamic Field.

In the book, I have tried to present physiology as a dynamic discipline with numerous unanswered questions that merit further investigation and research. It is essential that students appreciate that many of the "facts" they are learning are really only our current theories, especially as new information emerges from the union of the human genome project and basic research.

There is a tremendous need for systems level physiologists to come back into the research field to pull together clinical observations and the information uncovered by cellular and molecular research. I came across a perfect example the other day as I was doing research on guanylin and uroguanylin for this edition (see p. 583).

Leonard Forte *et al.* began their recent review[*] of guanylin and uroguanylin by describing the whole-

---

[*] Leonard R. Forte, Roslyn M. London, Ronald H. Freeman, and William J. Krause. (2000) Guanylin peptides: Renal actions mediated by cyclic GMP. *Am. J. Physiol.* 278 (2): F180-F191.

animal research from the 1970s that suggested that Na^+ sensors in the digestive system monitor salt ingestion and cause release of a blood-borne substance that enhances renal salt excretion. The bulk of their review then goes on to describe what we know about guanylin and uroguanylin, from the genes that encode the guanylin family peptides to their receptors and second messengers. But the key physiological experiments that confirm the relationship between salt intake and salt excretion have yet to be done.

The physiology students of today are the next generation of scientists and healthcare providers, the people who will try to answer the unanswered questions. It is my hope that this book will provide them with an integrated view of physiology so that they enter their chosen professions with respect for the complexity of the human body and a clear vision of the potential of physiological and biomedical research.

▶ SPECIAL FEATURES

This textbook has some special features that are designed to make the study of physiology easier. Please take a few minutes to look at the Owner's Manual on page vii. That section explains the symbols, boxes, and features that make this book unique.

More than a Textbook

As we begin the 21st century, computers provide us with an additional dimension to use in teaching. The book itself remains the single most important component of any physiology course, but it takes more than a textbook to improve teaching and learning.

I truly believe that HOW we teach is almost as important as WHAT we teach, particularly in this age of the Internet when students have so many options for gathering information in addition to coming to lecture and reading their textbook. What is the instructor's role in today's classroom? What can we do for our students besides be "talking heads" that convey content? I believe that we should be there to inspire our students and get them excited about physiology. We need to help them understand why the "facts" they memorize today may be "wrong" tomorrow. We are there to bridge the gap between systems physiology, which is what we teach, and physiology as it is practiced in the clinics and the research laboratory. With this in mind, we have assembled an array of innovative supplements for both students and instructors.

The *Human Physiology* Student Workbook   One of the most common reasons that faculty give for not including active learning in the classroom is that they are afraid that they will not be able to cover all the content their students need to know. The Student Workbook that accompanies *Human Physiology* is my solution to this problem, and it is derived from one that I use in my own teaching.

The *Teach Yourself the Basics* section of the workbook provides guided notetaking as students read the text. It is organized with the same headers as the chapter and it asks students to answer basic knowledge questions, such as "List the functions..." or "Define..." The workbook also has a complete list of vocabulary words for each chapter (*Talk the Talk*). Students can test their understanding with the questions in *Practice Makes Perfect* and *Quantitative Thinking*. A section on *Maps* provides additional lists of terms for students to compile into concept and reflex maps.

For students who want to extend their learning, each workbook chapter includes additional reading and information about the pathology discussed in the chapter's Running Problem. In *Beyond the Pages,* students and faculty alike will find suggestions for mini-demonstrations (*Try It*) and a list of articles about topics covered in the chapter. [0-13-019554-5]

The Silverthorn Physiology WebSite   The Companion Website for this text can be found at www.prenhall.com/silverthorn. This interactive site contains study questions and the answers to all end-of-chapter questions in the book. Additionally, there are links to other interesting physiology-related sites. Because the World Wide Web is such a fluid medium, you never know what you might find on this site, so I encourage you to be a regular visitor.

Problem Solving in Physiology   Written by Joel Michael and Allen Rovick, this workbook contains a wealth of real-world problems designed to help students truly understand physiology. Utilizing clinical data and realistic scenarios, students develop active learning skills. Instructor's Edition: [0-13-959933-9]

The New York Times "Themes of the Times"   **Themes of the Times** is a program sponsored jointly by Prentice Hall and The New York Times. Physiology-related articles have been compiled into a free supplement that helps students make the connection between the classroom and the outside world. It is designed to enhance student access to current, relevant information.

Video Tutor for Anatomy and Physiology   This videotape gives students the opportunity to see the dynamics of the most difficult physiological processes from the comfort of a favorite couch or chair. This two-hour video focuses on concepts that instructors across the country have consistently identified as the most challenging. Physiological concepts are highlighted using three-dimensional animations and video footage. On-camera narration and built-in review questions insure that these concepts come to life for the viewer. [0-13-751843-9]

Instructor's Resource Guide   This Instructor's Resource Guide is currently the only one for human physiology that was written by the author of the text. I believe very strongly that active and inquiry learning should be incorporated into the lecture setting on a regular basis, so I created a resource to assist faculty with this task. The IRG contains annotated chapter outlines with class-tested demonstrations and activities. Each chapter also contains additional detail that instructors will find useful in writing their lectures, as well as references from the research and clinical literature. I have also included some higher-level questions and problems that faculty can use in the classroom or on tests. [0-13-019026-8]

Test Item File and Computerized Quiz Management System A printed bank of over 3,000 questions is available with this text. Written in the same configuration as the end-of-chapter questions in the text, the test bank is a valuable compliment to an instructor's own test/quiz files. Available in Macintosh and Microsoft Windows formats, this powerful software includes easy-to-use Wizards (the Wizard asks the user questions and offers prompts to make test creation simple), an editing function, and a grade book program. This software also includes a test item analysis program that generates helpful statistics on class performance. [Print version: 0-13-019028-4; Windows: 0-13-019029-2; Macintosh: 0-13-019020-9]

Transparency Acetates and Transparency Masters Prentice Hall has prepared 250 full-color acetates and 50 black-and-white masters containing key illustrations from the text. Labels and figures have been enlarged for optimal viewing in the classroom. [0-13-019031-4]

Image Bank and PowerPoint Gallery CDROM for the Lab or Classroom Think of this supplement as a bank of images from the text, which you can integrate into your lecture notes, handouts, homework, exams, etc..., along with a series of pre-assembled PowerPoint slides for easy inclusion with your current presentations. With this CD, you have the tools to visually enhance your curriculum. [0-13-019551-0]

For the Laboratory

Dicarlo/Sipe/Layshock/Rosian, *Experiments and Demonstrations in Physiology.* 1998. This laboratory manual is appropriate for courses in physiology or the physiology component of an Anatomy and Physiology lab where there is very little equipment. The emphasis of the manual is on critical thinking and problem-solving. [0-13-636457-8]

Gerald Tharp, *Experiments in Physiology, 7th edition.* 1997. This traditional laboratory manual includes new computer exercises for key physiology experiments. [0-13-575788-6]

In addition, Dee Silverthorn will publish two laboratory manuals in human physiology, including a selection of over 150 class-tested laboratory exercises which can be customized for individual adoptions. Both will be available in 2001. Please contact a Prentice Hall sales representative for further information.

▶ ACKNOWLEDGMENTS

No one could write a textbook of this scope without a lot of help from a lot of people.

Many people devoted time and energy to making this book a reality, and I would like to make an attempt at thanking them all. I apologize in advance to anyone whose name I've left out.

Reviews

I am particularly grateful to the instructors who reviewed the first edition and made a second edition possible.

Brenda Alston-Mills, North Carolina State University

Roland M. Bagby, University of Tennessee

Patricia J. Berger, University of Utah

Ronald Beumer, Armstrong State College

Sunny K. Boyd, University of Notre Dame

Timothy Bradley, University of California-Irvine

Alan P. Brockway, University of Colorado at Denver

David Byman, Pennsylvania State University Worthington-Scranton Campus

Robert G. Carroll, East Carolina University School of Medicine

Kim Cooper, Arizona State University

Charles L. Costa, Eastern Illinois University

Gordon Garrett, Utah Valley State College

Margaret M. Gould, Georgia State University

Janet L. Haynes, Long Island University-Brooklyn

Richard W. Heninger, Brigham Young University

Carol Hoffman, San Jose City College

Bruce Johnson, Cornell University

William C. Kleinelp, Middlesex County College

James Larimer, University of Texas

Daniel L. Mark, Penn Valley Community College

Paul Matsudaira, Massachusetts Institute of Technology

Wayne Meyer, Austin College

Joel Michael, Rush Medical College

Alice C. Mills, Middle Tennessee State University

Ron Mobley, Wake Technical Community College

Harold Modell, National Resource for Computers iLife Science Education

John G. Moner, University of Massachusetts at Amherst

Patricia L. Munn, Longview Community College

Dell Redding, Evergreen Valley College

Mary Anne Rokitka, State University of New York (SUNY) at Buffalo

Allen Rovick, Rush Medical College

Evelyn Schlenker, University of South Dakota

Norm Scott, University of Waterloo

Brian R. Shmaefsky, Kingwood College

Sabyasachi Sircar, Delhi, India (IUPS Teaching Fellow, 1995-6)

Judy Sullivan, Antelope Valley College

Deborah Taylor, Kansas City Kansas Community College

Richard L. Walker, University of Calgary

Donald Whitmore, University of Texas at Arlington

Ronald L. Wiley, Miami University of Ohio

Jack Wilmore, Texas A&M University

David A. Woodman, University of Nebraska-Lincoln

The following colleagues read multiple chapters and suggested changes for this edition. Through this process, they provided valuable insights and expertise in a field where it is impossible for any one individual to be an expert in everything. I appreciate their thoughtful input and comments, and hope they will forgive me for not always making the changes they suggested. Reviewers for the second edition were:

Annabelle Cohen, CUNY The College of Staten Island

Lori K. Garrett, Danville Area Community College

Tammy Greco, Eastern Michigan University

Tracy M. Hodgson, University of Pittsburgh

Katja Hoehn, Mount Royal College

Roberta Meehan, Denver, Colorado

Patricia Munn, Longview Community College

Joanne Nosky, Okanagan Valley College of Massage Therapy

Carolyn Rivard, Fanshawe College

David Thomas, Fanshawe College

Curt Walter, Dixie College

Content Specialists

No one can be an expert in every area of physiology, and I am deeply thankful for my friends and colleagues who did everything from answer frantic emails with questions to review entire chapters. Even with their help, there may be errors, for which I take full responsibility.

I would particularly like to acknowledge Bruce Johnson, Cornell University, Department of Neurobiology and Behavior, whose expertise and guidance shaped the first edition chapters on the central nervous system and motor control. Other colleagues who contributed to the revision are:

Leona Young, Emory University

Virginia Scofield, University of Southern California

William Percy, University of South Dakota

Tom Blevins, M.D., Austin, TX

Catherine Dallemagne, Queensland University, Australia

Douglas Eaton, Emory University

Barbara Goodman, University of South Dakota

Dan Hardy, M.D., Austin, TX

Charles Levinson, University of Texas Health Science Center at San Antonio

Harold Modell, National Resource for Computers in Life Science Education

Joel Michael, Rush Medical College

Jeffrey Sands, Emory University

James Stockand, Emory University

Roger Tanner-Theis, University of Oklahoma (retired)

Jack Wilmore, Texas A&M University

Photographs

The immunofluorescent micrograph on the cover of the first edition was so popular that we decided to include more "scientific art" in the chapter openers. I would like to thank the following colleagues who generously provided micrographs from their research:

Jane Lubisher, University of Texas

Flora M. Love, University of Texas

Carole L. Moncman, University of Texas

Young-Jin Son, University of Texas

Daniel Casellas, Groupe Rein & Hypertension IURC, Montpellier, France

L. Gabriel Navar, Tulane University

Supplements

Pat Munn at Longview Community College in Kansas helped write the end-of-chapter questions. Pat and Alice Mills from Middle Tennessee State University wrote the test bank questions. Damian Hill worked with me to revise the Instructor's Resource Guide and Student Workbook. Mary Pat Wenderoth from University of Washington and Kristina Stanfield of G.D. Searle/Momsanto wrote the fun and creative MediaLabs that appear in the text and Student Workbook. Karina Loyo-Garcia and Jan Melton Machart revised the web site and appendices.

Prentice Hall

Making a book is a complex process that requires the energy and talents of many people. Special thanks go to Anne A. Reid, my development editor, whose experience, insight, and gentle guidance made the revision a pleasure. Debra Wechsler, my production editor, helped me keep my sanity despite a grueling schedule. Andy Stull and Don O'Neal masterfully coordinated the wonderful group of colleagues who contributed to the supplements. Bill Ober and Claire Garrison took the art from the first edition and made it even better. Thanks to Heather Scott for working with a fussy author on the book's cover and interior design. And special recognition goes to Jennifer Welchans, Marty McDonald, and the wonderful sales team at PH who made it possible for me to visit with faculty around the country. Finally, a heartfelt thank you to my two editors, David K. Brake and Halee Dinsey, for the time and energy they devoted to seeing that the book had all the resources that they could muster.

Special Thanks

I would also like to thank my students who looked for errors and areas that needed improvement. The Spring 2000 ZOO 365N class put a lot of energy into writing multiple choice questions that could be used on the Companion Website. I would particularly like to thank the following students who volunteered their free time to proofread pages.

Jeff Duman	Jeff Eaton	Rachel Halbrook
Christina Michaelis	Mark Muir	Brian Schwartz
Mark Zafereo		

My graduate teaching assistants have played a huge role in my teaching ever since I arrived at the University of Texas, and their input has helped shape how I teach. Many of them are now faculty members themselves. I would particularly like to thank:

Lawrence Brewer, Ph.D.	Peter English, Ph.D.
Jan Melton-Machart	Kurt Venator, Ph.D.
Ari Berman	Karina Loyo-Garcia

Finally, special thanks go to a group of my colleagues for their continuing support: Ruth Buskirk, Judy Edmiston, Jeanne Lagowski, and Marilla Svinicki (University of Texas), Penelope Hansen (Memorial University, St. John's), Mary Anne Rokitka (SUNY-Buffalo), Ann McNeal (Hampshire College), and Rob Carroll (East Carolina University School of Medicine).

I also must thank my family, friends, and colleagues for their understanding whenever I said, "I can't; I'm working on the book." And, as always, the biggest thank you to my husband Andy, who is always encouraging and supportive.

A Work In Progress

One of the most rewarding aspects of writing a textbook is the opportunity it has given me to meet or communicate with other instructors and students. In the three years since the first edition was published, I have heard from people around the world, and have had the pleasure of hearing how the book has been incorporated into their teaching and learning.

Because science textbooks are revised every three or four years, they are always works in progress. I invite you to contact me or my publisher with any suggestions, corrections, or comments about this second edition. I am most reachable through E-mail at silverth@utxvms.cc.utexas.edu. You can reach my editor at the following address:

halee_dinsey@prenhall.com

—Dee Silverthorn
University of Texas
Austin, Texas

This book is dedicated to my
students, past, present, and future,
with whom and for whom it was
written.

1 Introduction to Physiology

■ "Physiology is not a science or a profession but a point of view."—*Ralph W. Gerard* ■

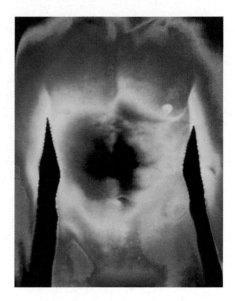

For most of recorded history, humans have been interested in how their bodies work. Early Egyptian, Indian, and Chinese writings describe attempts by physicians to treat various diseases and to restore health. Camel dung and powdered sheep horn may seem to us today to be bizarre therapies, but we must view them in light of what was known about the human body in those times. In order to appropriately treat disease and injury, we must first understand the human body in its healthy state. **Physiology** is the study of the functions of a living organism and its component parts, including all chemical and physical processes. The term *physiology* literally means "knowledge of nature." Aristotle (384–322 B.C.) used the word in this broad sense to describe the functioning of all living organisms, not just of the human body. However, Hippocrates (ca. 460–377 B.C.), considered the father of medicine, used the word *physiology* to mean "the healing power of nature," and the field became closely associated with medicine. By the sixteenth century in Europe, physiology had been formalized as the study of the vital functions of the human body, although today the term is again used to refer to the study of the functions of all animals and plants.

In contrast, **anatomy** is the study of structure, with minimum emphasis on function. Despite this distinction, anatomy and physiology cannot be truly separated. The function of a tissue or organ is closely tied to its structure, and the structure of an organism presumably evolved to provide an efficient physical base for its function.

There has never been a more exciting time to study human physiology. Today we benefit from centuries of work by physiologists who constructed a foundation of knowledge about how the human body functions. Since the 1970s, rapid advances in the fields of cellular and molecular biology have supplemented this work. A few decades ago we thought that we would find the key to the secret of life by sequencing the human genome. However, this deconstructionist view of biology has proved to have its limitations because living organisms are much more than the simple sum of their parts. Sequencing genes and proteins in the molecular biology laboratory does not always tell us the significance of these substances in the intact animal. As we begin the twenty-first century, molecular biologists are turning to physiologists to help them understand the function of molecules in the intact organism. The integration of function across all levels, from molecules to the living body, is the special skill of a physiologist. Today, as we try to reconcile the findings of the molecular biologists with what we know about how our bodies work, this skill is in more demand than ever.

In this chapter, we examine the fundamental principles of physiology and explore how we know what we know about the way the human body functions.

▶ LEVELS OF ORGANIZATION

To understand how the human body functions as a coordinated entity, we must first examine its component parts. Figure 1-1 ■ shows the different **levels of organization** of living organisms, ranging from the parts of a

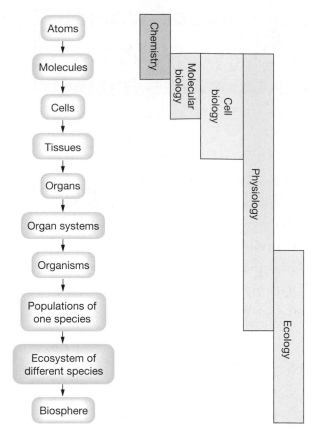

■ Figure 1-1   **Levels of organization and related fields of study**

> **The Physiome Project**   By 2002 scientists will successfully complete the Human Genome Project, having determined the human genome DNA sequence. However, a second goal of the project, to discover all the genes of humans, has been more elusive. The original idea that one segment of DNA contained only one gene had to be revised when it became clear that one segment of DNA may contain several different genes. In addition, scientists began to realize that applying those DNA sequences directly to the function of living organisms would be difficult. In 1997 an international group of physiologists met in St. Petersburg, Russia, to organize the Physiome Project, envisioned as the successor to the Genome Project. The Physiome Project will coordinate molecular, cellular, and physiological information about living organisms into an Internet database. Scientists around the world will be able to access this information and use it in their own research efforts to create better drugs or genetic therapies for curing and preventing disease. Some scientists plan to use the data to create mathematical models to explain how the body functions. The Physiome Project is an ambitious undertaking, but one that promises to integrate information from diverse areas of research so that we can improve our understanding of the complex processes that we call life.

single organism to groups of the same species (*populations*) and populations of different species living together in *ecosystems* and the *biosphere*. The different disciplines of chemistry and biology related to the study of each level are shown on the right side of the figure. There is considerable overlap between the different fields, and these artificial divisions will vary according to who is defining them. One distinguishing feature of physiology is that it encompasses many levels of organization.

At a fundamental level, atoms of elements link together to form molecules. One of the great, and possibly unsolvable, mysteries of science is how interactions between certain groups of molecules result in the unique properties of a living organism. The smallest unit of structure capable of carrying out all life processes is the **cell.** Cells are collections of molecules separated from the external environment by a cell membrane. Simple organisms are composed of only one cell, but complex organisms have many cells with different structural and functional specializations.

Collections of cells that carry out related functions are known as **tissues** [*texere*, to weave]. Tissues form structural and functional units known as **organs** [*organon*, tool], and groups of organs integrate their functions to create **organ systems.** The human body has 10 physiological organ systems (Table 1-1).

Figure 1-2 ■ is a schematic diagram showing the interrelationship of the systems in the human body. The

TABLE 1-1   Organ Systems of the Human Body

System Name	Representative Organs or Tissues	Functions
Circulatory	Heart, blood vessels, blood	Transport of materials between all cells of the body
Digestive	Stomach, intestines, liver, pancreas	Conversion of food into particles that can be transported into the body; elimination of some wastes
Endocrine	Thyroid gland, adrenal gland	Coordination of body function through synthesis and release of regulatory molecules
Immune	Thymus, spleen, lymph nodes	Defense against foreign invaders
Integumentary	Skin	Protection from external environment
Musculoskeletal	Skeletal muscles, bones	Support and movement
Nervous	Brain, spinal cord	Coordination of body function through electrical signals and release of regulatory molecules
Reproductive	Ovaries and uterus, testes	Perpetuation of the species
Respiratory	Lungs, airways	Exchange of oxygen and carbon dioxide between the internal and external environments
Urinary	Kidneys, bladder	Maintenance of water and solutes in the internal environment; waste removal

integumentary system, composed of the skin, forms a protective boundary that separates the body's internal environment from the external environment of the outside world. The **musculoskeletal system** provides support and body movement.

Four systems exchange material between the internal and external environments. The **respiratory system** exchanges gases, the **digestive system** takes up nutrients and water and eliminates wastes, the **urinary system** removes excess water and waste material, and the **reproductive system** produces eggs or sperm. The major organs of these four systems are hollow. Their interior

spaces, or **lumens** [*lumin,* window], are essentially extensions of the external environment. Material that enters the lumens of these organs is not truly part of the internal environment until it crosses the tissue wall of the organ.

An interesting illustration of the difference between the internal and external environment involves the bacterium *Escherichia coli.* This organism normally lives and reproduces in the lumen of the large intestine, an interior space that is continuous with the external environment. In this location, the organism does not harm the host. However, if the intestinal wall is punctured by disease or

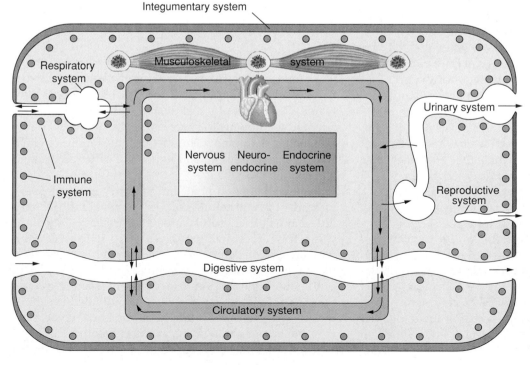

■ Figure 1-2 **The integration between systems of the body** This schematic figure shows the relationships between the 10 systems of the human body. The lumens of the hollow organs of the respiratory, digestive, reproductive, and urinary systems open to the external environment.

accident and *E. coli* enters the internal environment of the body, a serious infection can result.

The **immune system** of the body protects the internal environment from foreign invaders such as the *E. coli* bacterium. To carry out that function, the cells and tissues of the immune system are positioned to intercept material that may enter through the exchange surfaces or through a break in the skin. In addition, immune tissues are closely associated with the **circulatory system,** which distributes material throughout the body.

The **nervous** and **endocrine systems** coordinate body function. Note that the figure shows them as a continuum rather than as two distinct systems. Why? Because as we have learned more about the integrative nature of physiological function, the lines between these systems have blurred. How do we classify nerve cells that secrete hormones as endocrine cells do? How should we classify responses that begin in the nervous system and end with the endocrine system? One of the current challenges in physiology is integrating information from the different body systems into a cohesive picture of the living human body. One way that physiologists do this is by using maps or flow charts, schematic representations of structure and function (see Fig. 1-3 ■, Focus on Mapping).

▶ PHYSIOLOGY IS AN INTEGRATIVE SCIENCE

Physiology cuts across many levels of organization. At the systems level, there are relatively few unanswered questions about how the human body works. Most of these questions involve the nervous control of function, such as the nervous control mechanisms that govern breathing. The bulk of research in physiology today is focused on the cellular and molecular level because many questions about how the human body works at that level remain unanswered. Nevertheless, explaining what happens in test tubes or isolated cells can only partially answer questions about how the intact body works. We must also figure out how events in a single cell influence neighboring cells, tissues, organs, and systems throughout the body. In this respect, physiologists are like children throwing rocks into a pond. They want to see what kind of splash the rocks make where they enter the water, but they also want to know where the ripples go and what happens when they encounter other ripples.

The integration of function among systems is a special focus of physiology. Traditionally, students have studied the cardiovascular system and the regulation of blood pressure in one unit, then studied the kidneys and the control of body fluid volume in a separate unit. But body fluid volume plays a key role in determining blood pressure, so how can the two units be completely separated? Creating an integrated view that links together the different systems is one of the most interesting and most challenging aspects of physiology.

▶ FUNCTION AND PROCESS

What is the difference between function and process? Learning to make this distinction is an important skill for you to acquire as you study physiology. The **function** of a physiological system or event is the "why" of the system or event: why does the system exist? This way of thinking about a subject is called the **teleological approach** to science. For example, the teleological answer to the question of why red blood cells transport oxygen is "because red blood cells bring oxygen to the cells that need it." This answer explains the reason red blood cells transport oxygen but says nothing about *how* the cells transport oxygen.

In contrast, physiological **processes,** or **mechanisms,** are the "how" of a system. The **mechanistic approach** to physiology examines process. The mechanistic answer to the question of why red blood cells transport oxygen is "because there are hemoglobin molecules that combine reversibly with oxygen molecules." This very concrete answer explains exactly how oxygen transport occurs but says nothing about the significance of oxygen transport to the intact animal.

Although function and process seem to be two sides of the same coin, it is possible to study processes, particularly at the cellular and subcellular level, without understanding their function in the life of the organism. As biological knowledge becomes more complex, and as research focuses on the subcellular and molecular levels, scientists sometimes become so involved in the processes they study that they fail to step back and look at what those processes mean to cells, organ systems, or the intact animal. One role of physiology is to integrate function and process into a cohesive picture.

▶ THE EVOLUTION OF PHYSIOLOGICAL SYSTEMS

To understand why the systems of the human body work the way they do, note that human beings are large, mobile, terrestrial animals whose bodies are about 60% water. Our physiological systems and processes are adapted to allow us to survive in a dry, highly variable external environment. This fact becomes more obvious when we examine the adaptations of other animals to their habitats.

Life probably originated in tropical seas, a stable environment where salinity, oxygen content, and pH vary little and where light and temperature cycle in predictable ways. The earliest living organisms had an internal environment almost identical to that of seawater, just as today's marine invertebrates do. If external conditions changed, conditions inside the primitive organisms changed as well. Even today, marine invertebrates cannot tolerate significant changes in salinity and pH, as you know if you have ever maintained a saltwater aquarium. In both ancient and modern times, most marine organisms have relied on the

FOCUS ON...Mapping

■ **Figure 1-3** **Maps for physiology**

(a) A map showing structure/function relationships

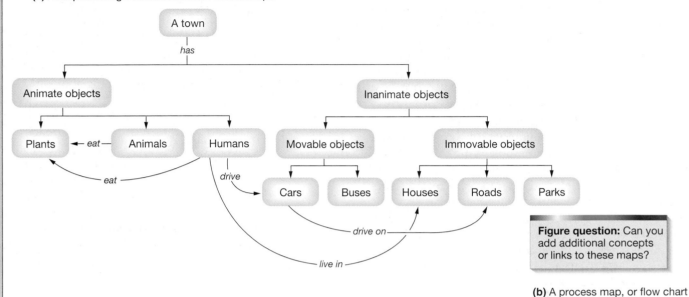

Figure question: Can you add additional concepts or links to these maps?

(b) A process map, or flow chart

Person working outside on a hot, dry day

↓

Loses body water by evaporation

↓

Body fluids become more concentrated

↓

Internal receptors sense change in internal concentration

↓

Thirst pathways stimulated

↓

Person seeks out and drinks water

↓

Addition of water to internal environment decreases its concentration

Figure question: Make a structure/function map of your home and a flow chart of one day in your physiology class.

Mapping is a nonlinear way of organizing material. It is a useful study tool because mapping requires active processing of information rather than memorization. Maps in physiology usually focus on the relationships between anatomical structures and physiological processes (structure/function), normal homeostatic control processes (flow charts), or response pathways to abnormal (pathophysiological) events. A map can take a variety of forms but often consists of terms linked by arrows. The connecting arrows can be labeled with explanatory phrases (for example, tissues "are composed of" cells). A map may include graphs or pictures.

The real benefit from mapping occurs when you prepare the maps yourself. By organizing the material on your own, you examine the relationships between terms, organize concepts into a hierarchical structure, and look for similarities and differences between items. Such interaction with the material ensures that you process it into long-term memory instead of simply memorizing it and forgetting it.

Often the most difficult part of mapping is deciding where to begin. Select the concepts to map and mentally arrange them in an organized fashion. If the map is a structure/function map, begin with the most general, important, or overriding concept from which all the others naturally stem. If you are making a process map or flow chart, start with the first event to occur. Next, break down the key idea into progressively more specific parts using the other concepts, or follow the event through its time course. Use arrows to point the direction of linkages, and include horizontal links to tie concepts together. Labeling the kind of linkage is useful. The downward development of the map will generally mean the passage of time or an increase in complexity. Color is very effective on concept maps. You can use colors for different types of links or for different sections.

These maps are examples of structure/function maps and flow charts.

constancy of their external environment to keep their internal environment in balance.

As animals migrated from the ancient seas into estuaries, where fresh and sea water mix, then into freshwater environments and onto the land, their external environment was no longer constant. Salinity varies when rains send fresh water into the estuaries or as the sun evaporates tidal pools among the rocks. Temperature fluctuates widely from day to night and from season to season. Animals that live in fresh water must cope with the influx of water into their body fluids, while animals that live on land constantly lose internal water to the dry air around them.

Coping with the demands of variable external environments required the evolution of a variety of physiological, structural, and behavioral mechanisms. For terrestrial animals, these mechanisms include anatomical, metabolic, and behavioral adaptations to reduce water loss (impermeable body surfaces, internalized surfaces for gas exchange, and highly efficient kidneys that conserve water). Fertilization and fetal development, two processes that require aqueous environments, often take place internally. In the absence of support from the buoyancy of water, terrestrial animals require rigid skeletons to support their weight. The result of all these adaptations is an efficient organism designed for survival in a hostile terrestrial world.

▌ HOMEOSTASIS

Although the human body as a whole is adapted to cope with a variable external environment, its individual cells are much less tolerant of change. Only a small minority of cells in a multicellular organism are actually in direct contact with the external environment (Fig. 1-4 ■). The vast majority of the cells are sheltered from the outside world by the buffer zone of the **extracellular fluid** [*extra-*, outside of], the body fluid that surrounds the cells and makes up one-third of the body's total volume. The bulk of the body's water is found in the **intracellular fluid** inside cells [*intra-*, within].

Extracellular fluid is the body's internal environment and serves as the interface between the external environment and the cells. When conditions outside the body change, the changes are reflected in the composition of the extracellular fluid, which in turn affects the cells. But our cells are not very tolerant of changes in their surroundings. As a result, a variety of mechanisms have evolved that maintain the composition of the extracellular fluid within a narrow range of values. The body's ability to maintain internal stability is known as **homeostasis** [*homeo-*, similar + *-stasis*, condition]. Homeostasis and the regulation of the internal environment are central precepts of physiology and create an underlying theme of each chapter in this book.

Failure to maintain homeostasis disrupts normal function and results in a disease state or **pathological** con-

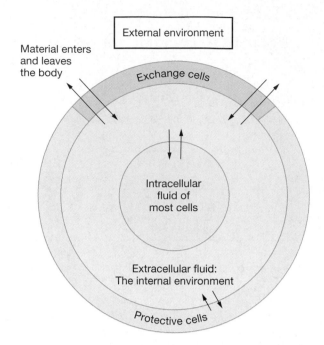

■ **Figure 1-4   The internal and external environments** Only a few cells in the body are able to exchange material with the external environment. Most cells are in contact with the internal environment, composed of the extracellular fluid.

dition [*pathos*, suffering]. Diseases generally can be divided into two groups according to their origin: those in which the problem arises from internal failure of some normal physiological process and those that originate from some outside source. Internal causes of disease include the abnormal growth of cells, which may cause cancer or benign tumors, the production of antibodies by the body against its own tissues (autoimmune diseases), and the premature death of cells or the failure of cell processes. Inherited disorders are also considered to have internal causes. External causes of disease include toxic chemicals, physical trauma, and foreign invaders such as viruses and bacteria. In both internally and externally caused diseases, when homeostasis is disturbed, the body attempts to compensate (Fig. 1-5 ■). If the compensation is successful, homeostasis is restored. If compensation fails, disease results. The study of body functions in a disease state is known as **pathophysiology**.

You will encounter many examples of pathophysiology as we study the various systems of the body. One very common pathological condition in the United States is **diabetes mellitus,** a metabolic disorder characterized by abnormally high blood glucose concentrations. Although we speak of diabetes as if it were a single disease, it is actually a whole family of diseases with varying causes and manifestations. You will learn more about diabetes in the focus boxes scattered throughout the chapters of this book. The influence of this one disease on many systems of the body makes it an excellent example of the integrative nature of physiology.

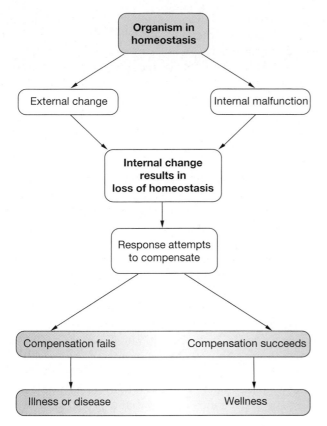

■ Figure 1-5   **Homeostasis**

▶ THEMES IN PHYSIOLOGY

Recall the quotation that opens this chapter: "Physiology is not a science or a profession but a point of view."* Physiologists pride themselves on relating the processes they study to the functioning of the animal as a whole. One key to critical thinking in physiology is the ability to recognize patterns. The amount of information you will be asked to understand can seem overwhelming at times, but once you realize that many specific processes are variations on a few basic themes, learning physiology is easier.

In this book we will focus on seven themes:

Homeostasis. As we have seen, one key theme of human physiology is homeostasis.

Integration of body systems. A related theme is the integration of body systems, the ability of the different organ systems to coordinate with each other, work together to maintain homeostasis, and carry out the functions of the body.

Cell-to-cell communication and coordination. Integration and homeostasis both require that the cells of the body communicate with one another rapidly and efficiently. A variety of sensing mechanisms monitor changes in the internal and external

environments. These sensors, which range from single cells to highly developed special sense organs such as the eyes, nose, and ears, must relay information about environmental changes to other cells. Most cell-to-cell communication uses chemical signals, although the nervous system adds speed by means of electrical signals. Information is sent between distant cells through the movement of blood in the circulatory system or along the specialized cells of the nervous system. Integration and coordination of responses take place in the brain and spinal cord and in endocrine and immune cells.

Movement of substances and information across cell membranes. Communication between the internal environment of the cells and the extracellular fluid requires information transfer across the cell membrane, another key theme. You will learn how some molecules can move across the cell membranes and how others remain outside the cell and simply pass information through the membrane.

Compartmentation of the body and cell. Cell membranes separate cells from each other and from the extracellular fluid. Similarly, within a cell, membranes divide the cell into discrete units or *organelles.* This **compartmentation,** or the presence of separate compartments within the body, allows different areas to specialize their functions.

Energy flow. Living processes require the continuous input of energy. Where does this energy come from, and how is it stored? As we answer those questions, you will learn how energy is used to do work. In the body, energy is used for synthesis and breakdown of molecules, to transport molecules across cell membranes, and to create movement.

The principles of mass balance and mass flow. Another common theme and important principle of physiology is the **law of mass balance.** This law says that if the amount of a substance in the body remains constant, any gain must be offset by an equal loss (Fig. 1-6 ■). For example, in order to maintain constant body temperature, heat gain from the external

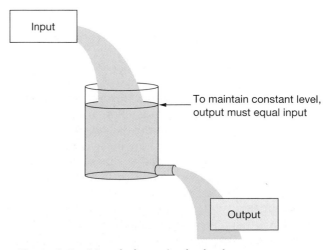

■ Figure 1-6   **Mass balance in the body**

*Ralph W. Gerard, *Mirror to Physiology: A Self Survey of Physiological Science* (Washington, D.C.: American Physiology Society, 1958).

environment and from metabolism must be offset by heat loss back to the external environment. Most substances enter the body from the outside environment, although some, like heat, can be produced internally. The major routes for loss from the body are the urinary and digestive systems (in the urine and feces), the respiratory system, and the integumentary system (skin). Some aspects of mass balance are monitored by sensors (receptors) that detect changes in the internal and external environments. When mass balance is disturbed, physiological and behavioral homeostatic reflexes bring the body back into balance. The law of mass balance is summarized by the following equation:

Total amount (or **load**) of substance x in the body = intake + production − output

Mass balance in the body is usually discussed as a function of time. We talk about sodium intake per day or carbon dioxide output per minute. Thus, the rate of intake, production, or output is an important consideration. This is expressed as **mass flow,** where

Mass flow (amount/min) = concentration (amount/vol) × volume flow (vol/min)

For example, a person is given an intravenous infusion of glucose containing 50 grams of glucose per liter. If the infusion is given at a rate of 2 milliliters per minute, the mass flow for glucose would be

50 g glucose/1000 mL solution × 2 mL solution/min = 0.1 g glucose/min

Substances whose concentrations are maintained through mass balance include oxygen and carbon dioxide, water, salts, and hydrogen ions. In addition, our bodies have regulatory mechanisms to keep temperature and energy stores within an acceptable range.

✔ If a person eats 12 milligrams of salt in a day and excretes 11 milligrams of it in the urine, what happened to the remaining 1 milligram?

▶ THE SCIENCE OF PHYSIOLOGY

How do we know what we know about the physiology of the human body? Observation and experimentation are the key elements of **scientific inquiry.** An investigator observes an event or phenomenon and, using prior knowledge, generates a **hypothesis** [*hypotithenai*, to assume], or logical guess, about how the event takes place. The next step is to test the hypothesis by designing an experiment in which the investigator manipulates some aspect of the phenomenon.

A common type of experiment removes or alters some element that the investigator thinks is an essential part of the observed phenomenon. That altered element is the **independent variable.** For example, a biologist notices that birds at a feeder seem to eat more in the winter than in the summer. She generates a hypothesis that cold temperatures cause birds to increase their food intake. To test her hypothesis, she designs an experiment in which she will keep birds at different temperatures and monitor how much food they eat. In her experiment, temperature, the manipulated element, is the independent variable. Food intake, which is hypothesized to be dependent on temperature, becomes the **dependent variable.**

An essential feature of any experiment is a **control.** A control group is usually a duplicate of the experimental group in every respect except that the manipulated variable is not changed from its normal value. For example, in the bird-feeding experiment, the control group would be a set of birds maintained at warm temperatures but otherwise treated exactly like the birds held at cold temperatures. The purpose of the control is to ensure that any observed changes are due to the experimental manipulation and not to an outside factor. For example, if the investigator in the bird experiment changed to a different food, and food intake increased, she could not determine if the increased food intake was due to temperature or to more palatable food unless she had a control group that was also fed the different food.

During an experiment, the investigator carefully collects facts, or **data** [plural; singular *datum,* a thing given], about the effect that the manipulated (independent) variable has on the observed (dependent) variable. Once the investigator feels that she has sufficient information to draw a conclusion, she begins to analyze the data. Analysis can take many forms and usually includes statistical analysis to determine if apparent differences are statistically significant. A common format for presenting data is a graph (see Fig. 1-7 ■, Focus on Graphs).

If one experiment supports the hypothesis that cold causes birds to eat more, then the experiment should be repeated to ensure that the results were not an unusual one-time event. This step is called **replication.** When data support a hypothesis in multiple experiments, the hypothesis may become a **scientific theory.** Most of the facts presented in textbooks like this one are really theoretical **models** that scientists have developed on the basis of the best available experimental evidence. On occasion, new experimental evidence that does not support a current model is presented. In that case, the model must be revised to fit the available evidence. Thus, although you may learn a physiological "fact" while using this textbook, in 10 years that fact may be inaccurate because of what scientists have discovered in the interval. For example, in 1970 students learned that the cell membrane was a "butter sandwich," composed of two layers of proteins with a layer of fats sandwiched between. But in 1972 scientists presented a very different model of the membrane, one in which globules of proteins float within a double layer of fats.

Where do our scientific models come from? We have learned much of what we know about human physiology from experiments on animals ranging from squid to rats. Experiments using human subjects are difficult to perform for many reasons. For one thing, there is tremendous variability within human populations. Although we talk in physiology about the *average* values for various parameters such as blood pressure, these average values simply represent a number that falls somewhere near the middle of a wide range of values. Thus, to show significant differences between experimental and control groups in a human experiment, an investigator would, ideally, have to include large numbers of identical subjects.

In human experiments, however, getting two groups of people who are *identical* in every respect is impossible. The researcher then must attempt to recruit subjects who are *similar* in as many aspects as possible. You may have seen advertisements in newspapers requesting volunteers: "Healthy males between 18 and 25, nonsmokers, within 10% of ideal body weight, to participate in a study. . . . " The variability inherent in even a select group of humans must be taken into account when doing experiments with human subjects, because variability may affect the researcher's ability to accurately interpret the significance of data collected on that group.

One way to reduce variability within a test population, whether human or animal, is to do a **crossover study.** In a crossover study, each animal acts both as experimental animal and control. Thus, the individual's response to the treatment can be compared with the individual's own control value. This method is particularly effective when there is wide variability within a population. For example, in a test of blood pressure medication, a group of subjects might be divided into two groups. Group A takes an inactive substance (**placebo,** from the Latin for "I shall be pleasing") for the first half of the experiment, then changes to the experimental drug for the second half. Group B starts with the experimental drug, then changes to the placebo. This scheme allows the researcher to assess the effect of the drug on each individual subject. Statistically, the resulting data can be analyzed using different methods that look at the changes in each individual rather than at changes in the collective group data.

Another significant variable in human studies is the psychological aspect of administering a treatment. If you give someone a pill and tell the person that it will have a certain effect, there is a strong possibility that it will have that effect, even if the pill contains only sugar or an inert substance. This is a well-documented phenomenon called the **placebo effect,** and it shows the ability of our minds to regulate the physiological functioning of our bodies.

In setting up an experiment with human subjects, we must try to control the placebo effect. The simplest way to do this is with a **blind study,** in which the subject does not know whether he or she is receiving the drug or the placebo. Even this can fail, however, if the researcher who is assessing the subject knows which type of treatment the subject is receiving. The researchers' expectations of what the treatment will or will not do may color their measurements or interpretations. To avoid this outcome, researchers often use **double-blind studies,** in which a third party, not involved in the experiment, is the only one who knows which group is receiving the experimental treatment and which group is receiving the control treatment. The most sophisticated experimental design for minimizing psychological effects is the **double-blind crossover study,** in which the control group in the first half of the experiment becomes the experimental group in the second half, and vice versa.

Finally, ethical questions arise when humans are used as experimental subjects, particularly when the subjects are people suffering from a disease or other illness. Is it ethical to withhold a new and promising treatment from the control group? A noteworthy example occurred a few years ago when researchers were testing the efficacy of a treatment for dissolving blood clots in heart attack victims. The survival rate among the treated patients was so much higher that testing was halted so that control groups could also be given the experimental drug. In contrast, tests on some anticancer agents have shown that the experimental treatments were less effective in stopping the spread of cancer than were the standard treatments used by the controls. Was it ethical to undertreat the experimental patients by depriving them of the more-effective current medical practice?

The difficulty of using human subjects is one reason that scientists use animals to develop many of our scientific models. However, not all results of studies done on other mammals such as mice or dogs can be applied to humans. For example, an antidepressant that had been used safely by Europeans for years was undergoing stringent testing required by the U.S. Food and Drug Administration before it could be sold in this country. When beagle dogs were given the drug for a period of months, the dogs started dying from heart problems. Scientists were alarmed until further research showed that beagles have a unique genetic makeup that causes them to break down the drug into a more toxic substance. The drug was perfectly safe in other kinds of dogs and in humans, and it was subsequently approved for human use.

Since the 1970s, physiological research has turned increasingly to techniques developed by cellular biologists and molecular geneticists. As we have come to understand the fundamentals of how chemical signals in the body are received and interpreted by the cells, we have unlocked the mysteries of many processes. In doing so, we also have come closer to being able to treat many diseases by correcting their cause rather than simply alleviating their symptoms.

FOCUS ON...Graphs

Graphs are pictorial representations of the relationship between two (or more) different variables, plotted in a rectangular region (Fig. 1-7a ■). Most graphs that you will encounter in physiology represent data as lines, bars, or dots. Three typical types of graphs are shown in Figure 1-7 ■. All three types have several features in common. The independent variable that was manipulated by the experimenter is graphed on the horizontal *x*-axis. Examples shown include (b) an experiment with measurements taken every day, (c) an experiment conducted at three different temperatures, and (d) an experiment in which the subjects lifted weights to increase their muscle mass. The dependent variable is the variable that was measured by the experimenter. Dependent variable values are plotted on the vertical *y*-axis. In graph (b), the dependent variable is body weight; in (c), it is the amount of food consumed each day, and in (d), it is the percentage of body fat in the subjects. If the experimental design is valid and the hypothesis is correct, changes in the dependent variable will vary with changes in the independent variable.

Each axis of the graph is divided into intervals represented by evenly spaced tick marks on the axis. A label on each axis tells what the axis represents (time, temperature, amount of food consumed) and in what units it is marked (days, degrees Celsius, grams per day). Each

■ Figure 1-7 Graphs

(a) The standard features of a graph include units and labels on the axes, a key, and a figure legend.

(b) Line graph. The *x*-axis frequently represents time; the points represent individual observations. The points may be connected by lines, in which case the slope of the line between two points shows the rate at which the variable changed.

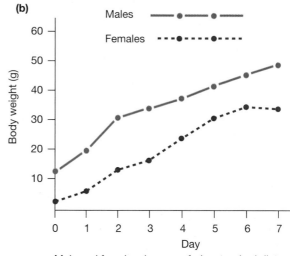

Male and female mice were fed a standard diet and weighed daily.

More and more, medicine is turning to therapies based on interventions at the molecular level. A classic example is the treatment of cystic fibrosis, an inherited disease in which the mucus of the lungs and digestive tracts is unusually thick. For many years, patients with this condition had few treatment options, and most died at a young age. However, basic research into the mechanisms by which cells move salt and water across their membranes provided clues to the underlying cause of cystic fibrosis: a defective protein in the membrane of certain cells. Once molecular geneticists found the gene that coded for that protein, the door was opened to the possibility of replacing the defective gene in cystic fibrosis patients with the gene for the normal protein. But without the basic research into how cells and tissues carry out their normal function, this cure would never have been developed.

As you read this book and learn what we know currently about how the human body works, keep in mind that many of the facts presented in it are just theories. There still are many questions in physiology waiting for investigators to find the answers.

graph has a legend that tells the reader what the graph represents. In graph (b), two study groups (males and females) are plotted on a single graph. A key shows which line corresponds to which group. The graphs show three different ways of plotting data: line graph, scatter plot, and bar graph.

Here are some questions to ask when you are trying to extract information from a graph:

1. What does each axis represent?
2. What is the relationship between the parameters represented by the axes? This can usually be expressed by substituting the labels on the axes into the fol-

lowing statement: *y* varies with *x*. For example, in graph (b), body weight varies with time. What are the relationships shown in graphs (c) and (d)?

3. Are any trends shown in the graph? For line graphs and scatter plots, is the line horizontal (no change in the dependent variable when the independent variable changes), or does it have a slope? For bar graphs, are the bars the same or different heights? Is there a trend in the direction of change if they are different heights? Draw a conclusion about each graph on the basis of what you observe.

(c) Bar graph. Each bar shows a fixed number or amount of a variable. The bars are lined up side by side along one axis so that they can be easily compared with one another. Scientific bar graphs traditionally have the bars running vertically.

(c)

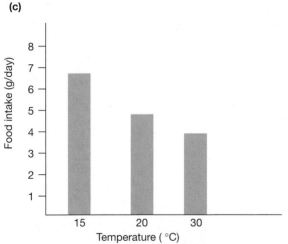

Canaries were acclimated to the temperatures shown, and their food intake was monitored for three weeks.

(d) Scatter plot. Each point represents one member of a test population. The individual points of a scatter plot are never connected by lines, but a best fit line may be estimated to show a trend in the data, or better yet, the line may be calculated by a mathematical equation.

(d)

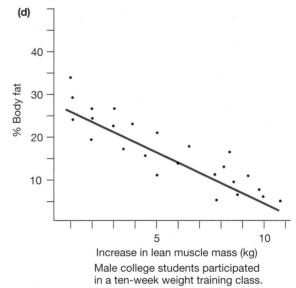

Male college students participated in a ten-week weight training class.

CHAPTER REVIEW

SUMMARY

1. **Physiology** is the study of the normal functioning of a living organism and its component parts. (p. 1)

Levels of Organization

2. Physiologists study many **levels of organization** of living organisms, from molecules to populations. (p. 2)

3. The **cell** is the smallest unit of structure capable of carrying out all life processes. (p. 2)

4. Collections of cells that carry out related functions group into **tissues** and **organs.** (p. 2)

5. The human body has 10 physiological organ systems: **integumentary, musculoskeletal, respiratory, digestive, urinary, immune, circulatory, nervous, endocrine,** and **reproductive.** (p. 2)

6. Material in the **lumens** of organs opening to the external environment is not part of the **internal environment** until it crosses the tissue wall of the organ. (p. 3)

Physiology Is an Integrative Science

7. Physiology is an integrative science that studies how the different systems of the body coordinate in the intact animal. (p. 4)

Function and Process

8. The **function** of a physiological system or event is the "why" of the system; the **process** by which events occur is the "how" of a system. A **teleological approach** to physiology explains why events happen; the **mechanistic approach** explains how they happen. (p. 4)

The Evolution of Physiological Systems

9. The human body as a whole is adapted to cope with a variable external environment, but most individual cells of the body are much less tolerant of change. For this reason, the body attempts to keep the **extracellular fluid** within a narrow range of values. (p. 4)

Homeostasis

10. **Homeostasis** is the body's coordinated response to change in order to maintain internal stability. (p. 6)

11. Failure to maintain homeostasis results in disease. (p. 6)

Themes in Physiology

12. Key themes in physiology include homeostasis, the integration of body systems, cell-to-cell communication, information transfer across the cell **membrane, compartmentation,** energy use, and the **law of mass balance.** (p. 7)

The Science of Physiology

13. Observation and experimentation are the key elements of **scientific inquiry.** A **hypothesis** is a logical guess about how an event takes place. (p. 8)

14. In scientific experimentation, the factor manipulated by the investigator is the **independent variable.** The observed factor is the **dependent variable.** All well-designed experiments have a **control** to ensure that observed changes are due to the experimental manipulation and not to an outside factor. (p. 8)

15. **Data,** the facts collected during the experiment, are analyzed and presented, often as a graph. (p. 8)

16. A **scientific theory** is a hypothesis that has been supported by data on multiple occasions. Sometimes new experimental evidence is presented that does not support a theory or **model,** and the model must be revised. (p. 8)

17. Animal experimentation is an important part of learning about human physiology because there is a tremendous amount of **variability** within human populations and because it is difficult to control human experiments. In addition, ethical questions arise when using humans as experimental animals. (p. 9)

18. In a **crossover study,** each subject acts both as experimental subject and control. For half the experiment, the subject takes an inactive substance known as a **placebo.** One difficulty in human experiments is the **placebo effect,** in which changes take place even if the treatment is inactive. (p. 9)

19. In a **blind study,** the subjects do not know whether they are receiving the experimental treatment or a placebo. In a **double-blind study,** a third party removed from the ex-

periment is the only one who knows which group receives the experimental treatment and which group receives the control treatment. In a **double-blind crossover study,** the control group in the first half of a double-blind experiment becomes the experimental group in the second half, and vice versa. (p. 9)

QUESTIONS

LEVEL ONE   Reviewing Facts and Terms

1. Define physiology. Describe the relationship between physiology and anatomy.

2. Name the different levels of organization in the biosphere and the disciplines that study different levels.

3. Name the 10 systems of the body and give their major function(s).

4. What is meant by the phrase "Physiology is an integrative science"?

5. Define homeostasis. What aspects of human function are maintained homeostatically?

6. Name six major themes in physiology.

LEVEL TWO   Reviewing Concepts

7. **Mapping exercise:**

   (a) Make a map of the law of mass balance using the following phrases. You may use terms more than once.

body contents	external environment	lungs
body synthesis	input	output
digestive system	kidneys	skin

   (b) Make a large map showing the organization of the human body. Use all levels of organization in the body (see Fig. 1-1 ■) and all 10 organ systems. Try to include functions of all components on the map and remember that some structures may share functions. (Hint: Start with the human body as the most important term. You may also draw the outline of a body and make your map using it as the basis.)

8. Distinguish between the following groups of terms:

   (a) tissues and organs

   (b) x-axis and y-axis on a graph

   (c) dependent and independent variables

   (d) teleological and mechanistic approaches

   (e) the internal and external environments for a human

   (f) blind, double-blind, and crossover studies

9. Name as many organs or body structures that have a lumen as you can. Do all of these lumens connect directly with the external environment? Which ones do not?

10. Which systems are responsible for coordinating body function? For protecting the body from outside invaders? Which systems exchange material with the external environment and what do they exchange?

LEVEL THREE   Problem Solving

11. A group of biology majors went to a mall and asked passersby, "Why does blood flow?" These are some of the answers they received. Which answers are teleological and which are mechanistic? (Not all answers are correct, but they can still be classified.)

    (a) Because of gravity

    (b) To bring oxygen and food to the cells

    (c) Because if it didn't flow, we would die.

    (d) Because of the pumping action of the heart

12. Although dehydration is one of the most serious physiological obstacles that animals on land must overcome, there are others. Think of as many as you can, and think of different ways that different terrestrial animals have to overcome these obstacles. (Hint: Think of humans, insects, and amphibians; also think of as many different terrestrial habitats as you can.)

LEVEL FOUR   Quantitative Problems

13. A group of students wanted to see what effect a diet deficient in vitamin D would have on the growth of baby guppies. They fed the guppies a special diet with low vitamin D and measured their length every third day for three weeks. The data they collected are in the table below:

Day	0	3	6	9	12	15	18	21
Average length	6 mm	7 mm	9 mm	12 mm	14 mm	16 mm	18 mm	21 mm

(a) What is the dependent variable and what is the independent variable in this experiment?

(b) What was the control in this experiment?

(c) Make a fully labeled graph with a legend, using the data in the table.

14. You performed an experiment in which you measured the volumes of slices of potato, then soaked the slices in solutions of varying salinities for 30 minutes. At the end of 30 minutes, you again measured the volumes of the slices. The results are shown below.

% change in volume after 30 minutes

Solution	*Sample 1*	*Sample 2*	*Sample 3*
Distilled water	+10%	+8%	+11%
1% salt (NaCl)	0%	+0.5%	−1%
9% salt (NaCl)	−8%	−12%	−11%

(a) What is the independent variable in this experiment? What is the dependent variable?

(b) Was there a control in this experiment or can you tell from the information given? If there was a control, what was it?

(c) Graph the results of the experiment using the most appropriate type of graph.

15. If a woman weighs 121 pounds and 51% of her total body weight is water, what is her total body water volume in liters? Assume that 2.2 pounds = 1 kilogram, and that 1 kilogram of water = 1 liter. Use the correct abbreviations for these units in your answer.

E X P L O R E <MediaLab>

Introduction

This is the first in a series of MediaLabs that you will find at the end of each chapter. Two Web Explorations are included within each MediaLab. To carry out the Explorations, you will need to go to the Companion Website for the text at

http://www.prenhall.com/silverthorn

These Web Explorations will help you enhance, expand, and assess your understanding of the material presented in the chapter. A major part of learning is building and testing mental models of the information you study. This is particularly relevant for physiology because physiology deals with the dynamics of how the body works. Some of the Web Explorations will provide visual reinforcement for your mental models. Others will ask you to use your model to critically analyze and assess short scientific papers, while others will provide historical perspective on the topic.

Web Exploration 1

Estimated time for completion = 5 minutes
The first Web Exploration is an example of how we will use visual models on the Web to help reinforce and refine your mental models of basic themes in physiology. Select the keyword **SCIENTIFIC METHOD** on the Website. This site shows you a flow diagram depicting the process involved in the scientific method. You will use this same method to develop your mental models of physiological systems.

Let's use the skeletal muscle system as an example. Write down three basic assumptions you currently have about how a muscle works—this will be your hypothesis. You have not yet studied the skeletal muscle system but you have some idea of how muscles work from your everyday experiences. The ideas you currently hold on this topic can either help or impede your future understanding of the subject.

Now, list three resources that you might use to gather information about the function of muscles. Use these resources to develop a new hypothesis to explain how a muscle works. Next, list two ways that you can test your hypothesis. Based on the data input you get from these sources, modify your understanding.

As you will see, effective learning uses the same processes as the scientific method. Learning requires constant assessment and refinement of your knowledge. To complete this exercise, visit MediaLab Web Exploration 1 in Chapter 1 of your Companion Website.

Web Exploration 2

Estimated time for completion = 10 minutes
Physiology consists of many recurring themes. One of the major themes you will see in every chapter on physiological systems is that of homeostasis. What does the word **homeostasis** mean? Select the keyword **WORDSMYTH** on your Companion Website and look up the definition of "homeo" and "stasis." (Rather than just memorizing the words, break the word down to its root words and understand their meanings.)

Homeostasis refers to the process of "maintaining an internal stability in the body." You use the process of homeostasis to maintain your speed while driving a car. On a piece of paper write down what you use as a sensor in this driving example. What do you do if you are going too fast? What if you are going too slowly? How often are you traveling exactly at the speed limit? Select the keyword **HOMEOSTASIS** on the Website and compare your answers with those used by scientists. To complete this exercise, visit MediaLab Web Exploration 2 in Chapter 1 of your Companion Website.

2 Atoms, Ions, and Molecules

■ *"One of the things that distinguishes ours from all earlier generations is this, that we have seen our atoms."*
—*Karl Kelchner Darrow* ■

CHAPTER OUTLINE

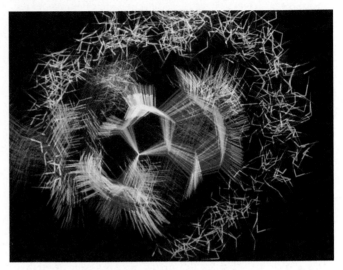

Protein motion measured using a fluorescent probe

Before we embark on our study of the human body, it is important that you understand its smallest particles. So we begin by looking at the atom and its three fundamental subunits: the proton, the neutron, and the electron. Then we examine how different atoms link together to form molecules. Because many of the molecules in the human body are dissolved in liquid to form solutions, we will study how to make solutions and express the concentration of the molecules dissolved in them. Finally, we will study the biologically important molecules or biomolecules: carbohydrates, lipids, proteins, and nucleotides.

▶ ELEMENTS AND ATOMS

The simplest kinds of matter are called **elements**. Some are substances that you encounter in everyday life: the oxygen that you breathe, the carbon in the soot of a barbecue grill. Elements are the building blocks of all matter, including the human body. Three of the elements—oxygen, carbon, and hydrogen—make up more than 90% of the body's mass. These elements and others bind together into more complex substances called molecules. To understand what distinguishes one element from another and how they combine into molecules, we will first look at atoms.

PROBLEM
Chromium Supplements

"Lose weight while gaining muscle," the ads promise. "Prevent heart disease." "Stabilize blood sugar." What is this miracle substance? It's chromium picolinate, a nutritional supplement being marketed to consumers looking for a quick fix. But does it work? Some athletes, like Stan—the star halfback on the college football team—swear by it. Stan takes 500 micrograms of chromium picolinate daily. Many researchers, however, are skeptical and feel that the necessity for and safety of chromium supplements has not been established.

...continued on page 18

• •

Atoms Are Composed of Protons, Neutrons, and Electrons

All elements are composed of small particles called **atoms,** once thought to be the smallest particles of matter [*atomos,* indivisible]. We now know, however, that an atom itself is composed of three types of even smaller particles: protons, neutrons, and electrons. These subatomic particles are distinguished from each other by their different electrical charges and masses. A **proton** has one positive charge (written +), an **electron** has one negative charge (written −), and a **neutron** has no charge, or zero charge—that is, it is neutral. Like the poles of a magnet, oppositely charged atomic particles (+ and −) attract, while those with like charge (+ and +, or − and −) repel. An atom usually has an equal number of protons and electrons, giving it an overall electrical charge of zero.

Protons and neutrons have nearly equal masses of about 1 **atomic mass unit** or **amu** (1 amu = 1.6605×10^{-27} kg). Electrons are much smaller: it takes the mass of 1836 electrons to equal the mass of one neutron. The arrangement of protons, neutrons, and electrons in an atom is always the same. The protons and neutrons of the atom, which account for almost all of its mass, are clustered into a dense body called the **nucleus** [*nuculeus,* little nut; plural, nuclei]. The space around the nucleus, almost all of the atom's volume, is occupied by the rapidly moving, lightweight, negatively charged electrons, held in their orbit by the positively charged protons (Fig. 2-1 ■).

Atoms are very tiny, having diameters in the range of 1 angstrom to 5 angstroms (1 angstrom = 1 Å = 10^{-10} m). Most of their volume is empty space. To get a feel for how little of an atom is occupied by the subatomic particles, imagine that the atom has a diameter equal to the length of a football field. The nucleus would then be the size of a large apple in the center of the field, and tiny electrons, smaller than a pea, would move through the remainder of the space.

If all atoms have protons, neutrons, and electrons, what makes an atom of one element different from that

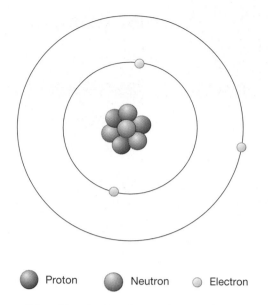

Proton   Neutron   Electron

■ **Figure 2-1   Atomic structure**   An atom has a central nucleus composed of protons and neutrons. The nucleus is surrounded by rapidly moving, lightweight electrons, one for each proton in the nucleus.

of another? For example, why is carbon so unlike oxygen? The next three sections will answer those questions by showing what happens to an atom when it gains or loses one of its subatomic particles.

An Element Is Distinguished by the Unique Number of Protons in Its Nucleus

There are more than 100 different elements in the world, about 90 of which occur naturally. The remainder are manufactured. All of the elements that we know of are listed in a **periodic table of the elements,** as shown in Figure 2-2 ■. Each box in the table contains the name of an element and its one- or two-letter symbol.

Only about one-fourth of the elements are commonly encountered in biological systems. Those that are necessary for life are called **essential elements.** Eleven **major essential elements** are found in the human body in fairly large amounts. Carbon (C), hydrogen (H), oxygen (O), and nitrogen (N) are by far the most common essential elements because they make up most biological molecules. The essential elements that are found in very low concentrations in living tissues are called **trace elements,** or minor essential elements. All of the trace elements are essential in some way to cell function, and the list of such elements continues to change as scientists discover more about how cells work. The need for trace elements is evident only when they are absent from a person's diet or present in insufficient quantities. One trace element that has received a lot of attention in recent years is chromium. A deficiency of chromium in the body has been postulated (but not proven) to play a role in the development of the metabolic disease diabetes mellitus ("sugar diabetes") in some people.

Periodic Table of the Elements

Key:
- Atomic number = number of protons — 6
- Name — Carbon
- Symbol — C
- Atomic mass — 12.0

Legend:
- ☐ Major essential elements
- ☐ Minor essential elements
- ☐ Not believed essential for life

Transitional metals (Groups 3–12)

Group 1	2	3	4	5	6	7	8	9	10	11	12	13	14	15	16	17	18
Period 1 1 Hydrogen **H** 1.0																	2 Helium **He** 4.0
Period 2 3 Lithium **Li** 6.9	4 Beryllium **Be** 9.0											5 Boron **B** 10.8	6 Carbon **C** 12.0	7 Nitrogen **N** 14.0	8 Oxygen **O** 16.0	9 Fluorine **F** 19.0	10 Neon **Ne** 20.2
Period 3 11 Sodium **Na** 23.0	12 Magnesium **Mg** 24.3											13 Aluminum **Al** 27.0	14 Silicon **Si** 28.1	15 Phosphorus **P** 31.0	16 Sulfur **S** 32.1	17 Chlorine **Cl** 35.5	18 Argon **Ar** 39.9
Period 4 19 Potassium **K** 39.1	20 Calcium **Ca** 40.1	21 Scandium **Sc** 45.0	22 Titanium **Ti** 47.9	23 Vanadium **V** 50.9	24 Chromium **Cr** 52.0	25 Manganese **Mn** 54.9	26 Iron **Fe** 55.8	27 Cobalt **Co** 58.9	28 Nickel **Ni** 58.7	29 Copper **Cu** 63.5	30 Zinc **Zn** 65.4	31 Gallium **Ga** 69.7	32 Germanium **Ge** 72.6	33 Arsenic **As** 74.9	34 Selenium **Se** 79.0	35 Bromine **Br** 79.9	36 Krypton **Kr** 83.8
Period 5 37 Rubidium **Rb** 85.5	38 Strontium **Sr** 87.6	39 Yttrium **Y** 88.9	40 Zirconium **Zr** 91.2	41 Niobium **Nb** 92.9	42 Molybdenum **Mo** 95.9	43 Technetium **Tc** (98)	44 Ruthenium **Ru** 101.1	45 Rhodium **Rh** 102.9	46 Palladium **Pd** 106.4	47 Silver **Ag** 107.9	48 Cadmium **Cd** 112.4	49 Indium **In** 114.8	50 Tin **Sn** 118.7	51 Antimony **Sb** 121.8	52 Tellurium **Te** 127.6	53 Iodine **I** 126.9	54 Xenon **Xe** 131.3
Period 6 55 Cesium **Cs** 132.9	56 Barium **Ba** 137.3	57* Lanthanum **La** 138.9	72 Hafnium **Hf** 178.5	73 Tantalum **Ta** 181.0	74 Tungsten **W** 183.9	75 Rhenium **Re** 186.2	76 Osmium **Os** 190.2	77 Iridium **Ir** 192.2	78 Platinum **Pt** 195.1	79 Gold **Au** 197.0	80 Mercury **Hg** 200.6	81 Thallium **Tl** 204.4	82 Lead **Pb** 207.2	83 Bismuth **Bi** 209.0	84 Polonium **Po** (209)	85 Astatine **At** (210)	86 Radon **Rn** (222)
Period 7 87 Francium **Fr** (223)	88 Radium **Ra** 226.0	89** Actinium **Ac** (227)	104 Dubnium **Db** (261)	105 Joliotium **Jl** (262)	106 Rutherfordium **Rf** (263)	107 Bohrium **Bh** (262)	108 Hannium **Hn** (265)	109 Meitnerium **Mt** (268)	110 (269)	111 (272)	112 (277)						

* Lanthanides:

58 Cerium **Ce** 140.1	59 Praseodymium **Pr** 140.1	60 Neodymium **Nd** 144.2	61 Promethium **Pm** (145)	62 Samarium **Sm** 150.4	63 Europium **Eu** 152.0	64 Gadolinium **Gd** 157.3	65 Terbium **Tb** 158.9	66 Dysprosium **Dy** 162.5	67 Holmium **Ho** 164.9	68 Erbium **Er** 167.3	69 Thulium **Tm** 168.9	70 Ytterbium **Yb** 173.0	71 Lutetium **Lu** 175.0

** Actinides:

90 Thorium **Th** 232.0	91 Protactinium **Pa** 231.0	92 Uranium **U** 238.0	93 Neptunium **Np** 237.0	94 Plutonium **Pu** (244)	95 Americium **Am** (243)	96 Curium **Cm** (247)	97 Berkelium **Bk** (247)	98 Californium **Cf** (251)	99 Einsteinium **Es** (252)	100 Fermium **Fm** (257)	101 Mendelevium **Md** (258)	102 Nobelium **No** (259)	103 Lawrencium **Lr** (260)

Note: Numbers in parentheses are mass numbers (the total number of protons and neutrons in the nucleus) of most stable or best-known isotope of radioactive elements.

Modern name	Latin name	Symbol
Copper	Cuprium	Cu
Iron	Ferrum	Fe
Potassium	Kalium	K
Sodium	Natrium	Na

■ Figure 2-2 **Periodic table of the elements**

The atoms of different elements are distinguished from one another by the fact that each element has a unique number of protons in its nucleus. For example, the lightest element, hydrogen, has only one proton in its nucleus. The next element, helium (He), has two protons, followed by lithium (Li) with three protons, and so on. The number of protons in one atom of an element is called the **atomic number** of the element. The periodic table is arranged by atomic number, starting at hydrogen, with an atomic number of 1, and ending with meitnerium (Mt), with an atomic number of 109.

The average **atomic mass** of an element is shown below the symbol of the element. Atomic mass is calculated by adding the mass of the protons and neutrons in one atom of the element. For example, one atom of carbon has six protons and six neutrons. Because the mass of each proton or neutron is one atomic mass unit, carbon's atomic mass is 12 amu. However, in the periodic table, not all atomic masses are whole numbers. For example, the box for chlorine (Cl), with 17 protons in its nucleus, shows an atomic mass of 35.5. This is because chlorine comes in two forms, one with 18 neutrons and one with 20. The atomic mass in the periodic table is a weighted average of the different forms of the element. In the next section we will discuss these different forms, which are known as isotopes.

Use the periodic table of the elements in Figure 2-2 ■ to answer the following questions:

✔ Which element has 30 protons?

✔ How many electrons does one atom of sodium (Na) have?

✔ How many electrons does one atom of calcium (Ca) have?

✔ What is the average atomic mass of nitrogen?

✔ What element is illustrated by the atom in Figure 2-1 ■?

...continued from page 16

What is chromium picolinate? Chromium (Cr) is an essential element that is linked to normal glucose metabolism. In the diet, chromium is found in brewer's yeast, broccoli, mushrooms, and apples. Chromium in food and in chromium chloride supplements is poorly absorbed from the digestive tract, so a scientist developed the compound chromium picolinate. Picolinate, derived from amino acids, enhances chromium uptake at the intestine. The recommended daily allowance (RDA) of chromium is 50 micrograms. As we've seen, Stan takes several times this amount.

Question 1:   *Locate chromium on the periodic table of the elements. What is its atomic number? Atomic mass? How many electrons does one atom of chromium have? Which elements close to chromium are also essential elements?*

Isotopes of an Element Are Atoms with Different Numbers of Neutrons

Different numbers of protons are what distinguishes one element from the next. But, though the number of protons in an atom of any given element is constant, the number of neutrons may differ (Fig. 2-3 ■). Atoms of one element that have different numbers of neutrons are known as **isotopes** of the element [*iso-*, same + *topos*, place (that is, the same place in the periodic table)]. The symbol for an isotope shows the atomic mass number written to the upper left of the chemical letter symbol. For example, ^1H, hydrogen-1, is the most common naturally occurring isotope of hydrogen. It has a single proton in its nucleus (Fig. 2-4 ■). But hydrogen has two other isotopes, hydrogen-2, or *deuterium* (^2H), and hydrogen-3, or *tritium* (^3H), which are formed by the addition of one or two neutrons to the nucleus. The different iso-

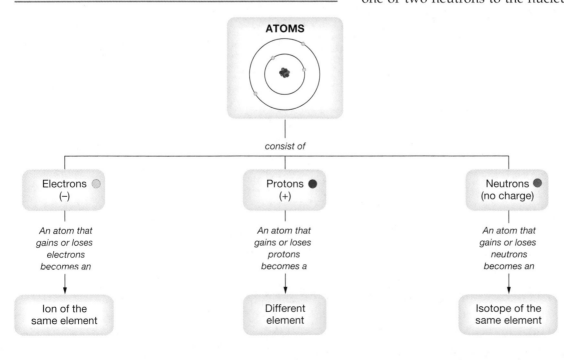

■ Figure 2-3 The relationship between elements, ions, isotopes, and atoms

 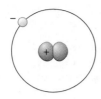 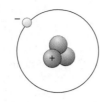

(a) Hydrogen-1, H   **(b)** Hydrogen-2, deuterium, ^{2}H   **(c)** Hydrogen-3, tritium, ^{3}H

■ **Figure 2-4   Hydrogen atoms**   (a) A typical hydrogen atom has one proton in the nucleus and one orbiting electron, but no neutrons. (b) A neutron has been added to the nucleus, creating the isotope hydrogen-2, or ^{2}H, also known as deuterium. (c) When a second neutron is added to the nucleus, the new isotope is hydrogen-3, (^{3}H), or tritium.

topes of a single element have the same chemical properties, because these properties are affected by the electron configuration and not the number of neutrons.

Isotopes are important in biology and medicine, because some are unstable and emit energy. These unstable isotopes are called **radioisotopes,** and the energy they emit is called **radiation.** Every element has at least one radioisotope, most of which are produced artificially by using high-energy particle accelerators to "shoot" extra neutrons into a stable nucleus. A small number of elements, such as carbon and uranium, have naturally occurring radioisotopes. Both artificial and natural radioisotopes have medical uses. For example, radioisotopes such as thallium-201 (^{201}Tl) produce images of the inside of the body for *diagnostic procedures* (those that identify disease). Radioisotopes are also used to treat cancer, an example of *therapeutic use.* The use of radioisotopes in the diagnosis and treatment of disease is a specialty known as *nuclear medicine.*

✔ What are the atomic masses (in amu) of the three hydrogen isotopes shown in Figure 2-4 ■?

Electrons Form Bonds Between Atoms and Capture Energy

Electrons, the third subatomic particle, play a significant role in physiology. The arrangement of electrons around the nucleus determines the ability of an atom to bind with other atoms and form molecules. The gain or loss of electrons changes the electrical charge on an atom and plays an essential role in communication throughout the body. Finally, electrons on certain atoms can capture energy from their environment and transfer it to other molecules, so that the energy can be used for movement and other life processes.

The electrons in an atom do not move around the nucleus at random. Instead, they are arranged in a series of **energy levels,** or **shells.** The lowest energy level, designated as level 1, is closest to the nucleus. As distance from the nucleus increases, energy levels increase. Each energy level is divided into sublevels called, for

Radioisotopes and Nuclear Medicine   Of the 2000 known isotopes of the elements, unstable isotopes that emit radiation outnumber the stable isotopes four to one. Some of these isotopes occur naturally, but many have been made artificially, including all those with atomic numbers of 84 or higher. Radioisotopes emit three types of radiation: alpha (α), beta (β), and gamma (γ). Alpha and beta radiation are composed of fast-moving particles (protons + neutrons, or electrons) ejected from an unstable atom. Gamma radiation is composed of high-energy waves rather than particles. Gamma radiation is able to penetrate matter much more deeply than α or β particles. X-rays are similar to gamma radiation and are used to create images of the body's interior.

Radioisotopes are widely used in medicine for diagnosis and treatment. In some instances, the radioisotope is used by itself. For example, ^{131}I, a radioactive isotope of ^{127}I, is useful for diagnosis and treatment of thyroid gland disorders. ^{131}I emits both α and β radiation. In small doses it does little damage, but larger doses can be administered to selectively destroy thyroid tissue when the gland is overactive. Another widely used radioisotope for diagnostic procedures is technetium-99, an element that can be bound to carrier molecules that target it to specific tissues.

historical reasons, *s, p, d,* and *f.* Each of these sublevels may be further subdivided into a certain number of **orbitals,** each of which holds a maximum of two electrons (Fig. 2-5 ■).

For example, the first energy level or shell, 1, has one sublevel, 1*s*, with one orbital. Thus, the first energy level can hold a total of two electrons. The second shell, 2, has two sublevels (2*s* and 2*p*) with a total of four orbitals, so it can hold a total of eight electrons.

If an energy sublevel has more than one orbital, each orbital is assigned one electron before any orbital gets two (Fig. 2-5b ■). The nitrogen atom is a good example of this principle: its highest energy level, 2*p*, has one electron in each of its three orbitals. The most stable electron configuration for an element is one in which all orbitals in the outer shell are filled with electrons. Neon, a completely unreactive element, is an example of an element whose outer electron shell is totally filled (Fig. 2-5b ■).

Electrons are able to absorb energy from their environment and "jump" up one or more energy levels to a higher orbital (Fig. 2-5c ■). High-energy electrons are not terribly stable, so they eventually fall back to their stable orbital position, emitting energy as they do so. The released energy can be captured by other atoms or may be emitted as radiation. The bioluminescence of fireflies is visible light emitted by high-energy electrons returning to their normal low-energy state. When biologists speak of "high-energy electrons" that capture and transfer energy in the cell, they mean the electron in a hydrogen atom that moves out of its 1*s* orbital into an unoccupied higher orbital.

(a)

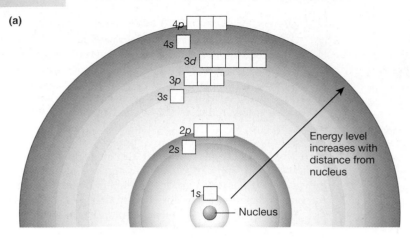

■ **Figure 2-5** **Energy levels of electrons in an atom** (a) The first eight energy levels of electrons in an atom. The lowest energy level, 1s, is closest to the nucleus. Within an energy level, the electrons move in orbitals, represented by boxes. (b) The electron configurations of some of the lighter elements. Each arrow represents one electron and each box represents an orbital. Within an energy level (e.g., 2p), each orbital gets one electron before any orbital gets two. In the most stable configuration, all occupied orbitals have two electrons. (c) The four possible energy levels for a hydrogen electron. When an electron absorbs energy from its environment, the energized electron "jumps" up one or more energy levels. The absorbed energy is released as light or mechanical energy when the electron falls back to a lower energy level.

(b)

Element	Total electrons	Orbital diagram

(c)

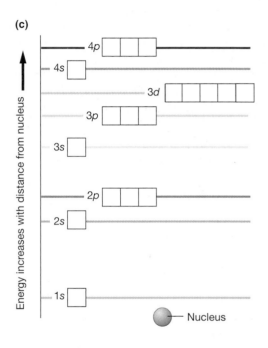

When an orbital is filled with two electrons, the electrons are said to be paired. Paired electrons are stable. An atom having an unpaired electron has a higher probability of losing the unpaired electron to another atom or gaining an unpaired electron from another atom. In some cases, two atoms with unpaired electrons share the electrons. When electrons are transferred or shared in that manner, each participating atom ends up with a stable orbital of paired electrons. The transfer and sharing of electrons brings us to our next topic: the bonds formed between atoms.

❱ MOLECULES AND BONDS

The transfer or sharing of electrons is a critical part of forming **bonds,** or links between atoms. When two or more atoms link by sharing electrons, they make units known as **molecules.** Some molecules, such as those of oxygen (O_2) and nitrogen (N_2) gas, are made of two atoms of the same element. But most molecules contain more than one element and are known as **compounds.** For example, although oxygen gas (O_2) is a molecule, it is not a compound because it has only oxygen atoms. In contrast, glucose, a molecule composed of carbon, hydrogen, and oxygen, is a compound.

What Does the Chemical Formula of a Molecule Tell Us?

The **chemical formula** of a molecule is a form of notation that describes the type and number of atoms in the molecule. Each element in the molecule is represented by its chemical symbol followed by a numeric subscript showing the number of atoms of the element. Thus water, which has two hydrogen atoms and one oxygen atom, is written as H_2O. Glucose is written $C_6H_{12}O_6$, meaning it has six atoms each of carbon and oxygen, and twelve atoms of hydrogen.

Imaging in Medicine When X-rays were discovered in 1895 by Wilhelm Roentgen, they created such a stir that they influenced all aspects of society, not just medicine. Marcel Duchamp's famous painting of *Nude Descending a Staircase* was probably inspired by the skeletal interior revealed by these high-energy emissions. X-rays were our only means of visualizing the interior structures of the body until the 1950s, when ultrasound was introduced. Ultrasound is a technique similar to sonar in the ocean, in which an image is created from sound waves that are reflected as they bounce off tissues and fluid. In the 1970s, computerized tomography (CT scans), positron emission tomography (PET scans [*tomos*, a section]), and magnetic resonance imaging (MRI) came into use. CT scans are series of X-rays that create images of very thin slices of the body. CT images are assembled by computer rather than projected onto a photographic plate as X-rays are, so CT can be used to visualize the three-dimensional structure of an organ. PET scans are made using radioisotopes that emit *positrons:* tiny positively charged particles released when a proton converts into a neutron. Because the body does not normally contain positron-emitting chemicals, these radioisotopes must be administered to a subject before a PET scan can be taken. For MRI scans, the subject is placed inside a strong magnetic field that increases the energy level of the hydrogen nucleus. When the nucleus "jumps" back to its lower energy level, it emits radiation that is captured to create an image of internal structure.

From the chemical formula of a molecule, we can calculate its **molecular weight.** This is the weight of one molecule, expressed in atomic mass units or, more often, in *daltons* (1 amu = 1 Da).* To calculate molecular weight, multiply the atomic mass of each element in the molecule by the number of atoms of the element, then add the results. For example, the molecular weight of glucose, $C_6H_{12}O_6$, would be calculated:

Element	Number of Atoms in the Molecule $\times$	Atomic Mass of the Element	
Carbon	6	12.0 amu	= 72.0
Hydrogen	12	1.0 amu	= 12.0
Oxygen	6	16.0 amu	= 96.0
		SUM	180.0 amu
		or	180 Da

✔ Calculate the molecular weight of water, H_2O.

Although the chemical formula gives the number of atoms in a molecule, it doesn't tell you which atom is linked to which. To understand that aspect of molecular structure, you need to know the different types of bonds

*In recognition of John Dalton, who proposed the modern theory of the atom around 1803.

that atoms can form with each other. There are four common bond types, two strong and two weak.

1. Covalent bonds are strong bonds that result when two atoms share a pair of electrons, one from each atom. Covalent bonds usually require the input of energy to break them apart.
2. Ionic bonds result when an atom has such a strong attraction for electrons that it pulls one or more electrons completely away from another atom.
3. Hydrogen bonds are weak attractive forces between hydrogen atoms and certain other atoms.
4. Van der Waals forces are another type of weak attractive force that acts between all molecules.

We will discuss each type of bond further in the sections below.

Covalent Bonds Are Formed When Adjacent Atoms Share Electrons

Covalent bonds are formed when atoms share one or more pairs of electrons. Each atom contributes one electron to each pair. Most of the bonds that link atoms together into molecules are covalent bonds. It is possible to predict how many covalent bonds an atom will form by knowing how many unpaired electrons the atom has in its outer shell. Figure 2-6 ■ shows the electron configurations of the three most plentiful elements in the body: hydrogen, carbon, and oxygen.

Hydrogen, the lightest element, has one proton in its nucleus and one unpaired electron in its outer shell (Fig. 2-6a ■). Because it has only one electron to share, it always forms a single covalent bond. Chemists represent the single electron by a single dot in the *electron-dot shorthand.*

Carbon, with an atomic number of 6, has six protons in the nucleus and therefore six electrons surrounding the nucleus. Two electrons fill the first electron shell, while the outermost second shell has an additional four electrons (Fig. 2-6b ■). There are four unoccupied places in the second shell, which has space for eight electrons. An atom is most stable when its outer electron shell is filled. This means that carbon, with four of eight possible outer electrons, binds to other atoms to find the four electrons it needs to complete its outer shell. When carbon is sharing all four of its outermost electrons with other elements, it forms four covalent bonds.

Oxygen has eight electrons: two electrons completely fill the first shell and six are found in the eight-place second shell (Fig. 2-6c ■). Because oxygen needs only two additional electrons to complete its outer shell, it can get them by sharing two of its outer electrons in covalent bonds. A water molecule is a simple example of covalent bonds formed by oxygen. In each molecule of water, two hydrogen atoms share their electrons with two electrons in the outer shell of the oxygen atom.

(a) Hydrogen

H·   1 proton

(b) Carbon

·C·   6 protons
             6 neutrons

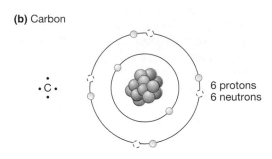

(c) Oxygen

:O: 8 protons
             8 neutrons

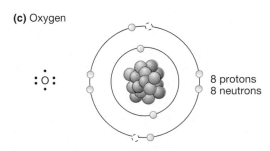

■ **Figure 2-6   Electron configuration of the three most common elements in the body**   An electron-dot model shows the number of electrons found in the outer shell of an atom. Each electron is represented by a single dot. (a) Hydrogen, the lightest element, has only one electron, which occupies the lowest energy level. (b) Carbon has six electrons, two in the first shell and four occupying half of the eight possible places in the second shell. (c) Oxygen has eight electrons, two in the first shell and six in the second. This leaves two unpaired electrons in the second shell. A shell is most stable when completely filled with electrons.

We can use electron-dot shorthand to show how atoms form covalent bonds in molecules. In the next example, one carbon combines with four hydrogens to form a molecule of methane, an odorless gas that is a principal constituent of natural gas. Each hydrogen fills its two-electron outer shell by sharing its one electron. At the same time, the carbon atom fills its eight-place outer shell with the four shared hydrogen electrons. When the outer electron shells are filled, even if it is with shared electrons, the atoms are more stable than if the shells are unfilled.

$$4 \ \text{H·} + 1 \ \text{·C·} \longrightarrow \text{H:C:H} \quad \text{or} \quad \text{H}-\overset{\text{H}}{\underset{\text{H}}{\text{C}}}-\text{H}$$

Methane CH_4

Another form of chemical shorthand shows the paired electrons of a covalent bond as a single line drawn between the chemical symbols for the elements of the molecule. Sometimes adjacent atoms share two pairs of electrons rather than just one. In this case the bond is called a *double bond* and is represented by a double line (═). The following example shows two carbon atoms sharing two pairs of electrons:

$$4 \ \text{H·} + 2 \ \text{·C·} \longrightarrow \text{C::C} \quad \text{or} \quad \text{C=C}$$

Ethylene C_2H_4

Because two of the four carbon electrons are shared in the carbon-carbon bond, this leaves only two electrons per carbon to bond with hydrogen. But if you count the bonds made by the carbons, once again each carbon forms four bonds and each hydrogen forms one. The arrangement of bonds is different but the number of electrons being shared has not changed. The chemical formula for this molecule, called ethylene, is C_2H_4.

✔ You know that the number of electrons present in the outer shell determines the number of covalent bonds an atom can form. Use this fact to decide whether a hydrogen atom can form a double bond. Why or why not?

✔ Why does carbon always form four bonds to other molecules?

✔ Write the chemical formula for this molecule:

Citric acid

✔ What is the molecular weight of the citric acid molecule above?

Functional Groups   There are several types of partial molecules, or **functional groups,** that are commonly found in biological molecules. The atoms in functional groups tend to move from molecule to molecule as a single unit. For example, hydroxyl groups, —OH, common on many biological molecules, are added and removed as a group rather than as single hydrogen or oxygen atoms. Functional groups usually attach to molecules by single covalent bonds. The most common groups are listed in Table 2-1.

Molecular Shape   Molecules can have many shapes. Molecular shape is important because it is closely related to molecular function. Probably the best-known mol-

TABLE 2-1   Common Functional Groups

Notice that oxygen, with two electrons to share, sometimes forms a double bond with another atom.

	Shorthand	Bond Structure
Carboxyl (acid)	—COOH	$-C\begin{smallmatrix}O\\ \\OH\end{smallmatrix}$
Hydroxyl	—OH	—O—H
Amino	—NH$_2$	—N$\begin{smallmatrix}H\\ \\H\end{smallmatrix}$
Phosphate	—H$_2$PO$_4$	$-O-\overset{OH}{\underset{OH}{P}}=O$

ecular shape is the double-helix of DNA, the molecule that stores genetic information. Figure 2-7 ■ shows how atoms are linked in two biological molecules. Notice the carbons linked in a straight chain in the leucine molecule, represented in Figure 2-7a ■. In the glucose molecule (Fig. 2-7b ■), the atoms join together to form a closed circle, or **ring.**

(a) Leucine

(b) Glucose: C$_6$H$_{12}$O$_6$

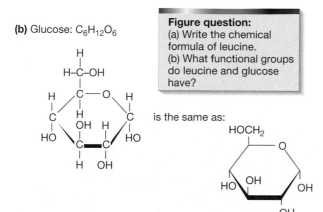

Figure question:
(a) Write the chemical formula of leucine.
(b) What functional groups do leucine and glucose have?

is the same as:

■ **Figure 2-7   Chemical structures and formulas of some biological molecules**   (a) Leucine, an amino acid, is a straight chain of carbon atoms with hydrogens and other functional groups attached. (b) Glucose has a ring of five carbon atoms and one oxygen atom. On the left, the glucose molecule is drawn with all of its carbons and hydrogens shown. In the shorthand version on the right, the carbons at each corner of the ring are omitted, as are the hydrogens attached to the carbons.

The three-dimensional structure of molecules is difficult to show on paper. The angles of the bonds between atoms may cause molecules to take on characteristic shapes. Two well-known examples are the double helix of the DNA molecule and the alpha helix of many proteins. These molecules are shown in Figures 2-16 ■ and 2-18 ■.

Polar and Nonpolar Molecules   Electron pairs in covalent bonds are not always evenly shared between the linked atoms. In some molecules, the nucleus of one atom has a stronger attraction for the shared electrons than the other nucleus does. The electron pair spends more time near the nucleus with the stronger attraction, giving that atom a slightly more negative charge. Meanwhile the atom that is farther from the electrons develops a slightly more positive charge. For both atoms, the change in charge is only a fraction of the charge carried by a single electron or proton. Molecules that develop regions of partial positive and negative charge are called **polar molecules** because they can be said to have positive and negative poles. Certain elements, particularly nitrogen and oxygen, have a strong attraction for electrons and are often found in polar molecules.

A good example of a polar molecule is water (H$_2$O). When the hydrogen atoms form covalent bonds with oxygen, their electrons are pulled away from the hydrogen nuclei toward the larger and stronger nucleus of the oxygen atom. This leaves the two hydrogen atoms of the molecule with a partial positive charge, and the single oxygen atom with a partial negative charge from the unevenly shared electrons (see Fig. 2-8 ■). These partial charges are represented by the lowercase Greek letter delta: δ^+ indicates a positive pole of the molecule, and δ^- indicates a negative pole. Note that the net charge for the entire water molecule is zero.

A **nonpolar molecule** is one whose electrons are distributed so evenly that there are no regions of partial positive or negative charge. For example, molecules composed mostly of carbon and hydrogen tend to form nonpolar bonds. This is because carbon does not attract electrons as strongly as oxygen does.

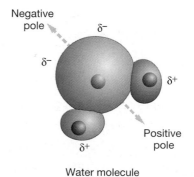

Water molecule

■ **Figure 2-8   Water is a polar molecule**   A water molecule develops separate regions, or poles, of partial positive (δ^+) and partial negative (δ^-) charge when the oxygen nucleus pulls the hydrogen electrons toward itself.

...continued from page 18

One advertising claim for chromium is that it helps glucose—the simple sugar that cells use to fuel all their activities—move from the bloodstream into cells. In the disease diabetes mellitus, cells are unable to take up glucose from the blood efficiently. It seemed logical, therefore, to see if addition of chromium to the diet could enhance glucose uptake in people with diabetes. In one Chinese study, diabetic patients receiving 500 micrograms of chromium picolinate twice a day showed significant improvement in their diabetes but patients receiving 100 micrograms or a placebo did not.

Question 2: *If people have a chromium deficiency, would you predict that their blood glucose level would be higher or lower than normal? From the results of the Chinese study, can you conclude that all people with diabetes have a chromium deficiency?*

The polarity of a molecule is important in determining whether it will dissolve in water. Polar molecules generally dissolve easily. Nonpolar molecules do not.

✔ Based on what you know about the tendency of the following substances to dissolve, predict whether they are polar or nonpolar molecules: table sugar, vegetable oil.

Ionic Bonds Form When Atoms Gain or Lose Electrons

When an atom or molecule completely gains or loses one or more electrons, it becomes an **ion.** Generally, the electrons that are gained have been released by another atom or molecule. Table 2-2 lists the most common ions that you will encounter in physiology. Notice that some of the ions are single atoms, and others are functional groups.

A molecule that gains electrons acquires one negative charge (-1) for each electron added. Such negatively charged ions are called **anions.** By convention, ions are indicated by a superscript notation following the symbol of the element, with the number of charges first and the sign of the charge second. For example, a chloride ion has one extra electron (Cl^-), and a phosphate ion has two (HPO_4^{2-}).

Any molecule that loses electrons has one positive charge ($+1$) for each electron lost. This is because of the protons that are left behind without matching electrons.

TABLE 2-2   Important Ions of the Body

Cations		Anions	
Na^+	Sodium	Cl^-	Chloride
K^+	Potassium	HCO_3^-	Bicarbonate
Ca^{2+}	Calcium	HPO_4^{2-}	Phosphate
H^+	Hydrogen	SO_4^{2-}	Sulfate
Mg^{2+}	Magnesium		

Positively charged ions are called **cations.** Hydrogen loses its only electron and ends up as a single proton, H^+. The ion formed by a calcium atom that loses two electrons is abbreviated as Ca^{2+}.

A chemical bond between a cation and an anion is called an **ionic bond.** Figure 2-9 ■ shows how sodium

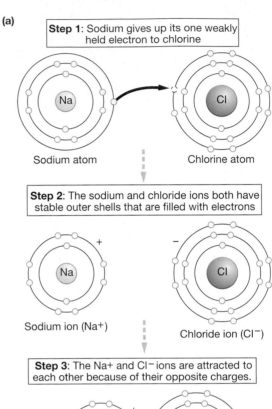

(a)

Step 1: Sodium gives up its one weakly held electron to chlorine

Sodium atom   Chlorine atom

Step 2: The sodium and chloride ions both have stable outer shells that are filled with electrons

Sodium ion (Na^+)   Chloride ion (Cl^-)

Step 3: The Na^+ and Cl^- ions are attracted to each other because of their opposite charges.

Sodium chloride (NaCl) molecule

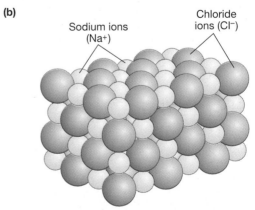

(b)

Sodium ions (Na^+)   Chloride ions (Cl^-)

Sodium chloride (NaCl) crystal

■ **Figure 2-9   Formation of ionic bonds and ions**   (a) Sodium gives up an electron to chlorine to form the sodium and chloride ions, Na^+ and Cl^-. (b) In the solid state, ionic bonds between Na^+ and Cl^- create a NaCl crystal.

and chlorine atoms become ions and how those ions form ionic bonds. A sodium (Na) atom has only one electron in its eight-place outer electron shell, while a chlorine (Cl) atom has seven of the eight electrons needed to fill its outer shell. The less stable sodium atom gives up its electron to chlorine, resulting in a positively charged sodium ion (Na^+) and a negatively charged chloride ion (Cl^-). In table salt, the solid form of NaCl, the opposite charges attract, creating a neatly ordered crystal of alternating Na^+ and Cl^- ions held together by ionic bonds.

✔ The H^+ ion is also called a proton. Why is it given that name?

Hydrogen Bonds Are Weak Bonds Between Molecules or Regions of a Molecule

A **hydrogen bond** is the weak attraction of a hydrogen atom to a nearby oxygen, nitrogen, or fluorine atom. Hydrogen bonds may occur between atoms in neighboring molecules or between atoms in different parts of the same molecule. For example, hydrogen bonds cause large biological molecules such as proteins to fold back on themselves, creating a three-dimensional shape that is essential for the function of the protein.

The hydrogen bonds of water are a good example of bonds between adjacent molecules. The positive hydrogen pole of one water molecule is attracted to the negative oxygen poles of as many as four other water molecules. As a result, the molecules line up with their neighbors in a somewhat ordered fashion (Fig. 2-10 ■). The weak interaction between the positive hydrogen regions and the negative oxygen regions is the hydrogen bond. Hydrogen bonding in water is responsible for the **surface tension** of water, the attractive force between water molecules that causes water to form spherical droplets when falling or to bead up when spilled onto a nonabsorbent surface (Fig. 2-10b ■). These weak forces of attraction also make it difficult to separate water molecules, as you may have noticed in trying to pick up a wet glass that has become "stuck" to a slick table top by a thin film of water.

(a)

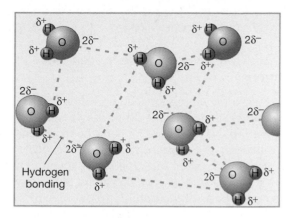

(b)

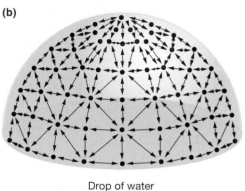

Drop of water

■ **Figure 2-10   Hydrogen bonds of water**   (a) The polar regions of adjacent water molecules allow them to form hydrogen bonds with one another. (b) Surface tension created by hydrogen bonds holds a drop of water in a hemispheric shape.

Van der Waals Forces Are Weak Attractions Between All Molecules

Van der Waals forces are even weaker bonds than hydrogen bonds. They are nonspecific attractions between the nucleus of any atom and the electrons of nearby atoms. Two atoms that are weakly attracted to each other by van der Waals forces move together until they are so close that their electrons begin to repulse each other. Consequently, van der Waals forces allow atoms to pack closely together and occupy a minimum amount of space. A single van der Waals attraction between atoms is very weak, but multiple van der Waals forces supplement the hydrogen bonds that hold proteins in their three-dimensional shapes.

✔ What do dishwashing detergents do that keeps water from beading up on glasses?

▶ SOLUTIONS AND SOLUTES

Substances that dissolve in liquid are collectively known as **solutes.** The liquids into which they dissolve are known as **solvents.** In biological solutions, water is the

...continued from page 24

Chromium is found in several ionic forms. The chromium usually found in biological systems and in dietary chromium supplements is the cation Cr^{3+}. This ion is called trivalent because it has a net charge of +3. The hexavalent cation, Cr^{6+}, with a charge of +6, is used in industries such as in the manufacturing of stainless steel and the chrome plating of metal parts.

Question 3:   How many electrons have been lost from the hexavalent ion of chromium? From the trivalent ion?

universal solvent, and the human body is about 60% water. The combination of solutes dissolved in solvent is called a **solution.** This section examines some of the properties of biological solutions.

Not All Molecules Dissolve in Aqueous Solutions

Life as we know it is established on water-based, or *aqueous,* solutions that resemble dilute seawater in their ionic composition. Na^+, K^+, and Cl^- are the main ions; other ions make up a lesser proportion. All molecules and cell components are either dissolved or suspended in these saline solutions. It is useful, therefore, to classify molecules on the basis of their ability to dissolve in water.

The degree to which a molecule is able to dissolve in a solvent is its **solubility:** the more easily it dissolves, the higher its solubility. Molecules that dissolve readily in water are **hydrophilic** [*hydro-*, water + *-philic,* loving]. Most are polar or ionic molecules whose positive and negative regions can form hydrogen bonds with water molecules (Fig. 2-11 ■). Molecules that do not dissolve readily in water are **hydrophobic** [*-phobic,* hating]. They tend to be nonpolar molecules that are unable to make hydrogen bonds with water molecules.

The lipids (fats and oils) are the most hydrophobic group of biological molecules. They cannot interact with water molecules when placed in an aqueous solution, so they tend to form separate layers, like oil floating on vine-

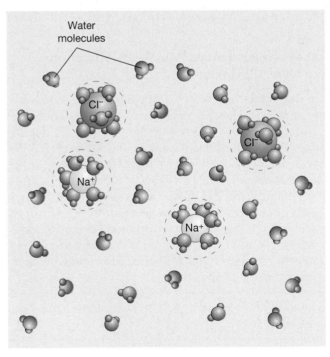

Sodium chloride in solution

■ **Figure 2-11   Sodium chloride dissolves in water** When salt is placed in water, the interaction of the ions with polar water molecules breaks the ionic bonds that held the NaCl crystal together, and the salt dissolves.

gar in a bottle of salad dressing. In order for hydrophobic molecules such as lipids to dissolve in body fluids, they must be combined with a hydrophilic molecule that can effectively carry them into solution. For example, cholesterol, a common animal fat, is a hydrophobic molecule. Fat from a piece of meat dropped into a glass of warm water will float to the top, undissolved. Similarly, cholesterol in the blood will not dissolve unless it is combined with water-soluble molecules called lipoproteins.

In addition to its role as medium for dissolved substances, water has other properties that affect the way our bodies work. The surface tension of water is important in lung function, as you will see in Chapter 17. Water also has the ability to capture and store large amounts of heat. The heat-retaining property of water is significant in heat loss from the body through sweating. Temperature control in the body is discussed in Chapter 21.

✔ Why does table salt (NaCl) dissolve in water?

There Are Several Ways to Express the Concentration of a Solution

The **concentration** of a solution is the amount of solute per unit volume of solution. There are many different ways to express concentration in physiology. The amount of solute can be described as the weight of the solute before it dissolves or as the number of molecules or ions of solute in the solution. Weight is usually given in grams (g) or milligrams (mg). Numbers of solute molecules are given in moles (mol), and numbers of solute ions are given in equivalents (eq). Volume is usually expressed as liters (L) or milliliters (mL).

Moles and Molarity   A **mole** [*moles,* a mass] is defined as 6.02×10^{23} atoms, ions, or molecules of a substance. Thus, one mole of a substance has the same number of particles as one mole of any other substance, just as a dozen eggs has the same number of items as a dozen roses. However, the weight of a mole depends on the weight of its particles.

* *

…continued from page 25

The hexavalent form of chromium used in industry is known to be toxic to humans. In 1992, officials at California's Hazard Evaluation System and Information Service warned that inhaling chromium dust, mist, or fumes placed chrome and stainless steel workers at increased risk for lung cancer. Officials found no risk to the public from normal contact with chrome surfaces or stainless steel. The first link between the biological trivalent form of chromium and cancer was presented in a controversial 1995 study. In these experiments, isolated cells exposed to moderate levels of Cr^{3+} developed cancerous changes.

Question 4:   *From this information, can you conclude that hexavalent and trivalent chromium are equally toxic?*

* *

The weight of one mole of a substance is equal to the atomic mass or molecular weight of the substance, expressed in grams. This value is called the **gram molecular weight.** For example, the molecular weight of glucose is 180 daltons. Therefore, one mole of glucose weighs 180 grams. One mole of NaCl, with the same number of molecules, weighs only 58.5 grams.

Solution concentrations in biology are frequently given as **molarity,** the number of moles of solute in a liter of solution. Molarity is abbreviated as mol/L or, simply, M. A one molar solution of glucose (1 mol/L, 1 M) contains 6.02×10^{23} molecules of glucose per liter. It is made by dissolving 180 grams (one mole) of glucose in enough water to make one liter of solution.

Typical biological solutions are so dilute that solute concentrations are usually expressed as **millimoles** per liter (mmol/L or mM). A millimole is 1/1000th of a mole [milli-, 1/1000].

✔ *How much does a mole of potassium chloride weigh?*

Equivalents   Concentrations of ions are sometimes expressed in equivalents per liter instead of moles per liter. An **equivalent** (eq) is equal to the molarity of an ion times the number of charges the ion carries. A sodium ion with a charge of +1 has one equivalent per mole. A calcium ion, Ca^{2+}, has two equivalents per mole. The same rule is true for negatively charged ions. One mole of chloride ions (Cl^-) has one equivalent per mole, one mole of phosphate ions (HPO_4^{2-}) has two equivalents. A **milliequivalent** (meq) is 1/1000th of an equivalent. Milliequivalents are useful for expressing the concentration of dilute biological solutions. Many clinical measurements, such as the concentration of sodium ions in the blood, are reported in milliequivalents per liter (meq/L).

Weight/Volume and Percent Solutions   In the laboratory or pharmacy, scientists cannot measure out solutes by the mole. Instead, they use the more conventional measurement of weight. The solute concentration of these solutions may then be expressed as a percentage of the total solution, or **percent solution.** A 10% solution means that there are 10 parts of solute per 100 parts *of total solution.* If the solute is normally a solid at room temperature, the percent solution is expressed as weight/volume. A 5% glucose solution would be made by weighing out 5 grams of glucose and adding water to make a final volume of 100 mL. If the solute is a liquid, such as concentrated hydrochloric acid, the solution is made using volume/volume measurements. To make 0.1% HCl, add 0.1 mL of concentrated acid to enough water to give a final volume of 100 mL.

Another convention is common in medicine. The concentrations of drugs and other chemicals in the body are often expressed as milligrams per deciliter (mg/dL). A **deciliter** (dL) is 1/10 of a liter, or 100 mL. A second way of expressing this, no longer commonly used, is mg%, where % means per 100 parts or 100 mL. A solution of 20 mg/dL could also be expressed as 20 mg%.

✔ *Which solution is more concentrated: a 100 mM solution of glucose or a 0.1 M solution of glucose?*

✔ *In making a 5% solution of glucose, why wouldn't you weigh out 5 grams of glucose and add it to 100 mL of water?*

The Concentration of Hydrogen Ions in the Body Is Expressed in pH Units

One of the important ionic solutes in the body is the hydrogen ion, H^+. The concentration of H^+ in body fluids determines the body's *acidity,* a physiological parameter that is closely regulated because H^+ in solution can interfere with hydrogen bonding and van der Waals forces. These bonds are responsible for the shapes of many important molecules, and when the bonds are disrupted, the shape of a molecule can change, thereby destroying the molecule's ability to function.

Where do hydrogen ions in body fluids come from? Some of them come from the separation of water (H_2O) into H^+ and OH^- ions. Others come from ionized molecules that release H^+ ions when they dissolve in water. A molecule that contributes H^+ to a solution is called an **acid.** For example, carbonic acid is an acid made in the body from CO_2 (carbon dioxide) and water. In solution, carbonic acid separates into a bicarbonate ion and a hydrogen ion:

$$CO_2 + H_2O \rightleftharpoons H_2CO_3 \text{ (carbonic acid)} \rightleftharpoons H^+ + HCO_3^-$$

Note that when the H^+ is part of the intact carbonic acid molecule, it does not contribute to acidity. *It is only free H^+ that affects H^+ concentration.*

Some molecules decrease the H^+ concentration of a solution by combining with free H^+. These molecules are called **bases.** Molecules that produce hydroxide ions, OH^-, in solution are bases because the hydroxide combines with H^+ to form water:

$$H^+ + OH^- \rightleftharpoons H_2O$$

Another molecule that acts as a base is ammonia, NH_3. It reacts with free H^+ to form an ammonium ion:

$$NH_3 + H^+ \rightleftharpoons NH_4^+$$

The concentration of H^+ in body fluids is measured in terms of **pH.** The expression pH stands for "power of hydrogen" and is calculated from the negative log of the hydrogen ion concentration: $pH = -\log [H^+]$ [∞ Appendix A]. The square brackets around the symbol for a solute are the shorthand notation for "concentration": $[H^+]$ = hydrogen ion concentration.

Another way to write the pH equation is $pH = \log(1/[H^+])$. This equation states that pH is inversely related to the H^+ concentration. In other words, as the H^+ concentration goes up, the pH goes down. The pH of a solution is measured on a numeric scale between 0 and 14. The H^+ concentration in pure water is 1×10^{-7} M, so pure water has a pH value of 7.0. It is considered neutral. Solutions that have gained H^+ from an acid have higher H^+ concentrations. Because $[H^+]$ is inversely proportional to pH, **acidic** solutions have a pH less than 7. Solutions that have lost H^+ to a base have a lower H^+ concentration than pure water; they are called **alkaline.** They have a pH value greater than 7. Remember, because pH is proportional to the reciprocal of the $[H^+]$, as the hydrogen ion concentration *increases*, the pH number *decreases*.

The pH scale is logarithmic, so that a change in pH value of 1 unit indicates a 10-fold increase or decrease in the hydrogen ion concentration. For example, if the pH of a solution is altered from a pH of 8 to a pH of 6, there has been a 100-fold (10 × 10) increase in the H^+ concentration.

Figure 2-12 ■ shows a pH scale and the pH values of various solutions. The normal pH of blood in the human body is 7.40, slightly alkaline. Regulation of the body's pH within a narrow range is critical because a blood pH more acidic than 7.00 (pH < 7.00) or more alkaline than 7.70 (pH > 7.70) is incompatible with life.

Buffers are a key factor in the body's ability to maintain normal pH. A **buffer** is any molecule that moderates changes in pH. Many buffers are anions that have a strong affinity for H^+ molecules. When free H^+ is added to a solution containing buffers, the buffers bond to the H^+, thereby diminishing any change in pH. The bicarbonate ion, HCO_3^-, is an important buffer in the human body. Sodium bicarbonate (baking soda) is a buffer used to adjust the pH of swimming pools and spas. The following equation shows how sodium bicarbonate buffers a solution to which hydrochloric acid (HCl) has been added. When placed in solution, HCl dissociates into H^+ and Cl^-, creating a high H^+ concentration and low pH. If sodium bicarbonate is present, some of the bicarbonate ions combine with some of the H^+ to form undissociated carbonic acid. The removal of free H^+ lowers the H^+ concentration and raises the pH:

$$H^+ + Cl^- + HCO_3^- + Na^+ \rightleftharpoons H_2CO_3 + Cl^- + Na^+$$

| Hydrochloric acid | Sodium bicarbonate | | Carbonic acid | Sodium chloride (table salt) |

✔ When the body becomes more acidic, does the pH increase or decrease?

✔ How can urine, stomach acid, and saliva all have pH values outside the pH range that is compatible with life?

▶ BIOMOLECULES

Most of the molecules of concern in human physiology have three elements in common: carbon (C), hydrogen (H), and oxygen (O). Molecules that contain carbon are

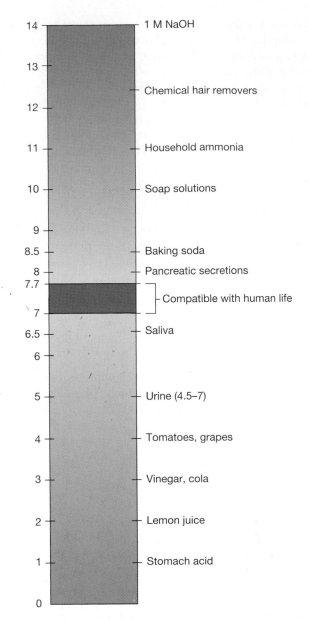

■ **Figure 2-12   pH scale**   This logarithmic scale is a measure of the hydrogen ion concentration. For example, at a pH of 14, the hydrogen ion concentration is 10^{-14} moles/liter.

...continued from page 26

To test the claim that chromium supplements enhance muscle development, a group of researchers gave high daily doses of chromium picolinate to football players during a two-month training period. Researchers found the players who took chromium supplements did not experience greater increase in muscle mass or strength than players who did not take the supplement.

Question 5: *Based on this study, which did not show enhanced muscle development from chromium supplements, and on the study that suggested that chromium supplements might cause cancer, do you feel that Stan should continue taking chromium?*

known as **organic molecules,** because it was once thought that they existed in or were derived from only plants and animals. Organic molecules associated with living organisms are also called **biomolecules.** There are four kinds of biomolecules: carbohydrates, lipids, proteins, and nucleotides. The first three groups are used by the body for energy and as the building blocks for cellular components. The fourth group, the nucleotides, includes DNA and RNA, the structural components of genetic material. Compounds that carry energy, such as adenosine triphosphate (ATP), or regulate metabolism, such as cyclic adenosine monophosphate (cAMP), are also nucleotides. Each group of biomolecules has a characteristic composition and molecular structure, which we describe briefly in the following sections.

Carbohydrates Are the Most Abundant Type of Biomolecule

Carbohydrates get their name from their structure: literally, carbons with water. The basic formula for a carbohydrate is $(CH_2O)_n$, showing that for each carbon there are two hydrogens and one oxygen, the same H:O ratio found in water. The n after the parentheses represents the number of repetitions of the CH_2O unit: $C_nH_{2n}O_n$. For example, glucose, $C_6H_{12}O_6$, has an n of 6. Glucose is an example of a simple sugar, the smallest type of carbohydrate.

The names of all the simple sugars end with the suffix *-ose*. There are two kinds of simple sugars: the **monosaccharides** [*mono-*, one + *sakcharon*, sugar] and the **disaccharides** [*di-*, two]. The monosaccharides are the building blocks of complex carbohydrates and have either five carbons, like **ribose,** or six carbons, like **glucose (dextrose), fructose,** and **galactose.**

When two monosaccharides bond to each other, they form a disaccharide molecule. Typical disaccharides are **maltose, lactose** (milk sugar), and **sucrose** (table sugar). Figure 2-13 ■ shows the composition of most of these simple sugars.

When many glucose molecules join together, they form very large molecules. These complex carbohydrate molecules are known as **polysaccharides** [*poly-*, many]. A large molecule made up of repeating units is called a **polymer.** Thus, the complex carbohydrates are all glucose polymers (Fig. 2-13 ■). Because these polymers have only one type of molecule, glucose, the differences in the polysaccharides come from the manner in which the glucose molecules are linked.

All living cells store glucose for energy in the form of a polysaccharide, and some cells produce polysaccharides for structural purposes as well. For example, yeasts and bacteria make a glucose storage polymer called *dextran.* Animal cells make a storage polysaccharide called **glycogen** that is found in all tissues of the body. Many invertebrate animals make a structural polysaccharide called *chitin.* Plants make two types of polysaccharides: the storage molecule **starch,** which humans can digest, and the structural molecule **cellulose,** which humans cannot digest. It is unfortunate that we are unable to digest cellulose and obtain its energy, because it is the most abundant organic molecule on earth.

Lipids Are the Most Diverse Biomolecules

Lipids are biomolecules made up of carbon, hydrogen, and oxygen, like the carbohydrates, but as a rule they contain much less oxygen. An important characteristic of this group is that they are not very soluble in water, because of their nonpolar structure. Technically, lipids are called fats if they are solid at room temperature and oils if they are liquid at room temperature. Most lipids derived from animal sources, like lard and butter, are fats, whereas most plant lipids are oils.

Lipids are the most diverse group of biomolecules. In addition to the true lipids, this category includes three lipid-related substances: phospholipids, steroids, and eicosanoids (Fig. 2-14 ■).

The true lipids and phospholipids are most similar to each other in structure. Both groups contain a simple 3-carbon molecule known as **glycerol** and long carbon-chain molecules known as **fatty acids.** Fatty acids are long chains of carbon atoms bound to hydrogens, with a carboxyl (—COOH) group at one end of the chain. Fatty acids are described as **saturated** if there are no double bonds between carbons; **monounsaturated** if there is one double bond in the molecule; or **polyunsaturated** if there are two or more double bonds in the molecule (Fig. 2-14 ■). You have probably heard these terms in commercials for margarine and cooking oil: manufacturers like to boast that their products are "High in polyunsaturates!!" A lot of research is being done trying to relate the degree of fat saturation in our diets to the development of heart disease. Saturated fats, such as the fat that marbles a piece of choice beef, are generally considered to contribute to the development of artery-clogging fatty deposits. For a time, it was thought that the more unsaturated fats were the least likely to be harmful, but recent studies suggest that monounsaturated fats such as olive oil may be the most healthful.

Fatty acids in the body link to glycerol to form mono-, di-, or triglycerides. **Triglycerides** (more correctly known as **triacylglycerol**) are the most important form of lipid in the body; more than 90% of our lipids are in this form. Triglyceride levels in the blood are also important as a predictor of artery disease; an elevated fasting triglyceride level is linked to a higher risk of disease.

The lipid-related molecules include phospholipids, steroids, and eicosanoids. Most **phospholipids** are diglycerides with a phosphate group attached to the single carbon that lacks a fatty acid. Phospholipids are important components of cell membranes.

Steroids are lipid-related molecules whose structure includes four linked carbon rings (Fig. 2-14 ■). Cholesterol

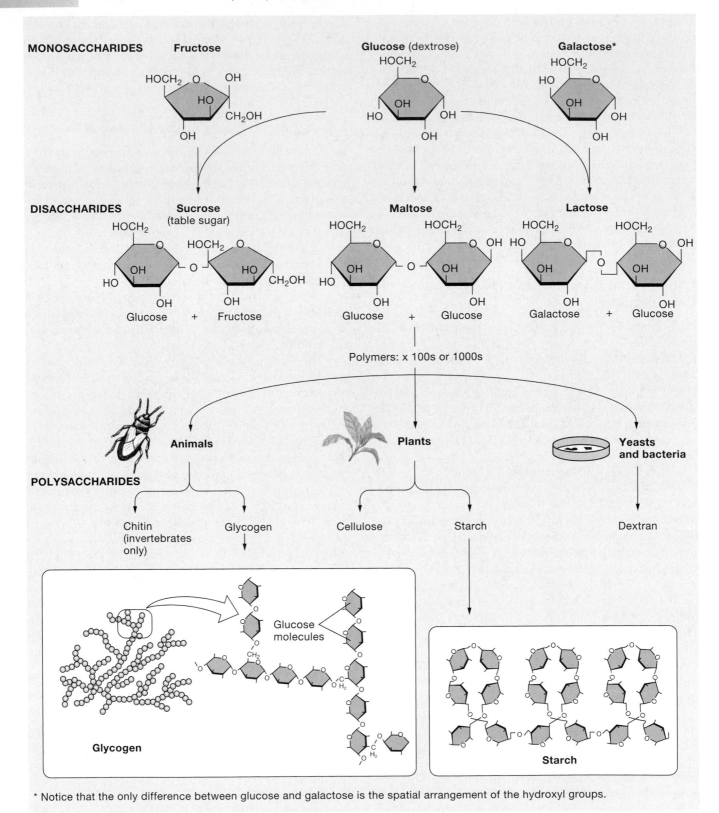

MONOSACCHARIDES

DISACCHARIDES

POLYSACCHARIDES

* Notice that the only difference between glucose and galactose is the spatial arrangement of the hydroxyl groups.

■ Figure 2-13   **Carbohydrates**

is the source of steroids in the human body. It is also an important component of animal cell membranes.

The **eicosanoids** are modified 20-carbon fatty acids that are found in animals [*eikosi*, twenty]. These molecules all contain one open five- or six-carbon ring that

makes the long carbon chain appear to double back on itself (Fig. 2-14 ■). The main eicosanoids are thromboxanes, leukotrienes, and prostaglandins. As you will learn in later chapters, eicosanoids are regulators of various physiological functions.

LIPIDS

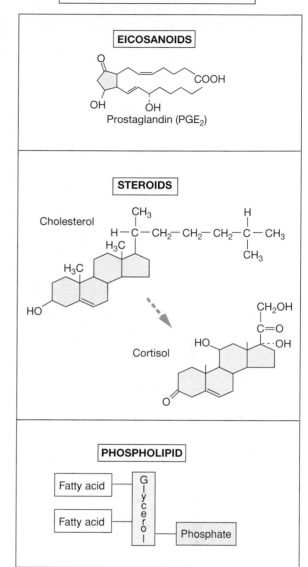

Palmitic acid, a saturated fatty acid

Glycerol

Oleic acid, a monounsaturated fatty acid

FATTY ACIDS

Glycerol | Fatty acid

MONOGLYCERIDE — Glycerol — Fatty acid

DIGLYCERIDE — Glycerol — Fatty acid / Fatty acid

TRIGLYCERIDE — Fatty acid — Glycerol — Fatty acid / Fatty acid

LIPID-RELATED MOLECULES

EICOSANOIDS

Prostaglandin (PGE₂)

STEROIDS

Cholesterol

Cortisol

PHOSPHOLIPID

Fatty acid / Fatty acid — Glycerol — Phosphate

■ **Figure 2-14 Lipids and lipid-related molecules** Palmitic acid is shown with all bonds and atoms. It has no double bonds between carbons and thus is a saturated fatty acid. Oleic acid, another fatty acid, is shown without symbols for the carbons and hydrogens in the middle of the molecule. Oleic acid has one double bond between carbons and thus is a monounsaturated fatty acid.

Proteins Are the Most Versatile of the Biomolecules

Proteins are large molecules made from smaller building-block molecules called **amino acids.** Twenty different amino acids commonly occur in natural proteins, and these building blocks may be assembled in an almost infinite number of combinations. All amino acids have a similar core structure: a central carbon atom is linked to a hydrogen atom, a nitrogen-containing amino group (—NH₂), a carboxyl group (—COOH), and a group of atoms designated "R" that is different in each of the amino acids (Fig. 2-15a ■). The R groups differ in their size, shape, and ability to form hydrogen bonds or ions. Because of the different R groups, each of the different amino acids reacts with other molecules in unique ways.

The human body can synthesize all but nine of the 20 protein-forming amino acids. Those nine, which must come from dietary proteins, are called the **essential amino acids** (Table 2-3). Because of the nitrogen in the amino group, proteins are the major dietary source of nitrogen.

There are also some amino acids that do not occur in proteins but that have important physiological functions. They include *homocysteine,* a sulfur-containing amino acid that occurs normally in the body but which in excess is linked to heart disease; γ-*amino butyric acid* (gamma-amino butyric acid or *GABA*), a chemical made by nerve cells; and *creatine,* a molecule that stores energy when it binds to a phosphate group.

When two amino acids link together, the amino group of one is joined to the carboxyl group of the other.

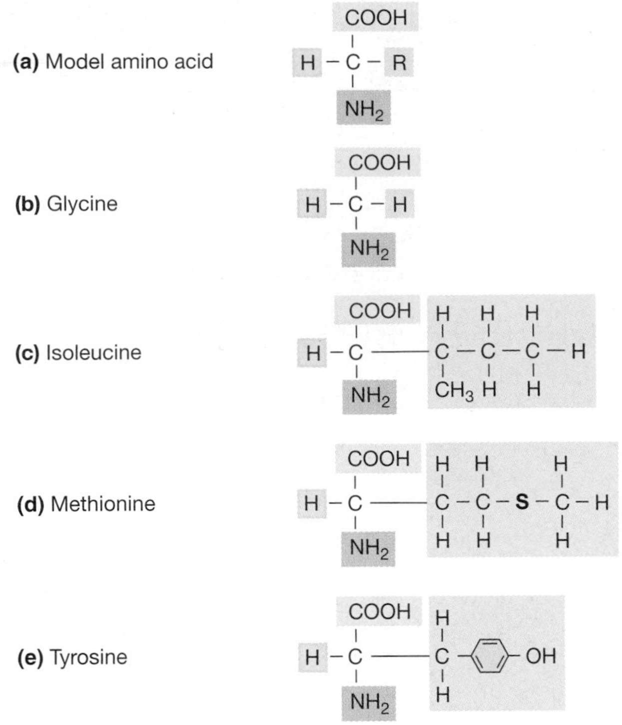

(a) Model amino acid

(b) Glycine

(c) Isoleucine

(d) Methionine

(e) Tyrosine

■ **Figure 2-15** **Amino acid structure** All amino acids have a carboxyl group (—COOH), an amino group (—NH₂), and a hydrogen attached to the carbon. The fourth carbon bond is to a variable "R" group. Notice the sulfur atom (—S—) in methionine.

TABLE 2-3 **The Amino Acids of the Human Body**

Each amino acid has both a three-letter abbreviation of its name and a single-letter abbreviation, used to spell out the sequence of amino acids in large proteins. The essential amino acids are marked with a star. These nine amino acids cannot be made by the body and must be obtained from the diet.

Amino Acid	Three-Letter Abbreviation	One-Letter Symbol
Alanine	Ala	A
Arginine	Arg	R
Asparagine	Asn	N
Asparagine or aspartic acid	Asx	B
Aspartic acid	Asp	D
Cysteine	Cys	C
Glutamic acid	Glu	E
Glutamine	Gln	Q
Glutamine or glutamic acid	Glx	Z
Glycine	Gly	G
*Histidine	His	H
*Isoleucine	Ile	I
*Leucine	Leu	L
*Lysine	Lys	K
*Methionine	Met	M
*Phenylalanine	Phe	F
Proline	Pro	P
Serine	Ser	S
*Threonine	Thr	T
*Tryptophan	Trp	W
Tyrosine	Tyr	Y
*Valine	Val	V

*Essential amino acids

By linking many molecules in this way, long chains of up to hundreds of amino acids can be formed. If there are two to nine amino acids in the amino acid chain, the molecule is called a **peptide.** A chain of 10 to 100 amino acids is called a **polypeptide.** A chain of more than 100 amino acids is called a **protein.**

The sequence of amino acids in the chain is called the **primary structure** of the protein (Fig. 2-16 ■). For example, the primary structure of thyrotropin-releasing hormone, a peptide made in the brain, is glutamic acid-histidine-proline. The primary structure of a protein is genetically determined and is essential to proper function. For example, a substitution of only one amino acid causes sickle cell disease, a condition in which red blood cells cannot function normally.

As the amino acid chain forms, it takes on its **secondary structure,** which is the spatial arrangement of the amino acids in the chain. Some amino acid chains curl into a spiral, or **α-helix,** whereas others create a **β-pleated sheet.** The secondary structure is determined by the angle of the bond between adjacent amino acids and is stabilized by hydrogen bonding between different parts of the molecule. Proteins that are destined for structural uses may be composed almost entirely of sheets, because this configuration is very stable. Other proteins combine sheets with helices.

The three-dimensional shape of an amino acid chain is its **tertiary structure.** Some long protein chains fold up into a ball or other compact shape when the R groups of amino acids in the α-helix are attracted to each other. The attraction may be between adjacent amino acids or between amino acids that are at different ends of the molecule. In some proteins, the tertiary structure is held in place by hydrogen bonds. In other proteins, sulfur atoms in two different *cysteine* molecules bond covalently to each other in a disulfide (S—S) bond.

Proteins are often classified as either globular or fibrous proteins based on their shape. **Globular proteins** have amino acid chains that are folded into ball-like shapes. **Fibrous proteins** are found as pleated sheets or in long chains that wind around each other.

A protein with more than one polypeptide chain is said to have **quaternary structure** (Fig. 2-16 ■). Hemoglobin, the oxygen-carrying pigment found in red blood cells, has a quaternary structure composed of four subunits: two polypeptide alpha chains of 141 amino acids each and two polypeptide beta chains of 146 amino acids

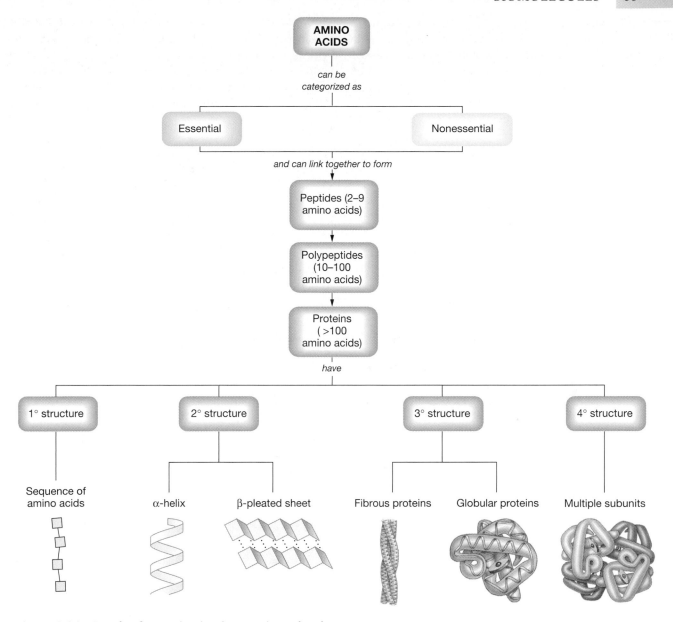

■ **Figure 2-16   Levels of organization in protein molecules**

each. The hemoglobin molecule is illustrated in Figure 2-16 under the label "Multiple subunits."

Proteins, peptides, and amino acids serve a number of important functions in the body. The fibrous proteins, which are insoluble in water, are important structural components of cells and tissues. Examples include *collagen,* a fibrous protein found in many types of connective tissue, and *keratin,* a fibrous protein found in hair and nails. The globular proteins are soluble in water. They act as *carriers* for insoluble lipids in the blood, binding to them and making them soluble. They may also serve as *enzymes* that increase the rate of chemical reactions. Soluble proteins also act as cell-to-cell messengers in the form of hormones and neurotransmitters and as defense molecules to help fight foreign invaders.

✔ Why must we include essential amino acids in our diet but not the other amino acids?

✔ What aspect of protein structure allows proteins to have more versatility than lipids or carbohydrates?

Some Molecules Combine Carbohydrates, Proteins, and Lipids

Not all biomolecules are pure protein, pure carbohydrate, or pure lipid. **Conjugated proteins** are protein molecules combined with another kind of biomolecule. Proteins combine with lipids to form **lipoproteins.** Lipoproteins are found mostly in cell membranes. They also carry hydrophobic molecules such as cholesterol in the blood: the lipid section of the molecule interacts with insoluble cholesterol while the protein section makes the entire lipoprotein-cholesterol complex soluble in water.

Proteins combine with carbohydrates to form **glycoproteins.** A "glyco-" molecule is composed of a carbo-

hydrate bonded to either a protein or a lipid (**glycolipid**). Glycoproteins and glycolipids, like lipoproteins, are important components of cell membranes (Chapter 5). The carbohydrate portions of these molecules are usually found on the external surface of the cell, where they can form hydrogen bonds with water and other molecules in

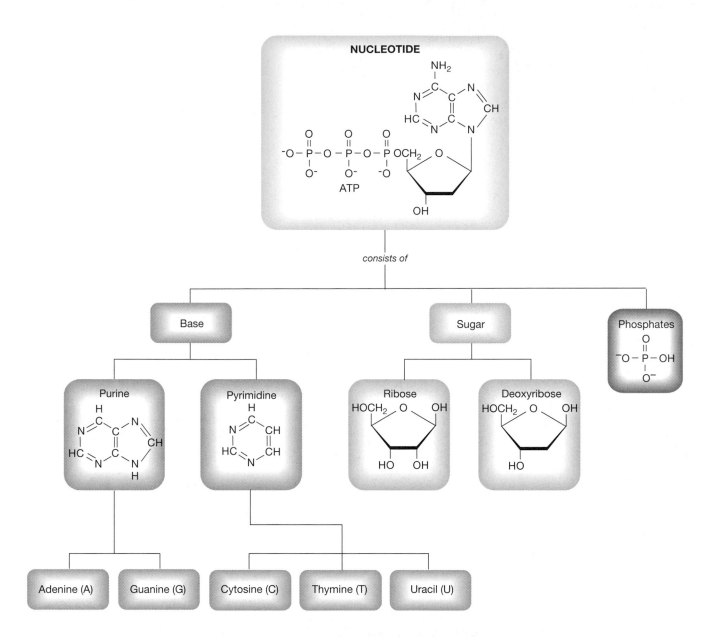

NUCLEOTIDES	Bases	Sugar	Phosphate groups	Other component
ATP	= Adenine	+ Ribose	+ 3	—
ADP	= Adenine	+ Ribose	+ 2	—
cAMP	= Adenine	+ Ribose	+ 1	—
NAD	= Adenine	+ 2 Ribose	+ 2	+ Nicotinamide
NUCLEIC ACIDS:				
DNA	= A,G,C,T	+ Deoxyribose	+ 1 per nucleotide	—
RNA	= A,G,C,U	+ Ribose	+ 1 per nucleotide	—

■ Figure 2-17   **Nucleotide structure**

body fluids. Glycolipids are believed to contribute to the stability of the cell membrane. Certain glycoproteins and glycolipids act as cell membrane receptors and cell markers. *Membrane receptors* bind to molecules outside the cell. The *cell markers* identify cells, like a label, so that the defense cells of the body can tell the difference between "self" and foreign invaders.

Nucleotides Transmit and Store Energy and Information

Nucleotides play an important role in many of the processes of the cell, including the storage and transmission of genetic information and the transfer of energy. A single **nucleotide** is a three-part molecule composed of (1) one or more phosphate groups, (2) a five-carbon sugar, and (3) a carbon-nitrogen ring structure called a **base** (Fig. 2-17 ■). Every nucleotide contains one of two possible sugars: **ribose** or **deoxyribose** [*de-*, without; *-oxy-*, oxygen]. There are also two types of bases: the purines and the pyrimidines. The **purines** have a double ring structure; the **pyrimidines** have a single ring. There are two different purines: **adenine** and **guanine.** There are three different pyrimidines: **cytosine, thymine,** and **uracil.**

Different nucleotides are made up of sugars, phosphate groups, and various combinations of bases. The smallest nucleotides include the energy-transferring compounds **ATP** (adenosine triphosphate) and **ADP** (adenosine diphosphate). These molecules contain the base adenine, the sugar ribose, and two or three phosphate groups (Fig. 2-17 ■). A related nucleotide is **cyclic AMP** (cyclic adenosine monophosphate, or cAMP), a molecule important in the transfer of signals between the extracellular fluid and the cell. Other energy-transferring nucleotides include **NAD** (nicotinamide adenine dinucleotide) and **FAD** (flavin adenine dinucleotide). Molecules that take part in energy transfer and storage are discussed in more detail in Chapter 4.

DNA and RNA are nucleotide polymers, or **nucleic acids.** They store genetic information within the cell and transmit it to future generations of cells. One complete DNA molecule with its associated proteins forms a *chromosome.*

DNA stands for *deoxyribonucleic acid;* **RNA** is short for *ribonucleic acid.* In each case, the name describes the type of sugar found in the nucleotides of the polymer. DNA contains the bases adenine, guanine, cytosine, and thymine. RNA also contains adenine, guanine, and cytosine but has uracil instead of thymine. Both polymers are formed by linking nucleotides into long chains, similar to the long amino acid chains of proteins. The sugar of one nucleotide links to the phosphate of the next, creating a chain or backbone of alternating sugar-phosphate-sugar-phosphate groups. The bases extend freely to the side (Fig. 2-18 ■).

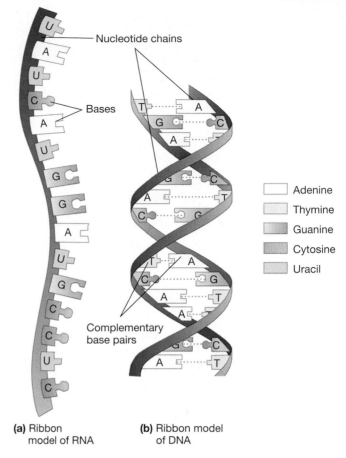

(a) Ribbon model of RNA   **(b)** Ribbon model of DNA

■ **Figure 2-18   RNA and DNA**

The DNA molecule is composed of two linked nucleotide chains that form a *double helix,* a three-dimensional structure that was first described by J. D. Watson and F. H. C. Crick, in 1953. The double helix is formed when two DNA chains link through complementary base pairing. Because of space limitations, the larger purines (adenine or guanine) must always link to a smaller pyrimidine (cytosine or thymine). Adenine (A) always pairs with thymine (T), and guanine (G) always pairs with cytosine (C). These base pairs (A—T and G—C) hold the double strand of DNA in its spiral shape. Base pairing and DNA replication are reviewed in Appendix B.

RNA usually takes the form of a single chain, although it is temporarily bound to DNA while it is being formed. RNA plays an important role in protein synthesis, discussed in Chapter 4.

In this chapter we have looked at how atoms combine to form molecules, how molecules dissolve to form solutions, and how different molecular building blocks combine into biologically important compounds (see Figure 2-19 ■ for a summary). The next chapter will consider the role of biomolecules in the structure of cells and tissues.

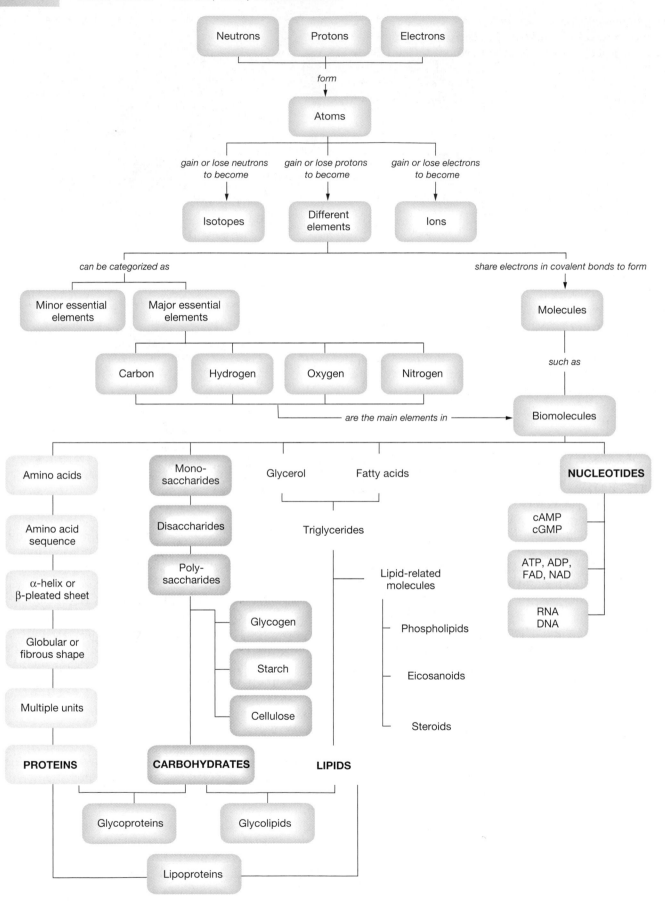

Figure 2-19   **Chemistry summary**

PROBLEM CONCLUSION

In this running problem, you have learned that the claims made for chromium picolinate have not been supported with evidence from controlled scientific experiments. You have also learned that hexavalent chromium is toxic, and that one study suggests that the biological trivalent form of chromium may be toxic in moderate doses.

Further check your understanding of this running problem by comparing your answers to those in the summary table.

	Question	Facts	Integration and Analysis
1a	Locate chromium on the periodic table of the elements.	The periodic table organizes the elements according to their atomic number.	N/A*
1b	What is chromium's atomic number? Atomic mass?	Reading from the table, chromium (Cr) has an atomic number of 24 and an average atomic mass of 52.	N/A
1c	How many electrons does one atom of chromium have?	The atomic number of an element is equal to the number of protons in one atom. An atom has equal numbers of protons and electrons.	The atomic number of chromium is 24, therefore one atom of chromium has 24 protons and 24 electrons.
1d	Which elements close to chromium are also essential elements?	Reading from the table, the essential elements close to chromium include molybdenum, manganese, and iron.	N/A
2a	If people have a chromium deficiency, would you predict that their blood glucose levels would be lower or higher than normal?	Chromium helps move glucose from the blood into cells.	If chromium is absent/lacking, less glucose will leave the blood and blood glucose would be elevated.
2b	From the results of the Chinese study, can you conclude that all people with diabetes suffer from chromium deficiency?	Higher doses of chromium supplements lowered elevated blood glucose levels but lower doses had no effect. This is only one study done in China and we have no information about similar studies elsewhere.	We have insufficient evidence from the information presented to draw a conclusion about the role of chromium deficiency in diabetes.
3	How many electrons have been lost from the hexavalent ion of chromium? From the trivalent ion?	For each electron lost from an ion, a positively charged proton is left behind with the ion.	The hexavalent ion of chromium, Cr^{6+}, has six unmatched protons and has lost six electrons. The trivalent ion, Cr^{3+}, has lost three electrons.
4	Can you conclude that hexavalent and trivalent chromium are equally toxic?	The hexavalent form of chromium is used in industry, and when inhaled, has been linked to an increased risk of lung cancer. Enough studies have shown an association that California's Hazard Evaluation System and Information Service has issued warnings to chromium workers. To date, only one study, performed on isolated cells rather than intact animals, has suggested that trivalent chromium may cause cancerous changes.	Although the toxicity of Cr^{6+} is well established, the toxicity of Cr^{3+} is questionable. One study is not sufficient proof of a link between Cr^{3+} and cancer. Additional studies need to be performed examining the effects of various doses of Cr^{3+} on biological systems.

*Not applicable

Continued on next page

Question	Facts	Integration and Analysis
5	Based on the study that did not show enhanced muscle development from chromium supplements, and on the study that suggested that chromium supplements might cause cancer, do you feel that Stan should continue taking chromium?	Researchers in one study found no increase in the muscle mass or strength in research subjects who took chromium supplements. Another research study found that chromium caused cancerous changes in isolated cells exposed to moderate doses.	Two isolated studies do not provide substantial evidence either in favor of or against taking chromium supplements for muscle development. Using risk-benefit analysis, no study showed benefit of taking chromium for muscle development, while one study suggested that there is possible risk. The decision to take chromium supplements is Stan's personal decision. He should remain informed of new developments that would change the risk-benefit analysis.

CHAPTER REVIEW

SUMMARY

Elements and Atoms

1. **Atoms** are composed of protons, neutrons, and electrons. A **proton** has a mass of about 1 **atomic mass unit** (amu), or **dalton** (Da), and an electrical charge of +1. A **neutron** has a mass of about 1 amu and no electrical charge. An **electron** has negligible mass, compared with protons and neutrons, and a charge of −1. (p. 16)

2. **Major essential elements** are necessary for life and are present in large quantities in the body. **Trace elements** are necessary elements found in tiny amounts. (p. 16)

3. **Elements** are distinguished from each other by their **atomic number,** which is the number of protons in the nucleus of one atom. (p. 18)

4. The **mass** of an atom is calculated by adding the number of protons and neutrons in its nucleus. (p. 18)

5. Atoms of one element that have different numbers of neutrons in their nuclei are known as **isotopes** of the element. Radioactive isotopes emit energy called **radiation.** (p. 18)

6. The electrons around an atom are arranged in **shells** centered on the nucleus. The shells are subdivided into **energy levels,** which are further subdivided into **orbitals.** An atom is most stable when all its outermost orbitals are occupied by pairs of electrons. (p. 19)

7. Electrons form **bonds** between atoms and capture energy. (p. 20)

Molecules and Bonds

8. Two or more atoms linked together create a **molecule.** Molecules that include atoms of more than one element are called **compounds.** (p. 20)

9. The **molecular weight** of a molecule is determined by adding the atomic masses of all its atoms. (p. 21)

10. **Covalent bonds** are strong bonds formed when adjacent atoms share one or more pairs of electrons. Each pair of shared electrons forms a single covalent bond. A **double bond** forms when two atoms are sharing two pairs of electrons. (p. 21)

11. Molecules whose atoms share electrons unevenly are called **polar molecules.** If electrons are shared evenly by the atoms in a molecule, the molecule is a **nonpolar molecule.** (p. 23)

12. An atom or molecule that gains or loses electrons acquires an electrical charge and is called an **ion.** (p. 24)

13. Atoms form **cations** when they lose electrons and **anions** when they gain electrons. **Ionic bonds** are strong bonds formed when oppositely charged ions are attracted to each other. (p. 24)

14. **Hydrogen bonds** form when hydrogens in polar molecules are attracted to oxygen, nitrogen, or fluorine atoms. Hydrogen bonds may form within a molecule or between adjacent molecules. Hydrogen bonding among water molecules is responsible for the **surface tension** of water. (p. 25)

15. **Van der Waals forces** are weak bonds that form when molecules are attracted to each other. (p. 25)

Solutions and Solutes

16. Molecules that dissolve in liquid are called **solutes.** The liquid in which they dissolve is called the **solvent.** The universal solvent for biological solutions is water. (p. 25)

17. The ease with which a molecule dissolves is called its **solubility. Hydrophilic molecules** dissolve easily in water, whereas **hydrophobic molecules** do not. (p. 26)

18. A **mole** is 6.02×10^{23} particles of an element or compound. The weight of a mole in grams is known as the **gram molecular weight** of the element or compound. (p. 26)

19. The **concentration** of a solution is the amount of solute per unit volume of solution. A one **molar** (1 M) solution has one mole of solute per liter of solution. The concentration of most biological solutions is expressed in **millimoles per liter** of solution (mM). (p. 26)

20. Ion concentrations may be expressed in milliequivalents, where one **equivalent** equals the molarity of the ion times its electrical charge. (p. 27)

21. The **pH** of a solution is a measure of its hydrogen ion concentration. **Acids** are molecules that contribute H^+ to a solution. **Bases** are molecules that remove H^+ from solution. The more acidic the solution, the lower the pH. (p. 27)

22. **Buffers** are molecules that moderate pH changes. (p. 28)

Biomolecules

23. There are four major groups of **biomolecules,** which all contain carbon, hydrogen, and oxygen: carbohydrates, lipids, proteins, and nucleotides. (p. 29)

24. **Carbohydrates** have the general formula $(CH_2O)_n$. **Monosaccharides** are the simplest building blocks of complex carbohydrates. Two monosaccharides join to form **disaccharides. Polysaccharides** are glucose **polymers** used for the storage of energy and as structural components of cells. (p. 29)

25. **Lipids** contain carbon and hydrogen with little oxygen. Most are nonpolar. **Fatty acids** are long chains of carbon atoms bound to hydrogens. **Triglycerides** are composed of glycerol and three fatty acids. (p. 29)

26. **Proteins** are the most versatile biomolecule because 20 different **amino acids** serve as their building blocks. (p. 31)

27. The **primary structure** of a protein is the sequence of its amino acids. The **secondary structure** is its spatial arrangement into helixes or sheets. The **tertiary structure** is the three-dimensional shape of the folded amino acid chain, and the **quaternary structure** is the arrangement of multiple amino acid chains in a single protein. (p. 32)

28. **Globular proteins** are soluble. They act as carriers, messengers, defense molecules, and enzymes. **Fibrous proteins** are not very soluble. They serve as structural components of cells and tissues. (p. 32)

29. Proteins, lipids, and carbohydrates combine to form **glycoproteins, glycolipids,** or **lipoproteins.** (p. 33)

30. Nucleotides are important in the transmission and storage of information and the transfer of energy. (p. 35)

31. A **nucleotide** is composed of phosphate groups, a sugar, and either a purine or a pyrimidine base. Small nucleotides include ATP, ADP, and cyclic AMP. Nucleotide polymers are the **nucleic acids** DNA and RNA. (p. 35)

QUESTIONS

LEVEL ONE   Reviewing Facts and Terms

1. List three major essential elements found in the human body.

2. When atoms of elements are bound to one another into compounds, such as H_2O or CO_2, one unit is called a _____.

3. The subatomic particle with a mass of about 1 amu and a charge of +1 is called the (a) _____. The subatomic particle with a mass of about 1 amu and a neutral charge is the (b) _____. The remaining subatomic particle is the electron, with a negligible mass but a charge of (c) _____.

4. Which subatomic particle in an element determines its atomic number?

5. Name the element associated with each of these symbols: Ca, C, O, Na, N, K, H, and P.

6. Isotopes of an element have the same number of _____ and _____, but differ in their number of _____.

7. Unstable isotopes emit energy called _____. Using this energy for the diagnosis or treatment of disease is a specialty called _____.

8. The smallest subdivision of an electron shell is called a(n) _____, and each holds _____ electrons.

9. When electrons in an atom or molecule are stable, are they paired or unpaired?

10. When an atom of an element gains or loses one or more electrons, it is called an _____ of that element.

11. Match each type of bond with its description:
    (a) covalent bond
    (b) ionic bond
    (c) hydrogen bond
    (d) van der Waals forces

    1. weak attractive forces between hydrogen and oxygen or nitrogen
    2. shares a pair of electrons
    3. weak attractive forces between molecules
    4. atoms gain or lose electrons

12. In mixing compounds with water, the polarity of the compounds helps to predict their solubility. When a compound's electrons are shared evenly, the compound is said to be _____; if shared unevenly, the compound is _____. Which group dissolves more readily in water?

13. The pH of a solution is an abbreviation conveying what information about that solution? A solution with a pH less than 7 is _____; a solution with a pH greater than 7 is _____.

14. A_____ is a molecule that moderates changes in the pH of a solution by binding to free H^+ or OH^- ions.

15. List the four kinds of biomolecules. Give an example of each kind that is relevant to physiology.

16. Match each carbohydrate with its description:
    (a) starch
    (b) chitin
    (c) glucose
    (d) cellulose
    (e) glycogen
    (f) lactose

    1. most abundant carbohydrate on earth
    2. disaccharide, found in milk
    3. storage form of glucose for animals
    4. storage form of glucose for plants
    5. structural polysaccharide of invertebrates
    6. monosaccharide, found in blood

17. Proteins may be combined with other types of molecules. They are called _____ when combined with fatty components. They are called _____ when combined with carbohydrates.

18. Match each lipid with its description:
    (a) triglycerides
    (b) eicosanoids
    (c) steroids
    (d) oils
    (e) phospholipids

    1. most common form of lipid in the body
    2. liquid at room temperature, usually from plants
    3. important component of cell membrane
    4. structure composed of carbon rings
    5. modified 20-carbon fatty acids

19. Match these terms pertaining to proteins and amino acids:
    (a) the building blocks
    (b) must be included in our diet
    (c) protein that speeds the rate of chemical reactions
    (d) sequence of amino acids in a protein
    (e) protein chains folded into a ball-shaped structure

    1. essential amino acids
    2. primary structure
    3. amino acids
    4. globular proteins
    5. enzymes
    6. tertiary structure
    7. fibrous proteins

20. List the components of a nucleotide.

LEVEL TWO   Reviewing Concepts

21. **Mapping exercises**: Make each list of terms into a separate map.

List 1	List 2
anion	concentration
atom	equivalent
atomic mass	hydrogen bond
atomic number	hydrophilic
cation	hydrophobic
covalent bond	molarity
electron	mole
electron shell	nonpolar molecule
element	polar molecule
ion	solubility
isotope	solute
neutron	solvent
nucleus	water
orbital	
proton	

22. Use the periodic table of the elements in Figure 2-2 ■ to answer the following questions: An atom of sodium has 11 protons in its nucleus. (a) How many electrons does the atom have? (b) What is the electrical charge of the atom? (c) How many neutrons does the atom have? (d) If this atom loses one electron, it would be called a(n) _____? (e) What would be the electrical charge of the substance formed in (d)? (f) Write the correct chemical symbol for the atom referred to in (d). (g) What does the sodium atom become if it loses a proton from its nucleus? (h) Write the correct chemical symbol for the atom referred to in (g).

23. A solution with $[H^+] \times 10^{-3}$ M is _____ (acidic/basic?), whereas a solution with an $[H^+] \times 10^{-10}$ M is _____ (acidic/basic?). Give the pH for each of these solutions.

24. Name three nucleotides, and tell why each one is important.

25. Explain the progression of protein structure from primary to quaternary.

26. Compare the structure of DNA and RNA.

27. Distinguish between purines and pyrimidines.

LEVEL THREE   Problem Solving

28. You have been summoned to assist with the autopsy of an alien being whose remains have been brought to your lab. The chemical analysis returns with 33% C, 40% O, 4% H, 14% N, and 9% P. From this information you conclude that the cells do contain nucleotides, possibly even DNA or RNA. Your assistant is demanding that you tell him how you knew this. What do you tell him?

29. A cell carrying out work normally produces CO_2. The harder a cell works, the more CO_2 it produces. CO_2 is carried in

the blood according to the following equation: $CO_2 + H_2O \rightleftarrows H_2CO_3 \rightleftarrows H^+ + HCO_3^-$. What effect does hard work by your muscle cells have on the pH of the blood?

30. When the radioactive isotope of iodine, ^{131}I, emits beta radiation, it converts a neutron to a proton and electron. What is the atomic mass, atomic number, and name of the resulting atom?

LEVEL FOUR   Quantitative Problems

31. Write the chemical formulas for each molecule depicted. Calculate the molecular weight of each molecule. Predict whether each molecule is polar or nonpolar.

(a)

(b) $O=C=O$

(c)

(d)

(e)

32. Calculate the amount of NaCl you would weigh out to make one liter of 0.9% NaCl. Explain how you would make a liter of this solution.

33. A 1.0 M NaCl solution has 58.5 g of salt per liter. (a) How many molecules of NaCl are present in this solution? (b) How many millimoles of NaCl are present? (c) How many equivalents of Na^+ are present? (d) Convert this measurement to a percent solution.

34. How would you make 200 mL of a 5% glucose solution? Calculate the molarity of this solution. How many millimoles of glucose are present in 500 mL of this solution?

E   X   P   L   O   R   E   <MediaLab>

Introduction

Biology is about combining atoms of elements into unique structures that can accomplish the many processes required for life. In this MediaLab we explore the bonds that hold elements together. You will have the opportunity to see how a particular sequence of amino acids can become an ion channel within the cell membrane. On the Website you will find an exercise to help you critically evaluate the benefits of chromium supplements to your body's health.

Web Exploration 1

Estimated time for completion = 5 minutes
In Chapter 1 you learned about the levels of organization of living organisms (Fig. 1-1 ■). Now you will use levels of organization to build your mental model of how molecules are assembled. What roles do biological molecules play in the human body? What forces keep molecules together (covalent, ionic, and hydrogen bonds, van der Waals forces)? What forces disrupt them?

One of the major molecules in your body is sodium chloride (NaCl). Select the keyword **MOLECULES** on your Compan-

ion Website to see a tutorial on atoms and molecules. To complete this exercise, visit MediaLab Web Exploration 1 in Chapter 2 of your Companion Website.

Web Exploration 2

Estimated time for completion = 5 minutes
As you continue your journey through the human body, you will come to realize that proteins are indeed "the most versatile of the biomolecules." Proteins serve as hormones in the endocrine system, enzymes in cellular metabolism, and channels, transporters, and receptors in the cell membrane. They help to regulate movement of substances into and out of the cell, make up the cellular components that give cells their shape, and are intimately involved in muscle function.

To learn more about these dynamic proteins, you must first know about the molecules that serve as their building blocks. Select the keyword **AMINO ACIDS** on your Website to review and compare the various amino acids that make up proteins. To complete this exercise, visit MediaLab Web Exploration 2 in Chapter 2 of your Companion Website.

3 Cells and Tissues

■ "Cells are organisms, and entire animals and plants are aggregates of these organisms arranged according to definite laws."
—*Theodor Schwann, 1839* ■

CHAPTER OUTLINE

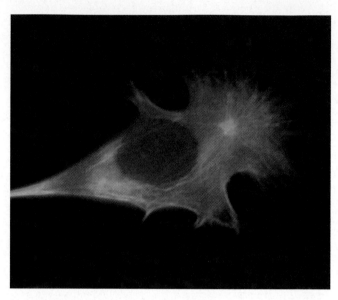

Developing cell in tissue culture showing microtubules (red), actin (green), and DNA (blue)

BACKGROUND BASICS

Units of measure (inside back cover)
Hydrophobic molecules (p. 26)
Proteins (p. 31)
Compound molecules (p. 20)
pH (p. 27)
DNA and RNA (p. 35)

Just as atoms combine to make molecules, a vast collection of molecules makes up the **cell,** the basic functional unit of most living organisms. Individual cells carry out the same life-sustaining functions as multicellular animals. They sense and respond to changes in their environment, and they communicate with neighboring cells by releasing chemicals or by generating electrical signals. Cells take in oxygen and nutrients, they extract energy from the nutrients for growth, repair, and reproduction, and they get rid of wastes such as carbon dioxide.

Much of what we know about cells comes from studies of simple organisms that consist of one cell. But humans are much more complex, with trillions of cells in their bodies. It has been estimated that there are more than 200 different types of cells within the human body, each cell with its own characteristic structure and function. In studying physiology, you must be aware both of how individual cells carry out their functions and of how they cooperate and communicate with each other in the human body.

Cytology [*kytos,* cell + *logia,* explaining] is the study of cells. In recent years, the word *cytology* has been used

PROBLEM
The Pap Smear

Dr. George Papanicolaou has saved the lives of millions of women by popularizing the Pap test, a screening method that detects the early signs of cervical cancer. In the 50 years the Pap test has been widely used, deaths from cervical cancer have dropped nearly 70%. If detected early, cervical cancer is one of the most treatable forms of cancer. Today, Jan Melton, who had an abnormal Pap test three months ago, returns to her family physician for a repeat test. The results will determine whether she should undergo treatment for cervical cancer.

...continued on page 45

• •

mostly to refer to the study of the structure of cells, whereas the more general term *cell biology* has referred mostly to the study of how cells work. The development of new research techniques has greatly expanded our understanding of how cells do what they do, and so we are better able to apply that knowledge to the study of tissues, organs, and the body as a whole.

Cooperation between cells is a central theme of physiology. This chapter looks first at the structure and function of individual cells. It then examines how groups of cells with similar functions join together to form the tissues and organs of the body.

▶ STUDYING CELLS AND TISSUES

The first recorded use of magnifying lenses to study the composition of biological material took place in the last half of the 1600s. In the Netherlands, Antoine van Leeuwenhoek discovered that pond water was teeming with "wee beasties" too small to be seen with the naked eye, while the English scientist Robert Hooke around 1665 used a simple light microscope to investigate the structure of cork, the

dead tissue layer under the bark of trees. It was Hooke who named the holes in cork "cells," from the Latin word *cella*, meaning a small room. But Hooke's cork cells, like the rooms of a building, were only walls and did not contain living material. Not until the early part of the 1800s, with improved microscopes, did we begin to understand that living cells were filled with a semifluid substance, and that the cell was the fundamental structural unit of life.

The light microscope makes it possible to examine both living cells and preserved cells. But even at the highest magnification and with the aid of dyes that stain different cellular components, it is possible to see only the general shape and size of cells or the larger internal structures such as the nucleus. Because most internal components are too small to be seen with the light microscope, the fine details of cell structure went undescribed until the electron microscope came into widespread use in the 1950s.

Figure 3-1 ■ shows a cell as seen under a light microscope, by transmission electron microscopy, and by scanning electron microscopy. The scanning electron microscope has greatly added to the understanding of cells and tissues by making it possible to view their fine structure in a realistic three-dimensional fashion rather than as thin flat slices that cut through internal structures. The next section of this chapter examines the details of cellular structure as seen with the aid of these microscopic techniques.

▶ CELLULAR ANATOMY

Animals begin life as a single cell, the **zygote,** or fertilized ovum (egg). As the zygote divides into two cells, then four cells, then eight, each daughter cell looks identical. But mature cells are not all alike. By a process known as **differentiation,** developing cells take on more than 200 different shapes and functions.

How does differentiation occur? Each cell inherits the same genetic information in its DNA, but no one cell uses

(a)   (b)   (c)

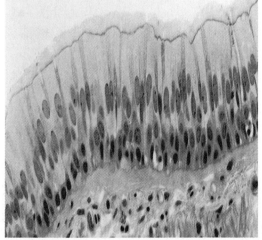
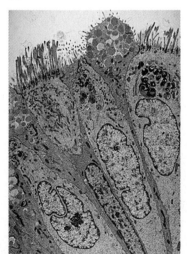

■ **Figure 3-1   One cell type viewed by light microscopy (a), transmission electron microscopy (b), and scanning electron microscopy (c)**   In (a), note that the fingerlike microvilli appear as a fuzzy border. Only the nucleus and cell membrane are easily seen.

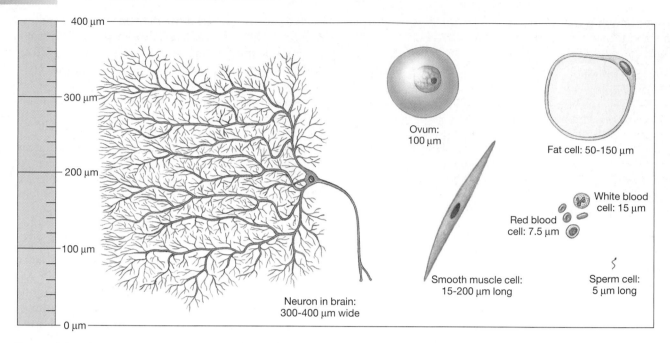

■ **Figure 3-2   Representative cell types in the human body**   Each cell has the genetic potential to become any type of cell, but selective activation of genes during differentiation determines the ultimate size, shape, and function of the cell.

all of it. During differentiation, only selected genes are activated, transforming the cell into a specialized unit. In most cases, the final shape and size of a cell and its contents reflect its function. Figure 3-2 ■ shows some of the different cells in the human body. Although these mature cells look very different from each other, they all started out alike and they retain many features in common. Let us look at the anatomical organization that is shared by most cells.

You can compare the structural organization of a cell to that of a medieval walled city. The city is separated from the surrounding countryside by a high wall, with entry and exit strictly controlled through gates that can be opened and closed. The city inside the walls is a diverse collection of houses and shops with varied functions. Within the city, a ruler in the castle oversees the everyday comings and goings of the city's inhabitants. Because the city depends on food and raw material from outside the walls, the ruler negotiates with the farmers in the countryside. Foreign invaders are always a threat, so the city ruler communicates and cooperates with the rulers of neighboring cities.

In the cell, the outer boundary is the cell membrane. Like the city wall, it controls the movement of material between the cell interior and the outside by opening and closing "gates," made of protein. The inside of the cell is divided into compartments rather than into shops and houses. Each of these compartments has a specific purpose that contributes to the function of the cell as a whole. In the cell, DNA in the nucleus is the "ruler in the castle," controlling both the internal workings of the cell and its interaction with other cells. Like the city, the cell depends on supplies from its external environment. It must also communicate and cooperate with other cells in order to keep the body functioning in a coordinated fashion.

Figure 3-3 ■ is a map of cell structure. Recall from Chapter 1 [∞ p. 6] that the cells of the body are surrounded by the dilute salt solution known as extracellular fluid (ECF) [*extra-*, outside]. This is the external

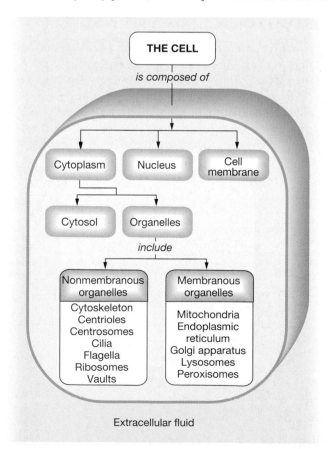

■ **Figure 3-3   A map for the study of cell structure**   The cells of the body are surrounded by the extracellular fluid.

environment of the cell from which the cell gains nutrients and into which it puts wastes and the chemicals that it uses to communicate with other cells. The cell membrane separates the inside environment of the cell from the extracellular fluid.

Internally the cell is divided into the cytoplasm and the nucleus. The cytoplasm consists of a fluid portion, called the cytosol, and assorted structures, collectively known as organelles. The organelles work in an integrated manner, each organelle taking on one or more of the cell's functions. A typical cell from the lining of the small intestine is shown in Figure 3-4d ■. It has most of the structures found in animal cells.

The Cell Membrane Separates the Cell from Its Environment

The **cell membrane** (also called the **plasma membrane** or *plasmalemma*) separates the inside of the cell from the surrounding extracellular fluid. Structurally, the membrane is a bilayer (double layer) of phospholipid molecules (Fig. 3-5 ■, p. 48). They are arranged so that the hydrophilic phosphate groups face the aqueous solutions inside and outside the cell, leaving the hydrophobic lipid portions hidden in the center of the membrane. The membrane is studded with protein molecules, like raisins in a slice of bread, and the extracellular surface has glycoproteins and glycolipids [∞ p. 34].

Functionally, the cell membrane serves as both a gateway and a barrier. All substances that enter or leave the cell must cross the cell membrane, which regulates exchange between the cell and the extracellular fluid. Because of the importance of membranes in physiology, we will cover the details of membrane structure and function separately in Chapter 5.

The Cytoplasm Includes Cytosol and Organelles

The **cytoplasm** includes all material inside the cell membrane, except for the nucleus. The cytoplasm has two components:

1. *Cytosol,* or intracellular fluid: The **cytosol** is a semigelatinous substance, separated from the extracellular fluid by the cell membrane. The cytosol contains dissolved nutrients, ions, waste products, and particles of insoluble materials known collectively as **inclusions.** Among the most common in-

• •

...*continued from page 43*

Cancer is a condition in which cells divide uncontrollably and fail to coordinate with the normal cells that surround them. Cancerous cells also fail to differentiate into specialized cell types.

Question 1: *Why is it necessary to remove or kill cancerous cells in order to treat cancer?*

• •

clusions are stored nutrients such as glycogen granules and lipid droplets. Also suspended in the cytosol are the:

2. *Organelles:* These "little organs" are like the organs of the body in that each has a specific role to play in the overall function of the cell. For example, mitochondria generate most of the cell's ATP, and lysosomes filled with enzymes act as the digestive system of the cell. Organelles are often classified as *nonmembranous* or *membranous* according to whether they have a membrane around them.

Nonmembranous Organelles Are in Direct Contact with the Cytosol

Nonmembranous organelles do not have boundary membranes and so are in direct contact with the cytosol. Movement of material between these organelles and the cytosol does not require transport across a membrane.

Nonmembranous organelles can be divided into two groups: those made from RNA and protein, and those made from insoluble protein fibers. The latter include the cytoskeleton, centrosomes and centrioles, cilia, and flagella.

Ribosomes Participate in Protein Synthesis   **Ribosomes** (Fig. 3-6 ■, p. 48) are small dense granules of RNA and protein that manufacture proteins under the direction of the cell's DNA (see Chapter 4). Most ribosomes are attached to the endoplasmic reticulum and are called **fixed ribosomes.** Ribosomes also float freely in the cytoplasm, where some form groups of 10 to 20 known as **polyribosomes.** Ribosomes are most numerous in cells that synthesize proteins, such as enzymes, for export out of the cell.

Vaults Are Newly Discovered Organelles   As scientists develop new staining techniques for looking at the fine structure of cells, they continue to find new organelles. One of the newest organelles, first described in 1986, is an eight-part particle of ribonucleoprotein called a **vault.** Although vaults are found in all cells with nuclei (eukaryotic cells), their function is still unclear. They appear to be located close to protein fibers in many cells. One of the vault proteins is found in increased amounts in some cancer cells and is associated with resistance to certain drugs in those cells.

There Are Three Sizes of Protein Fibers in the Cytoplasm   The protein fibers in the cytoplasm are classified by their diameter. The thinnest, **microfilaments,** are made of the protein *actin.* Somewhat larger **intermediate filaments** are made of several different types of protein, including *myosin* in muscle, *keratin* in hair and skin, and *neurofilament* in nerve cells. In muscle, intermediate myosin filaments combine to form **thick filaments.** The largest protein fiber organelles are the **microtubules,**

ANATOMY SUMMARY   Levels of Organization—Cells to Systems

■ Figure 3-4

System	Organ	Tissue

(a) The function of the digestive system is to take in food and water from the external environment, process it, and bring it into the internal environment of the body.

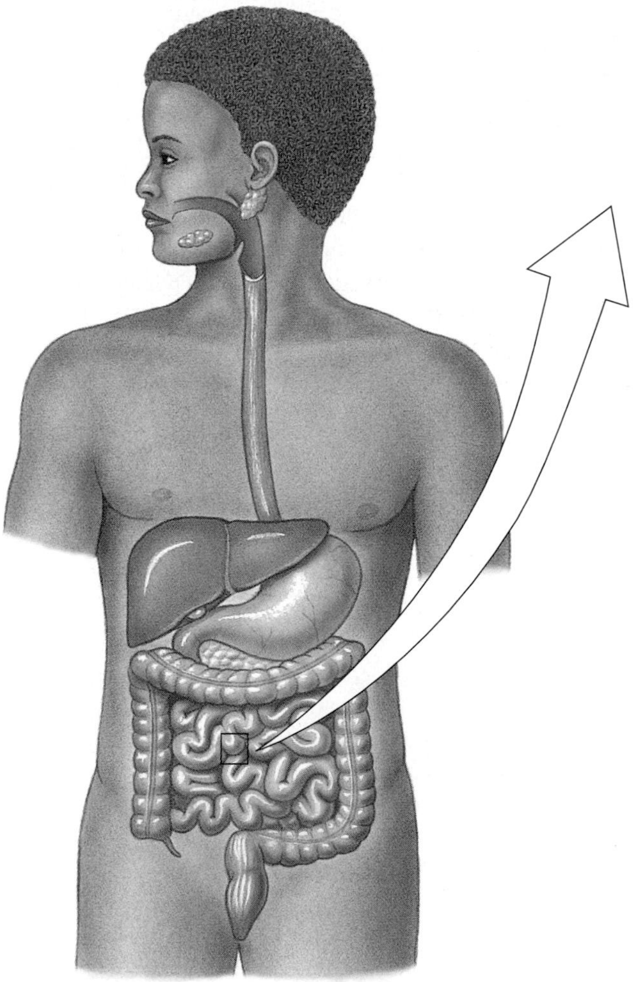

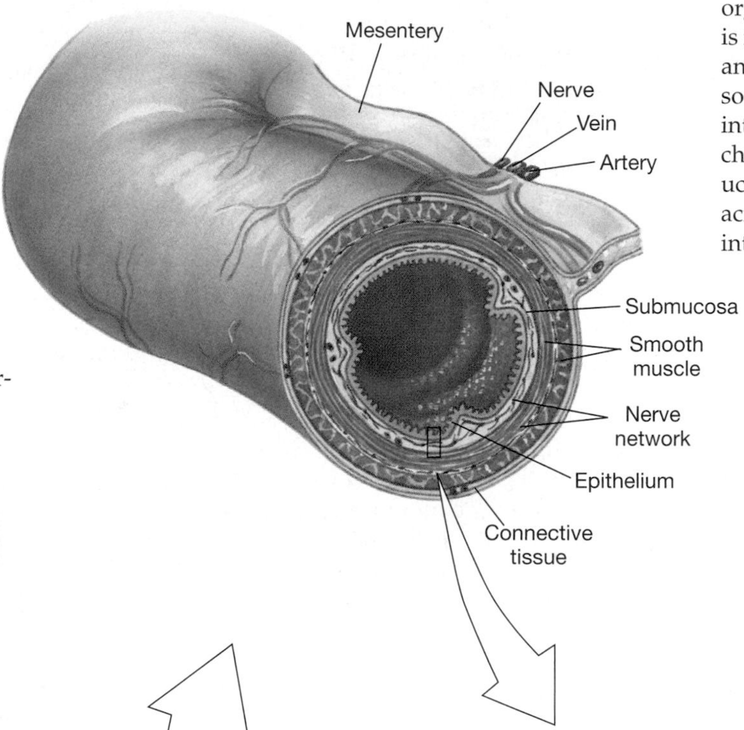

Mesentery

Nerve

Vein

Artery

(b) The small intestine, one organ of the digestive system, is responsible for mechanical and chemical digestion of food so that it can be absorbed. The intestine secretes enzymes for chemical digestion. The products of digestion are absorbed across the wall of the intestine into the blood.

Submucosa

Smooth muscle

Nerve network

Epithelium

Connective tissue

(c) A cross section of part of the small intestine, showing the different tissues that compose it. Nutrients and water are absorbed across the epithelium and enter either the blood vessels or the lymph.

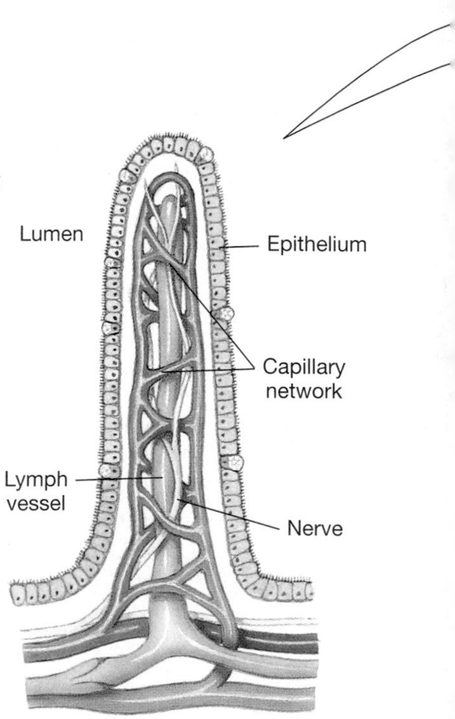

Lumen

Epithelium

Capillary network

Lymph vessel

Nerve

Cell	Sub-cellular structures (organelles)	Characteristics	Function
	(e) Microvilli	Extensive folding of the cell membrane	Increase surface area for absorption
	Cell membrane Cytosol	Bilayer of phospholipid molecules with inserted proteins	Acts as a gateway for and barrier to movement of material between the interior of the cell and the extracellular fluid
		Semi-gelatinous substance	Contains dissolved nutrients, ions, wastes, insoluble inclusions; suspends the organelles
	Centrioles	Bundles of microtubules	Direct movement of DNA during cell division
	Lysosomes and peroxisomes	Membrane-bound vesicles filled with enzymes	Digest bacteria, old organelles; metabolize fatty acids
	Mitochondrion	Double wall with central matrix	Produces most of cell's ATP
	Golgi apparatus	Hollow membranous sacs	Modify and package proteins
	Nucleus Nucleolus	Central lumen with a two-membrane outer envelope with pores	Contains DNA to direct all functions of the cell
		Region of DNA, RNA, and protein	Contains the genes that direct synthesis of ribosomal RNA
	Rough endoplasmic reticulum	Membrane tubules that are continuous with the outer nuclear membrane; edged with ribosomes	Site of protein synthesis
	Smooth endoplasmic reticulum	Same as rough ER without ribosomes	Synthesis of fatty acids, steroids, and lipids
	Ribosomes	Granules of RNA and protein	Assemble amino acids into proteins
	Microtubules and microfilaments	Protein fibers	Provide strength and support; enable cell motility; transport

(d) Each cell of the intestinal epithelium has structures that allow it to carry out the digestion and absorption of food. The microvilli on the side of the cell that faces the lumen of the intestine increase the surface area available for absorption. This epithelial cell also has most of the organelles of a typical cell.

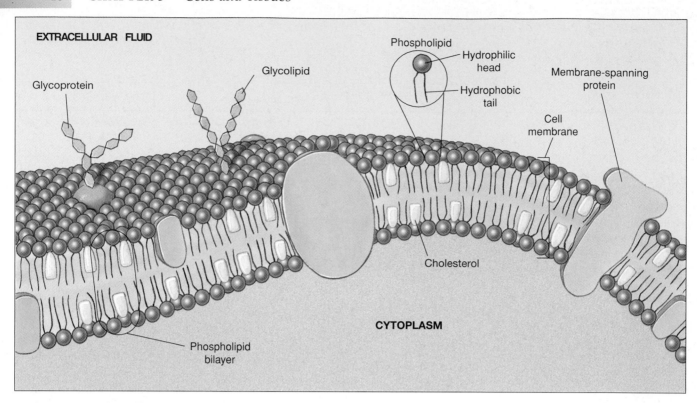

■ **Figure 3-5   The cell membrane**   The cell membrane is a double layer of phospholipid molecules. Proteins of various kinds are inserted into and through the phospholipid bilayer, and carbohydrates bind to proteins and lipids on the extracellular surface.

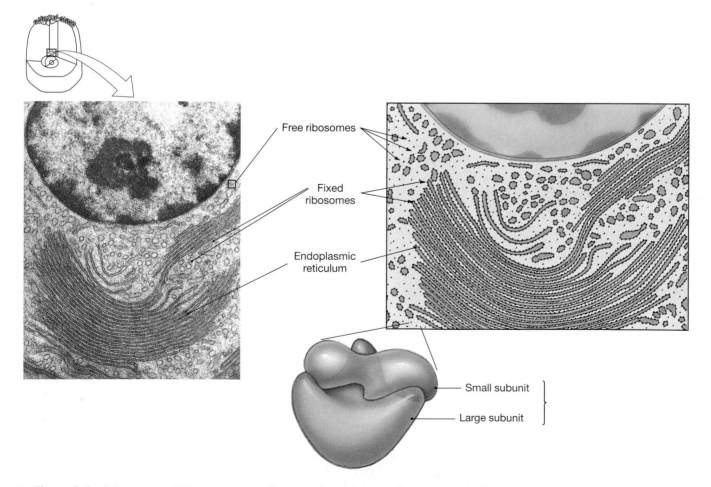

■ **Figure 3-6   Ribosomes**   Ribosomes are small nonmembranous organelles, composed of RNA and protein, organized into two subunits. Many ribosomes are attached to endoplasmic reticulum, but others are found free in the cytoplasm.

made of a protein called *tubulin*. Microtubules and other fibers link together to create the internal scaffolding of the cell, the cytoskeleton. In addition, microtubules combine to form the more complex structures of centrioles, cilia, and flagella.

The Cytoskeleton Is a Changeable Organelle The **cytoskeleton** [Greek *kytos*, cell + skeleton] is a flexible, changeable three-dimensional scaffolding of actin microfilaments, intermediate filaments, and microtubules that extends throughout the cytoplasm. Some cytoskeleton fibers are permanent, but others can be synthesized or disassembled according to the cell's needs. Because of the cytoskeleton's changeable nature, its organizational details are not fully understood.

The cytoskeleton has at least five important functions.

1. The protein scaffolding provides mechanical strength to the cell, and in some cells, plays an important role in determining the shape of the cell. Figure 3-7a ■ shows how the cytoskeleton helps support **microvilli** [*micros*, small + *villus*, tuft of hair], fingerlike extensions of the cell membrane that increase the surface area for absorption of material.

2. The fibers stabilize the positions of organelles. Figure 3-7b ■ is a diagram of a typical cell with organelles held in place by cytoskeletal fibers.

3. The cytoskeleton helps transport materials into the cell and within the cytoplasm, serving an intracellular "railroad track" system for moving vesicles and organelles. This function is particularly important in cells of the nervous system, where material must be transported over intracellular distances as long as a meter.

4. Fibers of the cytoskeleton connect with protein fibers in the extracellular space, linking cells to each other and to support material outside the cells. These linkages also allow the transfer of information from one cell to another.

5. Finally, the cytoskeleton enables certain cells of the human body to move. These mobile cells include white blood cells that squeeze out of blood vessels and travel to sites of infection, as well as growing nerve cells that send out long extensions as they elongate.

Centrosomes and Centrioles Are Associated with Microtubules **Centrosomes** are regions of darkly staining material that are often found close to the cell nucleus. Centrosomes usually act as the cell's *microtubule-organizing center*, where tubulin molecules are assembled into microtubules that lengthen outward through the cytoplasm. In most animal cells, the centrosome contains two **centrioles,** shown in the typical cell of Figure 3-4d ■. Each centriole is a cylindrical bundle of 27 microtubules, arranged in nine triplets (Fig. 3-8a ■). In cell division, the centrioles direct the movement of DNA strands (see Appendix B). Cells that have lost their ability to undergo cell division, such as mature nerve cells, lack centrioles.

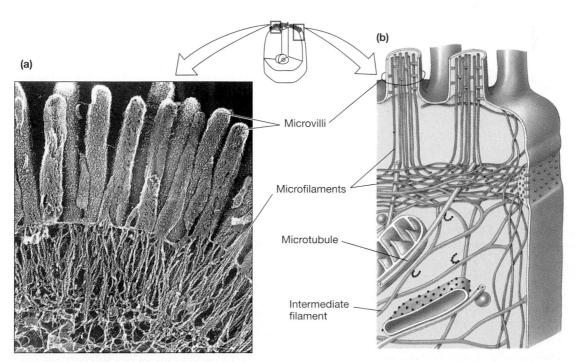

(a)

(b)

Microvilli

Microfilaments

Microtubule

Intermediate filament

■ **Figure 3-7** **The cytoskeleton and microfilaments** (a) Microfilaments form a network just under the cell membrane. They also extend into microvilli that project above the membrane surface. Microvilli greatly increase the surface area of the cell. (b) The cytoskeleton of a cell is composed of microfilaments, intermediate filaments, and microtubules. The cytoskeleton provides support for the cell membrane and anchors cell organelles in position.

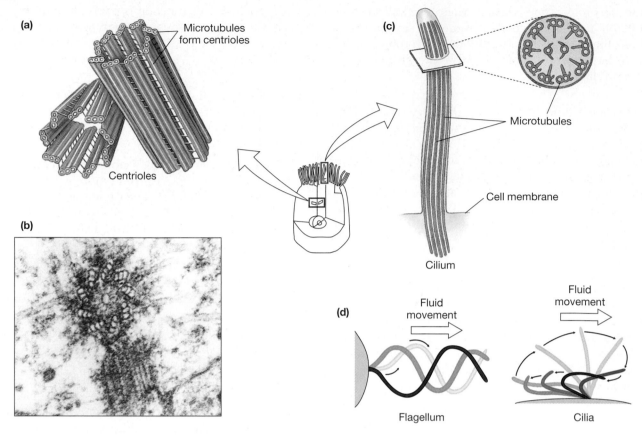

(a)
Microtubules form centrioles

Centrioles

(b)

(c)
Microtubules

Cell membrane

Cilium

(d)
Fluid movement

Fluid movement

Flagellum

Cilia

■ **Figure 3-8   Cilia and centrioles**   (a) Most cells contain a pair of centrioles that direct the movement of DNA during cell division. Each centriole is a cylindrical bundle of 27 microtubules, arranged in nine triplets. (b) A pair of centrioles. (c) The outer surface of a cilium or flagellum is a continuation of the cell membrane. Internally, cilia and flagella are composed of nine pairs of microtubules surrounding a central pair. (d) The beating of cilia and flagella creates fluid movement.

Cilia and Flagella Are Movable Hairlike Structures

Cilia [singular, *cilium,* Latin for eyelash] are short, hairlike structures projecting from the cell surface like the bristles of a brush. The surface of a cilium is a continuation of the cell membrane, and its core contains nine pairs of microtubules surrounding a central pair (Fig. 3-8c ■). The microtubules terminate just inside the cell at the *basal body.* Cilia beat rhythmically back and forth when the microtubule pairs in their core slide past each other in a process that consumes energy. This movement creates currents that sweep fluids or secretions across the cell surface.

Flagella [singular, *flagellum,* Latin for whip] have the same microtubule arrangement as cilia but are considerably longer. In addition, a flagellated animal cell usually only has one or two flagella, whereas ciliated cells may have one surface almost totally covered with cilia. Flagella are found on free-floating single cells; the only such cell in humans is the male sperm cell (see Fig. 3-2 ■). The function of a flagellum is to push the cell through fluid with wavelike movements of the flagellum, just as undulating contractions of a snake's body push it headfirst through its environment. Flagella bend and move by the same basic mechanism as cilia.

✔ How would the absence of a flagellum affect a sperm cell?

Membranous Organelles Create Compartments for Specialized Functions

Membranous organelles are separated from the cytosol by one or more phospholipid membranes similar in structure to the cell membrane. Many membranous organelles have hollow interiors. (The cavity of a hollow tube, organ, or organelle is known as its **lumen,** from the Latin word for light or window. Various anatomical structures have lumens.)

The membrane barrier between the inside of the organelle and the cytosol allows the organelle to contain substances that might be harmful to the cell if they were free. It also allows the cell to separate different functions. For example, proteins are synthesized in one organelle, modified in another, and stored in a third. Figure 3-4e ■ shows five types of membranous organelles: mitochondria, the endoplasmic reticulum, the Golgi apparatus, lysosomes, and peroxisomes.

Mitochondria Are the Powerhouse of the Cell   Mi-

tochondria [singular, *mitochondrion; mitos,* thread + *chondros,* granule] are small spherical to elliptical organelles with an unusual double wall (Fig. 3-9 ■). The outer membrane of the wall gives the mitochondrion its shape. The inner membrane is folded into leaflets or tubules called **cristae.** In the center of the mitochondrion, inside

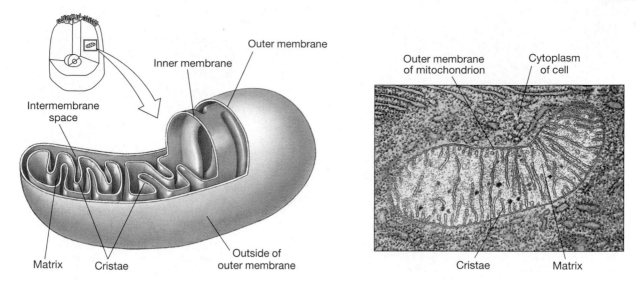

■ **Figure 3-9   Mitochondria**   Mitochondria have a double wall structure with a heavily folded inner membrane.

the inner membrane, is the mitochondrial **matrix** that contains enzymes, ribosomes, granules, and DNA. Between the outer and inner membranes is the **intermembrane space,** a region that plays an important role in the production of ATP by the mitochondria.

Mitochondria are where most ATP is generated, which is why they are nicknamed the cell's "powerhouse." The number of mitochondria in a particular cell depends on the cell's energy needs. Cells such as skeletal muscle cells, which use a lot of energy, have many more mitochondria than less active cells, such as adipose (fat) cells.

Mitochondria are unusual organelles in two ways. First of all, in their matrix they have their own unique DNA. This mitochondrial DNA, along with the ribosomes in the matrix, means that mitochondria can manufacture some of their own proteins. Why is this true of mitochondria and not other organelles? This question has been the subject of intense scrutiny. According to the *prokaryotic endosymbiont* theory, mitochondria are the descendants of bacteria that invaded cells millions of years ago. The bacteria developed a mutually beneficial relationship with their hosts and soon became an integral part of the host cell. Supporting evidence for this theory is the fact that our mitochondria contain DNA, RNA, and enzymes similar to the type found in bacteria but unlike those found in our own nuclei.

The second unusual characteristic of mitochondria is their ability to replicate themselves even when the cell to which they belong is not undergoing cell division. This process is aided by the presence of mitochondrial DNA that allows the organelles to direct their own duplication. Mitochondrial replication takes place by budding, with small daughter mitochondria pinching off an enlarged parent. Cells such as exercising muscle cells, which are subjected to increased energy demands over a period of time, may meet the demand for more ATP by increasing the number of mitochondria in their cytoplasm.

The Endoplasmic Reticulum Is the Site of Protein and Lipid Synthesis

The **endoplasmic reticulum,** or **ER,** is a network of interconnected membrane tubes (Fig. 3-6 and 3-10 ■) that are a continuation of the outer nuclear membrane. The name *reticulum* comes from the Latin word for *net* and refers to the netlike appearance of the tubules. Electron micrographs reveal that there are two forms of endoplasmic reticulum: rough endoplasmic reticulum and smooth endoplasmic reticulum. Rows of ribosomes dot the cytoplasmic surface of rough endoplasmic reticulum, giving it a granular or rough appearance. Smooth endoplasmic reticulum lacks the ribosomes and appears as smooth membrane tubes. Both types of endoplasmic reticulum have the same three major functions: synthesis, storage, and transport of biomolecules.

The **smooth endoplasmic reticulum** (SER) is the main site for the synthesis of fatty acids, steroids, and lipids [∞ p. 29]. Phospholipids for the cell membrane are produced here, and cholesterol is modified into steroid hormones, such as the sex hormones estrogen and testosterone. The smooth endoplasmic reticulum of liver and kidney cells detoxifies or inactivates drugs. In skeletal muscle cells, a modified form of smooth endoplasmic reticulum stores calcium ions (Ca^{2+}) to be used in muscle contraction.

The **rough endoplasmic reticulum** (RER) is the main site for the synthesis of proteins. Proteins are assembled on ribosomes attached to the cytoplasmic surface of the rough endoplasmic reticulum, then inserted into the lumen, where they undergo chemical modification. Most of these proteins are packaged into spherical membrane-bound **transport vesicles** [*vesicula,* bladder] that pinch off from the tips of the endoplasmic reticulum. The transport vesicles shuttle their contents across the cytosol to the Golgi apparatus.

The Golgi Apparatus Packages Proteins into Membrane-Bound Vesicles

The **Golgi apparatus** (Fig. 3-11 ■) was first described by Camillo Golgi in 1898. For years,

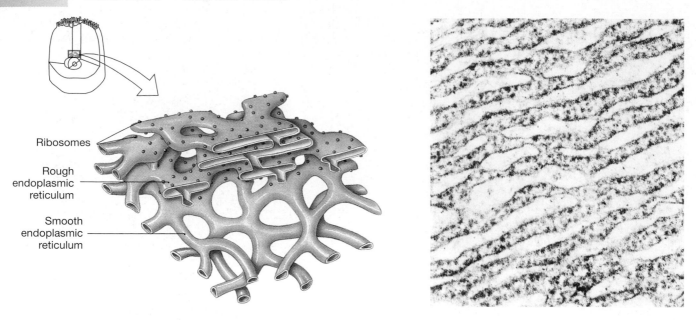

Ribosomes

Rough
endoplasmic
reticulum

Smooth
endoplasmic
reticulum

■ **Figure 3-10** **The endoplasmic reticulum** The endoplasmic reticulum is a series of hollow tubules formed by a continuation of the outer nuclear membrane. The appearance of the rough endoplasmic reticulum is due to the fixed ribosomes attached to it.

some investigators thought that this organelle was just a result of the fixation process needed to prepare tissues for viewing under the light microscope. However, we now know from electron microscope studies that the Golgi apparatus is indeed another membranous organelle. Its function is to modify proteins made by the rough endoplasmic reticulum and package them into membrane-bound vesicles.

The Golgi apparatus consists of five or six hollow curved sacs stacked on top of each other like a series of hot water bottles and connected so that they share a single lumen. The convex side of the stack faces the rough endoplasmic reticulum and receives the transport vesicles from it. These transport vesicles fuse with the membranes of the Golgi sac and discharge their contents into its lumen.

As the proteins move through the sac, they may be modified by enzymes. For example, long proteins may be cleaved into smaller proteins or carbohydrates may be attached to proteins to make glycoproteins. Finally, the

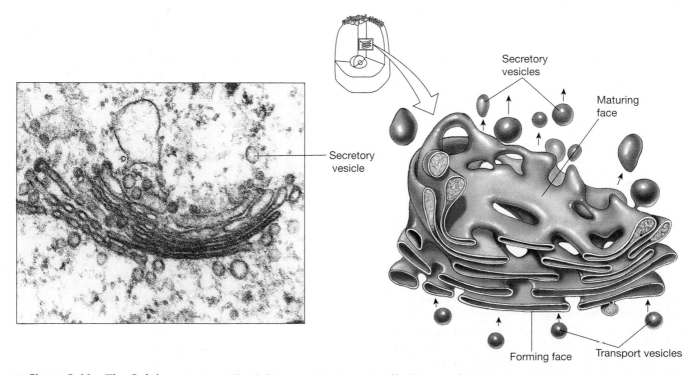

Secretory
vesicles

Maturing
face

Secretory
vesicle

Forming face

Transport vesicles

■ **Figure 3-11** **The Golgi apparatus** The Golgi apparatus is a series of hollow membranous sacs stacked on top of each other. Material arrives at the Golgi in transport vesicles from the endoplasmic reticulum and leaves the Golgi packaged in new vesicles.

processed proteins are enclosed in membrane-bound vesicles that pinch off from the concave face of the Golgi apparatus and move out into the cytosol.

Cytoplasmic vesicles are of two kinds: secretory vesicles and storage vesicles. **Secretory vesicles** contain proteins that will be exported to other parts of the body. (Secretion is the process by which a cell releases a substance into the extracellular space.) The contents of most **storage vesicles,** however, never leave the cytoplasm. Lysosomes are the major storage vesicles of the cell.

Lysosomes Are the Intracellular Digestive System

Lysosomes [*lysis*, dissolution + *soma*, body] are small, spherical storage vesicles that appear as membrane-bound granules in the cytoplasm (Fig. 3-12 ■). Lysosomes contain powerful enzymes and act as the digestive system of the cell. They take up bacteria or old organelles, such as mitochondria, and use the enzymes to break them down into their component molecules. Those molecules that can be reused are reabsorbed into the cytosol, while the rest are dumped out of the cell. As many as 50 types of enzymes have been identified from lysosomes of different cell types.

The digestive enzymes of lysosomes are not always kept isolated within the membranes of the organelle. Occasionally, lysosomes release their enzymes to dissolve extracellular support material, such as the hard calcium carbonate portion of bone. The inappropriate release of lysosomal enzymes has been implicated in certain disease states: for example, the inflammation and destruction of joint tissue in *rheumatoid arthritis.*

In other instances, cells allow the enzymes of their lysosomes to come in contact with the cytoplasm, leading to self-digestion of all or part of the cell. When mus-

cles *atrophy* (shrink) from lack of use or the uterus diminishes in size after pregnancy, the loss of cell mass is due to the action of lysosomes.

Because lysosomal enzymes are so powerful, one of the questions that puzzled researchers was why these enzymes do not normally destroy the cell that contains them. What they discovered was that these enzymes are activated only by very acid conditions, 100 times more acid than the normal cytoplasm. When lysosomes are first pinched off from the Golgi apparatus, their interior pH is about the same as that of the cytosol, 7.0–7.3. The enzymes are inactive at this pH. Their inactivity serves as a form of insurance. If the lysosome breaks or accidentally releases the enzyme, it will not harm the cell. However, as the lysosome sits in the cytoplasm, it accumulates H^+ ions in a process that uses energy. The pH inside the vesicle drops to 4.8–5.0, and the enzymes are activated. Once this has occurred, the enzymes are capable of breaking down various biomolecules.

In inherited conditions known as *lysosomal storage diseases,* lysosomes are not effective because they lack specific enzymes. One of the best-known lysosomal storage diseases is the fatal inherited condition known as Tay-Sachs disease. Infants with Tay-Sachs disease have defective lysosomes that fail to break down glycolipids in nerve cells. Accumulation of the glycolipids in the cells causes symptoms of nervous system dysfunction, including blindness and loss of coordination. Most of the afflicted infants die in early childhood.

Peroxisomes Contain Enzymes That Neutralize Toxins

Peroxisomes are storage vesicles that are even smaller than lysosomes. For years, they were thought to be a kind of lysosome, but we now know that they contain a different set of enzymes. Their main function appears to be to degrade long-chain fatty acids and potentially toxic foreign molecules. Peroxisomes get their name from the fact that the breakdown of fatty acids generates hydrogen peroxide (H_2O_2), a toxic molecule. The peroxisomes rapidly convert this peroxide to oxygen and water by using the enzyme *catalase.* Peroxisomal disorders disrupt the normal processing of lipids and can severely disrupt the normal function of the nervous system by altering the structure of nerve cell membranes.

✔ Microscopic examination of a cell reveals many mitochondria. What does this observation imply about the cell's energy requirements?

✔ You are examining tissue from a previously unknown species of fish. You discover a tissue with large amounts of smooth endoplasmic reticulum in its cells. What is one possible function of these cells?

✔ Which would have more rough endoplasmic reticulum: pancreatic cells that manufacture the protein hormone insulin, or adrenal cortex cells that synthesize the steroid hormone cortisol?

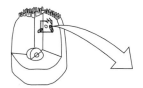

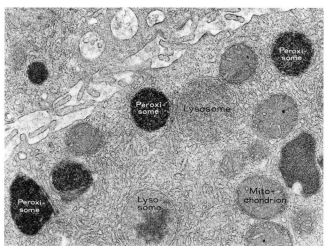

■ **Figure 3-12   Lysosomes and peroxisomes** Lysosomes and peroxisomes are membrane-bound vesicles filled with enzymes.

The Nucleus Is the Cell's Control Center

The **nucleus** of the cell contains DNA, the genetic material that ultimately controls all cell processes. How the nucleus receives information about conditions in the cell or elsewhere in the body and responds with fine-tuned control of the cell's protein synthesis is an active area of biological research. Figure 3-13 ■ illustrates the structure of a typical nucleus. The boundary, or **nuclear envelope,** is a two-membrane structure that separates the nucleus from the cytoplasmic compartment. The outer membrane of the envelope is connected with the endoplasmic reticulum, and both membranes of the envelope are pierced here and there by round holes, or **pores.**

Communication between the nucleus and cytosol occurs through the **nuclear pore complexes,** large protein complexes with a central channel. Ions and small molecules move freely through this channel when it is open, but proteins and RNA must be transported by a process that uses energy. This requirement allows the cell to restrict large molecules such as DNA to the nucleus and various enzymes to either the cytoplasm or the nucleus.

In electron micrographs of cells that are not dividing, most DNA in the nucleus appears as randomly scattered granular material, or **chromatin.** A nucleus usually also contains from one to four larger dark-staining bodies of DNA, RNA, and protein called **nucleoli** [singular, *nucleolus*, little nucleus]. Nucleoli contain the genes and proteins that control the synthesis of RNA for ribosomes (see Appendix B).

...continued from page 45

During a Pap test for cervical cancer, cell samples are swabbed from the cervix (neck) of the uterus. The cells are then smeared onto a glass slide and sent to a laboratory for examination by a trained cytologist. The cytologist looks for *dysplasia,* a change in the size and shape of the cells that is indicative of cancerous changes [*dys-,* abnormal + *-plasia,* growth or cell multiplication]. Cancer cells can usually be recognized by a large nucleus surrounded by a relatively small amount of cytoplasm. Jan's first Pap test showed all the hallmarks of dysplasia.

Question 2:   *What property of cancer cells explains the large size of their nucleus and the relatively small amount of cytoplasm?*

❱ TISSUES OF THE BODY

Despite the amazing variety of intracellular structures, no single cell can carry out all the processes of the mature human body. Instead, cells assemble into larger units called **tissues,** collections of cells held together by specialized connections called **cell junctions,** and by other support structures. Tissues range in complexity from simple tissues containing only one cell type, such as the lining of blood vessels, to complex tissues containing many cell types and extensive extracellular material, such as connective tissue. The cells of most tissues work together to achieve a common purpose.

The study of tissue structure and function is known as **histology** [*histos,* tissue]. Histologists describe tissues

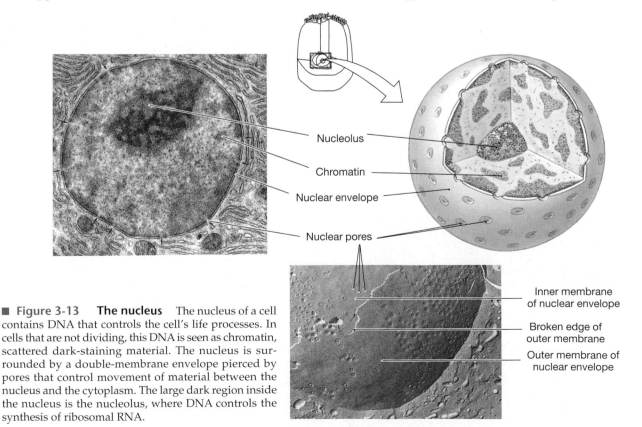

Nucleolus

Chromatin

Nuclear envelope

Nuclear pores

Inner membrane of nuclear envelope

Broken edge of outer membrane

Outer membrane of nuclear envelope

■ **Figure 3-13   The nucleus**   The nucleus of a cell contains DNA that controls the cell's life processes. In cells that are not dividing, this DNA is seen as chromatin, scattered dark-staining material. The nucleus is surrounded by a double-membrane envelope pierced by pores that control movement of material between the nucleus and the cytoplasm. The large dark region inside the nucleus is the nucleolus, where DNA controls the synthesis of ribosomal RNA.

by their physical features: (1) the shape and size of the cells, (2) how the cells are arranged in the tissue (in layers, scattered, and so on), (3) how the cells are connected to each other, and (4) the amount of extracellular material that is present in the tissue. There are four **primary tissue types** in the human body: epithelial, connective, muscle, and neural, or nerve (Table 3-1). Before we consider each tissue type specifically, let us look at how cells link together to form tissues.

Extracellular Matrix Helps Support Tissues

Matrix is extracellular material that is synthesized and secreted by the cells of a tissue. For years, we believed that its only role was to hold cells together, but experimental evidence now shows that matrix plays a key role in physiological function. When matrix proteins attach to proteins in the cell membrane, they provide a means of communication between the cell and its external environment. Indeed, a recent review article was entitled "Extracellular matrix: the central regulator of cell and tissue homeostasis." You will learn more about the physiological role of matrix in Chapter 5.

The amount of matrix in a tissue is highly variable. Nerve and muscle tissue have very little matrix, but the connective tissues, cartilage, bone, and blood, have extensive matrix that occupies as much volume as their cells. The consistency of matrix varies from watery (blood and lymph) to rigid (bone). In many tissues, matrix is composed of complex glycoprotein molecules mixed with insoluble protein fibers. The protein fibers provide strength and anchor cells to the matrix.

Cell Junctions Hold Cells Together to Form Tissues

How are individual cells within tissues connected to each other and to the extracellular matrix? Membrane proteins known as **cell adhesion molecules,** or **CAMs,** provide the linkages. There are three types of cell junctions: adhesive junctions, tight junctions, and gap junctions.

Adhesive Junctions   Adhesive junctions have been compared to buttons or zippers that link cells together and hold them in position within a tissue. They are recognizable in electron micrographs by the dense glycoprotein bodies, or *plaques,* that lie just inside the cell membranes in the region where the two cells connect (Fig. 3-14b ■). The protein linkage of adhesive cell junctions is very strong, and it allows sheets of tissue in skin and lining body cavities to resist damage from stretching and twisting. Even the tough protein fibers of adhesive junctions can be broken, however. If you have shoes that rub against your skin, the stress can shear the proteins connecting the different skin layers. When fluid accumulates in the resulting space and the layers separate, a *blister* results.

In vertebrates, adhesive junctions take the form of either desmosomes or adherens junctions. **Desmosomes** [*desmos,* band + *soma,* body] are created by membrane-spanning proteins called *cadherins* that connect with each

TABLE 3-1   Characteristics of the Four Tissue Types

	Epithelial	Connective	Muscle	Nerve
Matrix amount	Minimal	Extensive	Absent	Absent
Matrix type	Basement membrane	Varied—protein fibers in ground substance that ranges from liquid to gelatinous to firm to calcified	NA	NA
Unique features	No direct blood supply	Cartilage has no blood supply	Able to generate electrical signals, force, and movement	Able to generate electrical signals
Surface features of cells	Microvilli, cilia	NA	NA	NA
Locations	Covers body surface; lines cavities and hollow organs and tubes. Secretory glands	Supports skin and other organs. Cartilage, bone, and blood	Makes up skeletal muscles, hollow organs, and tubes	Throughout body. Concentrated in brain and spinal cord
Cell arrangement and shapes	Variable number of layers, from one to many; cells flattened, cuboidal, or columnar	Cells not in layers; usually randomly scattered in matrix; cell shape irregular to round	Cells linked in sheets or elongated bundles; cells shaped in elongated, thin cylinders. Heart muscle cells may be branched	Cells isolated or networked; cell appendages highly branched and/or elongated

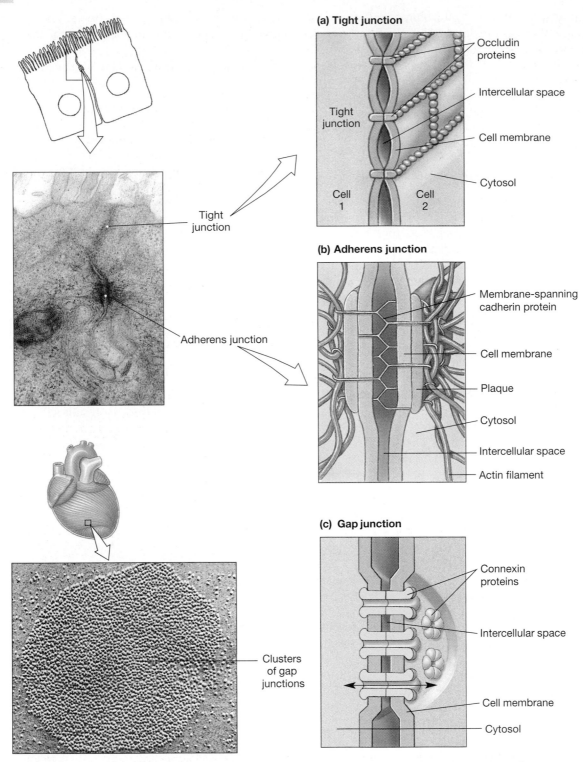

(a) Tight junction

Occludin proteins

Intercellular space

Cell membrane

Cytosol

Tight junction

Cell 1

Cell 2

(b) Adherens junction

Membrane-spanning cadherin protein

Cell membrane

Plaque

Cytosol

Intercellular space

Actin filament

(c) Gap junction

Connexin proteins

Intercellular space

Cell membrane

Cytosol

Tight junction

Adherens junction

Clusters of gap junctions

■ **Figure 3-14   Types of cell junctions**   A junction complex consists of a tight junction and an adherens junction. (a) Tight junctions prevent movement of material between the cells. (b) Adherens junctions anchor cells together using proteins called cadherins. (c) Gap junctions are formed when connexin proteins create a cytoplasmic bridge between adjacent cells.

other across the intercellular space. Inside the cells, cadherins link to intermediate filaments of the cytoskeleton. Desmosomes may be small points of contact between two cells (spot desmosomes) or bands that encircle the entire cell (belt desmosomes). **Hemidesmosomes** [*hemi-,* half]

use cell membrane proteins called *integrins* to anchor cells to matrix glycoproteins, such as *fibronectin.* **Adherens junctions** are similar to desmosomes but not quite as strong. They link actin microfilaments in adjacent cells together with the help of cadherins.

Tight Junctions **Tight junctions** serve to prevent the movement of material past the cells they link. In tight junctions, the cell membranes of adjacent cells partly fuse together with the help of proteins called *occludins,* thereby making a barrier (Fig. 3-14a ■). Tight junctions always occur next to adherens junctions; the combination is called a *junction complex.*

Tissues held together with tight junctions are like a solid brick wall: very little can pass from one side of the wall to the other between the bricks. By comparison, tissues held together with adhesive junctions are like a picket fence. The pickets form a continuous surface but the spaces between the individual pickets allow material to pass from one side of the fence to the other.

In physiology, tight junctions in the intestinal tract and kidney prevent substances from moving freely between the external and internal environments and thus allow cells to regulate what enters and leaves the body. Tight junctions also create the so-called blood-brain barrier that prevents potentially harmful substances in the blood from reaching the extracellular fluid of the brain.

Gap Junctions **Gap junctions** create cytoplasmic bridges between adjoining cells so that chemical and electrical signals pass rapidly from one cell to the next (Fig. 3-14c ■). Interlocking cylindrical proteins, called *connexins,* resemble hollow rivets with narrow channels through their centers. Small molecules and ions move through the channels from cell to cell. Gap junctions were once thought to be found only in certain muscle and nerve cells, but we now know they are important in cell-to-cell communication in many tissues, including the liver, pancreas, ovary, and thyroid gland.

Now that you understand how cells are held together into tissues, we will look at the four different tissue types in the body: (1) epithelial, (2) connective, (3) muscle, and (4) nervous tissues.

Epithelia Provide Protection and Regulate Exchange

The **epithelial tissues,** or **epithelia,** protect the internal environment of the body and regulate the exchange of material between the internal and external environments. They cover exposed surfaces, such as the skin, and line internal passageways, such as the digestive tract. **Any substance that enters or leaves the internal environment of the body must cross an epithelium.**

Some epithelia, such as those of the skin and mucous membranes of the mouth, act as a barrier to keep water in the body and invaders such as bacteria out. Other epithelia, such as those in the kidney and intestinal tract, control the movement of material between the external environment and the extracellular fluid of the body. Indeed, nutrients, gases, and wastes must often cross several different epithelia in their passage between cells and the outside world.

Matrix, Junctions, and Cancer The disappearance of cell junctions is a characteristic of cancer and probably contributes to the spread, or *metastasis,* of cancer throughout the body. Cancer cells lose the adherens junctions that tie them to adjacent cells because they have fewer of the cadherin molecules that make up the junctions. Once the cancer cell is released from its moorings, it secretes enzymes known as *proteases.* These enzymes, especially those called *matrix metalloproteinases* (MMPs), dissolve the extracellular matrix so that the cancer cells can invade adjacent tissues or enter the bloodstream. Researchers are investigating ways to block these enzymes to see if they can prevent metastasis.

Another type of epithelium is specialized to manufacture and secrete chemicals into the blood or to the external environment. Sweat and saliva are examples of substances secreted by epithelia into the environment. Hormones, signal molecules used to maintain homeostasis, are secreted into the blood.

Structure of Epithelia Epithelia typically have a thin layer of extracellular matrix that lies between the cells and their underlying tissues (Fig. 3-15 ■). This layer, called the **basal lamina** [*bassus,* low; *lamina,* a thin plate], or **basement membrane,** is composed of a network of fine protein filaments embedded in glycoprotein. The filaments hold the epithelial cells to the underlying cell layers, just as adhesive junctions hold the individual cells in the epithelium to each other.

The cell junctions in epithelia are variable. Physiologists classify epithelia either as "leaky" or "tight." In a leaky epithelium, adhesive junctions leave gaps or pores that allow molecules to pass across the epithelium between

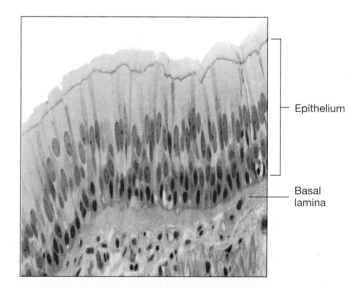

■ **Figure 3-15 Epithelial tissue** Epithelial tissue is supported by a layer of extracellular matrix known as the basal lamina. This matrix is secreted by the cells of the epithelium.

the cells. One leaky epithelium is the wall of the capillary blood vessels, where all dissolved molecules except for large proteins can pass between the cells. In a tight epithelium, such as the kidney, adjacent cells are bound to each other by tight junctions and infolded membranes that create a barrier to movement between the cells. In order to cross a tight epithelium, most substances must enter the cells and go *through* them. The tightness of an epithelium is directly related to how selective it is about what can move across it. Some epithelia, such as those of the intestine, have the ability to alter the tightness of their junctions according to the body's needs.

Types of Epithelia   Epithelial tissues can be divided into two general types: (1) sheets of tissue that lie on the surface of the body or that line the inside of tubes and hollow organs and (2) secretory epithelia that synthesize and release substances into the extracellular space. Histologists classify sheet epithelia by the number of cell layers in the tissue and by the shape of the cells in the surface layer. This classification scheme recognizes two types of layering—**simple** (one cell thick) and **stratified** (multiple cell layers) [*stratum,* layer + *facere,* to make]—and three cell shapes—**squamous** [*squama,* flattened plate or scale], **cuboidal,** and **columnar.** However, physiologists are more concerned with the functions of these tissues, so we will divide epithelia into five groups according to their function.

There are five functional types of epithelia: exchange, transporting, ciliated, protective, and secretory. **Exchange epithelia** permit rapid exchange of material, in particular, gases. **Transporting epithelia** are selective about what can cross them and are found primarily in the intestinal tract and the kidney. **Ciliated epithelia** are located in the airways of the respiratory system and in the female reproductive tract. **Protective epithelia** are found on the surface of the body and just inside the openings of body cavities. The **secretory epithelia** synthesize and release secretory products into the external environment or into the blood. Figure 3-16 ■ shows the

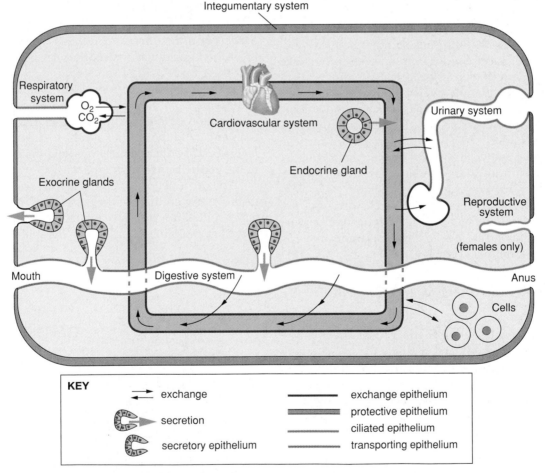

■ **Figure 3-16**   **Distribution of epithelia in the body**   The outer surface of the body and the openings of the body cavities are covered with protective epithelium (brown). The exchange surface of the lungs and the blood vessels are lined with exchange epithelium (red). Transporting epithelia (purple) are found primarily in the digestive tract and kidney, where they regulate the movement of substances into and out of the extracellular fluid. Ciliated epithelia (blue) are found in the airways of the respiratory tract and in part of the female reproductive system. The secretory epithelia (green) come in two forms. Endocrine glands, without ducts, secrete their hormones into the blood. Exocrine glands, with ducts, secrete their products outside the body—onto the surface of the skin or into the lumen of an organ that opens into the environment outside the body (∞ Fig. 1-2, p. 3).

TABLE 3-2   Types of Epithelia

	Number of Cell Layers	Cell Shape	Special Features	Where Found
Exchange	One	Flattened	Pores between cells to permit easy passage of molecules	Lungs, lining blood vessels
Transporting	One	Columnar or cuboidal	Tight junctions prevent movement between cells; surface area increased by folding of cell membrane into villi, or fingers	Intestine, kidney, some exocrine glands
Protective	Many	Flattened in surface layers; polygonal in deeper layers	Cells tightly connected by many desmosomes	Skin, lining cavities, such as the mouth, that open to the environment
Ciliated	One	Cuboidal to columnar	One side covered with cilia to move fluid across surface	Nose, trachea, and upper airways; female reproductive tract
Secretory	One to many	Columnar to polygonal	Protein-secreting cells filled with membrane-bound secretory granules and extensive RER. Steroid-secreting cells contain lipid droplets and extensive SER	Exocrine glands, including pancreas, sweat glands, salivary glands; endocrine glands, such as thyroid, gonads

distribution of these types of epithelia in the different systems of the body. Notice that most epithelia face the external environment on one surface and the extracellular fluid on the other. The only exception to this rule is the endocrine glands. Table 3-2 summarizes the types of epithelia.

Exchange epithelia The **exchange epithelia** are composed of very thin, flattened cells that allow gases (CO_2 and O_2) to pass rapidly through the cells. This type of epithelium lines the blood vessels and the lungs, the two major sites of gas exchange in the body. In the smallest

blood vessels, called capillaries, gaps or pores in the epithelium also allow molecules smaller than proteins to pass *between* the cells, making this a leaky epithelium (Fig. 3-17a ■). Histologists classify thin exchange tissue as *simple squamous epithelium* because it has a single layer of thin, flattened cells. The simple squamous epithelial lining of the heart and blood vessels is also called an **endothelium.**

Transporting epithelia The **transporting epithelia** actively and selectively regulate the exchange of nongaseous material, such as ions and nutrients, between the inter-

(a) Leaky exchange epithelium

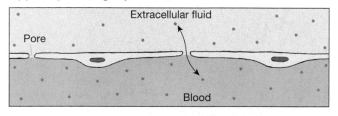

(b) Tight junction in a transporting epithelium

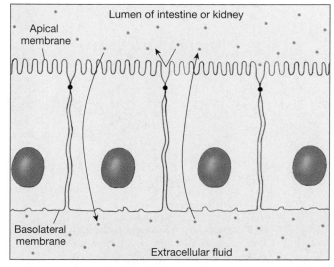

■ **Figure 3-17   Movement of substances across tight and leaky epithelia**   (a) In leaky epithelia, such as the capillary shown, substances can move through gaps or pores between the cells. (b) In transporting epithelia with tight junctions, substances crossing the epithelium must move through the epithelial cell.

nal and external environments. These epithelia are found lining the hollow tubes of the digestive system and the kidney, where lumens open into the external environment [∞ p. 3].

There are several characteristics of transporting epithelia:

1. The cells of transporting epithelia are much thicker than those of exchange epithelia, so they act as a barrier as well as an entry point. The cell layer is only one cell thick, but the cells are cuboidal or columnar in shape (Fig. 3-17b ■).

2. The **apical membrane,** the surface of the cell that faces the lumen, has **microvilli,** fingerlike projections that increase the surface area available for transport. A cell with microvilli has at least 20 times the surface area of a cell without them. In addition, the **basolateral membrane,** the side of the cell facing the extracellular fluid, may also have folds that increase the cell's surface area (Fig. 3-17b ■).

3. The cells of transporting epithelia are firmly attached to adjacent cells by moderately tight to very tight junctions. This means that material must move into an epithelial cell on one side of the tissue and out of the cell on the other in order to cross the epithelium (Fig. 3-17b ■).

4. Most cells that transport materials have numerous mitochondria to provide energy for transport processes (discussed further in Chapter 5).

The properties of transporting epithelia differ. For example, glucose can be moved from the lumen of the small intestine to the extracellular fluid, but cannot be moved across the epithelium of the large intestine. Furthermore, the transport properties of an epithelium can be regulated and modified in response to various stimuli. Hormones, for example, affect the transport of ions by the epithelium of the kidney. You will learn more about transporting epithelia when you study the kidney and the digestive systems.

Ciliated epithelia   **Ciliated epithelia** are nontransporting tissues found lining the respiratory system and parts of the female reproductive tract. The surface of the tissue facing the lumen is covered with cilia that beat in a coordinated, rhythmic fashion, moving fluid and particles across the surface of the tissue. Injury to the cilia or to their epithelial cells can stop ciliary movement. Paralysis of the ciliated epithelium lining the respiratory tract is one effect of smoking. It is considered to contribute to the higher incidence of respiratory infection in smokers, because the mucus that traps bacteria is no longer swept up out of the lungs by the cilia. Figure 3-18 ■ shows the ciliated columnar epithelium of the trachea.

Protective epithelia   The **protective epithelia** are designed to prevent any exchange between the internal and external environments. They are stratified tissues, composed of many stacked layers of cells. The epidermis [*epi,* upon + *derma,* skin] and linings of the mouth, pharynx,

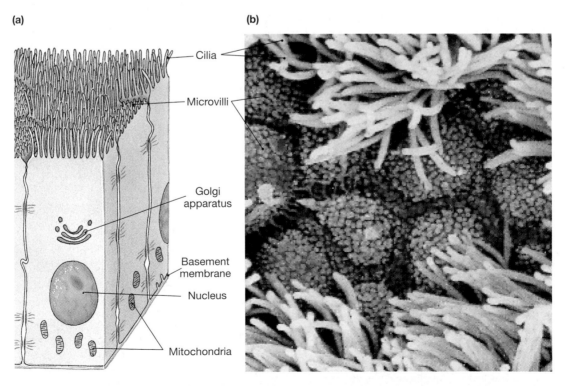

(a)   (b)

Cilia

Microvilli

Golgi apparatus

Basement membrane

Nucleus

Mitochondria

■ **Figure 3-18   Ciliated epithelia**   (a) This is a drawing of the ciliated epithelium lining the trachea. (b) This scanning electron micrograph shows ciliated cells interspersed with mucus-secreting cells whose surfaces are covered with shorter microvilli.

esophagus, urethra, and vagina are all protective epithelia. The multiple cell layers of this tissue type protect areas subject to mechanical or chemical stresses. Some of these tissues are toughened further by the secretion of *keratin* [*keras*, horn], the same insoluble protein abundant in hair and nails.

Because the protective epithelia are subjected to irritating chemicals, bacteria, and other destructive forces, the cells in them have a short life span. In the deeper layers, new cells are produced continually, displacing the older cells at the surface. Each time you wash your face, you scrub off dead cells on the surface layer. As the skin ages, the rate of cell turnover declines. Retin-A, a drug derived from vitamin A, speeds up cell division and surface shedding so that the treated skin develops a more youthful appearance.

Secretory epithelia **Secretory epithelia** are composed of cells that produce a substance inside the cell and release it into the extracellular space in a process is known as **secretion.** Secretory cells may be scattered among other epithelial cells, or they may group together to form a multicellular **gland.** There are two types of secretory gland: exocrine glands and endocrine glands.

Exocrine glands release their secretions into the external environment [*exo-*, outside + *krinein*, to secrete]. This may be onto the surface of the skin or onto an epithelium lining one of the internal passageways, such as the airways of the lung or the lumen of the intestine. In effect, an exocrine secretion leaves the body. This is important to know, if you consider that some exocrine secretions, like stomach acid, have a pH that is incompatible with life. Most exocrine glands release their products through open tubes known as **ducts.** Sweat glands, mammary glands found in the breast, salivary glands, the liver, and the pancreas are all exocrine glands.

...continued from page 54

Many kinds of cancer develop in epithelial cells that are subject to damage or trauma. The cervix consists of two types of epithelia. Secretory epithelium with mucus-secreting glands lines the inside of the cervix, while a protective epithelium covers the outside of the cervix. At the opening of the cervix, these two types of epithelia come together. As Jan lies on the examining table, her physician uses a small spatula and a little brush to take samples of cells from both inside and outside her cervix. She then smears the cells on a slide and sprays them with fixative. The slides will now be sent to a lab for evaluation.

Question 3: *What kind of damage or trauma are cervical epithelial cells normally subjected to? Which of its two types of epithelia is more likely to be affected by trauma?*

Exocrine gland cells produce two types of secretions. **Serous secretions** are watery solutions, and many of them contain enzymes. Tears, sweat, and digestive enzyme solutions are all serous exocrine secretions. **Mucous** secretions (also called **mucus**) are sticky solutions containing glycoproteins and proteoglycans. **Goblet cells,** shown in Figure 3-19 ■, are single exocrine cells that produce mucus. Mucus acts as a lubricant for food to be swallowed, as a trap for foreign particles and microorganisms inhaled or ingested, and as a protective barrier between the epithelium and the environment. Some exocrine glands contain more than one type of secretory cell, and they produce both serous and mucous secretions. For example, the salivary glands release mixed secretions.

Endocrine secretions, or **hormones,** are released by ductless glands or single cells directly into the extracellular space. Hormones then enter the blood for distribution to other portions of the body, where they regulate or coordinate the activities of various tissues, organs,

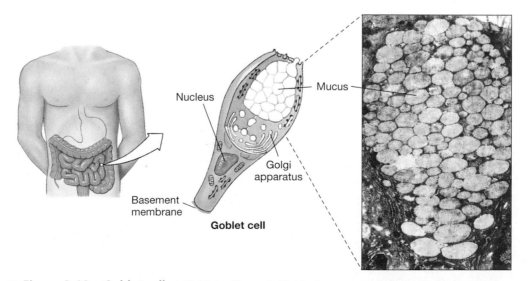

Nucleus

Mucus

Golgi apparatus

Basement membrane

Goblet cell

■ **Figure 3-19 Goblet cells** Goblet cells are individual mucus-secreting cells that are interspersed among other cells in an epithelium, such as that lining the intestine. Their apical ends, close to the lumen, are filled with large, clear droplets of mucus.

and organ systems. Some of the best-known endocrine glands are the pancreas, the thyroid gland, the gonads, and the pituitary gland. For years, it was thought that all hormones were produced by cells grouped together into endocrine glands. In recent years, however, isolated endocrine cells have been found scattered in the epithelial lining of the digestive tract, in the tubules of the kidney, and in the walls of the heart.

Figure 3-20 ■ shows the epithelial origin of endocrine and exocrine glands. During development, epithelial cells grow downward into the supporting tissues. Exocrine glands retain a link to the parent epithelium in the form of a hollow passageway or duct that transports the secretion to its destination. Endocrine glands lose the connecting cells and secrete their hormones into the bloodstream.

✔ You look at a tissue under a microscope and see a simple squamous epithelium. Can it be a sample of the skin surface? Explain.

✔ A cell of the intestinal epithelium secretes a substance into the extracellular space, where it is picked up by the blood and carried to the pancreas. Is the intestinal epithelium cell an endocrine or an exocrine cell?

Connective Tissues Provide Support and Barriers

Connective tissues, the second major tissue type, provide structural support and sometimes a physical barrier that, along with specialized cells, helps defend the body from foreign invaders such as bacteria. The distinguishing characteristic of connective tissues is the presence of extensive extracellular matrix containing widely scattered cells. The cells secrete and modify the matrix of the tissue. Connective tissues range from blood, to the support tissues for the skin and internal organs, to cartilage and bone.

Structure of Connective Tissue   The matrix of connective tissue is a **ground substance** of glycoproteins and water in which insoluble fibrous protein fibers are arranged, much like suspended pieces of fruit in a gelatin salad. The consistency of ground substance is highly variable, depending on the type of connective tissue. At one extreme is the watery matrix of blood, and at the other extreme is the hardened matrix of bone. In between are solutions of glycoproteins that vary in consistency from syrupy to gelatinous. The term *ground substance* is sometimes used interchangeably with *matrix*.

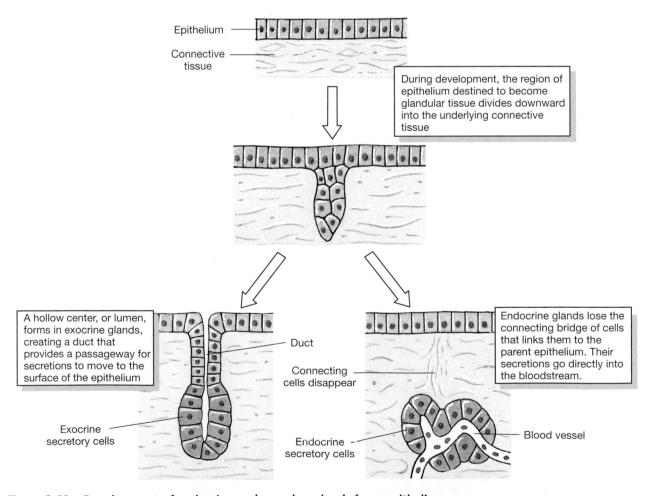

During development, the region of epithelium destined to become glandular tissue divides downward into the underlying connective tissue

A hollow center, or lumen, forms in exocrine glands, creating a duct that provides a passageway for secretions to move to the surface of the epithelium

Endocrine glands lose the connecting bridge of cells that links them to the parent epithelium. Their secretions go directly into the bloodstream.

Epithelium

Connective tissue

Duct

Connecting cells disappear

Exocrine secretory cells

Endocrine secretory cells

Blood vessel

■ Figure 3-20   **Development of endocrine and exocrine glands from epithelium**

DIABETES Diabetes and the Bioartificial Pancreas When organs such as the kidney, heart, or liver fail, the solution is often to transplant a replacement organ from a donor. But in one form of diabetes mellitus, in which the insulin-producing beta cells (β-cells) of the pancreas fail, the solution is not so easy. Most of the diabetic patient's pancreas is functioning normally, and only the beta cells need to be replaced. A variety of solutions have been proposed and are in the trial stage. For patients who have kidney failure due to their diabetes, combined human kidney-pancreas transplants have been performed. A second technique being investigated encloses isolated beta cells from humans or pigs in a semipermeable capsule. This bioartificial pancreas allows glucose and nutrients to enter and insulin secreted by the beta cells to leave. Both types of transplantation are complicated by rejection reactions in which the body recognizes the transplants as foreign and tries to destroy them, so research is continuing.

The actual cells of connective tissue lie embedded within the extracellular matrix. Connective tissue cells are described as *fixed* if they remain in one place, or *mobile* if they can move from place to place. **Fixed cells** are responsible for local maintenance, tissue repair, and energy storage. The **mobile cells** of connective tissue are responsible mainly for defense. The distinction between fixed and mobile cells is not absolute, because at least one cell type is found in both fixed and mobile forms.

Although matrix itself is nonliving, the connective tissue cells constantly modify it by adding, deleting, or rearranging molecules. The suffix *-blast* [*blastos*, sprout] on a connective tissue cell name indicates a cell that is either growing or actively secreting extracellular matrix. Cells that are actively breaking down matrix are identi-

fied by the suffix *-clast* [*klastos*, broken]. Cells that are neither growing, secreting, nor breaking down matrix are often identified by the suffix *-cyte*, meaning "cell." By remembering these suffixes you will be able to tell the functional difference between cells with similar names, such as the osteoblast, osteocyte, and osteoclast, three cell types found in bone.

In addition to secreting glycoprotein ground substance, connective tissue cells produce the fibers of the matrix. Four types of fiber proteins are found in the matrix of connective tissue, aggregated into insoluble protein fibers. **Collagen** [*kolla*, glue + *-genes*, produced] is the most abundant protein in the human body, almost one-third of its dry weight. It is also the most diverse of the four protein types, with at least 12 different variations. Collagen is found almost everywhere connective tissue is found, from the skin to muscles and bones. Individual collagen molecules pack together to form collagen fibers, flexible but inelastic fibers whose strength per unit weight exceeds that of steel. The amount and arrangement of collagen fibers is one of the distinguishing characteristics of different types of connective tissue.

Three other protein fibers in connective tissue are elastin, fibrillin, and fibronectin. **Elastin** is a coiled, wavy protein that returns to its original length after being stretched. This property is known as *elasticity*. Elastin combines with the very thin, straight fibers of **fibrillin** to form filaments and sheets of elastic fibers. These two fibers are important in elastic tissues such as the lungs, blood vessels, and skin. As discussed earlier, fibronectin connects cells to their extracellular matrix. Fibronectins also play an important role in wound healing and in blood clotting.

Types of Connective Tissue   Table 3-3 describes in detail the different types of connective tissue. The most common types of connective tissue are loose and dense

TABLE 3-3   Types of Connective Tissue

Tissue Name	Ground Substance	Fiber Type and Arrangement	Main Cell Types	Where Found
Loose connective tissue	Gel; more ground than fibers and cells	Collagen, elastic, reticular; random	Fibroblasts	Skin, around blood vessels and organs, under epithelia
Dense, irregular connective tissue	More fibers than ground	Mostly collagen; random	Fibroblasts	Muscle and nerve sheaths
Dense, regular connective tissue	More fibers than ground	Collagen, parallel	Fibroblasts	Tendons and ligaments
Adipose	Very little	None	Brown fat, white fat	Depends on age and sex
Blood	Aqueous	None	Blood cells	In blood and lymph vessels
Cartilage	Firm but flexible; hyaluronic acid	Collagen	Chondroblasts	Joint surfaces, spine, ear, nose, larynx
Bone	Rigid due to calcium salts	Collagen	Osteoblasts and osteoclasts	Bones

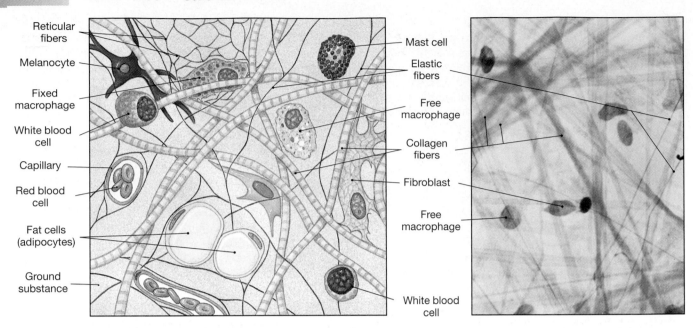

Reticular fibers
Melanocyte
Fixed macrophage
White blood cell
Capillary
Red blood cell
Fat cells (adipocytes)
Ground substance
Mast cell
Elastic fibers
Free macrophage
Collagen fibers
Fibroblast
Free macrophage
White blood cell

■ **Figure 3-21 Cells and fibers of connective tissue** This figure represents a typical loose connective tissue with extensive extracellular matrix. The cells labeled fibroblasts are the connective tissue cells that secrete the fibers and glycoproteins of the matrix. Macrophages and mast cells are two types of cells that defend against foreign invaders, such as bacteria.

connective tissue, adipose tissue, cartilage, bone, and blood.

Loose connective tissues (Fig. 3-21 ■) are the elastic tissues that underlie skin and provide support for small glands. **Dense connective tissues** are tissues whose primary function is strength or flexibility. Important examples of this tissue type are tendons, ligaments, and the sheaths that surround muscles and nerves. In these tissues, collagen fibers are the dominant type.

Tendons (Fig. 3-22 ■) attach skeletal muscles to bones. **Ligaments** connect one bone to another. Because ligaments contain elastic fibers in addition to collagen

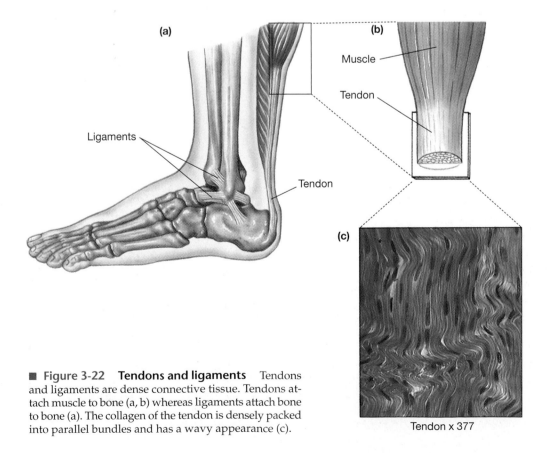

(a)
(b)
Muscle
Tendon
Ligaments
Tendon
(c)

Tendon x 377

■ **Figure 3-22 Tendons and ligaments** Tendons and ligaments are dense connective tissue. Tendons attach muscle to bone (a, b) whereas ligaments attach bone to bone (a). The collagen of the tendon is densely packed into parallel bundles and has a wavy appearance (c).

fibers, they have a limited ability to stretch, whereas tendons cannot stretch.

Adipose tissue is made up of **adipocytes,** or fat cells (Fig. 3-23 ■). An adipocyte of **white fat** typically contains a single enormous lipid droplet that occupies most of the volume of the cell. This is the most common form of adipose tissue in adults. **Brown fat** is composed of adipose cells that contain multiple lipid droplets rather than a single large droplet. This type of fat is almost completely lacking in adults, but it plays an important role in temperature regulation in infants.

Blood is an unusual connective tissue that is characterized by its watery matrix, consisting of a dilute solution of ions and dissolved organic molecules. The matrix lacks insoluble protein fibers but contains a large variety of soluble proteins. We will discuss these in Chapter 16.

Cartilage and bone together are considered supporting connective tissues. These tissues have a dense ground substance that contains closely packed fibers. **Cartilage** is found in structures such as the nose, ears, knee, and windpipe. It is solid, flexible, and notable for its lack of blood supply. Without a blood supply, nutrients and oxygen must reach the cells of cartilage by diffusion. This is a slow process, which means that damaged cartilage heals slowly.

The fibrous matrix of **bone** is said to be *calcified* because it contains mineral deposits, primarily calcium salts, such as calcium phosphate. These minerals give the bone strength and rigidity. We will examine the structure and formation of bone along with calcium metabolism in Chapter 21. Figure 3-24 ■ is a study map showing the components of connective tissues.

✔ Blood is a connective tissue with two components: plasma and cells. What is the matrix in this connective tissue?

✔ Why does torn cartilage heal more slowly than a cut in the skin?

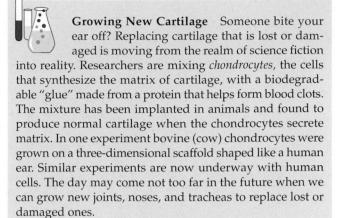

Growing New Cartilage Someone bite your ear off? Replacing cartilage that is lost or damaged is moving from the realm of science fiction into reality. Researchers are mixing *chondrocytes,* the cells that synthesize the matrix of cartilage, with a biodegradable "glue" made from a protein that helps form blood clots. The mixture has been implanted in animals and found to produce normal cartilage when the chondrocytes secrete matrix. In one experiment bovine (cow) chondrocytes were grown on a three-dimensional scaffold shaped like a human ear. Similar experiments are now underway with human cells. The day may come not too far in the future when we can grow new joints, noses, and tracheas to replace lost or damaged ones.

Muscle and Neural Tissues Are Excitable

Muscle and neural tissues are collectively called the *excitable tissues* because of their ability to generate electrical signals called *action potentials.* Both of these tissue types have little or no extracellular matrix.

Muscle tissue has the ability to contract and produce force and movement. There are three types of muscle tissue in the body: cardiac muscle, in the heart; smooth muscle, which makes up most internal organs; and skeletal muscle. Most skeletal muscles are attached to bone and are responsible for gross movement of the body. We will discuss muscle in Chapter 12.

Neural tissue has two types of cells. **Neurons,** or nerve cells, carry information in the form of chemical and electrical signals from one part of the body to another. They are concentrated in the brain and spinal cord but also include a network that extends to virtually every part of the body. **Glial cells,** or neuroglia, are the support cells for neurons. We will discuss neural tissue in Chapter 8.

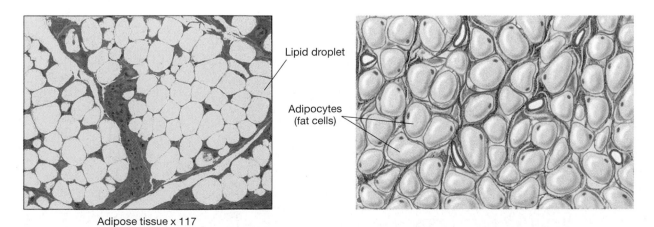

Lipid droplet

Adipocytes (fat cells)

Adipose tissue x 117

■ **Figure 3-23   Adipose tissue**   The cells of adipose tissue are almost filled with lipid and have very little cytosol.

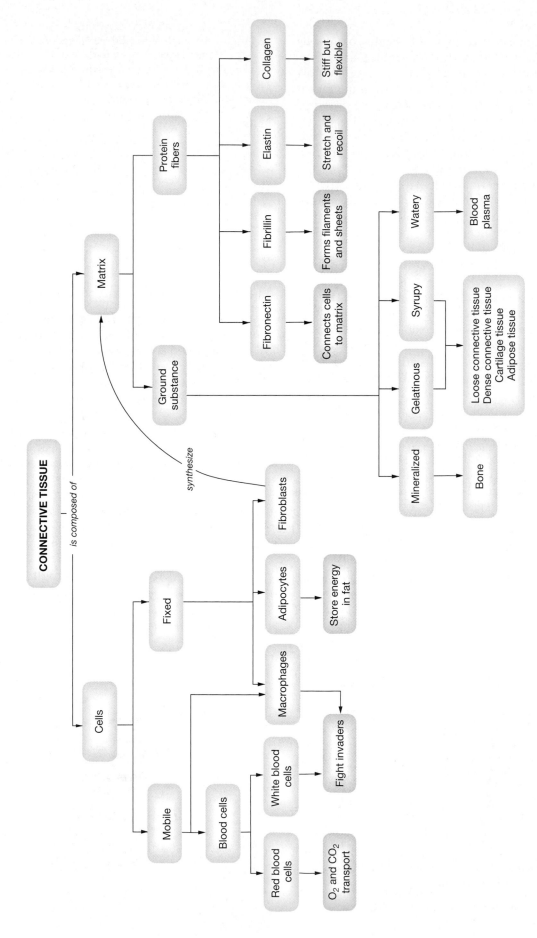

■ **Figure 3-24   Components of connective tissue**   This map shows the relationship of cells and the components of matrix in connective tissue.

...continued from page 61

The day after Jan's visit, a cytologist stains her slide with dye and inserts it into PAPNET, a computerized analysis system that assists cytologists by identifying abnormal cells in cervical samples. The robotic arm of PAPNET picks up Jan's slide and places it under a camera. A computer rapidly scans the cells on the slide, looking for abnormal size or shape. The computer is programmed to record the 128 cells that appear most abnormal. When the computer finds an abnormal cell, it is projected on a large screen.

Question 4:   *Has Jan's dysplasia improved or worsened? What evidence do you have to support your answer (Fig. 3-25 ■)?*

▶ TISSUE REMODELING

Growth is a process that most people associate with the period from birth to adulthood. But cell birth, growth, and death continue throughout a person's life. The tissues of the body are constantly changing as cells die and are replaced.

Apoptosis Is a Tidy Form of Cell Death

Cell death occurs two ways, one messy and one tidy. In **necrosis,** cells die from physical trauma, toxins, or lack of oxygen when their blood supply is cut off. Necrotic cells swell, their organelles deteriorate, and finally the cells rupture. The cell contents thus released include digestive enzymes that damage adjacent cells and trigger an inflammatory response. You see necrosis when you have a red area of skin surrounding a scab.

In contrast, cells that undergo programmed cell death, or **apoptosis** [app-oh-TOE-sis, with the second *p* silent; *apo-* apart, away + *ptosis,* a falling], do not disrupt their neighbors when they die. Apoptosis, also called cell suicide, is a complex process regulated by multiple chemical signals. Some signals keep apoptosis from occurring while other signals tell the cell to destruct. When the suicide signal wins out, chromatin in the nucleus condenses, the cell pulls away from its neighbors, shrinks, and finally breaks up into tidy membrane-bound blebs that are gobbled up by neighboring cells or by wandering cells of the immune system.

Apoptosis is a normal event in the life of an organism. During development, apoptosis removes unneeded cells, such as half the cells in the developing brain, and the webs of skin between fingers and toes. In adults, cells that are subject to wear and tear from exposure to the outside environment may live only a day or two before undergoing apoptosis. For example, it has been estimated that the intestinal epithelium is completely replaced with new cells every two to five days.

Stem Cells Can Create New Specialized Cells

If cells in the adult body are constantly dying, where do their replacements come from? In most cases, they come from mature cells of the same type that are still able to undergo the cell division process known as **mitosis** (see Appendix B). The exceptions to this rule are blood cells and nerve and muscle cells. Blood cells are formed from undifferentiated precursor cells, called **stem cells,** that are found in the bone marrow.

We used to believe that nerve and muscle cells, which are highly specialized in their mature forms, could not be replaced when they died. But exciting new research suggests that stem cells for these tissues do exist in the body. Given the right combination of chemical signals, they can replace nervous and muscle tissue. These findings have tremendous potential for helping us treat *degenerative* diseases in which neurons and muscle cells degenerate and die.

▶ ORGANS

Groups of tissues that carry out related functions may form structures known as **organs.** The organs of the body contain the four types of tissue in various combinations.

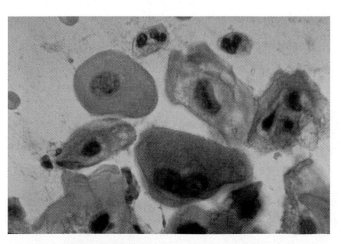
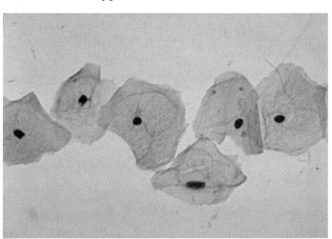

■ **Figure 3-25   First and second Pap smears**   Cells from Jan's previous Pap test are shown on the left, and cells from her second Pap test are shown on the right.

FOCUS ON… The Skin

■ **Figure 3-26**

(a) The layers of the skin   Skin has a number of important functions that are reflected in its structure. It is waterproof to prevent evaporation and tough to keep bacteria out. Sweat glands secrete fluid to cool the body, while hair and an underlying layer of fat insulate against cold. Sensory receptors monitor external conditions.

The outer layer of skin is called the **epidermis.** It consists of multiple cell layers that create a protective barrier. The inner **dermis** is loose connective tissue that contains exocrine glands, tiny blood vessels, muscles, and nerve endings. The **hypodermis** lies beneath the skin. It contains adipose tissue and the major blood vessels and nerves whose branches extend upward into the dermis.

The exocrine glands in the dermis are derived from the epidermal layer (as shown in Figure 3-20 ■, p. 62). **Sebaceous glands** secrete a lipid mixture, the secretions that, in excess, cause oily skin. **Sweat glands** secrete a dilute salt solution. **Apocrine glands** are located in the skin of the genitalia, anus, axillae [*axilla,* armpit], and eyelids. They release waxy or viscous milky secretions in response to fear or sexual excitement. The function of these secretions in humans is not known, but, in other mammals, similar secretions are used to mark territories or for sexual attraction. The chemical *musk,* commonly used in perfumes, comes from the anal apocrine glands of musk deer.

Hair follicles in the skin secrete the nonliving keratin shaft of hair. Tiny muscles attached to the follicle pull the follicle into a vertical position when the muscle contracts, creating "goose bumps."

(b) Epidermis   The outermost layer of skin is composed of the fibrous protein keratin. The thickness of the keratin layer varies from thick on the soles of feet and palms of hands to thin, as on the scalp. Keratin fibers are produced by **keratinocytes,** which are tied to each other by an extensive system of desmosomes. These cells also secrete a hydrophobic matrix that acts as the skin's main waterproofing agent. By the time older keratinocytes are pushed to the surface of the epidermis by newer cells, their cytoplasm is thick with keratin fibers. At this point the cells die, and their nuclei and organelles disappear, leaving behind a protective mat of linked keratin fibers.

(c) Basement membrane and the connection between epidermis and dermis   An acellular basement membrane lies between the epidermis and the underlying layer of connective tissue. The cells of the epidermis and fibers of the dermis are anchored to each other by their connection to the protein fibers that run throughout the basement membrane.

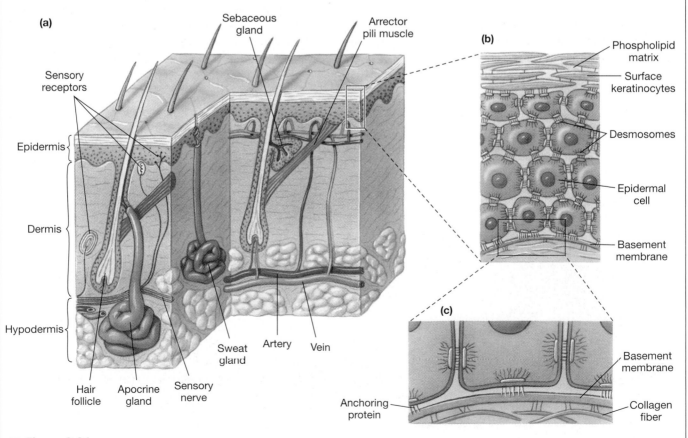

■ **Figure 3-26**

The skin is an excellent example of an organ that incorporates all four types of tissue into an integrated whole. We think of skin as a thin layer that covers the external surfaces of the body, but in reality it is the heaviest single organ, at about 16% of an adult's total body weight! If it were flattened out, it would cover a surface of between 1.2 and 2.3 square meters, about the size of a couple of card-table tops. Its size and weight make skin one of the most important organs of the body. The structure of this organ is highlighted in *Focus on the Skin,* Figure 3-26 ∎.

There are other organs, like the skin, whose functions do not fit neatly into any one chapter of this book. We will highlight several of them in special *Organ Focus* *Features* throughout the book. The illustrated boxes discuss structure and functions of these versatile organs so that you can gain an appreciation for the way different tissues combine for a united purpose.

As we consider the systems of the body in the succeeding chapters, you will see how diverse cells, tissues, and organs carry out the processes of the living body. Although the cells have different structures and different functions, they have one need in common: a continuous supply of energy. Without energy, cells cannot survive, let alone carry on all the other processes of daily living. The next chapter looks at cellular metabolism: how cells capture and harness the energy released by chemical reactions.

PROBLEM CONCLUSION

In this running problem, you have learned how the Pap test can detect the early changes that herald cancer.

Further check your understanding of this running problem by checking your answers against those in the summary table.

	Question	Facts	Integration and Analysis
1	Why is it necessary to remove or kill cancerous cells in order to treat cancer?	Cancerous cells divide uncontrollably and fail to coordinate with other cells. Cancerous cells fail to differentiate into specialized cells.	Unless removed, cancerous cells will displace normal cells. This may cause destruction of normal tissues. In addition, because cancerous cells do not become specialized, they cannot carry out the same functions as the specialized cells that they displace.
2	What property of cancer cells explains the large size of their nucleus and the relatively small amount of cytoplasm?	Cancerous cells divide uncontrollably. Dividing cells must duplicate their DNA prior to cell division [∞ Appendix B].	Actively reproducing cells are likely to have more DNA in their nucleus as they prepare to divide, so their nuclei tend to be larger. Each cell division splits the cytoplasm between the two daughter cells. If division is occurring rapidly, the daughter cells may not have time to synthesize new cytoplasm, so the amount of cytoplasm is less than in a normal cell.
3a	What kind of damage or trauma are cervical epithelial cells normally subjected to?	The cervix is the neck of the uterus.	The cervix is subject to trauma or damage during sexual intercourse and childbirth.
3b	Which of the two types of epithelium is more likely to be affected by trauma?	The cervix consists of secretory epithelium with mucus-secreting glands lining the inside, and protective epithelium that covers the outside.	Protective epithelium is composed of multiple layers of cells, and is designed to protect areas from mechanical and chemical stress [∞ p. 60]. Therefore, the secretory epithelium with its single cell layer is more easily damaged.
4	Has Jan's dysplasia improved or worsened? What evidence do you have to support your answer?	The slide from Jan's first Pap smear shows abnormal cells with large nuclei and little cytoplasm. These abnormal cells do not appear in the second smear.	The disappearance of the abnormal cells indicates that Jan's dysplasia has resolved.

CHAPTER REVIEW

SUMMARY

Cellular Anatomy

1. The **cell** is the basic functional unit of most living organisms. (p. 43)

2. The **cell membrane** is a phospholipid bilayer with embedded proteins that separates the inside of the cell from the **extracellular fluid.** (p. 44)

3. The **cytoplasm** consists of semigelatinous **cytosol** with dissolved nutrients, ions, wastes, insoluble **inclusions,** and **organelles** to carry out specific functions. (p. 45)

4. Nonmembranous organelles include **ribosomes,** which take part in protein synthesis, **vaults** of uncertain function, and insoluble protein fibers such as actin, myosin, keratin, and tubulin. (p. 45)

5. The **cytoskeleton** is made of **microfilaments, intermediate filaments,** and **microtubules.** It provides support, aids transport of materials within the cell, links cells together, and enables motility in certain cells. (p. 45)

6. **Centrosomes** are the microtubule-organizing center of the cell. **Centrioles** aid the movement of chromosomes during cell division. (p. 49)

7. **Cilia** move fluid or secretions across the cell surface. **Flagella** propel sperm through body fluids. (p. 50)

8. **Mitochondria** generate most ATP for the cell. (p. 50)

9. The **smooth endoplasmic reticulum** is the primary site of lipid synthesis. The **rough endoplasmic reticulum** is the primary site of protein synthesis. (p. 51)

10. The **Golgi apparatus** packages proteins into vesicles. **Secretory vesicles** secrete their contents into the extracellular fluid. (p. 51)

11. **Lysosomes** and **peroxisomes** are **storage vesicles** that contain digestive enzymes. (p. 53)

12. The **nucleus** contains DNA, the genetic material that ultimately controls all cell processes, in the form of **chromatin.** The **nuclear envelope** has **nuclear pore complexes** that allow controlled chemical communication between the nucleus and cytosol. **Nucleoli** are nuclear areas that control the synthesis of RNA for ribosomes. (p. 54)

Tissues of the Body

13. There are four **primary tissue types** in the human body: epithelial, connective, muscle, and neural. (p. 54)

14. Extracellular **matrix** secreted by cells provides support and a means of cell-cell communication. (p. 55)

15. Cells link together with the aid of **cell adhesion molecules (CAMs). Desmosomes** and **adherens junctions** anchor cells. **Tight junctions** prevent the movement of material between cells. **Gap junctions** allow chemical and electrical signals to pass directly from cell to cell. (p. 55)

16. **Epithelial tissues** protect the internal environment, regulate the exchange of material, or manufacture and secrete chemicals. (p. 57)

17. **Exchange epithelia** permit rapid exchange of material, particularly gases. **Transporting epithelia** actively regulate the selective exchange of nongaseous material between the internal and external environments. **Ciliated epithelia** move fluid and particles across the surface of the tissue. **Protective epithelia** help prevent exchange between the internal and external environments. The **secretory epithelia** release secretory products into the external environment or the blood. (p. 58)

18. **Exocrine glands** release their secretions into the external environment through **ducts. Endocrine glands** are ductless glands that release their secretions, called **hormones,** directly into the extracellular fluid. (p. 61)

19. **Connective tissues** provide structural support and a physical barrier. Connective tissues are notable for extensive extracellular matrix. (p. 62)

20. **Loose connective tissues** are the elastic tissues that underlie skin. **Dense connective tissues,** including **tendons** and **ligaments,** have strength or flexibility due to collagen. **Adipose tissue** stores fat. **Blood** is characterized by a watery matrix. **Cartilage** is solid and flexible and has no blood supply. The fibrous matrix of **bone** is hardened by deposits of calcium salts. (p. 64)

21. Muscle and neural tissues are called excitable tissues because of their ability to generate electrical signals called action potentials. **Muscle** has the ability to contract and produce force and movement. There are three types of muscle: cardiac, smooth, and skeletal. (p. 65)

22. **Neural tissue** includes **neurons** that use electrical and chemical signals to transmit information from one part of the body to another and support cells known as **glia.** (p. 65)

Tissue Remodeling

23. Cell death occurs by **necrosis,** which adversely affects neighboring cells, and by **apoptosis,** programmed cell death that does not disturb the tissue. (p. 67)

24. Most mature cells can reproduce themselves. Nerve, muscle, and blood cells must be replaced by cells that develop from unspecialized **stem cells.** (p. 67)

Organs

25. **Organs** are formed by groups of tissues that carry out related functions. The organs of the body contain the four types of tissues in various ratios. (p. 67)

QUESTIONS

LEVEL ONE   Reviewing Facts and Terms

1. What are the two primary types of biomolecules found in the cell membrane?

2. Define and distinguish between inclusions, membranous organelles, and nonmembranous organelles. Give an example of each.

3. Define cytoskeleton. List five functions of the cytoskeleton.

4. Match the terms with their descriptions:
   (a) cilia
   (b) centriole
   (c) flagellum
   (d) centrosome

   1. in human cells, appears as single, long, whiplike tail
   2. short, hairlike structures that beat to produce currents in fluids
   3. a bundle of microtubules that aids in mitosis
   4. the microtubule-organizing center

5. Match the organelle with its function:
   (a) endoplasmic reticulum
   (b) Golgi apparatus
   (c) lysosomes
   (d) mitochondria
   (e) peroxisomes

   1. "powerhouse" of the cell where most ATP is produced
   2. these degrade long-chain fatty acids and toxic foreign molecules
   3. network of membranous tubules that synthesize biomolecules
   4. digestive system of cell, degrading or recycling components
   5. modifies and packages proteins into vesicles

6. What process activates the enzymes inside lysosomes?

7. Exocrine glands produce secretions that are either watery, called _____, such as tears and sweat, or stickier solutions called _____.

8. _____ glands release hormones, which enter the blood and are carried throughout the body to regulate and coordinate the various activities of organs or systems.

9. List the four major tissue types. Give an example and location of each.

10. The largest and heaviest organ in the body is the _____.

11. Match the proteins to their functions. Functions may be used more than once.
   (a) cadherin
   (b) CAM
   (c) collagen
   (d) connexin
   (e) elastin
   (f) fibrillin
   (g) fibronectin
   (h) integrin
   (i) occludin

   1. membrane protein used to form cell junctions
   2. matrix glycoprotein used to anchor cells
   3. protein found in gap junctions
   4. matrix protein found in connective tissue

12. What types of glands can be found within the skin? Name the secretion of each type.

13. The term *matrix* is used in reference to an organelle and tissues. Compare the meanings of the term in these two contexts.

LEVEL TWO   Reviewing Concepts

14. **Mapping exercise:** Compile this list of terms into a map and add functions where appropriate.

actin	microtubule
cell	microtubule-organizing center
cell membrane	mitochondria
centriole	myosin
centrosome	neurofilament
cilia	nonmembranous organelle
cytoplasm	nucleus
cytoskeleton	peroxisome
cytosol	RER
extracellular matrix	ribosome
flagella	secretory vesicle
Golgi apparatus	SER
intermediate filament	storage vesicle
keratin	thick filament
lysosome	tubulin
membranous organelle	vault
microfilament	

15. List, compare, and contrast the three types of intercellular junctions and their subtypes. Give an example of where each type can be found in the body and describe its function in that location.

16. A number of organelles can be considered vesicles. Define *vesicle* and describe at least three examples.

17. Explain why a stratified epithelium offers more protection than a simple epithelium.

18. Sketch a short series of columnar cells. Label the apical and basolateral borders of these typical epithelial cells. Briefly explain the different kinds of activities that may take place at each cell border, as well as how the structures and junctions found on each side of the cell support these functions.

19. Compare and contrast the structure, locations, and functions of bone and cartilage.

20. Differentiate between the terms in each set below:
   (a) lumen/wall   (b) cytoplasm/cytosol
   (c) histology/cytology (d) myosin/keratin

21. When a tadpole turns into a frog, its tail shrinks and is re-absorbed. Is this an example of necrosis or apoptosis? Defend your answer.

22. Define the term *organ*. Explain the specific traits exhibited by the following body parts that allow them to qualify as organs: skin, stomach, uterus, heart. What similarities do they share?

LEVEL THREE   Problem Solving

23. One result of cigarette smoking is paralysis of the cilia that line the respiratory passageways. What function do these cilia serve? Based on what you have read in this chapter, why is it harmful when they no longer beat? What health problems would you expect to arise? How does this explain the hacking cough common among smokers?

24. Cancer is abnormal, uncontrolled cell division. What property of epithelial tissues might (and does) make them more prone to develop this?

E X P L <MediaLab>

Introduction

This chapter introduced you to the structure and function of individual cells. Cells are comprised of many small organelles, each with a specialized function. You also learned how cells form units called tissues. The cells of a tissue must coordinate functions and share products. One way cells coordinate shared tasks is through cell-to-cell junctions that allow the exchange of material

The following Web Explorations will help you gain an appreciation of the function of cellular organelles as well as the cell-to-cell junctions that allow cells to function as tissues. After reading the descriptions below, visit the MediaLab for Chapter 3 in your Companion Website and select the appropriate keyword.

Web Exploration 1

Estimated time for completion = 15 minutes
A cell contains many organelles that must work together to support the proper function and survival of that cell. If a cell is damaged by mechanical or toxic injury, it may die prematurely. But not all cell death is unwanted. Sometimes controlled or programmed cell death is necessary for the successful development and/or function of a tissue. Apoptosis (app-oh-TOE-sis) is programmed cell death that is important in a variety of tissue types. It is also the topic of a fast-growing field of research.

Select the keyword **APOPTOSIS** on the Website to read an article about cell death. Use the information you gather from this site to answer the following questions. What are some features of apoptosis that distinguish it from cell death due

to injury? Which of these features could best be visualized via light microscopy and which by electron microscopy? According to the article, what might influence whether a cell undergoes apoptosis instead of attempting to repair itself? How might understanding the mechanisms of apoptosis be of benefit to the treatment of cancer? To complete this exercise, visit MediaLab Web Exploration 1 in Chapter 3 of your Companion Website.

Web Exploration 2

Estimated time for completion = 10 minutes
Extracellular matrix (ECM) is a substance of varying composition that provides support to surrounding cells. At times, the ECM can be quite rigid. How then do developing and expanding tissues cope with a complex glycoprotein matrix to make space for themselves?

A web article, written by Eugene Russo, introduces us to a special set of proteases (enzymes that degrade protein) called **matrix metalloproteases** (MMPs). These proteases specifically degrade proteins commonly found in the ECM. Select the keyword **MMP** on your Website to learn more about MMPs and their potential medical benefits.

Can you envision how understanding cell migration can aid in the treatment of cancer? How might MMPs be involved in diseases such as arthritis that affect the ECM? Can MMPs be involved in blood vessel migration? What might happen to normal physiological function if MMPs are inhibited by drugs? To compete this exercise, visit MediaLab Web Exploration 2 in Chapter 3 of your Companion Website.

4 Cellular Metabolism

■ "There is no good evidence that in any of its manifestations life evades the second law of thermodynamics, but in the downward course of the energy-flow it interposes a barrier and dams up a reservoir which provides potential for its own remarkable activities."—*F. G. Hopkins, 1933* ■

CHAPTER OUTLINE

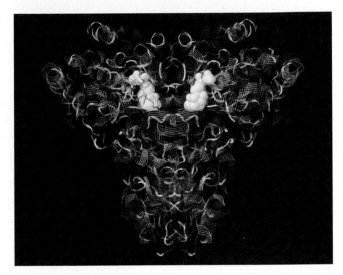

Transketolase, an enzyme for sugar metabolism, with cofactors (white)

BACKGROUND BASICS

Forms of energy and work (Appendix A)
DNA and RNA (Appendix B)
Hydrogen bonds (p. 25)
Covalent bonds (p. 21)
ATP (p. 35)
Organelles (p. 45)
Electrons (p. 16)
Protein structure (p. 32)
Carbohydrates (p. 29)
Graphing (p. 10)
Lipids (p. 29)

The structure of a cell is the framework for processes that are essential to living things: growth, movement, reproduction, and the maintenance of a relatively constant internal environment. Without these processes, the cell is a ghost town, filled with buildings but lacking the energy and organization brought to it by living people. A deserted town left to itself slowly crumbles into ruin because there is no one to work on repairs. People provide the energy for growth and maintenance of a town, for building new buildings and repairing old ones. Similar energy-dependent processes take place within a cell. Cells import raw materials, make new molecules, and repair or recycle aging parts. Their ability to extract energy from the external environment and use that energy to construct and maintain themselves as organized, functioning units is

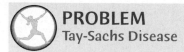

PROBLEM
Tay-Sachs Disease

In many ultra-orthodox Jewish communities of the United States—in which arranged marriages are the norm—the rabbi is entrusted with an important, life-saving task. He keeps a confidential record of individuals known to carry the gene for Tay-Sachs disease, a fatal, inherited condition that strikes one in 3600 American Jews of Eastern European descent. Babies born with this disease rarely live beyond age four, and there is no cure. Based on the family trees he constructs, the rabbi can avoid pairing two individuals who carry the deadly gene.

Sarah and David, who met while working on their college newspaper, are not orthodox Jews. But both are aware that their Jewish ancestry might put any children they might have at risk for Tay-Sachs disease. Six months before their wedding, they decide to see a genetic counselor to determine whether they are carriers of the gene for Tay-Sachs disease.

...continued on page 80

one of their most outstanding characteristics. In this chapter, we look at the cell processes through which the human body obtains energy and maintains its ordered systems.

▶ ENERGY IN BIOLOGICAL SYSTEMS

The cycling of energy in the environment and its use by living organisms is one of the fundamental concepts of biology. All cells use energy from their environment to grow, make new parts, and reproduce. Through photosynthesis, plants trap radiant energy from the sun and store it as chemical bond energy (Fig. 4-1 ■). They use carbon and oxygen from carbon dioxide, nitrogen from the soil, and hydrogen and oxygen from water to make glucose, amino acids, and other biomolecules. Thus, energy from the sun trapped by photosynthesis is the ultimate energy source for all animals, including humans.

Animals cannot trap energy from the sun or use carbon and nitrogen from the air and soil to synthesize biomolecules. They must import energy and biomolecules

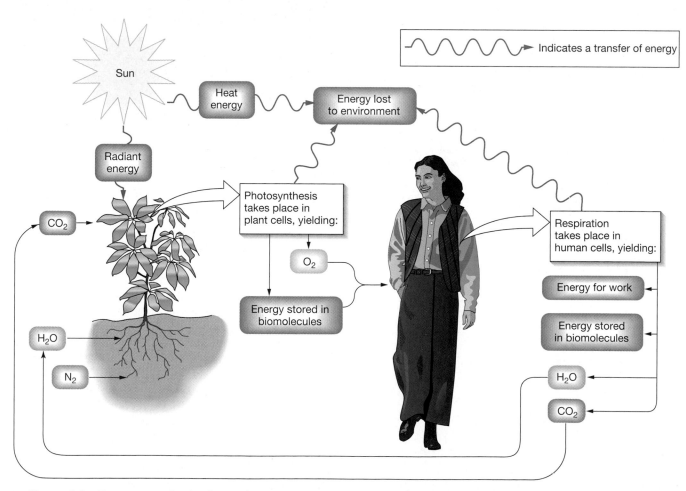

■ **Figure 4-1 Energy transfer in the environment** Plants use photosynthesis to trap radiant energy from the sun in chemical bonds. Animals eat the plants and use the energy from chemical bonds for growth, maintenance, reproduction, and movement. Some energy is lost to the environment as heat. The rest is stored as the chemical bond energy of glycogen and lipids.

synthesized by plants or obtained from plants by other animals. Animals extract energy from these biomolecules through respiration, consuming oxygen and producing carbon dioxide and water that recycle to be used again by plants. This process of energy extraction is never totally efficient. As a result, some chemical bond energy is returned into the environment as heat.

If animals ingest more energy than they need for immediate use, the excess energy is stored in chemical bonds, just as it is in plants. Glycogen (a glucose polymer) and lipid molecules are the main energy stores in animals [∞ p. 29]. These storage molecules are available for use at times when an animal's energy needs exceed its food intake.

Energy Is Used to Perform Work

As the preceding discussion suggests, a cell can obtain, store, and use energy to fuel its activities. Energy can be defined as the capacity to do work, but what is work? We use the word in everyday life to mean various things, from hammering a nail to sitting at a desk writing a paper. In biological systems, work has three basic forms: chemical work, transport work, and mechanical work.

Chemical work enables cells and organisms to grow, maintain a suitable internal environment, and store information needed for reproduction and normal activities. Protein synthesis for wound repair is an example of chemical work.

Transport work enables cells to move ions, molecules, and larger particles through the cell membrane and through membranes of organelles within the cell. Transport work is particularly useful for creating **concentration gradients,** in which the concentration of a molecule is higher on one side of a membrane than on the other. For example, certain types of endoplasmic reticulum use energy to import calcium ions from the cytosol. This creates a high calcium concentration inside the organelle and a low concentration in the cytosol. If calcium is then released back into the cytosol, it creates a "calcium signal" that causes the cell to perform an action such as muscle contraction.

Mechanical work in animals is used for movement. At the cellular level, this movement includes organelles that move around within the cell, and the beating of cilia and flagella [∞ p. 50]. In whole animals, movement usually involves muscle contraction. Most mechanical work, whether in cells or entire animals, is mediated by the movement of intracellular fibers and filaments of the cytoskeleton [∞ p. 49].

Energy Comes in Two Forms: Kinetic and Potential

Energy can be classified in various ways.* We often think of energy in terms that we deal with daily: thermal energy, electrical energy, mechanical energy. We speak of

---

*To review the common forms of energy and their standard units, see Appendix A.

energy stored in chemical bonds. Each type of energy has its own characteristics. However, all share an ability to appear in two forms: as potential energy or as kinetic energy.

Kinetic energy is the energy of motion [*kinetikos,* motion]. A ball rolling down a hill, perfume spreading through the air, electric current flowing through power lines, heat warming a frying pan, and molecules moving across biological membranes are all examples of kinetic energy.

Potential energy is stored energy. Potential energy may be stored in the position of a molecule with respect to its concentration gradient. In chemical bonds, potential energy is stored in the position of the electrons that form the bond [∞ p. 19]. A key feature of all forms of energy is the ability of potential energy to become kinetic energy and vice versa.

Energy Can Be Converted from One Form to Another

Recall that a general definition of energy is the capacity to do work. Work always involves movement and therefore always involves kinetic energy. Potential energy also can be used to perform work, but it must be converted to kinetic energy, and the conversion is never 100% efficient. A certain amount of energy is lost to the environment, usually as heat. The amount of energy lost in the transformation depends on the efficiency of the process. Many physiological processes in the human body are not very efficient. For example, 70% of the energy used in physical exercise is lost as heat rather than transformed into the work of muscle contraction.

Figure 4-2 ■ summarizes the relationship of kinetic energy, potential energy, and work. In Figure 4-2a ■, work is performed by pushing the ball up a ramp. Some of the kinetic energy of the moving ball is transformed into potential energy as it goes up the ramp. Figure 4-2b ■ shows potential energy, stored in the stationary ball at the top of the ramp. No work is being performed, but the capacity to do work is stored in the position of the ball. In Figure 4-2c ■, the potential energy of the ball becomes kinetic energy when the ball rolls down the ramp. Some of this kinetic energy is lost to the environment as heat resulting from friction between the ball and the air and ramp.

In biological systems, potential energy is stored in concentration gradients and chemical bonds. It is transformed into kinetic energy in order to do chemical, transport, or mechanical work.

Thermodynamics Is the Study of Energy Use

Two basic rules govern the transfer of energy in biological systems and in the universe as a whole. The **first law of thermodynamics** states that the total amount of energy in the universe is constant. The universe is called a "closed system"—nothing enters and nothing leaves.

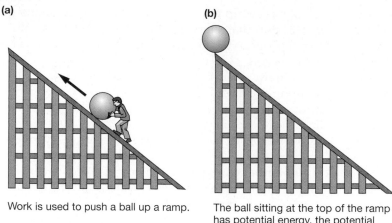

(a) Work is used to push a ball up a ramp.

(b) The ball sitting at the top of the ramp has potential energy, the potential to do work.

(c) The ball rolling down the ramp is converting the potential energy to kinetic energy. However, the conversion is not totally efficient, and some energy is lost as heat due to friction between the ball, ramp, and air.

■ **Figure 4-2   Potential energy, kinetic energy, and work**

Energy can be converted from one form to another, but the total amount of energy in a closed system never changes. The human body is not a closed system, however. As an open system, it exchanges material and energy with its surroundings. Our bodies cannot create energy, but must import it from outside in the form of food. By the same token, our bodies lose energy, especially in the form of heat, to the environment. Energy that stays within the body can be changed from one form to another or used to do work.

The **second law of thermodynamics** states that natural spontaneous processes move from a state of order (nonrandomness) to a condition of randomness or disorder, also known as **entropy.** To create and maintain order in an open system such as the body requires the input of energy. Disorder occurs when open systems only lose energy to their surroundings. The analogy of the deserted town illustrates the second law. When people put all their energy into activities outside the town, the town slowly falls into disrepair and becomes less organized. Without constant input of energy, a cell is unable to maintain its ordered internal environment. As the cell loses organization, its ability to carry out normal functions disappears, and it dies.

In the remainder of this chapter you will learn how cells obtain energy from and store energy in the chemical bonds of biomolecules. The chemical processes through which this takes place are collectively called *metabolism.* Through metabolism, cells transfer the potential energy of chemical bonds into kinetic energy for growth, maintenance, reproduction, and movement.

✔ How do animals store energy in their bodies?

✔ What is the difference between potential energy and kinetic energy?

▶ CHEMICAL REACTIONS

Living organisms are characterized by their ability to extract energy from the environment and use it to support life processes. The study of energy flow through biological systems is a field known as **bioenergetics** [*bios*, life + *en-*, in + *ergon*, work]. In a biological system, chemical reactions are a critical means of transferring energy from one part of the system to another. This section explains chemical reactions, discusses different types of reactions, and presents the general mechanisms for trapping, releasing, or transferring energy during reactions.

Energy Is Transferred Between Molecules During Reactions

In chemical reactions, a substance becomes a different substance, usually by breaking and/or making covalent bonds. A reaction begins with one or more molecules called **reactants** (or *substrates*) and ends with one or more **products** (Table 4-1). In this discussion, we will consider a reaction that begins with two reactants and ends with two products:

$$A + B \rightarrow C + D$$

The speed with which a reaction takes place, the **reaction rate,** is the disappearance rate of the reactants (A and B) or the appearance rate of the products (C and D). Reaction rate is measured as change in concentration during a certain time period and is often expressed as molarity per second (M/sec).

Energy Transfer in Reactions   The purpose of chemical reactions in cells is to transfer energy from one molecule to another or to use energy stored in the reactants to do work. The potential energy stored in the chemical bonds of a molecule is known as the **free energy** of the molecule. Generally, complex molecules have more

TABLE 4-1 Types of Chemical Reactions

Reaction Type	Reactants (Substrates)		Products
Combination	A + B	→	C
Decomposition	C	→	A + B
Single displacement*	L + MX	→	LX + M
Double displacement*	LX + MY	→	LY + MX

*X and Y represent atoms, ions, or chemical groups.

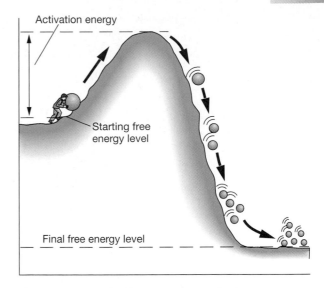

■ **Figure 4-3 Activation energy** This figure is a cartoon representation of a chemical reaction. The rock represents the reactants that are pushed over the crest of a hill. As the rock rolls to the bottom, it breaks into pieces that represent the products. All reactions require an initial input of energy in order to get them started, just as the rock in the figure must be pushed up the little hill before it can release its potential energy as kinetic energy.

chemical bonds and therefore higher free energies. For example, a large glycogen molecule has more free energy than a single glucose molecule, which in turn has more free energy than the carbon dioxide and water from which it was synthesized. The high free energy of complex molecules such as glycogen is the reason these molecules are used to store energy in cells.

To understand how chemical reactions transfer energy between molecules, we should answer two questions. First, how do reactions get started? The energy required to initiate a reaction is known as the *activation energy*. Second, what happens to the free energy of the products and reactants during a reaction? The difference in free energy between reactants and products is known as the *net free energy change of the reaction*.

Activation Energy **Activation energy** is the initial input of energy required to bring reactants into a position that allows them to react with each other. This "push" needed to start the reaction is shown in Figure 4-3 ■ as the little hill up which the rock must be pushed before it can roll by itself down the slope. A reaction with low activation energy will proceed spontaneously when the reactants are brought together. You can demonstrate a spontaneous reaction by pouring a little vinegar onto some baking soda and watching the two react to form carbon dioxide. Reactions with high activation energies will not proceed spontaneously or will proceed too slowly to be useful. For example, if you pour vinegar over a pat of butter, no observable reaction will take place.

Endergonic and Exergonic Reactions One characteristic property of any chemical reaction is its net free energy change. In other words, is the free energy of the products higher or lower than the free energy of the reactants? If the free energy of the products is greater than that of the reactants, some of the activation energy was trapped in the chemical bonds of the products. If the free energy of the products is lower than the free energy of the reactants, the reaction released energy. Energy released by a reaction can be used by other molecules to do work or may be given off as heat. In a few cases, released energy is stored as potential energy in a concentration gradient.

A reaction that releases energy is called an **exergonic reaction** [*ex-*, out + *ergon*, work]. A plot of an exergonic

reaction is shown in Figure 4-4a ■. Note that the products C and D have a lower free energy than the reactants A and B. This indicates the release of energy during the reaction.

An important biological example of an exergonic, or *energy-producing*, reaction is the combination of ATP and water to form ADP, inorganic phosphate (P_i), and H^+. Energy is released during this reaction when the high-energy phosphate bond of the ATP molecule is broken:

$$ATP + H_2O \rightarrow ADP + P_i + H^+ + energy$$

A reaction that requires net input of energy from an outside source is called an **endergonic reaction** [*end(o)*, within + *ergon*, work]. A plot of net free energy change for an endergonic reaction, $E + F \rightarrow G + H$, is shown in Figure 4-4b ■. During endergonic, or *energy-utilizing*, reactions, part of the activation energy added to the reactants is retained by the products. Compare this endergonic reaction with the energy-producing reaction of Figure 4-4a ■, in which all activation energy added to the reactants was released during the reaction.

Some of the energy added to an endergonic reaction remains trapped in the chemical bonds of the products. Thus, energy-utilizing reactions are often *synthesis* reactions in which complex molecules are made from smaller molecules. For example, many glucose molecules link together to create the glucose polymer glycogen. The complex glycogen molecule has higher free energy than the simple glucose molecules used to make it.

If a reaction is energy utilizing in one direction, it will be energy producing in the reverse direction. For example, the energy trapped in the bonds of glycogen

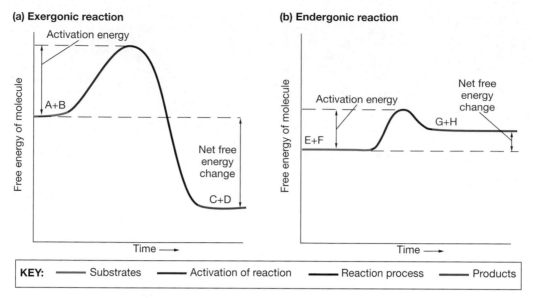

(a) Exergonic reaction

(b) Endergonic reaction

KEY: ⸻ Substrates   ⸻ Activation of reaction   ⸻ Reaction process   ⸻ Products

■ **Figure 4-4   Exergonic and endergonic reactions**   (a) In an exergonic reaction, the products C and D have lower free energy than the starting reactants A and B. The energy released during the reaction can be trapped in other molecules or given off as heat. (b) In an endergonic reaction, the products G and H have higher free energy than the starting reactants E and F. In this type of reaction, some of the activation energy is trapped by the products, increasing their free energy.

during its synthesis will be released when glycogen is broken back down into glucose.

Coupled Reactions   Where does the activation energy for metabolic reactions come from? The simplest source of activation energy is to couple an exergonic reaction that produces energy to an endergonic reaction that requires energy. Some of the most familiar coupled reactions are those that use the energy released by breaking the high-energy bond of ATP to drive an endergonic reaction:

$$\text{E + F} \xrightarrow{\quad \text{ATP} \quad \text{ADP + P}_i \quad} \text{G + H}$$

In this type of coupled reaction, the two reactions take place simultaneously and in the same location, so that the energy from ATP can be used immediately to drive the endergonic reaction of E and F.

However, it is not always practical for reactions to be directly coupled like this. Consequently, living cells have developed ways to trap and save energy released by exergonic reactions. The most common method for trapping energy is in the form of high-energy electrons carried on nucleotides [∞ p. 35]. The nucleotide molecules NADH, $FADH_2$, and NADPH all capture energy in the electrons of their hydrogen atoms (Fig. 4-5 ■). NADH and $FADH_2$ usually transfer most of that energy to ATP, which can be used to drive endergonic reactions.

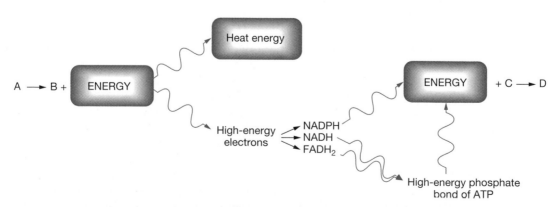

■ **Figure 4-5   Energy transfer and storage in biological reactions**   Some of the energy released by exergonic reactions can be trapped in the high-energy electrons of NADH, $FADH_2$, or NADPH. NADH and $FADH_2$, in turn, transfer energy to the high-energy phosphate bond of ATP. The stored energy of ATP and NADPH can be used to run endergonic reactions. Energy that is not trapped is given off as heat.

Reversible and Irreversible Reactions   A chemical reaction that can proceed in both directions is called a **reversible reaction.** In a reversible reaction, the forward reaction A + B → C + D and its reverse reaction (C + D → A + B) are both likely. If a reaction proceeds in one direction but not the other, it is an **irreversible reaction.**

The net free energy change of a reaction plays an important role in determining whether that reaction can be reversed, because the net free energy change of the forward reaction contributes to the activation energy of the reverse reaction. For example, look at the activation energy of the reaction C + D → A + B in Figure 4-6 ■. This reaction is the reverse of the reaction shown in Figure 4-4a ■. Because there was a large release of energy in the forward reaction A + B → C + D, the reverse reaction (Fig. 4-6 ■) has a large activation energy. As you will recall, the larger the activation energy, the less likely it is that the reaction will proceed spontaneously. Theoretically all reactions could be reversed with enough energy input, but some reactions have such large activation energies that they are essentially irreversible.

In your study of physiology, you will encounter a few irreversible reactions. However, most biological reactions are reversible: if the reaction A + B → C + D is possible, then so is the reaction C + D → A + B. Reversible reactions are shown with arrows that point in both directions: A + B ⇌ C + D. One of the main reasons that many biological reactions are reversible is that they are aided by the proteins known as enzymes.

✔ If you mix baking soda and vinegar together in a bowl, the mixture reacts and foams up, releasing carbon dioxide gas. Name the reactant(s) and product(s) in this reaction.

✔ Do you think this reaction is endergonic or exergonic? Do you think it is reversible? Defend your answers.

▶ ENZYMES

Enzymes are biological *catalysts,* molecules that speed up the rate of chemical reactions without themselves being changed. Without enzymes, most chemical reactions in the cell would go so slowly that the cell would be unable to live. Because an enzyme is not permanently changed or used up in the reaction that it catalyzes, we might write it in a reaction equation this way:

$$A + B + \text{enzyme} \rightarrow C + D + \text{enzyme}$$

This way of writing the reaction shows that the enzyme participates with reactants A and B but is unchanged at the end of the reaction. A more common shorthand for enzymatic reactions shows the name of the enzyme above the reaction arrow, like this:

$$A + B \xrightarrow{\text{enzyme}} C + D$$

In enzymatically catalyzed reactions, the reactants are called **substrates.**

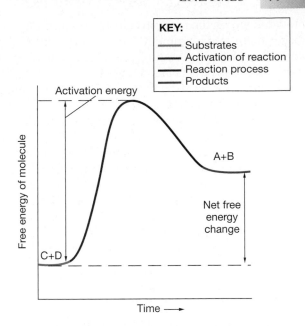

KEY:
— Substrates
— Activation of reaction
— Reaction process
— Products

Activation energy

Free energy of molecule

A+B

Net free energy change

C+D

Time →

■ **Figure 4-6   Some reactions have large activation energies**   This endergonic reaction, which is the reverse of the exergonic reaction shown in Fig. 4-4a, has a large activation energy.

Enzymes Lower the Activation Energy of Reactions

How does an enzyme increase the rate of a reaction? In thermodynamic terms, it lowers the activation energy, making it more likely that the reaction will start (Fig. 4-7 ■). Enzymes accomplish this by binding to the reactant molecules and bringing them together into the best position for reacting with each other. Without enzymes, the reaction

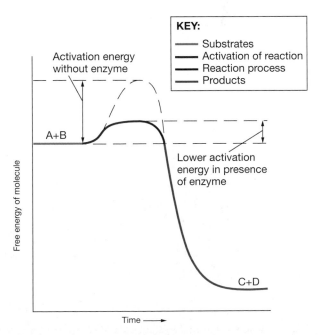

KEY:
— Substrates
— Activation of reaction
— Reaction process
— Products

Activation energy without enzyme

A+B

Free energy of molecule

Lower activation energy in presence of enzyme

C+D

Time →

■ **Figure 4-7   Enzymes lower the activation energy of reactions**   Enzymes speed up reactions by lowering the activation energy required to start the reaction.

would depend on the random collision of the molecules to bring them into alignment. Notice in the figure that neither the starting nor ending free energy content of the substrates and products has changed.

The rate of a reaction catalyzed by an enzyme is much more rapid than the same reaction taking place in the absence of the enzyme. For example, consider the enzyme *carbonic anhydrase,* which converts CO_2 and water to carbonic acid. A single molecule of carbonic anhydrase takes one second to catalyze the conversion of one million molecules of CO_2 and water to carbonic acid. In the absence of enzyme, it takes nearly 100 seconds for only one molecule of CO_2 and water to be converted to carbonic acid. Without enzymes in cells, biological reactions would go so slowly that the cell would be unable to obtain enough energy to live.

Enzymes Bind to Their Substrates

How does an enzyme bring the substrates into the best position for them to react? Most enzymes are large protein molecules with complex three-dimensional shapes [∞ p. 32].* On each enzyme molecule is a region known as the

---

*Recently, researchers discovered that RNA and DNA can sometimes act as enzymes. However, their significance in human physiology is unclear.

...continued from page 74

Tay-Sachs disease is a particularly devastating condition. Normally, lysosomes within cells contain enzymes that digest old, worn-out parts of the cell. In Tay-Sachs, hexosaminidase A, an enzyme that digests specific glycolipids called gangliosides, is absent. As a result, gangliosides accumulate in nerve cells in the brain, causing them to swell and function abnormally. Infants with Tay-Sachs disease slowly lose muscle control and brain function. They usually die before age four.

Question 1: *Hexosaminidase A is also required to remove gangliosides from the light-sensitive cells of the eye. Based on this information, what is another symptom of Tay-Sachs disease besides loss of muscle control and brain function?*

binding site, part of the protein molecule that actually binds to the substrates. When enzyme-substrate binding takes place, the substrate molecules are brought close to each other and to the enzyme's **active site,** the region that promotes reaction of the substrates with each other.

For many years, it was thought that the binding site was shaped to fit the substrates exactly, the way a key fits into a lock. A representation of this *lock-and-key model* of enzyme activity is shown in Figure 4-8a ∎. In recent

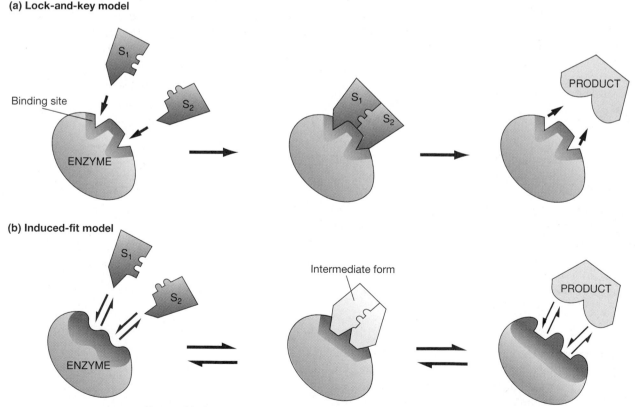

(a) Lock-and-key model

(b) Induced-fit model

Enzyme binds most tightly to an intermediate form of substrate or product.

∎ **Figure 4-8   Enzymes bind substrate at the binding site**   (a) In the older, lock-and-key model, the enzyme binding site was an exact match for the substrates (S_1 and S_2). (b) In the current, induced-fit model, the binding site is not an exact match for the substrates. Instead, the enzyme binds most tightly to a shape intermediate between the substrates and the products. This model is more compatible with reversible reactions that are catalyzed by the same enzyme.

decades, however, scientists have discovered that the binding site and the substrates do not need to fit one another exactly.

The binding site needs only to attract the substrates to that region of the enzyme. Then, as the binding site and the substrates begin to interact, the binding site changes shape to fit more closely to the substrates. This **induced-fit model** of enzyme-substrate interaction is shown in Figure 4-8b ■. According to this model, the binding site has an intermediate shape that can change to fit either the substrate or the product molecules. This trait would enable an enzyme to bind to either reactants or products and thus be able to catalyze a single reaction in both directions. In fact, enzymes can do this. Induced fit explains reversibility better than the older lock-and-key model, in which the binding site was matched to either the substrates or the products, but not to both.

Enzyme Specificity   Most enzymes react with only one set of substrates or with a group of similar substrates. The ability of an enzyme to catalyze a certain reaction or a group of closely related reactions is called **specificity.** Some enzymes are very specific about the reactions that they catalyze. For example, the sole function of the enzyme *glucokinase* is to attach a phosphate group onto glucose molecules when they enter a cell. Other enzymes act on whole groups of molecules. The enzymes known as *peptidases* act on peptide bonds of polypeptides without regard for which two amino acids are joined by those bonds. Peptidases therefore are not very specific in their action.

An Enzyme's Name Gives Information About Its Function   Most enzyme names are important clues to their function in a cell. The names of most enzymes are instantly recognizable by the suffix *-ase.* The first part of the enzyme's name (everything that precedes the suffix) usually refers to the type of reaction, the substrate upon which the enzyme acts, or both. For example, glucokinase, mentioned previously, has glucose as its substrate and as a kinase adds a phosphate group [∞ p. 24] to the substrate. Addition of a phosphate group is called **phosphorylation.** The peptidases mentioned above break up peptides into smaller peptides or amino acids by cleaving the peptide bond that links the amino acids.

Note that a few enzymes have two names. These enzymes were discovered before 1972, when the current standards for naming enzymes were first adopted. As a result, they have both a new name and a commonly used older name. Pepsin and trypsin, two different digestive enzymes, are examples of older enzyme names.

A few enzymes come in a variety of related forms; these are known as isozymes [*iso-*, equal]. **Isozymes** are enzymes that catalyze the same reaction but under different conditions or in different tissues. Their structures are slightly different, which causes the variability in their activity. Lactate dehydrogenase, for example, has sever-

TABLE 4-2   Diagnostically Important Enzymes

Elevated blood levels of these enzymes are suggestive of the pathologies listed.

Enzyme	*Related Diseases*
Acid phosphatase*	Cancer of the prostate
Alkaline phosphatase	Diseases of bone or liver
Amylase	Pancreatic disease
Creatine kinase (CPK)	Myocardial infarction (heart attack), muscle disease
Glutamate dehydrogenase (GDH)	Liver disease
Lactate dehydrogenase (LDH)	Myocardial infarction, liver disease, excess breakdown of red blood cells

*A newer test for a molecule called prostate specific antigen (PSA) has replaced the test for acid phosphatase in the diagnosis of prostate cancer.

al isozymes, including one found primarily in the heart and a second found in skeletal muscle and the liver.

Enzymes and their isozymes have an important role in the medical diagnosis of certain conditions. For example, in the hours following a heart attack, damaged heart muscle cells release enzymes into the blood. One way to determine whether a person's chest pain was indeed due to a heart attack is to look for elevated levels of heart isozymes in the blood. Some diagnostically important enzymes and the diseases of which they are suggestive are listed in Table 4-2.

Separation of Proteins by Electrophoresis Many enzymes that have isozymes are complex structures with multiple protein chains. For example, lactate dehydrogenase has two kinds of subunits, named H and M, that are assembled into *tetramers*—groups of four. The isozymes include H_4, H_2M_2, and M_4. One way to determine which isozymes are present in a tissue sample is to use a technique known as electrophoresis. In this technique, a solution of proteins is placed at one end of a container filled with a polyacrylamide polymer gel. An electric current is passed through the gel, causing the negatively charged proteins to move toward the positively charged end of the gel. The rate at which the proteins move depends upon their size, shape, and the electrical charge on the amino acids that make up the protein. The proteins move along the gel at different rates and become separated from each other, causing each of them to appear as an individual band of color when stained with a dye called Coomassie blue or with silver. Electrophoresis can be used to separate mixtures of charged macromolecules and is also used to identify DNA.

✔ Name the substrate for each of the following enzymes: lactase, peptidase, lipase, sucrase.

✔ How does the induced-fit model of enzyme action differ from the lock-and-key model?

✔ What would be a biological advantage of having isozymes of one enzyme?

Some Enzymes Must Be Activated

Some enzymes are not ready to catalyze reactions when they are first synthesized. These enzymes are produced instead as inactive molecules called *proenzymes* or *zymogens*. When the enzymes are needed, one or more portions of the molecule are chopped off, and the inactive protein becomes an active enzyme (Fig. 4-9 ■). This process is known as *proteolytic activation* [*proteo-*, protein + *lysis*, to dissolve]. The enzymes involved in blood clotting and digestion are among those that are produced as proenzymes. Protein hormones as well as enzymes undergo proteolytic activation.

If an enzyme has both an active and an inactive form, the inactive enzyme is often named by adding the suffix *-ogen* to the name of the active enzyme. For ex-ample, *pepsinogen* is the inactive form of the digestive enzyme *pepsin*.

Some Enzymes Require Cofactors or Coenzymes

Sometimes the activation of an enzyme requires the presence of an additional molecule or ion called a **cofactor.** Cofactors may be either inorganic molecules or nonprotein organic molecules. Inorganic cofactors are ions, such as Ca^{2+} or Mg^{2+}. They must attach to the enzyme in order for the substrates to bind to the binding site. This process is shown in Figure 4-10 ■.

Organic cofactors are called **coenzymes.** Coenzymes do not alter the enzyme's binding site as inorganic cofactors do. Instead, coenzymes act as receptors and carriers for atoms or functional groups that are removed from the substrates during the reaction. Although coenzymes are needed for metabolic reactions to take place, they are not required in large amounts.

Many chemicals that we call **vitamins** are the precursors of coenzymes. The water-soluble vitamins, such as the B vitamins, vitamin C, folic acid, biotin, and pantothenic acid, become coenzymes required for various metabolic reactions. For example, vitamin C is needed for adequate collagen synthesis.

Modulators Alter Enzyme Activity

The ability of an enzyme to speed up a reaction can be altered by various factors, including temperature, pH, or molecules that interact with the enzyme. A factor that influences enzyme activity is known as a **modulator.** If a modulator *activates* an enzyme, the reaction rate catalyzed by the enzyme will increase. If a modulator *inactivates* the enzyme, the reaction rate will decrease and may even stop completely.

There are two basic mechanisms by which modulation takes place. The modulator either (1) changes the ability of the substrate to bind to the binding site or (2) changes the ability of the enzyme to alter the activation energy of the reaction. The different types of modulation are discussed in the following sections and are summarized in Table 4-3.

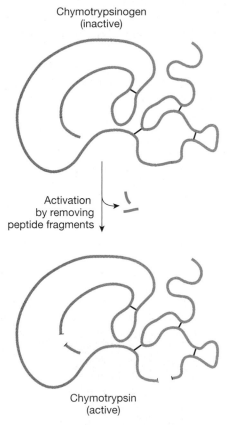

Chymotrypsinogen
(inactive)

Activation
by removing
peptide fragments

Chymotrypsin
(active)

■ **Figure 4-9   Proteolytic activation of an enzyme**   The inactive digestive enzyme chymotrypsinogen becomes active when two small peptide fragments are removed from its long peptide chain. The three peptide chains created by the activation (shown in different shades of blue) are held together by sulfur-sulfur bonds (orange) formed by sulfur-containing amino acids.

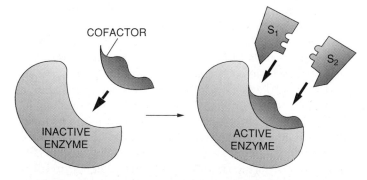

COFACTOR

S₁

S₂

INACTIVE
ENZYME

ACTIVE
ENZYME

■ **Figure 4-10   Cofactors bind to the binding site**   In the absence of the cofactor, the binding site is unable to bind to the substrates.

TABLE 4-3   Factors That Affect Reaction Rate

Essential for enzymatic reaction*

Cofactors	Required for substrate binding at active site.
Coenzymes	Act as receptors for atoms or groups.
Proteolytic activation	Converts inactive to active form by removal of part of molecule. Examples: digestive enzymes, blood-clotting enzymes.

Modulators that alter reaction rate

1. Changes substrate binding at active site

Competitive inhibitor	Competes directly with substrate by binding to active site.
Allosteric modulator (activator or inhibitor)	Binds to enzyme away from active site and changes active site. Reversible. Examples: Feedback inhibition, regulatory proteins.
pH	H^+ alters three-dimensional shape of enzyme by disrupting hydrogen or S-S bonds. May be irreversible if protein denatures.
Temperature	Alters three-dimensional shape of enzyme by disrupting hydrogen or S-S bonds. May be irreversible if protein denatures.

2. Changes ability of enzyme to decrease activation energy of the reaction

Covalent modulation	Binds covalently to enzyme and changes its ability to alter activation energy. Example: addition or removal of phosphate groups.

*Not all enzymes require these, but those that do must have these factors in order to catalyze a reaction.

Acidity and Temperature   Both temperature and pH can have dramatic effects on protein structure and function. Changes in temperature and pH can disrupt the bonds that hold a protein in its tertiary configuration, causing the protein to lose its shape (Fig. 4-11 ■). If you have ever fried an egg, you have watched this transformation happen to the egg white protein albumin as it changes from a slithery clear state to a firm white state.

Hydrogen ions in high enough concentration to be called acids have a similar effect on proteins. During preparation of ceviche, the national dish of Ecuador, raw fish is marinated in lime juice. The acidic lime juice contains hydrogen ions that disrupt hydrogen bonds in the muscle proteins of the fish, causing the proteins to lose their configuration. As a result, the meat becomes more firm and opaque, just as it would if it were cooked with heat.

Small changes in pH and temperature can increase or decrease the activity of enzymes (Fig. 4-12 ■). However, once the changes exceed some critical point, the structure of the enzyme is altered so much that its activity is destroyed. When this occurs, the enzyme is said to be *denatured*. In a few cases, activity can be restored if the modulator is removed. The protein then resumes its original shape as if nothing had happened. Usually, however, denaturation is a permanent loss of activity. There is certainly no way to unfry an egg or uncook a piece of fish. The potentially disastrous influence of temperature and pH on enzymes and other proteins is the reason that these factors are so closely regulated by the body.

Chemical Modulators   Chemical modulators are molecules that bind to enzymes and alter their catalytic ability. There are three types of chemical modulators: competitive inhibitors, allosteric modulators, and covalent modulators.

Competitive inhibitors bind to the binding site of the enzyme, blocking the site and thus preventing the enzyme from binding with the substrate (Fig. 4-13 ■). Some competitive inhibitors are alternative substrates that can also be acted upon by the enzyme. Others are

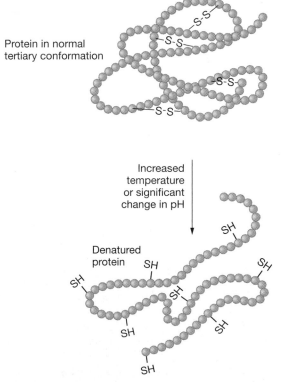

■ Figure 4-11   **Denaturation of proteins results when intramolecular bonds are broken**   Heat and acid can break the sulfur-sulfur or hydrogen bonds that hold a protein in its three-dimensional shape. When this happens to an enzyme, it loses its activity.

(a) Effect of changing temperature on enzyme activity.

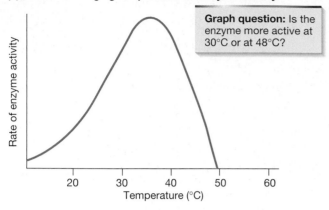

> **Graph question:** Is the enzyme more active at 30°C or at 48°C?

(b) Effect of changing pH on enzyme activity.

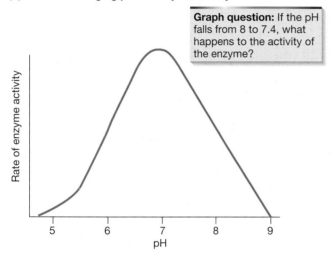

> **Graph question:** If the pH falls from 8 to 7.4, what happens to the activity of the enzyme?

■ **Figure 4-12   Effect of pH and temperature on enzyme activity**   (a) This enzyme denatures above temperatures of 50° C. Activity slows to almost zero in the cold. (b) Most enzymes in humans have optimal activity near the body's internal pH of 7.4. Extremes of pH also denature enzymes.

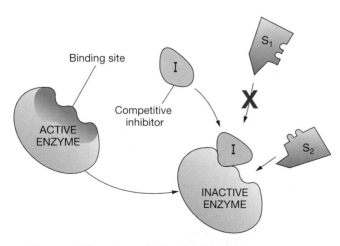

■ **Figure 4-13   Competitive inhibition**   Competitive inhibitors bind to the binding site of the enzyme, preventing the normal substrate from binding.

simply molecules that bind to the enzyme and block the binding site without being acted upon themselves. They are like the guy who slips into the front of the movie ticket line to chat with his girlfriend, the cashier. He has no interest in buying a ticket, but he prevents the people in line behind him from getting their tickets for the movie.

The next group, **allosteric modulators** [*allo-*, other + *stereos*, solid (as a shape)], bind to the enzyme away from the binding site and change the shape of the binding site (Fig. 4-14 ■). This type of modulation can either (1) increase the probability of enzyme-substrate binding and enhance enzyme activity or (2) decrease the affinity of the binding site for the substrate and inhibit enzyme activity.

Allosteric modulation also occurs in proteins other than enzymes. For example, you will learn how the ability of the respiratory protein hemoglobin to bind oxygen changes with allosteric modulation by carbon dioxide, H^+, and several other factors.

Covalent modulators are atoms or functional groups that bind by covalent bonds to enzymes and affect their ability to decrease the activation energy of the reaction. Like allosteric modulators, covalent modulators may either increase or decrease the activity of the enzyme. One of the most common covalent modulators is the phosphate group. Many enzymes in the cell can be either activated or inactivated when a phosphate group forms a covalent bond with them. The ability to turn reactions on and off or up and down allows the cell to regulate the flow of biomolecules through synthetic and energy-producing pathways.

One of the best known chemical modulators is the antibiotic penicillin. Alexander Fleming discovered in this compound in 1928, when he noticed that *Penicillium* mold inhibited bacterial growth. By 1938, researchers had extracted the active ingredient penicillin from the mold and used it to treat infections in humans. Yet it was not until 1965 that researchers figured out exactly how the antibiotic works. Penicillin is a competitive inhibitor that binds to a key bacterial enzyme by mimicking the normal substrate. In doing so, it forms bonds that are unbreakable and thereby irreversibly inhibits the enzyme. Without the enzyme, the bacterium is unable to make a rigid cell wall. Without the rigid wall, the bacterium swells, ruptures, and dies.

✔ Name two ways an enzyme modulator can work to affect the rate of an enzymatic reaction.

Enzyme and Substrate Concentration Affect Reaction Rate

We measure the rate of an enzymatic reaction by monitoring either how fast the products are synthesized or how fast the substrates are consumed. If the amount of enzyme present changes or if the substrate concentra-

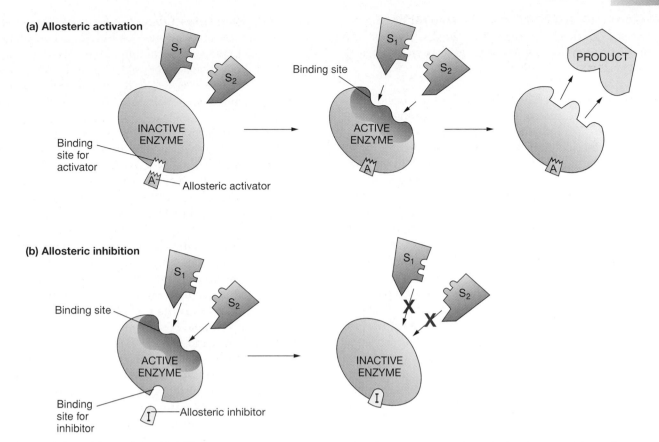

(a) Allosteric activation

(b) Allosteric inhibition

■ **Figure 4-14   Allosteric modulation**   Allosteric modulators bind to the enzyme away from the binding site. (a) An allosteric activator enhances enzyme-substrate binding, so the reaction proceeds at a faster rate. (b) An allosteric inhibitor decreases enzyme-substrate binding by changing the binding site so that it has a lower attraction for the substrates.

tion changes, the reaction rate will change. Now let's examine the relationships between enzyme and substrate concentration and reaction rate.

Reaction Rate Is Directly Related to the Amount of Enzyme Present   A major determinant of the rate of an enzymatic reaction is the amount of enzyme present. As we have seen, if there is no enzyme, most metabolic reactions go very slowly. If enzyme is present, the rate of the reaction will be proportional to the amount of enzyme.

The graph in Figure 4-15 ■ shows the results of a typical experiment in which the amount of substrate remains constant while the amount of enzyme is varied. As the graph shows, an increase in the amount of enzyme present causes an increase in the rate of the reaction. This is like the checkout lines in a supermarket. Imagine that each cashier is an enzyme, the waiting customers are substrates, and those leaving the store with their purchases are products. One hundred people will get checked out faster when there are 25 lines open than when there are only 10 lines. Likewise, in a reaction, the presence of more enzyme molecules means that more binding sites are available to interact with the substrate molecules. As a result, reactants are converted into products more rapidly.

The relationship between enzyme concentration and reaction rate is an important way that cells regulate their physiological processes. Cells control the amount of an enzyme by regulating both its synthesis and its breakdown. If enzyme synthesis exceeds breakdown, enzyme accumulates and the reaction speeds up. If enzyme

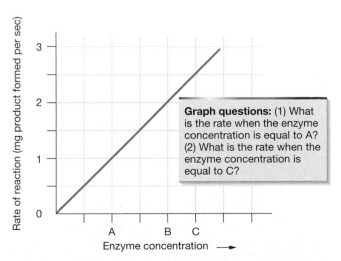

Graph questions: (1) What is the rate when the enzyme concentration is equal to A? (2) What is the rate when the enzyme concentration is equal to C?

■ **Figure 4-15   Effect of changing enzyme concentration on rate**   In this experiment, the substrate concentration remained constant while the enzyme concentration was varied.

breakdown exceeds synthesis, the amount of enzyme decreases, as does the reaction rate. Even when the amount of enzyme is constant, there is still a steady turnover of enzyme molecules.

Reaction Rate Can Reach a Maximum
Besides the amount of enzyme present, what else determines the rate of a reaction? If the concentration of the enzyme is constant, then the reaction rate will vary according to the concentration of substrate. Figure 4-16 ■ shows the results of a typical experiment in which the enzyme concentration is constant, but the concentration of substrate varies. At low substrate concentrations, the reaction rate is directly proportional to the substrate concentration. But when the concentration of the substrate molecules increases beyond a certain point, the limited number of enzyme molecules have no more binding sites free to bind substrate molecules. The enzyme is catalyzing reactions as fast as it can, and the rate of reaction reaches a maximum value. This condition is known as **saturation.**

An analogy to saturation appeared in the early days of television on the *I Love Lucy* show. Lucille Ball was working on the conveyor belt of a candy factory, loading chocolates into the little paper cups of a candy box. Initially, the belt was moving slowly, and she had no difficulty removing the candy and putting it into the box. Gradually, the belt brought candy to her more rapidly, and she had to increase her packing speed to keep up. Finally, the belt was bringing candy to her so fast that she couldn't pack it all in the boxes because she was working at her maximum rate. That was Lucy's saturation point. (Her solution was to stuff the candy into her mouth as well as into the box!)

✔ What happens to the rate of an enzymatic reaction as the amount of enzyme present increases?

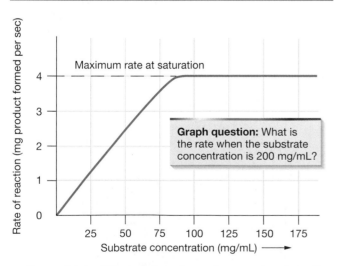

Graph question: What is the rate when the substrate concentration is 200 mg/mL?

■ **Figure 4-16   Effect of changing substrate concentration on rate**   In this experiment, the enzyme concentration remained constant while the substrate concentration was varied. The rate of the reaction increases as the substrate concentration increases, up to a maximum rate. At this rate, the enzyme is said to have reached saturation.

Reversible Reactions Obey the Law of Mass Action
The study of reaction rates in cells is complicated by the fact that some reactions are reversible. This means that the forward reaction A + B → C + D can reverse, so that C + D → A + B. In the reverse reaction, the products become the reactants and the reactants become the products.*

What then determines in which direction the reaction is to go? The answer is that reversible reactions go to a state of **equilibrium,** where the rate of the reaction in the forward direction (A + B → C + D) is exactly equal to the rate of the reverse reaction (C + D → A + B), as shown in Figure 4-17a ■. At equilibrium, there is no net change in the amount of reactant or product. As fast as A and B convert to C and D, the reverse reaction takes place: C and D turn back into A and B at an equal rate.

But what happens if some factor disturbs this equilibrium by increasing the amount of A and B within the system? You learned in the previous section that reaction rate is proportional to substrate concentration. If the concentration of A and B in the system goes up, so does the rate of the forward reaction (Fig. 4-17b ■). More A and B are converted into products C and D. But since the reaction is reversible, as the product concentration increases, so does the rate of the reverse reaction. The system swings back and forth until finally it reaches equilibrium again (Fig. 4-17c ■).

Reversible reactions like this obey the **law of mass action,** a simple relationship that holds for chemical reactions whether in a test tube or in a cell. You may have learned this law in chemistry as *Le Châtelier's principle.* In very general terms, the law of mass action says that when a reaction is at equilibrium, the ratio of the substrates to the products is always the same.

If the concentration of a substrate (A or B) or product (C or D) changes, the equilibrium will be disturbed. The system then adjusts the substrate and product concentrations until the equilibrium ratio is restored. The law of mass action is important in physiology because as the concentration of substrates changes, the change is reflected in the concentrations of the products.

Enzymatic Reactions Can Be Categorized

Most reactions catalyzed by enzymes can be classified into one of four categories: oxidation-reduction, hydrolysis-dehydration, exchange-addition-subtraction, and ligation reactions. Table 4-4 summarizes these categories.

Oxidation-Reduction Reactions
Oxidation-reduction reactions are the most important reactions in energy extraction and transfer within the cell. These reactions transfer electrons or protons (H^+) between atoms, ions, or molecules. A molecule that gains electrons or loses H^+ is said to be **reduced.** One way to think of this is to

---

*The naming of forward and reverse directions is arbitrary.

(a) Reaction at equilibrium

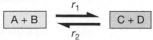

Rate of reaction in forward direction (r_1) = rate of reaction in reverse direction (r_2)

(b) Equilibrium disturbed

Add A and B to system

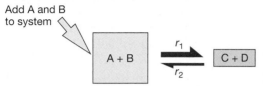

Reaction rate r_1 increases: $r_1 > r_2$

(c) Equilibrium restored

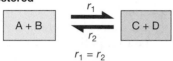

$$r_1 = r_2$$

The ratio of substrates to products is always the same at equilibrium

■ **Figure 4-17 Law of mass action** The law of mass action states that when a reaction is at equilibrium, the ratio of the products and substrates will remain constant. Although this figure shows A + B equal to C + D at equilibrium, the actual ratio can vary, depending on the reaction. (a) When the reaction is at equilibrium, the rate of the forward reaction, r_1, equals the rate of the reverse reaction, r_2. (b) When substrate is added to one side, some of the added substrate is converted into product. (c) The reaction returns to equilibrium when the substrate-to-product ratio returns to the constant value for this reaction.

...continued from page 80

Tay-Sachs disease is a recessive genetic disorder caused by a defect in the gene that directs the synthesis of hexosaminidase A. For a baby to be born with Tay-Sachs disease, it must inherit two defective genes, one from each parent. People with one Tay-Sachs gene and one normal gene are called carriers of the disease. Carriers will not develop the disease but can pass the defective gene on to their children. About 1 in 27 people of Eastern European Jewish descent in the United States carries the Tay-Sachs gene.

People who have two normal genes have normal amounts of hexosaminidase A in their blood. Carriers have lower-than-normal levels of the enzyme, but this amount is enough to prevent excessive accumulation of gangliosides in cells.

Question 2:   *How would you test Sarah and David for the presence of the Tay-Sachs gene?*

remember that the gain of negatively charged electrons *reduces* the electric charge on the molecule. Likewise, the loss of a positive proton leaves behind an electron with a negative charge. Conversely, molecules, atoms, or ions that lose electrons or gain H^+ are said to be **oxidized.**

Hydrolysis-Dehydration Reactions   Hydrolysis and dehydration reactions are important in the breakdown and synthesis of large biomolecules. In **dehydration reactions,** a water molecule is one of the products. In many dehydration reactions, two molecules combine into one, losing water in the process. For example, the monosaccharides glucose and fructose join to make one sucrose molecule [∞ p. 29]. In the process, one molecule loses a hydroxyl group (—OH) and the other molecule

TABLE 4-4   Types of Chemical Reactions

Reaction Type	What Happens	Representative Enzymes
1. Oxidation-reduction	Add or subtract electrons or H^+	
a. Oxidized	Transfer electrons from donor to oxygen	Oxidase
	Remove electrons and H+	Dehydrogenase
b. Reduced	Gain electrons	Reductase
2. Hydrolysis-dehydration	Add or subtract a water	Hydrolase
a. Hydrolysis	Splits large molecules by adding water	Protease, lipase
b. Dehydration	Removes water. Used to make large molecules from several smaller ones	
3. Transfer chemical groups	Add, subtract, or exchange groups between molecules	
a. Exchange reaction	Phosphate	Kinase
	Amino group	Transaminase
b. Add a group	Phosphate	Phosphorylase
	Amino group	Aminase
c. Subtract a group	Phosphate	Phosphatase
	Amino group	Deaminase
4. Ligation	Join two substrates using energy from ATP	Synthetase

loses a hydrogen to create water, H_2O. When a dehydration reaction results in the synthesis of a new molecule, the process is known as *dehydration synthesis.*

In a **hydrolysis reaction** [*hydro,* water + *lysis,* to loosen or dissolve], a substrate changes into one or more products by adding water. In these reactions, the covalent bonds of the water molecule are broken so the water reacts as a hydroxyl group (—OH) and a hydrogen (—H). For example, an amino acid can be removed from the end of a peptide with a hydrolysis reaction (see Fig. 4-27 ■, p. 96).

When an enzyme name consists of the substrate name with the suffix *-ase,* the enzyme causes a hydrolysis reaction. An example is lipase, an enzyme that breaks up large lipids into smaller lipids by hydrolysis. The enzyme that removes the amino acid from the peptide is called a peptidase.

Addition-Subtraction-Exchange Reactions An **addition reaction** adds a functional group to one or more of the reactants. A **subtraction reaction** removes a functional group from one or more of the reactants. Functional groups are exchanged between or among reactants during an **exchange reaction.**

For example, phosphate groups may be transferred between molecules during addition, subtraction, or exchange reactions. The transfer of phosphate groups is an important means of covalent modulation, turning reactions on, off, up, or down. Several types of enzymes catalyze reactions that transfer phosphate groups. The enzymes named *kinases* transfer a phosphate group from a substrate to an ADP molecule to create ATP, or from an ATP molecule to a substrate. For example, creatine kinase transfers a phosphate group from creatine phosphate to ADP, forming an ATP and leaving behind creatine.

The addition, subtraction, or exchange of amino groups [∞ p. 23] is also important in metabolism. Removal of an amino group from an amino acid or peptide is a **deamination** reaction, illustrated in Figure 4-27 ■. Addition of an amino group is **amination,** and the transfer of an amino group from one molecule to another is called **transamination.**

Ligation Reactions Ligation reactions join two molecules together using enzymes known as *synthetases* and energy from ATP. An example of a ligation reaction is the synthesis of acetyl coenzyme A from fatty acids and coenzyme A. Acetyl coenzyme A is an important molecule in metabolism, as you will learn in the next section.

▶ METABOLISM

Metabolism refers to all chemical reactions that take place within an organism. These reactions extract energy from nutrient biomolecules (such as proteins, carbohydrates, and lipids) and synthesize or break down molecules. Metabolism is often divided into **catabolism,** reactions that produce energy through the breakdown of large biomolecules, and **anabolism,** energy-utilizing

reactions that result in the synthesis of large biomolecules. Anabolic and catabolic reactions take place simultaneously in cells throughout the body, so that at any given moment, some biomolecules are being synthesized while others are being broken down.

The energy released from or stored in the chemical bonds of biomolecules during metabolism is commonly measured in kilocalories (kcal). A **kilocalorie** (or Calorie, with a capitol C) is the amount of energy needed to raise the temperature of 1 liter of water by 1 degree Celsius.

Much of the energy released during catabolism is trapped in the high-energy phosphate bonds of ATP or in the high-energy electrons of NADH, $FADH_2$, or NADPH [∞ p. 35]. From these temporary carriers, energy is transferred to ATP or the covalent bonds of biomolecules through anabolic reactions.

Metabolism is a highly coordinated process in which activities at any given moment are matched to the needs of the cell. The enzymatic reactions of metabolism form a network of interconnected chemical reactions, or **pathways.** Each step of a pathway is a different enzymatic reaction, and the reactions of a pathway proceed in sequence. Starting substrate A is changed into product B, which then becomes the substrate for the next reaction. In that reaction, B is changed into C, and so forth:

$$A \rightarrow B \rightarrow C \rightarrow D$$

We call the molecules of the pathway **intermediates** because the products of one reaction become the substrates for the next. You will sometimes hear metabolic pathways called *intermediary metabolism.* Certain intermediates, called *key intermediates,* participate in more than one pathway and act as the branch points for channeling substrate in one direction or another. Glucose is a key intermediate in several metabolic pathways.

In many ways, a network of metabolic pathways is similar to a detailed road map (Fig. 4-18 ■). Just as a map shows a network of roads that connect various cities and towns, metabolism can be thought of as a network of chemical reactions connecting various intermediate products. Each city or town is a different chemical intermediate. One-way roads are irreversible reactions, and big cities with roads to several destinations are key intermediates. Just as there may be more than one way to get from one place to another, there can be several pathways between chemical intermediates.

Cells Regulate Their Metabolic Pathways

How do cells regulate the flow of molecules through their metabolic pathways? They do so in five basic ways:

1. By controlling the enzyme concentration
2. By producing allosteric and covalent modulators
3. By using two different enzymes to catalyze reversible reactions
4. By isolating enzymes within intracellular organelles
5. By maintaining an optimum ratio of ATP to ADP

(a) Section of a road map

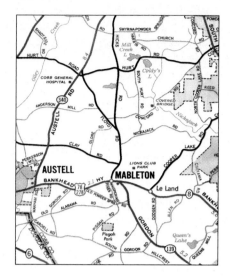

(b) Metabolic pathways drawn like a road map

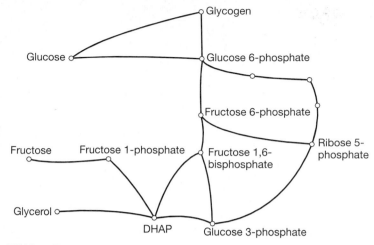

DHAP = dihydroxyacetone phosphate

■ **Figure 4-18   A group of metabolic pathways resembles a road map**   Cities on the map are equivalent to intermediates in metabolism. In metabolism, there may be more than one way to go from one intermediate to another, just as on the map there may be many ways to get from one city to another.

We discussed the effect of changing enzyme concentration in an earlier section of this chapter. The other four methods are described below.

Enzyme Modulation   Enzyme modulation, described on page 82, is frequently controlled by hormones and other signals coming from outside the cell. This type of outside regulation is a key element in the integrated control of the body's metabolism following a meal or during periods of fasting.

Metabolic pathways may also have their own built-in form of modulation, called **feedback inhibition.** In this form of modulation, the end product of a pathway, shown as Z in Figure 4-19 ■, acts as an inhibitory modulator of the pathway. As the pathway proceeds and Z accumulates, the enzyme catalyzing the conversion of A to B is inhibited. This slows down production of Z until the cell can use it up. Once the levels of Z drop, the inhibition is removed and the pathway starts to run again. Because Z is the end product of the pathway, this type of inhibition is sometimes called *end-product inhibition.*

Enzymes and Reversible Reactions   Cells can use reversible reactions to regulate the rate and direction of

metabolism. If a single enzyme can run the reaction in either direction (Fig. 4-20a ■), the reaction will go to a state of equilibrium as determined by the law of mass action. These reactions therefore cannot be closely regulated except by modulators and by control of the amount of enzyme. But if a reaction requires two different enzymes, one for the forward and one for the reverse reaction, the cell can regulate the reaction more closely (Fig. 4-20b ■). If there is no enzyme for the reverse reaction, the reaction is irreversible (Fig. 4-20c ■).

Isolation of Enzymes Within the Cell   Many enzymes of metabolism are localized in specific parts of the cell. Some, like the enzymes of carbohydrate metabolism, are found dissolved in the cytosol, whereas others are isolated within specific organelles. Mitochondria, endoplasmic reticulum, Golgi apparatus, and lysosomes all contain enzymes that are not found in the cytosol. This separation of pathways allows the cell to control metabolism by regulating the movement of substrate between the different compartments. The isolation of enzymes within organelles is an example of *compartmentation* [∞ p. 7].

Ratio of ATP to ADP   The energy status of the cell is one final mechanism that can influence metabolic pathways. Through complex regulation, the ratio of ATP to ADP in the cell determines whether pathways that re-

•••
…continued from page 87

In 1989, researchers discovered three genetic mutations responsible for Tay-Sachs disease. This discovery paved the way for a new carrier screening test that directly tests for the presence of the defective gene in blood cells, rather than testing for lower-than-normal hexosaminidase A levels. David and Sarah will undergo this new genetic test.

Question 3:   *Why is the new test for the Tay-Sachs gene more accurate than the old test, which detects decreased amounts of hexosaminidase?*

•••

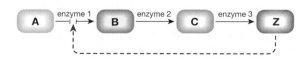

■ **Figure 4-19   Feedback inhibition**   The accumulation of product Z inhibits the first step of the pathway. As the cell consumes Z in another metabolic reaction, the inhibition is removed and the pathway resumes.

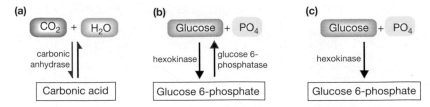

■ **Figure 4-20   Reversible and irreversible metabolic reactions**   (a) Some enzymes catalyze a reaction in both directions. (b) Other enzymes catalyze a reaction in only one direction. (c) If a cell has only one of the two enzymes needed for a reaction, the reaction is irreversible. For example, most cells of the body can phosphorylate glucose to glucose 6-phosphate but cannot make glucose by dephosphorylation, because they lack the enzyme glucose 6-phosphatase.

sult in ATP synthesis are turned on or off. When ATP levels are high, production of ATP drops. When ATP levels are low, the cell sends substrates through pathways that result in more ATP synthesis. In the next section, we look further into the role of ATP in cellular metabolism.

ATP Transfers Energy Between Reactions

The usefulness of metabolic pathways as suppliers of energy is often measured in terms of the net amount of adenosine triphosphate, or ATP, that the pathways can yield. ATP is a nucleotide molecule containing three phosphate groups [∞ p. 23]. The third phosphate group is attached by a covalent bond in an energy-requiring reaction. Energy is stored in this *high-energy phosphate bond* and then released when the bond is broken during removal of the phosphate group. This relationship is shown by the following reaction:

$$ADP + P_i + energy \leftrightarrow ADP \sim P (= ATP)$$

The squiggle indicates a high-energy bond and P_i is the abbreviation for an inorganic phosphate group. Estimates of the amount of free energy released when a high-energy phosphate bond is broken range from 7 to 12 kcal per mole of ATP.

ATP is more important as a carrier of energy than as an energy storage molecule. For one thing, cells have a limited amount of ATP. A resting adult human would require 40 kg (88 pounds) of ATP to support one day's worth of metabolic activity, far more than our cells could store. Instead, the body acquires most of its daily energy requirement from the chemical bond energy stored in complex biomolecules. Metabolic reactions transfer that energy to the high-energy bonds of ATP, or in a few cases, into high-energy bonds of the related nucleotide, *guanosine triphosphate*, or **GTP.**

The metabolic pathways that yield the most ATP molecules are those that require oxygen, the **aerobic,** or *oxidative,* pathways. **Anaerobic** [*an-,* without + *aer,* air] pathways that can proceed without oxygen also produce ATP molecules, but in much smaller quantities. The lower ATP yield of anaerobic pathways means that most animals (including humans) are unable to survive for extended periods on anaerobic metabolism alone. In the next section we will consider how the three major groups of biomolecules are metabolized in order to transfer energy to ATP.

✔ Name five ways that cells regulate the movement of substrates through metabolic pathways.

▶ ATP PRODUCTION

The catabolic pathways that extract energy from biomolecules and transfer it to ATP are summarized in Figure 4-21 ■. Aerobic production of ATP follows two common pathways: glycolysis and the citric acid (tricarboxylic acid) cycle. Both pathways produce small amounts of ATP directly, but their most important contributions to ATP synthesis are high-energy electrons carried by NADH and $FADH_2$ to the electron transport system in the mitochondria. The electron transport system, in turn, transfers energy from the electrons to the high-energy phosphate bond of ATP. At various points, the process produces carbon dioxide and water. The water can be used by the cell, but carbon dioxide is a waste product and must be removed from the body.

All three types of biomolecules used for energy (carbohydrates, proteins, and lipids) use the enzymes of glycolysis, the citric acid cycle, and the electron transport system to produce ATP. Carbohydrates enter the pathways in the form of glucose. Lipids are broken down into fatty acids and glycerol [∞ p. 29], which enter the pathways at different points. Proteins are broken down into assorted amino acids, which also enter at various points. Because glucose is the only molecule that follows the entire pathway, we look first at glucose catabolism. We will then see how protein and lipid catabolism differ from glucose catabolism.

The pathways for ATP production are a good example of compartmentation. The enzymes of glycolysis are located in the cytosol while the enzymes of the citric acid cycle are in the mitochondria.

Glycolysis Converts Glucose and Glycogen into Pyruvate

The main catabolic pathway of the cytosol is the series of reactions known as **glycolysis** [*glyco-,* sweet + *lysis,* dissolve]. During glycolysis, a molecule of glucose is converted by a series of enzymatically catalyzed reactions into two pyruvate molecules, producing a net gain in energy.

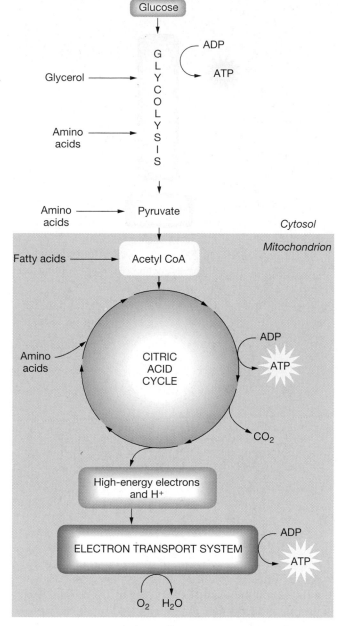

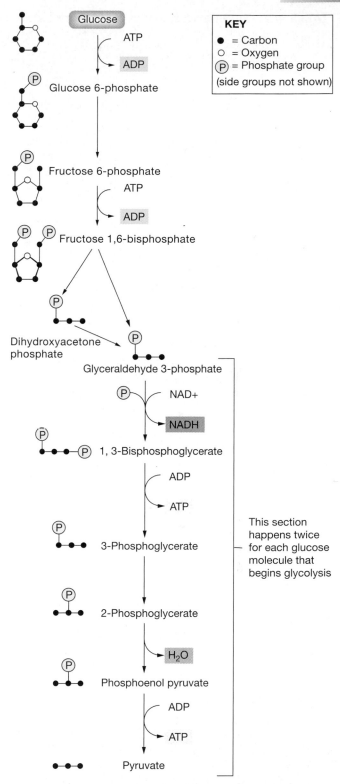

KEY

● = Carbon
○ = Oxygen
Ⓟ = Phosphate group
(side groups not shown)

This section
happens twice
for each glucose
molecule that
begins glycolysis

■ **Figure 4-21   Overview of aerobic pathways for ATP production**   ATP can be made through aerobic pathways using the chemical bond energy from glucose, amino acids, or glycerol and fatty acids.

Thus, glycolysis is considered to be an exergonic pathway even though some steps of the pathway require energy input. A portion of the energy produced is used to phosphorylate ADP molecules, yielding ATP molecules. The events of glycolysis can be summarized as follows:

$$\text{Glucose} + 2\,NAD^+ + 2\,ADP + 2\,P_i \rightarrow$$
$$2\,\text{Pyruvate} + 2\,ATP + 2\,NADH + 2\,H^+ + 2\,H_2O$$

Notice that glycolysis does not require oxygen. It therefore serves as a common pathway for aerobic and anaerobic catabolism of glucose.

Figure 4-22 ■ shows the steps of glycolysis. In the first step, glucose is phosphorylated with the aid of ATP

■ **Figure 4-22   Glycolysis**   Glycolysis is the pathway that converts glucose to pyruvate. Glucose, a six-carbon sugar, is metabolized into two three-carbon pyruvate molecules. The steps of glycolysis consume two ATP and make four ATP and two NADH. Pyruvate is the branch point for aerobic and anaerobic metabolism of glucose.

TABLE 4-5   Summary of Energy Yield from Catabolic Pathways

Aerobic metabolism

Pathway	Uses	Makes	Net Energy Yield
Glucose			
glucose → 2 pyruvate	2 ATP	4 ATP 2 NADH	+2 ATP +2 NADH*
Aerobic metabolism			
2 pyruvate → 2 acetyl CoA	—	2 NADH 2 CO_2	+2 NADH
2 acetyl CoA through citric acid cycle		2 ATP 6 NADH 2 $FADH_2$	+2 ATP +6 NADH +2 $FADH_2$
Total yield for 1 glucose through aerobic metabolism:			+4 ATP +2 NADH cytoplasm* +8 NADH mitochondria +2 $FADH_2$

Maximal ATP yield after mitochondrial oxidative phosphorylation:

4 ATP	4 ATP
2 cytoplasmic NADH × 2–3 ATP/NADH	4–6 ATP
8 mitochondrial NADH × 3 ATP/NADH	24 ATP
2 $FADH_2$ × 2 ATP/$FADH_2$	4 ATP
	= 36–38 ATP

Anaerobic metabolism

Pathway	Uses	Makes	Net Energy Yield
Glucose → 2 lactate	2 ATP 2 NADH	4 ATP 2 NADH	+2 ATP 0 NADH

*Cytoplasmic NADH molecules cannot enter mitochondria. Their ATP yield depends on whether their electrons go to NADH or to $FADH_2$ in the mitochondria.

to glucose 6-phosphate.* Glucose 6-phosphate is a key intermediate that leads into glycolysis as well as to other pathways we will not consider. Note that by the time the six-carbon glucose molecule reaches the end of glycolysis, it has been split into two molecules of pyruvate, each containing three carbons. Two steps of glycolysis are endothermic and require energy input from two ATP. Other steps are exothermic and produce four ATP and two NADH for each glucose that enters the pathway. The net energy yield from each glucose is therefore two ATP and two NADH. (See Table 4-5 for a summary of energy yield of various catabolic pathways.)

Anaerobic Metabolism Converts Pyruvate into Lactate

Pyruvate is a branch point for metabolic pathways, like the hub cities on the road map. Depending on the cell's needs and condition, pyruvate can be shuttled off into one of two pathways (Fig. 4-23 ■). If the cell contains adequate oxygen, pyruvate continues into the citric acid cycle.

If the cell lacks sufficient oxygen for aerobic pathways (a condition that may be caused by many factors, including strenuous exercise), pyruvate is converted into lactate* with the assistance of the enzyme lactate dehydrogenase mentioned earlier:

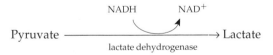

The conversion of pyruvate to lactate changes the NADH produced earlier in glycolysis back to NAD^+ when a hydrogen atom is transferred to the lactate molecule. As a result, the net energy yield for anaerobic metabolism of one glucose molecule is two ATP and no NADH.

---

*The "6" in glucose 6-phosphate tells on which carbon of glucose the phosphate group has been attached.

---

*Lactate is the anion of lactic acid. At normal cell conditions, many organic acids such as pyruvic acid, lactic acid, and glutamic acid are ionized and are therefore named as the corresponding anion: pyruvate, lactate, and glutamate. The suffix -ate identifies the anions of organic and some inorganic acids.

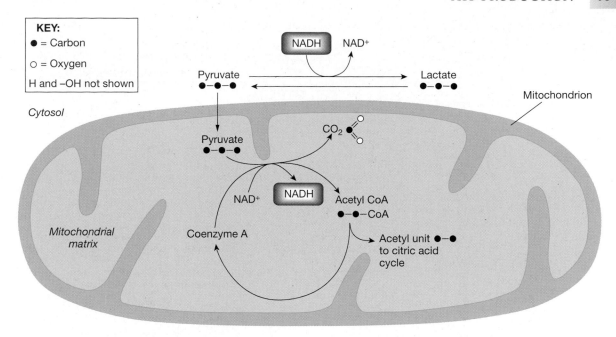

■ **Figure 4-23** **Pyruvate metabolism**   Under anaerobic conditions, pyruvate is made into lactate (lactic acid). If adequate amounts of oxygen are present in the cell, pyruvate is transported into the mitochondrial matrix and converted to acetyl CoA. The acetyl unit of acetyl CoA feeds into the citric acid cycle, while the coenzyme is released to react with another pyruvate.

✔ Using Fig. 4-22 ■, identify the steps of glycolysis that are:
  (a) obviously exergonic
  (b) obviously endergonic
  (c) catalyzed by kinases
  (d) catalyzed by dehydrogenases (see Table 4-4)
  (e) dehydration reactions

Pyruvate Enters the Citric Acid Cycle in Aerobic Metabolism

If the cell has adequate oxygen for aerobic metabolism, then pyruvate molecules formed from glucose during glycolysis are transported into the mitochondria (Fig. 4-23 ■). Once in the mitochondrial matrix, pyruvate is converted into another key intermediate called *acetyl coenzyme A*, or **acetyl CoA** for short. As its name suggests, this molecule has two parts: a two-carbon acetyl (acyl) unit, derived from pyruvate, and a coenzyme.

Coenzyme A is made from the vitamin pantothenic acid. Like enzymes, it is not consumed during metabolic reactions and can be reused. The synthesis of acetyl CoA from pyruvate and coenzyme A is an exergonic reaction that produces one NADH. During the reaction, one of pyruvate's three carbon atoms combines with oxygen and is released as CO_2, leaving acetyl CoA with only two carbons.

Acetyl CoA releases its two-carbon acetyl unit into a cyclical pathway known as the **citric acid cycle,** or *tricarboxylic acid cycle* (Fig. 4-24 ■). This pathway was first described by Sir Hans A. Krebs, so it is also known as the *Krebs cycle.* Because Sir Hans Krebs described other metabolic cycles, we will avoid confusion by using the term citric acid cycle. The citric acid cycle makes a never-ending circle, adding carbons from an acetyl CoA with each turn of the cycle and producing ATP, high-energy electrons, and carbon dioxide.

The two-carbon acetyl unit enters the cycle by combining with a four-carbon oxaloacetate molecule that is the last intermediate in the cycle. The resulting six-carbon citrate molecule then goes through a series of reactions until it completes the cycle as another oxaloacetate molecule. Most of the energy released by the reactions of the cycle is captured as high-energy electrons on three NADH and one $FADH_2$. However, some energy is used to make the high-energy phosphate bond of one ATP, and the remainder is given off as heat. In two of the reactions, carbon and oxygen are removed in the form of carbon dioxide. By the end of the cycle, the four-carbon oxaloacetate molecule that remains is ready to begin the cycle again.

In the next step of aerobic metabolism, NADH and $FADH_2$ transfer their high-energy electrons to the electron transport system and return to the citric acid cycle as NAD^+ and FAD^+.

The Electron Transport System Transfers Energy from NADH and $FADH_2$ to ATP

The final step in aerobic ATP production is the transfer of energy from the electrons of NADH and $FADH_2$ to ATP. This transfer is made possible by a group of mitochondrial proteins known as the **electron transport system,** located in the inner mitochondrial membrane [∞ p. 50]. The protein complexes of the electron transport

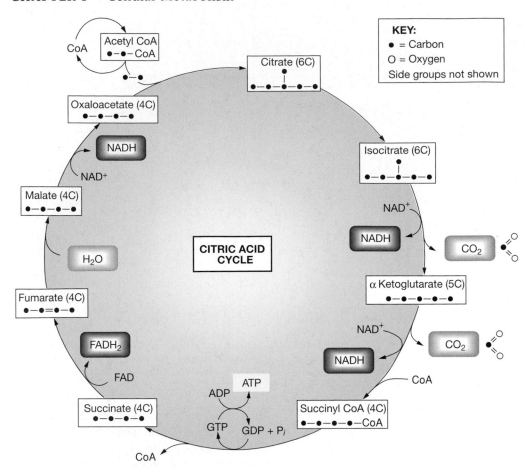

■ **Figure 4-24   The citric acid cycle**   Two-carbon acetyl units from acetyl CoA combine with four-carbon oxaloacetate molecules to make citrate. Each full turn of the cycle releases two carbons as carbon dioxide and produces 1 ATP, 3 NADH, and 1 FADH₂.

system include enzymes and the iron-containing proteins known as **cytochromes.**

According to the **chemiosmotic theory,** as high-energy electrons pass through the electron transport system, their energy is used to move H^+ from the mitochondrial matrix into the intermembrane space. The energy stored in this concentration gradient is then transferred to the high-energy bond of ATP when the H^+ move back across the membrane. The synthesis of ATP using the electron transport system is called **oxidative phosphorylation,** because the electron transport system requires oxygen to act as the final acceptor of electrons and H^+.

The process of oxidative phosphorylation is shown in detail in Figure 4-25 ■. Mitochondrial NADH and FADH₂ produced in the citric acid cycle give up high-energy electrons. Two at a time, the electrons pass through the protein complexes, losing energy with each transfer. During three of the transfers, enough energy is released to perform the work of transporting H^+ ions into the intermembrane space. During the other transfers, the released energy is given off as heat.

By the end of the electron transport chain, electrons have given up the usable portion of their stored energy.

At that point, each two electrons combine with two H^+ from the pool of H^+ ions in the matrix. The resulting hydrogen atoms then combine with an oxygen atom, creating a molecule of water, H_2O.

ATP Production Is Coupled to Movement of Hydrogen Ions Across the Inner Mitochondrial Membrane

As we have seen, the energy released by electrons moving through the electron transport system is stored as potential energy by H^+ ions concentrated in the intermembrane space. This stored energy is converted into chemical energy when the ions move back into the mitochondrial matrix through a protein known as the F_1F_0 *ATPase,* or **ATP synthase.** As H^+ ions move back into the matrix through a pore in the enzyme, the synthase transfers their kinetic energy to the high-energy phosphate bond of ATP. Because energy conversions are never completely efficient, a portion of the energy is released as heat. For each three H^+ that shuttle through the enzyme, a maximum of one ATP is formed.

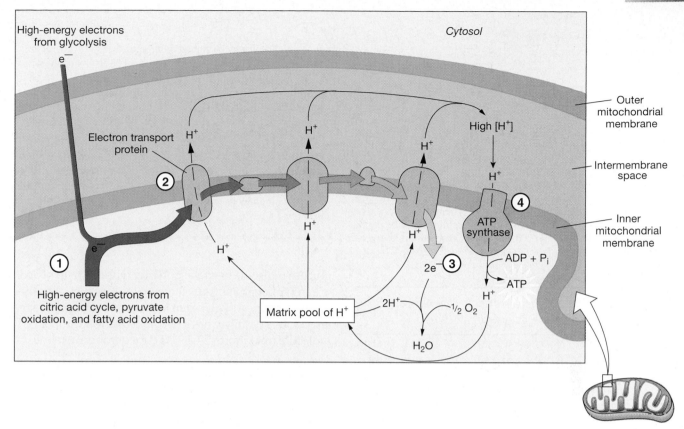

■ **Figure 4-25   The electron transport system and ATP synthesis**   (1) Chemical energy released during metabolism is captured by high-energy electrons. (2) When these high-energy electrons move through the electron transport system, some of their energy is used to transport H^+ from the matrix into the intermembrane space. (3) By the end of the electron transport system, the electrons are back to their normal energy state. They combine with H^+ and an oxygen atom to form water. (4) Potential energy from the H^+ ions concentrated in the intermembrane space is converted to kinetic energy when H^+ pass through the protein known as ATP synthase. The kinetic energy is captured in the high-energy bond of ATP.

The Net Energy Yield of One Glucose Molecule Is 36 to 38 ATP

If we tally up the maximum potential energy yield for the catabolism of one glucose molecule through aerobic pathways, the total comes to either 36 or 38 ATP (Table 4-5). From studies on mitochondria, scientists have determined that each mitochondrial NADH produces a maximum of three ATP. In contrast, $FADH_2$ electrons enter the electron transport system after the first protein complex, pumping fewer H^+ into the intermembrane space and resulting in a potential yield of only two $ATP/FADH_2$. Note that we say potential yield. This is because often the mitochondria do not work up to capacity. There are various reasons for this, including the fact that there is a certain amount of leakage of H^+ ions back into the matrix.

Another source of variability in the ATP produced per glucose comes from the two cytosolic NADH molecules produced during glycolysis. These NADH molecules are unable to enter the mitochondria. In order to use their high-energy electrons in oxidative phosphorylation, they must transfer the electrons to carrier molecules that take them through the mitochondrial membrane. Once in the mitochondria, if the electrons from the two cytosolic NADH pass to mitochondrial NADH, an additional six ATP molecules will be produced. If they go to mitochondrial $FADH_2$, only four ATP molecules will be made.

Even when oxidative phosphorylation is not working at maximum efficiency, the ATP yield from aerobic metabolism of glucose far exceeds the puny yield of two ATP from the anaerobic metabolism of glucose to lactate. If oxygen is not present to act as the final electron acceptor, the electron transport system stops. NADH and $FADH_2$ cannot give up their high-energy electrons and return to the citric acid cycle as NAD^+ and FAD^+, so the citric acid cycle runs out of electron acceptors. The entire aerobic pathway comes to a halt and glucose metabolism is diverted to the anaerobic production of lactate.

The low efficiency of anaerobic metabolism severely limits its usefulness in most vertebrate cells, whose metabolic energy demand is greater than can be met in this way. Some cells, such as exercising muscle, can tolerate anaerobic metabolism for a limited period of time, but they must then shift back to aerobic metabolism to get rid of the lactate that builds up.

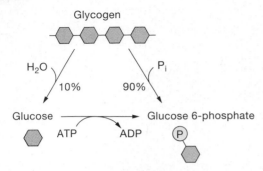

■ **Figure 4-26   Glycogen catabolism**   When glycogen is broken down for ATP production, 90% of the glycogen molecules are converted directly to glucose 6-phosphate. The remaining 10% of the glycogen is made into glucose molecules that must be phosphorylated with the help of ATP. The direct conversion of glycogen to glucose 6-phosphate saves one ATP per glucose.

✔ Name two differences between aerobic metabolism of glucose and anaerobic metabolism of glucose.

Large Biomolecules Can Be Used to Make ATP

We have now examined how glucose is used for ATP production, but how are large biomolecules such as glycogen, protein, and fat used to make ATP? The answer is a series of pathways that converts these biomolecules into intermediates that enter glycolysis and the citric acid cycle.

Glycogen Converts to Glucose   Glycogen, a glucose polymer, is the primary carbohydrate storage molecule in animals. Glycogen can be broken down through the pathways of **glycogenolysis** into glycolysis intermediates. About 10% of stored glycogen is hydrolyzed back to plain glucose molecules (Fig. 4-26 ■). These molecules must then be phosphorylated to glucose 6-phosphate with the help of an ATP.

The other 90% of glycogen molecules are converted directly to glucose 6-phosphate in a reaction that splits a

…continued from page 89

David and Sarah had their blood drawn for the genetic test several weeks ago and have been anxiously awaiting the results. Today, they returned to the hospital to hear the news. The tests show that Sarah carries the gene for Tay-Sachs disease but David does not. This means that although some of their children may be carriers of the Tay-Sachs gene like Sarah, none of the children will develop the disease.

Question 4:   *The Tay-Sachs gene is a recessive gene (t). If Sarah is a carrier of the Tay-Sachs gene (Tt) but David is not (TT), what is the chance that any child of theirs will be a carrier? (Consult a general biology or genetics text if you need help solving this problem.)*

glucose molecule from the polymer with the help of an inorganic phosphate from the cytosol. Because the direct phosphorylation of glycogen to glucose 6-phosphate saves an ATP, this pathway has a net yield of one additional ATP molecule (maximum 37–39 ATP for aerobic metabolism).

Proteins Can Be Catabolized to Produce ATP   The first step in protein catabolism is the digestion of proteins into smaller polypeptides by enzymes called proteases. Once that has been accomplished, other enzymes known as peptidases break the peptide bonds at the ends of the polypeptide, freeing individual amino acids (Fig. 4-27 ■). Most amino acids used for ATP synthesis come not from breaking down protein in cells, however, but from surplus amino acids in the diet.

Amino acids can be converted into intermediates for either glycolysis or the citric acid cycle. The first step in this conversion is *deamination,* or removal of the amino group, creating an ammonia molecule and an organic acid. Some of the organic acids created by deamination include pyruvate, acetyl CoA, and several intermediates of the citric acid cycle. The former amino acids thus enter

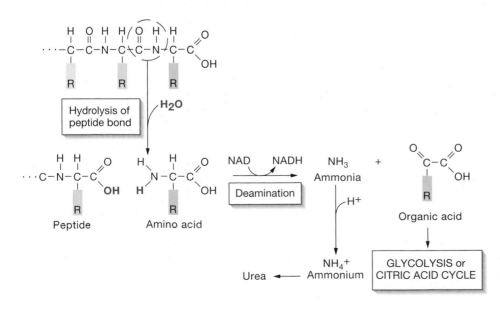

■ **Figure 4-27   Protein catabolism** Proteins are broken into amino acids by hydrolysis of their peptide bonds. Removal of the amino group from the amino acid creates an organic acid and ammonia, which is ultimately detoxified by conversion to urea.

the pathways of aerobic metabolism and become part of the ATP production process previously described.

The amino groups of amino acids are removed as ammonia (NH_3). They then rapidly pick up hydrogen ions to become ammonium ions (NH_4^+). Because both ammonia and ammonium are toxic, liver cells rapidly convert them into urea (CH_4N_2O). Urea is the main nitrogenous waste of the body and is excreted by the kidneys.

Lipids Yield More Energy per Unit Weight than Glucose or Proteins Lipids are the primary fuel storage molecule of the body, because they have a higher energy content than proteins or carbohydrates. In the reactions known as **lipolysis,** lipids are broken down by lipases into glycerol and fatty acids. Glycerol feeds into glycolysis about halfway through the pathway and goes through the same reactions as glucose from that point on (Fig. 4-28 ■).

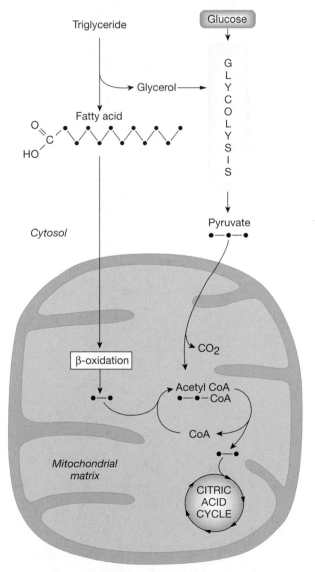

■ **Figure 4-28   Lipolysis**   Triglycerides are broken down into glycerol and long-chain fatty acids. Glycerol becomes a glycolysis substrate. Fatty acids lose two-carbon units, one at a time, in the process known as beta-oxidation. The two-carbon units become acetyl CoA, which can be used in the citric acid cycle.

The long carbon chains of fatty acids are not as easy to turn into ATP as glucose or amino acids. A fatty acid must first be transported from the cytosol into the mitochondria. There it is slowly disassembled by chopping two-carbon units off the end of the chain, one unit at a time, in the process of **beta-oxidation.** In most cells, the two-carbon units from fatty acids are converted into acetyl CoA, which feeds directly into the citric acid cycle. Because of the many acetyl CoA molecules that can be produced from a single fatty acid, lipids contain 9 kcal of stored energy per gram, compared with 4 kcal per gram for either proteins or carbohydrates.

The liver is one tissue that is unable to convert the two-carbon units into acetyl CoA and oxidize them to CO_2 and water. For this reason, liver cells must use alternative pathways to break down fatty acids. One result of these alternative pathways is the formation of **ketone bodies,** strong metabolic acids that can seriously disrupt the body's pH balance. We will cover the details of fatty acid metabolism in Chapter 21.

✔ Is the digestion of proteins to polypeptides more likely to be a dehydration reaction or a hydrolysis reaction?

▶ SYNTHETIC PATHWAYS

The metabolic reactions discussed to this point have been catabolic processes that break up large molecules into smaller ones in order to trap energy in the high-energy phosphate bond of ATP. We now examine anabolic reactions, those that synthesize polysaccharides, lipids, and proteins. In order to carry out synthetic reactions, the cell must have an adequate supply of organic nutrients, as well as energy in the form of ATP.

Glycogen Can Be Made from Glucose

Glycogen, the main storage form of glucose in the body, is found in all cells, but the liver and skeletal muscles contain especially high concentrations. Glycogen in skeletal muscles provides a ready energy source for muscle contraction. Glycogen in the liver acts as the main source of glucose for the body in periods between meals. It is estimated that the liver keeps about a four-hour supply of glucose stored as glycogen. Once that glycogen is exhausted by catabolism, the cells must turn to fats and proteins as their source of energy.

The synthesis of glycogen is essentially the reverse of glycogen breakdown (discussed in the previous section). Individual glucose molecules can be linked together into glycogen, or glucose 6-phosphate from glycolysis can be synthesized into the branching glucose chains of glycogen by removal of the phosphate group (see Fig. 4-26 ■).

Glucose Can Be Made from Glycerol or Amino Acids

Glucose is one of the key metabolic intermediates in the body for several reasons. As you have just learned, the aerobic metabolism of glucose is the most efficient way to make ATP, and so glucose is the primary substrate for energy production in all cells. More importantly, it is normally the only substrate for ATP synthesis in nervous tissue. If the brain is deprived of glucose for any period of time, its cells begin to die. For this reason, the body has multiple metabolic pathways that it can use to make glucose, ensuring a continuous supply for the brain. The simplest source of glucose is glycogen breakdown, or glycogenolysis, as previously discussed. But what happens when all the glycogen stores in the body are used up? In that case, cells can turn to fats and proteins for the intermediates needed to make glucose.

The production of glucose from nonglucose precursors such as proteins or glycerol is a process known as **gluconeogenesis,** or, literally, "birth of new glucose." The pathway is similar to glycolysis run in reverse, but uses different enzymes for some steps. Gluconeogenesis can start with glycerol, various amino acids, or lactate (Fig. 4-29 ■). In all cells, the process can go as far as glucose 6-phosphate (G-6-P). However, only liver and kidney cells have significant amounts of glucose 6-phosphatase, the enzyme that removes the phosphate group from G-6-P to make glucose. During periods of fasting, the liver is the primary source of glucose syn-

thesis, producing about 90% of the glucose made from the breakdown of fats and proteins.

✔ What is the difference between hexokinase and glucose 6-phosphatase? (Hint: See Table 4-4.)

Acetyl CoA Is an Important Precursor for Lipid Synthesis

Lipids are so diverse that generalizing about their synthesis is difficult. Generally, lipids are synthesized by enzymes in the smooth endoplasmic reticulum and in the cytosol. Fatty acids are made in the cytosol by an enzyme called *fatty acid synthetase,* which links the two-carbon acetyl groups of acetyl CoA together (Fig. 4-30 ■). The process also uses hydrogens and high-energy electrons from NADPH. Glycerol can be made from glucose or glycolysis intermediates. The combination of glycerol and fatty acids into triglycerides takes place in the endoplasmic reticulum. The phosphorylation steps that turn triglycerides into phospholipids also take place in the endoplasmic reticulum.

Cholesterol is an important component of cell membranes, and it forms the skeleton of steroid hormones. Cholesterol is usually obtained from the diet, but it is such an important molecule for cells that the body will synthesize it if the diet is deficient. Even vegetarians who eat no animal products (vegans) have substantial amounts of cholesterol in their cells. The entire structure

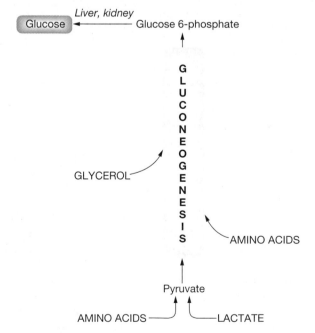

■ **Figure 4-29   Gluconeogenesis**   The production of glucose from nonglucose precursors is known as gluconeogenesis. Glycerol, amino acids, and lactate can all be used to make glucose by a series of reactions that are similar to the glycolysis pathway in reverse. Only the liver and kidneys are able to convert glucose 6-phosphate to glucose.

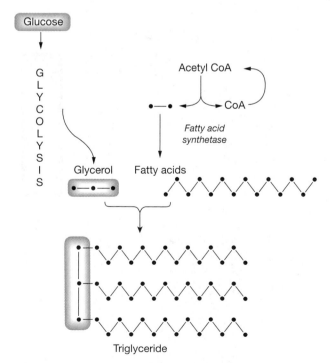

■ **Figure 4-30   Lipid synthesis**   Glycerol can be made from glucose through the glycolysis pathway. Two-carbon units from acetyl CoA are linked together by an enzyme called fatty acid synthetase to make long-chain fatty acids.

DIABETES Insulin and Metabolism Why do people with diabetes have elevated levels of blood glucose ("blood sugar")? Part of the answer lies with the disruption of glucose metabolism when insulin is absent or when cells do not respond to it. In a normal person, insulin promotes anabolic pathways by stimulating the appropriate enzymes. Glucose absorbed from the digestive system goes through glycolysis to make ATP or is stored as glycogen. Fatty acids are stored as fats, and amino acids are used to make proteins. In the absence of insulin, the enzymes associated with catabolic pathways are more active. Fats undergo beta-oxidation, amino acids go through gluconeogenesis to become glucose, and glycogen stores are turned back into glucose. The production of glucose within cells coupled to the inability of many cells to absorb glucose creates the elevated "blood sugar" levels that are a hallmark of diabetes.

of cholesterol can be made from acetyl CoA. From there, it is a simple matter to add or subtract functional groups to the carbon rings to change cholesterol into various hormones and other steroids. As you will see in Chapter 21, all the steroid hormones are very similar in structure.

Proteins Are the Key to Cell Function

Proteins are the molecules that run the cell from day to day. As enzymes, they control the synthesis and breakdown of carbohydrates, lipids, structural proteins, and signal molecules. Proteins in the cell membrane and organelle membranes regulate the movement of molecules into and out of the cell, or within the cell. Thus, protein synthesis is critical to cell function.

The power of proteins arises from their tremendous variability and specificity. Protein synthesis can be compared to creating a language with an alphabet of 20 letters (amino acids) and words up to hundreds of letters in length (peptides and proteins). The language is a rich one, with thousands of different words with different meanings. Like a language, a "spelling error" during the synthesis of a protein can change the function of the protein, just as changing the last letter turns the word "foot" into "food." In many inherited diseases, the substitution of a single amino acid alters the structure of a protein dramatically. The classic example is sickle cell anemia. In this inherited disease, a change of one amino acid alters the shape of hemoglobin enough so that red blood cells take on a sickle shape, which causes them to get tangled up and block small blood vessels.

The Protein "Alphabet" Begins with DNA One of the mysteries of biology until the 1960s was the question of how only four bases in the DNA molecule—adenine (A), guanine (G), cytosine (C), and thymine (T)—could code for more than 20 different amino acids. If each base controlled the synthesis of one amino acid, a cell could only make four different amino acids. If pairs of bases represented different amino acids, it could make 4^2, or 16, possible combinations. Because we have 20 amino acids, this is still not satisfactory. But if triplets of bases were the code for different molecules, DNA could create 4^3 or 64 different amino acids. These triplets, called **codons,** are indeed the way information is encoded in DNA (Fig. 4-31 ■).

Of the 64 possible triplet combinations, one DNA triplet (TAC) acts as the initiator or *"start codon"* that signifies the beginning of a coding sequence. Three codons are recognized as terminator or *"stop codons"* that show where the sequence ends. The remaining 60 triplets all code for amino acids. Methionine and tryptophan have only one codon each, but the other amino acids have between two and six different codons each. Thus, the information contained in the DNA sequence determines the amino acid sequence of proteins.

How is information translated from DNA into a chain of linked amino acids? First, the codon sequence of DNA is transferred to **messenger RNA (mRNA)** in the process known as **transcription.** The mRNA leaves the nucleus and enters the cytosol, where it directs **translation,** the assembly of amino acids into proteins. Two other forms of RNA, transfer RNA and ribosomal RNA, are needed to help mRNA complete the process of amino acid assembly.

During Transcription, DNA Guides the Synthesis of a Complementary mRNA Molecule Because DNA is a very large molecule, it cannot pass through the nuclear envelope to the cytosol where protein synthesis takes place. Instead, a smaller single chain of mRNA is made by using DNA as a template during transcription

First base of codon	Second base of codon: U	C	A	G	Third base of codon
U	UUU Phe	UCU Ser	UAU Tyr	UGU Cys	U
U	UUC Phe	UCC Ser	UAC Tyr	UGC Cys	C
U	UUA Leu	UCA Ser	UAA Stop	UGA Stop	A
U	UUG Leu	UCG Ser	UAG Stop	UGG Trp	G
C	CUU Leu	CCU Pro	CAU His	CGU Arg	U
C	CUC Leu	CCC Pro	CAC His	CGC Arg	C
C	CUA Leu	CCA Pro	CAA Gln	CGA Arg	A
C	CUG Leu	CCG Pro	CAG Gln	CGG Arg	G
A	AUU Ile	ACU Thr	AAU Asn	AGU Ser	U
A	AUC Ile	ACC Thr	AAC Asn	AGC Ser	C
A	AUA Ile	ACA Thr	AAA Lys	AGA Arg	A
A	AUG Met Start	ACG Thr	AAG Lys	AGG Arg	G
G	GUU Val	GCU Ala	GAU Asp	GGU Gly	U
G	GUC Val	GCC Ala	GAC Asp	GGC Gly	C
G	GUA Val	GCA Ala	GAA Glu	GGA Gly	A
G	GUG Val	GCG Ala	GAG Glu	GGG Gly	G

■ **Figure 4-31 The genetic code** This table shows the genetic code as it appears in the codons of mRNA. The three-letter abbreviations to the right of the brackets indicate the amino acids that these codons represent. The start and stop codons are also marked.

[*trans,* over + *scribe,* to write]. Messenger RNA is formed by **base pairing** in the same way that the double strand of DNA is formed (see Appendix B for a review of DNA synthesis). During transcription, part of the DNA chain unwinds and the complementary RNA base pairs come in to make a new RNA strand (Fig. 4-32 ■). The main difference between DNA and RNA synthesis is that RNA uses the base uracil (U) in place of thymine (T). Each adenine on the DNA pairs with a uracil on the RNA; the DNA thymine continues to pair with adenine.

The synthesis of mRNA from the DNA template requires an enzyme known as RNA polymerase, plus magnesium or manganese ions, and energy in the form of high-energy phosphate bonds:

DNA template + nucleotides A, U, C, G

$$\xrightarrow[\text{and energy}]{\begin{array}{c}\text{RNA polymerase,}\\ \text{Mg}^{2+}\text{ or Mn}^{2+},\end{array}}$$

mRNA + DNA template

How does the cell know which of the thousands of base pairs of DNA to use to make a strand of mRNA? It turns out that the information a cell needs to make a protein is contained within a segment of DNA known as a gene. A **gene** is a region of DNA that contains all the information needed to make a functional piece of mRNA. We used to believe that a gene coded for only one protein (the one gene-one protein theory), but we now know that one gene can create many proteins.

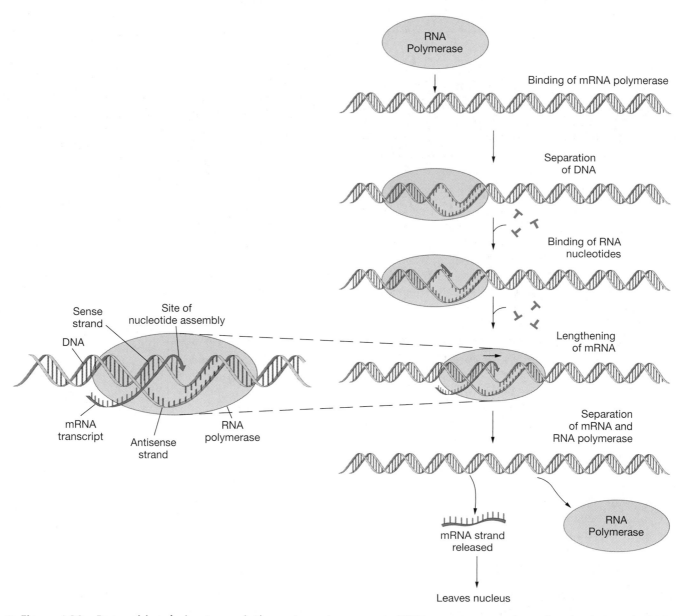

■ **Figure 4-32   Base pairing during transcription**   A gene is a segment of DNA necessary to produce a functional piece of mRNA. During transcription, the section of DNA that contains the gene unwinds. The enzyme RNA polymerase attaches and creates a single strand of mRNA corresponding to one of the DNA strands. Base pairing is the same as in DNA synthesis, except that the base uracil (U) substitutes for thymine (T). The mRNA strand is released from the DNA template and leaves the nucleus.

Each gene is preceded by a **promoter** region that must be activated in order for transcription to begin. The promoter region is not transcribed into mRNA. The gene itself is a nucleotide string that includes the codon sequence for the polypeptide to be synthesized, as well as some noncoding segments. Those segments of the gene that encode proteins are called **exons.** Noncoding segments between the start and stop codons are called **introns.** The designation of sequences of nucleotides as introns and exons is not fixed for a given gene. This allows a single piece of DNA to create more than one protein (see below).

The process of transcription begins when the gene is activated. Activation requires the presence of protein **initiation factors** that bind to the promoter region. This, in turn, enables the DNA to bind to **RNA polymerase.** This enzyme moves along a segment of the double helix strand of the DNA molecule and "unwinds" it by breaking the hydrogen bonds between the paired nucleotide bases. One strand of DNA, called the *sense strand,* serves as the guide for mRNA synthesis, while the other *antisense strand* sits idly by.

With the aid of RNA polymerase, nucleotide bases in the sense strand of the DNA molecule pair with the corresponding RNA bases, which are then linked together. For example, a DNA segment with the bases AGTAC will pair with the complementary RNA bases UCAUG. During transcription of mRNA, nucleotides are linked at an average rate of 40 per second. In humans, the largest mRNAs may contain as many as 5000 nucleotides, and their transcription may take over a minute—a long time for a cellular process. When RNA polymerase reaches the stop codon for the gene, it stops adding nucleotides and releases the mRNA.

The initial chain of mRNA synthesized has the same noncoding sequences as the sense strand of DNA. These sequences must be edited out before the mRNA leaves the nucleus. This **mRNA processing** is accomplished by enzymes that clip sections out of the middle or off the ends of the mRNA strand. Other enzymes then splice the remaining pieces back together. The result is a smaller piece of mRNA than was initially made from the DNA template (Fig. 4-33 ■).

One advantage of mRNA processing is that it allows a single nucleotide sequence on the DNA to code for more than one protein. Segments that are removed one time can be left in the next time, producing a finished mRNA with a different sequence. This phenomenon is known as **alternative splicing.** The closely related forms of a single enzyme known as *isozymes* are probably made by alternative splicing of a single gene.

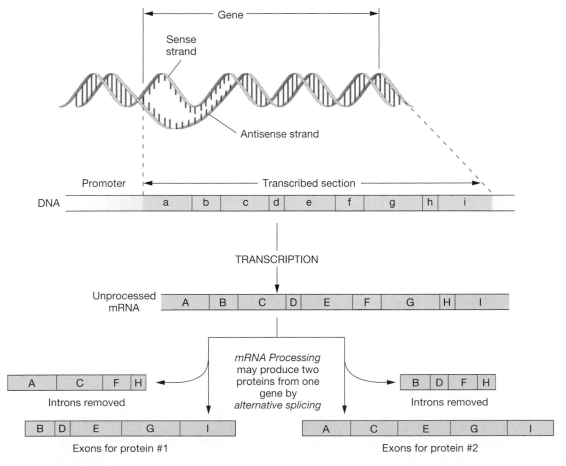

■ **Figure 4-33 Genes and mRNA processing** The promoter segment of DNA is needed to start transcription but is not itself transcribed into mRNA. Before the mRNA leaves the nucleus, it is processed by enzymatically removing segments of nucleotides that are not part of the code for the desired protein. This process, known as alternative splicing, allows one piece of mRNA to code for more than one protein.

After mRNA has been processed, it exits the nucleus through nuclear pores and goes to ribosomes in the cytosol. There it directs the construction of protein.

Translation of the Nucleotide Sequence of mRNA Produces a String of Amino Acids

Protein synthesis is a coordinated effort that requires cooperation between mRNA, ribosomal RNA, and transfer RNA. Upon arrival in the cytosol, mRNA binds to ribosomes. Ribosomes, nonmembranous organelles, are two-part structures made of protein and several types of **ribosomal RNA (rRNA).** Each ribosome has two subunits, one large and one small, which come together when protein synthesis begins. The small ribosomal subunit binds the mRNA, then adds the large subunit so that the mRNA is sandwiched in the middle (Fig. 4-34 ■). Now the com-

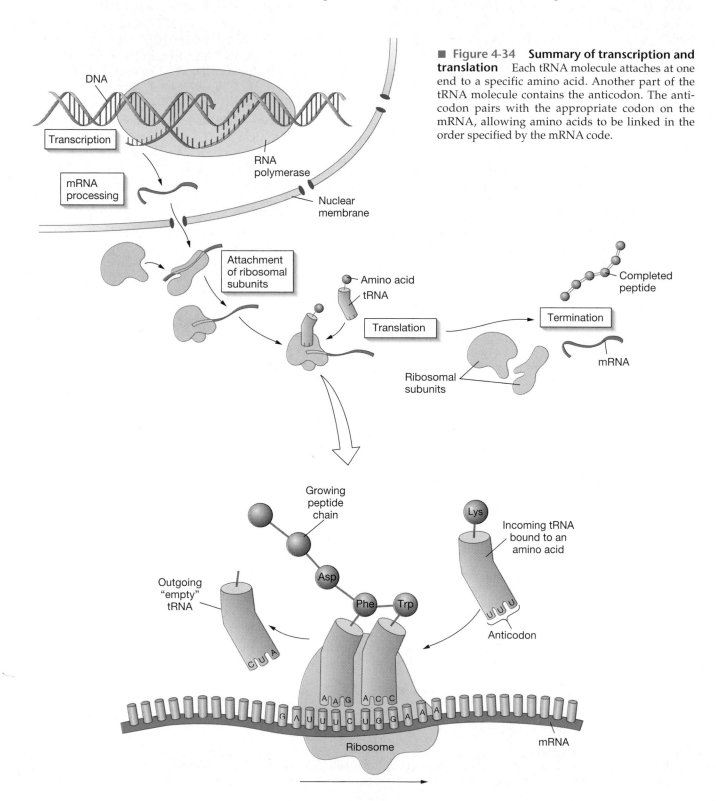

■ **Figure 4-34** **Summary of transcription and translation** Each tRNA molecule attaches at one end to a specific amino acid. Another part of the tRNA molecule contains the anticodon. The anticodon pairs with the appropriate codon on the mRNA, allowing amino acids to be linked in the order specified by the mRNA code.

plex of ribosome and mRNA is ready to begin the process of translation.

In translation, the triplet code of the mRNA codons must be matched to the proper amino acid. This is done with the assistance of a third form of RNA, **transfer RNA (tRNA)** (Fig. 4-34b ■). With the aid of a cytosolic enzyme and ATP, one portion of the tRNA binds to the amino acid that matches the type of tRNA. Another region of the molecule contains the **anticodon** for the amino acid.

The anticodon is the tRNA base triplet that pairs with the mRNA codon for the amino acid. For example, an mRNA codon for the amino acid lysine is AAA. The corresponding anticodon is UUU, found on the tRNA carrying lysine.

As translation begins, tRNAs carrying amino acids move to the ribosomes. The mRNA codons bind to the corresponding anticodons of the arriving tRNAs. This puts the amino acids carried by the tRNAs into the correct orientation. As each amino acid moves into position, it links to the preceding amino acid by a peptide bond [∞ p. 31]. This process is repeated to make a peptide chain.

When the last amino acid is joined, the mRNA, the polypeptide, and the ribosomal subunits separate. The ribosomes are ready for a new round of protein synthesis, but the mRNA is broken down by enzymes known as **ribonucleases.** Some forms of mRNA are broken down quite rapidly, while others may linger in the cytosol and be translated many times.

The polypeptide chain released from the ribosome is now free to take on its final three-dimensional shape. Research studies show that much of this structural change takes place spontaneously. The chain forms its secondary structure, which may be an α-helix or a β-pleated sheet [∞ p. 32]. The molecule curls into its final shape as hydrogen bonds, sulfur-sulfur bonds, and ionic bonds form between amino acids in the chain.

Protein Sorting Directs Proteins to Their Final Destination

One of the amazing aspects of protein synthesis is the way that specific proteins go from the ribosomes directly to where they are needed within the cell. It is as if each protein has an address label on it that tells the cell where it should go. Proteins destined to be dissolved in the cytosol are simply released directly from the ribosomes after they have been synthesized. Those that have other roles are identified by a special initial segment of amino acids known as a **signal sequence.** This signal directs the protein to the proper organelle, so that it can be processed, packaged, and delivered to its final destination.

Many peptides are directed through the membrane and into the lumen of the endoplasmic reticulum as they are being made. Within the endoplasmic reticulum, the signal sequence is enzymatically removed. The protein is then ready to be modified by the addition of sugars if it is to become a glycoprotein.

Once proteins are modified in the endoplasmic reticulum, they are sent to the adjacent Golgi apparatus by small membrane-bound transport vesicles that bud off the endoplasmic reticulum (Fig. 4-35 ■). The transport vesicles fuse with the Golgi membranes and dump their contents into the lumen of the Golgi apparatus. Once inside, the proteins may be modified further by addition of sugars or phosphate groups, combination with lipids to make lipoproteins, or activation by removal of segments.

When the protein reaches the end of the Golgi apparatus, it is packaged for delivery to its final destination. Vesicles that bud off the Golgi apparatus may become lysosomes, intracellular storage vesicles, or secretory vesicles, depending on their contents. How the contents of vesicles leave the cell is the subject of the next chapter, Movement Across Membranes.

✔ If the protein-coding portion of a piece of edited mRNA is 450 nucleotides long, how many amino acids will be in the corresponding polypeptide? (Hint: The start codon is translated into an amino acid but the stop codon is not.)

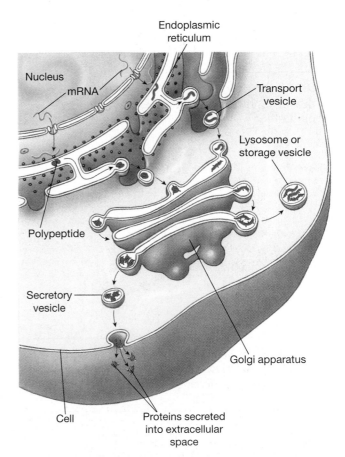

■ **Figure 4-35   Protein synthesis, modification, and packaging by the endoplasmic reticulum and Golgi apparatus** Many newly formed proteins are directed into the lumen of the endoplasmic reticulum. From there they go in transport vesicles to the Golgi complex for processing. Modified proteins leave the Golgi in vesicles that either remain in the cytoplasm or are released from the cell.

PROBLEM CONCLUSION

In this running problem you have learned that Tay-Sachs disease is an incurable, recessive genetic disorder in which the enzyme that breaks down gangliosides in cells is missing. One in 27 Americans of Eastern European Jewish descent carries the gene for this disorder. You have also seen how a blood test can detect the presence of the deadly gene in a person's blood.

Further check your understanding of this running problem by checking your answers against those in the summary table.

Question	Facts	Integration and Analysis
1 What is another symptom of Tay-Sachs disease?	Hexosaminidase A breaks down gangliosides. In Tay-Sachs disease, the enzyme is absent and gangliosides accumulate in cells, including light-sensitive cells of the eye, and cause them to function abnormally.	Damage to light-sensitive cells of the eye could cause vision problems and even blindness.
2 How would you test for the presence of the defective gene?	Carriers of the gene have lower-than-normal levels of hexosaminidase A.	Run tests to determine the average enzyme levels in known carriers of the disease (i.e., people who have had children with Tay-Sachs disease) and in people who have little likelihood of being carriers. Compare the enzyme levels of suspected carriers to the averages for the known carriers and noncarriers.
3 Why is the new test for Tay-Sachs disease more accurate than the old test?	The new test detects the defective gene. The old test analyzed the levels of the enzyme produced by the gene.	The new test is a direct way to test if a person is a carrier. The old test was an indirect way. It is possible for factors other than a defective gene to alter a person's enzyme level. If enzyme levels are decreased due to some factor other than the Tay-Sachs gene, a false-positive test result will occur, and the subject may be erroneously told that he or she is a carrier. In addition, the variability in enzyme levels among the general population may be so broad that there is no distinct difference in the enzyme levels between carriers and noncarriers.
4 The Tay-Sachs gene is a recessive gene (t). What is the chance that any child of a carrier and a noncarrier will be a carrier?	Tt × TT ↓ TT Tt TT Tt	Each child has a 50% chance of being a carrier (Tt).

CHAPTER REVIEW

SUMMARY

Energy in Biological Systems

1. **Energy** is the capacity to do work. **Chemical work** enables cells and organisms to grow, reproduce, and carry out normal activities. **Transport work** enables cells to move molecules to create concentration gradients. **Mechanical work** is used for movement. (p. 74)

2. **Kinetic energy** is the energy of motion. **Potential energy** is stored energy. (p. 75)

Chemical Reactions

3. A **chemical reaction** begins with one or more **reactants** and ends with one or more **products. Reaction rate** is measured as the change in concentration of products with time. (p. 76)

4. The energy stored in the chemical bonds of a molecule and available to perform work is the **free energy** of the molecule. (p. 76)

5. **Activation energy** is the initial input of energy required to begin a reaction. (p. 77)

6. **Exergonic reactions** are energy producing. **Endergonic reactions** are energy utilizing. (p. 77)

7. Metabolic pathways couple exergonic reactions to endergonic reactions. (p. 78)

8. Energy for driving endergonic reactions is stored in ATP. (p. 78)

9. A reaction that can proceed in both directions is called a **reversible reaction.** If a reaction can proceed in one direction but not the other, it is an **irreversible reaction.** The net free energy change of a reaction determines whether that reaction can be reversed. (p. 79)

Enzymes

10. **Enzymes** are biological catalysts that speed up the rate of chemical reactions without themselves being changed. Enzymes bind to reactant molecules, bringing them together into the best position for reacting with each other. In reactions catalyzed by enzymes, the reactants are called **substrates.** (p. 79)

11. The part of the enzyme that binds to the substrates is called the **binding site.** The **induced-fit model** of enzyme-substrate interaction says that the binding site has an intermediate shape that can change to fit either the substrate or the product molecules. (p. 80)

12. **Specificity** is the ability of an enzyme to catalyze a certain reaction or a group of closely related reactions. (p. 81)

13. Some enzymes are produced as inactive precursors and must be activated. This may require the presence of a **cofactor.** Organic cofactors are called **coenzymes.** (p. 82)

14. Enzyme activity is altered by temperature, pH, and modulator molecules. A **modulator** changes the ability of the enzyme to (1) bind the substrate or (2) alter activation energy of the reaction. (p. 82)

15. **Competitive inhibitors** block the binding site of the enzyme directly. **Allosteric modulators** bind to the enzyme away from the binding site and change the shape of the binding site. **Covalent modulators** affect the ability of an enzyme to decrease the activation energy of the reaction. (p. 83)

16. Factors that affect the rate of enzymatic reactions include modulation, the enzyme concentration, and the ratio of the concentrations of substrates and products. (p. 84)

17. When an enzyme is **saturated** with substrate, it works at its maximum possible rate. (p. 86)

18. Reversible reactions go to a state of **equilibrium,** where the rate of the reaction in the forward direction is exactly equal to the rate of the reverse reaction. Reversible reactions obey the **law of mass action:** when a reaction is at equilibrium, the ratio of substrates to products is always the same. If the concentration of a substrate or product changes, the equilibrium will be disturbed. (p. 86)

19. Most reactions can be classified as **oxidation-reduction, hydrolysis-dehydration, addition-subtraction-exchange,** or **ligation** reactions. (p. 86)

Metabolism

20. All the chemical reactions in the body are known collectively as **metabolism. Catabolic reactions** release energy and break down large biomolecules. **Anabolic reactions** require a net input of energy and synthesize large biomolecules. (p. 88)

21. Cells regulate the flow of molecules through their metabolic pathways by (1) controlling the enzyme concentration, (2) producing allosteric and covalent modulators, (3) using different enzymes to catalyze reversible reactions, (4) isolating enzymes within intracellular organelles, or (5) maintaining an optimum ratio of ATP to ADP. (p. 88)

22. **Aerobic pathways** require oxygen and yield the most ATP. **Anaerobic pathways** can proceed without oxygen, but produce ATP in much smaller quantities. (p. 90)

ATP Production

23. During **glycolysis,** one molecule of glucose is converted into two pyruvate molecules, 2 ATP, 2 NADH, and 2 H$^+$. Glycolysis does not require the presence of oxygen. (p. 90)

24. In **anaerobic metabolism,** pyruvate is converted into lactate with a net yield of 2 ATP for each glucose molecule. (p. 92)

25. Aerobic metabolism of pyruvate through the **citric acid cycle** yields ATP, carbon dioxide, water, and high-energy electrons captured by NADH and FADH$_2$. (p. 93)

26. **High-energy electrons** from NADH and FADH$_2$ give up their energy as they pass through the **electron transport system.** Their energy is trapped in the high-energy bonds of ATP. (p. 93)

27. Glycogen and lipids are the primary energy storage molecules in animals. Lipids are broken down for ATP production in the process of **beta-oxidation.** (p. 97)

Synthetic Pathways

28. Protein synthesis is controlled by **genes** made of DNA in the nucleus. The code from a gene is transcribed into a matching code on messenger RNA (**mRNA**). The mRNA leaves the nucleus and, with the assistance of **transfer RNA** and **ribosomal RNA,** assembles amino acids into the sequence designated by the gene. This process is called **translation** (p. 99)

29. Newly synthesized proteins may be modified in the rough endoplasmic reticulum or the Golgi apparatus. In the Golgi, they are packaged into membrane-bound vesicles that become lysosomes, storage vesicles, or secretory vesicles. (p. 103)

QUESTIONS

LEVEL ONE   Reviewing Facts and Terms

1. List the three basic forms of work and give a physiological example of each.

2. Explain the difference between potential and kinetic energy.

3. State the two laws of thermodynamics in your own words.

4. The sum of all chemical processes through which cells obtain and store energy is called _____.

5. In the reaction CO$_2$ + H$_2$O → H$_2$CO$_3$, water and carbon dioxide are the _____, and H$_2$CO$_3$ is the product. Because this reaction is catalyzed by an enzyme, it is also appropriate to call water and carbon dioxide _____. The speed at which this reaction occurs is called the reaction _____, often expressed as molarity/second.

6. _____ are protein molecules that speed up chemical reactions by (increasing or decreasing?) the activation energy of the reaction.

7. Match these terms with their definitions (not all terms are used):
   (a) a reaction that can run in either direction
   (b) the part of a protein molecule that actually binds the substrates
   (c) a reaction that releases energy
   (d) the ability of an enzyme to catalyze one reaction but not another
   (e) boost of energy needed to get a reaction started
   (f) the ability of a protein to alter its shape to fit more closely with that of the substrate

   1. exergonic
   2. endergonic
   3. activation energy
   4. reversible
   5. irreversible
   6. induced fit
   7. binding site
   8. specificity
   9. free energy
   10. saturation

8. Since 1972, enzymes have been designated by adding the suffix _____ to their name.

9. An ion, such as Ca^{2+} or Mg^{2+}, that must be present in order for an enzyme to work is called a _____. Organic molecules that must be present for an enzyme to function are called _____. The precursors of these organic molecules come from _____ in our diet.

10. A protein whose structure is altered to the point that its activity is destroyed is said to be _____.

11. In an oxidation-reduction reaction, where electrons are moved between molecules, the molecule that gains an electron is said to be _____, and the one that loses an electron is _____.

12. The removal of H$_2$O from reacting molecules is called _____. Using H$_2$O to break down polymers, such as starch, is called _____.

13. The removal of an amino group (NH$_2$) from a molecule (such as an amino acid) is called _____. Transfer of an amino group from one molecule to the carbon skeleton of another molecule (to form a different amino acid) is called _____. What happens to the amino group removed from an amino acid?

14. In metabolism, _____ reactions release energy and result in the breakdown of large biomolecules, and _____ reactions require a net input of energy and result in the synthesis of large biomolecules. In what units do we measure the energy of metabolism?

15. Metabolic regulation where the last product of a metabolic pathway (the end product) accumulates and slows or stops reactions earlier in the pathway is called _____.

16. Explain how H$^+$ movement across the inner mitochondrial membrane results in ATP synthesis.

17. List the two carrier molecules that deliver high-energy electrons to the electron transport system.

18. The breakdown of lipids for energy is termed _____. The long chains of fatty acids are broken into two-carbon acetyl CoA molecules by a pathway called _____.

Which pathway do these acetyl CoA molecules join in order to produce ATP?

19. How many kilocalories per gram are stored in carbohydrates, proteins, and fats, respectively?

20. Which cellular organelles are involved in lipid synthesis?

LEVEL TWO   Reviewing Concepts

21. Create concept maps using the following terms.

List 1:
acetyl CoA
amino acids
ATP
citric acid cycle
CO_2
cytosol
electron transport system
$FADH_2$
fatty acids
gluconeogenesis
glucose
glycerol
glycogen
glycolysis
high-energy electrons
lactate
mitochondria
NADH
oxygen
pyruvate
water

List 2:
alternative splicing
base pairing
DNA
exon
gene
initiation factors
intron
mRNA processing
mRNA
nucleotides (A, C, G, T, A, U)
promoter
RNA polymerase
sense strand
start codon
stop codon
transcription

22. When bonds are broken during a chemical reaction, what are the three possible fates for the potential energy found in those bonds?

23. Define, compare, and contrast the following terms: competitive inhibition, allosteric modulation, covalent modulation.

24. Explain why it is advantageous for a cell to store or secrete an enzyme in an inactive form.

25. Compare the energy yields from the aerobic breakdown of one glucose to CO_2 and H_2O to the yield from one glucose going through anaerobic glycolysis ending with lactate. What are the advantages each pathway offers?

26. What is the advantage of converting glycogen to glucose 6-phosphate rather than to glucose?

27. Briefly describe the processes of transcription and translation. Which organelles are involved in these processes?

28. On what molecule does the anticodon appear? Explain the role of this molecule in protein synthesis.

LEVEL THREE   Problem Solving

29. Transporting epithelial cells in the intestine do not use glucose as their primary energy source. What molecules might they use instead?

30. Given the following strand of DNA, list the sequence of bases that would appear in the matching mRNA. For the underlined triplets, underline the corresponding mRNA codon and list the appropriate anticodon.

DNA: CGC<u>TAC</u>A<u>AGT</u>C<u>AGG</u>TAC<u>CGT</u>AACG

mRNA:

Anticodons:

LEVEL FOUR   Quantitative Problems

31. This graph shows the free energy change for the reaction $A + B \rightarrow D$. Is this an endothermic or exothermic reaction?

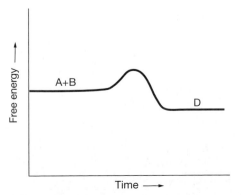

E X P L O R E <MediaLab>

Introduction

In this chapter you learned how cells create and use energy through chemical reactions, a process known as metabolism. Specialized proteins called *enzymes* catalyze the work done by reactions inside the cell. Enzymes are necessary for metabolism of carbohydrates and proteins as well as for the creation of ATP. The following Web Explorations will help you gain an appreciation of the metabolic reactions needed for energy and cellular function. After reading the description below, visit the MediaLab for Chapter 4 in your Companion Website and select the appropriate keyword.

Web Exploration 1

Estimated time for completion = 10 minutes
Imagine you are designing a working cell. To sustain life, your custom-made cell must extract, utilize, transfer, and store energy in the chemical bonds of biomolecules such as carbohydrates, proteins, and lipids. Now envision which molecules you would like to use for each type of energy exchange. If a molecule like ATP releases energy to drive a chemical reaction, would you suspect the activation energy of that reaction to be large or small? What is the rationale for your choice?

Select the keyword **ATP** on the Website to learn more about why this very special molecule is the energy currency for the cell. Is ATP an example of potential or kinetic energy? Then select **ACTIVATION ENERGY** to visualize the difference between the breakdown of sugar during cellular metabolism versus burning sugar with a match! Why is the rate of a reac-

tion important when one molecule provides the energy to drive another reaction? To complete this exercise, visit MediaLab Web Exploration 1 in Chapter 4 of your Companion Website.

Web Exploration 2

Estimated time for completion = 10 minutes
Many chemical reactions in glycolysis and the citric acid cycle are coupled. A coupled reaction is one in which an energy-utilizing (endergonic) reaction is paired with an energy-producing (exergonic) reaction. In this way, the energy produced by one reaction can be used to power the other reaction. In many metabolic cycles, a coupled reaction includes the conversion of NAD^+ to NADH.

Refer back to Figures 4-21 and 4-23 of the text. As you review the pathways, think of what is necessary for the cycle to proceed. What is the role of each molecule in the pathway? What would happen to these metabolic cycles if NAD^+ were depleted?

Select **CELLULAR RESPIRATION** on the Website to aid you with this process. What effect do the enzymes that catalyze these reactions have upon the reaction rate? What effect would a deficiency of an enzyme in the citric acid cycle have upon the total number of ATP produced? These concepts of feedback will continually arise as you delve deeper into your understanding of physiology! To complete this exercise, visit MediaLab Web Exploration 2 in Chapter 4 of your Companion Website.

5 Membrane Dynamics

■ "Since the cell membrane is not visible, evidence for its existence is indirect, deriving from several types of investigation."
—*Bradley T. Sheer, in his textbook* General Physiology, 1953 ■

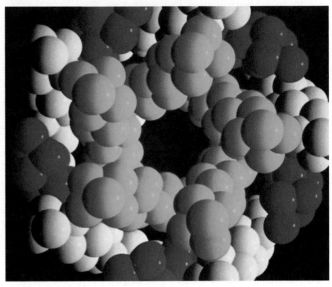

Acetylcholine receptor-channel

BACKGROUND BASICS

Up to this point, we have been concerned with the structures and processes of individual cells. However, the human body is made of trillions of cells that must communicate and cooperate with each other while maintaining their separate identities. The cell membrane plays a key role in this process because it is the barrier that separates the intracellular compartment from signals outside the cell.

Cell membranes are increasingly the focus of physiological and medical research. Almost daily, you can read in the newspaper about new medical treatments, such as drugs for high blood pressure, depression, gout, and stomach ulcers, that target membrane processes. Membrane research is also helping us understand how drugs we have used for years exert their effect at the cellular level. A good example of this is the sulfonylurea drugs used for the treatment of diabetes. In 1942 researchers accidentally discovered that sulfonylureas lower blood glucose levels by stimulating insulin release from the pancreatic beta cells. The drugs went into clinical use by the 1950s, but it took nearly 40 more years before we learned how they act on the beta cell. The mechanisms involved in insulin release exemplify basic aspects of membrane biology and are described in the last section of this chapter.

In this chapter and the next, you will learn more about four of the fundamental themes that are the basis of physiology [∞ p. 7]. Chapter 5 discusses the division of the body into intracellular and extracellular compartments by the cell membrane. You will learn how molecules cross cell membranes and how the selectivity of the cell membrane allows the intracellular and extracellular compartments to differ in their chemical and electrical properties. In Chapter 6 we examine how signals outside the cell are communicated to the cytoplasm, and we consider the role of cell-cell communication in maintaining homeostasis.

▶ MEMBRANES IN THE BODY

The word *membrane* has come to have two different meanings in physiology. The simplest membrane, the **plasma membrane** or **cell membrane,** is the membrane that surrounds the cell. But confusion sometimes arises because the word *membrane* is also used to describe epithelial tissues that line a cavity or separate two compartments [∞ p. 7]. Anatomically, we speak of mucous membranes in the mouth and vagina, the peritoneal membrane that lines the inside of the abdomen, the pleural membrane that covers the surface of the lungs, and the pericardial membrane that surrounds the heart. These visible membranes are actually tissues: thin, translucent layers of cells. To confuse matters further, epithelial membranes are often depicted in book illustrations as a single line, leading students to think of them as if they were similar in structure to the cell membrane (Fig. 5-1 ■).

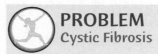

PROBLEM
Cystic Fibrosis

Over 100 years ago, midwives performed an unusual test on the infants they delivered: the midwife would lick the infant's forehead. A salty taste meant that the child was destined to die of a mysterious disease that withered the flesh and robbed the breath. Today, a similar "sweat test" will be performed in a major hospital—this time with state-of-the-art techniques—on Daniel Biller, a two-year-old with a history of weight loss and respiratory problems. The name of the mysterious disease? Cystic fibrosis.

...continued on page 115

Cell membranes, on the other hand, are a double layer (*bilayer*) of phospholipids [∞ p. 29] with protein molecules inserted in them. Material moving into and out of cells must cross the cell membrane, and anything that enters and leaves the body through an epithelial "membrane" passes through two cell membranes. In addition to forming the outer boundary of the cell, phospholipid-protein membranes also enclose the membranous organelles within the cytoplasm [∞ p. 45].

▶ CELL MEMBRANES

Phospholipid membranes surround the contents of the cytoplasm and divide the interior of the cell into compartments such as the nucleus, mitochondria, endoplasmic reticulum, and Golgi complex. In this chapter we will be concerned with the external cell boundary called the cell membrane. (We will use the term *cell membrane* rather than *plasma membrane* to avoid confusion with blood plasma.) The general functions of the cell membrane include:

1. **Physical isolation**: The cell membrane is a physical barrier that separates the inside of the cell (the

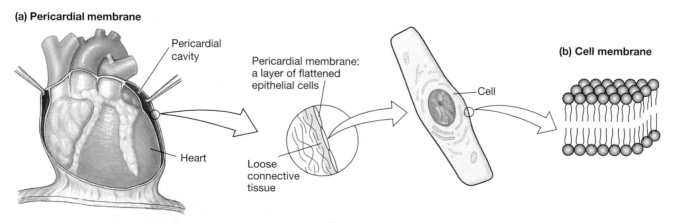

(a) Pericardial membrane

Pericardial cavity

Pericardial membrane: a layer of flattened epithelial cells

Cell

(b) Cell membrane

Heart

Loose connective tissue

■ **Figure 5-1   Membranes in the body**   (a) The pericardial membrane, depicted as a single line just like the cell membrane, is enlarged to show that it is a layer of flattened cells. Substances crossing the pericardial membrane must cross two phospholipid bilayers. (b) The phospholipid bilayer of a cell membrane.

cytosol or intracellular fluid) from the surrounding extracellular fluid.

2. **Regulation of exchange with the environment**: The cell membrane controls the entry of ions and nutrients, the elimination of wastes, and the release of secretory products.

3. **Communication between the cell and its environment**: The cell membrane is in direct contact with both the cytosol and the extracellular fluid [∞ p. 45]. It contains receptors that allow the cell to recognize and respond to molecules or changes in its external environment. Any alteration in the cell membrane may affect the cell's activities.

4. **Structural support**: Cell membrane proteins hold proteins of the cytoskeleton in place to maintain cell shape. They also create specialized junctions between adjacent cells or between cells and the extracellular matrix. These junctions stabilize the structure of tissues [∞ p. 55].

To understand how the cell membrane can carry out such diverse functions, we will first examine the current model of cell membrane structure.

Membranes Are Mostly Lipid and Protein

The structure of cell membranes has been a matter of great interest ever since scientists first observed cells with a microscope. By the 1890s, scientists had concluded that the outer membrane of cells was made of a thin layer of lipids that acted as a barrier between the aqueous interior and the watery environment outside the cell. However, this simple and uniform structure did not account for the highly variable properties of membranes found in different types of cells. How could water cross the cell membrane to enter a red blood cell but not be able to enter certain cells of the kidney tubule? The explanation had to lie in the structure of the cell membrane.

To isolate and analyze cell membranes, researchers in the first decades of the twentieth century ground up cells. They discovered that the cell membrane and all membranes within the cell consist of a combination of lipids and proteins with a small amount of carbohydrate. The ratio of protein to lipid varies widely, depending on the source of the membrane (Table 5-1). Generally, the more metabolically active a membrane is, the more pro-

teins it contains. For example, the inner mitochondrial membrane, which contains enzymes for oxidative phosphorylation [∞ p. 94], is three-quarters protein.

The chemical analysis of membranes was useful but it still did not explain how lipids and proteins were structurally arranged in the membrane. Studies in the 1920s suggested that there was enough lipid in a given area of membrane to create a double layer. This model was further modified in the 1930s to account for the presence of proteins. With the introduction of electron microscopy, scientists saw the membrane for the first time. The 1960s model of the membrane was a "butter sandwich"—a double layer of lipids sandwiched between two layers of proteins.

By the early 1970s, freeze-fracture electron micrographs had revealed the actual three-dimensional arrangement of lipids and proteins within cell membranes. Because of what scientists learned from looking at freeze-fractured membranes, S. J. Singer and G. L. Nicolson in 1972 proposed the **fluid mosaic model** of the membrane. Figure 5-2 ∎ highlights the major features of this contemporary model of membrane structure. The phospholipids of the membrane are arranged in a bilayer, with a variety of proteins inserted partly or completely through the bilayer. Carbohydrates on the extracellular surface are attached to both the proteins and the lipids, forming a "sugar coating." All cell membranes have similar components and are of uniform thickness, about 8 nm.

Membrane Lipids Form a Barrier Between the Cytoplasm and Extracellular Fluid

Two main types of lipids make up the cell membrane: phospholipids and cholesterol. **Phospholipids** are made of a glycerol backbone (Fig. 5-3a ∎) with two fatty acid chains extending to one side and a phosphate group extending to the other [∞ p. 29]. The glycerol-phosphate "head" of the molecule is polar and thus hydrophilic. The fatty acid "tail" is nonpolar and thus hydrophobic (Fig. 5-3b ∎).

When placed in an aqueous solution, phospholipids orient themselves so that the polar side of the molecule interacts with the water molecules while the nonpolar fatty acid tails hide by putting the polar heads between themselves and the water. This arrangement can be seen in three different structures: the micelle, the liposome, and the phospholipid bilayer (Fig. 5-3b ∎). **Micelles** are small droplets of phospholipid, arranged so that the interior is filled with hydrophobic fatty acid tails. Micelles are important in the digestion and absorption of fats in the digestive tract.

Liposomes are larger hollow spheres with phospholipid bilayer walls. This arrangement leaves a hollow center with an aqueous core that can be loaded with water-soluble molecules. Liposomes are being tested as a medium for the delivery of drugs through the skin.

TABLE 5-1 Composition of Selected Membranes

Membrane	Protein	Lipid	Carbohydrate
Red blood cell membrane	49%	43%	8%
Myelin membrane around nerve cells	18%	79%	3%
Inner mitochondrial membrane	76%	24%	0%

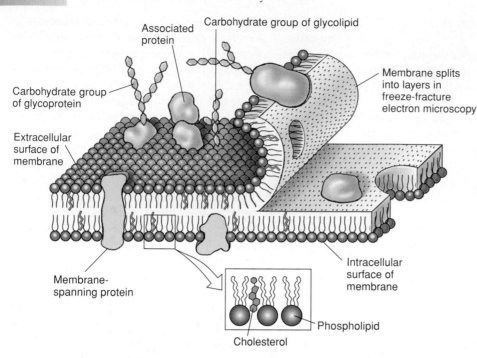

Carbohydrate group of glycolipid

Associated protein

Carbohydrate group of glycoprotein

Extracellular surface of membrane

Membrane splits into layers in freeze-fracture electron microscopy

Intracellular surface of membrane

Membrane-spanning protein

Cholesterol

Phospholipid

■ **Figure 5-2   The fluid mosaic model of the membrane**   The phospholipid molecules orient themselves so that the polar glycerol-phosphate ends (purple) face the aqueous intracellular and extracellular fluid, and the nonpolar lipid tails (yellow) are sandwiched in the middle. Cholesterol molecules insert themselves into the lipid layer. Integral proteins are tightly bound to the membrane. Associated proteins are more loosely attached at the surface. Carbohydrates attach to both membrane proteins and lipids.

The **phospholipid bilayer** is a sheetlike structure with two layers of phospholipids arranged so that the hydrophobic tails are sandwiched between two layers of hydrophilic heads. The major role of the bilayer is structural: to form a barrier across which only lipid-soluble molecules can migrate and to provide a framework for the membrane proteins. A few membrane lipids participate in communication between the extracellular environment and the cell.

Although phospholipids are the major lipid of membranes, cholesterol is also a significant part of many cell membranes. Because cholesterol molecules do not have

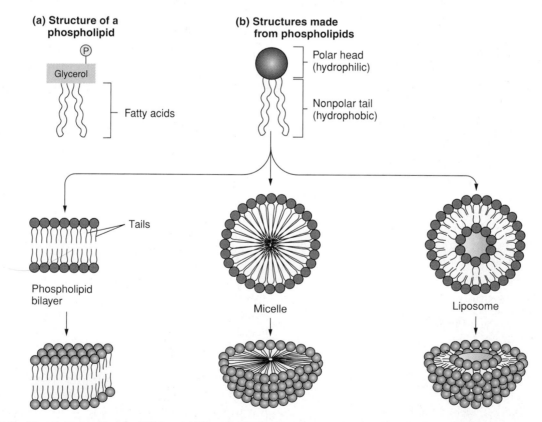

(a) Structure of a phospholipid

P

Glycerol

Fatty acids

(b) Structures made from phospholipids

Polar head (hydrophilic)

Nonpolar tail (hydrophobic)

Tails

Phospholipid bilayer

Micelle

Liposome

■ **Figure 5-3   Membrane phospholipids**   (a) Phospholipid molecules have polar glycerol-phosphate heads and nonpolar fatty acid tails. (b) Structures made from phospholipids include the phospholipid bilayer, the micelle, and the liposome.

Liposomes for Beauty and Health Many people first hear the term liposome in connection with cosmetic skin creams that promise to deliver special ingredients to the cells that need them. But the cosmetic manufacturers have adopted a medical technique developed to enhance the delivery of drugs. In medicine, the centers of liposomes are filled with drugs or with fragments of DNA for gene therapy. Then the liposomes are applied to the skin or injected into the bloodstream. The newest generation of liposomes has attached proteins that bind to receptors on specific cells. In this way, toxic drugs designed for anticancer chemotherapy can be targeted specifically to diseased tissues, enhancing their effectiveness and reducing their side effects on normal tissues. In one study, liposomes loaded with antibodies against the potent anticancer drug doxorubicin (DXR) were applied to the scalp of mice receiving DXR treatments. The liposomes were absorbed into hair follicles, where the antibodies protected the rapidly dividing follicle cells from DXR. This prevented the drug-induced hair loss usually associated with chemotherapy.

a polar region like that of phospholipids, they are hydrophobic. Consequently, they insert themselves into the central lipid portion of the bilayer (Fig. 5-2 ■). Cholesterol helps make membranes impermeable to small water-soluble molecules and keeps the membrane flexible over a wide range of temperatures.

Membrane Proteins May Be Loosely or Tightly Bound to the Membrane

Each cell has between 10 and 50 different types of proteins inserted into its phospholipid bilayer. Anatomically, membrane proteins are described as being either associated or integral. **Associated proteins,** also called *peripheral* or *extrinsic* proteins, attach themselves loosely to membrane-spanning proteins or to the polar regions of phospholipids. They can be removed without disrupting the integrity of the membrane itself. Associated pro-

teins include enzymes and some structural proteins that tie the cytoskeleton to the membrane (Fig. 5-4 ■).

Integral proteins, also called *intrinsic* proteins, are tightly bound into the phospholipid bilayer. The only way an integral protein can be removed is by disrupting the membrane structure with detergents or other harsh methods that destroy the membrane's integrity. Some integral proteins extend only partway into the lipid core of the membrane, while most span the entire membrane.

Membrane-Spanning Proteins **Membrane-spanning proteins** are integral proteins that extend all the way across the membrane. They can be classified into families according to how many transmembrane segments they have. Many physiologically important membrane proteins have 7 transmembrane segments, while others may have as few as 1 or as many as 12 or more.

When proteins cross the membrane more than once, amino acid chains between the membrane-spanning sections protrude into the cytoplasm and the extracellular fluid (Fig. 5-5 ■). Carbohydrates attach to the extracellular loops. Phosphate groups may attach to the intracellular loops, a method cells use to alter the functions of the proteins.

According to the fluid mosaic model of the cell membrane, membrane proteins may move laterally from location to location, directed by fibers of the cytoskeleton that run just under the membrane surface. On the other hand, some membrane-spanning proteins are immobile, held in position by cytoskeleton proteins (Fig. 5-4 ■). The ability of the cytoskeleton to restrict the movement of membrane proteins allows cells to develop *polarity,* in which different faces of the cell have different proteins and therefore different properties. This is particularly clear in the cells of the transporting epithelia, as you will learn later in this chapter.

✔ Name two types of lipids found in cell membranes.

✔ Describe two ways that membrane proteins are physically associated with the cell membrane.

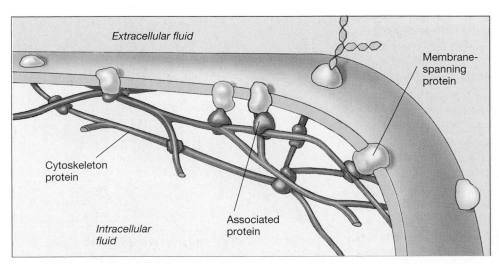

■ Figure 5-4 **Cytoskeleton proteins anchor to the cell membrane using associated proteins**

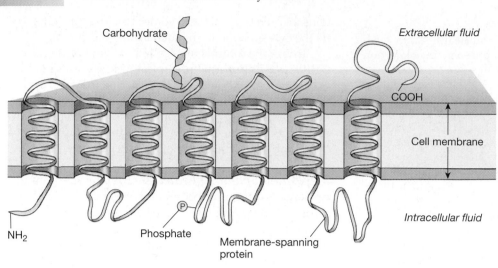

■ **Figure 5-5 Membrane-spanning proteins** Membrane-spanning proteins cross the membrane several times with loops of peptide chains that extend into cytoplasm and extracellular fluid.

Membrane Proteins Function as Structural Proteins, Enzymes, Receptors, and Transporters

For physiologists, classifying membrane proteins by their function is more useful than classifying them by their structure. Cell membrane proteins function as (1) structural elements, (2) enzymes, (3) receptors, and (4) transporters that move molecules into and out of the cell. Many physiological processes you will be studying involve membrane enzymes, receptors, and transporters.

Structural Proteins The structural proteins have two major roles. The first is to connect the membrane and the cytoskeleton to maintain the shape of the cell. The microvilli of transporting epithelia [∞ p. 58] are one example of membrane shaping by the cytoskeleton; the flattened disk shape of the red blood cell is another example.

The second role of structural membrane proteins is to create cell junctions that hold tissues together, such as tight junctions and gap junctions [∞ p. 55]. Membrane-spanning proteins also link cytoskeleton fibers to collagen and other fibers in the extracellular matrix [∞ Fig. 3-14, p. 56].

Enzymes Membrane-associated enzymes catalyze chemical reactions that take place on the cell's external surface or just inside the cytoplasm. For example, enzymes on the luminal surface of cells in the small intestine are responsible for digesting peptides and carbohydrates. Enzymes attached to the intracellular surface of many cell membranes play an important part in transferring signals from the extracellular environment to the cytoplasm of the cell.

Receptors Receptor proteins of the cell membrane are part of the body's chemical signaling system. Each receptor is *specific* for a certain molecule or family of related molecules. It will combine only with those molecules, just as enzymes recognize and bind only to specific substrates. The molecule that binds to a receptor is called the **ligand** [*ligare,* to bind or tie]. The binding of receptor and ligand usually triggers another event at

the membrane, such as activation of a membrane enzyme (Fig. 5-6 ■). An example of a ligand is the hormone insulin, which combines with an insulin receptor on a cell membrane to exert its effects. You will learn more about receptors in Chapter 6.

Transporters The fourth major function of membrane proteins is their role as transporters. Many molecules enter or leave the cell using transport proteins. We will classify transport proteins into two categories: channels and carriers. **Channel proteins** create water-filled passageways that link the intracellular and extracellular compartments. **Carrier proteins** bind to the substrates that they carry but never form a direct connection between the intracellular and extracellular fluid.

Why do cells need both channels and carriers? The answer lies in the different properties of the two transporters. Channel proteins allow more rapid transport across the membrane but are not as selective about what they transport. Carriers, while slower, are better at discriminating between closely related molecules. Carriers can also move larger molecules than channels can.

Channel proteins Channel proteins are like narrow doorways into the cell. If the door is closed, nothing can go

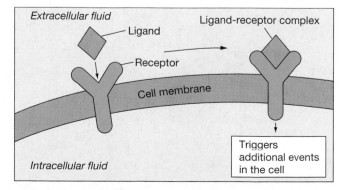

■ **Figure 5-6 Cell membrane receptor** Cell membrane receptors combine with molecules known as ligands. The receptor-ligand complex triggers additional events within the cell.

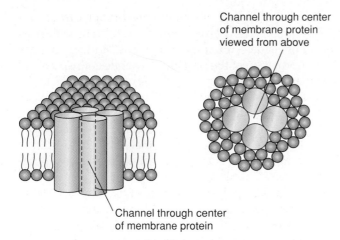

Channel through center of membrane protein viewed from above

Channel through center of membrane protein

■ **Figure 5-7   Structure of channel proteins**   Many membrane channel proteins are composed of multiple subunits that assemble in the membrane to form a unit. Hydrophilic amino acids facing the center of the channel create a water-filled passage that allows ions and very small molecules, such as water and urea, to pass through.

through. If the door is open, there is a continuous passage between the two rooms connected by the doorway. In cells, channels are made of protein segments that zigzag back and forth across the membrane, creating a cluster of cylinders that surround a water-filled pore (Fig. 5-7 ■). The diameter of the channel is quite narrow, so movement through it is restricted primarily to water and ions. Because water-filled channels are open to both sides of the membrane at once, tens of millions of ions per second can whisk through them unimpeded.

Channels are named according to the substance(s) they allow to pass. For example, most cells have water channels made from a protein called *aquaporin*. Ion channels may be specific for one ion or may allow ions of similar size and charge to pass. There are Na^+ channels, K^+ channels, and nonspecific *monovalent* ("one charge") cation channels that transport Na^+, K^+, and lithium ions (Li^+).

The selectivity of a channel is determined by its diameter and by the electrical charge of the amino acids that line the channel. If the channel amino acids are positively charged, positive ions will be repelled while negative ions will pass through the channel. For example, the nonspecific cation channel mentioned above must have a negative charge that attracts cations but prevents the passage of Cl^- or other anions.

Membrane channel proteins have regions that act like swinging "gates" to open and close the channel. According to current models, many "gates" are part of the

…continued from page 110

Cystic fibrosis is a debilitating disease caused by a defect in a gated channel protein that normally transports chloride ions (Cl^-). The channel protein—called the cystic fibrosis transmembrane regulator, or CFTR—is located in the cell membranes of epithelia lining the airways, sweat glands, and pancreas. The open state of the channel is regulated through the binding of nucleotides to specific regions of the channel. In people with cystic fibrosis, the CFTR channel is absent or defective. As a result, chloride transport (either secretion or reabsorption) across the epithelium is impaired.

Question 1:   Is CFTR a chemically gated, voltage-gated, or mechanically gated channel protein?

cytoplasmic side of the membrane protein (Fig. 5-8 ■). Such a gate can be envisioned as a ball on a chain that swings up and blocks the mouth of the channel. A few channels have a gate in the middle of the protein, and one type of channel in nerve cells has two gates.

Channels can be classified according to whether their gates are usually open or closed. **Open channels** spend most of their time in the open state, allowing ions to move back and forth across the membrane without regulation. Their gates may occasionally flicker closed, but for the most part, these channels behave as if they have no gates. Open channels are sometimes called *leak channels* or *pores*, as in *water pores*.

Gated channels spend most of their time in the closed state, which allows these channels to regulate the movement of molecules through them. If a gated channel is open, molecules move through the channel just as they move through open channels. If a gated channel is closed, which it may be much of the time, it allows no movement between the intracellular and extracellular fluid.

What controls the opening and closing of gated channels? The gating of these channels is controlled by intracellular messenger molecules or extracellular ligands (**chemically gated channels**), by the electrical state of the cell (**voltage-gated channels**), or by a physical change such as increased temperature or a blow that puts tension on the membrane and pops the channel open (**mechanically gated channels**). You will encounter many variations of these channel types as you study physiology.

Carrier proteins   The second type of transport protein is the carrier protein. If channels are like doorways, then carrier proteins are like revolving doors that allow

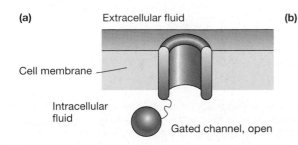

(a)   Extracellular fluid

Cell membrane

Intracellular fluid

Gated channel, open

(b)

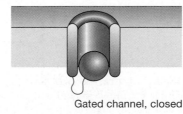

Gated channel, closed

■ **Figure 5-8   Gating of channel proteins**   (a) Channels that spend most of their time in the open state freely allow selected molecules through the cell membrane. (b) Gated channels that are usually closed can open in response to a signal, which may be chemical, mechanical, or electrical, depending on the channel.

movement between inside and outside without creating an open hole. Carrier proteins can transport molecules across a membrane in both directions, like a revolving door at a hotel, or they can restrict their transport to one direction, like the revolving doors at amusement parks that allow you out of the park but not back in.

Carrier proteins bind with specific molecules, called substrates, and carry them across the membrane by changing conformation or shape. Small organic molecules, such as glucose and amino acids, cross membranes only on carriers. On the other hand, ions such as Na^+ and K^+ may move by carriers as well as through channels. The conformation change makes carrier transport much slower, and only 1000 to 1,000,000 molecules per second can be moved by each protein.

Carrier proteins differ from channel proteins in another way. Carrier proteins never create a continuous passage between the inside and outside of the cell. They do form channels, but the channels have two gates, one on the extracellular side and one on the intracellular side. One gate is always closed, preventing free exchange between inside and outside.

In this respect, carrier proteins are like the Panama Canal (Fig. 5-9a ■). Picture the canal with only two gates, one on the Atlantic side and one on the Pacific side. Only one gate at a time is open. When the Atlantic gate is closed, the canal opens into the Pacific. A ship enters the canal from the Pacific and the gate closes behind it. Now the canal is isolated from both oceans with the ship trapped in the middle. Then the Atlantic gate opens, making the canal continuous with the Caribbean Sea. The ship sails out of the gate, having crossed the barrier of the land without the canal ever forming a direct connection between the two bodies of water.

Movement across the membrane through carrier proteins is similar (Fig. 5-9b ■). The substrate binds to the carrier on one side of the membrane. The binding changes the conformation of the protein so that one gate closes and the other one opens. As a result, the channel opens to the opposite side of the membrane. The carrier then releases the substrate into the other side, having brought it through the membrane without creating a direct connection between the extracellular fluid and the intracellular fluid.

The mechanisms by which molecules cross membranes, either on transporters or by moving through the phospholipid bilayer, are the subject of a later section in this chapter, Movement Across Membranes.

✔ Name four functions of membrane proteins.

Membrane Carbohydrates Attach to Both Lipids and Proteins

Most membrane carbohydrates are glucose polymers attached either to membrane proteins (*glycoproteins*) or to membrane lipids (*glycolipids*). They are found exclusively on the external surface of the cell, where they form a protective layer known as the **glycocalyx** [*glykys*, sweet + *kalyx*, husk or pod]. The glycoproteins on the surface of cells play a key role in the body's immune response.

Figure 5-10 ■ is a summary map organizing the structure of the cell membrane. Figure 5-11 ■ is a map showing the structure and function of membrane proteins. In the next two sections, we will see how membranes divide the body into compartments, and how membranes regulate the movement of molecules between these compartments.

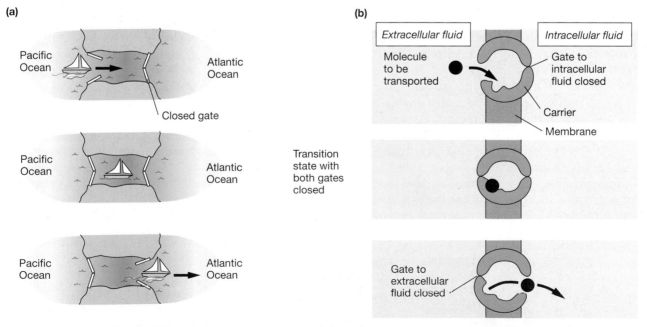

(a)

Pacific Ocean — Atlantic Ocean
Closed gate

Pacific Ocean — Atlantic Ocean
Transition state with both gates closed

Pacific Ocean — Atlantic Ocean

(b)

Extracellular fluid | Intracellular fluid

Molecule to be transported — Gate to intracellular fluid closed
Carrier
Membrane

Gate to extracellular fluid closed

■ **Figure 5-9** **Facilitated diffusion by means of a carrier protein** (a) Carrier proteins, like the canal illustrated, never form a continuous passageway between the extracellular and intracellular fluids. (b) Carrier proteins bind to the molecule to be transported on one side of the cell membrane, then change conformation, opening the channel to the other side and releasing the molecule.

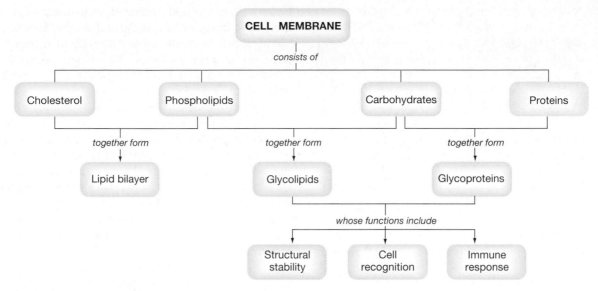

■ Figure 5-10   Cell membrane structure

▶ BODY FLUID COMPARTMENTS

Cell membranes divide the inside of the body into compartments, just as the membranes of organelles divide the inside of the cell. Most cells of the body are not in direct contact with the outside world. Instead they are surrounded by extracellular fluid [∞ Fig. 1-4, p. 6]. If we think of all the cells of the body together as a single unit, we can divide the body into two main compartments: (1) the intracellular fluid (ICF) within the cells and (2) the extracellular fluid (ECF) outside the cells. These compartments are separated by the barrier of the cell membrane.

The extracellular fluid can be further subdivided. The dividing "wall" in this case is the wall of the circulatory system. **Plasma,** the fluid portion of the blood, lies within

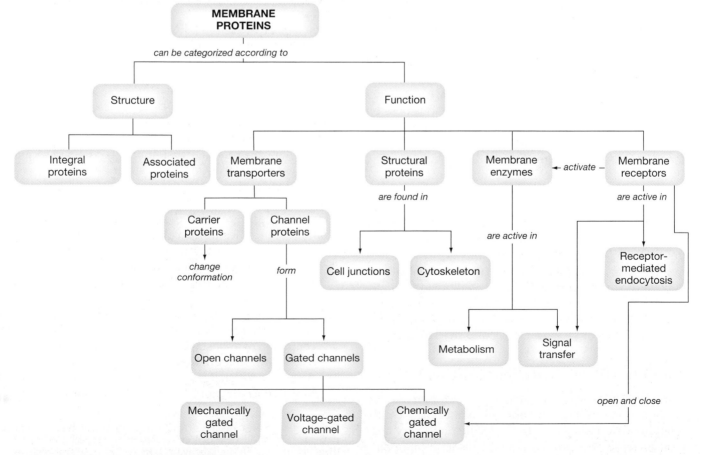

■ Figure 5-11   Membrane proteins

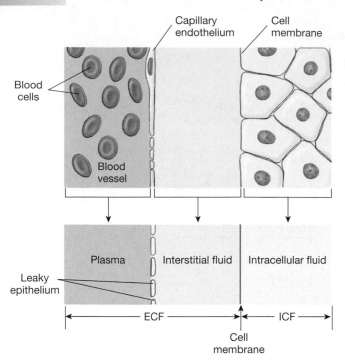

■ **Figure 5-12** **Body fluid compartments** The body is divided into two major compartments, intracellular and extracellular. Movement between these compartments requires crossing the cell membrane. The extracellular compartment is subdivided into the plasma and the interstitial fluid, divided by the epithelial layer of the capillary endothelium.

the circulatory system and forms one extracellular compartment. The fluid that lies outside the circulatory system and directly bathes the cells is known as **interstitial fluid** [*inter-*, between + *stare*, to stand]. Figure 5-12 ■ shows the relationships between these compartments.

In many parts of the circulatory system, no exchange between the plasma and the interstitial fluid is possible due to the thick walls of the blood vessels. However, in the smallest blood vessels, called *capillaries,* the walls are made of leaky epithelium. The capillary walls act like a sieve, retaining blood cells and large proteins but allowing the remainder of the plasma to cross freely through spaces between the cells [∞ Fig. 3-17, p. 59].

Although many materials move freely between the blood and the interstitial fluid, exchange between the intracellular and extracellular compartments is restricted by the cell membrane. Whether or not a substance enters a cell depends on the properties of the cell membrane and of the substance itself. To understand how cells can be selective about the materials that enter them, we must look in detail at the mechanisms by which molecules cross membranes.

◗ MOVEMENT ACROSS MEMBRANES

Some molecules, such as water, oxygen, carbon dioxide, and lipids, move easily across cell membranes. On the other hand, ions, larger polar molecules, and very large molecules, such as proteins, enter cells with more difficulty or may not enter at all. If a molecule crosses a membrane by any method, we say that the membrane is **permeable** to the molecule. If the membrane does not allow the molecule to cross, we say that the membrane is **impermeable** to the molecule. Membrane permeability is variable and can be changed by altering the proteins or lipids of the membrane.

Two properties of a molecule will influence its movement across cell membranes: the size of the molecule and its lipid-solubility or polarity [∞ p. 23]. Very small or lipid-soluble substances can cross directly through the phospholipid bilayer. Larger or less lipid-soluble molecules are excluded from crossing the cell membrane unless a cell has a specific process for getting them across. Usually, membrane proteins are the helpers or mediators that cells use to transport molecules across their membranes.

There are two ways to categorize how molecules move across membranes. One scheme describes movement according to whether it takes place through the phospholipid bilayer or with the aid of a membrane protein (Fig. 5-13 ■). A different scheme classifies movement according to its energy requirements. **Passive transport** does not require the input of energy. **Active transport** requires the input of energy from another source, such as the high-energy phosphate bond of ATP. In this section we examine all the different ways that substances move into and out of cells.

Diffusion Uses Only the Energy of Molecular Movement

All molecules are constantly in motion. Gas molecules and molecules in solution constantly move from one place to another, bouncing off other molecules or off the sides of a container (Fig. 5-14 ■). When molecules start out concentrated in one area of an enclosed space, their motion causes them to spread out gradually until they are evenly distributed throughout the available space. This process is known as diffusion. **Diffusion** may be defined as the movement of molecules from an area of higher concentration of the molecule to an area of lower concentration of the molecule.* If you leave a bottle of cologne open and later notice its fragrance across the room, it is because the aromatic molecules in the cologne have diffused from where they are more concentrated (in the bottle) to where they are less concentrated (across the room).

Diffusion has the following properties:

1. *Molecules move from an area of higher concentration to an area of lower concentration.* A difference in the concentration of a substance between two places is called a **concentration gradient.** We say that molecules diffuse *down the gradient,* from higher concentration to lower concentration. The rate of diffusion depends

---

*Some texts use the term *diffusion* to mean any random movement of molecules. They call molecular movement along a concentration gradient *net diffusion.* To simplify matters, we will use the term *diffusion* to mean movement of molecules down a concentration gradient.

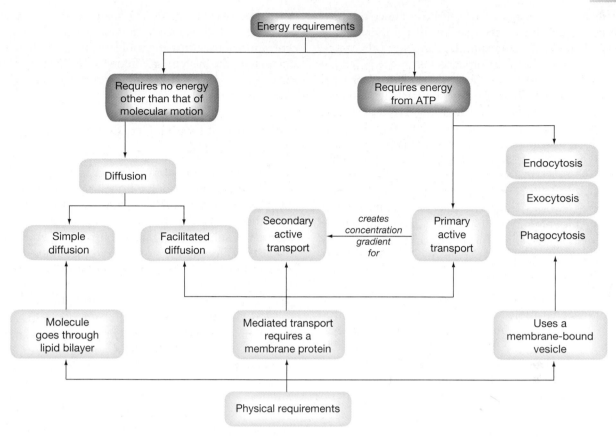

■ **Figure 5-13   How molecules move across cell membranes**   Movement of substances across cell membranes can be classified by energy source (orange boxes) or according to whether transport uses a membrane protein or a vesicle (blue boxes).

on the magnitude of the concentration gradient. The larger the concentration difference, the faster diffusion takes place. For example, when you open a bottle of ammonia, the rate of diffusion is most rapid as the ammonia molecules first escape from the bottle into the air. Later, when the ammonia has spread evenly throughout the room, the rate of diffusion has dropped to zero. The molecules are still in motion but diffusion has ceased because there is no longer a concentration gradient (Fig. 5-14 ■).

2. *Diffusion is a passive process.* Diffusion does not require the input of energy from another source, such as the energy stored in ATP. Diffusion uses only the kinetic energy possessed by all molecules.

3. *There will be net movement of molecules until the concentration is equal everywhere.* Once the molecules have distributed themselves evenly through the system, **equilibrium** occurs [*aequus*, equal + *libra*, balance] and diffusion stops. Individual molecules are still moving at equilibrium, but for each molecule that exits one area, another one enters. The equilibrium state in diffusion means that the concentration has equalized throughout the system.

4. *Diffusion is rapid over short distances but much slower over long distances.* Albert Einstein studied the diffusion of molecules in solution and found that the time required for a molecule to diffuse from point A to point B is proportional to the square of the distance

■ **Figure 5-14   Diffusion**   A crystal of dye, such as potassium permanganate, is placed in a beaker of water. As the crystal dissolves, dye molecules spread outward by random molecular motion, moving from the concentrated area into less concentrated areas within the beaker. In the last drawing, the dye molecules have distributed themselves evenly throughout the container and the system has reached equilibrium. Remember that individual molecules are still moving, although diffusion has stopped.

from A to B. In other words, if the distance doubles from 1 to 2, the time needed for diffusion increases from 1^2 to 2^2. In this example, diffusion over twice the distance takes four times as long!

What does the slow rate of diffusion over long distances mean for biological systems? In humans, nutrients take five seconds to diffuse from the blood to a cell that is 100 μm from the nearest capillary. At that rate, it would take years for nutrients to diffuse from the small intestine to cells in the big toe, and the cells would starve to death. To overcome the limitations of diffusion over distance, organisms have developed various transport mechanisms that speed up the movement of molecules. Most multicellular organisms have some form of circulatory system to bring oxygen and nutrients rapidly from the point at which they enter the body to the cells.

5. *Diffusion is directly related to temperature.* At higher temperatures, molecules move faster. Because diffusion is a result of molecular movement, as temperature increases so does the rate of diffusion. Generally, changes in temperature do not significantly affect diffusion in humans because we maintain a relatively constant body temperature.

6. *Diffusion is inversely related to molecular size.* The larger a molecule is, the more slowly it will diffuse. Albert Einstein showed that friction between the surface of a particle and the medium through which it diffuses is a source of resistance to movement. He calculated that diffusion is inversely proportional to the radius of the molecule: the larger the molecule, the slower its diffusion through a given medium.

7. *Diffusion can take place in an open system or across a partition that separates two systems.* Diffusion of ammonia or cologne within a room is an example of diffusion taking place in an open system. There are no barriers to molecular movement, so the molecules spread out to fill the entire system. Diffusion can take place between two different systems, such as the intracellular and extracellular compartments, but only if the partition dividing the two compartments allows the diffusing molecules to cross.

For example, if you close the top of an open bottle of ammonia, diffusion cannot spread molecules into the room because the bottle is not permeable to the ammonia. However, if you cover the open mouth of the bottle with a "vegetable-keeper" plastic bag with tiny holes in it, you will begin to smell the ammonia in the room because the bag is permeable to the ammonia. Similarly, if a cell membrane is permeable to a molecule, that molecule can enter or leave the cell by diffusion. If the membrane is not permeable, the molecule cannot cross.

In summary, diffusion is the passive movement of molecules down their concentration gradient due to random molecular movement. Diffusion is slower over long distances and slower for large molecules. When the concentration of the diffusing molecule is equal throughout the system, the system has come to equilibrium and diffusion stops, although the random movement of molecules continues.

Lipophilic Molecules Can Diffuse Through the Phospholipid Bilayer

The ability of cell membranes to act as barriers is a direct result of their hydrophobic lipid core. Water is the primary solvent of the body, and many vital nutrients, ions, and other molecules dissolve in it because of its polar nature. But substances that are hydrophilic and dissolve in water are lipophobic as a rule: they do not readily dissolve in lipids. For this reason, the lipid layer between the cytoplasm and the cell's exterior acts as a boundary that prevents hydrophilic molecules from crossing.

Substances that can pass through the lipid center of the cell membrane move by diffusion. Diffusion directly across the phospholipid bilayer of a membrane is called **simple diffusion** and has the following properties in addition to the properties of diffusion listed earlier.

1. *The rate of diffusion depends on the ability of the molecule to dissolve in the lipid layer of the membrane.* Most molecules in solution can mingle with the polar phosphate-glycerol heads of the membrane's phospholipid bilayer, but only nonpolar molecules that are lipid-soluble (lipophilic) can traverse the central lipid core of the membrane. As a rule, only lipids, steroids, and small lipophilic molecules can move across the membrane by simple diffusion.

Water, although a polar molecule, can diffuse across some phospholipid membranes. For years, it was thought that the polar nature of the water molecule prevented it from moving through the lipid center of the bilayer, but experiments done with artificial membranes have shown that the small size of the water molecule allows it to slip between the lipid tails in many membranes. How readily water passes through the lipid layer depends on the composition of the phospholipid bilayer. Membranes with high cholesterol content are less permeable to water than those with low cholesterol content, presumably because the cholesterol molecules fill the spaces between the fatty acid tails and thus exclude water. For example, the cell membranes of some sections of the kidney are essentially impermeable to water unless the cells insert special water channel proteins into the phospholipid bilayer.

2. *The rate of diffusion across a membrane is directly proportional to the surface area of the membrane.* In other words, the larger the surface area, the more molecules can diffuse across per unit time. This may seem obvious, but it has important implications in physiology. One striking example of how a change in surface

area affects diffusion is the lung disease emphysema. As lung tissue is destroyed, the surface area available for diffusion of oxygen decreases. Consequently, less oxygen can move into the body. In severe cases, the oxygen that reaches the cells is not enough to sustain any muscular activity and the patient is confined to bed.

3. *The rate of diffusion across a membrane is inversely proportional to the thickness of the membrane.* The thicker the membrane, the slower diffusion will take place. This factor comes into play in certain lung conditions in which the exchange epithelium of the lung is thickened with scar tissue. This slows diffusion so that the oxygen entering the body is not adequate to meet metabolic needs.

The rules for simple diffusion across membranes are summarized in Table 5-2. They can be combined mathematically into an equation known as Fick's law of diffusion. In an abbreviated form, Fick's law says that:

Rate of diffusion is proportional to

$$\frac{\text{available surface area} \times \text{concentration gradient}}{\text{resistance of membrane} \times \text{membrane thickness}}$$

The resistance of the membrane is the most complex of the four terms because it varies according to (1) the size of the diffusing molecule, (2) the lipid-solubility of the molecule, and (3) the composition of the lipid layer across which it is diffusing. Figure 5-15 ■ illustrates the principles of Fick's law.

Lipophobic molecules cannot enter a cell by simple diffusion, therefore cells must make special provisions for their transport. Smaller lipophobic molecules and ions cross cell membranes through protein carriers. Ions also cross membranes through open water-filled channels. However, the movement of ions through channels is influenced by electric gradients as well as concentration gradients. This means that ion movement cannot be predicted by using Fick's law and the diffusion rules in Table 5-2. We will discuss ion movement in the last sections of this chapter. Macromolecules and particles such as bacteria are too large to be transported by proteins. They move into and out of cells by way of membrane vesicles.

✔ Two compartments are separated by a membrane that is permeable only to water and yellow dye molecules. Compartment A is filled with a solution of yellow dye, compartment B is filled with a solution of equal concentration of blue dye. If the system is left undisturbed for a long period of time, what color will compartment A be: yellow, blue, or green? (Remember, yellow + blue makes green.)

✔ What keeps atmospheric oxygen from diffusing into our bodies across our skin? (Hint: What kind of epithelium is skin?)

TABLE 5-2 Rules for Simple Diffusion

1. Molecules diffuse from an area of higher concentration to an area of lower concentration. The larger the concentration difference, the faster diffusion takes place.
2. Diffusion is a passive process that does not require an outside energy source. It uses only the kinetic energy of molecular movement.
3. There will be net movement of molecules until the concentration is equal. Molecular movement continues, however, so this is called a dynamic equilibrium.
4. Diffusion is rapid over short distances but much slower over long distances.
5. Diffusion is directly related to temperature. At higher temperatures, molecules move faster and diffusion is more rapid.
6. Diffusion is inversely related to molecular size. The larger a molecule is, the slower it will diffuse.
7. Diffusion can take place in an open system or across a partition that separates two systems.

Diffusion across a membrane

8. The rate of diffusion through a cell membrane depends on the ability of the molecule to dissolve in the lipid layer of the membrane. Only lipid-soluble (lipophilic) molecules can diffuse across the membrane by simple diffusion.
9. The rate of diffusion through a membrane is directly proportional to the surface area of the membrane. The larger the surface area, the more molecules can diffuse across per unit time.
10. The rate of diffusion across a membrane is inversely proportional to the thickness of the membrane. The thicker the membrane, the slower diffusion will take place.

Carrier-Mediated Transport Exhibits Saturation, Specificity, and Competition

Movement of substances across cell membranes with the aid of a carrier protein is known as **mediated transport.** If mediated transport is passive, moves molecules down their concentration gradient, and net transport stops when concentrations are equal inside the cell and out, the process is known as **facilitated diffusion.** If protein-mediated transport requires energy from ATP or another outside source and moves a substance against its concentration gradient, the process is known as **active transport.**

Both passive and active forms of mediated transport demonstrate three properties: specificity, competition, and saturation. You have already encountered these concepts in the discussion on enzymes [∞ p. 81]. Mediated transport shares these properties with enzymes because both processes depend on the interaction of a substrate with a protein.

Specificity   **Specificity** refers to the ability of a carrier to move only one molecule or a group of closely related molecules. One example of specificity is the family

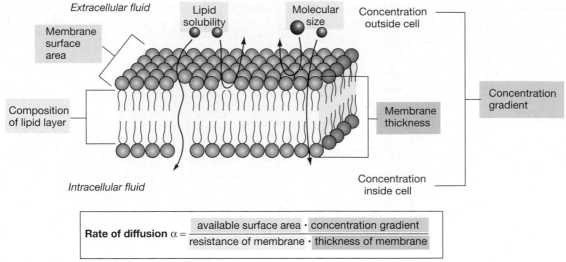

Rate of diffusion $\alpha = \dfrac{\text{available surface area} \cdot \text{concentration gradient}}{\text{resistance of membrane} \cdot \text{thickness of membrane}}$

■ **Figure 5-15** **Fick's law of diffusion** What factors affect the rate of simple diffusion across a membrane? They include the surface area available for diffusion, the resistance of the membrane (determined by the composition of the lipid layer and the size and lipid solubility of the diffusing molecule), the thickness of the membrane, and the concentration gradient (concentration outside minus concentration inside the cell) across the membrane. Fick's law of diffusion relates these factors mathematically.

of transporters known as the GLUT transporters. These membrane proteins move six-carbon sugars (*hexoses*), such as glucose, mannose, galactose, and fructose [∞ p. 29] across cell membranes. But GLUT transporters will not transport the disaccharide maltose or a form of glucose that is not normally found in nature. Thus we can say that GLUT transporters are specific for the naturally occurring six-carbon monosaccharides.

The GLUT Transporter Family Facilitated diffusion carriers for the hexose sugars are found in every cell of the body. For many years, scientists assumed that there were different variants of the carrier. In some cells, glucose transport was regulated by hormones, yet in other cells, it was independent of hormonal action. It was not until the 1980s that the first glucose transporter was isolated. To date, six glucose transporters have been identified. The first transporter, isolated from red blood cells, was named GLUT1. This protein is the facilitated diffusion carrier that moves glucose and other hexoses into most cells of the body. The GLUT2 transporter is found in the hepatocytes of the liver and on the basolateral membrane of transporting epithelial cells in the kidney and intestine. GLUT3 transports glucose and other hexoses into neurons. The GLUT4 transporter is found in adipose tissue and muscle. It is the glucose transporter regulated by the hormone insulin. GLUT5 is actually a fructose transporter and is the transporter present on the apical membrane of the intestinal epithelial cells. The protein named GLUT6 turned out not to be a transporter, but by that time GLUT7 had already been named. GLUT7 is an intracellular transporter found on the endoplasmic reticulum membrane. The restriction of different GLUT transporters to different tissues is an important feature in the metabolism and homeostasis of glucose.

Competition **Competition** is a property closely related to specificity. A transporter may move several members of a related group of substrates, but those substrates will compete with each other for the binding sites on the transport proteins. For example, the GLUT transporters move the family of hexose sugars, but each different GLUT transporter has a "preference" for one or more hexoses, based on its binding affinity.

The results of an experiment demonstrating competition are shown in Figure 5-16 ■. In the initial part of this graph, GLUT transporters are moving galactose into a cell at a constant rate. At the arrow, glucose is added to the extracellular fluid. Because the transporters have a higher binding affinity for glucose than for galactose, glucose will replace galactose on some of the binding sites. When galactose is displaced from some of the

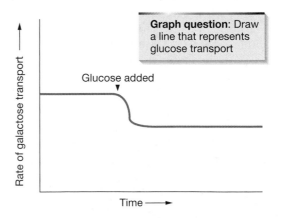

■ **Figure 5-16** **Graph of competition** Glucose and galactose are transported into cells on the same membrane protein. The carrier has a higher affinity for glucose than for galactose. When glucose is added to a system that has been transporting galactose, some galactose molecules are displaced as glucose binds to the carriers, and the rate of galactose transport decreases.

(a) Glucose transport

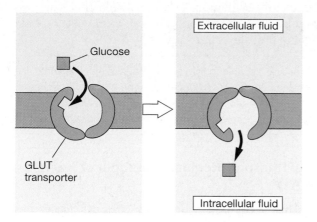

(b) Competitive inhibition by maltose

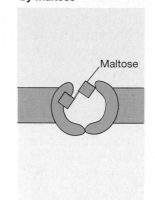

■ **Figure 5-17 Competitive inhibition** (a) Glucose is transported into cells on the GLUT transporter. (b) Maltose binds to the GLUT protein, preventing glucose transport, but not being transported itself. Maltose therefore is a competitive inhibitor of glucose transport.

transporters, the rate of galactose transport into the cell decreases. In this example, glucose and galactose are competing for the transporter.

Sometimes the competing molecule is not transported but merely blocks the transport of another substrate. In this case, the competing molecule is a *competitive inhibitor* [∞ p. 83]. In the glucose transport system, the disaccharide maltose is an inhibitor (Fig. 5-17 ■). It competes with glucose for the binding site, but, once bound, is too large to be moved across the membrane.

Competition between transported substrates has been put to good use in medicine. For example, gout is a disease caused by elevated uric acid in the plasma. One method of decreasing plasma uric acid is to enhance its excretion in the urine. Normally, the kidney's organic acid transporter reclaims uric acid from the urine. However, if an organic acid called probenecid is administered, the carrier binds to probenecid instead of to uric acid, preventing the reabsorption of uric acid. As a result, more uric acid leaves the body in the urine, lowering the uric acid concentration in the plasma.

Saturation   **Saturation** occurs when a group of membrane carriers are transporting substrate at their maximum rate. The rate of carrier activity depends on both the substrate concentration and the number of carrier molecules, just as the rate of enzyme activity depends on both enzyme and substrate concentration [∞ p. 84]. For a fixed number of carriers, however, as substrate concentration increases, the rate increases up to the point at which all carriers are filled with substrate. At this point, the carriers are said to be saturated with substrate. They are working at their maximum rate, and a further increase in substrate concentration will have no effect. Figure 5-18 ■ shows saturation represented graphically.

For an analogy, think of the carriers as doors into a concert hall. Each door has a maximum number of people that it can allow to enter the hall in a given period of time. Let us say that all the doors will allow 100 people per minute to enter the hall. This is the maximum transport rate, also called the *transport maximum*. When the

concert hall is empty, three maintenance people enter the doors every hour. The transport rate is 3 people/60 minutes, or 0.05 people/minute, well under the maximum. For the local dance recital, about 50 people per minute go through the doors, still well under the maximum. But when the most popular rock group of the day appears in concert, thousands of people gather outside. When the doors open, thousands of people are clamoring to get in. But the doors will only allow 100 people/minute into the hall. They are working at the maximum rate, so it does not matter whether there are 1000 or 3000 people trying to get in. The transport rate is saturated at 100 people per minute.

How can cells avoid having transport reach saturation? One way is to increase the number of carriers in the membrane. This would be like opening more doors into the concert hall. Under some circumstances, cells are able to insert additional carriers into their membranes. Under other circumstances, a cell may withdraw carriers to decrease movement of a molecule into or out of the cell.

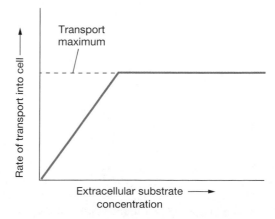

■ **Figure 5-18 Graph showing saturation of carrier-mediated transport** With a limited number of carrier proteins in the membrane, the rate of transport is proportional to the concentration of substrate available until all of the available carriers are working at maximum capacity. The maximum rate of transport, or transport maximum, is an indication that the carriers have been saturated with substrate.

All forms of mediated transport show saturation, specificity, and competition. But they differ in other ways. Passive mediated transport, better known as facilitated diffusion, requires no input of energy. Active transport requires energy input from ATP, either directly or indirectly.

Facilitated Diffusion Is Diffusion That Uses Membrane Proteins

Some polar molecules appear to move into and out of cells by diffusion, even though we know from their chemical properties that they are unable to pass easily through the lipid core of the cell membrane. Most of these polar molecules travel by facilitated diffusion, in which carrier proteins transport them across the cell membrane. Glucose and amino acids are examples of molecules that enter or leave cells with the help of facilitated diffusion carriers.

Facilitated diffusion has the same properties as simple diffusion. The molecules move down their concentration gradient, the process requires no input of energy, and net movement stops at equilibrium, when concentration inside the cell equals concentration outside the cell (Fig. 5-19 ■). Cells can avoid having diffusion reach

equilibrium by keeping the concentration of substrate in the cell low. With glucose, for example, this is accomplished by phosphorylating glucose to glucose 6-phosphate, the first step of glycolysis [∞ p. 90].

✔ Liver cells are able to convert glycogen to glucose and make intracellular glucose concentrations higher than the extracellular glucose concentration. What do you think happens to facilitated diffusion of glucose when this occurs?

Active Transport Requires the Input of Energy from ATP

Active transport is a process that moves molecules *against* their concentration gradient, that is, from areas of lower concentration to areas of higher concentration. Rather than creating an equilibrium state, where the concentration of the molecule is equal throughout the system, active transport creates a state of *dis*equilibrium by making concentration differences more pronounced.

Moving molecules against their concentration gradient requires the input of outside energy, just as pushing a ball up a hill requires energy [∞ Fig. 4-2, p. 76].

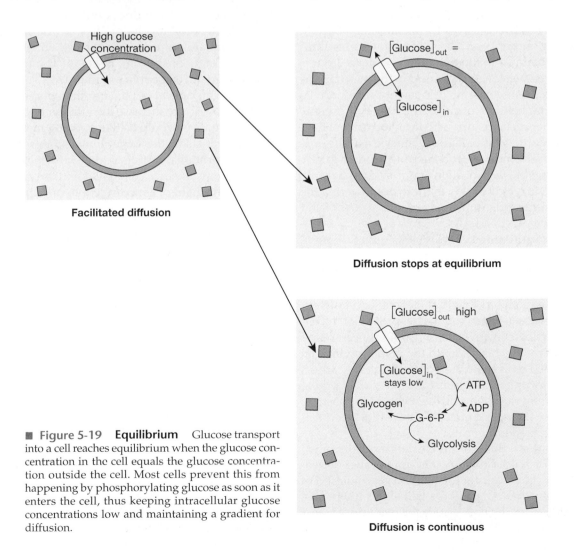

Facilitated diffusion

Diffusion stops at equilibrium

Diffusion is continuous

■ **Figure 5-19** **Equilibrium** Glucose transport into a cell reaches equilibrium when the glucose concentration in the cell equals the glucose concentration outside the cell. Most cells prevent this from happening by phosphorylating glucose as soon as it enters the cell, thus keeping intracellular glucose concentrations low and maintaining a gradient for diffusion.

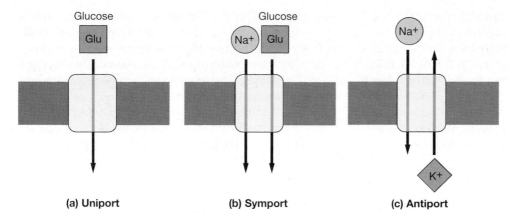

(a) Uniport   (b) Symport   (c) Antiport

■ **Figure 5-20   Carrier-mediated transport**   A carrier that transports one substrate is called a uniporter. When a carrier cotransports two or more molecules in the same direction across the cell membrane, the process is called symport. In antiport, molecules are exchanged in opposite directions, with one moving into the cell as the other moves out.

The energy for active transport comes directly or indirectly from the high-energy phosphate bond of ATP. In biological systems, active transport always involves the movement of molecules across cell membranes.

Some active transport carriers move only one kind of molecule, a process known as **uniport** (Fig. 5-20a ■). However, it is common to find carriers that move two or even three different kinds of molecules. A protein that moves more than one kind of molecule at one time is called a **co-transporter.** If the molecules are moving in the same direction, whether into or out of the cell, the process is called **symport** [*sym-*, together + *portare*, to carry]. If the molecules are being carried in different directions, the process is called **antiport** [*anti*, opposite + *portare*, to carry]. Symport and antiport are shown in Figure 5-20b, c ■.

Active transport can be divided into primary active transport and secondary active transport. In **primary (direct) active transport,** the energy for transport comes from the high-energy phosphate bond of ATP. If the energy to push molecules against their gradient comes from potential energy stored in a concentration gradient [∞ p. 75], the process is called **secondary (indirect) active transport.** All secondary active transport ultimately depends on primary active transport because the concentration gradients that drive it are created using energy from ATP.

The mechanism for active transport carriers appears to be similar to that for facilitated diffusion carriers. The substrate(s) binds to the protein carrier, which then changes conformation, releasing the substrate into the opposite compartment. Active transport differs from facilitated diffusion because these conformation changes require energy input.

Primary Active Transport   **Primary active transport** uses ATP as its energy source. Consequently, many of the primary active transporters are known as **ATPases.** You will recall from Chapter 4 that the suffix *-ase* signifies an enzyme, and the stem (ATP) is the substrate upon which the enzyme is acting [∞ p. 81]. ATPases hydrolyze ATP to ADP and inorganic phosphate (P_i), releasing usable energy in the process. Most of the ATPases you will encounter in your study of physiology are listed in Table 5-3. ATPases are sometimes called *pumps*, as in the sodium-potassium pump.

The sodium-potassium pump, or **Na^+-K^+-ATPase**, is probably the single most important transport protein in animal cells, because it maintains the concentration gradients of Na^+ and K^+ across the cell membrane. The transporter is arranged in the cell membrane so that it pumps 3 Na^+ out of the cell and 2 K^+ into the cell for each ATP consumed (Fig. 5-21 ■). In some cells, the energy needed to move these ions uses 30% of all the ATP

TABLE 5-3 Primary Active Transporters

Names	Type of Transport
Na^+-K^+-ATPase or sodium-potassium pump	Antiport
Ca^{2+}-ATPase	Uniport
H^+-ATPase or proton pump	Uniport
H^+-K^+-ATPase	Antiport

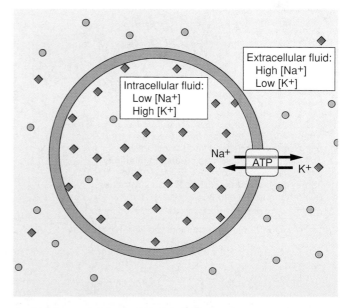

■ **Figure 5-21   The Na^+-K^+-ATPase**   The Na^+-K^+-ATPase pumps Na^+ out of the cell and K^+ into the cell against their concentration gradients by utilizing the energy stored in the high-energy phosphate bond of ATP.

produced by the cell. The current model of how the Na^+-K^+-ATPase works is illustrated in Figure 5-22 ∎.

The sodium concentration gradient is a source of potential energy that the cell can harness for other functions. For example, nerve cells use the sodium gradient to transmit electrical signals. Epithelial cells use it to drive the uptake of nutrients, ions, and water. Membrane transporters that use potential energy stored in concentration gradients are called secondary active transporters.

Secondary Active Transport **Secondary active transport** couples the kinetic energy of one molecule moving down its concentration gradient to the movement of another molecule against its concentration gradient. The cotransported molecules may go in the same direction across the membrane (symport) or one may leave the cell in exchange for the other (antiport). The most common secondary active transport systems are driven by the sodium concentration gradient. As Na^+

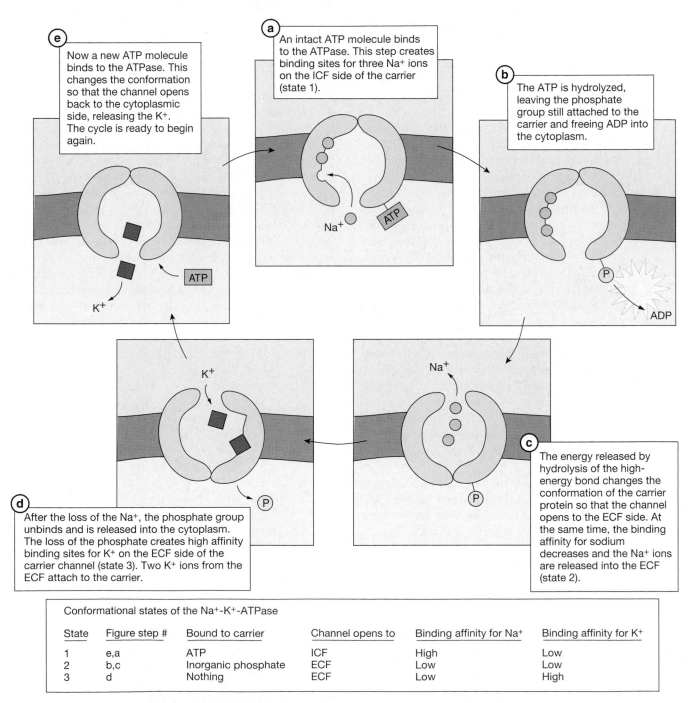

e Now a new ATP molecule binds to the ATPase. This changes the conformation so that the channel opens back to the cytoplasmic side, releasing the K^+. The cycle is ready to begin again.

a An intact ATP molecule binds to the ATPase. This step creates binding sites for three Na^+ ions on the ICF side of the carrier (state 1).

b The ATP is hydrolyzed, leaving the phosphate group still attached to the carrier and freeing ADP into the cytoplasm.

c The energy released by hydrolysis of the high-energy bond changes the conformation of the carrier protein so that the channel opens to the ECF side. At the same time, the binding affinity for sodium decreases and the Na^+ ions are released into the ECF (state 2).

d After the loss of the Na^+, the phosphate group unbinds and is released into the cytoplasm. The loss of the phosphate creates high affinity binding sites for K^+ on the ECF side of the carrier channel (state 3). Two K^+ ions from the ECF attach to the carrier.

Conformational states of the Na^+-K^+-ATPase

State	Figure step #	Bound to carrier	Channel opens to	Binding affinity for Na^+	Binding affinity for K^+
1	e,a	ATP	ICF	High	Low
2	b,c	Inorganic phosphate	ECF	Low	Low
3	d	Nothing	ECF	Low	High

∎ **Figure 5-22   Mechanism of the Na^+-K^+-ATPase**   How do active transporters use the energy from ATP to move molecules against their concentration gradient? This figure presents one model that explains how the Na^+-K^+-ATPase couples ion movement to the release of energy and inorganic phosphate from a molecule of ATP.

moves into the cell, it either brings one or more molecules with it or trades places with molecules exiting the cell. The major Na$^+$-dependent transporters are listed in Table 5-4. Notice that the cotransported molecules may be either other ions or uncharged molecules, such as glucose.

The mechanism of the Na$^+$-glucose symporter is illustrated in Figure 5-23 ■. Notice its similarity to both the facilitated diffusion carrier (Fig. 5-9 ■) and the Na$^+$-K$^+$-ATPase (Fig. 5-22 ■). The Na$^+$ and glucose bind to the carrier on the extracellular fluid side. Sodium binds first and causes a conformational change in the protein that creates a high-affinity binding site for glucose. When glucose binds, the protein changes conformation again and opens its channel to the intracellular fluid side. Sodium is released as it moves down its concentration gradient. The loss of Na$^+$ from the protein changes the binding site for glucose back to a low-affinity site, so glucose is released and follows Na$^+$ into the cytoplasm. The net result is the entry of glucose into the cell against its concentration gradient, coupled to the movement of Na$^+$ into the cell down its concentration gradient.

Why does the body have both a facilitated diffusion GLUT carrier and a Na$^+$-glucose symporter for glucose uptake? All cells use GLUT transporters to bring glucose into cells from the extracellular fluid. In addition, facilitated diffusion carriers can move molecules in either direction across a membrane, depending on the concentration gradient. For example, when blood glucose levels are high, the GLUT transporters on liver cells bring glucose into the cell. During times of fasting, when blood glucose levels fall, the liver cell converts its glycogen stores to glucose. When the glucose concentration inside the liver cell exceeds the glucose concentration in the plasma, glucose leaves the cell on the reversible GLUT transporters. In contrast, the Na$^+$-glucose transporter can move substrate only into the cell because it must follow the Na$^+$ gradient. Consequently, the Na$^+$-glucose transporter is found only on transporting epithelial cells

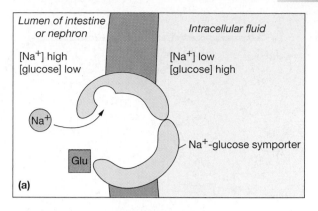

(a)

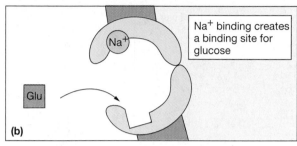

Na$^+$ binding creates a binding site for glucose

(b)

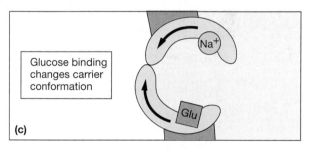

Glucose binding changes carrier conformation

(c)

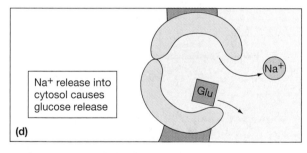

Na$^+$ release into cytosol causes glucose release

(d)

■ **Figure 5-23   Sodium-glucose symporter**   This transporter uses the potential energy of the Na$^+$ concentration gradient to move glucose against its concentration gradient. When the carrier is open to the extracellular fluid, it has a high-affinity binding site for Na$^+$ and a low-affinity site for glucose.

TABLE 5-4 Typical Secondary Active Transporters

Symport Carriers	*Antiport Carriers*
Sodium-dependent transporters	
Na$^+$-glucose	Na$^+$-H$^+$
Na$^+$-amino acids (several types)	Na$^+$-Ca^{2+}
Na$^+$-K$^+$-2 Cl$^-$	
Na$^+$-bile salts (small intestine)	
Na$^+$-choline uptake (nerve cells)	
Na$^+$-neurotransmitter uptake (nerve cells)	
Nonsodium-dependent transporters	
	HCO$_3^-$-Cl$^-$
	H$^+$-K$^+$

that use it to bring glucose into the body from the external environment.

The use of energy to drive secondary active transport is summarized in Figure 5-24 ■. The organism imports energy from the environment in the chemical bonds of nutrients such as glucose. This energy is transferred to the high-energy bonds of ATP through oxidative phosphorylation in the mitochondria [∞ p. 94]. ATP is used to create ion gradients. For example, the Na$^+$-K$^+$-ATPase

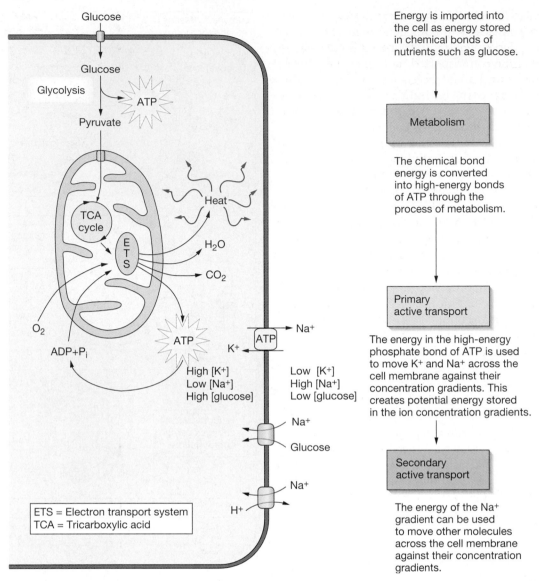

■ **Figure 5-24**   **Energy transfer in living systems**

pushes sodium out of the cell and potassium into it. Now the energy from ATP has been transformed into potential energy stored in the ion concentration gradients. This potential energy can be harnessed by cotransporters such as the Na^+-glucose transporter. The kinetic energy of Na^+ moving into the cell down its concentration gradient is linked to the uphill movement of glucose into the cell. As you study the different systems of the body, you will find these secondary active transporters taking part in many physiological processes.

Vesicles Move Large Molecules Across Membranes

What happens to the many macromolecules that are too large to enter or leave cells through protein channels or membrane transporters? They move in and out of the cell with the aid of bubblelike vesicles created from the cell membrane. Cells use two basic mechanisms to im-

port large molecules and particles: phagocytosis and endocytosis. Phagocytosis can be considered a type of endocytosis, but as scientists learned more about the mechanisms behind the two processes, they decided that phagocytosis was fundamentally different. Material leaves cells by the process known as exocytosis, similar to endocytosis in reverse.

Phagocytosis   If you studied *Amoeba* in your biology laboratory, you may have watched these one-cell creatures ingest their food by surrounding it and enclosing it into a vesicle that is brought into the cytoplasm. **Phagocytosis** is the process by which a cell engulfs a particle into a vesicle [*phagein*, to eat + *cyte*, cell + *-sis*, process]. In humans, this process occurs only in certain types of white blood cells called *phagocytes* that specialize in "eating" bacteria and other foreign particles (Fig. 5-25 ■). When a phagocyte encounters a bacterium, the bacterium binds to the phagocyte membrane ①. The

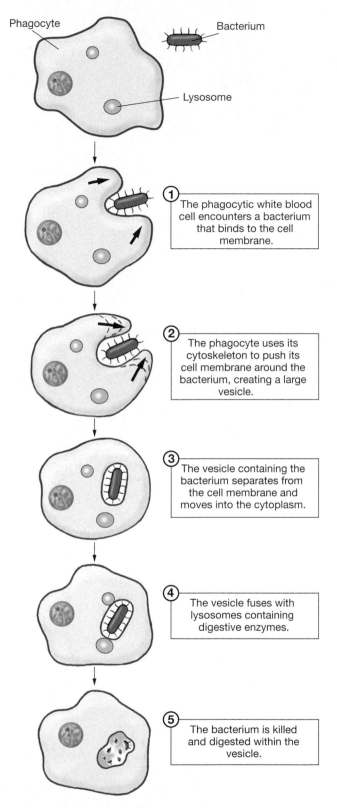

Phagocyte

Bacterium

Lysosome

① The phagocytic white blood cell encounters a bacterium that binds to the cell membrane.

② The phagocyte uses its cytoskeleton to push its cell membrane around the bacterium, creating a large vesicle.

③ The vesicle containing the bacterium separates from the cell membrane and moves into the cytoplasm.

④ The vesicle fuses with lysosomes containing digestive enzymes.

⑤ The bacterium is killed and digested within the vesicle.

■ **Figure 5-25   Phagocytosis**

phagocyte then uses the fibers of its cytoskeleton to push its cell membrane out around the particle so that the bacterium becomes completely engulfed ②. Once the bacterium is surrounded, the arms of the membrane fuse, leaving the bacterium contained within a membrane-

bound **vesicle** ③. The vesicle pinches off from the cell membrane and moves to the interior ④, where it fuses with a lysosome [∞ p. 53], so that digestive enzymes in the lysosome can destroy the bacterium ⑤. Phagocytosis requires energy from ATP for the movement of the cytoskeleton and for the intracellular transport of the vesicles.

Endocytosis   Another means by which molecules or particles can move into cells is called **endocytosis.** It differs from phagocytosis in that the membrane surface indents rather than pushes out, and the vesicles formed are much smaller. In addition, endocytosis is *constitutive,* that is, it is an essential function that is always taking place. Phagocytosis must be triggered by the presence of a substance to be ingested.

Endocytosis is an active process that requires ATP. It can be nonselective, allowing extracellular fluid to enter the cell (a process called **pinocytosis** [*pino-,* drink]) or it can be highly selective, allowing only specific molecules to enter the cell. Two types of endocytosis require a ligand to bind to a membrane receptor: receptor-mediated endocytosis and potocytosis.

Receptor-mediated endocytosis   **Receptor-mediated endocytosis** takes place in regions of the cell membrane known as *clathrin-coated pits* (Fig. 5-26 ■). In the first step of the process, membrane proteins recognize and bind to the substances (ligands) that will be brought into the cell ①. The receptor-ligand complex migrates along the cell surface until it encounters a coated pit, where the cytoplasmic side of the membrane has high concentrations of the protein clathrin ②. Once the receptor-ligand complex is in the coated pit, the membrane draws inward, or *invaginates* ③, then pinches off from the cell membrane, and becomes a cytoplasmic vesicle ④. The clathrin molecules are released and recycle back to the membrane. In the vesicle, the receptor and ligand part company ⑤. The ligand is sent to a lysosome if it is to be destroyed or to the Golgi apparatus for processing ⑥. The section of vesicle with the receptors ⑦ may then return to the cell membrane ⑧, allowing the receptors to be reused ⑨. The return of the vesicle membrane to the cell surface is a process known as **membrane recycling.**

Receptor-mediated endocytosis transports a variety of substances into the cell including protein hormones, growth factors, antibodies, and plasma proteins that serve as carriers for iron and cholesterol. Abnormalities in receptor-mediated removal of cholesterol from the blood are associated with elevated plasma cholesterol levels and cardiovascular disease.

Potocytosis and caveolae   **Potocytosis** is distinguished from receptor-mediated endocytosis because it uses **caveolae** ("little caves") rather than clathrin-coated pits to bring receptor-bound molecules into the cell. Caveolae are small, flask-shaped invaginations of the cell membrane

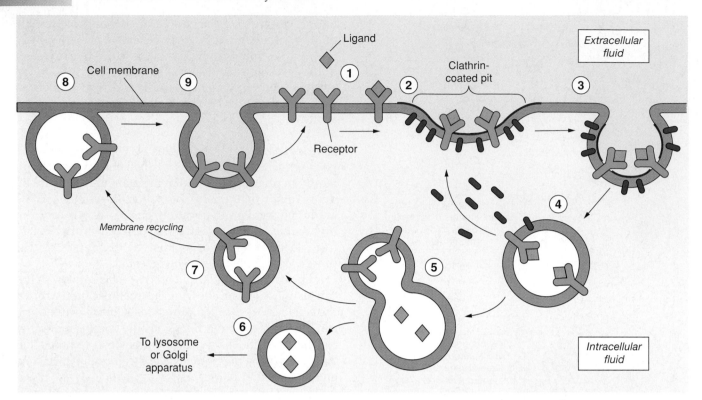

■ **Figure 5-26   Receptor-mediated endocytosis and exocytosis**   ① Protein receptors on the cell membrane bind to their specific ligand. ② The receptor-ligand complex migrates along the surface of the cell membrane to a clathrin-coated pit. ③ The coated pit invaginates to create a small membrane-bound vesicle. ④ The vesicle separates from the cell membrane. The clathrin molecules return to the cell membrane. ⑤ The vesicles move into the cell and the receptors and ligands separate, sometimes forming two distinct vesicles. ⑥ The vesicles with ligands may be processed through lysosomes or the Golgi apparatus. ⑦ Vesicles with the receptors move back to the cell membrane and fuse with it ⑧, a process called membrane recycling. ⑨ When the fused section opens during exocytosis, the vesicle membrane becomes part of the cell membrane.

that contain a variety of membrane proteins, including a unique protein named *caveolin*. The receptors in caveolae are linked to the membrane by a glycolipid anchor and are not integral proteins.

Caveolae have three functions: to concentrate and internalize small molecules for potocytosis, to help in the transfer of macromolecules across the capillary endothelium (see Fig. 5-29 ■), and to participate in cell-to-cell signaling. Scientists are currently trying to discover more details about the role of caveolae in these processes.

Exocytosis   Exocytosis is the opposite of endocytosis. In exocytosis, intracellular vesicles move to the cell membrane, fuse with it (Figure 5-26 ■, step 8), and then release their contents to the extracellular fluid (step 9). Cells use exocytosis to export large lipophobic molecules, such as proteins synthesized within the cell, and to get rid of wastes left in lysosomes from intracellular digestion.

The process by which the cell and vesicle membranes fuse is similar in a variety of cell types, from neurons to endocrine cells. Exocytosis involves a series of proteins with names such as SNARE, SNAP, rab, and

LDL—The Lethal Lipoprotein   Cholesterol is a molecule that is essential for membrane structure and for the synthesis of steroid hormones, such as the sex hormones. On the other hand, elevated cholesterol can lead to heart disease. Why is it that some people have too much cholesterol in their blood (*hypercholesterolemia*)? Cholesterol is an insoluble nonpolar molecule, which means it must be bound to a lipoprotein carrier molecule for transport in the blood. There are several forms of the carrier, the most common of which is *low-density lipoprotein* (*LDL*). LDL-cholesterol is taken into cells by endocytosis when LDL receptors on the cell surface bind to the protein portion of the molecule. People who inherit a genetic defect that decreases the number of LDL receptors on their cell membranes cannot transport cholesterol normally into their cells. As a result, LDL-cholesterol remains in the plasma. Abnormally high blood levels of LDL-cholesterol predispose these people to the development of **atherosclerosis,** commonly known as hardening of the arteries [*atheroma,* a tumor + *skleros,* hard + -*sis,* condition]. In this condition, the accumulation of cholesterol in blood vessels blocks blood flow and contributes to heart attacks.

VAMP. Although the details are still not fully understood, the process usually begins with an increase in intracellular Ca^{2+} concentration. The Ca^{2+} ions interact with one of the proteins, causing the secretory vesicles to attach to the membrane, then fuse with it. The fused area opens, and the vesicle membrane becomes part of the cell membrane as the vesicle contents diffuse into the extracellular space. Exocytosis, like endocytosis, requires energy in the form of ATP.

Exocytosis takes place continually in some cells but only intermittently in others. For example, goblet cells [∞ p. 61] in the intestine continually release mucus by exocytosis, and fibroblasts in connective tissue release collagen [∞ p. 63]. In many endocrine cells, hormones are stored in secretory vesicles in the cytoplasm and released in response to a signal from outside the cell. Exocytosis is also the means by which membrane proteins can be inserted into the cell membrane, as shown in Figure 5-26 ■. You will encounter many examples of exocytosis in your study of physiology.

✔ How does phagocytosis differ from endocytosis?

✔ How does receptor-mediated endocytosis differ from potocytosis?

Molecules Move Across Epithelia Using Passive and Active Transport

The transport processes described in the previous sections all deal with the movement of molecules across a single membrane, that of the cell. However, all molecules entering and leaving the body across an epithelium must cross two cell membranes. Molecules cross the first membrane when they move into an epithelial cell from the external environment and the second when they leave the epithelial cell to enter the extracellular fluid. Movement across epithelial cells, *transepithelial transport*, uses a combination of active and passive transport.

...continued from page 115

The "sweat test" that Daniel will undergo analyzes levels of salt (NaCl) in sweat. Sweat—a mixture of ions and water—is secreted into sweat ducts by the epithelial cells of sweat glands. As sweat moves toward the skin's surface through the sweat ducts, the CFTR channel allows chloride ions to move out of the sweat and back into the epithelial cells. Na^+ is also reabsorbed, following the Cl^-. This epithelium is not permeable to water, so normal reabsorption of NaCl creates sweat with a low salt content. However, without CFTR channels in the sweat gland epithelium, salt is not reabsorbed. "Normally, sweat contains about 120 millimoles of salt per liter," says Beryl Rosenstein, M.D., director of the Cystic Fibrosis Center at the Johns Hopkins Medical Institutions. "In cystic fibrosis, salt concentrations in the sweat can be four times the normal amount."

Question 2: Based on the information given, is the CFTR channel on the apical or basolateral surface of the sweat gland epithelium?

The transporting epithelia of the intestine and kidney are specialized to transport molecules into and out of the body. Recall that epithelial cells are connected to each other by adhesive junctions and tight junctions [∞ p. 55]. These act as barriers to minimize the unregulated diffusion of material through the space between the cells. They also mark the division of the cell membrane into two regions, or poles. The surface of the epithelial cell that faces the lumen of an organ is called the **apical** (or *mucosal*) membrane (Fig. 5-27 ■). It is often folded into microvilli that increase its surface area. The sides of the cell below the tight junctions and the surface of the cell membrane in contact with the extracellular fluid are collectively called the **basolateral** (or *serosal*) membrane.

Transporting epithelial cells are said to be *polarized* because their apical and basolateral membranes have very different properties. Some proteins, such as the Na^+-K^+-ATPase, are found on the basolateral membrane,

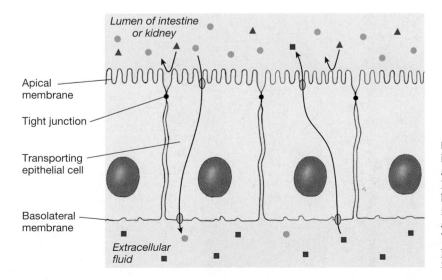

Apical membrane

Tight junction

Transporting epithelial cell

Basolateral membrane

Lumen of intestine or kidney

Extracellular fluid

■ **Figure 5-27 Polarized cells of transporting epithelia** The cells of transporting epithelia are polarized into an apical membrane that faces the lumen of the organ lined by the epithelium and a basolateral membrane that faces the extracellular fluid. These two membranes have different proteins and, consequently, different transporting properties. The polarization of cells allows directional transport of material from lumen to extracellular fluid or from the extracellular fluid into the lumen.

whereas others, like the Na⁺-glucose symporter, are found only on the apical membrane. This polarized distribution of transporters results in the one-way movement of certain molecules across the epithelium.

The cells of transporting epithelia can alter their permeability by selectively inserting or withdrawing membrane proteins. Transporters pulled out of the membrane may be destroyed in lysosomes or they may be stored in vesicles inside the cell, ready to be reinserted into the membrane in response to a signal (another example of membrane recycling).

The transepithelial movement of molecules requires that the molecules first enter the epithelial cell and then leave it to enter the extracellular fluid. If the molecule can be transported through a protein channel or on transport carriers, then the two-step process usually has one uphill step that requires energy and one downhill step in which the molecule moves passively down its concentration gradient. Molecules that are too large to be moved by membrane proteins can be transported across the cell in vesicles.

Transepithelial Transport Using Membrane Proteins
The movement of glucose from the lumen of the kidney tubule or intestine to the extracellular fluid is an important example of directional movement across a transporting epithelium. Transepithelial movement of glucose involves three different transport systems: the secondary active transport of glucose with Na⁺ from the lumen into the cell at the apical membrane, followed by the active transport of Na⁺ and the facilitated diffusion of glucose out of the cell and into the extracellular fluid on the basolateral side.

Figure 5-28 ■ shows the process in detail. The glucose concentration in the transporting epithelium cell is higher than the glucose concentration in either the extracellular fluid or the lumen of the intestine ①. To move glucose from the lumen into the cell is an uphill process and requires the input of energy—in this case, the energy stored in the Na⁺ concentration gradient. Sodium ions in the lumen bind to the symporter, as previously described, and bring glucose with them into the cell ②. The energy needed to move glucose against its concentration gradient comes from the kinetic energy of Na⁺ moving down its gradient.

Once glucose is in the cell, it leaves by moving down its concentration gradient through the facilitated diffusion GLUT transporter in the basolateral membrane ③. The Na⁺ is pumped out of the cell on the basolateral side using the Na⁺-K⁺-ATPase. Sodium is more concentrated in the extracellular fluid than in the cell, so this step requires energy provided by ATP.

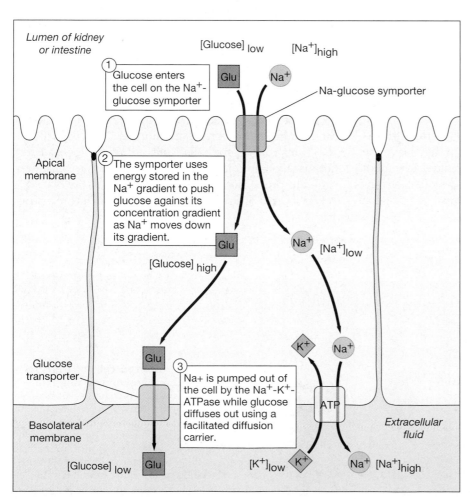

■ Figure 5-28   **Transepithelial transport of glucose**

The removal of Na^+ is essential if glucose is to continue to be absorbed, because the Na^+-glucose symporter depends on low intracellular concentrations of Na^+. If the basolateral Na^+-K^+-ATPase is poisoned with *ouabain* ("wah-bane"— a compound related to the heart drug digitalis), Na^+ that enters the cell cannot be pumped out. The Na^+ concentration inside the cell then increases until it is equal to that outside the cell. Without a sodium gradient, there is no energy source to run the Na^+-glucose symporter, and the movement of glucose across the epithelium stops.

Transepithelial transport can use ion movement through channels in addition to carrier-mediated transport. For example, the apical membrane of a transporting epithelium may have the Na^+-K^+-$2\,Cl^-$ symporter to bring K^+ into the cell against its concentration gradient, as Na^+ and Cl^- move in down their gradient. Because the K^+ concentration inside the cell is higher than in the extracellular fluid, the K^+ brought in by cotransport can move out of the cell on the basolateral side through open K^+ leak channels. The Na^+ leaves by the Na^+-K^+-ATPase. By this simple mechanism the body can absorb Na^+ and K^+ at the same time from the lumen of the intestine or the kidney.

✔ Ouabain, an inhibitor of the Na^+-K^+-ATPase, cannot pass through cell membranes. What would happen to the transepithelial glucose transport shown in Figure 5-28 ■ if ouabain were applied to the apical side of the epithelium? To the basolateral side of the epithelium?

✔ Which GLUT transporter is illustrated in Figure 5-28 ■? (Hint: See the box on GLUT transporters.)

Transcytosis and Vesicular Transport   Some molecules, such as proteins, are too large to cross epithelia on membrane transporters. Instead they are moved across epithelia by **transcytosis,** which is a combination of endocytosis, vesicular transport across the cell, and exocytosis (Fig. 5-29 ■). In this process, the molecule is brought into the cell using receptor-mediated endocytosis or potocytosis. The resulting vesicle attaches to microtubules in the cytoskeleton and is transported across the cell by a process known as **vesicular transport.** At the opposite side of the epithelium, the contents of the vesicle are expelled into the extracellular fluid by exocytosis.

Transcytosis makes it possible for large proteins to move across an epithelium and remain intact. It is the means by which infants absorb maternal antibodies in breast milk. The antibodies are absorbed on the apical surface of the infant's intestinal epithelium and then released into the extracellular fluid.

Caveolar transcytosis of macromolecules may turn out to play a significant role in the development of certain diseases. For example, cholesterol-lipoproteins are transported across blood vessel endothelium using caveolae. Problems with this transport may contribute to the development of atherosclerosis, the accumulation of cho-

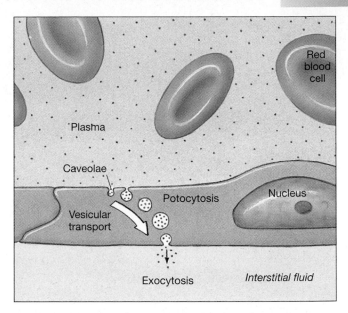

■ **Figure 5-29   Transcytosis across the capillary endothelium**   Receptors in caveolae on the plasma side of the endothelial cell bind to molecules. The caveolae form vesicles by potocytosis. The vesicles move across the cell with help from the cytoskeleton. Vesicle contents are released into the interstitial fluid by exocytosis.

lesterol in blood vessels that blocks blood flow and contributes to heart attacks.

✔ If a poison that disassembles microtubules is applied to a capillary endothelial cell, what happens to transcytosis?

▶ THE DISTRIBUTION OF WATER AND SOLUTES IN THE BODY

The distribution of solutes in the body depends on whether a substance can cross the cell membrane by simple diffusion or whether it needs a membrane protein or vesicle to move across. Cell membranes are **selectively permeable,** that is, a cell can select which molecules will enter and which will leave by varying the lipid composition and protein transporters of its membrane. In the remainder of this chapter, we will discuss how the selective permeability of cell membranes creates a body in which the intracellular and extracellular compartments are chemically and electrically different but have the same total concentration of solutes.

Living Cells Use Energy to Maintain a State of Chemical Disequilibrium

Earlier in this chapter, you learned that diffusion proceeds until the concentration of the diffusing molecule is equal in all parts of the system, a condition known as *chemical equilibrium.* However, few substances in the body are maintained in a state of chemical equilibrium. This is partly because the cell membrane and the capillary

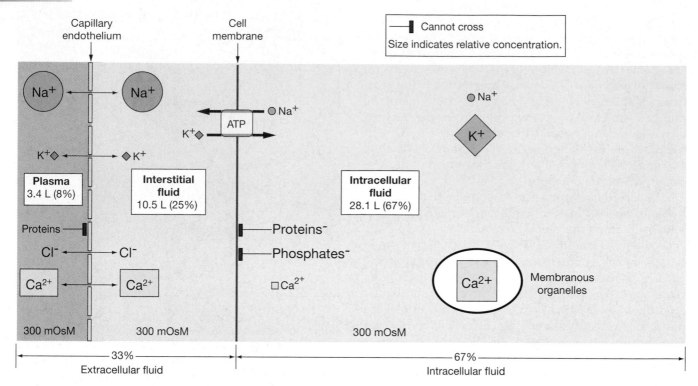

■ **Figure 5-30   Distribution of solutes in the body fluid compartments**   The compartments of the body are in a state of chemical disequilibrium. Intracellular fluid has high protein, phosphate, and K^+ concentrations but low Na^+, Cl^-, and free Ca^{2+}. In contrast, extracellular fluid K^+ concentration is low, but Na^+, Ca^{2+}, and Cl^- concentrations are high. In the extracellular space, proteins are found only in the plasma. Water moves freely between compartments, so that the body stays in osmotic equilibrium. This figure shows the compartment volumes for a 70-kg man.

endothelium act as selective barriers that prevent solutes from diffusing freely throughout the body. In addition, the active transport of solutes helps create or maintain differences in solute concentrations.

The chemical disequilibrium of the body is illustrated in Figure 5-30 ■. Phosphate ion and K^+ concentrations are higher inside the cell, while Na^+ and Cl^- are concentrated in the extracellular fluid. Proteins are found primarily in intracellular fluid and in the plasma. Calcium is more concentrated in the extracellular fluid than in the cytosol, although many cells store calcium ions inside membranous organelles such as the endoplasmic reticulum and mitochondria.

Only the constant input of energy keeps the body in this state of chemical disequilibrium, with active transporters returning molecules to compartments they have left by diffusion. For example, the Na^+-K^+-ATPase returns K^+ that leaks out of the cell to the cytoplasm and pumps Na^+ that leaks into the cell back to the extracellular fluid. The concentration differences of chemical disequilibrium are a hallmark of the living organism. If the cells die and cannot use energy, they obey the second law of thermodynamics [∞ p. 76] and return to a state of randomness.

The Body Is Mostly Water

Water is the most important molecule in the human body. It is the solvent for all living matter and moves freely between the cells and the extracellular fluid. If we look at the relative volumes of the body compartments (Fig. 5-30 ■), we find that the intracellular compartment contains about 67% of the body's water. The remaining 33% is split between the interstitial fluid (which contains three-quarters of the extracellular water) and the plasma (which contains about one-quarter of the extracellular water).

How much water is in the body? Because one individual differs from the next, there is no single answer. However, in human physiology we often speak of standard values for physiological functions, based on "the 70-kg man." These standard values are derived from data obtained by studying 21-year-old white males who weighed 70 kg. Thus, when we speak of standard or average values in physiology, remember that these numbers need to be adjusted for an individual's age, gender, weight, and ethnic origin.

Estimating Body Water   Clinicians estimate a person's fluid loss in dehydration by equating weight loss to water loss. Because 1 liter of pure water weighs 1 kilogram, a decrease in body weight of 1 kilogram is considered the equivalent of loss of 1 liter of body fluid. A baby with diarrhea can easily be weighed to estimate its fluid loss. A decrease of 1.1 pounds (0.5 kg) of body weight is assumed to mean loss of 500 mL of fluid. This calculation provides a quick estimate of how much fluid needs to be replaced.

TABLE 5-5 Water Content as Percentage of Total Body Weight by Age and Sex

Age	Male	Female
Infant	65%	65%
1–9	62%	62%
10–16	59%	57%
17–39	61%	51%
40–59	55%	47%
60+	52%	46%

Adapted from Edelman and Leibman, *Am. J. Med.* 27: 256–277, 1959.

The "average" 70-kilogram, or 154-pound, male has 60% of his total body weight, or 42 kg (92.4 lbs), in the form of water. Each kilogram of water equals one liter, so his *total body water* is 42 liters. This is the equivalent of fourteen 3-liter soft drink bottles! Women have less water per kilogram of body weight than men because women have more adipose tissue. Look back at Figure 3-23 [∞ p. 65] and note how the large fat droplets in adipose tissue occupy most of the cell, displacing the more aqueous cytoplasm. Age also influences body water content. Infants have relatively more water than adults, and water content decreases as people grow older than 60.

Table 5-5 shows water content as a percentage of total body weight in people of various ages and gender. In clinical practice, it is necessary to allow for the variability of body water content when prescribing drugs. Because women and older people have less body water, they will have a higher concentration of a drug in the plasma than young men will if all are given an equal dose per kilogram of body weight.

✔ A mother brings her baby to the Emergency Room because he has had diarrhea and vomiting for two days. The staff weighs the baby and finds that he has lost 2 pounds. If you assume that all the weight loss is water loss, what volume of water has the baby lost? (2.2 pounds = 1 kilogram)

The Body Is in Osmotic Equilibrium

Water is essentially the only molecule that moves freely between cells and the extracellular fluid and thus can reach a state of equilibrium. The equal distribution of water across body compartments is known as **osmotic equilibrium.** When water moves across a semipermeable membrane in response to a concentration gradient, the process is called **osmosis.** In osmosis, water always moves into the compartment with more concentrated solute.

Look at the example shown in Figure 5-31a ■, where two compartments of equal volume are separated by a membrane that is freely permeable to water but does not allow glucose to cross. Compartment A is partly filled with a dilute glucose solution, while compartment B is filled with an equal volume of concentrated glucose solution. This creates a concentration gradient for both glucose and water. However, the membrane is not permeable to glucose, so glucose cannot diffuse to equalize its distribution. Water, on the other hand, can cross the membrane freely. It will move by osmosis from compartment A, which contains the dilute glucose solution, to compartment B, which contains the more concentrated glucose solution. Thus, water moves to dilute the region of more concentrated solute. Osmosis will continue until the glucose concentrations in the two compartments are equal (Fig. 5-31b ■). At this point the system has reached osmotic equilibrium, and net water movement stops.

The same situation exists in the body as in the example just discussed. If you drink a glass of water, the water is absorbed through the intestinal epithelium and enters the extracellular fluid. The result is the dilution of solutes in the extracellular fluid. Now the cells have a higher solute concentration than the extracellular fluid. Because there is a concentration gradient, water flows by osmosis across the cell membranes into the cells. Cell volume increases and concentration decreases until the extracellular and intracellular fluid compartments are at equilibrium.

Some people like to think of osmosis as the diffusion of water because water moves from an area of higher *water* concentration to an area of lower *water* concentration.

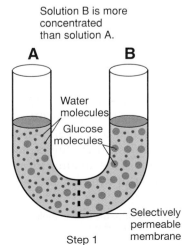

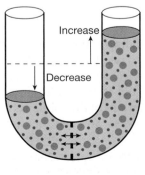

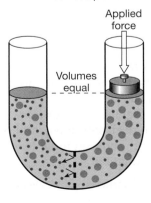

Solution B is more concentrated than solution A.

A   B

Water molecules
Glucose molecules

Selectively permeable membrane

Step 1

Water moves by osmosis into the more concentrated solution until concentrations are equal.

Increase

Decrease

Step 2a

Pressure exerted = osmotic pressure

Applied force

Volumes equal

Step 2b

■ **Figure 5-31 Osmosis and osmotic pressure** (a) Two compartments are separated by a membrane that is permeable to water but not to glucose. (b) Water moves by osmosis until the concentrations of the two compartments are equal. (c) The pressure that must be applied to the solution in compartment B to exactly oppose osmosis is the osmotic pressure of solution B.

particle (ion or molecule) can enter the cell, we call it a **penetrating solute.** We call particles that cannot cross the cell membrane **nonpenetrating solutes.**

The most important nonpenetrating solute in physiology is NaCl. If a cell is placed in a solution of NaCl, the Na^+ and Cl^- will not cross the membrane into the cell. (In reality, a few Na^+ ions may leak across, but they are immediately transported back to the extracellular fluid by the Na^+-K^+-ATPase. NaCl is therefore considered a *functionally* nonpenetrating solute.) If a *saline* solution of water and NaCl is put into the plasma, the Na^+ and Cl^- will stay in the extracellular fluid because they cannot enter the cells.

A key point to remember is that tonicity tells you NOTHING about the osmolarity of the solution. This is where tonicity becomes tricky. Suppose you know the composition and osmolarity of a solution. How can you figure out the tonicity of the solution without putting a cell in it? *The key lies in knowing the relative concentrations of nonpenetrating solutes in the cell and in the solution.*

To determine tonicity of a solution relative to a cell, you must consider the relative concentrations of nonpenetrating solutes in the solution and in the cell. Here are the rules:

1. *If the cell has a higher concentration of nonpenetrating solutes than the solution,* there will be net movement of water into the cell. The cell swells, and the solution is *hypotonic.*
2. *If the cell has a lower concentration of nonpenetrating solutes than the solution,* there will be net movement of water out of the cell. The cell shrinks, and the solution is *hypertonic.*
3. *If the concentrations of nonpenetrating solutes are the same in the cell and the solution,* there is no net movement of water at equilibrium. The solution is *isotonic* to the cell.

The starting osmolarities of the cell and solution cannot be used to determine whether the solution is hypotonic, hypertonic, or isotonic. And the presence of penetrating solutes must be ignored when determining tonicity, because these solutes move freely into the cell, as if the cell membrane did not exist. To understand why this is true, look at the following example.

In Figure 5-32a ■, the cell contains 6 particles of nonpenetrating solute in 1 liter of volume. The 1 liter of solution also contains 6 particles of solute per liter: 3 nonpenetrating particles and 3 penetrating particles. Because cell and solution have the same concentrations (6 particles per liter), they are isosmotic.

To determine tonicity of the solution, compare the concentrations of nonpenetrating solutes in the solution and cell. The solution has 3 nonpenetrating particles per liter, but the cell has 6 nonpenetrating particles per liter. By rule 1 above, this means that water will move from the solution into the cell. The net movement of water

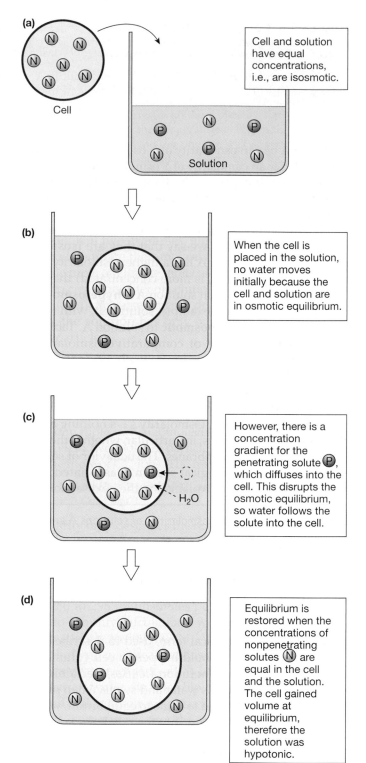

(a) Cell and solution have equal concentrations, i.e., are isosmotic.

Cell

Solution

(b) When the cell is placed in the solution, no water moves initially because the cell and solution are in osmotic equilibrium.

(c) However, there is a concentration gradient for the penetrating solute (P), which diffuses into the cell. This disrupts the osmotic equilibrium, so water follows the solute into the cell.

H_2O

(d) Equilibrium is restored when the concentrations of nonpenetrating solutes (N) are equal in the cell and the solution. The cell gained volume at equilibrium, therefore the solution was hypotonic.

Figure question: Using the same cell, give the relative osmolarity and the tonicity of the following solutions, if the solution has (a) 6 nonpenetrating particles, (b) 6 nonpenetrating and 3 penetrating particles, or (c) 3 nonpenetrating and 6 penetrating particles.

■ **Figure 5-32 Tonicity** A cell and a solution have equal concentrations of solute. Water and penetrating solutes (P) can cross the membrane but nonpenetrating solutes (N) cannot cross. Water will distribute until the concentrations of nonpenetrating solutes are equal on both sides of the membrane. At equilibrium, the cell has gained volume, therefore, the solution was hypotonic to the cell.

into the cell will increase its volume, so the solution is hypotonic to the cell, even though the solution was isosmotic to the cell before they were put together.

If the solution was isosmotic to the cell, what caused osmosis? The answer lies with the penetrating solutes. When the cell was first placed in the solution, there was no osmotic gradient (and no water movement) across the cell membrane because cell and solution both had concentrations of 6 particles per liter. However, there were chemical concentration gradients: the nonpenetrating solutes were more concentrated in the cell (6 versus 3), while the penetrating solutes were more concentrated in the solution (3 versus 0). The nonpenetrating solutes could not cross the membrane, but the penetrating solutes were free to diffuse down their concentration gradient and move into the cell. As soon as one particle moved in, an osmotic gradient was created (Figure 5-32b ■). The cell contained $6 + 1$ particles, while the solution had $6 - 1$ particles. Once an osmotic gradient was created, water followed the solute into the cell (Figure 5-32c ■) and the cell volume increased (solute movement has no significant effect on cell volume). Thus, in this example, an isosmotic solution is hypotonic.

Table 5-8 lists some rules for dealing with osmolarity and tonicity. Understanding the difference between the two properties is critical to making good clinical decisions about intravenous (IV) fluid therapy. In medicine, the tonicity of a solution is an important consideration.

The purpose of IV fluids is usually either to get water into dehydrated cells (in which case, a hypotonic solution is used) or to keep fluid in the extracellular fluid to replace blood loss (in which case, an isotonic solution is used). The choice of fluid depends on how the clinician wants the solutes and water to distribute between the extracellular and intracellular fluid compartments. Table 5-9 shows the most common IV solutions and their osmolarity and tonicity relative to the normal human cell.

✔ You have a patient who has lost 1 liter of blood and you need to restore volume quickly while waiting for a blood transfusion to arrive from the blood bank. Would it be better to administer 5% dextrose (glucose) in water or 0.9% NaCl in water? (Hint: Think about how these solutes distribute in the body.) Defend your choice.

✔ How much of your solution of choice would you have to administer to return blood volume to normal?

The Body Is in a State of Electrical Disequilibrium

Thus far, we have seen that the compartments of the body are in osmotic equilibrium but in chemical disequilibrium. Because many body solutes are ions that carry either positive or negative electrical charges, we

---

TABLE 5-8 Rules for Osmolarity and Tonicity

1. When a cell is exposed to a solution, water and penetrating solutes will move and the system will go to equilibrium. At equilibrium the total concentration of solute particles in the cell and the solution will be equal; that is, the osmolarities will be the same.
2. Compare the osmolarity of a cell and a solution before they are put together, because at equilibrium, the osmolarities will be the same.
3. The tonicity of a solution is determined by the volume of the cell at equilibrium (see Table 5-7).
4. A solution that is hyposmotic to a cell will always be hypotonic, no matter what the nature of the solutes in the solution.
5. You cannot predict the tonicity of isosmotic and hyperosmotic solutions unless you know the concentration of nonpenetrating solutes relative to the cell. Net water movement will be into the compartment that has the higher concentration of nonpenetrating solutes.

TABLE 5-9 Intravenous Solutions

Solution	Also Known as	Osmolarity	Tonicity
0.9% saline*	Normal saline	Isosmotic	Isotonic
D_5—0.9% saline	5% dextrose** in normal saline	Hyperosmotic	Isotonic
D_5W	5% dextrose in water	Isosmotic	Hypotonic
0.45% saline	Half-normal saline	Hyposmotic	Hypotonic
D_5—0.45% saline	5% dextrose in half-normal saline	Hyperosmotic	Hypotonic

*Saline = NaCl.
**Dextrose = glucose.

...continued from page 137

Three days after Daniel's sweat test, the lab returns the grim results: salt levels in his sweat are over twice the normal concentration. Daniel is diagnosed with cystic fibrosis. Now, along with antibiotics to prevent lung infections and therapy to loosen the mucus in his airways, Daniel must begin a regimen of enzymes to be taken whenever he eats, for the rest of his life. In cystic fibrosis, thick mucus in the pancreatic ducts blocks the secretion of digestive enzymes into the intestine. Without artificial enzymes, he would starve. Yet even with these treatments, Daniel will probably die before the age of 30.

Question 4: Why will Daniel starve if he does not take artificial pancreatic enzymes?

must also consider the distribution of electrical charge between the intracellular and extracellular compartments. Although the body as a whole is electrically neutral, it has a few excess negative ions in the intracellular fluid, while their matching positive ions are found in the extracellular fluid.

Potassium (K^+) is the major cation of the cells, and sodium (Na^+) dominates the extracellular fluid (Table 5-10). Chloride ions (Cl^-) remain with Na^+ in the extracellular fluid, whereas phosphate ions and negatively charged proteins are the anions of the intracellular fluid. However, there are some protein anions inside the cell that do not have matching cations and some K^+ ions in the extracellular fluid that do not have matching anions. One consequence of this uneven distribution of ions is that the intracellular and extracellular compartments are not in electrical equilibrium. The electrical disequilibrium in the body is commonly known as the *resting membrane potential difference.*

Electrical Signals Are Generated in Many Types of Cells
The concept of resting membrane potential has traditionally been taught in the chapters on nerve and

TABLE 5-10 Typical Distribution of Major Ions in the Intracellular and Extracellular Compartments (mmol/L)*

Ion	Intracellular	Extracellular	Normal Plasma Value
K^+	150	5	3.5–5.0
Na^+	2	140	135–145
Cl^-	10	105	100–108
Organic anions	65	0	**

*Clinically these values are usually given in milliequivalents per liter. However, because the ions shown all have only a single charge, mEq = mmol.
**Plasma proteins (unionized) = 60–80 g/L.
The totals of the intracellular and extracellular compartments are not equal because of other ions, such as calcium, bicarbonate, and phosphate, that are not shown in this table.

muscle function because those tissues generate electrical signals known as action potentials. Yet one of the most exciting recent discoveries in physiology is the realization that other kinds of cells also use electrical signals for communication. After we review basic principles of electricity and discuss what creates the resting membrane potential, we will examine how the beta cells of the pancreas monitor blood glucose concentration and use changes in their membrane potential to trigger insulin secretion.

Electricity review Atoms are electrically neutral [∞ p. 16]. They are composed of positively charged protons, negatively charged electrons, and uncharged neutrons, but in balanced proportions, so that an atom is neither positive nor negative. The removal or addition of electrons to an atom creates a charged particle known as an ion. We have discussed several ions that are important in the human body, such as Na^+, K^+, and H^+. For each of these positive ions, somewhere there is a matching electron, usually found as part of a negative ion. For example, when Na^+ in the body enters in the form of NaCl, the "missing" electron from Na^+ can be found on the Cl^-.

The following principles are important to remember when dealing with electricity in physiological systems:

1. The **law of conservation of electric charge** states that the net amount of electric charge produced in any process is zero. This means that, for every positive charge on an ion, there is an electron on another ion. Overall, the human body is electrically neutral.

2. Opposite charges (+ and −) are attracted to each other, but two charges of the same type (+ and + or − and −) repel. The protons and electrons in an atom exhibit this attraction.

3. To separate positive and negative charges, it is necessary to use energy. For example, energy is needed to separate the protons and electrons of an atom.

4. If separated positive and negative charges can move freely toward each other, the material through which they are moving is called a **conductor.** Water is a good conductor of electricity. If separated charges are unable to move through the material that separates them, the material is known as an **insulator.** The phospholipid bilayer of the cell membrane is a good insulator, as is the plastic coating on electrical wires.

The word *electricity* comes from the Greek word *elektron,* for amber, the fossilized resin of trees. The Greeks discovered that if they rubbed a rod of amber with cloth, the amber acquired the ability to attract hair and dust. This attraction (called static electricity) arises from the separation of electrical charge that occurs when electrons move from the amber atoms to the cloth. To separate these charged particles, energy (work) must be put into the system. In the case of the amber, work was done by rubbing the rod. In the case of biological systems, the work is usually done by energy stored in ATP.

The Cell Membrane Allows Separation of Electrical Charge in the Body

In the body, separation of electrical charge takes place across the cell membrane. This process is shown in Figure 5-33 ■. The diagram shows an artificial cell filled with molecules that dissociate into positive and negative ions, represented by the plus and minus signs. Because the molecules were electrically neutral to begin with, there are equal numbers of positive and negative ions inside the cell. The cell is placed in a similar solution, also electrically neutral. The phospholipid bilayer of the artificial cell, like the bilayer of a real cell, is not permeable to ions. Water can freely cross this cell membrane, so the extracellular and intracellular ion concentrations are equal. In Figure 5-33a ■, the system is at osmotic, chemical, and electrical equilibrium.

In the next step (Figure 5-33b ■), an active transport protein is inserted into the membrane. This carrier uses energy from ATP to move positive ions (+) out of the cell against their concentration gradient. The negative ions (−) in the cell attempt to follow the positive ions because of the attraction of positive and negative charges. But the membrane is impermeable to negative ions, so they remain trapped within the cell. Positive ions outside the cell might try to move into the cell, attracted by the negative charge of the intracellular fluid, but the membrane does not allow ions to leak across it.

As soon as the first positive ion leaves the cell, the electrical equilibrium between the extracellular fluid and intracellular fluid is disrupted: the cell's interior has a net charge of −1, while the cell's exterior has a net charge of +1. The input of energy to transport ions across the membrane has created an electrical gradient: in this example, the inside of the cell became negative relative to the outside.

The active transport of positive ions out of the cell also creates a concentration, or chemical, gradient: there are more positive ions outside the cell than inside. The combination of electrical and concentration gradients is called an **electrochemical gradient.** The cell remains in osmotic equilibrium because water can move freely across the membrane in response to solute movement.

(a) Cell and solution are electrically and chemically at equilibrium

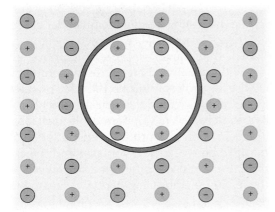

(b) Cell and solution in chemical and electrical disequilbrium

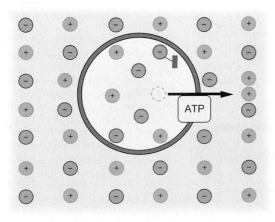

■ **Figure 5-33   Separation of electrical charge**   (a) The cell and the solution in this figure are in electrical and chemical equilibrium. The solutions inside and outside the cell are electrically neutral. (b) Energy from ATP is used to pump one positive charge out of the cell, disturbing the equilibrium. Now the system is in a state of chemical and electrical disequilibrium, with more positive charges outside and more negative charges inside. (c) On an absolute charge scale, the extracellular fluid (ECF) would be shown at +1 and the intracellular fluid (ICF) at −1. However, physiological measurements are always on a relative scale, on which the extracellular fluid is assigned a value of zero. This shifts the scale to the left and gives the inside of the cell a relative charge of −2.

(c) Comparison of absolute and relative charge scales

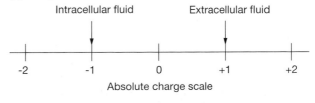

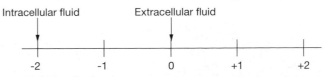

An electrical gradient between the extracellular fluid and the intracellular fluid is known as the **resting membrane potential difference** or **membrane potential,** for short. Although the name sounds intimidating, we can break it apart to see what it means.

1. The *resting* part of the name comes from the fact that this electrical gradient is seen in all living cells, even those that appear to be without electrical activity. The membrane potential has reached a steady state and is not changing.

2. The *potential* part of the name comes from the fact that the electrical gradient created by active transport of ions across the cell membrane is a source of stored, or potential, energy, just as the concentration gradient is a form of potential energy. When oppositely charged molecules come back together, they release energy that can be used to do work, in the same way that molecules moving down their concentration gradient can perform work. The work done by electrical energy includes opening voltage-gated membrane channels and sending electrical signals.

3. The *difference* part of the name is to remind you that the membrane potential represents a difference in the electrical charge inside and outside the cell. The word *difference* is often dropped.

In living systems, we measure electrical gradients on a relative scale rather than an absolute scale. Figure 5-33c ■ compares the two scales. On the absolute scale, the extracellular fluid in our simple example has a charge of +1 from the positive ion it gained, while the intracellular fluid has a charge of −1 for the negative ion that was left behind. However, in real life we cannot measure the charges as numbers of electrons gained or lost. Instead we use a device that measures the difference in electrical charge between two points. This device artificially sets one side of the membrane at zero and measures the second side relative to the first. In our example, resetting the extracellular fluid to zero gives the intracellular fluid a charge of −2.

The actual setup for measuring a cell's membrane potential difference is shown in Figure 5-34 ■. *Electrodes* are created from hollow glass tubes drawn to very fine points. These *micropipets* are filled with fluid that conducts electricity. They are connected to a *voltmeter* that measures the electrical difference between two points in volts (V) or millivolts (mV). A *recording electrode* is inserted through the cell membrane into the cytoplasm of the cell. A *reference electrode* is placed in the external fluid bath, which represents the extracellular fluid.

In living systems, by convention, the extracellular fluid is designated as the *ground* and assigned a charge of 0 mV. When the recording electrode is placed inside a living cell, the voltmeter measures the membrane potential (V_m), the electrical difference between the intracellular fluid and the extracellular fluid. If a chart recorder is connected to the voltmeter, you can make a recording of the membrane potential difference versus time.

For nerve and muscle cells, the voltmeter will record a resting membrane potential difference between −40 and −90 mV, indicating that intracellular fluid is negative relative to the extracellular fluid (0 mV). As you saw in Figure 5-31c ■, the extracellular fluid is not really neutral because it has excess positive charges that exactly balance the excess negative charges inside the cell.

The Resting Membrane Potential Is Due Mostly to Potassium

Which ions create the resting membrane potential difference in animal cells? The artificial cell shown in Figure 5-33b ■ used an active transport protein to move an unspecified ion with a positive charge across a membrane that was otherwise impermeable to ions. But what processes go on in living cells to create an electrical gradient?

Real cells are not completely impermeable to all ions. They have open channels and protein transporters that allow ions to move between the cytoplasm and the extracellular fluid. We can use a different artificial cell to show how the resting membrane potential is created in living cells.

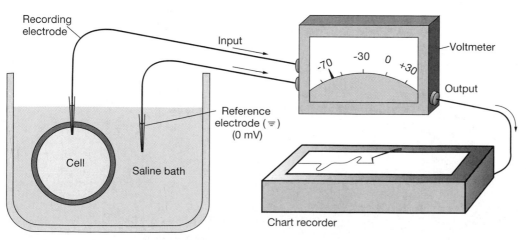

■ **Figure 5-34 Measuring membrane potential difference** In the laboratory, a voltmeter measures the difference in electrical charge between the inside of a cell and the surrounding solution. The ground, or reference electrode, is placed in the bath and given a value of 0 millivolts (mV). A recording electrode is placed inside the cell. The voltmeter measures the electrical potential difference between the two electrodes. This value is the membrane potential difference, or V_m.

Recording electrode
Input
Voltmeter
−70 −30 0 +30
Output
Reference electrode (⊽) (0 mV)
Cell
Saline bath
Chart recorder

The artificial cell shown in Figure 5-35a ■ contains K^+ ions and large negatively charged proteins, represented by Pr^-. The cell is placed in a solution of Na^+ and Cl^-. There is no uneven distribution of charge across the membrane, and the system is in electrical equilibrium. However, it is not in chemical equilibrium. There are concentration gradients for all four types of ions in the

system, and they would all diffuse down their respective concentration gradients if they could cross the cell membrane.

In Figure 5-35b ■, the membrane suddenly becomes permeable only to K^+. Because there is no K^+ in the extracellular fluid initially, some K^+ ions will leak out of the cell, moving down their concentration gradient (Fig. 5-33b ■). As the K^+ leave, the negatively charged proteins, Pr^-, are unable to follow. The proteins gradually build up a negative charge inside the cell as K^+ diffuses out of the cell.

If the only force acting on K^+ were the concentration gradient, K^+ would leak out of the cell until the K^+ concentration inside the cell equaled the K^+ concentration outside. But the loss of positive ions from the cell creates an electrical gradient. Because opposite charges attract, the negative Pr^- inside the cell try to pull K^+ back into the cell. Thus, when K^+ ions move out of the cell down their concentration gradient, the negative charge left inside the cell causes other K^+ to move into the cell.

At some point in this process the electrical force exerted by the negative charge inside the cell becomes equal in magnitude to the chemical concentration gradient driving K^+ out of the cell. At that point, net movement of K^+ across the membrane stops (Fig. 5-35c ■). The rate at which K^+ move out of the cell down the concentration gradient is exactly equal to the rate at which K^+ move into the cell down the electrical gradient.

In a cell that is permeable to only one ion, the membrane potential difference that exactly opposes the concentration gradient of the ion is known as the **equilibrium potential**, or E_{ion}. For example, when the concentration gradient is 150 mM K^+ inside and 5 mM K^+ outside the cell, the equilibrium potential for potassium, or E_K, is -90 mV. The equilibrium potential for any ion at 37° C (human body temperature) can be calculated using the Nernst equation:

$$E_{ion} = \frac{61}{z} \log \frac{[ion]_{out}}{[ion]_{in}}$$ where 61 is a combination of several constants plus temperature

z is the electrical charge on the ion ($+1$ for K^+)
[ion] are the ion concentrations inside and outside the cell

Now we will use the same artificial cell (K^+ and Pr^- inside, Na^+ and Cl^- outside), but this time we will make the membrane permeable only to Na^+ (Fig. 5-36 ■). Because Na^+ is more concentrated outside the cell, it will begin to move into the cell. As it does so, it causes positive charges to accumulate inside the cell. Meanwhile Cl^- ions left behind in the extracellular fluid make that compartment negative. This creates an electrical gradient that moves Na^+ back out of the cell. When the Na^+ concentration is 150 mM outside and 15 mM inside, the equilibrium potential for Na^+ (E_{Na}) is $+60$ mV. In other words, the concentration gradient moving Na^+ into the cell (150 mM → 15 mM) is exactly opposed by a positive intracellular charge of $+60$ mV.

(a) Membrane is impermeable

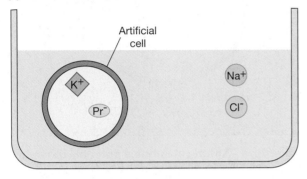

(b) Membrane permeable to K^+ only

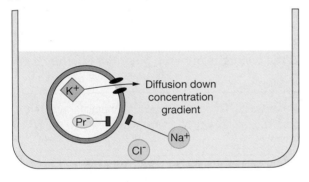

(c) Inside of cell develops negative membrane potential

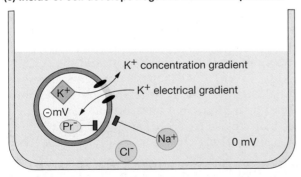

■ **Figure 5-35   Potassium equilibrium potential** (a) An artificial cell filled with K^+ and large proteins (Pr^-) is placed in a solution of Na^+ and Cl^-. The cell membrane is permeable only to K^+. (b) Some K^+ diffuse out of the cell because of the concentration gradient. As a result, the extracellular fluid gains K^+, while the cell has excess Pr^-. This creates a negative membrane potential difference. (c) Because opposite charges attract, some extracellular K^+ move back into the cell due to the electrical gradient. The membrane potential difference at which the movement down the concentration gradient is exactly equal to and opposite of the movement down the electrical gradient is called the equilibrium potential for the ion.

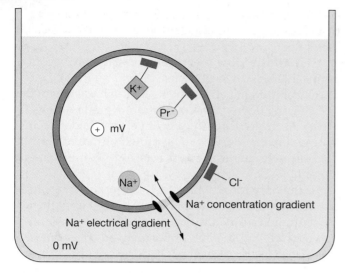

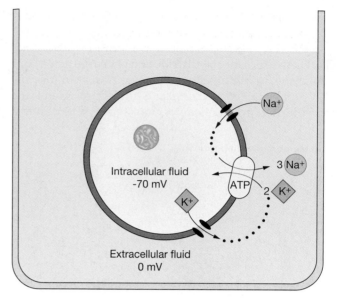

■ **Figure 5-36   Sodium equilibrium potential**   An artificial cell filled with K^+ and large proteins (Pr^-) is placed in a solution of Na^+ and Cl^-. This cell is permeable to Na^+ but not to any other ions. As sodium moves into the cell down its concentration gradient, the inside of the cell becomes positive. This continues until the cell reaches the equilibrium potential for Na^+, where movement of Na^+ into the cell down the concentration gradient is exactly opposed by the movement of Na^+ out of the cell down the electrical gradient.

■ **Figure 5-37   Resting membrane potential difference in an actual cell**   Most cells in the human body are about 40 times more permeable to K^+ than to Na^+, and the resting membrane potential difference is about -70 mV. The Na^+ leaking in and the K^+ leaking out are returned to their respective compartments by the Na^+-K^+-ATPase.

But living cells are not permeable to only one ion. The situation in real cells is similar to a combination of the two artificial systems just described. If a cell is permeable to several ions, we cannot use the Nernst equation to calculate membrane potential. Instead we must use a related equation called the Goldman equation that considers both the concentration gradients and the relative permeability of the cell to each ion. For more detail on the Nernst and Goldman equations, see Appendix A.

The cell illustrated in Figure 5-37 ■ has a resting membrane potential of -70 mV. Most cells are about 40 times more permeable to K^+ than to Na^+, and as a result, a cell's resting membrane potential is closer to the E_K of -90 mV than to the E_{Na} of $+60$ mV. A small amount of Na^+ leaks into the cell, making the inside of the cell less negative than it would be if Na^+ were totally excluded. The Na^+ that leaks in is promptly pumped out by the Na^+-K^+-ATPase as described earlier. At the same time, K^+ ions that leak out of the cell are pumped back in. The pump contributes to the membrane potential by pumping 3 Na^+ out for every 2 K^+ pumped in. Because the Na^+-K^+-ATPase helps maintain the electrical gradient, it is called an *electrogenic* pump.

Not all ion transport creates an electrical gradient. Some transporters, like the Na^+-K^+-2 Cl^- symporter, move one negative charge for each positive charge and are therefore electrically neutral. Other transporters make an even exchange of ions: for each charge that enters the cell, one of the same charge leaves. An example is the HCO_3^--Cl^- antiporter of red blood cells, which transports these ions in a one-for-one, electrically neu-

tral exchange. Electrically neutral transporters will not affect the resting membrane potential difference of the cell.

✔ What would happen to the resting membrane potential of a cell poisoned with ouabain, an inhibitor of the Na^+-K^+-ATPase?

Changes in Ion Permeability Change the Membrane Potential

As you have just learned, two factors influence the cell's membrane potential: (1) the concentration gradients of different ions across the membrane and (2) the permeability of the membrane to those ions. If the cell's permeability to an ion changes, the cell's membrane potential changes. We monitor changes in membrane potential using the same intracellular recording electrodes that we use to record resting membrane potential (see Fig. 5-34 ■).

Terminology for Changes in Membrane Potential

Figure 5-38 ■ shows a recording of membrane potential plotted against time. The membrane potential (V_m) begins at a steady resting value of -70 mV. When the trace moves upward (becomes less negative), the potential difference between the inside of the cell (-70 mV) and the outside (0 mV) decreases, and the cell is said to have **depolarized.** If the trace moves downward, the membrane potential becomes more negative, the potential difference has increased, and the cell has **hyperpolar-**

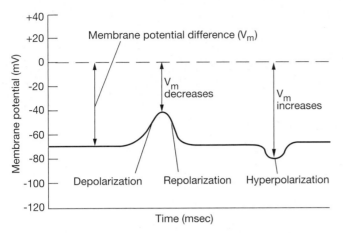

■ **Figure 5-38 Terminology associated with changes in membrane potential** If the membrane potential becomes less negative, the cell depolarizes. If the membrane potential becomes more negative, the cell hyperpolarizes.

ized. A return to the resting membrane potential from either direction is termed **repolarization.**

A major point of confusion when talking about changes in membrane potential is the use of the phrases "the membrane potential decreased" or "the membrane potential increased." Normally, in English we associate "increase" with becoming more positive and "decrease" with becoming more negative—the opposite of what is happening in those cases. One way to avoid confusion is to add the word *difference* after *membrane potential.* If the membrane potential difference is *increasing,* the V_m must be moving away from the ground value of zero and becoming *more negative.* If the membrane potential difference is *decreasing,* the V_m is moving closer to the ground value of 0 mV and is becoming *less negative.*

Changes in Ion Permeability Change Membrane Potential Four ions usually contribute the most to changes in membrane potential: Na^+, Ca^{2+}, Cl^-, and K^+. The first three are more concentrated in the extracellular fluid than in the cytosol, and the resting cell is minimally permeable to them. If the cell suddenly becomes more permeable to any one of these ions, ions will move across the membrane into the cell. Entry of Ca^{2+} or Na^+ will depolarize the cell; entry of Cl^- will hyperpolarize the cell.

Resting cells are fairly permeable to K^+, but making them more permeable will allow a cell to hyperpolarize until it reaches the equilibrium potential for K^+. Making the cell *less* permeable to K^+ will allow less K^+ to leak out of the cell. When the cell retains K^+, it becomes more positive and depolarizes. You will encounter instances of all these ion changes as you study physiology.

It is important to remember that a significant change in membrane potential occurs with the movement of very few ions. *You do not need to reverse the concentration gradient to change the membrane potential.* For example, to change

the membrane potential by 100 mV, only one out of every 100,000 K^+ must enter or leave the cell. This is such a tiny fraction of the total number of K^+ in the cell that the concentration of K^+ remains essentially unchanged.

▶ INTEGRATED MEMBRANE PROCESSES: INSULIN SECRETION

The movement of Na^+ and K^+ across cell membranes has been known to play a role in generating electrical signals in excitable tissues for many years. You will study these processes in detail in the chapters on the nervous and muscular systems. Recently, however, we have come to understand how small changes in membrane potential act as signals in nonexcitable tissues such as endocrine cells. One of the best-studied examples of this process is the beta cell of the pancreas. The release of insulin by beta cells demonstrates how membrane processes, such as facilitated diffusion, exocytosis, and the opening and closing of ion channels by ligands and membrane potential, work together to regulate cell function.

The beta cells of the pancreas synthesize the protein hormone insulin and store it in cytoplasmic secretory vesicles [∞ p. 53]. When blood glucose levels increase, such as after a meal, the beta cell releases insulin by exocytosis. Insulin then directs other cells of the body to take up and use glucose, bringing blood concentrations down to pre-meal levels. A key question about the process that went unanswered until recently was "How does the beta cell 'know' that glucose levels have gone up and that it needs to release insulin?" The answer, we have now learned, links the beta cell's metabolism to its electrical activity.

Figure 5-39a ■ shows a beta cell at rest. Recall from earlier sections in this chapter that the gates of membrane channels can be opened or closed by chemical or electrical signals. The beta cell has two such channels that help control insulin release. One is a **voltage-gated Ca^{2+} channel.** This channel is closed at the cell's resting membrane potential (Fig. 5-39a, ⑥ ■). The other is a K^+ leak channel (i.e., the channel is usually open) that closes when ATP binds to it. It is called an **ATP-gated K^+ channel,** or **K_{ATP} channel.** In the resting cell, when glucose concentrations are low, the cell makes less ATP (Fig. 5-39a, ① – ③ ■). There is little ATP to bind to the K_{ATP} channel, so the channel remains open, allowing K^+ to leak out of the cell (Fig. 5-39a, ④ ■).

Figure 5-39b ■ shows a beta cell secreting insulin. Following a meal, plasma glucose levels increase as glucose is absorbed from the intestine ①. Glucose reaching the beta cell diffuses into the cell with the aid of a GLUT transporter. Increased glucose in the cell stimulates the metabolic pathways of glycolysis and the citric acid cycle [∞ p. 90], and ATP production increases ②, ③. When ATP binds to the K_{ATP} channel, the gate to the channel closes, preventing K^+ from leaking out of the

(a) Beta cell at rest

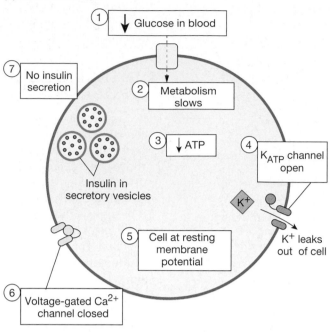

(b) Beta cell secretes insulin

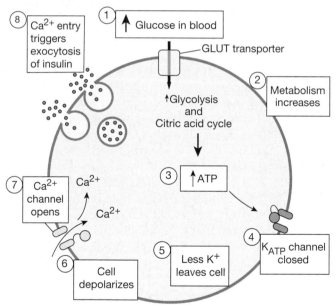

■ **Figure 5-39   Insulin secretion**   (a) In a resting beta cell, insulin is stored in secretory vesicles. ATP production is low, K_{ATP} channels are open, and the cell is at its resting membrane potential. (b) When plasma glucose concentrations increase, glucose enters the beta cell on GLUT transporters. Metabolism increases, and increased concentrations of ATP cause the K_{ATP} channels to close. Less K^+ leaks out of the cell, so the cell depolarizes. The change in membrane potential opens voltage-gated Ca^{2+} channels. Ca^{2+} enters the cell and triggers exocytosis of insulin from the secretory vesicles.

cell ④, ⑤. Retention of K^+ depolarizes the cell ⑥, which then causes the voltage-sensitive Ca^{2+} channels to open ⑦. Calcium ions enter the cell from the extracellular fluid, moving down their electrochemical gradient. The Ca^{2+} binds to proteins that initiate exocytosis of the insulin-containing vesicles ⑧, and insulin is released into the extracellular space.

The discovery that cells other than nerve and muscle cells use changes in membrane potential as signals for physiological responses has been an exciting one that promises to change our traditional thinking about the role of the resting membrane potential. In the next chapter, we will describe some other types of signals that the body uses for communication and coordination.

PROBLEM CONCLUSION

In this running problem, you learned that cystic fibrosis is caused by a defect in the CFTR channel, which regulates the transport of chloride ions out of epithelial cells. Because the CFTR channel is found in the epithelial cell membranes of several organs—the sweat glands, lungs, and pancreas—you have seen that this defect affects many different body processes.

Further check your understanding of this running problem by checking your answers against those in the summary table.

Question	Facts	Integration and Analysis
1 Is CFTR a chemically gated, voltage-gated, or mechanically gated channel protein?	Chemically gated channels open when a ligand bind to them. Voltage-gated channels open with a change in the cell's membrane potential. Mechanically gated channels open when a physical force opens the channel. CFTR channels open in response to the binding of nucleotides.	Nucleotides are chemical ligands, therefore the CFTR channel is a chemically gated channel.
2 Based on the information given, is the CFTR channel on the apical or basolateral surface of the sweat gland epithelium?	In normal people, the CFTR channels move Cl⁻ from sweat into epithelial cells.	The epithelial surface that faces the lumen of the sweat gland, which contains sweat, is the apical membrane. Therefore the CFTR channel is on the apical surface.
3 Why would failure to transport chloride ions into the lumen of the airways cause the secreted mucus to be thick? (Hint: Remember that "water moves to dilute the more concentrated region.")	If Cl⁻ (followed by Na⁺) is secreted into the lumen of the airways, the solute concentration of the airway fluid increases. Water moves in response to concentration gradients.	Normally, the movement of Cl⁻ and Na⁺ creates an osmotic gradient so that water also enters the airway lumen, creating a saline solution that thins the thick mucus. If Cl⁻ cannot be secreted into the airways, there will be no fluid movement to thin the mucus.
4 Why will Daniel starve if he does not take artificial pancreatic enzymes?	The pancreas secretes mucus and digestive enzymes into small ducts that empty into the small intestine. In cystic fibrosis, the mucus in the ducts will be thick due to lack of Cl⁻ and fluid secretion. This blocks the ducts and prevents the digestive enzymes from reaching the small intestine.	Without digestive enzymes, Daniel cannot digest the food he eats. His weight loss over the past six months suggests that this has already been a problem. Taking artificial enzymes will allow him to digest his food.

CHAPTER REVIEW

SUMMARY

Membranes in the Body

1. The word **membrane** is used for both cell membranes and for epithelial tissues that line a cavity or separate two compartments. (p. 110)

Cell Membranes

2. The cell membrane is a barrier between the intracellular and extracellular fluid. (p. 110)

3. The cell membrane regulates exchange and communication between the cell and its environment. It also provides structural support to the cell. (p. 111)

4. The **fluid mosaic model** of membranes says that membranes are **phospholipid bilayers** with proteins inserted into the bilayer. Carbohydrates attach to the extracellular surface. (p. 111)

5. **Membrane-spanning** and other **integral proteins** are tightly bound to the phospholipid bilayer. **Associated proteins** attach less tightly to either side of the membrane. (p. 113)

6. **Structural proteins** maintain cell shape and form cell junctions that hold tissues together. (p. 114)

7. **Membrane-associated enzymes** catalyze chemical reactions and help transfer signals across the membrane. (p. 114)

8. **Receptor proteins** on the cell membrane are part of the body's chemical signaling system. A molecule that binds to a receptor is called its **ligand.** (p. 114)

9. **Transport proteins** are the means by which many molecules enter or leave the cell. **Channel proteins** form water-filled channels that link the intracellular and extracellular compartments. **Gated channels** regulate movement of substances through them by opening and closing. Gated channels are regulated by ligands, by the electrical state of the cell, or by physical changes such as pressure. (p. 115)

10. **Carrier proteins** never form a direct connection between the intracellular and extracellular fluid. They bind to substrates, then change shape. (p. 115)

Body Fluid Compartments

11. Most cells of the body are surrounded by **extracellular fluid** (ECF), which can be subdivided into **interstitial fluid** bathing the cells and **plasma,** the fluid portion of the blood. (p. 117)

Movement Across Membranes

12. The cell membrane is a barrier that restricts free exchange between the cell and the interstitial fluid. The movement of a substance across the membrane depends on the **permeability** of the membrane to the substance. (p. 118)

13. Lipid-soluble substances can diffuse through the phospholipid bilayer. Less lipid-soluble molecules require the assistance of a membrane protein to cross the membrane. (p. 118)

14. **Passive transport** does not require the input of energy. (p. 118)

15. **Diffusion** is the passive movement of molecules from an area of higher concentration to an area of lower concentration. Diffusion stops when the system reaches **equilibrium,** although molecular movement continues. (p. 118)

16. Diffusion rate depends on the magnitude of the concentration gradient. Diffusion is slow over long distances, is directly related to temperature, and is inversely related to molecular size. (p. 119)

17. **Simple diffusion** is diffusion across the phospholipid bilayer of a membrane. The rate of simple diffusion is directly proportional to the surface area of the membrane, inversely proportional to the thickness of the membrane, and depends on the lipid solubility of the molecule that is diffusing. (p. 120)

18. **Mediated transport** is movement of molecules with the aid of a carrier protein. Passive mediated transport is called **facilitated diffusion.** (p. 121)

19. All mediated transport demonstrates **specificity, competition,** and **saturation.** Specificity refers to the ability of a transporter to move only one molecule or a group of closely related molecules. Related molecules will compete for a single transporter. Saturation occurs when a group of membrane transporters are working at their maximum rate. (p. 121)

20. **Active transport** moves molecules against their concentration gradient and requires an outside source of energy. In **primary (direct) active transport,** the energy comes directly from ATP. **Secondary (indirect) active transport** uses the potential energy stored in a concentration gradient and is indirectly driven by energy from ATP. (p. 124)

21. The most important primary active transporter is the **sodium-potassium ATPase** (Na$^+$-K$^+$-ATPase), which pumps Na$^+$ out of the cell and K$^+$ into the cell. (p. 125)

22. Most secondary active transport systems are driven by the sodium concentration gradient. (p. 126)

23. Large macromolecules and particles are brought into cells by **phagocytosis** and **endocytosis**. Material leaves cells by **exocytosis.** When vesicles that come into the cytoplasm by endocytosis are returned to the cell membrane, the process is called **membrane recycling.** (p. 128)

24. In **receptor-mediated endocytosis**, ligands bind to membrane receptors that concentrate in **clathrin-coated pits,** the site of endocytosis. In **potocytosis,** receptors are located in **caveolae** that have a nonclathrin protein coating. (p. 129)

25. In exocytosis, the vesicle membrane fuses with the cell membrane in a complex process that involves a series of proteins. Exocytosis requires ATP and an increase in cytosolic Ca^{2+} concentration. (p. 130)

26. Transporting epithelia in the intestine and kidney have different membrane proteins on their **apical** and **basolateral** surfaces. This polarization allows one-way movement of molecules across the epithelium. Larger molecules cross epithelia by **transcytosis,** which includes **vesicular transport.** (p. 131)

The Distribution of Water and Solutes in the Body

27. The body is in a state of **chemical disequilibrium** because cell membranes prevent solutes from diffusing freely and because of the active transport of solutes. (p. 133)

28. Water moves freely between body compartments, so the body is in a state of **osmotic equilibrium.** (p. 135)

29. The movement of water across a membrane in response to a concentration gradient is called **osmosis.** (p. 135)

30. To compare solutions, we express concentration in terms of **osmolarity,** the number of particles (ions or intact molecules) per liter of solution, expressed as milliosmoles (mosmol) per liter. **Osmolality** is concentration expressed as milliosmoles solute per kilogram of water. (p. 136)

31. **Tonicity** describes what a solution would do to cell volume if the cell were placed in the solution. Cells swell in **hypotonic solutions** and shrink in **hypertonic solutions.** If the cell does not change size at equilibrium, the solution is **isotonic.** (p. 137)

32. The starting osmolarities of a cell and solution cannot be used to determine the tonicity of a solution. The relative concentrations of **nonpenetrating solutes** in the cell and the solution determine the tonicity of the solution. **Pene-**trating solutes contribute to the osmolarity of a solution but not to its tonicity. (p. 138)

33. Although the total body is electrically neutral, it is in a state of **electrical disequilibrium** because diffusion and active transport of ions across the cell membrane create an **electrical gradient,** with the inside of the cell negative relative to the outside. (p. 139)

34. The electrical gradient between the extracellular fluid and the intracellular fluid is known as the **resting membrane potential difference.** (p. 140)

35. The movement of an ion across the cell membrane is influenced by the **electrochemical gradient** for that ion. (p. 141)

36. The membrane potential difference that exactly opposes the concentration gradient of an ion is known as the **equilibrium potential,** or E_{ion}. The equilibrium potential for any ion can be calculated using the Nernst equation. (p. 143)

37. In living cells, K$^+$ is the primary ion that determines the resting membrane potential. (p. 144)

38. Changes in membrane permeability to ions such as K$^+$, Na$^+$, Ca^{2+}, or Cl$^-$ will alter membrane potential and create electrical signals. (p. 144)

Integrated Membrane Processes: Insulin Secretion

39. We have recently learned that the use of electrical signals to initiate a cellular response is not limited to the excitable tissues of nerve and muscle. The beta cell of the pancreas releases insulin in response to a change in membrane potential. (p. 145)

QUESTIONS

LEVEL ONE   Reviewing Facts and Terms

1. List the four general functions of the cell membrane.

2. In 1972, Singer and Nicolson proposed a model of the cell membrane known as the _____ model. According to this model, the membrane is composed of a bilayer of _____ and a variety of embedded _____, with _____ on the extracellular surface.

3. List the four functions of membrane proteins and give an example of each.

4. Arrange these compartments in the order a glucose molecule entering the body at the intestine would encounter them: interstitial fluid, plasma, intracellular fluid. Which of these fluids is/are considered to be extracellular fluid(s)?

5. Distinguish between active and passive transport.

6. Which of the following processes are examples of active transport and which of passive transport? Simple diffusion, phagocytosis, facilitated diffusion, exocytosis, osmosis, endocytosis, potocytosis.

7. List four factors that will improve the rate of diffusion.

8. Match the membrane channels with the appropriate description(s). Answers may be used once, more than once, or not at all.

   (a) chemically gated channel
   (b) open pore
   (c) voltage-gated channel
   (d) mechanically gated channel

   1. channel that spends most of its time in the open state
   2. channel that opens in response to a signal
   3. channel opens when resting membrane potential changes
   4. channel that opens when a ligand binds to it
   5. channel that opens in response to membrane stretch
   6. channel through which water can pass

9. List the three physical methods by which materials enter cells.

10. A cotransporter is a protein that moves more than one molecule at a time. If the molecules are moved in the same direction, the process is called _____ ; if the molecules are transported in opposite directions, it is called _____. A transport protein that moves only one substrate is called a _____.

11. The two types of active transport are _____, which derives energy directly from ATP, and _____, which couples the kinetic energy of one molecule moving down its concentration gradient to the movement of another molecule against its concentration gradient.

12. A molecule that moves freely between the intracellular and extracellular compartments is said to be _____. A molecule that is not able to enter cells is called a _____ solute.

13. Based on percentage of body water, rank these individuals from highest to lowest: (a) a 25-year-old 70-kg male, (b) a 25-year-old 50-kg female, (c) a 65-year-old 50-kg female, and (d) a 1-year-old male toddler.

14. What determines the osmolarity of a solution? In what units is body osmolarity usually given?

15. What does it mean if we say that a solution is hypotonic to a cell? Hypertonic to the same cell? What determines the tonicity of a solution relative to a cell?

16. In your own words, state the four principles of electricity important in physiology.

17. Match each structure with its role in cellular activity.

   (a) Na^+-K^+-ATPase
   (b) proteins
   (c) unit of measurement for membrane potential
   (d) K^+
   (e) Cl^-
   (f) ATP
   (g) Na^+

   1. ion channel
   2. extracellular cation
   3. source of energy
   4. intracellular anion
   5. intracellular cation
   6. millivolts
   7. electrogenic pump
   8. extracellular anion
   9. milliosmoles

18. The membrane potential difference at which the electrical gradient exactly opposes the concentration gradient for an ion is known as the _____.

19. A material that allows free movement of electrical charges is called a(n) _____, while one that prevents this movement is called a(n) _____.

LEVEL TWO   Reviewing Concepts

20. Create a map using the following terms. You may add additional terms if you wish.

   active transport
   clathrin-coated pit
   carrier
   caveolae
   channel
   concentration gradient
   electrochemical gradient
   exocytosis
   facilitated diffusion
   glucose
   GLUT
   ion
   large polar molecule
   receptor-mediated endocytosis
   secondary active transport

   ligand
   Na^+-K^+-ATPase
   osmosis
   passive transport
   phospholipid bilayer
   potocytosis
   receptor
   simple diffusion
   small polar molecule
   transcytosis
   vesicle
   vesicular transport
   water

21. Phospholipids and cholesterol are found within the cell membrane. Explain their anatomical arrangement and the physiological results of this arrangement.

22. Compare and contrast chemically, mechanically, and voltage-gated channels.

23. What six factors influence the rate or speed of diffusion across a membrane? Briefly explain each one.

24. Define the following terms and explain how they differ: specificity, competition, saturation. Apply these terms in a short explanation of facilitated diffusion of glucose.

25. Red blood cells are suspended in a solution of NaCl. The cells have an osmolarity of 300 mOsM and the solution has an osmolarity of 250 mOsM. (a) The solution is (hypertonic, isotonic or hypotonic) to the cells? (b) Water would move (into the cells, out of the cells, or not at all)?

26. Two compartments are separated by a membrane that is permeable to glucose. Each compartment is filled with 1 M glucose. After 6 hours, compartment A contains 1.5 M glucose and compartment B contains 0.5 M glucose. What kind of transport occurred?

27. A 2 M NaCl solution and a 2 M glucose solution are separated by a membrane that is permeable to the water, but not to the NaCl or glucose. Are the following statements true or false? Defend your answer with a short statement. (a) The salt solution is isosmotic to the glucose solution. (b) The salt solution is hyperosmotic to the glucose solution.

(c) Water will move from the salt solution to the sugar solution. (d) The volume will increase on the glucose side of the membrane. (e) If the salt solution were inside a cell, the cell would swell.

28. Referring to the previous question, change the molarity of the salt and sugar solutions so that each false statement would become true.

29. Explain the difference between a chemical gradient, an electrical gradient, and an electrochemical gradient.

LEVEL THREE   Problem Solving

30. Sweat glands secrete a fluid into their lumen that is identical to interstitial fluid. As the fluid moves through the lumen on its way to the surface of the skin, the cells of the sweat gland's epithelium make the fluid hypotonic by removing Na^+ and leaving water behind. Design an epithelial cell that will reabsorb Na^+ but not water. You may place water pores, Na^+ leak channels, K^+ leak channels, and the Na^+-K^+-ATPase in the apical membrane, basolateral membrane, or both.

31. Insulin is a hormone that promotes the movement of glucose into many types of cells, thereby lowering blood glucose concentration. Propose a mechanism that explains how this occurs, using your knowledge of cell membrane transport.

32. The following terms have been applied to membrane transport molecules: specificity, competition, saturation. These terms can also be applied to enzymes. How has the application of these terms changed in the two situations? What chemical characteristics do enzymes and transport molecules share that allow these terms to be applied to both?

33. NaCl is a nonpenetrating solute and urea is a penetrating solute. Red blood cells are placed in each of the solutions below. The intracellular concentration of nonpenetrating solute is 300 mOsM. What will happen to the cell volume in each solution? Label each solution with all the terms that apply: hypertonic, isotonic, hypotonic, hyperosmotic, hyposmotic, isosmotic.

    (a) 150 mM NaCl plus 150 mM urea
    (b) 100 mM NaCl plus 50 mM urea
    (c) 100 mM NaCl plus 100 mM urea
    (d) 150 mM NaCl plus 100 mM urea
    (e) 100 mM NaCl plus 150 mM urea

LEVEL FOUR   Quantitative Problems

34. The addition of dissolved solutes to water lowers the freezing point of water. A 1 OsM solution depresses the freezing point of water 1.86°. If a patient's plasma shows a freezing-point depression of 0.550° C, what is her plasma osmolarity? (Assume that 1 kg water = 1 L.)

35. The patient in the previous question is found to have total body water volume of 42 L, ECF volume of 12.5 L, and plasma volume of 2.7 L.

    (a) What is her intracellular fluid (ICF) volume? Her interstitial fluid volume?
    (b) How much solute (osmoles) exists in her whole body? ECF? ICF? plasma? (Hint: concentration = solute amount/volume)

36. What is the osmolarity of half-normal saline (= 0.45% NaCl)? [∞ p. 27] Assume that all NaCl molecules dissociate into two ions.

37. If you give 1L of half-normal saline to the patient in question 35, what happens to each of the following at equilibrium? (Hint: NaCl is a nonpenetrating solute.)

    (a) her total body volume
    (b) her total body osmolarity
    (c) her ECF and ICF volumes
    (d) her ECF and ICF osmolarities

E X P L O R E   <MediaLab>

Introduction

In this chapter you learned about the intricacies and importance of cell membranes. Although you may think it strange, cells are often thought of as factories because they function much as a factory does. Within the cell, membranes create organelles, each with a dedicated function. The outer cell membrane acts both as a barrier and a line of communication between the cell and its external environment. The following Web Explorations will help you gain an appreciation of these membranes and their importance to life. After reading the description below, visit the MediaLab for Chapter 5 in your Companion Website and select the appropriate keyword.

Web Exploration 1

Estimated time for completion = 10 minutes

The lipid composition of the membrane and its specific proteins determine what and when material crosses a cell membrane. When building your mental model of transport processes, realize two criteria must be met before a substance can cross the cell membrane: 1) there must be a way across the membrane and 2) there must be a driving force. Select the keyword **PHOSPHOLIPID** on the Website to learn more about the types of lipids that can be found in the cell membrane. As you look at these different lipid molecules, try to determine how the lipid composition of the membrane influences fluidity and permeability. Based on the structure of the phospholipid, what can you predict about the orientation of

this molecule in the lipid bilayer? What is the basis of your prediction? How would the number of unsaturated bonds in the fatty acid tail alter the packing and fluidity of the membrane? You should have noticed how cholesterol fits into the lipid bilayer. How would increased cholesterol content alter water permeability of the membrane? To complete this exercise, visit MediaLab Web Exploration 1 in Chapter 5 of your Companion Website.

Web Exploration 2

Estimated time for completion = 10 minutes

What is the driving force for the movement of Na^+ and K^+ ions across the cell membrane? The Na-K pump is key to maintaining the ion gradient found between the intracellular and extracellular compartments. Select the keyword **Na-K PUMP** on the Website to see an animation of the Na-K pump. The animation also shows how secondary active transport of glucose is coupled to the pump activity. Based on your model of transport across the cell membrane (a way across and a driving force—refer to Web Exploration 1), what would happen to the movement of glucose into the cell via a Na-glucose cotransporter if the Na-K pump was inhibited? Why? How could the Na-K pump be inhibited? Search the Web with the keywords *Foxglove* and/or *Digitalis* to get more information. Discuss your results with members of your class to see what they discovered. To complete this exercise, visit MediaLab Web Exploration 2 in Chapter 5 of your Companion Website.

6 Communication, Integration, and Homeostasis

■ "Many chronic diseases represent disorders in communication between cells."
—*Howard Rasmussen, in* Biology and Medicine into the 21st Century, *1991* ■

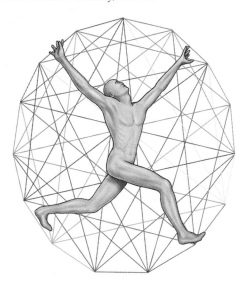

BACKGROUND BASICS

Membrane structure (p. 111)
Homeostasis (p. 6)
Exocytosis (p. 130)
Endocrine glands (p. 61)
Membrane proteins (p. 114)
Cell junctions (p. 55)
Extracellular matrix (p. 55)
Nucleotides (p. 35)

Complex living organisms are collections of molecules arranged into organelles, cells, tissues, and organs. The sheer size of the human body means that individual cells may be within micrometers of each other or meters apart. So how do the cells communicate and coordinate with one another? Diffusion is adequate for information exchange between neighboring cells, but as you learned in the last chapter, diffusion is ineffective over long distances. Multicellular organisms require additional methods for moving signals from one part of the body to another.

To meet the challenge of coordinating body functions, various anatomical and physiological mechanisms evolved. Blood flowing through our efficient circulatory system makes a complete circuit about once a minute, moving material from one part of the body to another. The nervous system takes care of more rapid communication, with electrical signals traveling as fast as 120 m/sec

PROBLEM
Diabetes Mellitus

It is 8:00 A.M., and Marvin Garcia, age 20, is hungry. As part of his routine physical examination before school starts, he is having a fasting blood glucose test. In this test, blood is drawn after an overnight fast, and the glucose concentration in the blood is measured. Marvin knows that he is in good physical condition, so he doesn't worry about the results. He is surprised, then, when the nurse practitioner at his family physician's office calls two days later. "Your fasting blood sugar is a bit elevated, Marvin. It is 150 milligrams per deciliter, and normal is 126 or less. Does anyone in your family have diabetes?" "Well, yeah—my dad has it. What exactly is diabetes?"

…continued on page 156

(268 mi/hr). The combination of simple diffusion across small distances, widespread distribution of materials through the circulatory system, and rapid, specific delivery of messages by the nervous system enables each cell in the body to communicate with most other cells.

In the first section of this chapter we examine the mechanisms of cell-to-cell communication. In the second half of the chapter we look at why efficient cell-to-cell communication is important to the body as a whole. We return to the central theme introduced in the first chapter, *homeostasis,* the process by which the body maintains a relatively stable internal environment in the face of either an external or an internal disturbance [∞ p. 6]. In this chapter and those that follow, we will see how the control of homeostasis resides primarily in the nervous and endocrine systems, with their combination of chemical and electrical signals.

▶ CELL-TO-CELL COMMUNICATION

By most estimates the human body is composed of about 75 *trillion* cells. Those cells face a daunting task—to communicate with each other in a manner that is rapid yet also conveys a tremendous amount of information. Surprisingly, there are only two basic types of physiological signals: electrical and chemical. Electrical signals are changes in a cell's membrane potential [∞ p. 34]. Chemical signals are molecules secreted into the extracellular fluid (ECF) by cells, and they are responsible for most communication within the body. The cells that receive electrical or chemical signals are called **target cells,** or **targets** for short.

Our bodies use three basic methods of cell-to-cell communication: (1) direct cytoplasmic transfer of electrical and chemical signals through gap junctions that connect adjacent cells; (2) local communication by chemicals that diffuse through the extracellular fluid, and (3) long-distance communication through a combination of electrical signals carried by nerve cells and chemical signals transported in the blood. A given molecule can func-

tion as a signal by more than one method. For example, a molecule can act close to the cell that releases it as well as act on cells in distant parts of the body.

Gap Junctions Transfer Chemical and Electrical Signals Directly Between Cells

The simplest form of cell-to-cell communication is the direct transfer of electrical and chemical signals through **gap junctions,** protein channels that create cytoplasmic bridges between adjacent cells (Fig. 6-1a ■) [∞ p. 57]. A gap junction forms when membrane-spanning proteins, called *connexins,* on two adjacent cells unite. The connexins create a protein channel (*connexon*) that can open and close. When the channel is open, the connected cells function like a single cell with multiple nuclei (a *syncytium*).

When gap junctions are open, ions and small molecules such as amino acids, ATP, and cyclic AMP pass directly from the cytoplasm of one cell to the cytoplasm of the next. As with other membrane channels, larger molecules are excluded. In addition, gap junctions are the only means by which electrical signals can pass *directly* from cell to cell. In humans, gap junctions are found in many different tissue types, including heart muscle, some types of smooth muscle, liver, and neurons of the brain.

Paracrines and Autocrines Are Chemical Signals Distributed by Diffusion

Local communication is accomplished by paracrines and autocrines. A **paracrine** [*para-,* beside + *krinein,* to secrete] is a chemical that is secreted by a cell to act on cells in its immediate vicinity. If this signal molecule acts on the cell that secreted it, the chemical is called an **autocrine** [*auto-,* self]. In some cases a chemical may act both as an autocrine and as a paracrine.

Paracrines and autocrines reach their target cells by diffusing through the interstitial fluid (Fig. 6-1b ■). Because distance is a limiting factor for diffusion, the effective range of the chemical signal is restricted to adjacent cells. A good example of a paracrine is histamine, a chemical released from damaged cells. When you scratch yourself with a pin, the red, raised wheal that results is due in part to the local release of histamine from the injured tissue. The histamine acts as a paracrine, diffusing over to the capillaries in the immediate area of the injury and making them more permeable. Fluid leaves the blood vessels and collects in the interstitial space, causing swelling around the area of injury.

Several important groups of molecules act as paracrines. **Neuromodulators** are paracrines and autocrines secreted by neurons. **Cytokines** are regulatory peptides that usually act close to the site where they are secreted. Cytokines are discussed further below.

Eicosanoids [∞ p. 30] are lipid-derived paracrines that play important roles in inflammation and allergic responses. This group includes *prostaglandins, thromboxanes,* and *leukotrienes.* Leukotrienes were discovered recently to play a significant role in asthma, a lung con-

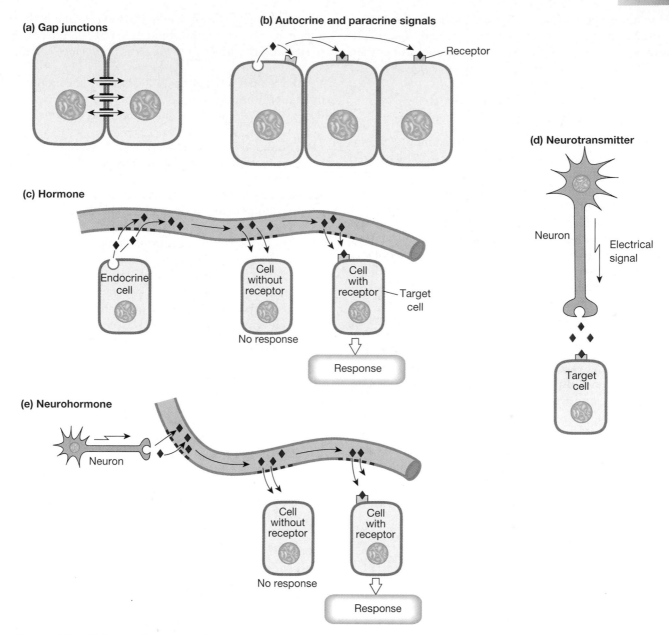

■ **Figure 6-1   Cell-to-cell communication**   (a) Gap junctions form direct cytoplasmic connections between adjacent cells. (b) Paracrines are secreted by one cell and diffuse to adjacent cells. Autocrines act on the same cell that secreted them. (c) Hormones are secreted by endocrine glands or cells into the blood. Only target cells, with receptors for the hormone, will respond to the signal. (d) Electrical signals travel long distances through neurons. Neurons release neurotransmitters that diffuse across a small gap to the target cell. (e) Neurons also release neurohormones into the blood. As with hormones, only cells with receptors respond to the neurohormone.

dition where the smooth muscle of the airways constricts, making it difficult to breathe. Some of the newest drugs to treat asthma either inhibit leukotriene synthesis or block leukotriene signal pathways.

Electrical Signals, Hormones, and Neurohormones Carry Out Long-Distance Communication

All cells in the body can release paracrines, but most long-distance communication between cells is the responsibility of the nervous and endocrine systems. The endocrine system communicates using **hormones** [*hormon*, to excite], chemical signals that are secreted into the blood and distributed all over the body by the circulation. Hormones come in contact with most cells of the body but only those cells with receptors for the hormone are target cells (Fig. 6-1c ■).

The nervous system uses a combination of chemical signals (**neurocrines**) and electrical signals to communicate. An electrical signal travels along a nerve cell (*neuron*) until it reaches the very end of the cell, where it is translated into a chemical signal secreted by the neuron. If the chemical signal diffuses from the neuron across a narrow extracellular space to a target cell, it is called a **neurotransmitter** (Fig. 6-1d ■). If the chemical released by the neuron diffuses into the blood for distribution, it is called a **neurohormone** (Fig. 6-1e ■). The similarities between neurohormones and hormones secreted by the endocrine system blur the distinction between the nervous and endocrine systems, making them a continuum rather than two distinct systems (see Fig. 6-26 ■, p. 178).

Cytokines Act as Both Local and Long-Distance Signals

Cytokines are the most recently identified family of communication molecules. Initially the term **cytokine** referred only to proteins that modulate immune responses, but in the past few years it has been broadened to include a variety of regulatory peptides. All nucleated cells synthesize and secrete cytokines in response to stimuli. Cytokines control cell development, differentiation, and the immune response. In development and differentiation, cytokines function as autocrines or paracrines. In stress and inflammation, cytokines act on relatively distant targets and are transported through the circulation just as hormones are.

How do cytokines differ from hormones? In general, they act on a broader spectrum of target cells. In addition, cytokines are not produced by specialized glands and are made on demand. In contrast, most protein or peptide hormones are made in advance and stored in the endocrine cell until needed. However, the distinction between cytokines and hormones is sometimes blurry. For example, erythropoietin, the molecule that controls synthesis of red blood cells, is by tradition considered a hormone but functionally fits the definition of a cytokine.

✔ Match the physiological signals on the left with the signal's properties on the right.

(a) autocrine
(b) cytokine
(c) gap junctions
(d) hormone
(e) neurohormone
(f) neurotransmitter
(g) paracrine

1. electrical signals
2. chemical signals
3. both electrical and chemical signals

✔ Which methods of cell-to-cell communication on the list above involve transport through the circulatory system?

✔ A cat sees a mouse and pounces on it. Do you think the internal signal to pounce could have been transmitted by a paracrine? Give two reasons to explain why or why not.

▶ SIGNAL PATHWAYS

Chemical signals in the form of paracrines, autocrines, and hormones are released from cells into the extracellular compartment. This is not a very specific way to ensure that they find their targets because substances that travel through the blood reach nearly every cell in the body. Yet cells do not respond to every signal that reaches them. Why do some cells respond to a chemical while others ignore it? The answer lies in the **receptor** proteins to which chemical signals bind [∞ p. 114]. *A cell cannot respond to a chemical signal if it lacks the appropriate receptor for that signal.*

If a cell has a receptor for a signal *ligand*, binding of the ligand to the receptor will initiate a response pathway. All signal pathways share common features:

1. The signal ligand, called a **first messenger** because it brings information to its target cell, binds to its receptor.
2. Activation of the receptor changes one or more intracellular **effectors** that direct the cell response [*effectus*, the carrying out of a task].
3. Many effectors are **protein kinases,** enzymes that transfer a phosphate group from ATP to a protein [∞ p. 88]. The *phosphorylation* of a protein can change its configuration and create a cellular response. Examples of changes that occur with phosphorylation include increased or decreased enzyme activity or changes in the open state of gated ion channels.

In the following sections, we will describe basic signal-receptor-response pathways. They may seem complex at first, but they follow patterns that you will encounter over and over as you study the systems of the body. Most physiological processes, from learning and memory to the beating of your heart, use some variation of these pathways. One of the wonders of physiology is the fundamental importance of these signal pathways and the way they have been conserved in animals ranging from worms to humans.

Receptors Are Located Inside the Cell or on the Cell Membrane

Receptors for chemical signal molecules have two functions: to bind ligands and to translate the message of the signal ligand into a cellular response. Target cell receptors are found in three locations: in the nucleus, in the cytosol, or as integral proteins of the cell membrane (Fig. 6-2 ■). *Lipophilic signal molecules* diffuse through the phospholipid

---

...continued from page 154

Later that day in the physician's office, the nurse practitioner explains diabetes to Marvin. Diabetes mellitus is a family of metabolic diseases caused by defects in the homeostatic pathways that regulate glucose metabolism. Several forms of diabetes exist, and some can be inherited. One form, called type 1 diabetes mellitus, is caused by deficient production of insulin, a protein hormone made in the pancreas. In another form, called type 2 diabetes mellitus, insulin is often present in normal or above normal levels. However, the insulin-sensitive cells of the body do not respond normally to the hormone.

Question 1:   *In which type of diabetes is the signal pathway for insulin more likely to be defective?*

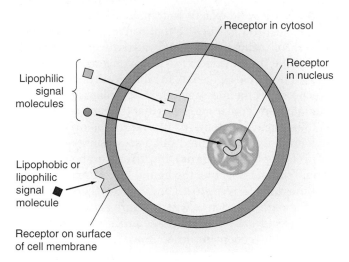

■ Figure 6-2   Target cell receptors   Lipophilic signal molecules diffuse through the cell membrane and bind to receptors in the cytosol or nucleus. Lipophobic signal molecules, which cannot diffuse through the cell membrane, and some lipophilic molecules bind to receptors on the cell membrane.

bilayer of cell membranes [∞ p. 120] and bind to cytosolic or nuclear receptors. In these cases, receptor activation usually turns on a gene and directs the nucleus to make new mRNA (*transcription*, [∞ p. 99]). The mRNA then provides a template for synthesis of new proteins (*translation*, [∞ p. 102]). This is a relatively slow process and the cell's response may not be noticeable for an hour or longer.

Lipophobic signal molecules are unable to enter the cell, so they bind to receptors on target cell membranes. Some lipophilic signal molecules have cell membrane receptors in addition to intracellular receptors. In general, the response time for membrane receptor-associated pathways is very rapid, and responses can be seen within milliseconds to minutes.

We can group membrane receptors and their effectors into four major categories, illustrated in Figure 6-3 ■. The simplest of these receptors are chemically gated (*ligand-gated*) ion channels [∞ p. 115]. Signal ligand binding opens or closes the channel and alters ion flow across the membrane.

For the other three receptor types shown in Figure 6-3 ■, information from the signal molecule must be passed across the membrane to initiate an intracellular response. This transmission of information from one side of a membrane to the other using membrane proteins is known as **signal transduction**, described in detail below.

The Most Rapid Signal Pathways Change Ion Flow Through Channels

As you have just learned, **ligand-gated ion channels** are the simplest receptors. When a signal molecule binds to the channel protein, a channel gate opens or closes, changing the cell's permeability to an ion. For example, if the signal opens a calcium receptor-channel, Ca^{2+} enters the cell. Cellular responses to calcium signals are discussed later in this chapter.

If the channel is selective for Na^+, K^+, or Cl^-, increased or decreased ion permeability changes the cell's membrane potential [∞ p. 145]. A change in membrane potential is an electrical signal that alters voltage-sensitive proteins (Fig. 6-4 ■). For example, at the end of Chapter 5 you learned that closure of an ATP-gated K^+ channel slows the leak of K^+ out of pancreatic beta cells, so they depolarize. The depolarization opens voltage-gated Ca^{2+} channels and ultimately leads to insulin secretion [∞ p. 145].

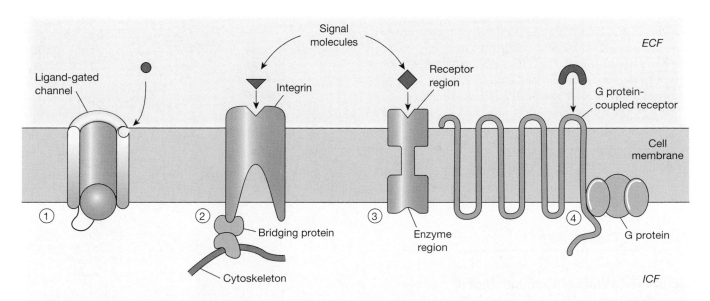

■ Figure 6-3   Four classes of membrane receptors   Membrane receptors can be ① ligand-gated ion channels; ② integrins linked to the cytoskeleton; ③ receptor-enzymes, where ligand binding activates an intracellular enzyme; or ④ G protein-coupled receptors with seven membrane-spanning regions.

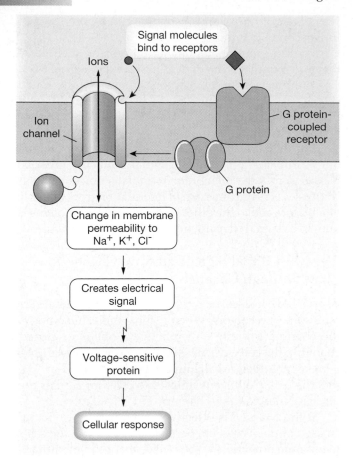

Figure 6-4　Ions create electrical signals　When signal molecules open or close ion channels, the membrane permeability for an ion changes. The resultant electrical signal alters voltage-sensitive proteins to create a response.

An example of a receptor-channel is the acetyl-choline-gated cation channel of skeletal muscle. The neurotransmitter *acetylcholine* opens the channel to initiate muscle contraction. Ligand-gated ion channels are particularly important in nerve and muscle.

Not all ligand-gated ion channels are receptors. Some are controlled by intracellular signal molecules called **second messengers,** such as cAMP. Others are linked to receptors by the membrane effectors called G proteins. We will discuss both second messengers and G proteins below.

✔ Will a cell depolarize or hyperpolarize in each of the following situations?

(a)　Cl^- channel opens

(b)　K^+ channel opens

(c)　Na^+ channel opens

Membrane Proteins Facilitate Signal Transduction

How can receptors that are not ion channels get information from an extracellular ligand into the cell? The answer is signal transduction, a process in which the ac-

tivated receptor alters intracellular molecules to create a response. But what exactly is transduction?

A **transducer** is a device that converts a signal from one form into a different one [*trans,* across + *ducere,* to lead]. For example, the transducer in a radio converts radio waves into sound waves (Fig. 6-5 ■). In biological systems, transducers convert the message of extracellular signal molecules into intracellular messages that trigger a response.

In biological systems, as in a radio, the original signal is not only transformed but also amplified [*amplificare,* to make larger]. **Signal amplification** begins with a single signal molecule, the ligand in Figure 6-6 ■, that combines with a receptor. The receptor then activates several molecules, each of which in turn activates several more molecules. By the end of the process, the effects of the ligand have been amplified much more than if there were a 1:1 ratio between each step. Amplification gives the body "more bang for the buck" by enabling a small amount of signal to create a large effect.

The basic pattern of a biological signal transduction pathway is shown in Figure 6-7 ■ and can be broken down into the following steps.

1. An extracellular *signal molecule* binds to and activates a protein or glycoprotein *membrane receptor*.
2. The receptor is linked to *effector proteins* that either
   a. activate *protein kinases*, or
   b. activate **amplifier enzymes** that create intracellular signal molecules known as **second messengers.** The most common amplifier enzymes and second messengers are listed in Table 6-1.

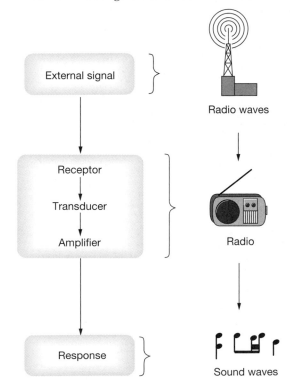

Figure 6-5　Signal transduction　In signal transduction, one form of signal is converted into a different form by a transducer. An example of signal transduction in the physical world is the radio, a transducer that converts radio waves into sound waves.

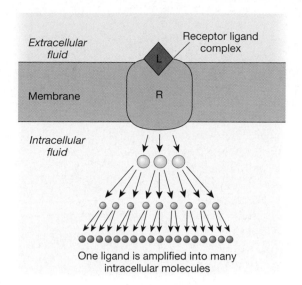

■ **Figure 6-6   Signal amplification**   One extracellular ligand (L) binds to a membrane receptor (R) and initiates a signal cascade that amplifies the message.

3. Second messenger molecules
   a. change enzyme activity, especially of protein kinases,
   b. increase intracellular calcium, or
   c. alter the open state of ion channels.
   Calcium binding to proteins or phosphorylation of proteins changes their function, creating a cellular response.
4. The proteins modified by calcium binding and phosphorylation control one or more of the following:
   a. metabolic enzymes,
   b. motor proteins for muscle contraction and cytoskeletal movement,
   c. proteins that regulate gene activity and protein synthesis, and
   d. membrane transport and receptor proteins.

If you think this list includes almost everything a cell does, you're right!

You can see from the pathway above that the steps of a signal transduction pathway form a **cascade** that starts when a stimulus (the signal molecule) converts inactive molecule A (the receptor) to an active form. Activated A then converts inactive molecule B into active B, active molecule B in turn converts inactive molecule C into active C, and so on, until at the final step a substrate is converted into a product (Fig. 6-8 ■). Many intracellular signaling pathways are cascades. Blood clotting is an important example of an extracellular cascade.

✔ Why do steroid hormones not require signal transduction and second messengers to exert their action? (Hint: Are steroids lipophobic or lipophilic? [∞ p. 23])

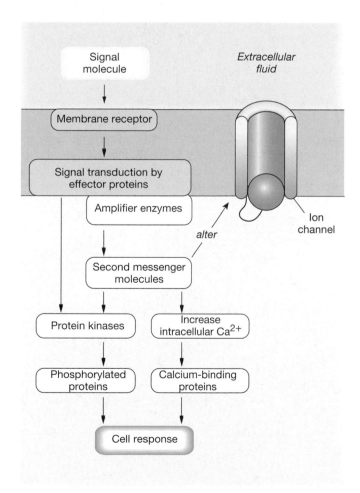

■ **Figure 6-7   Biological signal transduction**   Extracellular signal molecules bind to membrane receptors. The receptor activates effector proteins that mediate signal transduction. The process produces second messengers or active protein kinases, which in turn mediate the cellular response.

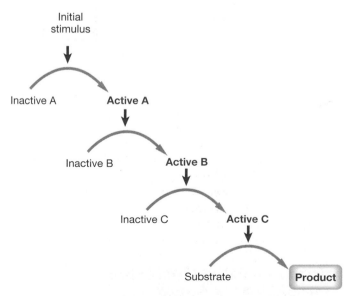

■ **Figure 6-8   Steps of a cascade**   An initial stimulus activates the first step in the cascade. In each step, the activated molecule from the previous step converts an inactive molecule into an active molecule. The final step is the conversion of a substrate into a product.

TABLE 6-1 Signal Transduction Pathways

Amplifier Enzyme	Cellular Location	Activated by	Converts	To
Adenylyl cyclase	Membrane	G protein-coupled receptor	ATP	cAMP
Guanylyl cyclase	Membrane Cytosol	Receptor-enzyme Nitric oxide (NO)	GTP	cGMP
Phospholipase C	Membrane	G protein-coupled receptor	Membrane phospholipids	IP$_3$ and DAG*

Second Messenger	Action	Effects
Ions		
Ca^{2+}	Binds to calmodulin	Alters enzyme activity
	Binds to other proteins	Exocytosis, muscle contraction, cytoskeleton movement
Nucleotides		
cAMP	Activates protein kinases, especially protein kinase A	Phosphorylates proteins
	Binds to ion channels	Alters channel opening
cGMP	Activates protein kinases, especially protein kinase C	Phosphorylates proteins
	Binds to ion channels	Alters channel opening
Lipid-derived		
IP$_3$	Releases Ca^{2+} from intracellular stores	See Ca^{2+} effects above
DAG	Activates protein kinase C	Phosphorylates proteins

* IP$_3$ = Inositol triphosphate and DAG = diacylglycerol

Integrin Receptors Transfer Information from the Extracellular Matrix

The membrane-spanning proteins called **integrins** (see Fig. 6-3 ■) mediate blood clotting, wound repair, cell adhesion and recognition in the immune response, and cell movement during development. On the extracellular side of the membrane, integrin receptors bind to proteins of the extracellular matrix [∞ p. 55] or to ligands such as antibodies and molecules involved in blood clotting. Inside the cell, integrins attach to the cytoskeleton via *bridging proteins*. Ligand binding to the receptor causes integrins to activate intracellular enzymes or alter the organization of the cytoskeleton.

The importance of integrin receptors is illustrated by two inherited conditions. In the first, absence of β2 integrin receptors results in white blood cells that cannot release bacteria-destroying chemicals, leading to recurrent bacterial infections. In the second condition, platelets, the cell fragments that play a key role in blood clotting, lack an integrin receptor. As a result, blood clotting is defective in these patients.

Receptor-Enzymes Have Protein Kinase or Guanylyl Cyclase Activity

Receptor-enzymes have receptors on the extracellular side of the membrane that activate enzymes on the cytoplasmic side (see Fig. 6-3 ■). In some instances, the receptor and enzyme are parts of the same protein molecule. In other cases, the enzyme is a separate protein. The enzymes of receptor-enzymes are either protein kinases, such as the tyrosine kinase illustrated in Figure 6-9 ■, or

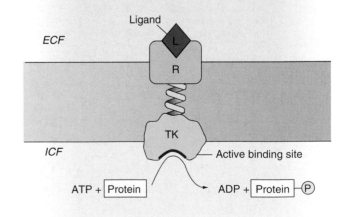

■ **Figure 6-9 Tyrosine kinase, an example of a receptor-enzyme** The receptor is a membrane-spanning protein whose cytoplasmic region has protein kinase activity. Receptor-ligand binding activates the enzyme, which in turn transfers a phosphate group from ATP to a tyrosine (amino acid) in the protein.

guanylyl cyclase. *Guanylyl cyclase* is the amplifier enzyme that converts GTP to *cyclic GMP* (cGMP) [∞ p. 35]. Ligands for receptor-enzymes include many growth factors and cytokines as well as the hormone insulin.

Most Signal Transduction Uses G Proteins

The **G protein-coupled receptors** are a large and complex family of membrane-spanning proteins that cross the phospholipid bilayer seven times. The cytoplasmic tail of the receptor protein is linked to a membrane transducer molecule known as a *G protein*. Hundreds of G protein-coupled receptors have been identified, and the list is growing daily. The types of ligands that bind to G protein-coupled receptors include hormones, growth factors, olfactory molecules, visual pigments, and neurotransmitters. In 1994 Alfred G. Gilman and Martin Rodbell received a Nobel prize for the discovery of G proteins and their role in cell signaling.

G proteins get their name from the fact that they bind guanosine nucleotides. Inactive G proteins are bound to guanosine diphosphate (GDP). Exchanging the GDP for guanosine triphosphate (GTP) activates the G protein. When G proteins are activated, they either (1) open an ion channel in the membrane or (2) alter enzyme activity on the cytoplasmic side of the membrane. G proteins linked to amplifier enzymes make up the bulk of all known signal transduction mechanisms.

DIABETES: Insulin's Signal Transduction Pathway In people with type 2 diabetes mellitus, insulin binds to its receptor but the cell fails to respond normally to the signal. As a result, researchers have been investigating the insulin receptor and its pathways in an effort to uncover the causes of type 2 diabetes. Insulin does not use the well-studied cAMP second messenger system, so for many years the insulin receptor-pathway was a "black box." In 1982 scientists discovered that the insulin receptor is a tyrosine kinase receptor-enzyme. Their next step was to discover what substrates the tyrosine kinase was phosphorylating. This search is still going on. At least two different substrates for the insulin receptor kinase have been identified, but the downstream steps of their cascades are still being investigated. Although we know what effects insulin has on glucose transport and metabolism, we still do not fully understand the pathways through which these effects take place.

The **G protein-coupled adenylyl cyclase-cAMP system** was the first identified signal transduction system (Fig. 6-10 ■). Dr. Earl Sutherland discovered it in the 1950s when he was studying the effects of hormones on carbohydrate metabolism. His work proved so significant to our understanding of signal transduction that Sutherland was awarded a Nobel prize in 1971. Cyclic AMP is the second messenger for many protein hormones.

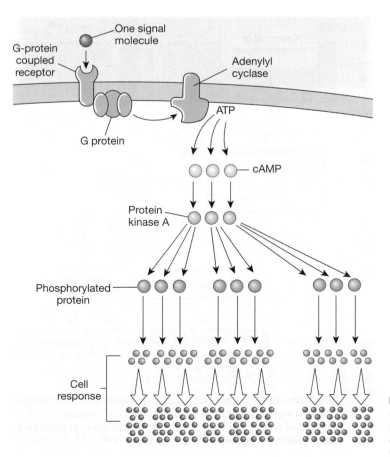

■ Figure 6-10   **The G protein-coupled adenylyl cyclase-cAMP system** G protein-coupled receptors activate adenylyl cyclase, an amplifier enzyme that converts ATP to cAMP. The second messenger cAMP activates protein kinase A, which phosphorylates proteins to continue the signal cascade.

Measuring Calcium Signals If you've ever run your hand through a tropical ocean at night and seen the glow of bioluminescent jellyfish, you've seen a calcium signal. Aequorin, a protein complex isolated from jellyfish, is one of the molecules that scientists use to indicate the presence of calcium during a cellular response. When aequorin combines with calcium, it releases light that can be measured by electronic detection systems. Since the first use of aequorin in 1967 researchers have been designing better and better indicators that allow them to follow calcium signals in cells. With the help of molecules called fura, Oregon green, BAPTA, and chameleons, we can now watch calcium diffuse through gap junctions and flow out of intracellular organelles.

half-life of only 2–30 seconds. In tissues, NO is synthesized by the action of the enzyme *nitric oxide synthase* (NOS) on the amino acid *arginine*. NO diffuses into target cells where it activates the cytosolic form of guanylyl cyclase and causes formation of the second messenger cGMP.

Nitric oxide in the brain acts as a neurotransmitter and neuromodulator. In blood vessels, NO is produced by endothelial cells lining the vessel. It then diffuses into adjacent smooth muscle cells, causing them to relax and dilate the blood vessel. In 1998 the Nobel prize for physiology and medicine was awarded jointly to Robert Furchgott, Louis Ignarro, and Ferid Murad for their work on NO as a signal molecule in the cardiovascular system.

✔ What is the difference between a first messenger and a second messenger?

✔ Organize the following terms into the correct order for a signal transduction cascade:

   (a) cell response, receptor, second messenger, signal ligand

   (b) amplifier enzyme, cell response, phosphorylated protein, protein kinase, second messenger

---

▶ MODULATION OF SIGNAL PATHWAYS

As you have just learned, signal pathways in the cell can be very complex. To complicate matters, different cells may respond differently to a single signal molecule. How can one molecule trigger response A in tissue 1 and response B in tissue 2? *For most signal molecules, the target cell response is determined by the receptor and its associated intracellular pathways, not by the ligand.* Because of the importance of signaling pathways, cells use receptors to maintain flexibility in their responses.

Receptors Exhibit Saturation, Specificity, and Competition

Receptors are proteins, so receptor-ligand binding exhibits the characteristics of specificity, competition, and saturation, just like the enzymes we discussed in Chap-

ter 4 [∞ p. 81] and the transporters in Chapter 5 [∞ p. 121].

Specificity and Competition: Multiple Ligands for One Receptor Receptors have binding sites for their ligands, just as enzymes and transporters do. As a result, different molecules with similar structures may be able to bind to the same receptor. A classic example of this principle involves two neurocrines: the neurotransmitter *norepinephrine* and its cousin *epinephrine* (also called *adrenaline*), a neurohormone. Both molecules bind to a class of receptors called *adrenergic receptors*. (*Adrenergic* is the adjective relating to adrenaline.) The ability of adrenergic receptors to bind these neurocrines but not others demonstrates *specificity* of the receptors.

Epinephrine and norepinephrine also exhibit *competition* for a single receptor type. Both neurocrines bind to subtypes of adrenergic receptors designated alpha (α) and beta (β). However, α receptors have a higher binding affinity for norepinephrine. The β receptor subtype designated β_2 has a higher affinity for epinephrine. And a third adrenergic receptor, designated β_1, has equal affinity for both signal molecules, so they compete for its binding sites on equal footing.

Agonists and Antagonists When a ligand combines with a receptor, one of two events follows. The ligand either turns on the receptor and elicits a response, or the ligand occupies the binding site and prevents the receptor from responding (Fig. 6-14 ■). Ligands that turn on receptors are known as **agonists** and ligands that block receptor activity are called **antagonists**.

Agonists may occur in nature. *Nicotine*, the chemical found in tobacco, mimics the activity of the neurotransmitter *acetylcholine*. Agonists can also be synthesized using what scientists learn from the study of receptor and ligand binding sites. The ability of agonists to mimic the activity of naturally occurring ligands has led to the development of many drugs.

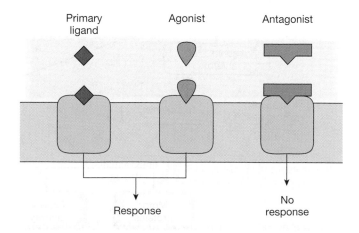

■ **Figure 6-14** **Agonists and antagonists** Molecules that bind to the same receptor and elicit a response are agonists. A molecule that binds and prevents a response is an antagonist.

As pharmacologists learn how particular ligands interact with their receptors and with the enzymes that break them down, they are able to design drugs that are longer acting and more resistant to enzymatic degradation. One example is the family of estrogens (female sex hormones) that are found in birth control pills. These drugs are agonists of naturally occurring estrogens but have chemical groups added to protect them from breakdown and extend their active life.

Antagonist ligands behave in the same fashion as competitive inhibitors of enzymes [∞ p. 83]. They bind to the receptor and block the normal ligand from binding without turning the receptor on themselves. As a result, the cell cannot respond. Antagonist drugs have proven useful for treating many conditions. For example, *tamoxifen*, an antagonist to the estrogen receptor, and *flutamide*, an androgen-receptor blocker, are used in the treatment of hormone-dependent cancers of the breast or prostate gland, respectively.

Multiple Receptors for One Ligand

For many years physiologists were unable to explain their observation that a single signal molecule could have different effects in different tissues. For example, epinephrine, the neurohormone described above, dilates blood vessels in skeletal muscle but constricts blood vessels in the intestine. How can one chemical have opposite effects?

The answer became clear when scientists discovered that many receptors come in variants known as *isoforms* [*iso-*, equal]. Receptor isoforms are equivalent to *isozymes*, enzymes that come in related forms [∞ p. 81]. When a signal molecule binds to different isoforms of its receptor, the cellular response will depend on the receptor isoform.

For example, the α and β adrenergic receptors for epinephrine described earlier are examples of receptor isoforms. When epinephrine binds to α-receptors on blood vessel smooth muscle, blood vessels constrict (Fig. 6-15 ■). When epinephrine binds to β-receptors on the muscle, blood vessels dilate. In other words, the response of the blood vessel to epinephrine depends on the receptor isoform, not on the ligand that activates the receptor. Many drugs now are designed so that they are specific for one receptor isoform and not others (See Emerging Concepts box: Selective Estrogen Receptor Modulators).

Selective Estrogen Receptor Modulators For many years scientists believed that estrogen, one of the female sex hormones, had only one receptor isoform and that all estrogen agonists had similar tissue-specific effects. But as pharmacologists developed the drug tamoxifen, they noticed that it was an estrogen agonist in some tissues but an estrogen antagonist in others. Tamoxifen belongs to a class of drugs known as *selective estrogen receptor modulators,* or SERMs. The action of SERMs is tissue specific. For example, SERMs are estrogen agonists in bone, where they are used to prevent osteoporosis. In other tissues, such as breast and uterus, SERMs are antagonists and are used for the treatment of hormone-dependent cancers. The observation that SERMs have tissue-specific effects is consistent with the 1996 discovery that there are two estrogen receptor isoforms, designated ERα and ERβ.

✔ Insulin increases glucose transport across the cell membrane of an adipocyte but not across the membrane of a liver cell. Think of at least two possible mechanisms that would allow this one hormone to have these two different effects.

Up- and Down-Regulation of Receptors Enables Cells to Modulate Cellular Response

Saturation of enzymes or transport proteins refers to the fact that enzymatic or transport activity can be "maxed out" with the presence of excess substrate because cells have limited numbers of protein molecules. The same phenomenon can occur with receptors. A cell's ability to respond to a chemical signal can be limited by the finite number of receptors for that signal.

A single cell contains between 500 and 100,000 receptors on the surface of its cell membrane, with additional receptors in the cytosol and nucleus. In any given cell, the number of receptors may change over time. Old receptors are withdrawn from the membrane by endocytosis and are broken down in lysosomes. New receptors are inserted into the membrane by exocytosis. Intracellular receptors are also made and broken down. This flexibility permits a cell to vary its responses to

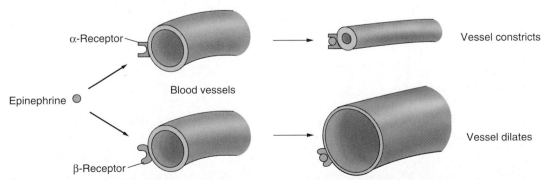

■ **Figure 6-15   Target response depends on the target receptor**   The response of a cell depends on the receptor. In this example, epinephrine will either dilate or constrict blood vessels depending on the receptor found on the blood vessel.

chemical signals depending on the extracellular conditions and the internal needs of the cell.

What happens if a signal molecule is present in the body in abnormally high concentrations for a sustained period of time? Initially the increased signal creates an enhanced response. As this continues, the target cells may attempt to bring their response back to normal by **down-regulation** of the receptors for the signal.

Down-regulation takes two forms, either a decrease in receptor number or a decrease in binding affinity. In both cases, the result is a lessened response of the target cell even though the concentration of the signal molecule remains high. Down-regulation is partially responsible for the development of *drug tolerance*, a condition in which the response to a given dose decreases despite continuous exposure to the drug. The development of tolerance to opiates such as morphine and codeine occurs when the receptors for these drugs down-regulate.

Up-regulation is the opposite of down-regulation. If the concentration of a ligand decreases, the target cell may insert more receptors into the cell membrane in an attempt to keep its response at a normal level. For example, if a nerve cell is damaged and unable to release normal amounts of neurotransmitter, its target cell will up-regulate its receptors. This up-regulation makes the target cell more sensitive to whatever neurotransmitters are present.

Cells Must Be Able to Terminate Signal Pathways

Signals turn on and off, so cells must be able to tell when a signal is over. This requires that signaling processes have built-in termination mechanisms. For example, to stop the response to a calcium signal, the cell removes Ca^{2+} from the cytosol by pumping it back into the endoplasmic reticulum or out into the extracellular fluid.

Receptor activity can be stopped in a variety of ways. The extracellular first messenger can be degraded by enzymes in the extracellular space. An example of

...continued from page 156

"My dad takes insulin shots for his diabetes," Marvin says. "What does insulin do?" The nurse practitioner replies that normally, insulin helps many cells take up and utilize glucose. But in all types of diabetes, the cells are not taking up and using glucose normally, so fasting blood glucose concentrations are elevated. If people with type 1 diabetes are given shots of insulin, their blood glucose levels decline. If people with type 2 diabetes are given insulin, blood glucose levels may change very little.

Question 2: *In which form of diabetes are the insulin receptors more likely to be up-regulated?*

this is the breakdown of the neurotransmitter acetylcholine. Other first messengers, particularly neurotransmitters, can be removed from the extracellular fluid by transporting them into neighboring cells. A widely used class of antidepressant drugs called *selective serotonin reuptake inhibitors,* or SSRIs, extends the active life of the neurotransmitter serotonin by slowing its removal from the extracellular fluid.

Once a receptor is bound to its ligand, activity can also be terminated by endocytosis of the entire receptor-ligand complex. This process was illustrated in Figure 5-26 [∞ p. 130]. After the vesicle is in the cell, the ligand is removed and the receptors can be returned to the membrane by exocytosis.

Many Diseases and Drugs Target the Proteins of Signal Transduction

As we learn more about cell signaling, scientists are realizing how many diseases are linked to problems with signaling pathways. Disease processes can be caused by alterations in receptors or by problems with G proteins or second messenger pathways (see Table 6-2 for some examples). A single change in the amino acid sequence of

TABLE 6-2 Diseases or Conditions Linked to Abnormal Signaling Mechanisms

Genetically inherited abnormal receptors		
Receptor	*Physiological Alteration*	*Disease or Condition That Results*
Vasopressin receptor (X-linked defect)	Shortens half-life of the receptor	Congenital diabetes insipidus
Calcium sensor in parathyroid gland	Fails to respond to increase in plasma Ca^{2+}	Familial hypercalcemia
Rhodopsin receptor in retina of eye	Improper protein folding	Retinitis pigmentosa
Toxins affecting signal pathways		
Toxin	*Physiological Effect*	*Condition That Results*
Bordetella pertussis	Blocks inhibition of adenylate cyclase (i.e., keeps it active)	Whooping cough
Cholera toxin	Blocks enzyme activity of G proteins; cell keeps making cAMP	Ions secreted into lumen of intestine cause massive diarrhea

G Proteins and Disease A variety of diseases have now been traced to disturbances in cell-signaling mechanisms (Table 6-2). Abnormalities in G protein function have been linked to certain oncogenes [*onkos*, mass]. An **oncogene** is an abnormal gene that plays a role in the growth and development of cancerous cells. One oncogene known as *ras* codes for G proteins linked to the phospholipase C pathway. When *ras* proteins are produced in tumor cells, the phosphorylated products that result apparently affect cell division, leading to the abnormal growth and division of the cancerous cells. Many researchers are studying oncogenes and the signal paths that they control, hoping that a clearer understanding of these mechanisms will lead them to better treatments for various forms of cancer.

a receptor protein can alter the shape of the receptor's binding site and destroy or modify its activity.

Pharmacologists are using information about signaling mechanisms to design drugs to treat disease. Some of the alphabet soup of drugs in widespread use include ARBs (angiotensin receptor blockers), SERMs (selective estrogen receptor modulators), beta-blockers that block adrenergic β-receptors, calcium-channel blockers, H_1 (histamine type 1) receptor antagonists, and COX2 (cycloxygenase 2) inhibitors that prevent the production of inflammatory prostaglandins. We will encounter many of these drugs again when we study the systems where they are effective.

▶ HOMEOSTASIS

In Chapter 1, you learned that homeostasis is the ability of the body to maintain a relatively stable internal environment [∞ p. 6]. In this section we examine homeostatic control mechanisms and compare the role of the nervous and endocrine systems in maintaining homeostasis.

The Development of the Concept of Homeostasis

The idea of an internal environment was developed by the French physician Claude Bernard in the mid-1800s. During his studies of experimental medicine, he noted the stability of various physiological **parameters*** such as body temperature, heart rate, and blood pressure. As chairman of physiology at the University of Paris, he wrote of "la fixité du milieu intérieur" (the constancy of the internal environment). This was an idea that applied to many of the experimental observations of his day, and it became the subject of discussion among physiologists and physicians.

In 1929, an American physiologist named Walter B. Cannon created the name *homeostasis* to describe the reg-

ulation of this internal environment. In his essay[†], Cannon explained that he selected the prefix *homeo-* meaning *like* or *similar* rather than the prefix *homo-* meaning *same* because the internal environment is maintained within a range of values rather than at an exact fixed value. He also pointed out that the suffix *-stasis* in this instance means *a condition*, not a state that is static and unchanging. Thus, Cannon's homeostasis is a state of maintaining "a similar condition," also described as "a relatively constant internal environment."

Cannon proposed a list of parameters that are under homeostatic control based on observations made by numerous physiologists and physicians during the nineteenth and early twentieth centuries. We now know that his list was both accurate and complete. Cannon divided his parameters into what he called environmental factors that affect cells (osmolarity, temperature, and pH) and "materials for cell needs"—nutrients, water, sodium, calcium, other inorganic ions, oxygen, and "internal secretions having general and continuous effects." Cannon's "internal secretions" are the hormones and other chemicals that our cells use to communicate with each other.

Cannon also postulated a number of properties of homeostasis that were validated in the succeeding years. You will encounter these properties repeatedly as you study the various systems of the body. Cannon's four postulates are:

1. **The role of the nervous system in preserving the "fitness" of the internal environment.** Fitness in this instance means conditions that are compatible with normal function. The nervous system coordinates and integrates blood volume, osmolarity, blood pressure, and body temperature, among other parameters.

2. **The concept of *tonic level* of activity [*tonos*, tone].** To quote Cannon, "an agent may exist which has … a moderate activity which can be varied up and down." This is like the volume control on a radio, which enables you to make the sound level louder or softer by turning a single knob. A physiological example of a tonically active control system is the nervous regulation of diameter in certain blood vessels, where an increase in input from the nervous system decreases diameter, while a decrease in input from the nervous system causes the diameter to increase (Fig. 6-16 ■). Tonic control is one of the more difficult concepts in physiology because we have a tendency to think of things as off or on rather than as more or less.

3. **The concept of antagonistic controls.** Cannon wrote, "When a factor is known which can shift a homeostatic state in one direction, it is reasonable to look for … a factor or factors having an opposing effect." Systems that are not under tonic control are usually under antagonistic control, either by hormones or the nervous

---

*A parameter [*para-*, beside + *meter*, measure] is one of the variables in a system.

[†]Walter B. Cannon, "Organization for Physiological Homeostasis," *Physiological Reviews* 9: 399–443, 1929.

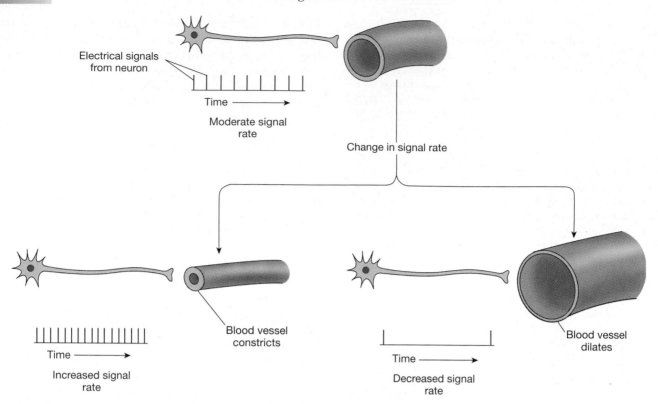

■ **Figure 6-16   Tonic control**   Tonic control regulates physiological parameters in an up-down fashion. In this example, a moderate rate of signaling from the controlling neuron results in a blood vessel of intermediate diameter. If the signal rate decreases, the blood vessel dilates. If the signal rate increases, the blood vessel constricts.

system. For example, insulin and glucagon are antagonistic hormones. Insulin decreases the concentration of glucose in the blood, while glucagon increases it. In pathways controlled by the nervous system, the sympathetic and parasympathetic divisions often have opposing effects. For example, chemical signals from a sympathetic neuron increase heart rate, while chemical signals from a parasympathetic neuron decrease it (Fig. 6-17 ■).

> ...*continued from page 166*
>
> "Why is an elevated blood glucose concentration bad?" Marvin asks. "The elevated blood glucose itself is not bad," says the nurse practitioner. "But when it is high after an overnight fast, it suggests that there is something wrong with the way your body is regulating its glucose metabolism." When a normal person absorbs a meal with carbohydrates, blood glucose levels increase and stimulate insulin release. Hours after the meal, when blood glucose levels fall as the cells metabolize glucose, secretion of another pancreatic hormone, glucagon, increases. Glucagon raises blood glucose and helps keep the level within the homeostatic range.
>
> **Question 3:**   *The homeostatic regulation of blood glucose levels by the hormones insulin and glucagon is an example of which of Cannon's postulates?*

4. **The concept that chemical signals can have different effects in different tissues of the body.** Cannon postulated that "Homeostatic agents, antagonistic in one region of the body, may be cooperative in another region." This theory was subsequently shown to be true. However, it was not until we learned about cell receptors that the basis for the seemingly contradictory actions of some hormones or nerves became clear. As you learned in the first part of this chapter, a single chemical signal can have different effects depending on the receptor at the target cell. An example of this is epinephrine constricting or dilating blood vessels depending on whether the vessel has alpha or beta adrenergic receptors (see Fig. 6-15 ■).

▶ **CONTROL PATHWAYS: RESPONSE AND FEEDBACK LOOPS**

Homeostasis is a continuous process that involves monitoring multiple parameters, then coordinating appropriate responses to minimize any disturbance. Homeostatic responses may take place in small, localized regions of the body or they may be bodywide, or *systemic*, responses. In either case, the process always has three components: a stimulus or change in condition, a cell or tissue that evaluates the stimulus and initiates a response, and the cells or tissues that carry out the response.

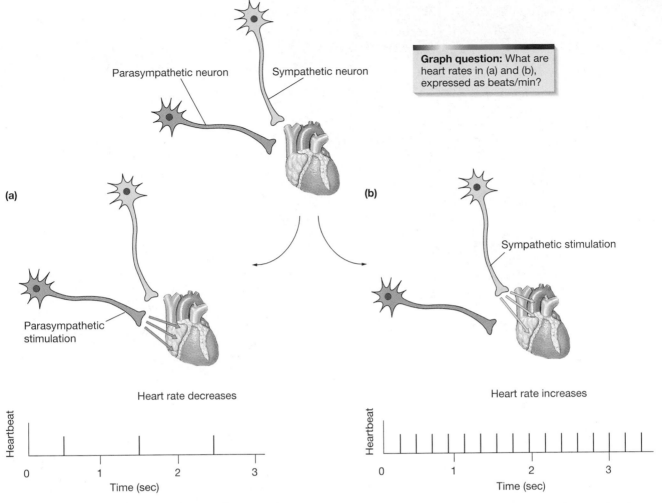

Graph question: What are heart rates in (a) and (b), expressed as beats/min?

■ **Figure 6-17   Antagonistic control of heart rate**   The heart is controlled by antagonistic neurons: some that speed up heart rate, some that slow down heart rate.

Homeostasis May Be Maintained by Local or Long-Distance Pathways

The simplest control takes place strictly at the tissue or cell involved. In **local control,** a relatively isolated change occurs in the vicinity of a cell or tissue and evokes a paracrine or autocrine response (Fig. 6-18 ■). More complicated **reflex control pathways** respond to changes that are more widespread or systemic in nature. In a reflex control pathway, a control center located away from the affected cell or tissue receives information and makes the decision to send a chemical or electrical signal to initiate a response.

Long-distance reflex pathways are traditionally considered to involve two control systems, the nervous system and the endocrine system. However, cytokines [∞ p. 156] are now considered the third control system. These proteins, secreted by all nucleated cells, can work by either local or reflex pathways depending on the circumstances. Acting as paracrines and autocrines, cytokines control the white blood cells that defend the body from outside invaders. During stress and systemic inflammatory responses, cytokines work together with the nervous and endocrine systems to integrate information from all over the body into coordinated reflex responses.

Local Control   Paracrines and autocrines are responsible for the simplest control systems. In local control a cell or tissue senses a change in its immediate vicinity and responds. The response is restricted to the region where the change took place, hence the term local control. One example of this type of response occurs when oxygen concentration in a tissue drops. The cells lining the small blood vessels bringing blood to that area sense the drop in oxygen concentration and respond by secreting a paracrine. The paracrine relaxes muscles in the blood vessel wall, dilating the blood vessels and bringing more blood and oxygen to the area. Paracrines that are probably involved in this response include carbon dioxide and metabolic products such as lactic acid.

Reflex Control   In a reflex pathway, control of the reaction lies outside the organ that carries out the response. We will use the term **reflex** to mean any long-distance pathway that uses the nervous system, endocrine system,

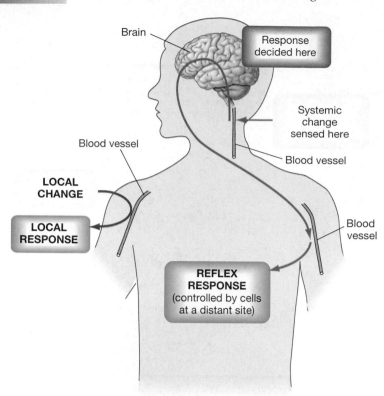

■ **Figure 6-18** **Comparison of local and reflex control** In local control, the response is initiated by cells in the vicinity of the change. In reflex control, a response is controlled by cells at a distant site. In the example shown, a change in systemic blood pressure is sensed in blood vessels in the neck. A signal is sent to the brain. The brain then sends a signal to create a response in blood vessels all over the body, represented here by a blood vessel in the arm.

or both to receive input about a change, integrate the information, and react appropriately. Reflex pathways can be broken down into response loops and feedback loops.

A **response loop** has three primary components: an *input signal, integration of the signal,* and an *output signal.* These three components can be broken down into the following sequence to form a pattern that is found with slight variations in all reflexes:

stimulus → receptor → afferent pathway → integrating center → efferent pathway → effector → response

The beginning of any homeostatic reflex pathway is (1) a **stimulus,** the disturbance or change that sets the pathway in motion. The stimulus may be a change in temperature, oxygen content, blood pressure, or any one of myriad parameters. The stimulus is sensed by (2) a *sensor,* or **receptor,** that is continually monitoring its environment. When alerted to a change, the receptor sends out a signal. (3) The signal, or **afferent** (incoming) **pathway,** links the receptor to (4) an integrating center. The **integrating center** is the control center that evaluates the incoming signal, compares it with the **setpoint,** or desired value, and decides on an appropriate response. From the integrating center, (5) an outgoing signal, the **efferent** (outgoing) **pathway,** travels to an effector. (6) The **effector** is the cell or tissue that carries out (7) the appropriate **response** to bring the situation back to within normal limits.

Sensory receptors The first step in a biological response loop is the receptor. NOTICE! This is a new and different use of the word *receptor.* Like many terms in physiology,

receptor can have different meanings (Fig. 6-19 ■). The receptors of a biological reflex are not membrane proteins like those involved in signal transduction. Reflex, or *sensory,* receptors are specialized cells, parts of cells, or complex multicellular receptors such as the eye. There are many sensory receptors in the body, each located anatomically where it is in the best position to monitor the parameter it detects. The eyes, ears, and nose are specialized receptors that sense light, sound, motion, and odors. Your skin is covered with less complex receptors that sense touch, temperature, vibration, and pain. Other receptors are internal: receptors in the joints of the skeleton, pressure receptors in blood vessels, osmolarity receptors, and receptors for oxygen and carbon dioxide. Biological reflex receptors are divided into *central receptors,* located in or closely linked to the brain, and *peripheral receptors* that reside elsewhere in the body.

All receptors have a **threshold,** a minimum stimulus that must be achieved to set the reflex response in motion. If a stimulus is below the threshold, no response loop will be initiated. You can demonstrate threshold easily by touching the back of your hand with a sharp, pointed object such as a pin. If you touch the point to your skin lightly enough, you can see the contact between the point and your skin even though you do not feel anything. In this case, the stimulus (pressure from the point of the pin) is below threshold and the pressure receptors of the skin are not responding. As you press harder, the stimulus reaches threshold, and the receptors respond by sending a signal through the afferent pathway, causing you to feel the pin.

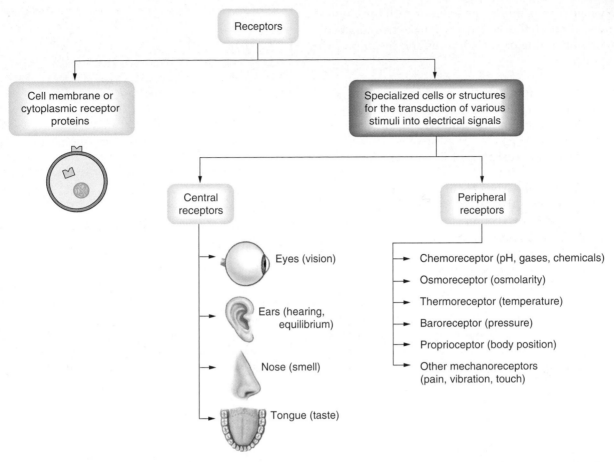

■ **Figure 6-19   Multiple meanings of the word receptor**   In biology, the word receptor may mean a protein that binds to a ligand. It may also refer to sensory receptors, the specialized cells or structures for transduction of stimuli into electrical signals. Sensory receptors can be divided into peripheral receptors scattered throughout the body and central receptors located in or close to the brain.

Afferent pathway   The afferent pathway in a reflex varies depending on the type of reflex. In a nervous reflex, such as the pin touch above, the afferent pathway is the electrical and chemical signal carried by a nerve cell. In endocrine reflexes, there is no afferent path because the stimulus comes directly into the endocrine cell.

Integrating center   The integrating center is the key to a reflex because it is the location that receives information about the change and decides on an appropriate response. If information is coming from a single stimulus, it is a relatively simple task for the integrating center to compare that information with the setpoint. In endocrine reflexes, the integrating center is the endocrine cell. In nervous reflexes, the integrating center lies within the central nervous system, which is composed of the brain and the spinal cord.

Integrating centers really "earn their pay" on occasions when two or more conflicting signals come in from different receptors. The center must evaluate each signal on the basis of its strength and importance and must come up with an appropriate response that integrates information from all the receptors. This is similar to the kind of decision making you must do when in one evening your parents want to take you to dinner, your friends are having a party, there is a television program you want to watch, and you have a major physiology test in three days. It is up to you to rank those items in order of importance and decide how you will act upon them.

Efferent pathway   Efferent pathways are relatively simple. In the nervous system, the efferent path is always the electrical and chemical signal transmitted by an efferent neuron. Because electrical signals traveling through the nervous system are identical, the distinguishing characteristic of the signal is the anatomical route taken by the nerve cell through which the signal goes. For example, the vagus nerve carries a nervous signal to the heart, and the phrenic nerve carries one to the diaphragm. Because the nature of the electrical message is always the same and there are relatively few types of neurotransmitters, nervous system efferent pathways are named using the anatomical name of the nerve that carries the signal.

In the case of a hormonal signal, the anatomical routing is always the same because all hormones travel in the blood to get to their target. Hormonal efferent pathways are distinguished by the chemical nature of the signal and are therefore named for the hormone that carries the message. For example, the efferent path for a reflex integrated through the pancreas will be either insulin or glucagon depending on the stimulus and the appropriate response.

Effectors The effectors of reflex pathways are the cells or tissues that carry out the response. The targets of nervous reflexes are muscles, glands, and some adipose tissue. The targets of endocrine pathways are any cells that have the proper receptor for the hormone.

Responses There are two levels of response that can be given for any reflex. One is the very specific *cell* or *tissue response* that results from combination of the signal ligand with the receptor: opening a channel, initiating protein synthesis, or modifying enzyme activity. The more general *systemic response* describes what those specific events mean to the tissue or the organism as a whole. For example, when the hormone epinephrine combines with β₂-adrenergic receptors on the walls of certain blood vessels, the cellular response is relaxation of the smooth muscle. The systemic response to relaxation of the blood vessel wall is increased blood flow through the vessel.

...continued from page 168

Marvin is fascinated by the ability of the body to keep track of glucose. "How does the pancreas know which hormone to secrete?" Special cells in the pancreas called beta cells sense an increase in blood glucose concentrations after a meal, and they release insulin in response. Insulin then acts on many tissues of the body so that they take up and utilize glucose.

Question 4: *In the insulin reflex that regulates blood glucose levels, what is the stimulus? the receptor? the integrating center? the efferent pathway? the effector(s)? the response(s)?*

Response Loops Begin with a Stimulus and End with a Response

Now that you have learned the parts of a reflex, we will apply them to nonbiological and biological examples. A simple nonbiological analogy to a homeostatic reflex pathway is an aquarium whose heater is set to maintain water temperature at 30° C in a room whose temperature is 25° C (Fig. 6-20 ■). The temperature at which the heater is set is the *setpoint* for the parameter.

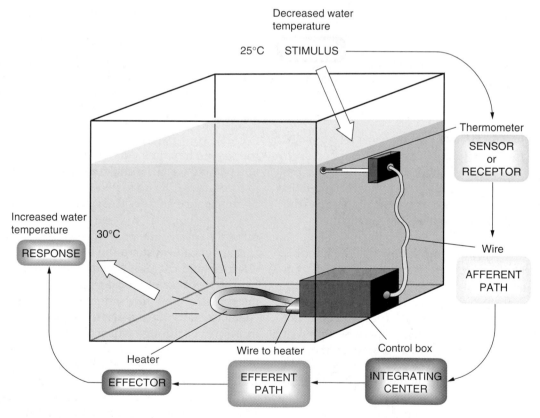

■ **Figure 6-20 Response and feedback loops** The control box of the aquarium is set to maintain a temperature of 30° ± 1° C in the water. The sensor (or receptor) sends a signal via the afferent path (a wire) to the control box that acts as the integrating center. The control box sends a message over the efferent path (another wire) to the effector, the heater. The heater turns on, completing the response loop. As the water heats, information about the water temperature is sent continuously to the control box. When the water temperature reaches the upper limit of the desired range, the control box turns off the heater. The increase in temperature that stops the response is a negative feedback signal.

The components of the aquarium are listed in the left column below, with the corresponding part of the response loop in the right column.

Component	Biological Reflex Equivalent
1. thermometer	1. receptor
2. communication wire	2. afferent pathway
3. control box	3. integrating center
4. communication wire	4. efferent pathway
5. heater	5. effector

Assume that initially the aquarium water is at room temperature, 25° C. When you turn the heater on, you set the response loop in motion. The thermometer (receptor) in the water senses a temperature of 25° C. It sends information about the current temperature via a wire (afferent path) to the control box (integrating center). The control box evaluates the incoming temperature signal, compares it with the setpoint for the system (30°), and decides that a response is needed to bring the water temperature up to the setpoint. The control box sends a message via another wire (efferent path) to the heater (effector), which turns on and starts heating the water (the response). This sequence from stimulus to response is the response loop.

The simple example given above demonstrates a parameter (temperature) that is under *tonic control* (see p. 167) by a single control system (the heater). We can also create a nonbiological system that obeys Cannon's postulate of *antagonistic control.* For example, think of a house that has both heating and air-conditioning. The owner would like the house to remain at 70° F (about 21° C). During the autumn when the mornings are chilly, the heater turns on to warm the house. Then as the day warms up, the heater is no longer needed. As sun heats the house and the temperature reaches 80° F, the air-conditioner turns on to cool the house back to 70°. The heater and air-conditioner have antagonistic control over the house temperature. A similar physiological example would be the hormones insulin and glucagon, which exert antagonistic control over glucose metabolism.

Setpoints Can Be Varied

In physiological systems, the **setpoint,** or normal value for any given parameter, can vary from person to person, or even for the same individual over a period of time. Factors that influence an individual's setpoint for a given parameter include inheritance as well as the conditions to which the person has become accustomed. The adaptation of physiological processes to a given set of environmental conditions is known as **acclimatization** if it occurs naturally or **acclimation** if the process is induced artificially in a laboratory setting. Each winter, northerners go south in February, hoping to escape the bitter subzero temperatures and snows of the northern climate. As they

walk around in 40° F weather in short-sleeve shirts, the southerners think they are crazy: the weather is cold! The locals are bundled up in coats and gloves. The difference in behavior is due to different temperature acclimatization, a difference in the setpoint for body temperature regulation that is a result of prior conditioning.

Physiological setpoints also vary within individuals in response to external cues such as the daily light-dark cycles and the seasons. These changing setpoints cause certain parameters to vary in predictable ways over a period of time, forming patterns of change known as biorhythms (see Biological Rhythms).

The response loop begins with a stimulus and ends with the response of the target cell. But it is only half of a reflex. The reflex is completed when the response becomes part of the stimulus and feeds back into the system, creating a feedback loop.

Feedback Loops Modulate the Response Loop

In the aquarium example above, the sensor sends a signal to the control box, telling it that the water is too cold. The response of the control box is to turn on the heater and warm up the water. But once the response starts, what keeps the heater from sending the temperature up to 50° C? The answer is a **feedback loop.** The response of the system, turning on the heater, alters the temperature of the water and changes the stimulus. The receptor monitoring the water sends a continuous signal back to the control box so that it receives constant input or feedback about the effect the heater is having. When the temperature warms up to the maximum acceptable, the control box shuts off the heater, thus completing the loop.

Negative Feedback Loops Are Homeostatic   For most reflexes, feedback loops are homeostatic, that is, designed to keep the system at or near a setpoint so that the parameter will be relatively stable. How well an integrating center succeeds in maintaining stability depends on the *sensitivity* of the system, the range of normal values. In the case of our aquarium, the control box is programmed to have a sensitivity of plus or minus one degree Celsius. If the water temperature drops from 30° to 29.5°, it is still within the acceptable range and no response is triggered. If the water temperature drops below 29° (30° − 1°), the control box turns the heater on (Fig. 6-21 ■). As the water heats up, the control box constantly receives information about the water temperature from the sensor. When the water reaches 31° (30° + 1°), the upper limit for the acceptable range, the feedback loop causes the control box to turn the heater off. The water then gradually cools off until the cycle starts all over again. The end result is a physiological variable that *oscillates* around the setpoint.

In biological systems, some receptors are more sensitive than others. For example, the sensors for osmolarity

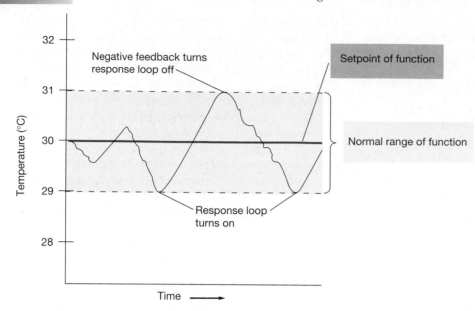

■ **Figure 6-21** **Oscillation around the setpoint** Most functions that are controlled homeostatically have a setpoint, or normal value for that function. The reflexes that control the function will operate when the function gets outside a predetermined range. As a result, the function oscillates around the setpoint.

trigger reflexes to conserve water when blood concentration increases only 3% above normal, but the sensors for low oxygen in the blood will not respond until oxygen has decreased by 40%.

A pathway where the response opposes or removes the signal is known as **negative feedback.** Negative feedback loops *stabilize* the physiological variable that is being regulated. In the aquarium example, the heater warms up the water (the response) and removes the stimulus (low water temperature). With the loss of the stimulus for the pathway, the response loop shuts off. All homeostatic reflexes are controlled by negative feedback so that the parameter being controlled will stay within a normal range. *Negative feedback loops can restore the normal state but cannot prevent the initial disturbance out of the normal range.*

Positive Feedback Loops Are Not Homeostatic A few reflexes are not homeostatic. In a **positive feedback loop,** the response *reinforces* the stimulus rather than decreasing it or removing it. In positive feedback, the response destabilizes the variable, triggering a vicious cycle of ever-increasing response and sending the system temporarily out of control (Fig. 6-22 ■). Because the response enhances the loop in positive feedback, these reflexes require some intervention or event outside the loop to stop them.

One example of a positive feedback loop involves the hormonal control of uterine contractions during childbirth (Fig. 6-23 ■). When the baby is ready to be delivered, it drops lower in the uterus and begins to put pressure on the *cervix,* the opening of the uterus. Sensory signals from the cervix to the brain cause release of the hormone *oxytocin,* which causes the uterus to contract

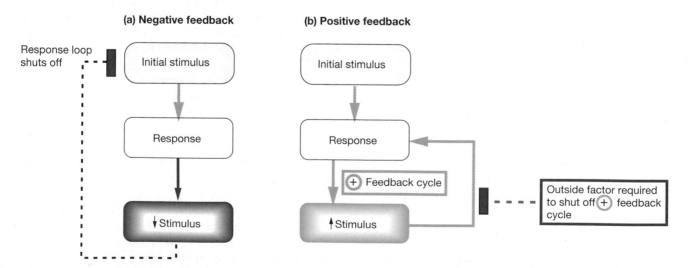

■ **Figure 6-22** **Negative and positive feedback** (a) In negative feedback, the response offsets the stimulus and the response loop shuts off. (b) In positive feedback, the response sends a message that reinforces the response loop, sending the parameter under control farther away from the setpoint. The response will continue to increase until some outside intervention stops the reflex loop.

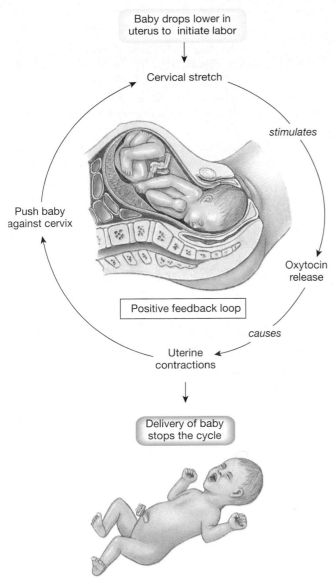

Baby drops lower in
uterus to initiate labor

Cervical stretch

stimulates

Push baby
against cervix

Oxytocin
release

Positive feedback loop

causes

Uterine
contractions

Delivery of baby
stops the cycle

■ **Figure 6-23   A positive feedback loop**   When labor begins, pressure created by the baby pushing against the cervix triggers release of the hormone oxytocin. Oxytocin causes strong uterine contractions that push the baby harder against the cervix, which reinforces oxytocin release. This positive feedback loop continues until the baby is delivered and the cervical stretch is relieved.

and push the baby's head even harder against the cervix, stretching it. The increased stretch causes more oxytocin release, which causes more contractions that push the baby harder against the cervix. This cycle continues until finally the baby is delivered, releasing the stretch on the cervix and stopping the positive feedback loop.

Feedforward Control Allows the Body to Anticipate Change and Maintain Stability

Negative feedback loops can stabilize a function and maintain it within a normal range but are unable to prevent the change that triggered the reflex in the first place. A few reflexes have evolved that allow the body to pre-

dict that a change is about to occur and start the response loop in anticipation of the change. These anticipatory responses are called **feedforward control.**

An easily understood physiological example of feedforward control is the reflex of salivation. The sight, smell, or even the thought of food is enough to start our mouths watering. The saliva is present in expectation of the food that will soon be eaten. This reflex extends even further, because the same stimuli can start the secretion of hydrochloric acid as the stomach anticipates food on the way. One of the most complex feedforward reflexes appears to be the body's response to exercise, to be discussed in Chapter 23.

Biological Rhythms Result from Changes in the Setpoint

In many reflex pathways, the triggers or stimuli are obviously related to the function of the reflex. In the aquarium example, a change in temperature is the trigger to ensure that temperature is maintained within the desired range. But this is not true of all reflexes. Many hormones, for example, are secreted continuously, with levels that rise and fall throughout the day. Most examples of these apparently spontaneous reflexes occur in a predictable manner and are often timed to coincide with a predictable environmental change such as light-dark cycles or the seasons. All animals exhibit some form of daily biological rhythm, called a **circadian rhythm** [*circa,* about + *dies,* a day]. Humans have circadian rhythms for many body functions, including blood pressure, body temperature, and metabolic processes. Body temperature peaks in the late afternoon and declines dramatically in the wee hours of the morning (Fig. 6-24a ■). Have you ever been studying late at night and noticed that you feel cold? This is not because of a drop in environmental temperature but because your thermoregulatory reflex has turned down your internal thermostat.

Many hormones in humans are secreted so that their concentration in the blood fluctuates predictably through a 24-hour cycle as their setpoints change. Cortisol, growth hormone, and the sex hormones are among the most noted examples. If an abnormality in hormone secretion is suspected, it is important to know the time of day that the test for hormone level is made. A level that is normal in a morning sample may be abnormally high or low in an afternoon sample (Fig. 6-24b ■). One method used to avoid this error involves collecting all urine for a 24-hour period and getting an average value for the hormone metabolites in the sample.

What is the adaptive significance of functions that vary with a circadian rhythm? Our best answer is that biological rhythms create an anticipatory response to a predictable environmental variable. There are seasonal rhythms of reproduction in many nonmammalian vertebrates and invertebrates, timed so that the offspring have food and other favorable conditions to maximize survival.

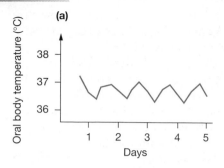

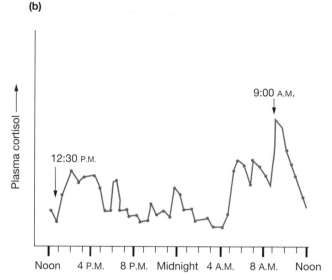

■ **Figure 6-24** **Circadian rhythms** (a) Body temperature rises and falls in a regular pattern on a daily basis. (b) Blood levels of the hormone cortisol change during a 24-hour period. Cortisol levels in the blood vary significantly between 9:00 A.M. and 12:30 P.M., so analysis of hormone levels must take into account the time of day that a blood sample was drawn.

Circadian rhythms cued by the light-dark cycle correspond to our rest-activity cycles. These rhythms allow our bodies to anticipate behavior and coordinate body processes accordingly. You may hear someone who is accustomed to eating dinner at 6 P.M. say that he cannot digest his food if he waits until 10 P.M. to eat because his digestive system has "shut down" in anticipation of going to bed.

One of the interesting correlations between circadian rhythms and behavior involves body temperature. Researchers found that self-described "morning people" have temperature rhythms that begin to climb before they awake in the morning, so that they get out of bed prepared to face the world. On the other hand, "night people" may be forced by school and work schedules to get out of bed while their body temperatures are still at their lowest point, before their bodies are prepared for activity. These same night people are still going strong and working productively into the wee hours of the morning when the morning people's body temperature is dropping and they are ready for bed.

...continued from page 172

"OK, just one more question," says Marvin. "You said that people with diabetes have high blood glucose levels. If glucose is so high, why can't it just seep into the cells?"

Question 5: *Why can't glucose always diffuse into cells when the blood glucose concentration is higher than the intracellular glucose concentration?*

Question 6: *What do you think happens to the rate of insulin secretion when blood glucose levels fall? What kind of feedback pathway is taking place?*

Circadian rhythms arise from special groups of cells in the brain and are reinforced by information about the light-dark cycle that comes in through the eyes. We will discuss the cellular basis for circadian rhythms further in Chapter 21.

Now that we have discussed response loops and feedback loops, let's look at the various patterns of biological reflexes.

Control Systems Vary in Their Speed and Specificity

Biological reflexes are mediated by the nervous system, the endocrine system, or a combination of the two (Fig. 6-25 ■). A reflex mediated solely by the nervous system or solely by the endocrine system is relatively simple, but combination reflexes can be quite complex. In the most complex reflex, signals pass through three different integrating centers before finally reaching the target tissue. With so much overlap between reflexes controlled by the nervous and endocrine systems, it makes sense to consider them a continuum rather than two discrete systems (Fig. 6-26 ■).

Why does the body need different types of control systems? To answer that question, let us compare endocrine control with nervous control and see what the differences are. These differences are summarized in Table 6-3 (p. 179).

Specificity Nervous control is very specific because each nerve cell has a specific target cell or cells to which it sends its message. Anatomically, we can isolate a neuron and trace it from its origin to where it terminates on its target cell(s). Endocrine control is more general because the chemical messenger is released into the blood and can reach virtually every cell in the body. As you learned in the first half of this chapter, the body's response to a specific hormone will depend on which cells have receptors for that hormone. Multiple tissues in the body can respond to a hormone simultaneously.

Nature of the Signal The nervous system uses both electrical and chemical signals to send information throughout the body. Electrical signals travel long

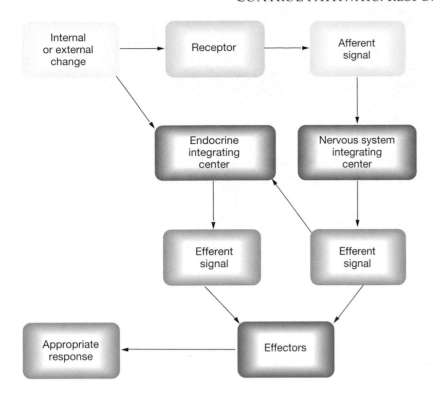

■ **Figure 6-25**   **The steps of a reflex**   Reflex response loops may be strictly nervous, strictly endocrine, or a combination of the two.

distances through nerve cells while chemical signals, the neurotransmitters, diffuse across the small gap between nerve cells or between nerve cell and target (Fig. 6-26 ■, pattern 1). In a limited number of instances, electrical signals pass directly from cell to cell through gap junctions.

The endocrine system uses only chemical signals: hormones secreted by endocrine glands or cells into the blood [∞ p. 155]. Neurohormones represent a hybrid of the two systems. A nerve cell creates an electrical signal, but the chemical released in response to the electrical signal goes into the blood for general distribution rather than to a specific target (Fig. 6-26 ■, patterns 6 and 2).

✔ In Figure 6-25 ■, which boxes represent the following specific parts of a reflex: afferent neuron, brain and spinal cord, central and peripheral sense organs, efferent neuron, hormone, neurohormone? In the figure, add a dotted line and arrow connecting boxes to show a negative feedback loop to complete the reflex.

Speed   Nervous reflexes are much faster than endocrine reflexes. The electrical signals of the nervous system cover great distances very rapidly, with speeds of up to 120 m/sec. Neurotransmitters also create very rapid responses, on the order of milliseconds.

Hormones are much slower than nervous reflexes. Their distribution through the circulatory system and diffusion from the capillary to their receptor takes considerably longer than signals through nerve cells. In addition, hormones have a slower onset of action. In target

tissues, the response may take minutes to hours before it can be measured.

Why do we need the speedy reflexes of the nervous system? Consider this example. A mouse ventures out of his hole and sees a cat ready to pounce on him and eat him. A signal must go from the eyes and brain of the mouse down to his feet, telling him to run back into the hole. If his brain and feet were only 5 micrometers (μm; 1/200 millimeter) apart, it would take a chemical signal 20 milliseconds (msec) to diffuse across the space. The mouse could escape. If the cells were 50 μm (1/20 millimeter) apart, diffusion would take 2 seconds: the mouse might get caught. But because the head and tail of a mouse are centimeters apart, at this rate of diffusion, it would take a chemical signal *three weeks* to diffuse from the head to the feet of the mouse. Poor mouse! Even if the distribution of the chemical were speeded by help from the circulatory system, the chemical message would still take 10 seconds to get to the feet, and the mouse would become cat food. The moral of this tale is that reflexes requiring a speedy response are mediated by the nervous system because it is so much more rapid.

Duration of Action   Nervous control is of shorter duration than endocrine control. The neurotransmitter released by a nerve cell combines with a receptor on the target cell and initiates a response. But the response is usually very brief because the neurotransmitter is rapidly removed from the vicinity of the receptor by various mechanisms. To get a sustained response, multiple repeating signals must be sent through the nerve cell.

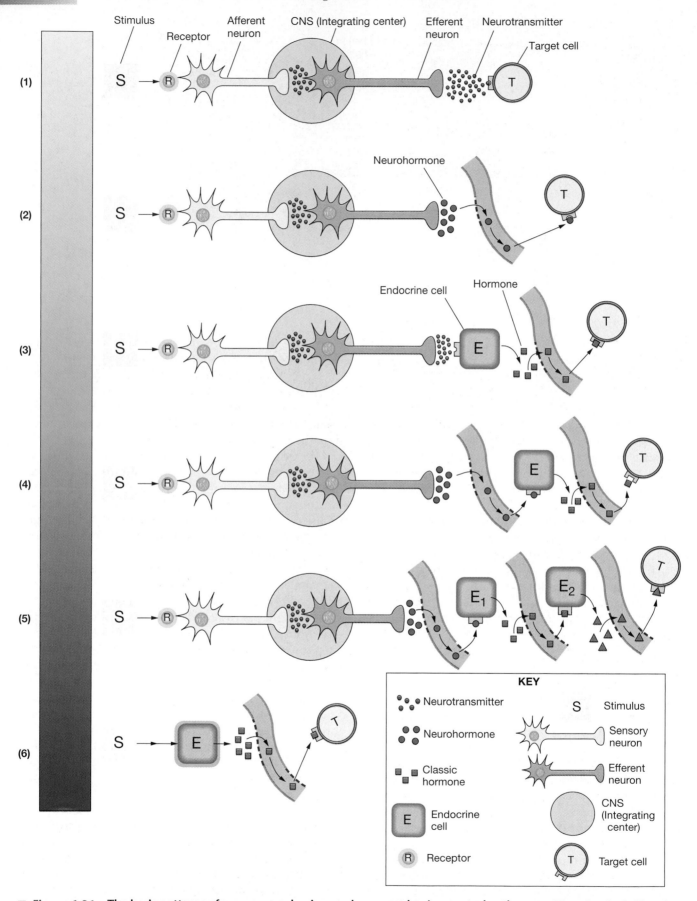

■ **Figure 6-26   The basic patterns of nervous, endocrine, and neuroendocrine control pathways**   The color-shaded bar along the left shows that there is no clear distinction between adjoining pathways. The nervous and endocrine systems are linked in a continuum.

TABLE 6-3 Comparison of Nervous and Endocrine Control

Property	Nervous Reflex	Endocrine Reflex
Specificity	Each neuron terminates on a single target cell or on a limited number of adjacent target cells.	Most cells of the body are exposed to a hormone. The response depends on which cells have receptors for the hormone.
Nature of the signal	Electrical signal through neuron with chemical neurotransmitters to pass signal from cell to cell. In a few cases, cell-to-cell communication takes place through gap junctions.	Chemical signals that are secreted in the blood for distribution throughout the body.
Speed	Very rapid.	Distribution of the signal and onset of action are much slower than in nervous responses.
Duration of action	Usually very rapid. Slower responses are mediated by neuromodulators.	Duration of action is usually much longer.
Coding for stimulus intensity	Each signal is identical in strength. Stimulus intensity reflected by increased frequency of signaling.	Stimulus intensity reflected by amount of hormone secreted.

Endocrine reflexes are slower to start but they are of longer duration. This means that most of the ongoing, long-term functions of the body such as metabolism and reproduction fall under the control of the endocrine system.

Coding for Stimulus Intensity As a stimulus increases in intensity, control systems must have a mechanism for conveying this information to the integrating center. The signal from any one neuron is constant in magnitude and therefore cannot reflect stimulus intensity. Instead, the frequency of signaling through the afferent neuron increases. In the endocrine system, stimulus intensity is reflected by the amount of hormone that is released: the stronger the stimulus, the more hormone.

Pathways for Nervous, Endocrine, and Neuroendocrine Reflexes May Be Complex, with Several Integrating Centers

Figure 6-26 ■ summarizes the different variations of endocrine, neuroendocrine, nervous, and mixed reflexes. The simplest reflexes are the pure nervous reflex (pattern 1), the pure endocrine reflex (pattern 6), and the simple neuroendocrine reflex (pattern 2).

In a classic endocrine reflex (pattern 6), the endocrine cell acts both as receptor and integrating center; there is no afferent pathway. Pure endocrine reflexes usually do not have a distinct sensory receptor in the same way that neural and neuroendocrine reflexes do. Instead, the endocrine cell itself acts as the receptor, monitoring the stimulus parameter directly. The efferent pathway is the hormone, and the target is any cell with the appropriate receptor.

An example of an endocrine reflex is the secretion of the hormone insulin in response to changes in blood glucose level. The endocrine cells that secrete insulin are able to directly monitor blood glucose concentrations. As blood glucose increases above the threshold level, the endocrine cells sense the change and respond by secreting insulin into the blood. Any cell in the body with insulin receptors will respond to the hormone and initiate processes that take glucose out of the blood. The removal of the stimulus acts in a negative feedback manner, and the response loop shuts off as blood glucose levels drop below a certain concentration.

In a pure nervous reflex, all the steps of the reflex are present, from receptor to target. This reflex is represented in its simplest form by the knee jerk reflex (see Fig. 13-7 ■, p. 392). A blow to the knee (the stimulus) activates a stretch receptor. An electrical and chemical signal travels through an afferent neuron to the spinal cord (the integrating center). If the blow is strong enough, a signal travels from the spinal cord through an efferent neuron to the muscles of the thigh. In response, the muscles contract, causing the lower leg to kick outward (the knee jerk).

The simple neuroendocrine reflex (pattern 2) is identical to the nervous reflex except that the chemical released by the neuron travels in the blood to its target, just like a hormone. A simple neuroendocrine reflex is the release of breast milk in response to a baby's suckling. The baby's mouth on the nipple stimulates sensory signals that travel to the brain (integrating center). An electrical signal in the efferent neuron triggers the release of the neurohormone oxytocin from the brain into the circulation. Oxytocin is carried to the breast, where it causes contraction of smooth muscles in the breast (effectors) with resultant ejection of milk.

In complex pathways, there may be more than one integrating center. Patterns 3, 4, and 5 represent combinations of the three basic pathways. Pattern 3 stacks an endocrine reflex on top of a nervous reflex. The target of the initial nervous reflex is an endocrine cell that releases

TABLE 6-4 Comparison of Nervous, Neuroendocrine, and Endocrine Reflexes

	Nervous	Neuroendocrine	Endocrine
Receptor	Special and somatic sensory receptors	Special and somatic sensory receptors	Endocrine cell
Afferent path	Afferent sensory neuron	Afferent sensory neuron	None
Integrating center	Brain or spinal cord	Brain or spinal cord	Endocrine cell
Efferent path	Efferent neuron (electrical signal and neurotransmitter)	Efferent neuron (electrical signal and neurohormone)	Hormone (via the circulation)
Effector(s)	Muscles and glands	Most cells of the body	Most cells of the body
Response	Contraction and secretion	Change in enzymatic reactions, membrane transport, or cell proteins	Change in enzymatic reactions, membrane transport, or cell proteins

a hormone. An example of this can be found in the control of insulin release by the nervous system. The endocrine cells of the pancreas have both stimulatory and inhibitory neurons terminating on them, and they integrate the information coming in through the nervous system with their direct detection of blood glucose levels.

Pattern 4 combines a neurosecretory reflex with an endocrine reflex. The secretion of growth hormone is an example of this pathway. Pattern 5 is the most complex pathway, starting with a neuroendocrine reflex and continuing with two sequential endocrine reflexes. This pattern is typical of most of the hormones released by the anterior pituitary, an endocrine gland located just below the brain (see Chapter 7 for details).

In describing all these complex pathways, it is simplest to mention only one receptor and afferent path, those that start the reflex. The brain is the first integrating center and the neurohormone is the first efferent path. The en-

docrine gland that is the target of the neurohormone becomes the second integrating center, and its hormone becomes the second efferent path. Finally, the endocrine target of efferent hormone 2 is the third integrating center, and its hormone is the third efferent path. The target of the last signal in the sequence is the effector.

Table 6-4 compares the various steps of reflexes in the nervous, neuroendocrine, and endocrine systems. There are some other combinations of endocrine and nervous reflexes that you may encounter in physiology. We will be using these general patterns throughout the book as an aid to simplifying endocrine pathways and as a way to classify them.

This chapter has described how the cells of the body communicate with each other to maintain a state of homeostasis. In the next chapter we look in detail at the endocrine system and the role it plays in these control processes.

PROBLEM CONCLUSION

Marvin Garcia underwent further tests and was diagnosed with type 2 diabetes. With careful attention to his diet and with a regular exercise program, he has managed to keep his blood glucose levels under control.

In this running problem, you have learned about how glucose homeostasis is maintained by insulin and glucagon, and how the disease diabetes mellitus is an indication that glucose homeostasis has been disrupted.

Further check your understanding of this running problem by comparing your answers to the questions against those in the summary table.

	Question	Facts	Integration and Analysis
1	In which type of diabetes is the signal transduction mechanism for insulin more likely to be defective?	Insulin is a peptide hormone that uses membrane receptors linked to second messengers to transmit its signal to cells [∞ p. 166]. People with type 1 diabetes lack insulin; people with type 2 have normal to elevated levels of insulin.	Normal or high insulin levels suggest that the problem is not with amount of insulin but with the action of the insulin at the cell. The problem in type 2 diabetes could be a defective signal transduction mechanism.

Question	Facts	Integration and Analysis
2 In which form of diabetes are the insulin receptors most likely to be up-regulated?	Up-regulation of receptors usually occurs if a signal molecule is present in unusually low concentrations [∞ p. 166]. In type 1 diabetes, insulin is not secreted by the pancreas.	In type 1 diabetes, insulin levels are low. Therefore, type 1 is most likely to cause up-regulation of the insulin receptors.
3 The homeostatic regulation of blood glucose levels by the hormones insulin and glucagon is an example of which of Cannon's postulates?	Cannon's postulates describe the role of the nervous system in maintaining homeostasis, and the concepts of tonic activity, antagonistic control, and different effects of signals in different tissues.	Insulin decreases blood glucose levels while glucagon increases them. Therefore, the two hormones are an example of an antagonistic control.
4 In the insulin reflex that regulates blood glucose levels, what is the stimulus? the receptor? the integrating center? the efferent pathway? the effector(s)? the response(s)?	See the steps of reflex pathways [∞ p. 170].	*Stimulus:* increase in blood glucose levels. *Receptor:* beta cells of the pancreas that sense the change. *Integrating center:* beta cells. *Efferent pathway* (the signal): insulin. *Effectors:* any tissues of the body that respond to insulin. *Responses:* cellular uptake and use of glucose.
5 Why can't glucose always diffuse into cells when the blood glucose concentration is higher than the intracellular glucose concentration?	Glucose is lipophobic. Simple diffusion goes across the phospholipid bilayer. Facilitated diffusion uses protein carriers [∞ p. 124]	Because glucose is lipophobic, it must cross the membrane by facilitated diffusion. If a cell lacks the necessary protein carriers, facilitated diffusion cannot take place.
6 What do you think happens to the rate of insulin secretion when blood glucose levels fall? What kind of feedback pathway is taking place?	The stimulus for insulin release is an increase in blood glucose levels. In negative feedback, the response offsets the stimulus. In positive feedback, the response enhances the stimulus.	An increase in blood glucose concentration stimulates insulin release; therefore a decrease in blood glucose should decrease insulin release. In this example, the response (lower blood glucose) offsets the stimulus (increased blood glucose), so the pathway exhibits negative feedback.

CHAPTER REVIEW

SUMMARY

Cell-to-Cell Communication

1. There are two basic physiological signals: chemical and electrical. Chemical signals are the basis for most communication within the body. (p. 154)

2. There are three methods of cell-to-cell communication: (1) direct cytoplasmic transfer through gap junctions, (2) local chemical communication, and (3) long-distance communication. (p. 154)

3. **Gap junctions** are protein channels that connect two adjacent cells. When they are open, ions and small molecules pass directly from one cell to the next. (p. 154)

4. Local communication is accomplished by **paracrines**, chemicals that act on cells in the immediate vicinity of the cell that secreted the paracrine. A chemical that acts on the cell that secreted it is called an **autocrine.** The action of

paracrines and autocrines is limited by diffusion distance. **Neuromodulators,** cytokines, and **eicosanoids** are paracrines. (p. 154)

5. Long-distance communication is accomplished by chemical **neurocrines** and electrical signals in the nervous system and by **hormones** in the endocrine system. Only cells that possess receptors for the hormone will be **target cells.** (p. 155)

6. **Cytokines** are regulatory molecules that control cell development, differentiation, and the immune response. They function as both local and long-distance signals. (p. 156)

Signal Pathways

7. Chemical signals released by cells are called **first messengers.** First messengers bind to **receptors** and change intracellular **effectors** that direct the response. (p. 156)

8. Protein phosphorylation by **protein kinases** is an important way to create a cellular response. (p. 156)

9. Lipophilic signal molecules enter the cell and combine with cytoplasmic or nuclear receptors. Lipophobic signal molecules and some lipophilic molecules combine with membrane receptors. (p. 157)

10. **Ligand-gated ion channels** open or close to create electrical signals. (p. 157)

11. **Signal transduction** pathways use membrane proteins and intracellular second messenger molecules to translate information from the first messenger into an intracellular response. (p. 158)

12. Many signal transduction pathways activate **amplifier enzymes** that create **second messenger** molecules. (p. 158)

13. Signal pathways create intracellular **cascades** that amplify the original signal. (p. 159)

14. **Integrin** receptors link the extracellular matrix to the cytoskeleton. (p. 160)

15. **Receptor-enzymes** activate protein kinases, such as **tyrosine kinase,** or the amplifier enzyme **guanylyl cyclase** that produces the second messenger **cGMP.** (p. 160)

16. **G proteins** linked to amplifier enzymes are the most prevalent signal transduction system. G protein-coupled receptors also alter ion channels. (p. 161)

17. The **G protein-adenylyl cyclase-cAMP-protein kinase A** pathway is the most common pathway for protein and peptide hormones. (p. 161)

18. The amplifier enzyme **phospholipase C** creates two second messengers, **IP$_3$** and **diacylglycerol.** IP$_3$ causes Ca^{2+} release from intracellular stores. Diacylglycerol activates **protein kinase C.** (p. 162)

19. Calcium is an important signal molecule that binds to **calmodulin** to alter enzyme activity. It also binds to other cell proteins to alter movement and initiate exocytosis. (p. 162)

20. **Nitric oxide** is a short-lived gaseous signal molecule that activates guanylyl cyclase directly. (p. 163)

Modulation of Signal Pathways

21. The response of a cell to a signal molecule is determined by the cell's receptor for the signal. (p. 164)

22. Receptor proteins exhibit specificity, competition, and saturation. (p. 164)

23. A receptor may have multiple ligands. **Agonists** mimic the action of a signal molecule. **Antagonists** block the signal pathway. (p. 164)

24. Receptors come in related forms called **isoforms.** One ligand may have different effects when binding to different isoforms. (p. 165)

25. Cells exposed to abnormally high concentrations of a signal for a sustained period of time attempt to bring their response back to normal by decreasing the number of receptors or decreasing the binding affinity of the receptors. This is known as **down-regulation** of the receptors. **Up-regulation** is the opposite of down-regulation. (p. 165)

26. Cells have mechanisms for terminating signal pathways. (p. 166)

27. Many diseases have been linked to defects with various aspects of signal pathways, such as missing or defective receptors. (p. 166)

Homeostasis

28. **Walter Cannon** first described the four basic postulates of homeostasis. (1) The nervous system plays an important role in maintaining homeostasis. (2) Some parameters are under **tonic control,** which allows the parameter to be increased or decreased by a single signal. (3) Other parameters are under **antagonistic control,** in which one hormone or neuron increases the parameter while another decreases it. (4) Chemical signals can have different effects in different tissues of the body, depending on the type of receptor present at the target cell. (p. 167)

Control Pathways: Response and Feedback Loops

29. The simplest homeostatic control takes place at the tissue or cell level and is known as **local control.** (p. 169)

30. In **reflex** control pathways, the decision that a response is needed is made away from the cell or tissue. A chemical or electrical signal is sent to the cell or tissue to initiate a response. Long-distance reflex pathways involve the nervous system, the endocrine system, and cytokines. (p. 169)

31. Reflex pathways can be broken down into **response loops**

and **feedback loops.** A response loop has an input signal, integration of the signal, and an output signal. A **stimulus** begins the reflex when it is sensed by a **receptor.** The receptor is linked by an **afferent pathway** to an **integrating center** that decides on an appropriate response. An **efferent pathway** travels from the integrating center to an **effector** that carries out the appropriate **response.** These steps create a response loop. (p. 170)

32. In **negative feedback,** a homeostatic response is turned off when the response of the system opposes or removes the original stimulus. (p. 173)

33. In **positive feedback** loops, the response reinforces the stimulus rather than decreasing it or removing it. This destabilizes the system until some intervention or event outside the feedback loop stops the response. (p. 174)

34. **Feedforward control** allows the body to predict that a change is about to occur and start the response loop in anticipation of the change. (p. 175)

35. Apparently spontaneous reflexes that occur in a predictable manner are called **biological rhythms.** Those that coincide with light-dark cycles are called **circadian rhythms.** (p. 175)

36. Nervous control is faster and more specific than endocrine control but is usually of shorter duration. Endocrine control is less specific and slower to start but is longer lasting and is usually amplified. (p. 176)

37. Many reflex pathways are combinations of nervous and endocrine control mechanisms. (p. 179)

QUESTIONS

LEVEL ONE   Reviewing Facts and Terms

1. What are the two routes for long-distance signal delivery in the body?

2. Which two body systems are charged with maintaining homeostasis by responding to changes in the environment?

3. What two types of physiological signals are used to send messages through the body? Of these two types, which is available to all cells?

4. The process of maintaining a relatively stable internal environment is called _____.

5. List at least three parameters maintained by homeostasis.

6. Distinguish between the "target" and the "receptor" in physiological systems.

7. In a signal pathway, the signal ligand, also called a _____ _____, binds to a _____ which activates and changes intracellular _____.

8. The three main amplifier enzymes are (a) _____, which forms cAMP; (b) _____, which forms cGMP, and (c) _____, which converts a phospholipid from the cell's membrane into two different second messenger molecules.

9. An enzyme known as protein kinase adds the functional group _____ to its substrate, by transferring it from a(n) _____ molecule.

10. Put these parts of a reflex in the correct order for a physiological response loop: efferent path, afferent path, effector, stimulus, response, integrating center.

11. Distinguish between central and peripheral receptors.

12. Match each term with its description:

(a) threshold	1. the desired target value for a parameter
(b) setpoint	2. the distance allowed away from the setpoint before a response starts
(c) effector	3. the minimum stimulus to trigger a response
(d) oscillation	4. the organ or gland that performs the change
(e) sensitivity	5. movement of a parameter within the desired range

13. Receptors for signal pathways may be found in the _____, _____, or _____.

14. The name for the daily fluctuations of body functions, including blood pressure, temperature, and metabolic processes, is _____ _____. These cycles arise in special cells in the _____.

15. Down-regulation results in a(n) (increased or decreased?) sensitivity to a prolonged signal, resulting in a condition called _____, often seen when a drug is administered over a long term.

16. List the two ways that down-regulation may be achieved.

17. In a negative feedback loop, the effector moves the system in the (same/opposite) direction as the stimulus.

LEVEL TWO   Reviewing Concepts

18. Explain the relationship among the terms in each of the following sets of terms. Give a physiological example or location if applicable.
    (a) gap junctions, connexins, syncytium, connexon

    (b) autocrine, paracrine, cytokine, neurocrine, hormone

    (c) agonist, antagonist

    (d) transduction, amplification, cascade

19. List and compare the four types of membrane receptors for signal pathways. Give an example of each.

20. Who was Walter Cannon? Restate his four postulates in your own words.

21. Briefly define each term in question 10 and give an anatomical example where applicable.

22. Explain the differences among positive feedback, negative

feedback, and feedforward mechanisms. Under what circumstances would each one be advantageous?

23. Compare and contrast the advantages and disadvantages of nervous versus endocrine control mechanisms.

24. Label each system below as positive or negative feedback.
    (a) glucagon secretion in response to declining blood glucose
    (b) increasing milk letdown and secretion in response to more suckling
    (c) urgency in emptying one's urinary bladder
    (d) sweating in response to rising body temperature

25. Go back to the preceding question and identify the effector organ for each example.

26. Now go back to question 24 again and try to identify the integrating center for each example.

LEVEL THREE Problem Solving

27. In each of the following situations, identify the components of the reflex.

    (a) You are sitting quietly at your desk, studying, aware of the bitterly cold winds blowing about at 30 mph, and you begin to feel a little chilly. You start to turn up the thermostat, remember last month's bill, and reach for an afghan to pull around you instead. Pretty soon you are toasty warm again.

    (b) You are strolling through the shopping district and the aroma of cinnamon sticky buns reaches you. You inhale appreciatively, but reassure yourself you're not hungry, since you just had lunch an hour ago. You go about your business, but 20 minutes later you're back at the bakery, sticky bun in hand, ravenously devouring its sweetness, saliva moistening your mouth.

E X P L <MediaLab>

Introduction

In this chapter you learned about the methods and mechanisms by which the body carries out internal communication, coordinates body functions, and maintains a consistent environment. Without these three functions—communication, integration, and homeostasis—large complicated organisms such as we are could not survive. The following Web Explorations will help you gain an appreciation of these coordinating functions. After reading the description below, visit the MediaLab for Chapter 6 in your Companion Website and select the appropriate keyword.

Web Exploration 1

Estimated time for completion = 5 minutes
Imagine yourself as a cell somewhere in your body. You are neighbors with a trillion other cells that must work together to maintain an internal environment (homeostasis) in the face of constant perturbation. As a cell, you neither have ears, eyes, or voice by which to send or receive messages. So how do you communicate? One of the most direct means of cell-to-cell communication is via "doorways" called gap junctions. Select the keyword **GAP JUNCTIONS** on the Website to see an electron micrograph (EM) of the intramembrane proteins of a gap junction. After you've viewed these micrographs and read the material, predict which molecules can freely pass between neighboring cells and which can not. As you begin to develop your model of the various ways cells communicate with

each other, it is important to factor in temporal (how fast does it work?) and spatial (over what distance does it work?) considerations as well as the nature of the substance being transported. What types of molecules can pass between cells? How rapid is this movement? What will be the consequence of the movement of this substance? As you learned in Chapter 5, molecules cross cell membranes only if there is a way across and if there is a driving force. What is the driving force for each molecule? To complete this exercise, visit MediaLab Web Exploration 1 in Chapter 6 of your Companion Website.

Web Exploration 2

Estimated time for completion = 5 minutes
The same chemical signal can cause very different responses throughout the body. How is this possible? One way is to use different receptors, while another is to use the same receptor but to couple that receptor to different second messenger cascades. A critical intermediate in many second messenger pathways is a G-protein. Select the keyword **G-PROTEIN** on your Website and read about the role of G-proteins in cellular signal transduction. As an interesting point of context, select the keyword **Rodbell** to read an example of how real scientific discoveries are made. The article describes how the actual discovery was worked out on a tablecloth in a bar. To complete this exercise, visit MediaLab Web Exploration 2 in Chapter 6 of your Companion Website.

7 Introduction to the Endocrine System

■ "The separation of the endocrine system into isolated subsystems must be recognized as an artificial one, convenient from a pedagogical point of view but not accurately reflecting the interrelated nature of all these systems."
—*Howard Rasmussen, in* Williams' Textbook of Endocrinology, 1974 ■

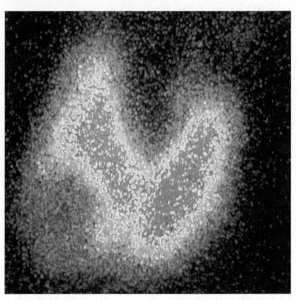

Radioactive scan of the thyroid gland

BACKGROUND BASICS

Receptors (p. 114)
Peptides and proteins (p. 32)
Comparison of endocrine and nervous systems (p. 176)
Signal transduction (p. 158)
Steroids (p. 29)
Specificity (p. 121)

David was seven years old when the symptoms first appeared. His appetite at meals increased, and he always seemed to be in the kitchen looking for food. But despite eating more, he was losing weight. When he started asking for water instead of soft drinks, David's mother became concerned. And when he wet the bed three nights in a row, she knew something was wrong. The doctor confirmed the suspected diagnosis after running tests to determine the concentration of glucose in David's blood and urine. David had diabetes mellitus. In David's case, the disease was due to lack of insulin, a hormone produced by the pancreas. David was placed on insulin injections, a treatment that he would continue for the rest of his life.

One hundred years ago, David would have died not long after the onset of symptoms. The field of endocrinology, the study of hormones, was then in its infancy. Most hormones had not been discovered, and the functions of known hormones were not well understood. There was no treatment for diabetes, no birth control pill

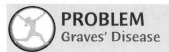

PROBLEM
Graves' Disease

Is hyperthyroidism—a condition in which the thyroid overproduces thyroid hormone—contagious? That's what some people wondered when, in 1991, both President and Mrs. Bush were diagnosed with Graves' disease, the most common form of hyperthyroidism. But Graves' disease is not caused by a virus or bacterium, so it can't be transmitted from one person to another. Still, many physicians believe that something in the White House precipitated this double diagnosis of thyroid trouble. That something was probably stress.

...continued on page 195

for contraception. Babies born with inadequate secretion of thyroid hormone did not grow or develop normally. Today, all that has changed. We have identified a long and growing list of hormones. The endocrine diseases that once killed or maimed are under control with synthetic hormones and sophisticated medical procedures. Although physicians do not hesitate to use these treatments, in many instances we still do not understand exactly how the hormones act on their target cells. This chapter provides an introduction to the basic principles of hormone structure and function. You will learn more about individual hormones as you encounter them in your study of the various systems.

▶ HORMONES

As you learned in Chapter 6, hormones are chemical messengers secreted into the blood by endocrine cells or by specialized neurons. Hormones are responsible for many functions that we think of as long-term, ongoing functions of the body. Processes that fall mostly under hormonal control include growth and development, metabolism, regulation of the internal environment (temperature, water balance, ions), and reproduction. Hormones act on their target cells in one of three basic ways. They (1) control the rates of enzymatic reactions, (2) control transport of molecules across cell membranes, or (3) control gene expression and the synthesis of proteins.

The effects of hormones on their target cells can be very complex. A single hormone may have different effects on various tissues. It may affect one tissue differently at different stages of development, or it may have no effect at all. Insulin is an example of a hormone with varied effects. It controls glucose transport into many cells of the body but not into the cells of the brain or liver.

The variable responsiveness of a cell to a hormone depends primarily on the cell's receptor and signal transduction pathways for the hormone [∞ p. 164]. If there are no hormone receptors in a tissue, its cells cannot respond. If tissues have different receptors and receptor-linked pathways for the same hormone, they will respond differently.

Traditionally, the field of endocrinology has focused on the **endocrine glands** [∞ p. 61], discrete and readily identifiable tissues that secrete hormones. But as you learned in the last chapter, the distinction between the nervous and endocrine systems is no longer clear-cut. Chemicals that act as hormones are secreted not only by endocrine glands but also by nerve cells. Furthermore, we have discovered that one hormone may come from a variety of sources. For example, *cholecystokinin* (CCK) was discovered in the 1920s in extracts of the intestine that caused contraction of the gall bladder. For many years, CCK was known only as an intestinal hormone. Then in the mid-1970s, CCK was found in neurons of the brain, where it acts as a neurotransmitter or neuromodulator. In recent years, it has become famous for its possible role in controlling hunger, a behavior controlled by the brain.

The multiple sources and roles of hormones have led some authors to group the hormones by name rather than by their anatomical sources. In this book, however, we will group the hormones anatomically.

Hormones Have Been Known Since Ancient Times

Although the scientific field of endocrinology is relatively young, diseases of the endocrine system have been documented for more than a thousand years. Evidence of endocrine abnormalities can even be seen in ancient art. For example, one pre-Colombian statue of a woman shows a mass on the front of her neck (Fig. 7-1 ■). This is an enlarged thyroid gland, or *goiter,* a common condition high in the Andes, where there was a lack of the dietary iodine needed to make thyroid hormones.

The first association of endocrine structure and function undoubtedly was the link between the testes and

Immunocytochemistry   Traditionally, physiologists thought that each hormone came from a single gland (although some glands secrete multiple hormones). But new techniques like immunocytochemistry have turned that view of endocrinology upside down. Immunochemical laboratory tests rely on antibodies made when an animal is injected with a foreign molecule (antigen) such as a hormone. The animal's immune system recognizes the molecule as "not self" and produces specific antibodies to it. The antibodies are extracted from the host animal and combined with a radioactive, fluorescent, or enzymatic marker. The tagged antibody is then applied to cultured cells or sections of tissue and examined under the microscope. If the original antigen is present in the tissue, the antibody binds to it and becomes visible under the microscope. By screening multiple tissues with the antibodies for different hormones, investigators discovered that many hormones are made in tissues far from the site where they were originally isolated.

■ **Figure 7-1   Endocrine disorders in ancient art**   This pre-Colombian stone carving of a woman shows a mass at her neck. This mass is an enlarged thyroid gland, a condition known as goiter. It was considered a sign of beauty among the people who lived high in the Andes mountains.

male sexuality. Castration of animals and men was a common practice in both Eastern and Western cultures because it was known to decrease the sex drive and render the male infertile.

In 1849, A. A. Berthold performed the first classic experiment in endocrinology. He removed the testes from roosters and observed that the castrated birds had smaller combs and less aggressiveness or sex drive than normal males. If the testes were surgically placed back into the donor cock or into another castrated male, normal male behavior and comb development resumed. Because the reimplanted testes were not connected to nerves, Berthold concluded that the glands must be secreting something into the blood that affected the entire body.

Experimental endocrinology did not receive much attention, however, until 1889, when the 72-year-old French physician Charles Brown-Séquard made a dramatic announcement of his sexual rejuvenation after injecting himself with extracts made from bull testes ground in water. An international uproar followed, and physicians on both sides of the Atlantic began to inject their patients with extracts of many different endocrine organs, a practice known as *organotherapy.*

We now know that the increased virility Brown-Séquard reported was most likely a placebo effect because testosterone is a hydrophobic steroid that cannot be extracted by an aqueous preparation. But the door to hormone therapy was opened, and in 1891 organotherapy had its first true success: a woman was treated for low thyroid hormone levels with glycerin extracts of sheep thyroid glands.

As the study of "internal secretions" grew, Berthold's experiments became a template for endocrine research. Once a gland or structure was suspected of secreting hormones, the classic steps for identifying an endocrine gland became:

1. Remove the suspected gland and monitor the animal for anatomical, behavioral, or physiological abnormalities. This is equivalent to inducing a state of *hormone deficiency.*
2. Replace the gland or an extract of the gland and see if the abnormalities disappear. *Replacement therapy* should eliminate the symptoms of hormone deficiency.
3. Implant the gland or administer an extract from the gland to a normal animal, and see if symptoms characteristic of *hormone excess* appear.
4. Once a gland is identified as a potential source of hormones, take extracts and purify them to separate the active substance. The test for hormone activity is usually a biological assay in which an animal is injected with the purified extract and monitored for a response.

The hormones identified by this technique are sometimes called the *classic* hormones. They include the hormones of the pancreas, adrenal glands, pituitary, and gonads, all discrete endocrine glands that could be easily identified and surgically removed. Not all hormones come from identifiable glands, however, and we have been slower to discover them.

The Anatomy Summary (Fig. 7-2 ■) shows the major hormones of the body and the glands or cells that secrete them, along with the major properties of each hormone.

What Makes a Chemical a Hormone?

In 1905, the term *hormone* was coined from the Greek verb meaning "to excite or arouse." A **hormone** is a chemical secreted by a cell or group of cells into the blood for transport to a distant target, where it exerts its effect at very low concentrations (10^{-9} to 10^{-12} M). Not all chemical signals transported in the blood to distant targets are hormones because some of them must be present in relatively high concentration before an effect is noticed.

All hormones bind to target cell receptors and initiate biochemical responses. These responses are known as the cellular mechanism of action of the hormone. As you can see from the table in Figure 7-2 ■, hormones often act on multiple tissues that may be far from the gland that secreted them.

ANATOMY SUMMARY   The Endocrine System

■ **Figure 7-2**

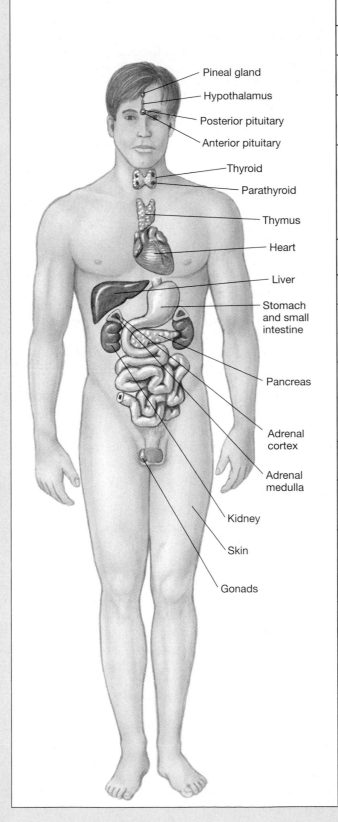

Location	Gland or Cell?	Chemical Class
Pineal gland	Gland	Amine
Hypothalamus	Clusters of neurons	Peptides
Posterior pituitary	Extensions of hypothalamic neurons	Peptides
Anterior pituitary	Gland	Peptides
Thyroid	Gland	Iodinated amines Peptide
Parathyroid	Gland	Peptide
Thymus	Gland	Peptides
Heart	Cells	Peptide
Liver	Cells	Peptides
Stomach and small intestine	Cells	Peptides
Pancreas	Gland	Peptide
Adrenal cortex	Gland	Steroids
Adrenal medulla	Gland	Amines
Kidney	Cells	Peptide Steroid
Skin	Cells	Steroid
Testes (male)	Glands	Steroids Peptide
Ovaries (female)	Glands	Steroids Peptide
Adipose tissue	Cells	Peptide
Placenta (pregnant females only)	Gland	Steroids Peptide

Hormone	Target	Main Effect
Melatonin	Unclear in humans	Circadian rhythms?
Trophic hormones (see Fig. 7-15) See posterior pituitary	Anterior pituitary	Release or inhibit pituitary hormones
Oxytocin	Breast and uterus	Milk ejection; labor and delivery; behavior
Vasopressin (ADH)	Kidney	Water reabsorption
Prolactin	Breast	Milk production
Growth hormone (somatotropin)	Many tissues	Growth and metabolism
Corticotropin (ACTH)	Adrenal cortex	Cortisol release
Thyrotropin (TSH)	Thyroid gland	Thyroid hormone synthesis and release
Follicle stimulating hormone (FSH)	Gonads	Egg or sperm production; sex hormone production
Luteinizing hormone (LH)	Gonads	Sex hormone production; egg or sperm production
Triiodothyronine and thyroxine	Many tissues	Metabolism, growth and development
Calcitonin	Bone	Plasma calcium levels (minimal effect in humans)
Parathyroid hormone (PTH)	Bone, kidney	Plasma calcium and phosphate levels
Thymosin, thymopoietin	Lymphocytes	Lymphocyte development
Atrial natriuretic peptide	Kidneys	Increase sodium excretion
Angiotensinogen	Adrenal cortex, blood vessels, brain	Aldosterone, blood pressure
Insulin-like growth factors	Many tissues	Growth
Gastrin, cholecystokinin, secretin, and others	GI tract and pancreas	Assist digestion and absorption of nutrients
Insulin, glucagon, somatostatin, pancreatic polypeptide	Many tissues	Metabolism of glucose and other nutrients
Aldosterone	Kidney	Electrolyte balance
Cortisol	Many tissues	Stress response
Androgens	Many tissues	Sex drive in females
Epinephrine, norepinephrine	Many tissues	Fight-or-flight response
Erythropoietin	Bone marrow	Red blood cell production
1,25 Dihydroxy-vitamin D_3 (calciferol)	Intestine	Increase calcium absorption
Vitamin D_3	Intermediate form of hormone	Precursor of 1,25 dihydroxy-vitamin D_3
Androgen	Many tissues	Sperm production, secondary sex characteristics
Inhibin	Anterior pituitary	FSH secretion
Estrogens and progesterone	Many tissues	Egg production, secondary sex characteristics
Inhibin	Anterior pituitary	FSH secretion
Relaxin (pregnancy)	Uterine muscle	Relaxes muscle
Leptin	Hypothalamus, other tissues	Food intake, metabolism, reproduction
Estrogens and progesterone	Many tissues	Fetal and maternal development
Chorionic somatomammotropin	Many tissues	Metabolism
Chorionic gonadotropin	Corpus luteum of ovary	Hormone secretion

Some hormone molecules never reach their targets, however. They are broken down into inactive *metabolites* by enzymes found primarily in the liver and kidney. The metabolites are then excreted in the bile or urine. The rate of hormone breakdown is indicated by a hormone's **half-life** in the circulation, the amount of time required to reduce the concentration of hormone by one-half. Half-life indicates how long a hormone is active in the body. By limiting hormone activity, the body can closely regulate hormonally controlled processes.

To be classified as a hormone, a molecule must be transported by the blood for action at a distant site. Experimentally, this property is sometimes difficult to demonstrate. Molecules that are suspected to be hormones but not fully accepted as such are called *candidate hormones*. They are usually identified by the word *factor*. In the early 1970s, the hypothalamic regulating hormones were known as "releasing factors" and "inhibiting factors" rather than releasing and inhibiting hormones.

Currently, a large group of substances that influence cell growth and division are being studied to determine if they meet all the criteria for hormones. Many of these **growth factors** have been shown to act as *paracrines* [∞ p. 154] but not as hormones. Evidence is lacking for their widespread distribution via the circulation.

The boundary between hormones and nonhormonal signal molecules is blurring as we learn more about signal molecules, just as the distinction between the nervous and endocrine systems has blurred. Many *cytokines* [∞ p. 156] seem to meet the definition of a hormone as stated above. However, experts in cytokine research do not consider them hormones since peptide cytokines are synthesized and released on demand, in contrast to classic peptide hormones, which are made in advance and stored in the parent endocrine cell. A few cytokines, such as **erythropoietin,** the molecule that controls red blood cell production, were considered to be hormones before the term *cytokine* was coined, so there is some overlap between the two groups of signal molecules.

✔ Based on what you know about the organelles involved in protein and steroid synthesis [∞ p. 51], what would be the major differences between the organelle composition of a steroid-producing cell and a protein-producing cell?

▶ THE CLASSIFICATION OF HORMONES

Chemically, there are three main classes of hormones: peptide hormones, steroid hormones, and amine hormones (Table 7-1). The peptide hormones are composed of amino acids. The steroid hormones are all derived from cholesterol [∞ p. 31]. The third group of hormones, the amine hormones, are derivatives of a single type of amino acid.

TABLE 7-1   Comparison of Peptide, Steroid, and Amine Hormones

			Amines	
	Peptide Hormones	*Steroid Hormones*	*Catecholamines*	*Thyroid Hormones*
Synthesis and storage	Made in advance, stored in secretory vesicles	Synthesized on demand from precursors	Made in advance; stored in secretory vesicles	Made in advance; precursor stored in secretory vesicles
Release from parent cell	Exocytosis	Simple diffusion	Exocytosis	Simple diffusion
Transport in blood	Freely dissolves in plasma	Bound to carrier proteins	Freely dissolves in plasma	Bound to carrier proteins
Half-life	Short	Long	Short	Long
Location of receptor	On cell membrane	Cytoplasm or nucleus; some have membrane receptors also	On cell membrane	Nucleus
Response to receptor-ligand binding	Activation of second messenger systems. May activate genes	Activate genes for transcription and translation. May have nongenomic actions	Activation of second messenger systems	Activate genes for transcription and translation
General target response	Modification of existing proteins and induction of new protein synthesis	Induction of new protein synthesis	Modification of existing proteins	Induction of new protein synthesis
Examples	Insulin, parathyroid hormone	Estrogen, androgens, cortisol	Epinephrine, norepinephrine	Thyroxine (T_4)

Most Hormones Are Peptides or Proteins

The peptide hormones range from small peptides of only three amino acids to large proteins and glycoproteins. Despite the variability among hormones in this group, for simplicity, they are usually called peptide hormones. You can remember which hormones fall into this category by exclusion: if the hormone is not a steroid and not an amine, then it must be a peptide or protein.

Peptide Hormone Synthesis, Storage, and Release

The synthesis and packaging of peptide hormones into membrane-bound secretory vesicles is similar to that of other proteins [∞ p. 99]. The initial peptide that comes off the ribosome is a large inactive protein known as a **preprohormone** (Fig. 7-3 ■). Preprohormones contain one or more copies of a peptide hormone, a *signal sequence* that directs the protein into the lumen of the rough endoplasmic reticulum, and other peptide sequences that may or may not have biological activity. As the inactive peptide moves through the endoplasmic reticulum and Golgi apparatus, the signal sequence is removed, creating a smaller, still inactive molecule called a **prohormone.** In the Golgi apparatus, the prohormone is packaged into secretory vesicles along with *proteolytic* [*proteo-*, protein + *lysis*, rupture] enzymes that chop the prohormone into active hormone and other fragments. This process is called *post-translational processing.*

The secretory vesicles containing peptides are stored in the cytoplasm of the endocrine cell until the cell receives a signal for secretion. At that time, the vesicles move to the cell membrane and release their contents by calcium-dependent exocytosis [∞ p. 130]. All of the peptide fragments created from the prohormone are released together into the extracellular fluid, in a process known as *co-secretion.*

Post-Translational Processing of Prohormones

Studies of prohormone processing have led to some interesting discoveries. Some prohormones, such as that for thyrotropin-releasing hormone (TRH), have multiple copies of the hormone (Fig. 7-4a ■). Another interesting prohormone is called **pro-opimelanocortin** (Fig. 7-4b ■). This prohormone processes into three active peptides plus an inactive fragment. In some instances, even the fragments are clinically useful. For example, proinsulin is cleaved into active insulin and an inactive fragment known as C-peptide (Fig. 7-4c ■). Clinicians measure the levels of C-peptide in the blood of diabetics to monitor how much insulin the patient's pancreas is producing.

Transport in the Blood and Half-Life of Peptide Hormones

Peptide hormones are water-soluble and therefore dissolve easily in the extracellular fluid for transport throughout the body. The half-life for peptide hormones is usually quite short, in the range of several minutes. If the response to a peptide hormone must be sustained for an extended period of time, the hormone must continue to be secreted.

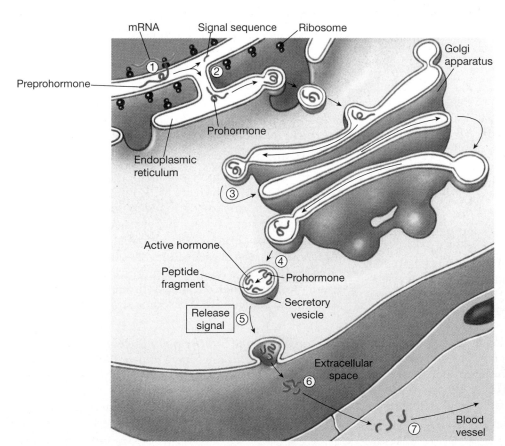

■ **Figure 7-3   Peptide hormone synthesis, packaging, and release** ① Messenger RNA on the ribosomes of the endoplasmic reticulum (ER) binds amino acids into a peptide chain called a preprohormone. The chain is directed into the ER lumen by a signal sequence of amino acids. ② Enzymes in the ER chop off the signal sequence, creating an inactive prohormone. ③ The prohormone passes from the ER through the Golgi apparatus. ④ Secretory vesicles containing enzymes and prohormone bud off the Golgi. The enzymes chop the prohormone into one or more active peptides plus additional peptide fragments. ⑤ The secretory vesicle releases its contents by exocytosis into the extracellular space ⑥. The hormone moves into the circulation for transport to its target ⑦.

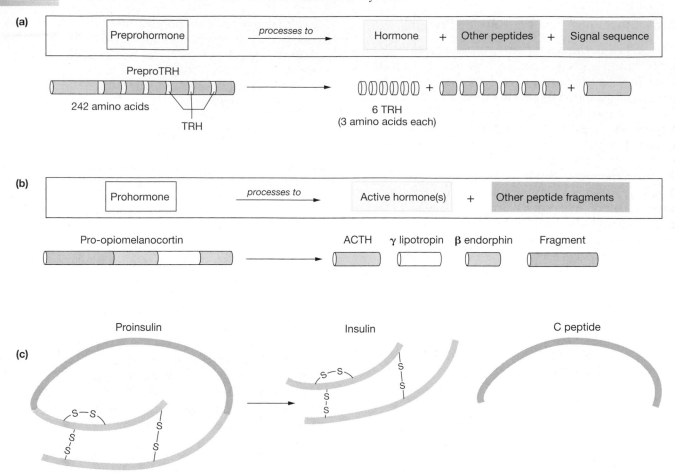

■ **Figure 7-4** **Peptide hormone processing** Peptide hormones are made as large, inactive preprohormones that include a signal sequence, one or more copies of hormone, and additional peptide fragments. (a) PreproTRH (thyrotropin-releasing hormone) has six copies of the 3-amino acid hormone TRH. (b) Prohormones, such as pro-opiomelanocortin, the prohormone for ACTH, may contain several peptide sequences with biological activity. (c) The peptide chain of insulin's prohormone folds back on itself with the help of disulfide (S-S) bonds. The prohormone cleaves to insulin and C-peptide.

Cellular Mechanism of Action of Peptide Hormones Because peptide hormones are lipophobic, they are usually unable to enter the target cell. Instead, they bind to surface membrane receptors. The hormone-receptor complex initiates the cellular response by means of a *signal transduction* system (Fig. 7-5 ■). Many peptide hormones work through cAMP second messenger systems [∞ p. 161]. A few peptide hormone receptors, such as that of insulin, have tyrosine kinase activity [∞ p. 160] or work through other signal transduction pathways.

The response of cells to peptide hormones is usually rapid because second messenger systems modify existing proteins. Some changes triggered by peptide hormones include opening or closing membrane channels and modulating metabolic enzymes or transport proteins. We have recently discovered that some peptide hormones also have longer-lasting effects when their second messenger systems activate genes and direct the synthesis of new proteins.

Steroid Hormones Are Derived from Cholesterol

Steroid hormones have similar chemical structures because they are all derived from cholesterol (Fig. 7-6 ■). Unlike peptide hormones, which are made in tissues all over the body, steroid hormones are made in only a few organs. Three types of steroid hormones are made in the **adrenal cortex,** the outer portion of the adrenal glands. One adrenal gland sits on top of each kidney [*ad-*, upon + *renal*, kidney; *cortex*, bark]. The gonads produce the sex steroids (estrogens, progesterone, and androgens). In pregnant women, the placenta is also a source of steroid hormones.

Steroid Hormone Synthesis and Release Cells that secrete steroid hormones have unusually large amounts of smooth endoplasmic reticulum, the organelle where steroids are synthesized. Steroids are lipophilic, so they diffuse easily across membranes, both out of their parent cell and into their target cell. This property also means

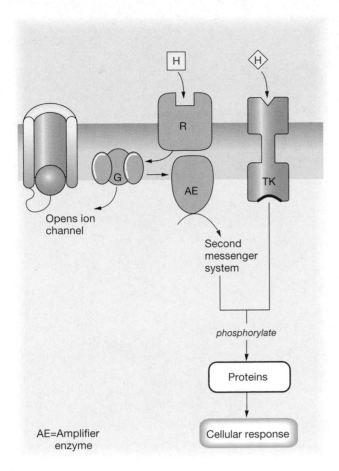

■ Figure 7-5   Receptors for peptide hormones   Because peptide hormones (H) cannot enter their target cells, they combine with membrane receptors (R) that activate signal transduction. Many hormones use a G protein-linked-adenylyl cyclase-cAMP second messenger, but some receptors have tyrosine kinase (TK) activity or use second messengers such as cGMP.

that steroid-secreting cells cannot store hormones in secretory vesicles. Instead, they synthesize their hormones as hormone is needed. When a stimulus activates the endocrine cell, precursors in the cytoplasm are rapidly converted to active hormone. The hormone concentration in the cytoplasm rises, and the hormones move out of the cell by simple diffusion.

Transport in the Blood and Half-Life of Steroid Hormones   Like their parent cholesterol, steroid hormones are not very soluble in plasma and other body fluids. For this reason, most of the steroid hormone molecules found in the blood are bound to protein carrier molecules [∞ p. 130]. Some hormones have specific carriers, such as corticosteroid-binding globulin. Others simply bind to general plasma proteins such as albumin.

The binding of steroid hormone to a carrier protein protects the hormone from enzymatic degradation and results in an extended half-life. For example, *cortisol*, a hormone produced by the adrenal cortex, has a half-life

of 60 to 90 minutes. (Compare this with epinephrine, an amine hormone whose half-life is measured in seconds.)

Although binding steroid hormones to protein carriers extends their half-life, it also blocks their entry into target cells. The carrier-steroid complex remains outside the cell since the carrier proteins are lipophobic and cannot diffuse through the membrane. Only unbound hormone molecules can diffuse into the target cell (Fig. 7-7 ■).

Fortunately, hormones are active in minute concentrations, so a tiny amount of unbound steroid is enough to produce a response. As unbound hormone leaves the blood and enters cells, the carriers release bound steroid so that a small fraction of hormone remaining in the blood is always unbound.

Cellular Mechanism of Action of Steroid Hormones
The best-studied steroid hormone receptors are found within cells, either in the cytoplasm or in the nucleus. The ultimate destination of steroid receptor-hormone complexes is the nucleus, where the complex acts as a *transcription factor*, binding to DNA and turning on one or more genes (Fig. 7-7 ■). The activated genes create mRNA that directs the synthesis of new proteins. A hormone that alters gene activity is said to have a *genomic effect* on the target cell.

When steroid hormones activate genes to direct the production of new proteins, there is usually a lag time between hormone-receptor binding and the first measurable biological effects. This lag can be as much as 90 minutes. Consequently, reflex pathways that require rapid responses are not mediated by steroid hormones.

In recent years researchers have discovered that several steroid hormones, including estrogen and aldosterone, have cell membrane receptors linked to signal transduction pathways, just as peptide hormones do. These receptors enable those steroid hormones to initiate rapid *nongenomic* responses in addition to their slower genomic effects. With the discovery of nongenomic effects of steroid hormones, the differences between steroid and peptide hormone mechanisms of action seem to have almost disappeared.

✔ The steroid hormone aldosterone has a short half-life for a steroid hormone, only about 20 minutes. What would you predict about the degree to which aldosterone is bound to blood proteins?

Amine Hormones Are Derived from a Single Type of Amino Acid

There are three groups of amine hormones in which the hormone molecule is created from a single amino acid. **Melatonin** is derived from the amino acid tryptophan (see Focus on the Pineal Gland, Fig. 7-21 ■). The other amine hormones, the catecholamines and thyroid hormones, are derived from the amino acid tyrosine, notable

■ **Figure 7-6   Steroid hormones of
the adrenal cortex**   Cholesterol is the
parent compound for all steroid hormones.
Its basic ring structure is modified by addition or subtraction of functional groups.
This figure shows three steroid hormones
secreted by the adrenal cortex and ovary:
aldosterone, cortisol, and estradiol.

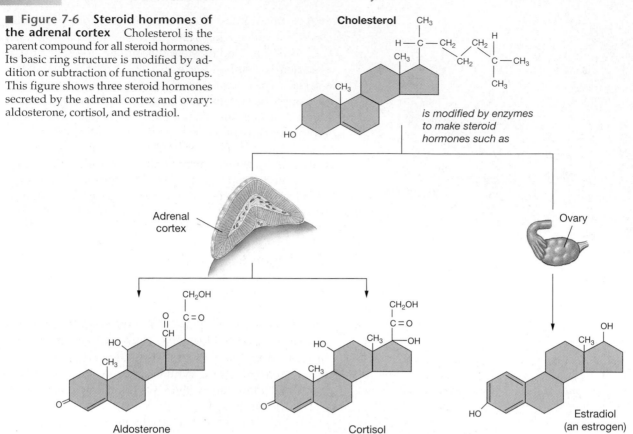

■ **Figure 7-7   Steroid hormone action**   Steroids are hydrophobic and must be bound to protein molecules in the blood. These same hormone-protein complexes are unable to cross cell membranes, and only unbound hormone can diffuse into the target cell. The primary receptors for steroid
hormones are located in the cytoplasm or nucleus of the target cell. The receptor-hormone complex
binds to DNA and activates one or more genes to produce mRNA. The resultant mRNA directs the
synthesis of new proteins in the cytoplasm. Some steroid hormones also have surface receptors that
activate second messenger systems and create rapid cellular responses.

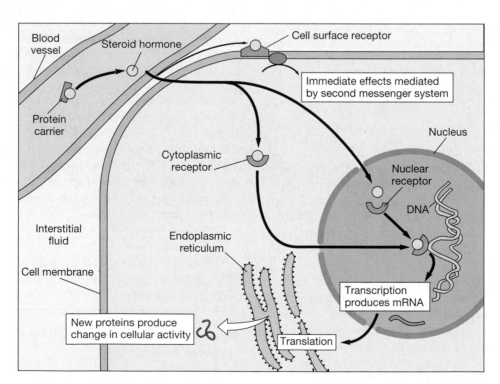

for its ring structure (Fig. 7-8 ■). Despite their common precursor, these amines have little in common. The **catecholamines** (epinephrine, norepinephrine, and dopamine) are neurohormones that have cell membrane receptors like typical peptide hormones. The **thyroid hormones,** produced by the butterfly-shaped thyroid gland in the neck, behave more like steroid hormones, with intracellular receptors that activate genes. Thyroid hormones will be discussed in detail in Chapter 21.

▶ CONTROL OF HORMONE RELEASE

You learned in Chapter 6 that homeostasis is maintained by local and reflex control pathways. All reflex pathways have similar components: an input signal, integration of the signal, and an output signal. In endocrine and neuroendocrine reflexes, a hormone or neurohormone is the output signal. Some complex endocrine pathways involve as many as three different hormones.

The Target of Most Trophic Hormones Is Another Endocrine Gland or Cell

Complex reflex control pathways have multiple integrating centers and involve more than one hormone [∞ p. 178]. In these complex pathways, any hormone that

…continued from page 186

Shaped like a butterfly, the thyroid gland straddles the trachea just below the Adam's apple. Responding to hormonal signals from the hypothalamus and anterior pituitary, the thyroid gland concentrates iodine, an element found in food (most notably as an added ingredient to salt) and combines it with the amino acid tyrosine to make two thyroid hormones, thyroxine and triiodothyronine. These thyroid hormones perform many important functions in the body, including the regulation of growth and development, oxygen consumption, and the maintenance of body temperature.

Question 1
a. *To which of the three classes of hormones do the thyroid hormones belong?*
b. *If a person's diet is low in iodine, predict what happens to thyroxine production.*

controls the secretion of another hormone is known as a **trophic (tropic)** hormone. The adjective *trophic* comes from the Greek word *trophikós*, which means "pertaining to food or nourishment" and refers to the manner in which the trophic hormone "nourishes" the target cell. Trophic hormones that act on endocrine cells may be recognized by the modifiers "releasing hormone" or "inhibiting hormone."

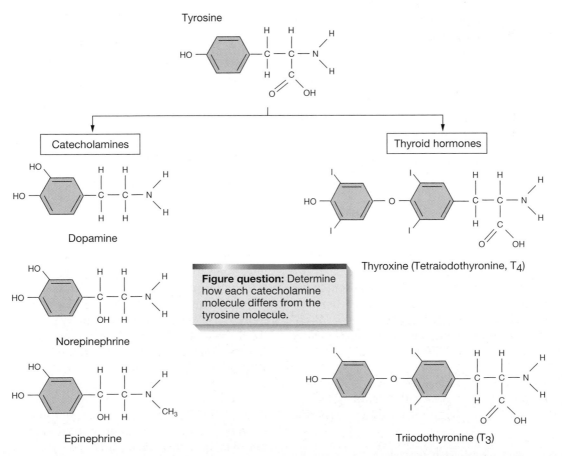

Figure question: Determine how each catecholamine molecule differs from the tyrosine molecule.

■ **Figure 7-8   Tyrosine-derived hormones**   The amino acid tyrosine has its side groups modified to create the catecholamines. Thyroid hormones are synthesized from two tyrosines and iodine (I) atoms.

• •

...continued from page 195

Thyroid hormone production is regulated by thyroid-stimulating hormone, or TSH, a hormone secreted by the anterior pituitary. The production of TSH is in turn regulated by the neurohormone TRH, or thyroid-releasing hormone, from the hypothalamus. Normally, when thyroid hormone levels in the blood increase, negative feedback decreases the secretion of TRH and TSH.

Question 2:   *If thyroid hormones are needed to regulate oxygen consumption and maintain body temperature, predict possible symptoms of hyperthyroidism.*

• •

Their names often end with the suffix *-tropin,* as in *gonadotropin.** The root word to which the suffix is attached is the target tissue: The gonadotropins are hormones that are trophic to the gonads.

Negative Feedback Turns Off Endocrine Reflexes

In many reflex pathways, the response of the pathway also serves as a negative feedback signal that turns off the reflex. For example, an increase in blood glucose concentration initiates an endocrine reflex to decrease blood glucose levels. Decreased blood glucose concentration then acts as a negative feedback signal to turn off the pathway and stop hormone secretion.

But in endocrine control pathways with multiple integrating centers, the hormones themselves act as negative feedback signals, as shown in Figure 7-9 ■. Each hormone in the pathway feeds back to suppress hormone secretion by integrating centers earlier in the reflex. This suppression decreases the output of trophic hormones and, in turn, means that less of the final hormone is produced. With this system of negative feedback, the system stays within the appropriate response range.

Hormones Can Be Classified by Endocrine Reflex Pathways

Endocrine reflex pathways are a convenient method by which you can classify hormones and simplify learning the control pathways that regulate their secretion. In the sections that follow, we will discuss the major reflex control pathways for hormones. This discussion is not all-inclusive, and you will encounter a few hormones in later chapters that do not fit into these pathways.

Classic Hormone Pathways Are the Simplest Endocrine Reflexes   The simplest reflex control pathway in the endocrine system is one in which an endocrine

*A few hormones whose names end in *-tropin* do not have endocrine cells as their targets. For example, melanotropin acts on pigment-containing cells in many animals.

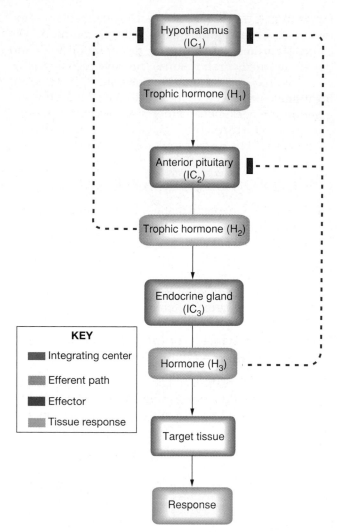

■ **Figure 7-9   Negative feedback loops in the hypothalamic-anterior pituitary pathway**   In complex endocrine pathways, the hormones of the pathway serve as negative feedback signals. Each hormone (H) feeds back to suppress hormone output from integrating centers (IC) higher in the control pathway.

cell directly senses a stimulus and responds by secreting its hormone [∞ Fig. 6-26, pattern 6, p. 178]. The hormone then travels through the blood to act on its target tissues. In this type of pathway, the endocrine cell acts both as sensor (receptor) and as integrating center.

An example of a classic hormone is parathyroid hormone (PTH), secreted by four small glands that lie behind the thyroid gland. The parathyroid endocrine cells monitor the concentration of Ca^{2+} in the blood (Fig. 7-10 ■). When the concentration of Ca^{2+} in the plasma falls below a certain point, the parathyroid cells secrete parathyroid hormone in response. Parathyroid hormone travels through the blood to act on its target tissues, initiating responses that increase the concentration of calcium in the plasma. The increase in plasma calcium is a negative feedback signal that turns off the reflex, ending the release of parathyroid hormone.

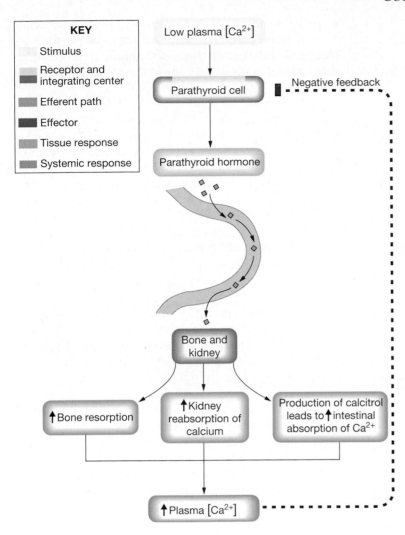

KEY

Stimulus

Receptor and integrating center

Efferent path

Effector

Tissue response

Systemic response

Low plasma [Ca^{2+}]

Parathyroid cell

Negative feedback

Parathyroid hormone

Bone and kidney

↑Bone resorption

↑Kidney reabsorption of calcium

Production of calcitrol leads to ↑intestinal absorption of Ca^{2+}

↑Plasma [Ca^{2+}]

■ **Figure 7-10   Endocrine reflex: parathyroid hormone**   Low plasma Ca^{2+} levels stimulate secretion of parathyroid hormone from the parathyroid glands. The hormone acts on various tissues to increase plasma Ca^{2+} concentrations. The increase in Ca^{2+} concentration acts as a negative feedback signal and shuts off the pathway.

✔ In Chapter 5 you learned how the beta cells of the pancreas sense an increase in blood glucose concentration [∞ p. 145]. Use what you know about ion channels, membrane transporters, and peptide hormone storage and secretion to design a parathyroid cell that secretes PTH when plasma Ca^{2+} concentrations decrease below a critical threshold. Draw a diagram of the process.

Some Endocrine Cells Respond to Multiple Stimuli

The secretion of some hormones is controlled by multiple stimuli. For example, insulin secretion can be triggered by an increase in the concentration of glucose in the blood, by stimulation from the nervous system, or by a hormone secreted from the digestive tract. The first two of these reflex pathways are diagramed in Figure 7-11 ■. No matter what the stimulus, insulin travels through the blood to act on its target tissues, whose responses ultimately remove glucose from the blood. The resultant decrease in blood glucose concentration acts as a negative feedback signal and turns off the reflex, ending release of insulin.

✔ Which three pathways in Figure 6-26 (p. 178) match insulin release by the three different mechanisms described above?

Neurohormones Are Secreted by Neurons into the Blood

There are three major groups of neurohormones: (1) the posterior pituitary neurohormones synthesized in the hypothalamus of the brain, (2) neurohormones released from the hypothalamus to control hormone release by the anterior pituitary, and (3) catecholamines made by the modified neurons that make up the adrenal medulla. Neurohormones are released when the nerve cell that synthesizes them receives a signal from the nervous system [∞ Fig. 6-26, pattern 2]. Many neurohormones are associated with the pituitary gland, so we will first describe that important endocrine structure.

The pituitary gland is actually two fused glands The **pituitary gland** has been called the master gland of the body because its hormones control so many vital functions.

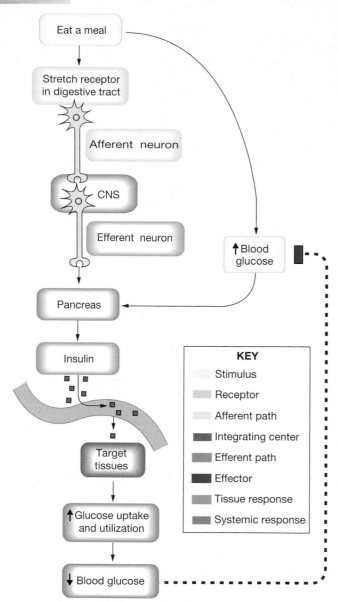

■ **Figure 7-11   Multiple stimuli for hormone release**   Insulin can be released directly by an increase in blood glucose levels or through nervous stimulation triggered by ingestion of a meal.

The pituitary is a lima bean-sized structure that extends downward from the brain, connected to it by a thin stalk and cradled in a protective pocket of bone (Fig. 7-12 ■). The first accurate description of the function of the pituitary gland came from Richard Lower (1631–1691), an experimental physiologist at Oxford University. Using observations and some experiments, he theorized that substances produced in the brain passed down the stalk into the gland and from there into the blood.

Lower did not realize that the pituitary gland is actually two different tissue types that merged during embryonic development. The **anterior pituitary** is a true endocrine gland of epithelial origin derived from the embryonic tissue that formed the roof of the mouth [∞ Fig. 3-20, p. 62]. It is also called the *adenohypophysis* [*adeno-*, gland + *hypo-*, beneath + *phyein*, to grow]. The anterior

pituitary synthesizes and releases peptide hormones. The **posterior pituitary** gland, or *neurohypophysis,* is an extension of the neural tissue of the brain. It secretes neurohormones made in neurons in the hypothalamus.

Neurohormones of the posterior pituitary   The posterior pituitary is the storage and release site for two neurohormones. These hormones are synthesized in neurons whose cell bodies lie in the hypothalamus, a region of the brain that controls many homeostatic functions (Fig. 7-13 ■). Each hormone is made in a separate cell type and packaged into secretory vesicles that are transported down the neuron for storage in the posterior pituitary. When a stimulus reaches the hypothalamus, the message is relayed to the posterior pituitary, and the vesicle contents are released into the circulation.

The two posterior pituitary neurohormones, vasopressin and oxytocin, are both small peptide hormones composed of nine amino acids each. **Vasopressin** (also known as **antidiuretic hormone**) regulates water balance in the body. In women, **oxytocin** released from the posterior pituitary controls the ejection of milk during breast-feeding and contractions of the uterus during labor and delivery.

Not all of the neurons that manufacture oxytocin send the peptide to the posterior pituitary for release as a neurohormone. Instead these neurons release oxytocin as a neurotransmitter or neuromodulator onto neurons in other parts of the brain. A number of animal and a few human experiments suggest that oxytocin plays an important role in social, sexual, and maternal behaviors. Some investigators have suggested that *autism,* a developmental disorder in which patients are unable to form normal social relationships, may be related to defects in the normal oxytocin-modulated pathways of the brain.

Complex Endocrine Pathways Include Two or More Hormones   In complex endocrine pathways, hypothalamic neurohormones act on the anterior pituitary gland, stimulating or inhibiting the synthesis and release of its hormones. In these pathways, the hypothalamic releasing and inhibiting hormones are acting as trophic hormones.

The hypothalamic-hypophyseal portal system   The hypothalamic trophic hormones that regulate secretion of anterior pituitary hormones are transported to the pituitary via a special set of blood vessels known as the **hypothalamic-hypophyseal portal system** (Fig. 7-14 ■). A **portal system** is a specialized region of the circulation consisting of two sets of capillaries directly connected by a set of blood vessels. There are three portal systems in the body, one in the kidneys, one in the digestive system, and this one in the brain.

The function of the hypothalamic-hypophyseal portal system is to transport trophic hormones from the hypothalamus directly to cells in the anterior pituitary.

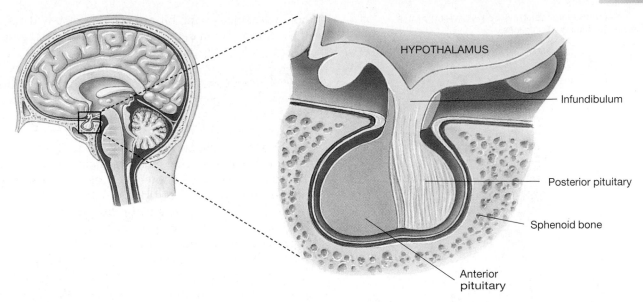

■ **Figure 7-12   Pituitary gland anatomy**   The pituitary gland sits in a protected pocket of bone, connected to the brain by a thin stalk, the infundibulum. The posterior pituitary is an extension of the neural tissue of the brain. The anterior pituitary is true endocrine tissue of epithelial origin.

Hormones secreted into a portal system have a distinct advantage over hormones secreted into the general circulation because a much smaller amount of hormone can be secreted to elicit the same response. A dose of hormone secreted into the general circulation will be rapidly diluted by the total blood volume, which is typically more than 5 L. The same dose secreted into the tiny volume of blood flowing through the portal system remains concentrated while it is taken directly to its target. Thus, a small number of neurosecretory neurons in the hypothalamus can effectively control the anterior pituitary.

The minute amounts of hormone secreted into the hypothalamic-hypophyseal portal system posed a great challenge to the researchers who first isolated these hormones. Because such tiny quantities of hypothalamic releasing hormones are secreted, Roger Guillemin and Andrew Shalley had to work with tremendous amounts of tissue to obtain enough hormone to analyze. Guillemin and his colleagues processed more than 50 tons of sheep hypothalami, and a major meat packer donated more than one million pig hypothalami to Shalley and his associates. For the final analysis, they needed 25,000 hypothalami to isolate and determine the amino acid sequence of just 1 mg of thyrotropin-releasing hormone, a tiny peptide made of three amino acids (see Fig. 7-4 ■). For their discovery, Guillemin and Shalley shared a Nobel Prize in 1977.

Hormones of the anterior pituitary As late as 1889, it was being said in reviews of physiological function that the pituitary was of little or no use to higher vertebrates! But by the early 1900s, researchers had discovered that animals with their anterior pituitary glands surgically removed were unable to survive more than a day or two. This observation, combined with the clinical symptoms associated with pituitary tumors, made scientists realize that the anterior pituitary is a major endocrine gland that secretes not one but six physiologically significant hormones.

The anterior pituitary hormones, their hypothalamic trophic hormones, and their targets are all illustrated

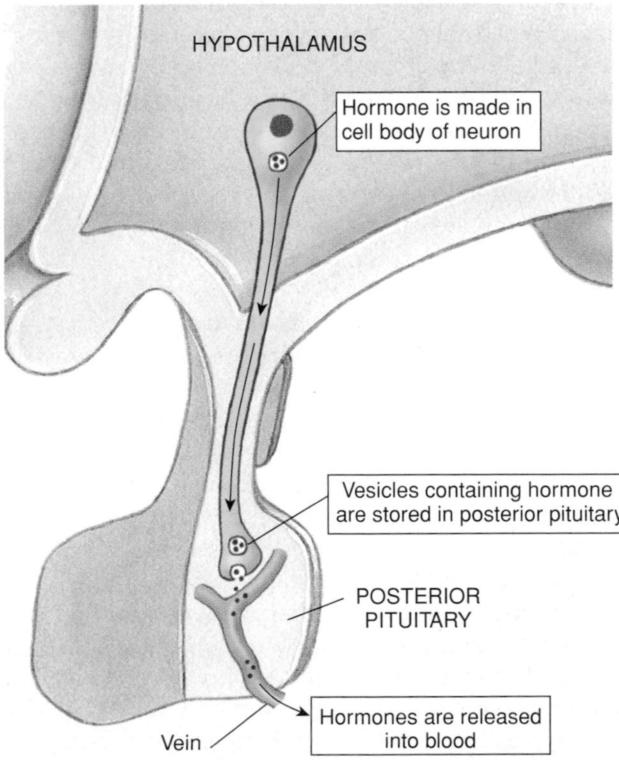

■ **Figure 7-13   Synthesis, storage, and release of posterior pituitary hormones**   Peptide neurohormones are synthesized in hypothalamic neurons and packaged into secretory vesicles. The secretory vesicles are transported down the neuron and stored in the posterior pituitary. At the appropriate signal, the neurohormones are released into the blood by exocytosis.

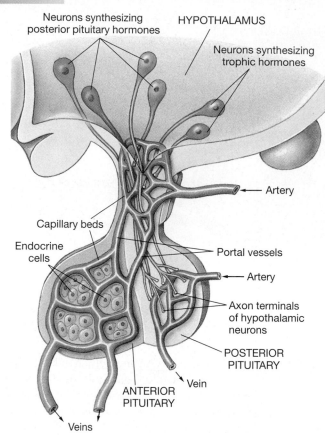

■ **Figure 7-14 Hypothalamic and pituitary hormones**
Neurons in the hypothalamus (green) make peptide hormones
that are released from the posterior pituitary. Other neuroendocrine neurons in the hypothalamus (purple) release their hormones into the capillaries of the portal system. These hormones are
carried directly to the endocrine cells of the anterior pituitary
where they exert their effects.

in Figure 7-15 ■. Notice that all but one of the anterior pituitary hormones have another endocrine gland or cell as
one of their targets. The pathways in which anterior pituitary hormones act as trophic hormones are among the
most complex reflexes because they have three different
integrating centers (the hypothalamus, the anterior pituitary, and the endocrine target of the pituitary hormone).

✔ Which pathway in Figure 6-26 (p. 178) fits the hypothalamic trophic hormone-anterior pituitary hormone-
endocrine target of the pituitary hormone pattern
described above?

One anterior pituitary hormone, **prolactin,** controls
milk production in the female breast. In both sexes, prolactin appears to play a role in regulation of the immune
system. **Growth hormone** (also called **somatotropin**) affects metabolism of many tissues in addition to stimulating hormone production by the liver (Fig. 7-16 ■).
Prolactin and growth hormone are the only two anteri-
or pituitary hormones whose secretion is controlled by

both releasing and inhibiting hormones. We will discuss
these hormones in detail in Chapters 24 and 21 respectively.

✔ Which pathway in Figure 6-26 (p. 178) fits the hypothalamic trophic hormone-prolactin-breast pattern described
above?

The remaining four anterior pituitary hormones all
have another endocrine gland as their primary target.
The two hormones known collectively as the **gonadotropins** act on the gonads. **Follicle-stimulating hormone** (FSH) and **luteinizing hormone** (LH) were
originally named for their effects on the ovaries, but both
hormones are trophic on the male testes as well. **Thyroid-stimulating hormone** (TSH, or **thyrotropin**) controls hormone synthesis and secretion in the thyroid
gland. **Adrenocorticotrophic hormone** (ACTH, or **corticotropin**) acts on certain cells of the adrenal cortex to control synthesis and release of the steroid hormone **cortisol.**

The nervous system plays an important role in controlling release of hypothalamic and anterior pituitary
hormones. One of the most fascinating links between the
brain and the anterior pituitary is the influence of emotions over the endocrine system and its functions. Physicians for centuries have recorded instances in which
emotional state has influenced health or normal physiological functioning. Women today know that the timing
of their menstrual periods may be altered by stress such
as travel or final exams. The condition known as "failure
to thrive" in infants can often be linked to environmental or emotional stress that increases secretion of some pituitary hormones and decreases production of others.
The interaction between stress, the endocrine system,
and the immune system is receiving intense study and
we will discuss it further in Chapter 22.

Feedback loops in the hypothalamic-pituitary pathway When
secretion of one hormone in a complex pathway increases or decreases, secretion of other hormones
changes as well because of feedback loops that link the
hormones. In pathways with one or two trophic hormones, the "downstream" hormone usually feeds back
to suppress the hormone(s) that controlled its secretion.
(A major exception to this is feedback by ovarian hormones, as you will learn in Chapter 24.)

In endocrine pathways that involve hypothalamic
releasing hormones, anterior pituitary hormones, and a
peripheral endocrine gland, the hormone secreted by the
peripheral gland will inhibit secretion of the two trophic hormones. This relationship is called **long-loop negative feedback.** For example, secretion of cortisol from
the adrenal cortex will suppress secretion of the trophic
hormones CRH and ACTH (see Fig. 7-15 ■). In **short-
loop negative feedback,** pituitary hormones feed back
to decrease hormone secretion by the hypothalamus.

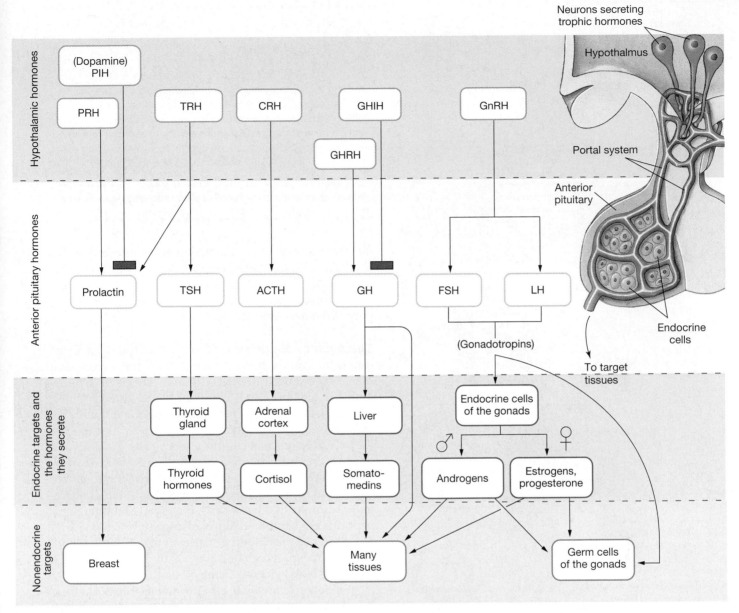

■ **Figure 7-15   Hormones of the hypothalamic-anterior pituitary pathway**   The hypothalamus secretes releasing and inhibiting hormones into the hypothalamic-hypophyseal portal system. These trophic hormones act on the endocrine cells of the anterior pituitary to influence secretion of their hormones. Those hormones may act on additional endocrine glands or may act directly on target cells. PIH = prolactin-inhibiting hormone; PRH = prolactin-releasing hormone; TRH = thyrotropin-releasing hormone; CRH = corticotropin-releasing hormone; GHIH = growth hormone-inhibiting hormone (somatostatin); GHRH = growth hormone-releasing hormone; GnRH = gonadotropin-releasing hormone; TSH = thyroid-stimulating hormone (thyrotropin); ACTH = adrenocorticotropic hormone (corticotropin); GH = growth hormone; FSH = follicle-stimulating hormone; LH = luteinizing hormone.

Feedback patterns are important in the diagnosis of endocrine pathologies, to be discussed later in the chapter.

✔ What is the target tissue of a hypothalamic hormone secreted into the hypothalamic-hypophyseal portal system?

✔ Look at the pathway for insulin release as the result of eating a meal (Fig. 7-11 ■). What event serves as the negative feedback signal to shut off insulin release?

Multiple Hormones Can Simultaneously Affect a Single Target

One of the most complicated and confusing aspects of endocrinology is the way that hormones interact at their target cells. It would be simple if each endocrine reflex were a separate entity and if each cell were under the influence of only a single hormone. But in many instances, cells and tissues are controlled by multiple hormones that may be present at the same time. Complicating the picture is the

KEY
Stimulus
Receptor
Afferent path
Integrating center
Efferent path
Effector
Tissue response

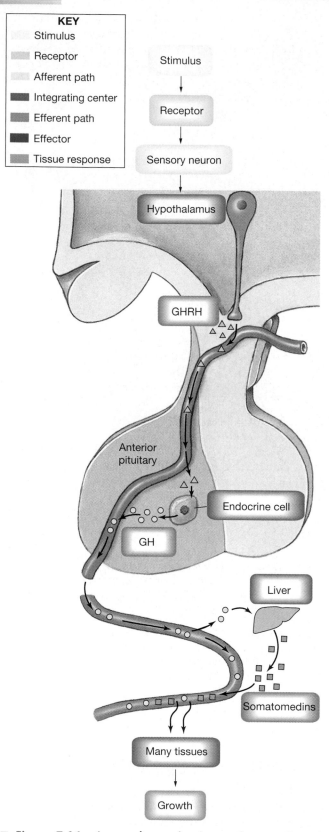

■ **Figure 7-16 A complex endocrine pathway** Growth hormone secreted by the anterior pituitary acts directly on many tissues of the body. It also influences liver production of somatomedins, also called insulin-like growth factors, another class of hormones that regulate growth.

...continued from page 196

Graves' disease is one form of hyperthyroidism. Therefore, the goal of treatment is to reduce thyroid hormone activity. One treatment involves the use of drugs that prevent the thyroid gland from using iodine. Another treatment is a single dose of radioactive iodine that results in the destruction of thyroid tissue. A third treatment is the surgical removal of all or part of the thyroid gland. Sometimes these treatments completely destroy thyroid gland function. In those cases, patients must take artificial thyroid hormones for the rest of their lives.

Question 3: *Why is radioactive iodine rather than some other radioactive element such as cobalt used to destroy thyroid tissue?*

fact that multiple hormones acting on a single cell can interact in ways that cannot be predicted by knowing the individual effects of the hormone. In this section, we examine three types of hormone interaction: synergism, permissiveness, and antagonism.

Synergism Sometimes different hormones have the same effect on the body, although they may accomplish that effect through different cellular mechanisms. One example is the hormonal control of blood glucose levels. Glucagon from the pancreas is the hormone primarily responsible for elevating blood glucose levels. But it is not the only hormone that has that effect. Cortisol raises blood glucose concentration as does epinephrine. What happens if two hormones are present, or if all three hormones are secreted at the same time? You might expect their effects to be additive. In other words, if a given amount of epinephrine elevates blood glucose 5 mg/100 mL blood, and glucagon elevates blood glucose 10 mg/100 mL blood, you would expect both hormones acting at the same time to elevate blood glucose 15 mg/100 mL blood (5 + 10).

Frequently, however, two hormones interact at their targets so that the combination of the two yields a result that is more than additive. This type of interaction is called **synergism.** Illustrated with the same example, a synergistic reaction would be:

epinephrine	elevates blood glucose 5 mg/100 mL blood
glucagon	elevates blood glucose 10 mg/100 mL blood
epinephrine + glucagon	elevate blood glucose 22 mg/100 mL blood

In other words, the combined effect of the two hormones is greater than the sum of the effects of the two hormones taken individually. An example of synergism between epinephrine, glucagon, and cortisol is shown in Figure 7-17 ■. The cellular mechanisms that underlie synergistic effects are not always clear, but with peptide hormones, synergism is often linked to overlapping effects on second messenger systems.

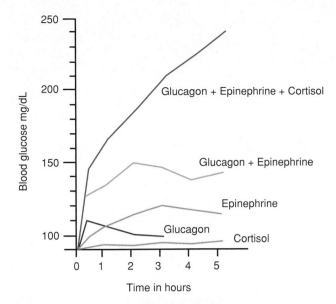

■ **Figure 7-17 Synergism** This graph shows the effect of hormone infusions on blood glucose levels. Cortisol has very little effect by itself. Glucagon raises glucose levels transiently, and epinephrine causes a sustained increase. The effects of glucagon and epinephrine together are greater than the summed effects of the two individual hormones, indicating a synergistic relationship. When cortisol is added to the other two hormones, blood glucose levels increase yet again, showing additional synergism. (Adapted from L. R. Johnson (ed.) *Essential Medical Physiology*: p. 626. New York: Raven Press, 1992.)

Synergism is not limited to hormones. It can occur with any two or more chemicals in the body. Pharmacologists have developed drugs with synergistic components. For example, the effectiveness of the antibiotic penicillin is enhanced by the presence of clavulanic acid in the same pill.

Permissiveness In **permissiveness,** one hormone cannot fully exert its effects unless a second hormone is present. For example, maturation of the reproductive system is controlled by gonadotropin-releasing hormone from the hypothalamus, gonadotropins from the anterior pituitary, and steroid hormones from the gonads. However, if thyroid hormone is not present in sufficient amounts, maturation of the reproductive system is delayed. Because thyroid hormone by itself cannot stimulate maturation of the reproductive system, thyroid hormone is considered to have a permissive effect on sexual maturation.

The results of this interaction can be summarized as follows:

thyroid hormone alone	no development of the reproductive system
reproductive hormones alone	delayed development of the reproductive system
reproductive hormones with adequate thyroid hormone	normal development of the reproductive system

Antagonism In some situations, two molecules work against each other, one diminishing the effectiveness of the other. This tendency of one substance to oppose the action of another is called **antagonism.** Recall from Chapter 6 that antagonism may result when two molecules compete for the same receptor [∞ p. 164]. When one molecule binds to the receptor but does not activate it, that molecule acts as a competitive inhibitor or antagonist to the other molecule. This type of antagonism has been put to use in the development of pharmaceutical compounds, such as the estrogen receptor antagonist tamoxifen.

In endocrinology, two hormones are considered antagonists if they have opposing physiological actions. For example, glucagon and growth hormone, both of which raise the concentration of glucose in the blood, are antagonistic to insulin, which lowers the concentration of glucose in the blood. Hormones with antagonistic actions do not necessarily compete for the same receptor. Instead, they may act through different metabolic pathways. Or one hormone may decrease the number of receptors for the opposing hormone. For example, evidence suggests that growth hormone decreases the number of insulin receptors, providing part of its antagonistic effects on blood glucose concentration.

The synergistic, permissive, and antagonistic interactions of hormones make the study of endocrinology both challenging and intriguing. With this brief survey of hormone interactions, you have built a solid foundation for learning more about hormone interactions in later chapters.

▶ **ENDOCRINE PATHOLOGIES**

As one endocrinologist said, "There are no good or bad hormones. A balance of hormones is important for a healthy life. ... Unbalance leads to diseases."* We can learn much about the normal functions of a hormone by studying the diseases caused by hormone imbalances. There are three basic patterns of endocrine pathology: hormone excess, hormone deficiency, and abnormal responsiveness of target tissues to a hormone.

To illustrate the endocrine pathologies, we will use a single example, that of cortisol production by the adrenal cortex (Fig. 7-18 ■). This is a complex reflex that starts with the secretion of corticotropin-releasing hormone (CRH) from the hypothalamus. CRH stimulates release of adrenocorticotrophic hormone (ACTH) from the anterior pituitary. ACTH in turn controls the synthesis and release of cortisol from the adrenal cortex. As in all homeostatic reflexes, negative feedback will shut off the pathway. As cortisol increases, it acts as a negative feedback signal, instructing the pituitary and hypothalamus to decrease their output of ACTH and CRH, respectively.

---

*W. König, preface to *Peptide and Protein Hormones* (New York: VCH Publishers, 1993).

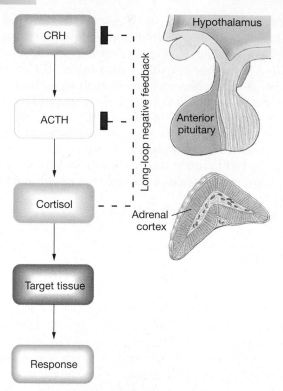

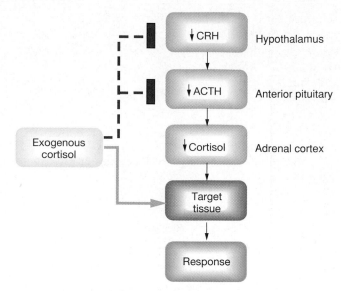

■ **Figure 7-19** **Negative feedback by exogenous cortisol** Hormone that is administered to a person for medical reasons will have the same negative feedback effect as the endogenous hormone. This figure shows that exogenous cortisol will shut down the hypothalamus and pituitary, just as cortisol from the adrenal cortex would. Without ACTH, the adrenal cortex also shuts down.

■ **Figure 7-18** **Control pathway for cortisol secretion** Cortisol is a steroid hormone secreted by the adrenal cortex. Its release is regulated by the trophic hormone ACTH (adrenocorticotropic hormone, or corticotropin) from the anterior pituitary. ACTH in turn is regulated by the CRH (corticotropin-releasing hormone) from the hypothalamus. Cortisol has a negative feedback effect on ACTH and CRH.

Hypersecretion Exaggerates a Hormone's Effects

If a hormone is present in excessive amounts, the normal effects of the hormone are exaggerated. Most instances of hormone excess are due to **hypersecretion.** There can be multiple causes for hypersecretion including benign tumors (adenomas) and cancerous tumors of the endocrine glands. Occasionally, nonendocrine tumors secrete hormones.

Sometimes an individual may exhibit symptoms of hypersecretion as the result of medical treatment with a hormone or agonist. In this case, the condition is said to be *iatrogenic,* or physician-caused [*iatros,* healer + *-gen,* to be born]. It seems simple enough to correct the hormone imbalance by stopping the *exogenous* [*exo-,* outside + *-gen,* to be born] hormone, but this is not always the case. In our example, exogenous cortisol in the body acts as a negative feedback signal, just as cortisol produced within the body would, shutting off the production of CRH and ACTH (Fig. 7-19 ■). Without the trophic "nourishing" influence of ACTH, the body's own cortisol production shuts down. If the pituitary remains suppressed and the adrenal cortex is deprived of ACTH long enough, the cells of both glands shrink and lose their

ability to manufacture ACTH and cortisol. This condition is known as **atrophy** [*a-,* without + *trophikós,* nourishment]. If the cells of an endocrine gland atrophy because of exogenous hormone administration, they may be very slow or totally unable to regain normal function when the treatment with exogenous hormone is stopped. As you may know, steroid hormones are often used to treat poison ivy and severe allergies. However, when treatment is complete, the dosage must be tapered off gradually to allow the pituitary and adrenal gland to work back up to normal hormone production. As a result, packages of steroid pills direct patients ending treatment to take six pills one day, five the day after that, and so on. Low-dose, over-the-counter steroid creams usually do not pose a risk of feedback suppression when used as directed.

Hyposecretion Diminishes or Eliminates a Hormone's Effects

Symptoms of hormone deficiency occur when too little hormone is secreted (**hyposecretion**). Hyposecretion may occur anywhere along the endocrine control pathway, in the hypothalamus, pituitary, or other endocrine glands. The most common cause of hyposecretion syndromes is atrophy of the gland due to some disease process. Hyposecretion of thyroid hormone may occur if there is insufficient dietary iodine for the thyroid gland to manufacture the iodinated hormone.

Negative feedback pathways are affected in hyposecretion, but in the opposite direction from hypersecretion. The absence of feedback causes trophic hormone levels to rise as the trophic hormones attempt to make the defective gland increase its hormone output. For example, if the adrenal cortex atrophies as a result of tuberculosis, cortisol production diminishes. The hypothalamus and anterior pituitary sense that cortisol levels are below normal, so they increase secretion of CRH and ACTH, respectively, in an attempt to stimulate the adrenal gland into making more cortisol.

Abnormal Tissue Responsiveness Can Be Due to Problems with Hormone Receptors or Second Messenger Pathways

Endocrine diseases do not always arise from problems with endocrine glands. They may also be triggered by changes in the responsiveness of the target tissues to the hormones. In these situations, the target tissues show abnormal responses although the hormone levels are within normal range. Changes in the target tissue response are usually caused by abnormal interaction between the hormone and its receptor or by alterations in signal transduction pathways. These concepts have been covered for signal molecules in general in Chapter 6 [∞ p. 165], so we will restrict this discussion to some typical examples of abnormal tissue responsiveness in the endocrine system.

Down-Regulation
If hormone secretion is abnormally high for an extended period of time, target cells will *down-regulate* (decrease the number of) their receptors in an effort to diminish their responsiveness to excess hormone. Hyperinsulinemia [*hyper-*, elevated + insulin + *-emia*, in the blood] is a classic example of down-regulation in the endocrine system. In this disorder, sustained high levels of insulin cause target cells to remove insulin receptors from the cell membrane.

Receptor and Signal Transduction Abnormalities
Many forms of inherited endocrine pathologies can be traced to problems with the hormone receptor in the target cell. If a mutation alters the structure of the protein receptor, the cellular response to receptor-hormone binding may be altered. In other cases the receptors may be absent or completely nonfunctional. For example, in testicular feminizing syndrome, androgen receptors are nonfunctional in the male fetus because of a genetic mutation. As a result, androgens produced by the developing fetus are unable to influence development of the genitalia, resulting in a child who appears to be female but who lacks a uterus and ovaries.

Genetic alterations in signal transduction pathways can lead to symptoms of hormone excess or deficiency. In the disease called *pseudohypoparathyroidism* [*pseudo-*, false + *hypo-*, decreased + parathyroid + *-ism*, condition or state of being], patients show signs of low parathyroid hormone even though blood levels of the hormone are normal to elevated. These patients have inherited a defect in the G protein that links the hormone receptor to the cAMP amplifier enzyme, adenylyl cyclase. Because the signal transduction pathway does not function, the cells are unable to respond to parathyroid hormone, and symptoms of hormone deficiency appear.

Diagnosis of Endocrine Pathologies Depends on the Complexity of the Reflex

Diagnosis of endocrine pathologies may be simple or complicated, depending on the complexity of the reflex. In the simplest endocrine reflex such as that for parathyroid hormone (see Fig. 7-10 ■), if there is too much or too little hormone, there is only one location where the problem can arise, the parathyroid glands. However, with complex hypothalamic-pituitary-endocrine gland reflexes, the diagnosis can be much more difficult.

If a pathology (deficiency or excess) arises in the last endocrine gland in a reflex, the problem is considered to be a **primary pathology.** For example, if a tumor in the adrenal cortex begins to produce excessive amounts of cortisol, the resulting condition is called *primary hypersecretion.* If dysfunction occurs in one of the tissues producing trophic hormones, the problem is a **secondary pathology.** For example, if the pituitary is damaged because of head trauma and ACTH secretion diminishes, the resulting cortisol deficiency is considered to be *secondary hyposecretion* of cortisol.

The diagnosis of pathologies in complex endocrine pathways depends on understanding negative feedback in the control pathway. Figure 7-20 ■ shows three possible causes of excess cortisol secretion. To determine which is the correct *etiology* (cause) of the disease in a particular patient, the clinician must assess the levels of the three hormones in the control pathway. If the problem is overproduction of CRH by the hypothalamus (Fig. 7-20a ■), CRH levels will be higher than normal. High CRH in turn causes high ACTH, which in turn causes high cortisol. This is, therefore, secondary hypersecretion arising from a problem in the hypothalamus.

…continued from page 202

Graves' disease causes the thyroid gland to produce too much thyroxine. People with Graves' disease therefore have elevated thyroxine levels in the blood. Their TSH levels are very low.

Question 4: If levels of TSH are low, but thyroxine levels are high, is Graves' disease a secondary disorder (one that arises as a result of a problem with the anterior pituitary or hypothalamus)? Explain your answer.

(a)

HYP

FOCUS ON... The Pineal Gland

Theories about the pineal gland, a pea-sized structure buried deep in the brain of humans (Fig. 7-21 ■), have alternated between giving it a prominent role in human physiology and relegating it to obscurity. Nearly 2000 years ago, the pineal was thought to act as a valve that regulated the flow of vital spirits and knowledge into the brain, so it came to be known as "the seat of the soul." But by 1950, scientists had decided that the human pineal was an evolutionary remnant with no known function. Then one of the wonderful coincidences of scientific research occurred. An investigator heard about a factor in beef pineal glands that could lighten the skin of amphibians. Using the classical methodology of endocrinology, he obtained pineal glands from a slaughterhouse and started making extracts. His biological assay consisted of dropping pineal extracts into bowls of live tadpoles to see if their skin color blanched. Several years and hundreds of thousands of pineal glands later, he had isolated a small amount of melatonin, an amine molecule synthesized by the pineal gland. This was the first evidence that the pineal gland produced a hormone. But what was its function?

Forty years later, we still do not know the role of the pineal gland in humans. We do know that melatonin is the "darkness hormone," secreted at night as we sleep. We know that in many seasonally breeding mammals, melatonin helps set the body clock that controls the onset of sexual activity. But the best we can say of the pineal gland in humans is that it transmits information about light-dark cycles to the center in the brain that governs the body's biological clock, using melatonin as the chemical message. Scientists and the popular press have tried to link melatonin to sexual function, the onset of puberty, seasonal affective depressive disorder (SADD) in the darker winter months, and sleep-wake cycles. To date, the only human function for melatonin that has been supported with scientific evidence is the hormone's ability to help shift the timing of the body's internal clock, making it useful in overcoming jet lag. As interest in melatonin grows, the pineal gland has come full cycle. No longer considered a vestigial structure of unknown function, the pineal may turn out to hold the key to the biological rhythms that provide an underlying pattern to our lives.

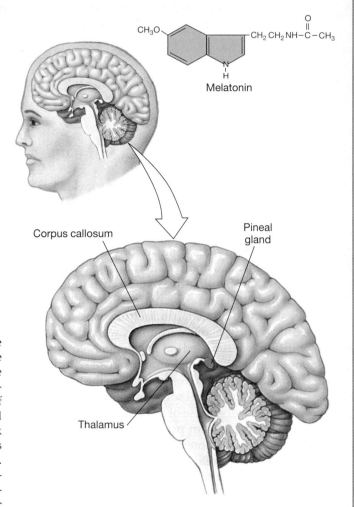

Melatonin

Corpus callosum

Pineal gland

Thalamus

■ **Figure 7-21   The pineal gland**

■ Figu
ondary
the pitu
a nega
The hig

Fi
cortiso
Once
produ
ative
produ
high A
If
are lov
20c ■).
[endo-
or the
peutic
els of c

humans. For example, melanocyte-stimulating hormone (MSH) from the intermediate lobe of the pituitary controls pigmentation in reptiles and amphibians. But adult humans have only a rudimentary intermediate lobe and normally do not have measurable levels of MSH in their blood.

In this chapter we have examined how the endocrine system with its hormones helps regulate the slower processes in the body. In the next chapter, you will learn how the nervous system can take care of the more rapid responses needed to maintain homeostasis.

PROBLEM CONCLUSION

In this running problem, you learned that the thyroid gland concentrates iodine to use for synthesis of thyroid hormones. You also learned that TRH and TSH control the secretion of thyroid hormones. In Graves' disease, thyroid hormone levels are high due to a immune system protein that mimics TSH.

Further check your understanding of this running problem by checking your answers against those in the summary table.

	Question	Facts	Integration and Analysis
1a	To which of the three classes of hormones do the thyroid hormones belong?	The three classes of hormones are peptides, steroids, and amines.	Thyroid hormones are made from the amino acid tyrosine, therefore they are amines.
1b	If a person's diet is low in iodine, predict what happens to thyroxine production.	In order to make thyroid hormones, the thyroid gland concentrates iodine and combines it with the amino acid tyrosine.	Iodine is an essential part of thyroid hormones. If it is lacking in the diet, a person will be unable to make thyroid hormone.
2	If thyroid hormones are needed to regulate oxygen consumption and maintain body temperature, predict possible symptoms of hyperthyroidism.	Oxygen consumption is a result of oxidative phosphorylation in the mitochondria. In this process, some of the energy released from nutrients, such as glucose, is released as heat while the rest is trapped in the high-energy bond of ATP [∞ p. 93].	People who are hyperthyroid will have an increased rate of oxygen consumption. Because oxidative phosphorylation requires nutrients as substrates, these people will have increased appetite and/or will lose weight. In addition, the extra heat they generate will tend to make them intolerant of hot environmental conditions.
3	Why is radioactive iodine rather than some other radioactive element such as cobalt used to destroy thyroid tissue?	In order to make thyroid hormones, the thyroid gland concentrates iodine.	Radioactive iodine will be concentrated in the thyroid gland and therefore selectively destroys that tissue. Other radioactive elements would distribute more widely throughout the body and might harm normal tissues.
4	If levels of TSH are low, but thyroxine levels are high, is Graves' disease a secondary disorder (one that arises as a result of a problem with the anterior pituitary or hypothalamus)? Explain your answer.	In secondary hypersecretion disorders, you would expect the levels of the hypothalamic and/or anterior pituitary trophic hormone levels to be elevated.	In Graves' disease, the level of the anterior pituitary trophic hormone TSH is very low. Therefore, the hypersecretion of thyroid hormones is not the result of hypersecretion of TSH. This means that Graves' disease is not a secondary disorder.
5	In Graves' disease, why doesn't negative feedback shut off thyroid hormone production before it becomes excessive?	In the normal negative feedback loop, as thyroid hormone levels increase, they shut off TSH. Consequently, without TSH stimulation, the thyroid stops producing thyroid hormone.	In Graves' disease, the high levels of thyroid hormone have shut off TSH production. However, the thyroid gland is producing hormone in response to the abnormal immune protein, not in response to TSH. Therefore, thyroid hormone production continues despite negative feedback.

CHAPTER REVIEW

SUMMARY

Hormones

1. The specificity of a hormone depends on its receptors and their associated signal transduction pathways. (p. 186)

2. A **hormone** is a chemical, secreted by a cell or group of cells into the blood, that acts on a distant target and is effective at very low concentrations. (p. 187)

3. The rate of hormone breakdown is indicated by a hormone's **half-life.** (p. 190)

The Classification of Hormones

4. There are three types of hormones: **peptide** and **protein hormones** composed of three or more amino acids, **steroid hormones** derived from cholesterol, and **amine hormones** derived from one type of amino acid. (p. 190)

5. Peptide hormones are made as inactive **preprohormones** and processed to **prohormones.** Prohormones are chopped into active hormone and peptide fragments that are **co-secreted.** (p. 191)

6. Peptide hormones dissolve in the plasma and have a short half-life. They bind to surface receptors and initiate cellular responses through signal transduction. As a result, cells usually respond rapidly to peptide hormones. In some instances, peptide hormones also initiate synthesis of new proteins. (p. 191)

7. Steroid hormones are synthesized as they are needed. They are hydrophobic, and most steroid hormones in the blood are bound to protein carriers. Steroids have an extended half-life. (p. 192)

8. The traditional receptors for steroid hormones are inside the target cell. Steroid hormones enter the nucleus, turn on genes, and direct the synthesis of new proteins. Cell response, as a result, is slower than with peptide hormones. Recently scientists have discovered that steroid hormones may bind to membrane receptors and have *nongenomic effects.* (p. 193)

9. Amine hormones are derived from a single type of amino acid. They may behave like typical peptide hormones or like a combination of steroid and peptide hormone. (p. 193)

Control of Hormone Release

10. **Trophic hormones** control the secretion of other hormones. (p. 195)

11. In many endocrine reflex pathways, the hormones of the pathway act as **negative feedback** signals. (p. 196)

12. Classic endocrine cells act as both receptor and integrating center in the reflex pathway. (p. 196)

13. Hormone secretion by some endocrine cells is controlled by multiple stimuli. (p. 197)

14. The pituitary gland has two parts. The **anterior pituitary** is a true endocrine gland, and the **posterior pituitary** is an extension of the brain. (p. 197)

15. The posterior pituitary releases two neurohormones, **oxytocin** and **vasopressin,** that are made in the **hypothalamus.** (p. 198)

16. The hormones of the anterior pituitary are controlled by **releasing** and **inhibiting hormones** from the hypothalamus. (p. 198)

17. The hypothalamic trophic hormones reach the pituitary through the **hypothalamic-hypophyseal portal system.** (p. 198)

18. There are six anterior pituitary hormones: prolactin, growth hormone, follicle-stimulating hormone, luteinizing hormone, thyroid-stimulating hormone, and adrenocorticotrophic hormone. (p. 199)

19. Hormones interact at their target cells. If the combination of two or more hormones yields a result that is more than additive, the interaction is **synergism.** If one hormone cannot exert its effects fully unless a second hormone is present, the second hormone is said to be **permissive** to the first. If one hormone opposes the action of another, they are **antagonistic.** (p. 201)

Endocrine Pathologies

20. Diseases of hormone excess are usually due to **hypersecretion.** Symptoms of hormone deficiency occur when too little hormone is secreted (**hyposecretion**). Changes in target tissue response may also result from abnormal interaction between the hormone and its receptor or from alterations in signal transduction pathways. (p. 204)

21. **Primary pathologies** arise in the last endocrine gland in a reflex. A **secondary pathology** is a problem with one of the tissues producing trophic hormones. (p. 205)

Hormone Evolution

22. Many hormones that are active in humans are also found in other vertebrate animals. (p. 207)

QUESTIONS

LEVEL ONE   Reviewing Facts and Terms

1. The study of hormones is called _____.

2. List the three basic ways hormones act on their target cells.

3. List five endocrine glands and give one example of a hormone each secretes. Give one effect of each hormone you listed.

4. Match the researchers with their experiments:
   (a) Lower
   (b) Berthold
   (c) Guillemin and Shalley
   (d) Brown-Séquard
   (e) Banting and Best

   1. isolated trophic hormones from the hypothalami of pigs and sheep
   2. claimed sexual rejuvenation after he injected himself with testicular extracts
   3. isolated insulin
   4. accurately described the function of the pituitary gland
   5. studied comb development in roosters that had been castrated

5. Put these steps for identifying an endocrine gland in order:
   (a) Purify the extracts and separate the active substances.
   (b) Perform replacement therapy with the gland or its extracts and see if the abnormalities disappear.
   (c) Implant the gland or administer the extract from the gland to a normal animal and see if symptoms characteristic of hormone excess appear.
   (d) Put the subject into a state of hormone deficiency by removing the suspected gland. Monitor the development of abnormalities.

6. In order for a chemical to be defined as a hormone, it must be secreted into the _____ for transport to a _____ and take effect at _____ concentrations.

7. What is meant by the term *half-life* in connection with the activity of hormone molecules?

8. Metabolites are inactivated hormone molecules, broken down by enzymes found primarily in the _____ and _____, to be excreted in the _____ and _____, respectively.

9. Candidate hormones often have the word _____ as part of their name.

10. List and define the three chemical classes of hormones. Give an example of one hormone in each class.

11. Decide if each statement below applies best to peptide hormones, steroid hormones, amine hormones, all of these, or none of these.
    (a) are lipophobic and must use a signal transduction system
    (b) short half-life, measured in minutes
    (c) often have a lag time of 90 minutes before effects are noticeable
    (d) water-soluble, therefore easily dissolved in the extracellular fluid for transport
    (e) most hormones belong to this class
    (f) all hormones in this group are derived from cholesterol
    (g) consist of three or more amino acids linked together
    (h) released into the blood to travel to a distant target organ
    (i) transported in the blood bound to protein carrier molecules
    (j) all are lipophilic, so diffuse easily across membranes

12. Why do steroid hormones take so much longer to act than peptide hormones?

13. When steroid hormones act on the nucleus, the hormone-receptor complex acts as a/an _____ factor, binds to DNA, and activates one or more _____, which create mRNA to direct the synthesis of new _____.

14. Researchers have discovered that some cells have additional steroid hormone receptors on their _____, enabling a faster response.

15. The amino acids that give rise to the amine hormones are _____, the basis for melatonin, and _____, from which the catecholamines and thyroid hormones are made.

16. A hormone that controls the secretion of another hormone is known as a _____ hormone.

17. In reflex control pathways involving trophic hormones and multiple integrating centers, the hormones themselves act as _____ _____ signals, suppressing trophic hormone secretion earlier in the reflex.

18. What characteristic defines neurohormones?

19. List the two hormones secreted by the posterior pituitary gland. To what chemical class do they belong?

20. What is the hypothalamic-hypophyseal portal system? Why is it important?

21. List the six hormones of the anterior pituitary gland; give an action of each. Which one(s) are considered to be trophic hormones?

22. How do long- and short-loop negative feedback differ? Give an example.

23. When two hormones work together to create a result that is greater than additive, that interaction is called _____. When two hormones must both be present to achieve full expression of an effect, that interaction is called _____. When hormone activities oppose each other, that effect is called _____.

LEVEL TWO   Reviewing Concepts

24. Map the following groups of terms. Add additional terms if you like.

    List 1
    co-secretion
    endoplasmic reticulum
    exocytosis
    Golgi apparatus
    hormone receptor
    peptide hormone
    preprohormone
    prohormone
    secretory vesicle
    signal sequence
    synthesis
    target cell response

    List 2
    ACTH
    anterior pituitary
    blood
    endocrine cell
    gonadotropins
    growth hormone
    hypothalamus
    inhibiting hormone
    neurohormone
    neuron
    oxytocin
    peptide/protein
    peripheral endocrine gland
    portal system
    posterior pituitary
    prolactin
    releasing hormone
    trophic hormone
    TSH
    vasopressin

25. Compare and contrast the following sets of terms:

    (a) paracrine, hormone, cytokine

    (b) primary and secondary endocrine pathologies

    (c) hypersecretion and hyposecretion

    (d) anterior and posterior pituitary

26. Compare and contrast the three chemical classes of hormones.

LEVEL THREE   Problem Solving

27. You have encountered the terms *specificity, receptors,* and *down-regulation* in other chapters in this text. Have their meanings changed? What chemical and physical characteristics do hormones, enzymes, transport proteins, and receptors have in common that makes specificity important?

28. Dexamethasone is a drug that is used to suppress the secretion of adrenocorticotrophic hormone (ACTH) from the anterior pituitary (see Fig. 7-18 ■). Two patients with hypersecretion of cortisol are given dexamethasone. Patient A's cortisol secretion level falls to normal levels as a result, but patient B's cortisol secretion remains elevated. Which patient has primary hypercortisolism, also known as Cushing's syndrome? Explain your reasoning.

29. Some early experiments for male birth control pills used drugs that suppressed gonadotropin (FSH and LH) release. However, men given these drugs stopped taking them because they decreased testosterone secretion and caused impotence. Then researchers suggested that a better treatment would be to give men extra testosterone. This would suppress sperm production by germ cells of the gonads without the side effect of impotence. Using the hypothalamus-gonadotropin-gonad pathway illustrated in Figure 7-15 ■ and what you know about feedback pathways, can you explain how the extra testosterone pathway would work?

LEVEL FOUR   Quantitative Problems

30. This graph represents the disappearance of a drug from the blood as the drug is metabolized and excreted. Based on the graph, what is the half-life of the drug?

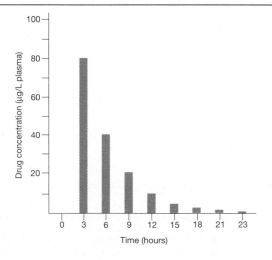

E X P L O R E <MediaLab>

Introduction

In this chapter you learned how tissues of the body communicate over long distances. Hormones are chemical messengers that travel in the blood. They regulate growth, development, and salt and water balance, as well as many other physiological processes. Disruptions of hormone secretion or action can lead to very serious medical problems. The following Web Explorations will help you gain an appreciation of the endocrine system and how feedback helps keep hormone secretion within a normal range. After reading the description below, visit the MediaLab for Chapter 7 in your Companion Website and select the appropriate keyword.

Web Exploration 1

Estimated time for completion = 10 minutes
The *half-life* of a hormone is the amount of time required to reduce the circulating concentration of a given hormone by one-half. This time provides important information about how long the hormone will be available to act on target tissues, i.e., its bioavailability. What influences the amount of time (half-life) that hormones stay in circulation? (Hint: think of how production and elimination are balanced and about which organs are involved in the elimination of hormones.)

The thyroid hormones, thyroxine (T_4) and triiodothyronine (T_3), are both secreted from the thyroid gland. T_3 is 3–4 times more potent than T_4, yet it has a shorter half-life. Select the keyword **THYROID HORMONE** on the Website to learn about the role of the blood protein known as thyroid-binding globulin (TBG). Is it a hormone? Why or why not? T_4 primarily circulates bound to TBG. What influence do you think this may have on the half-life of T_4, if any? Could a binding protein protect a circulating hormone? Would this be an advantage or disadvantage? To complete this exercise, visit MediaLab Web Exploration 1 in Chapter 7 of your Companion Website.

Web Exploration 2

Estimated time for completion = 10 minutes
The regulation of hormone secretion can happen via many chemical signals. However, feedback loops, in which a hormone regulates its *own* secretion, are a recurring theme in endocrinology. Most feedback loops are negative feedback loops, which serve to reinforce homeostasis. An example of an endocrine feedback loop would be a hormone that "feeds back" on the cell that synthesizes it. When hormone levels go up, the endocrine gland inhibits additional synthesis, thereby keeping the hormone concentration relatively constant. But there are also positive feedback loops that play a role in physiological processes where the maintenance of homeostasis is not needed or desired.

Select the keyword **POSITIVE FEEDBACK** on the Website to view a pathway image. What stops a positive feedback loop? To help you answer this question, think about the hormone oxytocin. It is secreted by the posterior pituitary and stimulates the contraction of smooth muscle in the uterus during labor. An increase in oxytocin production increases the uterine contractions, which in turn stimulate the secretion of more oxytocin. This is a good example of a positive feedback loop. What is the end result of increasing oxytocin levels in the blood of a woman in labor? What stops a positive feedback loop? To complete this exercise, visit MediaLab Web Exploration 2 in Chapter 7 of your Companion Website.

8 The Nervous System

■ *"The future of clinical neurology and psychiatry is intimately tied to that of molecular neural science." Eric R. Kandel, James H. Schwartz, and Thomas M. Jessell, in the preface to their book, Principles of Neural Science, 1991.* ■

CHAPTER OUTLINE

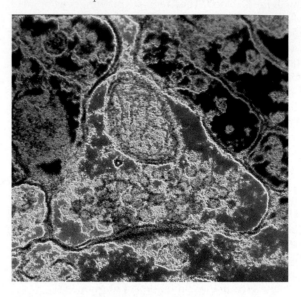

Vesicles filled with neurotransmitter (green) cluster near the end of a neuron as it synapses on its target.

BACKGROUND BASICS

In an eerie scene from a science fiction movie, white-coated technicians move quietly through a room filled with bubbling cylindrical fish tanks. As the camera zooms in on one tank, no fish are seen darting through aquatic plants. The lone occupant of the tank is a gray mass with a convoluted surface like a walnut and a long tail that appears to be edged with beads. Floating off the beads are hundreds of fine fibers, waving softly as the oxygen bubbles weave through them on their way to the surface. This is no sea creature ... it is a brain and spinal cord, removed from its original owner and awaiting transplantation into another body. Can this be real? Is this scenario possible? Or is it just the creation of an imaginative movie screenwriter?

PROBLEM
Mysterious Paralysis

"Like a polio ward from the 1950s" is how Guy McKhann, M.D., a neurology specialist at the Johns Hopkins School of Medicine, describes a section of Beijing Hospital that he visited on a trip to China in 1986. Dozens of paralyzed children—some attached to respirators to assist their breathing—filled the ward to overflowing. The Chinese doctors were convinced that the children had Guillain-Barré syndrome (GBS), an unusually rare paralytic condition. But Dr. McKhann immediately sensed that the diagnosis was wrong. There were simply too many stricken children for the illness to be the rare Guillain-Barré syndrome. Was it polio—as some of the Beijing staff feared? Or was it another illness, perhaps one that had not yet been discovered?

...continued on page 216

TABLE 8-1   Synonyms in Neuroscience

Term Used in This Book	Synonym
Action potential	Spike, nerve impulse, conduction signal
Autonomic nervous system	Visceral nervous system
Axon	Nerve fiber
Axonal transport	Axoplasmic flow
Axon terminal	Synaptic knob, synaptic bouton, presynaptic terminal
Axoplasm	Cytoplasm of the axon
Cell body	Cell soma
Cell membrane of the axon	Axolemma
Glial cells	Neuroglia, glia
Interneuron	Association neuron
Rough endoplasmic reticulum	Nissl substance, Nissl body
Sensory neuron	Afferent neuron, afferent

The brain is regarded as the seat of the soul, the mysterious source of those traits that we think of as setting humans apart from animals. The brain and spinal cord are also integrating centers for homeostasis, movement, and many other body functions. They are the control center of the **nervous system,** a network of billions or trillions of nerve cells linked together in a highly organized manner to form the rapid control system of the body.

Nerve cells, or **neurons,** are designed to carry electrical signals rapidly over long distances. They are uniquely shaped cells, and most have long, thin extensions, or *processes*, that can extend up to a meter in length. Neurons link together to form reflex pathways. In most pathways, neurons communicate with chemical signals called **neurotransmitters** across a small gap known as a *synapse*. In a few pathways, neurons are linked by *gap junctions* [∞ p. 57], allowing electrical signals to pass directly from cell to cell.

Reflex pathways in the nervous system do not necessarily follow a straight line from one neuron to the next. One neuron may influence multiple neurons, or many neurons may affect the function of a single neuron. Because of the complexity of the nervous system, we still do not fully understand how it functions. As a result, neurophysiology is one of the most active research areas in physiology today. Before you begin your study of the nervous system, you may want to become familiar with some specialized language used by neuroscientists. Table 8-1 lists some terms used in this book along with some common synonyms.

▶ ORGANIZATION OF THE NERVOUS SYSTEM

Information flow through the nervous system follows the basic pattern of a reflex described in Chapter 6 [∞ p. 169]. Sensory receptors throughout the body continuously monitor conditions in both the internal and external environments (Fig. 8-1 ■). These receptors send information along **afferent,** or **sensory, neurons** to the central nervous system. (Afferent pathways carry incoming information.) The **central nervous system,** or **CNS,** is composed of the **brain** and **spinal cord.** The neurons of the CNS integrate incoming information and determine if a response is needed. Signals directing an appropriate response (if any) are sent back to the effector cells of the body through **efferent neurons.** The afferent and efferent neurons together form the **peripheral nervous system** (sometimes called the PNS).

The efferent neurons of the peripheral nervous system are subdivided into the **somatic motor division,**[*] which controls skeletal muscles, and the **autonomic division,** which controls smooth and cardiac muscles, exocrine and some endocrine glands, and some types of adipose tissue.

The autonomic division is also called the *visceral nervous system* because it controls contraction and secretion in the various internal organs [*viscera,* internal organs]. Autonomic neurons are further divided into **sympathetic** and **parasympathetic branches,** which can be distinguished by their anatomical organization and by the chemicals that they use to communicate with their target cells. Many internal organs receive innervation from both types of autonomic neurons, and it is a common pattern to find that the two divisions exert *antagonistic control* over a single target [∞ p. 164].

In recent years, a third division of the nervous system has been receiving attention. The **enteric nervous system** is a network of neurons in the walls of the digestive

---

[*]The expression *motor neuron* can be used to refer to all efferent neurons. However, clinically, the term *motor neuron* (or *motoneuron*) is used only to describe somatic motor neurons that control skeletal muscles.

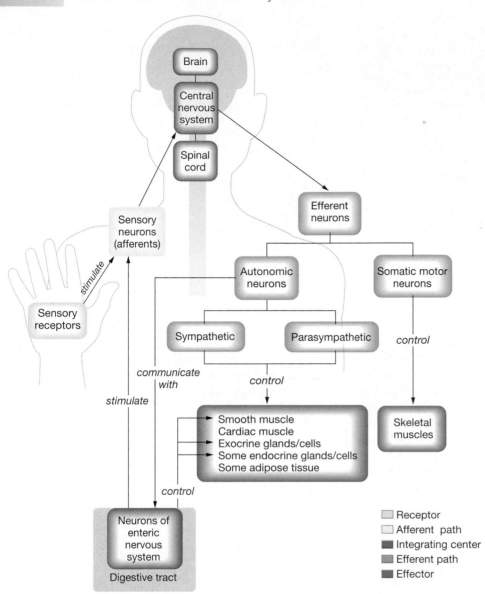

■ **Figure 8-1 Organization of the nervous system** The peripheral nervous system sends information to the central nervous system (CNS) through afferent (sensory) neurons and takes information from the CNS to target cells via efferent neurons. The enteric nervous system can act autonomously or can be controlled by the CNS through the autonomic division.

tract. It is frequently controlled by the autonomic division of the nervous system, but it is also able to function autonomously. We will discuss the enteric nervous system further in Chapter 20 on the digestive system.

▶ CELLS OF THE NERVOUS SYSTEM

The nervous system is composed primarily of two cell types: support cells known as glial cells (or glia or neuroglia) and nerve cells, or neurons, the basic signaling units of the nervous system.

Neurons Are Excitable Cells That Generate and Carry Electrical Signals

The neuron is the functional unit of the nervous system. (A *functional unit* is the smallest structure that can carry out all of the functions of a system.) Neurons are uniquely shaped cells with long appendages that extend outward

from the cell body. These appendages, or processes, are usually classified as either *dendrites* (which receive incoming signals) or *axons* (which carry outgoing information). The shape, number, and length of axons and dendrites vary from one neuron to the next, but they are an essential feature that allows neurons to communicate with each other and with other cells. Many neurons are

…continued from page 215

Guillain-Barré syndrome is a rare paralytic condition that strikes after a viral infection or immunization. There is no cure, but the illness usually resolves spontaneously. In Guillain-Barré, patients can neither feel sensations nor move their muscles.

Question 1: *Which division(s) of the nervous system may be involved in Guillain-Barré syndrome?*

similar to the model neurons shown in Figure 8-2 ■. In some neurons, processes shown in the model may be missing or modified. However, if you understand these model neurons, you will understand how all neurons work.

Neurons in the nervous system may be classified either structurally or functionally. Structurally, neurons are classified by the number of processes that originate from the cell body. They may be described as *pseudounipolar* (axon and dendrites fuse during development to create one long process; Fig. 8-3a ■), *bipolar* (single axon and single dendrite; Fig. 8-3b ■), or *multipolar* (many dendrites and branched axons; Fig. 8-3c,d ■). But, because physiology is concerned chiefly with function, we will classify neurons according to their function: (afferent) sensory neurons, interneurons, and efferent neurons.

Sensory neurons carry information about temperature, pressure, light, and similar stimuli from sensory receptors to the CNS. Structurally, they vary from the model neuron. For example, somatic sensory neurons have cell bodies that are found close to the CNS, with very long processes that extend to receptors in the limbs. In these neurons, the cell body is out of the direct path of signals originating at the dendrites (Fig. 8-3a ■). In contrast, sensory neurons for smell and vision have two long extensions, one an axon and one a dendrite (Fig. 8-3b ■). Signals that begin at the dendrites travel through the cell body to the axon.

Neurons that lie entirely within the CNS are known as **interneurons** (short for *interconnecting neurons*). They come in a variety of forms but often have quite complex branching processes that allow them to form synapses with many other neurons (Fig. 8-3c ■).

The **efferent neurons** of the somatic motor and autonomic divisions are generally very similar to the model neuron in Figure 8-2 ■. In the autonomic division, some neurons have enlarged regions along the axon called **varicosities.** The varicosities store and release neurotransmitter.

The Cell Body Is the Control Center of the Neuron

The nerve **cell body** (*cell soma*) resembles the typical cell described in Chapter 3, with a nucleus and all organelles needed to direct cellular activity [∞ p. 44]. An extensive cytoskeleton extends outward into the axon and dendrites. The position of the cell body varies in different types of neurons, but in most neurons the cell body is small, generally making up one-tenth or less of the total cell volume.

Despite its small size, the cell body with its nucleus is essential to the well-being of the cell. If a neuron is cut apart, any sections separated from the cell body are likely to degenerate slowly and die. With injured somatic motor neurons, the degeneration of the distal (distant) portions can cause permanent paralysis of the muscles that are *innervated* by the neuron. (The term *innervated* means "controlled by a neuron.") If the damaged neuron is a sensory neuron, the person may experience numbness or tingling in the region innervated by the neuron.

Dendrites Receive Incoming Signals

Dendrites [*dendron,* tree] are thin, branched processes whose main function is to receive incoming signals and transfer them to an integrating region within the neuron. Dendrites increase the surface area of a neuron so that it can communicate with multiple other neurons. Simple neurons may have only a single dendrite, whereas cells in the brain may have multiple dendrites with incredibly complex branching (Fig. 8-3c ■). A dendrite's surface area can be expanded even more by the presence of **dendritic spines** that vary from thin to mushroom-shaped.

Axons Carry Outgoing Signals to the Target

Most neurons have a single **axon,** a projection of the cell body that originates from a special region of the cell body

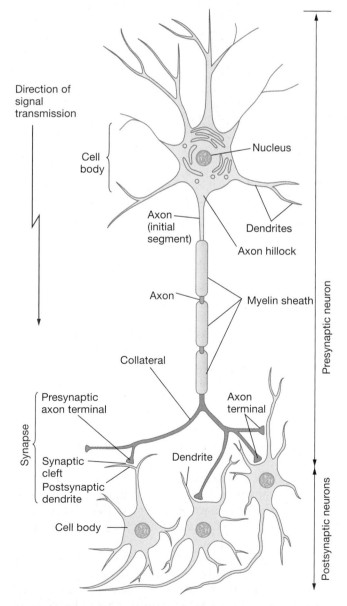

■ **Figure 8-2   Model neurons**   The neurons illustrated are typical neurons, consisting of a cell body and numerous extensions. The region where axon terminals communicate with their target cells is known as a synapse.

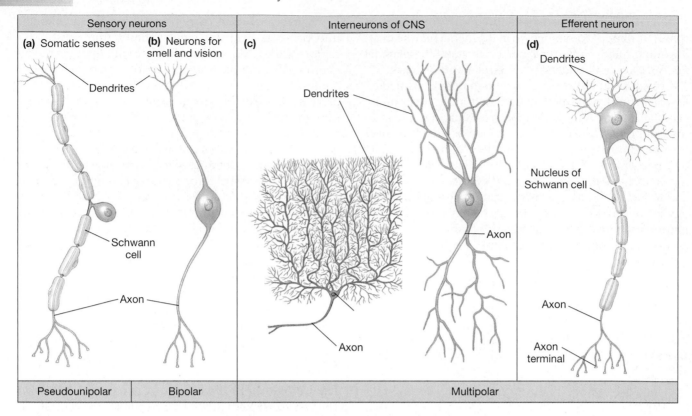

Sensory neurons		Interneurons of CNS	Efferent neuron
(a) Somatic senses	**(b)** Neurons for smell and vision	**(c)**	**(d)**
Pseudounipolar	Bipolar	Multipolar	

■ **Figure 8-3   Three functional types of neurons**   (a) Pseudounipolar sensory neurons appear to have only a single process. However, during development, the axon and dendrite fused, creating a single process called an axon. (b) Bipolar sensory neurons have two relatively equal fibers extending off the central cell body. These sensory neurons are found in the nose and eyes. (c) The interneurons of the CNS are highly branched (multipolar) but without long extensions. (d) The typical multipolar efferent neuron has five to seven dendrites coming off the cell body, each branching four to six times. The single long axon may branch several times. Somatic motor neurons such as the one illustrated terminate at swollen regions called axon terminals or synaptic boutons.

called the **axon hillock** (Fig. 8-2 ■). Axons vary from over a meter in length to only a few micrometers long. (A **nerve** is a collection of axons that carry information between the CNS and receptors or target cells.) Functionally, axons transmit outgoing electrical signals from the integrating center of the neuron to the end of the axon. At the distal end of the axon, the electrical signal is usually translated into a chemical message with the secretion of a *neurocrine* molecule [∞ p. 155].

Axons often branch sparsely along their length, forming **collaterals** [*col-*, with + *lateral*, something on the side]. In our model neuron, each collateral ends in a swelling called an **axon terminal.** The axon terminal con-

tains mitochondria and membrane-bound vesicles filled with neurocrine molecules.

Neurons that secrete neurotransmitters and neuromodulators terminate near their target cells, which are other neurons, muscles, or glands. The region where an axon terminal meets its target cell is called a **synapse** [*syn-*, together + *hapsis*, to join]. The neuron that delivers the signal to the synapse is known as the **presynaptic cell,** and the cell that receives the signal is called the **postsynaptic cell.** Neurons that secrete neurohormones terminate close to blood vessels so that neurohormones can enter the circulation.

The cytoplasm of axons is filled with many types of fibers and filaments but lacks ribosomes and endoplasmic reticulum. Therefore, any proteins destined for the axon, even the distant axon terminal, must be synthesized on the rough endoplasmic reticulum in the cell body. The proteins are then packaged into vesicles and moved down the axon by a process known as **axonal transport.**

Slow axonal transport moves material by **axoplasmic** (cytoplasmic) **flow** from the cell body to the axon terminal. Material moves only at a rate of 0.2–2.5 mm/day, therefore slow transport is used for components that are not consumed rapidly by the cell, such as enzymes and cytoskeleton proteins.

Fast axonal transport (Fig. 8-4 ■) moves organelles at rates of up to 400 mm (about 15.75 in.) per day. The

...continued from page 216

In Guillain-Barré syndrome, the disease affects both sensory and somatic motor neurons. Dr. McKhann observed that, although the Beijing children could not move their muscles, they could feel a pin prick.

Question 2:   *Do you think that the paralysis found in the Chinese children affected both sensory and somatic motor neurons? Why or why not?*

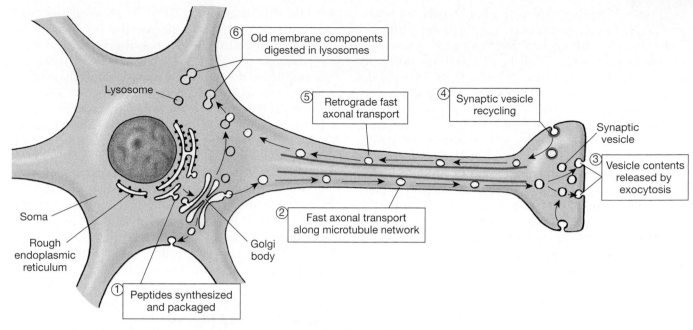

■ **Figure 8-4** **Axonal transport of membranous organelles**

neuron uses stationary microtubules as tracks along which transported particles "walk" with the aid of attached footlike proteins. These motor proteins, called **kinesins** [*kinesis,* motion], alternately bind and unbind to the microtubules with the help of ATP, stepping their attached organelles down the axon in a stop-and-go fashion. The role of motor proteins in axonal transport is similar to their role in muscle contraction and in the movement of chromosomes during cell division.

Fast axonal transport goes in two directions. Forward, or *anterograde,* transport takes synaptic and secretory vesicles and mitochondria from the cell body to the axon terminal. Backward, or *retrograde,* transport returns old membrane components from the axon terminal to the cell body for recycling. There is evidence that nerve growth factors and some viruses also reach the cell body by fast retrograde transport.

Glial Cells Are the Support Cells of the Nervous System

Glial cells [*glia,* glue] are the "unsung heroes" of the nervous system, outnumbering neurons in the nervous system by 10–50:1 (Fig. 8-5 ■). Although these neural

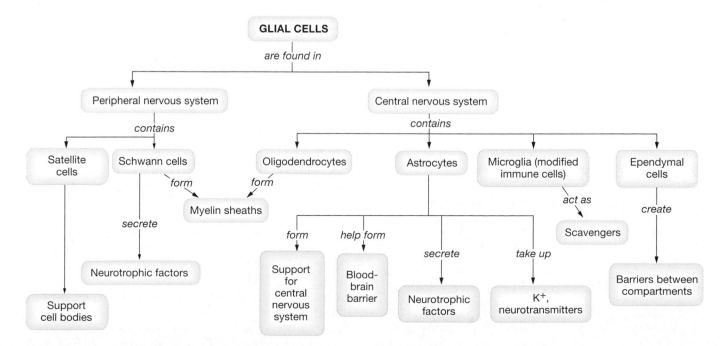

■ **Figure 8-5** **Glial cells and their functions**

support cells do not participate directly in the transmission of electrical signals over long distances, they do communicate with neurons and with each other using electrical and chemical signals. Glial cell function is an active area of neurophysiology research, and we still do not fully understand all the roles these important cells play in the nervous system.

Glial cells provide physical support for neurons since neural tissue has very little extracellular matrix [∞ p. 55]. They also direct the growth of neurons during repair and development. Recent research indicates that some glial cells provide metabolic support to neurons, help maintain the composition of the extracellular fluid, and even participate in information transfer. The peripheral nervous system has two types of glial cells: Schwann cells and satellite cells. The CNS has four types of support cells: oligodendrocytes, astrocytes, microglia, and ependymal cells.

Astrocytes [*astron,* a star] are highly branched cells that contact neurons and blood vessels and may transfer nutrients between the two. In addition, they take up K^+ and neurotransmitters from the extracellular fluid. **Microglia** are specialized immune cells that reside permanently within the CNS. When activated, they remove damaged cells and foreign invaders. **Ependymal cells** are epithelial cells that create a selectively permeable barrier between compartments in the brain.

Schwann cells in the peripheral nervous system and **oligodendrocytes** in the CNS support and insulate axons by creating **myelin,** multiple concentric layers of phospholipid membrane (Fig. 8-6a,c ■). Myelin forms when these glial cells wrap around an axon, squeezing out the glial cytoplasm so that each wrap becomes two membrane layers (Fig. 8-6b ■). As an analogy, think of wrapping a deflated balloon tightly around a pencil. Some neurons have as many as 150 wraps (300 membrane layers) in the myelin sheath that surrounds their axons.

In the CNS, one oligodendrocyte forms myelin around portions of several axons (Fig. 8-6a ■). In the peripheral nervous system, each Schwann cell associates with a single axon. One axon may have as many as 500 Schwann cells, each wrapped around a 1–1.5 mm segment of the axon (Fig. 8-6c ■). Between the insulated areas, a tiny region of axon membrane remains in direct contact with the extracellular fluid. These gaps, called the **nodes of Ranvier,** play an important role in the transmission of electrical signals along the axon.

The final type of glial cell is a nonmyelinating Schwann cell known as a **satellite cell.** These cells form supportive capsules around nerve cell bodies that are located in ganglia, outside the CNS.

▶ ELECTRICAL SIGNALS IN NEURONS

The unique property of nerve and muscle cells that characterizes them as *excitable tissues* is their ability to generate and propagate electrical signals. Recall from Chapter 5 that all living cells have a resting membrane potential difference [∞ p. 142]. The *membrane potential* (V_m) is the electrical disequilibrium that results from the uneven distribution of ions across the cell membrane.

Normally, sodium (Na^+) and calcium (Ca^{2+}) are more concentrated in the extracellular fluid than in the cytosol. Potassium (K^+) is more concentrated in the cells than in the extracellular fluid. At the same time, the resting cell membrane is much more permeable to K^+ than to Na^+ or Ca^{2+}, making K^+ the major determinant of the resting membrane potential. An average value for the V_m of neurons is −70 mV (inside the cell relative to outside).

Two factors influence the membrane potential: (1) the concentration gradients of different ions across the membrane and (2) the permeability of the membrane to those ions. If either of these factors changes, the membrane potential changes, creating an electrical signal that can be used to transmit information.

Ion Movement Across the Cell Membrane Creates Electrical Signals

Although Na^+ contributes minimally to the resting membrane potential, it plays a key role in generating electrical signals in excitable tissues. At rest, the cell membrane of a neuron is almost impermeable to Na^+. However, if the membrane suddenly increases its Na^+ permeability, Na^+ will enter the cell, moving down its electrochemical gradient [∞ p. 141]. The addition of positive Na^+ to the intracellular fluid *depolarizes* the cell membrane [∞ p. 144] and creates an electrical signal.

The movement of ions across the membrane can also *hyperpolarize* a cell. If the cell membrane suddenly becomes more permeable to K^+, positive charge is lost from inside the cell and the cell becomes more negative (hyperpolarizes). A cell may also hyperpolarize if negatively charged ions such as Cl^- enter the cell from the extracellular fluid.

✔ Would a cell with a resting membrane potential of −70 mV depolarize or hyperpolarize in the following cases? You must consider both the concentration and electrical gradients of the ion to determine net ion movement.

  (a)   Cell becomes more permeable to Ca^{2+}

  (b)   Cell becomes less permeable to K^+

✔ Would the cell membrane depolarize or hyperpolarize if a small amount of Na^+ leaked into the cell?

A significant change in membrane potential occurs with the movement of very few ions. For example, to change the membrane potential by 100 mV, only one out of every 100,000 K^+ must enter or leave the cell. This is such a tiny fraction of the total number of K^+ in the cell that the concentration of K^+ remains essentially unchanged. An equivalent situation is removing one grain of sand from a beach. There are so many grains of sand on the beach that the loss of one grain is not significant,

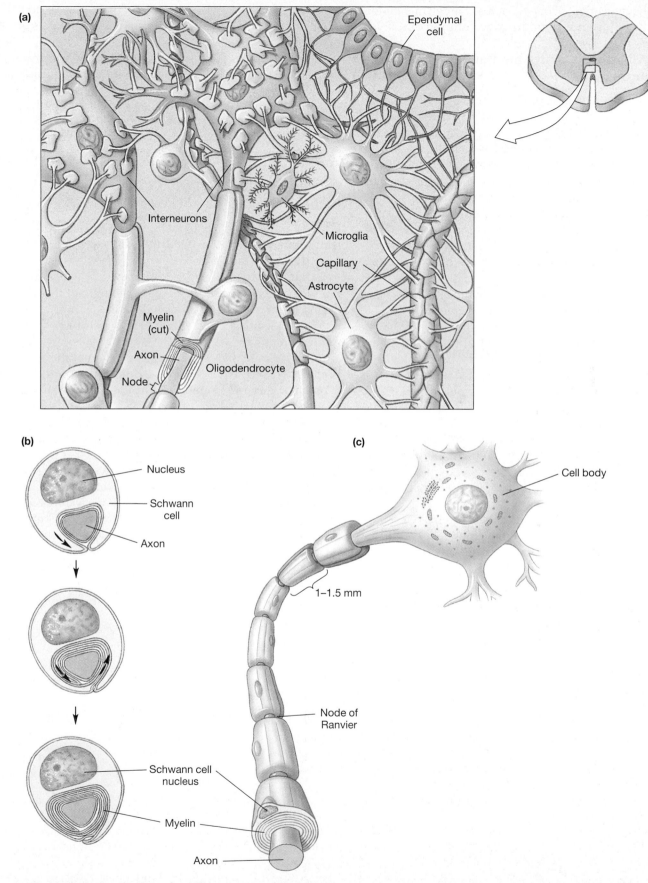

(a)

Ependymal cell

Interneurons

Microglia

Capillary

Astrocyte

Myelin (cut)

Axon

Node

Oligodendrocyte

(b)

Nucleus

Schwann cell

Axon

Schwann cell nucleus

Myelin

Axon

(c)

Cell body

1–1.5 mm

Node of Ranvier

■ **Figure 8-6   Formation of myelin**   (a) In the CNS, oligodendrocytes form myelin around portions of several interneuron axons. Astrocytes contact both neurons and blood vessels, but do not form myelin. (b) During myelin formation in the peripheral nervous system, the Schwann cell wraps around the axon many times while its nucleus is pushed to outside of the myelin sheath. (c) A Schwann cell forms myelin around a small segment of one axon.

just as the movement of one K^+ across the cell membrane does not significantly alter the concentration of K^+.

Gated Channels Control the Ion Permeability of the Neuron

How does a cell change its ion permeability? The simplest way is to open or close existing channels in the membrane. Neurons contain a variety of gated ion channels that alternate between open and closed states, depending on the intracellular and extracellular conditions [∞ p. 134]. A slower method is for the cell to insert or remove ion channels in the membrane.

Ion channels are usually named according to the primary ion(s) that they allow to pass through them. For example, Na^+ channels permit Na^+ to move across the membrane. There are four major types of selective ion channels in the neuron: (1) Na^+ channels, (2) K^+ channels, (3) Ca^{2+} channels, and (4) Cl^- channels. Other channels are less selective, such as the *monovalent cation channels* that allow both Na^+ and K^+ pass.

Ion channels may spend most of their time in an open state (*leak channels*) or they may have gates that open or close in response to particular stimuli [∞ p. 115]. Mechanically gated Na^+ channels are found in sensory neurons and open in response to physical forces such as pressure or light. Chemically gated ion channels in most neurons respond to a variety of ligands, such as extracellular neurotransmitters and neuromodulators or intracellular signal molecules. Voltage-gated ion channels play an important role in the initiation and conduction of electrical signals.

When ion channels open, ions move into or out of the cell, with their direction of movement depending on the concentration and electrical gradients for the ion. Potassium ions usually move out of the cell. Na^+, Cl^-, and Ca^{2+} usually flow into the cell. The net movement of electrical charge across the membrane depolarizes or hyperpolarizes the cell, creating an electrical signal.

These electrical signals can be classified into two basic types: graded potentials and action potentials (Table 8-2). **Graded potentials** are variable-strength signals that travel over short distances and lose strength as they travel through the cell. **Action potentials** are large, uniform depolarizations that can travel rapidly for long distances through the neuron without losing strength.

Graded Potentials Reflect the Strength of the Stimulus That Initiates Them

Graded potentials are depolarizations or hyperpolarizations that occur in the dendrites, cell body, or less frequently, near the axon terminals. They are called "graded" because their size or *amplitude* [*amplitudo*, large] is directly proportional to the strength of the triggering event. A large stimulus will cause a strong graded potential, and a small stimulus will result in a weak graded potential.

Graded potentials begin on the cell membrane at the point where ions enter from the extracellular fluid (Fig. 8-7 ■). For example, a neurotransmitter combines with receptors on the dendrite, opening Na^+ channels. Sodium ions move into the neuron, bringing in electrical energy. The positive charge carried by the Na^+ spreads as a wave of depolarization through the cytoplasm, just as a stone thrown into water creates ripples or waves that spread outward from the point of entry. The wave of depolarization that moves through the cell is known as **local current flow.** By convention, current flow in biological systems is the net movement of positive electrical charge.

The strength of the initial depolarization in a graded potential is determined by how much charge enters

TABLE 8-2 Comparison of Graded Potential and Action Potential

	Graded Potential	*Action Potential*
Type of signal	Input signal	Conduction signal
Where occurs	Usually dendrites and cell body	Trigger zone through axon
Types of gated ion channels involved	Mechanically, chemically, or voltage-gated channels	Voltage-gated channels
Ions involved	Usually Na^+, Cl^-, Ca^{2+}	Na^+ and K^+
Type of signal	Depolarizing (e.g., Na^+) or hyperpolarizing (e.g., Cl^-)	Depolarizing
Strength of signal	Depends on initial stimulus; can be summed	Is always the same (all-or-none phenomena); cannot be summed
What initiates the signal	Entry of ions through channels	Above-threshold graded potential at the trigger zone
Unique characteristics	No minimum level required to initiate	Threshold stimulus required to initiate
	Two signals coming close together in time will sum	Refractory period; two signals too close together in time cannot sum
		Initial stimulus strength is indicated by frequency of a series of action potentials

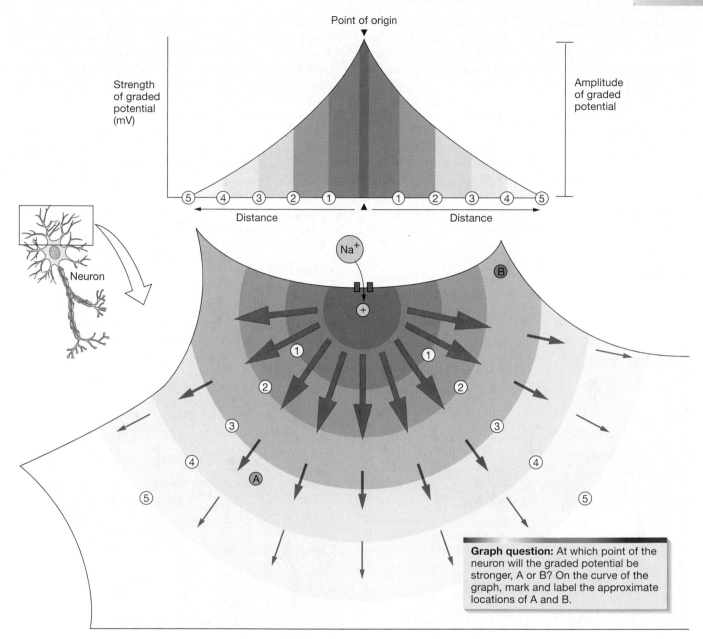

Graph question: At which point of the neuron will the graded potential be stronger, A or B? On the curve of the graph, mark and label the approximate locations of A and B.

■ Figure 8-7   **Graded potentials decrease in strength as they spread out from the point of origin**

the cell, just as the size of waves in the water is determined by the size of the stone. If more Na^+ channels open, more Na^+ enters and the graded potential has a higher initial amplitude. The stronger the initial amplitude, the farther the graded potential can spread through the neuron before it dies out.

Graded potentials may be hyperpolarizing as well as depolarizing. For example, opening Cl^- or K^+ channels in the neuron membrane will create a hyperpolarizing graded potential.

Graded potentials typically occur in the cell body and dendrites of a neuron. In neurons of the central and efferent nervous systems, graded potentials occur when chemical signals from other neurons open chemically gated ion channels.

Graded potentials travel through neurons until they die out or they reach the region known as the **trigger zone.** In efferent neurons and interneurons, the trigger zone is the axon hillock and the very first part of the axon, a region known as the **initial segment** (see Fig. 8-2 ■). In sensory neurons, the trigger zone is immediately adjacent to the receptor, where the dendrites join the axon.

✔ Identify the trigger zones of the neurons illustrated in Figure 8-3 ■.

The trigger zone is the integrating center of the neuron. If graded potentials reaching the trigger zone depolarize the membrane to a minimum level known as the **threshold** voltage, an *action potential* is initiated. If the depolarization does not reach threshold, no action potential is begun, and the graded potential simply dies out.

Because depolarization is necessary to *excite* the cell to fire an action potential, a depolarizing graded potential is known as an **excitatory postsynaptic potential** (**EPSP**). Any stimulus that makes a neuron more likely to fire an action potential is considered to be *excitatory*.

A hyperpolarizing graded potential moves the membrane potential farther from the threshold value and makes the neuron less likely to fire an action potential. Consequently, hyperpolarizing graded potentials are called **inhibitory postsynaptic potentials** (**IPSPs**). Any stimulus that makes a cell less likely to fire an action potential is considered to be *inhibitory*.

Figure 8-8 ■ shows a neuron that has several recording electrodes placed at intervals along the cell body and trigger zone. In the first example (Fig. 8-8a ■), a single stimulus triggers a *subthreshold* excitatory postsynaptic potential, one that is below threshold by the time it reaches the trigger zone. Although the cell is depolarized to −40 mV at the site where the graded potential begins, the

current decreases as it travels through the cell body. As a result, the graded potential is below threshold by the time it reaches the axon hillock. (For the average mammalian neuron, threshold is about −55 mV.) The stimulus is not strong enough to depolarize the cell to threshold at the trigger zone, so the graded potential dies out without triggering an action potential.

The second example (Fig. 8-8b ■) shows a stronger initial stimulus that initiates a stronger EPSP and ultimately causes an action potential. Although this excitatory graded potential also diminishes with distance through the neuron, its higher initial strength ensures that it is above threshold at the axon hillock. In this example, an action potential is triggered.

Action Potentials Travel Long Distances Without Losing Strength

Action potentials, also known as *spikes,* differ from graded potentials in several ways: (1) all action potentials are identical, and (2) they do not diminish in strength as they travel through the neuron.

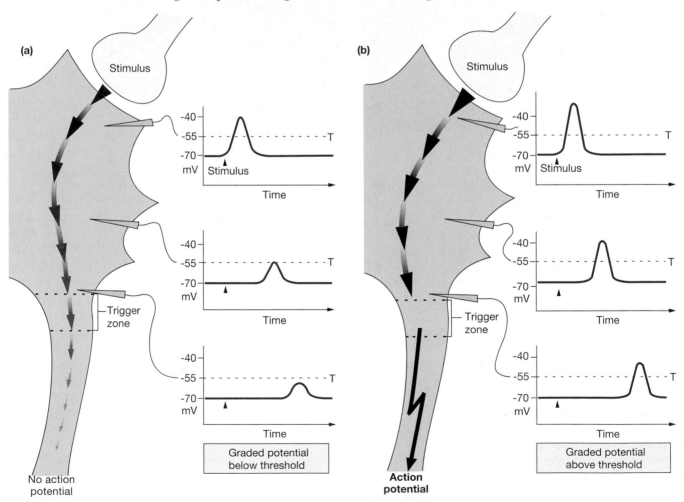

■ **Figure 8-8**   **Subthreshold and suprathreshold graded potentials in a neuron**   (a) A graded potential starts above threshold at the point where it is initiated but decreases in strength as it travels through the cell body. At the trigger zone it is below threshold (T) and therefore does not initiate an action potential. (b) A stronger stimulus at the same point on the cell body creates a graded potential that is still above threshold by the time it reaches the trigger zone, so an action potential results.

Recordings of action potentials show them to be identical depolarizations of about 100 mV amplitude. The strength of the graded potential that initiates an action potential has no influence on its amplitude. Action potentials are sometimes called **all-or-none** phenomena because they either occur as a maximal depolarization (if the stimulus reaches threshold) or do not occur at all (if the stimulus is below threshold).

The mechanism through which action potentials are generated and conducted along the axon allows them to stay constant. An action potential measured at the distal end of an axon is identical to the action potential that started at the trigger zone. This property is essential for the transmission of signals over long distances, such as from a fingertip to the spinal cord. Table 8-2 compares the key properties of action potentials and graded potentials.

Action Potentials Represent Movement of Na^+ and K^+ Across the Membrane

Action potentials are changes in membrane potential that occur when voltage-gated ion channels open, altering membrane permeability to Na^+ and K^+. Figure 8-9 ∎ shows the voltage and ion permeability changes that take place during an action potential. The graph can be divided into three phases: the rising phase of the action potential, the falling phase, and the after-hyperpolarization phase. Before and after the action potential, ① and ⑨, the neuron is at its resting membrane potential of −70 mV.

The rising phase of the action potential is due to a temporary increase in the cell permeability to Na^+. Action potentials begin when a graded potential reaching the trigger zone ② depolarizes the membrane to threshold (−55 mV) ③. As the cell depolarizes, voltage-gated Na^+ channels open and the membrane suddenly becomes much more permeable to Na^+. Because Na^+ is more concentrated outside the cell, and because the negative membrane potential inside the cell attracts these positively charged ions, Na^+ flows into the cell.

The addition of positive charge to the intracellular fluid depolarizes the cell membrane, making it progressively more positive ④. On the graph, this is shown by the steep rising phase of the action potential. In the top third of the rising phase, the membrane potential has reversed polarity; that is, the inside of the cell has become more positive than the outside. This reversal is represented on the graph by the **overshoot,** that portion of the action potential above 0 mV.

As soon as the cell membrane potential becomes positive, there is no longer an electrical driving force moving Na^+ into the cell (remember, like charges repel). However, the Na^+ concentration gradient remains, so Na^+ continues to move in. As long as Na^+ permeability remains high, the membrane potential moves toward the Na^+ *equilibrium potential* (E_{Na}) of +60 mV. (Recall from Chapter 5 that E_{Na} is the membrane potential at

which the movement of Na^+ into the cell down its concentration gradient is exactly opposed by the positive membrane potential [∞ p. 143].) However, before the E_{Na} is reached, the Na^+ channels in the axon close. Sodium permeability decreases dramatically, and the action potential peaks at +30 mV ⑤.

The falling phase of the action potential is due primarily to an increase in K^+ permeability. Voltage-gated K^+ channels, like Na^+ channels, open in response to depolarization. But the K^+ channel gates are slower to open, and peak K^+ permeability occurs later than peak Na^+ permeability (Fig. 8-9 ∎). By the time the K^+ channels are open, the membrane potential of the cell has reached +30 mV because of Na^+ influx. When the Na^+ channels close, the concentration and electrical gradients for K^+ favor movement of K^+ out of the cell.

As K^+ moves out of the cytoplasm, the membrane potential becomes more negative, creating the falling phase of the action potential and sending the cell toward its resting potential ⑥. The K^+ channels close slowly, so instead of repolarizing to exactly −70 mV, the cell hyperpolarizes, approaching the E_K of −90 mV. The additional loss of K^+ creates an **after-hyperpolarization** (also called the *undershoot*) ⑦. Once the voltage-gated K^+ channels close, K^+ leakage into the cell exceeds movement out, bringing the membrane potential back to its normal resting value of −70 mV ⑧.

To summarize, the action potential is a change in membrane potential that occurs when gated ion channels in the membrane open, increasing the cell's permeability first to Na^+ and then to K^+. The *influx* (movement into the cell) of Na^+ depolarizes the cell. This depolarization is followed by K^+ *efflux* (movement out of the cell), which restores the cell to the resting membrane potential.

Na^+ Channels in the Axon Have Two Gates

One question that puzzled scientists for many years was how the voltage-gated Na^+ channels could close when the cell was depolarized, because depolarization was the stimulus for Na^+ channel opening. After many years of study, they found the answer. The voltage-gated Na^+ channel has two gates to regulate ion movement rather than a single gate. These two gates, known as the *activation* and *inactivation gates,* flip-flop back and forth to open and shut the Na^+ channel.

When the neuron is at its resting membrane potential, the activation gate is closed and no Na^+ moves through the channel (Fig. 8-10a ∎). At the same time, the inactivation gate, apparently a ball-and-chain–like amino acid sequence on the cytoplasmic side of the channel protein, is open. When the cell membrane close to the channel depolarizes, the activation gate opens (Fig. 8-10b ∎). The Na^+ channel opens and allows Na^+ to move into the cell down its electrochemical gradient (Fig. 8-10c ∎). The addition of positive charge further depolarizes the inside of the cell and causes a positive feedback loop [∞ p. 174]

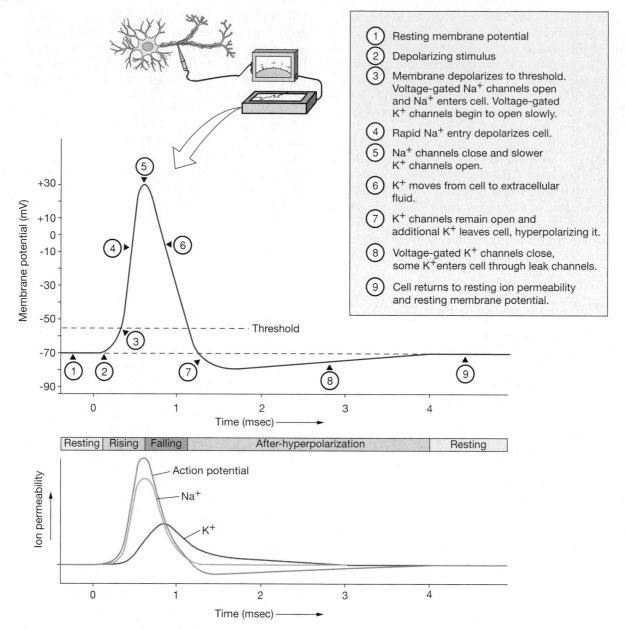

1. Resting membrane potential

2. Depolarizing stimulus

3. Membrane depolarizes to threshold. Voltage-gated Na⁺ channels open and Na⁺ enters cell. Voltage-gated K⁺ channels begin to open slowly.

4. Rapid Na⁺ entry depolarizes cell.

5. Na⁺ channels close and slower K⁺ channels open.

6. K⁺ moves from cell to extracellular fluid.

7. K⁺ channels remain open and additional K⁺ leaves cell, hyperpolarizing it.

8. Voltage-gated K⁺ channels close, some K⁺ enters cell through leak channels.

9. Cell returns to resting ion permeability and resting membrane potential.

■ Figure 8-9   **The action potential**

(Fig. 8-11 ■). More Na⁺ channels open and more Na⁺ enters, further depolarizing the cell. As long as the cell remains depolarized, the activation gates remain open.

As happens in all positive feedback loops, outside intervention is needed to stop the escalating depolarization of the cell. This outside intervention is the role of the second gate. Both gates respond to depolarization, but the inactivation gate delays for 0.5 msec. During that time the Na⁺ channel is open, allowing enough Na⁺ influx to create the rising phase of the action potential. When the slower inactivation gate of the Na⁺ channel finally closes (Fig. 8-10d ■), Na⁺ influx stops, and the action potential peaks.

As the neuron repolarizes during K⁺ efflux, the Na⁺ channel gates reset to their original positions so that they can respond to the next depolarization (Fig. 8-10e ■). Thus, the voltage-gated Na⁺ channel uses a two-step process for opening and closing rather than a single gate that swings back and forth. This is an important property that allows electrical signals along the axon to be conducted in only one direction, as you will see in the next section.

✔ If you could disable the inactivation gates of the Na⁺ channels so that the channels remained open, what would happen to the membrane potential? Explain your answer.

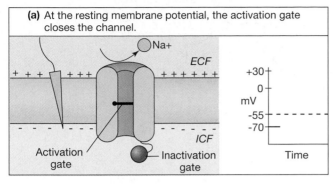

(a) At the resting membrane potential, the activation gate closes the channel.

Na+

ECF

+30
0
mV
-55
-70

ICF

Activation gate

Inactivation gate

Time

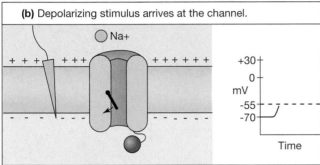

(b) Depolarizing stimulus arrives at the channel.

Na+

+30
0
mV
-55
-70

Time

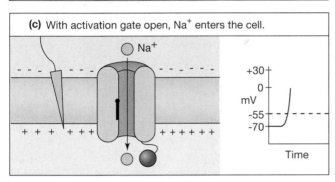

(c) With activation gate open, Na$^+$ enters the cell.

Na$^+$

+30
0
mV
-55
-70

Time

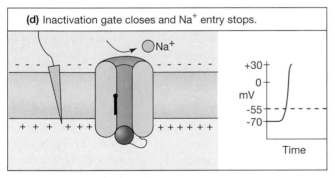

(d) Inactivation gate closes and Na$^+$ entry stops.

Na$^+$

+30
0
mV
-55
-70

Time

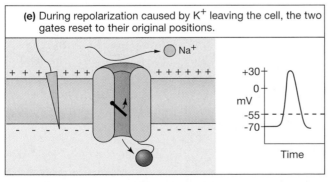

(e) During repolarization caused by K$^+$ leaving the cell, the two gates reset to their original positions.

Na$^+$

+30
0
mV
-55
-70

Time

■ **Figure 8-10   Model of the voltage-gated Na$^+$ channel**

Action Potentials Will Not Fire During the Refractory Period

The double gating of the Na$^+$ channel plays a major role in the phenomenon known as the **refractory period.** The adjective *refractory* comes from a Latin word meaning "stubborn." The "stubbornness" of the neuron refers to the fact that once an action potential has begun, for about 1 msec, a second action potential cannot be triggered, no matter how large the stimulus. This period is called the **absolute refractory period** (Fig. 8-12 ■). The absolute refractory period ensures that a second action potential will not occur before the first has finished. *Action potentials cannot overlap because of their refractory periods.*

After the Na$^+$ channel gates have reset to their original positions, but before the membrane is restored to its resting potential, a higher-than-normal graded potential can start another action potential. During this time, the neuron is said to be in its **relative refractory period.** A threshold-level depolarization during the relative refractory period will open those Na$^+$ channels that have returned to their resting position. However, Na$^+$ entry through these channels is offset by K$^+$ loss through still-open K$^+$ channels. The opposing flow of charge balances out, and the membrane potential does not reach threshold. For that reason, a stronger-than-normal depolarizing graded potential is needed to bring the cell up to the threshold for an action potential.

The refractory period is a key characteristic that distinguishes action potentials from graded potentials. If two stimuli reach the dendrites of a neuron within a short time, the successive graded potentials can be added to each other. But if two *suprathreshold* (above-threshold) graded potentials reach the action potential trigger zone within the absolute refractory period, the second stimulus will be ignored because the Na$^+$ channels are inactivated and are incapable of being stimulated again so soon.

Refractory periods limit the rate at which signals can be transmitted down a neuron. The absolute refractory period also assures one-way travel of an action potential from cell body to axon terminal by preventing backward conduction of the action potential.

Stimulus Intensity Is Coded by the Frequency of Action Potentials

One distinguishing characteristic of action potentials is that every action potential in a given neuron will be identical to every other action potential in that neuron. But if this is so, then how does the neuron transmit information about the strength and duration of the stimulus that started the action potential? The answer lies not in the amplitude of the action potential, but in the frequency of action potential propagation.

A stimulus in the form of a graded potential reaching the trigger zone does not usually trigger a single action

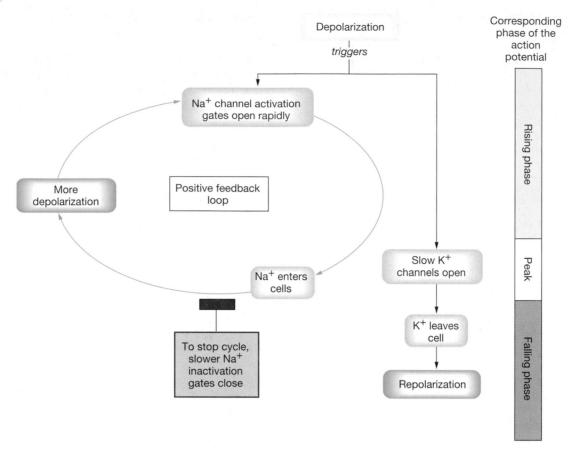

Figure 8-11   Ion movements during the action potential   Sodium entry creates a positive feedback loop that stops when the Na$^+$ channel inactivation gates close.

potential. Instead, even a minimal graded potential that is above threshold will trigger a burst of action potentials (Fig. 8-13a ■). If the graded potential increases in strength (amplitude), the frequency of action potentials fired (in action potentials per second) increases (Fig. 8-13b ■).

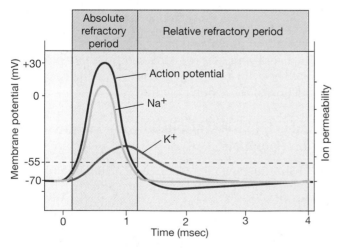

Figure 8-12   Refractory periods   During the absolute refractory period, no stimulus can trigger another action potential. During the relative refractory period, a larger-than-normal stimulus can initiate a new action potential. During this period, the Na$^+$ channel gates are resetting to their resting positions, and the K$^+$ channels are open.

The amount of neurotransmitter released at the axon terminal is directly related to the total number of action potentials that arrive at the terminal per unit time. An increase in signal strength will increase the neurotransmitter output, which in turn changes the magnitude of the graded potential in the postsynaptic cell.

The Na$^+$-K$^+$-ATPase Plays No Direct Role in the Action Potential

As you just learned, the action potential results from ion movements across the neuron membrane. First, Na$^+$ moves into the cell and then K$^+$ moves out. However, it is important that you understand that very few ions move across the membrane in a single action potential, so that *the relative Na$^+$ and K$^+$ concentrations remain essentially unchanged.* If a second action potential is triggered immediately after the first, it will be identical to the first one. The tiny amounts of Na$^+$ and K$^+$ that cross the membrane during action potentials do not disrupt the normal concentration gradients until a thousand or more action potentials have been fired.

The concentration gradients for Na$^+$ and K$^+$ in all cells are maintained by the *Na$^+$-K$^+$-ATPase* (Na$^+$-K$^+$ pump) in the cell membrane. The pump removes Na$^+$ that enters the cell and recaptures K$^+$ that has left it [∞ p. 125]. *But the Na$^+$-K$^+$-ATPase plays no direct role in the ac-*

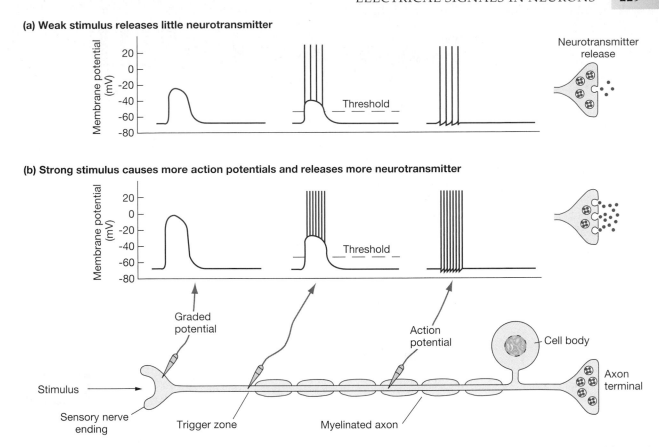

(a) Weak stimulus releases little neurotransmitter

(b) Strong stimulus causes more action potentials and releases more neurotransmitter

■ **Figure 8-13   Coding for stimulus intensity**   Stimuli of different intensity change the frequency of action potential firing along the axon. Since all action potentials in a neuron are identical, the strength of the stimulus is indicated by the frequency of action potential firing. (a) A graded potential that is barely above threshold causes a series of action potentials to pass down the axon and release neurotransmitter. (b) A stronger graded potential increases the frequency of action potential firing in the axon and releases more neurotransmitter.

tion potential. It treats the Na^+ that enters the cell during an action potential like any other Na^+ that leaks in, pumping it back out to the extracellular fluid. Since the number of Na^+ that enters the cell with each action potential is very small, a neuron poisoned with the pump inhibitor ouabain will continue to fire action potentials for an extended time before a change in the Na^+ concentration gradient can be noticed.

Action Potentials Are Conducted from the Trigger Zone to the Axon Terminal

The movement of an action potential through the axon at high speed is called **conduction.** Conduction represents the flow of electrical energy from one part of the cell to another in a process that ensures that any energy lost to friction or leakage out of the cell is immediately replenished. Thus, an action potential does not lose strength over distance as a graded potential does. To understand this, let's examine conduction at the cellular level to see why the action potential that reaches the end of the axon is identical to the action potential that started at the trigger zone.

When we talk about the conduction of an action potential, it is important to realize that there is no single ac-

tion potential that moves through the cell. The action potential that occurs at the trigger zone is like the movement in the first domino of a series of dominos standing on end. As the first domino falls, it strikes the next, passing on its kinetic energy. As the second domino falls, it passes kinetic energy to the third domino, and so on. If you could take a snapshot of the line of falling dominos, you would see that as the first domino is coming to rest in the fallen position, the next one is almost down, the third one most of the way down, etc., until you reach the domino that has just been hit and is starting to fall (Fig. 8-14a ■).

In the same way, a wave of action potentials moves down the axon. An action potential is simply a representation of the membrane potential in a given segment of cell membrane at any given moment in time. As the electrical energy of the action potential passes from one part of the axon to the next, the energy state is reflected in the membrane potential of that region. If we were to insert a series of recording electrodes along the length of an axon and start an electrical signal at the trigger zone, we would see a series of overlapping action potentials at different parts of the wave form, just like the dominos that are frozen in different positions (Fig. 8-14b ■).

Once a graded potential reaches threshold in the trigger zone, voltage-gated Na^+ channels open and the

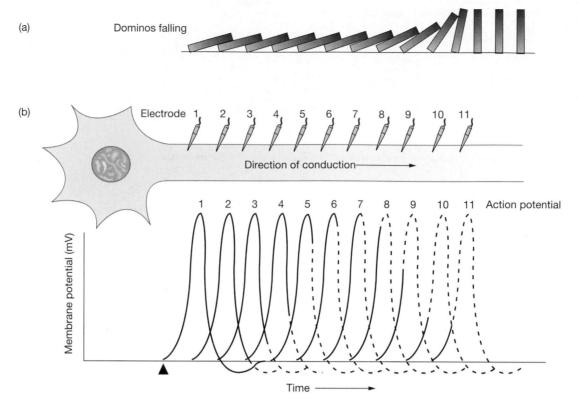

■ **Figure 8-14   Action potentials along an axon**   There are many action potentials taking place along an axon at any point in time. (a) The transmission of action potentials can be compared to a "snapshot" of dominos falling, where each domino is in a different position. (b) Similarly, simultaneous recordings of membrane potentials along an axon will show that each section of membrane is experiencing a different phase of the action potential. Twelve electrodes have been placed along the axon; the recordings from them are shown below the axon.

cell depolarizes (Fig. 8-15a ■). The positive current from Na$^+$ entry spreads through the cytoplasm in all directions, just as the depolarization of a graded potential spreads through the cell by local current flow (Fig. 8-15b ■). The current flow in the forward direction, down into the initial portions of the axon, is critical for conduction of the action potential.

The axon membrane is lined by the same type of voltage-sensitive Na$^+$ channels found in the membrane of the trigger zone. Positive charge from the depolarization of the trigger zone spreads into adjacent sections of membrane, attracted by the negative charge of the resting membrane potential. When the wave of depolarization reaches the voltage-sensitive Na$^+$ channels adjacent to the trigger zone, they open, allowing Na$^+$ into the cell (Fig. 8-15c ■). The membrane in the next segment of the axon is depolarized, and the positive feedback loop of depolarization begins: Na$^+$ channels open, Na$^+$ enters, causing depolarization, which opens more Na$^+$ channels in the adjacent membrane. The continuous entry of Na$^+$ means that the strength of the signal does not diminish as the action potential propagates itself. (Contrast this with graded potentials [Fig. 8-7 ■], in which Na$^+$ enters only at the point of stimulus, resulting in a membrane potential change that died out over distance.)

As each segment of axon reaches the peak of the action potential, its Na$^+$ channels inactivate. During the falling phase, K$^+$ channels open, allowing K$^+$ to leave the cytoplasm. Finally the K$^+$ channels close and the membrane returns to its resting potential.

Although positive charge from distal, depolarizing segments of membrane may flow back up into these sections by local current flow, the depolarization has no effect. This part of the axon is in the absolute refractory period, with its Na$^+$ channel inactivated, and the action potential cannot move backwards.

What happens to current that flows backward from the trigger zone into the cell body? We used to believe that there were few voltage-gated ion channels in that region, so it could be ignored. However, we now know that the cell body and dendrites may respond to the retrograde flow of current from action potentials by altering their electrical properties. The functional significance of these retrograde signals is unclear.

✔ If you place an electrode in the middle of an axon and artificially depolarize the cell above threshold, in which direction will an action potential travel: to the axon terminal, to the cell body, or to both? Explain your answer.

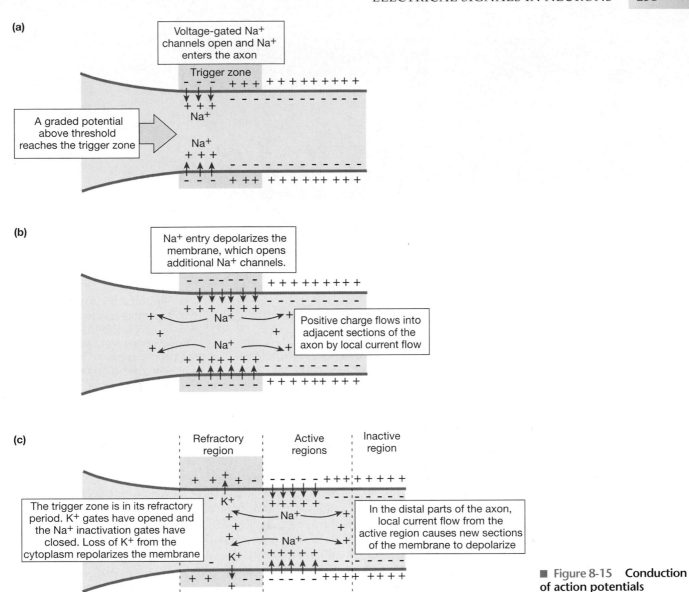

(a)

Voltage-gated Na⁺ channels open and Na⁺ enters the axon

Trigger zone

A graded potential above threshold reaches the trigger zone

(b)

Na⁺ entry depolarizes the membrane, which opens additional Na⁺ channels.

Positive charge flows into adjacent sections of the axon by local current flow

(c)

Refractory region   Active regions   Inactive region

The trigger zone is in its refractory period. K⁺ gates have opened and the Na⁺ inactivation gates have closed. Loss of K⁺ from the cytoplasm repolarizes the membrane

In the distal parts of the axon, local current flow from the active region causes new sections of the membrane to depolarize

■ Figure 8-15 **Conduction of action potentials**

The Diameter and Resistance of the Neuron Influence the Speed of Conduction

Two key physical parameters influence the speed of action potential conduction in mammalian neurons: (1) the diameter of the neurons and (2) the resistance of the neuron membrane to current leak out of the cell. The larger the diameter of the axon, the faster an action potential will move. To understand this relationship, think of a water pipe with water flowing through it. The water that touches the walls of the pipe encounters resistance due to friction between the flowing water molecules and the stationary walls. The water in the center of the pipe meets no direct resistance from the walls and therefore flows faster. In a large diameter pipe, a smaller fraction of the water flowing through the pipe is in contact with the walls, making the total resistance lower. In the same way, current flowing inside an axon meets resistance from the membrane. Thus, the larger the diameter of the axon, the lower its resistance to current flow.

The connection between axon diameter and speed of conduction is especially evident in the nerve cells of invertebrate animals such as the squid and the earthworm. The giant axons of these organisms may be up to 1 mm in diameter and have been very important to the development of our understanding of electrical signaling (Fig. 8-16a ■). The evolution of larger-diameter axons for rapid conduction works only in animals with simple nervous systems, however. If you compare a cross section of a squid giant axon with a cross section of a mammalian nerve, you will find that the mammalian nerve contains about 400 fibers in the same cross-sectional area (Fig. 8-16b ■). If each of those 400 axons were as large as a squid giant axon, the nerve would have to be 1.6 cm in diameter, much too large for our bodies to have many of them! Because vertebrates cannot rely on large axons for rapid transmission, another mechanism evolved to increase conduction: wrapping the axons in the insulating membranes of the myelin sheath.

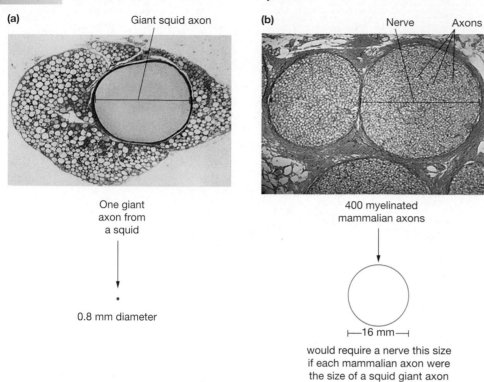

(a)

Giant squid axon

One giant
axon from
a squid

↓

·

0.8 mm diameter

(b)

Nerve   Axons

400 myelinated
mammalian axons

↓

⊙

⊢16 mm⊣

would require a nerve this size
if each mammalian axon were
the size of a squid giant axon

■ **Figure 8-16   Axon diameter
and speed of conduction**   Larger
axons offer less resistance to current
flow but occupy space. Myelinated
vertebrate axons can squeeze many
more axons into the same amount of
space.

The multiple layers of membrane in myelin create a high-resistance sheath that prevents current flow between the cytoplasm of the neuron and the extracellular fluid. The membranes are analogous to heavy coats of plastic surrounding electrical wires, increasing the effective thickness of the axon membrane by as much as 100-fold. In regions where the axon lacks the myelin sheath, the cell membrane has low resistance to current flow. In those areas, current leaks out of the cell in addition to moving forward into the next section of axon. Consequently, forward current flow is slower in regions of the axon that lack myelin.

When an action potential starts at the trigger zone, current flows rapidly down the myelinated axon to the first unmyelinated node of Ranvier (Fig. 8-17a ■). Each node has a high concentration of voltage-gated ion channels. When depolarization reaches the node, Na^+ channels open and Na^+ enters the axon, as described for the action potential. The Na^+ entry reinforces the depolarization, keeping the amplitude of the action potential constant, but it also slows the current flow.

When local current flow at the node reaches the next myelinated region, flow speeds up, zipping along until slowed at the next node of Ranvier. The apparent leapfrogging of the action potential from node to node has given this mode of current flow the name **saltatory conduction,** from the Latin word *saltare,* meaning "to leap." Saltatory conduction is an effective alternative to large-diameter axons and allows rapid action potentials through small axons. A myelinated frog axon 10 μm in diameter conducts action potentials at the same speed as an unmyelinated 500-μm squid axon. Thus, electrical

current flows through different axons at different rates, depending on the two parameters of axon diameter and myelination.

In *demyelinating diseases,* the loss of myelin from vertebrate neurons can have devastating effects on neural signaling. In the central and peripheral nervous systems, the loss of myelin slows the conduction of action potentials. The strength of the action potentials decreases when current leaks out of the uninsulated regions of membrane between the channel-rich nodes of Ranvier (Fig. 8-17b ■). Multiple sclerosis is the most common and best-known demyelinating disease. It is characterized by a variety of neurological complaints, including fatigue, muscle weakness, difficulty walking, and loss of vision. At this time, we can treat some of the symptoms but not the causes of demyelinating diseases, which are either inherited or autoimmune disorders. Currently, researchers are using recombinant DNA technology to study demyelinating disorders in mice.

…continued from page 218

Like multiple sclerosis, Guillain-Barré syndrome is an illness in which the myelin that insulates axons is destroyed. One way that multiple sclerosis and other demyelinating illnesses are diagnosed is through the use of a nerve conduction test. This test measures the strength of action potentials and the rate at which they are conducted as they travel down neurons.

Question 3:   *In Guillain-Barré syndrome, what would you expect the results of a nerve conduction test to be?*

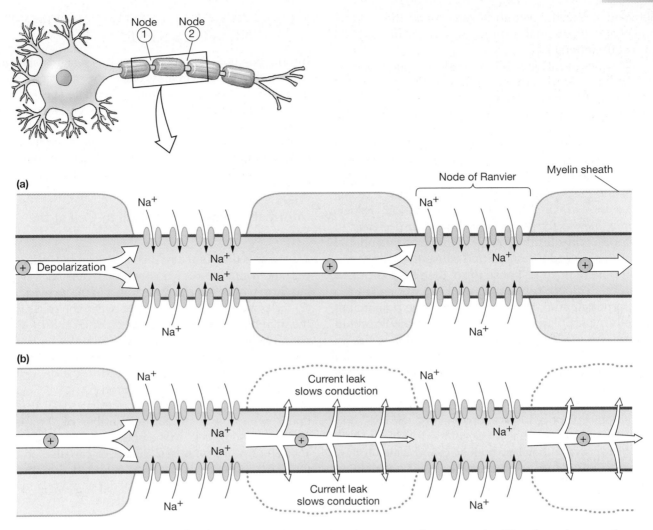

■ **Figure 8-17   Saltatory conduction**   (a) Action potentials appear to jump from one node of Ranvier to the next when local current flow moves rapidly through the myelinated sections between the nodes. Voltage-gated Na^+ channels are scarce between the nodes. (b) In demyelinating diseases, conduction slows down as current leaks out of the previously insulated regions between the nodes. Since there are few Na^+ channels in those regions, the action potentials are not conducted as strongly between the nodes.

Mutant Mouse Models   One technique that scientists use to study human diseases is to identify animals with a similar disease. The afflicted animal is then used as a model to explore the cause of the disease and to test potential treatments. Sometimes, natural mutations produce diseases that resemble human diseases. Two examples of such mutants are the twitcher mouse, in which normal myelin degenerates owing to an inherited metabolic problem, and the wobbler mouse, in which somatic motor neurons controlling the limbs die. In other cases, scientists have used biotechnology techniques to create mice that lack specific genes ("knock-out mice") or to breed mice that contain extra genes that were inserted artificially (transgenic mice). By comparing the genetically engineered mice with normal mice, we have expanded our understanding of how the nervous system responds during normal development and to disease.

Electrical Activity Can Be Altered by a Variety of Chemical Factors

A large variety of chemicals alter the conduction of action potentials by binding to Na^+, K^+, or Ca^{2+} channels in the neuron membrane. For example, some *neurotoxins* bind to and inactivate Na^+ channels. Local anesthetics such as procaine, which block sensation, function the same way. If the Na^+ channel is inactive, Na^+ cannot enter the axon. Thus, a depolarization that begins at the trigger zone loses strength as it moves down the axon, much like a normal graded potential. If the wave of depolarization manages to reach the axon terminal, it is too weak to release neurotransmitter. As a result, the message of the presynaptic neuron is not passed on to the postsynaptic cell.

Alterations in the extracellular fluid concentrations of K^+ and Ca^{2+} are also associated with abnormal electrical activity in the nervous system. The relationship between extracellular fluid K⁺ levels and the conduction of action potentials is the most straightforward and easiest to understand.

The concentration of K⁺ in the blood and interstitial fluid is the major determinant of the resting potential of all cells. If K⁺ concentrations in the blood move out of the normal range of 3.5–5 millimoles/liter (mmol/L), the result will be a change in the resting membrane potential of the cells. This change is not important to most cells, but it can have serious consequences to the body as a whole because of the relationship between resting potential and the excitability of nervous and muscle tissue.

An increase in blood K⁺ concentration, **hyperkalemia** [*hyper-*, above + *kalium*, potassium + *-emia*, in the blood], will shift the resting membrane potential of a neuron closer to threshold and cause the cells to fire action potentials in response to smaller graded potentials (Fig. 8-18c ■). If blood K⁺ concentrations fall too low (**hypokalemia**), the resting membrane potential of the cells hyperpolarizes, moving farther from threshold and requiring a larger-than-normal stimulus to fire an action potential (Fig. 8-18d ■). This condition shows up as muscle weakness because the neurons that control skeletal muscles are not firing normally.

Hypokalemia and its resultant muscle weakness are one reason that sport drinks supplemented with Na⁺ and K⁺, such as Gatorade®, were developed. When people sweat excessively, they lose both salts and water. If they replace this fluid loss with pure water, the K⁺ remaining in the blood will be diluted, causing hypokalemia. By replacing sweat loss with a dilute salt solution, they can prevent potentially dangerous drops in blood K⁺ levels.

Because of the importance of K⁺ to normal function of the nervous system, the body regulates blood K⁺ levels within a narrow range. We will discuss the important role of the kidneys in maintaining ion balance in Chapter 19.

▶ CELL-TO-CELL COMMUNICATION IN THE NERVOUS SYSTEM

Information flow through the nervous system using electrical and chemical signals is one of the most active areas of neurophysiology research today. The specificity of neural communication depends on several factors: neurotransmitters and neuromodulators secreted by neurons, the target cell receptors for these chemicals, and the anatomical connections between neurons and their targets, which occur in regions known as synapses.

Information Passes from Cell to Cell at the Synapse

A *synapse* is the point at which a neuron meets its target cell (another neuron or a non-neuronal cell). Each synapse has three parts: (1) the axon terminal of the *presynaptic cell*, (2) the **synaptic cleft**, or space between the cells, and (3) the membrane of the *postsynaptic cell* (Fig. 8-19 ■). In a neural reflex, information moves from presynaptic cell to postsynaptic cell. In most neuron-to-neuron synapses, the presynaptic axon terminals are next to the dendrites or cell body of the postsynaptic neuron.

In general, cells with many dendrites will also have many synapses. A moderate number of synapses is 10,000, but some cells in the brain are estimated to have 150,000 or more synapses! Synapses can also occur on the axon and even at the axon terminal. Synapses are classified as electrical or chemical depending on the type of signal that passes between cells.

Electrical Synapses   **Electrical synapses** pass an electrical signal or current directly from the cytoplasm of one cell to another through gap junctions. Information can flow in both directions through gap junctions. Electrical synapses are uncommon and occur mainly in the

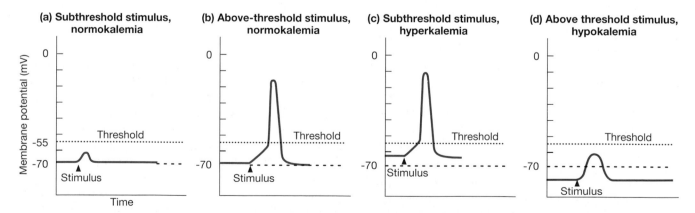

■ **Figure 8-18   Effect of changing extracellular potassium concentration on the excitability of neurons**   If the extracellular concentration of K⁺ changes, the resting membrane potential of cells changes. (a) A subthreshold graded potential does not fire an action potential when blood K⁺ concentration is in the normal range (normokalemia). (b) An above-threshold (suprathreshold) stimulus will fire an action potential when K⁺ concentration is normal. (c) Hyperkalemia, increased blood K⁺ concentration, brings the membrane closer to threshold. Now a subthreshold stimulus can trigger an action potential. (d) Hypokalemia, decreased blood K⁺ concentration, hyperpolarizes the membrane and makes the neuron less likely to fire an action potential in response to a stimulus that would normally be above threshold.

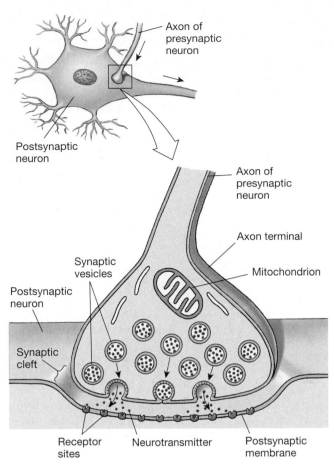

■ **Figure 8-19   A chemical synapse**   The axon terminal contains mitochondria and synaptic vesicles filled with neurotransmitter. The postsynaptic membrane has receptors for neurotransmitter that diffuses across the synaptic cleft.

CNS. Electrical synapses are also found in glial cells, in cardiac and smooth muscle, and in nonexcitable cells that use electrical signals, such as the pancreatic beta cell. The primary advantage of electrical synapses is rapid conduction of signals from cell to cell.

Chemical Synapses   The vast majority of synapses in the nervous system are **chemical synapses** that use neurotransmitters to carry information from one cell to the next. Much of what we know about chemical synapses comes from the study of the *neuromuscular junction,* formed by the synapse of a somatic motor neuron on a skeletal muscle fiber.

Neurotransmitter synthesis can take place either in the nerve cell body or in the axon terminal. However, axon

…continued from page 232

Dr. McKhann decided to perform nerve conduction tests on some of the paralyzed children. He found that, although the rate of conduction along the children's neurons was normal, the strength of the action potentials was greatly diminished.

Question 4:   *Is the paralytic illness that affected the Chinese children a demyelinating condition? Why or why not?*

terminals do not have the organelles needed for protein synthesis. Consequently, polypeptide neurotransmitters and protein enzymes needed for metabolism in the axon terminal must be made in the cell body. The dissolved enzymes are brought to axon terminals by slow axonal transport but the neurotransmitters, which are consumed more rapidly than enzymes, are moved in vesicles by fast axonal transport.

When we examine the axon terminal of a presynaptic cell with an electron microscope, we find many small **synaptic vesicles** and large mitochondria in the cytoplasm (Fig. 8-19 ■). Some vesicles are "docked" at active zones along the membrane closest to the synaptic cleft, waiting for a signal to release their contents. Other vesicles act as a reserve pool, clustering close to the docking sites. Each vesicle contains a certain amount of neurotransmitter that is released on demand.

✔ Which organelles are needed to synthesize proteins and package them into vesicles?

✔ How do mitochondria get to the axon terminals?

Calcium Is the Signal for Neurotransmitter Release at the Synapse

The release of neurotransmitters into the synapse takes place by exocytosis. From what we can tell, the basic process is similar in neurons and in other types of secretory cells, such as the pancreatic beta cell described in Chapter 5 [∞ p. 145]. Neurotoxins that block neurotransmitter release, including tetanus and botulinum toxins, exert their action by inhibiting specific proteins of the cell's exocytosis apparatus.

Figure 8-20 ■ shows how neurotransmitters are released at a synapse. When the depolarization wave of an action potential reaches the axon terminal ①, it sets off a sequence of events. The axon terminal membrane has voltage-gated Ca^{2+} channels that open in response to depolarization ②. Calcium ions are more concentrated outside the cell than in the cytosol, so they move into the cell. The Ca^{2+} ions bind to regulatory proteins and initiate exocytosis. ③ Exocytosis of synaptic vesicle contents releases neurotransmitter into the synaptic cleft ④. The molecules diffuse across the gap to bind with membrane receptors on the postsynaptic cell. When neurotransmitter molecules bind to their receptors, a response is initiated in the postsynaptic cell.

✔ In an experiment on synaptic transmission, a synapse was bathed in a Ca^{2+}-free medium that was otherwise equivalent to extracellular fluid. An action potential was triggered in the presynaptic neuron. Although the action potential reached the axon terminal at the synapse, the usual response of the postsynaptic cell did not occur. What conclusion did the researchers make on the basis of these results?

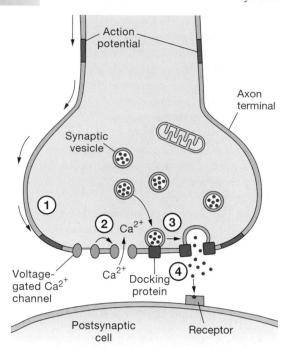

■ **Figure 8-20   Events at the synapse**   ① An action potential coming down the axon depolarizes the axon terminal. ② The depolarization opens voltage-gated Ca^{2+} channels and Ca^{2+} enters the cell. ③ Calcium entry triggers exocytosis of synaptic vesicle contents. ④ Neurotransmitter diffuses across the synaptic cleft and binds with receptors on the postsynaptic cell.

Neurocrines Convey Information from Neurons to Other Cells

What are these molecules that neurons release? Their chemical composition is varied, and they may function as neurotransmitters, neuromodulators, or neurohormones. Neurotransmitters and neuromodulators act as *paracrines*, with target cells located close to the neuron that secretes them. Neurohormones, on the other hand, are secreted into the blood and distributed throughout the body.

Generally, neurotransmitters act at a synapse. Neuromodulators act at nonsynaptic sites. Some neuromodulators and neurotransmitters also act on the cell that secretes them, making them *autocrine* signals as well as paracrines. The number of molecules identified as neurotransmitters and neuromodulators is large and growing daily.

The Nervous System Secretes a Variety of Neurotransmitters

The array of neurotransmitters in the body is truly staggering. They can be informally grouped into seven classes according to their structure: (1) acetylcholine, (2) amino acids, (3) amino acid–derived amines, (4) polypeptides, (5) purines, (6) gases, and (7) lipids. The largest variety of neurotransmitters is found within the CNS (Table 8-3). These neurotransmitters include many

Of Snakes, Snails, Spiders, and Sushi   What do snakes, marine snails, and spiders have to do with neurophysiology? They all provide neuroscientists with compounds for studying synaptic transmission, extracted from the neurotoxic venoms that they use to kill their prey. The Asian snake *Bungarus multicinctus* provides us with α-bungarotoxin, a long-lasting poison that binds tightly to nicotinic acetylcholine receptors. The fish-hunting cone snail, *Conus geographus*, and the funnel web spider, *Agelenopsis aperta*, use toxins that block different types of voltage-gated Ca^{2+} channels. But one of the most potent poisons known comes from the Japanese puffer fish, a highly prized delicacy whose flesh is consumed as *sushi*. The puffer has tetrodotoxin, or TTX, in its gonads. This neurotoxin blocks Na^+ channels on axons and prevents the transmission of action potentials, so ingestion of only a tiny amount can be fatal. The chefs who prepare the puffer fish, known as *fugu*, for consumption are carefully trained to avoid contaminating the fish's flesh as they remove the toxic gonads. But there's always some risk involved in eating *fugu*, one reason that the youngest person at the table is the first to sample the dish.

polypeptides known mostly for their hormonal activity. In contrast, three primary neurotransmitters are used within the peripheral nervous system: acetylcholine, norepinephrine, and epinephrine.

Acetylcholine   Acetylcholine (ACh), in a chemical class by itself, is synthesized from choline and acetyl coenzyme A (acetyl CoA). Choline is a small molecule also found in membrane phospholipids. Acetyl CoA is the metabolic intermediate that links glycolysis to the citric acid cycle [∞ p. 93]. The synthesis of ACh from these two precursors is a simple enzymatic reaction that takes place in the axon terminal (Fig. 8-21 ■). Neurons that secrete ACh and receptors that bind ACh are described as **cholinergic.**

Amines   The amine neurotransmitters, like the amine hormones [∞ p. 193], are derived from single amino acids. The amino acid tyrosine is converted to **dopamine, norepinephrine,** and **epinephrine.** All three of these neurocrines also function as neurohormones when secreted by the adrenal medulla [∞ p. 188].

Neurons that secrete norepinephrine are called **adrenergic neurons,** or, more properly, noradrenergic neurons. The adjective *adrenergic* does not have the same obvious link to its neurotransmitter as *cholinergic* does to *acetylcholine*. Instead, the adjective derives from the British name for epinephrine, *adrenaline*. In the early part of the twentieth century, British researchers thought that sympathetic neurons secreted adrenaline (epinephrine), hence the modifier *adrenergic*. Although our understanding changed, the name persists. Whenever you see reference to "adrenergic control" of a function, you must make the connection to a neuron secreting norepinephrine.

TABLE 8-3   Major Neurocrines

Chemical	Receptor	Type*	Receptor Location	Key Receptor Agonists/Antagonists
Acetylcholine (ACh)	Cholinergic			
	Nicotinic	ICR (Na$^+$, K$^+$)	Skeletal muscles, autonomic neurons, CNS	Nicotine—agonist Curare—antagonist
	Muscarinic	GPCR	Smooth and cardiac muscle, endocrine and exocrine glands, CNS	Muscarine—agonist Atropine—antagonist
Amines				
Norepinephrine (NE)	Adrenergic (α, β)	GPCR	Smooth and cardiac muscle, endocrine and exocrine glands, CNS	α—Prazosin (Minipress®) β—propranolol
Dopamine (DA)	Dopamine (D)	GPCR	CNS	Antipsychotic drugs—antagonists Bromocriptine—agonist
Serotonin (5-hydroxytryptamine, 5-HT)	Serotonergic (5-HT)	ICR (Na$^+$, K$^+$) GPCR	CNS	Sumatriptan—agonist LSD—antagonist
Histamine	Histamine (H)	GPCR	CNS	Ranitidine (Zantac®), cimetidine (Tagamet®)—antagonists
Amino acids				
Glutamate	Glutaminergic			
	AMPA	ICR (Na$^+$, K$^+$)	CNS	
	NMDA	ICR (Na$^+$, K$^+$, Ca^{2+})	CNS	
GABA (γ-aminobutyric acid)	GABA	ICR (Cl$^-$) GPCR	CNS	
Glycine	Glycine	ICR (Cl$^-$)	CNS	
Purines				
Adenosine	Purine (P)	GPCR	CNS	
Gases				
Nitric oxide (NO)	None	N/A	N/A	

*ICR = ion channel-receptor, GPCR = G protein–coupled receptor. AMPA = α-amino-3-hydroxy-5-methyl-4 isoxazole proprionic acid. NMDA = N-methyl-D-aspartate. LSD = lysergic acid diethylamine. N/A = Not applicable.

Other amine neurotransmitters include **serotonin** (also called *5-hydroxytryptamine* or **5-HT**), made from the amino acid tryptophan, and **histamine,** made from histadine. The amine neurotransmitters are all active in the CNS. In addition, norepinephrine is the major neurotransmitter of the peripheral autonomic sympathetic division.

Amino Acids   At least four amino acids function as neurotransmitters in the CNS. **Glutamate** is the primary excitatory neurotransmitter of the CNS, and **aspartate** serves the same function in selected regions of the brain. The main inhibitory neurotransmitter is **gammaaminobutyric acid (GABA)**. The amino acid **glycine** is also considered inhibitory, although it can bind to one type of glutamate receptor and be excitatory.

Polypeptides   The nervous system secretes a variety of peptides that act as neurotransmitters and neuromodulators in addition to functioning as neurohormones. These peptides include **substance P,** involved in some pain pathways, and the *opioid peptides* (**enkephalins** and **endorphins**) that mediate *analgesia* [*an-*, without + *algos*, pain]. Peptides that function as both hormones and neurotransmitters include *cholecystokinin (CCK), vasopressin,* and *atrial natriuretic peptide.* Many peptide neurotransmitters are cosecreted with other neurotransmitters.

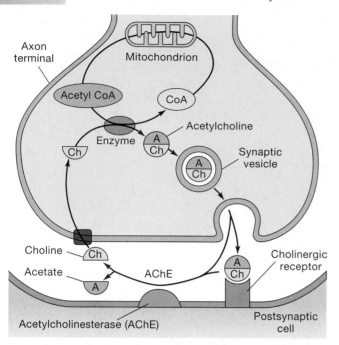

■ Figure 8-21   Synthesis and recycling of acetylcholine at the synapse   Acetylcholine (ACh) is made from choline and acetyl CoA in the axon terminal. In the synaptic cleft ACh is rapidly broken down by the enzyme acetylcholinesterase. Choline is transported back into the axon terminal and is used to make more ACh.

Purines   *Adenosine, adenosine monophosphate* (AMP), and *adenosine triphosphate* (ATP) can all act as neurotransmitters. These molecules, known collectively as *purines* [∞ p. 35], bind to *purinergic* receptors in the CNS and on other excitable tissues such as the heart.

Gases   One of the most interesting neurotransmitters is *nitric oxide* (NO), an unstable gas synthesized from oxygen and the amino acid arginine. Nitric oxide is an unusual neurotransmitter that diffuses freely into the target cell rather than binding to a membrane receptor [∞ p. 163]. Once inside the target cell, nitric oxide binds to proteins. With a half-life of only 2–30 seconds, nitric oxide is elusive and difficult to study. It is also released from cells other than neurons and often acts as a paracrine.

Recent work suggests that *carbon monoxide* (CO), best known as a toxic gas, may also be produced by the body in tiny amounts to serve as a neurotransmitter.

…continued from page 235

Dr. McKhann then asked to see autopsy reports on some of the children who had died of their paralysis at Beijing Hospital. In the reports, pathologists noted that the patients had normal myelin but damaged axons. In some cases, the axon had been completely destroyed, leaving only a hollow shell of myelin.

Question 5:   *Do the results of Dr. McKhann's investigation suggest that the Chinese children had Guillain-Barré syndrome? Why or why not?*

Multiple Receptor Types Amplify the Effects of Neurotransmitters

All neurotransmitters except nitric oxide have one or more receptor types with which they bind. Each receptor type may have multiple subtypes, a property that allows one neurotransmitter to have different effects in different tissues. Receptor subtypes are distinguished by combinations of letter and number subscripts. For example, serotonin (5-HT) has at least 14 different receptor subtypes that have been identified, such as 5-HT_{1A} or 5-HT_4.

Neurotransmitter receptors fall into two of the membrane receptor categories that we discussed in Chapter 6 [∞ p. 157]: ligand-gated ion channels and G protein–linked receptors. Receptors that alter ion channel function are called *inotropic* receptors, while those that exert their actions through second messenger systems are called *metabotropic* receptors. Some metabotropic G protein-linked receptors regulate the opening or closing of ion channels.

The study of neurotransmitters and their receptors has been greatly simplified by two advances in molecular biology. The genes for many receptor subtypes have been cloned, allowing researchers to create mutant receptors and study their properties. In addition, researchers have discovered or synthesized a variety of agonists and antagonist molecules that mimic or inhibit neurotransmitter activity by binding to the receptors [∞ p. 164]. Table 8-3 includes descriptions of receptor types and some of their agonists or antagonists.

Cholinergic Receptors   **Cholinergic receptors** come in two main subtypes: **nicotinic,** named because *nicotine* is an agonist, and **muscarinic,** for which *muscarine,* a compound found in some fungi, is an agonist. **Cholinergic nicotinic receptors** are found on skeletal muscle, in the autonomic division, and in the CNS. Nicotinic receptors are monovalent cation channels through which both Na^+ and K^+ can pass. Sodium entry into cells exceeds K^+ exit because the electrochemical gradient for Na^+ is stronger. As a result, net Na^+ entry depolarizes the postsynaptic cell and makes it more likely to fire an action potential.

Cholinergic muscarinic receptors come in five related subtypes. They are all coupled to G proteins and are linked to second messenger systems. The tissue response to activation of a muscarinic receptor varies with the receptor subtype. These receptors occur in the CNS and in the autonomic parasympathetic division.

Adrenergic Receptors   **Adrenergic receptors** come in two varieties: α (alpha) and β (beta). Like cholinergic muscarinic receptors, adrenergic receptors are linked to G proteins and initiate second messenger cascades. Indeed, it was the action of epinephrine on β receptors in dog liver that led E. W. Sutherland to the discovery of cyclic AMP and the concept of second messenger systems

§ **Myasthenia Gravis**   Myasthenia gravis is an autoimmune disease in which the body fails to recognize the acetylcholine (ACh) receptors on skeletal muscle as part of "self" and therefore produces molecules to attack the receptors. The molecules bind to the receptors at a site that is distinct from the ACh binding site. In doing so, they change the receptor protein in some way that causes the muscle cell to speed up turnover of the receptor. The bound receptors are withdrawn from the membrane by endocytosis and destroyed inside lysosomes. Their destruction leaves the muscle with fewer ACh receptors in the membrane and therefore a diminished response to ACh from the controlling somatic motor neuron. Assume for the moment that you are a researcher trying to help people who suffer from myasthenia gravis. On the basis of what you know about membrane receptors and synapses, think of possible methods for treating this disease.

as transducers of extracellular messengers [∞ p. 158]. The two types of adrenergic receptors work through different second messenger pathways.

Glutaminergic Receptors   Glutamate is the main excitatory neurotransmitter in the CNS and also acts as a neuromodulator. The action of glutamate at a particular synapse depends on which of the two receptor types occur on the cell. The NMDA receptors are named for the glutamate agonist *N-methyl-D-aspartate*, and the AMPA receptors are named for their agonist α-*amino-3-hydroxy-5-methyl-4 isoxazole proprionic acid.*

AMPA receptors are ligand-gated monovalent cation channels, similar to nicotinic acetylcholine channels. Glutamate binding opens the channel and the cell depolarizes due to net Na^+ influx.

NMDA receptors are unusual for several reasons. First, they are cation channels that allow Na^+, K^+, and Ca^{2+} to pass through the channel. Second, channel opening requires both glutamate binding and a change in membrane potential. The NMDA receptor channel is blocked by a magnesium ion (Mg^{2+}) at resting membrane potentials. Glutamate binding opens the ligand-activated gate but ions cannot flow past the Mg^{2+} ion. However, if the cell depolarizes, the Mg^{2+} blocking the channel is expelled and ions flow through the pore (see Fig. 8-29 ■).

Not All Postsynaptic Responses Are Rapid and of Short Duration

The combination of a neurotransmitter with its receptor sets in motion a series of responses in the postsynaptic cell (Fig. 8-22 ■). In the simplest response, the neurotransmitter opens a chemically gated ion channel, leading to ion movement between the cell and the extracellular fluid. The resulting change in membrane potential is called a **fast synaptic potential** because it begins quickly and lasts only a few milliseconds. These synaptic potentials are the excitatory (depolarizing) and inhibitory (hyperpolarizing) postsynaptic potentials (EPSPs and IPSPs) previously described.

In slow responses, neurotransmitters bind to receptors linked to G proteins and activate second messenger systems. The second messengers may act from the cytoplasmic side of the cell membrane to open or close ion channels. (Fast synaptic potentials always open ion channels.) Membrane potentials resulting from this process are called **slow synaptic potentials** because the second messenger method takes longer to create a response. In addition, the response itself lasts longer, usually seconds to minutes.

Slow responses are not limited to altering the open state of ion channels. Neurotransmitter activation of second messenger systems may also modify existing cell proteins or regulate the production of new cell proteins. This type of slow response has been linked to the growth and development of neurons and to the mechanisms underlying long-term memory.

Neurotransmitter Activity Is Rapidly Terminated

A key feature of nervous signaling is its short duration, which is achieved by the rapid removal or inactivation of neurotransmitter in the synaptic cleft. This can be accomplished various ways (Fig. 8-23 ■). Some neurotransmitter molecules simply diffuse away from the synapse, separating them from their receptors. Other neurotransmitters are inactivated by enzymes in the synaptic cleft. Many neurotransmitters are removed from the extracellular fluid by transport back into the presynaptic cell or into adjacent neurons or glial cells.

Acetylcholine   ACh in the extracellular fluid is rapidly broken down by the enzyme **acetylcholinesterase** (AChE) in the extracellular matrix and in the membrane of the postsynaptic cell (see Fig. 8-21 ■). Choline from degraded ACh is actively transported back into the presynaptic axon terminal and used to make new acetylcholine to refill recycled synaptic vesicles.

Norepinephrine   Norepinephrine action at the target tissue is terminated when the neurotransmitter is actively transported back into the presynaptic axon terminal. Once back in the axon terminal, norepinephrine is either repackaged into vesicles or broken down by intracellular enzymes such as *monoamine oxidase (MAO)*, found in mitochondria.

CNS Neurotransmitters   Once released into the synaptic cleft, these amine, polypeptide, amino acid neurotransmitters move into the circulation to be carried away or are actively transported back into the axon terminal to be metabolized or used again.

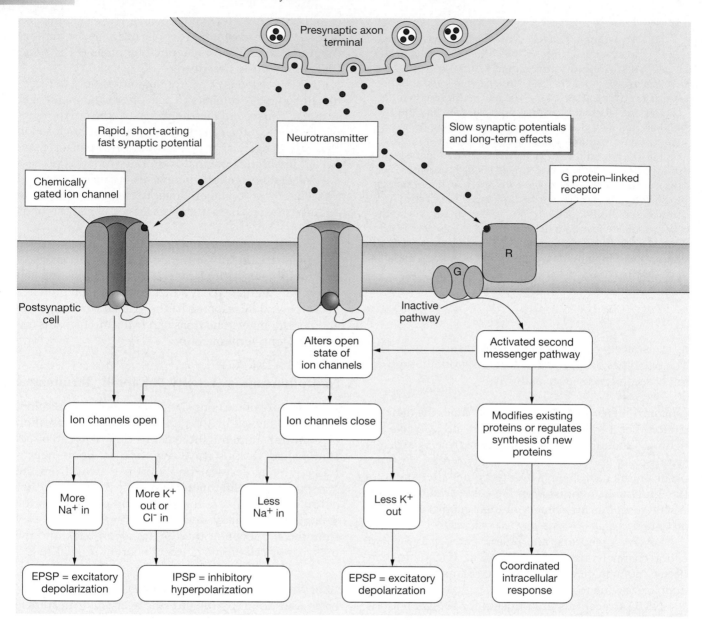

■ **Figure 8-22** **Fast and slow responses in postsynaptic cells** Many neurotransmitters create rapid, short-acting responses by opening ion channels. Some neurotransmitters create slower, longer-lasting responses by activating second messenger systems.

❱ INTEGRATION OF NEURAL INFORMATION TRANSFER

Traditionally, scientists viewed chemical synapses as sites of one-way communication, with all messages moving from a presynaptic cell to a postsynaptic cell. We now know that this is not always the case. In the brain, there are some synapses where cells on both sides of the synaptic cleft release neurotransmitter that acts on the opposite cell. Perhaps more important, we have learned that many postsynaptic cells "talk back" to their presynaptic neurons by sending neuromodulators that bind to presynaptic receptors. Let's examine some of the ways that complex neural pathways communicate.

Neural Pathways May Involve Many Neurons Simultaneously

Communication between neurons is not always a one-to-one event. Sometimes a single neuron branches, and its collaterals synapse on multiple target neurons. This pattern is known as **divergence** (Fig. 8-24a ■). On the other hand, a single postsynaptic neuron may have synapses with as many as 10,000 presynaptic neurons (Fig. 8-25 ■). If a larger number of presynaptic neurons provides input to a smaller number of postsynaptic neurons, the pattern is known as **convergence** (Fig. 8-24b ■).

One advantage of convergence in the nervous system is that input from multiple sources can influence the

(a) Divergence

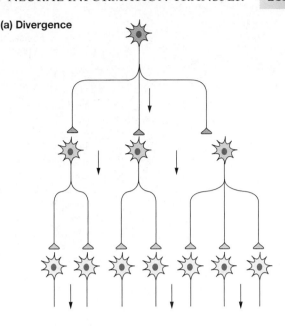

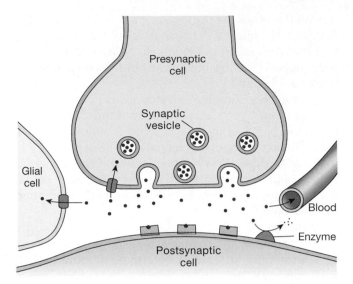

■ **Figure 8-23   Inactivation of neurotransmitters**   Neurotransmitter activity is stopped by enzymes that break down the molecule, transport of neurotransmitter into cells, or diffusion away from the synapse.

(b) Convergence

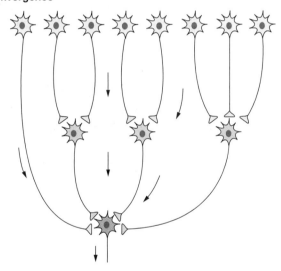

■ **Figure 8-24   Convergence and divergence**   (a) In a divergent pathway, a presynaptic neuron branches to affect a larger number of postsynaptic neurons. (b) In convergent pathways, a large number of presynaptic neurons converge to influence a smaller number of postsynaptic neurons.

output of a single postsynaptic cell. When two or more presynaptic neurons converge on the dendrites or cell body of a single postsynaptic cell, the response of the postsynaptic cell will be determined by the summed input from the presynaptic neurons.

For example, what happens if several stimuli arrive simultaneously at the dendrites or cell body of a postsynaptic neuron? If the stimuli are all subthreshold excitatory graded potentials (EPSPs), they can add together to create a *suprathreshold* potential [*supra-*, above] at the trigger zone. The initiation of an action potential from several simultaneous graded potentials is an example of **spatial summation.** The word *spatial* refers to the fact that the graded potentials originate simultaneously at different locations on the neuron.

Figure 8-26a ■ illustrates three presynaptic neurons converging on one postsynaptic neuron. The graded potentials from the presynaptic neurons are too weak to trigger an action potential by themselves, but if the neurons fire together, the sum of the three stimuli is above threshold and creates an action potential.

Postsynaptic inhibition occurs when a neuron releases an inhibitory neurotransmitter onto a postsynaptic cell and alters its response. Figure 8-26b ■ shows three neurons converging on a postsynaptic cell. The neurons fire, creating one inhibitory graded potential (IPSP) and two excitatory graded potentials (EPSPs) that sum as they reach the trigger zone. The IPSP counteracts the EPSPs, creating an integrated signal that is below threshold. As a result, no action potential leaves the trigger zone.

Summation of graded potentials does not always require input from more than one presynaptic neuron. Two

subthreshold graded potentials from the same presynaptic neuron can be summed if they arrive at the trigger zone close enough together in time. Let's see how this can happen.

Figure 8-27a ■ shows recordings from an electrode placed in the trigger zone of a neuron. A stimulus (X) starts a subthreshold graded potential on the cell body at time X_1. The graded potential reaches the trigger zone and depolarizes it as shown on the graph (A_1). A second stimulus occurs at time X_2, and its subthreshold graded

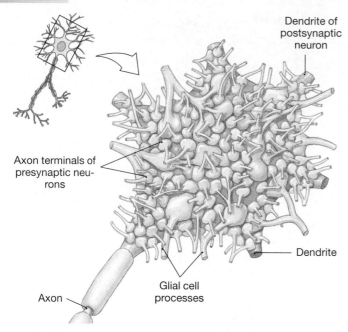

Dendrite of postsynaptic neuron

Axon terminals of presynaptic neurons

Dendrite

Glial cell processes

Axon

■ **Figure 8-25** **Locations of synapses on a postsynaptic neuron** The cell body and dendrites of a somatic motor neuron are nearly covered in hundreds of axon terminals from other neurons.

potential reaches the trigger zone a short time after the first (A$_2$). The interval between the two stimuli is so long that the two graded potentials do not overlap. Neither potential by itself is above threshold, so no action potential is triggered.

In Figure 8-27b ■, the two stimuli are given closer together in time. As a result, the two subthreshold graded potentials arrive at the trigger zone at almost the same time. The second graded potential adds its depolarization to that of the first, allowing the trigger zone to depolarize to threshold. Summation that occurs from graded potentials overlapping in time is called **temporal summation** [*tempus,* time].

In many situations, graded potentials in a neuron will incorporate both temporal and spatial summation. The summation of graded potentials demonstrates a key property of neurons: *postsynaptic integration.* When multiple signals reach a neuron, postsynaptic integration allows the neuron to evaluate the strength and duration of the signals. If the resultant integrated signal is above threshold, the neuron fires an action potential.

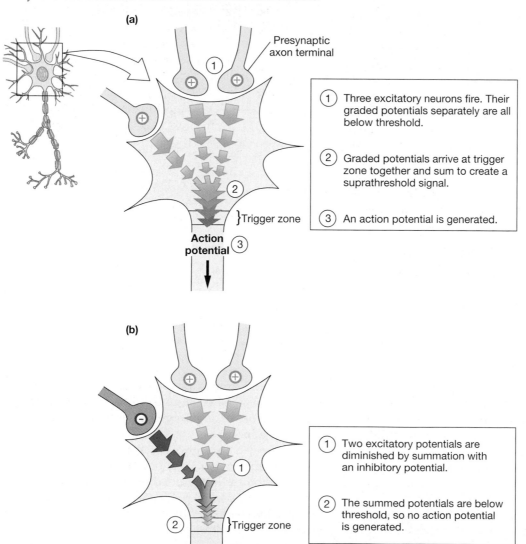

(a)

Presynaptic axon terminal

1. Three excitatory neurons fire. Their graded potentials separately are all below threshold.

2. Graded potentials arrive at trigger zone together and sum to create a suprathreshold signal.

} Trigger zone

3. An action potential is generated.

Action potential 3

(b)

1. Two excitatory potentials are diminished by summation with an inhibitory potential.

2. The summed potentials are below threshold, so no action potential is generated.

} Trigger zone

No action potential

■ **Figure 8-26** **Spatial summation**

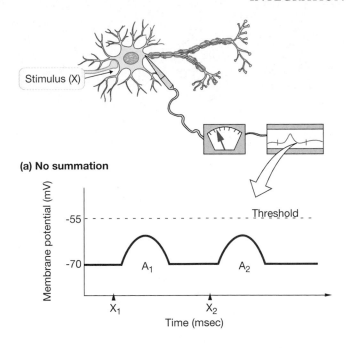

(a) No summation

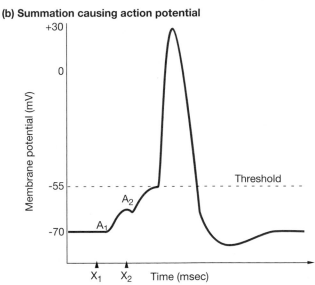

(b) Summation causing action potential

■ **Figure 8-27   Temporal summation**   (a) Two graded potentials will not cause an action potential if they are far apart in time. (b) If the same two subthreshold potentials arrive at the trigger zone within a short period of time, they may sum and create an action potential.

✔ In Figure 8-26b ■, assume the postsynaptic neuron has a resting membrane potential of −70 mV and threshold of −55 mV. If the inhibitory presynaptic neuron creates an IPSP of 5 mV and the two excitatory presynaptic neurons have EPSPs of 10 and 12 mV respectively, will the postsynaptic neuron fire an action potential?

✔ In the graphs of Figure 8-27 ■, why doesn't the membrane potential (A_1, A_2) change at the same time as the stimulus (X_1, X_2)?

Synaptic Activity Can Be Modulated at the Axon Terminal

Modulation of neuronal pathways can also occur at the axon terminal. Modulation at axon terminals, called **presynaptic modulation,** occurs when a modulatory neuron (inhibitory or excitatory) terminates on or close to an axon terminal of a presynaptic cell (Fig. 8-28a ■). The modulatory neuron influences the amount of neurotransmitter released by the presynaptic cell. If activity in the modulatory neuron decreases neurotransmitter release, the modulation is called *presynaptic inhibition*. In *presynaptic facilitation,* modulatory input increases neurotransmitter release by the presynaptic cell.

Presynaptic modulation provides a more precise means of control than postsynaptic modulation. If the responsiveness of a neuron is changed at the dendrites and cell body (postsynaptic modulation), all target cells of the neuron will be affected equally (Fig. 8-28b ■). In contrast, presynaptic inhibition on a divergent neuron allows selective modulation of collaterals and their targets (Fig. 8-28a ■). One collateral can be inhibited while others remain unaffected.

Synaptic activity can also be altered by changing the responsiveness of the target cell to neurotransmitter. This may be accomplished by changing the identity, affinity, and number of neurotransmitter receptors. Modulators can alter all of these parameters by influencing the synthesis of enzymes, membrane transporters, and receptors. Most neuromodulators act through second messenger systems that alter existing proteins, and their effects last much longer than do those of neurotransmitters.

✔ Why are axon terminals sometimes called "biological transducers"?

Long-Term Potentiation Alters Synaptic Communication

One of the "hot topics" in neurobiology today is **long-term potentiation** (LTP) (*potentia,* power), a process in which activity at a synapse induces sustained changes in the quality or quantity of synaptic connections. Long-term potentiation is believed to be related to the neural processes for learning and memory. Many scientists are working to discover the molecular details of the mechanism.

A key element in long-term potentiation is the amino acid glutamate, the main excitatory neurotransmitter in the CNS. As you learned above, glutamate has two types of receptors: NMDA receptors and AMPA receptors. When presynaptic neurons release glutamate, the chemical binds to both types of receptors on the postsynaptic cell (Fig. 8-29 ■). Activation of the AMPA receptor opens a Na^+ channel and depolarizes the cell. The gate of the NMDA receptor opens with glutamate binding but the channel is blocked by a Mg^{2+} ion. When the cell depolarizes, the

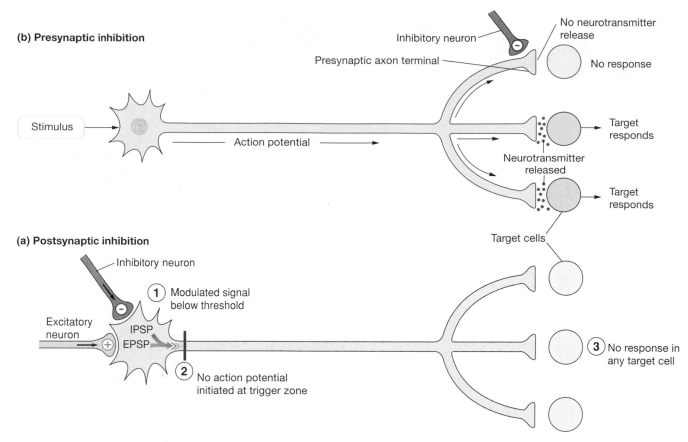

(b) Presynaptic inhibition

Inhibitory neuron

No neurotransmitter release

Presynaptic axon terminal

No response

Stimulus

Action potential

Neurotransmitter released

Target responds

Target responds

Target cells

(a) Postsynaptic inhibition

Inhibitory neuron

① Modulated signal below threshold

Excitatory neuron

IPSP

EPSP

② No action potential initiated at trigger zone

③ No response in any target cell

■ **Figure 8-28   Presynaptic and postsynaptic inhibition**   (a) In presynaptic inhibition, a modulatory neuron synapses on one collateral of the presynaptic neuron. In this way, one target of the neuron can be selectively inhibited. (b) In postsynaptic inhibition, all targets of the postsynaptic cell will be inhibited equally.

Mg^{2+} is knocked out of the channel, allowing Ca^{2+} to enter the cytosol.

The Ca^{2+} signal initiates second messenger pathways. As a result of these pathways, the postsynaptic cell releases a paracrine that acts on the presynaptic cell to enhance neurotransmitter release. The postsynaptic cell also becomes more sensitive to glutamate, possibly by inserting more glutamate receptors in the postsynaptic membrane.

Disorders of Synaptic Transmission Are Responsible for Many Diseases

Synaptic transmission is the most vulnerable step in the process of signaling through the nervous system. It is the point at which many things go wrong, leading to disruption of normal function. Yet, at the same time, the receptors at synapses are exposed to the extracellular fluid, making them more accessible to drugs than intracellular receptors. In recent years, we have discovered that a variety of nervous system disorders are related to problems with synaptic transmission. These include Parkinson's disease, schizophrenia, and depression. The best understood diseases of the synapse are those that involve the neuromuscular junction. Diseases resulting from synap-

tic transmission problems within the CNS have proved more difficult to study because, anatomically, they are more difficult to isolate.

Drugs that act on synaptic activity, particularly synapses in the CNS, are the oldest known and most widely used of all pharmacological agents. Caffeine, nicotine, and alcohol are common drugs in many cultures. Medically, some drugs that we use to treat conditions such as schizophrenia, anxiety, and epilepsy act by influencing events at the synapse. In many disorders arising in the CNS, we do not yet fully understand either the cause of the disorder or the mechanism of action of the drug. This subject is one major area of pharmaco-

…continued from page 238

Currently, Dr. McKhann believes that the disease afflicting the Chinese children—which he has named acute motor axonal polyneuropathy (AMAN)—may be caused by a bacterial infection. He also believes that the disease initiates its damage of axons at the neuromuscular junctions.

Question 6:   *Based on information provided in this chapter, can you name another disease that alters synaptic transmission?*

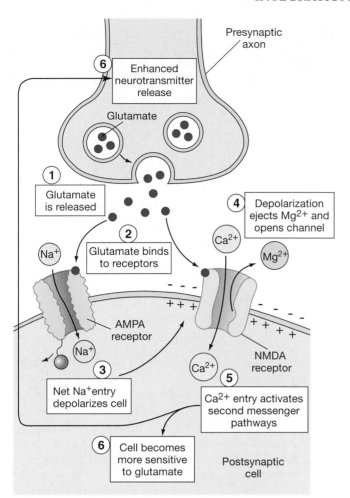

Presynaptic axon

6 Enhanced neurotransmitter release

Glutamate

1 Glutamate is released

2 Glutamate binds to receptors

Na+

AMPA receptor

Na+

3 Net Na+ entry depolarizes cell

6 Cell becomes more sensitive to glutamate

4 Depolarization ejects Mg2+ and opens channel

Ca2+

Mg2+

+ + +

- - - -
+ + + +

NMDA receptor

Ca2+

5 Ca2+ entry activates second messenger pathways

Postsynaptic cell

■ **Figure 8-29 Long-term potentiation** The NMDA receptors for glutamate are linked to long-term potentiation. As a result of this process, the presynaptic cell enhances glutamate release and the postsynaptic cell becomes more sensitive to glutamate.

logical research, and new classes of drugs are being formulated and approved every year.

Development of the Nervous System Depends on Chemical Signals

In a system as complex as the nervous system, how can more than 100 billion neurons in the brain find their correct targets and make synapses among more than 10 times that many glial cells? How can a somatic motor neuron in the spinal cord find the correct pathway to form a synapse with its target muscle in the big toe? The answer is found in the chemical signals used by the developing embryo.

Embryonic nerve cells send out axons that can grow until they find their target cell. In experiments in which the target cells are moved to an unusual location in the embryo, the axons in many instances are still able to find their targets by "sniffing out" their chemical scent. Once the axon reaches the target cell, a

synapse forms. But synapse formation must be followed by electrical and chemical activity, or the synapse will disappear.

This "use it or lose it" scenario is most dramatically reflected by the fact that the infant brain is only about one-fourth the size of the adult brain. Further brain growth is due not to an increase in cell number but to an increase in size and number of axons, dendrites, and synapses. Development is dependent on electrical activity in the brain, on action potentials moving through sensory pathways and interneurons.

Babies who are neglected or deprived of sensory input may have delayed development because of the lack of stimulation of the nervous system. On the other hand, there is no evidence that extra stimulation in infancy will enhance intellectual development, despite a popular movement to expose babies to art, music, and foreign languages before they can even walk. Once synapses form, they are not fixed for life. Variations in electrical activity can cause rearrangement of the connections. This process, known as the **plasticity** (changeability) of synapses, continues throughout life. It is one reason that older adults are urged to keep learning new skills and information.

When Neurons Are Injured, Segments Separated from the Cell Body Die

We can grow neurons when young, but what happens when neurons are injured? The responses of mature neurons to injury are similar in many ways to the growth of neurons during development, relying on a combination of chemical and electrical signals. This is another area of active research because spinal cord injuries and brain damage from illnesses and accidents disable many people each year. The loss of one neuron from a reflex can have drastic consequences on an entire reflex pathway.

When a neuron is injured, if the cell body dies, the entire neuron dies. If the cell body is intact and only the axon is severed, most of the neuron will survive. At the site of injury, the cytoplasm leaks out until membrane is recruited to seal the opening. The proximal segment (closest to the cell body) swells as organelles and filaments carried into the axon by axonal transport accumulate. Chemical factors produced by Schwann cells near the injury site move by retrograde transport to the cell body, telling it that an injury has occurred. The distal segment of axon, deprived of its source of protein, begins to degenerate slowly (Fig. 8-30 ■). The death of this part of the neuron may take a month or longer, although synaptic transmission ceases almost immediately. The myelin sheath around the distal axon begins to unravel and the axon itself collapses. The fragments are cleared away by scavenger microglia or phagocytes that ingest and digest the debris.

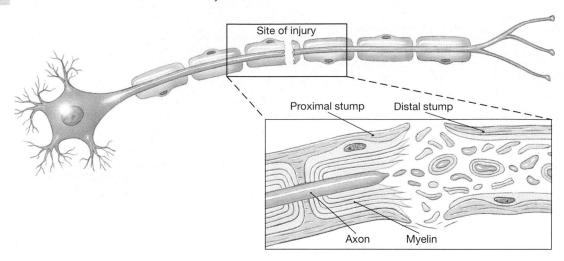

■ **Figure 8-30** **Injury to neurons** When an axon is cut, the section attached to the cell body continues to live, but the section of the axon distal to the cut will begin to disintegrate and die. Under some circumstances, the stump of the axon may regrow through the existing sheath of Schwann cells and reform a synapse with the proper target.

Under some conditions, axons in the peripheral nervous system can regenerate and reestablish their synaptic connections. The Schwann cells of the damaged neuron secrete certain neurotrophic factors that keep the cell body alive and that stimulate regrowth of the axon. The growing tip of a regenerating axon behaves much like that of a developing axon, following chemical signals in the extracellular matrix along its former path until the axon rejoins its target cell. But sometimes the loss of the distal axon is permanent, and the pathway is destroyed. Many scientists are studying these mechanisms of axon growth and inhibition in the hopes of finding treatments that will restore function to victims of spinal cord injury and degenerative neurological disorders.

PROBLEM CONCLUSION

In this running problem, you learned about a baffling paralytic illness that appears to be a new disease. Although it bears some resemblance to Guillain-Barré syndrome, upon closer investigation, it is actually quite different.

Further check your understanding of this running problem by checking your answers against those in the summary table.

	Question	Facts	Integration and Analysis
1	Which division(s) of the nervous system may be involved in Guillain-Barré syndrome (GBS)?	The nervous system is divided into peripheral divisions (sensory and efferent), and the central nervous system. The efferent division is composed of somatic motor neurons that control skeletal muscles, and autonomic neurons that control smooth and cardiac muscles, as well as glands.	Patients with GBS can neither feel sensations nor move their muscles. This suggests a problem with the sensory and somatic motor divisions. However, there could also be a problem within the central nervous system, which integrates information flow between the sensory and efferent divisions. You do not have enough information to determine which division is affected.
2	Do you think that the paralysis found in the Chinese children affected both sensory and efferent neurons? Why or why not?	The Chinese children can feel a pin prick but cannot move their muscles.	Because the children can feel the pin prick, their sensory neurons are normal. Inability to move their muscles indicates a problem with the motor neurons controlling the muscles, with the portions of the brain controlling movement, or with the muscles themselves.

	Question	Facts	Integration and Analysis
3	In Guillain-Barré syndrome, what would you expect the results of a nerve conduction test to be?	Nerve conduction tests measure the strength and conduction speed of action potentials. GBS is a condition in which myelin around the axons is destroyed.	Without myelin, electrical current flowing down the axon leaks out through the axon membrane, and saltatory conduction cannot occur. Thus, demyelinated axons conduct action potentials with reduced strength and slower speed.
4	Is the paralytic illness that affected the Chinese children a demyelinating condition? Why or why not?	Nerve conduction tests showed a normal speed of action potential conduction but a decrease in the strength of the action potentials.	Loss of myelin should decrease both strength and conduction speed of action potentials. Because the conduction speed was normal in these patients, their disease is probably not a demyelinating disease.
5	Do the results of Dr. McKhann's investigation suggest that the Chinese children had Guillain-Barré syndrome? Why or why not?	In addition to the nerve conduction tests, autopsy reports on children who died from the disease showed that they had damaged axons but undamaged myelin.	GBS is a demyelinating disease that affects both sensory and motor neurons. The Chinese children had normal sensory neurons, and nerve conduction tests and histological studies indicated normal myelin. Therefore, it is reasonable to conclude that the Chinese children did not have GBS.
6	Based on information provided in this chapter, can you name another disease that alters synaptic transmission?	Synaptic transmission can be altered by blocking neurotransmitter release, binding at the target cell, and removal from the synapse.	Parkinson's disease, depression, schizophrenia, and myasthenia gravis are related to problems with synaptic transmission (p. 235). Some neurotoxins also block neurotransmitter release (p. 236).

CHAPTER REVIEW

SUMMARY

1. The **nervous system** is a complex network of neurons that form the rapid control system of the body. (p. 215)

Organization of the Nervous System

2. The nervous system is divided into the **central nervous system (CNS),** composed of the **brain** and **spinal cord,** and the **peripheral nervous system.** (p. 215)

3. The peripheral nervous system has **afferent (sensory) neurons** that bring information into the CNS and **efferent neurons** that carry information away from the CNS back to various parts of the body. (p. 215)

4. The efferent neurons include **somatic motor neurons** that control skeletal muscles and **autonomic neurons** that control smooth and cardiac muscles, glands, and some adipose tissue. (p. 215)

5. Autonomic neurons are subdivided into **sympathetic** and **parasympathetic** neurons. (p. 215)

Cells of the Nervous System

6. Neurons have a **cell body** with a nucleus and organelles to direct cellular activity, **dendrites** to receive incoming signals, and an **axon** to transmit electrical signals from the cell body to the **axon terminals.** (p. 216)

7. **Interneurons** are neurons that lie entirely within the CNS. (p. 217)

8. The region where an axon terminal meets its target cell is called a **synapse.** The target cell is called the **postsynaptic cell,** and the neuron that releases the chemical signal is known as the **presynaptic cell.** (p. 218)

9. Material is transported between the cell body and axon terminal by **axonal transport.** (p. 218)

10. **Glial cells** provide physical support and direct the growth of neurons during repair and development. **Schwann cells** and **satellite cells** are associated with the peripheral nervous system. **Oligodendrocytes, astrocytes,** and **ependymal cells** are found in the CNS. **Microglia** are modified immune cells that act as scavengers. (p. 219)

11. Schwann cells and oligodendrocytes form insulating **myelin sheaths** around neurons. The **nodes of Ranvier** are sections of uninsulated membrane between Schwann cells. (p. 220)

Electrical Signals in Neurons

12. Membrane potential is influenced by the concentration gradients of ions across the membrane and by the permeability of the membrane to those ions. (p. 220)

13. The permeability of a cell to ions changes when ion channels in the membrane open and close. Movement of only a few ions will significantly change the membrane potential. (p. 220)

14. Gated ion channels in neurons open in response to chemical or mechanical signals or in response to depolarization of the cell membrane. (p. 222)

15. **Graded potentials** are depolarizations or hyperpolarizations whose strength is directly proportional to the strength of the triggering event. Graded potentials lose strength as they move through the cell. (p. 222)

16. The wave of depolarization that moves through the cell with a graded potential is known as **local current flow.** (p. 222)

17. **Action potentials** are rapid electrical signals that travel undiminished in amplitude from the cell body to the axon terminals. (p. 223)

18. Action potentials begin in the **trigger zone** if a single graded potential or the sum of multiple graded potentials exceeds a minimum depolarization known as the **threshold.** (p. 224)

19. **Excitatory postsynaptic potentials (EPSPs)** are depolarizing graded potentials that make a neuron more likely to fire an action potential. **Inhibitory postsynaptic potentials (IPSPs)** are hyperpolarizing graded potentials. (p. 224)

20. Action potentials are uniform, **all-or-none** depolarizations. (p. 225)

21. The rising phase of the action potential is due to increased Na^+ permeability. The falling phase of the action potential is due to increased K^+ permeability. (p. 225)

22. The voltage-gated Na^+ channels of the axon have a fast **activation gate** and a slower **inactivation gate.** (p. 225)

23. Once an action potential has begun, there is a brief period of time known as the **absolute refractory period** during which a second action potential cannot be triggered, no matter how large the stimulus. Because of this, action potentials cannot be summed. (p. 227)

24. During the **relative refractory period,** a higher-than-normal graded potential is required to trigger an action potential. (p. 227)

25. Information about the strength and duration of a stimulus is conveyed by the frequency of action potential propagation. (p. 227)

26. The Na^+-K^+-ATPase plays no direct role in the action potential because very few ions cross the membrane during an action potential. (p. 228)

27. The movement of an action potential through the axon at high speed is called **conduction.** (p. 229)

28. The diameter of the neuron and the resistance of the neuron membrane influence the speed of action potential conduction. The **myelin sheath** speeds up conduction by preventing current leakage. (p. 231)

29. The apparent jumping of action potentials from node to node is called **saltatory conduction.** (p. 232)

30. Changes in blood K^+ concentration affect resting membrane potential and the conduction of action potentials. (p. 234)

Cell-to-Cell Communication in the Nervous System

31. In **electrical synapses,** an electrical signal passes directly from the cytoplasm of one cell to another through gap junctions. **Chemical synapses** use neurotransmitters to carry information from one cell to the next. (p. 234)

32. **Neurotransmitters** are synthesized in the cell body or in the axon terminal. They are stored in **synaptic vesicles** and are released by exocytosis when an action potential reaches the axon terminal. Neurotransmitters combine with receptors on target cells. (p. 235)

33. Neurotransmitters come in a variety of forms. **Cholinergic** neurons secrete **acetylcholine. Adrenergic** neurons secrete **norepinephrine. Glutamate, GABA, serotonin, adenosine,** and **nitric oxide** are other major neurotransmitters. (p.236)

34. Neurotransmitter receptors are receptor-channels or G protein–coupled receptors. Ion channels create **fast synaptic potentials.** G protein–coupled receptors create **slow synaptic potentials** or modify cell metabolism. (p. 239)

35. Neurotransmitter action is rapidly terminated by reuptake into cells, diffusion away from the synapse, or enzymatic breakdown. (p. 239)

Integration of Neural Information Transfer

36. If a presynaptic neuron synapses on a larger number of postsynaptic neurons, the pattern is known as **divergence.** If several presynaptic neurons provide input to a smaller number of postsynaptic neurons, the pattern is known as **convergence.** (p.240)

37. The summation of simultaneous graded potentials from different neurons is known as **spatial summation.** The summation of graded potentials that follow each other sequentially is called **temporal summation.** (p. 241)

38. **Presynaptic modulation** of an axon terminal allows selective modulation of collaterals and their targets. **Postsynaptic modulation** occurs when a modulatory neuron, usually inhibitory, synapses on a postsynaptic cell. (p. 243)

39. **Long-term potentiation** is one mechanism by which neurons change the quality or quantity of their synaptic connections. (p. 243)

40. Developing neurons find their way to their targets by using chemical signals. (p. 245)

QUESTIONS

LEVEL ONE   Reviewing Facts and Terms

1. List the three functional classes of neurons, and explain how they differ from one another structurally and functionally.

2. Match the term with its description:

(a) axon	1. process of a neuron that receives incoming signals
(b) dendrite	
(c) afferents	2. sensory neurons, transmit information to CNS
(d) efferents	
(e) axon hillock	3. long process that transmits signals to the target cell
	4. region of neuron where action potential begins
	5. neurons that transmit information from CNS to the rest of the body

3. Somatic motor neurons control _____. _____ neurons control smooth and cardiac muscles, exocrine and some endocrine glands, and some types of adipose tissue.

4. Autonomic neurons are classified as either _____ or _____ neurons.

5. Name the two primary cell types found in the nervous system.

6. Draw a typical neuron and label the cell body, axon, dendrites, nucleus, trigger zone, axon hillock, collaterals, and axon terminals. Draw mitochondria, rough endoplasmic reticulum, Golgi bodies, and vesicles in the appropriate sections of the neuron.

7. Axonal transport refers to:

   (a) the release of neurotransmitter molecules into the synaptic cleft
   (b) the use of microtubules to send secretions from the cell body to the axon terminal
   (c) the transport of vesicles containing proteins down the length of the axon
   (d) the movement of the axon terminal to synapse with a new postsynaptic cell
   (e) none of these

8. Match the numbers of the appropriate characteristics with the two types of potentials. Characteristics may apply to one or both types.

(a) action potential	1. all-or-none
(b) graded potential	2. can be summed
	3. amplitude decreases with distance
	4. exhibits a refractory period
	5. amplitude depends on strength of stimulus
	6. has no threshold

9. A typical neuron synapses with how many other neurons?

   (a) 1   (b) fewer than 10
   (c) 1000   (d) 10,000
   (e) more than 100,000

10. List the four major types of ion channels found in neurons.

11. Arrange these events in the proper sequence:

    (a) Efferent neuron reaches threshold and fires an action potential.
    (b) Afferent neuron reaches threshold and fires an action potential.
    (c) Effector organ responds by performing output.
    (d) Integrating center reaches decision about response.
    (e) Sensory organ detects change in the environment.

12. An action potential is (circle all correct answers):

    (a) a reversal of the Na^+ and K^+ concentrations inside and outside the neuron.
    (b) the same size and shape at the beginning and end of the axon.
    (c) initiated by inhibitory postsynaptic graded potentials.
    (d) transmitted to the distal end of a neuron and causes release of neurotransmitter.

    In questions 13–17, choose the correct ion to fill in the blanks: Na^+, K^+, Ca^{2+}, Cl^-.

13. The resting cell membrane is more permeable to _____ ions than to _____ ions. Although _____ ions contribute little to the resting membrane potential, they play a key role in generating electrical signals in excitable tissues.

14. _____ is 12 times more concentrated outside the cell than inside.

15. _____ is 30 times more concentrated inside the cell than outside.

16. An action potential occurs when _____ enters the cell.

17. The resting membrane potential is due to the high _____ permeability of the cell.

18. What is the myelin sheath?

19. List two factors that enhance conduction speed.

20. List three ways neurotransmitters are removed from the synapse.

21. Draw an action potential and draw the positioning of the K^+ and Na^+ channel gates during each phase.

LEVEL TWO   Reviewing Concepts

22. Create a map using the following terms plus any terms you choose to add:

afferent signals	glial cells	peripheral division
astrocyte	integration	satellite cell
autonomic division	interneuron	Schwann cell
brain	microglia	sensory division
CNS	muscles	somatic motor division
efferent neuron	neuron	spinal cord
efferent signals	neurotransmitter	stimulus
ependymal cell	oligodendrocyte	sympathetic division
glands	parasympathetic division	

23. What causes the depolarization phase of the action potential? (Circle all that apply.)

    (a) K^+ leaving the cell through voltage-gated channels
    (b) K^+ being pumped into the cell by the Na^+-K^+-ATPase
    (c) Na^+ being pumped into the cell by the Na^+-K^+-ATPase
    (d) Na^+ entering the cell through voltage-gated channels
    (e) opening of the Na^+ channel inactivation gate

24. List four different types of glial cells, and briefly explain the role of each type.

25. Arrange the following terms to describe the sequence of events that happens after a neurotransmitter binds to a receptor on a postsynaptic neuron. Terms may be used more than once or not at all.

    (a) action potential fires at axon hillock
    (b) axon hillock reaches threshold
    (c) cell depolarizes
    (d) exocytosis
    (e) graded potential occurs
    (f) ligand-gated ion channel opens
    (g) local current flow
    (h) saltatory conduction
    (i) voltage-gated Ca^{2+} channels open
    (j) voltage-gated K^+ channels open
    (k) voltage-gated Na^+ channels open

26. Match the best term (hyperpolarize, depolarize, repolarize) to the events described. The cell in question has a resting membrane potential of −70 mV.

    (a) membrane potential changes from −70 mV to −50 mV
    (b) membrane potential changes from −70 mV to −90 mV
    (c) membrane potential changes from +20 mV to −60 mV
    (d) membrane potential changes from −80 mV to −70 mV

27. A neuron has a resting membrane potential of −70 mV. Will the neuron hyperpolarize or depolarize when each event below occurs? More than one answer may apply; list all those that are correct. What would happen if:

    (a) Na^+ enters the cell   (b) K^+ leaves the cell
    (c) Cl^- enters the cell   (d) Ca^{2+} enters the cell

28. Define, compare, and contrast the following concepts:

    (a) threshold, subthreshold, suprathreshold, spatial summation, temporal summation, all-or-none, overshoot, undershoot
    (b) graded potential, EPSP, IPSP, absolute refractory period, relative refractory period
    (c) afferent neuron, efferent neuron, interneuron
    (d) sensory, somatic motor, sympathetic, autonomic, parasympathetic neuron
    (e) fast and slow synaptic potentials
    (f) temporal and spatial summation
    (g) convergence and divergence

29. If all action potentials within a given neuron are identical, how does the neuron transmit information about the strength and duration of the stimulus?

30. The presence of myelin allows an axon to

    (a) produce more frequent action potentials
    (b) conduct impulses more rapidly
    (c) produce action potentials of larger amplitude
    (d) produce action potentials of longer duration

LEVEL THREE   Problem Solving

31. If human babies' muscles and neurons are fully developed and functional at birth, why can't they focus their eyes, sit up, or learn to crawl within hours of being born? (Hint: Muscle strength is not the problem.)

32. The voltage-gated Na^+ channels of the neuron open when the neuron depolarizes. If depolarization opens the channels, why do they close when the neuron is maximally depolarized?

33. One of the pills that Jim takes for high blood pressure caused his blood K^+ to decrease from 4.5 mM to 2.5 mM. What happens to the resting membrane potential of his liver cells? (Circle all that are correct)

    (a) it decreases   (b) it increases
    (c) it does not change   (d) it becomes more negative
    (e) it becomes less negative (f) it fires an action potential
    (g) it depolarizes   (h) it hyperpolarizes
    (i) it repolarizes

34. Match the example to the proper type of stimulus:

    (a) mechanical   1. bath water at 106°F
    (b) chemical   2. acetylcholine
    (c) thermal   3. a hint of perfume
                        4. epinephrine
                        5. lemon juice
                        6. a punch on the arm

LEVEL FOUR   Quantitative Problems

35. In each scenario below, will an action potential be produced? The postsynaptic neuron has a resting membrane potential of −70 mV.

    (a) Fifteen neurons synapse on one postsynaptic neuron. At the trigger zone, 12 of the neurons produce EPSPs of 2 mV each, and the other three produce IPSPs of

3 mV each. The threshold for the postsynaptic cell is −50 mV.

(b) Fourteen neurons synapse on one postsynaptic neuron. At the trigger zone, 11 of the neurons produce EPSPs of 2 mV each, and the other three produce IPSPs of 3 mV each. The threshold for the postsynaptic cell is −60 mV.

(c) Fifteen neurons synapse on one postsynaptic neuron. At the trigger zone, 14 of the neurons produce EPSPs of 2 mV each, and the other one produces an IPSP of 9 mV. The threshold for the postsynaptic cell is −50 mV.

E   X   P   L   O   R   E   < MediaLab >

Introduction

In this chapter you learned how the nervous system is organized and how the brain communicates with the rest of the body. Neurons carry electrical signals over long distances. These electrical signals cause release of a transmitter chemical that affects a target tissue or another neuron. This type of communication is very rapid.

To understand how nerve cells work, you must understand the action potential. The following Web Explorations will help you gain an appreciation of how electrical signals are generated, propagated, and passed from cell to cell. After reading the description below, visit the MediaLab for Chapter 8 in your Companion Website and select the appropriate keyword.

Web Exploration 1

Estimated time for completion = 10 minutes
Why are action potentials regarded as "all-or-none" stimuli? To help you with this exercise, think about what determines whether or not an action potential occurs. If many stimuli acted on a neuron, would the action potential be stronger or would there be an increase in the number of action potentials? Since graded potentials are also neuronal signals, what are the major differences between the action potentials and graded potentials? What might be the benefit of having one type of signal self-propagating and the other type of signal one that diminishes in intensity with time and distance?

Select the keyword **NEURONAL SIGNALS** on the Website. When you reach the page, scroll down to item #3 to see the differences between action potentials and graded potentials. Now, visualize an electric signal along an axon by selecting the keyword **ELECTRIC POTENTIALS.** Can you identify which is the graded potential and which is the action potential? To complete this exercise, visit MediaLab Web Exploration 1 in Chapter 8 of your Companion Website

Web Exploration 2

Estimated time for completion = 10 minutes
We can often learn a lot about the normal functioning of a system when we can study a perturbation of it! Normally, the chemical messenger acetylcholine (ACh) is released by efferent neurons to travel across the synapse. ACh binds to receptors on the postsynaptic cell and the receptors transduce the chemical signal into a mechanical signal. This is the signal for skeletal muscle contraction.

Now, think about ACh traveling across the synapse and binding to receptors. How might the resultant skeletal muscle contraction change if there were less ACh? What would occur if ACh was in excess but with fewer receptors to bind it?

A skeletal muscle disease known as myasthenia gravis is an interesting condition that can help you think about this system. Select the keyword **MYASTHENIA GRAVIS** on the Website to learn about this disease and to visualize this condition. Were your predictions correct? If you could inhibit the removal of ACh from its receptor on the skeletal muscle, how might muscle contraction be affected? To complete this exercise, visit MediaLab Web Exploration 2 in Chapter 8 of your Companion Website.

9 The Central Nervous System

■ "Neuronal assemblies have important properties that cannot be explained by the additive qualities of individual neurons."
- *O. Hechter, in* Biology and Medicine into the 21st Century, *1991.* ■

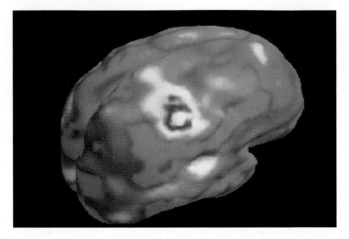

PET scan of brain

BACKGROUND BASICS

Mary Lou awoke to the sound of her 76-year-old husband crying in the darkness. When she turned on the bedside light, she saw Harold standing in his closet, facing the wall. Concerned, Mary Lou got out of bed and went to him. "Harold, what's wrong?" "I can't find my way back to bed!" her husband wailed. Harold suffers from Alzheimer's disease, a progressive neurodegenerative disease that robs people of their memories and leaves them unable to recognize familiar places and people. We still do not fully understand the cause of Alzheimer's disease, but we do know that normal communication among brain neurons is lost.

Neurons within the nervous system link together to form circuits with specific functions. The most complex

PROBLEM
Transient Ischemic Attacks

Today, Dorothy's brain sent an ominous warning. During lunch at a restaurant with her daughter, Dorothy suddenly dropped her fork and stared into space, unresponsive to her daughter's questions. Frightened, her daughter called Emergency Medical Services. By the time they reached the hospital, Dorothy was fine. The emergency room doctor, after listening to a description of the incident and Dorothy's history of high blood pressure, rendered his diagnosis: Dorothy had experienced a transient ischemic attack (TIA), or "mini-stroke." Transient ischemic attacks can be a warning that a stroke is imminent.

...continued on page 259

• •

circuits are those of the brain, in which thousands of millions of neurons are linked into intricate networks that converge and diverge, creating an infinite number of possible pathways. Signaling within these pathways creates thinking, language, feeling, learning, and memory, the complex behaviors that make us human. Some neuroscientists have proposed that the functional unit of the nervous system be changed from the individual neuron to groups of neurons because even the most basic functions require circuits of neurons.

▶ EVOLUTION OF NERVOUS SYSTEMS

How is it that combinations of neurons linked together into chains or networks collectively possess properties that are not found in any single neuron? This is a question to which we do not yet have the answer. Some scientists seek to answer it by looking for parallels between the nervous system and the integrated circuits of computers.

Computer programs have been written that attempt to mimic the thought processes of humans. This field of study, called *artificial intelligence,* has created some interesting programs such as the "psychiatrist" that is programmed to respond to typed complaints with appropriate comments and suggestions. But we are nowhere near creating a brain as complex as that of a human, or even as complex as that of Hal, the computer in the movie *2001.* Probably one reason is that computer models of the brain do not have **plasticity,** the ability to change circuit connections and function in response to sensory input and past experience. Although some computer programs can change their output under specialized conditions, they cannot begin to approximate the plasticity of human brain networks that easily restructure themselves as the result of sensory input, learning, emotion, and creativity.

How can linking neurons together create **affective behaviors,** which are related to feeling and emotion, and **cognitive behaviors** related to thinking? In their search for the answer to this perplexing question, scientists seek clues in the simplest animal nervous systems. Jellyfish and sea anemones, members of the phylum Cnidaria, have no central integrating center such as a brain, but they do possess neural elements that allow them to sense and respond to their environment. Their nervous system is a **nerve net** composed of afferent sensory neurons, connective interneurons, and efferent neurons. Stimuli sensed by receptors create a signal that is conducted in a diffuse manner throughout the network, and acted upon without input from an identifiable control center such as a brain.

In the primitive flatworms, we see the beginnings of a nervous system as we know it in higher animals, although in these animals the distinction between central nervous system and peripheral nervous system is not clear. Flatworms have a rudimentary brain, a cluster of nerve cell bodies called **ganglia** [singular *ganglion,* a swelling] concentrated in the head (*cephalic*) region. A large nerve called a nerve cord comes off the primitive brain formed by the cephalic ganglia and leads to a nerve network that innervates distal regions of the flatworm body.

The segmented worms, or annelids, have a more advanced central nervous system. Ganglia are no longer restricted to the head region but occur in fused pairs along a nerve cord. Because each segment of the worm contains a ganglion, simple reflexes can be integrated within that segment without input from the brain. Reflexes that do not require integration in the brain are also seen in higher animals and are called **spinal reflexes** in humans and other vertebrates.

Annelids have complex reflexes controlled through networks of neurons. Researchers use leeches (a type of annelid) and certain mollusks whose neurons can be easily seen and traced to study neural networks and synapse formation The neural function of these animals provides a basis upon which we can model more complex vertebrate networks.

Cephalic ganglia clustered into brains persist throughout the more advanced phyla and become increasingly more complex. One advantage to cephalic brains is that in most animals the head is the part of the body that contacts the environment first as the animal moves. In the higher arthropods, such as insects, specific regions of the brain begin to be associated with particular functions and to receive input from specific sensory structures. More sophisticated brains are associated with specialized cephalic receptors such as eyes or chemoreceptors for smell and taste.

The most dramatic change in vertebrate brain evolution is seen in the brain region known as the cerebrum. In primitive fish, the cerebrum is a small bulge, dedicated mainly to processing olfactory information about odors in the environment. In humans, the cerebrum is the largest and most distinctive part of the brain and is probably involved in all brain functions. More than anything else, the cerebrum is what makes us human. All evidence indicates that it is the part of the brain that allows reasoning and cognition.

Tracing Neurons in a Network One of the challenges facing neurophysiologists and neuroanatomists is tracing the networks of neurons that control specific functions, a task that can be likened to following one tiny thread through a tangled mass the size of a beach ball. In 1971, K. Kristensson and colleagues introduced the use of horseradish peroxidase (HRP), an enzyme that acts in the presence of its substrate to produce a visible product. When injected into the extracellular fluid near axon terminals, HRP is brought into the neuron by endocytosis. The vesicles containing HRP are then transported by fast retrograde (backward) axonal transport to the cell body and dendrites [∞ p. 218]. Once the enzyme-substrate reaction is completed, the neuron becomes visible, allowing the researcher to trace the entire neuron from its target back to its origin. Now, with the help of fluorescent antibodies, scientists can even tell what neurotransmitters and receptors the neuron uses.

Let's begin with an overview of the central nervous system and its functions. Then we will look at how neural networks create the higher brain functions of thought and emotion.

▶ ANATOMY OF THE CENTRAL NERVOUS SYSTEM

The **central nervous system (CNS)** consists of the cerebral ganglia (brain) and the associated nerve or spinal cord. As we follow the evolution of central nervous systems, we find that brains increase in complexity and degree of specialization as we move up the phylogenetic tree. Protection of these control centers becomes increasingly important, and, in vertebrates, the brain and spinal cord are encased in connective tissue and bone.

Many Layers of Protection Surround the Central Nervous System

The individual neurons and glial cells of the central nervous system have highly organized internal cytoskeletons that maintain cell shape and orientation, but the brain and spinal cord are soft and jellylike. Nervous tissue receives little support from its minimal extracellular matrix and must rely on external support to help protect it from trauma. This support comes in the form of an outer casing of bony tissue, three layers of connective tissue membrane, and fluid between the membranes.

In vertebrates, the brain is held in a protective bony case, the **skull,** or **cranium** (Fig. 9-1a ■), and the spinal cord runs through a canal in the **vertebral column** (Fig. 9-1c ■). The segmentation that is characteristic of many invertebrates can still be seen in the bony vertebrae, which are stacked on top of one another and separated by disks of connective tissue. The nerves of the peripheral nervous system enter and leave the spinal cord by passing through notches between the stacked vertebrae.

Between the bones and tissues of the central nervous system are three layers of membrane, collectively called the **meninges** [singular *meninx*, membrane]. These membranes help stabilize the neural tissue and protect it from bruising against the bones of the skeleton. Starting from the bones and moving toward the nervous tissue, the membranes are (1) the *dura mater*, (2) the *arachnoid membrane*, and (3) the *pia mater* (Fig. 9-1b ■).

Cerebrospinal Fluid and the Ventricular System

The cranium has an internal volume of 1.4 L, of which about 1 L is occupied by the cells. The remaining volume of the cranium is divided into two distinct extracellular compartments: the blood (100–150 mL) and the cerebrospinal fluid and interstitial fluid (250–300 mL). **Cerebrospinal fluid,** or **CSF,** is a salty solution that is continuously secreted into hollow cavities in the brain known as **ventricles** (Fig. 9-2 ■, p. 258). There are two lateral ventricles (the first and second) and two descending ventricles (the third and fourth).

Most cerebrospinal fluid is secreted into the ventricles by the **choroid plexus,** a specialized tissue found on the walls of the ventricles (Fig. 9-3b ■, p. 258). The choroid plexus is remarkably similar to kidney tissue and consists of capillaries and a transporting epithelium [∞ p. 59] called the **ependyma.** The ependymal cells actively pump sodium and other solutes into the ventricles, creating an osmotic gradient that draws water along with the solutes.

From the ventricles, cerebrospinal fluid flows into the subarachnoid space between the pia mater and the arachnoid membrane, surrounding the entire brain and spinal cord in fluid (Fig. 9-3a ■). The cerebrospinal fluid flows around the nervous tissue and is finally absorbed back into the blood by special areas of the arachnoid membrane in the cranium (Fig. 9-3c ■). The rate of fluid flow through the central nervous system is sufficient to replenish the entire volume of cerebrospinal fluid three or four times a day.

The cerebrospinal fluid serves two purposes: physical and chemical protection. The brain and spinal cord actually float in the thin layer of fluid between the membranes. The buoyancy of cerebrospinal fluid reduces the weight of the brain nearly 30-fold, to about 50 g. Lighter weight translates into less pressure on blood vessels and nerves attached to the central nervous system.

The cerebrospinal fluid also provides protective padding. When there is a blow to the head, the fluid must be compressed before the brain can hit the inside of the cranium. For a dramatic example of the protective power of cerebrospinal fluid, shake a block of tofu representing the brain in an empty jar. Then shake another block of tofu in a jar filled with water to see how the cerebrospinal fluid safeguards the brain.

In addition to physically protecting the delicate tissues of the central nervous system, cerebrospinal fluid creates a closely regulated extracellular environment for the neurons. The transporting epithelial cells of the choroid plexus are selective about what substances they allow to pass into the ventricles, and, as a result, the composition of cerebrospinal fluid is different from that of the plasma. The concentrations of K^+, Ca^{2+}, HCO_3^-, and glucose are lower in the cerebrospinal fluid, and the concentration of H^+ is higher. Only the concentration of Na^+ in cerebrospinal fluid is similar to that in the blood. In addition, cerebrospinal fluid contains very little protein and no blood cells.

By sampling the cerebrospinal fluid, one can get an indication of the chemical environment within the brain. This sampling procedure, known as a *spinal tap* or *lumbar puncture,* is generally done by withdrawing fluid from the subarachnoid space between vertebrae at the lower end of the spinal cord. The presence of proteins or blood cells in the cerebrospinal fluid is suggestive of infection.

✔ If the concentration of H^+ in the cerebrospinal fluid is higher than that in the blood, what can you say about the pH of the CSF?

✔ If a blood vessel running between the meninges is ruptured as a result of a blow to the cranium, what do you predict might happen?

✔ Is cerebrospinal fluid more like plasma or interstitial fluid? Defend your answer.

The Blood-Brain Barrier Protects the Brain from Harmful Substances in the Blood

Nervous tissue has special metabolic requirements. Neurons consume oxygen at high rates that reflect their need for ATP to conduct active transport [∞ p. 124]. Their only fuel source under normal circumstances is glucose. Consequently, the body closely regulates blood glucose concentrations and blood supply to the brain to ensure normal function.

About 15% of the blood pumped by the heart goes to the brain to provide adequate glucose and oxygen. Interruption of the blood supply to the brain can have devastating results, and brain death occurs after only a few minutes without oxygen. Progressive *hypoglycemia* (low blood glucose levels) leads to confusion, unconsciousness, and eventually death.

Because the brain is the main control center for the body, its cells must be protected from potentially harmful substances in the blood. Capillaries in the brain are much less permeable than most capillaries in the body, and they exclude many molecules that cross capillary walls elsewhere. These relatively impermeable capillaries form what is known as the **blood-brain barrier** (Fig. 9-4 ■, p. 259). The limited permeability of the blood-brain barrier is a protective mechanism that shelters the brain from fluctuations in the blood concentration of hor-

DIABETES Hypoglycemia and the Brain Neurons are picky about their food. Under most circumstances, the only biomolecule that neurons use for energy is glucose. Surprisingly, this can present a problem for diabetic patients, whose problem is too much glucose in the blood. In the face of sustained *hyperglycemia* (elevated blood glucose), the cells of the blood-brain barrier down-regulate [∞ p. 165] their glucose transporters. Then, if the patient's blood glucose level decreases due to insulin or other therapy, the neurons of the brain may be out of luck. They will be unable to transport glucose fast enough to sustain their electrical activity. The result is impaired CNS function, including confusion, irritability, and slurred speech. Unless the neurons are provided promptly with glucose, they can sustain permanent damage. In extreme cases, *hypoglycemia* can cause coma or even death.

mones, ions, and neuroactive substances such as neurotransmitters. However, lipid-soluble molecules can enter the interstitial fluid bathing the neurons by simply diffusing through the cell membranes [∞ p. 120]. This is one reason some antihistamines make you sleepy but others do not. The older antihistamines were lipid-soluble amines that readily crossed the blood-brain barrier and acted on brain centers controlling alertness. The newer drugs are much less lipid-soluble and as a result do not have the same sedative effect.

The blood-brain barrier, like the choroid plexus, is able to transport nutrients and other useful material from the blood into the brain interstitial fluid by using specific membrane carriers. One interesting illustration of how the blood-brain barrier works is seen in Parkinson's disease, a neurological disorder in which brain levels of the neurotransmitter dopamine are too low. Dopamine administered therapeutically is ineffective because it is unable to cross the blood-brain barrier. But the dopamine precursor, L-DOPA, is transported across the cells of the blood-brain barrier on an amino acid transporter [∞ p. 121]. Once in the cerebrospinal fluid, L-DOPA is metabolized to dopamine, thereby correcting the deficiency.

Why are capillaries in the brain so much less leaky than capillaries elsewhere in the body? The endothelial cells that form the walls of the brain capillaries are the same as those in other capillaries. However, the cells are connected to each other with tight junctions rather than leaky junctions and pores [∞ p. 55]. The formation of these tight junctions is induced by paracrines released by adjacent astrocytes. Thus, it is the brain tissue itself that creates the blood-brain barrier.

In a few areas of the brain, the blood-brain barrier is missing. In these areas, the function of adjacent neurons depends in some way on direct contact with the blood. For instance, the posterior pituitary releases neurosecretory hormones that must pass into the capillaries for

ANATOMY SUMMARY The Central Nervous System

■ **Figure 9-1**

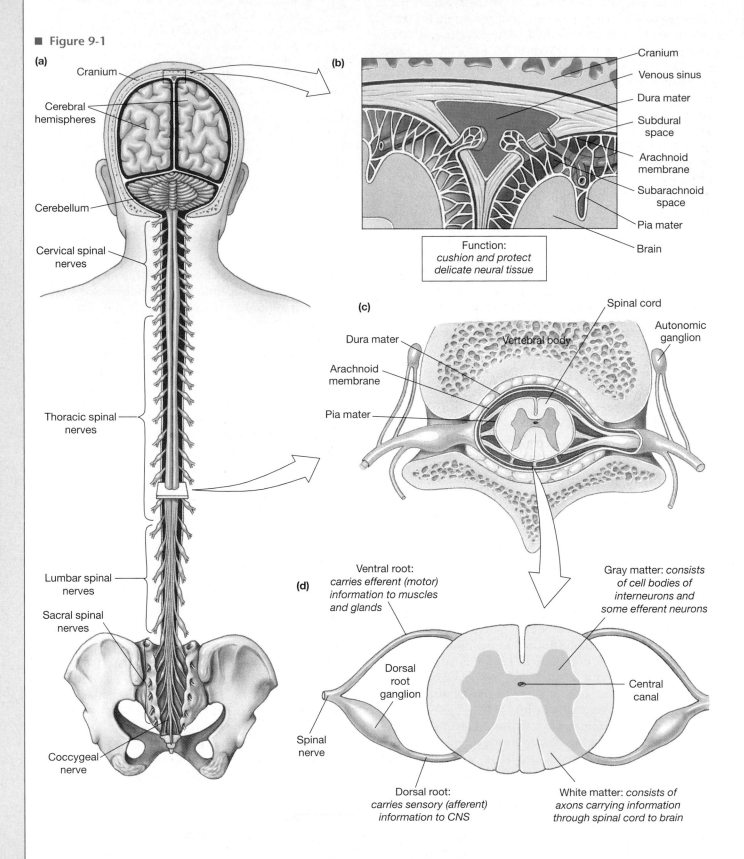

(a)

Cranium

Cerebral hemispheres

Cerebellum

Cervical spinal nerves

Thoracic spinal nerves

Lumbar spinal nerves

Sacral spinal nerves

Coccygeal nerve

(b)

Cranium

Venous sinus

Dura mater

Subdural space

Arachnoid membrane

Subarachnoid space

Pia mater

Brain

Function: cushion and protect delicate neural tissue

(c)

Spinal cord

Autonomic ganglion

Dura mater

Vertebral body

Arachnoid membrane

Pia mater

(d)

Ventral root: *carries efferent (motor) information to muscles and glands*

Dorsal root ganglion

Spinal nerve

Dorsal root: *carries sensory (afferent) information to CNS*

Gray matter: *consists of cell bodies of interneurons and some efferent neurons*

Central canal

White matter: *consists of axons carrying information through spinal cord to brain*

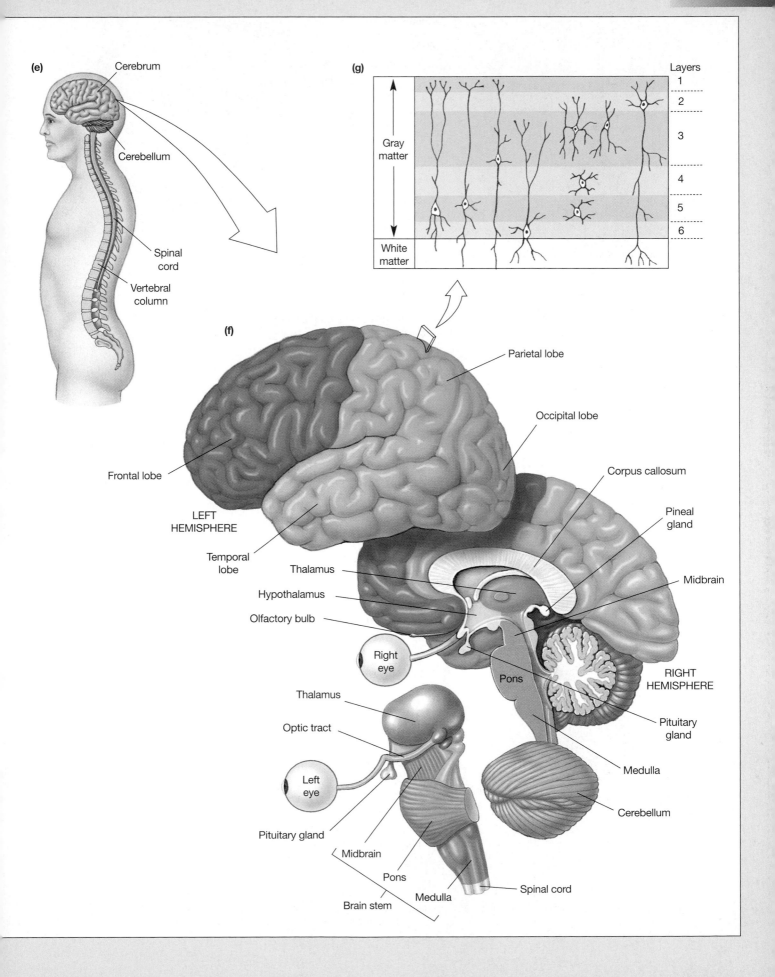

(e)

Cerebrum

Cerebellum

Spinal cord

Vertebral column

(g)

Layers

1
2
3
4
5
6

Gray matter

White matter

(f)

Parietal lobe

Occipital lobe

Corpus callosum

Pineal gland

Midbrain

Frontal lobe

LEFT HEMISPHERE

Temporal lobe

Thalamus

Hypothalamus

Olfactory bulb

Right eye

Pons

RIGHT HEMISPHERE

Pituitary gland

Medulla

Cerebellum

Spinal cord

Thalamus

Optic tract

Left eye

Pituitary gland

Midbrain

Pons

Medulla

Brain stem

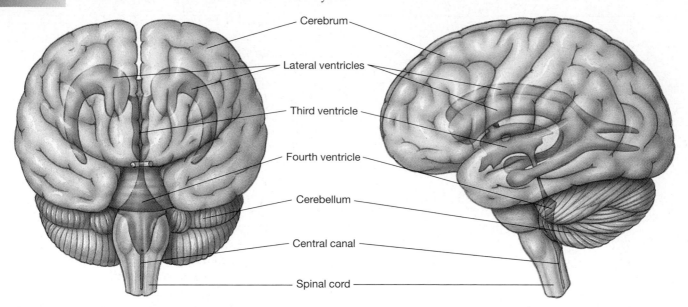

■ **Figure 9-2   Ventricles of the brain**   The first and second ventricles together form the lateral ventricles. The third and fourth ventricles extend through the brain stem and connect to the central canal that runs through the center of the spinal cord.

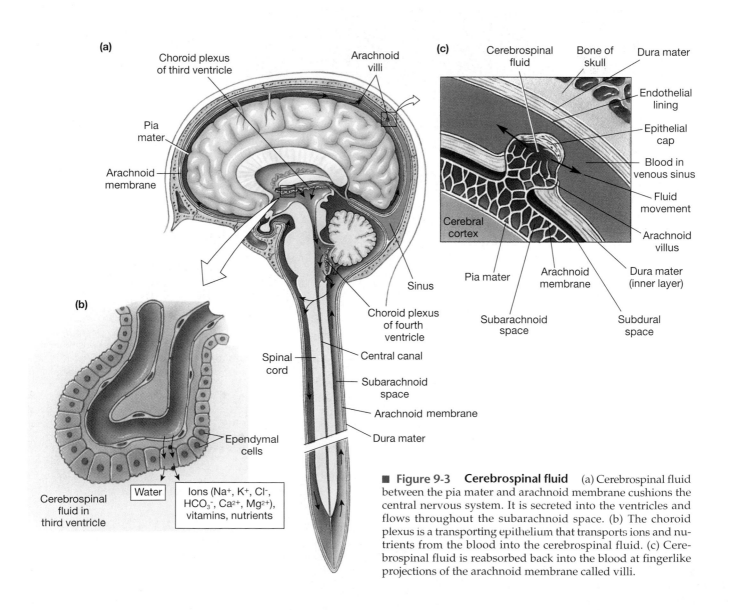

■ **Figure 9-3   Cerebrospinal fluid**   (a) Cerebrospinal fluid between the pia mater and arachnoid membrane cushions the central nervous system. It is secreted into the ventricles and flows throughout the subarachnoid space. (b) The choroid plexus is a transporting epithelium that transports ions and nutrients from the blood into the cerebrospinal fluid. (c) Cerebrospinal fluid is reabsorbed back into the blood at fingerlike projections of the arachnoid membrane called villi.

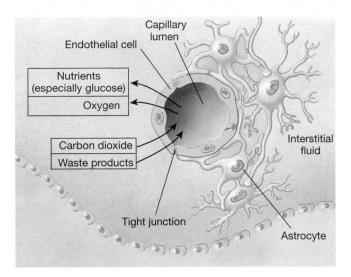

■ Figure 9-4   The blood-brain barrier   Paracrines secreted by astrocytes stimulate the development of tight junctions in cerebral capillaries and create the blood-brain barrier. The movement of material across the blood-brain barrier is regulated to protect the neurons from potentially harmful substances in the blood.

distribution to the body [∞ p. 198]. Another example is the vomiting center in the medulla (Fig. 9-1f ■). These neurons sense the presence of possibly toxic substances in the blood, such as drugs. They respond by inducing vomiting, thereby removing the contents of the digestive system in case the substance was ingested.

Now that we have seen how the central nervous system is protected, let's look at the general anatomical characteristics of the brain and spinal cord.

Neurons of the Central Nervous System Are Grouped into Nuclei and Tracts

The central nervous system, like the peripheral nervous system, is composed of neurons and supportive glial cells. Interneurons do not extend outside the central nervous system. Sensory and efferent neurons link interneurons in the central nervous system to peripheral receptors and effectors.

•••••••••••••••••••••••••••••••••••••••

...continued from page 253

A **stroke** is a form of cerebrovascular disease in which a blood vessel to the brain either ruptures (hemorrhagic stroke) or becomes blocked by a blood clot and fatty substance called plaque (ischemic stroke) [*ischein*, to suppress + *-emia*, blood]. The loss of blood flow to an area of brain tissue deprives the cells of oxygen, and they lose function or die. Often, strokes occur without warning. Sometimes, however, a transient ischemic attack (TIA)—or a series of them—heralds the onset of a stroke. In fact, a recent study found that TIAs occurred in 30% of people who later experienced full-fledged strokes.

Question 1:   *Do you think Dorothy's brain cells died as a result of her TIA? Explain.*

•••••••••••••••••••••••••••••••••••••••

When viewed on a macroscopic level, the tissues of the central nervous system are described by their color. The unmyelinated **gray matter** consists of nerve cell bodies, dendrites, and axon terminals (see Fig. 9-1d, g ■). The cell bodies are assembled in an organized fashion within both the brain and the spinal cord. They form layers in some parts of the brain or cluster into groups of neurons with a similar function. These clusters of cell bodies in the brain and spinal cord are known as **nuclei,** and they are usually identified by specific names.

White matter is made up mostly of axons with very few cell bodies. Its pale color comes from the myelin sheaths that surround the axons. Bundles of axons that connect different regions of the central nervous system are known as **tracts. Ascending tracts** mainly carry sensory information from the cord to the brain. **Descending tracts** carry mostly efferent (motor) signals from the brain to the cord. **Propriospinal tracts** remain within the cord [*proprius,* one's own]. *Tracts* in the central nervous system are equivalent to *nerves* in the peripheral nervous system.

▶ THE SPINAL CORD

The spinal cord is the major pathway of information flowing back and forth between the brain and the skin, joints, and muscles of the body. In addition, the spinal cord contains neural networks responsible for locomotion. If the spinal cord is severed, there is loss of sensation from the skin and muscles as well as loss of the ability to voluntarily control muscles (paralysis).

The interneurons of the spinal cord route sensory information from peripheral receptors to the brain and commands from the brain to the muscles and glands of the body. In many cases, the interneurons also modify information as it passes through. Spinal interneurons play a critical role in the coordination of movement (see Chapter 13).

The spinal cord is divided into four regions (*cervical, thoracic, lumbar,* and *sacral*), named to correspond to the adjacent vertebrae (Fig. 9-1a ■). Each region is further subdivided into segments, and each segment gives rise to a bilateral pair of *spinal nerves.* Just before a spinal nerve joins the spinal cord, it divides into two branches called **roots** (Fig. 9-1d ■). The **dorsal root** of each spinal nerve is specialized to carry incoming sensory information. The **dorsal root ganglia,** swellings found on the dorsal roots just before they enter the cord, contain cell bodies of sensory neurons. The **ventral root** carries information from the central nervous system to the muscles and glands.

In cross section, the spinal cord has a butterfly- or H-shaped core of gray matter and a surrounding rim of white matter (Fig. 9-5a ■). The color of the gray matter comes from neuron cell bodies. Interneurons in the **dorsal horn** of the gray matter synapse with sensory fibers from the dorsal roots. The cell bodies of these interneurons are organized into two distinct nuclei, one for somatic information

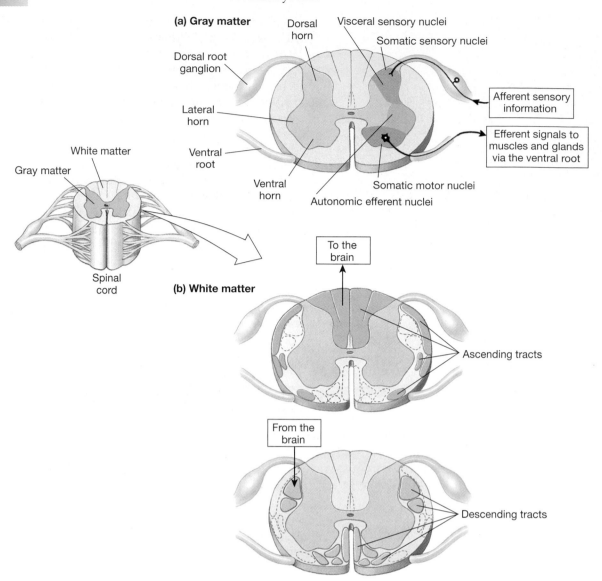

■ **Figure 9-5   Specialization in the spinal cord**   The two figures represent different regions of the spinal cord. (a) Cell bodies (nuclei) in the dorsal horn receive somatic and visceral sensory information. Efferent neurons that send signals to muscles and glands are organized into somatic motor and autonomic nuclei. The areas are shown here on only one side but exist on both sides. (b) The white matter consists of ascending tracts that take sensory information to the brain and descending tracts that carry commands to effector organs.

and one for visceral information. Similarly, the cell bodies of neurons sending efferent signals to muscles and glands are organized into somatic motor and autonomic nuclei. Efferent fibers leave the gray matter via the **ventral horns.**

The white matter of the spinal cord is the biological equivalent of the fiber-optic cables that phone companies use to carry our communications systems. The white matter can be divided into a number of **columns** composed of tracts of axons that transfer information up and down the cord. Ascending tracts that take sensory information to the brain occupy the dorsal and external lateral portions of the spinal cord (Fig. 9-5b ■). Descending tracts that carry commands to effector organs occupy the ventral and interior lateral portions of the white matter.

Not all information from peripheral receptors must go to the brain. The spinal cord itself functions as a self-contained integrating center for the simplest reflexes.

✔ What are the differences between horns, roots, columns, and tracts of the spinal cord?

✔ If a dorsal root of the spinal cord is cut, what function will be disrupted?

▶ THE BRAIN

Thousands of years ago, Aristotle declared that the heart was the seat of the soul. However, most people now agree that the brain is the organ that gives the human species its unique attributes. The adult human brain

...continued from page 259

Dorothy was surprised when her doctor prescribed a daily dose of half an aspirin. Acetylsalicylic acid, the active ingredient in aspirin, is a antiplatelet agent. "In other words," he said, "it helps keep blood cells from sticking together." He also asked Dorothy about her controllable risk factors for stroke: high blood pressure, high cholesterol levels, heavy alcohol consumption, and smoking. Dorothy did not smoke or drink, but she did have high blood pressure and elevated cholesterol levels. She was started on treatment to lower her blood pressure and given instructions for a cholesterol-reducing diet.

Question 2: *Explain the rationale for taking aspirin to prevent stroke.*

weighs about 1400 g and contains an estimated 10^{12} neurons. When you consider that each one of these millions of neurons may receive as many as 200,000 synapses, the number of possible neuronal connections is mind-boggling. Yet, despite this tremendous potential for variability, the central nervous system is remarkably orderly.

The human brain is compartmentalized so that different regions have different functions. It also has built-in backup mechanisms in case of failure. A single function may be carried out in more than one region (a feature known as *parallel processing*), and there is a certain degree of plasticity in the networks so that the functions of a damaged region can sometimes be taken over by other sections.

Microscopically, the brain contains two neural elements: (1) neuron cell bodies, some of which are specialized for neurohormone secretion, and (2) nerve **fibers** (axons) in bundles.

When viewed intact in profile, the brain of an adult can be grossly divided into the brain stem, the cerebellum, and the cerebrum (Fig. 9-1f ■). The **cerebrum** [*cerebrum,* brain; adjective *cerebral*] is the largest and most obvious structure we see when looking at a human brain. In the early embryo, the area that will become the cerebrum is not much larger than the other regions of the brain (Fig. 9-6a ■). But, as development proceeds, the growth of the cerebrum greatly outpaces that of the other regions (Fig. 9-6b ■). When viewed in sagittal section, the

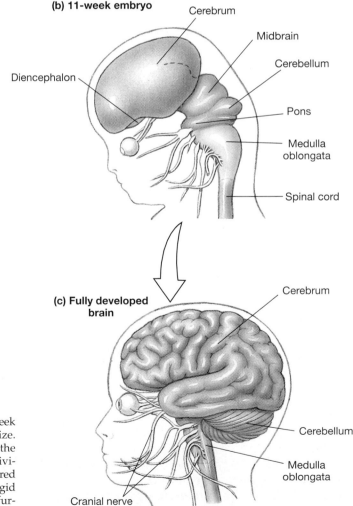

(a)

Brain (6-week embryo)	Brain Regions at Birth
	Cerebrum
	Diencephalon
	Midbrain
	Cerebellum and Pons
	Medulla oblongata

(b) 11-week embryo

Cerebrum
Midbrain
Cerebellum
Diencephalon
Pons
Medulla oblongata
Spinal cord

(c) Fully developed brain

Cerebrum
Cerebellum
Medulla oblongata
Cranial nerve

■ **Figure 9-6   Development of the brain**   (a) In the 6-week embryo, the different regions of the brain are fairly equal in size. (b) By 11 weeks of embryonic development, the growth of the cerebrum is noticeably more rapid than that of the other divisions of the brain. (c) In the adult brain, the cerebrum has covered most of the other brain regions. Its rapid growth within the rigid confines of the cranium forces it to develop a convoluted, furrowed surface.

adult cerebrum surrounds the hollow ventricles and many of the structures of the brain stem and diencephalon (Fig. 9-1f ■). Table 9-1 is a summary of the brain areas and their major functions.

The Brain Stem Contains Centers for Many Involuntary Functions

The brain stem is an extension of the spinal cord and is divided into three main sections: the medulla oblongata, the pons, and the midbrain, or mesencephalon (Fig. 9-6b ■). The brain stem also contains the fourth ventricle. Most of the cranial nerves emerge from the brain stem (Table 9-2).

The cranial nerves carry sensory and motor information for the head and neck. In addition, cranial nerve X, the **vagus nerve** [*vagus,* wandering], carries both sensory and motor fibers for many internal organs. The cranial nerves are described according to whether they include sensory fibers, efferent fibers, or both (mixed nerves).

✔ Using the information from Table 9-2, describe activities you might ask a patient to perform if you wished to test cranial nerve function.

The **medulla oblongata,** frequently just called the *medulla* [*medulla,* marrow; adjective *medullary*], is the transition from the spinal cord into the brain proper. Anatomically, it contains the *corticospinal fiber tracts* that convey information between the cerebral cortex and the spinal cord. Many of the corticospinal tracts cross the midline to the opposite side of the body in a region of the medulla known as the **pyramids.** As a result of this crossover, each side of the brain controls the opposite side of the body. The medulla also contains control centers for many involuntary functions such as blood pressure, breathing, swallowing, and vomiting.

The **pons** [*pons,* bridge; adjective *pontine*] is a bulbous protrusion on the ventral side of the brain stem below the midbrain. Because its primary function is to act as a relay station for information transfer between the cerebellum and cerebrum, the pons is often grouped with the cerebellum. The pons also coordinates the control of breathing along with centers in the medulla.

The **midbrain,** or **mesencephalon** [*mesos,* middle], is a relatively small area between the lower brain stem and the diencephalon. Its primary function is the control of eye movement, but it also relays signals for auditory and visual reflexes.

Running throughout the brain stem is the **reticular formation** (also known as the *reticular activating system*), diffuse groups of neurons that probably constitute the oldest part of the brain phylogenetically. The name *reticular* literally means "network" and comes from the redundant crisscrossed axons that branch profusely into superior sections of the brain and into the spinal cord. The reticular formation is best known for its role in arousal and sleep, but in recent years it has also been shown to be involved in muscle tone and stretch reflexes, coordination of breathing, blood pressure regulation, and modulation of pain.

TABLE 9-1   Parts of the Brain and their Major Functions

Cerebrum	
■ Cerebral cortex	
Sensory fields	Perception
Motor areas	Skeletal muscle movement
Association areas	Integration of information and direction of voluntary movement
■ Basal ganglia	Movement
■ Limbic system	
Amygdala	Emotion and memory
Hippocampus	Learning and memory
Diencephalon	
■ Thalamus	Integrating center and relay station
■ Hypothalamus	Homeostasis and behavioral drives (see Table 9-3)
■ Pineal gland	Melatonin secretion
Cerebellum	Movement coordination
Brain stem	
■ Midbrain	Eye movement
■ Pons	Relay station; coordination of breathing
■ Medulla oblongata	Control of involuntary functions
■ Reticular formation	Arousal, sleep, muscle tone, pain modulation

TABLE 9-2   The Cranial Nerves

Number	Name	Type	Function
I	Olfactory	Sensory	Olfactory (smell) information from nose
II	Optic	Sensory	Visual information from eyes
III	Oculomotor	Motor	Eye movement
IV	Trochlear	Motor	Eye movement
V	Trigeminal	Mixed	Sensory information from face; motor signals for chewing
VI	Abducens	Motor	Eye movement
VII	Facial	Mixed	Sensory for taste; efferent signals for tear and salivary glands, facial expression
VIII	Vestibulocochlear	Sensory	Hearing and equilibrium
IX	Glossopharyngeal	Mixed	Sensory from oral cavity, baro- and chemoreceptors in blood vessels; efferent for swallowing, parotid salivary gland
X	Vagus	Mixed	Sensory and efferents to many internal organs, muscles, and glands
XI	Accessory	Motor	Muscles of oral cavity, some muscles in neck and shoulder
XII	Hypoglossal	Motor	Tongue muscles

The Cerebellum Coordinates Movement

The name **cerebellum** [adjective *cerebellar*] means "little brain," and, indeed, most of the nerve cells in the brain are in the cerebellum. The specialized function of the cerebellum is to process sensory information and coordinate the execution of movement. Sensory input into the cerebellum comes from somatic receptors in the periphery of the body and from receptors for equilibrium and balance located in the inner ear. The cerebellum also receives motor input from neurons in the cerebral cortex. Cerebellar function is described in more detail in Chapter 13.

The Diencephalon Contains the Centers for Homeostasis

The **diencephalon,** or "between-brain," lies between the brain stem and the cerebrum (Fig. 9-7 ■). It is composed of two main sections, the thalamus and the hypothalamus. The diencephalon also contains the **pineal gland,** a small endocrine gland that secretes the hormone **melatonin** [∞ p. 208]. Most of the diencephalon is occupied by many small nuclei that make up the **thalamus** [*thalamus,* bedroom; adjective *thalamic*].

The thalamus is often described as a relay station because almost all sensory information from lower parts of the central nervous system is transmitted through the thalamus on its way to the cerebral cortex. But, like the spinal cord, the thalamus can shape information passing through it, making it an integrating center as well as a relay station. The thalamus receives sensory fibers from the optic tract, ears, and spinal cord, and it projects fibers to the cerebrum.

The **hypothalamus** lies beneath the thalamus. The stalk of the pituitary gland and the neuroendocrine tissue of the posterior pituitary are downgrowths of this region of the diencephalon [∞ p. 198]. The hypothalamus contains various centers for behavioral drives, such as hunger and thirst, and plays a key role in homeostasis. Its neural output controls many functions of the autonomic division of the nervous system as well as a variety of endocrine functions (Table 9-3). For example, neuroendocrine cells in the hypothalamus secrete neurohormones to control the release of anterior pituitary hormones.

The hypothalamus receives input from multiple sources including the cerebral cortex, the reticular formation, and various sensory receptors. Its output goes to the thalamus and eventually to multiple effector pathways. Most pathways in the hypothalamus are bidirectional, with two exceptions. A descending pathway takes neurohormones from the hypothalamus to the posterior pituitary. An ascending pathway leads from the retina to the **suprachiasmatic nucleus.** This region of the hypothalamus appears to be the center for the biological clock that creates circadian rhythms in the body [∞ p. 175].

The Cerebrum Is the Site of Higher Brain Functions

The **cerebrum** (Fig. 9-8 ■) fills most of the cranial cavity. It is composed of two hemispheres connected at the **corpus callosum** (see Fig. 9-12 ■). This connection ensures that the two hemispheres communicate and cooperate with each other. Each cerebral hemisphere is divided into four lobes, named for the bones of the skull under which they are located: **frontal, parietal, temporal,** and **occipital.**

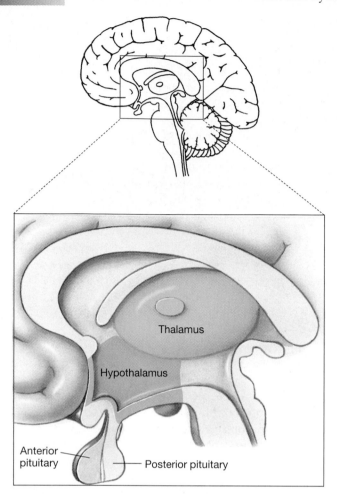

■ **Figure 9-7**   **The diencephalon**   The diencephalon lies between the brain stem and the cerebrum. It is composed of the thalamus, a relay station for information going to and from the cerebrum, and the hypothalamus, a center for many homeostatic reflexes. The pituitary appears to be an extension of the hypothalamus, but the anterior pituitary is actually an endocrine gland that developed from the same epithelium as the roof of the mouth.

TABLE 9-3   Functions of the Hypothalamus

1. Activation of sympathetic nervous system
   - Control of catecholamine release from adrenal medulla (as in fight-or-flight reaction)
   - Helps maintain blood glucose concentrations through effects on endocrine pancreas
2. Maintenance of body temperature
   - Shivering and sweating
3. Control of body osmolarity
   - Thirst and drinking behavior
   - Secretion of vasopressin
4. Control of reproductive functions
   - Secretion of oxytocin (uterine contractions and milk release)
   - Trophic hormone control of anterior pituitary hormones FSH and LH
5. Control of food intake
   - Satiety center
   - Feeding center
6. Interacts with limbic system to influence behavior and emotions
7. Influences cardiovascular control center in medulla oblongata
8. Secretes trophic hormones that control release of hormones from anterior pituitary gland

acts as the link between higher cognitive functions such as reasoning and more primitive emotional responses such as fear. The **amygdala** is linked to emotion and memory; the **hippocampus,** to learning and memory.

Organization of the Cerebral Cortex   The outer layer of neurons in the cerebrum, only a few millimeters thick, is called the **cerebral cortex** [*cortex,* bark or rind, adjective *cortical*]. Within this layer our higher brain

The surface of the cerebrum of humans and other primates has a furrowed walnut-like appearance. During development, the cerebrum grows faster than the surrounding cranium, causing the tissue to fold back on itself to fit into a smaller volume. The degree of folding is directly related to the level of processing of which the brain is capable; less advanced mammals, such as rodents, have brains with a relatively smooth surface. The human brain is so convoluted that if it were inflated enough to smooth the surfaces, it would be three times as large and would need a head the size of a beach ball.

The interior of the cerebrum contains three clusters of nuclei (cell bodies): the basal ganglia, the amygdala, and the hippocampus. The **basal ganglia** (also known as the *corpus striatum*) are involved in the control of movement. The amygdala and the hippocampus are part of the **limbic system,** a region of cerebrum that surrounds the brain stem (Fig. 9-9 ■). The limbic system probably represents the most primitive region of the cerebrum. It

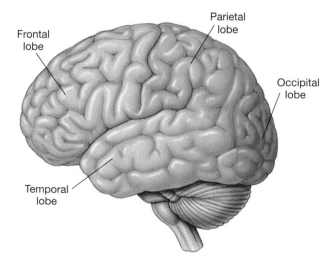

■ **Figure 9-8**   **The cerebrum**   The cerebrum has four lobes in each half, or hemisphere. The lobes are named for the bones of the cranium under which they lie.

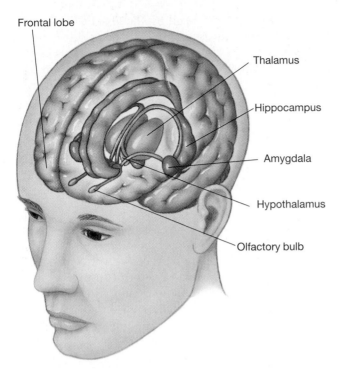

Frontal lobe

Thalamus

Hippocampus

Amygdala

Hypothalamus

Olfactory bulb

■ **Figure 9-9   The limbic system**   The limbic system is a section of the interior of the cerebrum that surrounds the brain stem. Structures in the limbic system have been linked to learning, memory, and emotion.

brain are not as obvious morphologically but have been identified by electrophysiological recordings. Information reaching a single fiber is transferred vertically as each neuron synapses on the next layer. The lateral spread of information is limited to a few millimeters, thus creating vertical columns of information transfer. Some of the best descriptions of cortical column function come from studies of the visual system.

The cerebral cortex contains three functional specializations: sensory areas (*fields*) that direct perception, motor areas that direct movement, and regions known as *association areas* (association cortices) that integrate information and direct voluntary behaviors (Fig. 9-10 ■). These functional areas do not necessarily correspond to the anatomical lobes but have been identified through experimental studies and observations of patients with various brain lesions.

Sensory areas include regions for perception of sensations from peripheral receptors as well as regions devoted to vision, hearing, and olfaction (smell). The **primary somatic sensory cortex** in the parietal lobe receives sensory information from the skin, musculoskeletal system, viscera, and taste buds. Damage to this part of the brain leads to reduced sensitivity of the skin on the opposite side of the body because sensory fibers cross to the opposite side of the midline as they ascend through the medulla.

The **visual cortex,** located in the occipital lobe, receives information from the eyes, and the **auditory cortex,** located in the temporal lobe, receives information from the ears. The **olfactory cortex,** a small region in the temporal lobe, receives input from chemoreceptors in the nose. The **gustatory cortex** in the parietal lobe receives sensory information from taste buds.

functions such as reasoning arise. The neurons of the cortex are arranged in anatomically distinct horizontal layers and functionally distinct vertical columns. If you look at a microscopic section of the cortex, you can see that the neurons are arranged into six layers parallel to the surface (Fig. 9-1g ■). The functional columns of the

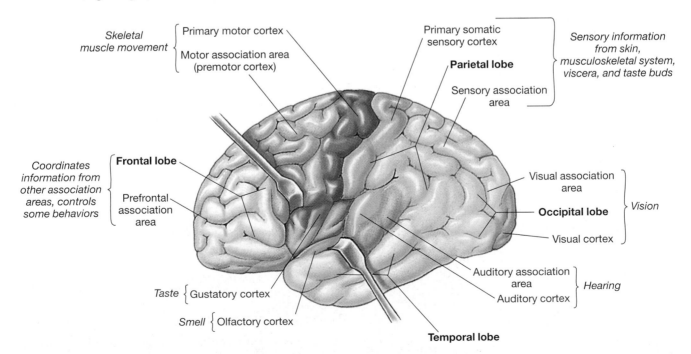

Skeletal muscle movement { Primary motor cortex
Motor association area (premotor cortex)

Primary somatic sensory cortex
Parietal lobe
Sensory association area

Sensory information from skin, musculoskeletal system, viscera, and taste buds

Coordinates information from other association areas, controls some behaviors { **Frontal lobe**
Prefrontal association area

Visual association area
Occipital lobe   } Vision
Visual cortex

Taste { Gustatory cortex

Smell { Olfactory cortex

Auditory association area
Auditory cortex   } Hearing

Temporal lobe

■ **Figure 9-10   Functional areas of the cerebral cortex**   The cerebral cortex is specialized into sensory areas for perception, motor areas that direct movement, and association areas that integrate information.

The **motor areas,** or **primary motor cortices,** in the frontal lobes process information about skeletal muscle movement and play a key role in voluntary movements. If part of one motor area is damaged by a stroke, paralysis on the opposite side of the body can result. The left side of the brain controls movement on the right side of the body and vice versa. Descending pathways from one motor cortex cross to the opposite side of the body as they pass through the medulla.

Association areas integrate sensory information such as somatic, visual, and auditory stimuli into perception. **Perception** is the brain's interpretation of sensory stimuli, and the perceived stimulus may be very different from the actual stimulus. For instance, the photoreceptors of the eye receive light waves of different frequencies, but we perceive the different wave energies as different colors. Similarly, the brain translates pressure waves hitting the ear into sound or interprets chemicals binding to chemoreceptors as taste or smell.

One interesting aspect of perception is the way that our brain fills in missing information to create a complete picture or translates a two-dimensional drawing into a three-dimensional shape (Fig. 9-11 ■). Our perceptual translation of sensory stimuli allows the information to be acted upon and used in voluntary motor control or complex cognitive functions such as language.

The distribution of different areas of functional specialization across the cerebral cortex is not symmetrical: each lobe has developed special functions not shared by the other lobe. This **cerebral lateralization** of function is sometimes referred to as cerebral dominance, more popularly known as left brain–right brain dominance. Language and verbal skills tend to be concentrated on the left side of the brain, the dominant hemisphere for right-handed people. Spatial skills are concentrated on the right side (Fig. 9-12 ■).

Neural connections in the cerebrum, like those in other parts of the nervous system, exhibit a certain degree of plasticity, an ability to change neuronal connections on the basis of experience. For example, if a person loses a finger, the regions of motor and sensory cortex previously devoted to control of the finger do not go dormant. Instead, adjacent regions of the cortex extend their functional fields and take over the parts of the cor-tex that are no longer used by the absent finger. Similarly, skills normally associated with one side of the cerebral cortex can be developed in the other hemisphere, as when a right-handed person with a broken hand learns to write with the left hand.

✔ Name the anatomical location in the brain where neurons from one side of the body cross to the opposite side.

✔ Name the divisions of the brain in anatomical order, starting from the spinal cord.

❱ BRAIN FUNCTION

The brain is an information processor, much like a computer. It receives informational input from the internal and external environments, integrates and processes the information, and, if appropriate, creates a response. What is amazing is that the brain is capable of generating information and output signals *in the absence of external input.* Another interesting aspect of higher brain function is the ability of this internally generated information to significantly affect tangible functions such as motor output or immune function. For example, you may be familiar with visual imaging, a technique sometimes used as an adjunct to traditional chemotherapy treatments for cancer. The patient is instructed to imagine the immune cells of the body attacking and destroying the malignant cells. According to some studies, the efficiency of the immune cells is improved by visual imaging. This field is currently receiving a lot of attention as scientists seek the physical links between the brain and the body that allow an internally derived mental image to enhance the effectiveness of the immune system.

For many years, studies of brain function were restricted to anatomical descriptions. As we developed techniques for monitoring electrical activity in neurons, research turned to physiological function. Studies on learning and memory can be performed using a variety of animal models, ranging from the mollusk *Aplysia* to rodents and primates. Brain functions dealing with perception are the most difficult to study because they require communication between the subject and the investigator.

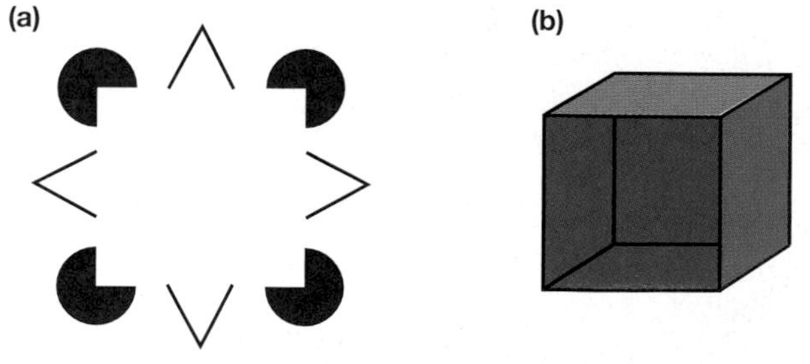

■ **Figure 9-11   Perception**   The brain has the ability to interpret sensory information to create the perception of shapes (a) or a three-dimensional object (b).

...continued from page 261

Dorothy took aspirin for about three months. Then one day she collapsed at home, the left side of her body paralyzed. At the hospital, tests showed that she had a blood clot causing a stroke. To further test the extent of damage caused by the ischemia, the doctor drew a tree and asked her to copy it. Dorothy drew a cluster of lines that did not even resemble a tree.

Question 3:   *Based on her symptoms, do you think Dorothy's stroke occurred in the right or left hemisphere?*

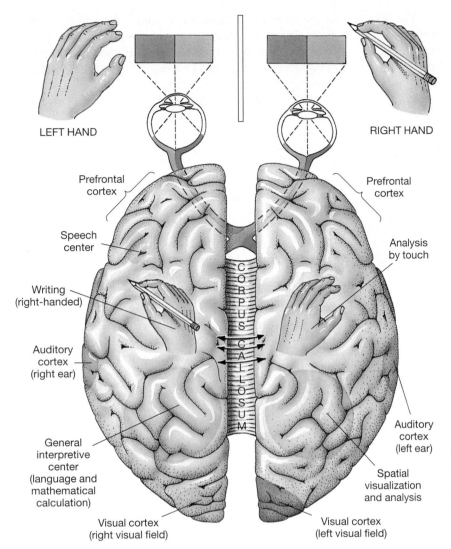

LEFT HAND

RIGHT HAND

Prefrontal cortex

Prefrontal cortex

Speech center

Analysis by touch

Writing (right-handed)

Auditory cortex (right ear)

CORPUS CALLOSUM

Auditory cortex (left ear)

General interpretive center (language and mathematical calculation)

Spatial visualization and analysis

Visual cortex (right visual field)

Visual cortex (left visual field)

■ **Figure 9-12** **Cerebral lateralization** The functional areas in the two cerebral hemispheres are not symmetrical. Language and verbal skills are concentrated on the left side of the brain, and spatial skills are concentrated on the right. Information is exchanged between the two hemispheres through the fibers of the corpus callosum.

Much of what we know today about human perception comes from the study of patients with inherited neurological defects, wounds from accidents or war, or surgical lesions that have been made to treat some medical condition such as uncontrollable epilepsy. These studies are enhanced by the fact that we can now see the human brain at work using noninvasive techniques such as positive emission tomography (PET) scans and magnetic resonance imaging (MRI) (Fig. 9-13 ■).

As a result of modern methods in neurobiology, we have discovered that some mental illnesses once thought to be untreatable have physiological origins. In many instances, they appear to result from abnormalities of neurotransmitter release or reception in different brain regions. For example, depression, bipolar disorder (manic-depressive disease), and schizophrenia are now treated as chemical imbalances in the brain. The search for a biological basis to disturbances of brain function has led to a whole battery of new drugs that work with varying degrees of effectiveness.

Use of some of these drugs raises moral and ethical questions, however, because they may alter personality in the process of treating disease. One example of such

a drug is the antidepressant drug Prozac®, a drug that prevents the normal reuptake of the neurotransmitter serotonin by presynaptic neurons. As a result of uptake inhibition, serotonin lingers in the synaptic cleft longer than usual, so serotonin-dependent activity within the

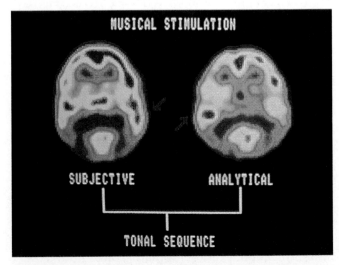

MUSICAL STIMULATION

SUBJECTIVE

ANALYTICAL

TONAL SEQUENCE

■ **Figure 9-13** **PET scan of the brain at work**

§ **Depression** **Depression** and other mood disorders are estimated to be the fourth leading cause of illness in the world today. Once thought to be a purely psychological problem, depression is now known to be a CNS dysfunction whose causes are not well understood. It is characterized by sleep and appetite disturbances and alterations of mood and libido. The drug therapy for depression has changed in recent years, but the major categories of antidepressant drugs alter some aspect of synaptic transmission. The older tricyclic antidepressants, such as amitriptyline, block reuptake of norepinephrine into the presynaptic neuron, thus extending the active life of the neurotransmitter. The newer selective serotonin reuptake inhibitors, or SSRIs, prevent serotonin (and possibly also norepinephrine) removal from the synapse. Interestingly, patients need to take these drugs for several weeks before they experience their full effect. This suggests that the changes taking place in the brain are long-term modulation of pathways rather than simply enhanced fast synaptic responses.

central nervous system increases. Although Prozac may effectively alleviate depression, some patients stop taking it because they feel that it alters their personality so that they are no longer themselves.

Other research into brain function has become quite controversial, particularly that dealing with sexuality and the degree to which behavior in general is genetically determined in humans. We will not delve deeply into any of these subjects because they are complex and would require lengthy explanations to do them justice. Instead, we will look briefly at some of the recent models proposed to explain the mechanisms that are the basis for higher brain functions.

Neurotransmitters and Neuromodulators Influence Communication in the Central Nervous System

In the last chapter, you learned how neurons communicate with each other at synapses [∞ p. 234]. In a neural reflex, the central nervous system acts as the integrating center, receiving sensory information from various receptors, processing it through networks of interacting neurons, and sending commands to effectors. The meaning of signals in the nervous system depends on two factors: (1) where the signals originate and terminate and (2) the physical and chemical connections between neurons in the pathway. Chemical communication has two components: the neurotransmitters and neuromodulators released by the presynaptic cells and the target cell receptors for these molecules. Various combinations of receptor and ligand create a large number of possible communication signals.

Neurotransmitters are responsible for fast synaptic communication. Neuromodulators act more slowly, often by using second messenger systems [∞ p. 158]. A

single chemical can be both a neurotransmitter and a modulator. For example, acetylcholine (ACh) acts as a neurotransmitter in the peripheral divisions of the nervous system but as a modulator or neurotransmitter in the central nervous system.

The neurotransmitters and neuromodulators in the central nervous system include amino acids, ACh, an assortment of peptides, and amines (Table 9-4). Two of the amino acids, gamma-aminobutyric acid (GABA) and glycine, are the most common inhibitory neurotransmitters of the central nervous system. They act by opening Cl^- channels, hyperpolarizing their postsynaptic targets, and making them less likely to fire an action potential. GABA acts primarily in the brain; glycine acts chiefly in the spinal cord. The amino acid glutamate is the primary excitatory neurotransmitter in the central nervous system; it depolarizes cells by opening channels to allow sodium ions to enter.

Many of the central nervous system neuromodulators are released from small but organized collections of neurons known as the **diffuse modulatory systems.** These neurons are clustered mostly in the brain stem and project their axons so that they affect large areas of the brain (Table 9-5). The diffuse modulatory systems influence such things as attention, motivation, wakefulness, memory, motor control, mood, and metabolic homeostasis.

In medicine, the pharmacological use of neuromodulators to treat various disorders is a hot topic. Neuromodulatory drugs that enhance serotonin activity are being used to treat depression and other mood disorders. The dopamine precursor L-DOPA is administered to alleviate the symptoms of Parkinson's disease.

The Reticular Formation Influences States of Arousal

One aspect of brain function that is difficult to define is consciousness: a state of arousal or awareness of self and environment. What distinguishes the awake state from

TABLE 9-4 Neurotransmitters and Neuromodulators in the CNS

Acetylcholine (ACh)
Amino acids
Gamma-aminobutyric acid (GABA)
Glycine
Glutamate
Amines
Dopamine
Norepinephrine
Epinephrine
Serotonin
Histamine

TABLE 9-5   The Diffuse Modulatory Systems

System (Neuromodulator)	Neurons Originate	Neurons Innervate	Functions Modulated by the System
Noradrenergic (norepinephrine)	Locus coeruleus of the pons	Cerebral cortex, thalamus, hypothalamus, olfactory bulb, cerebellum, midbrain, spinal cord	Attention, arousal, sleep-wake cycles, learning, memory, anxiety, pain, and mood
Serotonergic (serotonin)	Raphe nuclei along brain stem midline	Lower nuclei project to spinal cord	Pain; locomotion
		Upper nuclei project to most of brain	Sleep-wake cycle, mood and emotional behaviors such as aggression and depression
Dopaminergic (dopamine)	Substantia nigra in midbrain	Cortex	Motor control
	Ventral tegumentum in midbrain	Cortex and parts of limbic system	"Reward" centers linked to addictive behaviors
Cholinergic (acetylcholine)	Base of cerebrum; pons and midbrain	Cerebrum, hippocampus, thalamus	Sleep-wake cycles, arousal, learning, memory, sensory information passing through thalamus

various stages of sleep, sleep from coma, and coma from death? One way to define arousal states is by the pattern of electrical activity created by the cortical neurons. In awake states, many neurons are firing but not in a coordinated fashion. In the deepest stages of sleep, waves of synchronous depolarization sweep across the cerebral cortex. The desynchronization of electrical activity in waking states is produced by ascending signals from the reticular formation, the diffuse network of neurons that run through the brain stem. If connections between the reticular formation and the cortex are disrupted surgically, the animal becomes comatose for long periods of time. Other evidence for the importance of the reticular formation in states of arousal comes from studies showing that general anesthetics depress synaptic transmission in that region of the brain. Presumably, blocking the ascending pathways between the reticular formation and the cortex creates a state of unconsciousness.

The measurement of brain activity is recorded by a procedure known as **electroencephalography.** Surface electrodes placed on the scalp pick up depolarizations of the cortical neurons in the region just under the electrode. In the awake, aroused state, an electroencephalogram, or **EEG,** shows a rapid, irregular pattern with no dominant waves. In awake-resting (eyes closed) states, sleep, or coma, electrical activity of the neurons synchronizes into series of waves with characteristic patterns. The more synchronous the firing of cortical neurons, the larger the amplitude of the waves. As the person's state of arousal lessens, the frequency of the waves slows down. Thus, deep sleep is marked by large-amplitude, slow-frequency waves known as **delta waves;** the awake-resting state is characterized by small waves with high frequency

known as **alpha waves.** The significance of EEG wave patterns and why we need to sleep is one of the unsolved mysteries in neurophysiology.

Sleep   All organisms, even plants, have alternating daily patterns of rest and activity. When an organism is placed in conditions of constant light or dark, these activity rhythms persist, apparently cued by an internal clock. In mammals, the clock appears to be located in the suprachiasmatic nucleus of the hypothalamus, where it receives sensory information about light cycles from the eyes. Sleep-wake rhythms, like other biological cycles, generally follow a 24-hour light-dark cycle like our day and are known as **circadian rhythms** [*circa*, about + *dies*, a day].

In humans, our major rest period is marked by a behavior known as sleep. **Sleep** is an easily reversible state of inactivity, unlike coma. It is characterized by lack of interaction with the external environment. During some periods of sleep, sensory input to and motor output from the brain are interrupted.

Until the 1960s, sleep was thought to be a passive state that resulted from withdrawal of stimuli to the brain. Then, experiments showed that neuronal activity in ascending tracts from the brain stem to the cortex was *required* for sleep. As a result, we now consider sleep to be an active state. Each stage of sleep is marked by identifiable, predictable events associated with characteristic EEG patterns and somatic changes (Fig. 9-14 ■).

Sleep is divided into two major phases known as **REM (rapid eye movement) sleep** and **slow-wave sleep** (nonREM or deep sleep). During a sleep cycle, a person alternates between these two phases, spending variable periods of time in each phase, with transition stages in

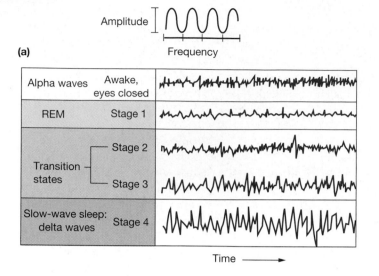

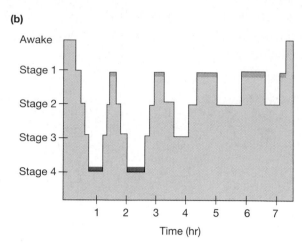

■ **Figure 9-14   Electroencephalograms (EEGs) and the sleep cycle**   (a) Recordings of electrical activity in the brain during waking and sleep periods show characteristic patterns. Neuronal depolarizations in the awake state are unsynchronized and appear as low-amplitude, high-frequency (many waves per second) waves. (b) During sleep periods, a person cycles between slow-wave sleep (dark blue), characterized by high-amplitude, low-frequency waves, and REM sleep (light green), characterized by low-amplitude, high-frequency waves. The transition states between slow-wave sleep and REM sleep (purple) show waves with intermediate characteristics.

between. Slow-wave sleep is marked on the EEG by the presence of delta waves, low-frequency waves of long duration. During this phase of the sleep cycle, sleepers adjust body position without conscious commands from the brain to do so.

In contrast, REM sleep is marked by an EEG pattern that resembles that of an awake person, with rapid, low-amplitude waves. During REM sleep, brain activity inhibits motor neurons to skeletal muscles, thus paralyzing them. Exceptions to this pattern are the muscles that move the eyes and those that control breathing. The control of homeostatic functions is depressed during REM sleep, and body temperature falls toward ambient temperature. REM sleep is the period of time during which most dreaming takes place. The eyes move behind closed lids, as if following the action of the dream. Sleepers are most likely to wake up spontaneously from periods of REM sleep.

If sleep is a neurologically active process, what is it that makes us sleepy? The possibility of a sleep-inducing factor was first proposed back in 1913, when scientists found that cerebrospinal fluid from sleep-deprived dogs could induce sleep in normal animals. Since then, a variety of sleep-inducing factors have been identified. Curiously, many of them are also substances that enhance the immune response, such as interleukin-1, interferon, serotonin, and tumor necrosis factor. As a result of this finding, some investigators have suggested that one answer to the puzzle of the biological reason for sleep is that we need to sleep to enhance our immune response. Whether or not that is a reason for why we sleep, the link between the immune system and sleep induction may help explain why we tend to sleep more when we are sick.

Sleep disorders are relatively common, as you can tell by looking at the variety of sleep-promoting agents available over the counter in drugstores. Among the more common sleep disorders are *insomnia*, the inability to go to sleep or remain asleep long enough to awake refreshed, and *sleep apnea* [*apnoos*, breathless], a condition in which sleepers awake when their airway muscles relax to the point of obstructing normal breathing.

The Hypothalamus Is the Primary Integrating Center for Many Homeostatic Reflexes

Although the hypothalamus occupies less than 1% of the total brain volume, it is the center for homeostasis. Output signals from this region of the brain influence

Sleepwalking   Sleepwalking, or somnambulism [*somnus*, sleep + *ambulare*, to walk], is a sleep behavior disorder that for many years was thought to represent the acting out of dreams. However, most dreaming occurs during REM sleep, whereas sleepwalking takes place during deep sleep. During sleepwalking episodes, which may last from 30 seconds to 30 minutes, the subject's eyes are open and registering the surroundings. The subject is able to avoid bumping into objects, can negotiate stairs, and in some cases is reported to perform tasks such as preparing food or folding clothes. The subject usually has little if any conscious recall of the sleepwalking episode upon awakening. Sleepwalking is most common in children, and the frequency of episodes declines with age. There is also a genetic component, as the tendency to sleepwalk runs in families.

endocrine and autonomic reflexes, as well as behaviors such as eating and drinking. In coordinating homeostasis, the hypothalamus receives sensory input from central and peripheral receptors. In addition, the limbic system provides the hypothalamus with emotional input that may affect homeostatic responses.

Probably the best-known hypothalamic pathway is the *fight-or-flight response,* a generalized reaction that occurs with fear or anger. When something occurs to threaten the well-being of the body, the hypothalamus takes over, preparing the body to fight or flee. All of the physical manifestations of fear, such as pounding heart, sweaty palms, and increased blood pressure, can be traced to increased sympathetic output directed by the hypothalamus. Sympathetic innervation of various organs is reinforced by sympathetically induced release of epinephrine (adrenaline) from the adrenal medulla.

In addition to regulation of the fight-or-flight response, the hypothalamus contains centers for temperature regulation, eating, and control of body osmolarity. The responses to stimulation of these centers may be neural or hormonal reflexes or a behavioral response. Stress, reproduction, and growth are also mediated by the hypothalamus by way of multiple hormones. You will encounter these homeostatic reflexes in later chapters as we discuss the various systems of the body.

Emotion and Motivation Are Complex Neural Pathways

Emotion and motivation are two aspects of brain function that are linked together through the hypothalamus, limbic system, and cerebral cortex. The pathways are complex and form closed circuits that cycle information between various parts of the brain.

Emotions are difficult to define. We know what they are and can name them, but in many ways they defy description. One characteristic of emotions is that they cannot be voluntarily turned on or off. The most commonly described emotions that arise in different parts of the brain are anger, aggression, sexual feelings, fear, pleasure, and contentment or happiness.

The limbic system, particularly the region known as the amygdala, is the center of emotion in the human brain. We learned about the role of this brain region through experiments in humans and animals. If the amygdala is artificially stimulated in humans, they report experiencing feelings of fear and anxiety. Experimental lesions that destroy the amygdala in animals cause the animals to become tamer and to display hypersexuality. As a result, neurobiologists feel that the amygdala is the center for basic instincts such as fear and for learned emotional states.

The pathways for emotions are complex (Fig. 9-15 ■). Sensory stimuli feeding into the cerebral cortex are constructed within the brain to create a representation (perception) of the world. After that information is integrated by the association areas, it is passed on to the limbic system. Feedback from the limbic system to the cortex creates awareness of the emotion, while descending pathways to the hypothalamus and brain stem initiate voluntary behaviors and unconscious responses mediated by autonomic, endocrine, immune, and somatic motor systems. The physical result of emotions can be as dramatic as the pounding heart of a fight-or-flight reaction or as insidious as the development of a gastric ulcer. The links between mind and body are difficult to study, and we may never completely understand them.

Motivation is defined as internal signals that shape voluntary behaviors. Some of these behaviors, such as eating, drinking, and sex, are related to survival; others, such as curiosity and sex (again), are linked to emotions. Motivational states are known as **drives** and generally have three properties in common: (1) they create an increased state of central nervous system arousal or alertness; (2) they create goal-oriented behavior; and (3) they are capable of coordinating disparate behaviors to achieve that goal.

Motivated behaviors often work in parallel with autonomic and endocrine responses in the body, as you might expect from behaviors originating in the hypothalamus. For example, if you eat salty popcorn, your body osmolarity increases. This stimulus acts on the thirst center of the hypothalamus, motivating you to seek something to drink. Increased osmolarity also acts on an endocrine center in the hypothalamus, releasing a hormone that increases water retention by the kidneys. Thus, one stimulus triggers both a motivated behavior and a homeostatic endocrine response.

Some motivation states can be activated by internal stimuli that may not even be obvious to the person in whom they are occurring. Many motivated behaviors stop when the person has reached a certain level of satisfaction, or **satiety.** Eating, curiosity, and sex drive are three examples of behaviors that have complex stimuli underlying their onset.

Pleasure is a motivational state that is being intensely studied because of its relationship to addictive behaviors

...continued from page 266

After running some additional tests, Dorothy's doctor administered an intravenous solution containing a drug called *tissue plasminogen activator* (t-PA). t-PA is a natural substance that the body produces to dissolve blood clots during the process of wound repair. The drug that Dorothy received is a genetically engineered version of t-PA. For maximum effect, it must be given within three hours of the onset of a stroke.

Question 4: *Explain the rationale for giving t-PA to people who have had a stroke. Why is it most effective when given shortly after the stroke occurs? Is there any type of stroke for which t-PA would not be appropriate? (Hint: What are the two types of strokes?)*

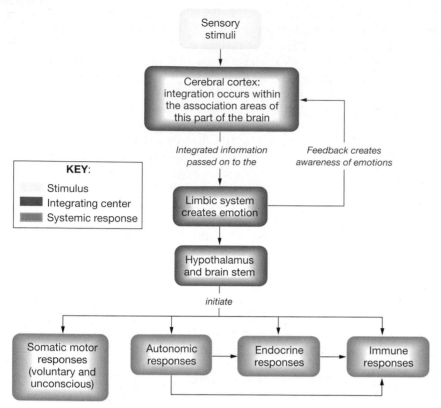

■ Figure 9-15 The link between emotions and physiological functions Sensory stimuli are perceived and integrated within the cerebral cortex, which passes the information on to the limbic system. The limbic system creates emotional reactions in response to the sensory information. It passes the emotions back to the cerebral cortex, where they are perceived, and on to the hypothalamus and brain stem. Physiological responses mediated through those parts of the brain can involve multiple systems, including skeletal muscle activity (somatic motor division), endocrine and autonomic responses, and even immune responses. The well-known link between stress and increased susceptibility to viruses is an example of an emotionally linked immune response.

such as drug use. Animal studies have shown that pleasure is a physiological state that is accompanied by increased activity of the neurotransmitter dopamine in certain parts of the brain. Drugs that are addictive, such as cocaine and possibly nicotine, act by enhancing the effectiveness of dopamine, thus increasing the pleasurable sensations perceived by the brain. As a result, use of these drugs rapidly becomes a learned behavior. Interestingly, not all behaviors that are addictive are pleasurable. For example, there are a variety of compulsive behaviors that involve self-mutilation, such as pulling out hair by the roots. Fortunately, many behaviors that are learned can also be unlearned, given motivation.

Learning and Memory Change Synaptic Connections in the Brain

For many years, motivation, learning, and memory were considered to be in the realm of psychology rather than biology. Neurobiologists in decades past were more concerned with the cellular aspects of neuronal function than with the integrated action of the central nervous system. However, in recent years, the two fields have overlapped more and more. Scientists have discovered that the underlying basis for cognitive function seems to be explainable in terms of cellular events, such as long-term potentiation [∞ p. 243]. One of the newest ideas is the concept of *plasticity:* that neuronal connections are not fixed and have the ability to change with experience. Two aspects of cognitive ability that probably involve plasticity are learning and memory.

Learning **Learning** is the acquisition of knowledge about the world around us. Learning is demonstrated by behavioral changes, but behavioral changes are not required for learning to occur. Learning can be internalized and is not always reflected by overt behavior.

Learning can be classified into two broad types: associative and nonassociative. **Associative learning** occurs when two stimuli are associated with each other, such as Pavlov's classic experiment in which he simultaneously presented dogs with food and rang a bell. After a period of time, the dogs came to associate the sound of a bell with food and began to salivate in anticipation of food whenever the bell was rung. Another form of associative learning occurs when an animal associates a stimulus with a given behavior. An example would be a mouse that gets a shock each time it touches a certain part of its cage. It soon associates that part of the cage with an unpleasant experience and avoids that area.

Nonassociative learning includes imitative behaviors such as learning a language. This type of learning includes habituation and sensitization, two adaptive behaviors that allow us to filter out and ignore background stimuli while responding more sensitively to potentially disruptive stimuli. In **habituation,** an animal shows a decreased response to an irrelevant stimulus that is repeated over and over. For example, a sudden loud noise may startle you, but if the noise is repeated over and over again, your brain begins to ignore it. Habituated responses allow us to filter out stimuli that we have evaluated and found to be insignificant.

Sensitization is the opposite of habituation and helps to increase an organism's chances for survival. In sensitization learning, exposure to a noxious or intense stimulus causes an enhanced response upon subsequent exposure. For example, people who become ill while eating certain foods may find that they lose their desire to eat that food again. Sensitization is adaptive since it helps us to avoid potentially harmful stimuli.

Memory   In humans, the hippocampus seems to be an important structure in both learning and memory. For example, patients who had part of the hippocampus destroyed to relieve a certain type of epilepsy also had trouble remembering new information to which they were exposed. When given a list of words to repeat, they could remember the words as long as their attention stayed focused on the task. But if they were distracted, the memory of the words disappeared and they had to learn the list again. Information stored in long-term memory before the operation was not affected. This inability to remember newly acquired information is a defect known as **anterograde amnesia** [*amnesia,* oblivion].

Memory has multiple levels of storage, and our memory bank is constantly changing. When a stimulus comes into the central nervous system, it first goes into **short-term memory,** a limited store of information that can hold only about 7 to 12 pieces of information at a time. Items in short-term memory will disappear unless an effort, such as repetition, is made to put them into a more permanent form (Fig. 9-16 ■).

Working memory is a special form of short-term memory. One region of the cerebral cortex is devoted to keeping track of bits of information long enough to put them to use in a task that takes place after the information has been acquired. Working memory in these regions is linked to long-term memory stores, so that newly acquired information can be integrated with stored information and acted upon.

For example, you are trying to cross a busy road. You look to the right and see that there are no cars coming for several blocks. You then look left and see that there are no cars coming from that direction either. Working memory has stored the information that the road to the right is clear, so using stored knowledge about safe-

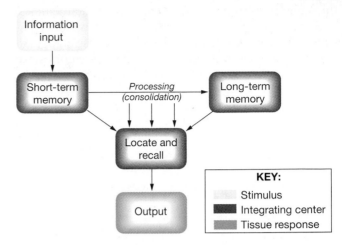

■ **Figure 9-16   Memory processing**   New information goes first into short-term memory, from which it can be recalled and acted upon. The amount of information that can be stored in short-term memory is limited, however, and information will be lost unless it is processed through consolidation and stored in long-term memory.

ty, you are able to conclude that there is no traffic from either direction and it is safe to cross the road.

In people with damage to the prefrontal lobes of the brain, this task becomes more difficult because they are unable to recall whether the road is clear from the right once they have looked away to assess traffic coming from the left. Working memory allows us to collect a series of facts from short- and long-term memory and connect them together in logical order to solve problems or plan actions.

The processing of information that converts short-term memory into **long-term memory** is known as **consolidation.** Consolidation can take varying periods of time, from seconds to minutes. Information passes through many intermediate levels of memory during consolidation, and in each of these stages, information can be located and recalled.

Long-term memory has been divided into two types that are consolidated and stored using different neuronal pathways (Table 9-6). **Reflexive (implicit) memory** is automatic and does not require conscious processes for its creation or recall. Information stored in this kind of memory is acquired slowly through repetition. Motor skills fall into the category of reflexive memory, as do procedures and rules. For example, you do not need to

TABLE 9-6   Types of Memory

Reflexive (Implicit) Memory	*Declarative (Explicit) Memory*
Recall is automatic and does not require conscious attention	Recall requires conscious attention
Acquired slowly through repetition	Depends on higher-level thinking skills such as inference, comparison, and evaluation
Includes motor skills and rules and procedures	Memories can be reported verbally
Memories can be demonstrated	

think about putting a period at the end of each sentence or about how to pick up a fork. Reflexive memory has also been called procedural memory because it generally concerns how to do things. Reflexive memories can be acquired through either associative or nonassociative learning processes.

Declarative (explicit) memory, on the other hand, requires conscious attention for its recall. Its creation generally depends on the use of higher-level cognitive skills such as inference, comparison, and evaluation. Declarative memories deal with knowledge about ourselves and the world around us that can be reported or described verbally.

Sometimes information can be transferred from declarative memory to reflexive memory. The quarterback on a football team is a good example. When he learned to throw the football as a small boy, he had to pay close attention to gripping the ball and coordinating his muscles to throw the ball accurately. At that point of learning to throw the ball, the process was in declarative memory and required conscious effort as the boy analyzed his movements. With repetition, however, the mechanics of throwing the ball were transferred to reflexive memory: they became a reflex that could be executed without conscious thought. That transfer allowed the quarterback to use his conscious mind to analyze the path and timing of the pass while the mechanics of the pass became automatic. Athletes often refer to this automaticity of learned body movements as "muscle memory."

Memory is an individual thing. We process information on the basis of our experiences and perception of the world. Because people have widely different experiences throughout their lives, it follows that no two people will process a piece of information in the same way. If you ask a group of people about what happened during a particular event such as a lecture, a robbery, or an accident, no two descriptions will be identical. Different people will have processed the event according to their own perceptions and experiences. Experiential processing is important to remember when studying in a group situation, because it is unlikely that all group members will learn or recall information using the same pathways.

Memory processing for reflexive and declarative memory appears to take place through different pathways. With noninvasive imaging techniques such as MRI and PET scans, researchers have been able to track brain activity as individuals learned to perform tasks. Memories are stored throughout the cerebral cortex in pathways known as **memory traces.** Declarative memories involve the temporal lobes of the brain; reflexive memory storage requires the amygdala and the cerebellum. Some components of memories are stored in the sensory cortices where they are processed. For example, pictures are stored in the visual cortex and sounds in the auditory cortex.

Learning or recall of a single task may involve multiple circuits in the brain that work in parallel. This parallel processing provides backup in case one of the circuits is damaged. It is also believed to be the means by which specific memories are generalized, allowing new information to be processed and matched to stored information. For example, a person who has never seen a volleyball will recognize it as a ball because the volleyball has the same general characteristics as all other balls the person has seen.

As scientists studied the consolidation of short-term memory into long-term memory, they discovered that the process involves changes in the synaptic connections of the circuits involved in learning. In some cases, new synapses form; in others, the effectiveness of synaptic transmission is altered. These changes are evidence of plasticity and show us that the brain is not "hard-wired," as we once had thought.

Language Is the Most Elaborate Cognitive Behavior

One of the hallmarks of an advanced nervous system is the ability of a species to exchange complex information with other members of the same species. Although found predominantly in birds and mammals, this ability also occurs in certain insects that convey amazingly detailed information by means of sound (crickets), touch and sight (bees), and odor (pheromones, also found in many other animals). In humans, communication takes place primarily through spoken and written language. Because language is considered the most elaborate cognitive behavior, it has received considerable attention from neurobiologists.

Language skills require the input of sensory information (primarily from hearing and vision), processing in various centers in the cerebral cortex, and the coordination of motor output for vocalization and writing. In most people the centers for language ability are found in the left hemisphere of the cerebrum. Even 70% of people who are left-handed (right-brain dominant) or ambidextrous use their left brain for speech. The ability to communicate through speech has been divided into two processes: the combination of different sounds to form words (vocalization) and the combination of words into grammatically correct and meaningful sentences.

The integration of spoken language in the human brain involves two regions in the cortex: **Wernicke's area** in the temporal lobe and **Broca's area** in the frontal lobe close to the motor cortex (Fig. 9-17 ■). Most of what we know about these areas comes from studies of people with brain lesions because nonhuman animals are not capable of speech. Even primates that communicate on the level of a small child through sign language and other visual means do not have the physical ability to vocalize the sounds of human language.

Input into the language areas comes from either the visual cortex (reading) or the auditory cortex (listening). Sensory input goes first to Wernicke's area, then to Broca's

(a) Speaking a written word

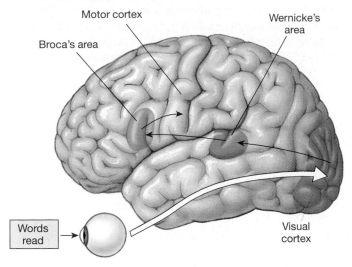

Broca's area

Motor cortex

Wernicke's area

Words read

Visual cortex

(b) Speaking a heard word

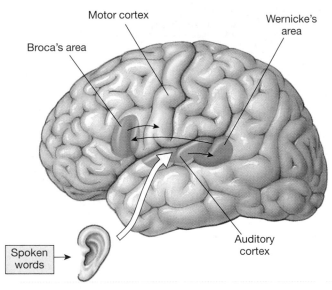

Broca's area

Motor cortex

Wernicke's area

Spoken words

Auditory cortex

■ **Figure 9-17   Cerebral processing of spoken and visual language**   Spoken and written language are processed through their respective sensory areas. The information is passed on to Wernicke's area, where it is interpreted, then to Broca's area for the coordination of speech and writing. People with damage to Wernicke's area do not understand spoken or written communication; those with damage to Broca's area understand, but are unable to respond appropriately.

area. After integration and processing, output to the motor cortex initiates a spoken or written response. If damage occurs to Wernicke's area, a person is unable to understand any spoken or visual information. The person's own speech, as a result, is nonsense, since the person is unaware of his or her own errors. This condition is known as **receptive aphasia** because the person is unable to understand sensory input [*a-*, not + *phatos*, spoken].

On the other hand, people with damage to Broca's area understand spoken and written language but are unable to express their response in normal syntax. Their response to a question often consists of appropriate words strung together in random order. These patients may have a difficult time dealing with their disability because they are aware of their mistakes but are powerless to correct them. Damage to Broca's area causes an **expressive aphasia.**

Mechanical forms of aphasia occur as a result of damage to the motor cortex. These patients find themselves unable to physically shape the sounds that make up words or unable to coordinate the muscles of their arm and hand to write.

Personality and Individuality Are a Combination of Experience and Inheritance

One of the most difficult aspects of brain function to translate from the abstract realm of philosophy into the physical circuits of neurobiology is the combination of attributes that we call personality. What is it that makes us individuals? The parents of more than one child will tell you that their offspring were different from birth, and even before that in the womb. If we all have the same brain structure, what makes us different?

This question is one that fascinates many people. The answer that is evolving as the result of neurobiology research is that we are a combination of our experiences and the genetic constraints that we inherit. What we learn or experience and what we store in memory create a unique pattern of neuronal connections in our brains. Sometimes these circuits malfunction, creating depression, schizophrenia, or any number of personality disturbances. Psychiatrists for many years attempted to treat these disorders as if they were due solely to events in the person's life, but now we know that there is a genetic component to many of these disorders.

One of the biggest advances in treating altered mental states was the discovery of diffuse modulatory pathways and the neuromodulatory action of serotonin and dopamine. The development of drugs that enhance the effectiveness of information exchange at serotonergic and dopaminergic synapses has created a major breakthrough in the treatment of depression and other conditions.

On the other hand, we still have much to learn about repairing damage to the central nervous system. One of the biggest tragedies in life is the personality change that sometimes accompanies head trauma. Physical damage to the delicate circuits of the brain can not only alter intelligence and memory storage but also can create a new personality. The person who exists after the injury may not be the same person who inhabited that body before the injury. Although the change may not be noticeable to the injured person, it can be devastating to the family and friends of the victim. Perhaps as we learn more about how neurons link to each other, we will be able to find a means of restoring damaged networks and preventing the lasting effects of head trauma.

One of the newest breakthroughs in this area is the discovery of neural *stem cells* [∞ p. 67], immature cells that can differentiate into neurons. Mature neurons are so highly specialized in their structure that they are unable to undergo mitosis and reproduce themselves. We used to believe that once a neuron died, it could never be replaced. However, scientists have now discovered neural stem cells in the brain. It remains to be seen what role these cells play in normal brain function and whether we can learn to use them to replace damaged cells.

PROBLEM CONCLUSION

Dorothy's paralysis slowly resolved and she spent the next three months in intensive speech and physical therapy. She was left with some residual weakness in her skeletal muscles, but her cognitive function returned to normal.

In this running problem, you have learned how a stroke that disrupts blood flow to the brain can lead to loss of body function. You have also learned about the risk factors and current treatments for stroke.

Further check your understanding of this running problem by comparing your answers against those in the summary table.

	Question	Facts	Integration and Analysis
1	Do you think Dorothy's brain cells died as a result of her TIA?	Without oxygen, brain cells lose function or die.	Dorothy regained function within a short period of time, indicating that her brain cells were not permanently damaged or killed.
2	Explain the rationale for taking aspirin to prevent stroke.	Aspirin helps prevent platelets from sticking together and forming blood clots.	Strokes may be caused by a blood clot that blocks a blood vessel. Therefore, a drug that prevents blood clot formation may help prevent a stroke.
3	Did Dorothy's stroke occur in the right or left hemisphere?	Dorothy lost function on the left side of her body and lost the ability to draw a tree. Motor functions on the left side of the body are controlled by the right motor cortex of the brain. Spatial visualization is controlled by the right side of the brain.	Dorothy's loss of function is consistent with damage to her right cerebral cortex.
4a	Explain the rationale for giving t-PA to people who have had a stroke.	t-PA is a natural substance produced by the body that dissolves blood clots.	If a stroke is caused by a blood clot, giving t-PA may help remove the blood clot and restore blood flow to the affected area of the brain
4b	Why is t-PA most effective when given shortly after the stroke occurs?	Brain cells have a high rate of oxygen consumption and die rapidly when deprived of oxygen.	Rapid administration of t-PA may help restore blood flow to ischemic areas of brain tissue and help minimize the damage caused by the stroke.
4c	Is there any type of stroke for which t-PA would not be appropriate?	Strokes are caused by rupture or blockage in a blood vessel. If a blood vessel ruptures, it causes bleeding. The body's normal response to bleeding is clot formation.	In strokes caused by blood vessel rupture, bleeding can damage brain tissue due to pressure buildup within the rigid skull. t-PA would prevent the normal clot formation that would stop bleeding, so it could actually worsen a hemorrhagic stroke.

CHAPTER REVIEW

SUMMARY

Anatomy of the Central Nervous System

1. The brain and spinal cord are encased in the membranes of the **meninges** and the bones of the **cranium** and vertebrae. (p. 254)

2. The **choroid plexus** secretes **cerebrospinal fluid (CSF)** into the **ventricles** of the brain. Cerebrospinal fluid helps cushion the brain and spinal cord and also creates a chemically controlled environment. (p. 254)

3. Because the normal fuel source for neurons is glucose, the body closely regulates blood glucose concentrations. (p. 255)

4. Relatively impermeable capillaries in the brain create a **blood-brain barrier** that prevents possibly harmful substances in the blood from entering the interstitial fluid. Paracrines released by astrocytes induce formation of tight junctions between capillary endothelial cells. (p. 255)

5. The **gray matter** of the CNS consists of nerve cell bodies, dendrites, and axon terminals. The cell bodies form layers in parts of the brain or cluster into groups of neurons known as **nuclei.** (p. 259)

The Spinal Cord

6. **White matter** in the CNS is made up mostly of myelinated axons with very few cell bodies. In the spinal cord, **ascending tracts** of white matter carry sensory information to the brain, and **descending tracts** carry efferent signals from the brain. **Propriospinal tracts** remain within the cord. (p. 259)

7. The **dorsal root** of each spinal nerve carries incoming sensory information, and the **ventral root** carries information from the central nervous system to the muscles and glands. (p. 259)

8. The nerve cell bodies of sensory neurons are found in the **dorsal root ganglia.** (p. 259)

The Brain

9. The brain can be divided into three areas: the brain stem, the cerebellum, and the cerebrum. (p. 261)

10. The brain stem is divided into the medulla oblongata, the pons, and the midbrain (mesencephalon). (p. 262)

11. The **medulla oblongata** contains the tracts that convey information between the cerebrum and the spinal cord or cranial nerves. Many of these tracts cross the midline in a region known as the **pyramids.** The medulla contains control centers for many involuntary functions. (p. 262)

12. The primary function of the **pons** is to act as a relay station for information transfer between the cerebellum and cerebrum. (p. 262)

13. The primary function of the **midbrain** is the control of eye movement. It also relays signals for auditory and visual reflexes. (p. 262)

14. The **reticular formation** is a diffuse group of neurons that play a role in arousal, sleep, muscle tone, stretch reflexes, coordination of breathing, blood pressure regulation, and modulation of pain. (p. 262)

15. The **cerebellum** processes sensory information and coordinates the execution of movement. (p. 263)

16. The **diencephalon** is composed of the thalamus and hypothalamus. The small nuclei that make up the **thalamus** serve as an integration and relay station for sensory information on its way to the cerebral cortex. (p. 263)

17. The **hypothalamus** contains centers for behavioral drives and plays a key role in homeostasis. It controls many functions of the autonomic division of the nervous system as well as a variety of endocrine functions. (p. 263)

18. The **cerebrum** is composed of two hemispheres connected at the **corpus callosum.** Each cerebral hemisphere is divided into frontal, parietal, temporal, and occipital lobes. (p. 263)

19. The interior of the cerebrum contains the **basal ganglia,** involved in the control of movement, and the **limbic system,** which acts as the link between higher cognitive functions and more primitive emotional responses. The **amygdala** of the limbic system is linked to emotion and memory, and the **hippocampus** is involved with learning and memory. (p. 264)

20. Our higher brain functions such as reasoning arise within the **cerebral cortex,** the outer layer of the cerebrum whose neurons are arranged in anatomically distinct horizontal layers and functionally distinct vertical columns. (p. 264)

21. The cerebral cortex contains three functional specializations: sensory areas, motor areas, and association areas. (p. 265)

22. Sensory areas direct perception of sensations. The **primary somatic sensory cortex** receives sensory information from the skin, musculoskeletal system, viscera, and taste buds. The **visual cortex, auditory cortex, gustatory cortex,** and **olfactory cortex** receive information about vision, sound, taste, and odors, respectively. (p. 265)

23. **Motor areas** direct skeletal muscle movement. A **primary motor cortex** is located in each hemisphere of the cerebrum. (p. 266)

24. **Association areas** integrate sensory information into perception. **Perception** is the brain's interpretation of sensory stimuli. (p. 266)

25. Each hemisphere of the cerebrum has developed special functions not shared by the other hemisphere. This division of labor is known as **cerebral lateralization.** (p. 266)

Brain Function

26. Neural connections in the CNS exhibit **plasticity,** the ability to change neuronal connections on the basis of past experience. (p. 266)

27. A variety of neurotransmitters and neuromodulators in the brain create complex pathways for the transmission and storage of information. Gamma-aminobutyric acid (GABA) and glycine are the most common inhibitory neurotransmitters of the central nervous system. Glutamate is the primary excitatory neurotransmitter. (p. 268)

28. Many of the central nervous system neuromodulators are released from the **diffuse modulatory systems.** These neurons influence attention, motivation, wakefulness, memory, motor control, mood, and metabolic homeostasis. (p. 268)

29. Neural activity in the reticular formation is linked to states of arousal. Varying levels of arousal are marked by different patterns of electrical activity in the neurons. Brain activity is recorded by **electroencephalography.** (p. 269)

30. Sleep is an easily reversible state of inactivity with characteristic stages marked by identifiable, predictable events. The two major phases of sleep are **REM (rapid eye movement) sleep** and **slow-wave sleep** (non-REM sleep). (p. 269)

31. The physiological reason for sleep is unknown but may be linked to maintenance of the immune system. (p. 270)

32. The hypothalamus interacts with the limbic system in a way that causes emotional events to influence physiological functions in the body. (p. 270)

33. The limbic system is the center of emotion in the human brain. The pathways for emotions are complex and difficult to study. (p. 271)

34. **Motivation** is defined as internal signals that shape voluntary behaviors related to survival or emotions. Motivational **drives** share three properties: (1) an increased state of central nervous system arousal, (2) goal-oriented behavior, and (3) the ability to coordinate disparate behaviors. (p. 271)

35. **Learning** is the acquisition of knowledge about the world around us. **Associative learning** occurs when two stimuli are associated with each other. **Nonassociative learning** includes imitative behaviors such as learning a language. (p. 272)

36. In **habituation,** an animal shows a decreased response to a stimulus that is repeated over and over. In **sensitization,** exposure to a noxious or intense stimulus creates an enhanced response upon subsequent exposure. (p. 273)

37. Memory has multiple levels of storage and is constantly changing. Information is first stored in **short-term memory,** but will disappear unless consolidated into long-term memory. (p. 273)

38. **Long-term memory** includes **reflexive memory,** which does not require conscious processes for its creation or recall, and **declarative memory,** which uses higher-level cognitive skills for formation and requires conscious attention for its recall. (p. 273)

39. The **consolidation** of short-term memory into long-term memory appears to involve changes in the synaptic connections of the circuits involved in learning. (p. 273)

40. Language is considered the most elaborate cognitive behavior. The integration of spoken language in the human brain involves information processing in **Wernicke's area** and **Broca's area.** (p. 274)

QUESTIONS

LEVEL ONE   Reviewing Facts and Terms

1. The ability of human brains to change circuit connections and function in response to sensory input and past experience is known as _____.

2. _____ behaviors are related to feeling and emotion. _____ behaviors are related to thinking.

3. The part of the brain called the _____ is what makes us human, allowing human reasoning and cognition.

4. In vertebrates, the central nervous system is protected by bony cases. The brain is found inside the _____, and the spinal cord is found in a canal inside the _____.

5. Name the meninges, beginning with the layer next to the bones.

6. List and explain the purposes of cerebrospinal fluid (CSF).

7. Compare the concentrations of these substances in CSF and plasma.
   (a) HCO_3^- (b) Ca^{2+}
   (c) glucose   (d) H^+
   (e) Na^+ (f) K^+

8. The only fuel source for neurons under normal circumstances is _____. Low concentrations of this fuel in the blood is termed _____ and if not treated can lead to confusion, unconsciousness, and eventually death. To synthesize enough ATP to continually transport ions, the neurons also exhibit high rates of _____ consumption. To supply these needs, about _____% of the blood pumped by the heart goes to the brain.

9. Match each area with its function.

(a) medulla oblongata
(b) pons
(c) midbrain
(d) reticular formation
(e) cerebellum
(f) diencephalon
(g) thalamus
(h) hypothalamus
(i) cerebrum

1. processes sensory information and coordinates execution of movement
2. composed of the thalamus and hypothalamus
3. composed of two hemispheres; this part fills most of the cranial cavity
4. control centers for blood pressure, vomiting, breathing, and swallowing
5. transmits and integrates sensory information between lower central nervous system and cerebral cortex
6. transfers information to cerebellum, coordinates breathing centers
7. contains centers for behavioral drives such as hunger and thirst; plays a key role in homeostasis
8. relays signals and visual reflexes, eye movement
9. arousal and sleep, muscle tone, and stretch reflexes

10. Relatively impermeable capillaries form the _____, sheltering the brain from what?

11. How are gray matter and white matter different, both anatomically and functionally?

12. Name the cerebral cortex areas that direct perception, direct movement, and integrate information and direct voluntary behaviors.

13. What does cerebral lateralization refer to? What functions tend to be centered in each hemisphere?

14. Name the two most common inhibitory neurotransmitters of the CNS. Where do they act? Which ion channels do they open?

15. List and define the two major phases of sleep. How are they different from each other?

16. List several homeostatic reflexes and behaviors influenced by output from the hypothalamus. What is the source of emotional input into this area?

17. The _____ region of the limbic system is believed to be the center for basic instincts such as fear and learned emotional states.

18. What are the broad categories of learning? Define habituation and sensitization. What anatomical structure (of the cerebrum) is important in both learning and memory?

19. What two centers of the cortex are involved in integrating spoken language?

LEVEL TWO   Reviewing Concepts

20. Map the following terms. You may add additional terms if you wish.

arachnoid membrane
ascending tracts
blood-brain barrier
brain
capillaries
cell bodies
cerebrospinal fluid
cervical nerves
choroid plexus
cranial nerves
descending tracts
dorsal root
dorsal root ganglion
dura mater

ependyma
gray matter
lumbar nerves
meninges
nuclei
pia mater
propriospinal tracts
sacral nerves
spinal cord
thoracic nerves
ventral root
ventricles
vertebral column
white matter

21. Trace the pathway that the cerebrospinal fluid follows through the nervous system.

22. How do the capillaries of the brain regulate their "leakiness"?

23. How can a single molecule act as a neurotransmitter at one time and a neuromodulator at another?

24. Compare and contrast the following concepts:

(a) the diffuse modulatory systems, the reticular formation, and the limbic system
(b) the different forms of memory
(c) nuclei and ganglia
(d) tracts, nerves, horns, nerve fibers, and roots

25. Fill in the question marks in the chart below:

Cerebral Area	Lobe	Functions
Primary somatic sensory cortex	?	Receives sensory information from peripheral receptors
?	Occipital	Processes information from the eyes
Auditory cortex	Temporal	?
?	Temporal	Receives input from chemoreceptors in the nose
Motor cortices	?	?
Association areas	NA	?

26. Given the wave below, draw examples with (a) lower frequency, (b) higher amplitude, (c) higher frequency. (Hint: see Figure 9-14, p. 270)

27. Define motivational states. What properties do they have in common?

28. What changes occur at synapses as memories are formed?

29. Mr. Andersen, a stroke patient, experiences expressive aphasia. His savvy therapist, Cheryl, teaches him to sing to communicate his needs. What symptoms did he exhibit before therapy? How do you know he did not have receptive aphasia? Using what you have learned about cerebral lateralization, hypothesize why singing worked for him.

30. A study was done in which 40 typical citizens were taught about using seatbelts in their cars. At the end of the presentations, all participants scored at least 90% on a comprehensive test. However, the people were secretly videotaped entering and leaving the parking lot. Twenty subjects entered wearing their seatbelts, 22 left wearing them. Did learning occur? What is the relationship between learning and actually buckling the seatbelts?

31. In 1913, Henri Pieron took a group of dogs and kept them awake for several days. Before allowing them to sleep, he withdrew cerebrospinal fluid from the sleep-deprived animals. He then injected this CSF into normal, rested dogs. The recipient dogs promptly went to sleep for periods ranging from two to six hours. What conclusion can you draw about the possible source of a sleep-inducing factor? What controls should Pieron have run?

E X P L O R E < MediaLab >

Introduction

This chapter introduced you to the organization of the brain and spinal cord. Regions of brain tissue that look the same can function quite differently. Their function depends on the origins and types of signals received. These complex arrangements of neurons in brain tissue are responsible for basic functions like breathing and for very intricate functions like learning and memory. The following Web Explorations will help you gain an appreciation of brain organization, anatomy, and function. After reading the description below, visit the MediaLab for chapter 9 in your Companion Website and select the appropriate keyword.

Web Exploration 1

Estimated time for completion = 5 minutes
Upon first glance the brain seems like an amorphous mass of jelly. However, first perceptions can be very misleading! Much of what we have learned about the discrete functional areas of the brain has been the result of studying the neurological manifestations of patients with head injuries, strokes, or surgical lesions of the cerebral cortex. Several methods exist that help physicians and scientists study brain activity. The PET (positive emission tomography) scan and MRI (magnetic resonance imaging) are often used to visualize brain function.

From what you have learned in this chapter, what area of the brain would be damaged in a patient who could only speak in a nonsensical string of words, i.e., has fluent aphagia? It might help to sketch a diagram of the brain and briefly map out the functional areas.

Select the keyword **FLUENT APHAGIA** on the Website to visualize the scans of a brain from a patient who has suffered a stroke and has this condition. [**Note:** in each brain scan, you are looking at the underside of the brain with the frontal lobes facing the top of the page.] Was your prediction correct? Now let's try to see if you can predict what symptoms might be present in a different patient (select the keyword **BRAIN SCAN**). What rationale did you use to justify your predictions? To complete this exercise, visit MediaLab Web Exploration 1 in Chapter 9 of your Companion Website.

Web Exploration 2

Estimated time for completion = 10 minutes
For a long time the adult brain was viewed as a fixed organ, incapable of changing. However, we are now uncovering evidence that supports the idea of the brain as a dynamic and changing organ. This plasticity of the adult brain can best be illustrated by understanding long-term potentiation (LTP). In LTP, the response of a postsynaptic cell to a constant stimulus is prolonged and enhanced.

Select the keyword **LTP** on the Website to learn how experiments have expanded our understanding of memory, learning, and the brain's plasticity. What do scientists think is the role of the hippocampus in learning? What experiment provided the evidence for this theory?

Take a look at the tracings of LTP in a mutant mouse that does not have normal calcium-calmodulin-dependent kinase II (CaMKII). Compare them to those of a normal mouse. Which parameters differ between the two mice? The amount of current flow? The duration of current flow? Both? Can you visualize what might be happening at the postsynaptic cell? Refer to Figure 9-17 ■ in Chapter 9 of your textbook for help with this.

Select the keyword **HIPPOCAMPUS** on the Website to view a preserved brain slice and the location of this very important functional area. To complete this exercise, visit MediaLab Web Exploration 2 in Chapter 9 of your Companion Website.

10 Sensory Physiology

■ *"Nature does not communicate with man by sending encoded messages."*
—*Oscar Hechter, in* Biology and Medicine into the 21st Century, *1991.* ■

CHAPTER OUTLINE

Stereocilia on hair cells of the ear

BACKGROUND BASICS

Imagine floating in the dark in a buoyant tank of salt water: there is no sound, no light, no breeze. The air and water are the same temperature as your body. You are in a sensory deprivation tank, and the only sensations you are aware of come from your own body. Your limbs are weightless, your breath moves in and out, and you feel your heart beating. In the absence of external stimuli, you turn your awareness inward to hear what your body has to say.

In recent decades, sensory deprivation tanks were a popular way to counter the stress of a busy world. These facilities are hard to find now, but they illustrate the role of the afferent division of the nervous system: to provide us with information about the environment outside and inside our bodies. Sometimes we perceive sensory signals when they reach a level of conscious awareness, but other times they are processed completely at the subconscious level (Table 10-1). Stimuli that your central nervous system (CNS) monitors without your conscious awareness include changes in muscle stretch and tension, as well as a variety of internal parameters that the body monitors to maintain homeostasis. The responses to these stimuli comprise many of the subconscious reflexes of the body.

In this chapter we are concerned primarily with the sensory stimuli that reach the conscious level of perception. We divide these stimuli into the **special senses** of vision, hearing, taste, smell, and equilibrium and the **somatic senses** of touch-pressure, temperature, pain, and proprioception. **Proprioception** is mediated by sensory receptors in the muscles and joints of the body that make us aware of our body position in space and of the relative location of various body parts to each other. If you close your eyes and raise your arm above your head, you are aware of its position in space because of the activation of various proprioceptors in the muscles and joints.

Let's first consider the general properties shared by the somatic and special senses. We will then look at the unique receptors and pathways that distinguish the different sensory systems from each other.

▶ GENERAL PROPERTIES OF SENSORY SYSTEMS

All sensory pathways have certain elements in common. They begin with a stimulus, internal or external, that acts on a sensory receptor. The receptor is a transducer that converts the stimulus into electrical graded potentials. If the graded potentials are above threshold, action potentials pass from the receptor along an afferent sensory neuron to the CNS, where the signals are integrated. Some stimuli pass upward to the cerebral cortex, where they reach conscious perception, but others are acted upon without our awareness. At each synapse along the

TABLE 10-1   Information Processing by the Sensory Division of the Nervous System

Conscious	Subconscious
Special senses	**Somatic stimuli**
Vision	Muscle length and tension
Hearing	**Visceral stimuli**
Taste	
Smell	Blood pressure
Equilibrium	pH and oxygen content of blood
	pH of cerebrospinal fluid
Somatic senses	Lung inflation
Touch-pressure	Osmolarity of body fluids
Temperature	Temperature
Pain	Blood glucose
Proprioception	Distension of gastrointestinal tract

PROBLEM
Ménière's Disease

After a visit to an Arles brothel on December 23, 1888, Vincent Van Gogh, the nineteenth-century French painter, returned to his room in a boardinghouse, picked up a knife, and cut off his own ear. A local physician, Dr. Felix Ray, examined Van Gogh that night and wrote that the painter had been "assailed by auditory hallucinations" and in an effort to relieve them, "mutilated himself by cutting off his ear." A few months later, Van Gogh committed himself to a lunatic asylum. By 1890, Van Gogh was dead by his own hand. Historians have assumed that Van Gogh suffered from epilepsy. But the painter's strange attacks of dizziness, nausea, and overwhelming tinnitus (ringing or other sounds in the ears), which he recorded in desperate letters to his relatives, are more consistent with Ménière's disease, a condition that affects the inner ear. Today, Anant Patel, a 20-year-old college student, will be examined by an otolaryngologist (ear-nose-throat specialist) to see if his periodic attacks of severe dizziness and nausea are caused by the same condition that drove Van Gogh to suicide.

...continued on page 289

way, the nervous system can modulate and shape the information. Let's first examine sensory receptors and pathways and see how sensory signals are transduced, coded, and processed.

Receptors Are Transducers That Convert Stimuli into Electrical Signals

Receptors in the sensory system vary widely in complexity, ranging from the branched endings of a single sensory neuron to multicellular *sense organs* such as the eye. Sense organs in humans can be very complex. The cochlea of the ear contains about 16,000 sensory receptors and more than a million associated parts, and one human eye has about 126 million sensory receptors.

The simplest receptors are **somatosensory** (somatic senses) receptors that consist of a neuron with naked ("free") nerve endings (Fig. 10-1a ■). In more complex somatosensory receptors, the nerve endings are encased in connective tissue capsules (Fig. 10-1b ■). The axons of somatosensory receptors may be myelinated or unmyelinated.

The special senses have **sense organs** with highly specialized receptor cells. The receptors for smell are neurons, but the other senses use non-neural receptor cells that synapse onto sensory neurons. The hair cell of the ear, shown in Figure 10-1c ■, is a non-neural receptor. When activated, the hair cell releases a neurotransmitter to create an action potential in the associated sensory neuron. Taste buds, derived from epithelial tissue, and photoreceptors are also non-neural receptors. Other non-neural body parts often act as accessories to sensory receptors. The lens and cornea of the eye and the hairs on our arms are examples of non-neural accessories.

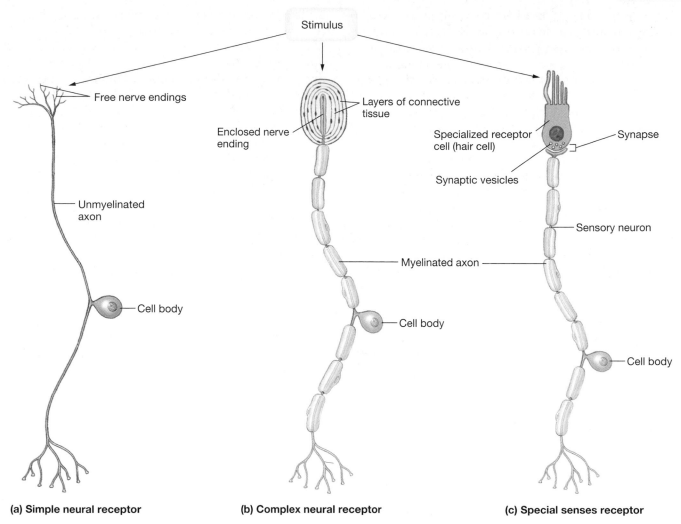

(a) Simple neural receptor   **(b) Complex neural receptor**   **(c) Special senses receptor**

■ **Figure 10-1   Sensory receptors**   (a) Simple receptors are neurons with free nerve endings. Axons may be either myelinated or unmyelinated. (b) Simple nerve endings enclosed in connective tissue capsules create complex neural receptors. This is a Pacinian corpuscle, which senses touch. (c) Specialized receptor cells release neurotransmitter onto sensory neurons, initiating an action potential. The cell illustrated in this figure is a hair cell, found in the ear.

Receptors are divided into five major groups, based on the type of stimulus to which they are most sensitive (Table 10-2). **Chemoreceptors** respond to chemical ligands that bind to the receptor. **Mechanoreceptors** respond to various forms of mechanical energy, including pressure, vibration, gravity, acceleration, and sound. **Thermoreceptors** respond to temperature, **photoreceptors** to

light, and **nociceptors** to noxious stimuli [*nocere,* to injure]. The specificity of a receptor for a particular type of stimulus is known as the **law of specific nerve energies.**

✔ What is the advantage of having myelinated axons?

✔ Look at each of the sensory modalities listed in Table 10-1 and decide which of the following receptor types is the appropriate transducer: mechano-, chemo-, photo-, thermo-, or noci-.

Sensory Pathways Carry Information to the Central Nervous System Integrating Centers

Sensory stimuli trigger action potentials in **primary sensory neurons** that travel to the CNS. In the spinal cord, most sensory neurons synapse onto interneurons called **secondary sensory neurons.** These interneurons may extend their axons (*project*) to other regions of the spinal cord or to the brain, where they in turn synapse onto **tertiary**

TABLE 10-2   Types of Sensory Receptors

Type of Receptor	Stimuli
Chemoreceptors	Oxygen, pH, various organic molecules such as glucose
Mechanoreceptors	Pressure (baroreceptors), cell stretch (osmoreceptors), vibration, acceleration, sound
Photoreceptors	Photons of light
Thermoreceptors	Varying degrees of heat
Nociceptors	Tissue damage interpreted as pain

sensory neurons. Tertiary neurons synapse onto quaternary neurons, and so on. (Primary, secondary, and tertiary neurons are also known as *first-order, second-order, third-order neurons.*)

Sensory information from the spinal cord ascends through the brain stem on its way to the sensory areas of the cerebrum. Each major division of the brain processes a particular type of sensory information (Fig. 10-2 ■). For example, the midbrain and thalamus receive visual information, and the medulla oblongata receives input for sound and taste. These pathways, along with those carrying somatosensory information, then project to the thalamus, which acts as a relay and processing station.

Only *olfactory* [*olfacere,* to sniff] information is not routed through the thalamus. Instead, information about odors travels from the nose directly to the cerebrum. Perhaps it is because of this direct link to the cerebrum that odors are so closely linked to memory and emotion. Most people have experienced encountering a smell that suddenly brings back a flood of memories of times or people from the past.

The cerebral cortex processes consciously perceived sensory information in ways that we still do not fully understand. One interesting aspect of CNS processing is the **perceptual threshold,** a level of stimulus intensity necessary for us to be aware of a particular sensation. Stimuli bombard our sensory receptors constantly, but our brain can "turn off" some stimuli to avoid being overwhelmed with information. This is accomplished by using inhibitory modulation [∞ p. 241] to decrease a suprathreshold stimulus until it is below the perceptual threshold. Inhibitory modulation often occurs in secondary and tertiary neurons of a sensory pathway. If it becomes necessary for us to pay attention to the stimulus, we can consciously focus our attention and overcome the inhibitory modulation.

You have experienced a change in perceptual threshold when you "tune out" the radio while studying or when you "zone out" during a lecture. In both cases, the noise stimulus is adequate to stimulate primary sensory neurons, but higher-order neurons dampen the perceived signal so that it does not reach the conscious brain. The threshold can be modified by an event such as a sudden question from the professor. At that point, the modulating influence is removed, and your conscious brain seeks to retrieve and recall recent sound input from the subconscious so that you can answer the question.

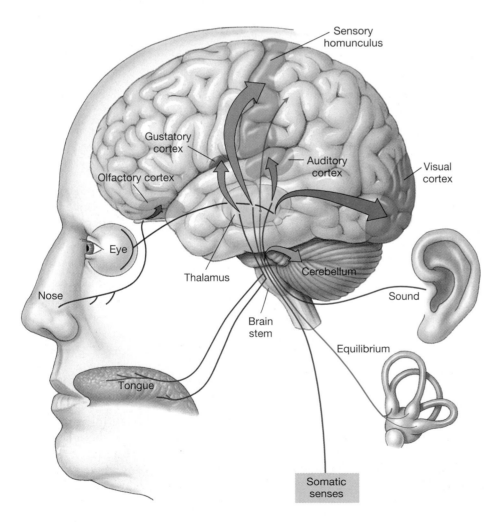

■ **Figure 10-2 Sensory pathways** Olfactory pathways from the nose project directly to the olfactory cortex. Equilibrium pathways project to the cerebellum. All other pathways as well as a branch of the equilibrium pathway pass through the thalamus on their way to cortical centers.

Receptive Fields Primary and secondary sensory neurons do not always exist in a 1:1 ratio. In many cases, multiple primary sensory neurons converge onto a single secondary sensory neuron (Fig. 10-3 ■). The receptor of each primary neuron picks up information from a specific area known as the **receptive field** of the receptor. These fields may be irregular in shape and frequently overlap with the receptive fields of neighboring receptors. If neighboring primary neurons converge on a single secondary neuron, their individual receptive fields merge into a single, large receptive field. Convergence of sensory neurons allows multiple simultaneous subthreshold stimuli to sum at the secondary neuron [∞ p. 241].

The size of secondary (neuron) receptive fields determines the sensitivity of an area to a stimulus. Sensitivity to touch is demonstrated by a **two-point discrimination test** (Fig. 10-4 ■). In some regions of skin, such as that on the arms and legs, two pins placed within 40 mm of each other will be interpreted by the brain as a single pinprick. In these areas, many primary neurons converge on a single secondary neuron, so the secondary receptive field is very large. On the other hand, more sensitive areas of skin have smaller receptive fields, with perhaps a 1:1 relationship between primary and secondary neurons. In these regions, two pins separated by as little as 2 mm can be perceived as two separate touches.

Sensory Transduction Converts Chemical and Mechanical Stimuli into Graded Potentials

How do receptors convert diverse physical stimuli into electrical signals? The first step is **transduction,** the conversion of stimulus energy into information that can be processed by the nervous system.* In many receptors, mechanical, chemical, thermal, or light energy is transformed into a change in membrane potential. Some nervous system transduction, particularly that mediated by neurotransmitters, involves second messenger systems.

Each receptor type has an **adequate stimulus,** a particular form of energy to which it is most responsive. For example, thermoreceptors are more sensitive to temperature than to pressure. Mechanoreceptors respond preferentially to stimuli that deform the cell membrane. Although receptors are specific for one form of energy, they can respond to most forms if the intensity is high enough. Photoreceptors of the eye respond most readily to light, but a blow to the eye may cause us to "see stars," an example of mechanical energy of sufficient force stimulating the photoreceptors.

Sensory receptors can be incredibly sensitive to their preferred form of stimulus. For example, a single photon of light stimulates certain photoreceptors, and a single *odorant* molecule may activate some olfactory chemoreceptors. The minimum stimulus required to activate a receptor is known as the **threshold,** just as the minimum depolarization required to trigger an action potential is called the threshold [∞ p. 223].

How is a physical or chemical stimulus converted into a change in membrane potential? The stimulus opens ion channels in the receptor membrane, either directly or indirectly (through a second messenger). In most cases, channel opening results in net influx of Na^+

---

*This is not to be confused with signal transduction that involves the transfer of information from an extracellular chemical messenger to an intracellular second messenger system [∞ p. 158].

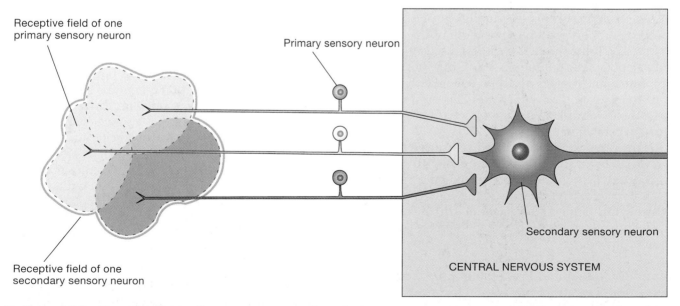

■ Figure 10-3 **Receptive fields of sensory neurons** Several primary sensory neurons converge and synapse onto a single secondary sensory neuron with one large receptive field. Convergence of primary sensory neurons allows simultaneous subthreshold stimuli to sum at the secondary sensory neuron and initiate an action potential. (In this illustration, only part of the secondary sensory neuron is shown.)

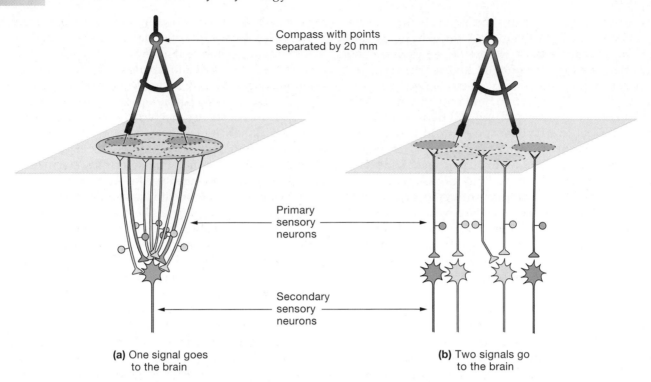

Compass with points separated by 20 mm

Primary sensory neurons

Secondary sensory neurons

(a) One signal goes to the brain

(b) Two signals go to the brain

■ **Figure 10-4** **Two-point discrimination**   (a) In certain regions of the body, such as the arms and legs, many primary sensory neurons converge onto a single secondary neuron, creating a very large receptive field. In these regions, two stimuli separated by as much as 40 mm will be perceived as a single point because the two stimuli fall within one secondary receptive field. (b) In more sensitive regions of the body, such as the fingertips, secondary receptive fields are much smaller since fewer neurons converge. In these areas two stimuli separated by as little as 2 mm will activate separate pathways and will be perceived as distinct stimuli.

into the receptor, depolarizing the membrane. In a few cases, the response to the stimulus is hyperpolarization when K^+ leaves the cell.

The change in receptor membrane potential is called the **generator potential** (somatic sense receptors) or **receptor potential** (special senses). These potentials are graded potentials because the amplitude of the voltage change is proportional to the strength of the stimulus [∞ p. 222]. If the receptor is part of a sensory neuron and a depolarizing stimulus is above threshold, the receptor potential triggers an action potential that travels along the sensory fiber to the CNS. This process is similar to a suprathreshold stimulus that reaches the trigger zone in other neurons [∞ p. 223].

If the receptor is a specialized cell such as the hair cell, variations in stimulus intensity change the amount of excitatory neurotransmitter released onto the primary sensory neuron. As more neurotransmitter is released, the firing rate of the sensory neuron increases, creating a code through which the CNS can interpret the stimulus intensity.

Stimulus Coding and Processing Are Used to Determine Location, Intensity, Duration, and Nature of a Stimulus

If all stimuli are converted to action potentials in the primary sensory neuron and all action potentials are identical, how can the body tell the difference between heat and pressure, or between a pinprick to the toe and one to the hand? The attributes of the stimulus must somehow be preserved once the stimulus enters the nervous system for processing. The body must distinguish four properties of a stimulus: the nature, or **modality,** of the stimulus, its location, its intensity, and its duration.

Sensory Modality   The modality of a signal is indicated by which sensory neurons are activated. Each receptor type is most sensitive to a particular type of stimulus. For example, some neurons respond most strongly to touch; others respond to changes in temperature. The brain thus associates a signal coming from a specific group of receptors with a specific modality. This 1:1 association of a receptor to a sensation is called **labeled line coding.** Stimulation of a cold receptor will always be perceived as cold, whether the actual stimulus was cold or whether an artificial depolarization of the receptor was the source of the stimulus. The blow to the eye that causes us to "see" a flash of light is another example of labeled line coding.

Location of the Stimulus   The location of a stimulus is also coded according to which group of neurons is active. The sensory regions of the cerebrum are highly organized in respect to the incoming signals, and input from adjacent sensory receptors is processed in adjacent columns of the cortex. This arrangement preserves the topographical organization of receptors on the skin, eye, or other regions in the processing centers of the brain.

For example, touch-pressure receptors in the hand project to a specific area of the cerebral cortex. Experimental stimulation of that area of the cortex during brain surgery is interpreted as a touch to the hand, although there may have been no actual touch. Similarly, the *phantom limb pain* reported by some amputees occurs when secondary sensory neurons in the spinal cord become hyperactive, resulting in the sensation of pain in a limb that is no longer there.

Auditory and olfactory information is the exception to the localization rule, however. Neurons in the ear and nose are sensitive to different frequencies of sound or different odors, but their activation provides no information about the location of the sound or odor. Instead, the brain uses the timing of receptor activation to compute a location for the sound or the odor, as shown in Figure 10-5 ■. A sound coming from directly in front of a person will reach both ears simultaneously. A sound placed off to one side will reach the closest ear several milliseconds before it reaches the other ear. The differ-

ence in the time it takes for the sound stimuli to reach the two sides of the auditory cortex is registered by the brain and used to compute the source of the sound.

Lateral inhibition is another way the location of a stimulus can be isolated. Figure 10-6 ■ shows this process for a pressure stimulus to the skin. The pin pushing on the skin activates three primary sensory neurons, each of which releases neurotransmitter onto its corresponding secondary neuron. However, the three secondary neurons do not all respond in the same fashion.

The secondary neuron closest to the actual stimulus (pathway B) suppresses the response of the secondary neurons to its sides (lateral), where the stimulus is weaker. It simultaneously allows its own pathway to proceed without interference. The inhibition of neurons farther from the stimulus enhances the contrast between the center and sides of the receptive field, so the sensation is more easily localized (Fig. 10-6 ■).

✔ What kind of ion channel do you think that neuron B (Fig. 10-6 ■) is opening in neurons A and C to depress their responsiveness: Na$^+$, K$^+$, Cl$^-$, or Ca^{2+}?

The pathway in Figure 10-6 ■ shows how multiple receptors can function together to send the CNS more information than would be possible from a single receptor. By comparing the input from multiple receptors, the CNS can make complex calculations about the spatial and temporal characteristics of a stimulus.

Intensity and Duration of the Stimulus   The intensity of a stimulus cannot be directly calculated from a single sensory neuron action potential because a single action potential is "all-or-none," with constant amplitude and duration. Instead, stimulus intensity is coded in two types of information: by the number of receptors activated (*population coding*) and by the frequency of action potentials coming from those receptors (*frequency coding*).

Population coding occurs because not all receptors have the same threshold to their preferred stimulus. Only the most sensitive receptors with the lowest thresholds will respond to a low-intensity stimulus. As a stimulus increases in intensity, additional receptors will be activated. The CNS then translates the number of active receptors into a measure of stimulus intensity.

For individual sensory neurons, intensity discrimination begins at the receptor. If a stimulus is below threshold, the primary sensory neuron does not respond. Once stimulus intensity exceeds threshold, the amplitude (strength) of the receptor potential increases in proportion to the intensity of the stimulus [∞ Figure 8-13, p. 229]. As the strength of the receptor potential increases, the frequency of action potentials in the primary sensory neuron increases.

Similarly, the duration of a stimulus is coded by the duration of action potentials in the sensory neuron. In general, the longer a stimulus persists, the longer a series of action potentials will be generated in the primary sen-

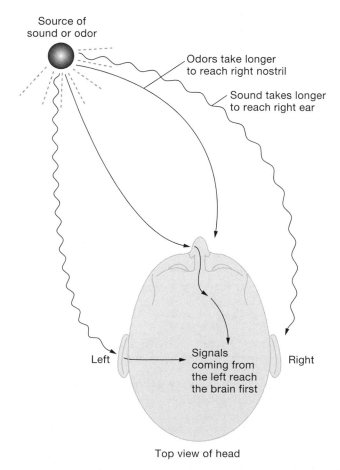

■ **Figure 10-5   Localization of sound and smell**   The brain uses timing differences rather than specific neurons to localize sound and smell. A stimulus located to the left side of the body will reach the left ear or nostril sooner than it reaches the right side of the body. Sensory input to the brain will show a similar time difference. This difference allows the brain to compute the location of the stimulus on the basis of the timing of stimulus arrival.

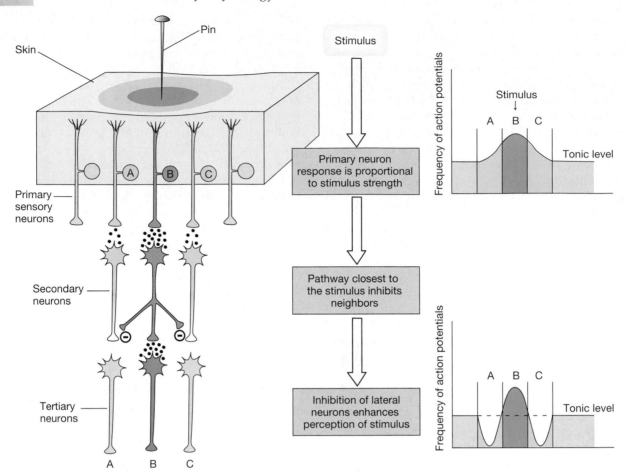

■ **Figure 10-6   Lateral inhibition**   The responses of primary neurons A, B, and C are proportional to the intensity of the stimulus in each receptive field. Secondary neurons A and C are inhibited by neuron B, creating greater contrast between B and its neighbors.

sory neuron (Fig. 10-7 ■). However, as a stimulus persists, some receptors turn off and cease to respond. Receptors fall into one of two classes, depending on how they react to continuous stimulation.

Tonic receptors are slowly adapting receptors that continue to transmit signals to the CNS as long as the stimulus is present (Fig. 10-8 ■). Pressure-sensitive baroreceptors, nociceptors, and some tactile and proprioreceptors are among the receptors that fall into this category. In general, the stimuli that activate tonic receptors are parameters that must be monitored continuously by the body.

In contrast, **phasic receptors** are rapidly adapting receptors that fire when they first receive a stimulus but cease firing if the strength of the stimulus remains constant. Phasic receptors are attuned to changing conditions. Once a stimulus reaches a steady intensity, phasic receptors **adapt** to the new steady state and turn off. This type of response allows the body to ignore information that has been evaluated and found not to threaten homeostasis or well-being.

Our sense of smell is an example of a sense that uses phasic receptors. For example, when you put on cologne in the morning, you can smell it, but as the day goes on, your olfactory receptors adapt and are no longer stimulated by the cologne molecules. You no longer smell the fragrance, yet others may comment on it. Adaptation of phasic receptors allows us to filter out extraneous sensory information and concentrate on what is new, different, or essential.

✔ How do sensory receptors tell the central nervous system about the intensity of a stimulus?

✔ What is the adaptive significance of having nociceptors that are tonic instead of phasic?

What happens at a receptor during **adaptation?** There is more than one explanation, depending on the receptor type. In some cases, additional ion channels in the receptor membrane open, causing the membrane to repolarize. In other cases, accessory structures decrease the amount of stimulus reaching the receptor. This occurs in the ear, for example, where muscle contraction decreases the vibration of small bones that occurs as a result of loud noises. In general, once adaptation of a phasic receptor has occurred, the only way to create a new signal is to change the intensity of the excitatory stimulus.

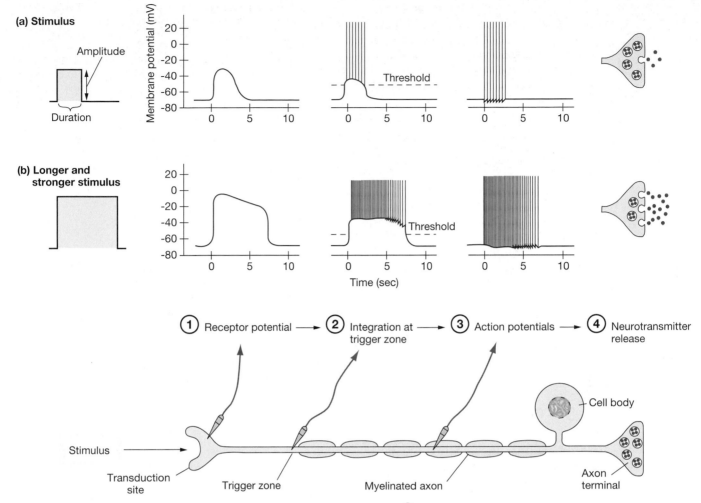

(a) Stimulus

(b) Longer and stronger stimulus

① Receptor potential → ② Integration at trigger zone → ③ Action potentials → ④ Neurotransmitter release

Cell body

Stimulus

Transduction site

Trigger zone

Myelinated axon

Axon terminal

■ **Figure 10-7   Sensory coding for stimulus intensity and duration**   ① The strength and duration of the receptor potential will vary according to the strength and duration of the stimulus. ② The frequency of action potentials initiated at the trigger zone will be proportional to the intensity of the stimulus. The duration of a series of action potentials will be proportional to the duration of the stimulus. ③ The same pattern of frequency and duration is seen in action potentials conducted down the axon. ④ The amount of neurotransmitter released depends on the pattern of action potentials arriving at the axon terminal.

To summarize, the specificity of sensory pathways is established in several ways. Each receptor is most sensitive to a particular type of stimulus. Activation of a receptor initiates action potentials in a primary sensory neuron that projects its axon to the CNS. Stimulus intensity and duration are coded in the pattern of action potentials reaching the CNS. Stimulus location and modality are coded according to which receptors are activated or, in the case of sound and smell, by the timing of receptor activation. Each sensory pathway projects to a region of the cerebral cortex dedicated to a particular receptive field. The brain can then tell the exact origin of each incoming signal.

▶ SOMATIC SENSES

There are four somatosensory modalities: touch-pressure, proprioception, temperature, and nociception. We will discuss details of proprioception in Chapter 13.

Pathways for Somatic Perception Project to the Somatosensory Cortex and Cerebellum

Primary sensory neurons bring information from somatic receptors to secondary sensory neurons in the CNS. The location of the synapse between primary and secondary neurons varies according to the type of receptor. Those neurons associated with receptors for nociception, temperature, and coarse touch synapse onto

...continued from page 282

Ménière's disease, named for its discoverer, the nineteenth-century French physician Prosper Ménière, is caused by a buildup of fluid within the inner ear. Symptoms of Ménière's disease include episodic attacks of vertigo, nausea, and tinnitus, accompanied by hearing loss. *Vertigo* is a false sensation of movement that patients often describe as dizziness.

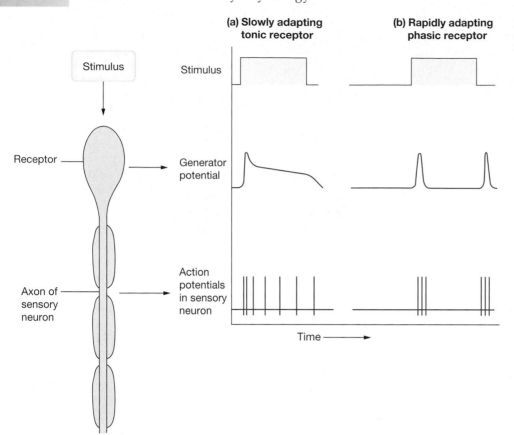

(a) Slowly adapting tonic receptor

(b) Rapidly adapting phasic receptor

Stimulus

Receptor

Axon of sensory neuron

Stimulus

Generator potential

Action potentials in sensory neuron

Time

■ **Figure 10-8 Tonic and phasic receptors** (a) Tonic receptors are slowly adapting receptors that continue to respond for the duration of a stimulus. (b) Phasic receptors are attuned to changing conditions. They rapidly adapt to a constant stimulus and turn off. When the stimulus turns off, the phasic receptors are activated briefly once more.

their secondary neurons shortly after entering the spinal cord. In contrast, most fine touch and proprioceptive neurons have very long axons that project up the spinal cord all the way to the medulla (Table 10-3).

All secondary sensory neurons cross the midline of the body, so that sensations from the left side of the body are processed in the right hemisphere of the brain and vice versa. In the thalamus, secondary sensory neurons synapse onto tertiary sensory neurons, which in turn project to the somatosensory region of the cerebral cortex. In addition, most pathways send branches to the cerebellum so that it can use the information to coordinate balance and movement.

The **somatosensory cortex** is the part of the brain that recognizes where ascending sensory tracts originate. Each sensory tract has a corresponding region of the cortex, as shown in Figure 10-9 ■, so that all sensory pathways for the left hand terminate in one area and so on. Within the different regions of the cortex, columns of

neurons are devoted to particular types of receptors. A cortical column activated by cold receptors in the left hand may be found next to a column activated by pressure receptors in the skin of the left hand. This columnar arrangement creates a highly organized structure that maintains the association between specific receptors and the sensory modality they transmit.

Some of the most interesting research about the somatosensory cortex has been done on patients during brain surgery. Because brain tissue has no pain fibers, this type of surgery can be performed with the patient awake under local anesthesia. The surgeon stimulates a particular region of the brain and asks the patient about sensations that occur. The ability of the patient to communicate with the surgeon during this process has expanded our knowledge of brain regions tremendously.

Experiments can also be done on animals by stimulating peripheral receptors and monitoring electrical activity in the cortex. We have learned from these

TABLE 10-3 Sensory Pathways

	Stimulus	
	Fine Touch, Proprioception	*Irritants, Temperature, Touch (Secondary)*
Primary sensory neuron terminates in:	Medulla	Dorsal horn of spinal cord
	Path crosses midline of body.	Path crosses midline of body.
Secondary sensory neuron terminates in:	Thalamus	Thalamus
Tertiary sensory neuron terminates in:	Somatosensory cortex	Somatosensory cortex

complex receptors (Table 10-4). Most are difficult to study because of their small size. However, the Pacinian corpuscle is one of the largest receptors in the body, and much of what we know about somatosensory receptors comes from studies on these structures.

Pacinian corpuscles are composed of nerve endings encapsulated in layers of connective tissue (Fig. 10-1b ■). They are found in the subcutaneous layers of skin, muscles, joints, and internal organs. The concentric layers of connective tissue in the corpuscle create large receptive fields. Pacinian corpuscles respond best to high-frequency vibrations whose energy is transferred through the connective tissue capsule to the nerve ending, where it opens mechanically gated ion channels. Pacinian corpuscles are rapidly adapting phasic receptors that turn off even if the stimulus continues (see Fig. 10-8 ■). This property allows them to respond to a change in touch but then ignore it.

Nociceptors and Pain   Nociceptors are activated by a variety of strong noxious stimuli that cause or have the potential to cause tissue damage. Nociceptors are sometimes called pain receptors, even though **pain** is a perceived sensation rather than a stimulus. Pain is an adaptive, protective response to environmental stress. For example, if we did not feel joint discomfort with overuse, we would damage our joints more quickly. In a way, pain protects us from wearing out our body.

The afferent signals for pain are carried in two different types of primary sensory fibers (see Table 10-5). **Fast pain,** described as sharp and localized, is transmitted rapidly by small, myelinated Aδ (A-delta) fibers. **Slow pain,** described as duller and more diffuse, is carried on small, unmyelinated C fibers. The timing distinction between the two is most obvious when the stimulus originates far from the CNS, such as when you stub your toe. You first experience a quick stabbing sensation (fast pain), followed shortly by a dull throbbing (slow pain).

The receptors for pain are free nerve endings that respond to chemical, mechanical, or thermal stimuli. One theory of nociceptor activation states that mechanical and thermal stimuli are mediated through local chemicals. We do know that a variety of molecules released with tissue injury either activate nociceptors or sensitize them by lowering their activation threshold. Chemical mediators of the pain response include K^+, histamine, and prostaglandins released from damaged cells, as well as serotonin released from platelets activated by injury. Primary sensory neurons release a peptide known as **substance P** that acts both as a neurotransmitter and as a modulator that enhances the pain pathway.

Painful stimuli are carried in ascending pathways to the cortex, where they become conscious sensation. However, not all irritant responses must go through the cortex. Subconscious withdrawal reflexes occur outside the cortex. **Withdrawal reflexes,** such as pulling back your hand when you accidentally touch a hot stove, automatically remove a stimulated area from the source of

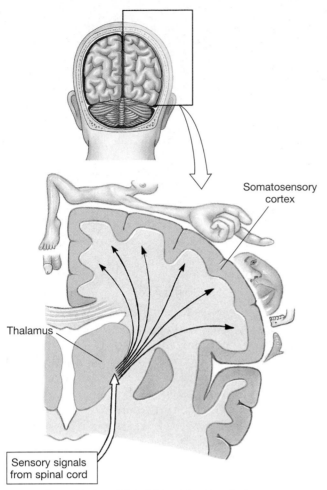

Cross section of the right cerebral hemisphere and sensory areas of the cerebral cortex

■ **Figure 10-9   The somatosensory cortex**   Somatosensory neurons from one side of the body project to the opposite side of the brain. The amount of space on the sensory cortex devoted to each part of the body is proportional to the sensitivity of that part. In this illustration, representations of body parts are shown adjacent to the area of the sensory cortex that processes their stimuli.

experiments that the more sensitive a region of the body is to tactile and other stimuli, the larger the corresponding region in the cortex. Interestingly, the size of the regions is not fixed. If a particular body part is used more extensively, its topographical region in the cortex will expand. For example, people who are visually handicapped and learn to read Braille with their fingertips develop an enlarged region of the somatosensory cortex devoted to the fingertips. This plasticity [∞ p. 253] shows the amazing versatility of the brain.

Touch-Pressure Receptors   Touch-pressure receptors are one of the most common receptors in the body. They are found both in superficial layers of the skin (Fig. 10-10 ■) and in deeper regions of the body such as the viscera. Some of these receptors are free nerve endings, such as those that wrap around the base of hairs; others are

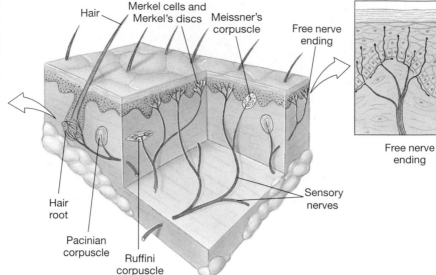

Hair
Merkel cells and Merkel's discs
Meissner's corpuscle
Free nerve ending

Free nerve ending of hair root

Hair root

Pacinian corpuscle

Ruffini corpuscle

Sensory nerves

Free nerve ending

■ Figure 10-10 **Touch-pressure receptors** Touch-pressure receptors may be either simple or complex receptors. They are scattered throughout the superficial and deep layers of the skin.

the stimulus. Reflexes that do not require integration in the brain are called **spinal reflexes.**

✔ What is the adaptive advantage of a spinal reflex?

The lack of brain involvement in many protective reflexes has been demonstrated in the classic "spinal frog" preparation, in which the animal's brain has been destroyed by pithing. If the frog's foot is placed in a beaker of hot water or strong acid, the withdrawal reflex causes the leg to contract and move the foot away from the stimulus, despite the absence of a functioning brain.

The ascending pathways for pain go to the limbic system and hypothalamus as well as to the sensory areas of the cortex. As a result, pain may be accompanied by

emotional distress and a variety of autonomic reactions such as nausea, vomiting, or sweating.

Our perception of pain is subject to modulation at several levels in the nervous system. It can be magnified by past experiences or suppressed in emergencies when survival depends on ignoring injury. In the latter situation, descending pathways travel through the thalamus and inhibit nociceptor neurons in the spinal cord. Artificial stimulation of these inhibitory pathways is one of the newer techniques being used to control chronic pain.

Pain can also be suppressed in the dorsal horn of the spinal cord, before the stimuli are sent to ascending spinal tracts. Normally, tonically active interneurons within the spinal cord inhibit ascending pathways for pain (Fig. 10-11a ■). C fibers from nociceptors synapse on

TABLE 10-4   Touch-Pressure Receptors

Receptor	Stimulus	Location	Structure	Adaptation
Free nerve endings	Various touch and pressure stimuli	Around hair roots and under surface of skin	Unmyelinated nerve endings	Variable
Meissner's corpuscle	Flutter	Superficial layers of skin	Encapsulated in connective tissue	Rapid
Pacinian corpuscle	Vibration	Deep layers of skin	Encapsulated in connective tissue	Rapid
Ruffini corpuscle	Steady pressure on skin	Deep layers of skin	Enlarged nerve endings	Slow
Merkel receptors	Steady pressure on skin	Superficial layers of skin	Enlarged nerve endings	Slow

TABLE 10-5   Classes of Somatosensory Nerve Fibers

Fiber Type	Fiber Characteristics	Speed of Conduction	Associated with
Aβ (beta)	Large, myelinated	30–70 m/sec	Mechanical stimuli
Aδ (delta)	Small, myelinated	12–30 m/sec	Cold, fast pain, mechanical stimuli
C	Small, unmyelinated	0.5–2 m/sec	Slow pain, temperature, mechanical stimuli

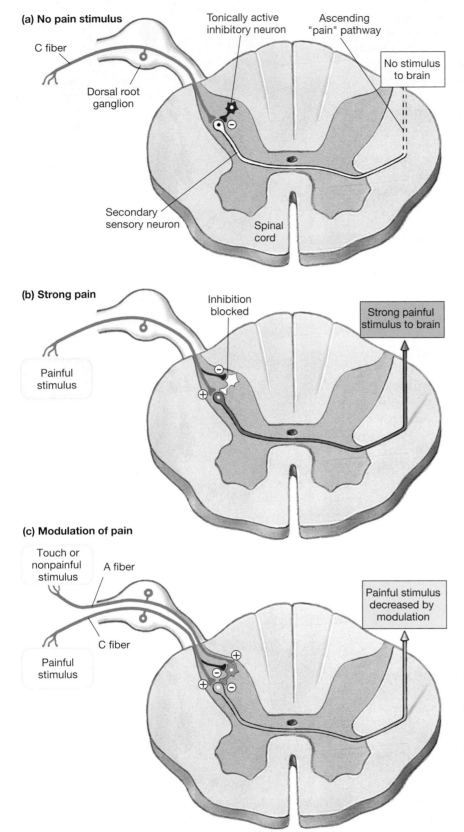

(a) No pain stimulus

C fiber

Dorsal root
ganglion

Tonically active
inhibitory neuron

Ascending
"pain" pathway

No stimulus
to brain

Secondary
sensory neuron

Spinal
cord

(b) Strong pain

Inhibition
blocked

Strong painful
stimulus to brain

Painful
stimulus

(c) Modulation of pain

Touch or
nonpainful
stimulus

A fiber

C fiber

Painful
stimulus

Painful stimulus
decreased by
modulation

■ **Figure 10-11   The gating theory of pain modulation**   (a) In the absence of stimulus, an inhibitory interneuron prevents ascending pain signals from going to the brain. (b) When a painful stimulus excites a nociceptor, the C fiber of the nociceptor blocks the inhibition, allowing a strong pain signal to be sent to the brain. (c) In the gating model, other non-nociceptor–mediated stimuli can diminish the pain signal by stimulating the inhibitory interneuron and overriding the effect of the C fiber.

the inhibitory neurons. When activated by a painful stimulus, the C fibers simultaneously excite the ascending path and block the tonic inhibition (Fig. 10-11b ■). This action allows the pain signal from the C fiber to travel unimpeded to the brain.

In the **gating theory** of pain modulation, Aβ and Aδ neurons from nonirritant somatic receptors also synapse on the inhibitory neurons, *enhancing* their inhibitory activity. If simultaneous stimuli reach the inhibitory neuron from the A and C fibers, the integrated response is partial inhibition of the ascending pain pathway. The pain perceived by the brain is lessened (Fig. 10-11c ■). You may have used the gating theory when you rubbed a bumped elbow or shin. The tactile stimulus activates A fibers and helps block the transmission of pain signals to the cortex.

Pain can be felt in skeletal muscles (*deep somatic pain*) as well as in the skin. Muscle pain during exercise is associated with the onset of anaerobic metabolism. It is often perceived as a burning sensation within the muscle (as in "go for the burn!"), apparently created by a metabolite released during anaerobic metabolism. Some investigators have suggested that the metabolite is K^+, known to enhance the pain response. Muscle pain from **ischemia** (lack of adequate blood flow that reduces oxygen supply) also occurs during *myocardial infarction* (heart attack).

Pain in the heart and other internal organs (*visceral pain*) is often poorly localized and may be felt in areas far removed from the site of the stimulus (Fig. 10-12a ■). For example, the pain of cardiac ischemia is often felt in the neck and down the left shoulder and arm. This **referred pain** apparently occurs because multiple primary sensory neurons converge onto a single ascending tract (Fig. 10-12b ■). According to this model, when painful stimuli arise in visceral receptors, the brain is unable to distinguish visceral signals from the more common signals arising from somatic receptors. As a result, it interprets the pain as coming from the somatic regions rather than the viscera.

Chronic pain of one sort or another affects millions of people in this country every year, and the alleviation of pain is of considerable interest to health professionals. **Analgesic drugs** [*analgesia,* painlessness] range from aspirin to potent opiates such as morphine. Aspirin inhibits prostaglandins and presumably slows the transmission of pain signals from the site of injury. The opiates act directly on opioid receptors in the brain, apparently activating descending pathways that inhibit incoming pain signals. In the case of severe chronic pain, it is sometimes possible to get surgical relief by severing afferent nerves at the dorsal root or by stimulating the inhibitor pathways of the brain. Acupuncture can also be effective, although the physiological reason for its effectiveness is not clear (see *Pain Control* box).

Temperature Receptors   Temperature receptors are free nerve endings that terminate in the subcutaneous layers of the skin. **Cold receptors** are sensitive primarily to temperatures lower than body temperature. **Warm receptors** are stimulated by temperatures from body temperature (37° C) to about 45° C. Above that temperature, pain receptors are activated, creating a sensation of painful heat.

The receptive field for a thermoreceptor is about 1 mm in diameter, and the receptors are scattered across the body. There are considerably more cold receptors

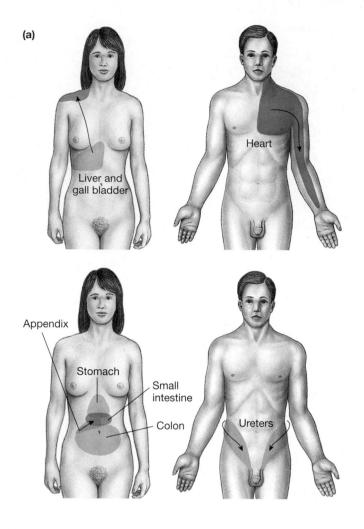

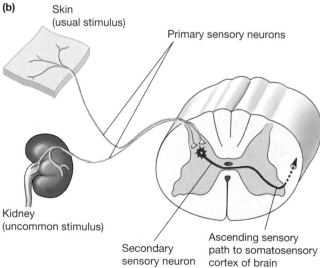

■ **Figure 10-12   Referred pain**   (a) Pain in the internal organs is often sensed on the surface of the body in the regions indicated. This sensation is known as referred pain. (b) One theory for the origin of referred pain says that the nociceptors for pain from several locations converge on a single ascending tract in the spinal cord. Pain signals from the skin are more common than pain from internal organs, so the brain learns to associate activation of the pathway with pain in a region of the skin, no matter where the pain originates.

§ **Pain Control**   Acupuncture is a centuries-old Chinese therapy that is based on the manipulation of needles inserted in special locations on the body. In the Chinese tradition, acupuncture enhances Qi (pronounced *chee*), a form of energy that promotes healing. Western medicine for many years scoffed at acupuncture, claiming that its results were due to the *placebo effect*: if the patient believes that a treatment will work, in many cases, it does. However, in recent years, evidence has been accumulating that acupuncture has a physiological basis. The leading theory on how acupuncture works proposes that properly placed acupuncture needles stimulate a type of sensory neuron and trigger the release of endorphins (morphine-like pain suppressants) by the brain. Several leading medical schools are now participating in controlled clinical trials on the effectiveness of acupuncture. Although much of this research is directed at pain relief, acupuncture may also be useful for repairing neural damage following strokes and for treating drug and alcohol addictions.

than warm. Temperature receptors adapt between 20° and 40° C, but outside that range, where the likelihood of tissue damage is greater, the receptors do not adapt. Temperature receptors play an important role in thermoregulation, which is discussed in Chapter 21.

✔ From an adaptive standpoint, why would touch and proprioception move through faster sensory pathways than would pain and temperature information?

✔ Our sense of smell uses phasic receptors. Can you think of other receptors (senses) that adapt to ongoing stimuli?

▶ CHEMORECEPTION: SMELL AND TASTE

The five special senses of smell, taste, hearing, equilibrium, and vision are concentrated in the head region. Like somatic senses, the special senses rely on receptors to transform information about the environment into patterns of action potentials that can be interpreted by the brain. Chemoreception is one of the oldest senses from an evolutionary perspective. Primitive animals without formalized nervous systems use chemoreception to locate food and mates. Research on bacteria even suggests that they use chemoreception as an integral part of sensing their environment. In humans, chemoreceptive senses have been refined into the special senses of smell (**olfaction**) and taste (**gustation**).

Olfaction Is One of the Oldest Senses

Imagine waking up one morning and discovering a whole new world around you, a world filled with odors that you never dreamed existed—scents that told you more about your surroundings than you ever imagined from looking at them. This is exactly what happened to a young patient of Dr. Oliver Sacks (an account is in *The Man Who Mistook His Wife for a Hat and Other Clinical Tales*). Or imagine skating along the sidewalk without a helmet, only to fall and hit your head. When you regain consciousness, the world has lost all odor: no smell of grass or perfume or garbage. Even your food has lost much of its taste, and you now eat only to survive because eating has lost its pleasure.

We do not realize the role that our sense of smell plays in our lives until a head cold or injury robs us of the ability to smell. Olfaction [*olfacere*, to sniff] allows us to discriminate between thousands of different odors. Even so, our noses are not nearly as sensitive as those of many other animals whose survival depends on olfactory cues. The **olfactory bulb,** the part of the brain that receives input from the primary olfactory neurons, is considered by many biologists to be one of the oldest parts of the cerebrum (Fig. 10-13a ■). Smell is probably not only the oldest sense but also the basis for chemical synaptic communication.

The olfactory receptor cells are concentrated in a 3-cm^2 patch of **olfactory epithelium** high in the nasal cavity (Fig. 10-13b ■). The receptors are neurons whose processes extend on one side to the surface of the olfactory epithelium and on the other to the olfactory bulb, located on the underside of the frontal lobe. Olfactory receptors, unlike other neurons in the body, are continuously dividing cells, with a turnover time of about two months (Fig. 10-13c ■). This means that the axon of each new neuron must find its way to the olfactory bulb and make the proper synaptic connection. Discovering how these neurons manage to repeat the same connection each time will give us insight on how developing neurons find their targets.

The surface of the olfactory epithelium is composed of the knobby terminals of the receptor cells, each knob sprouting multiple nonmobile cilia (Fig. 10-13c ■). The membranes of the cilia contain protein receptor molecules needed to start the signal transduction process. The cilia are embedded in a layer of mucus, so molecules must first penetrate the mucus before they can bind to a receptor. This process is aided by **olfactory binding proteins** in the mucus, soluble proteins that concentrate odorant molecules and present them to the receptors. Once the odorant combines with the receptor protein on the cilia, it activates a special G protein, G_{olf}. Activation of G_{olf} sets off a second messenger cascade that opens cation channels, depolarizing the cell and triggering a signal that travels to the brain.

✔ What part(s) of an olfactory receptor cell would be the dendrites?

✔ Are olfactory neurons pseudounipolar, bipolar, or multipolar? (Hint: Fig. 8-3 ■, p. 218)

Secondary and higher-order neurons project from the olfactory bulb through the first cranial nerve to the

ANATOMY SUMMARY Olfaction

■ Figure 10-13

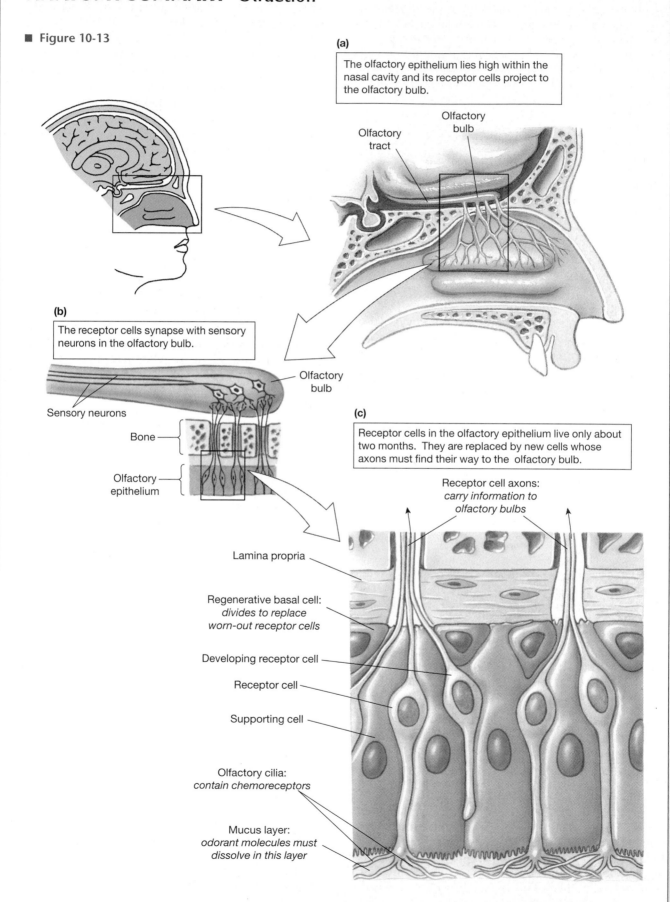

(a) The olfactory epithelium lies high within the nasal cavity and its receptor cells project to the olfactory bulb.

Olfactory tract

Olfactory bulb

(b) The receptor cells synapse with sensory neurons in the olfactory bulb.

Olfactory bulb

Sensory neurons

Bone

Olfactory epithelium

(c) Receptor cells in the olfactory epithelium live only about two months. They are replaced by new cells whose axons must find their way to the olfactory bulb.

Receptor cell axons: *carry information to olfactory bulbs*

Lamina propria

Regenerative basal cell: *divides to replace worn-out receptor cells*

Developing receptor cell

Receptor cell

Supporting cell

Olfactory cilia: *contain chemoreceptors*

Mucus layer: *odorant molecules must dissolve in this layer*

olfactory cortex. In addition, olfactory pathways lead to the amygdala and hippocampus, parts of the limbic system involved with emotion and memory.

One amazing aspect of olfaction is the link between smell, memory, and emotion. A special cologne or aroma of food can trigger memories and create a wave of nostalgia for the time, place, or people with whom the aroma is associated. In some way that we do not understand, the processing of odors through the limbic system creates deeply buried olfactory memories. Particular combinations of olfactory receptors become linked to other patterns of sensory experience so that stimulating one pathway stimulates them all.

An accessory olfactory structure in the nose, the **vomeronasal organ** (VNO), is known to be involved in behavioral responses to sex pheromones in rodents. A **pheromone** is an air- or waterborne signaling molecule used for communication between individual organisms. The existence of pheromones in humans has been postulated but never proved [∞ p. 207].

What is occurring at the cellular and molecular level that allows us to discriminate between thousands of different odors? Current research suggests olfactory coding is accomplished by receptors with sensitivity to a variety of substances. The brain uses information from hundreds of different receptors in different combinations to create the perception of many different smells.

Taste Is a Combination of Five Basic Sensations

Taste is closely linked to olfaction. Indeed, much of what we call the taste of food is actually the aroma, as you know if you have ever had a bad cold. Although smell is sensed by hundreds of different types of receptors, taste is a combination of five sensations: sweet, sour, salty, bitter, and *umami,* a taste associated with the amino acid glutamate and some nucleotides. Umami, a name derived from the Japanese word for "deliciousness," is a basic taste that enhances the flavor of foods. It is the reason that monosodium glutamate (MSG) is used as a food additive in some countries.

Interestingly, although we have only a limited number of taste sensations, each one is associated with an essential body function. Sour taste is triggered by H^+ and salty by Na^+, two ions whose concentrations in body fluids are closely regulated. The other three taste sensations result from organic molecules. Sweet and umami are associated with nutritious food. Bitter taste is recognized by the body as a warning of possibly toxic components. If something tastes bitter, our first reaction is often to spit it out.

The receptors for taste are located primarily on **taste buds** clustered together on the surface of the tongue. These sensory organs are composed of 50–150 receptor cells, support cells, and regenerative basal cells (Fig. 10-14b, c ■). Taste receptors are also scattered through other regions of the oral cavity, such as the palate. Receptors for sweet, sour, salty, and bitter taste sensations are localized to different sections of the tongue, with overlap (Fig. 10-14a ■). All taste buds appear to respond to all ligands, but with differing sensitivity.

Taste receptors are polarized epithelial cells [∞ p. 131], tucked down into the epithelium so that only a tiny tip protrudes. Tight junctions link the apical ends of the cells together so that movement of molecules into the taste bud between the cells is limited. The cell membrane that extends into the oral cavity is modified into microvilli to increase surface area in contact with the environment. The basal side of the cell forms a synapse with the primary sensory neuron.

For a substance to be tasted, it must first dissolve in the saliva and mucus of the mouth. There is now evidence for taste binding proteins that facilitate this process, similar to the olfactory binding protein for smell. Taste ligands bind to an apical membrane receptor (Fig. 10-14c ■). Binding either depolarizes the taste receptor, allowing Ca^{2+} influx from the extracellular fluid, or causes release of calcium from intracellular stores. The Ca^{2+} signal within the cell triggers exocytosis of neurotransmitter [∞ p. 235] into the synapse and initiates a series of action potentials in the primary sensory neuron.

The four primary taste sensations utilize different cellular mechanisms to activate the sensory neuron (Fig. 10-15 ■). Bitter and sweet ligands bind to membrane receptors and activate second messenger systems. Ionic salty (Na^+) and sour (H^+) ligands bind directly to membrane ion channels.

Bitter-tasting ligands activate a G protein named *transducin* linked to the IP_3 second messenger pathway. This cascade releases Ca^{2+} from intracellular stores. The Ca^{2+} signal then causes release of vesicle contents into the synapse, just as occurs in an axon terminal.

Sweet, salty, and sour ligands depolarize the taste receptor and open voltage-gated Ca^{2+} channels to create the calcium signal. The salt ligand Na^+ depolarizes

Satisfying Our Sweet Tooth   One challenge facing food scientists is to develop substances that will satisfy our desire for sweets without adding calories to the diet. Recent research has focused on two naturally occurring proteins, thaumatin and monellin, found in the fruits of West African shrubs. These proteins are intensely sweet-tasting to humans and other primates, nearly 100,000 times sweeter than table sugar (sucrose). For comparison, two artificially developed sweeteners, saccharin and aspartame (Nutrasweet) are merely 500 and 180 times sweeter, respectively. The problem with commercial use of thaumatin and monellin is the limited supply of the naturally occurring proteins. Consequently, the gene that codes for thaumatin has been introduced into various microbes to see if commercially viable yields can be produced. Other efforts under way include attempts to genetically engineer food crops that would produce large amounts of thaumatin.

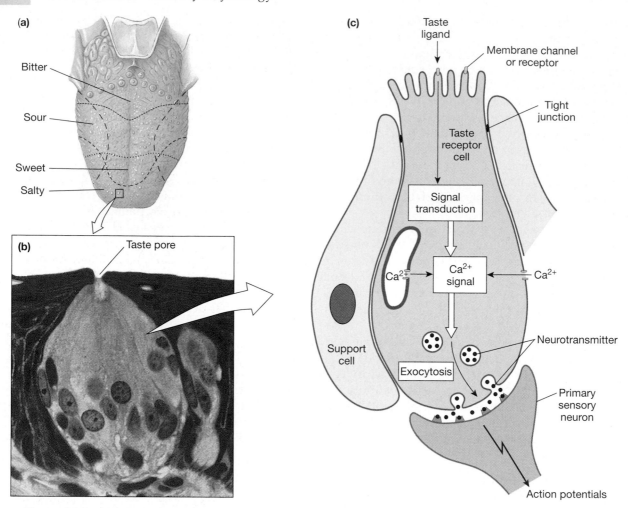

■ **Figure 10-14 Taste buds** (a) The most sensitive taste buds for different sensations are in specific regions on the dorsal surface of the tongue. The umami receptors (not shown) are located in the back of the pharynx. (b) A light micrograph of a taste bud. (c) Each taste bud is composed of taste receptor cells and support cells, joined near the apical surface with tight junctions. Taste ligands bind to the receptors and create calcium signals that release neurotransmitters onto primary sensory neurons.

the receptor cell by entering through open Na^+ channels. The sour ligand H^+ blocks K^+ channels and slows K^+ efflux. Retention of K^+ depolarizes the cell. Sweet ligands also decrease K^+ permeability but with the aid of a special G protein, *gustducin,* and cAMP. Many researchers are interested in sweet ligands as part of the search for sugar substitutes.

Sensory information from the taste buds passes to gustatory neurons that project to the medulla. Information then passes through the thalamus to the gustatory cortex (Fig. 10-2 ■). Central processing of sensory information compares the input from multiple taste buds and interprets the taste sensation based on which populations of neurons are responding most strongly.

An interesting psychological aspect of taste is the phenomenon named **specific hunger.** Humans and other animals who are lacking a particular nutrient may develop a craving for that substance. **Salt appetite,** representing a lack of Na^+ in the body, has been recognized for years. Hunters have used their knowledge of this

specific hunger to stake out salt licks because they know that animals will seek them out. Salt appetite is directly related to Na^+ concentration in the body and cannot be assuaged by ingestion of other cations such as Ca^{2+} or K^+. Other appetites are more difficult to relate to specific nutrient needs and are probably complex mixtures of physical, psychological, environmental, and cultural influences.

✔ With what essential body function is the umami taste sensation associated?

▶ THE EAR: HEARING

The ear is a sense organ that is specialized for two distinct functions: hearing and equilibrium. It can be divided into outer, middle, and inner sections, with the neurological structures housed in and protected by the structures of the inner ear. The vestibular complex of the

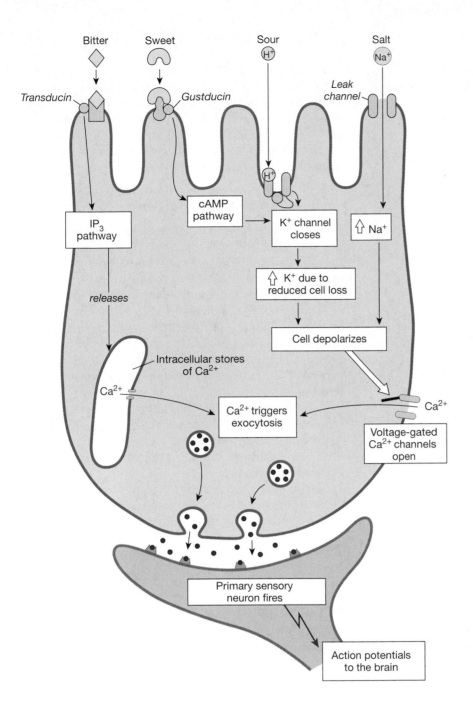

■ **Figure 10-15** **Taste transduction**
Bitter and sweet ligand signal transduction uses G protein–coupled membrane receptors. Transducin releases Ca^{2+} from intracellular stores. Gustducin activates a cAMP second messenger pathway that closes K^+ channels. Ionic ligands alter ion channels and depolarize the taste receptor, which allows Ca^{2+} entry from the extracellular fluid. For all taste ligands, the Ca^{2+} signal triggers exocytosis of neurotransmitter.

inner ear is the primary region involved in equilibrium. The remainder of the ear is used for hearing.

Sound Waves Vary in Their Pitch and Loudness

Hearing is our perception of the energy carried by sound waves. *Sound waves* are alternating pressure waves with peaks of compressed air (or water) and valleys where the molecules are farther apart (Fig. 10-16a ■). *Sound* is our interpretation of the amplitude, frequency, and duration of those waves. The classic question about hearing is, "If a tree falls in the forest with no one to hear, will it make a noise?" The answer is no. The falling tree will emit sound waves, but there is no noise unless some-

one or something is present to process and perceive the wave energy as sound.

We characterize sound waves by their pitch and their loudness, two properties that are psychophysiological interpretations of what we hear. We interpret **pitch** as high and low sounds, such as the screech of a fingernail on a blackboard (high pitch) or the rumble of distant thunder (low pitch). Physically, pitch is a function of the frequency of sound waves. Our brain interprets low-frequency waves as low-pitched sounds and high-frequency waves as high-pitched sounds.

Sound wave *frequency* is measured in waves per second, or **hertz (Hz)** (Figure 10-16b ■). The average human ear can hear sounds over the range of 20–20,000 Hz, with

(a)

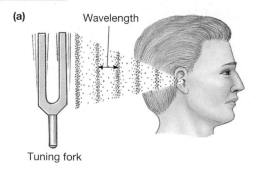

Wavelength

Tuning fork

(b) Frequency = 20 Hz (waves/sec)

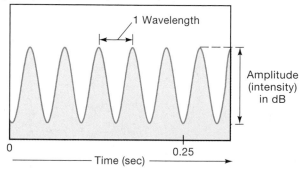

1 Wavelength

Amplitude (intensity) in dB

0

0.25

Time (sec)

(c) Frequency = 32 Hz

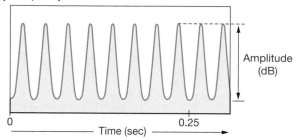

Amplitude (dB)

0

0.25

Time (sec)

■ **Figure 10-16   Sound waves**   (a) Sound waves are alternating pressure waves, with peaks of compressed air and valleys where the molecules are farther apart. (b) Sound waves are distinguished by their amplitude, measured in decibels (dB), and frequency, measured in hertz (Hz). (c) In this graph, the frequency of the sound wave is higher than that of the sound wave in (b).

the most acute hearing between 1000–3000 Hz. Our hearing is not as acute as that of many other animals, just as our sense of smell is less acute. Bats listen for ultra–high-frequency sound waves (in the kilohertz range) that bounce off objects in the dark. Elephants and some birds can hear sounds in the infrasound (very low frequency) range.

✔ What is a kilohertz?

Loudness is our interpretation of sound intensity and will be influenced by the sensitivity of an individual's ear. The *intensity* of a sound wave is a function of the wave *amplitude*. Intensity is measured with a logarithmic scale in units called **decibels (dB)** (Fig. 10-16b ■). Each

10-dB increase represents a 10-fold increase in intensity. Normal conversation has a typical noise level of about 60 dB. At a level of 80 dB or more, damage can occur to the sensitive hearing receptors of the ear, resulting in hearing loss. A typical heavy metal rock concert will have noise levels around 120 dB, an intensity that puts listeners in immediate danger of damage to their hearing. The amount of damage depends on the duration and frequency of the noise as well as its intensity.

Transduction of Sound Is a Multistep Process

Hearing is a complex sense that involves the transduction of energy from air waves into (1) mechanical vibrations, then (2) fluid waves, (3) chemical signals, and finally (4) action potentials. Sound waves striking the outer ear are directed into the ear canal by the outer ear, or **pinna** (Fig. 10-17 ■). The canal is sealed at its internal end by the **tympanic membrane,** or *eardrum,* a thin membranous sheet of tissue. Sound waves striking the eardrum create vibrations that are passed along the three bones of the middle ear to another membrane at the **oval window** of the **cochlea.**

Vibrations of the middle ear bones against the oval window create fluid waves in the channels of the cochlea. Because fluid is not compressible, energy of the fluid waves dissipates back into the air of the middle ear at the **round window.**

As fluid waves move through the cochlea, they activate the sensory hair cells, which release neurotransmitter (the chemical signal). Finally the primary sensory neurons send information coded by action potentials to higher brain centers, where it is decoded.

The Middle Ear Transfers Sound from the Eardrum to the Cochlea

The middle ear is an air-filled cavity that connects with the nasopharynx through the **eustachian tube.** The eustachian tube is normally collapsed but opens transiently to allow the pressure in the middle ear to equilibrate with atmospheric pressure during chewing, swallowing, and yawning. Colds or infections that cause swelling can block the eustachian tube and result in fluid buildup in the middle ear. If bacteria are trapped in the middle ear, the ear infection known as *otitis media* results.

Three small bones of the middle ear conduct sound: the **malleus** [hammer], the **incus** [anvil], and the **stapes** [stirrup] (see Figs. 10-17 and 10-19a ■). The three bones are connected to one another with the biological equivalent of hinges. One end of the malleus is attached to the tympanic membrane so that when sounds waves hit the eardrum, the vibrations are transferred to the malleus, the incus, and the stapes in sequence. The stirrup end of the stapes is attached to the thin tissue of the oval window. Vibrations of the oval window are converted into fluid waves in the next step of the transduction sequence.

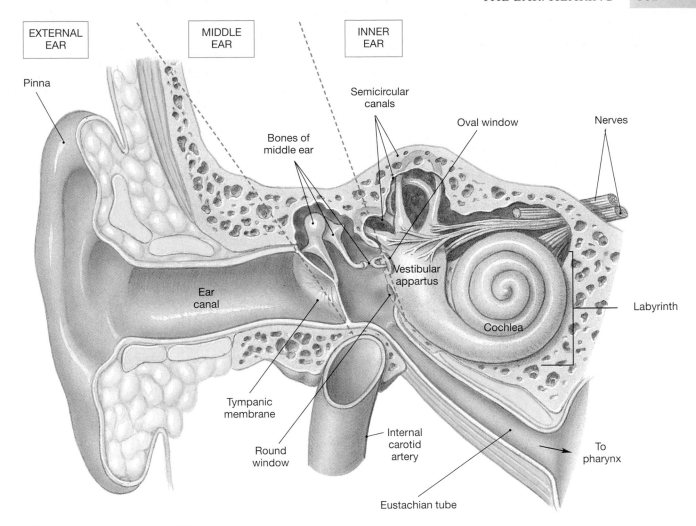

■ **Figure 10-17   Anatomy of the ear**

The arrangement of the three middle ear bones creates a lever that multiplies the force of the vibration (*amplification*) so that little energy is lost due to friction. If noise levels are so high that there is danger of damage to the inner ear, small muscles in the middle ear can pull on the bones to decrease their movement and dampen sound transmission to some degree.

The Cochlea of the Inner Ear Is Filled with Fluid

The **cochlea** of the inner ear is a membranous tube that lies coiled like a snail shell within a cavity of the bony **labyrinth** (Fig. 10-17 ■). Uncoiled, the cochlea can be seen to be composed of three parallel, fluid-filled channels: (1) the **vestibular duct** (*scala vestibuli*) [*scala*, stairway; *vestibulum*, entrance]; (2) the central **cochlear duct** (*scala media*) [*media*, middle]; and (3) the **tympanic duct** (*scala tympani*) [*tympanon*, drum] (Fig. 10-18 ■). The vestibular and tympanic ducts are continuous, and they connect at the tip of the cochlea through a small opening known as the **helicotrema** [*helix*, a spiral + *trema*, hole].

The fluid contained within the vestibular and tympanic ducts is similar in ion composition to plasma and is known as **perilymph.**

The cochlear duct is a dead-end tube filled with **endolymph,** the same fluid that fills the inner ear (see The Ear: Equilibrium). Endolymph is secreted by epithelial cells in the duct. It is unusual because it is more like intracellular fluid than extracellular fluid in its composition, with high concentrations of K^+ and low concentrations of Na^+.

The cochlear duct itself is formed from flexible tissue that moves in response to sound waves in the vestibular duct (see Fig. 10-19 ■). The cochlear duct contains the **organ of Corti,** composed of **hair cell** receptors and support cells. The organ of Corti sits on the **basilar membrane** and is partially covered by the flexible **tectorial membrane.**

Sound Transduction Through the Cochlea Depends on Movement of Hair Cell Stereocilia

Sound waves reach the cochlea as vibrations at the oval window. These vibrations set up fluid waves in the perilymph of the vestibular duct. As the fluid waves

ANATOMY SUMMARY The Cochlea

■ **Figure 10-18**

(a)

The cochlea is the region of the inner ear where sound waves are converted first into fluid waves, then into chemical signals, and finally into action potentials

(b) Note the oval window, the organ of Corti, and the round window. When the stapes vibrates against the oval window, it transfers sound into the cochlea. The organ of Corti contains hair cells that convert fluid wave energy into neurotransmitter release. Residual energy from the sound waves returns to the middle ear at the membrane known as the round window.

(c) The cochlear nerve transmits action potentials from the hair cells to the auditory cortex.

(d) The movement of the tectorial membrane with sound waves moves the cilia on the hair cells and affects neurotransmitter release by the hair cells.

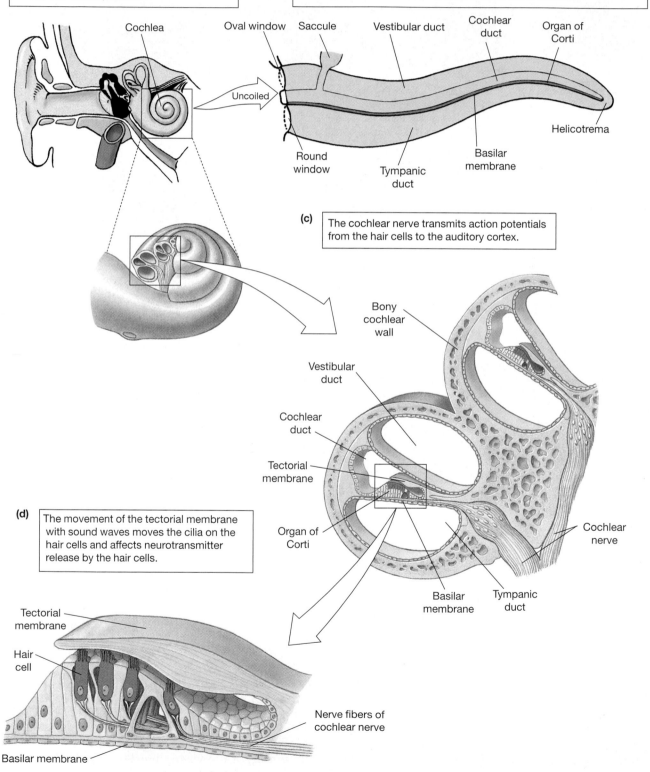

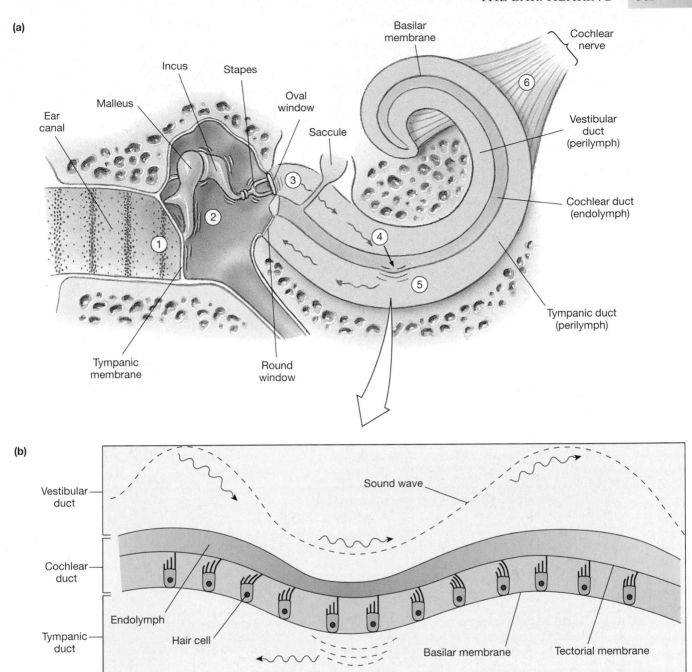

■ Figure 10-19   Sound transmission through the ear   (a) ① Sound waves in the air strike the tympanic membrane. ② The sound wave energy is transferred to bones of the middle ear, which vibrate. ③ The vibrations alternately compress and expand the fluid within the vestibular duct as the membrane of the oval window moves back and forth. This creates fluid waves within the cochlea. ④ The fluid waves push on the flexible membranes of the cochlear duct. ⑤ The waves transfer across the cochlear duct into the tympanic duct and are dissipated back into the air at the membrane of the round window. (b) Deformation of the cochlear duct by sound waves causes the tectorial membrane over the organ of Corti to move. This movement bends stereocilia of the hair cells.

travel down the duct, they displace the organ of Corti, creating up and down oscillations that bend the hair cells (Fig. 10-19 ■).

Hair cells, like taste receptors, are non-neural receptor cells. The apical surface of hair cells is modified into 50–100 stiffened cilia known as **stereocilia,** arranged in ascending height. The longest stereocilia of the hair cells are embedded in the overlying tectorial membrane. If the tectorial membrane moves, the underlying cilia do also.

When the hair cells move, their stereocilia flex, first one way, then the other. The stereocilia are attached to each other by protein bridges. These bridges act like little springs and are connected to "trap doors" that close off ion channels in the cilia membrane (Fig. 10-20a ■).

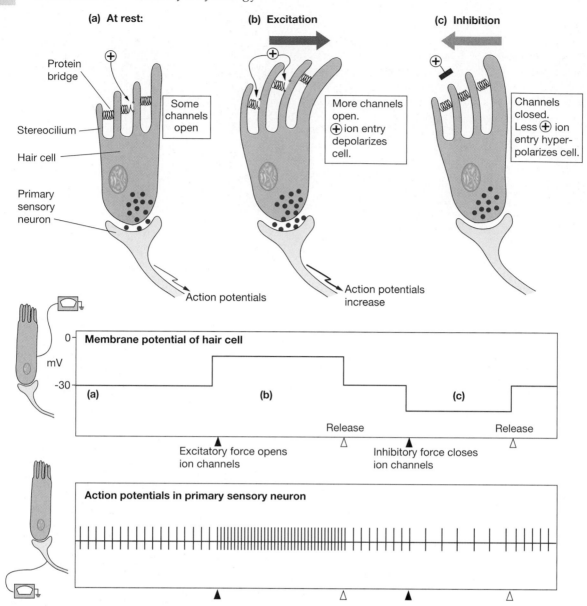

■ **Figure 10-20   Signal transduction in hair cells**   The stereocilia of hair cells have "trap doors" that close off ion channels. (a) At rest, about 10% of the ion channels are open and a tonic signal is sent by the sensory neuron. (b) When the hair cells are bent in one direction, more ion channels open, and the cell depolarizes. Depolarization increases the frequency of action potentials in the associated sensory neuron. (c) If the hair cells are bent back in the opposite direction, all ion channels close. The hair cell hyperpolarizes, and the signal rate in the sensory neuron decreases.

When the basilar membrane and cilia are in a neutral position, about 10% of the ion channels are open and there is a low level of tonic neurotransmitter release onto the primary sensory neuron.

When waves deflect the tectorial membrane so that the cilia bend toward the tallest members of a bundle, the bridges pop more channels open, cations enter the cell, and it depolarizes (Fig. 10-20b ■). Neurotransmitter release increases and the sensory neuron increases its firing rate.

When the basilar membrane pushes the cilia away from the tallest members, the springy bridges relax and all the ion channels close. Cation influx slows, the membrane hyperpolarizes, less transmitter is released, and the sensory neuron firing decreases.

The vibration pattern of waves reaching the inner ear is thus transformed into a pattern of action potentials going to the CNS. Because the tectorial membrane vibrations reflect the sound wave frequency, the hair cells and sensory neurons must be able to respond to sounds of nearly 20,000 waves per second, the highest frequency audible by a human ear.

Sounds Are Processed First in the Cochlea

Sound waves are processed by the auditory system so that they can be discriminated based on pitch, loudness, duration, and location. The initial processing for pitch, loudness, and duration takes place in the cochlea of each ear. Localization of sound is a higher function that re-quires sensory input from both ears coupled with so-phisticated computation by the brain.

Coding sound for pitch is primarily a function of the basilar membrane. This membrane is stiff and narrow near its attachment by the round and oval windows but widens and becomes more flexible near its distal end (Fig. 10-21 ■). High-frequency waves entering the vestibular duct create maximum displacement of the basilar membrane close to the oval window and are not transmitted very far along the cochlea. Low-frequency waves travel along the length of the basilar membrane and create their maximum displacement near the flexi-ble distal end.

This differential response to frequency transforms the temporal aspect of frequency (number of sound waves per second) into spatial coding by location along the basilar membrane. This spatial coding is preserved in the auditory cortex as neurons project from the basi-lar membrane to corresponding regions in the brain.

Loudness is coded by the ear in the same way that signal strength is coded in somatic receptors. The loud-er the noise, the more rapidly action potentials fire in the sensory neuron.

Auditory Pathways Project to the Auditory Cortex

Once sound waves have been transformed into electri-cal signals in the cochlea, the sensory neurons transfer their information to nuclei in the medulla. At that point, sound from each ear is projected to both *ipsilateral* (same side) and *contralateral* (opposite side) nuclei. As a result, each side of the brain gets information from both ears. The ascending tracts from the medulla divide, with col-laterals taking information to the reticular formation and the cerebellum. The main auditory pathway synapses in nuclei in the midbrain and thalamus before projecting to the auditory cortex (Fig. 10-2, p. 284).

The localization of a sound source is an integrative task for the CNS that requires simultaneous input from both ears. Unless sound is coming from directly in front

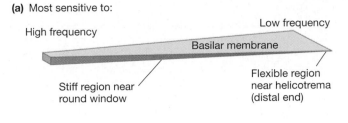

(a) Most sensitive to:

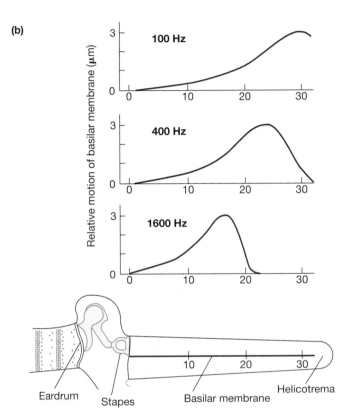

■ **Figure 10-21   Sensory coding for pitch**   (a) The basilar membrane of the cochlear duct is stiff and narrow close to oval and round windows. It widens and becomes more flexible near its distal end. (b) The frequency of sound waves determines the displacement of the basilar membrane. High-frequency sound displaces the mem-brane closer to the oval window than low-frequency sound. The lo-cation of active hair cells along the membrane creates a code that the brain translates as information about the pitch of sound.

...*continued from page 289*

Anant reports to the otolaryngologist that his attacks of dizziness strike without warning and last from 10 minutes to an hour. They often cause him to vomit. He also reports that he has a persistent low buzzing sound in one ear and that he does not seem to hear low tones as well as he could before. The buzzing sound (tinnitus) often gets worse during his dizzy attacks.

Question 1:   *Subjective tinnitus occurs when an abnormality somewhere along the anatomical pathway for hearing causes the brain to perceive a sound that does not exist outside the au-ditory system. Starting from the ear canal, name the auditory structures in which problems may arise.*

---

of a person, it will not reach both ears at the same time (see Fig. 10-5 ■). The brain records the time differential for sound arriving at the ears and uses complex com-putational skills to create a three-dimensional represen-tation of the sound source.

Hearing Loss May Result from Mechanical or Neural Damage

There are three forms of hearing loss: conductive, sen-sorineural, and central. In *conductive hearing loss*, sound is unable to be transmitted through the external or middle ear. The causes of conductive hearing loss range from an ear canal plugged with ear wax (*cerumen*), to fluid in the middle ear from an infection, to diseases or trauma that

Cochlear Implants One technique used to treat sensorineural hearing loss is the cochlear implant. The newest generation of cochlear implants has three components: a computerized speech processor, a transmitter, and 8–24 electrodes. The speech processor is a transducer that converts sound into electrical impulses. The smallest speech processors now fit behind the ear like a conventional hearing aid, thanks to microchip technology. The transmitter converts the processor's electrical impulses into radio waves that activate the electrodes. The electrodes are surgically placed under the skin and run directly into the cochlea where they stimulate the sensory nerves. Cochlear implants have been remarkably successful for many profoundly deaf people, allowing them to hear loud noises and modulate their own voices. In the most successful cases, individuals can even use the telephone.

impede vibration of the malleus, incus, or stapes. Correction of conductive hearing loss includes new microsurgical techniques in which the bones of the middle ear can be reconstructed.

Sensorineural hearing loss arises from damage to the structures of the inner ear, including degeneration of hair cells as a result of loud noises. This form of hearing loss is more difficult to treat, but amazing results have been obtained with cochlear implants attached to tiny computers. *Central hearing loss* results from damage to the neural pathways between the ear and cerebral cortex or from damage to the cortex itself. This form of hearing loss is relatively uncommon.

Hearing is probably our most important social sense. Suicide rates are higher among deaf people than among those who have lost their sight. Hearing connects us to other people and the world around us more than any other sense.

✔ Why is somatosensory information projected to only one hemisphere of the brain but auditory information projected to both hemispheres? (Hint: see Fig. 10-5 and 10-9.)

✔ Would a cochlear implant help a person who suffers from nerve deafness? From conductive hearing loss?

▶ THE EAR: EQUILIBRIUM

Equilibrium is a state of balance, whether the word is used to describe ion concentrations in body fluids or the positioning of the body in space. In body positioning, sensory information from the inner ear and from joint and muscle proprioceptors tells our brain the location of different body parts in relation to each other and to the environment. Visual information also plays an important role in equilibrium, as you know if you have ever gone to one of the 360° movie theaters where the scene tilts suddenly to one side and the audience tilts with it!

Our sense of equilibrium is mediated by hair cells lining the fluid-filled vestibular apparatus and semicircular canals of the inner ear. These receptors respond to changes in rotational, vertical, and linear acceleration. The hair cells function just like those of the cochlea (see Sound Transduction Through the Cochlea), but gravity and acceleration rather than sound waves provide the force that moves the stereocilia.

The Vestibular Apparatus Is Filled with Endolymph

The **vestibular apparatus** consists of two saclike **otolith organs,** the **utricle** and the **saccule,** along with three **semicircular canals** (Fig. 10-22a ■). The semicircular canals connect to the utricle at their bases. They are oriented at right angles to each other, like three planes that come together to form the corner of a box (Fig. 10-22b ■).

At one end of each canal is an enlarged chamber, the **ampulla** [bottle], that contains a sensory receptor known as a **crista** [a crest; plural *cristae*]. The crista consists of a gelatinous mass, the **cupula** [small tub], that stretches from floor to ceiling of the ampulla, closing it off (Fig. 10-22c ■). Embedded in the cupula are the cilia of hair cells. The basal membranes of the hair cells synapse on the sensory neurons of the **vestibular nerve** (a branch of the VIII cranial nerve).

The sensory receptors of the utricle and saccule, the **maculae,** are different from those of the cristae (Fig. 10-22d ■). Each macula consists of a gelatinous mass known as the **otolith membrane** in which small crystals of calcium carbonate called **otoliths** are embedded [*oto,* ear + *lithos,* stone]. Underneath the otolith membrane are hair cells whose cilia are connected to the membrane.

The vestibular apparatus, like the cochlear duct, is filled with fluid endolymph secreted by epithelial cells. Like cerebrospinal fluid, endolymph is secreted continuously and drains from the inner ear into the venous sinus in the dura mater of the brain. If endolymph production exceeds the drainage rate, buildup of fluid in the inner ear may cause *Ménière's disease*. This condition is marked by episodes of dizziness and nausea, apparently as a result of changes in fluid pressure within the vestibular apparatus. It may also cause hearing loss if the organ of Corti in the cochlear duct is damaged.

The Vestibular Apparatus Provides Information About Movement and Position in Space

The special sense of equilibrium has two components: a *dynamic component* that tells us about our movement through space and a *static component* that tells us if our head is displaced from its normal upright position. The three semicircular canals sense rotational acceleration in various directions. The otolith organs tell us about linear acceleration and head position.

The receptor cells of both sets of sense organs, the hair cells, function just like the hair cells of the organ of

ANATOMY SUMMARY Vestibular Apparatus

■ **Figure 10-22**

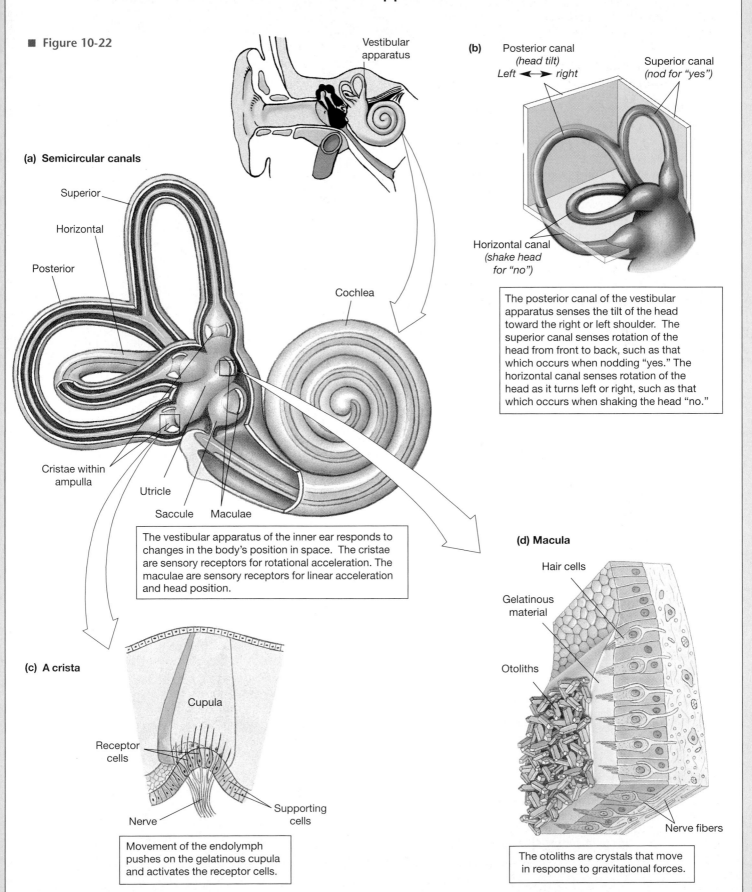

Vestibular apparatus

(a) Semicircular canals

Superior

Horizontal

Posterior

Cochlea

Cristae within ampulla

Utricle

Saccule   Maculae

(b) Posterior canal *(head tilt)*
Left ⟷ right

Superior canal *(nod for "yes")*

Horizontal canal *(shake head for "no")*

The posterior canal of the vestibular apparatus senses the tilt of the head toward the right or left shoulder. The superior canal senses rotation of the head from front to back, such as that which occurs when nodding "yes." The horizontal canal senses rotation of the head as it turns left or right, such as that which occurs when shaking the head "no."

The vestibular apparatus of the inner ear responds to changes in the body's position in space. The cristae are sensory receptors for rotational acceleration. The maculae are sensory receptors for linear acceleration and head position.

(c) A crista

Cupula

Receptor cells

Nerve

Supporting cells

Movement of the endolymph pushes on the gelatinous cupula and activates the receptor cells.

(d) Macula

Hair cells

Gelatinous material

Otoliths

Nerve fibers

The otoliths are crystals that move in response to gravitational forces.

Corti. When the cilia bend in one direction, they depolarize, and when they bend in the opposite direction, they hyperpolarize. Vestibular hair cells have one long cilium called a **kinocilium** that is located at one side of the bundle, creating a reference point for the direction of bending.

The Semicircular Canals Sense Rotational Acceleration

How is rotation sensed? As the head turns, the bony structure of the skull and the membranous labyrinth move, but the endolymph within the canals cannot keep up because of inertia. The drag of the fluid bends the cupula and its hair cells in the direction opposite to the way the head is turning.

For an analogy, think of pulling a paintbrush (the crista attached to the wall of the canal) through sticky wet paint (the endolymph) on a board. If you pull the brush to the right, the drag of the paint on the bristles bends them back to the left (Fig. 10-23 ■). In the same way, the inertia of the fluid in the semicircular canal pulls the cupula and the cilia of the hair cells to the left when the head turns right.

If rotation continues, the moving endolymph finally catches up. Then if head rotation stops suddenly, the fluid has built up momentum and cannot stop as fast. The fluid continues to rotate in the direction of the head rotation, leaving the person with a turning sensation. If the sensation is strong enough, the person may throw his or her body in the direction opposite the rotation in a reflexive attempt to compensate for the apparent loss of equilibrium.

The Otolith Organs Sense Linear Acceleration and Head Position

The otolith organs are arranged to sense linear forces. The maculae of the utricle are horizontal when the head is in its normal upright position. If the head tips back, the otoliths in the gelatinous membrane slide backward owing to gravity acting on the dense particles (Fig. 10-24 ■). The cilia of the hair cells bend and set off a signal.

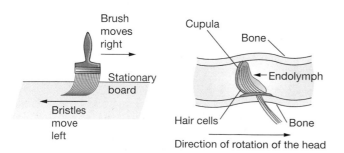

■ **Figure 10-23** **Transduction of rotational forces in the cristae** When the head turns, inertia keeps endolymph inside the ampulla from moving as rapidly as the surrounding cranium. When the head turns right, the endolymph pushes the cupula to the left, just as a brush dragged to the right causes the brushes' bristles to bend to the left.

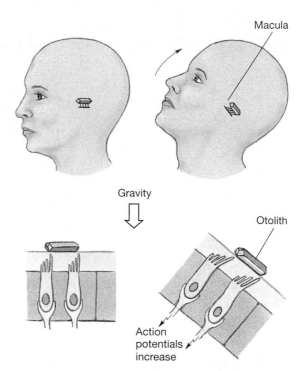

■ **Figure 10-24** **Otolith organs** The crystalline otoliths are attached to gelatinous material in the maculae. When the head tilts, gravity causes the otoliths to slide, pulling the stereocilia of the hair cells out of their vertical position and increasing the action potentials in the sensory neurons.

The maculae of the saccule are oriented vertically when the head is erect and are sensitive to vertical forces, such as dropping downward in an elevator. The brain analyzes the pattern of which hair cells are depolarized and which are hyperpolarized in order to compute head position and direction of movement.

Equilibrium Pathways Project Primarily to the Cerebellum

The hair cells of the vestibular apparatus stimulate primary sensory neurons in the vestibular nerve. Those neurons either synapse in the vestibular nuclei of the medulla or run without synapsing to the cerebellum (Fig. 10-25 ■). Collateral pathways run from the medulla to the cerebellum or upward through the reticular formation and thalamus. There are some poorly defined pathways to the cerebral cortex, but most integration for equilibrium comes from the cerebellum. Descending pathways from the vestibular nuclei go to certain motor neurons involved in eye movement. These pathways help keep the eyes locked on an object as the head turns.

✔ Why does hearing decrease if you have an ear infection with fluid buildup in the middle ear?

✔ When dancers perform multiple turns, they try to keep their vision fixed on a single point ("spotting"). How does spotting keep the dancer from getting dizzy?

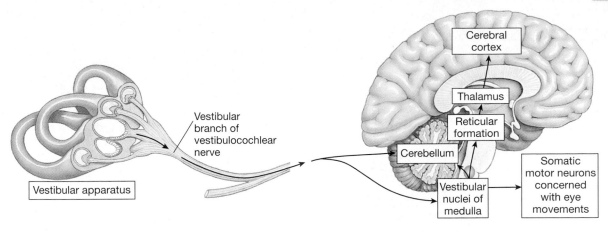

■ **Figure 10-25   Central nervous system pathways for equilibrium**

▶ THE EYE AND VISION

The eye is a sensory receptor that functions much like a camera. It focuses light on a light-sensitive surface (the retina) using a lens and an aperture or opening (the pupil), whose size can be adjusted to change the amount of entering light. **Vision,** the process through which light reflected from objects in our environment is translated into a mental image, can be divided into three steps. In the first step, light enters the eye and is focused by a lens on the retina. In the second step, the photoreceptors of the retina transduce light energy into an electrical signal. The third step involves neural pathways and processing of these electrical signals.

The electromagnetic energy that we call **visible light** is composed of waves with a frequency of 4.0–7.5×10^{14} waves per second (hertz) and a wavelength of 400–750 nanometers (nm). But these waves are only a small fraction of the electromagnetic spectrum that extends from high-energy, very short wavelength waves such as X-rays and gamma rays to low-energy, lower-frequency microwaves and radio waves (Fig. 10-26 ■).

The waves of ultraviolet and infrared light border the ends of the visible light spectrum. Although our eyes are unable to perceive these wavelengths, some animals can see light in the infrared and ultraviolet range. For example, bees use ultraviolet "runways" on flowers to guide them to pollen and nectar.

The Optic Tract Extends from the Eye to the Visual Cortex

The anatomy of the eye and the neural pathways for vision are shown in the Anatomy Summary, Figure 10-27 ■. The eye itself is a hollow sphere divided into two compartments separated by a lens. The compartment in front of the lens and its attachment is filled with the **aqueous humor,** a low-protein plasma-like fluid secreted by the ciliary epithelium supporting the lens (Fig. 10-27b ■). Behind the lens is a much larger chamber filled mostly with the **vitreous humor,** a clear, gelatinous matrix.

...continued from page 306

Although many vestibular disorders can cause the symptoms Anant is experiencing, two of the most common are benign positional vertigo (BPV) and Ménière's disease. In benign positional vertigo, the calcium crystals normally embedded in the otolith membrane of the maculae become dislodged and float toward the semicircular canals. The primary symptom of BPV is brief episodes of severe dizziness brought on by a change in position. People with this condition often say that they cannot go to sleep because they feel dizzy when they turn over in bed.

Question 2:   *When a person with BPV changes position, the displaced crystals float toward the semicircular canals. Why would this cause dizziness?*

Question 3:   *Compare the symptoms of BPV and Ménière's disease. On the basis of Anant's symptoms, which condition do you think he has?*

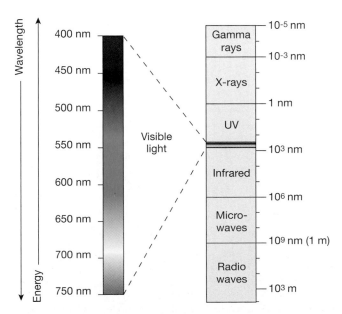

■ **Figure 10-26   The electromagnetic spectrum**

ANATOMY SUMMARY The Eye

■ Figure 10-27

(a) Neural pathway for vision

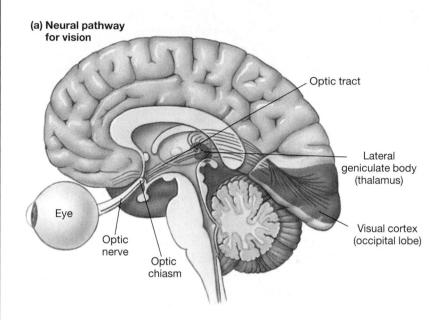

Optic tract

Lateral geniculate body (thalamus)

Eye

Visual cortex (occipital lobe)

Optic nerve

Optic chiasm

(b) Cross section of the eye

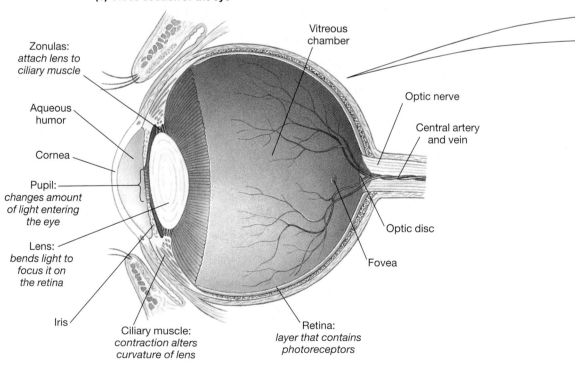

Zonulas: *attach lens to ciliary muscle*

Aqueous humor

Cornea

Pupil: *changes amount of light entering the eye*

Lens: *bends light to focus it on the retina*

Iris

Ciliary muscle: *contraction alters curvature of lens*

Vitreous chamber

Optic nerve

Central artery and vein

Optic disc

Fovea

Retina: *layer that contains photoreceptors*

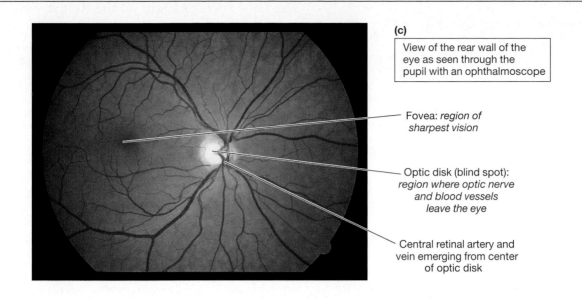

(c)

View of the rear wall of the eye as seen through the pupil with an ophthalmoscope

Fovea: *region of sharpest vision*

Optic disk (blind spot): *region where optic nerve and blood vessels leave the eye*

Central retinal artery and vein emerging from center of optic disk

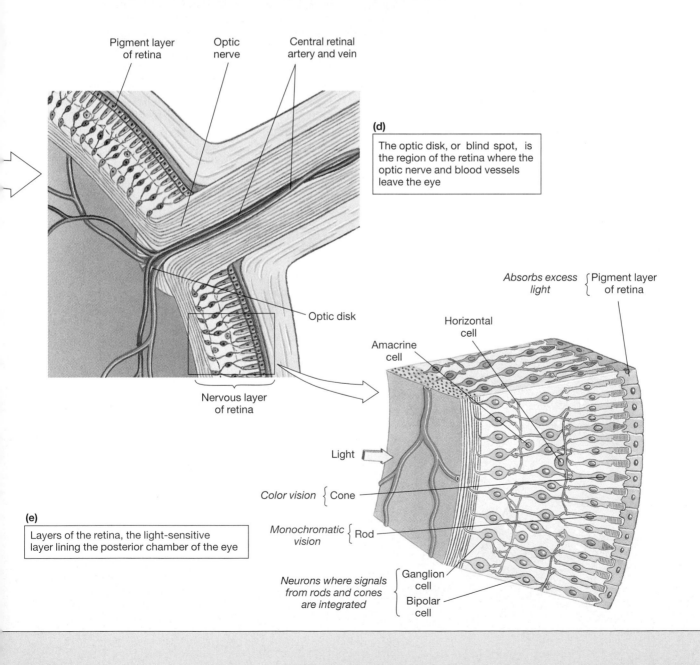

Pigment layer of retina

Optic nerve

Central retinal artery and vein

(d)

The optic disk, or blind spot, is the region of the retina where the optic nerve and blood vessels leave the eye

Optic disk

Nervous layer of retina

Absorbs excess light { Pigment layer of retina

Horizontal cell

Amacrine cell

Light

Color vision { Cone

Monochromatic vision { Rod

(e)

Layers of the retina, the light-sensitive layer lining the posterior chamber of the eye

Neurons where signals from rods and cones are integrated { Ganglion cell

Bipolar cell

DIABETES **Diabetic retinopathy** Diabetes mellitus is a metabolic disease whose complications can affect every system in the body. For example, did you know that diabetes is the leading cause of blindness in the younger population? *Diabetic retinopathy* is a condition in which the cells of the retina are damaged, leading to vision changes and eventual blindness. In the early stages, blood flow to the retina is decreased by fatty deposits in the blood vessels. In response, the small blood vessels (the *microvasculature*) of the retina and their supporting fibrous tissue proliferate. The overgrowth of the blood vessels can lead to retinal detachment and blindness.

Light enters the eye at the **cornea,** a transparent disk of tissue on the anterior surface. After crossing the aqueous humor, light passes through the opening of the pupil and strikes the **lens,** a transparent structure with two convex surfaces. The cornea and lens together bend incoming light rays so that they focus on the **retina,** the light-sensitive lining of the eye that contains the *photoreceptors.*

When viewed through the pupil of the eye with an ophthalmoscope [*ophthalmos,* eye], the retina is crisscrossed with small arteries and veins that radiate out from one spot, the **optic disk** (Fig. 10-27c ■). The optic disk is the location where neurons of the visual pathway form the **optic nerve** and exit the eye (Fig. 10-27d ■).

The optic nerves from each eye go to the **optic chiasm** in the brain, where some of the fibers cross to the opposite side. After synapsing in the **lateral geniculate body** (*lateral geniculate nucleus*) of the thalamus, the neurons of the tract terminate in the occipital lobe at the visual cortex (Fig. 10-27a ■).

External structures associated with the eye include six extrinsic eye muscles, skeletal muscles that attach to the outer surface of the sphere and control eye movements. The eyelids close over the anterior surface of the eye, and the *lacrimal apparatus,* a system of glands and ducts, keeps a continuous flow of tears washing across the cornea so that it remains moist and free of debris.

✔ What function(s) might the aqueous humor serve?

✔ What would happen to the eye if the normal drainage pathways for the aqueous humor became blocked?

The Lens Focuses Light on the Retina

Light striking the eye is modified two ways before it strikes the retina. First, the amount of light that reaches photoreceptors is modulated by changing the size of the pupil. Second, the light waves are focused by changing the shape of the lens. In this section we describe the principles of **optics,** the physical relationship between light and lenses.

Image Focusing and the Lens   Light waves passing from air into a medium of different density such as glass or water will bend, or **refract.** Light entering the eye is refracted twice, most strongly when it passes through the cornea and again when it passes through the lens. To simplify this discussion, we will consider only the refraction that occurs as light passes through the lens. The angle of refraction depends on two factors: (1) the difference in density of the two objects, and (2) the angle at which the light meets the face of the object into which it is passing. This angle depends on the curvature of the lens surface as well as on the angle of the beam of light.

If light waves strike a surface perpendicular to them, the light will pass through without bending. But if the surface is not perpendicular, light waves striking the surface will bend. Parallel light waves striking a *concave* lens such as that shown in Figure 10-28a ■ will be refracted into a wider beam. A light beam striking a spherical *convex* lens will be bent inward and focused into a point (Fig. 10-28b ■). You may have demonstrated the properties of a convex lens if you used a magnifying glass to focus sunlight onto a piece of paper or other surface.

The point where parallel light waves passing through a lens converge is known as the **focal point** of the lens. The distance from the center of the lens to the focal point is known as the **focal length** (or *focal distance*). These parameters are fixed for any given lens.

For us to see an object in focus, the focal point must fall precisely on the retina of the eye. In Figure 10-29a ■, light rays from a distant source are parallel as they strike the lens, resulting in a focal point that falls on the retina. The object is therefore in focus. For the human eye, any object that is 20 feet or more away from the eye will create parallel light rays.

But what happens if an object is closer than 20 feet to the lens? In that case, the light rays reflected from the

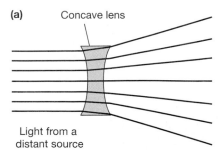

(a) Concave lens

Light from a distant source

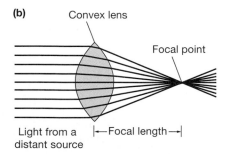

(b) Convex lens

Focal point

Light from a distant source |←Focal length→|

■ **Figure 10-28   Refraction of light**   (a) Light is refracted and scattered by a concave lens. (b) A spherical convex lens focuses a beam of light to a single point called the focal point. The distance between the center of the lens and the focal point is called the focal length (focal distance).

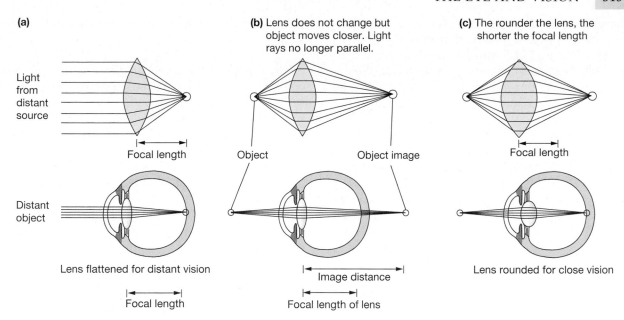

(a)

Light from distant source

Focal length

Distant object

Lens flattened for distant vision

Focal length

(b) Lens does not change but object moves closer. Light rays no longer parallel.

Object   Object image

Image distance

Focal length of lens

(c) The rounder the lens, the shorter the focal length

Focal length

Lens rounded for close vision

■ **Figure 10-29   Optics**   (a) Light reflecting off a distant object reaches the eye as nearly parallel rays. The lens is flattened so that the focal point falls on the retina. (b) If an object moves within 20 feet, the light rays from it are no longer parallel. The object is seen out of focus because the light beam is not focused on the retina. (c) To keep an object in focus as it moves closer, the lens becomes more rounded. This adjustment is known as accommodation.

object are not parallel and therefore strike the lens at an oblique angle (Fig. 10-29b ■). The light rays falling on the retina are not focused to a point, so the object appears fuzzy and out of focus.

To keep a near object in focus, the lens must round to increase the angle of refraction (Fig. 10-29c ■). The adjustment causes the light rays to converge *on* the retina instead of behind it, so the object comes into focus. The process by which the eye adjusts the shape of the lens to keep objects in focus is known as **accommodation.** If an object is so close to the lens that it falls within the focal length, no amount of accommodation can bring the object into focus. The closest distance at which you can focus on an object is known as the **near point of accommodation.**

You can demonstrate the accommodation reflex easily by closing one eye and holding your hand up about 8 in. in front of your open eye, fingers spread widely apart.

Focus your eye on some object in the distance that is between your fingers. Notice that when you do so, your fingers are visible but out of focus. Your lens is flattened for distance vision, so the focal point for near objects falls behind the retina. Those objects appear out of focus.

Now shift your gaze to your fingers and notice that they come into focus. The light waves coming off your fingers have not changed their angle, but your lens has become more rounded and the light rays now converge on the retina.

How can the lens, which is clear and does not have any muscle fibers in it, change shape? The lens is attached to the ciliary muscle of the eye by inelastic fibers known as **zonulas** (Fig. 10-30 ■). If no tension is placed on the lens by the zonulas, the lens assumes its natural rounded shape because of the elasticity of its capsule. If the zonulas pull on the lens, it flattens out and assumes the shape required for distance vision.

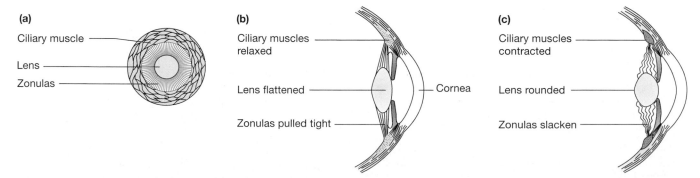

(a)

Ciliary muscle

Lens

Zonulas

(b)

Ciliary muscles relaxed

Lens flattened

Zonulas pulled tight

Cornea

(c)

Ciliary muscles contracted

Lens rounded

Zonulas slacken

■ **Figure 10-30   Accommodation**   (a) The lens is attached to the circular ciliary muscles by inelastic fibers known as zonulas. (b) When the ciliary muscle is relaxed, the zonulas pull tight and keep the elastic lens flattened. (c) When the ciliary muscle contracts, it releases tension on the zonulas, which go slack. The elastic lens can then assume a more rounded shape.

Tension on the zonulas is controlled by the **ciliary muscle,** a ring of smooth muscle that surrounds the lens. When the circular ciliary muscles contract, the muscle ring gets smaller, releasing tension on the zonulas so that the lens rounds. When the ciliary muscles are relaxed, the ring is more open and the lens is pulled into a flatter shape.

Young people can focus on items as close as 8 cm, but the accommodation reflex diminishes from the age of 10 on. By age 40, accommodation is only about half of what it was at age 10, and by age 60, many people lose the reflex completely because the lens has lost flexibility. The loss of accommodation, **presbyopia,** is the reason that most people begin to wear reading glasses in their 40s.

Two other common vision problems, nearsightedness (**myopia**) and farsightedness (**hyperopia**), occur when the focal point falls either in front of or behind the retina, respectively (Fig. 10-31 ■). These conditions can be corrected by placement of a lens with the appropriate curvature in front of the eye to change the focal length. A third common vision problem, **astigma-tism,** is usually caused by a cornea that is not a perfectly shaped dome.

Light Entering the Eye and the Pupil The human eye functions over a 100,000-fold range of light intensity. Most of this ability comes from the sensitivity of the photoreceptors, but the pupils of the eyes assist by regulating the amount of light that falls on the retina. In bright sunlight, the pupils constrict to about 1.5 mm. In the dark, the opening dilates to 8 mm, a 28-fold increase.

In addition to regulating the amount of light that hits the retina, the pupils create what is known as **depth of field.** A simple example comes from photography. Imagine a picture of a puppy sitting in the foreground of a field of wildflowers. If only the puppy and the flowers immediately around her are in focus, the picture is said to have a shallow depth of field. But if the puppy and the wildflowers all the way back to the horizon are in focus, the picture has full depth of field. Depth of field is created by constricting the pupil (or shutter on the camera) so that only a narrow beam of light enters the eye. In this way, more of the depth of the image is focused on the retina.

Light hitting the retina activates the **pupillary reflex,** a parasympathetic pathway that constricts the circular pupillary muscles and decreases entering light. To open the pupil, a set of radial muscles that lie perpendicular to the circular muscles contract under the influence of sympathetic neurons.

✔ An antagonist for what neurotransmitter could be used to dilate the pupils before an eye exam?

✔ What kind of lens is needed to correct myopic vision? Explain.

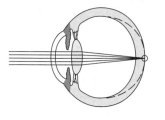
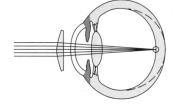

(a) Hyperopia   Hyperopia (corrected with a convex lens)

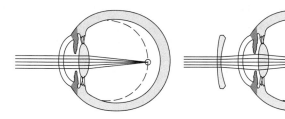

(b) Myopia   Myopia (corrected with a concave lens)

Figure question: Explain (1) what the corrective lenses do to the refraction of light, and (2) why each lens type corrects the visual defect.

■ **Figure 10-31** **Visual defects** (a) Hyperopia, or far-sightedness, occurs when the length of the eyeball is too short and the focal point falls behind the retina. It is corrected with a convex lens. (b) Myopia, or near-sightedness, occurs when the eyeball is too long and the focal point falls in front of the retina. It is corrected with a concave lens.

Phototransduction Occurs at the Retina

Light that enters the pupil hits the retina, the sensory organ of the eye. The retina develops from the same embryonic tissue as the CNS, and like the cortex of the brain, its neurons are organized into layers. **Photoreceptors** tran-

…continued from page 309

The otolaryngologist strongly suspects that Anant has Ménière's disease, with excessive endolymph within the vestibular apparatus and cochlea. Many treatments are available, from simple dietary changes to surgery. For now, the physician suggests that Anant limit his salt intake and take diuretics, drugs that cause the kidneys to remove excess fluid from the body.

Question 4: *Why is limiting salt (NaCl) intake suggested as a treatment for Ménière's disease? (Hint: Remember that water follows solute.)*

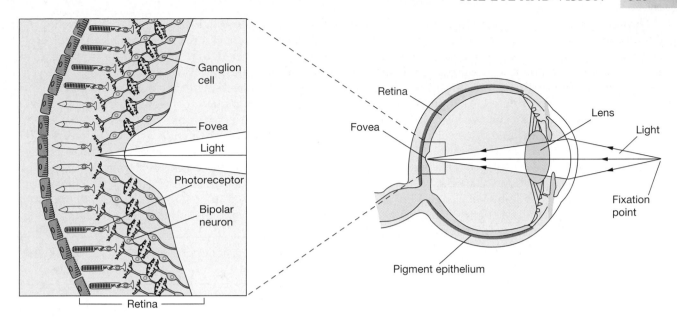

■ **Figure 10-32   Photoreceptors in the retina**   Light striking the retina usually must pass through several layers of retinal cells prior to reaching the photoreceptor layer. However, in the fovea the overlying layers of neurons are pushed aside and the light strikes the photoreceptors directly. For this reason, vision is sharpest in the fovea.

duce light energy into electrical energy that passes along a pathway consisting of **bipolar neurons,** then **ganglion cells.** The axons of ganglion cells form the *optic nerve* that leaves the eye at the optic disk. Because the optic disk has no photoreceptors, images projected onto this region will not be seen, creating the so-called **blind spot.**

Backing the photosensitive portion of the retina is a black-pigmented epithelium layer. Its function is to absorb any light rays that escape the photoreceptors, preventing distracting light from reflecting inside the eye and distorting the visual image. The black color of the epithelial cells comes from granules of the pigment *melanin.*

The layers of retinal neurons are in reverse order from what you might expect. Instead of the photoreceptors facing the pupil where photons of light will strike them first, they are at the back, with their photosensitive tips against the pigment epithelium (Fig. 10-32 ■). Most light entering the eye must pass through relatively transparent layers of cell bodies and processes before striking the photoreceptors.

The one exception is a small region of the retina known as the **fovea** [pit]. In this area, the photoreceptors receive light directly because the intervening neurons are pushed off to the side. The fovea is the region of most acute vision and is the point on which light is focused when you look at an object.

For example, look at Figure 10-33 ■. The eye is focused on the green bar so light from that section of the visual field falls on the fovea and will be in sharp focus. Notice that the image falling on the retina is upside down. Later visual processing reverses the image again so that we perceive it in the correct orientation.

Photoreceptors: Rods and Cones   There are two types of photoreceptors in the eye: rods and cones. Rods are the most numerous by a 20:1 ratio, except in the fovea, where cones are concentrated. **Rods** are responsible for monochromatic nighttime vision and function in low light levels. **Cones** are responsible for *high-acuity* vision and color vision during the daytime, when light levels are higher. *Acuity* means keenness and is derived from the Latin *acuere,* meaning "to sharpen."

The two types of photoreceptors have the same basic structure: (1) an outer segment whose tip touches the pigment epithelium of the retina, (2) an inner segment that contains the cell nucleus and organelles for ATP and protein synthesis, and (3) a base segment that synapses with the bipolar cells (Fig. 10-34 ■). In the outer segment, the cell membrane has deep folds that form disklike layers, like candy mints stacked in a wrapper. In rods, toward the tip of the outer segments, these layers actually separate from the cell membrane and form free-floating disks. In the cones, they stay attached.

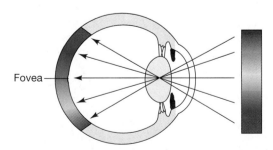

■ **Figure 10-33   Image projection onto the retina**   The image projected onto the retina is upside down. Visual processing in the brain reverses the image.

The light-sensitive **visual pigments** are bound to the cell membranes of the photoreceptors. These pigments are the transducers that convert light energy into a change in membrane potential. Rods have one type of visual pigment, **rhodopsin.** Cones have three different pigments that are closely related to rhodopsin.

Cones are responsible for color vision and contain pigments that are excited by different wavelengths of light. White light is a combination of colors, as you see when you separate white light by passing it through a prism. The eye contains cones for red, green, and blue light. Each cone type is stimulated by a range of light wavelengths but is most sensitive to a particular wavelength (Fig. 10-35 ■). Red, green, and blue are the three primary colors for light, just as red, blue, and yellow are the three primary colors for paints.

The color of objects in our environment depends on the wavelengths of light reflected by the object. Green leaves reflect green light and bananas reflect yellow light. White objects reflect most wavelengths. Black objects absorb most wavelengths, one reason they heat up in sunlight while white objects stay cool.

Our brain recognizes the color of an object by interpreting the combination of signals coming to it from the three different color cones. This seems like a simple process. However, the details of color vision are still not well understood and there is some controversy about how color is processed in the cerebral cortex. **Color-blindness** is a condition in which a person inherits a defect in one or more of the three types of cones and is therefore unable to distinguish certain colors.

✔ Why is our vision in the dark in black and white rather than color?

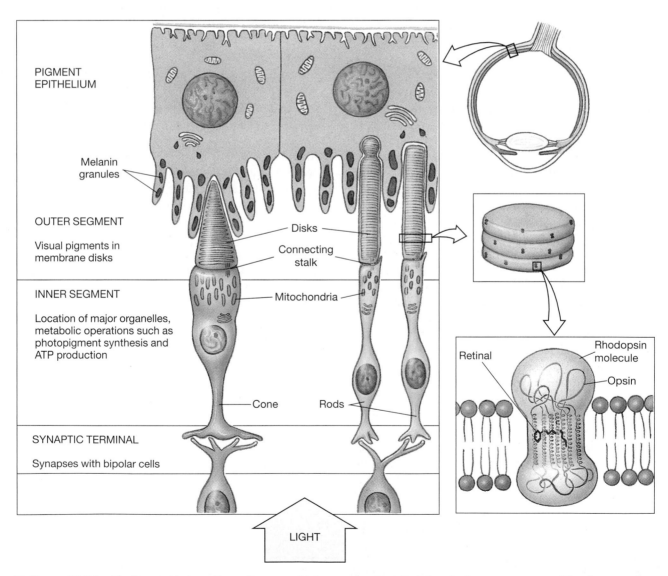

■ **Figure 10-34   Photoreceptors: rods and cones**   Light transduction takes place in the outer segment of the photoreceptor, where visual pigments are contained in the membranes of disklike structures. In rods, the visual pigment is rhodopsin, composed of opsin and an attached retinal molecule. Depolarization of the photoreceptor causes release of neurotransmitter from the synaptic terminal onto bipolar cells. The dark pigment layer absorbs extra light and prevents it from reflecting back and distorting vision.

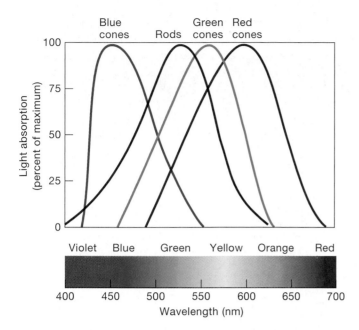

Violet   Blue   Green   Yellow   Orange   Red

Graph question: Which pigment absorbs light over the broadest spectrum of wavelengths? Over the narrowest? Which pigment absorbs the most light at 500 nm?

■ **Figure 10-35   Light absorption of visual pigments** There are three types of cone pigment, each with a characteristic light absorption spectrum.

Phototransduction   The process of phototransduction is similar for rhodopsin and the three color pigments. The visual pigment of the rods, rhodopsin, is composed of two sections: **opsin,** a protein embedded in the membrane of the rod disks, and **retinal,** a vitamin A derivative that is the light-absorbing portion of the pigment. In the absence of light, retinal binds snugly into a binding site on the opsin (Fig. 10-36a ■). When activated by as little as one photon of light, retinal changes shape to a new configuration. The activated retinal no longer binds to opsin and is released from the pigment in the process known as **bleaching.**

But how does rhodopsin bleaching lead to action potentials traveling through the optical pathway? In order to understand the link, we must look at other properties of the rods. As you learned in Chapters 5 and 8, electrical signals in cells occur as a result of the movement of ions between the intracellular and extracellular compartments. Rods contain many cation channels that allow Na^+ to enter the rod and K^+ channels that allow K^+ to leak out of the cell.

When the rods are in darkness and rhodopsin is not active, cyclic GMP levels in the rod are high and both sets of channels are open. Sodium ion influx is greater than K^+ efflux, so the rod stays depolarized to an average membrane potential of -40 mV, instead of the more usual -70 mV. At this membrane potential, there is tonic (continuous) release of neurotransmitter from the synaptic portion of the rod onto the adjacent bipolar cell.

When light activates rhodopsin, a second messenger cascade is initiated through **transducin,** the same G protein found in bitter taste receptors (Fig. 10-36b ■). The second messenger cascade decreases the concentration of cGMP, and, in turn, the cation channels close. As a result, Na^+ influx slows or stops. Less Na^+ influx but continued K^+ efflux causes the inside of the cell to hyperpolarize. Neurotransmitter release onto the bipolar neurons consequently decreases. Bright light will close all Na^+ channels and stop all neurotransmitter release. Dimmer light will cause a response that is graded in proportion to the light intensity.

Since a change in neurotransmitter release by rods is mediated through a second messenger system, the response of rods to changes in light intensity is much slower than you might expect for a process associated with the nervous system. When you move from bright light into the dark, or vice versa, the time required for your eyes to adapt to the new light level is partially determined by the slow change in rod membrane potential as ion channels open or close. The recovery of rhodopsin from bleaching also can take some time. After activation, retinal diffuses out of the rod and is transported into the pigment epithelium. There, it reverts to its inactive form before moving back into the rod and being reunited with opsin (Fig. 10-36c ■).

We now turn from the cellular mechanism of light transduction to the processing of light signals in the bipolar and ganglion cells of the retina.

Signal Processing in the Retina Occurs When Light Strikes Visual Fields

One hallmark of signal processing in the retina is *convergence* [∞ p. 240], in which multiple neurons synapse onto a single postsynaptic cell (Fig. 10-37 ■). Groups of 15 to 45 photoreceptors may synapse onto one bipolar neuron, although in the fovea, a few cones have a 1:1 relationship with their bipolar neurons. Multiple bipolar neurons in turn innervate a single ganglion cell, so that the information from hundreds of millions of photoreceptors is condensed down to a mere 1 million axons leaving the eye in each optic nerve.

...*continued from page 314*

Anant's condition does not improve with the low-salt diet and diuretics, and he continues to suffer from disabling attacks of vertigo with vomiting. As a last resort, surgery is sometimes performed to treat severe cases of Ménière's disease. In one surgical procedure for the disease, the vestibular nerve is severed. This surgery is difficult to perform, as the vestibular nerve lies near many other important nerves, including facial nerves and the auditory nerve. Patients who undergo this procedure are advised that the surgery can result in deafness if the auditory nerve is inadvertently severed.

Question 5:   *Why would severing the vestibular nerve alleviate Ménière's disease?*

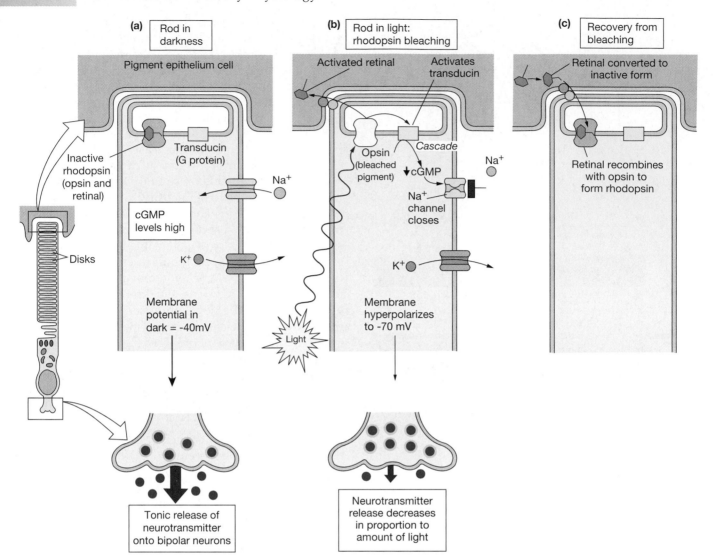

(a) Rod in darkness

Pigment epithelium cell

Inactive rhodopsin (opsin and retinal)

Transducin (G protein)

cGMP levels high

Na$^+$

Disks

K$^+$

Membrane potential in dark = -40mV

Tonic release of neurotransmitter onto bipolar neurons

(b) Rod in light: rhodopsin bleaching

Activated retinal

Activates transducin

Opsin (bleached pigment)

Cascade

↓cGMP

Na$^+$

Na$^+$ channel closes

K$^+$

Light

Membrane hyperpolarizes to -70 mV

Neurotransmitter release decreases in proportion to amount of light

(c) Recovery from bleaching

Retinal converted to inactive form

Retinal recombines with opsin to form rhodopsin

■ **Figure 10-36   Phototransduction in rods**   (a) In darkness, the visual pigment rhodopsin is in its inactive form. Cyclic GMP levels are high within the rod, and both Na$^+$ and K$^+$ channels are open. The net influx of Na$^+$ depolarizes the cell to −40 mV. At this membrane potential, there is tonic release of neurotransmitter onto the bipolar neurons. (b) When a photon of light strikes the rhodopsin molecule, it activates the retinal molecule, causing it to unbind from the rhodopsin. This process is known as bleaching. The resultant opsin molecule activates a second messenger cascade that decreases intracellular cGMP and closes the Na$^+$ channels. With less Na$^+$ influx, the cell hyperpolarizes and neurotransmitter release decreases. (c) In the recovery phase, active retinal is reduced to its inactive form in the cells of the pigment layer. It then returns to the rod and recombines with the opsin to form rhodopsin.

The release of the neurotransmitter glutamate from the photoreceptors onto bipolar neurons is the first step in the signal pathway. Glutamate excites some bipolar neurons but inhibits others, depending on the type of glutamate receptor on the bipolar neuron. In this fashion, one stimulus (light) creates two responses with a single neurotransmitter.

The second synapse in the pathway occurs between the bipolar neurons and the ganglion cells. These connections can also be either excitatory or inhibitory. We know more about ganglion cells because they lie on the surface of the retina, and their axons are the most accessible to researchers. Extensive studies have been done in which researchers stimulated the retina with carefully placed light and evaluated the response of the ganglion cells.

Each ganglion cell has a particular area of retina to which it is linked. These areas, known as **visual fields,** have two interesting properties. First of all, they are circular, un-

like the irregular shape of receptive fields in the somatic sensory system. The visual field of a ganglion cell near the fovea is quite small. Only a few receptors are associated with each ganglion, so visual acuity is greatest in these areas. At the edge of the visual field, a single ganglion cell may have multiple receptors, so vision is not as sharp.

The second property of visual fields is that they are divided into two sections: a round center and its doughnut-shaped **surround** (Fig. 10-38 ■). There are two basic types of visual fields. In the "on-center/off-surround" field, the associated ganglion cell will respond most strongly with a series of action potentials when light is brightest in the center of the field. If light is brightest in the "off" surround region of the field, the ganglion cell will be inhibited and stop firing action potentials.

The reverse happens with off-center/on-surround visual fields. If light is uniform across the visual field,

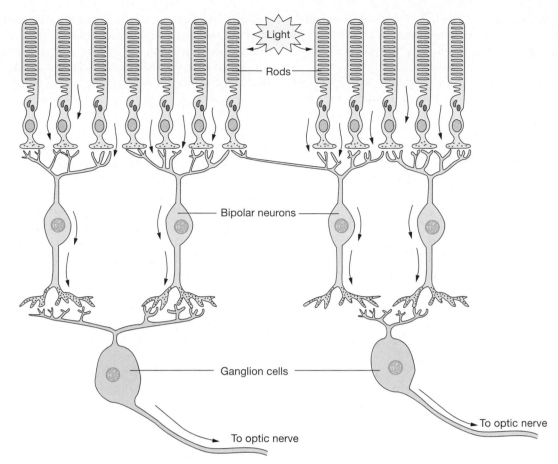

■ **Figure 10-37   Convergence in the retina**   Multiple photoreceptors (rods or cones) can converge onto a single bipolar neuron, although rods and cones do not converge onto the same cell. A single receptor can provide input into more than one bipolar neuron. Multiple bipolar neurons can converge onto a single ganglion cell.

the ganglion cell responds weakly. Thus, the retina uses *contrast* rather than absolute light intensity to recognize objects in the environment. One advantage of using contrast is that it allows better detection of weak stimuli.

Scientists have now identified two types of ganglion cells in the retina. Large *magnocellular* ganglion cells, or **M cells,** carry information about movement, location, and depth perception. Smaller *parvocellular* ganglion cells, or **P cells,** transmit signals that pertain to color, form, and texture of objects in the visual field. Once action potentials leave these cells, they travel along the optic nerves to the CNS for further processing.

✔ Why is the difference in visual acuity between the fovea and the edge of the visual field similar to the difference in touch discrimination between the fingertips and the skin of the arm?

Visual Processing in the Central Nervous System Takes Place in the Visual Cortex

The optic nerves enter the brain in a region known as the **optic chiasm.** This is also the point at which some fibers from each eye cross to the contralateral side of the

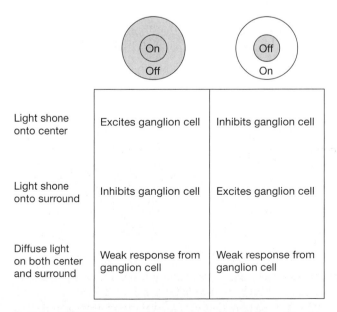

	On center / Off surround	Off center / On surround
Light shone onto center	Excites ganglion cell	Inhibits ganglion cell
Light shone onto surround	Inhibits ganglion cell	Excites ganglion cell
Diffuse light on both center and surround	Weak response from ganglion cell	Weak response from ganglion cell

■ **Figure 10-38   Ganglion cell receptive fields**   The shape of the visual fields allows the retina to use contrast rather than absolute light intensity for better detection of weak stimuli.

brain for processing. Figure 10-39 ■ shows how information from the right side of the visual field is processed on the left side of the brain, while information from the left side of the field is processed on the right.

The central portion of the visual field, where left and right visual fields overlap, is the **binocular zone** (Fig. 10-40 ■). The two eyes have slightly different views of objects in this region, so the brain processes and integrates the two views to create three-dimensional representations of the objects. Our sense of depth perception—that is, whether one object is in front of or

behind another—depends on binocular vision. Objects that fall within the visual field of only one eye are in the **monocular zone** and are viewed in two dimensions.

Once axons leave the optic chiasm, a few fibers project to the midbrain, where they participate in control of eye movement or coordinate with somatosensory and auditory information for balance and movement. Most axons, however, project to the *lateral geniculate body* of the thalamus, where the optic fibers synapse onto neurons leading to the visual cortex (see Fig. 10-27a ■).

The lateral geniculate body is organized in layers that correspond to the different parts of the visual field. Thus, information from adjacent objects is processed together. This **topographical organization** is maintained in the visual cortex, with the six layers of neurons grouped into vertical columns. Within each portion of the visual field, information is further sorted by form, color, and movement.

Visual processing and our perception of the world around us are exceedingly complex subjects whose details are beyond the scope of this book. The cortex merges monocular information from the two eyes to give us a binocular view of our surroundings. Information from on/off combinations of ganglion cells is translated into sensitivity to line orientation in the simplest pathways or into color, movement, and detailed structure in the most complex. Each of these attributes of visual stimuli is processed through a separate pathway, creating a network whose complexity we are just beginning to unravel.

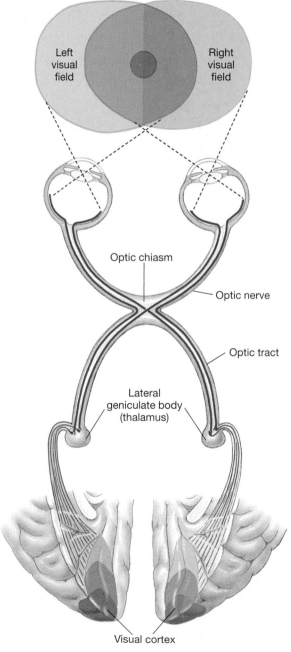

■ **Figure 10-39   Projection of visual fields**   The left side of the visual field in each eye is projected to the right side of the brain, while the right side of the field projects to the left visual cortex. Visual fibers from each eye meet at the optic chiasm. From there, they project through the lateral geniculate nucleus to the visual cortex. The cells of the cortex reflect the spatial organization of the visual fields.

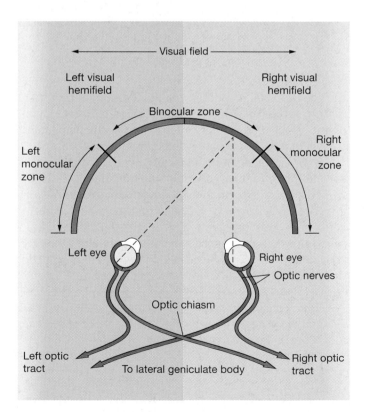

■ **Figure 10-40   Binocular vision**   The central portion of the visual field, known as the binocular zone, is seen by both eyes. Objects in the binocular zone will be perceived in three dimensions. Objects falling outside the binocular zone will only be perceived in two dimensions.

PROBLEM CONCLUSION

Anant was told about the surgical options but elected to continue medical treatment rather than risk deafness. Over the next two months, his Ménière's disease gradually resolved. The cause of it was never pinpointed.

In this running problem, you have learned that vestibular problems can result in severe dizziness. You have also learned about some treatments that are available to alleviate Ménière's disease. Further check your understanding of this running problem by comparing your answers to those in the summary table.

Question	Facts	Integration and Analysis
1 Subjective tinnitus occurs when the brain perceives a sound that does not exist outside the auditory system. Starting from the ear canal, name the auditory structures in which problems may arise.	The middle ear consists of the three bones that vibrate with sound. The hearing portion of the inner ear consists of the hair cells in the fluid-filled cochlea. The neural portion of the auditory system is the auditory nerve that leads to the brain.	Subjective tinnitus could arise from a problem with any of the structures named. Abnormal bone growth can affect the bones of the middle ear. Excessive fluid accumulation in the inner ear in Ménière's disease will affect the hair cells. Neural defects may cause the auditory nerve to fire spontaneously, creating the perception of sound.
2 When a person with BPV changes position, the displaced crystals float through the semicircular canals. Why would this cause dizziness?	The ends of the semicircular canals contain the sensory cristae, each consisting of a cupula with embedded hair cells. Displacement of the cupula creates a sensation of rotational movement.	If the floating crystals displace the cupula of the cristae, the brain will perceive movement that is not matched to sensory information coming from the eyes. The result is vertigo, an illusion of movement.
3 Compare the symptoms of BPV and Ménière's disease. On the basis of Anant's symptoms, which condition do you think he has?	The primary symptom of BP is brief dizziness following a change in position. Ménière's disease combines vertigo with tinnitus and hearing loss.	Anant complains of tinnitus and hearing loss. These are not typical of BPV. In addition, his dizzy attacks typically last up to an hour. BPV attacks are very brief because they are caused by a change in position. It is most likely that Anant has Ménière's disease.
4 Why is limiting salt (NaCl) intake suggested as a treatment for Ménière's disease? (Hint: Remember that the body regulates its osmolarity.)	Ménière's disease is a condition characterized by excessive endolymph in the inner ear. Endolymph is an extracellular fluid. However, its composition resembles intracellular fluid with high concentrations of K^+ and lower concentrations of Na^+.	Reducing salt intake should also reduce the amount of fluid in the extracellular compartment because the body will not retain water. This reduction may result in less fluid accumulation in the inner ear. On the other hand, the primary cation in endolymph is K^+, not Na^+, so decreasing NaCl intake may not be effective. Some diuretics remove K^+ along with water.
5 Why would severing the vestibular nerve alleviate Ménière's disease?	The vestibular nerve transmits information about balance and rotational movement from the vestibular apparatus to the brain.	Severing the vestibular nerve prevents false information about body rotation from reaching the brain, thus alleviating the vertigo of Ménière's disease.

to which a receptor responds is called its _____. The minimum stimulus required to activate a receptor is known as the _____.

8. When a sensory receptor membrane depolarizes (or hyperpolarizes in a few cases), the change in membrane potential is called the _____ or _____ potential. Is this a graded potential or an all-or-none potential?

9. Explain what is meant by adequate stimulus to a receptor.

10. The organization of sensory regions in the _____ of the brain preserves the topographical organization of receptors on the skin, eye, or other regions. However, there are exceptions to this rule. In what two senses does the brain rely on the timing of receptor activation to determine the location of the initial stimulus?

11. What is lateral inhibition?

12. Define tonic receptors and list some examples. Define phasic receptors and give some examples. Which type adapts? How is adaptation useful?

13. When cardiac ischemia is perceived as coming from the neck and down the left arm, this is an example of _____ pain, which often occurs when pain arises in visceral receptors.

14. What are the five basic tastes? Describe the hypothesis that attempts to account for the existence of each taste.

15. The unit of sound wave measurement is _____, which measures the frequency of the sound waves per second. The loudness or intensity of a sound is a function of the _____ of the sound waves and is measured in _____. The range of hearing for the average human ear is from _____ to _____ [units], with the most acute hearing in the range of _____ to _____ [units].

16. Which structure of the inner ear codes sound for pitch? Define spatial coding.

17. Loud noises cause action potentials to:
   (a) fire more frequently
   (b) have higher amplitudes
   (c) have longer refractory periods

18. Once sound waves have been transformed into electrical signals in the cochlea, sensory neurons transfer information to the _____, from which collaterals then take the information to the _____ and _____. The main auditory pathway synapses in the _____ and _____ before finally projecting to the _____ in the _____.

19. The parts of the vestibular apparatus that tell our brain about our movements through space are the _____, which sense rotation, and the _____, which responds to linear forces. Movement is the _____ component, whereas the otolith organs also function in telling us about head position while standing, the _____ component.

20. Put these structures in the correct order for a beam of light entering the eye: (a) pupil, (b) cornea, (c) retina, (d) lens, and (e) aqueous humor.

21. The three primary colors of vision are _____, _____, and _____. White light containing these colors stimulates photoreceptors called _____. Lack of the ability to distinguish some colors is called _____.

LEVEL TWO   Reviewing Concepts

22. Compare and contrast the following:
   (a) the special senses with the somatic senses
   (b) different types of touch-pressure receptors with respect to structure, size, and location
   (c) transmission of sharp localized pain with the transmission of dull and diffuse pain (include the particular fiber types involved as well as the presence or absence of myelin in your discussion)
   (d) the forms of hearing loss

23. Describe the neural pathways that link pain with emotional distress, nausea, and vomiting.

24. Trace the neural pathways involved in olfaction. Define olfactory binding protein, vomeronasal organ, pheromones, and G_{olf}.

25. Explain the process of signal transduction in the taste buds.

26. Put these structures in the order in which a sound wave would encounter them: (a) pinna, (b) cochlear duct, (c) stapes, (d) ion channels, (e) oval window, (f) hair cells/stereocilia, (g) tympanic membrane, (h) incus, (i) vestibular duct, and (j) malleus.

27. Sketch the structures and receptors of the vestibular apparatus for equilibrium. Label the components. Briefly describe how they function to notify the brain of movement.

28. Explain how accommodation by the eye occurs. What is the loss of accommodation called?

29. List four common visual problems and explain how they occur.

30. Explain how the intensity and duration of a stimulus are coded so that the stimulus can be interpreted by the brain. (Remember, action potentials are all-or-none phenomena.)

31. Map the following terms related to vision. Add additional terms if you wish.

accommodation reflex	lens
binocular vision	melanin
bipolar cells	opsin
bleaching	optic chiasm
blind spot	phototransduction
ciliary muscle	pupillary reflex
cones	retina
depth of field	retinal
field of vision	rhodopsin
focal point	rods
fovea	transducin
ganglion cells	visual field
lateral geniculate	zonulas

LEVEL THREE   Problem Solving

32. You are prodding your blindfolded lab partner's arm with two needle probes (with her permission). Sometimes she can tell you are using two probes, but when you probe less-sensitive areas, she thinks there is just one probe. What sense are you testing? Which receptors are being stimulated?

33. Make a table of the special senses. In the first row, start with these stimuli: sound, standing on the deck of a rocking boat, light, a taste, an aroma. Make rows for the location of the receptor for the sense and the structure or properties of each receptor. Last, name the cranial nerve(s) that convey(s) each sensation to the brain. [∞ p. 263]

34. Consuming alcohol depresses the nervous system and vestibular apparatus. In a sobriety check, police officers use this information to determine if an individual is inebriated. What kinds of tests can you suggest that would show evidence of this inhibition?

35. Often, children are brought to medical attention because of speech difficulties. If you were a clinician, which sense would you test first, and why?

36. Draw a diagram to explain the processes of phototransduction. Start by showing how bleaching occurs and continue with the release of the neurotransmitter from the rod. Explain how phototransduction differs in bright and low-level light. Continue to map the pathway to the visual cortex.

E X P L O R E   <MediaLab>

Introduction

We all live in the same world but different animals perceive the world differently. Dogs hear sounds we can't, nocturnal animals have better night vision than we do, and a stallion can smell if a mare is in estrus (ready to breed). An animal can only perceive stimuli for which it has sensory receptors. For example, radio waves fill the air around us, yet we only hear them if the radio is on and turned to the proper frequency.

In this chapter you explored sensory receptors in the human body and learned how each is exquisitely designed to allow us to perceive different aspects of the world around us. After reading the descriptions below, visit the MediaLab for Chapter 10 in your Companion Website and select the appropriate keyword.

Web Exploration 1

Estimated time for completion = 10 minutes
Throughout this chapter you should have been building your mental model of the basic characteristics shared by all sensory receptors. Do they all work the same at the molecular level? Take a minute and draw a basic model showing how a stimulus could generate a receptor potential. Now fill in the specifics for tasting a sour lemon ball and seeing the color red.

This chapter did not describe olfactory receptors in detail because scientists are still trying to determine how these receptors work. Based on your model of how taste and vision work, what would you look for if you were doing research in this area? First, you would need to categorize the odors you are sensing—to see if you can name five different categories of odor.

Select the keyword **SMELL PRISM** on the Website and see how close you came to the current categorization of odors.

Now that you have identified possible odors, can you identify their receptors and predict how they work? Select the keyword **OLFACTORY RECEPTOR** on the Website to read about how Doctor Axel and Buck are pursuing this line of research. What is Dr. Buck's model of how this receptor might work? What is the foundation of her model?

Web Exploration 2

Estimated time for completion = 15 minutes
Can you believe your eyes? In the chapter section on the eye and vision you learned how phototransduction occurs at the rods and cones. Draw a flow diagram showing what happens at the cellular level when a photon strikes a rod cell. To see an animation of this process, select **PHOTOTRANSDUCTION** on the Website.

Using the bar at the bottom of the animation, slowly step through the animation and compare the observed steps to those in your flow diagram. How accurate was your diagram? Light from an image strikes your rods and cones and causes them to release glutamate, but how is this message processed?

Optical illusions actually tell us a lot about the type of information the retina sends to the brain and how the brain processes this information. To learn more, select the keyword **LATERAL INHIBITION** on the Website. At the level of the bipolar and ganglion neuron, how does lateral inhibition work? What type of information about the image is being sent to the brain? How does your brain "fill in" images? Once you have developed your mental model of how lateral inhibition works, select the keyword **LATERAL INHIBITION SIMULATOR** and test your understanding by predicting the neuron output for the grid on the site.

11 Efferent Peripheral Nervous System: The Autonomic and Somatic Motor Divisions

■ "Because a number of cells in the autonomic nervous system act in conjunction, they have relinquished their independence to function as a coherent whole."—*Otto Appenzeller and Emilio Oribe, in* The Autonomic Nervous System, *1997.* ■

CHAPTER OUTLINE

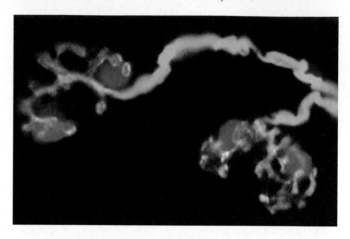

The neuromuscular junction, showing terminal Schwann cells covering axon terminals (red), somatic motor neuron (green), and ACh receptors in the postsynaptic muscle membrane (blue). Can you find Schwann cells associated with the axon and a node of Ranvier?

BACKGROUND BASICS

Neuron structure (p. 217)
Membrane receptors (p. 157)
Synapses (p. 234)
Neurotransmitters (p. 236)
Second messenger systems (p. 158)
Catecholamines (p. 190)
Action potentials (p. 224)
Slow synaptic potentials (p. 239)
Up- and down-regulation (p. 165)
Tonic and antagonist control (p. 167)

The picnic lunch was wonderful. You are now dozing in the warm spring sunlight on the grass as you let the meal digest. Suddenly you feel something moving across your lower leg. You open your eyes, and as they adjust to the bright light, you see a four-foot-long snake slithering over your foot. More by instinct than reason, you fling the snake into the grass while scrambling to a safe perch on top of the nearby picnic table. You are breathing heavily and your heart is pounding.

In less than a second, your body has gone from a state of quiet rest and digestion to a state of panic and frantic activity. This reflex reaction is integrated and coordinated through the central nervous system (CNS), then carried out by the efferent division of the peripheral nervous system. The fibers of efferent neurons are bundled together into nerves that carry commands from the

PROBLEM
A Powerful Addiction

Every day, more than 50 million Americans intentionally absorb a toxin that has been linked to the deaths of over 6 million people since 1964. Why would people knowingly poison themselves? If you've guessed that the toxin is nicotine, you already know part of the answer. One of over a thousand chemicals found in tobacco, nicotine is highly addictive. So powerful is this addiction that only 20% of tobacco users are able to quit smoking the first time they try. Shanika Thomas, a smoker for six years, is attempting for the second time to quit smoking. But the odds are in her favor because, this time, Shanika has sought help. She has decided to try the nicotine patch.

...continued on page 328

CNS to the muscles and glands of the body. Some nerves, called *mixed nerves,* carry sensory information through afferent fibers as well as efferent signals.

The efferent side of the peripheral nervous system has two divisions: (1) the **somatic motor neurons,** which control skeletal muscles, and (2) the **autonomic neurons,** which control smooth muscle, cardiac muscle, many glands, and some adipose tissue. The somatic and autonomic divisions are sometimes called the voluntary and involuntary divisions of the nervous system. Although it is true that movement controlled by somatic pathways usually requires conscious thought and that autonomic reflexes are mainly involuntary, this distinction does not always hold true.

For example, some skeletal muscle reflexes, such as swallowing and the knee jerk reflex, are involuntary. And a person can learn to modulate some involuntary autonomic functions, such as heart rate and blood pressure, through biofeedback training. Thus, the classification of efferent divisions into voluntary and involuntary branches is a broad generalization that is not accurate in all situations.

Let's begin by looking at the autonomic division. Then we will consider the somatic motor division, as preparation for learning about muscles in Chapter 12.

▸ THE AUTONOMIC DIVISION

The autonomic division of the efferent nervous system is also known in older writings as the *vegetative nervous system,* reflecting the observation that its functions were not under voluntary control. The name *autonomic* comes from the same roots as *autonomous,* meaning *self-governing.* You will also find the autonomic division called the *visceral nervous system* because of its control over the function of internal organs. Autonomic reflexes regulate blood pressure, heart rate, respiration, water balance, body temperature, and a host of other homeostatic functions.

The autonomic division is subdivided into **sympathetic** and **parasympathetic branches.** Some parts of the sympathetic branch were first described by the Greek physician Claudius Galen (ca. 130–200), who is famous for his compilation of anatomy, physiology, and medicine as they were known up to his time. As a result of his dissections, Galen proposed that "animal spirits" flowed from the brain through hollow nerves to the tissues, creating "sympathy" between the different parts of the body. Galen's "sympathy" later gave rise to the name for the sympathetic branch. The prefix *para-,* for the parasympathetic branch, means *beside,* or *alongside of.*

Although the sympathetic and parasympathetic branches can be distinguished anatomically, there is no simple way to separate the actions of the two types of neurons on their targets. The best generalization is to characterize the two branches according to the type of situation in which they are most active.

The picnic scene that began the chapter typifies the two extremes at which the sympathetic and parasympathetic branches function. If you are resting quietly after a meal, the parasympathetic branch is dominant, taking command of the routine, quiet activities of day-to-day living such as digestion. Consequently, parasympathetic neurons are sometimes said to control "rest and digest" functions.

In contrast, the sympathetic branch is dominant in stressful situations, such as the potential threat from the snake. One of the most dramatic examples of sympathetic action is the fight-or-flight response, in which the brain triggers massive simultaneous sympathetic discharge throughout the body. As the body prepares to fight or flee, the heart speeds up, blood vessels to muscles of the arms, legs, and heart dilate, and the liver starts to produce glucose to provide energy for muscle contraction. Digestion becomes a low priority when life and limb are threatened, so blood is diverted from the gastrointestinal tract to skeletal muscles. The massive sympathetic discharge that occurs in fight-or-flight situations is mediated through the hypothalamus and is a total-body response to a crisis. If you have ever been scared by the squealing of brakes or a sudden sound in the dark, you know how rapidly the nervous system can influence multiple systems of the body.

However, fight-or-flight is only an occasional job of the sympathetic nervous system. This autonomic branch also plays a significant role in daily life. One of the key functions of the sympathetic branch is its control over blood flow to the tissues, as you will learn in Chapter 15. Most blood vessels are controlled by the sympathetic branch, which directs blood to the organs and tissues where it is most needed. *Usually sympathetic responses are not the all-out response of a fight-or-flight reflex, and activating one sympathetic pathway does not automatically activate them all.*

The Sympathetic and Parasympathetic Branches Maintain Homeostasis

The maintenance of homeostasis within the body generally involves balancing sympathetic and parasympathetic control. Like the seesaw in Figure 11-1 ■, the

...continued from page 327

Nicotine is an addictive substance that binds to one subtype of acetylcholine (ACh) receptor. People who smoke have elevated nicotine levels in their blood.

Question 1:   *What is the usual response of cells that are chronically exposed to increased concentrations of a signal molecule? (Hint: This topic was discussed in Chapter 6 [⚬ p. 165]).*

nervous control of body function shifts back and forth between the two branches as they cooperate to fine-tune various processes. Only occasionally, as in the fight-or-flight example that opened the chapter, does the seesaw move to one extreme or the other.

The sympathetic and parasympathetic branches together display all of Walter Cannon's properties of homeostasis: (1) maintenance of the internal environment, (2) "up-down" regulation by varying tonic control, (3) antagonistic control, and (4) variable tissue responses that result from different membrane receptors [⚬ p. 157].

Most internal organs are innervated by both autonomic branches, which exhibit antagonistic control. One branch is excitatory while the other branch is inhibitory (see table in Fig. 11-4 ■). For example, the contraction rate of the heart is under tonic regulation by both autonomic branches. Sympathetic innervation increases heart rate and parasympathetic stimulation decreases it. Consequently, heart rate can be changed by altering the relative proportion of sympathetic and parasympathetic control.

Exceptions to dual antagonistic innervation include the sweat glands and smooth muscle in most blood vessels. These tissues are innervated only by the sympathetic branch and rely strictly on tonic (up-down) control.

Although the two autonomic branches are usually antagonistic in their control of a single target tissue, they sometimes work cooperatively on different tissues to achieve a common goal. For example, in the male sex act, erection of the penis is under the control of the parasympathetic branch but the ejaculation of sperm is directed by the sympathetic branch.

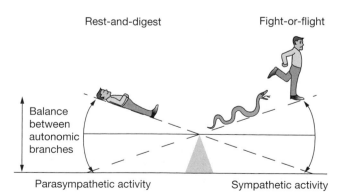

Rest-and-digest   Fight-or-flight

Balance between autonomic branches

Parasympathetic activity   Sympathetic activity

■ **Figure 11-1   Homeostasis and the autonomic division** A dynamic balance between sympathetic and parasympathetic activity maintains homeostasis in the body.

In some autonomic pathways, the membrane receptor determines the response of the target tissue. For instance, most blood vessels contain adrenergic receptors that cause smooth muscle contraction (*vasoconstriction*). But other blood vessels contain a different subtype of adrenergic receptor that causes the smooth muscle to relax (*vasodilation*). Both receptor types are activated by catecholamines. In this example, the receptor rather than the chemical signal determines the response.

The Autonomic Division Is Regulated by the Brain

The efferent neurons of the autonomic division are closely linked to homeostatic control centers in the hypothalamus, pons, and medulla of the brain (Fig. 11-2 ■). Autonomic functions controlled through the brain include heart rate, blood pressure, temperature regulation, and water balance. In addition, the cerebral cortex and limbic system can influence autonomic output through descending pathways. Blushing, fainting at the sight of a hypodermic needle, and "butterflies in the stomach" are all examples of emotional influences on autonomic functions. Autonomic pathways also influence the release of hormones such as insulin. Understanding the autonomic and hormonal control of organ systems is the key to understanding maintenance of homeostasis in virtually every system of the body.

Despite the link between autonomic efferent pathways and the brain, many autonomic reflexes, classified as spinal reflexes, are capable of taking place without input from the brain. These reflexes include urination, defecation, and penile erection, bodily functions that are normally influenced through pathways from the brain. People with spinal cord injuries that disrupt communication between the brain and spinal cord usually retain some spinal reflexes but lose the ability to sense or control them. More than half of the people who suffer spinal cord injuries each year are between the ages of 16 and 30, and over 80% of them are male.

✔ What do you think are the most common causes of spinal cord injuries?

...continued from above

Nicotinic cholinergic receptors are found throughout the peripheral nervous system and the CNS. Researchers speculate that nicotinic receptors in certain regions of the brain may play a key role in nicotine addiction.

Question 2:   *Cholinergic receptors are classified as either nicotinic or muscarinic, on the basis of the agonist molecules that bind to them. What happens to the postsynaptic cell when nicotine rather than ACh binds to a nicotinic cholinergic receptor?*

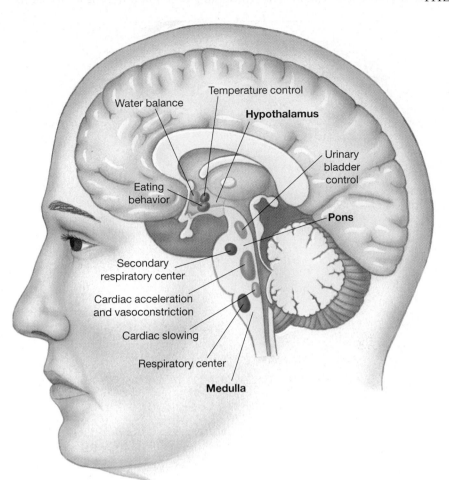

The Autonomic Division Consists of Two Efferent Neurons in Series

The sympathetic and parasympathetic branches of the autonomic division share the same general structure (Fig. 11-3 ■). All autonomic pathways consist of two neurons in series. They synapse in an **autonomic ganglion** located outside the CNS. (A *ganglion* is a cluster of nerve cell bodies that lie outside the CNS. The equivalent in the CNS is a *nucleus,* such as the lateral geniculate nucleus [∞ p. 312].)

Autonomic ganglia, where the two efferent neurons synapse, are more than simply clusters of axon terminals and nerve cell bodies. The ganglia also contain **intrinsic neurons** that lie completely within the ganglion, just as interneurons lie completely within the CNS. These intrinsic neurons modulate messages moving through the autonomic neurons.

For all autonomic pathways, the first neuron in the pathway, the **preganglionic neuron,** has its cell body in the CNS and its terminals in the ganglion. The second neuron in the pathway, the **postganglionic neuron,** has its cell body in the ganglion and projects its axon to the target tissue.

Divergence [∞ p. 240] is an important feature of autonomic pathways. On average, one neuron entering a ganglion synapses with eight or nine postganglionic neurons. Some synapse on as many as 32 neurons! Each postganglionic neuron may innervate a different target, meaning that a single signal from the CNS can affect a large number of target cells simultaneously.

Sympathetic and Parasympathetic Branches Exit the Spinal Cord in Different Regions

How then do the two autonomic branches differ anatomically? The main differences are where the pathways originate in the CNS and the location of the autonomic ganglia. As Figure 11-4 ■ shows, most sympathetic path-

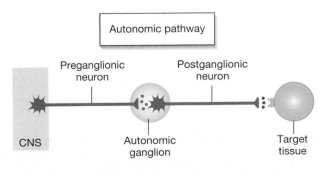

■ Figure 11-3   **Autonomic pathways**   All autonomic pathways consist of two neurons that synapse in an autonomic ganglion.

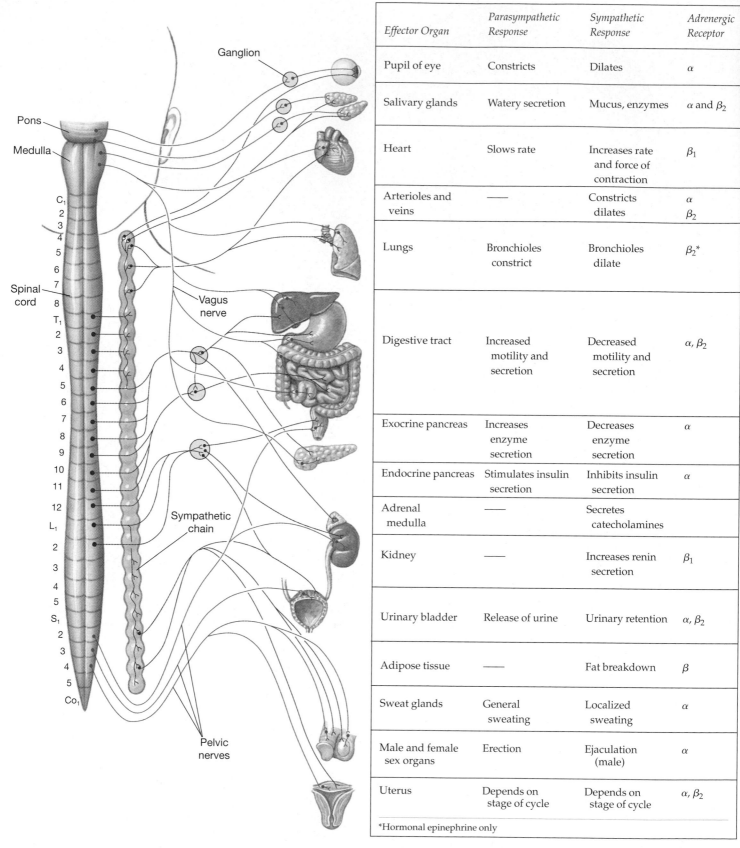

Effector Organ	Parasympathetic Response	Sympathetic Response	Adrenergic Receptor
Pupil of eye	Constricts	Dilates	α
Salivary glands	Watery secretion	Mucus, enzymes	α and β_2
Heart	Slows rate	Increases rate and force of contraction	β_1
Arterioles and veins	—	Constricts dilates	α β_2
Lungs	Bronchioles constrict	Bronchioles dilate	β_2*
Digestive tract	Increased motility and secretion	Decreased motility and secretion	α, β_2
Exocrine pancreas	Increases enzyme secretion	Decreases enzyme secretion	α
Endocrine pancreas	Stimulates insulin secretion	Inhibits insulin secretion	α
Adrenal medulla	—	Secretes catecholamines	
Kidney	—	Increases renin secretion	β_1
Urinary bladder	Release of urine	Urinary retention	α, β_2
Adipose tissue	—	Fat breakdown	β
Sweat glands	General sweating	Localized sweating	α
Male and female sex organs	Erection	Ejaculation (male)	α
Uterus	Depends on stage of cycle	Depends on stage of cycle	α, β_2

*Hormonal epinephrine only

■ **Figure 11-4 Autonomic sympathetic and parasympathetic pathways**   Most sympathetic neurons (red) originate in the thoracic and lumbar regions of the spinal cord. Sympathetic ganglia are located close to the spinal column or along the descending aorta. Parasympathetic neurons (blue) originate in the brain stem or sacral region of the spinal cord. Parasympathetic ganglia are located on or near their target organs.

Ganglia as Integrating Centers According to the traditional view of the autonomic division, a signal that originates in the CNS passes directly along a two-neuron chain to the target tissue. In that model, the ganglion is simply a way-station for transfer of signals from preganglionic neurons to postganglionic neurons. But in recent years, scientists have discovered that ganglia can act as mini-integrating centers, receiving sensory input from the periphery and modulating the outgoing responses. Chemical studies show that intrinsic neurons, particularly in sympathetic ganglia, secrete dopamine, which acts as an inhibitory neurotransmitter. The pathways modified by these neurons are still being investigated. It has recently been shown that some sensory receptors in the abdomen are linked by intrinsic neurons directly to postganglionic sympathetic neurons. Presumably this means that a sensory signal could influence the efferent output of the sympathetic neuron without integration by the CNS. If further research confirms this model, we must modify our view of the CNS as the only site of integration within the nervous system.

ways (red) originate in the thoracic and lumbar regions of the spinal cord. Sympathetic ganglia are found in a chain that runs close to the spinal column or along the descending aorta. Long nerves (axons) project to the target tissues. Because most sympathetic ganglia lie close to the spinal cord, sympathetic pathways generally have short preganglionic neurons and long postganglionic neurons.

Many parasympathetic pathways (shown in blue in Figure 11-4 ■) originate in the brain stem. Other parasympathetic pathways (also in blue) originate in the sacral region (near the lower end of the spinal cord) and control pelvic organs. In general, parasympathetic ganglia are located on or near their target organs. Consequently, parasympathetic preganglionic neurons have long axons and postganglionic neurons have short axons.

The axons of parasympathetic neurons in the brain stem leave the CNS in several cranial nerves [∞ p. 263]. The major parasympathetic tract is part of the **vagus nerve** (cranial nerve X), a large nerve carrying both afferent sensory neurons from and parasympathetic fibers to most of the internal organs (Fig. 11-5 ■). *Vagotomy*, a procedure in which the vagus nerve is surgically cut, was an experimental technique used in the nineteenth and early twentieth centuries to study the impact of the autonomic nervous system on various organs. For a time, vagotomy was the preferred treatment for stomach ulcers because removal of parasympathetic innervation to the stomach decreased the secretion of stomach acid. However, this procedure had many unwanted side effects and has been abandoned in favor of drug therapies that treat the problem more specifically.

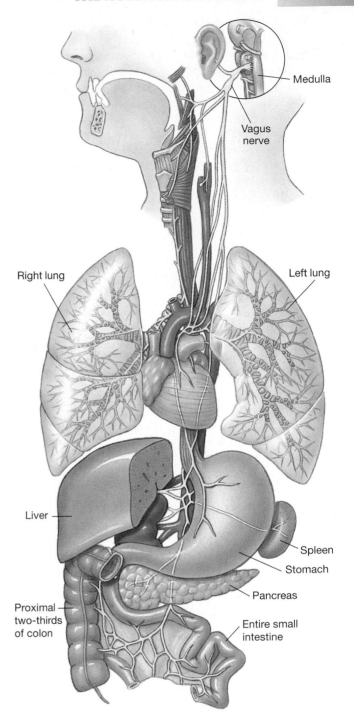

■ **Figure 11-5** **The vagus nerve** The major parasympathetic tract is the vagus nerve, which carries parasympathetic fibers to most of the internal organs.

Autonomic Pathways Control Smooth Muscle, Cardiac Muscle, and Glands

The targets of autonomic neurons are smooth muscle, cardiac muscle, many exocrine glands, a few endocrine glands, and some adipose tissue. The synapse between a postganglionic autonomic neuron and its target cell is called the **neuroeffector junction.**

The structure of an autonomic synapse differs from the model synapse shown in Chapter 8 [∞ p. 234]. Autonomic axons end with a series of swollen areas at their distal ends, like beads spaced out along a string (Fig. 11-6 ■). Each of these swellings, known as a **varicosity** [*varicosus*, abnormally enlarged or swollen], contains vesicles filled with neurotransmitter.

The branched ends of the axon lie across the surface of the target tissue, but the underlying membrane does not possess clusters of neurotransmitter receptors in specific sites. Instead, the neurotransmitter is simply released into the interstitial fluid to diffuse to wherever the receptors are, often some distance from the site of release. The result is a less-directed form of communication than that which occurs between a somatic motor neuron and a skeletal muscle. The diffuse release of autonomic neurotransmitter means that a single postganglionic neuron can affect a large area of target tissue.

The release of autonomic neurotransmitters is subject to modulation from a variety of sources. For example, the membranes of sympathetic varicosities contain receptors for hormones and for paracrines such as histamine. These modulators may either facilitate or inhibit the release of norepinephrine from the varicosity. Some preganglionic neurons co-secrete polypeptides along with acetylcholine. The peptides act as neuromodulators, producing slow synaptic potentials that modify the activity of postganglionic neurons [∞ p. 239].

The Adrenal Medulla Secretes Catecholamines

The **adrenal medulla,** part of the adrenal gland, is a neuroendocrine gland that is closely allied with the sympathetic branch of the autonomic division. During development, the neural tissue destined to secrete catecholamines (epinephrine, norepinephrine) splits into two functional entities: the sympathetic branch of the nervous system and the adrenal medulla, a neuroendocrine gland.

The adrenal glands are paired glands that sit on top of the kidneys (Fig. 11-7a ■). Like the pituitary gland, the adrenals are two glands of different embryological origin that fused during development (Fig. 11-7b ■). The outer portion, the adrenal cortex, is a true endocrine gland of epidermal origin that secretes steroid hormones [∞ p. 192].

The adrenal medulla [*ad-*, upon; *renal*, kidney; *medulla*, marrow] forms the small core of the gland. It is actually a modified sympathetic ganglion. Normal preganglionic neurons from the spinal cord terminate in the ganglion but its postganglionic neurons lack axons that would normally project to target cells (Fig. 11-7c ■). Instead, the cell bodies in the adrenal medulla secrete neurohormones directly into the blood. In response to alarm signals from the CNS, the adrenal medulla releases large amounts of catecholamines, primarily epinephrine, for general distribution throughout the body.

Neurotransmitter Activity at Target Tissues Is Limited

In the autonomic division, neurotransmitter synthesis takes place in the axon varicosities. The main autonomic neurotransmitters are acetylcholine (ACh) and norepinephrine, both small molecules that are easily synthesized by cytoplasmic enzymes. Neurotransmitter made in the varicosities is packaged into synaptic vesicles for storage [∞ p. 235].

When an action potential arrives at the varicosity, voltage-gated Ca^{2+} channels open, Ca^{2+} enters the neuron, and the synaptic vesicle contents are released by exocytosis (Fig. 11-8 ■). Once autonomic neurotransmitters are released into the synapse, they diffuse through the interstitial fluid until they encounter a receptor on the effector cell or drift away from the synapse (Fig. 11-8 ■).

The concentration of neurotransmitter in the synapse is a major factor in the control that an autonomic neuron exerts on its target: the more neurotransmitter, the longer or stronger the response. The concentration of neurotransmitter in a synapse is influenced by its rate of breakdown or removal. Neurotransmitter action terminates when the neurotransmitter is metabolized by enzymes in the extracellular fluid or actively transported into cells around the synapse. The return of neurotransmitter to varicosities allows neurons to reuse the chemicals.

Catecholamine activity at a synapse terminates when norepinephrine is transported back into neurons (Fig. 11-8 ■). Once returned to the axon terminal, norepinephrine is either repackaged into vesicles or broken down by intracellular enzymes. *Monoamine oxidase (MAO),* found in mitochondria, is the main enzyme responsible for degradation of catecholamines.

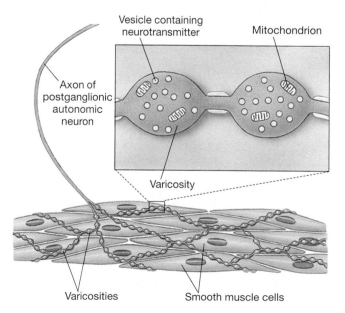

Vesicle containing neurotransmitter

Mitochondrion

Axon of postganglionic autonomic neuron

Varicosity

Varicosities

Smooth muscle cells

■ **Figure 11-6** **Varicosities on autonomic neurons** Autonomic neurons release their neurotransmitter from enlarged areas known as varicosities, found along the distal end of the postganglionic axon.

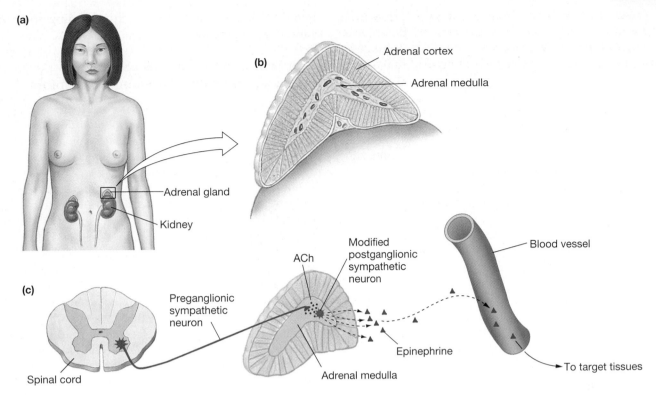

(a)

Adrenal gland

Kidney

(b)

Adrenal cortex

Adrenal medulla

(c)

Spinal cord

Preganglionic
sympathetic
neuron

ACh

Modified
postganglionic
sympathetic
neuron

Adrenal medulla

Epinephrine

Blood vessel

To target tissues

■ **Figure 11-7   The adrenal medulla**   The adrenal medulla is a modified sympathetic ganglion whose postganglionic neurons have no axons. Instead, they secrete epinephrine into the blood as a neurohormone.

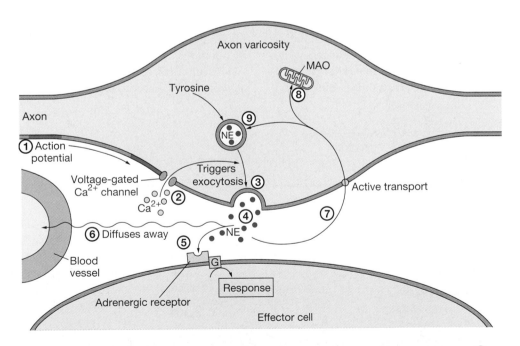

Axon varicosity

MAO

Tyrosine

Axon

⑨

NE

⑧

① Action
potential

Voltage-gated
Ca²⁺ channel

Triggers
exocytosis   ③

② Ca²⁺

Active transport

④

⑦

⑥ Diffuses away

⑤   ● NE

Blood
vessel

G

Response

Adrenergic receptor

Effector cell

■ **Figure 11-8   Norepinephrine release at a varicosity of a sympathetic neuron**   ① An action potential arriving at the varicosity opens voltage-gated Ca²⁺ channels ②, triggering exocytosis of synaptic vesicles ③. Norepinephrine (NE) in the synaptic cleft ④ can combine with an adrenergic receptor ⑤, diffuse away from the synapse ⑥, or be transported back into the axon ⑦. Once back in the axon, norepinephrine can be metabolized by monoamine oxidase (MAO) ⑧ or taken back into synaptic vesicles for re-release ⑨.

Acetylcholine action in the synapse is terminated primarily by the extracellular enzyme *acetylcholinesterase* [∞ p. 239]. The choline portion of the ACh molecule is transported back into neurons and used to make new ACh.

The Autonomic Nervous System Uses a Variety of Neurotransmitters and Modulators

The autonomic nervous system exerts its control through a combination of neurotransmitters and receptors. The sympathetic and parasympathetic branches can be distinguished by their neurotransmitters and receptors, using the following rules:

1. All autonomic preganglionic neurons release acetylcholine (ACh) onto cholinergic nicotinic receptors [∞ p. 238].
2. Postganglionic sympathetic neurons secrete norepinephrine onto adrenergic receptors.
3. Postganglionic parasympathetic neurons secrete acetylcholine onto cholinergic muscarinic receptors. These differences are illustrated in Figure 11-9 ■.

The two most common neurotransmitters secreted by autonomic neurons are acetylcholine and norepinephrine. A few autonomic neurons secrete neither norepinephrine nor acetylcholine and are therefore known as *non-adrenergic, non-cholinergic* neurons. Some of the chemicals they use as neurotransmitters include substance P, somatostatin, vasoactive intestinal peptide (VIP), adenosine, nitric oxide, and ATP. We will mention these unusual neurotransmitters in later chapters when they play a significant role. The non-adrenergic, non-cholinergic neurons are assigned to either the sympa-

thetic or parasympathetic branch according to their anatomical organization.

Most Sympathetic Pathways Secrete Norepinephrine onto Adrenergic Receptors

Most sympathetic postganglionic neurons are *adrenergic neurons* that release norepinephrine at their targets. A few sympathetic neurons, such as those that terminate on sweat glands, secrete ACh rather than norepinephrine. These neurons are therefore called *sympathetic cholinergic neurons.*

Adrenergic receptors for catecholamines are linked to G proteins. Adrenergic receptors come in two varieties: α (alpha) and β (beta). Alpha receptors are the most common sympathetic receptor. They respond strongly to norepinephrine and only weakly to epinephrine (Table 11-1). Both α and β receptors have subtypes.

Beta receptors have two main subtypes, β_1 and β_2. β_1 receptors are found on heart muscle and in certain cells of the kidney. They are innervated by sympathetic neurons and respond equally strongly to norepinephrine and epinephrine. β_2 receptors are found in locations not innervated by sympathetic neurons, such as smooth muscle in the airways and in certain blood vessels. Because β_2 receptors are not innervated (no sympathetic neurons terminate near them), they are more sensitive to epinephrine than to norepinephrine, which is primarily a neurotransmitter (Fig. 11-10 ■).

The different types of adrenergic receptors work through different second messenger pathways (Table 11-1). Catecholamine binding to α_1 receptors activates phospholipase C and IP_3. This creates intracellular Ca^{2+} signals by opening Ca^{2+} channels in the cell membrane, releasing Ca^{2+} from intracellular stores, or doing both. In general, activation of α_1 receptors causes muscle contraction or se-

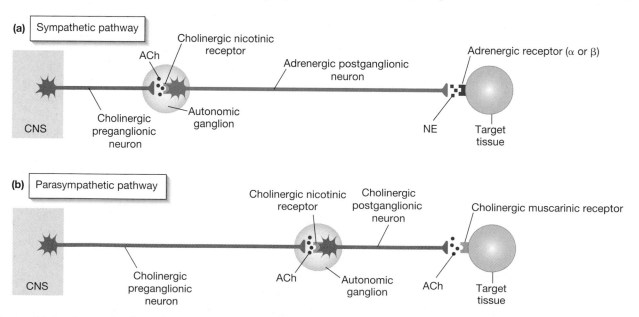

■ **Figure 11-9   Sympathetic and parasympathetic pathways**   (a) Sympathetic pathways are distinguished by ganglia close to the CNS. They use norepinephrine (NE) as the postganglionic neurotransmitter. (b) Parasympathetic pathways generally have their ganglia closer to their target tissues. They use acetylcholine (ACh) as the postganglionic neurotransmitter.

TABLE 11-1 **Sensitivity of Peripheral Adrenergic Receptors to Catecholamines**

Receptor	Found in	Sensitivity	Second Messenger
α_1	Most sympathetic target tissues	NE > E	Activates phospholipase C
α_2	Gastrointestinal tract and pancreas	NE > E	Inhibits cAMP
β_1	Heart muscle, kidney	NE = E	Activates cAMP
β_2	Certain blood vessels and smooth muscle of some organs	E > NE	Activates cAMP

NE = norepinephrine, E = epinephrine.

cretion by exocytosis. Activated α_2 receptors decrease cyclic AMP and cause smooth muscle relaxation (gastrointestinal tract) or decreased secretion (pancreas).

Catecholamine binding to β receptors activates cyclic AMP and triggers the phosphorylation of intra-cellular proteins. Activation of β_1 receptors enhances cardiac muscle contraction and speeds up electrical conduction in cardiac muscle. Activation of β_2 receptors by the hormone epinephrine results in smooth muscle relaxation in many tissues.

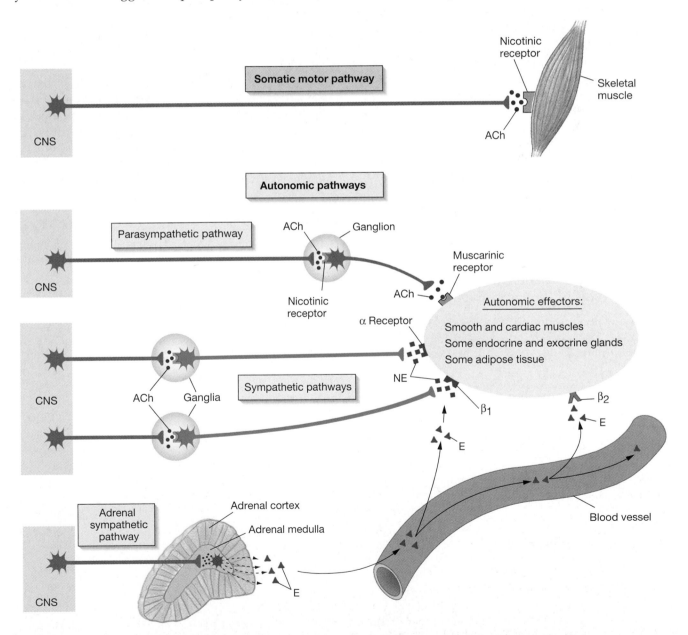

■ **Figure 11-10** **Summary of efferent pathways** ACh = acetylcholine, NE = norepinephrine, E = epinephrine.

…continued from page 328

One research study that examined brains at autopsy found that smokers have a greater number of receptors that bind nicotine than do nonsmokers. Normally, chronic exposure to an agonist would cause cells to down-regulate their receptors. Up-regulation, an increase in receptor numbers, is usually seen when cells are chronically exposed to receptor antagonists.

Question 3: *Although nicotine has been shown in short-term studies to be an agonist at nicotinic receptors, chronic exposure to nicotine causes up-regulation of the receptors. Speculate on why up-regulation might occur.*

For all adrenergic receptors, second messenger activity within the target tissue can persist for a longer time than is normally associated with the rapid action of the nervous system. These ongoing metabolic effects result from modification of existing proteins or from the synthesis of new proteins. We will discuss the specific effects of catecholamines on various tissues in subsequent chapters.

Parasympathetic Pathways Secrete Acetylcholine onto Muscarinic Receptors

As a rule, parasympathetic neurons release ACh at their targets. *Cholinergic muscarinic receptors* [∞ p. 238] are found at the neuroeffector junctions of the parasympathetic branch. Muscarinic receptors are all coupled to G proteins and linked either to second messenger systems or to gated K^+ channels. The tissue response to activation of a muscarinic receptor varies with the receptor subtype, of which there are at least five.

Autonomic Agonists and Antagonists Are Important Tools in Research and Medicine

The study of the autonomic branches has been greatly simplified by advances in molecular biology. The genes for many autonomic receptors and their subtypes have been cloned, allowing researchers to create mutant receptors and study their properties. In addition, researchers have discovered or synthesized a variety of agonists and antagonist molecules (Table 11-2). Some agonists and antagonists combine with the target receptor to mimic or block neurotransmitter action.

Other agonists and antagonists act by altering secretion, reuptake, or degradation of neurotransmitters. For example, cocaine blocks the reuptake of norepinephrine into adrenergic nerve terminals, thereby extending norepinephrine's stimulatory effect on the target. The drug neostigmine is a cholinesterase inhibitor that is used to treat myasthenia gravis [∞ p. 239], a condition in which skeletal muscles have fewer-than-normal ACh receptors. Neostigmine blocks ACh degradation and extends the active life of each ACh molecule. The toxic organophosphate insecticides are also cholinesterase inhibitors.

Many of the drugs used to treat depression act either on membrane transporters for neurotransmitters (tricyclic antidepressants and selective serotonin reuptake inhibitors) or on their metabolism (monoamine oxidase inhibitors). The older antidepressant drugs that act on norepinephrine transport and metabolism (tricyclics and MAO inhibitors) may have side effects related to their actions in the autonomic nervous system, including cardiovascular problems, constipation, urinary difficulty, and sexual dysfunction [*dys-*, abnormal or ill]. The newer serotonin reuptake inhibitors have fewer autonomic side effects because serotonin is primarily a CNS neurotransmitter.

Many new drugs have been developed from our studies of agonists and antagonists. The discovery of α and β adrenergic receptors led to the development of drugs that block only one of the two receptor types. Antagonists of α receptors proved to have little clinical significance, but the drugs known as beta-blockers have given physicians a powerful tool for treating high blood pressure, one of the most common disorders in the United States today.

TABLE 11-2 Agonists and Antagonists of Autonomic and Somatic Neurotransmitter Receptors

Receptor	Agonists	Antagonists	Other
Cholinergic	Acetylcholine		*AChE* inhibitors:* neostigmine, parathion *Inhibit ACh release:* botulinus toxin
Muscarinic	Muscarine	Atropine, scopolamine	
Nicotinic	Nicotine	α-bungarotoxin (muscle only), tetraethylammonium (TEA) (ganglia only), curare	
Adrenergic	Norepinephrine, epinephrine		*Stimulate NE release:* ephedrine, amphetamines *Prevent NE uptake:* cocaine
Alpha (α)	Phenylephrine	"Alpha-blockers"	
Beta (β)	Isoproterenol	"Beta-blockers": propranolol (β_1 and β_2), metoprolol (β_1 only)	

*AChE = acetylcholinesterase.

Primary Disorders of the Autonomic Nervous System Are Relatively Uncommon

Diseases and malfunction of the autonomic nervous system are relatively rare. Direct damage (trauma) to hypothalamic control centers may cause abnormalities in water balance or temperature regulation. Generalized sympathetic dysfunction may result from systemic diseases such as cancer and diabetes mellitus. There are also some conditions, such as *Shy-Drager syndrome,* in which the CNS control centers for autonomic functions degenerate. In all cases of sympathetic dysfunction, the symptoms are manifested most strongly in the cardiovascular system, when low sympathetic input to blood vessels results in abnormally low blood pressure. Other prominent symptoms of sympathetic pathology include *urinary incontinence* [*in-,* unable + *continere,* to contain], the loss of bladder control, and *impotence,* the inability to obtain or sustain a penile erection.

Occasionally, patients suffer from primary autonomic failure when sympathetic neurons degenerate. In the face of continuing diminished sympathetic input, target tissues up-regulate [∞ p. 165], putting more receptors into the cell membrane to maximize the cell's response to available norepinephrine. This increase in receptor number leads to *denervation hypersensitivity,* a state in which the administration of exogenous adrenergic agonists causes a greater-than-expected response.

As you have seen in this discussion, the branches of the autonomic nervous system share some features but are distinguished by others. Many of these features are summarized in Figure 11-10 ■ and compared in Table 11-3.

■ Both sympathetic and parasympathetic pathways are composed of two neurons. One exception to this rule is the adrenal gland, in which postganglionic sympathetic neurons have been modified into a neuroendocrine organ.

§ DIABETES Diabetic Neuropathy Primary disorders of the autonomic division are rare, but a very common autonomic problem is the condition known as *diabetic neuropathy.* This complication of diabetes can involve both sensory and autonomic divisions of the peripheral nervous system. It often begins with sensory loss in the hands and feet. In other patients, pain is the primary symptom. Autonomic neuropathies cause dysfunction of cardiovascular, gastrointestinal, urinary, and reproductive systems. The cause of these neuropathies is not clear, so most treatment is designed to treat the symptoms.

■ All preganglionic autonomic neurons secrete acetylcholine onto nicotinic receptors. Most postganglionic sympathetic neurons secrete norepinephrine onto adrenergic receptors. Postganglionic parasympathetic neurons secrete acetylcholine onto muscarinic receptors.

■ Most sympathetic neurons originate in the thoracic and lumbar regions of the spinal cord. The sympathetic ganglia are located close to the spinal cord (*paravertebral*). Parasympathetic neurons leave the CNS at the brain stem and in the sacral region of the spinal cord. Parasympathetic ganglia are located close to or in the target tissue.

■ The sympathetic branch controls functions that are useful in stress or emergencies (fight-or-flight). The parasympathetic branch is dominant during rest-and-digest activities.

◗ THE SOMATIC MOTOR DIVISION

Somatic motor pathways, which control skeletal muscles, differ from autonomic pathways both anatomically and functionally (Table 11-4). Somatic pathways have

TABLE 11-3   Comparison of Sympathetic and Parasympathetic Branches

	Sympathetic	*Parasympathetic*
Point of CNS origin	1st thoracic to 2nd lumbar segments	Midbrain, medulla, and 2nd–4th sacral segments
Location of peripheral ganglia	Primarily in paravertebral sympathetic chain; 3 outlying ganglia located alongside descending aorta	On or near target organs
Structure of region from which neurotransmitter is released	Varicosities	Boutons and varicosities
Neurotransmitter at target synapse	Norepinephrine (adrenergic neurons)	ACh (cholinergic neurons)
Inactivation of neurotransmitter at synapse	Uptake into varicosity, diffusion	Enzymatic breakdown, diffusion
Types of neurotransmitter receptors	α and β	Nicotinic and muscarinic
Ganglionic synapse	ACh on nicotinic receptor	ACh on nicotinic receptor
Neuron-target synapse	NE on α or β receptors	ACh on muscarinic receptor

TABLE 11-4 Comparison of Somatic and Autonomic Divisions

	Somatic	Autonomic
Number of neurons in efferent path	1	2
Neurotransmitter/receptor at neuron-target synapse	ACh (nicotinic)	ACh (muscarinic) or NE (α or β)
Target tissue	Skeletal muscle	Smooth and cardiac muscle; some endocrine and exocrine glands; some adipose tissue
Structure of axon terminal regions	Boutons	Boutons and varicosities
Effects on target tissue	Excitatory only: muscle contracts	Excitatory or inhibitory
Peripheral components found outside the CNS	Axons only	Preganglionic axons, ganglia, postganglionic neurons
Summary of function	Posture and movement	Visceral function, including movement in internal organs and secretion; control of metabolism

a single neuron that originates in the CNS and projects its axon to the target tissue, which is always a skeletal muscle. Unlike autonomic pathways, which may be either excitatory or inhibitory, somatic pathways are always excitatory.

A Somatic Motor Pathway Consists of One Neuron

The cell bodies of somatic motor neurons are located within the gray matter of the spinal cord or brain, with a long single axon projecting to the skeletal muscle target (Fig. 11-10 ■). These myelinated axons may be up to a meter or more in length, like the somatic motor neurons that innervate the muscles of the foot and hand.

Somatic motor neurons branch close to their targets. Each branch divides into a cluster of enlarged axon terminals, or *boutons* [from the French word for "knob" or "button"], that lie on the surface of the skeletal muscle cells (Fig. 11-11 ■). This branching structure allows a single motor neuron to control many muscle fibers at one time.

The synapse of a somatic motor neuron on a muscle fiber is called the **neuromuscular junction** (Fig. 11-12 ■). Like all synapses, this region has three components: (1) the motor neuron's presynaptic axon terminal filled with synaptic vesicles and mitochondria, (2) the synaptic cleft, and (3) the postsynaptic membrane of the skeletal muscle fiber.

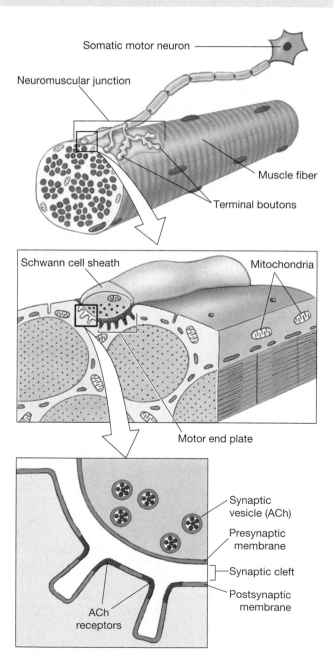

■ **Figure 11-11 Anatomy of the neuromuscular junction** A somatic motor neuron branches at its distal end. Each axon terminal (bouton) lies opposite a special region of the skeletal muscle membrane, the motor end plate, that contains a high density of ACh receptors.

Within the neuromuscular junction, the membrane of the muscle cell that lies opposite the axon terminal is modified into a **motor end plate,** a series of folds that look like shallow gutters. Along the upper edge of each gutter, ACh receptor channels cluster together into an active zone. Between the axon and the muscle, the synaptic cleft is filled with a fibrous matrix whose collagen fibers hold the axon terminal and the motor end plate in the proper alignment. The matrix also contains *acetylcholinesterase* (AChE), the enzyme that rapidly deactivates ACh by degrading it into acetyl and choline.

✔ Compare gating and ion selectivity of ion channels in the motor end plate with the ion channels along the axon of a somatic motor neuron.

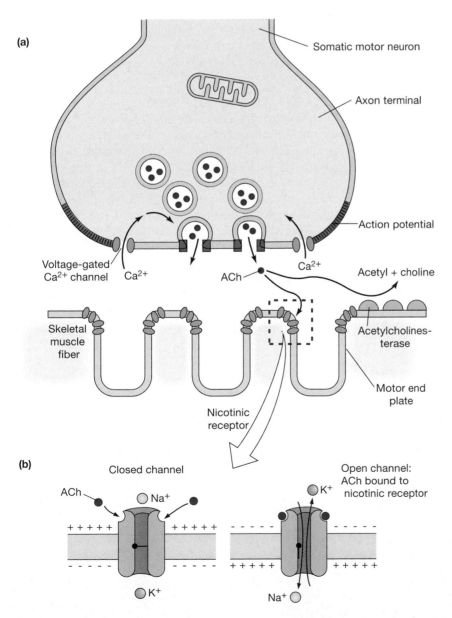

■ **Figure 11-12   Events at the neuromuscular junction**   (a) An action potential opens voltage-gated Ca^{2+} channels in the axon terminal. Calcium ions enter the terminal, triggering exocytosis of synaptic vesicles. Acetylcholine (ACh) in the synaptic cleft can combine with a nicotinic receptor on the motor end plate or be metabolized by acetylcholinesterase (AChE) to acetyl and choline. (b) The nicotinic cholinergic receptor binds two ACh molecules, opening a nonspecific monovalent cation channel. Sodium ion influx exceeds K^+ efflux, and the muscle fiber depolarizes.

● ●

...continued from page 336

One of the hallmarks of addiction is the withdrawal symptoms that occur when a person stops ingesting a substance. Nicotine withdrawal symptoms include lethargy, hunger, and irritability. To avoid withdrawal symptoms, people who are trying to stop smoking may use a nicotine patch. These adhesive patches deliver a small, continuous dose of nicotine that is usually enough to keep withdrawal symptoms in check. For Shanika, the first 24 hours without cigarettes have not been easy. Although she wears a nicotine patch, all she wants is a cigarette to smoke. When she reads the nicotine package insert, she notices a warning to keep the patches away from children. An overdose of nicotine (highly unlikely when the patch is used as directed) could result in complete paralysis of the respiratory muscles (the diaphragm and the skeletal muscles of the chest wall).

Question 4: *Why might excessive levels of nicotine cause respiratory paralysis?*

● ●

The Neuromuscular Junction Contains Nicotinic Receptors

As in all neurons, action potentials arriving at the axon terminal open voltage-gated Ca^{2+} channels in the membrane. Calcium diffuses into the cell down its electrochemical gradient, triggering the release of ACh-containing synaptic vesicles. Acetylcholine diffuses across the synaptic cleft and combines with nicotinic receptor channels on the skeletal muscle membrane (Fig. 11-12 ■). These receptors are similar but not identical to the nicotinic ACh receptors found in ganglia (and those found in the CNS). This difference is evidenced by the fact that the snake toxin α-bungarotoxin binds to nicotinic skeletal muscle receptors but not to those in autonomic ganglia.

Nicotinic receptors are chemically gated ion channels with two binding sites for ACh (Fig. 11-12b ■). When ACh binds to the receptor, the channel gate opens and allows monovalent cations to flow through. Net Na^+ entry into the muscle fiber depolarizes it, triggering an action potential that causes contraction of the skeletal muscle cell.

Acetylcholine acting on the motor end plate is always excitatory and creates muscle contraction. There is no dual innervation to relax skeletal muscles. Instead, relaxation occurs when the somatic motor neurons are inhibited, preventing ACh release. In Chapter 13 you will learn more about how inhibition of somatic motor pathways controls body movement.

✔ A nonsmoker who chews nicotine-containing gum might notice an increase in heart rate, a function controlled by sympathetic neurons. If postganglionic sympathetic neurons secrete norepinephrine, why would nicotine affect heart rate?

✔ The Huaorani Indians of South America use blowguns to shoot darts poisoned with curare at monkeys. Curare is a plant toxin that binds to and inactivates nicotinic ACh receptors. What happens to the monkey if struck by these darts?

✔ Patients with myasthenia gravis have a deficiency of ACh receptors on their skeletal muscles and have weak muscle function as a result. Why would administration of an anticholinesterase drug that inhibits acetylcholinesterase improve their muscle function?

Somatic motor neurons do more than simply create contractions: they are necessary for muscle health. "Use it or lose it" is a cliché that is very appropriate to the dynamics of muscle mass, because disrupting synaptic transmission at the neuromuscular junction has devastating effects on the entire body. Without communication between the motor neuron and the muscle, for whatever reason, the skeletal muscles for movement and posture weaken, as do the skeletal muscles used for breathing. In the severest cases, loss of respiratory function can be fatal unless the patient is placed on artificial ventilation. Myasthenia gravis, the loss of ACh receptors discussed earlier, is the most common disorder of the neuromuscular junction.

PROBLEM CONCLUSION

Shanika is determined to stop smoking this time because her grandfather, a smoker for many years, was just diagnosed with lung cancer. She finds that the patch alone does not stop her craving for a cigarette, so she attends behavioral modification classes as well. After six months, she proudly informs her family that she thinks she has kicked the habit.

Nicotine is an agonist for nicotinic ACh receptors, but with chronic exposure, nicotine inactivates the ACh receptors. With fewer active receptors, target cells up-regulate their receptor number. When smokers stop smoking, the removal of nicotine from the system allows all receptors to return to an active state. A larger than usual number of active receptors is probably responsible for some of the withdrawal symptoms noted by people who are trying to stop smoking. The nicotine patch allows the former smoker to gradually decrease nicotine levels in the body, preventing withdrawal symptoms during the time the cells are down-regulating their receptors back to the normal number.

In this running problem, you learned that nicotine is an addictive agonist of ACh. You also learned how cells respond to long-term exposure to nicotine. Further check your understanding of this running problem by comparing your answers to those in the summary table.

Question	Facts	Integration and Analysis
1 What is the usual response of cells that are chronically exposed to increased concentrations of a signal molecule? (Hint: This topic was discussed in Chapter 6.)	A cell that is exposed to increased concentrations of a signal molecule will decrease its receptors for that molecule [∞ p. 165].	Down-regulation of receptors allows a cell to respond normally even if the concentration of ligand is elevated.
2 What happens to the postsynaptic cell when nicotine rather than ACh binds to a nicotinic cholinergic receptor?	Nicotine is an agonist of ACh. Agonists mimic the activity of a ligand.	Nicotine binding to a nicotinic cholinergic receptor will open ion channels in the postsynaptic cell and the cell will depolarize. This is the same effect that ACh binding creates [∞ p. 164].
3 Although nicotine in short-term studies is a nicotinic receptor agonist, chronic exposure to nicotine causes up-regulation of the receptors. Speculate on why up-regulation might occur.	Chronic exposure to an agonist usually causes down-regulation. Chronic exposure to an antagonist usually causes up-regulation.	Although nicotine is a short-term agonist, it appears to be acting as an antagonist during long-term exposure. Researchers have found that nicotinic ACh receptors have several functional states: unbound and inactive, bound to ligand and active, and bound to ligand but inactivated. Long-term exposure to nicotine apparently places many nicotinic ACh receptors into the third state.
4 Why might excessive levels of nicotine cause respiratory paralysis?	Nicotinic receptors are found at the neuromuscular junction that controls skeletal muscle contraction. The diaphragm and chest wall muscles that regulate breathing are skeletal muscles.	The nicotinic receptors of the neuromuscular junction are not as sensitive to nicotine as are those of the CNS and autonomic ganglia. However, excessively high amounts of nicotine will activate the nicotinic cholinergic receptors of the motor end plate, causing the muscle fiber to depolarize and contract. The continued presence of nicotine keeps the ion channels open, and the muscle remains depolarized. In this state, it is unable to contract again, resulting in paralysis.

CHAPTER REVIEW

SUMMARY

The Autonomic Division

1. The efferent division of the nervous system is divided into **somatic motor neurons,** which control skeletal muscles, and **autonomic neurons,** which control smooth muscle, cardiac muscle, many glands, and some adipose tissue. (p. 327)

2. The autonomic division is subdivided into a **sympathetic branch** and a **parasympathetic branch.** (p. 327)

3. The maintenance of homeostasis within the body is a balance between sympathetic and parasympathetic control. (p. 327)

4. The two autonomic branches demonstrate all properties of homeostasis: maintenance of the internal environment, tonic control, antagonistic control, and variable tissue responses. (p. 328)

5. CNS control of the autonomic division is closely linked to centers in the hypothalamus, pons, and medulla of the brain. (p. 328)

6. Some autonomic reflexes are spinal reflexes. Many of these can be modulated by input from the brain. (p. 328)

7. An autonomic pathway is composed of a **preganglionic neuron** from the CNS that synapses with a **postganglionic neuron** in an **autonomic ganglion.** (p. 329)

8. Autonomic ganglia are capable of modulating information passing through them with the assistance of **intrinsic neurons** that lie completely within the ganglia. (p. 329)

9. Most sympathetic preganglionic neurons originate in the thoracic and lumbar regions of the spinal cord. Most sympathetic ganglia lie close to the spinal cord or along the descending aorta. (p. 331)

10. The cell bodies of parasympathetic preganglionic neurons are found either in the brain stem or in the sacral region of the spinal cord. Parasympathetic ganglia are located on or near their target organs. (p. 331)

11. The synapse between a postganglionic autonomic neuron and its target cells is called the **neuroeffector junction.** (p. 331)

12. Autonomic axons end with **varicosities** from which neurotransmitter is released. (p. 332)

13. The **adrenal medulla** is a neurosecretory endocrine gland that is controlled by sympathetic preganglionic neurons. It secretes catecholamines (primarily epinephrine). (p. 332)

14. The level of nervous control that an autonomic neuron exerts on its target depends on the concentration of neurotransmitter in the synapse as well as the identity and number of neurotransmitter receptors in the target tissues. (p. 332)

15. The two most common autonomic neurotransmitters are acetylcholine and norepinephrine. All preganglionic neurons secrete ACh. As a rule, sympathetic postganglionic neurons secrete norepinephrine, and postganglionic parasympathetic neurons secrete ACh. (p. 332)

16. **Adrenergic receptors** are G protein-linked receptors whose activation initiates a second messenger cascade. **Alpha receptors** are found on most target tissues of the sympathetic nervous system. β_1 **receptors** are found on heart muscle and in the kidney. β_2 **receptors** are found in locations not innervated by sympathetic neurons. (p. 334)

17. **Cholinergic receptors** come in two main types. **Nicotinic receptors** are found in autonomic ganglia. Binding of ACh to a nicotinic receptor opens cation channels and depolarizes the postsynaptic cell. **Muscarinic receptors** are found at the neuroeffector junctions of the parasympathetic branch. (p. 336)

The Somatic Motor Division

18. Somatic motor pathways, which control skeletal muscles, have a single neuron that originates in the CNS and projects its axon to a skeletal muscle. Somatic motor pathways are always excitatory and result in muscle contraction. (p. 338)

19. Somatic motor neurons branch at their distal end so that a single motor neuron controls many muscle fibers at one time. (p. 338)

20. The synapse of a somatic motor neuron on a muscle fiber is called the **neuromuscular junction.** The membrane of the muscle cell at the synapse is modified into a **motor end plate** that contains a high concentration of nicotinic ACh receptor channels. (p. 338)

21. Binding of ACh to the ACh receptor opens ion channels and allows net Na^+ entry into the muscle fiber, depolarizing it. Acetylcholine in the synapse is broken down by the enzyme acetylcholinesterase. (p. 340)

QUESTIONS

LEVEL ONE   Reviewing Facts and Terms

1. What are the two divisions of the efferent part of the peripheral nervous system? What effectors does each control?

2. The autonomic nervous system is sometimes called the _____ nervous system. Why is this an appropriate name? List some of the functions controlled by the autonomic nervous system.

3. What are the two branches of the autonomic nervous

system? How are these branches distinguished anatomically and physiologically?

4. Which neurosecretory endocrine gland is closely allied to the sympathetic branch of the autonomic division?

5. Neurons that secrete acetylcholine are described as _____, whereas those that secrete norepinephrine are called _____ or _____ neurons.

6. List four things that can happen to autonomic neurotransmitters after they are released into the synapse.

7. The main enzyme responsible for catecholamine degradation is _____, abbreviated _____.

8. Somatic motor pathways

 (a) are excitatory or inhibitory?
 (b) are composed of a single neuron or a preganglionic and a postganglionic neuron?
 (c) synapse with glands or with smooth, cardiac, or skeletal muscle?

9. What is acetylcholinesterase? Describe its action.

10. What kind of receptors are found on the postsynaptic cell in the neuromuscular junction?

LEVEL TWO   Reviewing Concepts

11. **Concept map:** Make a map comparing the somatic, sympathetic, and parasympathetic branches. Use the following terms and add any additional terms you like:

acetylcholine	muscarinic receptor
adipose tissue	nicotinic receptor
alpha receptor	norepinephrine
autonomic division	one-neuron pathway
beta receptor	smooth muscle
ganglion	parasympathetic branch
cardiac muscle	skeletal muscle
cholinergic receptor	somatic motor division
efferent division	sympathetic branch
endocrine gland	two-neuron pathway
exocrine gland	

12. What is the advantage of divergence of neural pathways in the autonomic nervous system?

13. Compare and contrast the following:

 (a) the neuroeffector junctions of the autonomic nervous system with the neuromuscular junction of the somatic nervous system
 (b) alpha, beta, muscarinic, and nicotinic receptors in the efferent side of the nervous system. Describe where each is found, and the ligands that bind to them.

14. Compare and contrast:

 (a) autonomic ganglia and CNS nuclei
 (b) the adrenal medulla and the posterior pituitary gland
 (c) boutons and varicosities

LEVEL THREE   Problem Solving

15. If the neurotransmitter receptor is _____ (use items in left column), the neuron(s) releasing neurotransmitter onto the receptor must be _____. (Use all appropriate items in right column)

 (a) cholinergic nicotinic
 (b) adrenergic α
 (c) cholinergic muscarinic
 (d) adrenergic β

 1. somatic motor neuron
 2. autonomic preganglionic neuron
 3. sympathetic postganglionic neuron
 4. parasympathetic postganglionic neuron

16. If nicotinic receptor channels allow both K^+ and Na^+ to flow through, why does Na^+ influx exceed K^+ efflux? (Hint: [∞ p. 145])

17. Little Danny was riding his tricycle in the driveway when a horrible thing happened. The parking brake on his Mom's car failed, and it rolled backward, flipping the tricycle and pinning Danny underneath, with the car on the handlebars, anchoring it down firmly. Sandy, his mother, was close by and was able to lift the car away so that she could get Danny out. Everybody was OK. Sandy weighs about 125 pounds. Is there a physiological explanation for what happened?

E X P L <MediaLab>

Introduction

In the previous chapters you learned about the parts of the nervous system, how individual neurons work, how your body senses the environment, and how it processes this information in the central nervous system. It is now time for action. What types of activities does the CNS control? In this Web Exploration, you will consider the question of neural dysfunction and how it can be treated. After reading the descriptions below, visit the MediaLab for Chapter 11 in your Companion Website and select the appropriate keyword.

Web Exploration 1

Estimated time for completion = 15 minutes

The previous chapter dealt with how we sense and process information from our internal and external environments. Chapter 9 helped us understand how this information is integrated to generate a response. But what are those responses, and how does your body regulate them?

Think of all the systems of the body. Then make a list of the tissues and organs in each system that perform an action and indicate their functions. For example, the stomach will increase its contractions after you have eaten but decrease its activity when you exercise. Based on your understanding of the nervous system, draw the neural circuit(s) that would accomplish each function you listed.

Select the keyword **NEURAL CIRCUIT** from the Website and see how close your model of the motor output of the nervous system came to the real thing. Select the keyword **INTERNAL ORGANS** to see which of the internal organs receive neural inputs. Predict how the function of each of these organs changes under a "fight or flight" situation. How will this same organ be regulated in a "rest and digest" condition? Select the keyword **ORGAN RESPONSES** to see if your predictions are correct.

Web Exploration 2

Estimated time for completion = 10 minutes

Suppose that one day you find that you have a hard time keeping your eyelids open (you begin to look like Sleepy of the seven dwarfs). As the day goes on, you develop double vision and have difficulty swallowing. Sometimes you have slurred speech, but you never drink liquor. On your worse days your muscles feel like you have been power-lifting for days, when all you've been doing is walking to and from class. What is going on? What could account for all these symptoms?

Draw out a basic model for the neural control of skeletal muscles. Where along this neural-muscular circuit could a problem cause these types of symptoms? After a visit to the doctor you are given a prescription of mestinon. Select the keyword **MESTINON** from the Website. Based on the mode of action of this drug, what is the cause of your muscle dysfunction? Select the keyword **MG** from the Website to see if your prediction is correct and to learn the molecular basis of your muscle dysfunction. Use this knowledge to explain how it is that this drug can increase your muscle strength.

12 Muscles

■ "A muscle is ... an engine, capable of converting chemical energy into mechanical energy. It is quite unique in nature, for there has been no artificial engine devised with the great versatility of living muscle."
— *Ralph W. Stacy and John A. Santolucito, in* Modern College Physiology, *1966.* ■

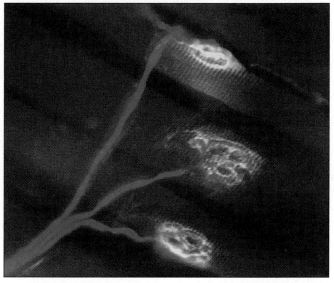

One somatic motor neuron branches to innervate several skeletal muscle fibers.

BACKGROUND BASICS

It was his first time to be the starting pitcher. As he ran from the bullpen onto the field, his heart was pounding and his stomach felt as if it were tied in knots. He stepped onto the mound and gathered his thoughts before throwing the first practice pitch. Gradually, as he went through the familiar routine of throwing and catching the baseball, his heart slowed and his stomach relaxed. It was going to be a good game.

The pitcher's pounding heart, queasy stomach, and movements as he runs and throws the ball are all the results of muscle contraction. Our muscles have two common functions: (1) to generate motion and (2) to generate force. Our skeletal muscles also generate heat and contribute significantly to the homeostasis of body temperature. When homeostasis is threatened by cold conditions, our brain may direct our muscles to shiver, creating additional heat.

Three types of muscle tissue occur in the human body: skeletal muscle, cardiac muscle, and smooth muscle. Most **skeletal muscles** are attached to the bones of the skeleton, enabling them to control body movement. **Cardiac muscle** is found only in the heart and is responsible for creating movement of blood through the circulatory system. Skeletal and cardiac muscles are classified as **striated muscles** [*stria*, groove] because of their

alternating light and dark bands under the light microscope (Fig. 12-1a, b ■).

Smooth muscle is the primary muscle of internal organs and tubes such as the stomach, urinary bladder, and blood vessels. Its primary function is to influence the movement of material into, out of, or within the body. An example is the passage of food through the gastrointestinal tract. Smooth muscle has a homogeneous appearance under the microscope, without obvious cross-bands (Fig. 12-1c ■). Its lack of banding results from the less organized arrangement of contractile fibers within the muscle cells.

Skeletal muscles are often described as the voluntary muscles, and smooth and cardiac muscle are described as

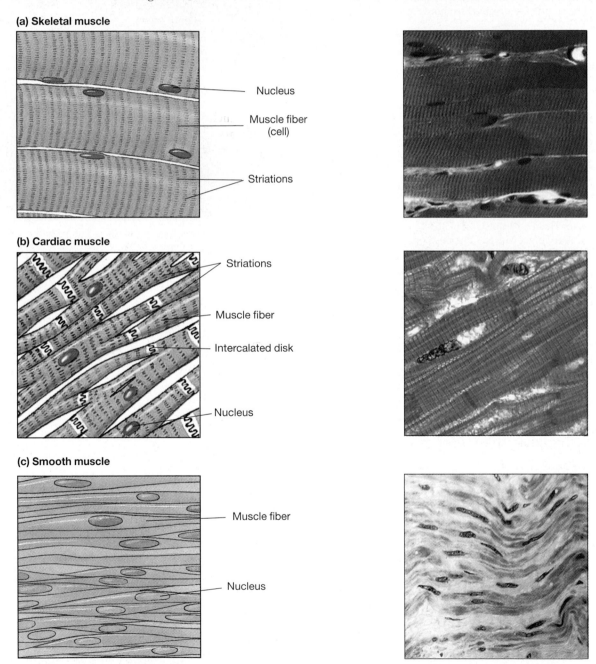

(a) Skeletal muscle

Nucleus

Muscle fiber (cell)

Striations

(b) Cardiac muscle

Striations

Muscle fiber

Intercalated disk

Nucleus

(c) Smooth muscle

Muscle fiber

Nucleus

■ **Figure 12-1   Three types of muscles**   (a) Skeletal muscle fibers are large multinucleate cells that appear striated under the microscope. (b) Cardiac muscle fibers are striated uninucleate cells that are branched. They connect to each other through specialized junctions known as intercalated disks. (c) Smooth muscle fibers are smaller, uninucleate cells without obvious banding patterns.

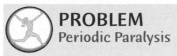
involuntary. However, this is not a precise classification. Skeletal muscles can contract without conscious direction, and we can learn a certain degree of conscious control over some smooth and cardiac muscle [∞ p. 327]

Skeletal muscles are unique in that they only contract in response to a signal from a somatic motor neuron. They cannot initiate their own contraction, nor is their contraction influenced directly by hormones.

In contrast, cardiac and smooth muscle have multiple levels of control. Although their primary extrinsic control arises through autonomic innervation, some types of smooth and cardiac muscle can contract spontaneously, without signals from the central nervous system. In addition, the activity of cardiac and some smooth muscle is subject to modulation by the endocrine system. Despite these differences, smooth and cardiac muscle share many properties of skeletal muscle.

▶ SKELETAL MUSCLE

Skeletal muscles make up the bulk of muscle in the body, about 40% of total body weight. They are responsible for positioning and movement of the skeleton, as their name

suggests. Skeletal muscles are usually attached to bones by collagenous **tendons** [∞ p. 64]. The **origin** of a muscle is the end of the muscle that is attached closest to the trunk or to the more stationary bone. The **insertion** of the muscle is the more *distal* [*distantia,* distant] or mobile attachment.

When the bones attached to a muscle are connected by a flexible **joint,** contraction of the muscle results in movement of the skeleton. If the centers of the connected bones are brought closer together when the muscle contracts, the muscle is called a **flexor.** If the bones move away from each other when the muscle contracts, the muscle is called an **extensor.**

Most joints in the body have both flexor and extensor muscles, since a contracting muscle can pull a bone in one direction but cannot push it back. Flexor-extensor pairs are called **antagonistic muscle groups** because they exert opposite effects. Figure 12-2 ■ shows a pair of antagonistic muscles in the arm: the *biceps brachii* [*brachion,* arm], which acts as the flexor, and the *triceps brachii,* which acts as the extensor. When the biceps muscle contracts, the hand and forearm move toward the body. Contraction of the triceps extends the flexed arm. In each case, when one muscle contracts and shortens, the antagonistic muscle must relax and lengthen.

✔ Identify as many pairs of antagonistic muscle groups in the body as you can. If you cannot name them, point out the probable location of the flexor and extensor of each group.

Skeletal Muscles Are Composed of Muscle Fibers

Muscles function together as a unit. A skeletal **muscle** is a collection of muscle cells, or **muscle fibers** (see Anatomy Summary, Fig. 12-3 ■), just as a nerve is a collection

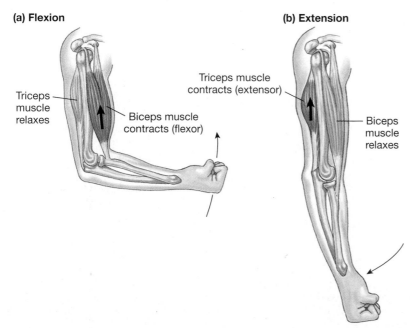

(a) Flexion

Triceps muscle relaxes

Biceps muscle contracts (flexor)

(b) Extension

Triceps muscle contracts (extensor)

Biceps muscle relaxes

■ **Figure 12-2** **Antagonistic muscle groups** Muscle contraction can pull on a bone, but cannot push a bone away. Muscles in the body are therefore often arranged in pairs or groups to provide movement in several directions.

ANATOMY SUMMARY   Skeletal Muscle

■ Figure 12-3

(a)

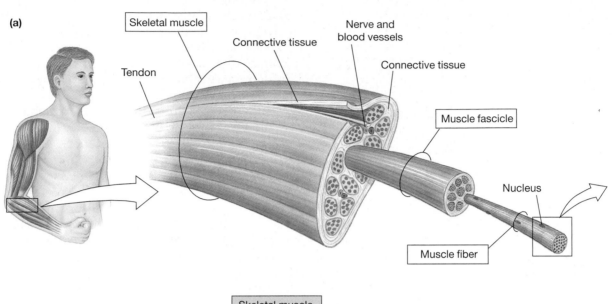

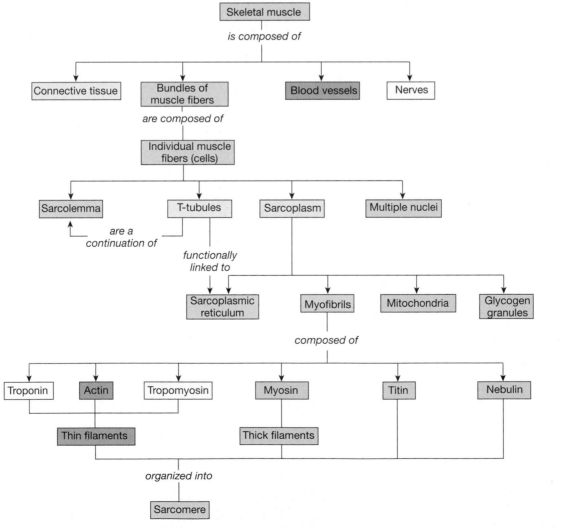

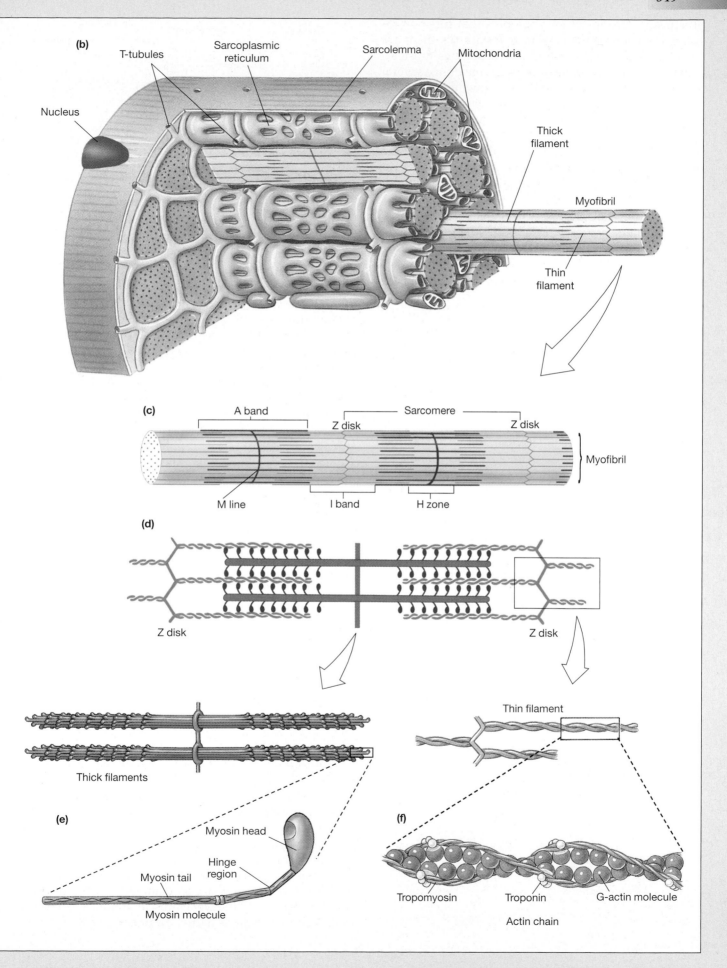

(b) Nucleus · T-tubules · Sarcoplasmic reticulum · Sarcolemma · Mitochondria · Thick filament · Myofibril · Thin filament

(c) A band · Sarcomere · Z disk · Z disk · Myofibril · M line · I band · H zone

(d) Z disk · Z disk

(e) Thick filaments · Myosin head · Hinge region · Myosin tail · Myosin molecule

(f) Thin filament · Tropomyosin · Troponin · G-actin molecule · Actin chain

of neurons. Groups of muscle fibers that function together and the motor neuron that controls them are called a *motor unit*. A small muscle, such as those attached to the bones of the middle ear, may have several hundred muscle fibers and a few motor units. Larger muscles, such as the quadriceps muscles of the thigh, may have several thousand fibers and many motor units.

Each skeletal muscle fiber is a long, cylindrical cell with up to several hundred nuclei on the surface of the fiber (Fig. 12-3b ■). Skeletal muscle fibers are the largest cells in the body, created by the fusion of many individual embryonic muscle cells.

The individual fibers of a skeletal muscle are bound with connective tissue and arranged with their long axes in parallel (Fig. 12-3a ■). Collagen, elastic fibers, nerves, and blood vessels are found between the bundles of muscle fibers. The entire muscle is enclosed in a connective tissue sheath that is continuous with the connective tissue around the muscle fibers and with the tendons holding the muscle to underlying bones.

Muscle Fiber Anatomy Muscle physiologists, like neurophysiologists, have adopted a specialized vocabulary (Table 12-1). The cell membrane of a muscle fiber is called the **sarcolemma** [*sarkos*, flesh + *lemma*, shell], and the cytoplasm is called the **sarcoplasm.** A muscle fiber contains little cytosol. The main intracellular structures are the **myofibrils,** bundles of contractile and elastic proteins that carry out the work of contraction.

Skeletal muscle fibers contain an extensive **sarcoplasmic reticulum** (SR), a form of modified endoplasmic reticulum that wraps around each myofibril like a piece of lace (Figs. 12-3b, 12-4 ■). The sarcoplasmic reticulum consists of longitudinal tubules, which release Ca^{2+}, and the **terminal cisternae,** which concentrate and sequester Ca^{2+} [*sequestrare,* to put in the hands of a trustee].

A branching network of **transverse tubules,** also known as **t-tubules,** is closely associated with the terminal cisternae. One t-tubule with its flanking terminal cisternae is known as a *triad.* The membranes of t-tubules are a continuation of the muscle fiber membrane. This makes the lumen of the t-tubules continuous with the extracellular fluid. To understand how this network of tubules in the heart of the muscle fiber communicates with the outside, take a lump of soft clay and stick your

TABLE 12-1   Muscle Terminology

General Term	Muscle Equivalent
Muscle cell	Muscle fiber
Cell membrane	Sarcolemma
Cytoplasm	Sarcoplasm
Modified endoplasmic reticulum	Sarcoplasmic reticulum

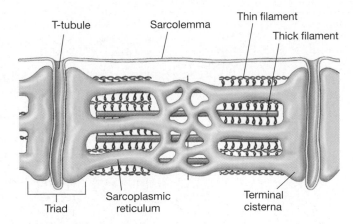

■ **Figure 12-4   T-tubules and the sarcoplasmic reticulum**
The sarcoplasmic reticulum wraps around each myofibril. The t-tubule system is closely associated with the sarcoplasmic reticulum.

finger into the middle of it several times. Notice how the outside surface of the clay (analogous to the cell membrane of the muscle fiber) is now continuous with the sides of the holes that you have poked in the clay (the membrane of the t-tubules).

T-tubules allow action potentials that originate at the neuromuscular junction on the cell surface to move rapidly into the interior of the fiber. Without t-tubules, the action potential could reach the center of the fiber only by diffusion of positive charge through the cytosol, a slower process that would delay the response time of the muscle fiber.

The cytosol between the myofibrils contains many glycogen granules and mitochondria. Glycogen, the storage form of glucose found in animals, is a reserve source of energy. Mitochondria provide much of the ATP for muscle contraction through oxidative phosphorylation of glucose and other biomolecules.

Myofibrils Are the Contractile Structures of a Muscle Fiber

Each muscle fiber contains a thousand or more myofibrils that occupy most of the intracellular volume, leaving little space for cytosol and organelles (Fig. 12-3b ■). Each myofibril is composed of several types of proteins: the contractile proteins *myosin* and *actin,* the regulatory proteins *tropomyosin* and *troponin,* and the giant accessory proteins *titin* and *nebulin.*

Myosin [*myo-,* muscle] is the protein that makes up the thick filaments of the myofibril. Various *isoforms* [*iso-,* equal], or different types, of myosin occur in different types of muscle. They are related to the muscle's speed of contraction.

Each myosin molecule is composed of two heavy protein chains that intertwine to form a long coiled tail and a pair of tadpolelike heads (Fig. 12-3e ■). Two lightweight protein chains are associated with the heavy chain of each head. In skeletal muscle, about 250 myosin molecules join to create a **thick filament.** The thick filament is arranged so that the myosin heads are clustered at the ends, and the central region of the filament is a bundle of myosin tails. The rodlike portion of the thick filament is stiff, but the protruding myosin heads have an elastic hinge region where the heads join the rods. This hinge region allows the heads to swivel around their point of attachment.

Actin [*actum,* to do] is the protein that makes up the thin filaments of the muscle fiber. One actin molecule is a globular protein (*G-actin*), represented in Figure 12-3f ▪ by a round ball. Usually, multiple globular actin molecules polymerize to form long chains or filaments (*F-actin*). In skeletal muscle, two F-actin polymers twist together like a double strand of beads, creating the **thin filaments** of the myofibril.

Most of the time, the parallel thick and thin filaments of the myofibril are connected by **crossbridges** that span the space between the filaments. The crossbridges are the myosin heads that loosely bind to the actin filaments. Each G-actin molecule has a single binding site for a myosin head.

Under a light microscope, the arrangement of thick and thin filaments in a myofibril creates a repeating pattern of alternating light and dark bands (Figs. 12-1a, 12-3c▪). One repeat of the pattern forms a **sarcomere** [*sarkos,* flesh + -*mere,* a unit or segment], which has the following elements (Fig. 12-5a ▪):

- **Z disks:** These zigzag structures are made of proteins that serve as the attachment site for the thin

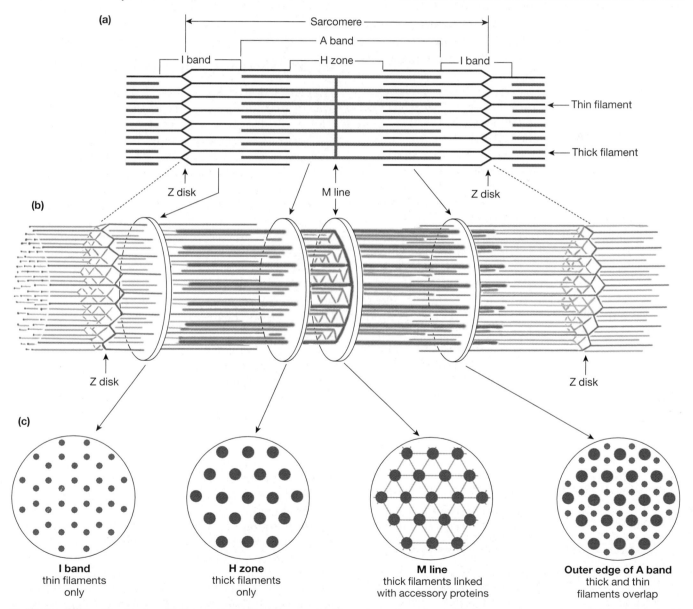

■ **Figure 12-5 The two- and three-dimensional organization of a sarcomere** (a) A sarcomere runs from Z disk to Z disk. (b) Within a myofibril, the thick and thin filaments overlap in a three-dimensional array. (c) The thick and thin filaments are arranged so that when they overlap, each thick filament is surrounded by six thin filaments (as seen in the A band) and each thin filament is surrounded by three thick filaments. The M line is the region where the thick filaments are linked together. The Z disk (not shown) has a similar arrangement of accessory proteins that link the thin filaments.

filaments. One sarcomere is composed of two Z disks and the filaments found between them. The abbreviation Z comes from *zwischen,* the German word for "between."

- **I band:** These are the lightest color bands of the sarcomere and represent a region occupied only by thin filaments. The abbreviation *I* comes from *isotropic,* a description from early microscopists meaning that this region reflects light uniformly under a polarizing microscope. A Z disk runs through the middle of an I band, so each half of the I band belongs to a different sarcomere.

- **A band:** This is the darkest of the bands in a sarcomere and encompasses the entire length of a thick filament. At the outer edges of the A band, the thick and thin filaments overlap. The center of the A band is occupied by thick filaments only. The abbreviation *A* comes from *anisotropic* [*an-,* not], meaning that the protein fibers in this region scatter light unevenly.

- **H zone:** This central region of the A band is lighter than the outer edges of the A band because the H zone is occupied by thick filaments only. The *H* comes from *helles,* the German word for "clear."

- **M line:** This band represents the attachment site for the thick filaments, equivalent to the Z disk for the thin filaments. One A band is divided in half by an M line. *M* is the abbreviation for *mittel,* the German word for "middle."

In three-dimensional array, the actin and myosin molecules form a lattice of parallel, overlapping thin and thick filaments (Fig. 12-5b ■). When viewed end-on, each thin filament is surrounded by three thick filaments, and six thin filaments encircle each thick filament (Fig. 12-5c ■).

The proper alignment of filaments within a sarcomere is ensured by two types of proteins, titin and nebulin (Fig. 12-6 ■). **Titin** is a huge elastic molecule and the largest known protein, with more than 25,000 amino acids. A single titin molecule stretches from one Z disk to the next M line. To get an idea of the immense size of titin, imagine that one titin molecule is an 8-foot-long piece of the very thick rope used to tie ships to a wharf. By comparison, a single actin molecule would be about the length and weight of a single eyelash.

Titin has two functions: it stabilizes the position of the contractile filaments, and its elasticity returns stretched muscles to their resting length. Titin is helped by **nebulin,** an inelastic giant protein that lies alongside thin filaments and attaches to the Z disk. Nebulin helps align the actin filaments of the sarcomere.

✔ Why are the ends of the A band the darkest region of the sarcomere when viewed under the light microscope?

✔ What is the function of t-tubules?

✔ Why are skeletal muscles called striated muscle?

Muscles Shorten When They Contract

The contraction of skeletal muscle fibers is a remarkable process that allows us to create force to move or resist a load. In muscle physiology, the force created by the contracting muscle is called the muscle **tension.** The **load** is a weight or force that opposes contraction of a muscle. The creation of tension in a muscle is an active process that requires energy input from ATP.

In previous centuries, scientists observed that when muscles move a load, they shorten. This observation led to early theories of contraction, which proposed that muscles were made of molecules that curled up and shortened when active, then relaxed and stretched at rest, like elastic in reverse. The theory received support when myosin was found to be a helical molecule that shortened upon heating (the reason meat shrinks when you cook it).

In 1954, however, Sir Andrew Huxley and Rolf Niedeigerke discovered that the length of the A band of a myofibril remains constant during contraction. Because the A band represents the myosin filament, they realized that shortening of the myosin molecule could not be responsible for contraction. Subsequently, they proposed an alternative model, the **sliding filament theory of contraction.** In this model, overlapping muscle fibers of fixed length (the thick and thin filaments) slide past each other in an energy-requiring process, resulting in muscle contraction.

The sliding filament theory takes into account the observation that when a muscle contracts, it does not always shorten. A muscle can contract and create force without creating movement. For example, if you push on a wall, you are creating tension in many muscles of your body without shortening the muscles and without moving the wall. According to the sliding filament theory, tension generated in a muscle fiber is directly proportional to interaction between the thick and thin filaments.

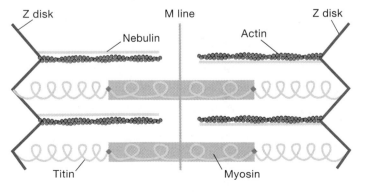

■ **Figure 12-6** **Titin and nebulin** Titin spans the distance from one Z disk to the next M line. Its elastic component helps return a stretched sarcomere to resting length. Titin and nonelastic nebulin stabilize the filaments within the sarcomere.

…continued from page 347

Paul had experienced mild attacks of muscle weakness in his legs before, usually in the morning. Twice, the weakness had come on after exposure to cold. Each attack had disappeared within minutes, and Paul seemed to suffer no lasting effects. On the advice of Paul's family doctor, Mrs. Leong took her son to see a specialist in muscle disorders. She diagnosed a condition called periodic paralysis. This condition is caused by a genetic defect in the Na^+ channels in the membranes of his skeletal muscle fibers. Under certain conditions, the defective Na^+ channels remain open and continuously admit Na^+ into the interior of the muscle cells.

Question 1:   What effect does the continuous influx of Na^+ have on the membrane potential of Paul's muscle fibers?

Sliding Filament Theory of Contraction

If you examined a myofibril at its resting length, you would see that within each sarcomere, the ends of the thick and thin filaments overlap slightly. As the muscle contracts, the thick and thin filaments slide past each other, moving the Z disks of the sarcomere closer together.

This phenomenon can be seen in light micrographs of relaxed and contracted muscle (Fig. 12-7 ■). In the relaxed state, a sarcomere has a large I band (thin filaments only) and an A band whose length is the length of the thick filament. As contraction occurs, the sarcomere shortens. The two Z disks at each end move closer to-

gether while the I band and H zone, regions where actin and myosin do not overlap in resting muscle, almost disappear. Despite the shortening of the sarcomere, the length of the A band remains constant. These changes are consistent with the sliding of thin actin filaments along the thick myosin filaments as they move toward the M line in the center of the sarcomere. It is from this process that the sliding filament theory of contraction derives its name.

The force that pushes the actin filament is the movement of myosin crossbridges that link actin and myosin. Each myosin head has two binding sites on it: one for an ATP molecule and one for actin. The actin molecules serve as the "road" along which the myosin heads walk.

During the **power stroke** that is the basis for muscle contraction, movement of the flexible myosin molecules pushes actin filaments toward the center of the sarcomere. At the end of a power stroke, the myosin head releases its bound actin, then swings back and binds to a new actin molecule, ready to start another contractile cycle.

This process repeats many times as a muscle fiber contracts. The myosin heads repeatedly bind and release actin molecules as myosin pushes the thin filaments toward the center of the sarcomere. But what causes movement of the myosin molecules? The answer is energy from ATP.

Myosin is a *motor protein* that converts the chemical bond energy of ATP into the mechanical energy of motion. Each myosin molecule is an ATPase (*myosin ATPase*)

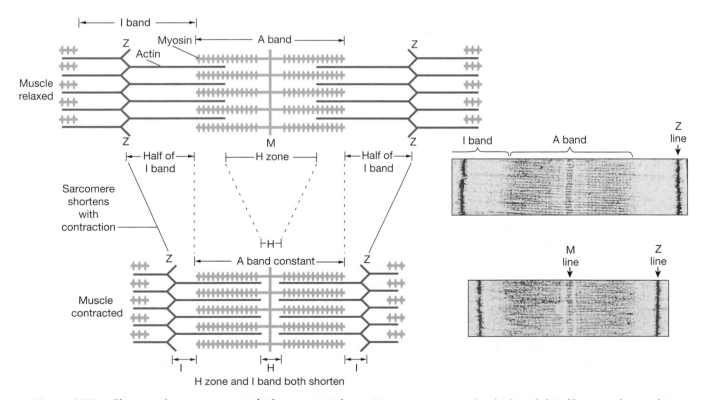

■ **Figure 12-7   Changes in a sarcomere during contraction**   During contraction the thick and thin filaments do not change length. The thin actin filaments slide over the thick myosin filaments, moving toward the M line in the center of the sarcomere. The A band does not change in length, but both the H zone and the I band get shorter as the filaments overlap.

The *In Vitro* Motility Assay One big step forward in understanding the power stroke of myosin was the development of the *in vitro* motility assay in the 1980s. In this assay, isolated myosin molecules are randomly bonded to a specially coated glass coverslip. A fluorescently labeled actin molecule is placed on top of the myosin molecules. With ATP as a source of energy, the myosin heads bind to the actin and move it across the coverslip, marked by a fluorescent trail as it goes. In even more ingenious experiments, developed in 1995, a single myosin molecule is bound to a tiny bead that elevates it above the surface of the cover slip. An actin molecule is placed on top of the myosin molecule, like the balancing pole of a tightrope walker. As the myosin "motor" moves the actin molecule, lasers measure the nanometer movements and piconewton forces created with each cycle of the myosin head. Because of this technique, researchers can now measure the mechanical work being done by a single myosin molecule!

that binds ATP and hydrolyzes it to ADP and inorganic phosphate (P_i). The energy released by hydrolyzing ATP is stored as potential energy in the angle between the myosin molecule and the long axis of the myosin filament. This potential energy is then used to create the power stroke that moves actin.

Figure 12-8 ■ shows the molecular events of a contractile cycle in detail. The starting point for this cycle is chosen arbitrarily.

1. Our contractile cycle starts with the myosin head of the crossbridge tightly bound to a G-actin molecule. In this state, no nucleotide (ATP or ADP) occupies the second binding site on the myosin head. This tight binding is known as the **rigor state** [*rigere*, to be stiff]. In living muscle, the rigor state occurs for only a very brief period.

2. Next, an ATP molecule binds within the pocketlike nucleotide-binding site on the myosin head. As ATP binds, it changes the affinity of the actin-binding site so that the myosin head releases from actin.

3. The nucleotide-binding site on the myosin closes around the ATP and hydrolyses it to ADP and inorganic phosphate. Both products remain bound to myosin.

4. The energy released from ATP rotates the free myosin molecule and it binds weakly to a new G-actin molecule, one or two positions away from the G-actin to which it was previously bound. At this point, the myosin has potential energy, like a stretched spring, and is ready to execute the power stroke that will move the actin filament past it.

5. The power stroke of myosin begins when inorganic phosphate is released from the myosin binding site. As myosin moves, it pushes the attached actin fila-

ment toward the center of the sarcomere. The power stroke is also called *crossbridge tilting* because the myosin head tilts from a 90° angle relative to the two filaments to a 45° angle. Our current model of the process also suggests that part of the long myosin arm bends, acting as a lever.

6. In the last step of the contractile cycle, myosin releases ADP, the second product of ATP hydrolysis. At this point, the myosin head is again tightly bound to actin. The cycle is ready to begin again when a new ATP binds to myosin.

Although our contractile cycle began with the rigor state in which no ATP or ADP was bound to myosin, relaxed muscle fibers remain mostly in stage 4 above. The rigor state in living muscle is normally brief because the muscle fiber has a sufficient supply of ATP that binds to myosin when ADP is released in step 6.

After death, however, when metabolism stops and ATP supplies are exhausted, muscles are unable to bind more ATP, so they remain in the tightly bound state described in step 1. In the condition known as *rigor mortis*, the muscles "freeze" owing to immovable crossbridges. The tight binding of actin and myosin persists for a day or so after death, until enzymes within the decaying fiber begin to break down the muscle proteins.

Actin-myosin binding and crossbridge tilting explain how actin filaments slide past myosin filaments during contraction. But what regulates this process? If ATP is always available in the living muscle fiber, what keeps the filaments from continuously interacting? The answer lies with the two regulatory proteins tropomyosin and troponin.

Contraction Is Regulated by Troponin and Tropomyosin

The thin actin filaments of the myofibril are associated with two regulatory proteins that prevent myosin heads from completing their power stroke, just as the safety latch on a gun keeps the cocked trigger from being pulled. **Tropomyosin** [*tropos*, to turn] is an elongated protein polymer that wraps around the actin filament. In doing so, tropomyosin blocks part of the myosin-binding site on each actin molecule (Fig. 12-9a ■). When tropomyosin is in this blocking, or "off," position, weak actin-myosin binding can take place, but myosin is blocked from completing its power stroke. For contraction to occur, tropomyosin must be shifted to an "on" position that uncovers the remainder of the binding site. This allows the power stroke to take place.

The "off-and-on" positioning of tropomyosin is regulated by troponin. **Troponin** is a complex of three proteins associated with tropomyosin. As contraction begins, one protein of the complex, **troponin C**, binds reversibly to Ca^{2+} (Fig. 12-9b ■). Calcium binding pulls tropomyosin toward the groove of the actin filament and

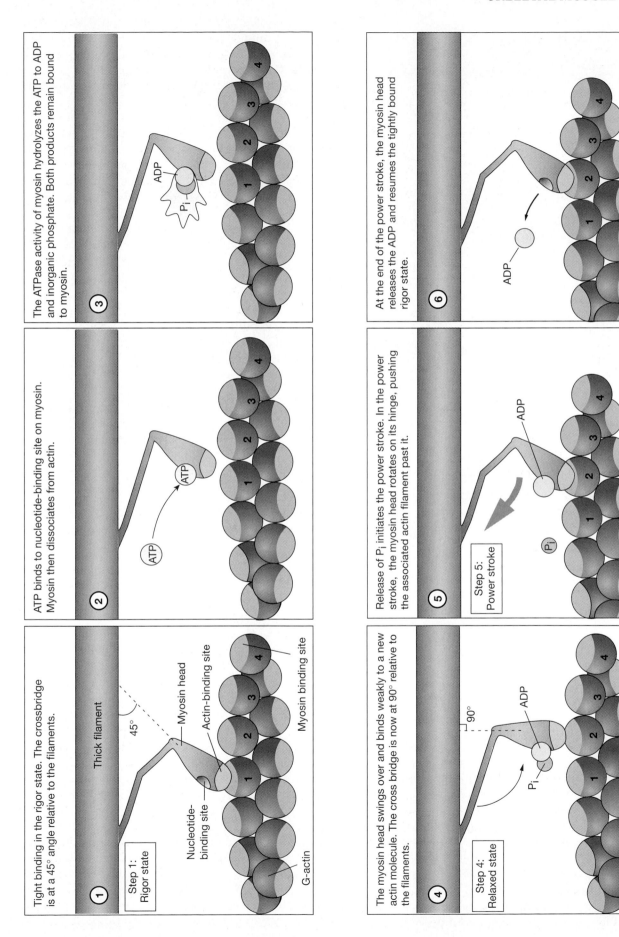

① **Step 1: Rigor state**

Tight binding in the rigor state. The crossbridge is at a 45° angle relative to the filaments.

Thick filament
45°
Myosin head
Actin-binding site
Nucleotide-binding site
Myosin binding site
G-actin

② ATP binds to nucleotide-binding site on myosin. Myosin then dissociates from actin.

ATP

③ The ATPase activity of myosin hydrolyzes the ATP to ADP and inorganic phosphate. Both products remain bound to myosin.

ADP
Pᵢ

④ **Step 4: Relaxed state**

The myosin head swings over and binds weakly to a new actin molecule. The cross bridge is now at 90° relative to the filaments.

90°
ADP
Pᵢ

⑤ **Step 5: Power stroke**

Release of Pᵢ initiates the power stroke. In the power stroke, the myosin head rotates on its hinge, pushing the associated actin filament past it.

ADP
Pᵢ
Actin filament moves

⑥ At the end of the power stroke, the myosin head releases the ADP and resumes the tightly bound rigor state.

ADP

■ **Figure 12-8   The molecular basis of contraction**

(a) Relaxed state

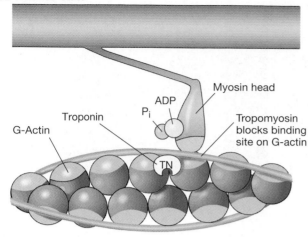

G-Actin

Troponin

ADP

P_i

Myosin head

Tropomyosin blocks binding site on G-actin

TN

(b) Initiation of contraction

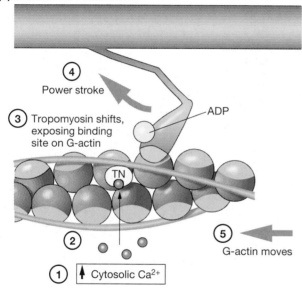

④ Power stroke

③ Tropomyosin shifts, exposing binding site on G-actin

ADP

TN

②

⑤ G-actin moves

① ↑ Cytosolic Ca²⁺

■ **Figure 12-9   Regulatory role of tropomyosin and troponin**   (a) In the relaxed state, troponin partially blocks the myosin-binding sites on actin. (b) Contraction begins when Ca^{2+} binds to troponin. This uncovers the myosin-binding site on actin so that myosin can complete its power stroke.

unblocks the myosin-binding sites. This "on" position enables the myosin heads to carry out their power strokes and move the actin filament. Contractile cycles repeat as long as the binding sites are uncovered.

For relaxation to occur, Ca^{2+} concentrations in the cytosol must decrease so that Ca^{2+} unbinds from troponin. Without Ca^{2+}, the troponin-tropomyosin complex returns to its "off" position, covering most of the myosin-binding site. During the portion of the relaxation phase when actin and myosin are not bound to each other, the filaments of the sarcomere slide back to their original positions with the aid of elastic connective tissue within the muscle.

Although the preceding discussion sounds as if we know everything there is to know about the molecular

basis of muscle contraction, in reality these are simply our current models for the process. Studying the movement of molecules within a myofibril has proved very difficult. Many research techniques rely on crystallized molecules, electron microscopy, and other tools that cannot be used with living tissues.

Often we can see filaments only at the beginning and end of contraction. This is like studying a baseball pitcher's delivery from two photos. The start of the pitch can be seen in a photo taken at the end of the windup, when the pitcher rears back, with his right arm cocked and left knee raised. The second photo shows the pitcher after the ball has been released: his right arm is extended forward toward home plate, his left foot is on the ground, and his right foot is in the air behind him. The dilemma is to reconstruct the motion that occurred between the two poses. This is one problem facing scientists who are trying to explain exactly how actin and myosin slide past each other. Progress is being made, however, and perhaps in the next decade you will see a "movie" of muscle contraction, constructed from photos of sliding filaments.

✔ In the sliding filament theory of contraction, what prevents the filaments from sliding back to their original position each time they release to bind to the next binding site? (Hint: What would happen if all crossbridges released simultaneously?)

Acetylcholine from Somatic Motor Neurons Initiates Excitation-Contraction Coupling

How does a muscle contraction begin? Signals for muscle contraction come from the central nervous system to skeletal muscles by way of somatic motor neurons. Acetylcholine from the somatic motor neuron initiates an action potential in the muscle fiber. The action potential in turn triggers calcium release, which is the intracellular signal for contraction. This combination of electrical and mechanical events in a muscle fiber is called **excitation-contraction coupling.** Let's look at this process in detail.

Acetylcholine released into the synapse at a neuromuscular junction binds to ACh receptor-channels on the motor end plate [∞ p. 338]. These channels allow both Na^+ and K^+ to cross the sarcolemma. When the channels open, Na^+ influx exceeds K^+ efflux because the electrochemical driving force is greater for Na^+ [∞ p. 143]. The addition of net positive charge to the muscle fiber depolarizes the membrane, creating an **end-plate potential (EPP).** Normally, end-plate potentials always reach threshold and initiate a muscle action potential.

The action potential is conducted across the surface of the muscle fiber and into the t-tubules by the opening of voltage-gated Na^+ channels (Fig. 12-10 ■). The process is similar to the conduction of action potentials in axons, although action potentials in skeletal muscle are con-

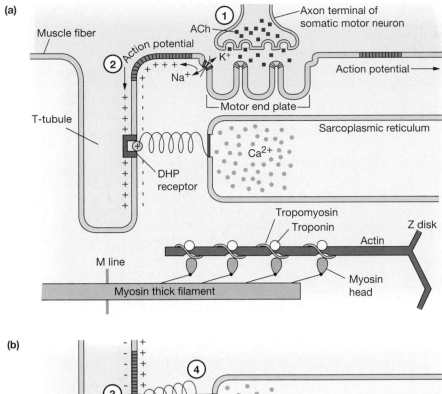

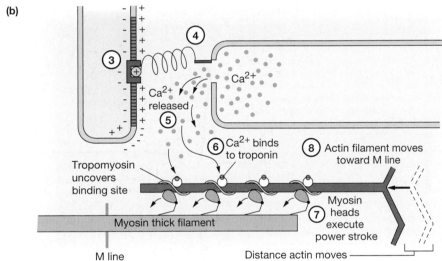

■ **Figure 12-10   Excitation-contraction coupling**   (a) Acetylcholine creates an action potential that moves across the fiber membrane and down the t-tubules. (b) Voltage-sensitive dihydropyridine (DHP) receptors open Ca^{2+} channels in the sarcoplasmic reticulum. Ca^{2+} combines with troponin to uncover the binding sites on actin and initiate contraction.

ducted more slowly than the action potentials of neurons [∞ p. 229].

The action potential that moves across the membrane and down the t-tubules causes Ca^{2+} release from the sarcoplasmic reticulum. The t-tubule membrane contains voltage-sensing receptors (the **dihydropyridine, or DHP, receptors**) that are mechanically linked to Ca^{2+} release channels in the adjacent sarcoplasmic reticulum (*ryanodine receptor-channels,* also called *junctional foot proteins*). When a wave of depolarization reaches a DHP receptor, its conformation changes, opening Ca^{2+} channels in the sarcoplasmic reticulum. Stored Ca^{2+} then diffuses into the cytosol, where it initiates contraction.

Free cytosolic Ca^{2+} levels in a resting muscle are normally quite low, but after an action potential, they increase about 100-fold. When cytosolic Ca^{2+} levels are high, Ca^{2+} binds to troponin, tropomyosin moves to the "on" position, and contraction occurs.

Relaxation of the muscle fiber occurs when the sarcoplasmic reticulum pumps Ca^{2+} back into its lumen using a Ca^{2+}-ATPase. As the free cytosolic Ca^{2+} decreases, Ca^{2+} releases from troponin, tropomyosin slides back to block the myosin-binding site, and the fiber relaxes. The events of excitation-contraction coupling and contraction are summarized in Table 12-2.

The discovery that Ca^{2+}, not the action potential, is the signal for contraction was the first piece of evidence suggesting that calcium acts as a messenger inside cells. For years, it was thought that calcium signals occurred only in muscles, but we now know that calcium is an almost universal second messenger [∞ p. 162].

Figure 12-11 ■ shows the temporal sequence of events during excitation-contraction coupling. The somatic motor neuron action potential is followed by the skeletal muscle action potential, which in turn is followed by contraction. A single contraction-relaxation cycle in a

TABLE 12-2   Muscle Excitation and Contraction

1. Action potential in somatic motor neuron arrives at axon terminal.
2. Voltage-gated Ca^{2+} channels open. Ca^{2+} entry triggers exocytosis of ACh-containing synaptic vesicles.
3. ACh diffuses into synaptic cleft and binds with nicotinic channel-receptors on muscle motor end plate.
4. Net influx of Na^+ through the ACh-gated channel depolarizes the muscle membrane, creating an end-plate potential.
5. The EPP always creates a muscle action potential.
6. The action potential spreads from the neuromuscular junction along the fiber membrane and down the t-tubules.
7. Voltage-sensitive dihydropyridine receptors open Ca^{2+} release channels in the sarcoplasmic reticulum.
8. Ca^{2+} diffuses into the cytosol and binds to troponin, pulling tropomyosin away from the myosin-binding site. This action allows myosin to release inorganic phosphate from ATP hydrolysis and complete its power stroke.
9. At the end of the power stroke, the myosin crossbridge releases ADP and remains tightly bound to actin. Myosin must bind an ATP molecule in order to release from this rigor state.
10. The muscle fiber relaxes when Ca^{2+} is transported out of the cytosol by a Ca^{2+}-ATPase. Ca^{2+} removal from troponin allows tropomyosin to again block the myosin-binding site of actin.
11. The ATPase activity of myosin hydrolyses ATP to ADP and inorganic phosphate. Both remain bound to myosin. The myosin head swivels and binds to a new actin molecule, ready to execute its next power stroke.

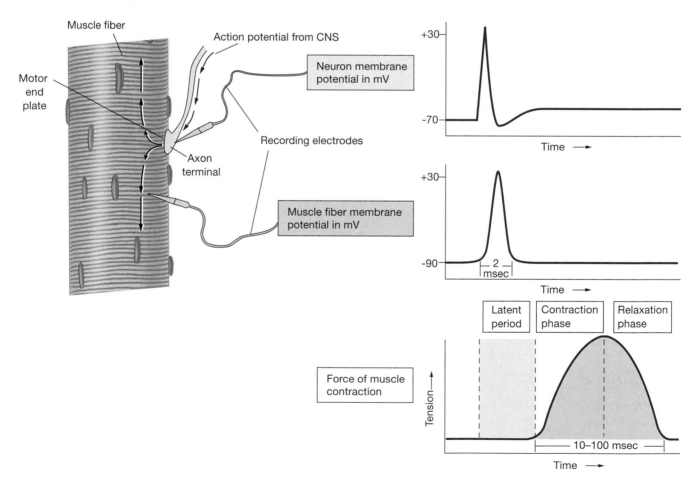

■ **Figure 12-11   Temporal sequence of electrical and mechanical events in muscle contraction**   Action potentials in the axon terminal and the muscle fiber are followed by the muscle twitch. The latent period represents the time required for Ca^{2+} release from the sarcoplasmic reticulum.

skeletal muscle fiber is known as a **twitch.** Notice that there is a short delay, the **latent period,** between the muscle action potential and the beginning of muscle tension development. This interval represents the time required for excitation-contraction coupling to take place.

Once contraction begins, muscle tension increases steadily to a maximum value as crossbridge interaction increases. Tension then decreases in the relaxation phase of the twitch. During relaxation, the elastic elements of the muscle return the sarcomeres to their resting length.

A single action potential in a muscle fiber evokes a single twitch. But muscle twitches vary from fiber to fiber in the speed with which they develop tension (the rising slope of the twitch curve), the maximum tension they achieve (the height of the twitch curve), and the duration of the twitch (the width of the twitch curve). We will discuss factors that affect these parameters in upcoming sections. First let's look at how muscles produce ATP to provide energy for contraction and relaxation.

✔ What aspect of contraction requires ATP? Why doesn't relaxation require ATP?

Skeletal Muscle Contraction Depends on a Steady Supply of ATP

The muscle fiber's use of ATP is a key feature of muscle physiology. Muscles require energy constantly: during contraction for crossbridge movement and release, during relaxation to pump Ca^{2+} back into the sarcoplasmic reticulum, and after excitation-contraction coupling to restore Na^+ and K^+ to the extracellular and intracellular compartments, respectively. Where do muscles get the ATP they need for this work?

● ●

…continued from page 353

Two forms of periodic paralysis exist. One form, called hyperkalemic periodic paralysis, is characterized by increased blood levels of K^+ during paralytic episodes. The other form, called hypokalemic periodic paralysis, is characterized by decreased blood levels of K^+ during episodes. Results of a blood sample revealed that Paul has the hyperkalemic form. It is not known why altered levels of K^+ trigger these attacks in people with the Na^+ channel defect.

Question 2: In people with hyperkalemic periodic paralysis, attacks tend to occur during a period of exercise (i.e., repeated muscle contractions). Why? (Hint: What ion is responsible for the repolarization phase of muscle contraction, and in which direction does this ion move across the muscle fiber membrane?)

Question 3: Why does the Na^+ channel defect cause paralysis? (Hint: The gated channels that release Ca^{2+} from the sarcoplasmic reticulum close in a time-dependent manner, even if the cell remains depolarized.)

● ●

The amount of ATP within the muscle fiber at any one time is sufficient for only about eight twitches. As a backup energy source, muscles contain **phosphocreatine,** a molecule whose high-energy phosphate bonds are created from creatine and ATP when muscles are at rest (Fig. 12-12 ■). When muscles become active, such as during exercise, the high-energy phosphate group of phosphocreatine is transferred to ADP, creating ATP.

The enzyme responsible for transferring the phosphate group is *creatine kinase (CK),* also known as *creatine phosphokinase (CPK).* Muscle cells contain large amounts of this enzyme. Consequently, elevated blood levels of creatine kinase usually indicate damage to skeletal or cardiac muscle. Because the two muscle types contain different isozymes [∞ p. 81], clinicians can distinguish tissue damage during a heart attack from skeletal muscle damage.

Energy stored in high-energy phosphate bonds is very limited, so muscle fibers must use metabolism to transfer energy from the chemical bonds of nutrients to ATP. Carbohydrates, particularly glucose, are the most rapid and efficient source of energy for ATP production. Glucose is metabolized through glycolysis to pyruvate [∞ p. 90]. In the presence of adequate oxygen, pyruvate goes into the citric acid cycle, producing about 30 ATP for each molecule of glucose.

When oxygen concentrations are too low to maintain aerobic metabolism, the muscle fiber shifts to *anaerobic glycolysis.* In this pathway, glucose is metabolized to lactic acid with a yield of only 2 ATP per glucose. Anaerobic metabolism of glucose is a quicker source of ATP but produces many fewer ATP per glucose. When muscle energy

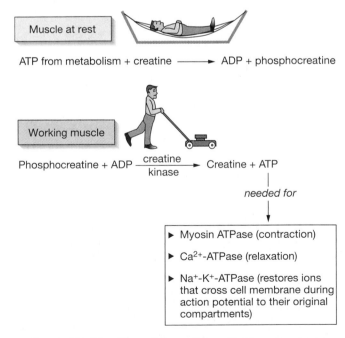

■ **Figure 12-12   Phosphocreatine**   Resting muscle stores energy from ATP in high-energy phosphate bonds of phosphocreatine. Working muscle transfers energy from phosphocreatine back to ATP.

demands outpace the amount of ATP that can be produced through anaerobic metabolism of glucose, muscles can function for only a short period of time without fatiguing.

Muscle fibers also obtain energy from fatty acids, although this process always requires oxygen [∞ p. 97]. During rest and light exercise, skeletal muscles burn fatty acids along with glucose, one reason that modest exercise programs of brisk walking are an effective way to reduce body fat. But beta-oxidation, the process by which fatty acids are converted to acetyl CoA, is a slow process and cannot produce ATP rapidly enough to meet the energy needs of muscle fibers during heavy exercise. Under these conditions, muscle fibers rely more on glucose.

Proteins are not normally a source of energy for muscle contraction. Most of the amino acids found within muscle fibers are used to synthesize proteins rather than to produce ATP.

Do muscles ever run out of ATP? You might think so if you have ever exercised to the point of fatigue, where you feel that you cannot continue or your limbs refuse to obey the commands from your brain. But most studies show that even intense exercise uses only 30% of the ATP within a muscle fiber. The condition we call fatigue must come from other changes in the exercising muscle.

Muscle Fatigue Has Multiple Causes

The term **fatigue** describes a condition in which a muscle is no longer able to generate or sustain the expected power output. Fatigue is highly variable. It is influenced by the intensity and duration of the contractile activity, by whether the muscle fiber is using aerobic or anaerobic metabolism, by the composition of the muscle, and by the fitness level of the individual.

Several factors have been proposed to play a role in fatigue. They include changes in the ionic composition of the muscle fiber after numerous contractions, depletion of muscle nutrients, and rarely, diminished neurotransmitter output from the somatic motor neuron. Most experimental evidence suggests that muscle fatigue arises from excitation-contraction failure within the muscle fiber itself rather than from failure of control neurons or neuromuscular transmission.

Fatigue within the muscle fiber itself can occur in any one of several sites. In extended submaximal exertion, fatigue is associated with the depletion of glycogen stores within the muscle. Because most studies show that lack of ATP is not a limiting factor, glycogen depletion may be affecting some other aspect of contraction, such as the release of Ca^{2+} from the sarcoplasmic reticulum.

In short-duration maximal exertion, fatigue has a different cause. It apparently results from increased production of inorganic phosphate from ATP and phosphocreatine breakdown. Elevated inorganic phosphate may slow P_i release from myosin and thereby alter the power stroke (Fig. 12-8, ⑤ ■).

Another factor implicated in fatigue is K^+. During maximal exercise, K^+ leaves the cell with each action potential, so K^+ concentrations rise in the extracellular fluid of the t-tubules. The shift in K^+ alters the membrane potential of the muscle fiber and is believed to decrease Ca^{2+} release from the sarcoplasmic reticulum. However, there appears to be no single cause for excitation-contraction coupling failure in fatigue, and the study of this phenomenon is quite complex.

Neural causes of fatigue could arise either from communication failure at the neuromuscular junction or from failure of the central nervous system command neurons. If ACh is not synthesized in the axon terminal fast enough to keep up with the firing rate of the neuron, neurotransmitter release at the synapse will decrease. Consequently, the muscle end-plate potential may fail to reach the threshold value needed to trigger a muscle fiber action potential, resulting in contraction failure. This type of fatigue is associated with some neuromuscular diseases, but it is probably not a factor in normal exercise.

Central fatigue includes subjective feelings of tiredness and a desire to cease activity. Several studies have shown that this fatigue actually precedes physiological fatigue in the muscles and therefore may be a protective mechanism. Lactic acid production from anaerobic metabolism is often mentioned as a possible cause for fatigue. Some evidence suggests that acidosis caused by lactic acid dumped into the blood may influence the sensation of fatigue perceived by the brain. However, because the homeostatic mechanisms for pH balance maintain blood pH at normal levels until exertion is nearly maximal, pH is not a factor in central fatigue in most instances of submaximal exertion.

Skeletal Muscle Fibers Are Classified by Speed of Contraction and Resistance to Fatigue

Skeletal muscle fibers can be classified into three broad groups on the basis of their speed of contraction and their resistance to fatigue with repeated stimulation. The three groups include **fast-twitch glycolytic fibers, fast-**

…continued from page 359

Paul's doctor explained to Mrs. Leong that the paralytic attacks associated with hyperkalemic periodic paralysis last only a few minutes to a few hours and generally involve only the muscles of the extremities. "Is there any treatment?" asked Mrs. Leong. The doctor replied that, although the condition is presently incurable, attacks can be prevented with several drugs. Diuretics, for example, increase the body's excretion of water and ions (including Na^+ and K^+), and they have been shown to prevent attacks of paralysis in people with this condition.

Question 4:   *Speculate on how diuretics prevent attacks in people with hyperkalemic periodic paralysis.*

twitch oxidative fibers, and **slow-twitch (oxidative) fibers.**

Fast-twitch muscle fibers develop tension two to three times as fast as slow-twitch fibers. The speed with which a muscle fiber contracts is determined by the isoform of myosin ATPase present in the fiber's thick filaments. Fast-twitch fibers split ATP more rapidly and can therefore complete multiple contractile cycles more rapidly than slow-twitch fibers. This speed translates into faster tension development in the muscle fiber.

The duration of contraction also varies according to the fiber type. Twitch duration is determined largely by how fast the sarcoplasmic reticulum removes Ca^{2+} from the cytosol. As cytosolic Ca^{2+} concentrations fall, Ca^{2+} unbinds from troponin, allowing tropomyosin to move back and partially block the myosin-binding sites. With the power stroke inhibited, the muscle fiber relaxes.

Fast-twitch fibers pump Ca^{2+} into their sarcoplasmic reticulum more rapidly than slow-twitch fibers, so fast-twitch fibers have quicker twitches. Fast-fiber twitches last only about 7.5 milliseconds (msec), making these muscles useful for fine, quick movements such as playing the piano.

Contractions in slow-twitch muscle fibers may last more than 10 times as long. These muscles are used for strong, sustained movements such as lifting heavy loads. Fast fibers are used occasionally, but the slow-fiber types are used almost constantly for maintaining posture, standing, or walking.

Another major difference between fast-twitch glycolytic and slow-twitch fibers is their ability to resist fatigue. Fast glycolytic muscle fibers rely primarily on anaerobic glycolysis to produce ATP. However, the accumulation of lactic acid contributes to acidosis, a condition implicated in the development of fatigue. As a result, fast-twitch fibers fatigue more easily than do slow-twitch fibers, which do not produce large amounts of lactic acid.

Slow-twitch fibers, on the other hand, depend primarily on oxidative phosphorylation for production of ATP, hence their descriptive name as oxidative fibers. Slow-twitch fibers have more mitochondria, the site of enzymes for the citric acid cycle and oxidative phosphorylation. They also have more blood vessels in their connective tissue to bring oxygen to the muscle fibers (Fig. 12-13 ■).

The efficiency with which muscle fibers obtain oxygen is a factor in their preferred route of glucose metabolism. Slow-twitch fibers are described as "red muscle" because large amounts of **myoglobin,** a red oxygen-binding pigment, give them their characteristic color.

Oxygen brought from the lungs to the muscles must diffuse from the interstitial fluid into the interior of the fiber to reach the mitochondria. In muscles with myoglobin, this diffusion process is faster because myoglobin has a high binding affinity for oxygen. This affinity allows myoglobin to act as a transfer molecule, bringing oxygen more rapidly to the interior of the fiber.

In addition, slow-twitch fibers have a smaller diameter, so the distance through which oxygen must pass before reaching the mitochondria is less. Because slow-twitch muscle fibers have more capillaries to bring blood to the cells, are smaller, and have a high myoglobin content, they maintain a better supply of oxygen and are more likely to use oxidative phosphorylation for ATP production.

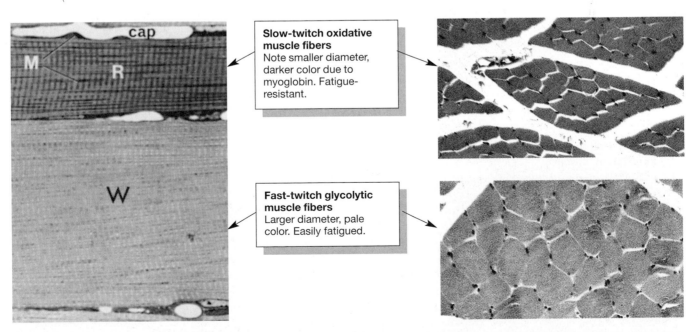

Slow-twitch oxidative muscle fibers Note smaller diameter, darker color due to myoglobin. Fatigue-resistant.

Fast-twitch glycolytic muscle fibers Larger diameter, pale color. Easily fatigued.

■ **Figure 12-13   Fast-twitch glycolytic and slow-twitch muscle fibers**   Large amounts of red myoglobin, numerous mitochondria (M), and extensive capillary blood supply (cap) distinguish slow-twitch oxidative muscle (R) from fast-twitch glycolytic muscle (W).

Fast-twitch fibers, on the other hand, are described as "white muscle" because of their lower myoglobin content. These muscle fibers are also larger in diameter. The combination of larger size, less myoglobin, and fewer blood vessels means that fast-twitch fibers are more likely to run out of oxygen with repeated contractions. They are therefore forced to use anaerobic metabolism for ATP synthesis.

Fast-twitch fibers have been divided into two subcategories based on their relative diameters and fatigue resistance. Fast-twitch glycolytic fibers are the largest in diameter and rely primarily on anaerobic metabolism. They fatigue most rapidly of the three fiber types. Fast-twitch oxidative fibers are smaller, contain some myoglobin, and use a combination of oxidative and glycolytic metabolism to produce ATP. Because of their intermediate size and the use of oxidative phosphorylation for ATP synthesis, fast-twitch oxidative fibers are more fatigue-resistant than their fast glycolytic cousins. Characteristics of the three muscle fiber types are compared in Table 12-3.

Tension Developed by Individual Muscle Fibers Is a Function of Fiber Length

Within a single muscle fiber, the tension of a twitch is a direct reflection of the length of individual sarcomeres before contraction begins (Fig. 12-14 ■). Each sarcomere will contract with optimum force if it is an optimum length, neither too long nor too short, before the contraction begins. Fortunately, the normal resting length of skeletal muscles usually ensures that sarcomeres within the individual muscle fibers are at optimum length when they begin a contraction.

At the molecular level, sarcomere length reflects the overlap between the thick and thin filaments. The sliding filament theory predicts that *the tension a muscle fiber can generate is directly proportional to the number of cross-*

bridges formed between the thick and thin filaments. If the fibers start a contraction at a very long sarcomere length, the thick and thin filaments are barely overlapped, forming few crossbridges (Fig. 12-14e ■). This means that in the initial part of the contraction, the sliding filaments can interact only minimally and therefore cannot generate much force.

At the optimum sarcomere length, the filaments begin contracting with more crossbridge linkage between the thick and thin filaments, allowing the fiber to generate optimum force in that twitch (Fig. 12-14c ■). If the sarcomere is shorter than optimum length at the beginning of the contraction, the thick and thin fibers were too well overlapped before the contraction began. Consequently, the thick filaments can move the thin filaments only a short distance before the thin actin filaments from opposite ends of the sarcomere start to overlap. This overlap prevents crossbridge formation.

Finally, if the sarcomere shortens too much, the thick filaments run into the Z disks at the ends of the sarcomere. Once this happens, the thick filaments are unable to find new binding sites for crossbridge formation, so tension decreases rapidly (Fig. 12-14a ■). Thus, the development of single-twitch tension within a muscle fiber is a passive property that depends on filament overlap and sarcomere length.

Force of Contraction Increases with Summation of Muscle Twitches

Although we have just shown that single-twitch tension is determined by the length of the sarcomere, it is important to note that a single twitch does not represent the maximum force that the muscle fiber can develop. The force generated by the contraction of a single muscle fiber can be increased by increasing the rate (frequency) at which muscle action potentials stimulate the muscle fiber.

TABLE 12-3   Characteristics of Muscle Fiber Types

	Slow-Twitch Oxidative; Red Muscle	Fast-Twitch Oxidative; Red Muscle	Fast-Twitch Glycolytic; White Muscle
Time to development of maximum tension	Slowest	Intermediate	Fastest
Myosin ATPase activity	Slow	Fast	Fast
Diameter	Small	Medium	Large
Contraction duration	Longest	Short	Short
Ca^{2+}-ATPase activity in SR	Moderate	High	High
Endurance	Fatigue-resistant	Fatigue-resistant	Easily fatigued
Use	Most used: posture	Standing, walking	Least used: jumping
Metabolism	Oxidative; aerobic; numerous large mitochondria	Glycolytic but becomes more oxidative with endurance training	Glycolytic; more anaerobic than fast-twitch oxidative type
Color	Dark red (myoglobin)	Red	Pale

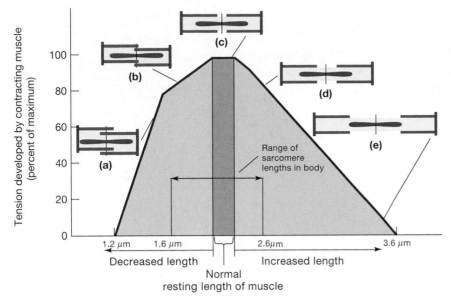

■ **Figure 12-14** **Length-tension relationships in contracting muscle** The graph compares tension generated by a muscle with its resting length before contraction starts. If the resting muscle is stretched, filaments in the sarcomere barely overlap and cannot form as many crossbridge links (e). If the muscle begins its contraction at a very short length, the sarcomere cannot shorten very much before the myosin filaments run into the Z disks at each end (a).

A typical muscle action potential lasts between 1 and 3 msec, while the muscle contraction itself may last 100 msec. If repeated action potentials are separated by long intervals of time, the muscle fibers have time to relax completely between stimuli (Fig. 12-15a ■). If the interval of time between action potentials is shortened, the muscle fiber will not have relaxed completely at the time of the second stimulus, resulting in a more forceful contraction (Fig. 12-15b ■). This process is known as **summation** and is similar to the summation that takes place in neurons [∞ p. 241].

If action potentials continue to stimulate the muscle fiber repeatedly at short intervals (high frequency), relaxation between contractions diminishes until the muscle fiber achieves a state of maximal contraction known as **tetanus.** There are two types of tetanus. In *incomplete,* or *unfused, tetanus,* the stimulation rate of the muscle fiber is slower, and the fiber relaxes slightly between stimuli (Fig. 12-15c ■).

In *complete,* or *fused, tetanus,* the stimulation rate is fast enough that the muscle fiber does not have time to relax. Instead, it reaches maximum tension and remains there (Fig. 12-15d ■). Thus, it is possible to increase the tension developed in a single muscle fiber by changing the rate at which action potentials occur in the fiber. Muscle action potentials are initiated by the somatic motor neuron that controls the muscle fiber.

✔ Summation in muscle fibers means that the _____ of the fiber increases with repeated action potentials.

✔ Temporal summation in neurons means that the _____ of the neuron increases when two depolarizing stimuli occur close together in time.

One Somatic Motor Neuron and the Muscle Fibers It Innervates Form a Motor Unit

The basic unit of contraction in an intact skeletal muscle is a **motor unit,** composed of a group of muscle fibers and the somatic motor neuron that controls them (see chapter opener photo and Fig. 12-16 ■). When the somatic motor neuron fires an action potential, all muscle fibers in the motor unit contract. Note that although one somatic motor neuron innervates multiple fibers, each muscle fiber is innervated by only a single neuron.

The number of muscle fibers in a motor unit varies. In muscles used for fine motor actions, such as the muscles that move the eyes or the muscles of the hand, a motor unit contains as few as three to five muscle fibers. If one motor unit is activated, only a few fibers contract, and the muscle response is quite small. If additional motor units are activated, the response increases by small increments since only a few more muscle fibers contract with the addition of each motor unit. This arrangement allows fine gradations of movement.

In muscles used for gross motor actions, such as standing or walking, a single somatic motor neuron may innervate hundreds or even thousands of muscle fibers. The gastrocnemius muscle in the calf of the leg, for example, has about 2000 muscle fibers in each motor unit. Each time an additional motor unit is activated in these muscles, many more muscle fibers contract, and the muscle response jumps by correspondingly greater increments.

All muscle fibers in a single motor unit are of the same fiber type. Thus, there are fast-twitch motor units and slow-twitch motor units. The determination of which kind of fiber associates with a particular neuron appears to lie with the neuron itself. During embryological development, the somatic motor neuron secretes a growth factor

(a) Single twitches

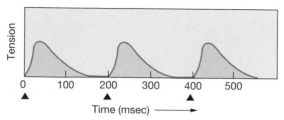

(b) Summation

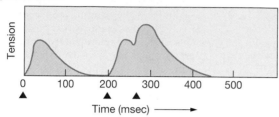

(c) Summation leading to unfused tetanus

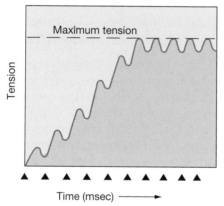

Maximum tension

Time (msec) →

(d) Summation leading to complete tetanus

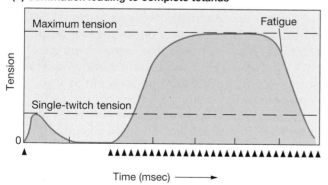

Maximum tension

Fatigue

Single-twitch tension

Time (msec) →

■ **Figure 12-15 Summation of contractions** (a) If stimuli are far enough apart in time, the muscle will relax completely between twitches. (b) As the stimuli move closer and closer together in time, the muscle fiber does not have time to relax and the contractions sum, creating a contraction with greater tension. (c) If the stimuli come very rapidly, the muscle will reach its maximum tension. If the muscle has a chance to relax slightly between stimuli, it has reached unfused tetanus. (d) If the muscle reaches a steady tension, it is said to be in complete tetanus. Once the muscle fiber starts to fatigue, the tension it creates diminishes even though the stimuli continue.

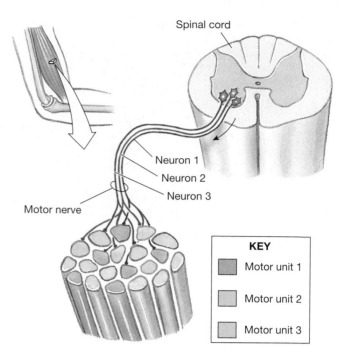

■ **Figure 12-16 A muscle may have many motor units of different fiber types**

that directs the differentiation of all muscle fibers in its motor unit so that they develop into the same fiber type.

Intuitively, it would seem that people who inherit a predominance of one fiber type over another would excel in certain sports. They do, to some extent. Endurance athletes such as distance runners and cross-country skiers have a predominance of aerobic slow-twitch fibers, whereas sprinters, ice hockey players, and weight lifters tend to have larger percentages of fast-twitch fibers.

But inheritance is not the only determining factor for fiber composition in the body. The metabolic characteristics of muscle fibers can be changed to some extent. With endurance training, the aerobic capacity of some fast-twitch fibers can be enhanced until they are almost as fatigue-resistant as a slow-twitch fiber. Since the conversion occurs only in those muscles that are being trained, a neuromodulator chemical is probably involved. In addition, endurance training increases the number of capillaries and mitochondria in the muscle tissue, allowing more oxygen-carrying blood to reach the contracting muscle and contributing to the increased aerobic capacity of the muscle fibers.

✔ Which type of runner would you expect to have more slow-twitch fibers, a sprinter or a marathoner?

Contraction in Intact Muscles Depends on the Types and Numbers of Motor Units in the Muscle

Within a skeletal muscle, each motor unit contracts in an all-or-none manner. How then can muscles create graded contractions of varying force and duration? The answer lies in the fact that intact muscles are composed of multiple motor units of different types (Fig. 12-16 ■). This diversity allows the muscle to vary contraction by (1) changing the types of motor units that are active or (2) changing the number of motor units that are responding at any one time.

The force of contraction within a skeletal muscle can be increased by recruiting additional motor units. **Recruitment** is controlled by the nervous system and proceeds in a standardized sequence. A weak stimulus directed onto a pool of somatic motor neurons in the central nervous system activates only the neurons with the lowest thresholds [∞ p. 223]. Studies have shown that these low-threshold neurons control fatigue-resistant slow-twitch fibers that generate minimal force.

As the stimulus onto the motor neuron pool increases in strength, additional motor neurons with higher thresholds begin to fire. These neurons in turn stimulate motor units composed of fatigue-resistant oxidative fast-twitch fibers. Because more motor units (and thus, more muscle fibers) are participating in the contraction, greater force is generated.

As the stimulus increases to even higher levels, somatic motor neurons with the highest thresholds begin to fire. These neurons stimulate motor units composed of glycolytic fast-twitch fibers. At this point, the muscle contraction is approaching its maximum force. Because of differences in myosin and crossbridge formation, fast-twitch fibers can generate more force than slow-twitch fibers. However, because fast-twitch fibers fatigue more rapidly, it is impossible to hold a muscle contraction at maximum force for an extended period of time. You can demonstrate this by clenching your fist as hard as you can: how long can you hold it before some of the muscle fibers begin to fatigue?

Sustained contractions within a muscle require a continuous train of action potentials from the central nervous system to the muscle. But, as you learned earlier, increasing the stimulation rate of a muscle fiber results in summation of its contractions. If the muscle fiber is easily fatigued, summation will lead to fatigue and diminished tension.

One way that the nervous system avoids fatigue in a sustained contraction is by **asynchronous recruitment** of motor units. The nervous system modulates the firing rates of the motor neurons so that different motor units take turns maintaining the muscle tension. The alternation of active motor units allows some of the motor units to rest between contractions, preventing fatigue.

Asynchronous recruitment will prevent fatigue only in submaximal contractions, however. In high-tension, sustained contractions, the individual motor units may reach a state of unfused tetanus, in which the muscle fibers cycle between contraction and partial relaxation. In general, we do not notice this cycling, because the different motor units in the muscle are contracting and relaxing at slightly different times. As a result, the contractions

and relaxations of the motor units average out and appear to be one smooth contraction. But, as different motor units fatigue, we notice that we are unable to maintain the same amount of tension in the muscle, and the force of the contraction gradually decreases.

✔ What is the response of a muscle fiber to an increase in the firing rate of the somatic motor neuron?

✔ How does the nervous system increase the force of contraction in a muscle composed of many motor units?

❱ MECHANICS OF BODY MOVEMENT

Because one main role of skeletal muscles is to move the body, we now turn to the mechanics of body movement. The term *mechanics* refers to how muscles move loads and how their anatomical relationship to the bones of the skeleton maximizes the work they can do.

Isotonic Contractions Move Loads but Isometric Contractions Create Force Without Movement

When we described the function of muscles, we noted that they create force to generate movement and that they can also create force without generating movement. You can demonstrate both properties with a pair of heavy weights. Pick up the weights and bend your elbows so that the weights touch your shoulders. You have just performed an **isotonic contraction** [*iso,* equal + *teinein,* to stretch]. Any contraction that creates force and moves a load is an isotonic contraction.

When you bent your arms at the elbows and brought the weights to your shoulders, the biceps muscles shortened in a **concentric action.** If you slowly extend your arms, resisting the tendency of the weights to pull them down, you are performing another type of isotonic contraction known as an **eccentric,** or lengthening, **action.** Eccentric exercise is thought to contribute most to cellular damage after exercise and to lead to delayed muscle soreness.

If you pick up the weights and hold them stationary in front of you, the muscles of your arms are creating tension (force) to overcome the load of the weights but are not creating movement. These contractions that create force without movement are called **isometric** (static) **contractions** [*iso,* equal + *metric,* measurement].

Isotonic and isometric contractions are illustrated in Figure 12-17 ■. To demonstrate an isotonic contraction experimentally, we hang a weight (the load) from the muscle and electrically stimulate the muscle to contract. The muscle contracts, lifting the weight. The graph on the right shows the development of force throughout the contraction.

To demonstrate an isometric contraction experimentally, we attach a heavier weight to the muscle.

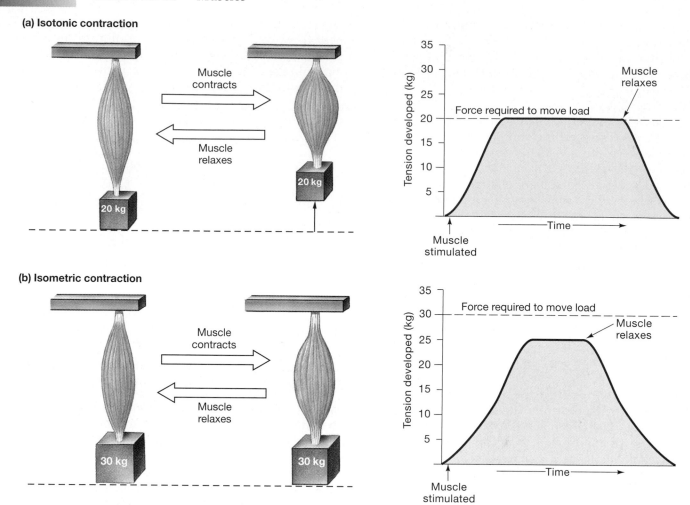

(a) Isotonic contraction

Muscle
contracts

Muscle
relaxes

20 kg

20 kg

Muscle
stimulated

Force required to move load

Muscle
relaxes

Time

(b) Isometric contraction

Muscle
contracts

Muscle
relaxes

30 kg

30 kg

Muscle
stimulated

Force required to move load

Muscle
relaxes

Time

■ **Figure 12-17** **Isotonic and isometric contractions** (a) In an isotonic contraction, the muscle creates enough tension to move the load. (b) In an isometric contraction, the muscle is unable to create enough tension to move the load. The muscle contracts but does not shorten.

When the muscle is stimulated, it develops tension, but the force created is not enough to move the load. In isometric contractions, muscles create force without shortening significantly. For example, when your exercise instructor yells at you to "tuck in your tummy," your response is isometric contraction of abdominal muscles.

How can an isometric contraction create force if the length of the muscle does not change? The elastic elements of the muscle provide the answer. All muscles contain elastic fibers in the tendons and other connective tissues that attach muscles to bone and in the connective tissue between muscle fibers. Within the muscle fibers, elastic cytoskeletal proteins occur between the myofibrils and as part of the sarcomere. Finally, the contractile filaments themselves apparently can stretch even while generating force. All of these elastic components behave collectively as if they were connected in series (one after the other) to the contractile elements of the muscle. Consequently, they are often called the **series elastic elements** of the muscle (Fig. 12-18a ■).

When the sarcomeres shorten in an isometric contraction, the elastic elements stretch. This stretching of the elastic elements allows the fibers to maintain constant length even though the sarcomeres are shortening and creating tension (Fig. 12-18c ■). Once the elastic elements have been stretched and the force generated by the sarcomeres equals the load, the muscle shortens in an isotonic contraction and lifts the load (Fig. 12-18d ■).

Bones and Muscles Around Joints Form Levers and Fulcrums

The anatomical arrangement of muscles and bones in the body is directly related to how muscles work. The body uses its bones and joints as levers and fulcrums on which muscles exert force to move or resist a load. A lever is a rigid bar that pivots around a point known as the fulcrum. In the body, bones form levers, flexible joints form the fulcrums, and muscles attached to bones create force when they contract. The work being performed by the muscle can be expressed as

$$\text{Work} = \text{force} \times \text{distance}$$

In a lever and fulcrum system, distance is measured from the fulcrum to the point along the lever at which

(a) Schematic of the series elastic elements

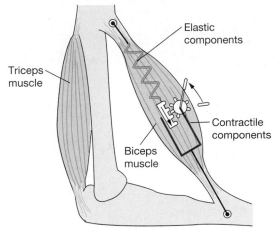

Triceps muscle

Elastic components

Contractile components

Biceps muscle

(b) Muscle at rest	**(c) Isometric contraction:** Muscle has not shortened. Sarcomeres shorten, generating force, but elastic elements stretch, allowing muscle length to remain the same.	**(d) Isotonic contraction:** Sarcomeres shorten more but, since elastic elements are already stretched, the entire muscle must shorten.

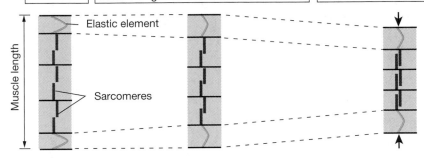

Muscle length

Elastic element

Sarcomeres

■ **Figure 12-18   Series elastic elements in muscle** (a) A muscle has both contractile components (sarcomeres, shown here as a gear and rachet) and elastic components (shown here as a spring). (b)–(d) Isometric contraction stretches the series elastic elements. Isotonic contractions shorten the entire muscle.

force is being exerted (Fig. 12-19a ■). Most lever systems in the body are similar to the one shown where the fulcrum is located at one end of the lever. This arrangement maximizes the distance and speed with which the lever can move the load but also requires that the muscles do more work. We will use the flexion of the forearm to illustrate how this lever system functions.

In the lever and fulcrum system of the forearm, the elbow joint acts as a fulcrum around which movement of the forearm, the lever, takes place (Fig. 12-19b ■). The biceps muscle is attached at its origin at the shoulder and inserts into the radius bone of the forearm (the lever) about 5 cm away from the elbow joint (the fulcrum). When the biceps contracts, it creates an upward force as it pulls on the bone.

If the biceps muscle is to hold the arm stationary and flexed at a 90° angle, it must do work (W_1) that exactly opposes the downward pull of the weight of the forearm and hand (W_2). For the arm illustrated in Figure 12-19b ■, the biceps muscle must create 6 kg of force simply to hold the arm at a 90° angle. Since the muscle is not shortening, this is an isometric contraction.

Now what happens if a 7-kg weight is placed in the hand? This places an additional load on the lever that is even farther from the fulcrum than the center of gravity. Unless the biceps muscle can create additional upward force to offset the downward force created by the

weight, the hand will drop. If the weight on the hand in Figure 12-19c ■ is 25 cm from the elbow, the biceps muscle must create an *additional* 35 kg of force to keep the arm from dropping the 7-kg weight!

What happens to the work required of the biceps muscle if the distance between the fulcrum and the muscle insertion changes? Genetic variability in the insertion point could have a dramatic effect on the force required to move or resist a load. For example, if the biceps muscle in Figure 12-19c ■ inserts 7 cm from the fulcrum instead of 5 cm, it would only need to generate 25 kg of force to offset the same 7-kg weight. Some studies have shown a correlation between the insertion points of muscles and success in certain athletic events.

In the example so far, we have assumed that the load is stationary and that the muscle is contracting isometrically. What happens if we want to flex the arm and lift the load? Then, the biceps muscle must exert a force that exceeds the force created by the load. The disadvantage of a lever system in which the fulcrum is positioned near one end of the lever is that the muscle is required to create large amounts of force to move or resist small loads.

However, this type of lever system maximizes speed and mobility. A small movement of the bone at the point where the muscle inserts becomes a much larger movement at the hand (Fig. 12-19d ■). In addition, the two movements occur in the same amount of time, so the

(a)

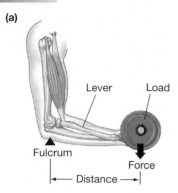

Lever

Load

Fulcrum

Force

Distance

The human forearm acts as a lever. The fulcurm is close to the elbow. The work required to move a load on the lever is equal to the force created by the load (*F*) times distance of the load from the fulcrum (*D*).

Work = force × distance

(a)

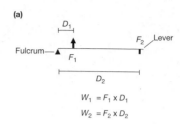

$W_1 = F_1 \times D_1$

$W_2 = F_2 \times D_2$

(b)

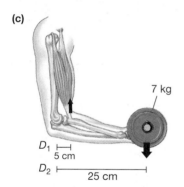

Biceps muscle

Center of gravity

F_1

D_1 ⊢⊣ 5 cm

D_2 ⊢——⊣ 15 cm

$F_2 = 2$ kg

The weight of the forearm exerts a downward force of 2 kg at its center of gravity, which is 15 cm from the fulcrum.

$$\text{Work}_{down} = 2 \text{ kg} \times 15 \text{ cm}$$

The biceps inserts into the lever 5 cm from the fulcrum. Work done by the biceps depends on the force created by the contracting muscle:

$$\text{Work}_{up} = \text{biceps force} \times 5 \text{ cm}$$

To hold the arm stationary at 90 degrees, the work done by the biceps pulling upward must equal the work created by the downward pull of the arm's weight.

$$\text{Work}_{up} = \text{Work}_{down}$$

$$\text{Biceps force} \times 5 \text{ cm} = 2 \text{ kg} \times 15 \text{ cm}$$

$$\text{Biceps force} = \frac{30 \text{ kg} \cdot \text{cm}}{5 \text{ cm}}$$

$$\text{Biceps force} = 6 \text{ kg}$$

(c)

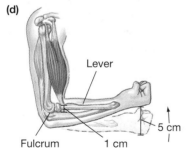

7 kg

D_1 ⊢⊣ 5 cm

D_2 ⊢————⊣ 25 cm

A 7-kg load is added to the hand 25 cm from the elbow.

$$\text{Work}_{down} = 7 \text{ kg} \times 25 \text{ cm}$$

$$\text{Biceps force} \times 5 \text{ cm} = 7 \text{ kg} \times 25 \text{ cm}$$

$$\text{Biceps force} = \frac{175 \text{ kg} \cdot \text{cm}}{5 \text{ cm}} = 35 \text{ kg}$$

The biceps muscle must exert an additional 35 kg of force to keep the hand from dropping.

(d)

Lever

Fulcrum   1 cm   5 cm

Because the insertion of the biceps is close to the fulcrum, a small movement of the biceps becomes a much larger movement of the hand. When the biceps contracts and shortens 1 cm, the hand moves upward 5 cm. If the muscle shortens in 1 second, the hand then moves at a speed of 5 cm/sec.

■ **Figure 12-19   The arm is a lever and fulcrum system**

speed of contraction at the insertion point is amplified at the end of the arm. For instance, if the biceps muscle in our example contracts and shortens 1 cm, the hand elevates 5 cm. If the contraction takes 1 second, the muscle shortens at a speed of 1 cm/sec, but the hand moves at a speed of 5 cm/sec. Thus, the lever system of the arm amplifies both the distance and the speed of movement of the load.

In muscle physiology, the speed with which a muscle contracts depends on the type of muscle fiber (fast-twitch versus slow-twitch) and on the load that is being

moved. Intuitively, you can see that you can flex your arm much faster with nothing in your hand than you can while holding a 7-kg weight in your hand. The relationship between load and velocity of contraction in a muscle fiber, determined experimentally, is graphed in Figure 12-20 ■. Contraction is fastest when the load on the muscle is zero. When the load on the muscle equals the ability of the muscle to create force, the muscle is unable to move the load and the velocity drops to zero. The muscle can still contract, but the contraction becomes isometric instead of isotonic. Because speed is a function

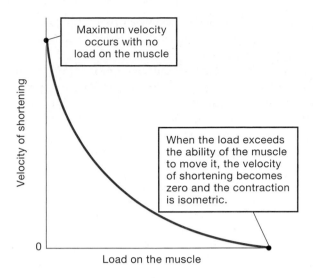

Figure 12-20 **Load-velocity relationship in skeletal muscle**

of load and muscle fiber type, it cannot be regulated by the body except through recruitment of faster muscle fiber types. However, the arrangement of muscles, bones, and joints allows the body to amplify speed so that regulation at the cellular level becomes less important.

✔ One study found that many world-class athletes have muscle insertions that are farther from the joint than in the average person. Why would this trait translate into an advantage for a weight lifter?

Muscle Disorders Have Multiple Causes

Dysfunctions of skeletal muscles can arise from a problem with the signal from the nervous system, from miscommunication at the neuromuscular junction, or from defects in the muscle itself. Unfortunately, in many muscle conditions, even the simple ones, we do not fully understand the mechanism of the primary defect. As a result, we can treat the symptoms but may not be able to cure the problem.

One common muscle disorder is a "charlie horse," or muscle cramp: a sustained painful contraction of skele-

tal muscles. As people age, they often develop nighttime cramps in the calves of their legs. Many muscle cramps are caused by hyperexcitability of the somatic motor neurons controlling the muscle. As the neuron fires repeatedly, the muscle fibers of its motor unit go into a state of painful sustained contraction. Sometimes muscle cramps can be relieved by forcibly stretching the muscle. Apparently, stretching sends sensory information to the central nervous system that inhibits the somatic motor neuron, relieving the cramp.

The simplest muscle disorders arise from overuse. Most of us have exercised too long or too hard and suffered from muscle fatigue or soreness as a result. Trauma to muscles can cause muscle fibers, the connective tissue sheath, or the union of muscle and tendon to tear.

Disuse of muscles can be as traumatic as overuse. With prolonged inactivity such as may occur when a limb is immobilized in a cast, the skeletal muscles will *atrophy*, or waste away [*a-*, without + *trephein*, nourishment]. The blood supply to the muscle diminishes and the muscle fibers get smaller. If activity is resumed in less than a year, the fibers usually will regenerate. Atrophy of longer than one year is usually permanent. If the atrophy is occurring because of somatic motor neuron dysfunction, therapists now try to maintain muscle function with the aid of electrical stimulation that directly stimulates the muscle fibers.

Acquired disorders that affect the skeletal muscle system include infectious diseases, such as influenza, that lead to weakness and aching, and poisoning by toxins, such as those produced in botulism (*Clostridium botulinus*) and tetanus (*Clostridium tetani*). Botulinum toxin acts by decreasing the release of acetylcholine from the somatic motor neuron. Recently, clinical investigators successfully used injections of botulinum toxin as a treatment for writer's cramp, a disabling cramp of the hand that apparently arises as a result of hyperexcitability in

...continued from page 360

Three weeks later, Paul had another attack of paralysis, this time at kindergarten after a game of tag. He was rushed to the hospital and given glucose by mouth. Within minutes, he was able to move his legs and arms and asked for his mother.

Question 5: Explain why oral glucose might help bring Paul out of his paralysis. (Hint: Cells use glucose to produce ATP for active transport. The Na^+-K^+-ATPase actively exchanges K^+ and Na^+ across the cell membrane. What happens to the extracellular K^+ level when Paul is given glucose?)

Muscular Dystrophies Skeletal muscle diseases that arise from dysfunction in the muscles themselves are known as myopathic diseases, or *myopathies*. Some of the best-known myopathies are the inherited diseases known collectively as muscular dystrophy. Although the name implies a single entity, the muscular dystrophies are a family of diseases that result from biochemical abnormalities in the muscle fiber. In Duchenne muscular dystrophy, the inherited defect is the absence of a cytoskeletal protein known as dystrophin. Muscle fibers that lack dystrophin have tiny tears in their cell membranes that allow extracellular Ca^{2+} to enter the fiber. Consequently, intracellular enzymes are activated, resulting in breakdown of the fiber components. The major symptom of Duchenne dystrophy is progressive muscle weakness, and patients usually die before age 30 from failure of the respiratory muscles.

the distal portion of the somatic motor neuron. You may also have heard of "bo-tox" injections for cosmetic wrinkle reduction. Botulinum toxin injected under the skin temporarily paralyzes facial muscles that pull the skin into wrinkles. (Would you do this?)

Inherited muscular disorders are the most difficult to treat. These conditions include various forms of muscular dystrophy as well as biochemical defects in glycogen and lipid storage. One way that we are trying to learn more about muscle diseases is by using animal models such as genetically engineered "knockout" mice. These mice lack the genes for certain muscle proteins, so researchers try to correlate the absence of protein with particular disruptions in function.

▶ SMOOTH MUSCLE

Although most muscle in the body is skeletal muscle, cardiac and smooth muscle are more important in the maintenance of homeostasis. Smooth muscle is found mostly in the walls of hollow organs and tubes, where its contraction will change the shape of the organ. Often smooth muscle generates force to move material through the lumen of the organ. For example, sequential waves of contraction in the intestinal tract move ingested material from the esophagus to the colon.

Smooth muscle is noticeably different from striated muscles in the way that it develops tension (Fig. 12-21 ■). A muscle twitch in a striated muscle fiber develops tension much more rapidly than does a twitch in a smooth muscle fiber. Striated muscle relaxes more rapidly as well.

Smooth muscle, although slower, can sustain contractions for extended periods without fatiguing. This ability allows the walls of organs to maintain tension with a continued load. For example, think of the urinary bladder filled with urine or the digestive tract filled with food.

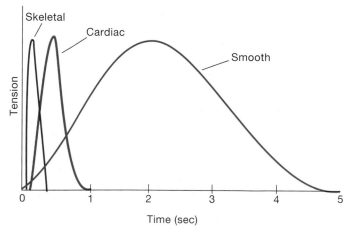

■ **Figure 12-21** **Duration of muscle contraction in three types of muscle** Skeletal muscle has the quickest contraction and relaxation time. Smooth muscle is much slower to contract and relax. Tension development in a cardiac muscle twitch is more like that in a skeletal muscle twitch than in a smooth muscle twitch.

Some smooth muscles are **tonically contracted** and maintain tension at most times. The esophageal and urinary bladder **sphincters** [*sphingein*, to close] are examples of tonically contracted muscles whose function is to close off the opening to a hollow organ. These sphincters relax when it is necessary to allow material to enter or leave the organ. A muscle that maintains a measurable level of tension is said to have **tone.**

Until recently, smooth muscle had not been studied as extensively as skeletal muscle for many reasons. One factor is its complexity. Many types of smooth muscle with widely differing properties are found throughout the animal kingdom, making a single model of smooth muscle function impossible.

A second factor is anatomy. The contractile fibers of a smooth muscle fiber are arranged in oblique bundles rather than in parallel sarcomeres so that a contraction pulls on the cell membrane in many directions. In addition, within an organ the layers of smooth muscle may run in several directions. For example, the intestine has one layer that encircles the lumen and a perpendicular layer that runs the length of the intestine. It is difficult to measure tension developing in both layers at once.

A third factor that complicates study of smooth muscle is its control. Smooth muscle fibers have electrical properties that make them difficult to stimulate directly. Also, unlike skeletal muscle, smooth muscle is controlled by hormones and paracrines in addition to a variety of neurotransmitters.

Finally, a fourth factor involves the response of smooth muscles to stimuli. Although depolarization of a smooth muscle fiber usually results in contraction, neurotransmitters and hormones acting on a smooth muscle fiber can inhibit contraction as well as stimulate it. All these factors make smooth muscle more difficult to work with in the laboratory. In recent years, however, the fact that smooth muscle contraction is regulated through signal transduction and second messenger systems has made it a popular tissue to study.

Smooth Muscle Fibers Are Much Smaller than Skeletal Muscle Fibers

Smooth muscle fibers are small spindle-shaped cells, about the same diameter as a single myofibril within a skeletal muscle fiber. Smooth muscle gets its name from its homogeneous appearance under the microscope (see Fig. 12-1c ■). The contractile fibers are not arranged in organized sarcomeres, so smooth muscle does not have distinct banding patterns as striated muscle does.

Instead, actin and myosin are arranged in long bundles that extend diagonally around the cell periphery, forming a lattice around the central nucleus (Fig. 12-22a ■). The oblique arrangement of contractile elements beneath the cell membrane causes smooth muscle fibers to become globular when they contract, rather than simply shortening as skeletal muscles do (Fig. 12-22b ■).

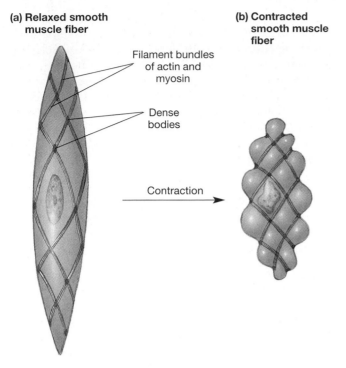

(a) Relaxed smooth muscle fiber

Filament bundles of actin and myosin

Dense bodies

Contraction

(b) Contracted smooth muscle fiber

■ **Figure 12-22   Anatomy of a smooth muscle fiber**   (a) Smooth muscle fibers have actin and myosin loosely arranged around the periphery of the cell, held in place by protein-dense bodies. (b) The arrangement of the fibers causes the cell to become globular when it contracts.

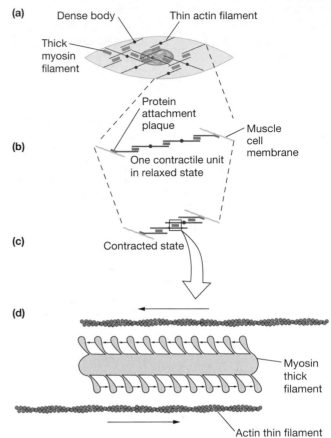

(a) Dense body   Thin actin filament

Thick myosin filament

(b) Protein attachment plaque

Muscle cell membrane

One contractile unit in relaxed state

(c) Contracted state

(d) Myosin thick filament

Actin thin filament

■ **Figure 12-23   Sliding filaments in smooth muscle**   (a) The actin and myosin filaments of smooth muscle are longer than in skeletal muscle. (b) The long actin filaments attach to dense bodies in the cytoplasm and terminate at protein plaques in the cell membrane. (c) Myosin can slide along actin for long distances without encountering the end of a sarcomere. (d) Smooth muscle myosin has hinged heads all along its length, in contrast to skeletal muscle myosin, which has no heads in the center of each filament.

In smooth muscle, as in skeletal muscle, actin is associated with tropomyosin. However, unlike skeletal muscle, smooth muscle lacks troponin. The long actin filaments attach to **dense bodies** of protein in the cytoplasm, structures that are analogous to the Z disks of the sarcomere (Fig. 12-23b ■). The ends of the actin filaments terminate at protein plaques in the cell membrane.

The ratio of actin to myosin in smooth muscles is 10–12:1, compared with 2–4:1 in striated muscle. The less numerous myosin filaments lie bundled between the long actin fibers and are arranged so that their entire surface is covered by myosin heads (Fig. 12-23d ■). (Recall that in the sarcomere, the center of the myosin filament is bare of myosin heads.) The continuous line of myosin heads allows the actin filaments to slide down the myosin without interruption, maintaining continuous tension from crossbridge interaction.

The myosin found in smooth muscles is a different isoform than that of skeletal muscle. Its ATPase activity is much slower, so the rate of crossbridge cycling is slower and the contraction phase of the twitch is longer. In addition, one of the smaller protein chains in the myosin head serves a regulatory role to control contraction and relaxation. The small protein chains are called *myosin light chains.*

Smooth muscle has relatively little sarcoplasmic reticulum compared with skeletal muscle, although the amount varies from one type of smooth muscle to another. The sarcoplasmic reticulum in smooth muscle is

chemically linked to the cell membrane rather than mechanically linked, with Ca^{2+} influx acting as the signal for sarcoplasmic Ca^{2+} release.

The calcium-storage function of the sarcoplasmic reticulum is supplemented by **caveolae** [∞ p. 128], small vesicles that cluster close to the cell membrane (Fig. 12-24 ■). The membranes of caveolae contain gated Ca^{2+} channels that open in response to either a change in membrane potential or the binding of a ligand. When the channels open, Ca^{2+} concentrated inside the caveolae enters the cell.

For years, it was believed that mature smooth muscle cells, like skeletal muscle, had lost the ability to undergo mitosis. Recent studies have shown otherwise however. Not only can smooth muscle fibers reproduce themselves, but their proliferation may also play a significant role in *atherosclerosis,* commonly known as hardening of the arteries.

There are two types of smooth muscle in the human body. Most smooth muscle is **single-unit smooth muscle.** This tissue is also called **visceral smooth muscle** because

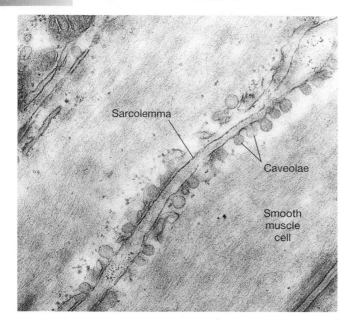

■ **Figure 12-24   Caveolae in smooth muscle**   Caveolae are small invaginations of the cell membrane. They concentrate Ca^{2+}.

it forms the walls of the internal organs (viscera) such as the blood vessels, the intestinal tract, and the ureters.

The muscle fibers of single-unit smooth muscle are electrically coupled by numerous gap junctions [∞ p. 57] that allow depolarizations to pass rapidly from cell to cell (Fig. 12-25a ■). This arrangement permits an action potential in one cell to be spread rapidly throughout a sheet of tissue so that many cells contract as a single unit. Because the cells of single-unit smooth muscle are electrically coupled, it is not necessary to stimulate each muscle fiber to make the entire sheet contract.

Multi-unit smooth muscle is found in the iris and ciliary body of the eye, in part of the male reproductive tract, and in the uterus, except just prior to labor and delivery. The cells of multi-unit smooth muscle are not linked electrically. Consequently, each individual mus-

cle fiber must be closely associated with an axon terminal or varicosity (Fig. 12-25b ■). This arrangement allows fine control of contractions in these muscles by selective activation of individual muscle fibers.

Interestingly, the multi-unit fibers of uterine muscle change and become single-unit just before labor and delivery. The addition of gap junctions to the cell membranes synchronizes the electrical signals and apparently allows the uterine muscle to contract more effectively while working to expel the baby.

Smooth Muscle Can Vary Its Force of Contraction

The unique organization of actin and myosin in smooth muscle enables this tissue to maintain tension over a wide range of fiber lengths. Smooth muscles have longer

(a) Single-unit smooth muscle

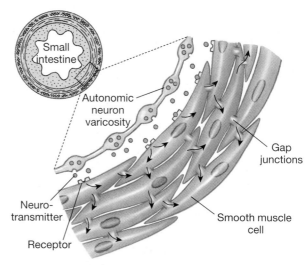

(b) Multi-unit smooth muscle

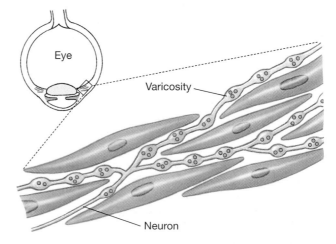

■ **Figure 12-25   Types of smooth muscle**   (a) Single-unit smooth muscle is connected by gap junctions so that the sheet of muscle contracts as a single unit. (b) In multi-unit smooth muscle, the cells are not electrically connected. Each cell must be stimulated independently.

Smooth Muscle and Atherosclerosis   Atherosclerosis is a disease of the blood vessels in which the vessel lumen narrows because of thickening of the vessel wall. Calcified fatty deposits are found in the extracellular matrix, leading to the hardened state that gives atherosclerosis its popular name, hardening of the arteries. As part of the disease process, smooth muscle cells in the blood vessel wall convert from a contractile state to a proliferative state. During proliferation, the muscle cells migrate toward the lumen, divide, and accumulate cholesterol. This change in smooth muscle function is regulated by growth factors and cytokines released by white blood cells in the region of the atherosclerotic lesion. If researchers can find a way to block the action of these growth factors and cytokines, they will have a powerful tool to help prevent atherosclerosis.

actin and myosin filaments. The extra length allows the fibers to be stretched more, yet still maintain enough overlap to create optimum tension.

In addition, once contraction begins, the actin filaments can slide along myosin for longer distances before the end is reached. This property is important for organs whose luminal volume may vary, such as the stomach, intestine, bladder, and uterus. The filling of the bladder with urine is a good example in which the smooth muscle must stretch as the lumen fills yet still maintain the ability to contract in order to expel the contents of the lumen.

Variable force of contraction in smooth muscle depends on the fiber type. Multi-unit smooth muscle resembles skeletal muscle because each cell responds to a stimulus independently. Increasing the force of contraction requires the recruitment of additional fibers. On the other hand, the fibers of single-unit smooth muscle are all electrically connected, so if one fiber fires, all fibers fire. No reserve units are left to be recruited. In single-unit smooth muscle, graded contractions vary with the amount of Ca^{2+} that enters the cell.

Smooth muscle uses less energy than skeletal muscle to generate a given amount of force. For example, it has been estimated that a smooth muscle cell can generate maximum tension with only 25%–30% of its crossbridges active. In addition, smooth muscle has low oxygen consumption rates yet does not fatigue easily. We still do not fully understand the mechanisms through which this is accomplished, but we will discuss some possibilities below.

✔ What is the difference between the way contraction force is varied in multi-unit and single-unit smooth muscle?

Phosphorylation of Proteins Plays a Key Role in Smooth Muscle Contraction

The molecular events of contraction in smooth muscle are similar in many ways to those in skeletal muscle, but some important differences exist. Here is a summary of the key points as we currently understand them:

1. In both smooth muscle and skeletal muscle, the signal to initiate contraction is an increase in cytosolic Ca^{2+}.
2. In smooth muscle, Ca^{2+} enters from the extracellular fluid in addition to being released from the sarcoplasmic reticulum.
3. In smooth muscle, Ca^{2+} binds to **calmodulin,** a binding protein found in the cytosol. In skeletal muscle, Ca^{2+} binds to troponin. (Smooth muscle lacks troponin.)
4. In smooth muscle, Ca^{2+} binding to calmodulin is only the first step in a cascade [∞ p. 159] that ends with contraction. In skeletal muscle, Ca^{2+} binding to troponin initiates contraction immediately.

5. In smooth muscle, the phosphorylation of proteins is an essential feature of the contraction process.
6. In smooth muscle, regulation of myosin ATPase is the primary control of contraction. In skeletal muscle, actin binding sites are regulated.

Let us follow these steps in sequence (Fig. 12-26 ■). Smooth muscle contraction begins when a stimulus (electrical or chemical) opens Ca^{2+} channels in the sarcolemma and sarcoplasmic reticulum. Calcium ions enter the cytosol and bind to calmodulin (CaM). The Ca^{2+}-calmodulin complex activates an enzyme called **myosin light chain kinase (MLCK)**. MLCK then activates myosin ATPase by phosphorylating the light protein chains in the myosin head (Fig. 12-27a ■). When myosin ATPase activity is high, actin binding and crossbridge cycling can take place.

Relaxation in a smooth muscle fiber is a multistep process (Fig. 12-26 ■). Calcium is removed from the cytosol partially by a Ca^{2+}-Na^+ antiport exchange and partially by a Ca^{2+}-ATPase. Removal of the phosphate group from myosin is accomplished with the aid of **myosin light chain phosphatase.** When the myosin light chain is dephosphorylated, ATPase activity and contraction are inhibited. Thus, smooth muscle contraction is primarily controlled through myosin-linked regulatory processes, a feature not seen in skeletal muscle. The steps of smooth muscle contraction and relaxation are summarized in Table 12-4.

Curiously, dephosphorylation of myosin does not automatically result in relaxation. Under conditions that we do not fully understand, dephosphorylated myosin may remain attached to actin for a period of time in what is known as a **latch state.** This condition maintains tension in the muscle fiber without consuming ATP. It is a significant factor in the ability of smooth muscle to sustain contractions without fatiguing. The hinge muscles of certain bivalve mollusks such as oysters can enter a similar latch state that allows them to remain tightly closed under anaerobic conditions.

In some types of smooth muscle, myosin regulation is supplemented by regulation of actin. In these cells, actin binds to a regulatory protein called **caldesmon,** inhibiting actin-myosin binding (Fig. 12-27b ■). If caldesmon is phosphorylated, it can no longer bind to actin, allowing actin and myosin to interact. Phosphorylation of caldesmon takes place with the aid of enzymes known as **protein kinases.**

✔ What happens to contraction if a smooth muscle is placed in a saline bath from which all calcium has been removed?

✔ Tetrodotoxin (TTX) is a poison that blocks sodium channels. When TTX is applied to certain types of smooth muscle, it does not affect the spontaneous generation of action potentials. What conclusion can you draw about the action potential of this smooth muscle on the basis of this observation?

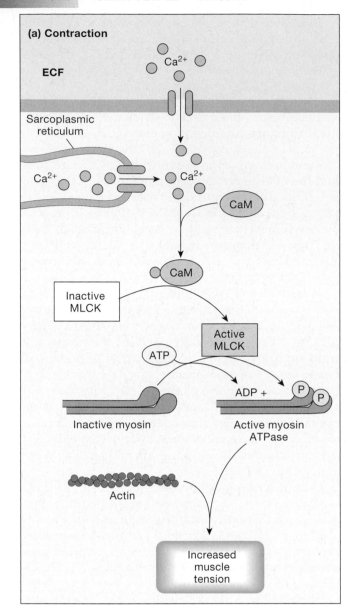

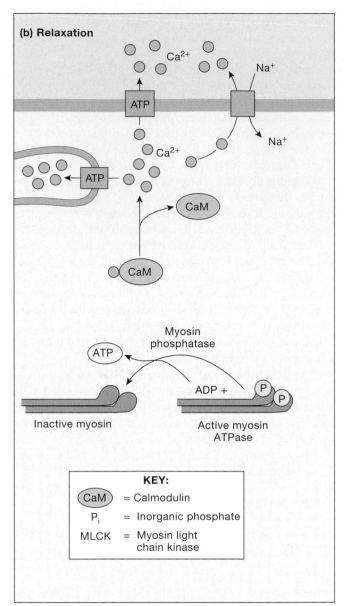

■ **Figure 12-26   The role of calcium in smooth muscle contraction**   Calcium and calmodulin (CaM) activate myosin light chain kinase (MLCK), which phosphorylates myosin to enhance myosin ATPase activity. For relaxation, calcium must be removed from the cytosol while myosin ATPase has a phosphate group removed.

Some Smooth Muscles Have Unstable Membrane Potentials

Many types of smooth muscle display unstable resting membrane potentials that vary between −40 and −80 mV. Cells that show cyclic depolarization and repolarization of their membrane potential are said to have **slow wave potentials** (Fig. 12-28a ■). Sometimes the cell simply cycles through a series of subthreshold slow waves. However, if the peak of the depolarization reaches threshold, action potentials fire, followed by contraction of the muscle.

Other types of smooth muscle have a regular depolarization that always reaches threshold and fires an ac-

tion potential (Fig. 12-28b ■). These depolarizations are called **pacemaker potentials** because they create regular rhythms of contraction. Pacemaker potentials are found in some cardiac muscles as well as in smooth muscle. Both slow wave and pacemaker potentials are due to ion channels in the cell membrane that spontaneously open and close.

Action potentials resulting from slow wave potentials and pacemaker potentials will spread to adjacent muscle cells in single-unit smooth muscle, allowing coordinated contractions in large masses of muscle. The same phenomenon can be observed in cardiac muscle of the heart.

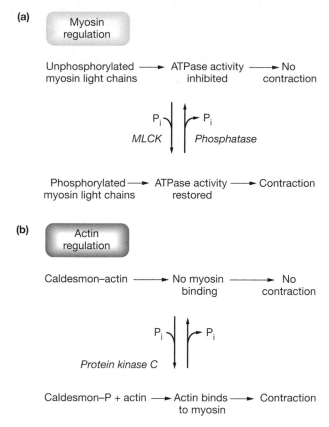

■ Figure 12-27   Regulation of contraction in smooth muscle   Both actin and myosin can be regulated in smooth muscle by the phosphorylation of proteins associated with the contractile filaments.

✔ How do pacemaker potentials differ from slow wave potentials?

Calcium Entry Is the Signal for Smooth Muscle Contraction

Action potentials in smooth muscle are different from action potentials of neurons and skeletal muscle because the depolarization phase is due to the entry of Ca^{2+} rather than Na^+. The Ca^{2+} influx also acts as the signal to initiate contraction. Interestingly, an action potential is not *required* to open voltage-gated Ca^{2+} channels in smooth muscle. Graded potentials open a few Ca^{2+} channels, allowing small amounts of Ca^{2+} into the cell. The entry of variable amounts of Ca^{2+} into the muscle fiber creates contractions whose force is graded according to the strength of the Ca^{2+} signal.

Calcium influx from the extracellular fluid initiates the release of Ca^{2+} from the sarcoplasmic reticulum, a process known as **Ca^{2+}-induced Ca^{2+} release**. However, since the Ca^{2+} stores in smooth muscle are limited, sustained contractions depend on continued influx of Ca^{2+} from the extracellular fluid.

Smooth muscle cell membranes contain stretch-activated and chemically gated Ca^{2+} channels in addition to voltage-gated Ca^{2+} channels. Stretch-activated channels open when pressure or other force distorts the cell membrane, allowing Ca^{2+} to enter and initiate contraction. Because this type of contraction originates from a property of the muscle fiber itself and not because of innervation or hormones, it is known as a **myogenic** contraction. Myogenic contractions are common in blood vessels that maintain a certain amount of tone at all times.

TABLE 12-4   Summary of Smooth Muscle Contraction and Relaxation

Contraction

1. Ca^{2+} enters smooth muscle fibers through:
   - ■ voltage-gated channels opened when cell depolarizes
   - ■ stretch-activated channels opened when cell membrane is deformed
   - ■ chemically gated channels opened by neurotransmitters, hormones, and paracrines
2. Ca^{2+} entry causes additional Ca^{2+} release from the sarcoplasmic reticulum.
3. Ca^{2+} binds to calmodulin.
4. Ca^{2+}-calmodulin activates myosin light chain kinase (MLCK).
5. Activated MLCK phosphorylates myosin light chains, using energy and P_i from ATP.
6. Phosphorylated myosin has ATPase activity that allows crossbridge cycling and contraction.
7. In some smooth muscle, caldesmon must be phosphorylated before actin can bind to myosin.

Relaxation

1. Myosin phosphatase removes phosphate from myosin, decreasing its ATPase activity.
2. Ca^{2+} is removed from the cytoplasm using Ca^{2+}-Na^+ antiport and Ca^{2+}-ATPase.
3. Calmodulin releases Ca^{2+}. MLCK inactivates.
4. In some smooth muscle, phosphate removal from caldesmon allows caldesmon to inactivate actin.

(a) Slow wave potential

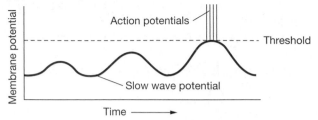

(b) Pacemaker potential

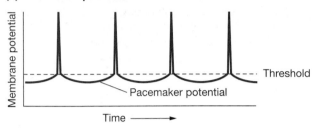

(c) Pharmacomechanical coupling

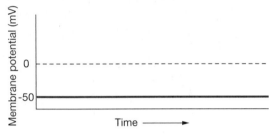

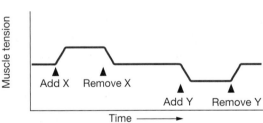

■ **Figure 12-28   Depolarizations in smooth muscle**   (a) Slow wave potentials do not always exceed threshold. (b) Pacemaker potentials steadily depolarize until they reach threshold. (c) In pharmacomechanical coupling, the muscle contracts or relaxes without a change in membrane potential. Chemical signal molecules open Ca^{2+} channels in the cell membrane, and Ca^{2+} entering the cell acts as a second messenger to influence muscle tension.

Chemically gated Ca^{2+} channels open when a ligand binds to them. In **pharmacomechanical coupling,** Ca^{2+} entry causes smooth muscle contraction without a significant change in membrane potential (Fig. 12-28c ■). Although Ca^{2+} entry would usually depolarize the cell, a simultaneous increase in Na^+ removal from the cell prevents a change in the membrane potential. As a result, contraction occurs without depolarization.

Smooth Muscle Contraction Is Regulated by Chemical Signals

Smooth muscle contraction is controlled by a variety of chemical signals that may be either excitatory or inhibitory. In general, if the stimulus depolarizes the cell, it is more likely to contract. Hyperpolarization of the cell decreases the likelihood that it will contract.

In neurally controlled smooth muscle, neurotransmitter is released from varicosities of autonomic neurons [∞ p. 332] onto the surface of the muscle fibers (see Fig. 12-25a ■). Smooth muscle lacks specialized receptor regions such as the motor end plates found in skeletal muscle synapses, so the neurotransmitter simply diffuses across the cell surface.

Many smooth muscles have dual innervation and are controlled by both sympathetic and parasympathetic neurons. However, other smooth muscles, such as those found in blood vessels, are innervated by only one of the two autonomic divisions. Single innervation is usually accompanied by tonic control and the response is graded by increasing or decreasing the amount of neurotransmitter [∞ p. 167].

As we have seen, neurotransmitters can have different effects in different tissues, depending upon the receptors to which they bind. Thus, both the neurotransmitter and the muscle receptor determine the response of a smooth muscle to nervous stimulation.

An amazing variety of neurotransmitters such as vasoactive intestinal peptide and substance P are active in smooth muscle, in addition to norepinephrine and acetylcholine. In many instances, we know how these chemicals affect smooth muscle contraction, but we do not understand the reflex pathways that trigger their release.

Hormones and paracrines also control smooth muscle contraction, unlike skeletal muscle which is controlled only by the nervous system. Smooth muscles in the cardiovascular, gastrointestinal, urinary, respiratory, and reproductive systems all respond to blood-borne or locally released chemicals. For example, asthma is a condition in which smooth muscle of the airways constricts in response to histamine release. This constriction can be reversed by the administration of a hormone, epinephrine, that relaxes smooth muscle and dilates the airway. Note that not all physiological responses are adaptive or favorable to the body: constriction of the airways triggered during an asthma attack, if untreated, can be fatal.

Another important paracrine that affects smooth muscle contraction is nitric oxide [∞ p. 163]. This gas is synthesized by the endothelial lining of blood vessels and relaxes adjacent smooth muscle that regulates the diameter of the blood vessels. For many years, the identity of this *endothelium-derived relaxing factor,* or *EDRF,* eluded scientists even though its presence could be demonstrated experimentally. We know now that EDRF is nitric oxide, an important paracrine in many systems of the body.

TABLE 12-5   Comparison of Three Muscle Types

	Skeletal	*Smooth*	*Cardiac*
Appearance under light microscope	Striated	Smooth	Striated
Fiber arrangement	Sarcomeres	Longitudinal bundles	Sarcomeres
Fiber proteins	Actin, myosin; troponin and tropomyosin	Actin, myosin, tropomyosin	Actin, myosin; troponin and tropomyosin
Control	■ Voluntary ■ Ca^{2+} and troponin ■ Fibers independent	■ Involuntary ■ Ca^{2+} and calmodulin ■ Fibers electrically linked via gap junctions	■ Involuntary ■ Ca^{2+} and troponin ■ Fibers electrically linked via gap junctions
Nervous control	Somatic motor neuron	Autonomic neurons	Autonomic neurons
Hormonal influence	None	Multiple hormones	Epinephrine
Location	Attached to bones; a few sphincters close off hollow organs	Forms the walls of hollow organs and tubes; some sphincters	Heart muscle
Morphology	Multinucleate; large, cylindrical fibers	Uninucleate; small spindle-shaped fibers	Uninucleate; shorter branching fibers
Internal structure	T-tubule and sarcoplasmic reticulum	No t-tubules; sarcoplasmic reticulum reduced or absent	T-tubule and sarcoplasmic reticulum
Contraction speed	Fastest	Slowest	Intermediate
Contraction force of single fiber	All-or-none	Graded	Graded
Initiation of contraction	Requires input from motor neuron	Can be autorhythmic	Autorhythmic

Sometimes, no chemical signal is needed to initiate smooth muscle contraction. In these cases, the smooth muscle fibers can be induced to contract by stretching the muscle and opening stretch-activated Ca^{2+} channels. Some smooth muscles show adaptation to stretch if the muscle fiber is stretched for a period of time. As the stretch continues, the Ca^{2+} channels begin to close in a time-dependent fashion. Then, as Ca^{2+} is pumped out of the cell, the muscle relaxes. This response explains why the bladder develops tension as it fills, then relaxes as it adjusts to the increased volume.

In skeletal muscle, contraction is controlled by the nervous system, and the muscle fiber always responds to an action potential with a twitch. In marked contrast, smooth muscle fibers can be directly inhibited by neurotransmitters, hormones, or paracrines. And, because multiple signals might reach the muscle fiber at one time, smooth muscle fibers also must act as integrating centers. For example, certain blood vessels may receive contradictory messages from two different sources: one message signals for contraction, the other for relaxation. The smooth muscle fibers must integrate the two signals and execute an appropriate response.

Although smooth muscles in the body do not have nearly the mass of skeletal muscles, they play a critical role in the function of most organ systems. You will learn more about smooth muscle physiology in the chapters to come.

▶ CARDIAC MUSCLE

Cardiac muscle, the specialized muscle of the heart, shares features with both smooth and skeletal muscle (Table 12-5). Like skeletal muscle fibers, cardiac muscle fibers are striated and have a sarcomere structure. However, cardiac muscle fibers are shorter than skeletal muscle fibers, may be branched, and have a single nucleus (unlike multinucleate skeletal muscle fibers).

Like single-unit smooth muscle, cardiac muscle fibers are electrically linked to each other. The gap junctions are contained in specialized cell junctions known as **intercalated disks.** Some cardiac muscle, like some smooth muscle, exhibits pacemaker potentials. In addition, cardiac muscle is under sympathetic and parasympathetic control as well as hormonal control. In Chapter 14, you will learn more about cardiac muscle and how it functions within the heart.

PROBLEM CONCLUSION

In this running problem, you learned about hyperkalemic periodic paralysis, a condition caused by a defect in the Na^+ channels that are present on muscle cell membranes. You also learned how the attacks that characterize the disease are triggered by increased levels of K^+ in the blood.

Further check your understanding of this running problem by comparing your answers with those in the summary table.

	Question	Facts	Integration and Analysis
1	What effect does the continuous influx of Na^+ have on the membrane potential of Paul's muscle fibers?	The resting membrane potential of cells is negative relative to the extracellular fluid.	The influx of positive charge will make the interior of the muscle fiber less negative (i.e., will depolarize the membrane potential).
2	In people with hyperkalemic periodic paralysis, attacks tend to occur during a period of rest after exercise. Why?	Each muscle twitch results from an action potential in the muscle fiber. In the repolarization phase of the action potential, K^+ leaves the cell.	During the repeated contractions of exercise, K^+ leaves the muscle fiber and accumulates in the t-tubules. The increase in K^+ concentration affects the Na^+ channels and triggers an attack.
3	Why does the Na^+ channel defect cause paralysis?	The gated channels that release Ca^{2+} from the sarcoplasmic reticulum are opened by the depolarization of an action potential. However, they close in a time-dependent manner, even if the cell remains depolarized.	During an attack, the Na^+ channels remain open and continuously admit Na^+ into the muscle cell, so the cell remains depolarized. The muscle fiber is unable to repolarize and fire additional action potentials. The first action potential causes a twitch, but the muscle then goes into a state of flaccid (uncontracted) paralysis.
4	Speculate on how diuretics prevent attacks in people with hyperkalemic periodic paralysis.	Diuretics cause the loss of water and ions from the body.	Because diuretics lower K^+ concentrations in the blood, they may prevent the higher than normal K^+ concentrations that trigger periodic attacks of paralysis.
5	Explain why oral glucose might help bring Paul out of his paralysis.	Cells use glucose to produce ATP, and they use ATP to run the Na^+-K^+-ATPase that actively exchanges K^+ and Na^+ across the cell membrane.	Providing glucose to cells may increase ATP available to run the Na^+-K^+-ATPase, which removes K^+ from the extracellular fluid. Insulin, a hormone that enhances glucose uptake into cells, also helps relieve the paralysis.

CHAPTER REVIEW

SUMMARY

1. Muscles generate motion, force, and heat. (p. 345)

2. The three types of muscle are **skeletal muscle, cardiac muscle,** and **smooth muscle.** Skeletal and cardiac muscles are **striated muscles.** (p. 346)

3. Skeletal muscles are controlled by somatic motor neurons. Cardiac and smooth muscle are controlled by autonomic innervation, paracrines, and hormones. Some smooth and cardiac muscles are autorhythmic and contract spontaneously. (p. 347)

Skeletal Muscle

4. Skeletal muscles are usually attached to bones by **tendons.** The **origin** is the end of the muscle attached closest to the trunk or to the more stationary bone. The **insertion** is the more distal or mobile attachment. (p. 347)

5. At a flexible **joint,** muscle contraction moves the skeleton. **Flexors** bring bones closer together; **extensors** move bones away from each other. Flexor-extensor pairs are called **antagonistic muscle groups.** (p. 347)

6. A skeletal **muscle** is a collection of **muscle fibers.** Skeletal muscle fibers are large cells with many nuclei. (p. 347)

7. **T-tubules** allow action potentials to move rapidly into the interior of the fiber and release calcium from the **sarcoplasmic reticulum.** (p. 350)

8. **Myofibrils** are intracellular bundles of contractile and elastic proteins. **Thick filaments** are made of **myosin. Thin filaments** are made mostly of **actin. Titin** and **nebulin** hold thick and thin filaments in position. (p. 350)

9. Myosin binds to actin, creating **crossbridges** between the thick and thin filaments. (p. 351)

10. One **sarcomere** is composed of two Z disks and the filaments between them. A sarcomere is divided into **I bands** (thin filaments only), an **A band** that runs the length of the thick filament, and a central **H zone** occupied by thick filaments only. The **M line** and **Z disks** represent attachment sites for thick filaments and thin filaments, respectively. (p. 351)

11. The force created by a contracting muscle is called muscle **tension.** The **load** is a weight or force that opposes contraction of a muscle. (p. 352)

12. The **sliding filament theory of contraction** states that during contraction, overlapping thick and thin filaments slide past each other in an energy-dependent manner as a result of actin-myosin crossbridge movement. (p. 352)

13. Myosin converts energy from ATP into motion. The myosin ATPase activity hydrolyses ATP to ADP and inorganic phosphate. Both products remain bound to myosin as myosin swivels and binds to a new actin molecule. (p. 353)

14. When myosin releases inorganic phosphate, movement of the myosin molecule forms the **power stroke.** Myosin pushes actin toward the center of the sarcomere. At the end of the power stroke, myosin releases ADP. The cycle ends with myosin tightly bound to actin. (p. 353)

15. **Tropomyosin** blocks the myosin-binding site on actin. As contraction begins, **troponin** associated with tropomyosin binds to Ca^{2+}. This calcium binding unblocks the myosin-binding sites and allows myosin to complete its power stroke. (p. 354)

16. During relaxation, the sarcoplasmic reticulum uses a Ca^{2+}-ATPase to pump Ca^{2+} back into its lumen so that crossbridge formation is blocked once again. (p. 356)

17. A somatic motor neuron releases ACh, which initiates a skeletal muscle action potential that leads to contraction. This combination of events is called **excitation-contraction coupling.** (p. 356)

18. Voltage-sensing **dihydropyridine receptors** in the t-tubules are linked to Ca^{2+} release channels in the sarcoplasmic reticulum. An action potential causes Ca^{2+} release. (p. 357)

19. A single contraction-relaxation cycle is known as a **twitch.** The **latent period** between the end of the muscle action potential and the beginning of muscle tension development represents the time required for Ca^{2+} release and binding to troponin. (p. 359)

20. Muscle fibers store energy for contraction in **phosphocreatine.** Additional energy comes from metabolism. Anaerobic metabolism of glucose is a rapid source of ATP but is not efficient. Aerobic metabolism is very efficient but requires an adequate supply of oxygen to the muscles. (p. 359)

21. Muscle **fatigue** is a condition in which a muscle is no longer able to generate or sustain the expected power output. Fatigue probably results from failure in the excitation-contraction coupling mechanism. (p. 360)

22. Skeletal muscle fibers can be classified on the basis of their speed of contraction and resistance to fatigue into **fast-twitch glycolytic fibers, fast-twitch oxidative fibers,** and **slow-twitch oxidative fibers.** Oxidative fibers are the most fatigue-resistant. (p. 360)

23. **Myoglobin** is a red oxygen-binding pigment that transfers oxygen to the interior of the muscle fiber. (p. 361)

24. The tension of a muscle twitch is determined by the length of the sarcomeres before contraction begins. (p. 362)

25. Increasing the stimulus frequency causes **summation** of twitches with an increase of tension. A state of maximal contraction is known as **tetanus.** (p. 363)

26. A **motor unit** is composed of a group of muscle fibers and the somatic motor neuron that controls them. The number of muscle fibers in a motor unit varies, but all fibers in a single motor unit are of the same fiber type. (p. 363)

27. The force of contraction within a skeletal muscle can be increased by **recruitment** of additional motor units. The nervous system avoids fatigue in sustained contractions through **asynchronous recruitment** of motor units. (p. 365)

Mechanics of Body Movement

28. An **isotonic contraction** creates force and moves a load. An **isometric contraction** creates force without movement. **Concentric actions** are shortening contractions. **Eccentric actions** are lengthening contractions. (p. 365)

29. Isometric contractions occur because **series elastic elements** allow the fibers to maintain constant length even though the sarcomeres are shortening and creating tension. (p. 366)

30. The body uses its bones and joints as levers and fulcrums. Most lever systems in the body maximize distance and speed but also require that muscles do more work. (p. 366)

31. Contraction speed is a function of muscle fiber type and load. Contraction is fastest when the load on the muscle is zero. (p. 367)

Smooth Muscle

32. Smooth muscle is slower than skeletal muscle but can sustain contractions for longer without fatiguing. Some smooth muscles are **tonically contracted** and are said to have **tone.** (p. 370)

33. Actin and myosin in smooth muscle are arranged diagonally in a lattice around the central nucleus. (p. 370)

34. Smooth muscle has relatively little sarcoplasmic reticulum. Calcium storage is supplemented by **caveolae.** (p. 371)

35. **Single-unit smooth muscle** contracts as a single unit when depolarizations pass from cell to cell through gap junctions. In **multi-unit smooth muscle,** individual muscle fibers are stimulated independently. (p. 371)

36. Smooth muscles exhibit summation and tetanus. Only multi-unit smooth muscle has recruitment. Contraction force in single-unit smooth muscle is a function of Ca^{2+} entry. (p. 373)

37. Smooth muscle uses less energy than skeletal muscle to generate a given amount of force. (p. 373)

38. Phosphorylation of myosin light protein chains by **myosin light chain kinase (MLCK)** activates myosin ATPase.

Myosin light chain phosphatase inhibits myosin ATPase activity. (p. 373)

39. MLCK is activated by Ca^{2+}**-calmodulin.** (p. 373)

40. In smooth muscle, both actin and myosin regulate contraction. Actin has no troponin and is associated with **caldesmon** instead. (p. 373)

41. During relaxation, Ca^{2+} is removed from the cytosol and the myosin light chains are dephosphorylated. (p. 373)

42. Unstable membrane potentials in smooth muscle take the form of **slow wave potentials** or **pacemaker potential.** (p. 374)

43. The rising phase of smooth muscle action potentials is due to Ca^{2+} entry rather than Na^+ entry as in neurons and skeletal muscles. Ca^{2+} influx triggers Ca^{2+}-induced Ca^{2+} release from the sarcoplasmic reticulum. (p. 375)

44. In **pharmacomechanical coupling,** Ca^{2+} entry causes smooth muscle contraction without a significant change in membrane potential. (p. 376)

45. Smooth muscle is controlled by a variety of chemical signals, and sympathetic and parasympathetic neurons. (p. 376)

Cardiac Muscle

46. Cardiac muscle fibers are striated, have a single nucleus, and are electrically linked through gap junctions. Cardiac muscle shares features with both skeletal and smooth muscle. (p. 377)

QUESTIONS

LEVEL ONE   Reviewing Facts and Terms

1. The three types of muscle tissue found in the human body are _____, _____, and _____. Which type is attached to the bones, enabling it to control body movement?

2. Which two muscle types are striated?

3. Which type of muscle tissue is controlled strictly by a signal from a somatic motor neuron?

4. Which statement is true about skeletal muscles?
   (a) They constitute about 40% of a person's total body weight.
   (b) They position and move the skeleton.
   (c) The insertion of the muscle is more distal or mobile than the origin.
   (d) They are often paired into antagonistic muscle groups called flexors and extensors.
   (e) All of these statements are true.

5. Arrange these skeletal muscle components in order, from outermost to innermost: sarcolemma, connective tissue sheath, myofilaments, myofibrils.

6. The modified endoplasmic reticulum of skeletal muscle is called the _____. Its role is to sequester _____ ions.

7. T-tubules allow _____ to move to the interior of the muscle fiber.

8. List the proteins that make up the myofibrils. Which one has heads that move to form the power stroke?

9. List the letters used to label the elements of a sarcomere. Which band has a Z disk in the middle? Which is the darkest band? Why? Which element forms the boundaries of a sarcomere? Name the line that divides the A band in half. What is the function of this line?

10. Briefly explain the functions of titin and nebulin.

11. During contraction, the _____ band remains a constant length. This band is composed primarily of _____ molecules. Which lettered components approach each other during contraction?

12. The current theory of muscle contraction is called _____. Why?

13. Match the characteristics below with the appropriate type(s) of muscle.

    (a) largest diameter
    (b) anaerobic metabolism, thus fatigues quickly
    (c) has the most blood vessels
    (d) has some myoglobin
    (e) used for quick, fine movements
    (f) also called red muscle
    (g) use a combination of oxidative and glycolytic metabolism
    (h) has the most mitochondria

    1. fast-twitch glycolytic fibers
    2. fast-twitch oxidative fibers
    3. slow-twitch oxidative fibers

14. Explain the roles of troponin, tropomyosin, and Ca^{2+} in muscle contraction.

15. Which neurotransmitter is released by somatic motor neurons?

16. What is the motor end plate, and what kinds of receptors are found there? Explain how neurotransmitter binding to these receptors creates an action potential.

17. A single contraction-relaxation cycle in a skeletal muscle fiber is known as a _____.

18. List the steps of the contraction cycle that require ATP.

19. The basic unit of contraction in an intact skeletal muscle is the _____. The force of contraction within a skeletal muscle is increased by _____ additional motor units.

20. The two types of smooth muscle are _____, also called visceral smooth muscle, and _____.

LEVEL TWO   Reviewing Concepts

21. **Concept map:** Make a map of muscle fiber structure using the following words or terms. Add additional terms if you like.

    actin
    Ca^{2+}
    cell membrane
    cell
    crossbridges
    cytoplasm
    contractile protein
    elastic protein
    glycogen
    mitochondria
    muscle fiber

    myosin
    nucleus
    regulatory protein
    sarcolemma
    sarcoplasm
    sarcoplasmic reticulum
    t-tubule
    titin
    tropomyosin
    troponin

22. How does an action potential in a muscle fiber trigger a Ca^{2+} signal inside the fiber?

23. Muscle cells depend on a continuous supply of ATP. How do the different types of muscle cells generate ATP? What is used for a backup energy source?

24. Define muscle fatigue. Summarize factors that could play a role in its development. How can muscle cells adapt to resist fatigue?

25. Explain how you vary the strength and effort made by your muscles in picking up a pencil versus a full gallon container of milk.

26. Compare and contrast the cellular anatomy and neural and chemical control contraction in skeletal, smooth, and cardiac muscle.

27. What is the role of the sarcoplasmic reticulum in muscular contraction? How can smooth muscle contract when it has so little sarcoplasmic reticulum?

28. Compare and contrast:

    (a) fast-twitch oxidative, fast-twitch glycolytic, and slow-twitch muscle fibers
    (b) a twitch and tetanus
    (c) action potentials in motor neurons and action potentials in skeletal muscles
    (d) temporal summation in motor neurons and summation in skeletal muscles
    (e) isotonic contraction, isometric contraction, concentric action, and eccentric action
    (f) slow-wave and pacemaker potentials

29. Explain smooth muscle contraction and relaxation, particularly the roles of MLCK and calmodulin.

LEVEL THREE   Problem Solving

30. One way that scientists study muscles is to put them into a state of rigor or stiffness by removing ATP. In this condition, the actin and myosin are strongly linked but unable to move. On the basis of what you know about muscle contraction, predict what would happen to these muscles in a state of rigor if you (a) added ATP but no free calcium ions; (b) added ATP with a substantial concentration of calcium ions.

31. When curare, a South American Indian arrow poison, is placed on a nerve-muscle preparation, the muscle will not contract when the nerve is stimulated, even though neu-

rotransmitter is still being released from the nerve. Name all possible explanations for the action of curare that you can think of.

32. On the basis of what you have learned about muscle fiber types and metabolism, predict what variations in structure you would find among these athletes.

    (a) a 7 foot, 2 inch tall, 325-pound basketball player
    (b) a 5 foot, 10 inch tall, 180-pound steer wrestler
    (c) a 5 foot, 7 inch tall, 130-pound female figure skater
    (d) a 4 foot, 11 inch tall, 89-pound female gymnast

LEVEL FOUR Quantitative Problems

33. Use the arm in Figure 12-19 ■ to answer the following questions.

(a) How much force would a biceps muscle inserted 4 cm from the fulcrum need to hold the arm stationary at a 90° angle? How does this compare to the arm with insertion at 5 cm?

(b) If a 7-kg weight band is placed around the wrist 20 cm from the fulcrum, how much force does the biceps muscle inserted 5 cm from the fulcrum need to hold the arm stationary at a 90° angle? How does this compare to placing the same weight in the hand?

E X P L O R E <MediaLab>

Introduction

Americans seem obsessed with their bodies, so it is not surprising that the business world is capitalizing on our obsessions. We now have different clothes, shoes, food and drink for each sport. Gyms are packed and having a good physique is in. To understand how to build a physique, you first need to know about muscles and how they work.

In this MediaLab you will consider the significance of muscle placement in our bodies and view the inside of a muscle cell. After reading the descriptions below, visit the MediaLab for Chapter 12 in your Companion Website and select the appropriate keywords.

Web Exploration 1

Estimated time for completion = 5 minutes
In the first "Check Your Knowledge" problem of this chapter, you were asked to list as many antagonistic pairs of muscles in your body as you could. Why are skeletal muscles found in antagonistic pairs? (Think about how muscles change their shape when they contract.)

Select the keyword **BODY MOVEMENTS** from the Website and look at the different types of movements that occur at each joint. Which muscles are responsible for these movements and which muscles are antagonistic pairs? Can you think of any muscles that do not have an antagonist? (Think about your head.) Keep the idea of antagonistic muscle pairs in mind as you study the respiratory system. You will learn the diaphragm is used for inspiration. What do you think is the antagonistic force used during expiration?

Web Exploration 2

Estimated time for completion = 15 minutes
Have you ever thought about how muscles actually work? Select the keyword **MYOFILAMENT ORGANIZATION** from the Website and look at the images of how myosin, actin, tropomyosin, and troponin are assembled in the three-dimensional structure of the myofibril. As you look at these images, ask yourself how a muscle cell builds a myofibril. How is it that all the sarcomeres have A and I bands with similar spacing? Does the cell have some type of "ruler" to measure out these distances?

Select the keyword **TITIN** from the Website and read about the role of titin (a.k.a. connectin) molecule in myofibril assembly and muscle mechanics. How do the protein's names relate to its size and function? Why was it so difficult to conduct research on titin? What is the functional role of titin in muscle contraction? What was Trinick's hypothesis? What disease is caused when titin is damaged? How were antibodies used to help determine the position of titin in the sarcomere?

13 Integrative Physiology I: Control of Body Movement

■ "Extracting signals directly from the brain to directly control robotic devices has been a science fiction theme that seems destined to become fact."—*Dr. Eberhard E. Fetz,* Science News *156: 142, 8/28/99.* ■

CHAPTER OUTLINE

BACKGROUND BASICS

Reflex pathways (p. 169)
Central nervous system (p. 215)
Summation of action potentials (p. 241)
Isometric contraction (p. 365)
Sensory pathways and receptors (p. 282)
Graded potentials (p. 222)
Tonic control (p. 167)
Tendons (p. 64)

Think back to the baseball pitcher in the last chapter. As he stands on the mound, watching the first batter, he receives sensory information from multiple sources: the sound of the crowd, the sight of the batter and the catcher, the feel of the ball within his hand, the alignment of his body as he prepares to throw. Sensory receptors code this information and send it to the central nervous system (CNS), where it is integrated.

The pitcher acts on some of the information consciously: he decides that it is time to throw a fast ball. But he processes other information at the subconscious level and acts upon it without conscious thought. As he thinks about starting his pitch, he shifts his weight to offset the impending movement of his arm. The integration of sensory information into an involuntary response is the hallmark of a *reflex* [∞ p. 169].

In this chapter we will first look at the characteristics of nervous reflexes. We then discuss how the CNS

PROBLEM
Tetanus

"She has not been able to talk to us. We're afraid she may have had a stroke." That is how her neighbors described 77-year-old Cecile Evans when they brought her to the emergency room. But when a neurological examination revealed no problems other than Mrs. Evans's inability to open her mouth and stiffness in her neck, emergency room physician Dr. Doris Ling began to consider other diagnoses. She noticed some scratches healing on Mrs. Evans's arms and legs and asked the neighbors if they knew what had caused them. "Oh, yes. She told us a few days ago that her dog jumped up and knocked her against the barbed wire fence." At that point, Dr. Ling realized she was probably dealing with her first case of tetanus.

...continued on page 386

uses reflexes to control muscle contraction and movement of the body.

▶ NERVOUS REFLEXES

All nervous reflexes begin with a stimulus that activates a sensory receptor. The receptor sends information in the form of action potentials through sensory neurons to the CNS [∞ p. 215]. The CNS is the integrating center that evaluates all incoming information and selects an appropriate response. It then initiates action potentials in efferent neurons to direct the response of muscles and glands, the effectors.

A key feature of many reflex pathways is **negative feedback**, a concept introduced in Chapter 6 [∞ p. 173]. Negative feedback signals from the muscles and joints of the body keep the CNS continuously informed of changing body position. Some reflexes have a **feedforward** component that allows the body to anticipate a stimulus and begin the response [∞ p. 175]. Bracing yourself in anticipation of a collision would be an example of a feedforward response.

Nervous Reflex Pathways Can Be Classified Different Ways

Reflex pathways in the nervous system consist of chains or networks of neurons that link sensory receptors to muscles or glands. Neural reflexes can be classified in several ways (Table 13-1):

1. *By the efferent division of the nervous system that controls the response.* Reflexes that involve somatic motor neurons and skeletal muscles are known as **somatic reflexes.** Reflexes whose responses are controlled by autonomic neurons are called **autonomic,** or **visceral, reflexes.**

2. *By the CNS location where the reflex is integrated.* **Spinal reflexes** are integrated within the spinal cord. These reflexes may be modulated by higher input from the brain, but they can occur without that input. Reflexes integrated in the brain are called **cranial reflexes.**

3. *By whether the reflex is inborn or learned.* Many reflexes are innate, that is, we are born with them and they are genetically determined. For example, the knee jerk reflex, in which the lower leg kicks out when the lower edge of the kneecap is tapped, is an innate response. Other reflexes are acquired through experience. The example of Pavlov's dogs salivating upon hearing a bell is the classic example of a learned (conditioned) reflex.

TABLE 13-1   Classification of Neural Reflexes

Neural reflexes can be classified by:

1. Efferent division that controls the effector
   a. Somatic motor neurons control skeletal muscles.
   b. Autonomic neurons control smooth and cardiac muscle, glands, and adipose tissue.
2. Integrating region within the central nervous system
   a. Spinal reflexes do not require input from the brain.
   b. Cranial reflexes are integrated within the brain.
3. Time at which the reflex develops
   a. Innate (inborn) reflexes are genetically determined.
   b. Learned (conditioned) reflexes are acquired through experience.
4. The number of neurons in the reflex loop
   a. Monosynaptic reflexes have only two neurons: one afferent (sensory) and one efferent. Only somatic motor reflexes can be monosynaptic.
   b. Polysynaptic reflexes add one or more interneurons between the afferent and efferent neurons. All autonomic reflexes are polysynaptic because they have three neurons: one afferent and two efferent.

4. *By the number of neurons in the reflex pathway.* The simplest reflex pathway is a **monosynaptic reflex** with only two neurons: an afferent sensory neuron and an efferent somatic motor neuron. These two neurons synapse within the spinal cord, allowing a signal from the receptor to go directly to the skeletal muscle effector (Fig. 13-1a ■). The monosynaptic reflex gets its name from the single synapse between the neurons. In this instance, the synapse at the neuromuscular junction is ignored.

Most reflexes have three or more neurons in the pathway, leading to their designation as **polysynaptic reflexes** (Fig. 13-1b ■). Polysynaptic reflexes may be quite complex and often branch within the CNS to form networks with multiple interneurons. *Divergence* of pathways allows a single stimulus to affect multiple targets. *Convergence* allows modulation of pathways by integrating the input from multiple sources. Recall from Chapter 8 that modulation can be either excitatory or inhibitory, as occurs with presynaptic inhibition [∞ p. 243].

 Visualization Techniques in Sports Presynaptic facilitation is now believed to be the physiological mechanism that underlies the success of visualization techniques in sports. Visualization enables athletes to maximize their performance by "psyching" themselves, picturing in their minds the perfect vault or the leaping catch for a touchdown. By pathways that we still do not understand, the visual image conjured up by the cerebral cortex is translated into signals that find their way to the muscles. This is only one example of many fascinating connections between the higher brain and the body.

▶ AUTONOMIC REFLEXES

Autonomic reflexes are also known as *visceral reflexes* because they often involve the internal organs of the body. Some visceral reflexes, such as urination and defecation, are spinal reflexes that can take place without input from the brain. However, spinal reflexes are often modulated by excitatory or inhibitory input from the brain, carried in descending tracts from higher brain centers.

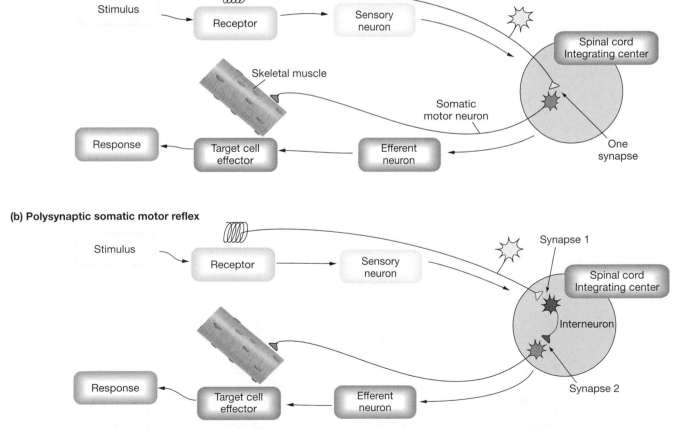

■ **Figure 13-1   Monosynaptic and polysynaptic somatic motor reflexes**   (a) A monosynaptic reflex has a single synapse between the afferent and efferent neurons. The synapse at the target organ is not counted. All monosynaptic reflexes are somatic motor reflexes. (b) Polysynaptic reflexes have two or more synapses. In somatic motor reflexes, the synapses all occur in the central nervous system, as illustrated.

● ●

...continued from page 384

Tetanus, also known as lockjaw, is a devastating disease caused by the bacterium *Clostridium tetani*. These bacteria are commonly found in soil and enter the human body through a cut or wound. As the bacteria reproduce in the tissues, they release a protein neurotoxin. This toxin, called tetanospasmin, is taken up by motor neurons at the peripheral nerve endings. Tetanospasmin then travels along the axons until it reaches the motor neuron cell body in the spinal cord.

Question 1: Tetanospasmin is a protein. By what process is it taken up into neurons? [Hint: ∞ p. 129] By what process does it travel up the axon to the nerve cell body? [Hint: ∞ p. 218]

● ●

For example, urination may be voluntarily initiated by conscious thought, or it may be inhibited by emotion or a stressful situation, such as the presence of other people (a syndrome known as "bashful bladder"). Often, the higher control of a spinal reflex is a learned response. Toilet training that we master as toddlers is an example of a learned reflex that our CNS uses to modulate the simple spinal reflex of urination.

Other autonomic reflexes are integrated in the brain, primarily in the hypothalamus, thalamus, and brain stem. These regions contain centers that coordinate body functions needed to maintain homeostasis, such as heart rate, blood pressure, breathing, eating, water balance, and maintenance of body temperature [∞ Fig. 11-9, p. 334]. The brain stem also contains the integrating centers for autonomic reflexes such as salivating, vomiting, sneezing, coughing, swallowing, and gagging.

An interesting type of autonomic reflex is the conversion of emotional stimuli into visceral responses. The limbic system, the site of primitive drives such as sex, fear, rage, aggression, and hunger, has been called the "visceral brain" because of its role in these emotionally driven reflexes. We speak of "gut feelings" and "butterflies in the stomach"—all transformations of emotion into somatic sensation and visceral function. Other emotion-linked autonomic reflexes include urination, defecation, blushing, blanching, and *piloerection,* in which tiny muscles in the hair follicles pull the shaft of the hair erect ("I was so scared that my hair stood on end!").

Autonomic reflexes are all polysynaptic, with at least one synapse in the CNS between the sensory neuron and the preganglionic autonomic neuron, and an additional synapse found between the two autonomic neurons (Fig. 13-2 ■). Many autonomic reflexes are characterized by tonic activity, a continuous stream of action potentials that create an ongoing response in the effector. For example, the tonic control of blood vessels discussed in Chapter 6 is an example of a continuously active autonomic reflex [∞ p. 167]. You will encounter many autonomic reflexes as you continue your study of the systems of the body.

✔ Name the general steps of a reflex pathway and the anatomical structures in the nervous system that correspond to each step.

✔ If the membrane potential hyperpolarizes, does it become more positive or more negative? Move closer to threshold or move farther away from threshold?

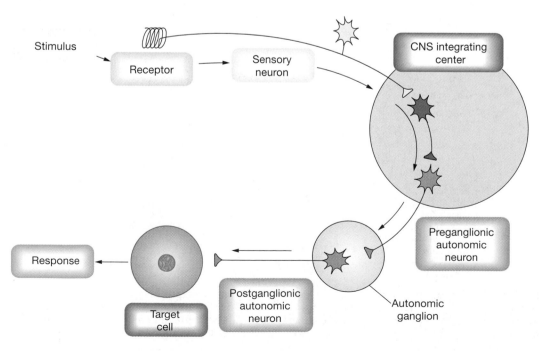

■ **Figure 13-2 Autonomic reflexes** All autonomic reflexes are polysynaptic, with at least one synapse in the CNS and another in the autonomic ganglion.

▶ SKELETAL MUSCLE REFLEXES

Although we are not always aware of them, skeletal muscle reflexes are involved in almost everything we do. Receptors that sense changes in muscle tension, stretch, and pressure feed information to the CNS, which responds in one of two ways. If muscle contraction is the appropriate response, action potentials pass along motor neurons to the muscle fibers. If muscles need to be relaxed to achieve the response, the CNS *inhibits* the motor neurons controlling the muscle.

Recall that somatic motor neurons are always excitatory and always cause contraction in skeletal muscle. For relaxation to occur, the excitatory somatic motor neuron must be inhibited. Thus, relaxation requires control at the level of the CNS, where output by the somatic motor neuron can be altered.

Receptors within the skeletal muscles themselves sense changes in muscle length and tension and activate muscle reflexes. Sensory neurons lead from these receptors to the CNS. Once the input signal has been integrated, output information is carried to skeletal muscles by two different types of motor neurons.

Alpha motor neurons are efferent neurons that innervate the normal contractile fibers of the muscle, also known as **extrafusal muscle fibers.** Action potentials in extrafusal muscle fibers cause muscle contraction. **Gamma motor neurons** are smaller neurons associated with specialized muscle fibers within sensory receptors.

The three types of sensory receptors found in skeletal muscle are *muscle spindles, Golgi tendon organs,* and *joint capsule mechanoreceptors* (proprioceptors). These receptors send information to the CNS about the relative positioning of bones linked by flexible joints. In the next two sections, we examine the function of muscle spindles and tendon organs, two interesting and unique receptors. We will not discuss joint capsule mechanoreceptors.

Muscle Spindles Respond to Muscle Stretch

The **muscle spindles** act as stretch receptors, sending information about muscle length to the CNS. These small, elongated structures are scattered among and arranged parallel to the contractile extrafusal fibers of the muscle. Each spindle consists of a connective tissue sheath that encloses a group of **intrafusal fibers** [*intra-,* within + *fusus,* spindle]. Intrafusal fibers are modified muscle fibers that lack myofibrils in their central portions (Fig. 13-3b ■). Sensory neurons wrap around the middle of the intrafusal fibers and project to the spinal cord. These neurons fire when the center of the intrafusal fiber stretches.

Although the centers of the intrafusal fibers are not contractile, the ends do contain fibers that contract when stimulated by gamma motor neurons. Although this contraction does not affect overall muscle tension, it does stretch the central portion of the intrafusal fiber.

The sensory neurons that wrap around the center of the spindle are tonically active, firing action potentials when the muscle is at its resting length (Fig. 13-4a ■). These signals travel to the spinal cord, where the afferent spindle neuron synapses with alpha motor neurons. The alpha motor neurons in turn create tonic excitation and contraction in the extrafusal fibers associated with the muscle spindle. Because of this tonic activity, even a muscle at rest maintains a certain level of tension, known as *muscle tone.*

The sensory neurons in the muscle spindle change their activity depending on the length of the intrafusal fibers of the spindle. If the muscle stretches, the intrafusal fibers of the spindle also stretch, so the sensory neurons fire more rapidly (Fig. 13-4b ■). This more rapid firing initiates reflex contraction of the muscle, relieving the stretch on the muscle and preventing damage from overstretching. Because contraction also decreases stretch in the spindle fibers, their firing rate declines. This reflex pathway, in which muscle stretch initiates a contraction response, is known as a **stretch reflex.**

What happens to the firing rate of tonically active spindle afferents when a resting muscle contracts and shortens? You might predict that the afferent neurons slow their firing rate or stop firing altogether (Fig. 13-5a ■, p. 390). But they neither slow nor stop firing, because of **alpha-gamma coactivation.**

When alpha motor neurons to extrafusal fibers are activated for muscle contraction, the gamma motor neurons to the spindle are activated at the same time (Fig. 13-5b ■). The gamma motor neurons innervate the contractile ends of the intrafusal fibers, which contract and shorten. Contraction at both ends of the fibers lengthens the central section and maintains stretch on the sensory nerve endings, counteracting the release of tension on the muscle spindle. Thus, when the extrafusal muscle fibers shorten during contraction, the intrafusal fibers remain stretched and continue to monitor tension in the muscle.

An example of how muscle spindles work during a stretch reflex is shown in Figure 13-6 ■, p. 391. You can try this yourself with a friend. The subject stands with eyes closed, elbows at 90°, and one hand extended with the palm up. A small book or other flat weight is placed in the hand. The assistant suddenly drops a heavier load, such as another book, onto the hand. The added weight sends the hand downward, stretching the biceps muscle and activating its muscle spindles. Sensory input into the spinal cord then activates the alpha motor neurons of the biceps muscle. The biceps contracts, bringing the arm back to its original position.

Golgi Tendon Organs Protect the Muscle

A second type of receptor for stretch reflexes is the **Golgi tendon organ.** These receptors are found at the junction of the tendons and muscle fibers, placing them in series with the muscle fibers. Golgi tendon organs respond to both stretch and contraction of the muscle by causing a

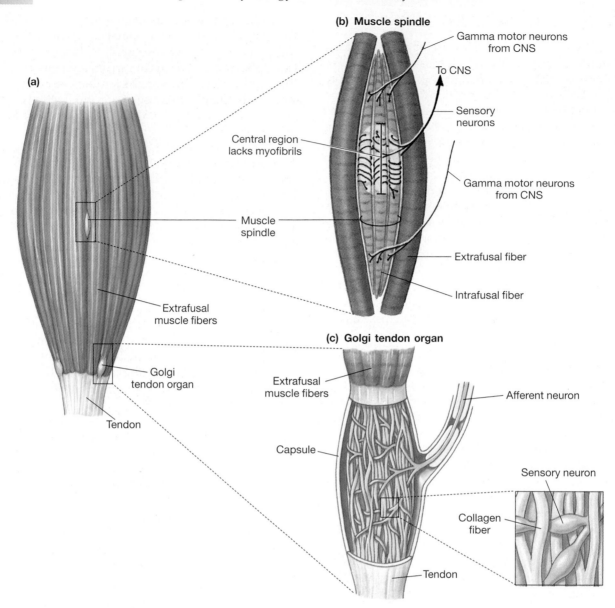

(a)

(b) Muscle spindle

Gamma motor neurons from CNS

To CNS

Sensory neurons

Central region lacks myofibrils

Gamma motor neurons from CNS

Muscle spindle

Extrafusal fiber

Intrafusal fiber

Extrafusal muscle fibers

(c) Golgi tendon organ

Golgi tendon organ

Extrafusal muscle fibers

Afferent neuron

Capsule

Tendon

Sensory neuron

Collagen fiber

Tendon

■ **Figure 13-3** **Sensory receptors in muscle** (a) Buried among the normal contractile fibers (extrafusal fibers) of the muscle are stretch receptors known as muscle spindles. Golgi tendon organs are receptors that link the muscle and the tendon. Contraction in extrafusal fibers is controlled by alpha motor neurons, and contraction of the muscle spindles is controlled by gamma motor neurons. The Golgi tendon organ does not contract. (b) The central region of the muscle spindle lacks myofibrils and cannot contract. Sensory nerve endings wrap around the central region and fire when the central section of the muscle spindle stretches. The ends of the muscle spindle contain myofibrils that contract in response to commands carried by gamma motor neurons. (c) The Golgi tendon organ consists of sensory neurons interwoven among collagen fibers. If the collagen fibers are stretched, they pinch the sensory neurons and trigger action potentials.

reflexive relaxation. This is the opposite of muscle spindles, which cause reflex contraction.

Golgi tendon organs are composed of free nerve endings that wind between collagen fibers inside a connective tissue capsule (Fig. 13-3c ■). Because Golgi tendon organs are in the elastic element of the muscle, they respond to both stretch and contraction of the muscle. If the muscle is stretched, the collagen fibers pull tight, pinching sensory endings of the afferent neurons and causing them to fire.

If the muscle contracts, the tendons act as an elastic component and are pulled taut during the isometric

phase of the contraction [∞ p. 365]. Again, the sensory neurons in the Golgi tendon organ are stimulated and fire. Although both stretch and contraction stimulate the neurons, they respond most strongly to contraction.

Activation of the Golgi tendon organ inhibits alpha motor neurons and decreases muscle contraction. Under most circumstances, this reflex slows muscle contraction as the force of contraction increases. In other instances, the Golgi tendon organs prevent excessive contraction that might injure the muscle. In the example of the book placed on the outstretched hand, if the added weight requires

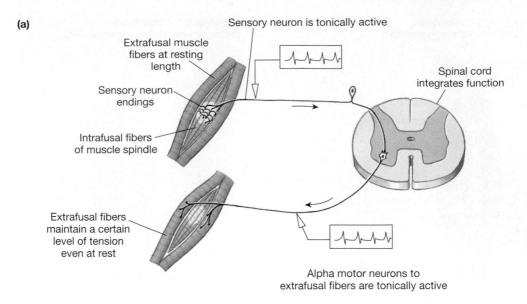

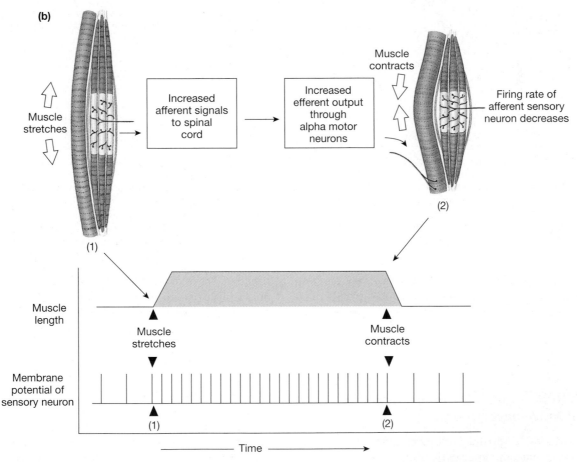

■ **Figure 13-4   Muscle spindle function**   (a) When a muscle is at its resting length, the muscle spindle is slightly stretched and its associated sensory neuron shows tonic activity. As a result of tonic activity of the reflex, the associated muscle maintains a certain level of tension, or tone, even at rest. (b) If a muscle is stretched, its muscle spindles are also stretched. This stretching increases the firing rate of the spindle afferents, and the muscle contracts. Contraction relieves the stretch on the spindle and acts as negative feedback to diminish the reflex.

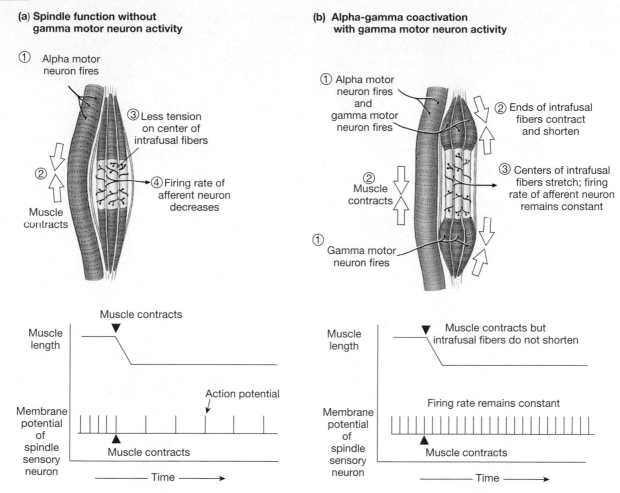

(a) Spindle function without gamma motor neuron activity

① Alpha motor neuron fires

③ Less tension on center of intrafusal fibers

② Muscle contracts

④ Firing rate of afferent neuron decreases

Muscle length — Muscle contracts

Membrane potential of spindle sensory neuron — Action potential — Muscle contracts

Time

(b) Alpha-gamma coactivation with gamma motor neuron activity

① Alpha motor neuron fires and gamma motor neuron fires

② Ends of intrafusal fibers contract and shorten

② Muscle contracts

③ Centers of intrafusal fibers stretch; firing rate of afferent neuron remains constant

① Gamma motor neuron fires

Muscle length — Muscle contracts but intrafusal fibers do not shorten

Membrane potential of spindle sensory neuron — Firing rate remains constant — Muscle contracts

Time

■ **Figure 13-5   Gamma motor neurons**   The gamma motor neurons allow the muscle spindle to maintain function when the muscle is contracting. (a) If the gamma motor neurons to the muscle spindle are cut, the spindle loses activity when the muscle contracts because the central region of the spindle loses its stretch. (b) Spindle function is maintained through alpha-gamma coactivation: when the alpha motor neurons are active, so are the gamma motor neurons to the spindle. Contraction of the ends of the spindle stretches the central region and maintains the tonic activity of the spindle.

more tension than the muscle can develop, the Golgi tendon organ will respond as muscle tension nears its maximum. It triggers reflex *inhibition* of the biceps muscle, causing the arm to relax. Thus, the person will drop the added weight before the muscle fibers can be damaged (Fig. 13-6d, e ■).

Stretch Reflexes and Reciprocal Inhibition Control Movement Around a Joint

Movement around most flexible joints in the body is controlled by groups of synergistic and antagonistic muscles that act in a coordinated fashion. Afferent sensory neurons from a muscle and efferent motor neurons that control the muscle are linked together by diverging and converging pathways of interneurons within the spinal cord. The collection of pathways controlling a single joint is known as a **myotatic unit.**

The simplest reflex in a myotatic unit is the monosynaptic stretch reflex that involves only two neurons: the sensory neuron from the muscle spindle and the so-matic motor neuron to the muscle. The knee jerk reflex is an example of a monosynaptic stretch reflex (Fig. 13-7 ■).

To demonstrate the knee jerk reflex, the subject sits on the edge of a table so that the lower leg hangs relaxed. When the patellar tendon below the kneecap is tapped with a small rubber hammer, the pressure stretches the quadriceps muscle that runs up the front of the thigh. This stretching activates the muscle spindle stretch receptors within the muscle, sending an action potential through the sensory neuron to the spinal cord. The sensory neuron synapses directly onto the motor neuron that controls contraction of the quadriceps muscle (a monosynaptic reflex). The action potential in the motor neuron causes the muscle fibers of the quadriceps to contract, and the lower leg swings forward.

In addition to muscle contraction that extends the leg, the relaxation of antagonistic flexor muscles must also occur (**reciprocal inhibition**). In the leg, this requires relaxation of the hamstring muscles that run up the back of the thigh. The single stimulus of the blow to the tendon accomplishes both contraction of the quadriceps

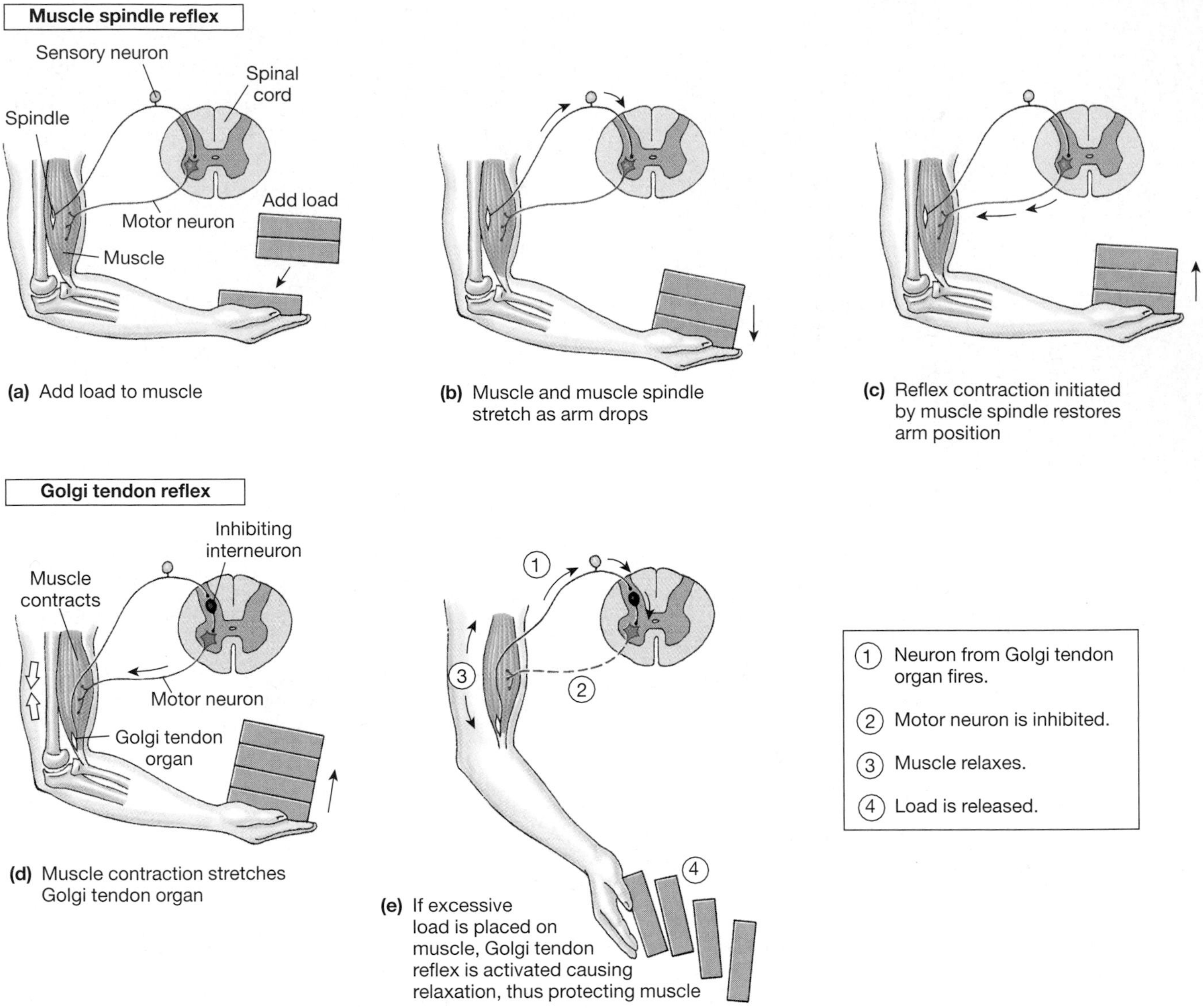

Muscle spindle reflex

(a) Add load to muscle

(b) Muscle and muscle spindle stretch as arm drops

(c) Reflex contraction initiated by muscle spindle restores arm position

Golgi tendon reflex

(d) Muscle contraction stretches Golgi tendon organ

(e) If excessive load is placed on muscle, Golgi tendon reflex is activated causing relaxation, thus protecting muscle

1. Neuron from Golgi tendon organ fires.
2. Motor neuron is inhibited.
3. Muscle relaxes.
4. Load is released.

■ **Figure 13-6   Muscle reflexes**   (a–c) The muscle spindle reflex: the addition of a load to a muscle stretches the muscle and the spindles, creating a reflex contraction. (d, e) The Golgi tendon reflex protects the muscle from excessively heavy loads. If the muscle contraction initiated by the spindle reflex approaches maximum tension, the Golgi tendon organs fire, causing the muscle to relax and drop the load.

muscle and reciprocal inhibition of the hamstrings. The sensory neuron branches upon entering the spinal cord. One collateral activates the motor neuron innervating the quadriceps, while the other collateral synapses with an inhibitory interneuron. This interneuron inhibits the motor neuron controlling the hamstrings, preventing it from firing an action potential (a polysynaptic reflex). The result is a relaxation of the hamstrings that allows contraction of the quadriceps to proceed unopposed.

Flexion Reflexes Pull Limbs Away from Painful Stimuli

Flexion reflexes are polysynaptic reflex pathways that cause an arm or leg to be pulled away from a painful stimulus such as a pinprick or a hot stove. These reflexes, like the reciprocal inhibition reflex just described, rely on divergent pathways in the spinal cord. The afferent fiber from the nociceptor (pain receptor) activates several excitatory interneurons (Fig. 13-8 ■). Some of these interneurons excite alpha motor neurons and cause contraction in all the flexor muscles in the affected limb. The other interneurons simultaneously activate inhibitory interneurons that cause relaxation of the antagonistic muscle groups. With reciprocal inhibition, the limb is flexed, withdrawing it from the painful stimulus. This type of reflex takes longer to happen than a stretch reflex such as the knee jerk, an indication that it is a polysynaptic rather than a monosynaptic reflex.

Flexion reflexes, particularly in the legs, are usually accompanied by the **crossed extensor reflex,** a postural reflex that helps maintain balance when one foot is lifted from the ground. In this reflex, also shown in Figure 13-8 ■, the quick withdrawal of the right foot from a

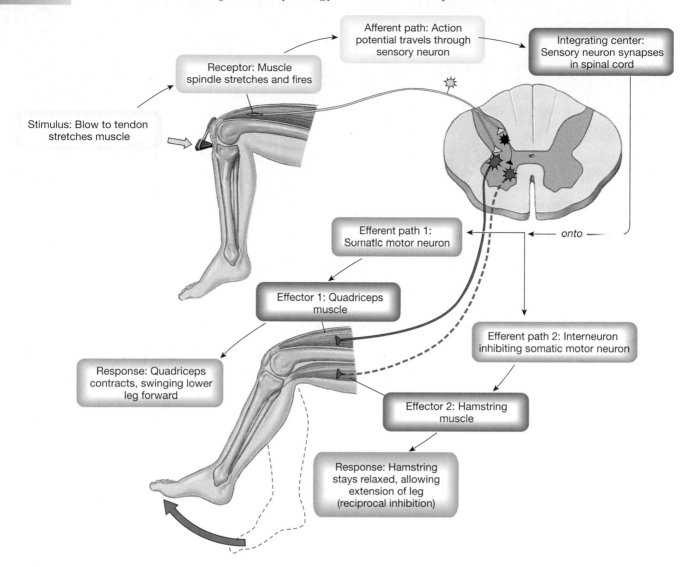

Afferent path: Action potential travels through sensory neuron

Integrating center: Sensory neuron synapses in spinal cord

Receptor: Muscle spindle stretches and fires

Stimulus: Blow to tendon stretches muscle

Efferent path 1: Somatic motor neuron

— *onto* —

Effector 1: Quadriceps muscle

Efferent path 2: Interneuron inhibiting somatic motor neuron

Response: Quadriceps contracts, swinging lower leg forward

Effector 2: Hamstring muscle

Response: Hamstring stays relaxed, allowing extension of leg (reciprocal inhibition)

■ **Figure 13-7** **The knee jerk reflex**   The knee jerk or patellar tendon reflex is a monosynaptic spinal reflex in which a blow to the patellar tendon causes contraction of the quadriceps muscle (pathway 1). An integral part of the response is the polysynaptic reflex inhibition of the hamstring muscles (pathway 2).

painful stimulus (a tack) is matched by extension of the left leg so that it can support the sudden shift in weight. The extensors contract in the supporting left leg and relax in the withdrawing right leg while the opposite occurs in the flexor muscles.

Note in the figure how the one sensory neuron synapses on multiple interneurons. Divergence of the sensory signal permits a single stimulus to control two sets of antagonistic muscle groups and also send sensory information to the brain. This type of complex reflex with multiple neuron interactions is more typical of our reflexes than the simple monosynaptic knee jerk stretch reflex.

The most complicated spinal reflex pathways are controlled by networks of neurons in the CNS called **central pattern generators.** Once activated, central pattern generators create spontaneous repetitive movement

without further sensory input. In humans, rhythmic movements controlled by central pattern generators include locomotion and the unconscious rhythmicity of quiet breathing.

In the next section of the chapter, we look at how the CNS controls movements that range from involuntary reflexes to complex, voluntary movement patterns such as dancing, throwing a ball, or playing a musical instrument.

✔ As you pick up a heavy weight, which of the following are active in your biceps muscle: alpha motor neuron, gamma motor neuron, muscle spindle afferent neurons, Golgi tendon organ afferent neurons?

✔ What distinguishes a stretch reflex from an extensor reflex?

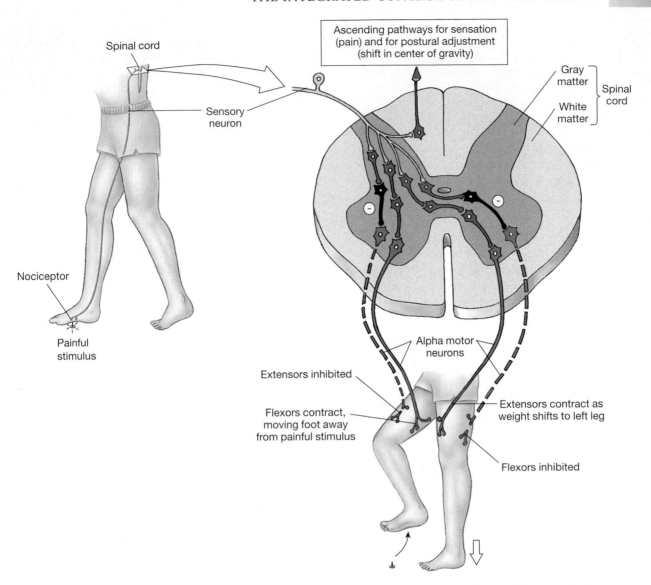

■ **Figure 13-8   Flexion reflex and the crossed extensor reflex**   Coordination of reflexes with postural adjustments is essential for maintaining balance. When the right foot encounters a painful stimulus, a polysynaptic withdrawal reflex causes the right flexor muscles to contract. Reciprocal inhibition keeps the right extensor muscles relaxed. The crossed extensor reflex, activated by the same sensory pathway, straightens the left leg so that it will support the sudden addition of weight.

Reflexes and Muscle Tone   Clinicians use reflexes to investigate the condition of the nervous system and the muscles. For a reflex to be normal, there must be normal conduction through all neurons in the pathway, normal synaptic transmission at the neuromuscular junction, and normal muscle contraction. A reflex that is absent, abnormally slow, or greater than normal (hyperactive) suggests the presence of a disease. Interestingly, not all abnormal reflexes suggest neuromuscular disorders. For example, slowed relaxation of the ankle flexion reflex suggests hypothyroidism. Besides testing reflexes, clinicians assess muscle tone. Even when relaxed and at rest, muscles have a certain resistance to stretch that is the result of continuous (tonic) output by alpha motor neurons. If muscle tone is absent or if a muscle resists being passively stretched by the examiner (increased tone), there is probably a problem with the pathways that control muscle contraction.

▶ THE INTEGRATED CONTROL OF BODY MOVEMENT

Most of us never think about how our body translates thoughts into action. Even the simplest movement requires proper timing so that antagonistic and synergistic muscle groups contract in the appropriate sequence. In addition, the body must continuously adjust its position to compensate for differences between the intended movement and the actual one. For example, the baseball pitcher steps off the mound as he reaches for a ground ball, but in doing so, he slips on a wet patch of grass. His brain quickly compensates for the unexpected change in position through reflex muscle activity, and he stays on his feet to intercept the ball.

Skeletal muscles cannot communicate with each other directly, so they send messages to the CNS, allowing that

...continued from page 386

Once in the spinal cord, tetanospasmin is released from the motor neuron. It then selectively blocks neurotransmitter release at inhibitory synapses. Patients with tetanus experience muscle spasms that begin in the jaw and may eventually affect the entire body. When the extremities become involved, the arms and legs may go into painful, rigid spasms.

Question 2: Using the reflex pathways diagrammed in Figures 13-7 ■ and 13-8 ■, explain why inhibition of inhibitory interneurons might result in uncontrollable muscle spasms.

integrating center to take charge and direct movement. Most body movement is a highly integrated, coordinated response that requires input from multiple regions of the brain. In this section we examine the integrating centers within the CNS that are responsible for control of body movement.

Movement Can Be Classified as Reflex, Voluntary, or Rhythmic

Movement can be loosely classified into three categories: reflex movement, voluntary movement, and rhythmic movement (Table 13-2). **Reflex movements** are the least complex and are integrated primarily in the spinal cord. However, like other spinal reflexes, reflex movements can be modulated by input from higher brain centers (Fig. 13-9 ■). In addition, the sensory input that initiates reflex movements, such as the input from muscle spindles and Golgi tendon organs, may go to the brain and participate in the coordination of voluntary movements or postural reflexes.

Postural reflexes that help us maintain body position as we stand or move through space are integrated in the brain stem. They require continuous sensory input from visual and vestibular (inner ear) sensory systems and from the muscles themselves [∞ p. 306]. Muscle, tendon, and joint receptors provide information about the positions of various body parts relative to one another. You can tell if your arm is bent even when your eyes are closed because these receptors provide information about body position to the brain.

Information from the vestibular apparatus of the ear and visual cues help us maintain our position in space. For example, we use the horizon to tell us our spatial orientation relative to the ground. In the absence of visual cues, we rely on tactile input. People trying to move in a dark room instinctively reach for a wall or piece of furniture to help orient themselves. Without visual and tactile cues, our orientation skills may fail. The lack of cues is what makes flying airplanes in clouds or fog impossible without instruments. The effect of gravity on the vestibular system is such a weak input when compared with visual or tactile cues that pilots may find themselves flying upside down relative to the ground.

Voluntary movements are the most complex type of movement. They require integration at the cerebral cortex, and they can be initiated at will without external stimuli. Learned voluntary movements improve with practice and some even become involuntary, like reflexes. Think about how difficult it was to learn to ride a bicycle. However, once you learned to pedal smoothly and to keep your balance, the movements became automatic. "Muscle memory" is the name dancers and athletes give the ability of the unconscious brain to reproduce voluntary, learned movements and positions.

Rhythmic movements such as walking or running are a combination of reflex movements and voluntary movements. Rhythmic movements must be initiated and terminated by input from the cerebral cortex, but once initiated, they can be sustained without further input from the brain. The rhythmic activity of skeletal muscles is maintained by groups of spinal interneurons. These act as central pattern generators, alternately con-

TABLE 13-2 Types of Movement

	Reflex	Voluntary	Rhythmic
Stimulus that initiates movement	Primarily external via sensory receptors; minimally voluntary	External stimuli or at will	Initiation and termination voluntary
Example	Knee jerk, cough, postural reflexes	Playing piano	Walk, run
Complexity	Least complex; integrated at level of spinal cord with higher center modulation	Most complex; integrated in cerebral cortex	Intermediate complexity; integrated in spinal cord with higher center input required
Comments	Inherent, rapid	Learned movements that improve with practice; once learned, may become subconscious ("muscle memory")	Spinal circuits act as pattern generators; activation of these pathways requires input from brain stem

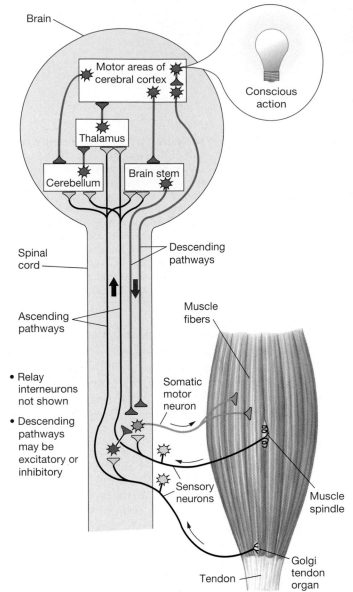

■ **Figure 13-9 Integration of muscle reflexes** Although many muscle reflexes are simple spinal reflexes, sensory information about them is transmitted to the brain through ascending pathways. In addition, the conscious and subconscious brain sends modulatory messages to the spinal integrating centers through descending pathways.

...continued from page 394

Dr. Ling admits Mrs. Evans to the intensive care unit. There Mrs. Evans is given tetanus antitoxin to deactivate any toxin that has not yet entered motor neurons. She also receives penicillin, an antibiotic that kills the bacteria, and drugs to help relax her muscles. Despite these treatments, by the third day Mrs. Evans is having difficulty breathing because of spasms in her chest muscles. Dr. Ling calls in the chief of anesthesiology to administer metocurine, a drug similar to curare. Curare and metocurine induce temporary paralysis of muscles by binding to ACh receptors on the motor end plate. Patients must be placed on respirators that breathe for them while receiving metocurine. For people with tetanus, however, metocurine can temporarily halt the muscle spasms and allow the body to recover.

Question 3: *Why does the binding of metocurine to ACh receptors on the motor end plate induce muscle paralysis? (Hint: What is the function of ACh in synaptic transmission?) Is metocurine an agonist or an antagonist of ACh?*

prepare for a voluntary movement, and feedback mechanisms are used to create a smooth, continuous motion. Coordination of movement thus requires cooperation from many parts of the brain.

The CNS Integrates Movement

Three levels of the nervous system control movement: the spinal cord, the brain stem, and the motor areas of the cerebral cortex (Fig. 13-10 ■). The brain stem is influenced by input from the cerebellum, while the basal ganglia, clusters of neurons surrounding the thalamus, and the cerebellum help cerebral cortical motor areas plan movement. The thalamus acts as a relay station for

tracting and relaxing muscles in a repetitive, rhythmic fashion until instructed to stop by signals from the brain. As an analogy, think of a battery-operated bunny. When the switch is thrown to "on," the bunny begins to hop. It continues its repetitive hopping until someone turns it off (or until the battery runs down).

The distinction between reflex, voluntary, and rhythmic movement is not always clear-cut. The precision of voluntary movements improves with practice, but so does that of some reflexes. Voluntary movements, once learned, can become reflexive. In addition, most voluntary movements require continuous input from postural reflexes. Feedforward reflexes allow the body to

Central Pattern Generators and Spinal Cord Injuries The ability of CNS pattern generators, or central pattern generators, to sustain rhythmic movement without continued input from the central nervous system has proved important for research on spinal cord injuries. Scientists have artificially stimulated rhythmic movements in experimental animals when the spinal cord between the brain and the motor neurons has been severed. For example, an animal paralyzed by a spinal cord injury that blocks the "start walking" signal from the brain will walk if supported on a moving treadmill and given an electrical stimulus to activate the spinal pattern generator. As the treadmill moves the animal's legs, the spinal pattern generator, reinforced by sensory signals from muscle spindles, drives contraction of the leg muscles. Researchers are trying to take advantage of these rhythmic reflexes in people with spinal cord injuries. Some of the newest therapeutic techniques involve artificially stimulating portions of the spinal cord so that movement is restored to formerly paralyzed limbs.

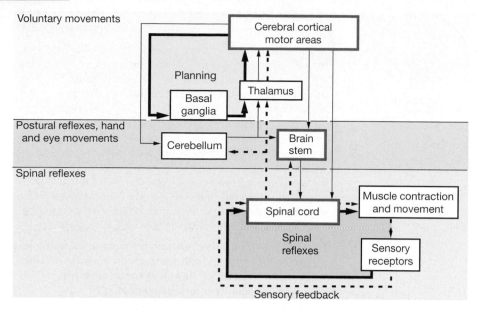

■ **Figure 13-10**   **Higher center control of movement**   The simplest control of movement takes place through spinal reflexes. Descending pathways from the brain stem and cerebral cortex can initiate or modify movement. The cerebellum receives feedback information from sensory receptors and combines it with information from the cerebral cortex to modify the descending commands. The basal ganglia assist the cerebral cortex in planning movement.

signals being sent to the cerebral cortex from the spinal cord, basal ganglia, and cerebellum (Table 13-3).

The simplest movements involve spinal reflexes integrated in the spinal cord. Sensory receptors such as muscle spindles, Golgi tendon organs, and joint proprioreceptors provide information to the spinal cord that can be acted upon without input from higher regions of the CNS. Some of this sensory information is sent through ascending pathways to the brain stem, along with sensory input from the eyes and vestibular apparatus. The brain stem is in charge of postural reflexes and hand and eye movements. It also gets commands from the cerebellum, the part of the brain responsible for "fine-tuning" movement.

Voluntary movements require the participation of the cerebral cortex, cerebellum, and the basal ganglia (Fig. 13-11 ■). The control of these behaviors can be divided into three steps: (1) decision making and planning,

(2) initiating the movement, and (3) executing the movement. The cerebral cortex plays a key role in the first two steps. Behaviors such as movement require knowledge of the body's position in space (where am I?), a decision on what movement should be executed (what shall I do?), a plan for executing the movement (how shall I do it?), and the ability to hold the plan in memory long enough to carry it out.

Let's return to our baseball pitcher and trace the process as he decides whether to throw a fast ball or a slow inside curve. Standing out on the mound, the pitcher is acutely aware of his surroundings: the other players on the field, the batter in the box, the dirt beneath his feet. With the help of visual and somatosensory input, he is aware of his body position as he steadies himself for the pitch. Deciding which type of pitch to throw and anticipating the consequences occupy many pathways in his cortical association areas. Once he makes the deci-

TABLE 13-3   Neural Control of Movement

Location	Role	Receives Input from:	Sends Integrative Output to:
Spinal cord	Spinal reflexes	Sensory receptors	Brain stem, cerebellum, thalamus/cerebral cortex
Brain stem	Posture, hand and eye movements	Cerebellum, visual and vestibular sensory receptors	Spinal cord
Motor areas of cerebral cortex	Planning and coordinating complex movement	Thalamus	Brain stem, spinal cord, cerebellum, basal ganglia
Cerebellum	Monitors output signals from motor areas and adjusts movements	Spinal cord (sensory), cerebral cortex (commands)	Brain stem, cerebral cortex (Note: All output is inhibitory.)
Thalamus	Contains relay nuclei that pass messages to cerebral cortex	Basal ganglia, cerebellum, spinal cord	Cerebral cortex
Basal ganglia	Motor planning	Cerebral cortex	Cerebral cortex

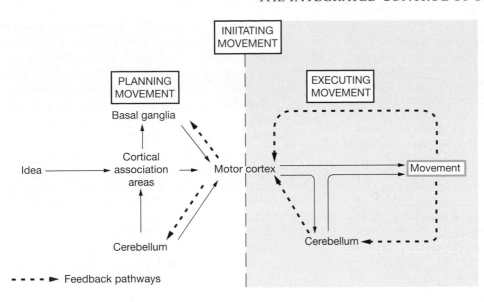

■ **Figure 13-11** **CNS control of voluntary movement** Voluntary movements can be divided into three phases: planning, initiation, and execution. Planning involves the exchange and coordination of information between the cortical association areas, the basal ganglia, and the cerebellum. Initiation is the responsibility of the motor cortex. Execution of voluntary movements is carried out through descending pathways to somatic motor neurons. Sensory information from the joints and muscles provides feedback to the brain that allows it to correct for any deviation between the planned movement and the actual movement.

sion to throw a fast ball, this information is sent to the motor cortex, the primary region in charge of planning and organizing complex movements.

Now it is time to initiate the movement. Information from the association areas is sent to the cerebellum, which makes postural adjustments by combining feedback from peripheral sensory receptors with information from the cerebral cortex. This information also goes to the basal ganglia, which assist the cortical motor areas in planning the pitch. Once this network completes a plan of action, the motor cortex begins the execution of the movement.

The pitcher's decision to throw a fast ball now is translated into action potentials that travel down through the corticospinal tract, a group of interneurons that connect the cortex to the spinal cord. At the same time, *feedforward postural reflexes* adjust the body position, shifting weight slightly in anticipation of the changes that are about to occur (Fig. 13-12 ■). Through divergent pathways, action potentials race down the interneurons to the somatic motor neurons that control the muscles used for the pitch: some are excited, others are

inhibited. Reciprocal inhibition allows precise control over antagonistic muscle groups as the pitcher flexes and retracts his right arm. His weight shifts onto his right foot as his arm moves back.

Each of these movements activates sensory receptors that feed back information to the brain stem, activating postural reflexes. These reflexes adjust his body position so that the pitcher does not lose his balance and fall over backward. Finally, the pitcher releases the ball, catching his balance on the follow-through: another example of postural reflexes mediated through sensory feedback. His head stays erect and his eyes stay on the ball as it reaches the batter. WHACK . . . home run. As the pitcher's eyes follow the ball and he evaluates the result of his pitch, his brain is preparing for the next batter, hoping to use what it has learned from these pitches to improve those to come.

. . . continued from page 395

Four weeks later, Mrs. Evans is ready to go home, completely recovered from her infection and showing no signs of lingering effects. Once she could talk, Mrs. Evans, who was born on the farm where she still lived, was able to tell Dr. Ling that she had never had immunization shots for tetanus or any other diseases. "Well, that made you one of only a handful of people in the United States who will develop tetanus this year," Dr. Ling told her. "You've been given your first two tetanus shots here in the hospital. Be sure to come back in six months for the last one so that this won't happen again." Because of national immunization programs begun in the 1950s, tetanus is now a rare disease in the United States. However, in developing countries without immunization programs, tetanus is still a common and serious condition.

Question 4: *On the basis of what you know about who receives immunization shots in the United States, predict the age and background of people who are most likely to develop tetanus this year.*

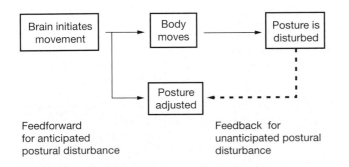

■ **Figure 13-12** **Feedforward reflexes and feedback of information during movement**

Parkinson's Disease and the Basal Ganglia   Our understanding of the role of the basal ganglia in the control of movement has been slow to develop because, for many years, animal experiments yielded little information. Randomly destroying portions of the basal ganglia did not affect research animals. However, research focusing on Parkinson's disease (Parkinsonism) in humans has been more rewarding. Parkinsonism is a condition characterized by abnormal movements and speech difficulties. These symptoms are associated with loss of the inhibitory neurotransmitter dopamine in neurons of the basal ganglia. The cause of Parkinson's disease is usually not known. However, a few years ago, a number of young drug users were diagnosed with Parkinsonism. Their disease was traced to the use of homemade heroin that contained a toxic contaminant that destroyed dopaminergic (dopamine-secreting) neurons. This contaminant has been isolated and now enables researchers to induce Parkinson's disease in experimental animals so that we have an animal model on which to test new treatments.

▶ CONTROL OF MOVEMENT IN VISCERAL MUSCLES

Movement created by contracting smooth and cardiac muscles is very different from that created by skeletal muscles, in large part because smooth and cardiac muscle are not attached to bone. In the internal organs, or viscera, muscle contraction usually changes the shape of an organ, narrowing the lumen of a hollow organ or shortening the length of a tube. In many hollow internal organs, muscle contraction pushes material through the lumen of the organ: the heart pumps blood, the digestive tract moves food, the uterus expels a baby.

Visceral muscle contraction is often reflexively controlled by the autonomic nervous system, but not always. Some types of smooth and cardiac muscle are capable of generating their own action potentials, independent of an external signal. Both the heart and digestive tract have spontaneously depolarizing muscle fibers (often called *pacemakers*) that give rise to regular, rhythmic contractions.

Reflex control of visceral smooth muscle varies from that of skeletal muscle. Skeletal muscles are controlled only by the nervous system, but in many types of visceral muscle, hormones are important in regulating contraction. In addition, some visceral muscle cells are connected to each other by gap junctions that allow electrical signals to pass directly from cell to cell.

Because smooth and cardiac muscle have such a variety of control mechanisms, we will discuss their control as we cover the appropriate organ system for each type of muscle. In the next chapter, we examine cardiac muscle and its function in the heart.

PROBLEM CONCLUSION

In this running problem, you learned that the tetanus toxin is transported from peripheral wounds to the spinal cord, where it blocks the release of inhibitory neurotransmitters. You also learned that drugs such as curare can cause paralysis of skeletal muscles.

Further check your understanding of this running problem by comparing your answers to those in the summary table.

	Question	Facts	Integration and Analysis
1a	By what process is tetanospasmin taken up into neurons?	Tetanospasmin is a protein.	Proteins are too large to cross cell membranes by mediated transport. Therefore, tetanospasmin must be taken up by endocytosis [∞ p. 129].
1b	By what process does tetanospasmin travel up the axon to the nerve cell body?	Movement of substances from the axon terminal to the cell body is called retrograde axonal transport [∞ p. 218].	Tetanospasmin is taken up by endocytosis, so it will be contained in endocytotic vesicles. These vesicles are "walked" up the axon along microtubules through retrograde axonal transport.

2	Using Figures 13-7 ■ and 13-8 ■, explain why inhibition of inhibitory interneurons might result in uncontrollable muscle spasms.	Muscles often occur in antagonistic pairs. When one muscle is contracting, its antagonist must be inhibited.	If the inhibitory interneurons are not functioning, both sets of antagonistic muscles can contract at the same time. This would lead to muscle spasms and rigidity because the bones attached to the muscles would be unable to move in any one direction.
3a	Why does the binding of metocurine to ACh receptors on the motor end plate induce muscle paralysis?	Acetylcholine (ACh) is the somatic motor neuron neurotransmitter that initiates skeletal muscle contraction.	If metocurine binds to ACh receptors, it prevents ACh from binding. Without ACh binding, the muscle fiber will not depolarize and cannot contract, resulting in paralysis.
3b	Is metocurine an agonist or an antagonist of ACh?	Agonists mimic the effects of a substance. Antagonists block the effects of a substance.	Metocurine blocks ACh action; therefore, it is an antagonist.
4	On the basis of what you know about who receives immunization shots in the United States, predict the age and background of people who are most likely to develop tetanus this year.	Immunizations are required for all children of school age. This practice has been in effect since about the 1950s. In addition, most people who suffer puncture wounds or dirty wounds are given tetanus booster shots when they are treated for those wounds.	Most cases of tetanus in the United States will occur in people over the age of 60 who have never been immunized, in immigrants (particularly migrant workers), and in newborn infants. Another source of the disease is contaminated heroin; injection of the drug under the skin may cause tetanus.

CHAPTER REVIEW

SUMMARY

Nervous Reflexes

1. A nervous reflex consists of the following elements: stimulus → receptor → sensory neurons → CNS → efferent neurons → effectors (muscles and glands) → response. (p. 384)

2. Nervous reflexes can be classified in several different ways. **Somatic reflexes** involve somatic motor neurons and skeletal muscles. **Autonomic,** or **visceral, reflexes** are controlled by autonomic neurons. (p. 384)

3. **Spinal reflexes** are integrated within the spinal cord. **Cranial reflexes** are integrated in the brain. (p. 384)

4. Many reflexes are inborn. Others are acquired through experience. (p. 384)

5. The simplest reflex pathway is a **monosynaptic reflex** with only two neurons. **Polysynaptic reflexes** have three or more neurons in the pathway. (p. 385)

Autonomic Reflexes

6. Some visceral reflexes are spinal reflexes that are modulated by input from the brain. Other autonomic reflexes are integrated in the brain, primarily in the hypothalamus, thalamus, and brain stem. The brain controls reflexes needed to maintain homeostasis. (p. 385)

7. Autonomic reflexes are all polysynaptic, and many are characterized by tonic activity. (p. 386)

Skeletal Muscle Reflexes

8. Skeletal muscle relaxation must be controlled by the CNS because activation of a somatic motor neuron always causes contraction in skeletal muscle. (p. 387)

9. The normal contractile fibers of a muscle are known as **extrafusal muscle fibers.** Their contraction is controlled by **alpha motor neurons.** (p. 387)

10. **Muscle spindles** send information about muscle length to the CNS. These receptors consist of **intrafusal fibers,** modified muscle fibers that lack myofibrils in their central portions. Sensory neurons wrap around the noncontractile center, and **gamma motor neurons** innervate the contractile ends of the intrafusal fibers. (p. 387)

11. Muscle spindles are tonically active stretch receptors. Their output is translated into tonic excitation of extrafusal muscle fibers by the alpha motor neurons. Because of this tonic activity, a muscle at rest maintains a certain level of tension, known as muscle tone. (p. 387)

12. If an intact muscle stretches, the intrafusal fibers of the spindle stretch and initiate reflex contraction of the muscle. The contraction prevents damage from overstretching. This reflex pathway is known as a **stretch reflex.** (p. 387)

13. When a muscle contracts, **alpha-gamma coactivation** ensures that the muscle spindle remains active. Activation of gamma motor neurons causes contraction of the ends of the intrafusal fibers. This contraction lengthens the central section of the intrafusal fiber and maintains stretch on the sensory nerve endings. (p. 387)

14. **Golgi tendon organs** are found at the junction of the tendons and muscle fibers. They consist of free nerve endings that wind between collagen fibers. Golgi tendon organs respond to both stretch and contraction of the muscle by causing a reflexive relaxation. (p. 387)

15. The synergistic and antagonistic muscles that control a single joint are known as a **myotatic unit.** When one set of muscles in a myotatic unit contracts, the antagonistic muscles must relax through a reflex known as **reciprocal inhibition.** (p. 390)

16. **Flexion reflexes** are polysynaptic reflexes that cause an arm or leg to be pulled away from a painful stimulus. Flexion reflexes are usually accompanied by the **crossed extensor reflex,** a postural reflex that helps maintain balance when one foot is lifted from the ground. (p. 391)

17. **Central pattern generators** are networks of neurons in the CNS that function spontaneously to control certain rhythmic muscle movements. (p. 392)

The Integrated Control of Body Movement

18. Movement can be loosely classified into three categories: reflex movement, voluntary movement, and rhythmic movement. **Reflex movements** are the least complex and are integrated primarily in the spinal cord. **Postural reflexes** help us maintain body position as we stand or move and are integrated in the brain stem. (p. 394)

19. **Voluntary movements** require integration at the cerebral cortex and can be initiated at will without external stimuli. Learned voluntary movements improve with practice and some even become involuntary, like reflexes. (p. 394)

20. **Rhythmic movements** such as walking are a combination of reflexes and voluntary movements. Rhythmic movements must be initiated and terminated by input from the cerebral cortex, but once initiated, they can be sustained by central pattern generators. (p. 394)

21. Feedforward reflexes allow the body to prepare for a voluntary movement; feedback mechanisms are used to create a smooth, continuous motion. (p. 395)

22. Three levels of the nervous system control movement: the spinal cord, the brain stem, and the motor areas of the cerebral cortex. The basal ganglia and the cerebellum help cortical motor areas plan movement. The thalamus acts as a relay station for signals being sent to the cerebral cortex. (p. 396)

23. Voluntary movements require the cerebral cortex, cerebellum, and basal ganglia. The control of voluntary behavior can be divided into decision making and planning, initiating the movement, and executing the movement. (p. 397)

Control of Movement in Visceral Muscles

24. Contraction in smooth and cardiac muscles may occur spontaneously or may be controlled by hormones or by the autonomic division of the nervous system. (p. 398)

QUESTIONS

LEVEL ONE   Reviewing Facts and Terms

1. All nervous reflexes begin with a _____ that activates a receptor.

2. Somatic reflexes involve _____ muscles; _____, or visceral, reflexes are controlled by autonomic neurons.

3. The pathway pattern that brings information from many neurons into a smaller number of neurons is known as _____.

4. When the axon terminal of a modulatory neuron (cell M) terminates close to the axon terminal of a presynaptic cell (cell P) and decreases the amount of neurotransmitter released by cell P, the resulting type of modulation is called _____. (Hint: see p. 243)

5. Autonomic reflexes are also called _____ reflexes. Why?

6. Some autonomic reflexes are spinal reflexes; others are integrated in the brain. List examples of each.

7. What part of the brain transforms emotions into somatic sensation and visceral function? List three autonomic reflexes that are linked to one's emotions.

8. How many synapses occur in the simplest autonomic reflexes? Where do the synapses occur?

9. List the three types of sensory receptors that convey information for muscle reflexes.

10. Because of tonic activity in neurons, a resting muscle maintains a low level of tension known as _____.

11. Stretching an intact skeletal muscle causes sensory neurons to (increase/decrease) their rate of firing, causing the muscle to contract, thereby relieving the stretch. Why is this a useful reflex?

12. Match the organ to all correct statements about it.

(a) muscle spindle
(b) Golgi tendon organ
(c) joint capsule mechanoreceptor

1. strictly a sensory receptor
2. has afferent neurons that bring information to the CNS
3. associated with two types of efferent neurons
4. conveys information about the relative positioning of bones
5. innervated by gamma motor neurons
6. eventually synapses with alpha motor neurons that innervate extrafusal muscle fibers.

13. The Golgi tendon organ responds to both _____ and _____, although _____ elicits the stronger response. Activation (increases/decreases) muscle contraction via the _____ neuron.

14. The simplest reflex requires a minimum of how many neurons? How many synapses? Give an example.

15. List and differentiate the three categories of movement. Give an example of each.

LEVEL TWO   Reviewing Concepts

16. What is the purpose of alpha-gamma coactivation? Explain how it occurs.

17. Modulatory neuron M synapses on the axon terminal of neuron P, just before P synapses with the effector organ. If M is an inhibitory neuron, what happens to neurotransmitter release by P? What effect does M's neurotransmitter have on the postsynaptic membrane potential of P?

LEVEL THREE   Problem Solving

18. There are several theories about how presynaptic inhibition works at the cellular level. Use what you have learned about membrane potentials and synaptic transmission to explain how each of the following mechanisms could result in presynaptic inhibition:

(a) Voltage-gated Ca^{2+} channels in axon terminal are inhibited.
(b) Cl^- channels in axon terminal open.
(c) K^+ channels in axon terminal open.

19. Andy is working on improving his golf swing. He must watch the ball, swing the club back, then forward, twist his hips, straighten the left arm, then complete the follow-through, where the club arcs in front of him. What specific parts of the brain are involved in adjusting how hard he hits the ball, keeping all his body parts moving correctly, watching the ball, and then repeating these actions once he has verified that this swing is successful?

20. You are playing the clarinet, reading the music, and fingering the proper notes as they are called for in the music. You are even tapping your toe to keep time. Diagram five different neural pathways involved in performing these actions, from the receptors to the muscles, and from the brain to the effectors.

21. Set up your own scenario involving a task familiar to you, such as driving a car and positioning it properly at the drive-through window, crocheting according to a printed pattern, playing tennis, or keyboarding and reproducing text from a page. Identify the types of movements involved (reflexive, voluntary, rhythmic, or a combination of these types) and sketch the neural pathways involved in completing the given task.

E X P L <MediaLab> E

Introduction

How many times have you heard someone say "I did it without thinking?" In effect, they were telling you that their action was just a reflex response. There are many ways to control the functions of muscles and glands of the body, but a neural reflex is the simplest and the fastest. Speed of response can mean the difference between damaged or undamaged cells, so neural reflexes can serve as a primary protective mechanism. In this MediaLab you will explore how your body responds to threatening situations and how health care providers use your reflex responses to assess the health of your nervous system. After reading the descriptions below, visit the Media-Lab for Chapter 13 in your Companion Website and select the appropriate keywords.

Web Exploration 1

Estimated time for completion = 5 minutes

It's Halloween and you are walking through the Haunted House. You've heard this is the scariest haunted house around. As you turn the corner and enter the dungeon, a skeleton reaches out and grabs your arm. You let out a scream. Your heart rate quickens and you feel the hairs on your arm stand on end. What has just happened to you? How does fear alter your heart rate and what is going on with the hair on your skin?

Draw a reflex are that shows what just happened. Where in the brain is fear processed? Select the keyword **FEAR** from the Website. What are the functions of this part of the brain? Which branch (somatic or autonomic) of the motor output does it control? What will be the target organs for this response? How is it possible for your hair to stand on end when hair is made of proteins that do not contract? Given that the autonomic nervous system is mediating this reflex response, what type of tissue do you expect to find attached to hair follicles?

Select the keyword **SKIN** on the Website. Look carefully at this picture. What is connected to the base of the hair follicles? Could this be used to alter the position of your hair? What purpose is served when hair "stands on end?" Think of what happens when a cat is frightened or cold.

Web Exploration 2

Estimated time for completion = 10 minutes

At your last physical, your physician checked your patellar tendon reflex by tapping just below your knee while you sat quietly on the edge of the table. What was she checking when she did this test? Draw the complete neural circuit that is involved in this pathway.

Select the keyword **PATELLAR REFLEX** from the Website and compare your circuit to the one in the diagram at the bottom of the Web page. What is the black neuron in the spinal cord doing? What would happen if you were worried about falling off the table during this test and were very tense? How would this affect the reflex test? Draw this additional input to the efferent motor neurons. Indicate where the cell bodies of these neurons are found (where the message is originating). Are these neurons causing EPSPs or IPSPs at the spinal motor neuron?

Your physician notices that you are tense, and she asks you to count backwards from 100 by 3's. By the time you get to 88, your patellar reflex is working. Explain what has happened. Draw a picture of your explanation.

14 Cardiovascular Physiology

■ *"Only in the 17th century did the brain displace the heart as the controller of our actions."*
—*Mary A. B. Brazier,* A History of Neurophysiology in the 19th Century, *1988* ■

CHAPTER OUTLINE

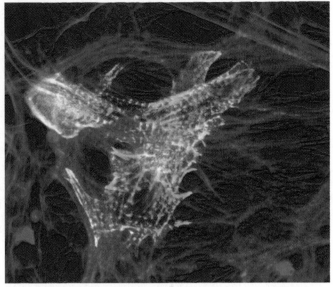

Immunofluorescent staining of cultured cardiac muscle cells. Actin is pink, nebulin is green.

BACKGROUND BASICS

Diffusion (p. 118)
Striated muscle (p. 346)
Desmosomes (p. 55)
Excitation-contraction coupling (p. 356)
Length-tension relationship in muscle (p. 362)
Tetanus in skeletal muscle (p. 363)
Portal system (p. 198)
Muscle contraction (p. 352)
Gap junctions (p. 57)
Catecholamines (p. 190)
Vagus nerve (p. 331)
Isometric contraction (p. 365)

As life evolved, simple one-celled organisms began to band together, first into cooperative colonies and then into multicelled organisms. In most multicellular animals, only the surface layer of cells is in direct contact with the environment. This body structure presents a problem for the transfer of oxygen and nutrients from the environment to the interior cells. Diffusion is severely limited by distance [∞ p. 118], and, in larger animals, oxygen consumption in interior cells exceeds the rate at which oxygen can diffuse from the body surface.

One solution to this problem was the evolutionary development of circulatory systems that move fluid between the body surface and its deepest parts. In simple animals, muscular activity creates fluid flow when the animal moves. More complex animals have muscular pumps called hearts to circulate internal fluid.

In the most efficient circulatory systems, the heart pumps blood through a closed system of vessels. This one-way circuit steers blood flow along a specific route and ensures efficient distribution of gases, nutrients, signal molecules, and wastes. A circulatory system with a heart and blood vessels is known as a **cardiovascular system** [*kardia,* heart + *vasculum,* little vessel].

Although the idea of a closed cardiovascular system that cycles blood in an endless loop seems intuitive to us today, it has not always been so. **Capillaries,** the microscopic vessels in which blood exchanges material with the interstitial fluid, were not discovered until Marcello Malpighi, an Italian anatomist, observed them with a microscope in the middle of the seventeenth century. At that time European medicine was still heavily influenced by the ancient belief that the cardiovascular system distributed both blood and air.

Blood was thought to be made in the liver and distributed throughout the body in the veins. Air went from the lungs to the heart, where it was "digested" and picked up "vital spirits." From the heart, air was distributed to the tissues through vessels called *arteries.* Anomalies such as the fact that a cut artery squirted blood rather than air were ingeniously explained by unseen links between arteries and veins that opened with injury.

According to this model of the circulatory system, the tissues consumed all blood delivered to them, and the liver had to continuously synthesize new blood. It took the calculations of William Harvey (1578–1657), court physician to King Charles I of England, to show that the weight of the blood pumped by the heart in a single hour is greater than the weight of the entire body! Once it became obvious that the liver could not make blood as rapidly as the heart pumped it, Harvey looked for an anatomical route that would allow the blood to recirculate rather than be consumed in the tissues. He

showed that valves in the heart and veins created a one-way flow of blood, and that veins carried blood back to the heart, not out to the limbs. He also showed that blood returning to the right side of the heart from the body had to go to the lungs before it could go to the left side of the heart.

The results of these studies created a furor among Harvey's contemporaries, leading Harvey to say in a huff that no one under the age of 40 could understand his conclusions. Ultimately, Harvey's work became the foundation of modern cardiovascular physiology. Today, we understand the structure of the cardiovascular system at microscopic and molecular levels that Harvey never dreamed existed. Yet some things have not changed. Even now, with our sophisticated technology, we are searching for "spirits" in the blood, although we call them by names such as "hormone" and "cytokine."

▶ OVERVIEW OF THE CARDIOVASCULAR SYSTEM

In the simplest terms, a cardiovascular system is a series of tubes (the blood vessels) filled with fluid (blood) and connected to a pump (the heart). Pressure generated in the heart propels blood through the system continuously. The blood picks up oxygen at the lungs and nutrients in the intestine, delivering them to cells, while simultaneously removing cellular wastes for excretion. In addition, the cardiovascular system plays an important role in cell-to-cell communication and in defending the body against foreign invaders. This chapter focuses on an overview of the cardiovascular system and on the heart as a pump. We will examine the properties of the blood vessels and the homeostatic controls that regulate blood flow and blood pressure in Chapter 15.

The Cardiovascular System Transports Material Throughout the Body

The primary function of the cardiovascular system is the transport of materials between all parts of the body. Substances transported by the cardiovascular system can be divided into nutrients, water, and gases that enter the body from the external environment, materials that move from cell to cell within the body, and wastes that the cells eliminate (Table 14-1).

Oxygen enters the blood in the lungs. Nutrients and water are absorbed across the intestinal epithelium. Once in the body, the circulation distributes these materials. A steady supply of oxygen for the cells is particularly important because many cells deprived of oxygen become irreparably damaged within a short period of time.

For example, about 5–10 seconds after blood flow to the brain is stopped, a person loses consciousness. If oxygen delivery stops for 5–10 minutes, permanent brain damage results. Neurons of the brain have a very high rate of oxygen consumption and are unable to meet their

PROBLEM
Myocardial Infarction

At 9:06 A.M., the blockage that had been silently growing in Walter Parker's left coronary artery made its sinister presence known. The 53-year-old advertising executive had arrived at the Dallas Convention Center feeling fine, but suddenly a dull ache started in the center of his chest and he became nauseated. At first he brushed it off as aftereffects of the convention banquet the night before. However, when it persisted, he made his way to the Aid Station. "I'm not feeling very well," he told the medic. "I think it may be indigestion." The medic, on hearing Walter's symptoms and seeing his pale, sweaty face, immediately thought of a heart attack. "Let's get you over to the hospital and get this checked out."

...continued on page 411

TABLE 14-1   Transport in the Cardiovascular System

Substance Moved	From	To
Materials entering the body		
Oxygen	Lungs	All cells
Nutrients and water	Intestinal tract	All cells
Materials moved from cell to cell		
Wastes	Some cells	Liver for processing
Immune cells, antibodies, clotting proteins	Present in blood continuously	Available for any cell that needs them
Hormones	Endocrine cells	Target cells
Stored nutrients	Liver and adipose tissue	All cells
Materials leaving the body		
Metabolic wastes	All cells	Kidneys
Heat	All cells	Skin
Carbon dioxide	All cells	Lungs

metabolic need for ATP by using anaerobic pathways. Because of the brain's sensitivity to *hypoxia* [*hypo-*, low + *-oxia*, oxygen], homeostatic controls do everything possible to maintain cerebral blood flow, even if it means depriving cells in other tissues of oxygen.

Cell-to-cell communication is a key function of the cardiovascular system. For example, hormones secreted by endocrine glands are carried to their targets. Nutrients such as glucose from the liver or fatty acids released by adipose tissue are transported to metabolically active cells. Finally, the defense team of white blood cells and antibodies patrols the circulation to intercept foreign invaders.

The cardiovascular system also picks up wastes released by cells. The blood transports carbon dioxide and metabolic wastes to the lungs and the kidneys for excretion. Some waste products are transported to the liver for processing before they are excreted in the urine and feces. Heat also circulates through the blood, moving from the body core to the surface, where it dissipates.

The Cardiovascular System Consists of the Heart and Blood Vessels

The cardiovascular system is composed of the heart, the blood vessels (also known as the *vasculature*), and the cells and plasma of the blood. Blood vessels that carry blood away from the heart are called **arteries.** Blood vessels that return blood to the heart are called **veins.**

As blood moves through the cardiovascular system, a system of valves in the heart and veins ensures that blood flows in one direction. Like the turnstiles at an amusement park, the valves keep the blood from reversing its direction of flow. Figure 14-1 ■ is a schematic diagram that shows these components and the route that blood flow follows through the body.

The heart is divided by a central wall, or **septum,** into left and right halves. Each half consists of an **atrium** [*atrium*, central room; plural *atria*], which receives blood returning to the heart from the blood vessels, and a **ventricle** [*ventriculus*, belly], which pumps blood out into the blood vessels. The right side of the heart receives blood from the tissues and sends it to the lungs for oxygenation. The left side of the heart receives the newly oxygenated blood from the lungs and pumps it to tissues throughout the body.

Starting in the right atrium, trace the path taken by blood as it flows through the cardiovascular system. Note that in Figure 14-1 ■, blood in the right side of the heart is colored blue. This is a convention used to show blood from which the tissues have extracted oxygen. Although this blood is often described as "deoxygenated," it is not completely devoid of oxygen. It simply has less oxygen than blood going from the lungs to the tissues. In living people, well-oxygenated blood is bright red; low-oxygen blood becomes darker red. Under some conditions, low-oxygen blood can impart a bluish color to certain areas of the skin, such as around the mouth and under the fingernails. This condition, known as *cyanosis* [*kyanos*, dark blue], is the reason blue is used in drawings to indicate blood with lower oxygen content.

From the right atrium, blood flows into the right ventricle of the heart. From there it is pumped through the **pulmonary arteries** to the lungs, where it is oxygenated. Note the color change from blue to red, indicating higher oxygen content. From the lungs, blood travels back to the left side of the heart through the **pulmonary veins.** The blood vessels that go from the right ventricle to the lungs and back to the left atrium are known collectively as the **pulmonary circulation.**

Blood from the lungs reenters the heart at the **left atrium** and passes into the **left ventricle.** Blood pumped out

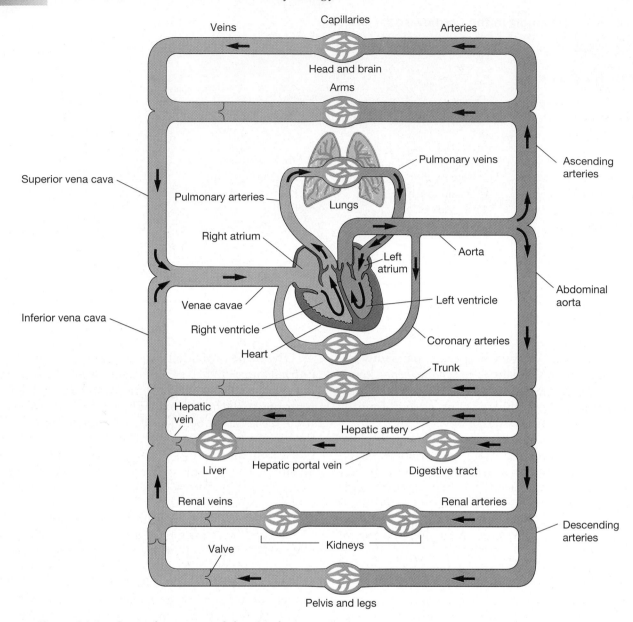

■ Figure 14-1   **General anatomy of the circulatory system**

of the left ventricle enters the large artery known as the **aorta.** The aorta branches into a series of smaller and smaller arteries that finally lead into networks of capillaries.

Capillaries are the tiny thin-walled vessels in which material is exchanged between the blood and the tissues. Notice the color change from red to blue in the capillaries, showing that oxygen has left the blood and diffused into the tissues.

Blood in capillaries flows into the *venous* side of the circulation, moving from small veins into larger and larger veins. The veins from the upper part of the body join together to form the **superior vena cava.** Those from the lower part of the body form the **inferior vena cava.** The two **venae cavae** empty into the right atrium. The blood vessels that go from the left side of the heart to the tissues and back to the right side of the heart are collectively known as the **systemic circulation.**

Return to the aorta and follow its divisions. The first branch represents the *coronary arteries* that nourish the heart muscle itself. Blood from coronary arteries flows into capillaries, then into *coronary veins* that empty directly into the right atrium at the *coronary sinus.*

Ascending branches of the aorta go to the head, arms, and brain. The abdominal aorta supplies blood to the trunk, the legs, and the internal organs such as the kidneys (*renal arteries*), liver (*hepatic artery*), and digestive tract.

Notice two special regions of the circulation. One is the blood supply to the digestive tract and liver. Both regions receive well-oxygenated blood through their own arteries, but, in addition, blood leaving the digestive tract goes directly to the liver by means of the *hepatic portal vein.* The liver is an important site for nutrient processing and plays a major role in the detoxification of foreign substances. Most nutrients absorbed in the intestine are routed directly to the liver, allowing that organ to process the material before it is released into the general circulation.

The two capillary beds of the digestive tract and liver, joined by the hepatic portal vein, are an example of a *portal system*. Another portal system, discussed earlier, is the hypothalamic-hypophyseal portal system that connects the hypothalamus and the anterior pituitary [∞ p. 198]. The third major portal system is in the kidneys, where two capillary beds are connected in series. You will learn more about this special vascular arrangement in Chapter 18.

▶ PRESSURE, VOLUME, FLOW, AND RESISTANCE

If you ask people why blood flows through the cardiovascular system, many of them respond, "So that oxygen and nutrients can get to all parts of the body." Although this is true, it is a teleological answer, one that describes the purpose of blood flow. In physiology, we are also concerned with how blood flows: the mechanisms or forces that create blood flow.

A simple mechanistic answer to "Why does blood flow?" is that liquids and gases flow down a **pressure gradient** (ΔP) from regions of higher pressure to regions of lower pressure. Therefore, blood can flow in the cardiovascular system only if one region develops higher pressure than other regions.

In humans, high pressure is created in the chambers of the heart when it contracts. Blood flows out of the heart (the region of highest pressure) into the closed loop of blood vessels (a region of lower pressure). As blood moves through the system, pressure is lost because of friction between the fluid and the blood vessel walls. Consequently, pressure falls continuously as blood moves farther from the heart (Fig. 14-2 ■). The highest pressure in the vessels of the circulatory system is found in the aorta and systemic arteries as they receive blood from the left ventricle. The lowest pressure is in the venae cavae, just before they empty into the right atrium.

In this section we review the laws of physics that explain the interaction of pressure, volume, flow, and resistance in the cardiovascular system. Many of these principles apply broadly to the flow of all types of liquids and gases, including that of air in the respiratory system. However, in this chapter we will focus on the flow of blood and its relevance to the function of the heart.

The Pressure of Fluid in Motion Decreases over Distance

Pressure in fluids is the force exerted by a fluid on the container holding the fluid. If the fluid is not moving, the pressure is called **hydrostatic pressure** (Fig. 14-3a ■), and force is exerted equally in all directions. For example, a

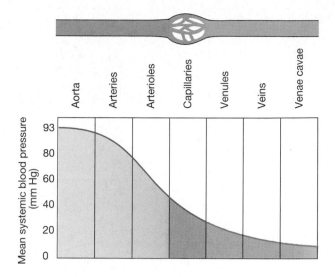

■ **Figure 14-2   Pressure gradient in the blood vessels**   The mean blood pressure of the systemic circulation ranges from a high of 93 mm Hg (millimeters of mercury) in the arteries to a low of a few mm Hg in the venae cavae.

column of fluid in a tube exerts hydrostatic pressure on the floor and sides of the tube. In the heart and blood vessels, pressure is commonly measured in millimeters of mercury (mm Hg, or torr).* One millimeter of mercury pressure is equivalent to the hydrostatic pressure exerted by a 1-mm-high column of mercury on an area 1 cm^2.

In a system in which fluid is flowing, pressure falls over distance as energy is lost because of friction (Fig. 14-3b ■). In addition, the pressure exerted by the moving fluid has two components: a dynamic, flowing component that represents the kinetic energy of the system and a lateral component that represents the hydrostatic pressure (potential energy) exerted on the walls of the system. Pressure within our cardiovascular system is usually called hydrostatic pressure, although it is a system in which fluid is in motion. Some textbooks are beginning to replace the term *hydrostatic pressure* with the term *hydraulic pressure*. Hydraulics is the study of fluid in motion.

Compressing a Fluid Raises its Pressure

If the walls of a fluid-filled container contract, the pressure exerted on the fluid in the container increases. You can demonstrate this principle by filling a balloon with water and squeezing it in your hand. Water is not very compressible, so the pressure applied to the balloon is transmitted throughout the fluid. As you squeeze, higher pressure in the fluid causes parts of the balloon to bulge. If the pressure becomes high enough, the stress on the balloon will cause it to pop. The water volume inside the balloon has not changed, but its pressure has increased.

---

*1 mm Hg = 1 torr. Some physiological literature reports pressures in centimeters of water: 1 cm H_2O = 0.74 mm Hg.

(a)

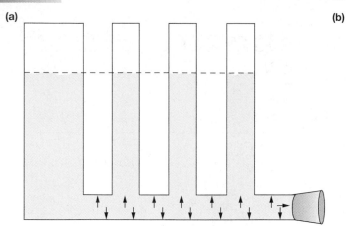

Hydrostatic pressure is the pressure exerted on the walls of the container by the fluid within the container. Hydrostatic pressure is proportional to the height of the water column.

(b)

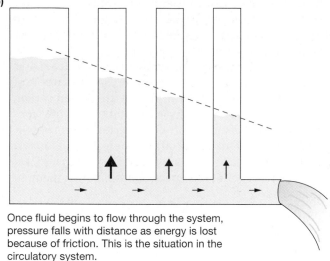

Once fluid begins to flow through the system, pressure falls with distance as energy is lost because of friction. This is the situation in the circulatory system.

■ **Figure 14-3   Hydrostatic pressure**

In the human heart, contraction of the ventricles on blood that fills them is like squeezing a water balloon: the pressure created by the contracting muscle is transferred to the blood. This high-pressure blood then flows out of the ventricle and into the blood vessels, displacing lower-pressure blood already in the vessels. The pressure created within the ventricles is called the **driving pressure** because it is the force that drives blood through the blood vessels.

When the muscular walls of a fluid-filled container relax, the pressure of the fluid inside the container decreases. Thus, when the heart relaxes and expands, pressure in the chambers falls.

Pressure changes can also take place in the blood vessels. If blood vessels dilate, blood pressure inside them falls. If blood vessels constrict, blood pressure increases. Volume changes of the blood vessels and heart are major factors that influence blood pressure within the cardiovascular system.

Blood Flows from an Area of Higher Pressure to One of Lower Pressure

As stated earlier, blood flow through the circulatory system requires a pressure gradient. This pressure gradient, ΔP, is the difference in pressure between two ends of a tube through which fluid flows (Fig. 14-4a ■). Flow through the tube is directly proportional to ($\propto$) the pressure gradient (ΔP):

(1) Flow $\propto \Delta P$ where $\Delta P = P_1 - P_2$

This relationship says that the higher the pressure gradient, the greater the flow of fluid.

A pressure gradient is not the same thing as the absolute pressure in the system. For example, the tube in Figure 14-4b ■ has an absolute pressure of 100 mm Hg at each end. However, because there is no pressure gradi-

ent between the two ends of the tube, there is no flow through the tube.

On the other hand, two identical tubes can have very different absolute pressures but the same flow (Fig. 14-4c ■). The first tube has a hydrostatic pressure of 100 mm Hg at one end and 75 mm Hg at the other end, with a pressure gradient between the ends of the tube equal to

(a)

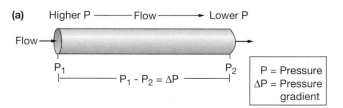

(b) No pressure gradient, so no flow

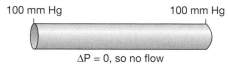

(c) Flow depends on ΔP, not absolute P

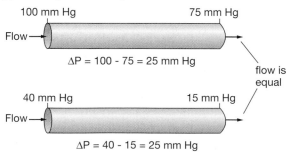

■ **Figure 14-4   Fluid flow through a tube**   (a) Fluid flows if there is a pressure gradient between two ends of the tube. (b) If there is no pressure gradient, there is no fluid flow. (c) The flow rate depends on the pressure gradient, not the absolute pressures in the tube. In the example shown, both tubes have the same pressure gradient and therefore the same flow.

25 mm Hg. The identical bottom tube has hydrostatic pressure of 40 mm Hg at one end and 15 mm Hg at the other end. This tube has lower absolute pressure all along its length, but the same pressure gradient as the top tube, 25 mm Hg. Because the pressure difference in the two tubes is identical, the fluid flow through the tubes will be the same.

Resistance Opposes Flow

In an ideal system, a substance in motion would remain in motion. However, no system is ideal because all movement creates friction. Just as a ball rolled across the ground loses energy to friction, blood flowing through blood vessels encounters friction from the walls of the blood vessels and from cells within the blood rubbing against each other as they flow.

The tendency of the cardiovascular system to oppose blood flow is called its **resistance** to flow. Resistance (R) is a term that most of us understand already from everyday life. We speak of people being resistant to change or taking the path of least resistance. This concept translates well to the circulatory system because blood flow will also take the path of least resistance. An increase in the resistance of a blood vessel means a decrease in flow through that vessel. We can express that relationship as

(2) Flow $\propto 1/R$

Flow is inversely proportional to resistance. If resistance increases, flow decreases. If resistance decreases, flow increases.

What parameters determine resistance? For fluid flowing through a tube, resistance is influenced by three components: the length of the tube (L), the radius of the tube (r), and the viscosity (thickness) of the fluid flowing through the tube (η, eta). The following equation, derived by the French physician Jean Leonard Marie Poiseuille, shows the relationship between these factors:

(3) $R = 8L\eta/\pi r^4$

Because the value of $8/\pi$ is a constant, the relationship can be rewritten as

(4) $R \propto L\eta/r^4$

This expression says that resistance increases as the length of the tube and the viscosity of the fluid increase, but it decreases as the radius increases.

To remember these relationships, think of drinking through a straw. You do not need to suck as hard on a short straw as on a long one (resistance increases with length). Drinking water through a straw is easier than drinking a thick milkshake (resistance increases with viscosity). And drinking the milkshake through a big fat straw is much easier than through a skinny cocktail straw (resistance increases as radius decreases).

How significant are length, viscosity, and radius to blood flow in a normal individual? The length of the sys-

temic circulation is determined by the anatomy of the system and is essentially constant. The viscosity of blood is determined by the ratio of red blood cells to plasma and by how much protein is in the plasma. Normally, viscosity is constant, and small changes in either length or viscosity have little effect on resistance. This leaves changes in the radius of the blood vessels as the main contributor to variable resistance in the systemic circulation.

Let's return to the example of the straw and the milkshake to illustrate how changes in radius affect resistance. If we assume that the length of the straw and the viscosity of the milkshake do not change, this system is similar to the circulatory system—the radius of the tube has the greatest effect on resistance. If we consider only resistance (R) and radius (r) from equation 4, the relationship between resistance and radius can be expressed as follows:

(5) $R \propto 1/r^4$

If the skinny straw has a radius of 1, its resistance is proportional to $1/1^4$, or 1. If the larger straw has a radius of 2, its resistance is $1/2^4$, or 1/16th, that of the skinny straw. Since flow is inversely proportional to resistance, flow increases 16-fold when the radius doubles (Fig. 14-5 ■).

As you can see from this example, a small change in the radius of a tube will have a large effect on the flow of a liquid through that tube. A small change in the radius of a blood vessel will have a large effect on the resistance to blood flow. A decrease in blood vessel diameter is known as **vasoconstriction** [*vas*, a vessel or duct]. An increase in blood vessel diameter is called **vasodilation.** Vasoconstriction will decrease blood flow through a vessel. Vasodilation will increase blood flow through a vessel.

In summary, by combining equations 1 and 2 above, we get the following equation:

(6) Flow $\propto \Delta P/R$

The flow of blood within the circulatory system is directly proportional to the pressure gradient in the system and inversely proportional to the resistance of the system to flow. If the driving pressure (ΔP) remains constant, then flow will vary inversely with resistance.

✔ The two identical tubes below have the pressures shown at each end. Which tube has the greater flow? Defend your choice.

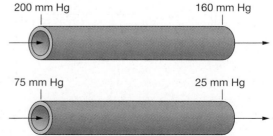

200 mm Hg   160 mm Hg

75 mm Hg   25 mm Hg

✔ All four tubes below have the same driving pressure. Which tube has the greatest flow? Which has the least flow? Defend your choices.

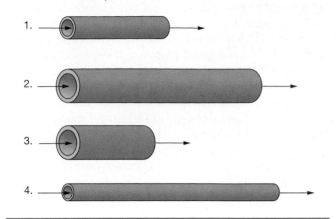

1.

2.

3.

4.

Velocity of Flow Depends on the Flow Rate and the Cross-Sectional Area

The word *flow* is often used imprecisely in cardiovascular physiology, leading to confusion. Flow usually means **flow rate,** the volume of blood that passes one point in the system per unit time. In the circulation, flow is expressed in liters per minute (L/min) or milliliters per minute (mL/min). For instance, blood flow through the aorta in a 70-kg man at rest is about 5 L/min.

Flow rate should not be confused with **velocity of flow,** the distance a fixed volume of blood will travel in a given period of time. Velocity of flow is a measure of *how fast* blood flows past a point. Flow rate measures *how much* (volume) blood flows past a point in a given period of time. For example, look through the open door at the hallway outside your classroom. The number of people passing the door in one minute is the flow rate of people through the hallway. How quickly those people are walking past the door is their velocity.

In a tube of fixed size, velocity of flow is directly related to flow rate. In a tube of variable diameter, if flow

rate is constant, blood velocity will vary inversely with the diameter. In other words, velocity will be faster through narrow sections and slower in wider sections. The relationship between flow rate (Q), cross-sectional area of the tube (A), and velocity of flow (v) is expressed by the following equation:

(7) Velocity = flow rate/cross-section area

Figure 14-6 ■ shows how the velocity of flow varies as the cross-sectional area of the tube changes. The vessel in the figure has two widths: narrow, with a small cross-sectional area of 1 cm², and wide, with a large cross-sectional area of 12 cm². The flow rate is identical in both parts of the vessel: 12 cm³ per minute.* This flow rate means that in one minute, 12 cm³ of fluid flows past point X in the narrow section, and 12 cm³ flows past point Y in the wide section.

But *how fast* does the fluid need to flow to accomplish that rate? According to equation 7, the velocity of flow at point X is 12 cm/min, but at point Y it is only 1 cm/min. Thus, fluid flows more rapidly through narrow sections of a tube than through wide sections.

To see this principle in action, watch a leaf as it floats down a stream or gutter. Where the stream is narrow, the leaf moves rapidly, carried by the fast velocity of the water. In sections where the stream widens into a pool, the velocity of the water drops and the leaf meanders more slowly.

Table 14-2 summarizes the factors that influence blood flow and velocity of flow. Now let's see how the heart creates the pressure gradients that create blood flow.

✔ Two canals in Amsterdam are identical in size, but the water flows faster through one than through the other. Which canal has the highest flow rate?

---

*1 cm³ = 1 mL

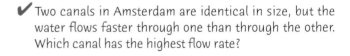

Resistance $\propto \dfrac{1}{radius^4}$	
Tube A	Tube B
$R \propto \dfrac{1}{1^4}$	$R \propto \dfrac{1}{2^4}$
$R \propto 1$	$R \propto \dfrac{1}{16}$

Flow $\propto \dfrac{1}{resistance}$	
Tube A	Tube B
Flow $\propto \dfrac{1}{1}$	Flow $\propto \dfrac{1}{\frac{1}{16}}$
Flow $\propto 1$	Flow $\propto 16$

Radius of B = 2

Radius of A = 1

Volume of A = 1

Volume of B = 16

■ **Figure 14-5 The role of radius in determining resistance to flow** As the radius (r) of a tube increases, the resistance (R) of the tube to fluid flow decreases dramatically because $R \propto 1/r^4$.

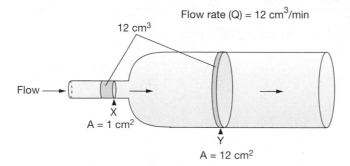

Flow rate (Q) = 12 cm³/min

12 cm³

Flow →

X
A = 1 cm²

Y
A = 12 cm²

Velocity (v) = $\dfrac{\text{Flow rate (Q)}}{\text{Cross-sectional area (A)}}$	
At point X	**At point Y**
$v = \dfrac{12 \text{ cm}^3/\text{min}}{1 \text{ cm}^2}$	$v = \dfrac{12 \text{ cm}^3/\text{min}}{12 \text{ cm}^2}$
v = 12 cm/min	v = 1 cm/min

■ **Figure 14-6   Flow rate versus velocity of flow** This figure illustrates the fact that the more narrow the vessel, the faster the velocity of flow.

TABLE 14-2   Rules for the Cardiovascular System

1. Blood flows if a pressure gradient is present. Blood flows from areas of higher pressure to areas of lower pressure. (Fig. 14-4 ■)
2. Blood flow is opposed by the resistance of the system. The three factors that affect resistance are length of the system, radius of the blood vessels, and viscosity of blood. (Fig. 14-5 ■)
3. Flow rate is the volume of blood that passes one point in the system per unit time. It is usually expressed in liters or milliliters per minute (L/min or mL/min). Flow rate is determined by the pressure gradient and by the resistance of the system.
4. Velocity of flow is how far a volume of blood will travel in a given period of time. It is usually expressed as centimeters per minute (cm/min) or millimeters per second (mm/sec). The primary determinant of velocity of flow (when flow rate is constant) is the total cross-sectional area of the vessel. (Fig. 14-6 ■)

▶ CARDIAC MUSCLE AND THE HEART

To ancient civilizations, the heart was more than a pump—it was "the seat of the mind." When ancient Egyptians mummified their dead, they removed most of the viscera, but left the heart in place so that the gods could weigh it as an indicator of the owner's worthiness. Aristotle characterized the heart as the most important organ of the body, as well as "the seat of intelligence." We can still find evidence of these ancient beliefs in modern expressions such as "heartfelt emotions." The link between the heart and mind is one that is still explored today.

The heart is the workhorse of the body, a muscle that contracts continually, resting only in the milliseconds pause between beats. By one estimate, the heart performs work equivalent to lifting a 5-pound weight up one foot, once a minute. The energy demands of this work require a continuous supply of nutrients and oxygen to the heart muscle.

The Heart Has Four Chambers

The heart is a muscular organ, about the size of a fist, that lies in the center of the *thoracic cavity* (see Anatomy Summary, Fig. 14-7a, b, d ■). The pointed *apex* of the heart angles down to the left side of the body, while the broader *base* lies just behind the breastbone, or *sternum*. Because we usually associate the word *base* with the bottom, remember that the base of a cone is the broad end and the apex is the pointed end. The heart can be thought of as an inverted cone with apex down and base up. Within the thoracic cavity, the heart lies on the ventral side, sandwiched between the two lungs, with its apex resting on the diaphragm (Fig. 14-7b ■).

The heart is encased in a tough membranous sac, the **pericardium** [*peri*, around + *kardia*, heart] (Fig. 14-7d, e ■). A thin layer of clear pericardial fluid inside the pericardium lubricates the external surface of the heart as it beats within the sac. Inflammation of the pericardium (*pericarditis*) may reduce this lubrication to the point that the heart rubs against the pericardium, creating a sound known as a *friction rub*.

...continued from page 404

The blockage in Walter's coronary artery had restricted blood flow to his heart muscle, and its cells were beginning to die from lack of oxygen. When someone has a heart attack, medical intervention is critical to prevent additional damage and possibly save the patient's life. While waiting for the ambulance, the medic gave Walter oxygen, hooked him to a heart monitor, and started an intravenous (IV) injection of normal (isotonic) saline. With an intravenous injection line in place, other drugs could be given rapidly if Walter's condition should suddenly get worse.

Question 1:   Why did the medic give Walter oxygen?

Question 2:   What effect would the injection of isotonic saline have on Walter's extracellular fluid volume? On his intracellular fluid volume? On his total body osmolarity? [Hint: ∞ p. 137]

ANATOMY SUMMARY The Cardiovascular System

■ Figure 14-7

(a)

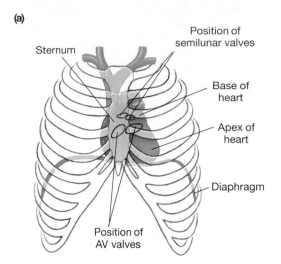

Sternum

Position of semilunar valves

Base of heart

Apex of heart

Diaphragm

Position of AV valves

(b)

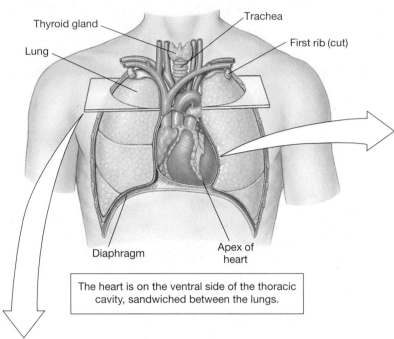

Thyroid gland

Lung

Trachea

First rib (cut)

Diaphragm

Apex of heart

The heart is on the ventral side of the thoracic cavity, sandwiched between the lungs.

(c)

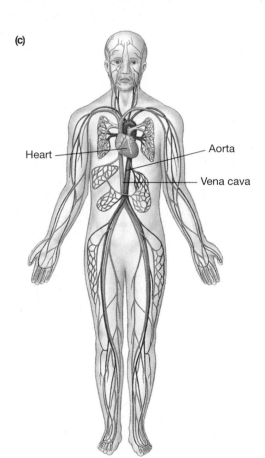

Heart

Aorta

Vena cava

Vessels that carry well-oxygenated blood are red; those with less well-oxygenated blood are blue.

(d) Superior view of transverse plane in (b)

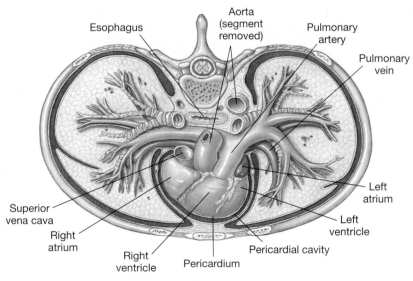

Esophagus

Aorta (segment removed)

Pulmonary artery

Pulmonary vein

Superior vena cava

Right atrium

Right ventricle

Pericardium

Left atrium

Left ventricle

Pericardial cavity

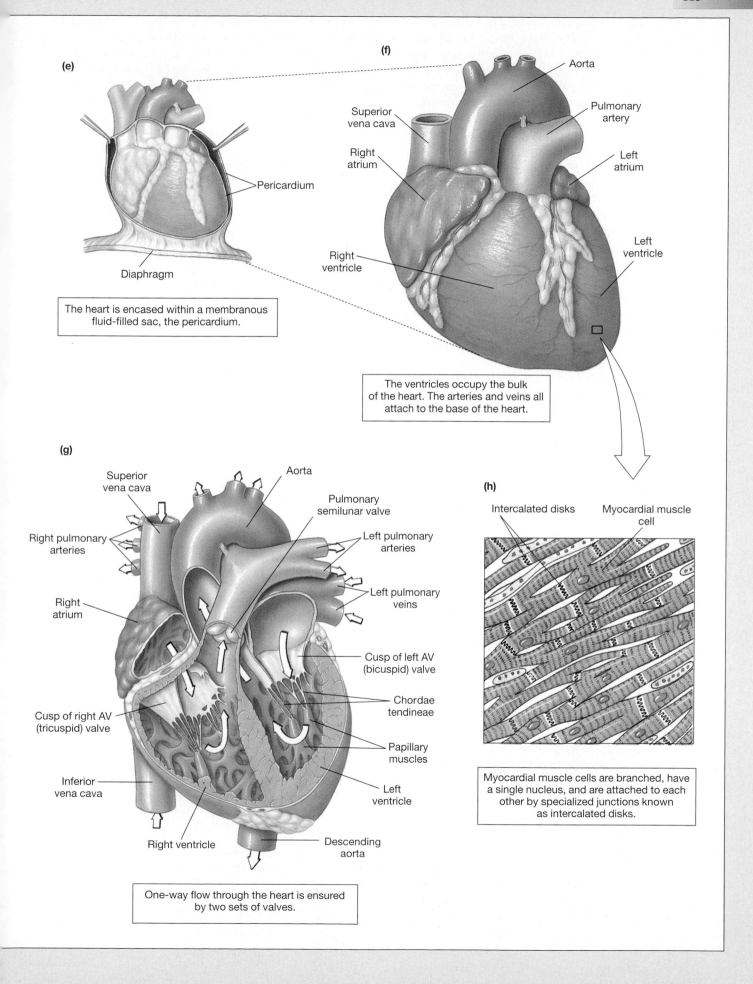

(e)

Pericardium

Diaphragm

The heart is encased within a membranous fluid-filled sac, the pericardium.

(f)

Aorta

Superior vena cava

Pulmonary artery

Right atrium

Left atrium

Right ventricle

Left ventricle

The ventricles occupy the bulk of the heart. The arteries and veins all attach to the base of the heart.

(g)

Superior vena cava

Aorta

Pulmonary semilunar valve

Right pulmonary arteries

Left pulmonary arteries

Left pulmonary veins

Right atrium

Cusp of left AV (bicuspid) valve

Cusp of right AV (tricuspid) valve

Chordae tendineae

Papillary muscles

Inferior vena cava

Left ventricle

Right ventricle

Descending aorta

One-way flow through the heart is ensured by two sets of valves.

(h)

Intercalated disks

Myocardial muscle cell

Myocardial muscle cells are branched, have a single nucleus, and are attached to each other by specialized junctions known as intercalated disks.

The heart itself is composed mostly of cardiac muscle, or **myocardium** [*myo,* muscle + *kardia,* heart], covered by thin outer and inner layers of epithelium and connective tissue. If seen from the outside, the bulk of the heart is the thick muscular walls of the ventricles, the two lower chambers (Fig. 14-7f ■). The thinner-walled atria lie above the ventricles.

The major blood vessels all emerge from the top of the heart. The aorta and pulmonary trunk (artery) take blood from the heart to the tissues and lungs, respectively. The venae cavae and pulmonary veins return blood to the heart (see Table 14-3). When the heart is viewed from the front (anterior view), as in Figure 14-7f ■, the pulmonary veins are hidden behind the other blood vessels. Running across the surface of the ventricles are three grooves containing the **coronary arteries** and **coronary veins** that supply blood to the heart muscle.

The relationship between the atria and ventricles can be seen in a cross-sectional view of the heart (Fig. 14-7g ■). Notice that in the figure the *right* side of the heart is on the *left* side of the page; that is, the heart is labeled as if you were viewing the heart of a person facing you. The left and right sides of the heart are separated from each other by the interventricular septum, so that blood on one side does not mix with blood on the other side. Although blood flow in the left and right heart is separated, the two sides of the heart contract in a coordinated fashion. First the atria contract, then the ventricles contract.

Blood flows from veins into the atria, thin-walled upper chambers that serve as a reception area. From there, blood passes through one-way valves into the ventricles, the pumping chambers. Blood leaves the heart via the pulmonary trunk from the right ventricle and via the aorta from the left ventricle. A second set of valves guards the exits of the ventricles so that blood cannot flow back into the heart once it has been ejected.

Notice that blood enters the ventricles at the top of the chamber but also leaves at the top (Fig. 14-7g ■). During development, the tubular embryonic heart twists back on itself. This puts the arteries (through which blood leaves) close to the top of the ventricles (Fig. 14-8 ■). Functionally, this means that the ventricles must contract from the bottom up so that blood is squeezed out of the top.

The heart contains four fibrous connective tissue rings that surround the openings to the major arteries and the openings between the chambers (Fig. 14-7a ■).

TABLE 14-3 The Heart and Major Blood Vessels

Blue type indicates structures containing blood with lower oxygen content; red type indicates well-oxygenated blood.

	Receives Blood from	*Sends Blood to*
Heart		
Right atrium	Venae cavae	Right ventricle
Right ventricle	Right atrium	Lungs
Left atrium	Pulmonary veins	Left ventricle
Left ventricle	Left atrium	Body except for lungs
Vessels		
Venae cavae	Systemic veins	Right atrium
Pulmonary trunk (artery)	Right ventricle	Lungs
Pulmonary vein	Veins of the lungs	Left atrium
Aorta	Left ventricle	Systemic arteries

(a) Age: embryo, day 25

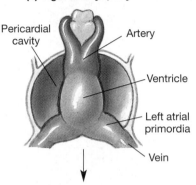

Pericardial cavity — Artery — Ventricle — Left atrial primordia — Vein

(b) Age: embryo, four weeks

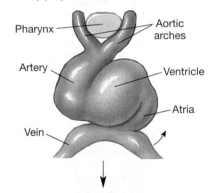

Pharynx — Aortic arches — Artery — Ventricle — Atria — Vein

(c) Age: one year (arteries not shown)

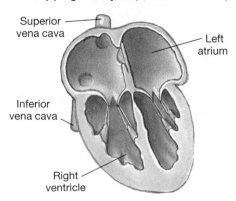

Superior vena cava — Left atrium — Inferior vena cava — Right ventricle

■ **Figure 14-8   Embryological development of the heart** (a) In the early embryo, the heart is a single tube. (b) By four weeks of development, the atria and ventricles can be distinguished. Eventually, the heart begins to twist so that the atria move on top of the ventricles. (c) By birth, veins and arteries both emerge from the base of the heart.

These rings form both the origin and insertion for the cardiac muscle, an arrangement that pulls the apex and base of the heart together when the ventricles contract.

In addition, the fibrous connective tissue acts as an insulator, blocking most transmission of electrical signals between the atria and the ventricles. This arrangement ensures that the electrical signal can be directed through special cells to the apex of the heart for the bottom-to-top contraction.

One-Way Flow in the Heart Is Ensured by the Heart Valves As the arrows in Figure 14-7g ■ indicate, blood flows through the heart in one direction. One-way flow is ensured by the heart valves, one set between the atria and ventricles and the second set between the ventricles and the arteries. Although the two sets of valves are very different in structure, they serve the same function, preventing the backward flow of blood.

The passageways between the atria and ventricles are guarded by the **atrioventricular, or AV, valves** (Fig. 14-7g ■). These valves are thin flaps of tissue attached at their base to a ring of connective tissue. The flaps are slightly thickened at the edge and are attached on the ventricular side to collagenous cords, the **chordae tendineae.** The opposite ends of the cords are tethered to moundlike extensions of ventricular muscle known as the **papillary muscles** [*papilla*, nipple]. Most of the chordae attach to the edges of the valve flaps, but some attach to the middle and the base.

When the ventricle contracts, blood pushes against the bottom side of the AV valve and forces it upward into a closed position (Fig. 14-9b ■). The tendons prevent the valve from being pushed back into the atrium, just as the struts on an umbrella keep the umbrella from turning inside out in a high wind. Occasionally, the tendons fail and the valve is pushed back into the atrium during ventricular contraction, an abnormal condition known as *prolapse.*

The AV valves are not identical. The valve that separates the right atrium and right ventricle has three flaps and is called the **tricuspid valve** [*cuspis*, point] (Fig. 14-9a ■). The valve between the left atrium and left ventricle has only two flaps and is called the **bicuspid valve.** The bicuspid is also called the **mitral valve** because of its resemblance to the tall headdress, known as a miter, worn by popes and bishops. You can match valves to the proper side of the heart by remembering that the right side has the valve with the "r"—the tricuspid.

The second set of valves lies at the openings from the ventricles into the arteries (aorta and pulmonary trunk). These two valves prevent blood pumped out of the heart from returning to it (Fig. 14-9d ■). They are called the **semilunar valves** because of their crescent-moon shape in cross section. They are referred to as the **aortic valve** or **pulmonary valve** according to their location. Both semilunar valves have three cuplike leaflets that fill with blood and snap closed when backward pressure is exerted on them (Fig. 14-9c ■). Because of their shape, the semilunar valves do not need connective tendons as the AV valves do.

✔ What prevents electrical signals from passing through the connective tissue within the heart?

✔ Trace a drop of blood from the superior vena cava to the aorta, naming all structures the drop of blood encounters.

Cardiac Muscle Cells Contract Without Nervous Stimulation

The bulk of the heart is composed of cardiac muscle cells, or myocardium. Most cardiac muscle is contractile, but about 1% of the myocardial cells are specialized to generate action potentials spontaneously. These cells are responsible for a unique property of the heart: its ability to contract without any outside signal. For example, Spanish explorers in the New World wrote of witnessing human sacrifices in which hearts that had been torn from the chests of living victims and held aloft continued to beat for minutes. The heart can contract without a connection to other parts of the body because the signal for contraction is *myogenic,* originating within the heart muscle itself.

The signal for myocardial contraction comes not from the nervous system but from specialized myocardial cells known as **autorhythmic cells.** The autorhythmic cells are also called **pacemakers** because they set the rate of the heartbeat. Myocardial autorhythmic cells are anatomically distinct from contractile cells: they are smaller and contain few contractile fibers. Because they do not have organized sarcomeres, the autorhythmic cells do not contribute to the contractile force of the heart.

Most myocardial cells in the heart are typical striated muscle, with contractile fibers organized into sarcomeres [∞ p. 351]. Cardiac muscle fibers are much smaller than skeletal muscle fibers and have a single nucleus. About one-third of the cell volume of a cardiac contractile fiber is occupied by mitochondria, a reflection of the high energy demand of these cells.

By one estimate, cardiac muscle extracts 70%–80% of the oxygen delivered to it by the blood, more than twice the amount removed by other cells in the body. This high oxygen demand does not leave much oxygen in the coronary blood to supply periods of increased activity. The only way to get more oxygen to the heart muscle is to increase its blood flow. Because the heart muscle depends on adequate blood flow, narrowing of a coronary vessel by a clot or fatty deposit that reduces myocardial blood flow can damage the myocardial cells.

Individual cardiac muscle cells branch and join neighboring cells end-to-end to create a complex network (Fig. 14-10b ■). The cell junctions are specialized regions known as **intercalated disks.** They consist of interdigitated membranes tightly linked by *desmosomes* that tie adjacent cells together [∞ p. 55]. These strong connections allow force created in one cell to be transferred to the adjacent cell.

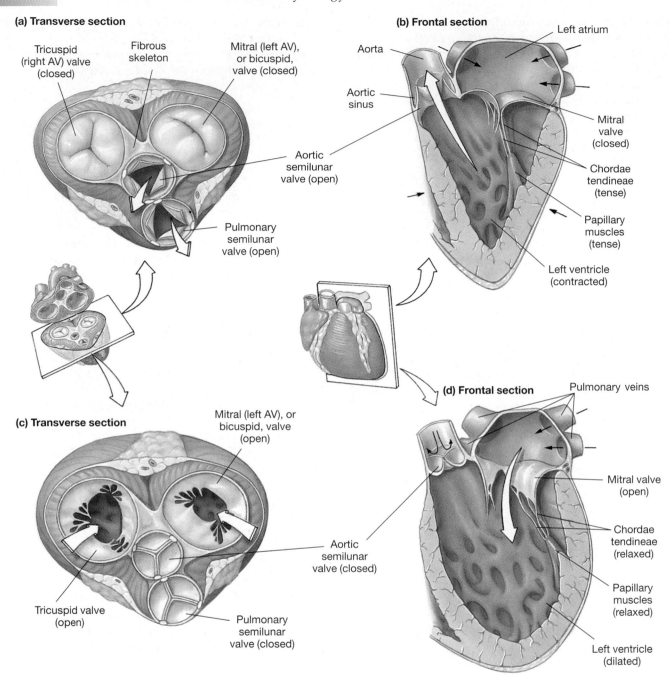

(a) Transverse section

Tricuspid (right AV) valve (closed)

Fibrous skeleton

Mitral (left AV), or bicuspid, valve (closed)

Aortic semilunar valve (open)

Pulmonary semilunar valve (open)

(b) Frontal section

Left atrium

Aorta

Aortic sinus

Mitral valve (closed)

Chordae tendineae (tense)

Papillary muscles (tense)

Left ventricle (contracted)

(c) Transverse section

Mitral (left AV), or bicuspid, valve (open)

Tricuspid valve (open)

Pulmonary semilunar valve (closed)

Aortic semilunar valve (closed)

(d) Frontal section

Pulmonary veins

Mitral valve (open)

Chordae tendineae (relaxed)

Papillary muscles (relaxed)

Left ventricle (dilated)

■ **Figure 14-9   Heart valves**   The atrioventricular (AV), or cuspid, valves separate the atria from the ventricles. During ventricular contraction, these valves remain closed to prevent blood flow backward into the atria. The semilunar valves prevent blood that has entered the arteries from flowing back into the ventricles during ventricular relaxation. Views (a) and (c) show the valves as viewed from the atria and from inside the arteries.

In addition, intercalated disks contain *gap junctions,* linked transmembrane proteins that allow direct movement of ions from the cytoplasm of one cell to the cytoplasm of the next cell [∞ p. 57]. Gap junctions electrically couple the heart muscle so that waves of depolarization spread rapidly from cell to cell, allowing them to contract almost simultaneously.

Although cardiac muscle is a striated muscle, it differs in several ways from skeletal muscle. The t-tubules of myocardial cells are larger, and they branch inside the myocardial cell. The sarcoplasmic reticulum is smaller than that of skeletal muscle, reflecting the fact that car-

diac muscle depends in part on extracellular Ca^{2+} to initiate contraction.

Excitation-Contraction Coupling in Cardiac Muscle Is Similar to Skeletal Muscle

Contraction in cardiac muscle takes place by the same type of sliding filament movement that occurs in skeletal muscle [∞ p. 352]. As in skeletal muscle, an action potential in the muscle cell is the signal to initiate contraction. But there the similarity ends.

(a)

(b)

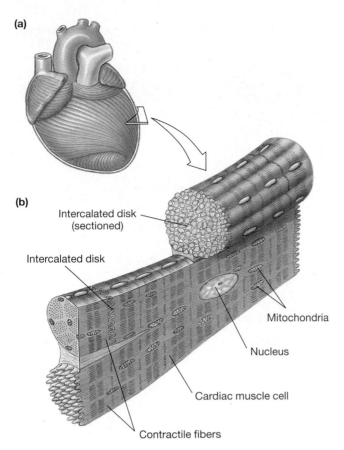

Intercalated disk
(sectioned)

Intercalated disk

Mitochondria

Nucleus

Cardiac muscle cell

Contractile fibers

■ **Figure 14-10   Cardiac muscle**   (a) The spiral arrangement of ventricular muscle allows ventricular contraction to squeeze the blood upward from the apex of the heart. (b) Intercalated disks contain desmosomes that transfer force from cell to cell and gap junctions that allow electrical signals to pass rapidly from cell to cell.

In cardiac muscle, the action potential itself does not release Ca^{2+} from the sarcoplasmic reticulum. Instead, the depolarization opens voltage-gated Ca^{2+} channels in the cell membrane and allows Ca^{2+} to enter the cell from the extracellular fluid, moving into the cell down

...continued from page 411

When Walter arrived at the University of Texas Southwestern Medical Center emergency room, one of the first tasks was to determine whether he had actually had a heart attack. Walter's vital signs (pulse and breathing rates, blood pressure, temperature) were taken. A resident physician drew blood for enzyme assays to determine the level of cardiac enzymes in Walter's blood. When heart muscle cells die, they release various enzymes that serve as markers of a heart attack. In addition, a second tube of blood was sent for an assay of its troponin I level. Troponin I is a good indicator of heart damage following a heart attack.

Question 3:   *Some of the enzymes that are used as markers for heart attacks are also found in related forms in skeletal muscle. What are related forms of an enzyme called?* [Hint: ∞ p. 81]

Question 4:   *What is troponin, and why would elevated blood levels of troponin indicate heart damage?* [Hint: ∞ p. 354]

its electrochemical gradient (Fig. 14-11 ■). Through a process known as **calcium-induced calcium release,** the influx of Ca^{2+} triggers the release of stored Ca^{2+} from the sarcoplasmic reticulum.

The Ca^{2+} released from the sarcoplasmic reticulum provides about 90% of the Ca^{2+} needed for muscle contraction. In a process similar to that in skeletal muscle, Ca^{2+} diffuses through the cytosol to the contractile elements, where the ions bind to troponin and initiate the cycle of crossbridge formation and movement.

Relaxation in cardiac muscle occurs when Ca^{2+} unbinds from troponin. Calcium ions are transported back into the sarcoplasmic reticulum with the help of a *Ca^{2+}-ATPase.* They can also be removed from the cell in exchange for sodium ions. The latter process uses a *Na^+-Ca^{2+} indirect active transport protein.* Ca^{2+} moves out of the cell against its electrochemical gradient in exchange for Na^+ entering the cell down its electrochemical gradient. The Na^+ that enters the cell during this transfer is removed by the Na^+-K^+-ATPase.

✔ If a myocardial contractile cell is placed in interstitial fluid and depolarized, the cell will contract. If Ca^{2+} is removed from the fluid surrounding the myocardial cell, and the cell is depolarized, it will not contract. If the experiment is repeated with a skeletal muscle fiber, the skeletal muscle will contract when depolarized, whether Ca^{2+} is present or not. What conclusion can you draw from the results of this experiment?

Cardiac Muscle Contraction Can Be Graded

A key property of cardiac muscle cells is the ability of a single muscle fiber to execute *graded contractions,* in which the fiber varies the amount of force it generates. (Recall that in skeletal muscle, contraction in a single

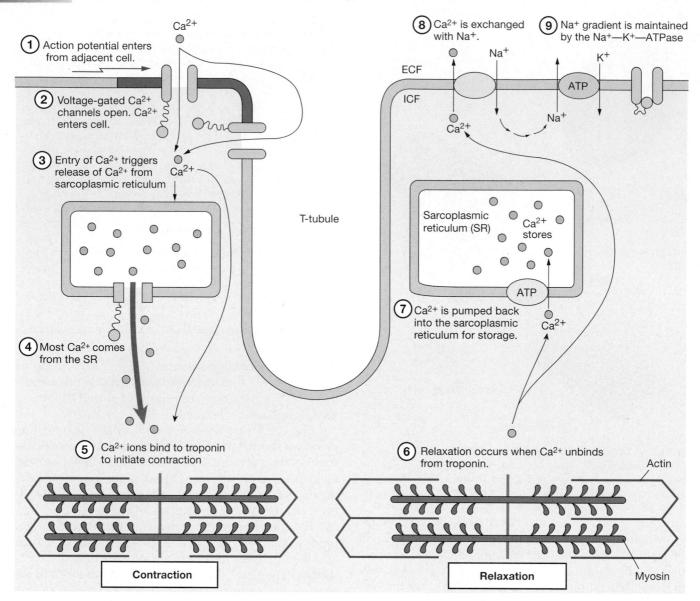

Figure 14-11   Role of calcium in cardiac muscle contraction

fiber is all-or-none at any given fiber length.) The force generated by cardiac muscle is proportional to the number of crossbridges that are active.

If cytosolic Ca^{2+} concentrations are low, some crossbridges will not be activated and force will be low. If additional Ca^{2+} enters the cell from the extracellular fluid, more Ca^{2+} is released from the sarcoplasmic reticulum. This additional Ca^{2+} release allows greater crossbridge formation and cycling, creating additional force.

The catecholamines epinephrine and norepinephrine are regulatory molecules that affect the amount of Ca^{2+} available for cardiac muscle contraction [∞ p. 352]. As you may recall, epinephrine is the "fight-or-flight" hormone released by the adrenal medulla. Norepinephrine is the neurotransmitter released by sympathetic neurons. Both molecules modulate contraction by binding to and activating β_1 adrenergic receptors [∞ p. 334] on the membrane of the cardiac muscle cell (Fig. 14-12 ■).

Once activated, β_1 receptors use a cyclic AMP second messenger system to phosphorylate specific intracellular proteins [∞ p. 161]. Phosphorylation of the voltage-gated Ca^{2+} channels increases the probability that they will open. More open channels means more Ca^{2+} enters the cell.

In addition, phosphorylation of a regulatory protein known as **phospholamban** enhances Ca^{2+}-ATPase activity in the sarcoplasmic reticulum. This ATPase concentrates Ca^{2+} in the sarcoplasmic reticulum, making more Ca^{2+} available for calcium-induced calcium release. Because the force of contraction increases as the number of active crossbridges increases, the net result of catecholamine stimulation is a stronger contraction.

Catecholamines shorten the duration of cardiac contraction in addition to increasing force. The enhanced Ca^{2+}-ATPase removes Ca^{2+} from the cytosol more rapidly. This in turn shortens the time that Ca^{2+} is bound to

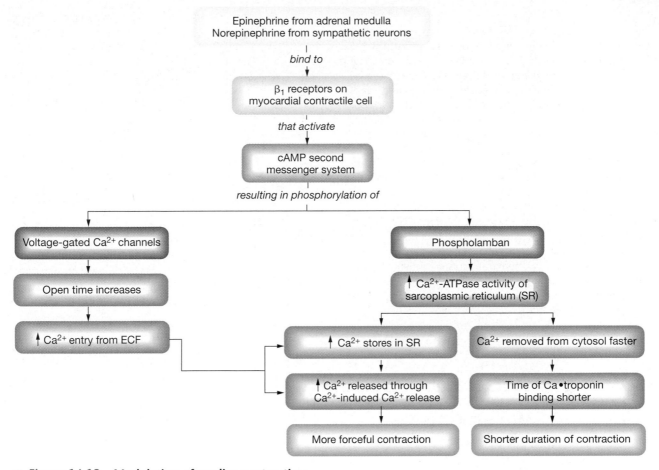

■ **Figure 14-12   Modulation of cardiac contraction**

troponin and the active time of the myosin crossbridges. The muscle twitch is therefore briefer.

When Cardiac Muscle Is Stretched, It Contracts More Forcefully

Another factor that affects the force of contraction in cardiac muscle is the sarcomere length at the beginning of contraction. For both cardiac and skeletal muscle, the tension generated is directly proportional to the initial length of the muscle fiber [∞ p. 362]. As muscle fiber length and sarcomere length increase, tension increases, up to a maximum (Fig. 14-13 ■).

For years, we believed that the length-tension relationship in cardiac muscle was strictly a function of the overlap between thick and thin filaments, as in skeletal muscle. But new evidence suggests that stretching the myocardial cell allows more Ca^{2+} to enter, contributing to a more forceful contraction, as discussed in the previous section. Thus, although it is traditional to separate changes in force due to fiber length from changes in force created by catecholamines, there is some overlap between the mechanisms underlying the two.

In the intact heart, stretch on the individual fibers depends on how much blood is in the chambers of the heart. The relationship between force and ventricular volume is an important property of the intact heart and is discussed in detail later in this chapter.

✔ A drug that blocks all Ca^{2+} channels in the myocardial cell membrane is placed in the solution around the cell. What will happen to the force of contraction in that cell?

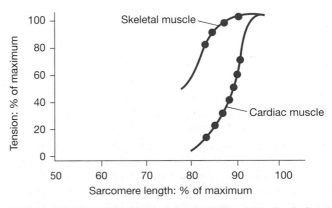

■ **Figure 14-13   Length-tension relationships in skeletal and cardiac muscle**   The lines represent the range of sarcomere length within which the muscles can operate. The dots indicate the normal range of operation.

Foxglove for a Failing Heart   Another mechanism for increasing Ca^{2+} in the cytosol is preventing Ca^{2+} removal from the cell. This mechanism is not active under normal physiological conditions. However, it can be triggered by the administration of cardiac glycosides, a class of molecules first discovered in the plant *Digitalis purpurea* (purple foxglove). The cardiac glycosides include digitoxin and the related compound ouabain, a molecule used to inhibit sodium transport in research studies. Because cardiac glycosides depress Na^+-K^+-ATPase activity, they are highly toxic. They poison the pumps on all cells, not just those of the heart. However, if administered in low doses, cardiac glycosides can partially block the removal of Na^+ from the myocardial cell. As the concentration of Na^+ builds up in the cytosol, the concentration gradient for Na^+ across the cell membrane diminishes. In turn, the potential energy available for indirect active transport diminishes. In the myocardial cell, cardiac glycosides decrease the cell's ability to remove Ca^{2+} by means of the Na^+-Ca^{2+} exchanger. The resultant increase in cytosolic Ca^{2+} causes stronger myocardial contractions. For this reason, cardiac glycosides have been used since the eighteenth century as a remedy for heart failure, a condition in which the heart is unable to contract forcefully.

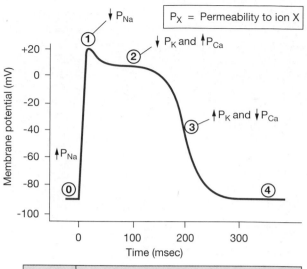

Phase	Membrane channels
⓪	Na^+ channels open
①	Na^+ channels close
②	Ca^{2+} channels open; fast K^+ channels close
③	Ca^{2+} channels close; slow K^+ channels open
④	Resting potential

■ **Figure 14-14   Action potential of a cardiac contractile cell**

Action Potentials in Myocardial Cells Vary According to Cell Type

Cardiac muscle, like skeletal muscle and neurons, is an excitable tissue with the ability to generate action potentials. Each of the two types of cardiac muscle cells has a distinctive action potential. In both types, Ca^{2+} plays an important role in the action potential, in contrast to the action potentials of skeletal muscle and neurons.

Myocardial Contractile Cells   The action potentials of myocardial contractile cells are similar in several ways to those of neurons and skeletal muscle [∞ p. 224]. The rapid depolarization phase of the action potential is due to Na^+ entry, and the steep repolarization phase is due to K^+ leaving the cell. The main difference between the action potential of the myocardial contraction cell and that of a skeletal muscle fiber or a neuron is a lengthening of the action potential due to Ca^{2+} entry. Let's take a look at these longer action potentials.

Myocardial contractile cells have a stable resting potential of about −90 mV (phase 4, Fig. 14-14 ■). When a wave of depolarization moves into a contractile cell through the gap junctions, the membrane potential becomes more positive. Voltage-gated Na^+ channels open, allowing Na^+ to enter the cell and rapidly depolarize it (phase 0). The membrane potential reaches about +20 mV before the Na^+ channels close. These are double-gated Na^+ channels, just like the voltage-gated Na^+ channels of the axon [∞ p. 225].

When the Na^+ channels close, the cell begins to repolarize as K^+ leaves through open K^+ channels (phase 1). But then the action potential flattens into a plateau (phase 2). Flattening occurs as the result of two events: a decrease in K^+ permeability and an increase in Ca^{2+} permeability. Voltage-gated Ca^{2+} channels activated by depolarization have been slowly opening during phases 0 and 1. When they finally open, Ca^{2+} enters the cell. At the same time, some K^+ channels close. The combination of Ca^{2+} influx and decreased K^+ efflux causes the action potential to flatten out into a plateau.

The plateau ends when the Ca^{2+} channels close and K^+ permeability increases once more. The K^+ channels responsible for this phase are similar to those in the neuron: they are activated by depolarization but are slow to open. When the delayed K^+ channels open, K^+ exits rapidly (phase 3), returning the cell to its resting potential (phase 4).

The influx of Ca^{2+} during phase 2 lengthens the duration of a myocardial action potential. A typical action potential in a neuron or skeletal muscle fiber lasts between 1 and 5 msec. In a contractile myocardial cell, the action potential is typically 200 msec or more.

The longer action potential in myocardial cells helps prevent the sustained contraction called tetanus. Prevention of tetanus in the heart is important because the muscles must relax between contractions so that the ventricles can fill with blood.

To understand why a longer action potential prevents tetanus, compare the relationship between action potentials and contraction in skeletal and cardiac muscle cells. As you may recall from Chapter 12, the skeletal muscle action potential is ending as contraction begins (Fig. 14-15a ■). Thus, a second action potential fired immediately after the refractory period will cause summation of the contractions. If a series of action potentials occurs in rapid succession, the sustained contraction known as tetanus results (Fig. 14-15b ■).

Tetanus cannot occur in cardiac muscle because the refractory period and the contraction end almost simultaneously due to the longer action potential (Fig. 14-15c ■). By the time a second action potential takes place, the myocardial cell has almost completely relaxed. Consequently, no summation will occur (Fig. 14-15d ■).

✔ Tetrodotoxin (TTX) is a molecule that blocks voltage-gated Na⁺ channels. What will happen to the action potential of a myocardial contractile cell if TTX is applied to it?

Myocardial Autorhythmic Cells   What gives myocardial autorhythmic cells their unique ability to generate action potentials spontaneously in the absence of

input from the nervous system? This property results from their unstable membrane potential that starts at -60 mV and slowly drifts upward toward threshold (Fig. 14-16 ■). Because the membrane potential never "rests" at a constant value, it is called a *pacemaker potential* rather than a resting membrane potential. Whenever the pacemaker potential depolarizes to threshold, the autorhythmic cell fires an action potential.

What causes the membrane potential of these cells to be unstable? Our current understanding is that the autorhythmic cells contain channels that are different from the channels of other excitable tissues. When the cell membrane potential is -60 mV, channels that are permeable to both K⁺ and Na⁺ open. These channels are called I_f **channels** because they allow current (I) to flow and because of their unusual properties. The researchers who first described these channels did not initially understand their behavior and named them "funny" channels, hence the subscript f.

When I_f channels open at negative membrane potentials, Na⁺ influx exceeds K⁺ efflux. (This is similar to

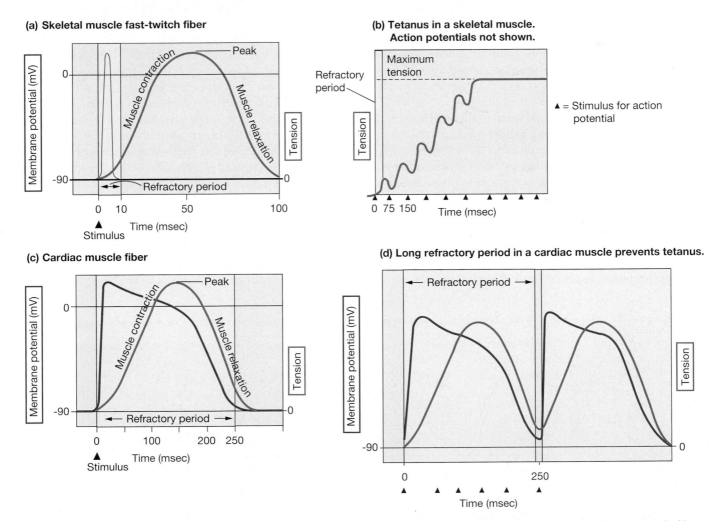

■ Figure 14-15   Refractory periods in skeletal and cardiac muscle   (a) The refractory period (yellow) of a skeletal muscle fiber is very short compared with the amount of time required for the development of tension. Compare this with the cardiac muscle (c), in which the refractory period lasts almost as long as the entire muscle twitch. (b) Skeletal muscles that are stimulated repeatedly will exhibit summation and tetanus. (d) Cardiac muscle will not show summation because of the long refractory period.

(a) Pacemaker and action potentials

(b) Ion movements during an action and pacemaker potential

(c) State of various ion channels

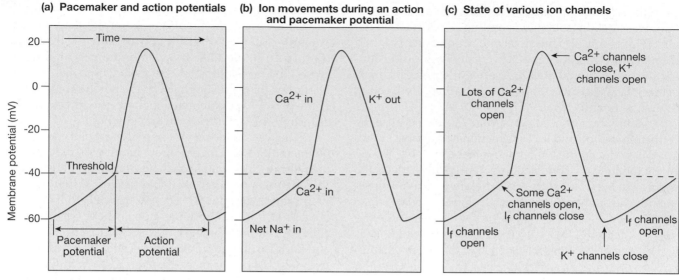

■ **Figure 14-16   Action potentials in cardiac autorhythmic cells**

what happens at the neuromuscular junction when non-specific cation channels open [∞ p. 338].) The net influx of positive charge slowly depolarizes the autorhythmic cell. As the membrane potential becomes more positive, the I_f channels gradually close and a set of Ca^{2+} channels opens. The subsequent influx of Ca^{2+} continues the depolarization, and the membrane potential moves steadily toward threshold.

When the membrane potential reaches threshold, many Ca^{2+} channels open. Calcium ions rush in, creating the steep depolarization phase of the action potential. Note that this process is different from that in other excitable cells, in which the depolarization phase is due to the opening of voltage-gated Na^+ channels.

When the Ca^{2+} channels close at the peak of the action potential, slow K^+ channels have opened. The repolarization phase of the autorhythmic action potential is due to the resultant efflux of K^+. This phase is similar to repolarization in other types of excitable cells.

The timing of autorhythmic action potentials can be modified by altering the permeability of the autorhythmic cells to different ions. Norepinephrine from sympathetic neurons and epinephrine from the adrenal medulla increase the flow of ions through both the I_f and Ca^{2+} channels. More rapid cation entry speeds up the rate of the pacemaker depolarization, causing the cell to reach threshold faster and increasing the rate of action potential firing. When the pacemaker fires action potentials more rapidly, heart rate increases (Fig. 14-17a ■). The catecholamines exert their effect by binding to and activating β_1 receptors on the autorhythmic cells. The β_1 receptors use a cAMP second messenger system to alter the transport properties of the ion channels.

The parasympathetic neurotransmitter acetylcholine (ACh) slows heart rate. Acetylcholine activates muscarinic receptors that influence K^+ and Ca^{2+} channels in the pacemaker membrane. Potassium permeability increases, hyperpolarizing the cell so that the pacemaker

(a) Sympathetic stimulation with SA node pacemaker activity

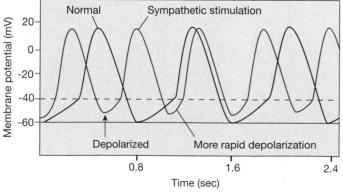

(b) Parasympathetic stimulation with SA node pacemaker activity

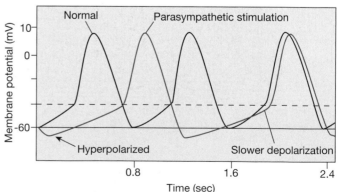

■ **Figure 14-17   Modulation of heart rate by the nervous system**   (a) Sympathetic stimulation and epinephrine depolarize the autorhythmic cell and speed up the depolarization rate, increasing the heart rate. (b) Parasympathetic stimulation hyperpolarizes the membrane potential of the autorhythmic cell and slows the depolarization rate, slowing down the heart rate.

TABLE 14-4   Comparison of Action Potentials in Cardiac and Skeletal Muscle

	Skeletal Muscle	Contractile Myocardium	Autorhythmic Myocardium
Membrane potential	Stable at −70 mV	Stable at −90 mV	Unstable pacemaker potential; usually starts at −60 mV
Events leading to threshold potential	Net Na⁺ entry through nonspecific channels opened by ACh	Depolarization of adjacent cells enters via gap junctions	Net Na⁺ entry through I_f channels that open at negative membrane potentials; reinforced with some Ca^{2+} entry
Rising phase of action potential	Na⁺ entry	Na⁺ entry	Ca^{2+} entry
Repolarization phase	Rapid; caused by K⁺ efflux	Extended plateau caused by Ca^{2+} entry; rapid phase caused by K⁺ efflux	Rapid; caused by K⁺ efflux
Hyperpolarization	Due to excessive K⁺ efflux at high K⁺ permeability; when K⁺ channels close, leak of K⁺ and Na⁺ restores potential to resting state	None; resting potential is −90 mV, the equilibrium potential for K⁺	None; when repolarization hits −60 mV, the I_f channels open and the pacemaker potential begins
Duration of action potential	Short: 1–2 msec	Extended: 200+ msec	Variable; generally 150+ msec
Refractory period	Results from time required to reset Na⁺ channel gates; generally brief	Resetting of Na⁺ channel gates delayed until end of action potential, leading to long refractory period	None

potential begins at a more negative value. At the same time, Ca^{2+} permeability of the pacemaker decreases. This decrease in Ca^{2+} permeability slows the depolarization rate of the pacemaker potential. The combination of the two effects causes the cell to take longer to reach threshold, delaying the onset of the action potential in the pacemaker and slowing the heart rate (Fig. 14-17b ■).

Table 14-4 compares the action potentials of the two types of myocardial muscle with those of skeletal muscle. Next we will see how the action potentials of autorhythmic cells spread throughout the heart to coordinate contraction.

✔ Do you think that the Ca^{2+} channels in the autorhythmic cells are the same as the Ca^{2+} channels in the contractile cells? Defend your answer.

✔ What will happen to the action potential of a myocardial autorhythmic cell if tetrodotoxin (TTX), which blocks voltage-gated Na⁺ channels, is applied to it?

✔ In an experiment, the vagus nerve to the heart was cut. The investigators noticed that heart rate increased. Why did this happen? [∞ p. 331]

▶ THE HEART AS A PUMP

We now turn from single myocardial cells to the intact heart. How can one tiny noncontractile autorhythmic cell cause the entire heart to beat? And why do those doctors on the TV shows shock people with electric pad-

dles when their hearts malfunction? You're about to learn the answers to these questions.

Electrical Conduction in the Heart Coordinates Contraction

The heart is like a group of people around a stalled car. One person can push on the car, but it's not likely to move very far unless everyone pushes together. In the

...continued from page 417

The results of the cardiac enzyme and troponin I assays do not come back for several hours. If Walter had a blockage in one of his cardiac blood vessels, damage to the heart muscle could be severe by that time. In Walter's case, an electrocardiogram (ECG) showed an abnormal pattern of electrical activity in the myocardium. "He's definitely had an MI," said the ER physician, referring to a myocardial infarction, or heart attack. "Let's start him on t-PA." t-PA is short for *tissue plasminogen activator*. t-PA activates plasminogen, a substance produced in the body that dissolves blood clots. When t-PA is given within 1–3 hours of a heart attack, it can help dissolve blood clots that are blocking blood flow to the heart muscle. This will help limit the extent of ischemic damage.

Question 5:   *How do electrical signals move from cell to cell in the myocardium?*

Question 6:   *What happens to contraction in a myocardial contractile cell if a wave of depolarization passing through the heart bypasses it?*

same way, individual myocardial cells must depolarize and contract in a coordinated fashion if the heart is to create enough force to circulate the blood.

Electrical communication in the heart begins with an action potential in an autorhythmic cell. The depolarization spreads rapidly to adjacent cells through gap junctions in the intercalated disks (Fig. 14-18 ■). The depolarization wave is followed by a wave of contraction that passes across the atria, then moves into the ventricles.

The **sinoatrial node (SA node),** a group of autorhythmic cells in the right atrium near the entry of the superior vena cava, is the main pacemaker of the heart (Fig. 14-19a ■). Other noncontractile autorhythmic fibers form a specialized conducting system that rapidly transmits action potentials through the heart. An **internodal pathway** connects the SA node to the **atrioventricular node (AV node),** a group of autorhythmic cells found near the floor of the right atrium.

From the AV node, action potentials move into fibers known as the **bundle of His** ("hiss"), or **atrioventricular bundle.** The bundle passes from the AV node into the wall (*septum*) between the ventricles. A short way down the septum, the bundle divides into *left and right bundle branches.* These fibers continue downward to the apex where they divide into many small **Purkinje fibers** that spread outward among the contractile cells.

The electrical signal that initiates contraction begins when the SA node fires an action potential (Fig. 14-19b ■). The depolarization of the pacemaker action potential spreads to adjacent cells through gap junctions. Conduction of the electrical signal is rapid through the internodal conducting pathways (Fig. 14-19c ■), but slower through the contractile cells of the atrium (Fig. 14-19d ■).

As action potentials spread outward across the atria, they encounter the fibrous skeleton of the heart at the junction of the atria and ventricles. This barricade prevents the transfer of electrical signals from the atria to the ventricles. Consequently, the AV node is the only pathway through which action potentials can reach the contractile fibers of the ventricles (Fig. 14-19d ■).

The electrical signal passes from the AV nodes through the bundle of His and the bundle branches to the apex of the heart (Fig. 14-19c ■). At that point, the Purkinje fibers transmit impulses very rapidly, with speeds up to 4 m/sec, so that all contractile cells in the apex contract nearly simultaneously (Fig. 14-19f ■).

Why is it necessary to direct the electrical signals through the AV node? Why not allow them to spread downward from the atria? The answer lies in the fact that blood being pumped out of the ventricles leaves through openings at the top of the chambers. If the electrical signals from the atria were conducted directly into the ventricles, the ventricles would start contraction at the top. Then blood would be squeezed downward and would be trapped in the bottom of the ventricle (think of squeezing toothpaste from a tube starting at the top). The apex-to-base contraction squeezes blood toward the arterial openings at the base of the heart.

The ejection of blood from the ventricles is aided by the spiral arrangement of the muscles in the walls (see Fig. 14-10a ■). As these muscles contract, they pull the apex and base of the heart closer together, squeezing blood out openings at the top of the ventricles.

A second function of the AV node is to delay the transmission of action potentials slightly, allowing the atria to complete their contraction before ventricular contraction begins. The **AV node delay** is accomplished by slowing conduction through the AV node cells. Action potentials here move at only 1/20 the rate of action potentials in the atrial internodal pathway.

Pacemakers Set the Heart Rate

The cells of the SA node set the pace of the heartbeat. Other cells in the conducting system, such as the AV node and the Purkinje fibers, have unstable resting po-

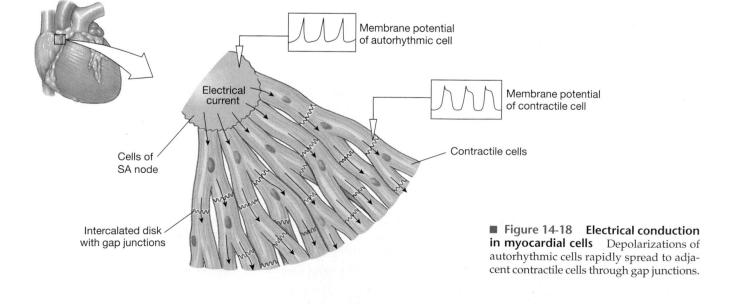

■ **Figure 14-18 Electrical conduction in myocardial cells** Depolarizations of autorhythmic cells rapidly spread to adjacent contractile cells through gap junctions.

Membrane potential of autorhythmic cell

Membrane potential of contractile cell

Electrical current

Cells of SA node

Contractile cells

Intercalated disk with gap junctions

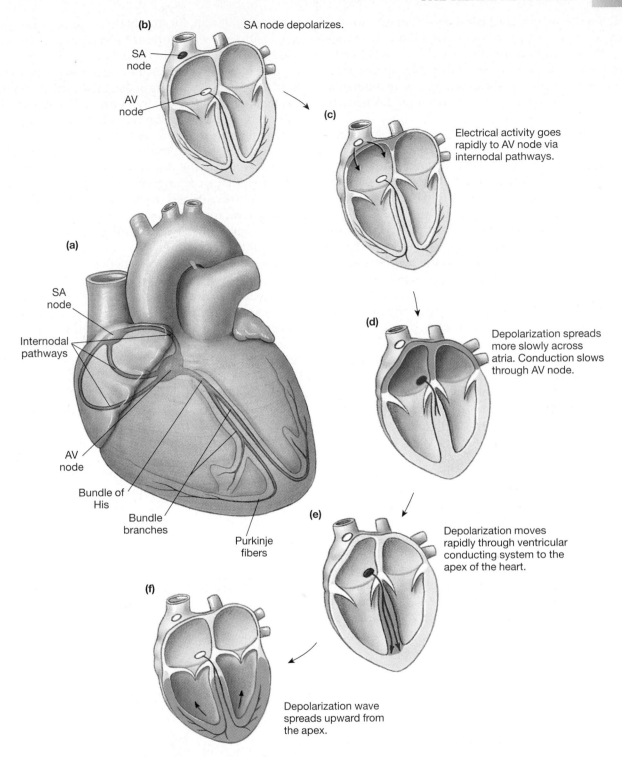

(b) SA node depolarizes.

SA node

AV node

(a)

SA node

Internodal pathways

AV node

Bundle of His

Bundle branches

Purkinje fibers

(c) Electrical activity goes rapidly to AV node via internodal pathways.

(d) Depolarization spreads more slowly across atria. Conduction slows through AV node.

(e) Depolarization moves rapidly through ventricular conducting system to the apex of the heart.

(f) Depolarization wave spreads upward from the apex.

■ **Figure 14-19   Electrical conduction in the heart**   (a) The conducting system of the heart consists of the SA node, internodal pathways, AV node, bundle of His, bundle branches, and Purkinje fibers. (b)–(f) Purple shading represents depolarization.

tentials and can also act as pacemakers under some conditions. However, because their rhythm is slower than that of the SA node, they do not usually have a chance to set the heartbeat. The Purkinje fibers, for example, can spontaneously fire action potentials, but their rate of firing is very slow, between 25 and 40 beats per minute.

Why does the fastest pacemaker determine the pace of the heartbeat? Consider the following analogy. A group of people are playing "Follow the Leader" as they walk. Initially, everyone is walking at a different pace: some fast, some slow. When the game starts, everyone must match his or her pace to the pace of the person who is walking the fastest. The fastest person in the group is the SA node, walking at 70 steps per minute. Everyone else in the group (autorhythmic and contractile cells) sees that the SA node is fastest, so they pick up their pace and follow the leader. In the heart, the cue to follow the leader is the electrical signal sent from the SA node to the other cells.

Now suppose that the SA node gets tired and drops out of the group. The role of leader defaults to the next fastest person, the AV node, who is walking at a rate of 50 steps per minute. The group slows to match the pace of the AV node, but they are still following the fastest leader.

What happens if the group divides? When they reach the corner, the AV node leader goes left, but a renegade Purkinje fiber decides to go right. Those people who follow the AV node continue to walk at 50 steps per minute, but the people who follow the Purkinje fiber slow down to match his pace of 35 steps per minute. Now there are two leaders, each walking at a different pace.

In the heart, the SA node is the fastest pacemaker and normally sets the heart rate. But if the SA node is damaged and cannot function, one of the slower pacemakers in the heart takes over. Heart rate then matches the rate of the new pacemaker. It is even possible for different parts of the heart to follow different pacemakers, just as the walking group split at the corner.

In a condition known as *complete heart block,* the conduction of electrical signals from the atria to the ventricles through the AV node is disrupted. The SA node fires at its rate of 70 beats per minute, but those signals never reach the ventricles. So the ventricles coordinate with their fastest pacemaker. Because ventricular autorhythmic cells discharge only about 35 times a minute, the ventricles contract at a totally different rate than the atria. In conditions such as complete heart block, in which the rate of ventricular contraction is too slow to maintain adequate blood flow, it may be necessary for the heart's rhythm to be set artificially by a surgically implanted mechanical pacemaker. These battery-powered devices artificially stimulate the heart at a predetermined rate.

✔ Name two functions of the AV node.

✔ Occasionally an *ectopic pacemaker* [*ektopos,* out of place] will develop in part of the conducting system. What happens to heart rate if an ectopic pacemaker in the atria depolarizes at a rate of 120 times per minute?

The Electrocardiogram Reflects the Electrical Activity of the Heart

In the beginning of the twentieth century, physiologists discovered that they could use electrocardiograms to obtain information about the function of the heart. An **electrocardiogram,** or **ECG,** is a recording of the electrical activity of the heart made from electrodes placed on the surface of the skin. By simply placing electrodes on the arms and legs, we can record electrical activity that is taking place deep within the body.

Salt solutions such as the NaCl-based extracellular fluid are good conductors of electricity and can transfer electrical activity to the skin's surface. The electrical signals are very weak by the time they get to the skin. A ventricular action potential, for example, has a voltage change of 110 mV, but the ECG signal has an amplitude of only 1 mV by the time it reaches the surface of the body. External recordings allow us to monitor electrical activity also in muscles (*electromyogram,* or *EMG*) and in the brain (*electroencephalogram,* or *EEG*).

The first human electrocardiogram was recorded in 1887, but the procedure was not refined for clinical use until the first years of the twentieth century. The father of the modern ECG was a Dutch physiologist named Walter Einthoven. He named the parts of the ECG as we know them today and created "Einthoven's triangle," a hypothetical triangle created around the heart when electrodes are placed on both arms and the left leg (Fig. 14-20 ■). The sides of the triangle are numbered to correspond with the three leads, or pairs of electrodes, used for a recording.

An ECG tracing shows the sum of the electrical potentials generated by all the cells of the heart at any moment. Each component of the ECG reflects depolarization or repolarization of a portion of the heart. Because depolarization is the signal for contraction, the *electrical events* (waves) of an ECG can be associated approximately to contraction or relaxation of the atria or ventri-

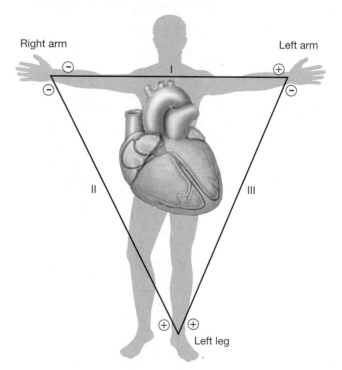

Right arm   Left arm

Left leg

■ **Figure 14-20 Einthoven's triangle** Electrodes are attached to the skin surface on both arms and the left leg, forming a triangle. A lead consists of two electrodes, one positive and one negative. Each lead gives a different electrical "view" of the heart.

cles (collectively referred to as the *mechanical events* of the cardiac cycle). Let's follow an ECG through a single contraction-relaxation, otherwise known as a **cardiac cycle.**

There are three major components of an ECG: the P wave, the combined waves of the QRS complex, and the T wave (Fig. 14-21 ■). The first wave is the **P wave,** which corresponds to depolarization of the atria (Fig.14-22 ■). The next trio of waves, the **QRS complex,** represents the progressive wave of ventricular depolarization. The final wave, the **T wave,** represents the repolarization of the ventricles. Atrial repolarization is not represented by a special wave but is incorporated into the QRS complex.

The mechanical events of the cardiac cycle lag slightly behind the electrical signals, just as the contraction of a single cardiac muscle cell follows its action potential (see Fig. 14-15c ■). Atrial contraction begins during the latter part of the P wave and continues during the P-R segment. Ventricular contraction begins just after the Q wave and continues through the T wave.

One thing that many people find puzzling is that downward deflection may correspond to periods of depolarization in the heart. It is important to remember that the ECG is different from a single action potential (Fig. 14-23 ■). The single action potential is an electrical event in a single cell, recorded using an intracellular electrode. The ECG represents multiple action potentials taking place in the heart muscle at a given time and recorded from the surface of the body.

When recording an ECG, one active surface electrode is the positive electrode, and the other active electrode is the negative electrode. (The third electrode is inactive.) If net electrical current in the heart moves toward the positive electrode, the tracing of the ECG goes up from the baseline. If net current moves toward the negative electrode, the tracing moves downward.

Another important point to remember is that the ECG is an electrical "view" of a three-dimensional object. This is one reason that we use multiple leads to assess heart function. Think of looking at an automobile. From the air, it looks like a rectangle, but from the side and front it has different shapes. Not everything that you see from the front of the car can be seen from its side, and vice versa.

In the same way, ECGs recorded from the different leads look different and give information about different regions of the heart. Clinically, using 12 leads is now common: the three limb electrodes plus nine more electrodes placed on the chest and trunk. These additional "views" provide even more information about electrical conduction in the heart. Electrocardiograms are important diagnostic tools in medicine because they are quick, painless, and noninvasive (that is, do not puncture the skin).

An ECG provides information on heart rate and rhythm, conduction velocity, and even the condition of tissues within the heart. Thus, although obtaining an ECG is simple, interpreting some of its subtleties can be quite complicated. The interpretation of an ECG begins with the following questions.

What is the heart rate? Heart rate is normally timed from the beginning of one P wave to the beginning of the next P wave, or from peak to peak of the QRS complexes. A normal resting heart rate is 60–100 beats per minute,

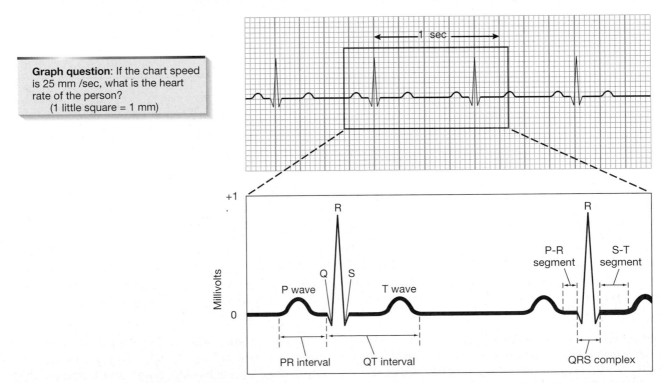

Graph question: If the chart speed is 25 mm /sec, what is the heart rate of the person?
(1 little square = 1 mm)

■ **Figure 14-21   The electrocardiogram**   The electrocardiogram (ECG) is divided into waves (P, Q, R, S, and T), segments between the waves, and intervals that include waves.

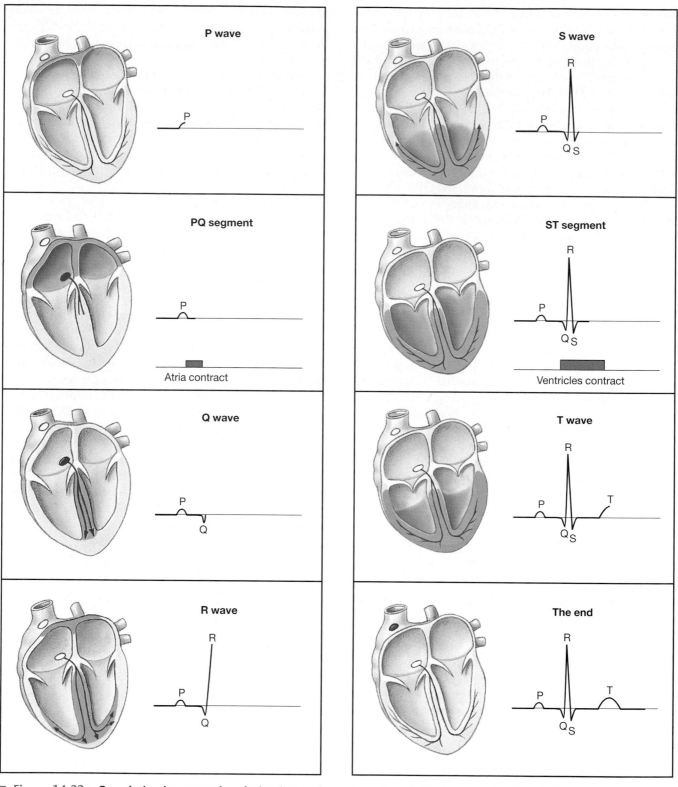

■ **Figure 14-22   Correlation between depolarization and repolarization in the heart and in the ECG**   The depolarized regions are shown in purple, along with the corresponding section of the ECG. The repolarizing regions are shown in peach.

although trained athletes often have slower heart rates at rest. A faster rate is known as *tachycardia,* and a slower rate is called *bradycardia* [*tachys,* swift; *bradys,* slow].

Is the rhythm of the heartbeat regular or irregular? That is, does it occur at regular intervals? An irregular rhythm, or *arrhythmia,* can result from a benign extra beat or from more serious conditions such as atrial fibrillation, in which the SA node has lost control of the pacemaking.

After determining heart rate and rhythm, the next step in analyzing an ECG is to look at the relationship of the

(a)

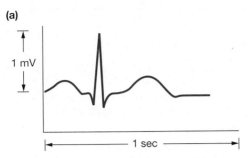

1 mV

|— 1 sec —|

The electrocardiogram represents the summed electrical activity of all cells recorded from the surface of the body.

(b)

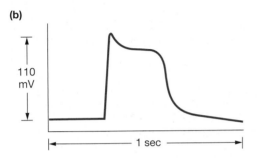

110 mV

|— 1 sec —|

The ventricular action potential is recorded from a single cell using an intracellular electrode. Notice that the voltage change is much greater when recorded intracellularly.

■ **Figure 14-23   Comparison of the ECG and a myocardial action potential**

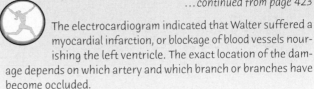

...continued from page 423

The electrocardiogram indicated that Walter suffered a myocardial infarction, or blockage of blood vessels nourishing the left ventricle. The exact location of the damage depends on which artery and which branch or branches have become occluded.

Question 7:   *If the ventricle of the heart is damaged, in which wave or waves of the electrocardiogram would you expect to see abnormal changes?*

The Heart Contracts and Relaxes Once During a Cardiac Cycle

A *cardiac cycle* is the period of time from the beginning of one heartbeat to the beginning of the next. Each cycle has two phases: **diastole,** the time during which cardiac muscle relaxes, and **systole,** the time during which the muscle is contracting [*diastole,* dilation; *systole,* contraction]. The atria and ventricles do not contract and relax at the same time, so we will discuss atrial and ventricular events separately.

In this discussion, the cardiac cycle is divided into five phases.

The Heart at Rest: Atrial and Ventricular Diastole

We begin our cardiac cycle with the brief moment during which both the atria and ventricles are relaxing (Fig. 14-25a ■). The atria are filling with blood from the veins, while the ventricles have just completed a contraction. As the ventricles relax, the AV valves between the atria and ventricles open, and blood flows from the atria into the ventricles. The relaxing ventricles expand to accommodate the entering blood.

Completion of Ventricular Filling: Atrial Systole

Although most blood enters the ventricles while the atria are relaxed, the last 20% of filling is accomplished when the atria contract (Fig. 14-25b ■). Atrial systole, or contraction, begins following depolarization of the autorhythmic cells in the SA node. A wave of contraction follows the electrical signal across the atria, pushing blood into the ventricles and completing ventricular filling.

Although the contribution of atrial contraction to ventricular filling is minor in someone at rest, it provides a margin of safety when heart rate increases or if problems develop. One condition in which atrial contraction aids filling is *stenosis* (narrowing) of the AV valves [*stenos,* narrow]. This condition may be inherited or may result from inflammation or other disease processes. Because the flow of blood from the atria to the ventricles is impaired in such conditions, atrial contraction helps force blood through the narrowed opening between the chambers.

Contraction of the atria forces blood into the ventricles. The pressure increase that accompanies contraction also pushes a small amount of blood backward into the veins. Although the openings of the veins narrow during

various waves. *Does a QRS complex follow each P wave, and is the P-R segment constant in length?* If not, a problem with conduction of signals through the AV node may exist.

In heart block, the conduction problem mentioned earlier, action potentials from the SA node sometimes fail to be transmitted through the AV node to the ventricles. In these conditions, one or more P waves may occur without initiating a QRS complex. In the most severe form of heart block (third-degree), the atria depolarize regularly at one pace while the ventricles contract at a much slower pace (Fig. 14-24b ■).

The more difficult aspects of interpreting an ECG include looking for subtle changes such as alterations in the shape or duration of various waves or segments. An experienced clinician can find signs pointing to changes in conduction velocity, enlargement of the heart, or tissue damage resulting from periods of ischemia. An amazing number of conclusions can be drawn about heart function simply by looking at alterations in the electrical activity of the heart as recorded on an ECG.

✔ Some abnormal ECGs are shown in Figure 14-24 ■. Study them and see if you can relate the ECG changes to disruption of the normal electrical conduction pattern in the heart.

(a) Normal ECG

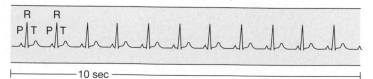

|———— 10 sec ————|

(b) Third-degree block

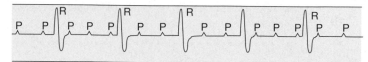

(c) Atrial fibrillation

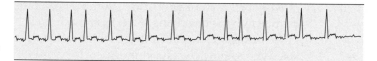

(d) Ventricular fibrillation

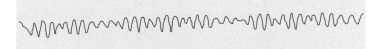

(e)

> **Graph question:** Identify the waves on the ECG below. Look at the pattern of their occurrence and describe what has happened to electrical conduction in the heart.

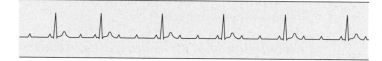

> **Graph questions about ECG tracings:**
>
> 1. What is the rate? Is it within the normal range of 50-100 beats per minute?
>
> 2. Is the rhythm regular?
>
> 3. Are all normal waves present in recognizable form?
>
> 4. Is there one QRS complex for each P wave? If yes, is the P-R segment constant in length?
>
> 5. If there is not one QRS complex for each P wave, count the heart rate using the P waves, then count it according to the R waves. Are the rates the same? Which wave would agree with the pulse felt at the wrist?

■ **Figure 14-24 Normal and abnormal electrocardiograms** All tracings represent 10-second recordings. To analyze, ask the above questions.

---

contraction, there are no one-way valves to block backward flow. Thus, the retrograde movement of blood back into the veins may be observed as a pulse in the jugular vein of a normal person who is lying with the head and chest elevated about 30°. (Look in the hollow formed where the sternocleidomastoid muscle runs under the clavicle.) An observable jugular pulse higher on the neck of a person sitting upright is a sign that pressure in the right atrium is higher than normal.

Early Ventricular Contraction and the First Heart Sound As the atria are contracting, the depolarization wave is moving slowly through the conducting cells of the AV node, then rapidly down the atrioventricular bundle to the apex of the heart. Ventricular systole begins at the apex of the heart as spiral bands of muscle squeeze the blood upward toward the base. Blood pushing upward on the underside of the AV valves forces them closed so that blood cannot flow back into the atria. Vi-

brations following closure of the AV valves create the first heart sound, the "lub" of "lub-dup."

With both sets of AV and semilunar valves closed, blood in the ventricles has nowhere to go. Nevertheless, the ventricles continue to contract, squeezing on the blood in the same way that you might squeeze a water balloon in your hand. This is similar to an isometric contraction, in which muscle fibers create force without movement [∞ p. 365]. To return to the toothpaste analogy, it is like squeezing the toothpaste tube with the cap still on: high pressure develops within the tube, but the toothpaste has nowhere to go. This phase is called **isovolumic ventricular contraction** [*iso-*, equal], to underscore the fact that the volume of blood in the ventricle is not changing (Fig. 14-25c ■).

As the ventricles begin their contraction, the atrial muscle fibers repolarize and relax. When the pressure within the atria falls below that in the veins, blood flow from the veins into the atria resumes. Closure of the AV

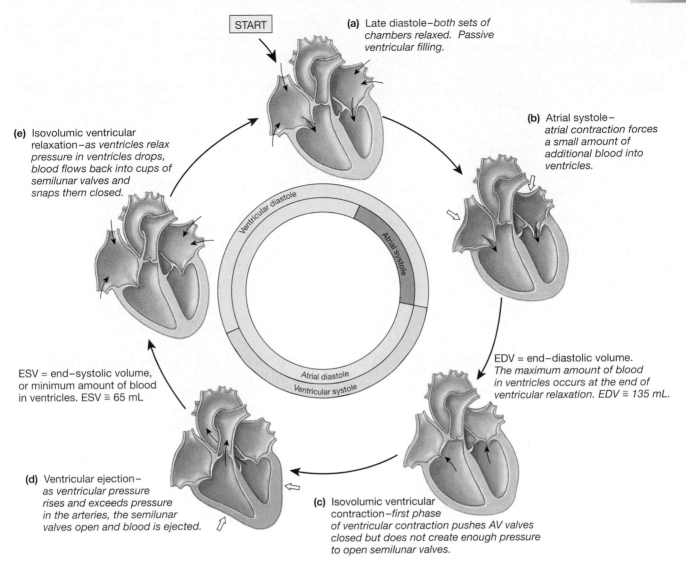

START

(a) Late diastole–*both sets of chambers relaxed. Passive ventricular filling.*

(b) Atrial systole– *atrial contraction forces a small amount of additional blood into ventricles.*

(e) Isovolumic ventricular relaxation–*as ventricles relax pressure in ventricles drops, blood flows back into cups of semilunar valves and snaps them closed.*

Ventricular diastole

Atrial systole

Atrial diastole

Ventricular systole

EDV = end–diastolic volume. *The maximum amount of blood in ventricles occurs at the end of ventricular relaxation. EDV ≅ 135 mL.*

ESV = end–systolic volume, or minimum amount of blood in ventricles. ESV ≅ 65 mL

(d) Ventricular ejection– *as ventricular pressure rises and exceeds pressure in the arteries, the semilunar valves open and blood is ejected.*

(c) Isovolumic ventricular contraction–*first phase of ventricular contraction pushes AV valves closed but does not create enough pressure to open semilunar valves.*

■ **Figure 14-25   The cardiac cycle**

valves isolates the upper and lower cardiac chambers, so atrial filling is independent of events taking place in the ventricles.

The Heart Pumps: Ventricular Ejection   As the ventricles contract, they generate enough pressure to open the semilunar valves, and blood is pushed into the arteries (Fig. 14-25d ■). The pressure created by ventricular contraction becomes the driving force for blood flow. High-pressure blood is forced into the arteries, displacing the low-pressure blood that fills them and pushing it farther into the vasculature.

Ventricular Relaxation and the Second Heart Sound
At the end of each ventricular contraction, the ventricles begin to relax. As they do so, ventricular pressure decreases. Once ventricular pressure falls below the pressure in the arteries, blood starts to flow backward into the heart. This backflow of blood fills the cuplike cusps of the semilunar valves, forcing them together into the closed

position. The vibrations created by semilunar valve closure are the second heart sound, the "dup" of "lub-dup."

Once the semilunar valves close, the ventricles again become a sealed chamber. The AV valves remain closed because ventricular pressure, although falling, is still higher than atrial pressure (Fig. 14-25e ■). When ventricular pressure falls to the point at which it is below atrial pressure, the AV valves open. Blood that has been accumulating in the atria during ventricular contraction rushes into the ventricles. The cardiac cycle has begun again.

Pressure-Volume Curves Represent One Cardiac Cycle

Another way to describe the cardiac cycle is with the use of a pressure-volume graph, shown in Figure 14-26 ■. This figure represents the changes in volume (*x*-axis) and pressure (*y*-axis) that occur during one cardiac cycle. The flow of blood through the heart is governed by the same principle that governs the flow of all liquids and gases:

flow proceeds from areas of higher pressure to areas of lower pressure. When the heart contracts, the pressure increases and blood flows out of the heart into areas of lower pressure.

The graph shown in Figure 14-26 ■ represents pressure and volume changes in the left ventricle, which sends blood into the systemic circulation. The left side of the heart creates higher pressures than the right side, which sends blood through the shorter pulmonary circuit.

The cycle begins at point A. The ventricle has completed a contraction and contains the minimum amount of blood that it will hold during the cycle. It has relaxed, and its pressure is also at its minimum value. Blood is flowing into the atrium from the pulmonary veins.

Once pressure in the atrium exceeds pressure in the ventricle, the mitral valve, between the atrium and ventricle, opens (Fig. 14-26 ■, point A). Atrial blood now flows into the ventricle, increasing its volume (point A to point B). As blood flows in, the relaxing ventricle expands to accommodate the entering blood. Consequently, the volume of the ventricle increases, but the pressure in the ventricle goes up very little.

The last portion of ventricular filling is completed by atrial contraction. The ventricles now contain the maximum volume of blood that they will hold during this cardiac cycle (point B). Because maximum filling occurs at the end of ventricular relaxation (diastole), this volume is called the **end-diastolic volume** (EDV).

In a 70-kg man at rest, end-diastolic volume is about 135 mL. The EDV will vary under different conditions. For instance, during periods of very high heart rate, when the ventricles do not have time to fill completely between beats, the end-diastolic value may be less.

When ventricular contraction begins, the mitral valve closes. With both AV and semilunar valves closed, blood in the ventricle has nowhere to go. Nevertheless, the ventricle continues to contract, causing pressure within the ventricle to increase rapidly during isovolumic contraction (B→C). Once ventricular pressure exceeds the pressure in the aorta, the aortic valve opens (point C). Pressure continues to increase as the ventricle contracts further, but ventricular volume decreases as blood is pushed out into the aorta (C→D).

The heart does not empty itself completely of blood each time the ventricle contracts. The amount of blood left in the ventricle at the end of contraction is known as the **end-systolic volume** (ESV). The ESV is the minimum amount of blood that the ventricle will contain during one cycle (point D). An average value for the ESV in a person at rest is 65 mL, meaning that nearly half of the 135 mL that was in the ventricle at the start of the contraction is still there at the end of the contraction.

At the end of each ventricular contraction, the ventricle begins to relax. As it does so, ventricular pressure decreases. Once pressure in the ventricle falls below aortic pressure, the semilunar valve closes and the ventricle again becomes a sealed chamber. The remainder of relaxation occurs without a change in blood volume, so this phase is called *isovolumic relaxation* (Fig. 14-26 ■, D→A). When ventricular pressure finally falls to the point at which atrial pressure exceeds ventricular pressure, the mitral valve opens and the cycle begins again.

The electrical and mechanical events of the cardiac cycle are summarized together in Figure 14-27 ■, known

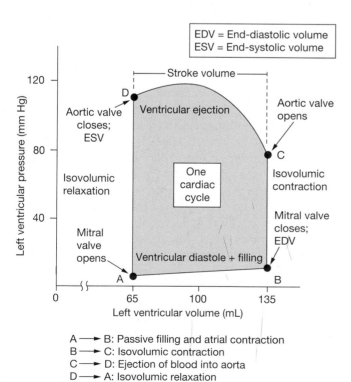

A ⟶ B: Passive filling and atrial contraction
B ⟶ C: Isovolumic contraction
C ⟶ D: Ejection of blood into aorta
D ⟶ A: Isovolumic relaxation

■ **Figure 14-26   Left ventricular pressure-volume changes during one cardiac cycle**

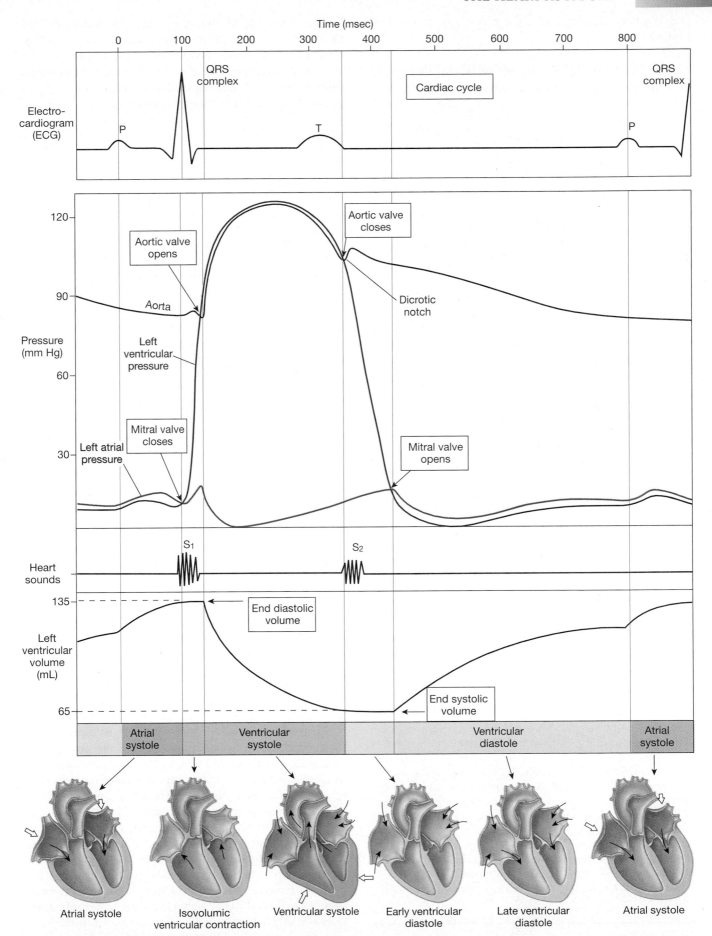

■ **Figure 14-27   The Wiggers diagram**   The Wiggers diagram shows the relationship between the electrocardiogram, heart sounds, and pressure and volume changes in the left heart and aorta.

as a Wiggers diagram, after the physiologist who first created it.

Stroke Volume Is the Volume of Blood Pumped by One Ventricle in One Contraction

What is the purpose of leaving blood in the ventricles at the end of each contraction? For one thing, the resting end-systolic volume of 65 mL provides a safety margin. With a more forceful contraction, the heart will decrease its ESV, sending additional blood to the tissues. Like many physiological functions, the heart does not usually work "all out."

The amount of blood pumped by one ventricle during a contraction is known as the **stroke volume.** It is measured in milliliters per beat and can be calculated as follows:

Volume of blood in the ventricles before contraction −

volume of blood after contraction = stroke volume

EDV − ESV = stroke volume

For the average contraction in a person at rest:

135 mL − 65 mL = 70 mL, the normal stroke volume

Stroke volume is not constant and can increase to as much as 100 mL during exercise. Stroke volume, like heart rate, is homeostatically regulated by mechanisms discussed later in this chapter.

Cardiac Output Is a Measure of Cardiac Performance

How can we assess the effectiveness of the heart as a pump? One way is to measure cardiac output, the amount of blood pumped by the heart in a given period of time. All blood that leaves the heart flows through the tissues, therefore cardiac output is an indicator of total blood flow

through the body. However, cardiac output does not tell us how blood is distributed to various tissues. That aspect of blood flow is regulated at the tissue level.

Cardiac output is defined as the amount of blood pumped per ventricle per unit time. It can be calculated by multiplying heart rate (beats per minute) by stroke volume (mL/beat, or contraction). For an average resting heart rate of 72 beats per minute and stroke volume of 70 mL per beat:

Cardiac output (CO) = heart rate × stroke volume

CO = 72 beats/min × 70 mL/beat

CO = 5040 mL/min ≈ 5 L/min

Average total blood volume is about 5 liters. This means that at rest, one side of the heart pumps all the blood in the body through it in only one minute!

During exercise, cardiac output may increase to 30–35 L/min. Homeostatic changes in cardiac output are accomplished by varying the heart rate, the stroke volume, or both. Normally, cardiac output is the same for both ventricles. If one side of the heart begins to fail for some reason and is unable to pump efficiently, cardiac output becomes mismatched. In that situation, blood will pool in the circulation behind the weaker side of the heart.

✔ If the stroke volume of the left ventricle is 250 mL/beat and the stroke volume of the right ventricle is 251 mL/beat, what will happen to the relative distribution of blood between the systemic and pulmonary circulation after 10 beats?

Heart Rate Is Varied by Autonomic Neurons and Catecholamines

An average resting heart rate in an adult is 70 beats per minute (bpm). The normal range for heart rate is highly variable, however. Trained athletes may have resting heart rates of 50 bpm or less. Someone who is excited or anxious may have a rate of 125 bpm or higher. Children have higher average heart rates than adults. Although heart rate is initiated by autorhythmic cells in the SA node, it is modulated by neural and hormonal input.

The sympathetic and parasympathetic branches of the autonomic division influence heart rate through antagonistic control (Fig. 14-28 ■). Parasympathetic activity slows heart rate, while sympathetic activity speeds it up. Normally, the tonic control of heart rate is dominated by the parasympathetic branch.

This can be shown experimentally by blocking all autonomic input to the heart. When all sympathetic and parasympathetic input is blocked, the spontaneous depolarization rate of the SA node is 90–100 times per minute. This means that to achieve a heart rate of 70 beats per minute, tonic parasympathetic activity must slow the intrinsic rate of 90–100 beats per minute.

Fibrillation Coordination of myocardial contraction is essential for normal cardiac function. In extreme cases in which the myocardial cells contract in an extremely disorganized manner, a condition known as **fibrillation** results. Ventricular fibrillation is a life-threatening emergency because without coordinated contraction of the muscle fibers, the ventricles cannot pump enough blood to supply adequate oxygen to the brain. One way to correct this problem is to administer an electrical shock to the heart. The shock creates a depolarization that triggers action potentials in all cells simultaneously, coordinating them again. You have probably seen this procedure on television hospital shows, when the doctors place flat paddles on the patient's chest and tell everyone to stand back ("Clear!") while they pass an electrical current through the body.

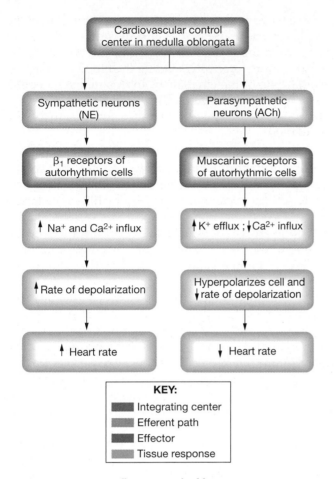

■ Figure 14-28   Reflex control of heart rate

...continued from page 429

Walter was in the cardiac care unit by 1 P.M., where the cardiologist visited him. "We need to keep an eye on you here for the next week. There is a good chance the damage from your heart attack could cause an irregular heartbeat." Once Walter is stable, he will have a coronary angiogram, a procedure in which an opaque dye visible on X-rays shows regions where the lumen of the vessel narrows. Depending on the results of that test, the physician may recommend either balloon angioplasty, in which a tube passed into the coronary artery is inflated to open up the blockage, or coronary bypass surgery, in which veins from other parts of the body are grafted onto the heart arteries to provide bypass channels around blocked regions.

Question 8: *If Walter's heart attack has damaged the muscle of his left ventricle, what do you predict will happen to his left cardiac output?*

An increase in heart rate can be achieved in two ways. The simplest method is to decrease the activity of the parasympathetic neurons. As parasympathetic influence is withdrawn from the autorhythmic cells, they resume their intrinsic rate of depolarization, and heart rate increases to 90–100 beats per minute. Sympathetic activity is required to increase heart rate above that value. As you learned earlier, norepinephrine (or epinephrine) on β_1 receptors speeds up the depolarization rate of the autorhythmic cells and increases heart rate.

Both autonomic branches also alter the rate of conduction through the AV node. Acetylcholine secreted by parasympathetic neurons slows the conduction of action potentials through the AV node, thus increasing AV node delay. In contrast, the catecholamines epinephrine and norepinephrine enhance conduction of action potentials through the AV node and through the conducting system.

Multiple Factors Influence Stroke Volume

Stroke volume, the amount of blood pumped per ventricle per contraction, is directly related to the force generated by cardiac muscle during a contraction. The greater the force of contraction, the greater the stroke volume. Force is affected by two parameters: the length

of the muscle fiber at the beginning of contraction and the contractility of the heart.

Contractility is the intrinsic ability of a cardiac muscle fiber to contract at any given fiber length. Changes in contractility are traditionally considered different from changes in force due to variations in muscle length, although we now know that Ca^{2+} plays a role in both.

Length-Tension Relationships and Starling's Law of the Heart   As you learned previously, the force created by myocardial muscle is directly related to the length of the sarcomere. As sarcomere length increases (up to an optimum length), tension created during contraction increases. In the intact heart, as tension increases, so does the stroke volume. If additional blood flows into the ventricles, the muscle fibers stretch and contract more forcefully, ejecting more blood.

This relationship between stretch and force in the intact heart was first described by a German physiologist, Otto Frank. A British physiologist, Ernest Starling, then expanded upon it. Starling used an isolated heart-lung preparation from a dog and hooked it to a reservoir so that he could regulate the amount of blood returning to the heart. In the absence of any nervous or hormonal control, he found that under normal conditions, the heart would pump all the blood that returned to it.

The relationship between stretch and force in the intact heart is plotted on a Starling curve (Fig. 14-29 ■). The *x*-axis represents the end-diastolic volume (EDV), the volume of blood in the ventricle at the beginning of contraction. Sarcomere length is proportional to the end-diastolic volume. The *y*-axis of the Starling curve represents the stroke volume, or amount of blood ejected from the ventricle per contraction.

The graph shows that stroke volume is proportional to force. As additional blood enters the heart, the heart contracts more forcefully. This relationship is known as

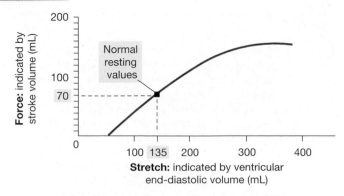

■ **Figure 14-29 Length-force relationships in the intact heart** As the end-diastolic volume increases, the stroke volume increases. In the intact heart, stroke volume is used as an indicator of contractile force.

the *Frank-Starling law of the heart*. It means that within physiological limits, the heart pumps all the blood that returns to it.

Stroke Volume and Venous Return According to the Frank-Starling law, the more blood there is in the ventricle at the beginning of contraction (increased end-diastolic volume), the greater the stroke volume is. End-diastolic volume in turn is determined by **venous return,** the amount of blood that enters the heart from the venous circulation. The degree of myocardial stretch created by venous return is sometimes called the *pre-load* on the heart because it represents the load placed on cardiac muscles before contraction.

What factors affect venous return? One is skeletal muscle contractions that squeeze the veins and push blood toward the heart (the *skeletal muscle pump*), and the so-called **respiratory pump** created by the movement of the thorax during inspiration (breathing in). As the chest expands and the diaphragm drops, the thoracic cavity, a sealed compartment, enlarges and develops a subatmospheric pressure. This low pressure creates lower pressure in the inferior vena cava as it passes through the thorax. The lower pressure helps draw more blood into the vena cava from veins in the abdomen.

The respiratory pump is aided by higher pressure placed on the outside of abdominal veins when the abdominal contents are compressed during inspiration. The combination of increased pressure in the abdominal veins and decreased pressure in thoracic veins enhances venous return during inspiration.

Constriction of veins by sympathetic activity is a third factor that affects venous return. When the veins constrict, their volume decreases. As a result, more blood flows out of the veins into the heart, and a larger vol-

ume of blood is in the ventricles at the beginning of the next contraction (increased end-diastolic volume). In turn, according to the Frank-Starling law, the ventricles contract more forcefully, sending the blood out into the arterial side of the circulation. In this manner, sympathetic innervation of veins allows the body to redistribute some venous blood to the arterial side of the circulation.

Reflex Control of Contractility Contractility of the heart is controlled by the nervous and endocrine systems. Any chemical that affects contractility is called an *inotropic agent* [*ino*, fiber]. For example, catecholamines and drugs such as digitalis enhance contractility and are therefore considered to have a positive inotropic effect.

Figure 14-30 ■ shows a normal Starling curve (the control curve) along with a curve showing how the stroke volume changes with increased contractility due to norepinephrine. Note that contractility is distinct from the length-tension relationship. A muscle can remain at one length (for example, one reflected by the end-diastolic volume marked A) but show increased contractility. Contractility increases as calcium available for contraction increases. As you learned earlier, catecholamines increase Ca^{2+} entry and storage as the basis for their positive inotropic effect.

The factors that affect cardiac output are summarized in Figure 14-31 ■. Cardiac output varies with both heart rate and stroke volume. Heart rate is modulated by the autonomic division of the nervous system and by

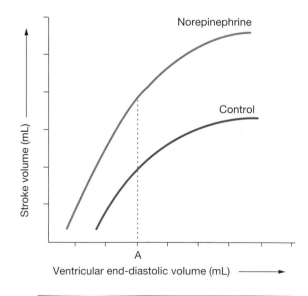

■ **Figure 14-30 The effect of norepinephrine on contractility of the heart** Norepinephrine is a positive inotropic agent.

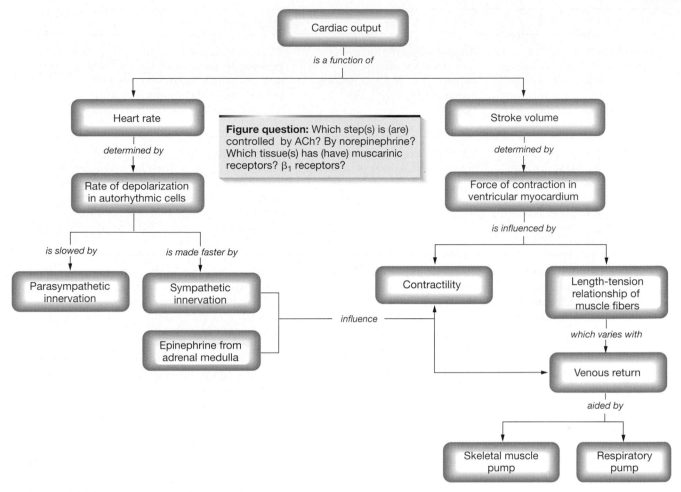

■ Figure 14-31   **Factors that affect cardiac output**

epinephrine. The determination of stroke volume is more complex, consisting of an intrinsic myocardial response to stretch (Starling's law), interacting with adrenergically mediated changes in contractility. Venous return is a major determinant of end-diastolic volume and stretch. In the next chapter, we will examine how cardiac output plays a key role in blood flow through the circulation.

 PROBLEM CONCLUSION

Walter's angiogram showed two blocked arteries, which were opened by balloon angioplasty. He returned home with instructions from his doctor for modifying his lifestyle to include a better diet, regular exercise, and no cigarette smoking.

In this running problem, you learned about some current techniques for diagnosing and treating heart attacks. You also learned that many of these treatments depend on speed to work effectively.

Further check your understanding of this running problem by comparing your answers to those in the summary table.

	Question	Facts	Integration and Analysis
1	Why did the medic give Walter oxygen?	The medic suspects that Walter has had a heart attack. Blood flow and oxygen supply to the heart muscle may be blocked.	If the heart is not pumping effectively, the brain may not receive adequate oxygen. Administration of oxygen will raise the amount of oxygen that reaches both the heart and the brain.

	Question	Facts	Integration and Analysis
2	What effect would the injection of isotonic saline have on Walter's extracellular fluid volume? On his intracellular fluid volume? On his total body body osmolarity?	An isotonic solution is one that does not change cell volume [∞ p. 137]. Isotonic saline is isosmotic to the body.	The extracellular volume will increase because all saline administered will remain in that compartment. The intracellular volume and total body osmolarity will not change.
3	Some of the enzymes that are used as markers for heart attacks are also found in related forms in skeletal muscle. What are related forms of an enzyme called?	Related forms of an enzyme are called isozymes [∞ p. 81].	Although isozymes are variants of the same enzymes, their activity may vary with different conditions, and their structures are slightly different. Cardiac and skeletal muscle isozymes can be distinguished by their different structures.
4	What is troponin, and and why would elevated blood levels of troponin indicate heart damage?	Troponin is the regulatory protein bound to tropomyosin [∞ p. 354]. Ca^{2+} binding to troponin uncovers the myosin-binding site of actin to allow contraction.	Troponin is part of the contractile apparatus of the muscle cell. If troponin escapes from the cell and enters the blood, this is an indication that the cell has been damaged or is dead.
5	How do electrical signals move from cell to cell in the myocardium?	Electrical signals can pass through gap junction in intercalated disks [∞ p. 57].	The cells of the heart are electrically linked by gap junctions.
6	What happens to contraction in a myocardial contractile cell if a wave of depolarization passing through the heart bypasses it?	Depolarization in a muscle cell is the signal for contraction.	If a myocardial cell is not depolarized, it will not contract. Failure to contract creates a nonfunctioning region of heart muscle and impairs the pumping function of the heart.
7	If the ventricles of the heart are damaged, in which wave or waves of the electrocardiogram would you expect to see abnormal changes?	The P wave of the ECG represents atrial depolarization. The QRS complex and T wave represent ventricular depolarization and repolarization, respectively.	The QRS complex and the T wave are mostly likely to show changes after a heart attack. Changes indicative of myocardial damage include enlargement of the Q wave, shifting of the S-T segment off the base-line (elevated or depressed), and inversion of the T wave.
8	If Walter's heart attack has damaged the muscle of his left ventricle, what do you predict will happen to his left cardiac output?	Cardiac output equals stroke volume (the amount of blood pumped by one ventricle per contraction) times heart rate.	If the ventricular myocardium has been weakened, stroke volume may decrease. Decreased stroke volume would in turn decrease cardiac output.

CHAPTER REVIEW

SUMMARY

Overview of the Cardiovascular System

1. The human **cardiovascular system** consists of a heart that pumps blood through a closed system of blood vessels. (p. 404)

2. The primary function of the cardiovascular system is the transport of nutrients, water, gases, wastes, and chemical signals among all parts of the body. (p. 404)

3. Blood vessels that carry blood away from the heart are called **arteries.** Blood vessels that return blood to the heart are called **veins.** Valves in the heart and veins ensure unidirectional blood flow. (p. 405)

4. The heart is divided into two **atria** and two **ventricles.** (p. 405)

5. The **pulmonary circulation** goes from the right side of the heart to the lungs and back to the heart. The **systemic circulation** goes from the left side of the heart to the tissues and back to the heart. (p. 405)

Pressure, Volume, Flow, and Resistance

6. Blood flows down a **pressure gradient** (ΔP), from the highest pressure in the arteries to the lowest pressure in the venae cavae and pulmonary veins. (p. 407)

7. In a system in which fluid is flowing, pressure decreases over distance. (p. 407)

8. The pressure created when the ventricles contract is called the **driving pressure** for blood flow. (p. 408)

9. **Resistance** of a fluid flowing through a tube increases as length of the tube and viscosity (thickness) of the fluid increase. Resistance increases as the radius (r) of the tube decreases. Radius has the greatest effect on resistance. (p. 409)

10. If resistance increases, flow decreases. If resistance decreases, flow increases. (p. 409)

11. Fluid flow through a tube is proportional to the pressure gradient (ΔP). A pressure gradient is not the same thing as the absolute pressure in the system. (p. 409)

12. **Flow rate** (flow) is the volume of blood that passes one point in the system per unit time. (p. 410)

13. **Velocity of flow** is the distance a volume of blood will travel in a given period of time. At a constant flow rate, blood will flow faster through a smaller tube than it will through a larger tube. (p. 410)

Cardiac Muscle and the Heart

14. The heart is composed mostly of cardiac muscle, or myocardium. Most cardiac muscle is typical striated muscle. (p. 414)

15. The signal for contraction originates in **autorhythmic cells** within the heart. Autorhythmic cells are noncontractile myocardium. (p. 415)

16. Myocardial cells are linked by **intercalated disks** that contain gap junctions. Gap junctions allow depolarization to spread rapidly from cell to cell. (p. 415)

17. In contractile cells, an action potential opens voltage-gated Ca^{2+} channels. Ca^{2+} enters the cell and triggers the release of Ca^{2+} from the sarcoplasmic reticulum through **calcium-induced calcium release.** (p. 417)

18. The force of cardiac muscle contraction can be graded according to how much Ca^{2+} enters the cell. (p. 417)

19. Epinephrine and norepinephrine increase the force of myocardial contraction when they bind to β_1 adrenergic receptors. They also shorten the duration of cardiac contraction. (p. 418)

20. As initial muscle fiber length increases, force of contraction increases. (p. 419)

21. The action potentials of myocardial contractile cells have a rapid depolarization phase created by Na^+ influx and a steep repolarization phase due to K^+ efflux. The action potential also has a plateau phase created by Ca^{2+} influx. (p. 420)

22. **Autorhythmic** myocardial cells have an unstable membrane potential called a **pacemaker potential.** The pacemaker potential is due to I_f **channels** that allow net influx of positive charge. (p. 421)

23. The steep depolarization phase of the autorhythmic cell action potential is caused by Ca^{2+} influx. The repolarization phase is due to K^+ efflux. (p. 422)

24. Norepinephrine and epinephrine act on β_1 receptors to speed up the rate of the pacemaker depolarization and increase heart rate. Acetylcholine activates muscarinic receptors and slows down heart rate. (p. 422)

The Heart as a Pump

25. Action potentials originate at the **sinoatrial node (SA node)** and spread rapidly from cell to cell. Action potentials are followed by a wave of contraction that passes across the atria, then moves into the ventricles. (p. 424)

26. The electrical signal moves from the SA node through the **internodal pathway** to the **atrioventricular node (AV node),** then into the **bundle of His,** bundle branches, **Purkinje fibers,** and myocardial contractile cells. (p. 424)

27. The SA node sets the pace of the heartbeat. If the SA node malfunctions, other autorhythmic cells in the AV node or ventricles will take control of the heart rate. (p. 424)

28. An **electrocardiogram,** or **ECG,** is a surface recording of the electrical activity of the heart. The **P wave** represents atrial depolarization. The **QRS complex** represents ventricular depolarization. The **T wave** represents ventricular repolarization. Atrial repolarization is incorporated in the QRS complex. (p. 426)

29. An ECG provides information on heart rate and rhythm, conduction velocity, and the condition of tissues within the heart. (p. 427)

30. One **cardiac cycle** includes one cycle of contraction and relaxation. **Diastole** is the relaxation phase. **Systole** is the contraction phase. (p. 429)

31. Most blood enters the ventricles while the atria are relaxed. Only 20% of ventricular filling is due to atrial contraction. (p. 429)

32. The AV valves prevent backflow of blood into the atria. Vibrations following closure of the AV valves create the first heart sound. (p. 430)

33. During **isovolumic ventricular contraction,** the ventricular blood volume does not change but pressure rises. When ventricular pressure exceeds arterial pressure, the semilunar valves open and blood is ejected into the arteries. (p. 430)

34. When the ventricles relax and ventricular pressure falls, the semilunar valves close, creating the second heart sound. (p. 431)

35. The amount of blood pumped by one ventricle during a contraction is known as the **stroke volume.** (p. 434)

36. **Cardiac output** is the amount of blood pumped per ventricle per unit time. It is equal to heart rate times stroke volume. The average cardiac output at rest is 5 L/min. (p. 434)

37. Homeostatic changes in cardiac output are accomplished by varying heart rate, stroke volume, or both. (p. 434)

38. Parasympathetic activity slows heart rate. Sympathetic activity speeds it up. (p. 434)

39. The Frank-Starling law of the heart says that an increase in end-diastolic volume will result in a greater stroke volume. (p. 435)

40. End-diastolic volume is determined by **venous return.** Factors that affect venous return include skeletal muscle contractions, the **respiratory pump,** and constriction of veins by sympathetic activity. (p. 436)

41. Contractility of the heart is enhanced by catecholamines and certain drugs. (p. 436)

QUESTIONS

LEVEL ONE   Reviewing Facts and Terms

1. What contributions to understanding the cardiovascular system did each of these people make?

    (a) William Harvey
    (b) Frank and Starling
    (c) Malpighi

2. List three functions of the cardiovascular system.

3. Put these structures in the order blood passes through them, starting and ending with the left ventricle:

    (a) left ventricle
    (b) systemic veins
    (c) pulmonary circulation
    (d) systemic arteries
    (e) aorta
    (f) right ventricle

4. The primary reason blood flows through the body is due to a _____ gradient. In humans, this value is highest at the _____ and in the _____. It is lowest in the _____. In a system in which fluid is flowing, pressure decreases over distance due to _____.

5. If vasodilation occurs in a blood vessel, pressure (increases/decreases).

6. The specialized cell junctions between myocardial cells are called _____. These areas also contain _____ that allow rapid conduction of electrical signals.

7. Trace an action potential from the SA node through the conducting system of the heart.

8. Distinguish between the following pairs:

    (a) end-systolic volume and end-diastolic volume
    (b) sympathetic and parasympathetic control of heart rate
    (c) diastole and systole
    (d) systemic and pulmonary circulation
    (e) AV node and SA node

9. Match the descriptions with the correct anatomic terms. Not all terms are used. Give a definition for the unused terms.

    (a) tough membranous sac that encases the heart
    (b) valves between ventricles and the main arteries
    (c) a vessel that carries blood away from the heart
    (d) lower chamber of the heart
    (e) valve between left atrium and left ventricle
    (f) primary artery of the systemic circulation
    (g) muscular layer of the heart
    (h) narrow end of the heart; points downward
    (i) valve with papillary muscles
    (j) the upper chambers of the heart

    1. aorta
    2. apex
    3. artery
    4. atria
    5. atrium
    6. AV valve
    7. base
    8. bicuspid valve
    9. endothelium
    10. myocardium
    11. pericardium
    12. semilunar valve
    13. tricuspid valve
    14. ventricle

10. What events cause the two principal heart sounds?

11. What is the proper term for each of the following?

    (a) number of heart contractions per minute
    (b) volume of blood in the ventricle before the heart contracts
    (c) volume of blood that enters the aorta with each contraction
    (d) volume of blood that leaves the heart in one minute
    (e) volume of blood in the entire body

LEVEL TWO   Reviewing Concepts

12. List the events of the cardiac cycle in sequence, beginning with atrial and ventricular diastole. Note where valves open and close. Describe what happens to pressure and blood flow in each chamber at each step of the cycle.

13. Compare and contrast the structure of a cardiac muscle cell with that of a striated muscle cell. What unique properties of cardiac muscle are essential to its function?

14. **Concept maps:** You may add any terms you like to the lists given.

    (a) Create a map showing the flow of blood through the heart and body. Label as many structures as you can.
    (b) Map the following terms to show how they influence cardiac output.

β_1 receptor	heart rate
ACh	length-tension relationship
adrenal medulla	muscarinic receptor
autorhythmic cells	norepinephrine
Ca^{2+}	parasympathetic neurons
Ca^{2+}-induced Ca^{2+} release	respiratory pump
cardiac output	skeletal muscle pump
contractile myocardium	stroke volume
contractility	sympathetic neurons
force of contraction	venous return

15. Explain why contractions in cardiac muscle cannot sum or exhibit tetanus.

16. Correlate the waves of the ECG with mechanical events in the atria and ventricles. Why are there only three electrical events but four mechanical events?

17. Match the ion movements with the appropriate phrase. More than one answer may apply to a single phrase. Answers may not all be used.

    (a) slow rising phase of autorhythmic cells
    (b) plateau phase of contractile cells
    (c) rapid rising phase of contractile cells
    (d) rapid rising phase of autorhythmic cells
    (e) rapid falling phase of contractile cells
    (f) falling phase of autorhythmic cells
    (g) cardiac muscle contraction
    (h) cardiac muscle relaxation

    1. K^+ from ECF to ICF
    2. K^+ from ICF to ECF
    3. Na^+ from ECF to ICF
    4. Na^+ from ICF to ECF
    5. Ca^{2+} from ECF to ICF
    6. Ca^{2+} from ICF to ECF

18. List and briefly explain four types of information that the ECG provides about the heart.

19. Define inotropic effect. Give examples of two drugs with a positive inotropic effect. How is calcium related?

LEVEL THREE   Problem Solving

20. Two drugs used to reduce cardiac output are calcium channel blockers and beta (receptor) blockers. What effect do these drugs have on the heart that explains how they decrease cardiac output?

21. Police Captain Jeffers has suffered a myocardial infarction.

    (a) Explain to his (nonmedically oriented) family what has happened to his heart.
    (b) When you analyzed his ECG, you referred to several different leads, such as lead I. What are leads?
    (c) Why is it possible to record an ECG on the body surface without direct access to the heart?

22. What might cause a longer-than-normal P-R interval in an ECG?

23. The following paragraph is a summary of a newspaper article:

    A new treatment for atrial fibrillation due to an excessively rapid rate at the SA node involves a high-voltage electrical pulse administered to the AV node to destroy its autorhythmic cells. A ventricular pacemaker is then implanted in the patient.

    Briefly explain the physiological rationale for this treatment. Why is a rapid atrial depolarization rate dangerous, why is the AV node destroyed, and why must a pacemaker be implanted?

LEVEL FOUR   Quantitative Problems

24. Police Captain Jeffers in question 21 above has an ejection fraction (SV divided by EDV) of only 25%. His stroke volume is 40 mL and his heart rate is 100 bpm. What are his EDV, ESV, and CO? Show your calculations.

25. If 1 cm water = 0.74 mm Hg:

    (a) Convert a pressure of 120 mm Hg to cm H_2O.
    (b) Convert a pressure of 80 mm Hg to cm H_2O.

26. Calculate cardiac output if stroke volume is 65 mL and heart rate is 80 bpm.

27. Calculate end-systolic volume if end-diastolic volume is 150 mL and stroke volume is 65 mL.

28. A person has a total blood volume of 5 liters. Of this total, assume 4 liters is contained in the systemic circulation and 1 liter is in the pulmonary circulation. If the person has a cardiac output of 5 L/min: (a) How long will it take for a drop of blood leaving the left ventricle to return to the left ventricle? (b) How long will it take for a drop of blood to go from the right ventricle to the left ventricle?

E X P L O R E <MediaLab>

Introduction

This chapter introduced you to cardiovascular physiology and explained how nutrients, oxygen, and hormones are transferred throughout the body. The cardiovascular system is a wonderful example of both electrical and mechanical physiology. The heart is made of a variety of tissues that function distinctly yet in cooperation with one another. This MediaLab will help you expand your understanding of the heart and circulatory system. After reading the description below, visit the MediaLab for Chapter 14 in your Companion Website and select the appropriate keyword.

Web Exploration 1

Estimated time for completion = 15 minutes

Your heart beats about 3,000,000,000 times during your life. Each of these beats arises from an action potential. Draw a diagram of a ventricular action potential and label the five distinct phases. Annotate this drawing to show which ions are moving, at what phase they are moving, and in what direction they are moving (into or out of the cell). Select the keyword **ACTION POTENTIAL** and compare your drawing to that of the diagram provided on the Website. Note the correlation between phases of the action potential and the electrocardiogram.

Web Exploration 2

Estimated time for completion = 10 minutes

You decide to start running five miles three times a week. Over time you notice your resting heart rate slowing. What is happening to resting stroke volume? How does this affect total cardiac output? What is happening at the cellular level to account for the change in stroke volume? How can a myocardial cell increase its force of contraction?

Based on your answers, how do you think the resting cardiac output of a trained athlete compares with that of a nonathlete? What is the hemodynamic factor (heart rate, stroke volume, or peripheral resistance) that contributes most to increased cardiac output during strenuous exercise? Select the keyword **CARDIAC OUTPUT** from your Website and read about how exercise changes cardiac dynamics. Were your answers correct?

15 Blood Flow and the Control of Blood Pressure

■ "59,700,000 Americans have one or more types of cardiovascular disease (CVD) according to current estimates. … Since 1900, CVD has been the No. 1 killer in the United States every year but 1918."—1999 Heart and Stroke Statistical Update, *American Heart Association* ■

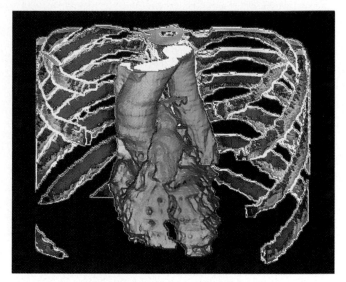

Colored scan of the heart enclosed in the rib cage

BACKGROUND BASICS

Basement membrane (p. 57)
Nitric oxide (p. 163)
Transcytosis (p. 133)
Fight-or-flight response (p. 327)
Exchange epithelium (p. 59)
Catecholamines and receptors (pp. 190, 334)
Caveolae (p. 129)
Diffusion (p. 118)
Tonic control (p. 167)

Anthony was always sure he was going to be a physician, until the day in physiology laboratory that they studied blood types. When the lancet pierced his fingertip and he saw the drop of bright red blood well up, the room started to spin, then everything went black. He awoke, much embarrassed, to the sight of his classmates and the teacher bending over him.

Anthony suffered an attack of *vasovagal syncope* (fainting), a benign and common reaction to blood, hypodermic needles, or other upsetting sights. Normally, homeostatic regulation of the cardiovascular system maintains adequate blood flow (*perfusion*) to the heart and brain. In vasovagal syncope, the signals from the nervous system cause a sudden fall in blood pressure, and the patient faints from lack of oxygen to the brain. In this chapter you will learn how the heart and blood vessels work together most of the time to prevent such problems.

PROBLEM
Essential Hypertension

"Doc, I'm as healthy as a horse," says Kurt English, age 56, during his annual physical examination. "I don't want to waste your time. Let's get this over with." But to Dr. Arthur Cortez, Kurt does not appear to be the picture of health: he is about 30 pounds overweight. When Dr. Cortez asks about his diet, Kurt replies, "Well, I like to eat." Exercise? "Who has the time?" replies Kurt. Dr. Cortez slips a blood pressure cuff around Kurt's arm and takes a reading. "Your blood pressure is 164 over 100," says Dr. Cortez. "We'll take it again in 15 minutes. If it's still high, we'll need to discuss it further." Kurt stares at his doctor, flabbergasted. "But how can my blood pressure be too high? I feel fine!" he protests.

...continued on page 450

A simplified model of the cardiovascular system (Fig. 15-1 ■) summarizes the key points that we will discuss in this chapter. This model shows the heart as two separate pumps that work in series (one after the other), with the right heart pumping blood to the lungs, then to the left heart. The left heart then pumps blood through the rest of the body.

Blood leaving the left heart enters the systemic arteries, shown here as an expandable, elastic region. Pressure produced by contraction of the left ventricle is stored in the elastic walls of the arteries and slowly released through *elastic recoil*. This mechanism maintains a continuous driving pressure for blood flow during the time when the ventricles are relaxing.

The arterioles create a high-resistance outlet for arterial blood flow. In addition, the arterioles direct distribution of blood flow to individual tissues by selectively constricting and dilating. Arteriolar diameter is regulated both by local factors, such as tissue oxygen, and by homeostatic control.

Once blood flows into the capillaries, a leaky epithelium allows exchange of material between the plasma, the interstitial fluid, and the cells of the body. At the distal end of the capillaries, blood flows into the venous side of the circulation and from there back to the right side of the heart.

Total blood flow through any level of the circulation is equal to the cardiac output. For example, if cardiac output is 5 L/min, blood flow through all the systemic capillaries is 5 L/min. In the same manner, blood flow through the pulmonary side of the circulation is equal to blood flow through the systemic circulation.

▶ THE BLOOD VESSELS

The walls of blood vessels are made from layers of smooth muscle, elastic connective tissue, and fibrous connective tissue (Fig. 15-2 ■). The inner lining of all blood vessels is a thin layer of **endothelium,** a type of epithelium. For years, the endothelium was thought to be simply a passive barrier. However, we now know that

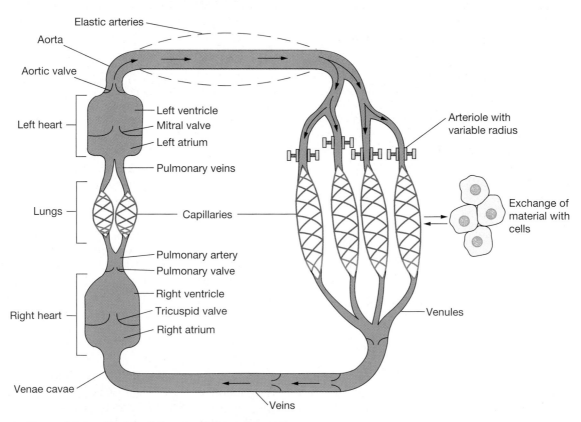

■ **Figure 15-1** **Model of the cardiovascular system**

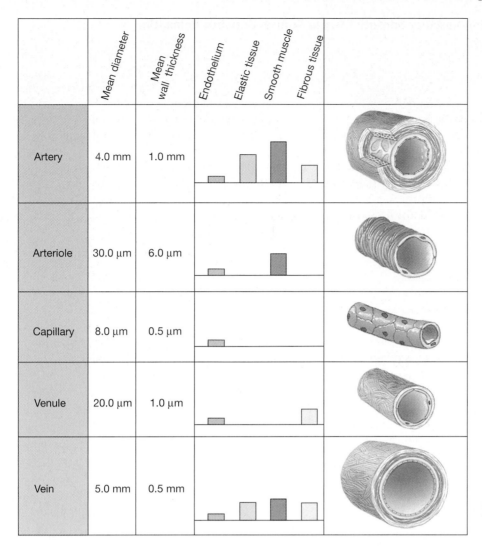

	Mean diameter	Mean wall thickness	Endothelium	Elastic tissue	Smooth muscle	Fibrous tissue	
Artery	4.0 mm	1.0 mm					
Arteriole	30.0 μm	6.0 μm					
Capillary	8.0 μm	0.5 μm					
Venule	20.0 μm	1.0 μm					
Vein	5.0 mm	0.5 mm					

■ **Figure 15-2   Blood vessels**   The walls of blood vessels vary in their diameter and composition.

endothelial cells play important roles in the regulation of blood pressure, blood vessel growth, and absorption of materials.

Surrounding the endothelium are layers of connective tissue and smooth muscle. The thickness and type of connective tissue vary with different vessels, as does the thickness of the smooth muscle layer. The descriptions below apply to the vessels of the systemic circulation, although those of the pulmonary circulation are generally similar.

Blood Vessels Contain Vascular Smooth Muscle

The smooth muscle of blood vessels is known as **vascular smooth muscle.** Most blood vessels contain smooth muscle, arranged in either circular or spiral layers. *Vasoconstriction* narrows the diameter of the vessel lumen and *vasodilation* widens it.

In most blood vessels, smooth muscle cells maintain a state of partial contraction at all times, creating the condition known as *muscle tone.* Contraction of smooth muscle, like that of cardiac muscle, depends on the entry of Ca^{2+} from the extracellular fluid through Ca^{2+} channels [∞ p. 417].

A variety of chemicals influences vascular smooth muscle tone, including neurotransmitters, hormones, and paracrines (Table 15-1). Paracrines [∞ p. 154] are a particularly important component in the local control of blood flow. Many vasoactive paracrines are secreted by endothelial cells lining the blood vessels or by the tissues surrounding the vessels.

Calcium Channel Blockers   The calcium channels of both vascular smooth muscle and cardiac muscle can be blocked by a class of drugs known as calcium channel blockers. These drugs bind to the Ca^{2+} channel proteins, making it less likely that the channels will open in response to depolarization. With diminished Ca^{2+} entry, vascular smooth muscle dilates, while in the heart, the depolarization rate of the SA node decreases. Vascular smooth muscle is more sensitive than cardiac muscle to certain classes of calcium channel blockers, and it is possible to get vasodilation at low doses that have no effect on heart rate. Other tissues with Ca^{2+} channels, such as neurons, are only minimally affected by calcium channel blockers because their Ca^{2+} channels are a different subtype.

TABLE 15-1   Chemicals Mediating Vascular Smooth Muscle Contraction and Relaxation

Chemical	Physiological Role	Source	Type
Contraction			
Norepinephrine (α receptors)	Baroreceptor reflex	Sympathetic neurons	Neural
Endothelin	Paracrine mediator	Vascular endothelium	Local
Serotonin	Platelet aggregation, smooth muscle contraction	Neurons, digestive tract, platelets	Local, neural
Substance P	Pain, increase capillary permeability	Neurons, digestive tract	Local, neural
Vasopressin	Increase blood pressure in hemorrhage	Posterior pituitary	Hormonal
Angiotensin II	Increase blood pressure	Plasma hormone	Hormonal
Prostacyclin	Minimize blood loss from damaged vessels before coagulation	Endothelium	Local
Relaxation			
Nitric oxide	Paracrine mediator	Endothelium	Local
Atrial natriuretic peptide	Reduce blood pressure	Atrial myocardium, brain	Hormonal
Vasoactive intestinal peptide	Digestive secretion, relax smooth muscle	Neurons	Neural, hormonal
Histamine	Increase blood flow	Mast cells	Local, systemic
Epinephrine (β₂ receptors)	Enhance local blood flow to skeletal muscle, heart, liver	Adrenal medulla	Hormonal
Acetylcholine (muscarinic receptors)	Erection of clitoris or penis	Parasympathetic neurons	Neural
Bradykinin	Increase blood flow via nitric oxide	Multiple tissues	Local
Adenosine	Enhance blood flow to match metabolism	Hypoxic cells	Local

Arteries and Arterioles Carry Blood Away from the Heart

The aorta and major arteries are characterized by walls that are both stiff and springy. Arteries have thick smooth muscle layers with large amounts of elastic and fibrous tissue (Fig. 15-2 ■). The stiffness of the fibrous tissue requires substantial amounts of energy to stretch the walls of an artery outward. This energy comes in the form of high-pressure blood ejected from the left ventricle. Once the artery is distended with blood, the energy stored by stretching elastic fibers is released through elastic recoil.

The arteries and arterioles are characterized by a divergent pattern of blood flow. As the major arteries divide into smaller and smaller arteries, the character of the wall changes, becoming less elastic and more muscular. Finally, the smallest arteries become **arterioles.** The walls of arterioles contain several layers of smooth muscle that contract and relax under the influence of various chemical signals.

Some arterioles branch into vessels known as **metarterioles** [*meta-*, beyond]. Metarterioles have only part of their wall surrounded by smooth muscle, in comparison with true arterioles, whose walls have a continuous muscle layer. Blood flowing through metarterioles either can be directed into adjoining capillary beds (Fig. 15-3 ■) or can bypass the capillaries and go directly to the

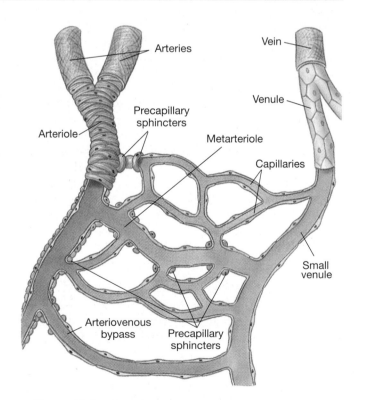

■ **Figure 15-3   Metarterioles**   The metarterioles act as a by-pass channel that allows white blood cells to go directly from the arterial to the venous side of the circulation. They also regulate flow into capillary beds.

venous circulation if precapillary sphincters (rings of muscle) are contracted.

In addition to regulating blood flow through the capillaries, metarterioles allow white blood cells to go directly from the arterial to the venous circulation. Capillaries are barely large enough to let red blood cells through, much less white blood cells that are twice as large.

The arterioles, along with the capillaries and venules, are called the *microcirculation.* The regulation of blood flow through the microcirculation is an active area of physiological research.

Exchange Between the Blood and Interstitial Fluid Takes Place in the Capillaries

Capillaries are the smallest vessels in the cardiovascular system. They and the postcapillary venules are the site of exchange between the blood and the interstitial fluid. To facilitate exchange of material, capillary walls lack smooth muscle and elastic or fibrous tissue reinforcement (Fig. 15-2 ■). Instead, they consist of a flat layer of endothelium, one cell thick, supported on an acellular matrix called the *basement membrane* or *basal lamina* [∞ p. 57]. Because the capillary endothelium is an exchange epithelium, it usually has leaky junctions between the cells. The notable exception to this pattern is the tight-junction endothelium of cerebral capillaries that creates the blood-brain barrier [∞ p. 255].

Blood Flow Converges in the Venules and Veins

Blood flows from the capillaries into small vessels called **venules.** The very smallest venules are similar to capillaries, with a thin exchange epithelium and little connective tissue (Fig. 15-2 ■). They are distinguished from capillaries by their convergent pattern of flow.

Smooth muscle begins to appear in the walls of larger venules. From venules, blood flows into veins that become larger in diameter as they travel toward the heart.

Pericytes   Although capillaries are generally described as having walls composed of a single layer of endothelium, they are closely associated with a type of cell known as the **pericyte** [*peri-*, around]. These elongated, highly branched cells surround the capillaries, forming a meshlike outer layer between the endothelium and the interstitial fluid. Pericytes apparently contribute to the "tightness" of capillary permeability: the more pericytes, the less leaky the capillary endothelium. Pericytes are thought to contribute to many aspects of capillary function including growth, repair following injury, and local control of blood flow through secretion of vasoactive agents. Pericytes may also play a role in several diseases that affect the microcirculation, including diabetes, high blood pressure, and Alzheimer's disease. It is likely that in the coming years we will begin to see new therapies for these conditions that target the pericytes.

Finally, the largest veins, the venae cavae, empty into the right atrium.

Veins are more numerous than arteries and have a larger diameter. As a result of their large volume, the veins hold more than half of the blood in the circulatory system. Veins lie closer to the surface of the body than arteries, forming the bluish blood vessels that you see running just under the skin. Veins have thinner walls than arteries, with less elastic tissue. As a result, they expand easily when they fill with blood.

When you have blood drawn from your arm (*venipuncture*), the technician uses a tourniquet to exert pressure on the blood vessels. Blood flow coming into the arm through high-pressure arteries is not affected, but pressure of the tourniquet stops outflow through the low-pressure veins. As a result, blood collects in the surface veins, making them stand out against the underlying muscle tissue.

Angiogenesis Creates New Blood Vessels

One topic of tremendous interest to researchers is **angiogenesis,** the process by which new blood vessels develop, especially after birth [*angeion*, vessel + *gignesthai*, to beget]. In children, capillary growth is necessary for normal development. In adults, angiogenesis takes place during wound healing and during growth of the uterine lining after menstruation. Angiogenesis also occurs with endurance exercise training, enhancing blood flow to the heart muscle and to skeletal muscles.

From studies of normal blood vessels and tumor cells, scientists learned that angiogenesis is controlled by a balance of angiogenic and antiangiogenic cytokines. A number of related growth factors, including *vascular endothelial growth factor* (VEGF) and *fibroblast growth factor* (FGF), promote angiogenesis. These growth factors are *mitogens*, meaning they promote mitosis, or cell division. They are normally produced by smooth muscle cells and pericytes. Cytokines that inhibit angiogenesis include *angiostatin,* made from the blood protein plasminogen, and *endostatin* [*stasis,* a state of standing still].

Scientists are currently using these cytokines to develop new treatments for two major illnesses: cancer and coronary artery disease. As cancer cells invade tissues and multiply, they must grow new blood vessels to maintain a supply of nutrients and oxygen. Without these vessels, the interior cells of a cancerous mass are unable to get adequate oxygen and nutrients, so they die. Angiostatin and endostatin are being used in clinical trials to see if they can block angiogenesis and literally starve tumors to death.

Coronary artery disease, in contrast, is a condition in which we would like to be able to selectively induce angiogenesis. In coronary artery disease, blood flow to the myocardium is decreased by fatty deposits that narrow the lumen of the coronary blood vessels. If we can induce the growth of new blood vessels to replace the vessels that are becoming blocked, we might be able to prevent the damage caused by lack of oxygen. Both VEGF and FGF are being tested in clinical trials to see how well they promote angiogenesis in people suffering from *myocardial ischemia* [∞ p. 417]. If successful, this same technique might be applied to people who have suffered *strokes* in which blood supply to the brain is diminished because of a blocked blood vessel.

▶ BLOOD PRESSURE

The pressure created by ventricular contraction is the driving force for blood flow through the vessels of the system [∞ p. 408]. As blood leaves the left ventricle, the aorta and arteries expand to accommodate it (Fig. 15-4a ■). When the ventricle relaxes and the semilunar valve closes, the elastic arterial walls recoil, propelling the blood forward into the smaller arteries and arterioles (Fig. 15-4b ■).

By sustaining the driving pressure for blood flow during ventricular relaxation, the arteries produce continuous blood flow through the blood vessels. Flow in the arterial side of the circulation is pulsatile, reflecting the changes in arterial pressure throughout a cardiac cycle. Once past the arterioles, pulse waves disappear.

Blood Pressure in the Systemic Circulation Is Highest in the Arteries and Lowest in the Veins

Blood pressure is highest in the arteries and falls continuously as blood flows through the circulatory system. The decrease in pressure occurs because energy is lost due to the resistance of the vessels to blood flow (Fig. 15-5 ■). Resistance to blood flow also comes from friction between the blood cells.

In the systemic circulation, the highest pressure occurs in the aorta and reflects the pressure created by the left ventricle. Aortic pressure reaches an average high of 120 mm Hg during ventricular systole (**systolic pressure**), then falls steadily to a low of 80 mm Hg during ventricular diastole (**diastolic pressure**). Notice that although pressure in the ventricle falls to nearly 0 mm Hg as the ventricle relaxes, the diastolic pressure in the large arteries remains relatively high. The high diastolic pressure in the arteries reflects the ability of those vessels to capture and store energy in their elastic walls.

The rapid pressure increase that occurs when the left ventricle pushes blood into the aorta can be felt as a **pulse,** or pressure wave, transmitted through the fluid-filled arteries of the cardiovascular system. The pressure wave travels about 10 times as fast as the blood itself.

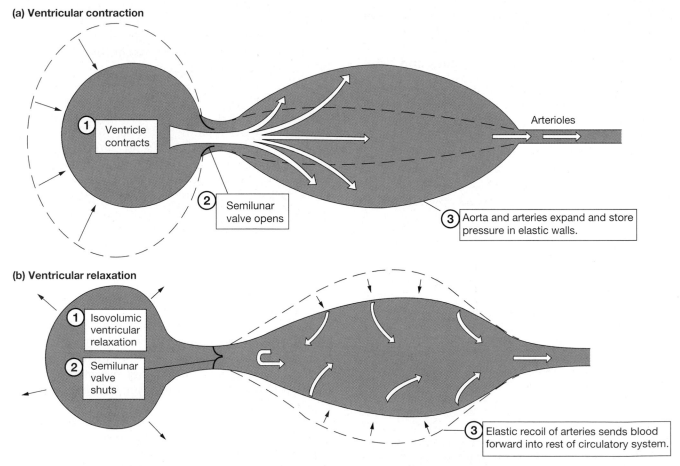

(a) Ventricular contraction

① Ventricle contracts

② Semilunar valve opens

③ Aorta and arteries expand and store pressure in elastic walls.

Arterioles

(b) Ventricular relaxation

① Isovolumic ventricular relaxation

② Semilunar valve shuts

③ Elastic recoil of arteries sends blood forward into rest of circulatory system.

■ **Figure 15-4** **Elastic recoil in the arteries** (a) During ventricular contraction, the elastic walls of the arteries expand as the arteries fill with high-pressure blood. (b) During ventricular relaxation, once the semilunar valve shuts, elastic recoil of the arterial walls pushes the arterial blood forward into the circulatory system.

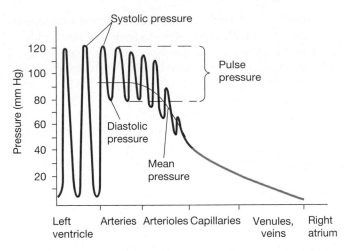

■ **Figure 15-5   Pressure throughout the systemic circulation**   Pressure waves created by ventricular contraction are reflected into the blood vessels. They diminish in amplitude with distance and disappear at the capillaries.

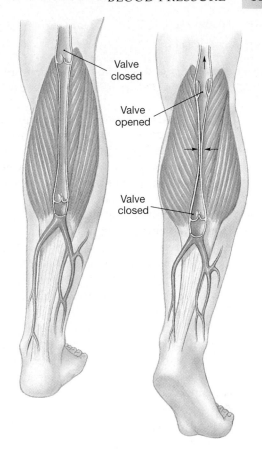

■ **Figure 15-6   Valves create one-way flow in the veins**   When the skeletal muscles compress the veins, they force blood toward the heart (the skeletal muscle pump). Valves in the veins prevent backflow of blood.

Even so, a pulse felt in the arm is occurring slightly after the ventricular contraction that created the wave.

The amplitude of the pulse decreases over distance because of friction, finally disappearing at the capillaries (Fig. 15-5 ■). **Pulse pressure,** a measure of the strength of the pressure wave, is defined as systolic pressure minus diastolic pressure:

Systolic pressure − diastolic pressure = pulse pressure

For example, in the aorta:

120 mm Hg − 80 mm Hg = 40 mm Hg pulse pressure

By the time blood flow reaches the veins, pressure has fallen because of friction, and a pulse wave no longer exists. Low-pressure blood in veins below the heart must flow "uphill," or against gravity, to return to the heart. Try holding your arm straight down without moving for several minutes and notice how the veins in the back of your hand begin to stand out as they fill with blood. (This effect may be more evident in older people, whose subcutaneous connective tissue has lost elasticity.) Then raise your hand so that gravity assists the venous flow and watch the bulging veins disappear.

To assist venous flow, some veins have internal one-way valves (Fig. 15-6 ■). These valves, like those in the heart, ensure that blood passing the valve cannot flow backward. Once blood reaches the vena cava, there are no valves. Blood flow is steady, pushed along by the continuous movement of blood into the venous circulation from the capillaries.

Venous return to the heart is aided by the *skeletal muscle* and *respiratory pumps* [∞ p. 436]. When muscles such as those in the calf of the leg contract, they compress the veins, forcing the blood upward past the valves. While your hand is hanging down, try clenching and unclenching your fist to see the effect muscle contraction has on distention of the veins.

✔ Do you think that veins from the brain to the heart have valves? Defend your answer.

✔ If you checked the pulse in a person's carotid artery and left wrist at the same time, would the pressure waves occur simultaneously? Explain.

Arterial Blood Pressure Reflects the Driving Pressure for Blood Flow

Arterial blood pressure, or simply "blood pressure," reflects the driving pressure created by pumping action of the heart. Because ventricular pressure is difficult to measure, it is customary to assume that arterial blood pressure reflects ventricular pressure. Arterial pressure is pulsatile, so we use a single value for arterial pressure that is representative of the driving pressure. This number is the **mean arterial pressure** (MAP), estimated as the diastolic pressure plus one-third of the pulse pressure:

MAP = diastolic P + 1/3(systolic P − diastolic P)

For a person whose systolic pressure is 120 and diastolic pressure is 80:

$$MAP = 80 \text{ mm Hg} + 1/3(120 - 80 \text{ mm Hg})$$

$$MAP = 93 \text{ mm Hg}$$

Mean arterial pressure is closer to diastolic pressure than to systolic pressure because diastole lasts twice as long as systole.

This formula for MAP applies to a person whose heart rate is in the range of 60–80 beats per minute, a typical resting heart rate. If heart rate increases, the relative amount of time the heart spends in diastole decreases. In that case, the contribution of systolic pressure to mean arterial pressure increases.

Abnormally high or low arterial blood pressure can be indicative of a problem in the cardiovascular system. If blood pressure drops too low (*hypotension*), the driving force for blood flow will be unable to overcome opposition by gravity. In this instance, blood flow and oxygen supply to the brain are impaired, and the subject may become dizzy or faint.

On the other hand, if blood pressure is chronically elevated (a condition known as *hypertension*, or high blood pressure), high pressure on the walls of blood vessels may cause weakened areas to rupture and bleed into the tissues. If a rupture occurs in the brain, it is called a *cerebral hemorrhage* and may cause the loss of neurological function commonly called a *stroke*. If a weakened area ruptures in a major artery such as the descending aorta, rapid blood loss into the abdominal cavity will cause blood pressure to fall below the critical minimum. Without prompt treatment, rupture of a major artery is fatal.

✔ Peter's systolic pressure is 112 mm Hg and his diastolic pressure is 68 mm Hg (written 112/68). What is his pulse pressure? His mean arterial pressure?

Blood Pressure Is Estimated by Sphygmomanometry

We estimate arterial blood pressure in the radial artery of the arm using a *sphygmomanometer*, an instrument consisting of an inflatable cuff and a pressure gauge [*sphyg-*

• •

...continued from page 444

Kurt's second blood pressure reading is 158/98. Dr. Cortez asks him to take his blood pressure at home daily for two weeks, then return to the office. When Kurt comes back with his diary, the story is the same: his blood pressure continues to average 160/100. After running some tests, Dr. Cortez concludes that Kurt is one of approximately 50 million adult Americans with high blood pressure, also called hypertension. If not controlled, hypertension can lead to congestive heart failure, stroke, and kidney failure.

Question 1: *Why are people with high blood pressure at higher risk for having a hemorrhagic, or bleeding, stroke?*

• •

mus, pulse + *manometer*, an instrument for measuring pressure of a fluid]. The cuff encircles the upper arm and is inflated until it exerts pressure higher than the systolic pressure driving arterial blood. When cuff pressure exceeds arterial pressure, blood flow into the lower arm stops (Fig. 15-7a ■).

Now pressure on the cuff is gradually released. When cuff pressure falls below systolic arterial blood pressure, blood begins to flow again. As blood squeezes through the still-compressed artery, the turbulent flow makes a noise called a **Korotkoff sound** that can be heard through a stethoscope placed just below the cuff (Fig. 15-7b ■).

A Korotkoff sound is heard with each heartbeat as long as the artery is compressed. Once the pressure exerted by the cuff no longer compresses the artery, blood flow is no longer turbulent and the sounds disappear (Fig. 15-7c ■).

The pressure at which blood flow is first heard represents the highest pressure in the artery and is recorded as the **systolic pressure.** The point at which all sound disappears is the lowest pressure in the artery and is recorded as the **diastolic pressure.** By convention, blood pressure is written as systolic pressure over diastolic pressure.

The average value for blood pressure is considered to be 120/80. Like many average physiological values, however, these numbers are subject to wide variability between people or within a single individual from moment to moment. A systolic pressure that is repeatedly over 140 mm Hg at rest or a diastolic pressure that is chronically over 90 mm Hg is considered to be a sign of hypertension.

Cardiac Output and Peripheral Resistance Are the Main Factors Influencing Mean Arterial Pressure

Mean arterial pressure, often called simply *blood pressure*, is a major determinant of blood flow. But what determines mean arterial pressure? Arterial pressure is determined by the balance between blood flow into the arteries and blood flow out of the arteries to the tissues. If flow in exceeds flow out, blood collects in the arteries and mean arterial pressure rises. If flow out exceeds flow in, mean arterial pressure falls.

Blood flow into the aorta, the main artery, is equal to the cardiac output of the left ventricle. Blood flow out of the arteries is influenced primarily by the resistance of the arterioles (**peripheral resistance**). We can express these relationships as

Mean arterial pressure ∝
cardiac output × arteriolar resistance

$$MAP \propto CO \times R_{arterioles}$$

Mean arterial pressure therefore is a function of cardiac output and the resistance of the arterioles (Fig. 15-8 ■).

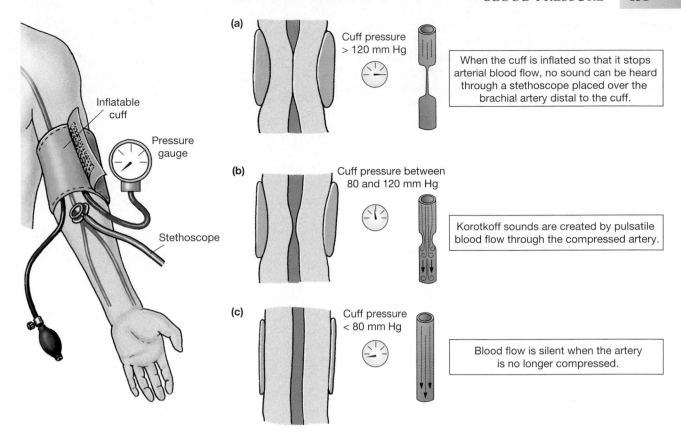

(a) Cuff pressure > 120 mm Hg

When the cuff is inflated so that it stops arterial blood flow, no sound can be heard through a stethoscope placed over the brachial artery distal to the cuff.

(b) Cuff pressure between 80 and 120 mm Hg

Korotkoff sounds are created by pulsatile blood flow through the compressed artery.

(c) Cuff pressure < 80 mm Hg

Blood flow is silent when the artery is no longer compressed.

Inflatable cuff

Pressure gauge

Stethoscope

■ **Figure 15-7   Measurement of arterial blood pressure**   Arterial blood pressure is measured with a sphygmomanometer, consisting of an inflatable cuff and a pressure gauge, and a stethoscope.

If the heart pumps more blood into the arteries, and resistance to blood flow out of the arteries does not change, arterial blood pressure will rise. Similarly, if cardiac output remains unchanged but peripheral resistance increases, blood will accumulate in the arteries, and the blood pressure will rise. Most cases of hypertension are associated with increased peripheral resistance without changes in cardiac output. Two additional factors influence mean arterial blood pressure: total blood volume and the distribution of blood in the systemic circulation.

Changes in Blood Volume Affect Blood Pressure

Although the volume of the blood within the circulation is usually relatively constant, changes in blood volume can affect arterial blood pressure. If blood volume in-

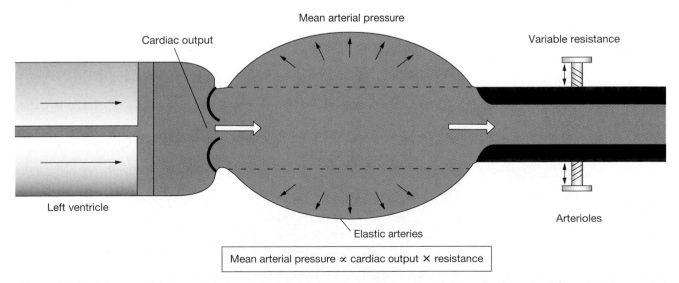

Mean arterial pressure

Cardiac output

Variable resistance

Left ventricle

Arterioles

Elastic arteries

Mean arterial pressure ∝ cardiac output × resistance

■ **Figure 15-8   Mean arterial pressure is a function of cardiac output and resistance in the arterioles**   In this model, the ventricle is represented by a syringe. The variable diameter of the arterioles is represented by a screw clamp.

creases, blood pressure increases. When blood volume decreases, blood pressure decreases.

To understand the relationship between blood volume and pressure, think of the circulatory system as an elastic balloon filled with water. If only a small amount of water is in the balloon, little pressure is exerted on the walls, and the balloon is soft and flabby. As more water is added to the balloon, more pressure is exerted on the elastic walls. If you fill a balloon close to the bursting point, you risk popping the balloon. The best way to reduce this pressure is to remove some of the water.

Small increases in blood volume occur throughout the day due to ingestion of food and liquids, but they usually do not create long-lasting changes in blood pressure because of homeostatic compensations. Adjustments for increased blood volume are primarily the responsibility of the kidneys. If the blood volume increases, the kidneys restore normal volume by excreting excess water in the urine (Fig. 15-9 ■).

Compensation for a decrease in blood volume is more difficult because it requires an integrated response from the kidneys and cardiovascular system in order to restore homeostasis. If blood volume decreases, *the kidneys cannot restore the lost fluid.* They can only conserve blood volume in order to prevent further decreases in blood pressure. The only way to restore lost fluid volume is through drinking or intravenous infusions.

...continued from page 450

Most hypertension is essential hypertension, or high blood pressure that cannot be attributed to any particular cause. "Since your blood pressure is only mildly elevated," Dr. Cortez tells Kurt, "let's see if we can control it with lifestyle changes. You need to reduce salt and fat in your diet, exercise, and lose some weight." "Looks like you're asking me to turn over a whole new leaf," says Kurt.

Question 2: *What is the rationale for reducing salt intake to control hypertension? (Hint: Salt causes water retention.)*

Cardiovascular compensation for decreased blood volume includes vasoconstriction and increased sympathetic stimulation of the heart [∞ p. 437]. However, there are limits to the effectiveness of cardiovascular compensation, and, if the fluid loss is too great, adequate blood pressure cannot be maintained. Typical events that cause major changes in blood volume include dehydration, hemorrhage, and ingestion of a large quantity of fluid. The integrated compensation for these events will be discussed in Chapter 19.

In addition to the absolute volume of blood in the cardiovascular system, the relative distribution of blood between the arterial and venous sides of the circulation can be an important factor in maintaining arterial blood

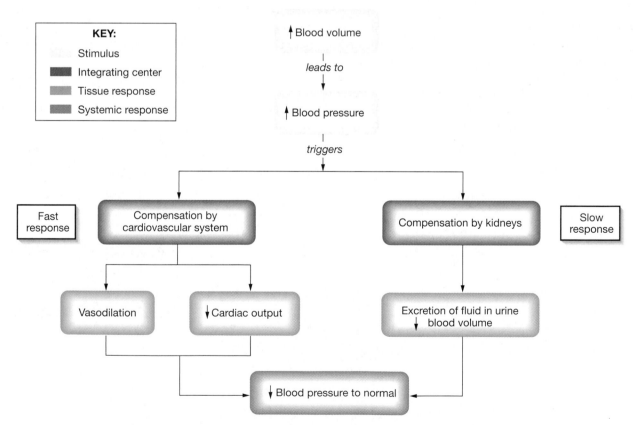

■ Figure 15-9 **Blood pressure control involves both the cardiovascular and renal systems** The cardiovascular system carries out rapid responses to changes in blood pressure, while the kidneys excrete excess fluid to decrease blood volume and blood pressure.

pressure. Arteries are low-volume vessels that usually contain only about 11% of the total blood volume. Veins, in contrast, are high-volume vessels that hold about 60% of the circulating blood volume. The blood volume in the venous side of the circulation serves as a reservoir that can be used to add blood to the arteries if arterial blood pressure falls. When arterial blood pressure falls, increased sympathetic activity constricts veins, decreasing their holding capacity and redistributing blood to the arterial side of the circulation.

Figure 15-10 ■ summarizes the four key factors that influence mean arterial blood pressure.

▶ RESISTANCE IN THE ARTERIOLES

As we saw in Chapter 14 [∞ p. 409], resistance to blood flow (R) is directly proportional to the length of the tubing through which the fluid flows (L) and the viscosity (η) of the fluid, and inversely proportional to the radius of the tubing (r) to the fourth power:

$$R \propto L\eta/r^4$$

Normally the length of the systemic circulation and the blood's viscosity are relatively constant. That leaves only the radius of the blood vessels as a factor in creating resistance to blood flow:

$$R \propto 1/r^4$$

The arterioles are the main site of variable resistance in the systemic circulation and contribute more than 60% of the total resistance to flow in the system. Resistance in

Shock *Shock* is a broad term that refers to generalized, severe circulatory failure. Shock can arise from multiple causes: failure of the heart to maintain normal cardiac output (*cardiogenic shock*), decreased circulating blood volume (*hypovolemic shock*), bacterial toxins (*septic shock*), and miscellaneous causes such as massive immune reactions that cause *anaphylactic shock*. No matter what the cause, the results are similar: low cardiac output and falling peripheral blood pressure. As tissue perfusion falls below the level needed to maintain adequate oxygen supply, the cells begin to sustain damage from inadequate oxygen and the buildup of metabolic wastes. Once this damage occurs, a positive feedback cycle begins. The shock becomes progressively worse until it becomes irreversible, and the patient dies. The management of shock includes administration of oxygen, fluids, and norepinephrine to stimulate vasoconstriction and increase cardiac output. If the shock arises from a cause that is treatable, such as a bacterial infection, measures must also be taken to remove the precipitating cause.

arterioles is variable because of the large amounts of smooth muscle in the arteriolar walls. When the smooth muscle contracts and relaxes, the radius of the arterioles changes.

Arteriolar resistance is influenced by both reflex and local control mechanisms. Autonomic reflexes mediated by the central nervous system maintain mean arterial blood pressure and govern blood distribution for certain homeostatic needs, such as temperature regula-

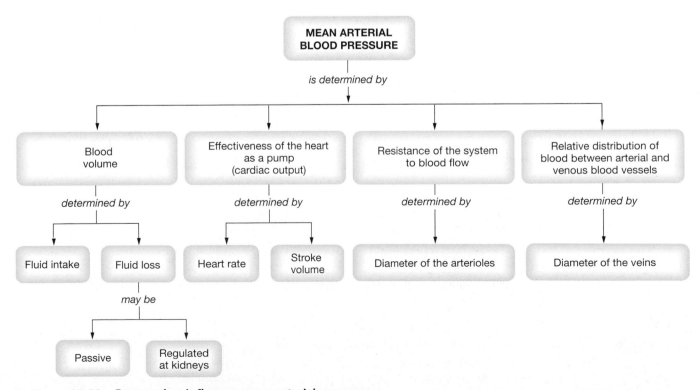

■ **Figure 15-10** **Factors that influence mean arterial pressure**

tion. Local control of arteriolar resistance matches tissue blood flow to the metabolic needs of the tissue. In the heart and skeletal muscle, these local controls actually take precedence over reflex control by the central nervous system. The following sections look at the factors that influence blood flow at the tissue level.

Myogenic Autoregulation Automatically Adjusts Blood Flow

Vascular smooth muscle has the ability to regulate its own state of contraction, a process called **myogenic autoregulation.** In general, an increase in blood pressure increases blood flow through an arteriole. However, when smooth muscle fibers in the wall of the arteriole stretch because of increased blood pressure, the arteriole constricts. This vasoconstriction increases the resistance of the arteriole, automatically decreasing its blood flow. With this simple and direct response to pressure, arterioles regulate their own blood flow.

The mechanism responsible for the intrinsic response of vascular smooth muscle to stretch is unclear. Some evidence suggests that stretch opens Ca^{2+} channels in the muscle membrane. Other researchers propose that vasoconstrictive paracrines are released from endothelial cells when the walls of the arteriole are stretched. The **endothelins,** a family of endothelial vasoconstrictors, may be the paracrines involved in myogenic autoregulation.

Paracrines Alter Vascular Smooth Muscle Contraction

Local control of arteriolar resistance is mediated by paracrines released from the vascular endothelium and from tissues. Paracrines that cause vasodilation include nitric oxide [∞ p. 163], H^+ from metabolically produced acids such as lactic acid, K^+, CO_2, prostaglandins, and histamine released by allergic reactions or tissue damage. Low dissolved oxygen levels in the tissues also relax vascular smooth muscle and cause vasodilation.

The concentrations of many paracrines change as cells become more or less metabolically active (Table 15-2). For

...continued from page 452

After a month, Kurt returns to the office for a checkup. He has lost five pounds and is walking at least a mile daily, but his blood pressure has not changed. "I swear, I'm trying to do better," says Kurt. "But it's difficult." Because lifestyle changes alone have not lowered Kurt's blood pressure, Dr. Cortez prescribes an antihypertensive drug. "This drug, called an ACE inhibitor, blocks production of a chemical called angiotensin II, a powerful vasoconstrictor. This medication should bring your blood pressure back to a normal value."

Question 3: Why would blocking the action of a vasoconstrictor lower blood pressure?

example, if aerobic metabolism increases, tissue O_2 levels fall while CO_2 production goes up. The arterioles dilate in response to low O_2 or high CO_2 in the interstitial fluid.

Vasodilation increases blood flow into the tissue, bringing in additional O_2 to match increased metabolism (Fig. 15-11a ■). This process is known as **active hyperemia:** an increase in blood flow accompanies an increase in metabolic activity [*hyper-*, above normal + *(h)aimia*, blood].

If blood flow to a tissue stops completely for a few seconds to a few minutes, metabolically produced paracrines such as CO_2 and H^+ accumulate in the interstitial fluid around the cells. When blood flow resumes, the increased concentration of these paracrines immediately triggers significant vasodilation. As the vasodilators are washed away by the restored tissue blood flow, the radius of the arteriole gradually returns to normal. An increase in tissue blood flow following a period of low perfusion is known as **reactive hyperemia** (Fig. 15-11b ■).

One vasodilator paracrine being studied is the nucleotide **adenosine.** In heart muscle, if oxygen consumption exceeds the rate of oxygen supply by the blood, myocardial hypoxia results. In response to low tissue oxygen, the myocardial cells release adenosine. Adenosine dilates coronary arterioles in an attempt to bring additional blood flow into the muscle.

TABLE 15-2   Control of Arteriolar Diameter

Type	Mechanism	Response
Myogenic activity	↑ Stretch of wall due to ↑ pressure	Vasoconstrict
Paracrines from metabolism	↓ O_2, ↑ CO_2, ↑ H^+, ↑ K^+	Vasodilate
Paracrine signal molecules	Nitric oxide, histamine, adenosine	Vasodilate
	Endothelins	Vasoconstrict
Reflex control		
Nervous	↑ Sympathetic (NE on α receptor)	Vasoconstrict
Hormonal	Epinephrine (adrenal medulla) on β_2 receptors	Vasodilate
	Angiotensin II	Vasoconstrict
	Atrial natriuretic peptide	Vasodilate

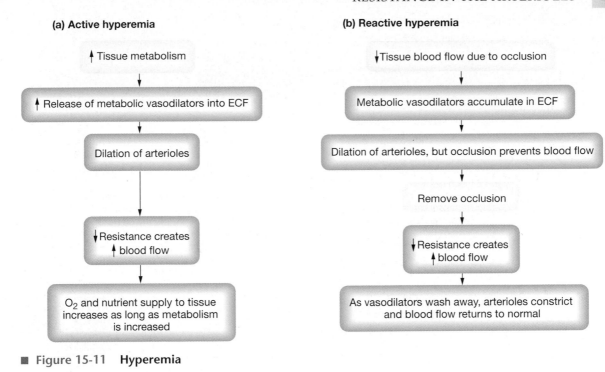

(a) Active hyperemia

↑ Tissue metabolism

↓

↑ Release of metabolic vasodilators into ECF

↓

Dilation of arterioles

↓

↓ Resistance creates
↑ blood flow

↓

O₂ and nutrient supply to tissue increases as long as metabolism is increased

(b) Reactive hyperemia

↓ Tissue blood flow due to occlusion

↓

Metabolic vasodilators accumulate in ECF

↓

Dilation of arterioles, but occlusion prevents blood flow

↓

Remove occlusion

↓

↓ Resistance creates
↑ blood flow

↓

As vasodilators wash away, arterioles constrict and blood flow returns to normal

■ **Figure 15-11   Hyperemia**

Not all vasoactive paracrines reflect changes in metabolism. For example, *kinins* and *histamine* are potent vasodilators that play a role in inflammation. *Serotonin*, previously mentioned as a CNS neurotransmitter [∞ p. 237], is a paracrine released by activated platelets. Serotonin triggers the constriction of arterioles, so if blood vessels are damaged, constriction helps slow blood loss. A serotonin agonist, *sumatriptan*, is used to treat migraine headaches that are caused by inappropriate cerebral vasodilation.

✔ Extracellular fluid concentration of K⁺ increases in exercising skeletal muscles. What effect will the increase in K⁺ have on blood flow in the muscle?

The Sympathetic Division Is Responsible for Most Reflex Control of Vascular Smooth Muscle

Smooth muscle contraction in arterioles is regulated by neural and hormonal signals in addition to locally produced paracrines. Among the hormones with significant vasoactive properties are *atrial natriuretic peptide* and *angiotensin II*. These hormones also have significant effects on the kidney's excretion of ions and water, as you will learn in Chapter 19.

Most systemic arterioles, except those in the brain, are innervated by the sympathetic branch of the nervous system. Another notable exception is arterioles involved in the erection reflex of the penis and clitoris. They are controlled by parasympathetic innervation.

Tonic discharge of norepinephrine from sympathetic neurons helps maintain myogenic tone (Fig. 15-12 ■).

Norepinephrine binding to α receptors on vascular smooth muscle causes vasoconstriction. If sympathetic release of norepinephrine decreases, the arterioles dilate. If sympathetic stimulation increases, arterioles constrict.

Epinephrine from the adrenal medulla travels through the blood to bind with α receptors and therefore reinforces vasoconstriction. However, α receptors have a lower affinity for epinephrine than for norepinephrine and do not respond as strongly to epinephrine [∞ p. 334].

Epinephrine also binds to β₂ receptors, found only in vascular smooth muscle of the heart, liver, and skeletal muscle. β₂ receptors are not innervated and therefore respond primarily to circulating epinephrine [∞ p. 335]. Activation of vascular β₂ receptors by epinephrine causes vasodilation.

One way to remember which arterioles have β₂ receptors is to think of a fight-or-flight response to a stressful event [∞ p. 327]. The fight-or-flight response includes a generalized increase in sympathetic activity, along with the release of epinephrine. Blood vessels that have β₂ receptors will respond to epinephrine by vasodilating. β₂-mediated vasodilation enhances blood flow to the heart, skeletal muscles, and liver, tissues that are active during the fight-or-flight response. (The liver produces glucose for muscle contraction.)

During fight-or-flight, increased sympathetic activity at arteriolar α receptors causes vasoconstriction. The increase in resistance diverts blood from nonessential organs, such as the gastrointestinal tract, to the muscles, liver, and heart. The nervous system's ability to selectively alter blood flow to organs is an important aspect of cardiovascular regulation.

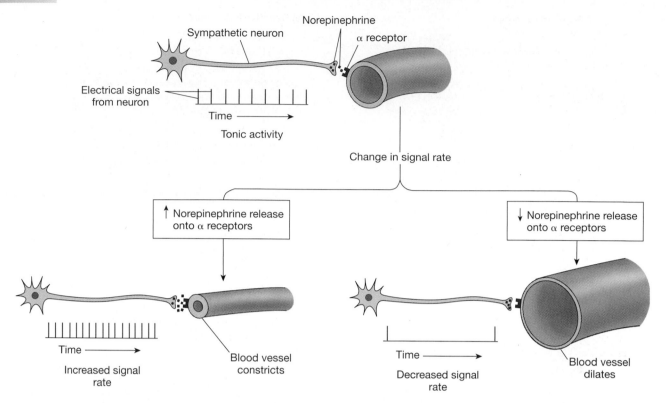

■ **Figure 15-12   Tonic control of arteriolar diameter**   Arteriolar diameter is regulated by norepinephrine from sympathetic neurons acting on α receptors.

✔ Draw a reflex map showing the vascular response to a fight-or-flight situation, such as Dr. Ian Malcolm fleeing from the *T. rex* in *Jurassic Park*. (Hint: Fear is integrated in the limbic system.)

▶ DISTRIBUTION OF BLOOD TO THE TISSUES

The distribution of systemic blood varies according to the metabolic needs of individual organs. Distribution is governed by a combination of local control mechanisms and homeostatic reflexes. For example, at rest, skeletal muscles receive about 20% of the cardiac output. During exercise, when the muscles use more oxygen and nutrients, they receive as much as 85%.

Blood flow to individual organs is set to some degree by the number and size of arteries feeding the organ. Figure 15-13 ■ shows how blood is distributed to various organs when the body is at rest. Usually, over two-thirds of the cardiac output is routed to the digestive tract, liver, muscles, and kidneys. On a per unit weight basis, the kidneys have the highest flow of all the major organs.

Variations in blood flow to individual tissues are possible because the arterioles in the body are arranged in parallel. That is, all arterioles receive blood at the same time from the aorta (see Fig. 15-1 ■). Total blood flow through *all* the arterioles of the body is always equal to the cardiac output.

However, the flow through individual arterioles depends on their resistance. The higher the resistance in the arteriole, the lower the blood flow through it. If an arteriole constricts and resistance goes up, blood flow through that arteriole goes down (Fig. 15-14 ■):

$$\text{Flow}_{\text{arteriole}} \propto 1/\text{resistance}_{\text{arteriole}}$$

In other words, blood is diverted from high-resistance arterioles to lower-resistance arterioles. You might say that blood traveling through the arterioles takes the path of least resistance.

Within a tissue, blood flow into individual capillaries can be regulated by **precapillary sphincters** [*sphingein*, to hold tight]. These are small bands of smooth muscle found at the junction of the metarteriole and capillary. If precapillary sphincters constrict, less blood flows through them into the capillaries (Fig. 15-15 ■). If the sphincters dilate, blood flow increases. This mechanism provides an additional site for local control of blood flow.

✔ Use Figure 15-13 ■ to answer these questions. (a) Which tissue has the highest blood flow per unit weight? (b) Which tissue has the least blood flow, regardless of weight?

▶ EXCHANGE AT THE CAPILLARIES

The transport of materials in the blood is only part of the function of the cardiovascular system. Once blood reaches the capillaries, the plasma and the cells exchange

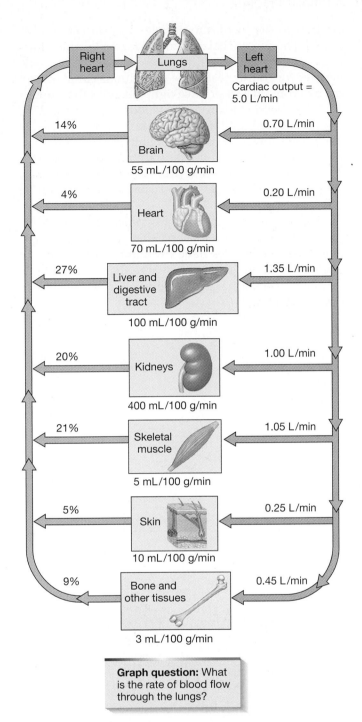

Graph question: What is the rate of blood flow through the lungs?

■ **Figure 15-13   Distribution of blood in the body at rest**
Blood flow to the major organs is represented as a percentage of total flow, on a per 100-g tissue-weight basis, and as an absolute rate of flow in liters per minute.

material. Molecules from the blood move into the interstitial space primarily by diffusion. Most cells are located within 0.1 mm of the nearest capillary, so diffusion over this short distance proceeds rapidly.

In addition, there is bulk flow of fluid across the capillary wall, caused by hydrostatic pressure that literally forces fluid out of the leaky capillary. As an analogy, think of the garden "soaker" hoses whose leaky walls allow water to ooze out.

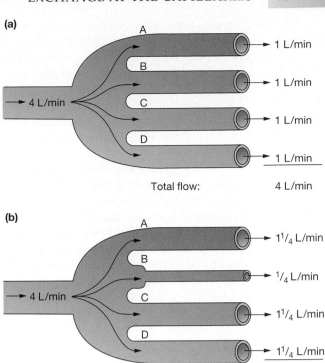

(a)

Total flow: 4 L/min

(b)

Total flow unchanged: 4 L/min

■ **Figure 15-14   Blood flow through individual blood vessels is determined by their resistance**   (a) Blood flow through four identical vessels (A–D) is equal. Total flow into the vessels equals total flow out. (b) When vessel B constricts, resistance of B increases and flow through B decreases. Since total flow must remain the same, flow diverted from B is divided among the lower-resistance vessels A, C, and D.

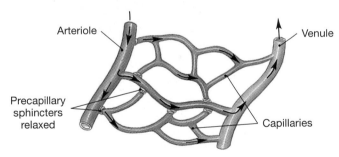

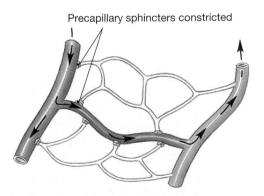

■ **Figure 15-15   Precapillary sphincters**   (a) When precapillary sphincters are relaxed, blood flows through all capillaries in the bed. (b) If precapillary sphincters constrict, blood flow bypasses capillaries completely and flows through metarterioles.

The capillary density in any given tissue is directly related to the metabolic activity of its cells. Tissues with a higher metabolic rate require more oxygen and nutrients. Those tissues have more capillaries per unit area. Subcutaneous tissue and cartilage have the lowest capillary density. Muscles and glands have the highest. The total capillary exchange surface area in an adult is more than 6300 m², nearly the surface area of two football fields.

Capillaries have the thinnest walls of all the blood vessels. They also have junctions that allow water, gases, and most dissolved solutes to pass freely. Proteins and blood cells cannot pass because of their size.

Capillaries can be divided into two categories based on structure. The most common is the **continuous capillary,** whose endothelial cells are closely joined (Fig. 15-16a ■). Some small molecules pass through the cell junctions, but other molecules, including some proteins,

are transported across in vesicles by *transcytosis* [∞ p. 133]. The endothelial cell surface facing the interstitial fluid appears dotted with numerous pits that become vesicles for transcytosis. The continuous capillaries of the blood-brain barrier have tight junctions that protect neural tissue from toxins in the bloodstream [∞ p. 255].

Fenestrated capillaries [*fenestra*, window] have large pores that allow high volumes of fluid to pass rapidly between the plasma and interstitial fluid (Fig. 15-16b ■). These capillaries are found primarily in the kidney and the intestine, where they are associated with absorptive transporting epithelia.

In the bone marrow, which synthesizes blood cells, and in the liver, where most plasma proteins are made, fenestrated capillaries have the ability to temporarily open gaps. These gaps are so large that both proteins and blood cells can squeeze between adjacent cells to enter the blood.

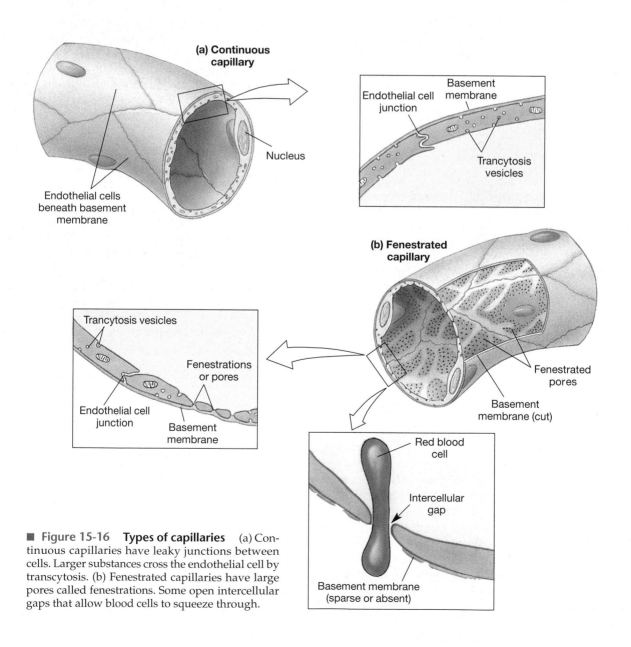

■ **Figure 15-16 Types of capillaries** (a) Continuous capillaries have leaky junctions between cells. Larger substances cross the endothelial cell by transcytosis. (b) Fenestrated capillaries have large pores called fenestrations. Some open intercellular gaps that allow blood cells to squeeze through.

The Velocity of Blood Flow Is Lowest in the Capillaries

In Chapter 14, you learned that at a constant flow rate, velocity of flow will be higher in a smaller vessel than it will be in a larger vessel [∞ p. 140]. From this, you might conclude that blood moves very rapidly through a capillary because it is the smallest blood vessel. But the primary determinant for velocity of flow is not the diameter of an individual capillary but the *total cross-sectional area* of all the capillaries.

What is total cross-sectional area? Imagine circles representing cross sections of all the capillaries placed edge to edge, and you have it. For the capillaries, those circles would cover an area much larger than the total cross-sectional areas of all the arteries and veins combined. Therefore, because total cross-sectional area of the capillaries is so large, the velocity of their blood flow is low.

Figure 15-17 ■ compares cross-sectional areas of different parts of the systemic circulation with the velocity of blood flow in each part. The fastest flow is in the relatively small-diameter arterial system. The slowest flow is in the capillaries and venules, which collectively have the largest cross-sectional area. The low velocity of blood flow through capillaries is a useful characteristic that allows diffusion to go to equilibrium [∞ p. 118].

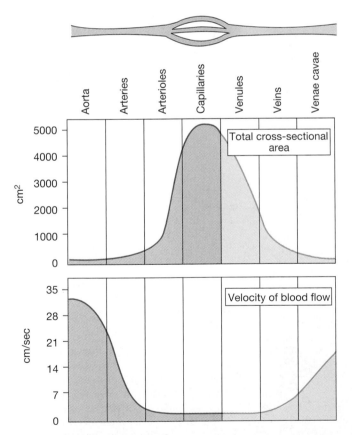

■ Figure 15-17   Vessel diameter, total cross-sectional area, and velocity of flow The velocity of flow depends on the total cross-sectional area. Flow is slowest in the capillaries and most rapid in the aorta and major arteries.

Most Capillary Exchange Takes Place by Diffusion and Transcytosis

Most exchange between the plasma and interstitial fluid takes place by simple diffusion, either through pores in the capillary wall or through the cells of the endothelium. The diffusion rate for permeable solutes is determined primarily by the concentration gradient between the plasma and the interstitial fluid. Oxygen and carbon dioxide diffuse freely across the thin endothelium. Their concentrations reach equilibrium with the interstitial fluid and cells by the time blood reaches the venous end of the capillary.

Blood cells and most plasma proteins are unable to pass between the capillary cells. The few proteins that escape from the blood are returned to the plasma by the lymphatic system. (We will look at the lymphatic system later in this chapter.)

If proteins cannot get through the pores, how do protein hormones and cytokines get from the plasma to the interstitial fluid? A few tissues such as the liver have capillary walls that can open gaps large enough for proteins to pass. In other tissues, transcytosis moves plasma proteins from one side of the endothelium cell to the other.

Capillary Filtration and Absorption Take Place by Bulk Flow

A third form of capillary exchange is the bulk flow of fluid into and out of the capillary. **Bulk flow** refers to the mass movement of water and dissolved solutes between the blood and the interstitial fluid as the result of hydrostatic or osmotic pressure. If the direction of flow is out of the capillary, the fluid movement is known as **filtration.** If the direction of flow is into the capillary, it is called **absorption.**

Most capillaries show a transition from net filtration at the arterial end to net absorption at the venous end. There are some exceptions to this rule. Capillaries in part of the kidney filter fluid along their entire length. Some capillaries in the intestine are only absorptive, picking up digested nutrients that have just been transported into the interstitial fluid from the lumen.

Two forces regulate bulk flow in the capillaries. One is *hydrostatic pressure,* the lateral pressure component of blood flow that pushes fluid out through the capillary pores [∞ p. 407]. The second factor is osmotic pressure [∞ p. 136]. These forces are sometimes called Starling forces, after the English physiologist E. H. Starling who first described them (the same Starling of Starling's law of the heart).

Osmotic pressure is determined by solute concentration differences between two compartments. The main solute difference between plasma and interstitial fluid is due to proteins that are present in the plasma but mostly absent from interstitial fluid. Osmotic pressure due to the presence of these proteins is known as *colloid osmotic pressure (π).* Colloid osmotic pressure is *not* equivalent to the total osmotic pressure. It is simply a measure of the osmotic pressure created by proteins. The capillary endothelium is freely permeable to other solutes in the plasma, and they therefore do not contribute to the osmotic gradient.

Colloid osmotic pressure is higher in the plasma (π_{CAP} = 25 mm Hg) than in the interstitial fluid (π_{IF} = 0 mm Hg). Therefore the osmotic gradient favors water movement by osmosis from the interstitial fluid into the plasma (Fig. 15-18a ■). For the purposes of our discussion, we will consider colloid osmotic pressure to be constant along the length of the capillary.

Capillary hydrostatic pressure (P_{CAP}), on the other hand, decreases along the length of the capillary as energy is lost to friction. Average values for capillary hy-

drostatic pressure are 32 mm Hg at the arterial end of a capillary and 15 mm Hg at the venous end. The hydrostatic pressure of the interstitial fluid (P_{IF}) is very low, so we will consider it to be essentially zero. This means that water movement due to hydrostatic pressure will always be directed out of the capillary, with the magnitude decreasing from the arterial end to the venous end.

Net fluid flow across the capillary is determined by the combined colloid osmotic pressure and hydrostatic pressure gradients. We will arbitrarily set pressures favoring filtration as positive and pressures favoring absorption as negative.

Net pressure = hydrostatic pressure gradient + colloid osmotic pressure gradient

Net pressure = $(P_{CAP} - P_{IF}) + (\pi_{CAP} - \pi_{IF})$

(a) Filtration in systemic capillaries

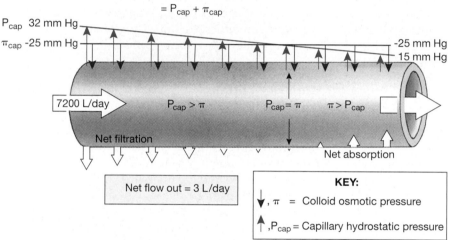

(b) Relationship between capillaries and lymph vessels

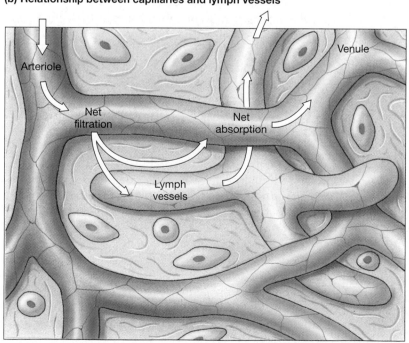

■ **Figure 15-18   Fluid exchange at the capillary**   (a) Hydrostatic pressure forces fluid out of the capillary. Colloid osmotic pressure of proteins within the capillary pulls fluid into the capillary. An average of 3 L/day of fluid filters out of the capillaries. (b) The excess water and solutes that filter out of the capillary are picked up by the lymph vessels and returned to the circulation.

If we assume that the interstitial hydrostatic and colloid osmotic pressures are zero, as discussed above, then we get the following values at the arterial end of a capillary:

$$\text{Net } P_{\text{Arterial end}} = (32 \text{ mm Hg} - 0) + (-25 \text{ mm Hg} - 0)$$

$$= 32 - 25 \text{ mm Hg}$$

$$\text{Net } P_{\text{Arterial end}} = 7 \text{ mm Hg, directed out of the capillary}$$
(filtration)

At the venous end, where capillary hydrostatic pressure is less:

$$\text{Net } P_{\text{Venous end}} = (15 \text{ mm Hg} - 0) + (-25 \text{ mm Hg} - 0)$$

$$= 15 - 25 \text{ mm Hg}$$

$$\text{Net } P_{\text{Venous end}} = -10 \text{ mm Hg, directed into the capillary}$$
(absorption)

Fluid movement down the length of a capillary is shown in Figure 15-18a ■. At the arterial end, there is net filtration, and at the venous end, there is net absorption. If the point at which filtration equals absorption occurred in the middle of the capillary, there would be no net movement of fluid. All volume that was filtered at the arterial end would be absorbed at the venous end.

However, filtration is usually greater than absorption, resulting in bulk flow of fluid out of the capillary into the interstitial space. By most estimates, that bulk flow amounts to about 3 liters per day, or the equivalent of the entire plasma volume! Unless this filtered fluid is returned to the plasma, the blood will turn into a sludge of blood cells and proteins. Restoring fluid lost from the capillaries to the circulatory system is one of the functions of the lymphatic system.

✔ Hydrostatic pressure in a capillary increases to 35 mm Hg on the arterial end and 20 mm Hg on the venous end. Will net filtration in this capillary decrease, increase, or stay the same?

▶ THE LYMPHATIC SYSTEM

The vessels of the lymphatic system interact with three physiological systems: the cardiovascular system, the digestive system, and the immune system. Functions of the lymphatic system include (1) returning fluid and proteins filtered out of the capillaries to the circulatory system, (2) picking up fat absorbed at the small intestine and transferring it to the circulatory system, and (3) serving as a filter to help capture and destroy foreign pathogens. In this discussion, we focus on the role of the lymphatic system in fluid transport. The other two functions will be discussed in connection with digestion (Chapter 20) and immunity (Chapter 22).

The lymph system is designed for the one-way movement of interstitial fluid from the tissues into the circulation. Closed-end lymph vessels called *lymph capillaries* lie close to all blood capillaries, except those in the kidney and central nervous system (Fig. 15-18b ■).

The lymph capillaries are a single layer of flattened endothelium, even thinner than that of the blood capillaries. The walls of the lymph capillaries are anchored to the surrounding connective tissue by fibers that hold the thin-walled vessels open. Large gaps between cells allow fluid, interstitial proteins, and particulate matter such as bacteria to be swept into the lymph capillary by bulk flow. Once inside the lymphatic system, this clear fluid is simply called **lymph.**

Lymph capillaries in the tissues join to form larger collecting vessels that progressively increase in size (Fig. 15-19 ■). These vessels have a system of semilunar

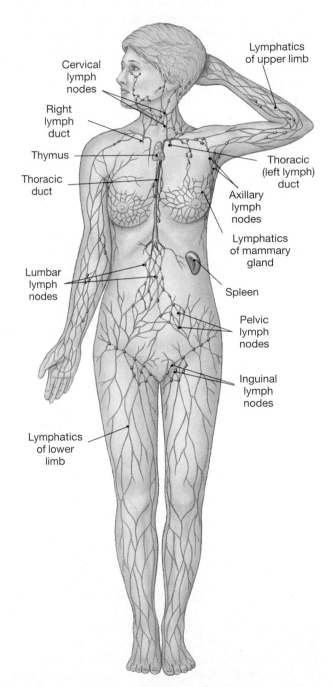

■ Figure 15-19 The lymphatic system The lymphatic system begins with blind-end lymph capillaries in the tissues and ends when lymph fluid empties into the venous circulation.

valves, similar to valves in the venous circulation. Finally, the largest lymph vessels empty into the venous circulation just under the collarbones, where the left and right subclavian veins join the internal jugular veins. At intervals along the way, vessels enter **lymph nodes,** bean-shaped nodules of tissue with a fibrous outer capsule and an internal collection of immunologically active cells, including lymphocytes and macrophages.

The lymphatic system has no single pump like the heart. Lymph flow depends primarily on waves of contraction of smooth muscle in the walls of larger lymph vessels. Flow is aided by contractile fibers in the endothelial cells, by the one-way valves, and by external compression created by skeletal muscles.

The skeletal muscle pump plays a significant role in lymph flow, as you know if you have ever injured a wrist or ankle. An immobilized limb frequently swells from the accumulation of fluid in the interstitial space, a condition known as *edema* [*oidema,* swelling]. Edema is the reason patients are told to elevate an injured limb above the level of the heart: gravity will assist lymph flow back to the blood.

An important reason for returning filtered fluid to the circulation is the recycling of plasma proteins. The body must maintain low protein concentration in the interstitial fluid because the colloid osmotic pressure gradient is the only significant force that opposes capillary hydrostatic pressure. If proteins move from the plasma to the interstitial fluid, the osmotic pressure gradient that opposes filtration decreases. With less opposition to capillary hydrostatic pressure, additional fluid moves into the interstitial space.

For example, inflammation is a situation in which the balance of colloid osmotic and hydrostatic pressures is disrupted. Histamine released in the inflammatory response makes capillary walls leakier and allows proteins to escape from plasma into the interstitial fluid. Thus, the local swelling that accompanies a region of inflammation is an example of edema caused by redistribution of proteins from the plasma to the interstitial fluid.

Edema Is the Result of Alterations in Capillary Exchange

Edema is a sign that normal capillary-lymph exchange has been disrupted. Edema usually arises from one of two causes: (1) inadequate drainage of lymph or (2) capillary filtration that greatly exceeds capillary absorption.

Inadequate lymph drainage occurs with obstruction of the lymphatic system, particularly at the lymph nodes. Parasites, cancer, or fibrotic tissue growth caused by therapeutic radiation can block the movement of lymph through the system. For example, *elephantiasis* is a chronic condition marked by gross enlargement of the legs and lower appendages when parasites block the lymph vessels. Lymph drainage may also be impaired if lymph nodes are removed during surgery, a common procedure in the diagnosis and treatment of cancer.

Factors that disrupt the normal balance between capillary filtration and absorption include:

1. *An increase in capillary hydrostatic pressure.* Increased capillary pressure is usually indicative of elevated venous pressure. An increase in arterial pressure is usually not noticeable at the capillaries because of the autoregulation of pressure in the arterioles.

   One common cause of increased venous pressure is *heart failure,* a condition in which the ventricle on one side of the heart is unable to pump all the blood sent to it by the other ventricle. As blood backs up, pressure rises first in the atrium, then in the veins and capillaries supplying blood to that side of the heart. When capillary hydrostatic pressure increases, filtration greatly exceeds absorption, leading to edema. If the left ventricle fails to pump normally, edema may occur in the lungs. Oxygen exchange is impaired, and breathing may become difficult. This condition is known as *pulmonary edema.*

2. *A decrease in plasma protein concentration.* Plasma protein concentrations may decrease as a result of severe malnutrition or liver failure. The liver is the main site for plasma protein synthesis.

3. *An increase in interstitial proteins.* As discussed earlier, excessive leakage of proteins out of the blood will decrease plasma colloid osmotic pressure and will increase net capillary filtration.

On occasion, changes in the balance between filtration and absorption help the body maintain homeostasis. For example, if arterial blood pressure falls, capillary hydrostatic pressure also falls. This change increases fluid absorption. If pressure falls low enough, there will be net absorption in the capillaries rather than net filtration. This passive mechanism helps maintain blood volume in situations in which blood pressure is very low, such as hemorrhage or severe dehydration.

✔ Malnourished children who have inadequate protein in their diet often have grotesquely swollen bellies. This condition is known as ascites and can be described as edema of the abdomen. Use the information you have just learned about capillary filtration to explain why malnutrition causes ascites.

▶ REGULATION OF BLOOD PRESSURE

The central nervous system coordinates the reflex control of blood pressure, with the main integrating center in the medulla oblongata of the brain. However, because of the difficulty of studying neural networks in the brain, we still know relatively little about the nuclei, neurotransmitters, and interneurons of the **medullary cardiovascular control center.**

The Baroreceptor Reflex Is the Primary Homeostatic Control for Blood Pressure

The primary goal of the cardiovascular control center is to maintain adequate blood flow to the brain and heart. Sensory input to this integrating center comes from a variety of peripheral sensory receptors. Stretch-sensitive mechanoreceptors known as **baroreceptors** are located in the walls of the carotid artery and aorta (Fig. 15-20 ■), where they monitor the pressure of blood flowing to the brain (carotid body receptors) and to the body (aortic body receptor). The carotid and aortic baroreceptors are tonically active stretch receptors that fire action potentials continuously at normal blood pressures.

The primary reflex pathway for homeostatic control of blood pressure is the **baroreceptor reflex.** When increased blood pressure in the arteries stretches the baroreceptor membrane, the firing rate of the receptor increases. If blood pressure falls, the firing rate of the receptor decreases.

Action potentials from the baroreceptors travel to the cardiovascular control center via sensory neurons. The cardiovascular control center integrates the sensory input and initiates an appropriate response. The response of

the baroreceptor reflex is quite rapid: changes in cardiac output and peripheral resistance occur within two heartbeats of the stimulus.

Efferent output from the medullary cardiovascular control center is carried via both sympathetic and parasympathetic autonomic neurons. Peripheral resistance is under tonic sympathetic control, with increased sympathetic discharge causing vasoconstriction.

Heart function is regulated by antagonistic control. Increased sympathetic activity increases heart rate at the SA node, shortens conduction time through the AV node, and enhances the force of myocardial contraction. Increased parasympathetic activity slows down heart rate but has no significant effect on ventricular contraction.

The baroreceptor reflex is summarized in Figure 15-21 ■. Baroreceptors increase their firing rate as blood pressure increases, activating the cardiovascular control center. In response, the cardiovascular control center increases parasympathetic activity and decreases sympathetic activity to slow down the heart. When heart rate falls, cardiac output falls. In the circulation, decreased sympathetic activity causes dilation of the arterioles, allowing more blood to flow out of the arteries. The combination of reduced cardiac output and decreased

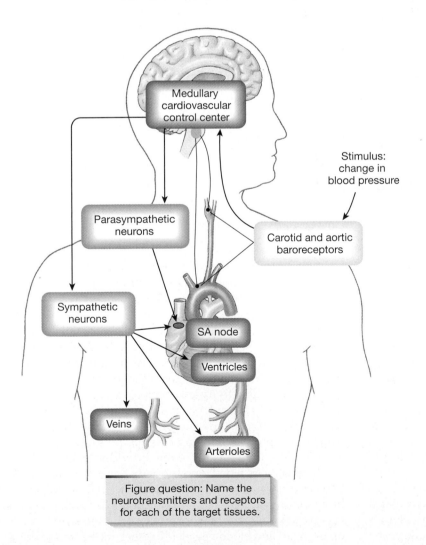

Figure question: Name the neurotransmitters and receptors for each of the target tissues.

■ **Figure 15-20** Components of the baroreceptor reflex

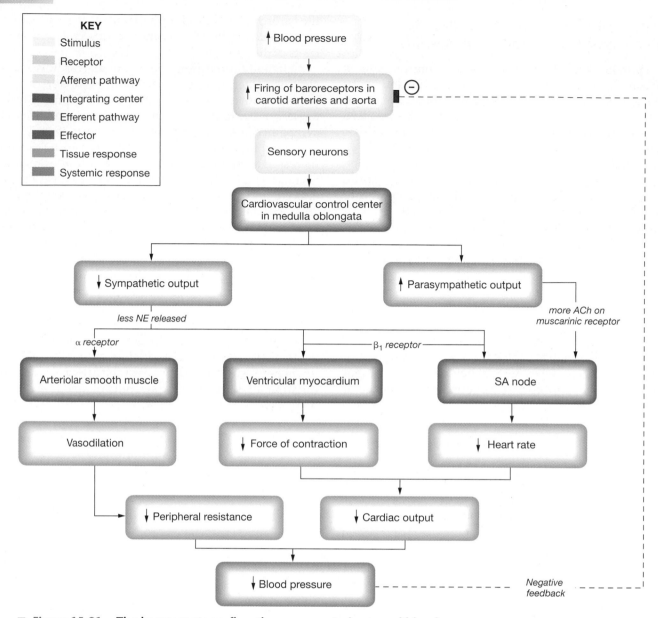

KEY
- Stimulus
- Receptor
- Afferent pathway
- Integrating center
- Efferent pathway
- Effector
- Tissue response
- Systemic response

■ Figure 15-21 **The baroreceptor reflex: the response to increased blood pressure**

peripheral resistance lowers the mean arterial blood pressure.

Cardiovascular function can be modulated by input from peripheral receptors other than the baroreceptors. For example, if arterial chemoreceptors are activated by low blood oxygen levels, they increase cardiac output. The cardiovascular control center also has reciprocal communication with centers in the medulla that control breathing.

The integration of function between the respiratory and circulatory systems is adaptive. If tissues require more oxygen, it is supplied by the cardiovascular system working in tandem with the respiratory system. Consequently, increases in breathing rate are usually accompanied by increases in cardiac output.

Blood pressure is also subject to modulation by higher brain centers such as the hypothalamus and cerebral cortex. The hypothalamus is responsible for vascular responses involved in body temperature regulation (see

Chapter 21) and for the fight-or-flight response. Learned and emotional responses may originate in the cerebral cortex and be expressed by cardiovascular responses such as blushing and fainting.

One such reflex is the *vasovagal response*, which may be triggered in some people by the sight of blood or a hypodermic needle. (Recall Anthony's experience at the beginning of this chapter.) In this pathway, increased parasympathetic activity and decreased sympathetic activity slow heart rate and cause widespread vasodilation. Cardiac output and peripheral resistance both fall, triggering a precipitous drop in blood pressure. With insufficient blood to the brain, the individual faints.

Regulation of blood pressure in the cardiovascular system is closely tied to the regulation of body fluid balance by the kidneys. Thus, certain hormones from the heart act on the kidneys, while hormones from the kidney act on the heart and blood vessels. Together, the heart and the kidney play a major role in maintaining the homeostasis of body fluids, the subject of Chapter 19. The overlap of these systems is an excellent example of the integration of organ system function that must take place in the body in order to maintain homeostasis.

✔ Baroreceptors have stretch-sensitive ion channels in their cell membrane. Increased pressure on the membrane opens the channels and allows ion flow that initiates action potentials. What kind of ion probably flows through these channels and in what direction, into or out of the cell?

Orthostatic Hypotension Triggers the Baroreceptor Reflex

The baroreceptor reflex functions every morning when you get out of bed. When you are lying flat, gravitational forces are distributed evenly up and down the length of the body, so blood also is distributed evenly throughout the circulation. When you stand up, gravity causes blood to pool in the lower extremities. This pooling creates an instantaneous decrease in venous return. As a result, less blood is in the ventricles at the beginning of the next contraction. Cardiac output falls from 5 L/min to 3 L/min, causing arterial blood pressure to decrease. This decrease in blood pressure upon standing is known as *orthostatic hypotension* [*orthos*, upright + *statikos*, to stand].

Orthostatic hypotension in a normal person triggers the baroreceptor reflex. The carotid and aortic baroreceptors respond to the drop in arterial blood pressure by decreasing their firing rate (Fig. 15-22 ■). Diminished sensory input into the cardiovascular control center increases sympathetic activity and decreases parasympathetic activity. As a result of autonomic changes, heart rate and force of contraction increase, while arterioles and veins constrict. The combination of increased car-

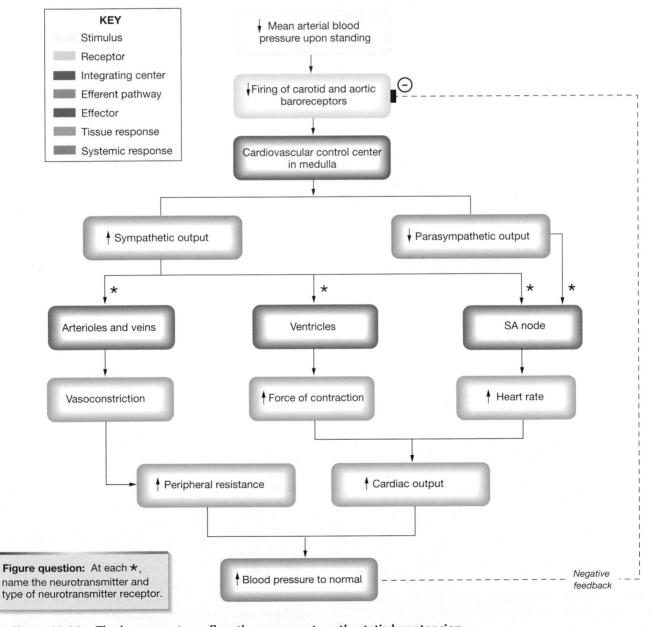

Figure question: At each ★, name the neurotransmitter and type of neurotransmitter receptor.

■ **Figure 15-22   The baroreceptor reflex: the response to orthostatic hypotension**

diac output and increased peripheral resistance raises mean arterial pressure and restores it to normal within two heartbeats. The skeletal muscle pump also contributes to the recovery by enhancing venous return when abdominal and leg muscles contract to maintain an upright position.

The baroreceptor reflex is not always effective, however. For example, during extended bed rest or in the zero-gravity conditions of space flights, blood from the lower extremities is redistributed more evenly throughout the body. This redistribution raises arterial pressure, triggering the kidneys to excrete what they perceive to be excess fluid. Over the course of three days, excretion of water leads to a 12% decrease in blood volume. When the person finally gets out of bed or returns to earth, gravity again causes blood to pool in the legs. Orthostatic hypotension occurs, and the baroreceptors attempt to compensate. But, in this instance, the cardiovascular system is unable to restore normal pressure because of the loss of blood volume. As a result, the subject may become dizzy or even faint from reduced delivery of oxygen to the brain.

▶ CARDIOVASCULAR DISEASE

Diseases of the heart and blood vessels, such as heart attacks and strokes, are responsible for nearly half of all deaths in the United States. Of these cardiovascular diseases, **coronary artery disease,** also known as *coronary heart disease,* is the leading cause of death in both men and women. By one estimate, this condition, in which the heart's blood vessels become blocked by cholesterol, calcium, plaques, and blood clots, costs people in the United States $50 billion to $100 billion per year in medical expenses and lost wages.

The prevalence of cardiovascular diseases is reflected in the tremendous amount of research that is being done worldwide. The scientific investigations range from large-scale clinical studies that track cardiovascular disease in thousands of people, such as the Framingham (Massachusetts) Heart Study, to experiments at the molecular level.

Much of the research at the cellular and molecular level is designed to expand our understanding of both normal and abnormal function in the heart and blood vessels. Scientists are studying a virtual alphabet soup of transporters and regulators. Some of these molecules, such as adenosine, endothelin, vascular endothelial growth factor (VEGF), phospholamban, and nitric oxide, you have studied in these chapters.

As we increase our knowledge of cardiovascular function, we begin to understand how drugs that have been used for centuries work. A classic example is the drug digitalis [∞ p. 420], whose mechanism of action we finally understand. It is a humbling thought to realize that for many therapeutic drugs, we know *what* they do without fully understanding *how* they do it.

…continued from page 454

Another few weeks go by, and Kurt again returns to Dr. Cortez for a checkup. Kurt's blood pressure is finally in the normal range and has been averaging 135/87. "But Doc, can you give me something for this dry, hacking cough I've been having? I don't feel bad, but it's driving me nuts." Dr. Cortez explains that a dry cough is an occasional side effect of taking ACE inhibitors. "It is more of a nuisance than anything else. But let's change your medicine. I'd like to try you on a calcium channel blocker this time."

Question 4:   *How do calcium channel blockers lower blood pressure?*

Risk Factors for Cardiovascular Disease Include Smoking, Obesity, and Inheritable Factors

We can predict the likelihood that a person will develop cardiovascular disease during his or her lifetime by examining the various risk factors that the person possesses. The list of risk factors is the result of many years following the medical histories of thousands of people in studies such as the famous Framingham Heart Study. As more data become available, additional risk factors may be added.

Risk factors are generally divided into those over which the person has no control and those that can be controlled. Medical intervention is aimed at reducing the risk from the controllable factors.

The risk factors that cannot be controlled include gender, age, and family history of early cardiovascular disease. Men have a 3–4 times higher risk of developing coronary artery disease through middle age than do women. After age 55, when most women have entered menopause, the death rate from coronary artery disease is nearly equal in men and in women who do not take estrogen replacement therapy. In general, the risk of coronary artery disease increases as people age.

Heredity also plays an important role. If a person has one or more close relatives with coronary artery disease, his or her risk is elevated. Diabetes mellitus, an inherited metabolic disorder that contributes to the development of fatty deposits in the blood vessels, also puts a person at risk for developing coronary artery disease.

Risk factors that can be controlled include cigarette smoking, obesity, sedentary lifestyle, elevated serum cholesterol and triglycerides, untreated high blood pressure (hypertension), and diabetes mellitus (to some extent). Most of these factors contribute directly to the development of *atherosclerosis,* also known as hardening of the arteries.

In atherosclerosis, macrophages ingest cholesterol and other lipids, then form fatty streaks in the extracellular matrix just under the endothelial lining of the larger arteries (Fig. 15-23 ■). As the streak grows, it forms a lipid core. Paracrines released by the macrophages attract smooth muscle cells that migrate from the underlying layer of vascular smooth muscle.

As the condition progresses, the lipid core grows and smooth muscle cells reproduce, forming bulging *plaques* that protrude into the lumen of the artery. In the advanced stages of atherosclerosis, the plaques develop hard, calcified regions and fibrous collagen caps.

We used to believe that the *occlusion* (blockage) of coronary blood vessels by large plaques was the primary cause of heart attacks, but that theory has been revised. The new model for the relationship between atherosclerosis and heart attacks divides plaques into two groups: stable plaques and vulnerable (unstable) plaques.

Stable plaques have thick collagen caps that separate the lipid core from the blood. Vulnerable plaques have thin caps that are likely to rupture and trigger the formation of blood clots (*thrombi*). One theory says that macrophages in some plaques release enzymes that dissolve collagen and convert stable plaques to vulnerable plaques. If a clot blocks blood flow to the heart muscle, a heart attack, or myocardial infarction, results.

Blocked blood flow decreases the heart muscle's oxygen supply. Without oxygen, the cells may be irreversibly damaged, and the normal conduction of electrical signals through the ventricle may be disrupted. This disruption can lead to an irregular heartbeat (*arrhythmia*) and potentially result in cardiac arrest and death.

The role of elevated blood cholesterol levels in the development of coronary artery disease is well established. Cholesterol, like other lipids, is not very soluble in aqueous solutions such as the plasma. Therefore, when cholesterol in the diet is absorbed from the digestive tract, it combines with a lipoprotein molecule so that it dissolves in the plasma. There are multiple types of lipoprotein carriers for cholesterol, but clinicians generally are concerned with two: **high-density lipoprotein, or HDL,** and **low-density lipoprotein, or LDL.** HDL-cholesterol is the more desirable form because high levels of HDL-cholesterol are associated with lower risk of heart attacks. (Remember, "H" stands for "healthy.")

LDL-cholesterol is sometimes called the bad cholesterol, because elevated plasma LDL levels are associated with coronary artery disease. LDL-cholesterol in normal levels is not bad, however, because the LDL carrier is necessary for uptake of cholesterol into cells. The LDL portion of the complex combines with an LDL receptor on the cell membrane. The entire receptor-LDL-cholesterol complex is brought into the cell by endocytosis. Once

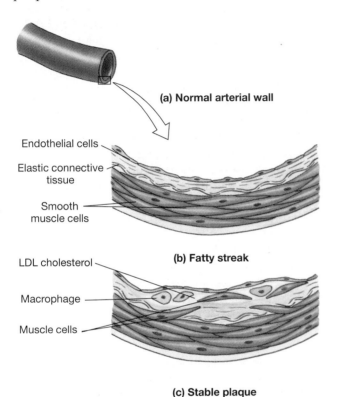

(a) Normal arterial wall

Endothelial cells
Elastic connective tissue
Smooth muscle cells

(b) Fatty streak

LDL cholesterol
Macrophage
Muscle cells

(c) Stable plaque

Fibrous scar tissue
Thickening of smooth muscle

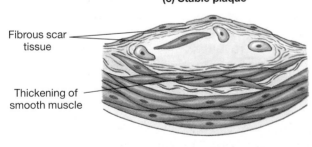

(d) Vulnerable plaque

Calcifications
Platelets

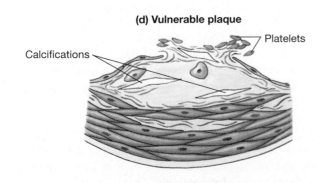

■ **Figure 15-23   The development of atherosclerotic plaques**   (a) The normal arterial wall consists of smooth muscle and connective tissue with a layer of endothelial cells lining the lumen of the artery. (b) In early stages, excess LDL-cholesterol accumulates in the acellular layer between the endothelium and the connective tissue. There, it is oxidized and phagocytosed by macrophages, white blood cells that patrol the area. When the macrophages begin to fill with oxidized LDL-cholesterol, they produce paracrines that attract smooth muscle cells to the region. At this stage, the lesion is called a fatty streak. (c) As cholesterol continues to accumulate, fibrous scar tissue forms around it. In addition, the migrating smooth muscle cells divide, thickening the arterial wall and narrowing the lumen of the artery. This stage is known as a fibrous plaque. (d) In the advanced stages of atherosclerosis, cells in the fatty deposit may die, and calcified scar tissue will form. The deposits of calcium carbonate in the scar tissue create the "hardening of the arteries." If the endothelium is damaged and collagen is exposed, platelets will stick to the damaged area and to each other. A blood clot (thrombus) may form on the wall of the vessel. If the clot completely blocks the lumen of the artery, blood flow to tissues beyond the clot will stop. If blood flow in the coronary blood vessel is stopped, a heart attack is the result.

DIABETES **Diabetes and Cardiovascular Disease**
Having diabetes is one of the major risk factors for developing cardiovascular disease, and almost two-thirds of people with diabetes will die from cardiovascular problems. In diabetes, cells that cannot use glucose for energy turn to fats and proteins. The body breaks down fat stores into fatty acids [∞ p. 97] and dumps them into the blood. Plasma cholesterol levels are also elevated. These are the very conditions that lead to development of atherosclerosis. Blockage of small and medium blood vessels in the lower extremities, combined with loss of sensation, can lead to gangrene (tissue death) in the feet. Atherosclerosis in larger vessels causes heart attacks and strokes.

freed from its carrier lipoprotein, cholesterol is used to make cell membranes or steroid hormones.

Problems arise when LDL-cholesterol is not taken into cells and remains in the blood. Under those conditions, excess LDL-cholesterol moves into the walls of the arteries. There, it is ingested by macrophages, starting a series of events that lead to atherosclerosis. Because of the pivotal role that LDL-cholesterol plays in atherosclerosis, many forms of therapy, ranging from dietary modification to drugs, are aimed at lowering LDL-cholesterol.

Hypertension Represents a Failure of Homeostasis

The last controllable risk factor for coronary artery disease is hypertension. **Hypertension** is defined as chronically elevated blood pressure, with either systolic pressures greater than 140 mm Hg or diastolic pressures greater than 90 mm Hg. Hypertension is a common disease in the United States and is the single most common reason for visits to physicians and for the use of prescription drugs.

More than 90% of all patients with hypertension are considered to have *essential* or *primary hypertension,* with no clear-cut cause. Cardiac output is usually normal in these people, and their elevated blood pressure appears to be associated with increased peripheral resistance. Some investigators have speculated that the increased resistance may be due to a lack of nitric oxide, the locally produced vasodilator that is produced by endothelial cells in the arterioles.

In the remaining 5%–10% of hypertensive cases, the cause is known, and the hypertension is considered to be secondary to an underlying pathology. For instance, the cause might be an endocrine disorder that causes fluid retention.

A key feature of hypertension from all causes is adaptation of the carotid and aortic baroreceptors to higher pressure, with subsequent down-regulation of their activity. Without input from the baroreceptors, the medullary cardiovascular control center interprets the high blood pressure as "normal," so no reflex reduction of pressure occurs.

Hypertension is a risk factor for atherosclerosis because high pressure in the arteries damages the endothelial lining of the vessels and promotes the formation of atherosclerotic plaques. In addition, high arterial blood pressure puts additional strain on the heart. When resistance in the arterioles is high, the myocardium must work harder to push the blood into the arteries.

To understand why this occurs, think of a swinging door (the semilunar valves) between the kitchen (the ventricles) and the dining room (the arteries). A waiter pushing through the door uses a certain amount of force to push the door open each time he passes through. The increased resistance of essential hypertension is equivalent to piling furniture against the back of the door. When that happens, the waiter must use a lot more force to push through the door because he must push not only the door but also the furniture. The additional resistance created by the weight of the furniture is analogous to the increased resistance of the arterioles to blood flow.

Amazingly, stroke volume in hypertensive patients remains constant, up to a mean blood pressure of about 200 mm Hg, despite the increasing amount of work that the ventricle must perform as blood pressure increases. The cardiac muscle of the left ventricle responds to chronic high systemic resistance in the same way that skeletal muscle responds to a weight-lifting routine. The heart muscle *hypertrophies,* increasing the size and strength of the muscle fibers.

However, if resistance remains high over time, the heart muscle is unable to meet the work load and begins to fail. At that point, cardiac output by the left ventricle decreases. If the cardiac output of the right heart remains normal while the output from the left side decreases, fluid collects in the lungs, creating *pulmonary edema.*

At this point, a detrimental positive feedback loop begins. Oxygen exchange in the lungs diminishes because of the pulmonary edema, leading to less oxygen in the blood. Lack of oxygen for aerobic metabolism further weakens the heart, so its pumping effectiveness diminishes even more. Unless treated, this condition, known as *congestive heart failure,* eventually leads to death.

Many of the treatments for hypertension have their basis in the cardiovascular physiology that you learned in these chapters. Calcium channel blockers decrease Ca^{2+} entry into vascular smooth muscle cells, causing arterioles to dilate and reducing peripheral resistance (see *Focus on Calcium Channel Blockers*, p. 445). Beta-blocking drugs that target β_1 receptors decrease catecholamine stimulation of cardiac output.

A third major group of antihypertensive drugs, the ACE inhibitors, act by blocking an enzyme that promotes production of a powerful vasoconstrictor substance. You will learn more about this enzyme later in the book, when you study the integrated control of blood pressure by the cardiovascular and renal systems. In the future, we may be seeing additional treatments for hypertension that are based on the molecular physiology of the heart and blood vessels.

PROBLEM CONCLUSION

Kurt remained on the calcium channel blocker, and, after several months, his blood pressure stabilized at 130/85—a significant improvement. Kurt's new diet also brought his total blood cholesterol down below 200 mg/dL plasma. By removing two of his controllable risk factors, Kurt decreased his chances of having a heart attack.

In this running problem, you learned about hypertension and some of the therapies currently used to treat it. Further check your understanding of this running problem by comparing your answers with those in the summary table.

	Question	Fact	Integration and Analysis
1	Why are people with high blood pressure at higher risk for having a hemorrhagic, or bleeding, stroke?	High blood pressure exerts force on the walls of the blood vessels.	If a blood vessel has a weakened or damaged wall, high blood pressure may cause that area to rupture, allowing blood to leak out of the vessel into the surrounding tissues.
2	What is the rationale for reducing salt intake to control hypertension?	Salt causes water retention.	Blood pressure will increase if the circulating blood volume increases. By restricting salt in the diet, a person can decrease retention of fluid in the extracellular compartment, which includes the plasma.
3	Why would blocking the action of a vasoconstrictor lower blood pressure?	Peripheral blood pressure is determined by cardiac output and peripheral resistance.	Resistance is inversely proportional to the radius of the blood vessels. Therefore, if blood vessels dilate by blocking action of vasoconstrictors, resistance and blood pressure will decrease.
4	How do calcium channel blockers lower blood pressure?	Calcium entry from the extracellular fluid plays an important role in both smooth and cardiac muscle contraction.	Blocking Ca^{2+} entry through Ca^{2+} channels will decrease force of cardiac contraction and decrease contractility of vascular smooth muscle. Both of these effects will lower blood pressure.

CHAPTER REVIEW

SUMMARY

1. Homeostatic regulation of the cardiovascular system is aimed at maintaining adequate blood flow to the brain and heart. (p. 443)

2. Total blood flow at any level of the circulation is equal to the cardiac output. (p. 444)

The Blood Vessels

3. Blood vessels are composed of layers of smooth muscle, elastic and fibrous connective tissue, and **endothelium.** (p. 444)

4. **Vascular smooth muscle** maintains a state of muscle tone. (p. 445)

5. The walls of the aorta and major arteries are both stiff and springy. This property allows them to absorb energy and release it through elastic recoil. (p. 446)

6. **Metarterioles** regulate blood flow through capillaries and allow white blood cells to go directly from arterioles to the venous circulation. (p. 446)

7. Capillaries and postcapillary **venules** are the site of exchange between blood and interstitial fluid. (p. 447)

8. Veins hold more than half of the blood in the circulatory system. Veins have thinner walls with less elastic tissue, so they expand easily when they fill with blood. (p. 447)

9. **Angiogenesis** is the process by which new blood vessels grow and develop, especially after birth. (p. 447)

Blood Pressure

10. The ventricles create high pressure that is the driving force for blood flow. (p. 448)

11. Blood pressure is highest in the arteries and falls continuously as blood flows through the circulatory system. At rest, average **systolic pressure** is 120 mm Hg, and average **diastolic pressure** is 80 mm Hg. (p. 448)

12. Pressure created by the ventricles can be felt as a **pulse** in the arteries. **Pulse pressure** equals systolic pressure minus diastolic pressure. (p. 448)

13. Blood flow against gravity in the veins is assisted by one-way valves and the respiratory and skeletal muscle pumps. (p. 449)

14. Arterial blood pressure is indicative of the driving pressure for blood flow. **Mean arterial pressure** (MAP) is defined as diastolic pressure + 1/3(systolic pressure − diastolic pressure). (p. 449)

15. Arterial blood pressure is usually measured in the brachial artery of the arm with a sphygmomanometer. When blood squeezes through the compressed artery, it makes a **Korotkoff sound** that can be heard through a stethoscope. (p. 450)

16. Arterial pressure is a balance between cardiac output and resistance of arterioles to blood flow (**peripheral resistance**). (p. 450)

17. If blood volume increases, blood pressure increases. When blood volume decreases, pressure decreases. (p. 451)

18. Venous blood volume can be shifted to the arteries if arterial blood pressure falls. (p. 452)

Resistance in the Arterioles

19. The arterioles are the main site of variable resistance in the systemic circulation. A small change in the radius of an arteriole creates a large change in resistance: $R \propto 1/r^4$. (p. 453)

20. Arterioles regulate their own blood flow through **myogenic autoregulation.** Vasoconstriction increases resistance of an arteriole and decreases its blood flow. (p. 454)

21. Arteriolar resistance is influenced by local control mechanisms that match tissue blood flow to the metabolic needs of the tissue. Vasodilator paracrines released from the vascular endothelium and from tissues include nitric oxide, H^+, K^+, CO_2, prostaglandins, adenosine, and histamine. Low dissolved oxygen levels also cause vasodilation. **Endothelins** are powerful vasoconstrictors. (p. 454)

22. **Active hyperemia** is a process in which increased blood flow accompanies increased metabolic activity. **Reactive hyperemia** is an increase in tissue blood flow following a period of low perfusion. (p. 454)

23. Most systemic arterioles, except those in the brain, are under tonic sympathetic control. Norepinephrine on vascular smooth muscle receptors causes vasoconstriction. Decreased sympathetic stimulation causes vasodilation. (p. 455)

24. Epinephrine binds to arteriolar α receptors and causes vasoconstriction. Epinephrine on β_2 receptors, found in the arterioles of the heart, liver, and skeletal muscle, causes vasodilation. (p. 455)

25. Changing the resistance of the arterioles affects mean arterial pressure and alters blood flow through the arteriole. (p. 455)

Distribution of Blood to the Tissues

26. The flow through individual arterioles depends on their resistance. The higher the resistance in the arteriole, the lower the blood flow in those arterioles: $Flow_{arteriole} \propto 1/R_{arteriole}$. (p. 456)

27. Blood flow into individual capillaries can be regulated by **precapillary sphincters.** (p. 456)

Exchange at the Capillaries

28. Exchange of material between the blood and the interstitial fluid is primarily by diffusion, aided by the hydrostatic pressure of the blood. (p. 457)

29. **Continuous capillaries** have leaky junctions between cells, but also transport material using transcytosis. Continuous capillaries with tight junctions form the blood-brain barrier. (p. 458)

30. **Fenestrated capillaries** have pores that allow large volumes of fluid to pass rapidly. (p. 458)

31. The velocity of blood flow through the capillaries is quite slow, so diffusion goes to equilibrium. (p. 459)

32. **Bulk flow** refers to the mass movement of water and dissolved solutes between the blood and the interstitial fluid. Fluid movement is called **filtration,** if the direction of flow is out of the capillary, or **absorption,** if the flow is directed into the capillary. (p. 459)

33. The osmotic pressure difference between plasma and interstitial fluid due to the presence of plasma proteins is the **colloid osmotic pressure.** (p. 460)

The Lymphatic System

34. About 3 liters of fluid filter out of the capillaries each day. The lymphatic system returns it to the circulatory system. (p. 461)

35. The lymphatic system also takes fat absorbed at the small intestine to the circulatory system and captures and destroys foreign pathogens. (p. 461)

36. Lymph capillaries accumulate fluid, interstitial proteins, and particulate matter by bulk flow. Flow in the lymphatic system depends on smooth muscle in vessel walls, one-way valves, and the skeletal muscle pump. (p. 462)

37. Excess fluid in the interstitial space is called **edema.** Factors that disrupt the normal balance between capillary filtration and absorption will cause edema. (p. 462)

Regulation of Blood Pressure

38. The reflex control of blood pressure resides in the medulla oblongata. **Baroreceptors** in the carotid and aortic bodies monitor arterial blood pressure and trigger the **baroreceptor reflex.** (p. 463)

39. Efferent output from the **medullary cardiovascular control center** goes to the heart and arterioles. Increased sympathetic activity increases heart rate and enhances the force of myocardial contraction. Increased parasympathetic activity slows down heart rate but has no significant effect on ventricular contraction. Increased sympathetic discharge at the arterioles causes vasoconstriction. There is no significant parasympathetic control of arterioles. (p. 463)

40. Cardiovascular function can be modulated by input from higher brain centers and from the respiratory control center of the medulla. (p. 464)

41. The baroreceptor reflex functions each time a person stands up, causing arterial blood pressure to drop. A decrease in blood pressure upon standing is known as orthostatic hypotension. (p. 465)

Cardiovascular Disease

42. **Coronary artery disease** is the leading cause of death in both men and women. It is possible to predict the likelihood that a person will develop cardiovascular disease during his or her lifetime by examining the various risk factors that the person possesses. (p. 466)

43. Atherosclerosis is a condition in which fatty deposits develop in large arteries. If these deposits are unstable, they may block the arteries by triggering blood clots. (p. 467)

QUESTIONS

LEVEL ONE   Reviewing Facts and Terms

1. The first priority of blood pressure homeostasis is to maintain adequate perfusion to what two organs?

2. Match the blood vessels with their function(s). Each vessel may have more than one match, and matching items may be used more than once.

   (a) arterioles
   (b) arteries
   (c) capillaries
   (d) veins
   (e) venules

   1. store pressure generated by the heart
   2. walls are both stiff and elastic
   3. carry low-oxygen blood
   4. thin walls of exchange epithelium
   5. act as a volume reservoir
   6. diameter can be altered by nervous input
   7. blood flow slowest through these vessels
   8. have the lowest blood pressure
   9. main site of variable resistance

3. List the components of blood vessel walls, in order from inner lining to outer covering. Briefly describe the importance of each component.

4. Blood flow to individual tissues is regulated by selective vasoconstriction and dilation of which vessels?

5. Aortic pressure reaches a typical high value of _____ (give units) during _____, or contraction of the heart. As the heart relaxes during the event called _____, aortic pressure declines to a typical low value of _____. This blood pressure reading would be written as _____ / _____.

6. The rapid pressure increase that occurs when the ventricles push blood into the aorta can be felt as a pressure wave, or _____. What is the equation used to calculate the strength of this pressure wave?

7. List the factors that aid venous return to the heart.

8. What is hypertension, and why is it a threat to a person's well-being?

9. What causes Korotkoff sounds, and when are you likely to hear them?

10. List at least three paracrines that cause vasodilation. What is the source of each one? In addition to paracrines, list two other ways to control smooth muscle contraction in arterioles.

11. What is hyperemia? How does active hyperemia differ from reactive hyperemia?

12. Most systemic arterioles are innervated by the _____ branch of the nervous system. List two exceptions.

13. Match the event at left with all appropriate neurotransmitter(s) *and* receptor(s) from the list on the right. [∞ p. 334]

    (a) vasoconstriction of intestinal arterioles
    (b) vasodilation of coronary arterioles
    (c) increased heart rate
    (d) decreased heart rate
    (e) vasoconstriction of coronary arterioles

    1. norepinephrine
    2. epinephrine
    3. acetylcholine
    4. β_1 receptor
    5. α receptor
    6. β_2 receptor
    7. nicotinic receptor
    8. muscarinic receptor

14. What organs receive over two-thirds of the cardiac output at rest? Which organs have the highest flow of blood, on a per unit weight basis?

15. By looking at the density of capillaries in a tissue, you can make assumptions about what property of the tissue? Which tissues have the lowest and highest capillary densities?

16. What type of transport is used to move each listed substance across the capillary endothelium?

    (a) oxygen
    (b) proteins
    (c) glucose
    (d) water

17. With which three physiological systems do the vessels of the lymphatic system interact?

18. Define edema. List some ways in which it can arise.

19. Define the following terms and explain their significance to cardiovascular physiology.

    (a) perfusion
    (b) colloid osmotic pressure
    (c) vasoconstriction
    (d) angiogenesis
    (e) metarterioles
    (f) pericytes

20. The two major lipoprotein carriers of cholesterol are _____ and _____. Which type is bad in elevated amounts?

LEVEL TWO   Reviewing Concepts

21. **Concept map:** Map all the following factors that influence mean arterial pressure. You may add additional terms.

aorta	parasympathetic neuron
arteriole	peripheral resistance
baroreceptor	SA node
blood volume	sensory neuron
cardiac output	stroke volume
carotid artery	sympathetic neuron
contractility	vein
heart rate	venous return
medulla oblongata	ventricle

22. Compare and contrast the following sets of terms:

    (a) lymphatic capillaries and systemic capillaries
    (b) roles of the sympathetic and parasympathetic branches in blood pressure control
    (c) lymph fluid and blood
    (d) continuous and fenestrated capillaries
    (e) hydrostatic and colloid osmotic pressure in systemic capillaries

23. Drugs known as calcium channel blockers prevent Ca^{2+} movement through Ca^{2+} channels. Explain two ways that this action can lower blood pressure. Why are neurons and other cells unaffected by these drugs?

24. Define myogenic autoregulation. What mechanisms have been proposed to explain it?

25. Left ventricular failure may be accompanied by edema, shortness of breath, and increased venous pressure. Explain how these symptoms develop.

LEVEL THREE   Problem Solving

26. Sylvia fell on the ice and sprained her knee, requiring three days of bed rest. When she got up on the third day, she experienced dizziness and loss of balance and had to sit on the side of the bed for a few minutes to avoid falling down. Why did she become dizzy?

27. A physiologist placed a section of excised arteriole in a perfusion chamber with saline. When the oxygen content of the saline perfusing (flowing through) the arteriole was reduced, the arteriole dilated. In a follow-up experiment, she placed an isolated piece of arteriolar smooth muscle stripped away from the other layers of the arteriole wall in saline. When the oxygen content of the saline was reduced as in the first experiment, the muscle showed no response. What do these two experiments suggest about how low oxygen exerts local control over arterioles?

28. Robert is a nonsmoker, 52 years old. He weighs 180 lbs, stands 5'9" tall, and his blood pressure averaged 145/95 on three successive visits to his doctor's office. His father, grandfather, and uncle all had heart attacks in their early 50s, and his mother died of a stroke at the age of 71.

    (a) Identify Robert's risk factors for coronary artery disease.
    (b) Does Robert have hypertension? Explain.
    (c) Robert's doctor prescribes a drug called a beta blocker. Explain the mechanism by which a beta-receptor-blocking drug may help lower blood pressure.

29. Draw a reflex map that would explain Anthony's vasovagal syncope at the sight of blood. Be sure to include all the steps of the reflex and explain if pathways are being stimulated or inhibited.

LEVEL FOUR   Quantitative Problems

30. Using the appropriate equation, mathematically explain what happens to blood flow if the diameter of a blood vessel increases from 2 mm to 4 mm.

31. Duplicate William Harvey's experiment that led him to believe that blood circulated in a closed loop.

    (a) Take your resting pulse.
    (b) Assume that your heart at rest pumps 70 mL/beat and that 1 mL of blood weighs one gram. Calculate how long it would take your heart to pump your weight in blood. (2.2 pounds = 1 kilogram)

32. Calculate the mean arterial pressure (MAP) and pulse pressure for a person with a blood pressure of 115/73.

33. According to the Fick principle, the oxygen consumption rate of an organ is equal to the blood flow through that organ times the amount of oxygen extracted from the blood as it flows through the organ:

> oxygen consumption rate = blood flow × systemic (arterial oxygen content − venous oxygen content)
>
> mL O_2 consumed/min = mL blood/min × mL O_2/mL blood

A person has a total body oxygen consumption rate of 250 mL oxygen consumed per minute. The oxygen content of blood in his aorta is 20 mL O_2/100 mL blood. The oxygen content of pulmonary artery blood is 16 mL O_2/100 mL blood. What is his cardiac output?

E   X   P   L   O   R   E   <MediaLab>

Introduction

In this chapter you learned about the factors that contribute to the maintenance of blood pressure and the details of blood flow throughout the body. A variety of mechanisms exist to monitor and alter the blood's movement through vessels. The driving pressure and the resistance of arterioles are the important factors that determine the extent to which different tissues are provided with oxygen and nutrients. This Media-Lab will challenge your knowledge of the mechanisms that regulate blood pressure. After reading the description below, visit the MediaLab for Chapter 15 in your Companion Website and select the appropriate keyword.

Web Exploration 1

Estimated time for completion = 15 minutes
During a physical exam, Dr. Cordis tells a 53-year-old man that he has a "right carotid artery bruit" (pronounced *brew-ee*). The carotid arteries, located deep within the neck, supply blood to the brain. Can you find your carotid artery pulse? Select the keyword **CAROTID** from your Website and read about the bruit. What causes the bruit? Visit the **CHOLESTEROL** website to find out.

What is Poiseuille's law for fluid dynamics? Using Poiseuille's law, diagram and detail why the presence of the carotid bruit is a problem in this patient. Dr. Cordis recommends surgery to repair the cause of the bruit. During the operation, the surgeon injects a drug, Nipride, locally into the carotid artery. Go to the **NIPRIDE** website and read about this drug's mechanism of action. How will Nipride affect your previous Poiseuille's law calculation?

Web Exploration 2

Estimated time for completion = 10 minutes
Why is "junk food" considered so 'bad' for you? Let's say you just walked out of the local convenience store with a cherry pie and a bag of nacho corn chips. What do you think the saturated fat and cholesterol content in these foods is? The terms *fat* and *cholesterol* should have you thinking about the parameters that affect arterial compliance. What is the definition of arterial compliance? How does it relate to arterial resistance?

As you place the junk food in your car, you suddenly recall the definition of arterial pulse pressure. Now your mind starts racing. You decide to contemplate the possible long-term effects on your cardiovascular system of eating cherry pies and nacho chips every single day. Select the keyword **HEALTHY HEART** from your Website to examine a healthy, normal artery. Picture how blood will flow through a vascular system composed of clean, "virgin" arteries such as this. What is the approximate pulse pressure in a normal individual? How does compliance affect arterial pulse pressure? Now visit the **ATHEROSCLEROSIS** websites. How does the presence of atherosclerosis throughout the cardiovascular system affect arterial compliance? How does it change pulse pressure? What do you expect to happen to pulse pressure as a person ages? Why? (Do you think most people have a decrease or an increase in arterial pulse pressure as they age?) Are you still hungry for that cherry pie?

16 Blood

■ "Who would have thought the old man to have had so much blood in him?"—Macbeth, V, i, 42, *by William Shakespeare* ■

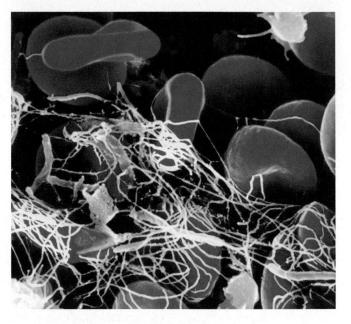

Red blood cells trapped in fibrin mesh

BACKGROUND BASICS

Blood, the fluid that circulates in the cardiovascular system, has occupied a prominent place throughout history as an almost mystical fluid. Humans undoubtedly made the association between blood and life by the time they began to fashion tools and hunt animals. A wounded animal that lost blood would weaken and die if the blood loss was severe enough. The logical conclusion was that blood was necessary for existence. This observation eventually led to the term *lifeblood,* meaning anything essential for existence.

Ancient Chinese physicians linked blood to energy flow in the body. They wrote about circulation of blood through the heart and blood vessels long before William Harvey described it in seventeenth-century Europe. In China, changes in blood flow were used as diagnostic clues to illness. Chinese physicians were expected to recognize some 50 variations in the pulse. Because blood was considered a vital fluid to be conserved and main-

PROBLEM
Von Willebrand's Disease

Emily Richards, 20, had heard horror stories about getting wisdom teeth pulled, but none could compare with what she was experiencing. In addition to the pain and swelling, the extraction sites would not stop bleeding. After three days of lying in bed with gauze stuffed in her mouth, she finally called her doctor. "I think something weird is going on," she told the receptionist. "Could I have hemophilia?" Emily's doctor ran blood tests and called her the next day with the results: "You do have a bleeding disorder, Emily, but it is not hemophilia. You have von Willebrand's disease."

...continued on page 477

• •

tained, bleeding patients to cure disease was not a standard form of treatment.

In contrast, Western civilizations came to believe that disease-causing evil spirits circulated in the blood. The way to remove these spirits was to remove the blood containing them. However, it was recognized that blood was an essential fluid, so bloodletting had to be done judiciously. Veins were opened with knives or sharp instruments (*venesection*), or blood-sucking leeches were applied to the skin. In ancient India, people believed that leeches could distinguish between healthy and infected blood.

There is no written evidence for venesection in ancient Egypt, but Galen of Pergamum in the second century A.D. influenced Western medicine for nearly 2000 years by advocating bleeding as treatment for many disorders. The location, timing, and frequency of the bleeding depended on the condition, and the physician was instructed to remove enough blood to bring the patient to the point of fainting. Over the years, this practice undoubtedly killed more people than it cured.

What is even more remarkable is the fact that as late as 1923, there was an American medical textbook that advocated bleeding for treating certain infectious diseases such as pneumonia! Now that we better understand the importance of blood in the immune response, it is doubtful that modern medicine will ever again turn to blood removal as a nonspecific means of treating disease. It is still used, however, for selected *hematological disorders* [*haima*, blood].

▶ PLASMA AND THE CELLULAR ELEMENTS OF BLOOD

What is this remarkable fluid that flows through the circulatory system? Blood makes up one-fourth of the extracellular fluid, the internal environment that bathes the cells and acts as a buffer between the cells and the external environment. Blood is the circulating portion of the extracellular fluid, responsible for carrying material from one part of the body to another.

Total blood volume in the 70-kg man is equal to about 7% of his total body weight. Thus, if we assume

that 1 kilogram of blood occupies a volume of 1 liter, then the 70-kg man has about 5 liters of blood ($0.07 \times 70\,kg = 4.9\,L$). Of this volume, about 2 liters is composed of blood cells, while the remaining 3 liters is composed of plasma, the fluid portion of the blood.

This chapter presents an overview of the components of blood, which are summarized in Figure 16-1 ■. We will consider the functions of plasma, red blood cells, and platelets in detail but will reserve detailed discussion of white blood cells for Chapter 22.

Plasma Is Composed of Water, Ions, Organic Molecules, and Dissolved Gases

Plasma is the fluid portion of the blood, within which cellular elements are suspended. Water is the main component of plasma, accounting for about 92% of its weight. Proteins account for another 7%. The remaining 1% is dissolved organic molecules (amino acids, glucose, lipids, nitrogenous wastes), ions (Na^+, K^+, Cl^-, H^+, Ca^{2+}, and HCO_3^-), trace elements and vitamins, and dissolved oxygen (O_2) and carbon dioxide (CO_2).

Plasma is identical in composition to interstitial fluid except for the presence of **plasma proteins.** The liver makes most plasma proteins and secretes them into the blood. **Albumins** are the most prevalent type of protein in the plasma. **Globulins** and the clotting protein **fibrinogen** are also common. Some globulins, known as *immunoglobulins* or *antibodies,* are synthesized and secreted by special cells rather than by the liver.

The presence of proteins in the plasma makes the osmotic pressure of the blood higher than that of the interstitial fluid. This gradient pulls water from the interstitial fluid into the capillaries, as you learned in the last chapter [∞ p. 459].

Plasma proteins participate in many functions, including blood clotting and defense against foreign invaders. In addition, they act as carriers for steroid hormones, cholesterol, certain ions such as iron (Fe^{2+}), and drugs. Finally, some plasma proteins act as hormones or as extracellular enzymes. Table 16-1 summarizes the functions of plasma proteins.

The Cellular Elements Include Red Blood Cells, White Blood Cells, and Platelets

Three main cellular elements are found in blood: **red blood cells,** also called **erythrocytes** [*erythros,* red]; **white blood cells,** also called **leukocytes** [*leukos,* white]; and **platelets.** Only the white blood cells are fully functional cells in the circulation. Red blood cells have lost their nuclei by the time they enter the bloodstream. Platelets are cell fragments that have split off a parent cell known as a **megakaryocyte.**

Red blood cells play a key role in transporting oxygen and carbon dioxide between the lungs and the tissues. Platelets are instrumental in *coagulation,* the process by which blood clots prevent blood loss in damaged

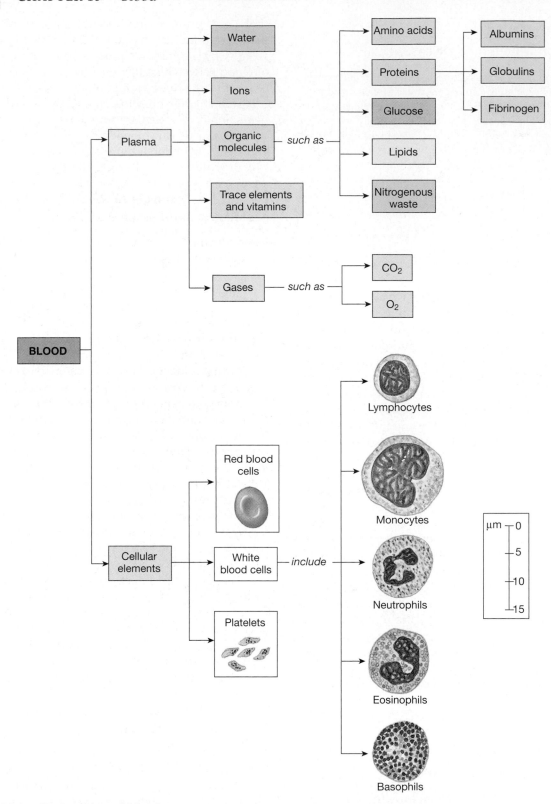

■ Figure 16-1   **Composition of blood**

vessels. White blood cells play a key role in defending the body against foreign invaders such as parasites, bacteria, and viruses. Although most white blood cells circulate through the body in the blood, their work is usually carried out in the tissues rather than inside the circulatory system.

Blood contains five types of mature white blood cells: (1) **lymphocytes,** (2) **monocytes,** (3) **neutrophils,** (4) **eosinophils,** and (5) **basophils.** Monocytes that leave the circulation and enter the tissues develop into **macrophages.**

The types of white blood cells may be grouped according to common morphological or functional char-

TABLE 16-1   Functions of Plasma Proteins

Name	Source	Function
Albumins (multiple types)	Liver	Major contributors to osmotic pressure of plasma; carriers for various substances
Globulins (multiple types)	Liver and lymphoid tissue	Clotting factors, enzymes, antibodies, carriers for various substances
Fibrinogen	Liver	Forms fibrin threads essential to blood clotting

acteristics. Neutrophils, monocytes, and macrophages are collectively known as **phagocytes** because they can engulf and ingest foreign particles such as bacteria (phagocytosis; ∞ p. 128). Lymphocytes are sometimes called **immunocytes** because they are responsible for specific immune responses directed against invaders. Basophils, eosinophils, and neutrophils are called **granulocytes** because they contain cytoplasmic inclusions that give them a granular appearance.

✔ On the basis of what you have learned about the origin and role of plasma proteins, explain why patients with advanced degeneration of the liver frequently suffer from edema [∞ p. 459].

▶ BLOOD CELL PRODUCTION

Where do these different blood cells come from? They are all descendants of a single precursor cell type known as the *pluripotent hematopoietic stem cell*. This cell type is found primarily in **bone marrow,** a soft tissue that fills the hollow centers of bones. Pluripotent stem cells have the remarkable ability to develop into an assortment of

The Differential White Cell Count When the body's defense system is called on to fight off foreign invaders, both the absolute number of white blood cells and the relative proportions of the different types of white blood cells in the circulation will change. Clinicians often rely on a differential white cell count to help them arrive at a diagnosis (Fig. 16-2 ■). For example, a person with a bacterial infection will usually have a high total number of white blood cells in the blood with an increase in the percentage that are neutrophils. A person with a viral infection may have a high, normal, or low total white cell count but often shows an increase in the percentage of lymphocytes. Identifying the different types of white blood cells is not as simple as you might think from doing it in the student laboratory. In abnormal smears, accurate identification may be complicated by the presence of immature blood cells that are difficult to distinguish from each other. Medical technologists undergo months of special training in hematology, and those who work in laboratories certified by the College of American Pathologists are tested on cell identification three times each year.

committed stem cells (Fig. 16-3 ■). These cell lines then differentiate into red blood cells, lymphocytes, other white blood cells, and megakaryocytes, the parent cells of platelets. It is estimated that only about one out of every 100 thousand cells in the bone marrow is an uncommitted stem cell, making the isolation and study of these cells very difficult.

Blood Cells Are Produced in the Bone Marrow

Hematopoiesis [*haima*, blood + *poiesis*, formation]**,** the synthesis of blood cells, begins early in embryonic development and continues throughout a person's life. In about the third week of fetal development, special cells in the yolk sac of the embryo form clusters. Some of these cell clusters are destined to become the endothelial lining of blood vessels, while others become blood cells. The common embryological origin of the endothelium and blood cells perhaps explains why many cytokines that control hematopoiesis are released by the vascular endothelium.

As the embryo develops, blood cell production spreads from the yolk sac to the liver, spleen, and bone marrow. By birth, the liver and spleen no longer produce blood cells. Hematopoiesis continues within all the bones of the skeleton until age five. However, as the child continues to age, the active regions of marrow decrease.

In adults, only the spine, ribs, pelvis, and proximal ends of long bones produce blood cells. Bone marrow that is actively producing blood cells retains a red color because of the presence of *hemoglobin*, the oxygen-binding protein of red blood cells. Inactive marrow is yellow

…continued from page 475

Von Willebrand's disease, named for the Finnish physician who discovered it in the late nineteenth century, is often diagnosed after an individual undergoes a dental or other surgical procedure. Besides prolonged bleeding, other symptoms include nosebleeds and bruising. The disease is caused by a deficiency in von Willebrand factor, a protein found in the blood that acts as a carrier for clotting factor VIII. Von Willebrand factor also promotes adherence of platelets to the endothelium and to each other during coagulation.

Question 1: *Why does von Willebrand's disease cause prolonged bleeding after a dental or surgical procedure?*

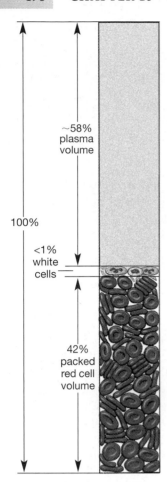

1. The **hematocrit** is the percentage (%) of the total blood volume that is occupied by packed (centrifuged) red blood cells (see figure at left).

2. The **hemoglobin** (Hb) content of erythrocytes. This value is measured as total hemoglobin content of blood (g Hb/dL).

3. **Mean red cell volume** (MCV). In some disease states, the red blood cells may be either abnormally large or abnormally small. For example, they are abnormally small in iron-deficiency anemia.

4. **Red cell count** in millions of cells per microliter. A machine counts the cells as they stream through a beam of light.

5. The **morphology** of the red blood cells also gives clues to diseases. Sometimes the cells lose their flattened disk shape and become spherical (spherocytosis). In sickle cell anemia, the cells are shaped like a sickle or crescent moon.

6. **Total white cell count** tells the total number of all types of white blood cells but does not distinguish between the different types.

7. The **differential white cell** count estimates the relative numbers of the five types of white cells. The differential count is made by a medical technologist from a thin smear of blood that is stained with biological dyes.

8. **Platelet count** is suggestive of the ability of the blood to clot.

	Males	Females
Hematocrit (% of whole blood that is packed red cells)	40%–54%	37%–47%
Hemoglobin (g/dL blood)	14–17	12–16
Red cell count (cells/μL)	$4.5–6.5 \times 10^6$	$3.9–5.6 \times 10^6$
Total white cell count (cells/μL)	$4–11 \times 10^3$	$4–11 \times 10^3$
Differential white cell count		
Neutrophils	50%–70%	50%–70%
Eosinophils	1%–4%	1%–4%
Basophils	<1%	<1%
Lymphocytes	20%–40%	20%–40%
Monocytes	2%–8%	2%–8%
Platelets (per μL)	$200–500 \times 10^3$	$200–500 \times 10^3$

■ **Figure 16-2 The blood count**   The table at the bottom shows the normal range of values. A deciliter is equal to 100 mL.

because of an abundance of adipocytes (fat cells). You can see the difference between red and yellow marrow the next time you look at bony cuts of meat in the grocery store. Although blood synthesis in adults is limited to small areas, the liver, spleen, and inactive (yellow) regions of marrow can resume blood cell production in times of need.

In the regions of marrow that are actively producing blood cells, about 25% of the cells are developing red blood cells, while 75% are destined to become white blood cells. The life span of white blood cells is considerably shorter than that of red blood cells, so they must

be replaced more frequently. For example, neutrophils have a six-hour half-life, and the body must make 100 billion neutrophils each day in order to replace those that die. Red blood cells, on the other hand, live for nearly four months in the circulation.

Hematopoiesis Is Controlled by Cytokines, Growth Factors, and Interleukins

What controls the production and development of blood cells? Chemical factors known as cytokines are responsible. *Cytokines* are peptides or proteins released from

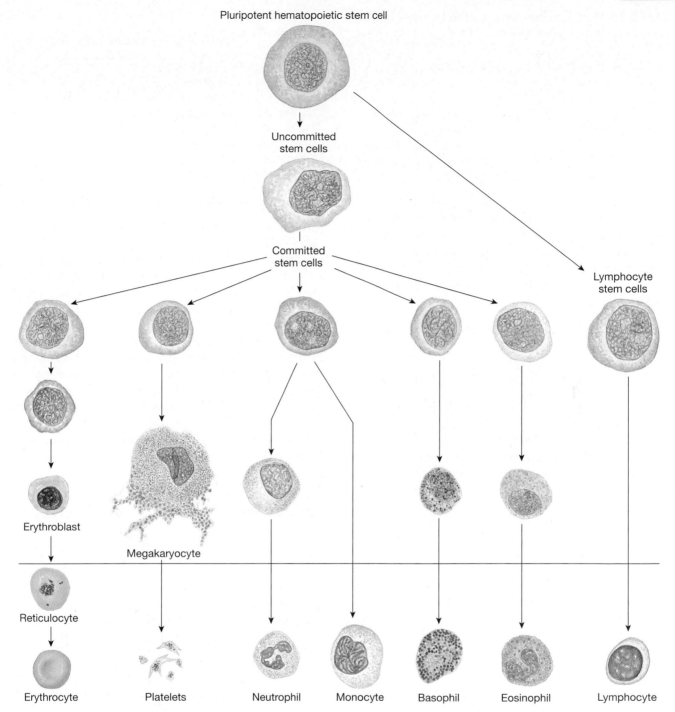

Pluripotent hematopoietic stem cell

Uncommitted
stem cells

Committed
stem cells

Lymphocyte
stem cells

Erythroblast

Megakaryocyte

Reticulocyte

Erythrocyte   Platelets   Neutrophil   Monocyte   Basophil   Eosinophil   Lymphocyte

■ **Figure 16-3   Hematopoiesis**   All blood cells are derived from a pluripotent stem cell. Cells below the horizontal line are the predominant forms found circulating in the blood. Cells above the line are found mostly in the bone marrow.

one cell that affect the growth or activity of another cell [∞ p. 156]. When cytokines are first discovered, they are often called *factors* and given a modifier that describes their actions: growth factor, differentiating factor, activating factor.

Some of the best-known cytokines in hematopoiesis are the **colony-stimulating factors,** molecules made by endothelial cells and white blood cells. Others are the **interleukins** [*inter-*, between + *leuko,* white], such as IL-3. The name *interleukin* was first given to cytokines

released by one white blood cell to act on another white blood cell. Numbered interleukin names are given to cytokines once their amino acid sequences have been determined. Interleukins also play important roles in the immune system, as we will discuss in Chapter 22.

Another hematopoietic cytokine is *erythropoietin,* which controls red blood cell synthesis. You will often see erythropoietin called a hormone, but technically it fits the definition of a cytokine because it is made on demand rather than stored in vesicles.

TABLE 16-2 Cytokines Involved in Hematopoiesis

Name	Cell That Produces Cytokine	Influences Growth or Differentiation of
Erythropoietin (EPO)	Kidney cells	Red blood cells
Granulocyte-macrophage colony-stimulating factor (GM-CSF)	Endothelial cells and fibroblasts of bone marrow; other white blood cells	Megakaryocytes, eosinophils, neutrophils, monocytes, red blood cells
Granulocyte colony-stimulating factor (G-CSF)	Endothelial cells and fibroblasts of bone marrow; monocytes	Neutrophils
Macrophage colony-stimulating factor (M-CSF)	Endothelial cells and fibroblasts of bone marrow; monocytes	Monocytes

Table 16-2 lists a few of the many cytokines linked to hematopoiesis. In fact, a recent review on the subject was titled "Regulation of hematopoiesis in a sea of chemokine family members with a plethora of redundant activities"!* Because of the complexity of the subject, we will not attempt to discuss hematopoietic cytokines in detail.

Colony-Stimulating Factors Regulate Leukopoiesis

Colony-stimulating factors (CSFs) were identified and named for their ability to stimulate the growth of white blood cell colonies in culture. These cytokines, made by endothelial cells, marrow fibroblasts, and white blood cells, regulate white blood cell production and development (**leukopoiesis**). They are required to induce both cell division (mitosis) and cell maturation. Once a leukocyte matures, it loses its ability to undergo mitosis.

One fascinating aspect of hematopoiesis is that white blood cell production is regulated in part by the existing white blood cells. This form of control allows white blood cell development to be very specific and tailored to the body's needs. For example, if you have a bacterial infection, cytokines released by active white blood cells fighting the infection stimulate the production of additional neutrophils and macrophages. The complex process by which leukocyte production is matched to need is still not well understood and is an active area of research.

Scientists hope to create a model for the control of hematopoiesis so that they can develop effective treatments for diseases characterized by either a lack or an excess of white blood cells. The *leukemias* are a group of diseases characterized by the abnormal growth and development of white blood cells. In *neutropenias* [*penia*, poverty], patients have too few white blood cells and are unable to fight off bacterial and viral infections as a result. Researchers hope to find better treatments for both leukemias and neutropenias by unlocking the secrets of how the body regulates cell growth and division.

*H. E. Broxmeyer and C. H. Kim, *Experimental Hematology* 27 (7): 1113–23, July 1999.

Thrombopoietin Regulates Platelet Production

Thrombopoietin (**TPO**) is a glycoprotein that regulates the growth and maturation of megakaryocytes, the parent cells of platelets. It is produced primarily in the liver but is also present in the kidney. The gene for this cytokine, whose activity was first described in 1958, was cloned in 1994. Within a year, TPO was widely available to researchers through the use of recombinant DNA techniques, and there has been an explosion of papers describing its effects on megakaryocytes and platelet production. Although we still do not understand everything about the basic biology of the process, thrombopoietin synthesis is a huge step forward in the search for answers.

Erythropoietin Regulates Red Blood Cell Production

Red blood cell production (**erythropoiesis**) is controlled by the glycoprotein hormone (cytokine) **erythropoietin** (**EPO**), assisted by several other cytokines. Erythropoi-

 Blood Cell Culture The successful laboratory culture of isolated blood cells was first reported around 1961. By 1966, researchers had recognized that isolated blood cells either must have growth factors added to the culture dish or be raised close to cells that secrete those factors. The fact that growth factors are secreted by other cells can be shown by the following experiment. In a petri dish, blood cells are grown in a semisolid layer of nutrient material, separated from a layer of "feeder cells" (macrophages or other cytokine-secreting cells) by a solid agar layer (Fig. 16-4 ■). The solid agar layer prevents direct contact between the blood cells and feeder cells. However, cytokine growth factors secreted by the feeder cells diffuse through the solid agar layer. These growth factors were named *colony-stimulating factors*. They are secreted in such small amounts that it seemed impossible to get enough for clinical use. But with technological advances, the factors were identified, and their genes were located and cloned. We now have commercially available recombinant factors in widespread clinical use. The cost of developing colony-stimulating factors for therapeutic use is extraordinary. It has been estimated that the development cost for one hematopoietic factor from laboratory to bedside is greater than $500 million.

...continued from page 477

"Is this disease serious?" Emily asks her doctor. "It can be," he replies. "Several different types of the disease exist, with varying degrees of severity. Since you have not had symptoms before this, you probably have one of the milder forms." Von Willebrand factor is present in megakaryocytes, endothelial cells, platelets, and plasma. In each location, the factor exists in three different "sizes," or multimers. The different types of von Willebrand disease vary according to the presence or absence of different multimers. The least serious type (type I) is characterized by decreased production of all von Willebrand multimers in all tissues. The most serious type (type III) is characterized by the complete absence of von Willebrand multimers in plasma.

Question 2: *Why is type I the least serious form of von Willebrand's disease?*

etin is made primarily in the kidney of adults. The stimulus for EPO synthesis and release is *hypoxia,* low oxygen levels in the tissues. This pathway, like other endocrine pathways, helps the body maintain homeostasis. By stimulating red blood cell synthesis, EPO puts more hemoglobin into the circulation to carry oxygen.

The existence of a substance controlling red blood cell production was first suggested in the 1950s, but two decades passed before scientists succeeded in purifying the hormone. One reason for the delay is that EPO is made on demand and not stored, as in an endocrine cell. Scientists took another nine years to sequence the hormone and to isolate and clone the gene for it. However, an incredible leap in progress was made after cloning the gene for EPO: only two years later, the hormone was produced by recombinant DNA technology and put into clinical use.

Today, physicians can prescribe not only genetically engineered EPO (epoetin alfa) but also several colony-stimulating factors (sargramostim and filgrastim) that

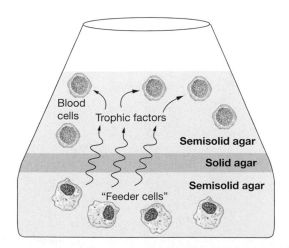

■ **Figure 16-4** **Blood cell culture** Cytokines secreted by feeder cells in the bottom layer diffuse through the solid agar and control the growth and development of blood cells in the top layer.

stimulate white blood cell synthesis. Other hematopoietic cytokines are being tested in clinical trials. Cancer patients in whom hematopoiesis has been suppressed by chemotherapy have benefited from injections of these hematopoietic hormones.

▶ RED BLOOD CELLS

Red blood cells, or *erythrocytes,* are the most abundant cell type in the blood. A microliter of blood contains about 5 million red blood cells, compared with only 4,000–11,000 white blood cells and 20,000–500,000 platelets. The primary function of red blood cells is to facilitate oxygen and carbon dioxide transport between the lungs and cells.

The ratio of red blood cells to plasma is indicated clinically by the **hematocrit** and is expressed as a percentage of the total blood volume (see Fig. 16-2 ■). Hematocrit is determined by drawing a blood sample into a narrow capillary tube and spinning it in a centrifuge so that the heavier red blood cells go to the bottom of the sealed tube, leaving the thin "buffy layer" of lighter white blood cells and platelets in the middle and the plasma on top.

The column of *packed red cells* is measured and compared to the total sample volume to find the hematocrit. The normal range of hematocrit is 40%–54% for a man and 37%–47% for a woman. This test provides a rapid and inexpensive way to estimate a person's red cell count because blood for a hematocrit can be collected by simply sticking a finger.

Mature Red Blood Cells Lack a Nucleus

In the bone marrow, committed stem cells differentiate through several stages into large, nucleated *erythroblasts.* As erythroblasts mature, the nucleus condenses and the cell shrinks in diameter from 20 μm to about 7 μm. In the last stage before maturation, the nucleus is pinched off and phagocytosed by marrow macrophages. At the same time, other membranous organelles such as mitochondria break down and disappear. The final immature cell form, called a *reticulocyte,* leaves the marrow and enters the circulation, where it matures into an erythrocyte in about 24 hours (see Fig. 16-5 ■).

Mature mammalian red blood cells are biconcave disks, shaped much like jelly doughnuts with the filling squeezed out of the middle (Fig. 16-6a ■). They lack a nucleus and membranous organelles and are a simple membranous "bag" filled with enzymes and hemoglobin, a red oxygen-carrying pigment.

Because red blood cells contain no mitochondria, they cannot carry out aerobic metabolism. Glycolysis is the primary source of ATP. Without a nucleus and endoplasmic reticulum to carry out protein synthesis, the cell is unable to make new enzymes or to renew its membrane components. This inability leads to an increasing

FOCUS ON... Bone Marrow

The bone marrow, hidden within the bones of the skeleton, is easily overlooked as a tissue, although collectively, it is nearly the size and weight of the liver! Bone marrow consists of blood cells in different stages of development and supporting tissue known as the **stroma** [mattress]. The stroma is composed of fibroblast-like reticular cells [∞ p. 64], collagenous fibers, and extracellular matrix. Fixed macrophages, a mature form of monocytes, are responsible for removing aged red blood cells. Other macrophages create centers around which stem cells cluster. Marrow is a highly vascular tissue, filled with **blood sinuses**, widened regions lined with endothelium. Mature blood cells squeeze through the endothelium to reach the circulation (Fig. 16-5 ■).

■ Figure 16-5   **Bone marrow**

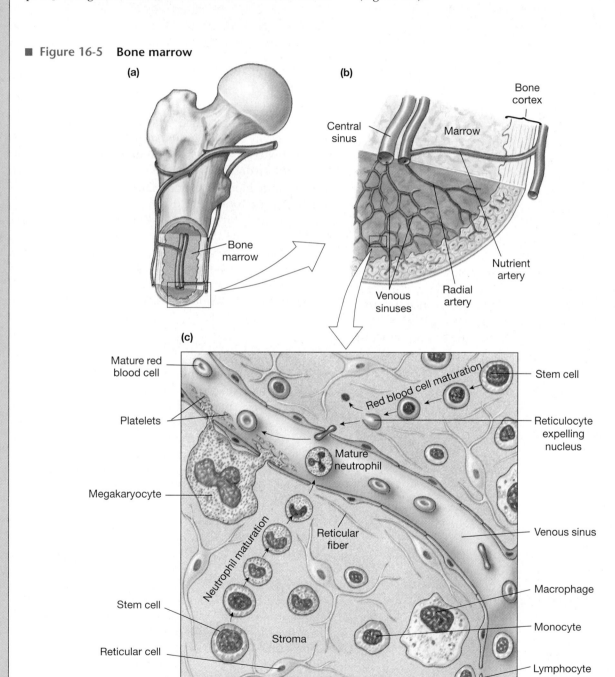

(a)

Bone marrow

(b)

Central sinus

Marrow

Bone cortex

Nutrient artery

Radial artery

Venous sinuses

(c)

Mature red blood cell

Platelets

Megakaryocyte

Stem cell

Reticular cell

Neutrophil maturation

Stroma

Mature neutrophil

Reticular fiber

Red blood cell maturation

Stem cell

Reticulocyte expelling nucleus

Venous sinus

Macrophage

Monocyte

Lymphocyte

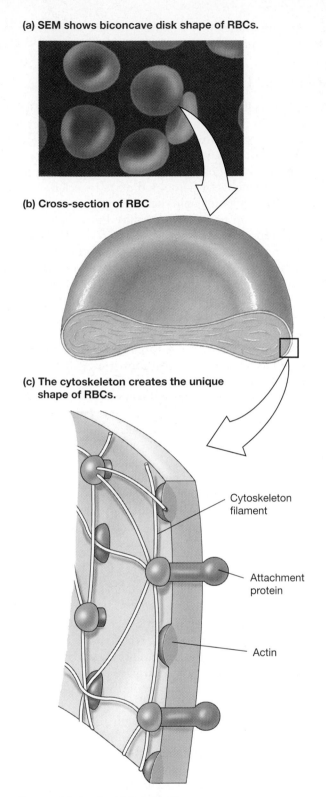

(a) SEM shows biconcave disk shape of RBCs.

(b) Cross-section of RBC

(c) The cytoskeleton creates the unique shape of RBCs.

Cytoskeleton filament

Attachment protein

Actin

■ **Figure 16-6   Red blood cells**

(Fig. 16-6c ■). Despite the cytoskeleton, red cells are remarkably flexible, like a partially filled water balloon that can compress into various shapes. This flexibility allows the cells to change shape as they squeeze through the narrow capillaries of the circulation.

The disklike structure of red blood cells also allows them to modify their shape in response to osmotic changes in the blood. A red cell placed in slightly hypotonic medium [∞ p. 137] will swell and form a sphere without disruption of its membrane integrity. In hypertonic media, red cells shrink up and develop a spiky surface where the membrane pulls tight against the cytoskeleton.

Hemoglobin Synthesis Requires Iron

Hemoglobin, the oxygen-carrying pigment of red blood cells, is a large complex molecule. Its quaternary structure has four globular protein chains, each of which is wrapped around an iron-containing **heme** group (Fig. 16-7 ■). There are several forms of the **globin** protein chain. The most common forms are designated *alpha* (α), *beta* (β), *gamma* (γ), and *delta* (δ), depending on the structure of the chain. Most adult hemoglobin has two alpha and two beta chains (designated *HbA*). However, a small portion of adult hemoglobin (about 2.5%) has two alpha and two delta chains (*HbA$_2$*).

loss of membrane flexibility, making older cells more fragile and likely to rupture.

The biconcave shape of the red blood cell is one of its most distinguishing features. The membrane is held in place by a complex cytoskeleton composed of filaments linked to transmembrane attachment proteins

(a) Hemoglobin molecule

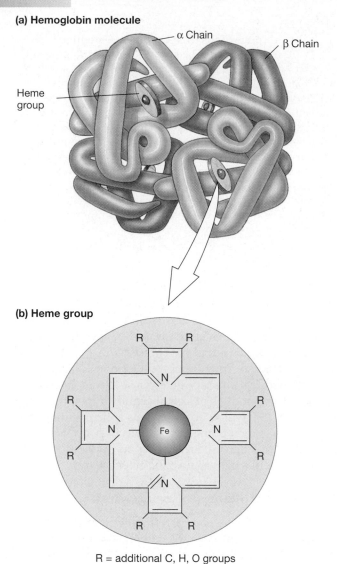

α Chain

β Chain

Heme
group

(b) Heme group

R = additional C, H, O groups

■ **Figure 16-7   Hemoglobin**   (a) A hemoglobin molecule is composed of four protein globin chains, each centered around a heme group. In most adult hemoglobin, there are two alpha chains and two beta chains as shown. (b) Each heme group consists of a porphyrin ring with an iron atom in the center.

The human fetus has a different form of hemoglobin adapted to attract oxygen from maternal blood in the placenta. Fetal hemoglobin (*HbF*) has two gamma chains in place of the two beta chains found in adult hemoglobin. Shortly after an infant's birth, fetal hemoglobin is replaced with the adult form as new red blood cells are made.

The four heme groups in a hemoglobin molecule are identical. Each consists of a carbon-hydrogen-nitrogen *porphyrin ring* with an iron atom (Fe) in the center (Fig. 16-7b ■). Some 70% of the iron in the body is found in the heme groups of red blood cells.

Hemoglobin synthesis requires an adequate supply of iron in the diet. Absorbed iron is transported in the blood on a carrier protein called *transferrin* (Fig. 16-8 ■). In the bone marrow, iron is taken into the mitochondria of developing red blood cells and used to make heme.

Excess iron in the body is stored, mostly in the liver, in the protein **ferritin** and its derivatives.

Red Blood Cells Live About Four Months

Red blood cells in the circulation live for 120 days (plus or minus 20 days). The increasingly fragile older red blood cells may rupture as they try to squeeze through narrow capillaries, or they may be captured by scavenging macrophages as they pass through the spleen. Many components of hemoglobin are recycled. Amino acids from the globin chains become new proteins, while some iron from heme groups may be reused in new heme groups.

The remainder of the old heme is converted by cells of the spleen and liver to a colored pigment called **bilirubin.** Bilirubin is carried by plasma albumin to the liver, where it is metabolized and incorporated into **bile** (Fig. 16-8 ■). Bile is secreted into the digestive tract, and the bilirubin metabolites leave the body in the feces. Small amounts of other bilirubin metabolites are filtered from the blood into the kidneys, where they contribute to the yellow color of the urine.

In some circumstances, bilirubin levels in the blood become elevated (*hyperbilirubinemia*). This condition, known as **jaundice,** causes the skin and whites of the eyes to take on a yellow cast. The accumulation of bilirubin can occur from multiple causes. Newborns whose fetal hemoglobin is being broken down and replaced with adult hemoglobin are particularly susceptible, so doctors monitor babies for jaundice in the first weeks of life. Another common cause of jaundice is liver disease, in which the liver is unable to process or excrete bilirubin.

Red Blood Cell Disorders Decrease Oxygen Transport

Hemoglobin plays a key role in oxygen transport, so the hemoglobin content of the body is important. If the hemoglobin content is too low, a condition known as **anemia,** the blood cannot transport enough oxygen to the tissues. People with anemia are usually tired and weak, especially during exercise. The major causes of anemia are summarized in Table 16-3. Anemia usually results either from blood loss, accelerated red blood cell destruction, or defective red blood cell or hemoglobin synthesis.

In the *hemolytic anemias* [*lysis,* rupture], the rate of red blood cell destruction exceeds the rate of red blood cell production. The hemolytic anemias are usually hereditary defects in which the body makes fragile cells. For example, in *hereditary spherocytosis,* the cytoskeleton does not link properly because of defective or deficient cytoskeletal proteins. Consequently, the cells are shaped more like spheres than like biconcave disks. This disruption in the cytoskeleton results in cells that rupture easily and that are not able to withstand osmotic changes as well as normal cells.

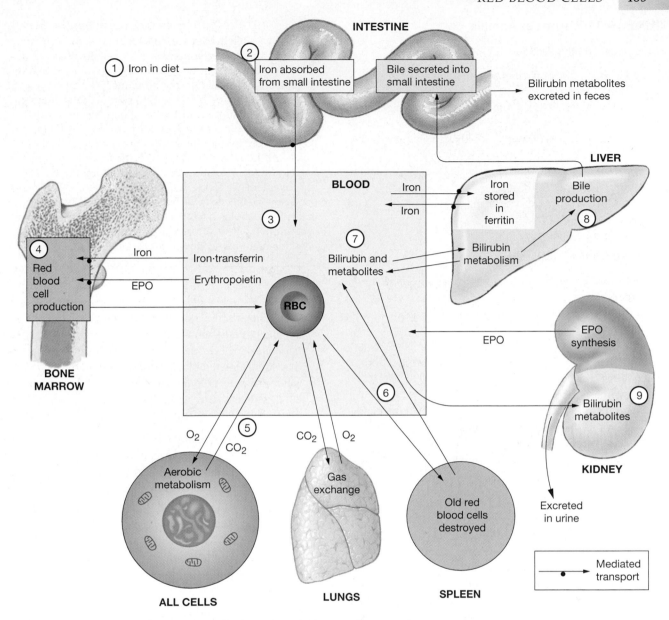

■ Figure 16-8   **Iron metabolism and the life of a red blood cell**

Other anemias result from the failure of the bone marrow to make adequate amounts of hemoglobin. One of the most common examples of an anemia that results from insufficient hemoglobin synthesis is **iron-deficiency anemia.** If iron loss exceeds iron intake, the marrow does not have adequate iron to make heme groups, and hemoglobin synthesis slows.

People with iron-deficiency anemia have a low red blood cell count (reflected in a low hematocrit) or a low hemoglobin content in their blood. Their red blood cells are often smaller than usual (*microcytic* red blood cells), and the lower hemoglobin content may cause the cells to be paler than normal (*hypochromic*). Women who menstruate are likely to suffer from iron-deficiency anemia because of iron loss in menstrual blood.

Sickle cell disease is a genetic defect in which glutamine, the number 6 amino acid in the 146-amino acid beta chain of hemoglobin, is replaced by valine. The result is abnormal hemoglobin (HbS) that crystallizes when it gives up its oxygen. This pulls the red blood cells into a sickle shape, like a crescent moon (Fig. 16-9 ■). The sickled cells get tangled with other sickled red cells as they pass through the smallest blood vessels, causing them to jam up and block blood flow to the tissues. This blockage creates tissue damage and pain from hypoxia.

Although the anemias are common, it is also possible to have too many red blood cells. *Polycythemia vera* [*vera,* true] is a stem cell dysfunction that produces too many blood cells, white as well as red. These patients may have hematocrits as high as 60%–70% (normal is 37%–54%). The increased number of cells causes the blood to become more viscous and thus more resistant to flow through the circulatory system [∞ p. 409].

TABLE 16-3   Causes of Anemia

Accelerated red blood cell loss

Blood loss: Cells are normal in size and hemoglobin content but low in number

Hemolytic anemias: Cells rupture at an abnormally high rate

Hereditary

Membrane defects (example: hereditary spherocytosis)

Enzyme defects

Abnormal hemoglobin (example: sickle cell anemia)

Acquired

Parasitic infections (example: malaria)

Drugs

Autoimmune reactions

Decreased red blood cell production

Defective red blood cell or hemoglobin synthesis in the bone marrow

Aplastic anemia: Can be caused by certain drugs or radiation

Inadequate dietary intake of essential nutrients

Iron deficiency (iron is required for heme production)

Folic acid deficiency (folic acid is required for DNA synthesis)

Vitamin B_{12} deficiency. May be due to lack of intrinsic factor for B_{12} absorption (vitamin B_{12} is required for DNA synthesis)

Inadequate production of erythropoietin

Sickle Cell Disease and Hydroxyurea   One new treatment developed for sickle cell disease is the administration of hydroxyurea, a compound that inhibits DNA synthesis. In the bone marrow, hydroxyurea causes immature red blood cells to produce fetal hemoglobin (HbF), with alpha and gamma globin chains, instead of adult hemoglobin. The theory for hydroxyurea therapy is that fetal hemoglobin interferes with the crystallization of sickle cell hemoglobin (HbS), so the red blood cells no longer develop a sickle shape. However, some studies show improvements in sickle cell symptoms before HbF levels increase. One theory of why this occurs is that hydroxyurea in rats is metabolized to nitric oxide (NO), which causes vasodilation. Inhaled nitric oxide is now being proposed as a new treatment for sickle cell disease symptoms.

✔ Erythropoietin (EPO) was first isolated from the urine of anemic patients who had high circulating levels of the hormone. However, despite the presence of the hormone that stimulates red blood cell production, these patients were unable to produce adequate amounts of hemoglobin or red cells. Give some possible reasons why the patients' own EPO was unable to correct their anemia.

✔ A person who goes from sea level to a city that is 5000 feet above sea level will show an increased hematocrit within two to three days. Draw the reflex pathway that links the hypoxia of high altitude to increased red blood cell production.

✔ An individual who becomes dehydrated will have an elevated hematocrit. Explain how dehydration leads to an elevated hematocrit although blood cell production is normal.

In *relative polycythemia,* the person's actual red blood cell number is normal, but the hematocrit is elevated due to loss of plasma volume. This might occur with dehydration, for example. The opposite problem can also occur. If an athlete overhydrates, the hematocrit may decrease temporarily because of increased plasma volume. In both of these situations, there is no actual pathology with the red blood cells.

▶ PLATELETS AND COAGULATION

Because of its fluid nature, blood flows freely throughout the circulatory system. However, if there is a break in the "piping" of the system, blood will be lost unless steps are taken. One of the challenges for the body is to plug holes in damaged blood vessels while still maintaining blood flow through the vessel.

It would be simple to block off a damaged blood vessel completely, like putting a barricade across a street full of potholes. But just as shopkeepers on that street lose business if traffic is blocked, cells downstream from the point of injury die from lack of oxygen and nutrients if the vessel is completely blocked. The body's task is to allow blood flow through the vessel while simultaneously repairing the damaged wall.

This challenge is complicated by the fact that blood in the system is under pressure. If the repair "patch" is too weak, it will simply be blown out by the blood pressure. Thus, stopping blood loss involves several steps. First, the

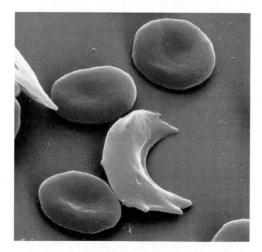

■ **Figure 16-9   Sickled red blood cells**

pressure in the vessel must be decreased long enough to create a secure mechanical seal in the form of a blood clot. Once the clot is in place and blood loss has been stopped, the body's repair mechanisms can take over. Then, as the wound heals, enzymes gradually dissolve the clot, while scavenger white blood cells ingest and destroy the debris.

Platelets Are Small Fragments of Cells

Platelets (*thrombocytes*) are cell fragments produced in the bone marrow from huge cells called megakaryocytes [*mega*, extremely large + *karyon*, kernel + *-cyte*, cell]. Megakaryocytes develop their formidable size by undergoing mitosis up to seven times without undergoing nuclear or cytoplasmic division. The result is a *polyploid* cell with a lobed nucleus (Fig. 16-10 ■).

The outer edges of marrow megakaryocytes extend through the endothelium into the lumen of marrow blood sinuses, where the cytoplasmic extensions fragment into disklike platelets (see Fig. 16-5 ■). Platelets are smaller than red blood cells, are colorless, and have no nucleus. Their cytoplasm contains mitochondria, smooth endoplasmic reticulum, and many granules filled with clotting proteins and cytokines.

The typical life span of a platelet is about 10 days. Platelets are always present in the blood, but they are not active unless damage has occurred to the walls of the circulatory system.

Hemostasis Prevents Blood Loss from Damaged Vessels

Hemostasis [*haima*, blood + *stasis*, stoppage] is the process of keeping blood within a damaged blood vessels. (The opposite of hemostasis is *hemorrhage* [*-rrhagia*, abnormal flow].) Hemostasis has three major steps: (1) vasoconstriction, (2) temporary blockage of a break by a platelet plug, and (3) blood coagulation or formation of a clot that seals the hole until tissues are repaired.

Let's take a quick look at these steps, shown in Figure 16-11 ■. The first step in hemostasis is immediate constriction of damaged vessels. The mechanism is unclear but may be due to vasoconstrictive paracrines released by the endothelium. Vasoconstriction temporarily decreases flow and pressure within the vessel, helping formation of the platelet plug. When you put pressure on a bleeding wound, you also decrease flow within the damaged vessel.

Vasoconstriction is rapidly followed by the second step, mechanical blockage of the hole by a **platelet plug.** Platelets stick to the exposed collagen (**platelet adhesion**) and become activated, releasing cytokines into the

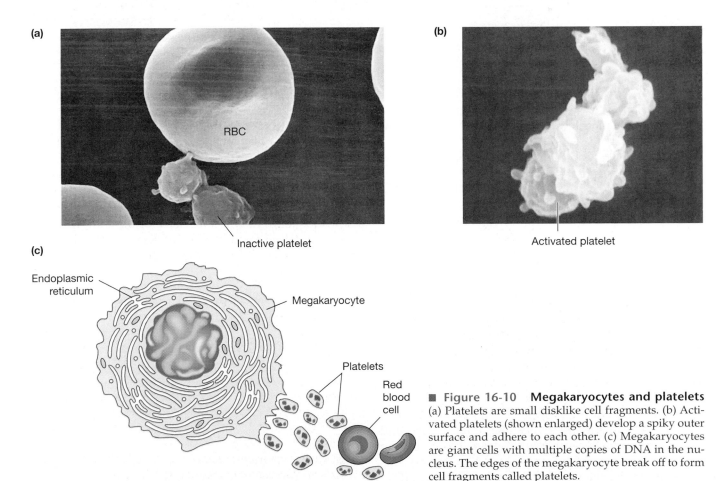

(a)

RBC

Inactive platelet

(b)

Activated platelet

(c)

Endoplasmic reticulum

Megakaryocyte

Platelets

Red blood cell

■ **Figure 16-10   Megakaryocytes and platelets** (a) Platelets are small disklike cell fragments. (b) Activated platelets (shown enlarged) develop a spiky outer surface and adhere to each other. (c) Megakaryocytes are giant cells with multiple copies of DNA in the nucleus. The edges of the megakaryocyte break off to form cell fragments called platelets.

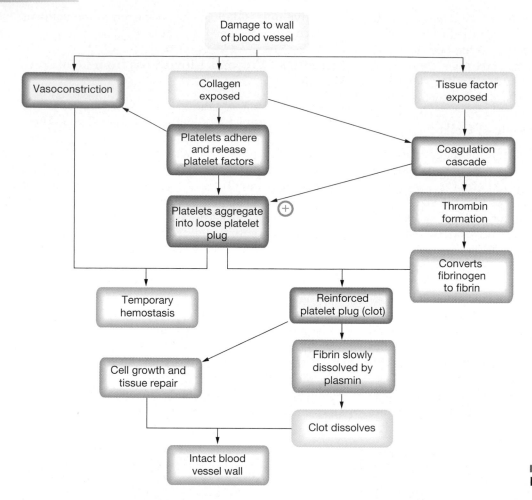

■ Figure 16-11   **Overview of hemostasis and tissue repair**

area around the injury. Platelet factors reinforce local vasoconstriction and activate more platelets, which stick to each other (**platelet aggregation**) to form a loose platelet plug.

Simultaneously, in the third step, exposed collagen and tissue factor initiate a series of reactions known as the **coagulation cascade.** In the cascade, inactive plasma proteins are converted into active enzymes. In the last step, the enzyme **thrombin** converts **fibrinogen** into **fibrin** fibers that intertwine with and reinforce the platelet plug. The reinforced platelet plug is called a **clot.** Finally, as cell growth and division repairs the damaged vessel, the clot retracts and is slowly dissolved by the enzyme **plasmin.**

The body must maintain the proper balance during hemostasis. Too little hemostasis allows excessive bleeding. Too much creates a **thrombus,** or blood clot, that adheres to the undamaged wall of a blood vessel [*thrombos,* a clot or lump]. A large thrombus can block the lumen of the vessel and stop blood flow.

Although hemostasis seems straightforward, there are still many unanswered questions at the cellular and molecular levels. Because inappropriate blood clotting plays an important role in strokes and heart attacks, this area of research is very active. Research has led to the development and use of "clot busters," enzymes that can dissolve clots in arteries after heart attacks and strokes (see running problem in Chapter 9, starting on p. 253).

A detailed study of hemostasis involves many chemical factors, some of which play multiple roles and have multiple names. Thus, learning about hemostasis may be especially challenging. For example, some factors participate in both platelet plug formation and coagulation. Another factor activates enzymes for both clot formation and clot dissolution. Because of the complexity of the coagulation cascade, we will discuss only a few aspects of hemostasis in additional detail.

…continued from page 481

Fortunately, Emily is diagnosed with type I disease. "From now on," explains her doctor, "you should take medication before any surgical procedure to prevent excessive bleeding." The medication consists of desmopressin and an antifibrinolytic agent. Desmopressin, a synthetic form of the hormone vasopressin (also called antidiuretic hormone), causes blood vessel constriction. The antifibrinolytic agent inhibits the action of plasmin in a blood clot.

Question 3:   *Explain the rationale for giving an antifibrinolytic agent to individuals with von Willebrand's disease.*

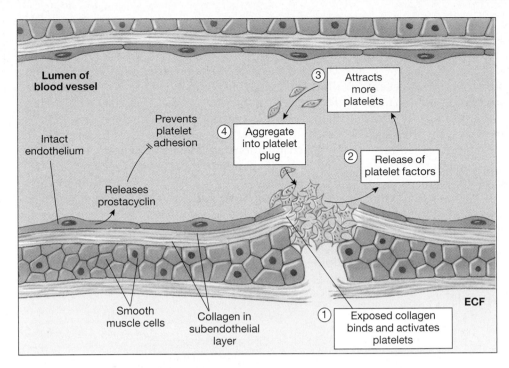

■ **Figure 16-12   Platelet plug formation**

Platelet Activation Begins the Clotting Process

When a blood vessel wall is first damaged, exposed collagen and chemicals from endothelial cells activate platelets (Fig. 16-12 ■; Table 16-4). Normally, the blood vessel endothelium separates the collagenous matrix fibers from the circulating blood. But if the vessel is damaged, collagen is exposed, and platelets rapidly begin to adhere to it.

Platelets adhere to collagen with the help of *integrins*, membrane receptor proteins that are linked to the cytoskeleton [∞ p. 157]. Binding activates platelets so they release the contents of their intracellular granules, including *serotonin* (5-hydroxytryptamine), ADP, and **platelet-activating factor** (**PAF**). PAF sets up a positive feedback loop by activating more platelets. It also initiates pathways that convert platelet membrane phospholipids into **thromboxane A$_2$**.

TABLE 16-4   Factors Involved in Platelet Function

Chemical factor	Source	Activated by or Released in Response to	Role in Platelet Plug Formation	Other Roles and Comments
Collagen	Subendothelial extracellular matrix	Injury exposes platelets to collagen	Binds platelets to begin platelet plug	NA
von Willebrand factor (vWF)	Endothelium, megakaryocytes	Exposure to collagen	Links platelets to collagen	Deficiency or defect causes prolonged bleeding
Serotonin	Secretory vesicles of platelets	Platelet activation	Platelet aggregation	Vasoconstrictor
Adenosine diphosphate (ADP)	Platelet mitochondria	Platelet activation, thrombin	Platelet aggregation	NA
Platelet-activating factor (PAF)	Platelets, neutrophils, monocytes	Platelet activation	Platelet aggregation	Plays role in inflammation; increases capillary permeability
Thromboxane A$_2$	Phospholipids in platelet membranes	Platelet activating factor	Platelet aggregation	Vasoconstrictor; eicosanoid
Platelet-derived growth factor (PDGF)	Platelets	Platelet activation	NA	Promotes wound healing by attracting fibroblasts and smooth muscle cells

Serotonin and thromboxane A_2 are vasoconstrictors. They also contribute to platelet aggregation, along with ADP and PAF. The net result is a growing platelet plug that seals the damaged vessel wall.

If platelet aggregation is a positive feedback event, what prevents the platelet plug from continuing to form and thus spreading beyond the site of injury to other areas of the blood vessel wall? The answer lies in the fact that platelets will not adhere to normal endothelium.

Intact vascular endothelial cells convert their membrane lipids into **prostacyclin,** an eicosanoid [∞ p. 30] that blocks platelet adhesion and aggregation. Nitric oxide (NO) released by normal, intact endothelium also inhibits platelets from adhering. The combination of platelet attraction to the injury site and repulsion from the normal vessel wall creates a localized response that limits the platelet plug to the area of damage.

Coagulation Converts the Platelet Plug into a More Stable Clot

The third major step in hemostasis, coagulation, is a complex process in which fluid blood forms a gelatinous clot. The coagulation cascade is a series of enzymatic reactions that ends with formation of a protein fiber mesh that stabilizes the platelet plug. Some chemical factors involved in the coagulation cascade also promote platelet adhesion and aggregation to the damaged region.

Coagulation is divided into two pathways (Fig. 16-13 ■). An **intrinsic pathway** (yellow) begins with collagen exposure and uses proteins already present in the plasma. An **extrinsic pathway** (blue) starts when damaged tissues expose **tissue factor,** a protein-phospholipid mixture, to the plasma proteins. The two pathways unite at the common pathway (green) to create thrombin,

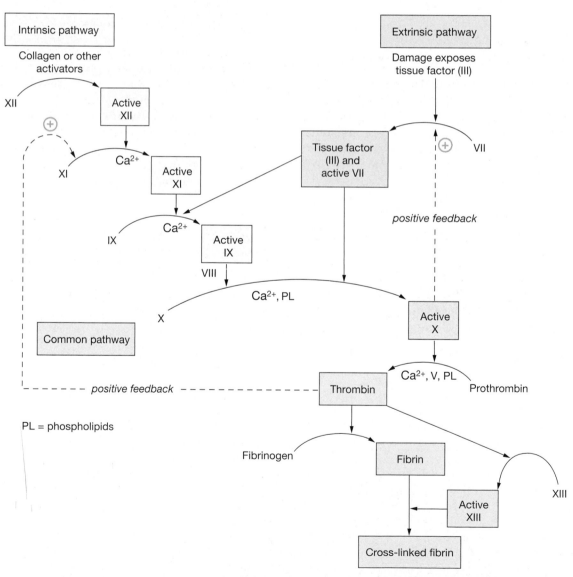

■ **Figure 16-13   The coagulation cascade**   Inactive plasma proteins are converted into active enzymes in each step of the pathway. In the final step, the enzyme thrombin converts fibrinogen to fibrin, stabilizing the platelet plug.

the enzyme that converts fibrinogen into insoluble fibrin fibers. These fibers become part of the clot.

Coagulation was initially regarded as a cascade similar to the second messenger cascades diagrammed in Chapter 6 [∞ p. 159]. At each step an enzyme converts an inactive precursor into an active enzyme, often with the help of Ca^{2+}, membrane phospholipids, or additional factors.

However, we now know that the process is more than a simple cascade. Factors in the intrinsic and extrinsic pathways interact with each other, making coagulation a network rather than a simple cascade. In addition, there are several positive feedback loops that sustain the cascade until one or more of the participating plasma proteins is completely consumed.

The end result of the coagulation process is the conversion of fibrinogen into fibrin, a reaction catalyzed by the enzyme thrombin (Figure 16-14 ■). The fibrin fibers weave through the platelet plug and trap red blood cells within their mesh (see chapter opening photo on p. 474). Active factor XIII converts fibrin into a cross-linked polymer that stabilizes the clot.

As the clot forms, it incorporates **plasmin** molecules, the seeds of its own destruction. Plasmin is an enzyme created from plasminogen by thrombin and **tissue plasminogen activator (t-PA)**. Plasmin breaks down fibrin polymers into fibrin fragments. The dissolution of fibrin by plasmin is known as **fibrinolysis.** As the damaged vessel wall slowly repairs itself, fibrinolysis helps remove the clot.

The large number of factors involved in coagulation and the fact that a single factor may have many different names can be confusing (Table 16-5). Scientists assigned numbers to the coagulation factors but the factors are not numbered in the order in which they participate in the coagulation cascade. Instead, they are numbered according to the order in which they were discovered.

✔ On Figure 16-11 ■, label the proper arrows for the intrinsic and extrinsic pathways of coagulation.

Anticoagulants Prevent Coagulation

Once coagulation begins, what keeps it from continuing until the entire circulation has clotted? Two different mechanisms limit the extent of blood clotting within a

Clot Busters and Antiplatelet Agents The discovery of the factors controlling coagulation and fibrinolysis was an important step in developing treatments for heart attacks, more properly called *myocardial infarctions* (MIs). An **infarct** is an area of tissue deprived of its blood (and oxygen) supply because the vessel is blocked by an inappropriate clot (thrombus) or by an atherosclerotic plaque. Unless the blockage is removed promptly, the tissue will die or be severely damaged. If the blockage occurs in a coronary artery, the heart may sustain so much damage that it can no longer function properly. Many clinical trials around the world are studying the best way to treat myocardial infarctions. One option is to use fibrinolytic drugs, such as streptokinase (from bacteria) and tissue plasminogen activator (t-PA), to dissolve the clots. These drugs are now being combined with other agents that prevent further platelet plug and clot formation. Radio ads for aspirin remind consumers that this old standby can help prevent heart attack damage. The newest drugs for treating heart attacks are antiplatelet agents, including some that are antagonists to platelet integrin receptors.

vessel: (1) inhibition of platelet adhesion and (2) inhibition of the coagulation cascade and fibrin production (Table 16-6). As mentioned earlier, factors such as prostacyclin in the blood vessel endothelium and plasma ensure that the platelet plug is restricted to the area of damage.

In addition, endothelial cells release chemicals known as **anticoagulants** that prevent coagulation from taking place. Most act by blocking one or more of the reactions in the coagulation cascade. The body produces two anticoagulants, **heparin** and **antithrombin III,** which work together to block active factors IX, X, XI, and XII. **Protein C,** another anticoagulant in the body, inhibits clotting factors V and VIII.

Anticoagulant drugs may be prescribed for people who are in danger of forming small blood clots that could block off critical vessels in the brain, heart, or lungs. Two of these drugs, *coumarin* and *warfarin*, block the action of vitamin K, a cofactor [∞ p. 82] in the synthesis of clotting factors II, VII, IX, and X.

When blood samples are drawn into glass tubes, clotting takes place very rapidly unless the tube contains an anticoagulant. Several of the anticoagulants that are used for this purpose remove free Ca^{2+} from the plasma. Ca^{2+}

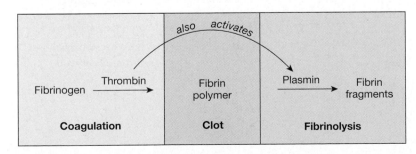

■ Figure 16-14 **Coagulation and fibrinolysis**

TABLE 16-5   Factors Involved in Coagulation

Chemical factor	Source	Activated by or Released in Response to	Role in Coagulation	Other Roles and Comments
Collagen	Subendothelial extracellular matrix	Injury exposes collagen to plasma clotting factors	Starts intrinsic pathway	NA
von Willebrand factor (vWF)	Endothelium, megakaryocytes	Exposure to collagen	Regulates level of factor VIII	Deficiency or defect causes prolonged bleeding
Kininogen and kallikrein	Liver and plasma	Cofactors normally present in plasma pathway	Cofactors for contact activation of intrinsic pathway	Mediate inflammatory response; enhance fibrinolysis
Tissue factor, or tissue thrombo-plastin (factor III)	Most cells except platelets	Damage to tissue	Starts extrinsic pathway	NA
Prothrombin and thrombin (factor II)	Liver and plasma	Platelet lipids, Ca^{2+}, and factor V	Fibrin production	NA
Fibrinogen and fibrin (factor I)	Liver and plasma	Thrombin	Forms insoluble fibers that stabilize platelet plug	NA
Fibrin-stabilizing factor (XIII)	Liver, megakaryocytes	Platelets	Cross-links fibrin polymers to make stable mesh	NA
Ca^{2+} (factor IV)	Plasma ions	NA	Required for several steps of coagulation cascade	Never a limiting factor
Vitamin K	Diet	NA	Needed for synthesis of factors II, VII, IX, X	NA

is an essential clotting factor, so with no Ca^{2+}, no coagulation can occur. In the living body however, plasma Ca^{2+} levels never decrease to levels that interfere with coagulation.

Acetylsalicylic acid (aspirin) is an agent that prevents platelet plug formation. People who are at risk of developing small blood clots are sometimes told to take one aspirin every other day "to thin the blood." The aspirin does not actually dilute the blood, but it does prevent clots from forming by blocking platelet aggregation.

Several inherited diseases affect the coagulation process. Patients with coagulation disorders bruise easily. In severe forms, spontaneous bleeding may occur throughout the body. Bleeding into the joints and mus-

TABLE 16-6   Endogenous Factors Involved in Fibrinolysis and Anticoagulation

Chemical factor	Source	Activated by or Released in Response to	Role in Anticoagulation or Fibrinolysis	Other Roles and Comments
Plasminogen and plasmin	Liver and plasma	t-PA and thrombin	Dissolves fibrin and fibrinogen	NA
Tissue plasminogen activator (t-PA)	Many tissues	Normally present; levels increase with stress, protein C	Activates plasminogen	Recombinant t-PA used clinically to dissolve clots
Antithrombin III	Liver and plasma	NA	Anticoagulant; blocks factors IX, X, XI, XII, thrombin, kallikrein	Facilitated by heparin; no effect on thrombin despite name
Prostacyclin (prostaglandin I, or PGI_2)	Endothelial cells	NA	Blocks platelet aggregation	Vasodilator

cles can be painful and disabling. If bleeding occurs in the brain, it can be fatal.

The best-known coagulation disorder is **hemophilia,** a name given to several diseases in which one of the factors in the coagulation cascade is defective or lacking. Hemophilia A, a factor VIII deficiency, is the most common form, occurring in about 80% of all cases. This disease is a recessive sex-linked trait that usually affects only males.

One exciting recent development in the treatment of hemophilia was a report on the first patients to be given gene therapy for hemophilia B, a deficiency in clotting factor IX. The patients who were injected with a virus engineered to carry the gene for factor IX started to produce some of the factor on their own, reducing their need for expensive injections of artificial factor IX.

PROBLEM CONCLUSION

"Can I live a normal life with von Willebrand disease?" asks Emily. "Yes," her doctor replies. "This coagulation disorder is different from hemophilia, in which a clotting factor is defective. Your clotting factor is normal." Interestingly, people with hemophilia have normal amounts of von Willebrand factor.

In this running problem, you learned about a coagulation disorder called von Willebrand's disease. You also learned how this disease is treated and how it differs from hemophilia. Further check your understanding of this running problem by comparing your answers with those in the summary table.

	Question	Facts	Integration and Analysis
1	Why does von Willebrand's disease cause prolonged bleeding after a dental or surgical procedure?	Von Willebrand's disease is caused by a deficiency in von Willebrand factor, a protein found in the blood that promotes adherence of platelets to the endothelium and to each other during coagulation.	Because of a deficiency in von Willebrand factor, platelets do not stick together during coagulation, and blood clots do not form easily.
2	Why is type I the least serious form of von Willebrand's disease?	Type I is characterized by decreased production of von Willebrand factor.	Because at least some von Willebrand factor is present, some clotting does take place. Therefore, this form is the least serious.
3	Explain the rationale for giving an antifibrinolytic agent to individuals with von Willebrand's disease.	Antifibrinolytic agents inhibit the action of plasmin in a blood clot.	Plasmin dissolves blood clots in the later stages of blood clotting. Inhibiting plasmin blocks this dissolution and allows clot formation to proceed.

CHAPTER REVIEW

SUMMARY

Plasma and the Cellular Elements of Blood

1. Blood is the circulating portion of the extracellular fluid. (p. 475)

2. **Plasma** is composed mostly of water, with dissolved proteins, organic molecules, ions, and dissolved gases. (p. 475)

3. The **plasma proteins** include **albumins, globulins,** and the clotting protein **fibrinogen.** They function in blood clotting, defense, and as hormones, enzymes, or carriers for different substances. (p. 475)

4. The cellular elements of blood are red blood cells (**erythrocytes**), white blood cells (**leukocytes**), and **platelets.** Platelets are fragments of cells called **megakaryocytes.** (p. 475)

5. Blood contains five types of white blood cells: (1) **lymphocytes,** (2) **monocytes,** (3) **neutrophils,** (4) **eosinophils,** and (5) **basophils.** (p. 476)

Blood Cell Production

6. All blood cells develop from a pluripotent hematopoietic stem cell. (p. 477)

7. **Hematopoiesis** begins early in embryonic development and continues throughout a person's life. Most hematopoiesis takes place in the **bone marrow.** (p. 477)

8. **Colony-stimulating factors** and other cytokines control white blood cell production. **Thrombopoietin** regulates the growth and maturation of megakaryocytes. Red blood cell production is regulated primarily by **erythropoietin.** (p. 478)

Red Blood Cells

9. Mature mammalian red blood cells are anucleate, biconcave disks. They contain hemoglobin, a red oxygen-carrying pigment. (p. 481)

10. One hemoglobin molecule consists of four **globin** chains, each centered around a **heme** group. (p. 483)

11. Hemoglobin synthesis requires iron in the diet. Iron is transported in the blood on transferrin and stored mostly in the liver, on **ferritin.** (p. 484)

12. When hemoglobin is broken down, some heme groups are converted into **bilirubin,** which is incorporated into **bile** and excreted. Elevated bilirubin concentrations in the blood cause **jaundice.** (p. 484)

Platelets and Coagulation

13. **Platelets** are cell fragments filled with granules containing clotting proteins and cytokines. Platelets are activated by damage to vascular endothelium. (p. 486)

14. **Hemostasis** begins with formation of a **platelet plug.** (p. 487)

15. Exposed collagen triggers **platelet adhesion** and **platelet aggregation.** The platelet plug is converted into a clot when reinforced by **fibrin.** (p. 487)

16. In the last step of the **coagulation cascade,** fibrin is made from **fibrinogen** through the action of **thrombin.** (p. 490)

17. As the damaged vessel is repaired, **plasmin** trapped in the platelet plug dissolves fibrin (**fibrinolysis**) and breaks down the clot. (p. 491)

18. Platelet plugs are restricted to the site of injury by **prostacyclin** in the membrane of intact vascular endothelium. **Anticoagulants** limit the extent of blood clotting within a vessel. (p. 491)

QUESTIONS

LEVEL ONE   Reviewing Facts and Terms

1. The fluid portion of the blood, called _____, is composed mainly of _____.

2. List the three types of plasma proteins in order of prevalence. Name at least one function of each type.

3. List the cellular elements found in blood and name at least one function of each.

4. Blood cell production is called _____. When and where does it occur?

5. What role do colony-stimulating factors, cytokines, and interleukins play in blood cell production? How are these chemical signal molecules different? Give two examples of each.

6. List the technical terms for production of red blood cells, platelets, and white blood cells respectively.

7. The hormone that directs red blood cell synthesis is called _____. Where is it produced and what is the stimulus for its production?

8. How are the terms *hematocrit* and *packed red cells* related? What are normal hematocrit values for men and women?

9. Distinguish between an erythroblast and an erythrocyte. Give three distinct characteristics of erythrocytes.

10. What chemical element in the diet is important for hemoglobin synthesis?

11. Define the following terms and explain their significance in hematology.

    (a) jaundice
    (b) anemia
    (c) transferrin
    (d) hemophilia

12. Chemicals that prevent blood clotting from occurring are called _____.

LEVEL TWO   Reviewing Concepts

13. **Concept map:** Combine each list of terms into a map. You may add other terms.

List 1	List 2	List 3
ADP	fibrin	hemoglobin
collagen	thrombin	iron
integrins	plasmin	ferritin
membrane phospholipids	infarct	transferrin
	fibrinolysis	erythropoietin
platelet activation	fibrinogen	bone marrow
platelet adhesion	clot	bile
platelet aggregation	coagulation	liver
	plasminogen	intestine
platelet plug	polymer	bilirubin
platelet-activating factor		heme
		globin
positive feedback		porphyrin
serotonin		hematocrit
thromboxane A_2		reticulocyte
vasoconstriction		

14. Distinguish between the intrinsic, extrinsic, and common pathways of the coagulation cascade.

15. Once platelets are activated to aggregate, what factors halt their activity?

LEVEL THREE   Problem Solving

16. Rachel is undergoing chemotherapy for breast cancer. She has blood cell counts at regular intervals; below is the computer printout of her results:

	Normal Count	10 Days Post-Chemotherapy	20 Days Post-Chemotherapy
WBC ($\times 10^3$)	4.6–10.2	2.6	4.9
RBC ($\times 10^6$)	4.04–5.48	3.85	4.2
Platelets	142–424	133	151

At the time of the first test, Jen, the nurse, notes that Rachel, although pale and complaining of being tired, does not have any bruises on her skin. Jen tells Rachel to eat foods high in protein, take a multivitamin tablet with iron, and stay home and away from crowds as much as possible. How are Jen's observations and recommendations related to the results of the first and second blood tests?

LEVEL FOUR   Quantitative Problems

17. If we estimate that total blood volume is 7% of body weight, calculate the total blood volume in a 200-lb man and in a 130-lb woman (2.2 lb/kg). What are their plasma volumes if the man's hematocrit is 52% and the woman's hematocrit is 41%?

E   X   P   L   <MediaLab>

Introduction

In this chapter you were introduced to the concept of blood as a tissue. Blood is composed of plasma and numerous cell types with diverse roles. Collectively these components serve as a functional unit that carries nutrients to tissues and provides protection from hemorrhage as well as protection from infection. The composition of blood and the regulation of hemostasis are important concepts you will need to understand other areas of physiology. This MediaLab will help you gain an appreciation of the various components of blood and their importance in the body. After reading the description below, visit the MediaLab for Chapter 16 in your Companion Website and select the appropriate keyword.

Web Exploration 1

Estimated time for completion = 15 minutes

Blood is a remarkable fluid that flows throughout our body. Can you remember each of the cellular elements that are present in the blood? Which of them do not have a nucleus?

One of the anucleate cellular elements is the erythrocyte, or red blood cell. What cytoplasmic molecule makes erythrocytes an indispensable component of blood? Select the keyword **HEMOGLOBIN** from your Website to visualize two complete hemoglobin molecules. Can you identify the four heme groups and the four iron groups? Are the four heme groups in a hemoglobin molecule identical? Why would increasing the synthesis of this molecule increase the amount of oxygen that is delivered to tissues? How important is hemoglobin to the overall structure of the red blood cell?

Web Exploration 2

Estimated time for completion = 10 minutes

Excessive blood loss can cause death. Why? Take a minute to list the components of blood. Which of those components are necessary to sustain life? What prevent minor injuries, like a cut on the finger, from resulting in hemorrhaging and eventually death? Select the keyword **HOMEOSTASIS** from your Website and view diagrams of how a blood vessel responds to injury.

Clotting mechanisms are always ready to prevent fatal loss of blood from our circulatory system. Inhibition of clotting can result in hemorrhage, but inappropriate clotting can prevent essential tissues such as heart muscle or brain neurons from getting oxygen. In Chapter 15 we learned about flood flow and blood pressure. Can turbulent blood flow cause damage to a vessel? What about high blood pressure?

17 Respiratory Physiology

■ "This being of mine, whatever it really is, consists of a little flesh, a little breath, and the part which governs."
—*Marcus Aurelius Antoninus (121—180 A.D.)* ■

Airways are lined with ciliated epithelium (green) interspersed with goblet cells covered in microvilli (orange). Two yellow mucus droplets sit on the microvilli.

Imagine covering the playing surface of a racquetball court (about 75 m²) with thin plastic wrap, then crumpling up the wrap and stuffing it into a 3-liter soft drink bottle. Impossible? Maybe so, if you use plastic wrap and a drink bottle. But the lungs of a 70-kg man have a gas exchange surface the size of that plastic wrap, compressed into a volume that is less than the size of the bottle. This tremendous area for gas exchange is needed to supply the trillions of cells in the body with adequate amounts of oxygen.

PROBLEM
Emphysema

"Diagnosis: COPD (blue bloater)," reads Edna Wilson's patient chart. COPD—chronic obstructive pulmonary disease—is a name given to diseases in which air exchange is impaired by narrowing of the airways. Most people with COPD have emphysema or chronic bronchitis or a combination of the two. Individuals in whom chronic bronchitis predominates are nicknamed "blue bloaters," owing to the bluish tinge of their skin and a tendency to be overweight. "Pink puffers" suffer more from emphysema. They tend to be thin, have normal (pink) skin coloration, and breathe shallow, rapid breaths. Because COPD is commonly caused by smoking, most people can avoid the disease simply by not smoking. Unfortunately, Edna has been a hard-core smoker for 35 of her 47 years.

...continued on page 502

• •

Aerobic metabolism in animal cells depends on a steady supply of oxygen and nutrients from the environment, coupled with the removal of carbon dioxide and other wastes. In small, simple aquatic animals, simple diffusion across the body surface meets these needs. Distance limits diffusion rate, however, and as a result, most multicelled animals use specialized respiratory organs associated with a circulatory system. Respiratory organs take a variety of forms, but all possess a large surface area compressed into a small space.

Besides needing a large exchange surface, humans and other terrestrial animals face an additional physiological challenge: dehydration. Their exchange surfaces must be thin and moist to allow gases to pass from air into solution, yet at the same time they must be protected from drying out due to exposure to air. Some terrestrial animals such as the slug, a shell-less snail, meet the challenge of dehydration with behavioral adaptations that restrict them to humid environments and nighttime activities.

However, a more common solution is anatomical: an internalized respiratory epithelium. Human lungs are enclosed within the chest cavity to control their contact with the outside air. Internalization creates a humid environment for the exchange of gases with the blood and protects the exchange surface from damage.

Internalized lungs create another problem, however: how to exchange air between the atmosphere and the exchange surface deep within the body. Air flow requires a muscular pump to create pressure gradients. Thus, in more complex animals, the respiratory system consists of two separate components: a muscle-driven pump and a thin, moist exchange surface. In humans, the pump is the musculoskeletal structure of the thorax. The lungs themselves consist of the exchange epithelium and associated blood vessels.

The primary functions of the respiratory system are:

1. **Exchange of gases between the atmosphere and the blood.** The body brings in oxygen for distribution to the tissues, and eliminates carbon dioxide waste produced by metabolism.
2. **Homeostatic regulation of body pH.** The lungs can alter body pH by selectively retaining or excreting CO_2.
3. **Protection from inhaled pathogens and irritating substances.** Like all epithelia that contact the external environment, the respiratory epithelium is well supplied with mechanisms to trap and destroy potentially harmful substances before they can enter the body.
4. **Vocalization.** Air moving across the vocal cords creates vibrations used for speech, singing, and other forms of communication.

In addition to serving these functions, the respiratory system is also a source of significant losses of water and heat from the body. These losses must be balanced using homeostatic compensations.

In this chapter you will learn how the respiratory system carries out these functions by exchanging air between the environment and the interior of the lungs. In addition, we will see how the circulatory and respiratory systems cooperate to move oxygen and carbon dioxide between the lungs and the cells of the body.

▶ THE RESPIRATORY SYSTEM

The word *respiration* can have several meanings (Fig. 17-1 ■). **Cellular respiration** refers to the intracellular reaction of oxygen with organic molecules to produce carbon dioxide, water, and energy in the form of ATP [∞ p. 94]. **External respiration,** the topic of this chapter, is the interchange of gases between the environment and the body's cells. External respiration can be subdivided into four integrated processes:

1. The exchange of air between the atmosphere and the lungs. This process is known as **ventilation,** or *breathing.* **Inspiration** is the movement of air into the lungs. **Expiration** is the movement of air out of the lungs.
2. The exchange of oxygen and carbon dioxide between the lungs and the blood.
3. The transport of oxygen and carbon dioxide by the blood.
4. The exchange of gases between blood and the cells.

External respiration requires the coordinated functioning of the respiratory and cardiovascular systems. The **respiratory system** is composed of the structures involved in ventilation and gas exchange (Fig. 17-2 ■). It consists of:

1. The **conducting system** of passages, or **airways,** that lead from the environment to the exchange surface of the **lungs.**
2. The *alveoli,* the exchange surface of the lungs, where oxygen and carbon dioxide transfer between air and blood.
3. The bones and muscles of the **thorax** (chest) that assist in ventilation.

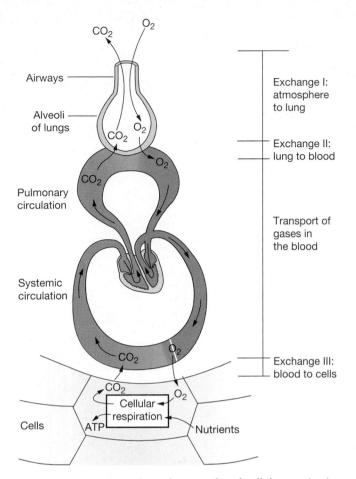

■ **Figure 17-1   Overview of external and cellular respiration**

The respiratory system can be divided into two parts. The **upper respiratory tract** consists of the mouth, nasal cavity, pharynx, and larynx. The **lower respiratory tract** consists of the trachea, two bronchi, their branches, and the lungs. It is also known as the *thoracic* portion of the respiratory system because it is enclosed in the thorax.

The Bones and Muscles of the Thorax Surround the Lungs

The thorax, or chest cavity, is bounded by the bones of the spine and rib cage and their associated muscles. Together the bones and muscles are called the *thoracic cage.* The ribs and spine (the *chest wall*) form the sides and top of the cage. A dome-shaped sheet of skeletal muscle, the **diaphragm,** forms the floor (Fig. 17-2b ■).

Two sets of **intercostal muscles,** internal and external, connect the 12 pairs of ribs (Fig. 17-2a ■). Additional muscles, the **sternocleidomastoids** and the **scalenes,** run from the head and neck to the sternum and first two ribs.

Functionally, the thorax is a sealed container filled with three membranous bags, or sacs. One, the *pericardial sac,* contains the heart. The other two bags, the *pleural sacs,* contain the lungs [*pleura,* rib, or side]. The esophagus and thoracic blood vessels and nerves pass between the pleural sacs (Fig. 17-2d ■).

Pleural Sacs Enclose the Lungs

The lungs (Fig. 17-2b ■) consist of light, spongy tissue whose volume is mostly occupied by air-filled spaces. These irregular cone-shaped organs nearly fill the thoracic cavity, with their bases resting on the curved diaphragm. Rigid conducting airways, the bronchi, connect the lungs to the main airway, the trachea.

Each lung is surrounded by a double-walled pleural sac whose membranes line the inside of the thorax and cover the outer surface of the lungs (Fig. 17-3 ■ p. 502). The pleural membranes, or **pleura,** contain several layers of elastic connective tissue and numerous capillaries. The opposing layers of pleural membrane are held together by a thin film of **pleural fluid** whose total volume is only a few milliliters. The result is similar to an air-filled balloon (the lung) surrounded by a water-filled balloon (the pleural sac). Most illustrations exaggerate the volume of the pleural fluid, but you can appreciate its thinness if you imagine spreading 3 mL of water evenly over the surface of a 3-liter soft drink bottle.

Pleural fluid serves several purposes. First, it creates a moist, slippery surface so that the opposing membranes can slide across one another as the lungs move within the thorax. The second important function of pleural fluid is to hold the lungs tight against the thoracic wall.

To visualize this arrangement, think of two panes of glass stuck together by a thin film of water. You can slide the panes back and forth across each other, but you cannot pull them apart because of cohesiveness of the water [∞ p. 25]. A similar fluid bond between the pleural membranes makes the lungs "stick" to the thoracic cage and holds them stretched in a partially inflated state, even at rest.

The Airways Connect Lungs to the Environment

Air enters the upper respiratory tract through the mouth and nose and passes into the **pharynx,** a common passageway for food, liquids, and air [*pharynx,* throat]. From the pharynx, air flows through the **larynx** into the **trachea,** or windpipe (Fig. 17-2b ■). The larynx contains the **vocal cords,** connective tissue bands that tighten to create sound when air moves past them.

The trachea is a semiflexible tube held open by 15 to 20 C-shaped cartilage rings (Fig. 17-2e ■). It extends down into the thorax, where it branches (division 1) into a pair of **primary bronchi,** one *bronchus* to each lung. Within the lungs, the bronchi branch repeatedly (divisions 2–11) into progressively smaller bronchi (Fig. 17-2b, e ■). Like the trachea, the bronchi are semirigid tubes supported by cartilage.

Within the lungs, the smallest bronchi branch to become **bronchioles,** small collapsible passageways with smooth muscle walls. The bronchioles continue branching (divisions 12–23) until the *respiratory bronchioles* form a transition between the airways and the exchange epithelium of the lung.

ANATOMY SUMMARY Respiratory System

■ **Figure 17-2**

(a) Muscles used for ventilation
The muscles of inspiration include the diaphragm, external intercostals, sternocleidomastoids, and scalenes. The muscles of expiration include the internal intercostals and the abdominals.

(b) The respiratory system
The respiratory system consists of the upper respiratory system (mouth, nasal cavity, pharynx, larynx) and the lower respiratory system (trachea, bronchi, lungs). The lower respiratory system is enclosed in the thorax, bounded by the ribs, spine, and diaphragm.

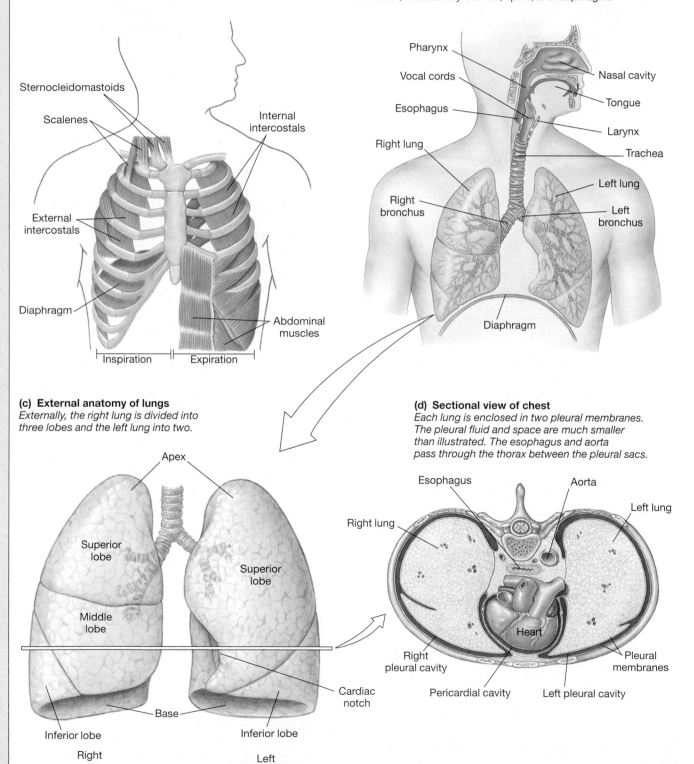

(c) External anatomy of lungs
Externally, the right lung is divided into three lobes and the left lung into two.

(d) Sectional view of chest
Each lung is enclosed in two pleural membranes. The pleural fluid and space are much smaller than illustrated. The esophagus and aorta pass through the thorax between the pleural sacs.

(e) Branching of airways
The trachea branches into two bronchi, one to each lung. Each bronchus branches 22 more times, finally terminating in a cluster of alveoli.

(f) Structure of lung lobule
Each cluster of alveoli is surrounded by elastic fibers and a network of capillaries.

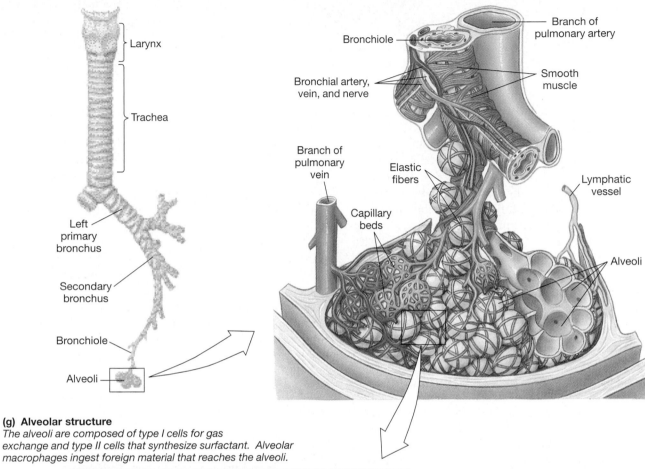

Larynx

Trachea

Left primary bronchus

Secondary bronchus

Bronchiole

Alveoli

Bronchiole

Bronchial artery, vein, and nerve

Branch of pulmonary vein

Capillary beds

Elastic fibers

Branch of pulmonary artery

Smooth muscle

Lymphatic vessel

Alveoli

(g) Alveolar structure
The alveoli are composed of type I cells for gas exchange and type II cells that synthesize surfactant. Alveolar macrophages ingest foreign material that reaches the alveoli.

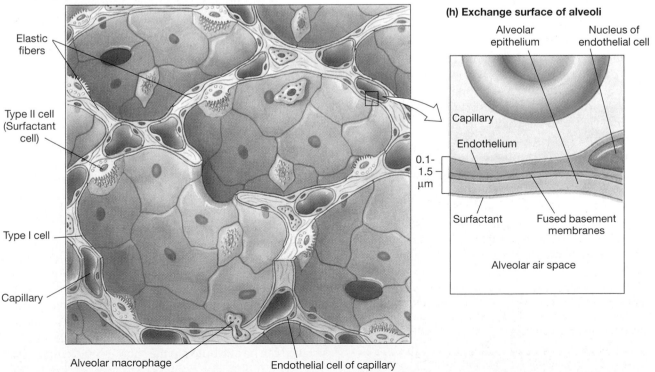

Elastic fibers

Type II cell (Surfactant cell)

Type I cell

Capillary

Alveolar macrophage

Endothelial cell of capillary

(h) Exchange surface of alveoli

Alveolar epithelium

Nucleus of endothelial cell

Capillary

Endothelium

0.1-1.5 µm

Surfactant

Fused basement membranes

Alveolar air space

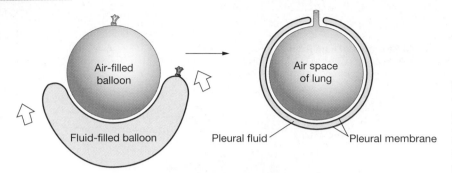

■ **Figure 17-3** **The relationship between the pleural sac and the lung** The pleural sac forms a double membrane surrounding the lung, similar to a fluid-filled balloon surrounding an air-filled balloon. The pleural fluid has a much smaller volume than is suggested by this illustration.

The diameter of the airways becomes progressively smaller from the trachea to the bronchioles, but as the individual airways get narrower, their numbers increase (Fig. 17-4 ■). As a result, the total cross-sectional area increases with each division of the airways. Total cross-sectional area is lowest in the upper respiratory tract and greatest in the bronchioles, analogous to the increase in cross-sectional area that occurs from the aorta to the capillaries. Velocity of air flow is therefore highest in the trachea and lowest in the terminal bronchioles [∞ p. 410].

The Alveoli Are the Site of Gas Exchange

The bulk of lung tissue consists of exchange sacs known as **alveoli** [*alveus*, a concave vessel; singular *alveolus*]. The alveoli are grapelike clusters at the ends of the terminal bronchioles (Fig. 17-2f, g ■). Their primary function is the exchange of gases between air in the alveoli and the blood.

Each tiny alveolus is composed of a single layer of thin exchange epithelium (Fig. 17-2g ■). Two types of epithelial cells are found in the alveoli, and they occur in roughly equal numbers. The larger **type I alveolar cells** are very thin so that gases can diffuse rapidly through them. The smaller but thicker **type II alveolar cells** synthesize and secrete a type of chemical known as *surfactant*. Surfactant mixes with the thin fluid lining of the alveoli to ease the expansion of the lungs during breathing, as we will see later in this chapter.

The thin walls of the alveoli do not contain muscle, because muscle fibers would block rapid gas exchange. As a result, lung tissue itself cannot contract. However, connective tissue between the alveolar epithelial cells

does contain many elastin fibers that create elastic recoil when lung tissue is stretched.

The close association of alveoli with an extensive network of capillaries demonstrates the intimate link between the respiratory and cardiovascular systems. Blood vessels cover 80%–90% of the alveolar surface, forming an almost continuous "sheet" of blood in close contact with the air-filled alveoli (Fig. 17-2f ■).

Gas exchange in the lungs occurs by diffusion through the thin alveolar type I cells to the capillaries (Fig. 17-2h ■). In much of the exchange area, the basement membrane underlying the alveolar epithelium has fused with that of the capillary endothelium, and only a small amount of interstitial fluid is present. The proximity of capillary blood to air in the alveolus is essential for the rapid exchange of gases.

The Pulmonary Circulation Is a High-Flow, Low-Pressure System

The pulmonary circulation begins with the pulmonary trunk that receives low-oxygen blood from the right ventricle. The pulmonary trunk then divides into two pulmonary arteries, one to each lung [∞ Fig. 14-1, p. 406]. Oxygenated blood from the lungs returns to the left atrium via the pulmonary veins.

At any given moment, the pulmonary circulation contains about 0.5 liters of blood, or 10% of the total blood volume. About 75 mL of this amount is found in the capillaries, where gas exchange takes place, with the remainder in the pulmonary arteries and veins. The rate of blood flow through the lungs is quite high when compared with other tissues [∞ p. 457] because the lungs receive the entire cardiac output of the right ventricle, 5 L/min. This means that as much blood flows through the lungs in one minute as flows through the rest of the body in the same amount of time!

Despite the high flow rate, blood pressure in the pulmonary circulation is low. Pulmonary arterial blood pressure averages 25/8 mm Hg, compared with the average systemic arterial blood pressure of 120/80 mm Hg. The right ventricle does not have to pump as forcefully to create blood flow through the lungs because the resistance of the pulmonary circulation is low. This low resistance can be attributed to the shorter total length of the

---

…continued from page 498

Patients with chronic bronchitis have excessive mucus production and general inflammation of the entire respiratory tract. The mucus narrows the airways and makes breathing difficult.

Question 1: *What does narrowing of the airways do to the resistance of the airways to air flow? (Hint: The relationship between radius and resistance is the same for air flow as it was for blood flow in the circulatory system;* ∞ *p. 409.)*

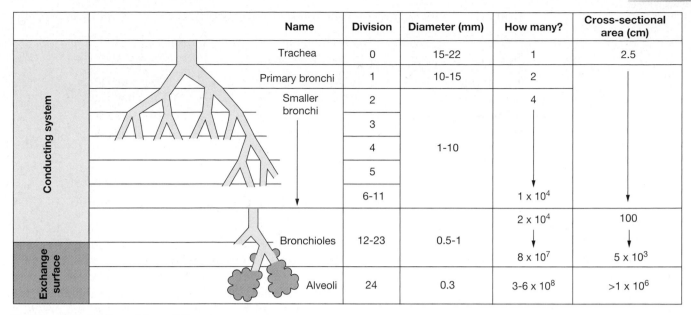

	Name	Division	Diameter (mm)	How many?	Cross-sectional area (cm)
Conducting system	Trachea	0	15-22	1	2.5
	Primary bronchi	1	10-15	2	
	Smaller bronchi	2		4	
		3			
		4	1-10		
		5			
		6-11		1 x 10⁴	
Exchange surface	Bronchioles	12-23	0.5-1	2 x 10⁴ ↓ 8 x 10⁷	100 ↓ 5 x 10³
	Alveoli	24	0.3	3-6 x 10⁸	>1 x 10⁶

■ **Figure 17-4** **Branching of the airways**

pulmonary blood vessels and the distensibility and large cross-sectional area of pulmonary arterioles.

Because mean blood pressure is low in the pulmonary capillaries, the net hydrostatic pressure forcing fluid out of the capillary into the interstitial space is also low [∞ p. 459]. The lymphatic system efficiently removes fluid filtered from the pulmonary capillaries, so normally lung interstitial fluid volume is small. As a result, the distance between alveolus and capillary is short, and gases diffuse rapidly between them.

✔ A person has left ventricular failure but normal right ventricular function. As a result, blood pools in the pulmonary circulation, doubling pulmonary capillary hydrostatic pressure. What happens to net fluid flow across the walls of the pulmonary capillaries?

▶ GAS LAWS

Air flow in the respiratory system is very similar to blood flow in the cardiovascular system in many respects, although blood is a noncompressible liquid and air is a compressible mixture of gases. Blood pressure and environmental air pressure (**atmospheric pressure**) are both reported in millimeters of mercury (mm Hg).* At sea level, atmospheric pressure is 760 mm Hg.

One convention that we will follow in this book is the designation of atmospheric pressure as 0 mm Hg. Because atmospheric pressure varies with altitude and because very few people live exactly at sea level, this convention allows us to compare pressure differences that occur during ventilation without correcting for al-

---

*Respiratory physiologists sometimes report gas pressures in units of centimeters of water: 1 mm Hg = 1.36 cm H₂O.

titude. Negative numbers designate subatmospheric pressures, and positive numbers denote higher-than-atmospheric pressures.

Table 17-1 summarizes the rules that govern the behavior of gases in air and in solution. These rules provide the basis for the exchange of oxygen and carbon dioxide between the environment and cells, as we'll see.

Air Is a Mixture of Gases

The atmosphere surrounding the earth is a mixture of gases and water vapor. **Dalton's law** states that the total pressure of a mixture of gases is the sum of the pressures of the individual gases. Thus, in dry air at an atmospheric pressure of 760 mm Hg, 78% of the total pressure is due to nitrogen molecules, 21% to oxygen molecules, and so on (Table 17-2).

In respiratory physiology, we are concerned not only with total atmospheric pressure but also with the individual pressures of oxygen and carbon dioxide. The pressure of a single gas is known as its **partial pressure,**

TABLE 17-1 Gas Laws

1. The total pressure of a mixture of gases is the sum of the pressures of the individual gases (Dalton's law).
2. Gases, singly or in a mixture, move from areas of higher pressure to areas of lower pressure.
3. If the volume of a container of gas changes, the pressure of the gas will change in an inverse manner (Boyle's law).
4. The amount of a gas that will dissolve in a liquid is determined by the partial pressure of the gas and the gas's solubility in the liquid (Henry's law).

TABLE 17-2   Partial Pressures of Some Atmospheric Gases at 25° C and 760 mm Hg

Gas	Partial Pressure in Atmospheric Air, Dry		Partial Pressure in Atmospheric Air, 100% Humidity	
Nitrogen (N_2)	593	mm Hg	575	mm Hg
Oxygen (O_2)	160	mm Hg	152	mm Hg
Carbon dioxide (CO_2)	0.25	mm Hg	0.24	mm Hg
Water vapor	0	mm Hg	23.8	mm Hg

abbreviated P_{gas}. To find the partial pressure of a gas, multiply the atmospheric pressure (P_{atm}) by the gas's relative contribution (%):

Partial pressure of an atmospheric gas = P_{atm} × % of gas in atmosphere

Partial pressure of oxygen = 760 mm Hg × 21%

P_{O_2} = 760 × 0.21 = 160 mm Hg

Thus, the P_{O_2}, or partial pressure of oxygen, in dry air at sea level is 160 mm Hg. The pressure of an individual gas is determined only by its relative abundance in the mixture and is independent of the molecular size or weight of the gas.

The actual partial pressure of gases varies slightly depending on how much water vapor is in the air. Water pressure "dilutes" the contribution of other gases to the total pressure. Table 17-2 compares the partial pressures of some atmospheric gases in dry air and at 100% humidity.

Gases Move from Areas of Higher Pressure to Areas of Lower Pressure

Air flow occurs whenever there is a pressure gradient. Air flow, like blood flow, moves from areas of higher pressure to areas of lower pressure. Meteorologists predict the weather by knowing that areas of high atmospheric pressure move in to replace areas of low pressure. In ventilation, flow of air down pressure gradients explains why air exchanges between the environment and the lungs. The movement of the thorax during breathing creates alternating conditions of high and low pressure within the lungs.

Movement down pressure gradients also applies to single gases. Oxygen moves from areas of higher oxygen partial pressure (P_{O_2}) to areas of lower oxygen partial pressure. The simple diffusion of oxygen and carbon dioxide between lung and blood, or between blood and cells, depends on pressure gradients for these gases. Respiratory physiologists talk about the partial pressure of oxygen or carbon dioxide in the body rather than the concentration of the gas in solution because this measure allows direct comparison with partial pressures of the gases in air.

Boyle's Law Describes Pressure-Volume Relationships of Gases

The pressure exerted by a gas or mixture of gases in a sealed container is created by the collisions of the moving gas molecules with the walls of the container and with each other. If the size of the container is reduced, the collisions will become more frequent and the pressure will rise. This relationship can be expressed by the following equation:

$P_1V_1 = P_2V_2$ where P represents pressure and V represents volume

For example, start with a 1-liter container (V_1) of a gas whose pressure is 100 mm Hg (P_1). One side of the container moves in to decrease the volume to 0.5 L (Fig. 17-5 ■). What happens to the pressure of the gas? According to the equation,

$P_1V_1 = P_2V_2$

100 mm Hg × 1 L = P_2 × 0.5 L

P_2 = 200 mm Hg

If the volume is reduced by one-half, the pressure doubles. If the volume were to double, the pressure would be reduced by one-half. This relationship between

Boyle's Law: $P_1V_1 = P_2V_2$

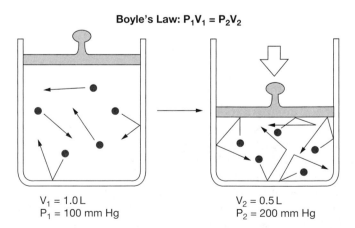

V_1 = 1.0 L
P_1 = 100 mm Hg

V_2 = 0.5 L
P_2 = 200 mm Hg

■ **Figure 17-5   Boyle's law**   Decreasing the volume of a gas causes more collisions between gas molecules and between gas molecules and the container. The higher number of collisions increases the pressure exerted by the gas. Boyle's law ($P_1V_1 = P_2V_2$) assumes that temperature and number of gas molecules remain constant.

pressure and volume was first noted by Robert Boyle in the 1600s and has been called **Boyle's law** of gases.

In the respiratory system, changes in volume of the chest cavity during ventilation cause pressure gradients that create air flow. When the chest increases in volume, the alveolar pressure drops and air flows into the respiratory system. When the chest decreases in volume, the pressure rises and air flows out into the atmosphere. This movement of air is called *bulk flow* because the entire gas mixture is moving rather than merely one or two gas species.

The Solubility of Gases in Liquids Depends on Pressure, Solubility, and Temperature

When a gas is placed in contact with water, if there is a pressure gradient, the gas molecules move from one phase to the other. If the gas pressure is higher in the water than in the gaseous phase, gas molecules will leave the water. If the gas pressure is higher in the gaseous phase than in the water, the gas will dissolve into the water.

The movement of gas molecules from air into solution is directly proportional to three factors: (1) the pressure gradient of the individual gas, (2) the solubility of the gas in the given liquid, and (3) temperature. Because temperature is relatively constant in mammals, we will ignore its contribution in this discussion.

The ease with which a gas dissolves in a solution is its **solubility.** If a gas is very soluble, large numbers of gas molecules will go into solution at low partial pressures. With less soluble gases, high partial pressures may cause only a few molecules of the gas to dissolve.

For example, imagine a container of water exposed to air with a P_{O_2} of 100 mm Hg (Fig. 17-6a ■). Initially, the water has no oxygen dissolved in it (P_{O_2} = 0 mm Hg). As the gas phase stays in contact with the water, the moving oxygen molecules in the gas diffuse into the water and dissolve (Fig. 17-6b ■). This process will continue until equilibrium is reached.

At equilibrium (Fig. 17-6c ■), the movement of oxygen into the water is equal to the movement of oxygen back into the air. We refer to the amount of oxygen that dissolves in the water at any given P_{O_2} as the *partial pressure of the gas in solution.* Thus, if the gaseous phase has a P_{O_2} of 100 mm Hg, at equilibrium the water will also have a P_{O_2} of 100 mm Hg.

Note that this does *not* mean that the concentration of oxygen is the same in the air and in the water! The concentration of dissolved oxygen also depends on the *solubility* of oxygen. For example, when the P_{O_2} of both air and water is 100 mm Hg, the concentration of oxygen in air is 5.2 mmoles O_2/L air, but the concentration of oxygen in water is only 0.15 mmoles O_2/L water. As you can see, oxygen is not very soluble in aqueous solutions.

(a) Initial state:
no O_2 in solution

(b) Oxygen dissolves

(c) At equilibrium, P_{O_2} in air and water is equal. Low O_2 solubility means concentrations are not equal.

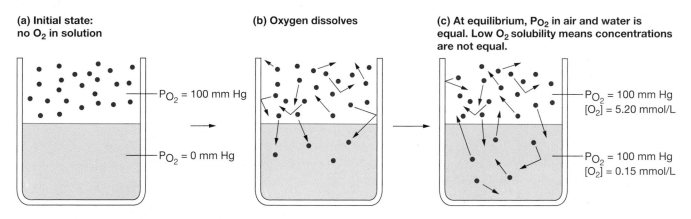

P_{O_2} = 100 mm Hg

P_{O_2} = 0 mm Hg

P_{O_2} = 100 mm Hg
$[O_2]$ = 5.20 mmol/L

P_{O_2} = 100 mm Hg
$[O_2]$ = 0.15 mmol/L

(d) When CO_2 is at equilibrium at the same partial pressure, more CO_2 dissolves.

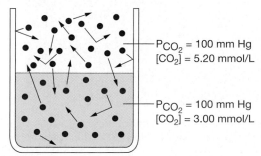

P_{CO_2} = 100 mm Hg
$[CO_2]$ = 5.20 mmol/L

P_{CO_2} = 100 mm Hg
$[CO_2]$ = 3.00 mmol/L

■ **Figure 17-6   Gases in solution**   The amount of a gas that dissolves in solution depends on the solubility and partial pressure of the gas if temperature is assumed to be constant.

Its insolubility is one reason for the evolution of oxygen-carrying molecules in blood.

Now compare oxygen to carbon dioxide (Fig. 17-6d ■). Carbon dioxide is about 20 times as soluble in water as oxygen is. At a P_{CO_2} of 100 mm Hg, the CO_2 concentration in air is 5.2 mmoles CO_2 /L air and its concentration in solution is 3.0 mmoles/L water.

✔ A saline solution is exposed to a gas mixture with equal partial pressures of O_2 and CO_2. What information do you need to know to predict if equal amounts of O_2 and CO_2 will dissolve in the saline?

✔ If nitrogen is 78% of atmospheric air, what is the partial pressure of nitrogen (P_{N_2}) when the dry atmospheric pressure is 720 mm Hg?

▶ VENTILATION

The first exchange in respiratory physiology is ventilation, or breathing, the movement of air between the environment and the alveoli.

The Airways Warm, Humidify, and Filter Inspired Air

The upper airways and the bronchi do more than simply serve as passageways for air. They play an important role in conditioning air before it reaches the alveoli. Conditioning has three components:

1. Warming air to 37° C so that core body temperature will not change and alveoli will not be damaged by cold air,
2. Adding water vapor until the air reaches 100% humidity so that the moist exchange epithelium will not dry out,
3. Filtering out foreign material so that viruses, bacteria, and inorganic particles will not reach the alveoli.

Inhaled air is warmed and moistened by heat and water evaporating from the mucosal lining of the airways. Under normal circumstances, by the time air reaches the trachea, it has been conditioned to 37° C and 100% humidity.

Breathing through the mouth is not nearly as effective at warming and moistening air as breathing through the nose. If you exercise outdoors in very cold weather, you may be familiar with the ache in your chest that results from breathing cold air through your mouth.

Filtration of air takes place in the trachea and bronchi as well. These airways are lined with a ciliated epithelium that secretes both mucus and a dilute saline solution (see chapter opening photo, p. 497). The cilia themselves are bathed in a watery saline layer covered by a sticky layer of mucus (Fig. 17-7 ■). The mucus is secreted by *goblet cells* in the epithelium [∞ p. 61].

Mucus traps most inhaled particles larger than 2 mm, and its immunoglobulins disable many inhaled microorganisms. The mucus layer is continuously moved toward the pharynx by the upward beating of the cilia,

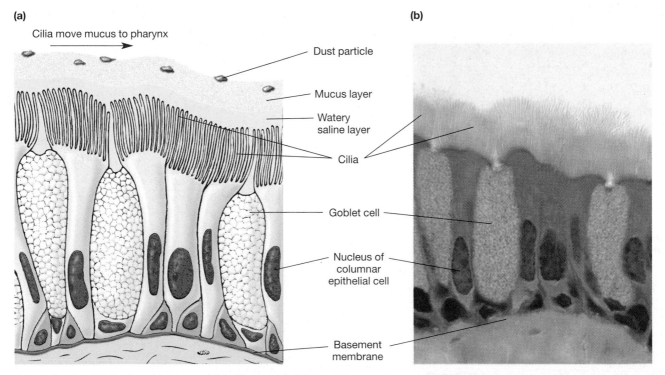

(a)

Cilia move mucus to pharynx

Dust particle

Mucus layer

Watery saline layer

Cilia

Goblet cell

Nucleus of columnar epithelial cell

Basement membrane

(b)

■ **Figure 17-7   Ciliated respiratory epithelium**   (a) Goblet cells in respiratory epithelium secrete a thick mucus layer that traps inhaled particles. The mucus floats on top of a watery saline that allows the cilia to push the mucus upward toward the pharynx, where it is swallowed. (b) This micrograph shows the ciliated epithelium of the trachea.

...continued from page 502

Smokers usually develop chronic bronchitis before they develop emphysema. Cigarette smoke paralyzes the cilia that sweep debris and mucus out of the airways. Without the action of cilia, mucus and debris pool in the airways, leading to a chronic cough. Eventually, breathing becomes difficult.

Question 2:   *Why do people with chronic bronchitis have a higher-than-normal rate of respiratory infections?*

a process called the *mucus escalator*. Once mucus reaches the pharynx, it is swallowed, and acid and enzymes in the stomach destroy any remaining microorganisms.

Secretion of the watery layer beneath the mucus is a critical step in the mucus escalator. Without the watery layer, the cilia would become trapped in the thick, sticky mucus and cease to function.

✔ Cigarette smoking paralyzes the cilia in the epithelial lining of the airways. Why would paralysis of the cilia cause smokers to develop a cough?

During Ventilation, Air Flows Because of Pressure Gradients

Air flows into the lungs because of pressure gradients created by a pump, just as blood flows in the cardiovascular system because of the pumping action of the heart. In the respiratory system, most lung tissue is thin exchange epithelium, so the muscles of the thoracic cage and diaphragm function as the pump. When these muscles contract, moving the rib cage and diaphragm, the lungs expand, held to the inside of the thoracic cage by the pleural fluid.

Breathing is an active process that uses muscle contraction to create a pressure gradient. The primary muscles involved in quiet breathing (breathing at rest) are the diaphragm, the intercostals, and the scalenes. During forced breathing, other muscles of the chest and abdomen may be recruited to assist. Examples of situations in which breathing is forced include exercise, playing a wind instrument, and blowing up a balloon.

Air flow in the respiratory tract obeys the same rule as blood flow in the cardiovascular system:

$$\text{Flow} \propto \Delta P / R$$

This equation means that (1) air flows in response to a pressure gradient (ΔP) and that (2) flow decreases as the resistance (R) of the system to flow increases. Before we discuss resistance, let's consider how the respiratory system creates a pressure gradient.

Pressures in the respiratory system can be measured either in the air spaces of the lungs (**alveolar pressure, P_A**) or within the pleural fluid (**intrapleural pressure**). Because atmospheric pressure is relatively constant, pressure in the lungs must be higher or lower than atmospheric pressure for air to flow between the environment and the alveoli.

The respiratory system ends in a dead end, so the direction of air flow reverses. Air flows into the lungs when you inhale (*inspiration*) and out of the lungs when you exhale (*expiration*). A single **respiratory cycle** consists of an inspiration followed by an expiration. The pressure-volume relationships of Boyle's law provide the basis for pulmonary ventilation.

Inspiration Occurs When Alveolar Pressure Decreases

During inspiration, somatic motor neurons trigger contraction of the diaphragm and the inspiratory muscles. When the diaphragm contracts, it loses its dome shape and drops down toward the abdomen. In quiet breathing, the diaphragm drops about 1.5 cm. This movement increases the volume of the thoracic cavity by flattening its floor (Fig. 17-8 ■). Contraction of the diaphragm causes between 60% and 75% of the inspiratory volume change during normal quiet breathing.

Movement of the rib cage creates the remaining 25% to 40% of the volume change. **External intercostal** and **scalene muscles** contract and pull the ribs upward and out. Rib movement during inspiration has been likened to a pump handle lifting up and away from the pump

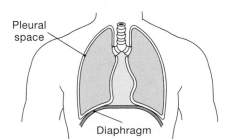

Pleural space

Diaphragm

(a) At rest, diaphragm is relaxed

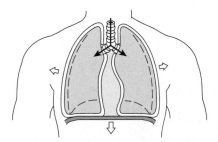

(b) Diaphragm contracts, thoracic volume increases.

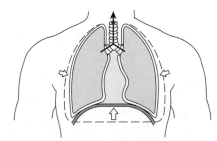

(c) Diaphragm relaxes, thoracic volume decreases.

■ **Figure 17-8   Movement of the diaphragm**

(the ribs moving up and away from the spine) and to the movement of a pair of bucket handles as they lift away from the sides of a bucket (ribs moving outward in a lateral direction). The combination of these two movements broadens the rib cage in all directions (Fig. 17-9 ■). As the volume of the thoracic cavity increases, pressure decreases and air flows into the lungs.

For many years, quiet breathing was attributed solely to the action of the diaphragm and the external intercostal muscles. It was thought that the scalenes and *sternocleido-mastoid muscles* were active only during deep breathing. But in recent years, studies of patients with neuromuscular disorders have changed our understanding of how these accessory muscles contribute to quiet breathing.

It now appears that without lifting of the sternum and upper ribs by the scalenes, contraction of the diaphragm during inspiration pulls the lower ribs inward. This action decreases the volume of the thoracic cage and works against inspiration. Because we know that the lower ribs move up and out in normal quiet breathing, the scalenes must assist quiet inspiration.

New evidence also downplays the role of the external intercostal muscles during quiet breathing. However, the external intercostals play an increasingly important role as respiratory activity increases. Because the exact contribution of the external intercostals and scalenes varies depending upon the type of breathing, we will group these muscles together and call them the *inspiratory muscles.*

(a) "Pump handle" motion increases anterior-posterior dimension of rib cage

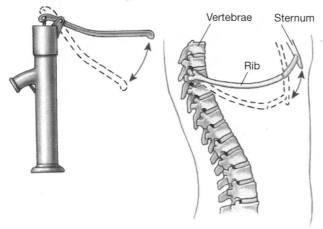

(b) "Bucket handle" motion increases lateral dimension of rib cage

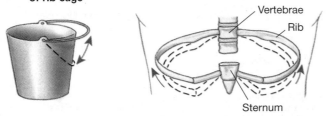

■ **Figure 17-9** **Movement of the rib cage during inspiration**

Now let's follow alveolar pressure as it changes during a single inspiration. At the start of an inspiration, the brief pause between breaths, alveolar pressure is equal to atmospheric pressure and there is no air flow (Fig. 17-10 ■, point A_1). As inspiration begins, the muscles of the thoracic cage contract and thoracic volume increases. With the increase in volume, alveolar pressure falls about 1 mm Hg below atmospheric pressure (-1 mm Hg) and air begins to flow into the alveoli.

The thoracic volume changes faster than air can flow, so alveolar pressure reaches its lowest value about halfway through inspiration (point A_2). As air continues to flow into the alveoli, pressure increases until the thoracic cage stops expanding, just before the end of inspiration. Air movement continues for a fraction of a second longer, until the pressure inside the lungs equalizes with atmospheric pressure (point A_3). At the end of inspiration, the volume of air in the lungs is at its maximum for the respiratory cycle (point C_2) and alveolar pressure is equal to atmospheric pressure.

You can demonstrate this phenomenon by taking a deep breath and stopping the movement of your chest at the end of inspiration. (Do not "hold your breath," because doing so causes the opening of the pharynx to close and prevents air flow.) If you do this correctly, you will notice that air flow stops after you freeze the inspiratory movement. This exercise shows that, at the end of inspiration, air pressure in the alveoli is equal to atmospheric pressure.

Expiration Occurs When Alveolar Pressure Exceeds Atmospheric Pressure

At the end of inspiration, impulses from somatic motor neurons to the inspiratory muscles cease and the muscles relax. Elastic recoil of the lungs returns the diaphragm and rib cage to their original relaxed positions, just as a stretched elastic waistband recoils when released. Because expiration during quiet breathing involves passive elastic recoil rather than active muscle contraction, it is called **passive expiration.**

As the volume of the lungs and thorax decreases during expiration, air pressure in the lungs increases, reaching about 1 mm Hg above atmospheric pressure at its maximum (Fig. 17-10 ■, point A_4). Alveolar pressure is now higher than atmospheric pressure, so air flow reverses and air moves out of the lungs.

At the end of expiration, air movement ceases when the alveolar pressure is again equal to atmospheric pressure (point A_5). The volume of air in the lungs reaches its minimum for the respiratory cycle (point C_3). At this point, the respiratory cycle has ended and is ready to begin again with the next breath.

The pressure differences shown in Figure 17-10 ■ apply to quiet breathing. During exercise or forced heavy breathing, these values will become proportionately larger. **Active expiration** occurs during voluntary exhalations and when ventilation exceeds 30–40 breaths per minute. (Nor-

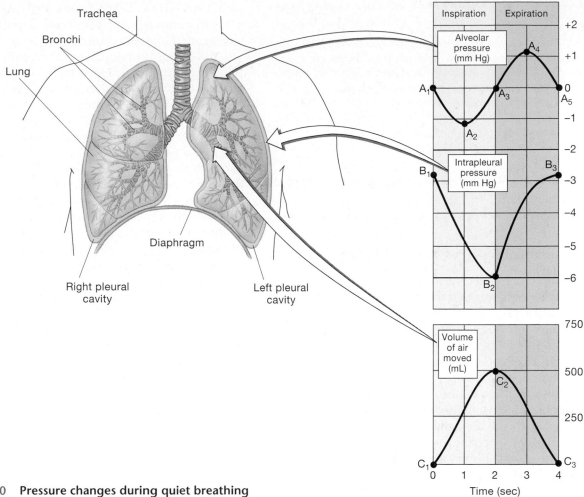

■ **Figure 17-10** **Pressure changes during quiet breathing**

mal resting ventilation rate is 12–20 breaths per minute for an adult.) Active expiration uses a different set of muscles from those used during inspiration, namely the internal intercostal muscles and the abdominal muscles. These muscles are collectively called the *expiratory muscles.*

The **internal intercostal muscles** line the inside of the rib cage. When they contract, they pull the ribs inward, reducing the volume of the thoracic cavity. To feel this action, place your hands on your rib cage. Forcefully blow as much air out of your lungs as you can, noting the movement of your hands as you do so.

The internal and external intercostals act as antagonistic muscle groups [∞ p. 347] to alter the position and volume of the rib cage during ventilation, but the diaphragm has no antagonistic muscles. Therefore, the *abdominal muscles* become active during active expiration to supplement the activity of the intercostals. The diaphragm stays relaxed during active expiration. Contraction of the diaphragm would move it downward, working against the abdominal muscles that are trying to push the diaphragm upward.

Contraction of abdominal muscles during active expiration pulls the lower rib cage inward and decreases abdominal volume. The displaced intestines and liver push the diaphragm up into the thoracic cavity, pas-

sively decreasing the chest volume even more. This is why when you are doing abdominal exercises in an aerobics class, the instructor tells you to blow air out as you lift your head and shoulders. The active process of blowing air out helps you contract your abdominals, the very muscles you are trying to strengthen.

Any neuromuscular disease that weakens skeletal muscles or damages their motor neurons can adversely affect ventilation. With decreased ventilation, less fresh air enters the lungs. In addition, loss of the ability to cough increases the risk of pneumonia and other infections. Examples of diseases that affect the motor control of ventilation include **myasthenia gravis** [∞ p. 239], an illness in which acetylcholine receptors of the motor end plate of skeletal muscles are destroyed, and **polio** (poliomyelitis), a viral illness that paralyzes skeletal muscles.

✔ Scarlett O'Hara was trying to squeeze herself into a corset with an 18-inch waist. Was she more successful when she took a deep breath and held it or when she blew all the air out of her lungs? Why?

✔ Why would loss of the ability to cough increase the risk of respiratory infections? (Hint: What does coughing do to mucus in the airways?)

...continued from page 507

Emphysema is characterized by a loss of elastin, the elastic fibers that help the alveoli recoil during expiration. Researchers believe that elastin is destroyed by proteases released by immune system cells, which work overtime in smokers to rid the lungs of irritants. People with emphysema have no difficulty inhaling air. However, because their alveoli have lost elastic recoil, expiration—normally a passive process—requires conscious effort. They literally have to work to push air out of their lungs.

Question 3: *Name the muscles that patients with emphysema use to exhale forcefully.*

Intrapleural Pressure Changes During Ventilation

Ventilation requires that the lungs, which are unable to expand and contract on their own, move in association with the contraction and relaxation of the thorax. As we noted earlier in this chapter, the lungs are "stuck" to the thoracic cage by the pleural fluid, the fluid between the two pleural membranes.

The *intrapleural pressure,* pressure within the fluid between the pleural membranes, is normally subatmospheric. This subatmospheric pressure arises during development, when the thoracic cage with its associated pleural membrane grows more rapidly than the lung with its associated pleural membrane. The two pleural membranes are held together by the pleural fluid bond, so the elastic lungs are forced to stretch to conform to the larger volume of the thoracic cavity. At the same time, however, elastic recoil of the lungs creates an inwardly directed force that attempts to pull the lungs away from the chest wall (Fig. 17-11a ■). The combination of the outward pull of the thoracic cage and inward recoil of the elastic lungs creates an intrapleural pressure of about −3 mm Hg.

You can create a similar situation with a syringe half-filled with fluid and capped with a plugged-up needle. Begin with the fluid inside the syringe at atmospheric pressure. Now pick up the syringe and hold the barrel (the chest wall) in one hand while you try to withdraw the plunger (the elastic lung pulling away from the chest wall). As you pull on the plunger, the volume inside the barrel increases slightly, but the cohesive forces between the water molecules cause the fluid inside the syringe to resist expansion. The pressure within the barrel, which was initially equal to atmospheric pressure, decreases slightly as you pull on the plunger. If you release the plunger, it snaps back to its resting position, restoring atmospheric pressure inside the syringe.

But what happens to subatmospheric intrapleural pressure if an opening is made between the sealed pleural cavity and the atmosphere? A knife thrust between the ribs, a broken rib that punctures the pleural membrane, or any event that opens the pleural cavity to the atmosphere will allow air to flow in down its pressure gradient, just as air enters when you break the seal on a vacuum-packed can.

Air in the pleural cavity breaks the fluid bond holding the lung to the chest wall. The chest wall expands outward while the elastic lung collapses to an unstretched state, like a deflated balloon (Fig. 17-11b ■). This condition, called **pneumothorax** [*pneuma*, air + *thorax*, chest], results in a collapsed lung that is unable to function normally. Pneumothorax can also occur spontaneously if a congenital *bleb*, or weakened section of lung tissue, ruptures, allowing air from inside the lung to enter the pleural cavity.

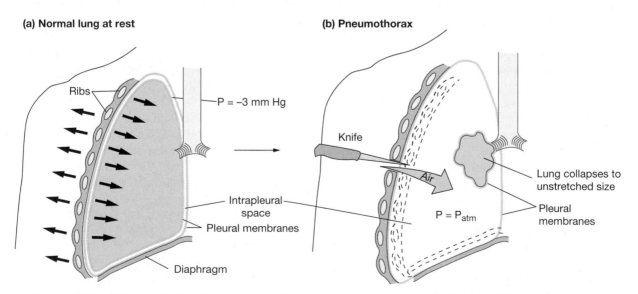

(a) Normal lung at rest

Ribs

P = −3 mm Hg

Intrapleural space

Pleural membranes

Diaphragm

(b) Pneumothorax

Knife

Air

P = P$_{atm}$

Lung collapses to unstretched size

Pleural membranes

■ **Figure 17-11** **Pressure in the pleural cavity** (a) Elastic recoil of the normal lung at rest creates an inward pull while the elastic recoil of the chest wall tries to pull the chest wall outward. Pressure in the fluid between the pleural membranes is subatmospheric. (b) If the sealed pleural cavity is opened to the atmosphere, air flows into the cavity. The lung collapses to its unstretched size, and the rib cage expands slightly. This condition is called pneumothorax.

Correction of a pneumothorax has two components: removing as much air from the pleural cavity as possible with a suction pump, and sealing the hole to prevent more air from entering. Any air remaining in the cavity will gradually be absorbed into the blood, restoring the pleural fluid bond and reinflating the lung.

Pressures in the pleural fluid vary during a respiratory cycle. At the beginning of inspiration, intrapleural pressure is about -3 mm Hg (Fig. 17-10 ■, point B_1). As inspiration proceeds, the pleural membranes and lungs follow the thoracic cage because of the pleural fluid bond. But the elastic lung tissue resists being stretched. The lungs attempt to pull farther away from the chest wall, causing the intrapleural pressure to become even more negative (Fig. 17-10 ■, point B_2).

Because this process is difficult to visualize, return to the analogy of the fluid-filled syringe with the blocked needle. You can pull the plunger out a small distance without much effort, but if you try to pull it out farther, it is harder to do because of the cohesiveness of the fluid. The increased amount of work you do trying to pull out the plunger is paralleled by the work your muscles must do when they contract for inspiration. The bigger the breath, the more work is required to "stretch" the pleural fluid and the elastic lung.

By the end of quiet inspiration, when lungs are fully expanded, intrapleural pressure drops to around -6 mm Hg (point B_2). During exercise or other powerful inspirations, intrapleural pressure may fall to -8 mm Hg.

Upon expiration, the thoracic cage returns to its resting position. The lungs are released from their extra-stretched position, and the intrapleural pressure returns to its normal value of -3 mm Hg (point B_3). Notice that intrapleural pressure never equilibrates with the atmosphere because the pleural cavity is a closed compartment.

The pressure gradients required for air flow are created by the work of skeletal muscle contraction. Normally, about 3%–5% of the body's energy expenditure is used for quiet breathing. During exercise, the energy required for breathing increases substantially. The two factors that have the greatest influence on the amount of work needed for breathing are (1) the stretchability, or compliance, of the lungs and (2) the resistance of the airways to air flow.

✔ A person has periodic spastic contractions of the diaphragm, otherwise known as hiccups. What happens to intrapleural and alveolar pressures when a person hiccups?

✔ A stabbing victim is brought into the emergency room with a knife wound between the ribs on the left side of his chest. What has probably happened to his left lung? To his right lung? Why does the left side of his rib cage seem larger than the right side?

Lung Compliance and Elastance May Change in Disease States

Adequate ventilation depends on the ability of the lungs to expand normally. Most of the work of breathing goes into overcoming the resistance of the elastic lungs and the thoracic cage to stretch. Clinically, the ability of the lung to stretch is called **compliance.** A high-compliance lung stretches easily, just as a compliant person is easy to persuade. A low-compliance lung requires more force from the inspiratory muscles to stretch it.

Compliance is different from **elastance** (elasticity). The fact that a lung stretches easily (high compliance) does not necessarily mean that it will return to its resting volume when the stretching force is released (elastance). You may have experienced this with old gym shorts. After many washings the elastic waistband is easy to stretch (high compliance) but lacking in elasticity, making the shorts impossible to keep around your waist.

Analogous problems occur in the respiratory system. For example, *emphysema* is a disease in which the elastin fibers normally found in lung tissue are destroyed. Destruction of the elastin fibers results in lungs that exhibit high compliance and stretch easily during inspiration. However, these lungs also have decreased elastance, and therefore they do not recoil to their resting position during expiration.

To understand the importance of elastic recoil to expiration, think of an inflated balloon and an inflated plastic bag. The balloon is similar to the normal lung. Its elastic walls squeeze on the air inside the balloon, thereby increasing the internal air pressure. When the neck of the balloon is opened to the atmosphere, elastic recoil causes air to flow out of the balloon. The inflated plastic bag, on the other hand, is like the lung of an individual with emphysema. It has high compliance and is easily inflated, but it has little elastic recoil. If the inflated plastic bag is opened to the atmosphere, most of the air remains inside the bag.

A decrease in lung compliance affects ventilation because more work must be expended to stretch a stiff lung. Pathological conditions in which compliance is reduced are called **restrictive lung diseases.** In these conditions, the energy expenditure required to stretch less-compliant lungs can far exceed the normal range. Two common causes of decreased compliance are inelastic scar tissue formed in *fibrotic lung diseases* and inadequate production of surfactant in the alveoli.

Surfactant Decreases the Work of Breathing

For years, physiologists assumed that elastin and other elastic fibers were the primary source of resistance to stretch in the lung. But studies comparing the work required to expand air-filled and saline-filled lungs showed that air-filled lungs are much harder to inflate.

From this, researchers concluded that lung tissue itself contributes less to resistance than they had thought. Some other property of the normal air-filled lung, not present in a saline-filled lung, must create most of the resistance to stretch.

This property is **surface tension** [∞ p. 25] created by the thin fluid layer between the alveolar cells and the air. At any air-fluid interface, the surface of a liquid behaves as if it is under tension, like a thin membrane being stretched. This surface tension arises because of the hydrogen bonds between water molecules. The surface molecules are attracted to other water molecules beside and beneath them, but not to the air.

If water is isolated in a drop, it will take on a rounded shape (∞ Fig. 2-10, p. 25). If water is in the shape of a bubble, as it is when it lines the alveoli, surface tension exerts a force directed toward the center of the bubble. The **law of LaPlace** is an expression of the force, or pressure, created by a fluid sphere or bubble.

In physiology, the sphere is analogous to a fluid-lined alveolus. The law of LaPlace states that the pressure within a fluid-lined alveolus depends on two factors: the surface tension of the fluid and the radius of the alveolus. This relationship is expressed by the equation

$P = 2 \times T/r$ where P is the pressure inside the alveolus

T is the surface tension of the fluid lining the alveolus

r is the radius of the alveolus

If two alveoli have different diameters but are lined by fluids with the same surface tension, the pressure inside the smaller alveolus will be greater (Fig. 17-12a ■). If the two alveoli are connected to each other, air will flow from the higher-pressure small alveolus to the lower-pressure large alveolus. This air movement causes the smaller alveolus to collapse, while the volume of the larger alveolus increases.

Thus, the presence of fluid lining the alveoli creates surface tension that increases the resistance of the lung to stretch. This surface tension also makes the alveoli behave like elastic bubbles or inflated balloons that, once expanded, tend to collapse. Consequently, surface tension has the potential to increase the work needed to expand the alveoli with each breath.

Normally, however, our lungs secrete a chemical surfactant that reduces surface tension. **Surfactants** are molecules that disrupt cohesive forces between water molecules. For example, in dishwasher rinses, surfactants

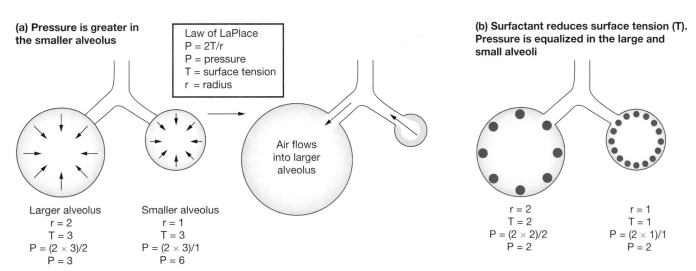

(a) Pressure is greater in the smaller alveolus

Law of LaPlace
$P = 2T/r$
P = pressure
T = surface tension
r = radius

Air flows into larger alveolus

Larger alveolus	Smaller alveolus
r = 2	r = 1
T = 3	T = 3
P = (2 × 3)/2	P = (2 × 3)/1
P = 3	P = 6

(b) Surfactant reduces surface tension (T). Pressure is equalized in the large and small alveoli

r = 2	r = 1
T = 2	T = 1
P = (2 × 2)/2	P = (2 × 1)/1
P = 2	P = 2

■ **Figure 17-12   Surfactant prevents the collapse of alveoli**   (a) According to the law of LaPlace, if two alveoli have the same surface tension, the small alveolus will have higher pressure and its air will flow into the larger alveolus. (b) Surfactant decreases surface tension in fluid lining the alveoli. Smaller alveoli have a higher concentration of surfactant so their surface tension is lower. This equalizes the pressure between smaller and larger alveoli.

keep rinse water from beading up on the dishes. In the lungs, surfactant decreases the surface tension of the fluid lining the alveoli and prevents small alveoli from collapsing. Human surfactant is a mixture containing lipoproteins such as *dipalmitoylphosphatidylcholine.*

Surfactant is more concentrated in smaller alveoli, making the surface tension in the smaller alveoli less than that in larger alveoli (Fig. 17-12b ■). Lower surface tension equalizes the pressure among different sizes of alveoli and keeps the smaller alveoli from collapsing when their air flows into larger, lower-pressure alveoli. With lower surface tension, the work needed to expand the alveoli with each breath is also greatly reduced.

Surfactant is manufactured and secreted into the alveolar air space by the type II alveolar cells (see Fig. 17-2g ■). Normally, surfactant synthesis begins about the twenty-fifth week of fetal development under the influence of various hormones. Production usually reaches adequate levels by the thirty-second week (about eight weeks before normal delivery). Babies who are born prematurely without adequate concentrations of surfactant in their alveoli develop *newborn respiratory distress syndrome (NRDS)*. In addition to "stiff" (low-compliance) lungs that require tremendous effort to expand with each breath, babies with NRDS also have alveoli that collapse.

✔ Coal miners who spend years inhaling fine coal dust have much of their alveolar surface area covered with scarlike tissue. What happens to their lung compliance as a result?

✔ When a baby takes its first breath, what happens to the alveoli if surfactant is not present?

---

§ Newborn Respiratory Distress Syndrome and Surfactant The clue to what creates additional resistance in an air-filled alveolus came from studies of premature babies with newborn respiratory distress syndrome (NRDS). From their first breath, NRDS babies have difficulty keeping their lungs inflated. With each breath, their lungs collapse, and they must use a tremendous amount of energy to expand them again for the next breath. Unless treatment is initiated rapidly, about 50% of these infants die. When President and Mrs. John Fitzgerald Kennedy's son was born with NRDS in the 1960s, the physicians could do little but administer oxygen and hope that the baby's lungs would mature before hypoxia and exertion overwhelmed him. Now the prognosis for NRDS babies is much better. Amniotic fluid can be sampled to assess if the baby's lungs are producing adequate amounts of surfactant. If they are not, and if delivery cannot be delayed, the NRDS baby can be treated by artificial ventilation that forces air into the lungs and keeps the alveoli open. The current treatment includes the aerosol administration of artificial surfactant until the baby's lungs mature enough to produce their own.

Airway Diameter Is the Primary Determinant of Airway Resistance

The other factor besides compliance that influences the work of breathing is the resistance of the respiratory system to air flow. Resistance in the respiratory system is similar in many ways to resistance in the cardiovascular system [∞ p. 409]. Parameters that contribute to resistance are:

1. The length of the system (L)
2. The viscosity of the substance flowing through the system (η)
3. The radius of the tubes in the system (r)

The equation that expresses the relation of those parameters is:

$$R \propto L\eta/r^4$$

Because the length of the respiratory system is constant, we can ignore that variable.

The viscosity of air is almost constant, although you may have noticed that it feels harder to breathe in a sauna filled with steam than in a room with normal humidity. Water droplets in the steam increase the viscosity of the steamy air, thereby increasing its resistance to flow.

Viscosity also changes slightly with atmospheric pressure, increasing as pressure increases. Divers breathing high-pressure compressed air may feel a little more resistance to air flow, while someone at high altitude may feel less resistance. Despite these exceptions, viscosity plays very little role in resistance to air flow.

Because length and viscosity are essentially constant for the respiratory system, the radius (or diameter) of the airways becomes the primary determinant of airway resistance. Normally, however, the work needed to overcome the resistance of the airways to air flow is small compared with the work needed to overcome the resistance of the lungs and thoracic cage to stretch.

Nearly 90% of airway resistance normally can be attributed to the trachea and bronchi, rigid structures with the smallest total cross-sectional area of the airways. These structures are supported by cartilage and bone, so their diameters normally do not change, and their resistance to air flow is constant. However, mucus accumulation from allergies or infections can dramatically increase resistance. If you have ever tried breathing through your nose when you have a cold, you can appreciate how the narrowing of an upper airway limits air flow!

The bronchioles normally do not contribute significantly to airway resistance because their total cross-sectional area is about 2000 times that of the trachea. But, because the bronchioles are collapsible tubes, a decrease in their diameter can suddenly turn them into a significant source of airway resistance. **Bronchoconstriction** increases resistance to air flow and decreases the amount of fresh air that reaches the alveoli.

Bronchioles, like arterioles, are subject to reflex control by the nervous system and by hormones. However, most minute-to-minute changes in bronchiolar diameter occur in response to paracrines. Carbon dioxide in the airways is the primary paracrine that affects bronchiolar diameter. Increased carbon dioxide in expired air relaxes bronchiolar smooth muscle and cause **bronchodilation.**

Histamine is a paracrine that acts as a powerful bronchoconstrictor. This chemical is released by mast cells [∞ p. 64] in response to tissue damage or allergic reactions. In severe allergic reactions, large amounts of histamine may lead to widespread bronchoconstriction and difficult breathing. Immediate medical treatment in these patients is imperative.

The primary neural control of bronchioles comes from parasympathetic neurons that cause bronchoconstriction, a reflex designed to protect the lower respiratory tract from inhaled irritants. There is no significant sympathetic innervation of the bronchioles in humans.

However, smooth muscle in the bronchioles is well supplied with β_2 receptors that respond to epinephrine. Stimulation of β_2 receptors relaxes airway smooth muscle and results in bronchodilation. This reflex is used therapeutically in the treatment of asthma and various allergic reactions characterized by histamine release and bronchoconstriction. Table 17-3 summarizes the factors that alter airway resistance.

✔ A cancerous lung tumor has grown into the walls of a group of bronchioles, narrowing their lumens. What has happened to the resistance to air flow in these bronchioles?

✔ Why does it feel more difficult to breathe in a steam room than outside in drier air?

✔ Name the neurotransmitter and receptor for parasympathetic bronchoconstriction.

Pulmonary Function Tests Measure Lung Volumes During Ventilation

Physiologists and clinicians assess a person's pulmonary function by measuring how much air the person moves during quiet breathing and with maximum effort. These

§ **Asthma** Asthma is an obstructive lung disease characterized by bronchoconstriction and airway edema, which increase airway resistance and decrease air flow. Patients complain of shortness of breath (*dyspnea*), and when they are asked to exhale forcefully, a wheezing sound is heard as air whistles through the narrowed lower airways. The severity of asthma attacks ranges from mild to life threatening. Asthma is an inflammatory condition that is often associated with allergies. It can also be triggered by exercise (exercise-induced asthma) and by rapid changes in the temperature or humidity of inspired air. At the cellular level, a variety of chemical signals may be responsible for inducing bronchoconstriction, including acetylcholine, histamine, substance P (a neuropeptide), and leukotrienes secreted by mast cells, macrophages, and eosinophils. *Leukotrienes* are lipid-like bronchoconstrictors that are released during the inflammatory response. Asthma is treated with inhaled and oral medications that include β_2-adrenergic agonists, anti-inflammatory drugs, and new leukotriene antagonists.

pulmonary function tests use a **spirometer,** an instrument that measures the volume of air moving with each breath (Fig. 17-13 ■). Most spirometers in clinical use today are small computerized machines rather than the traditional spirometer illustrated.

Although pulmonary function tests are relatively simple to perform, they have considerable diagnostic value. For example, in asthma, the bronchioles are constricted. They tend to collapse and close off before a forced expiration is completed, reducing both the amount and rate of air flow. Diseases in which air flow during expiration is diminished due to narrowing of the bronchioles are known as **obstructive lung diseases.** Emphysema and chronic bronchitis are sometimes called *chronic obstructive pulmonary disease,* or *COPD,* because of their ongoing, or *chronic,* nature.

Lung Volumes The air moved during breathing can be divided into four lung volumes: (1) tidal volume, (2) inspiratory reserve volume, (3) expiratory reserve volume, and (4) residual volume. These volumes are diagrammed in Figure 17-14 ■ and described below.

TABLE 17-3 Factors That Affect Airway Resistance

Factor	*Affected by*	*Mediated by*
Length of the system	Constant; not a factor	
Viscosity of air	Usually constant. Humidity and altitude may alter slightly.	
Diameter of airways		
Upper airways	Physical obstruction	Mucus and other factors
Bronchioles	Bronchoconstriction	Parasympathetic neurons (muscarinic receptors), histamine, leukotrienes
	Bronchodilation	Carbon dioxide, epinephrine (β_2 receptors)

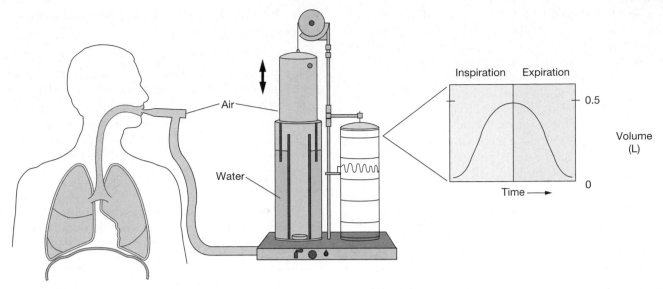

■ **Figure 17-13   The recording spirometer**   Lung volumes and air flow are recorded using a spirometer. (a) The subject inserts a mouthpiece that is attached to an inverted bell filled with air or oxygen. The volume of the bell and the volume of the subject's respiratory tract create a closed system. When the subject inhales, air moves into the lungs. The volume of the bell decreases, and the pen rises on the tracing. When the subject exhales, air moves into the bell, its volume increases, and the pen drops. Most clinical spirometers have been computerized, but the spirometer illustrated is still widely used in student laboratories.

The numerical values given represent average volumes for the 70-kg man. The volumes for women are typically about 20%–25% less. Lung volumes vary considerably with the age, sex, and height of the individual. Each paragraph below begins with the instructions that you would be given if you were being tested for these volumes.

"Breathe quietly." The volume of air that moves in a single normal inspiration or expiration is known as the tidal volume (V_T). Average tidal volume during quiet breathing is about 500 mL.

"Now, at the end of a quiet inspiration, take in as much additional air as you possibly can." The additional volume you inspire above the tidal volume represents your **inspiratory reserve volume** (IRV). In a 70-kg man, this volume is about 3000 mL, a sixfold increase over the normal tidal volume.

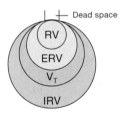

RV   = Residual volume
ERV = Expiratory reserve volume
V_T = Tidal volume
IRV = Inspiratory reserve volume

Pulmonary volumes		
	Males	Females
IRV	3000	1900 } Inspiratory capacity
Vital capacity V_T	500	500
ERV	1100	700 } Functional residual capacity
Residual volume	1200	1100
	5800 mL	4200 mL

■ **Figure 17-14   Lung volumes and capacities**   (a) The four lung volumes. (b) A spirometer tracing showing the lung volumes and capacities.

"Now stop at the end of a normal exhalation, then exhale as much air as you possibly can." The amount of air exhaled after the end of a normal expiration is the **expiratory reserve volume** (ERV), which averages about 1100 mL.

The fourth volume is one that cannot be measured directly. Even if you blow out as much air as you can, air still remains in the lungs and the airways. The volume of air in the respiratory system after maximal exhalation, about 1200 mL, is called the **residual volume** (RV). Most of this residual volume exists because the lungs are held stretched against the thoracic wall by the pleural fluid.

Lung Capacities   The sum of two or more lung volumes is called a **capacity**. The **vital capacity** (VC) is the sum of the inspiratory reserve volume, expiratory reserve volume, and tidal volume. Vital capacity represents the maximum amount of air that can be voluntarily moved into or out of the respiratory system with one breath. To measure vital capacity, you would instruct the person to take in as much air as possible, then blow it all out. Vital capacity decreases with age.

The vital capacity plus the residual volume yields the **total lung capacity** (TLC). Other capacities of importance in pulmonary medicine include the **inspiratory capacity** (tidal volume + inspiratory reserve volume) and the **functional residual capacity** (expiratory reserve volume + residual volume).

Auscultation of Breath Sounds   Auscultation of breath sounds is an important diagnostic technique in pulmonary medicine, just as auscultation of heart sounds is used in cardiovascular diagnosis [∞ p. 432]. But breath sounds are more complicated to interpret due to a wide range of normal variation.

In general, breath sounds are distributed evenly over the lungs and resemble a quiet "whoosh" made by flowing air. In conditions in which air flow is reduced, such as pneumothorax, breath sounds may be diminished or absent. Abnormal sounds include a variety of squeaks, pops, wheezes, or bubbling sounds caused by fluid and secretions in the airways or alveoli. Inflammation of the pleural membrane results in a crackling or grating sound known as a *friction rub*. It is caused by swollen, inflamed pleural membranes rubbing against each other, and it disappears when fluid again separates them.

✔ Restrictive lung disease decreases the compliance of the lung. How will the inspiratory reserve volume change in patients with a restrictive lung disease?

✔ Chronic obstructive lung disease causes patients to lose the ability to exhale fully. How does residual volume change in these patients?

✔ If vital capacity decreases with age but total lung capacity does not change, what volume must be changing? In which direction?

Rate and Depth of Breathing Determine the Efficiency of Breathing

You may recall that the efficiency of the heart is measured by the cardiac output, which is calculated by multiplying heart rate by stroke volume. Likewise, we can estimate the effectiveness of ventilation by calculating the **total pulmonary ventilation**, the volume of air moved into and out of the lungs each minute. Total pulmonary ventilation, also known as the *minute volume*, is calculated as follows:

Ventilation rate × tidal volume (V_T) = total pulmonary ventilation

The normal ventilation rate for an adult is 12–20 breaths per minute. Using the average tidal volume of 500 mL and the slowest ventilation rate, we get:

12 breaths/min × 500 mL/breath = 6000 mL/min
= 6 L/min

Total pulmonary ventilation represents the physical movement of air into and out of the respiratory tract. But is it a good indicator of how much fresh air reaches the alveolar exchange surface? Not necessarily. Some air that enters the respiratory system does not reach the alveoli because part of every breath remains in the conducting airways, such as the trachea and bronchi. Because the conducting airways do not exchange gases with the blood, they are known as the **anatomic dead space**. Anatomic dead space averages about 150 mL.

To illustrate the difference between air that enters the airways and fresh air that reaches the alveoli, let's consider a typical breath bringing 500 mL of fresh air into the respiratory system (Fig. 17-15 ■). The entering air displaces the 150 mL of stale air in the anatomic dead space, sending it into the alveoli. The first 350 mL of the fresh air also enters the alveoli. The last 150 mL of inspired fresh air remains in the dead space. This fresh air will never reach the alveoli because it will be replaced during expiration with 150 mL of stale air. That stale air then becomes the first air into the alveoli on the next inspiration.

Thus, although 500 mL of air entered the alveoli, only 350 mL of that volume was fresh air. The fresh air entering the alveoli equals the tidal volume minus the dead space volume.

Because a significant portion of inspired air never reaches an exchange surface, a more accurate indicator of the efficiency of ventilation is **alveolar ventilation**, the amount of fresh air that reaches the alveoli each minute. Alveolar ventilation is calculated by multiplying ventilation rate by the volume of fresh air that reaches the alveoli:

Ventilation rate × (tidal volume − dead space volume)
= alveolar ventilation

Using the same ventilation rate and tidal volume as above and a dead space volume of 150 mL yields an alveolar ventilation of:

12 breaths/min × (500 mL/breath − 150 mL/breath)
= 4200 mL/min

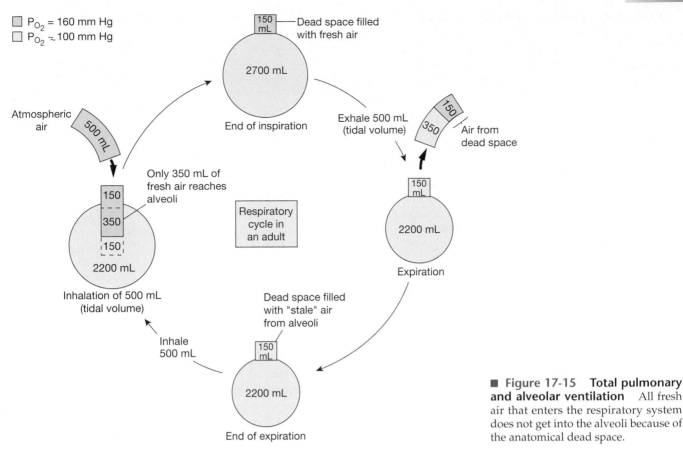

☐ $P_{O_2} = 160$ mm Hg
☐ $P_{O_2} \approx 100$ mm Hg

■ **Figure 17-15 Total pulmonary and alveolar ventilation** All fresh air that enters the respiratory system does not get into the alveoli because of the anatomical dead space.

Thus, at 12 breaths per minute, the alveolar ventilation is 4.2 L/min. Although 6 L/min of fresh air entered the respiratory system, only 4.2 L reached the alveoli.

Alveolar ventilation can be drastically affected by changes in the rate or depth of breathing. Table 17-4 shows that three people can have the same total pulmonary ventilation (minute volume) but dramatically different alveolar ventilation. *Maximum voluntary ventilation,* which involves breathing as deeply and quickly as possible, may increase total pulmonary ventilation to as much as 170 L/min. Table 17-5 describes various patterns of ventilation.

Gas Composition in the Alveoli Varies Little During Normal Breathing

How much can a change in alveolar ventilation affect the amount of fresh air and oxygen that reaches the alveoli? Figure 17-16 ■ shows how P_{O_2} and P_{CO_2} vary with hyper- and hypoventilation.

As alveolar ventilation increases above normal levels (**hyperventilation**), alveolar P_{O_2} rises to about 120 mm Hg and alveolar P_{CO_2} falls to around 20 mm Hg. During **hypoventilation,** when less fresh air enters the alveoli, alveolar P_{O_2} decreases and alveolar P_{CO_2} increases.

Although a dramatic change in alveolar ventilation pattern will affect gas partial pressures in the alveoli, the P_{O_2} and P_{CO_2} in the alveoli change surprisingly little during normal quiet breathing. Alveolar P_{O_2} is fairly constant at 100 mm Hg, and alveolar P_{CO_2} stays close to 40 mm Hg.

Intuitively, one might think that P_{O_2} would increase when fresh air first enters the alveoli, then decrease steadily as oxygen leaves to enter the blood. Instead, we find only very small swings in P_{O_2}. Why? The reasons are that (1) the amount of oxygen that enters the alveoli with each breath is roughly equal to the amount of oxygen that enters the blood, and (2) the amount of fresh air that enters

TABLE 17-4 Effects of Breathing Pattern on Alveolar Ventilation

Tidal Volume (mL)	Respiratory Rate (breaths/min)	Total Pulmonary Ventilation (mL/min)	Fresh Air to Alveoli (mL) (tidal volume − dead space volume*)	Alveolar Ventilation (mL/min)
500 (normal)	12 (normal)	6000	350	4200
300 (shallow)	20 (rapid)	6000	150	3000
750 (deep)	8 (slow)	6000	600	4800

*Dead space volume is assumed to be 150 mL.

TABLE 17-5 Types and Patterns of Ventilation

Name	Description	Examples
Eupnea	Normal quiet breathing	
Hyperpnea	Increased respiratory rate and/or volume in response to increased metabolism	Exercise
Hyperventilation	Increased respiratory rate and/or volume without increased metabolism	Emotional hyperventilation; blowing up a balloon
Hypoventilation	Decreased alveolar ventilation	Shallow breathing; asthma; restrictive lung disease
Tachypnea	Rapid breathing; usually increased respiratory rate with decreased depth	Panting
Dyspnea	Difficulty breathing (a subjective feeling sometimes described as "air hunger")	Various pathologies or hard exercise
Apnea	Cessation of breathing	Voluntary breath-holding; depression of CNS control centers

the lungs with each breath is only a little more than 10% of the total lung volume at the end of inspiration.

✔ If a person increased the depth of breathing (i.e., increased his tidal volume), what do you predict would happen to his alveolar P_{O_2}?

Ventilation and Alveolar Blood Flow Are Matched

Moving oxygen from the atmosphere to the alveolar exchange surface is only the first step in external respiration. Next there must be normal gas exchange across the

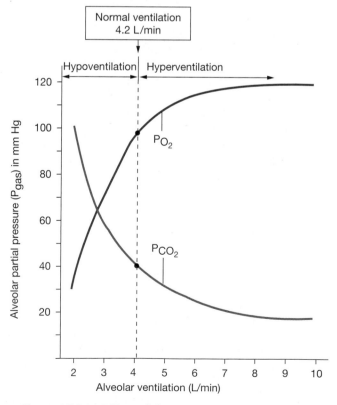

■ **Figure 17-16** **Effect of changing alveolar ventilation on P_{O_2} and P_{CO_2} in the alveoli**

alveolar-capillary interface. Finally, blood flow (*perfusion*) past the alveoli must be adequate to pick up the available oxygen. Matching ventilation into groups of alveoli with blood flow past those alveoli is a two-part process involving local regulation of both air flow and blood flow.

Alterations in pulmonary blood flow depend almost exclusively on properties of the capillaries and on local factors such as the concentrations of oxygen and carbon dioxide in the lung tissue. Capillaries in the lungs are unusual because they are collapsible. If the pressure of blood flowing through the capillaries falls below a certain point, the capillaries close off, diverting blood to other pulmonary capillary beds with higher pressure.

In a person at rest, some capillary beds found in the apex (top) of the lung are closed off because of low hydrostatic pressure. Capillary beds at the base of the lung have higher hydrostatic pressure because of gravity and remain open. Consequently, blood flow is diverted toward the base of the lung.

During exercise, when blood pressure rises, the closed apical capillary beds open, ensuring that the increased cardiac output will be fully oxygenated as it passes through the lungs. The ability of the lungs to recruit additional capillary beds during exercise is an example of the reserve capacity of the body.

At the local level, the body attempts to match air and blood flow in each section of the lung by regulating the diameters of the arterioles and the bronchioles. Bronchiolar diameter is mediated primarily by CO_2 levels in exhaled air passing through them (Table 17-6). An increase in the P_{CO_2} of expired air causes the bronchioles to dilate. A decrease in the P_{CO_2} of expired air causes bronchioles to constrict.

Although there is some autonomic innervation of the pulmonary arterioles, there is no evidence for neural control of pulmonary blood flow. The resistance of pulmonary arterioles to blood flow is regulated primarily by the oxygen content of the interstitial fluid around the arteriole (Fig. 17-17 ■).

TABLE 17-6 Local Control of Arterioles and Bronchioles by Oxygen and Carbon Dioxide

Gas Composition	Bronchioles	Pulmonary Arterioles	Systemic Arterioles
P_{CO_2} increases	Dilate	(Constrict)	Dilate
P_{CO_2} decreases	Constrict	(Dilate)	Constrict
P_{O_2} increases	(Constrict)	Dilate	Constrict
P_{O_2} decreases	(Dilate)	Constrict	Dilate

Note: Responses in parentheses indicate weak responses.

If the ventilation of alveoli in an area of the lung is diminished, the tissue P_{O_2} in that area decreases. The arterioles respond to lower P_{O_2} by constricting. Vasoconstriction diverts blood away from the underventilated region to better-ventilated parts of the lung.

On the other hand, if P_{O_2} in the lung becomes higher than normal, the alveoli in that region are being overventilated relative to the blood flow past them. The arterioles supplying those alveoli dilate to bring in additional blood that can pick up the extra oxygen.

Note that constriction of pulmonary arterioles in response to low P_{O_2} is exactly the opposite of what occurs in the systemic circulation [∞ p. 454]. In the systemic circulation, a decrease in the concentration of oxygen in the tissues causes the arterioles to dilate, bringing more blood to metabolically active tissues that are consuming oxygen.

One important point must be noted here. Local control mechanisms are not effective regulators of air and blood flow under all circumstances. If blood flow is blocked in one pulmonary artery, or if air flow is blocked at the level of the larger airways, local responses that shunt air or blood to other parts of the lung are ineffective because no place in the lung has normal ventilation or perfusion.

✔ If a tumor in the lung tissue decreases blood flow in one small section of the lung to a minimum, what happens to P_{O_2} in those alveoli and in the surrounding tissue? What happens to P_{CO_2} in the same region? What is the compensatory response of the bronchioles in that region? Will the compensation bring ventilation in that section of the lung back to normal? Explain.

▶ GAS EXCHANGE IN THE LUNGS

The diffusion of gases between the alveoli and the blood obeys the rules for simple diffusion introduced in Chapter 5 [∞ p. 121]:

1. The rate of diffusion across membranes is directly proportional to the partial pressure (concentration) gradient.
2. The rate of diffusion across membranes is directly proportional to the available surface area.

(a) Ventilation in alveoli is matched to perfusion through pulmonary capillaries

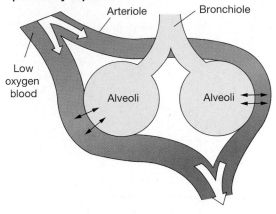

(b) If ventilation decreases in a group of alveoli (blue), P_{CO_2} increases and P_{O_2} decreases. Blood flowing past those alveoli does not get oxygenated.

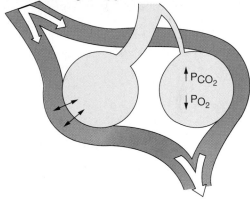

(c) Decreased tissue P_{O_2} around underventilated alveoli constricts their arterioles, diverting blood to better ventilated alveoli.

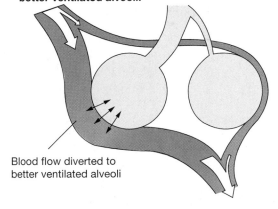

Blood flow diverted to better ventilated alveoli

■ **Figure 17-17 Local control matches ventilation and perfusion**

3. The rate of diffusion across membranes is inversely proportional to the thickness of the membrane.
4. Diffusion is most rapid over short distances.

Normally, gas exchange in the lungs is rapid and goes to equilibrium, but a change in any of the parameters listed above has the potential to limit diffusion significantly.

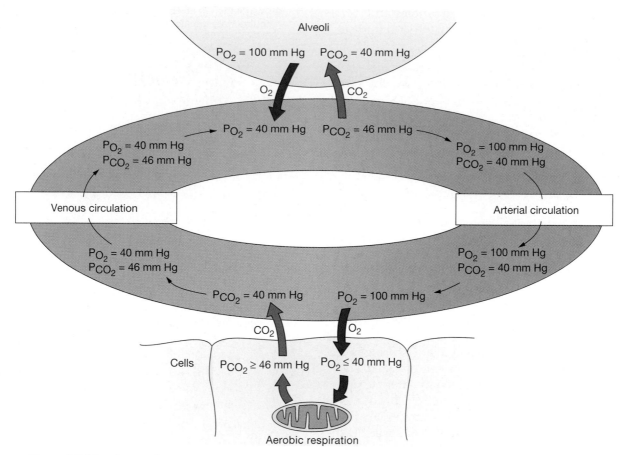

■ **Figure 17-18**   **Gas exchange at the alveoli and cells**

Pathological changes in these factors that affect gas exchange are discussed briefly below.

Gas Exchange Requires Pressure Gradients

As you learned earlier in the chapter, the amount of oxygen or carbon dioxide that dissolves in the plasma depends on both the pressure gradient and the solubility of the gas. Because solubility is normally constant, the primary determinant of gas exchange is the partial pressure gradient of the gas across the alveolar-capillary membrane.

The gas laws state that individual gases flow from regions of higher partial pressure to regions of lower partial pressure. This rule governs the exchange of oxygen and carbon dioxide at the lungs and tissues (Fig. 17-18 ■). Normal alveolar P_{O_2} is 100 mm Hg. The P_{O_2} of systemic venous blood arriving at the lungs is 40 mm Hg. Oxygen therefore moves from the alveoli into the capillaries.

The reverse process occurs with carbon dioxide. Alveolar P_{CO_2} is 40 mm Hg and venous blood P_{CO_2} is 46 mm Hg. Carbon dioxide thus moves from the blood into the alveoli. By the time blood leaves the alveoli, it has a P_{O_2} of 100 mm Hg and a P_{CO_2} of 40 mm Hg, identical to the partial pressures of those two gases in the alveoli (Table 17-7).

Although oxygen and carbon dioxide diffuse across two cell layers (capillary and alveolus) and through the

interstitial fluid, exchange is very rapid because (1) the distance is small and (2) the gases are both water- and lipid-soluble. Diffusion of carbon dioxide and oxygen reaches equilibrium in less than one second.

Any factor that decreases alveolar P_{O_2} decreases the pressure gradient and results in less oxygen entering the blood. If alveolar P_{O_2} is low, there are two questions to ask: (1) is the composition of the inspired air normal? and (2) is alveolar ventilation adequate?

The main factor that affects the oxygen content of inspired air is altitude. The partial pressure of oxygen in air decreases along with total atmospheric pressure as you move from sea level (760 mm Hg) to higher alti-

TABLE 17-7   Normal Values in Pulmonary Medicine

Total pulmonary ventilation	6 L/min
Total alveolar ventilation	4.2 L/min
Maximum voluntary ventilation	125–170 L/min
Respiration rate	12–20 breaths/min

Blood values		
	Arterial	*Venous*
P_{O_2}	95 mm Hg (85–100)	40 mm Hg
P_{CO_2}	40 mm Hg (35–45)	46 mm Hg
pH	7.4 (7.38–7.42)	7.37

tudes. At the summit of Mt. Everest, an altitude of 8848 m, atmospheric pressure is only 253 mm Hg, and atmospheric P_{O_2} is only 53 mm Hg. As a result, alveolar and arterial P_{O_2} fall from 100 mm Hg to 35 mm Hg, a value barely large enough to sustain life!

If the composition of the inspired air is normal but alveolar P_{O_2} is low, then the cause is a decrease in alveolar ventilation. Low alveolar ventilation is also known as *hypoventilation* and is characterized by less fresh air entering the alveoli. Pathological factors that cause alveolar hypoventilation include increased airway resistance (asthma; Fig. 17-19e ■), decreased lung compliance (fibrosis; Fig. 17-19c ■), and overdoses of drugs or alcohol that depress the central nervous system and slow the ventilation rate.

Changes in the Alveolar Membrane Alter Gas Exchange

In some situations, alveolar P_{O_2} is normal but the P_{O_2} of the arterial blood leaving the lungs is low. In these cases, there is a problem with the exchange of gases between the alve-

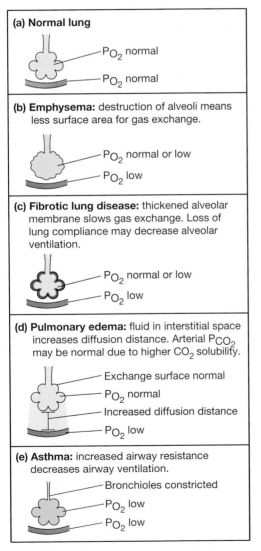

■ **Figure 17-19 Pulmonary pathologies that affect alveolar ventilation and gas exchange**

oli and the blood. Factors that adversely affect gas exchange include (1) a decrease in the amount of alveolar surface area that is available for gas exchange and (2) an increase in the diffusion distance between the alveoli and the blood.

Physical loss of alveolar surface area is demonstrated dramatically in emphysema, a degenerative lung disease most often caused by cigarette smoking. The irritating effect of smoke in the alveoli activates alveolar macrophages that release proteolytic enzymes. These enzymes destroy the elastic fibers of the lung, as discussed earlier, and kill cells, breaking down the walls of the alveoli. The result is a high-compliance/low-elastic recoil lung with fewer and larger alveoli and less surface area for gas exchange (Fig. 17-19b ■).

Any change in the alveolar membrane that alters its properties will slow gas exchange. For example, in fibrotic lung diseases, deposition of scar tissue thickens the alveolar membrane (Fig. 17-19c ■). Diffusion of gases through this scar tissue is much slower than normal. However, the lungs have a built-in reserve capacity. As a result, one-third of the exchange epithelium must be incapacitated before arterial P_{O_2} falls significantly.

A situation in which the diffusion distance between alveoli and blood increases is **pulmonary edema,** characterized by excessive interstitial fluid volume in the lungs. Normally, only small amounts of interstitial fluid are found in the lungs, due to low pulmonary blood pressure and effective lymph drainage. But if pulmonary blood pressure rises for some reason such as left ventricular failure or mitral valve dysfunction, capillary hydrostatic pressure increases.

Fluid filtration out of the capillary increases to the point that the lymphatics become unable to remove all the fluid (Fig. 17-19d ■). Excess fluid accumulates in the pulmonary interstitial space (edema) and may even leak across the alveolar membrane, collecting inside the alveoli.

The increased diffusion distance does not usually affect carbon dioxide exchange because carbon dioxide is relatively soluble in body fluids. Oxygen, however, is much less soluble and unable to cross the increased diffusion distance as easily. Subsequently, oxygen exchange diminishes. In these cases, it is not unusual to find normal arterial P_{CO_2} accompanying decreased arterial P_{O_2}.

If the diffusion of oxygen from alveolus to blood is significantly impaired, **hypoxia,** or too little oxygen in the cells, results. Hypoxia frequently goes hand in hand with **hypercapnia,** elevated concentrations of carbon dioxide. These two conditions are clinical signs, not diseases, and the clinician must gather additional information to pinpoint their cause (Table 17-8).

✔ Why does left ventricular failure or mitral valve dysfunction cause elevated pulmonary blood pressure? (Hint: ∞ p. 444.)

✔ If alveolar ventilation increases, what do you predict will happen to arterial P_{O_2}? To arterial P_{CO_2}? To venous P_{O_2} and P_{CO_2}? Explain.

TABLE 17-8   Causes of Hypoxia and Examples

- **Hypoxic hypoxia:** low arterial P_{O_2}
  - Physiological: high altitude
  - Alveolar hypoventilation
  - Decreased lung diffusion capacity
  - Abnormal ventilation-perfusion ratio
- **Anemic hypoxia:** decrease in total amount of O_2 bound to hemoglobin
  - Blood loss and anemia
  - Carbon monoxide prevents O_2 binding
- **Ischemic hypoxia:** due to reduced blood flow in tissue
  - General (heart failure), peripheral (shock), or to a single organ (coronary thrombosis)
- **Histotoxic hypoxia:** failure of cell to use O_2 due to poisoning
  - Cyanide and other metabolic poisons

GAS EXCHANGE IN THE TISSUES

Diffusion of gases between the blood and cells also depends on the pressure gradient between the two compartments. The cells are continuously using oxygen and producing carbon dioxide through cellular respiration, so at rest their P_{O_2} is 40 mm Hg and their P_{CO_2} is 46 mm Hg.

Arterial blood reaching the systemic capillaries has a P_{O_2} of 100 mm Hg and a P_{CO_2} of 40 mm Hg (Fig. 17-18 ■). The lower P_{O_2} in the cells sets up a gradient favoring diffusion of oxygen from plasma into the cells. Conversely, higher P_{CO_2} in the cells than in capillary blood allows carbon dioxide to diffuse out of cells into the capillary.

Just as at the alveoli, exchange is rapid and goes to equilibrium. Blood in the systemic venous circulation has an average P_{O_2} of 40 mm Hg and a P_{CO_2} of 46 mm Hg. Gas exchange at the tissue level may be affected slightly by tissue edema, but it usually proceeds without interference.

The Pulse Oximeter   One important clinical indicator of the effectiveness of gas exchange in the lungs is the amount of oxygen in arterial blood. Previously, obtaining an arterial blood sample was difficult and painful because it meant finding an accessible artery (most blood is drawn from superficial veins rather than from arteries, which lie deeper within the body). Recently, scientists developed instruments that quickly and painlessly measure blood oxygen levels through the surface of the skin on a finger or earlobe. The pulse oximeter clips onto skin and, within seconds, gives a digital reading of arterial hemoglobin saturation by measuring light absorbance of the tissue at two wavelengths. Transcutaneous oxygen sensors measure dissolved oxygen (P_{O_2}) using a variant of traditional gas-measuring electrodes. Both methods have their limitations but are popular because they give a rapid, noninvasive means of estimating arterial oxygen concentration.

GAS TRANSPORT IN THE BLOOD

Now that we have described gas exchange in the lungs, let's turn our attention to how oxygen and carbon dioxide are transported in the blood.

Hemoglobin Transports Most Oxygen to the Tissues

Hemoglobin (Hb), the oxygen-binding protein in red blood cells [∞ p. 483], is essential for normal gas transport. Oxygen is only slightly soluble in aqueous solutions, so without adequate amounts of hemoglobin, we cannot survive.

Oxygen is transported two ways in the blood: dissolved in the plasma and bound to hemoglobin (Hb). This fact can be summarized in the following statement:

Total blood oxygen content = amount dissolved in plasma + amount bound to hemoglobin

Because of the low solubility of oxygen, only 3 mL of O_2 will dissolve in the plasma fraction of 1 liter of arterial blood (Fig. 17-20a ■). Thus, with a typical cardiac output of 5 L blood/min, about 15 mL of dissolved oxygen reaches the systemic tissues each minute.

This amount cannot begin to meet the needs of the tissues, however. Oxygen consumption at rest is about 250 mL O_2 consumed/min, and that figure increases dramatically with exercise. The oxygen brought to cells in the plasma provides less than 10% of what the cells consume. Thus, the body is heavily dependent on oxygen carried by hemoglobin.

More than 98% of the oxygen in a given volume of blood is transported inside red blood cells bound to hemoglobin. At normal hemoglobin levels, oxygen content of red blood cells is about 197 mL/L blood (Fig. 17-20b ■). Thus:

Total blood oxygen content = 3 mL plasma O_2 + 197 mL HbO_2 = 200 mL O_2/L blood

If the cardiac output remains 5 L/min, oxygen delivery to cells jumps to almost 1000 mL/min, nearly four times the oxygen consumption of the tissues at rest.

The amount of oxygen that binds to hemoglobin depends on two factors: (1) the P_{O_2} of the plasma surrounding the red blood cells and (2) the number of potential oxygen-binding sites available within the red blood cells.

The P_{O_2} of plasma is the primary factor determining how many of the available hemoglobin binding sites are occupied by oxygen. Arterial plasma P_{O_2} is established by the composition of inspired air, the alveolar ventilation rate, and the efficiency of gas exchange between lung and blood, as you learned in previous sections.

The total number of binding sites depends on the number of hemoglobin molecules in the blood. Clinically, this number can be estimated either by counting the red blood cells and quantifying the amount of hemoglobin per red blood cell (*mean corpuscular hemoglobin*) or by simply determining the blood hemoglobin content

(a) Oxygen transport in blood without hemoglobin. Alveolar P_{O_2} = arterial P_{O_2}

P_{O_2} = 100 mm Hg

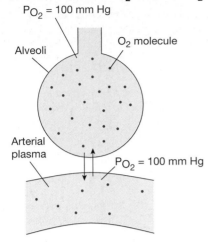

Alveoli

O_2 molecule

Arterial plasma

P_{O_2} = 100 mm Hg

O_2 content of plasma = 3 mL O_2/L blood	
O_2 content of red blood cells	= 0
Total O_2 carrying capacity	3 mL O_2/L blood

(b) Oxygen transport at normal P_{O_2} in blood with hemoglobin

P_{O_2} = 100 mm Hg

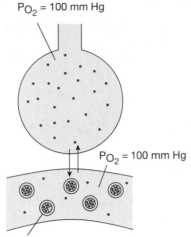

P_{O_2} = 100 mm Hg

Red blood cells with hemoglobin are carrying 98% of their maximum load of oxygen

O_2 content of plasma =	3 mL O_2/L blood
O_2 content of red blood cells	= 197 mL O_2/L blood
Total O_2 carrying capacity	200 mL O_2/L blood

(c) Oxygen transport at reduced P_{O_2} in blood with hemoglobin

P_{O_2} = 28 mm Hg

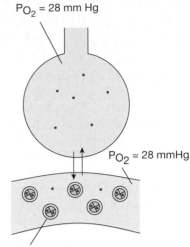

P_{O_2} = 28 mmHg

Red blood cells carrying 50% of their maximum load of oxygen

O_2 content of plasma =	0.8 mL O_2/L blood
O_2 content of red blood cells	= 99.5 mL O_2/L blood
Total O_2 carrying capacity	100.3 mL O_2/L blood

■ **Figure 17-20**   **Hemoglobin and oxygen transport**

(g Hb/dL whole blood). Any pathology that decreases the amount of hemoglobin in the cells or the number of red blood cells will adversely affect the blood's oxygen-transporting capacity.

People who have lost large amounts of blood need to replace hemoglobin for oxygen transport. A blood transfusion is the ideal replacement for blood loss, but in emergencies this is not always possible. Saline infusions can replace lost blood volume, but saline, like plasma, cannot transport sufficient quantities of oxygen to support cellular respiration.

Researchers are currently trying to find artificial oxygen carriers to replace hemoglobin. In times of large-scale disasters, these hemoglobin substitutes will eliminate the need to establish blood type before giving transfusions for blood loss.

Each Hemoglobin Molecule Binds Up to Four Oxygen Molecules

Why is hemoglobin an effective oxygen carrier? One hemoglobin molecule is composed of four globin subunits, each centered around a heme group whose central iron atom binds reversibly with an oxygen molecule [∞ p. 483]. Because there are four iron atoms per hemoglobin, one hemoglobin molecule has the potential to bind four oxygen molecules. The iron-oxygen interaction is a weak bond, and the Hb-O_2 link can be easily broken without altering either the hemoglobin or the oxygen.

Hemoglobin (Hb) bound to oxygen is known as *oxyhemoglobin*, abbreviated HbO_2. (It would be more accurate to show the actual number of oxygen molecules carried on each hemoglobin molecule: $Hb(O_2)_{1-4}$, but we will use the simpler abbreviation because the number bound varies from molecule to molecule.) The reversible

Blood Substitutes Physiologists have been attempting to find a substitute for blood ever since 1878, when an intrepid scientist named T. Gaillard Thomas transfused whole milk in place of blood. Although milk seems an unlikely replacement for blood, it has two important properties: proteins to provide colloid osmotic pressure and molecules (emulsified lipids) capable of binding to oxygen. In the development of hemoglobin substitutes, oxygen transport is the most difficult property to mimic. A hemoglobin solution would seem to be the obvious answer, but hemoglobin that is not compartmentalized in red blood cells behaves differently than hemoglobin that is. First, the quaternary structure changes so that the four-subunit (tetramer) form is in equilibrium with a smaller two-subunit (dimer) form. The smaller version of hemoglobin is easily excreted by the kidneys, so almost half a dose of hemoglobin solution disappears from the blood in 2–4 hours. Second, hemoglobin found outside the red blood cells does not release oxygen as easily in peripheral tissues. Investigators are making progress by polymerizing hemoglobin into larger, more stable molecules and loading these hemoglobin polymers into phospholipid liposomes [∞ p. 111]. By encapsulating hemoglobin into these artificial red blood cells, researchers hope to extend the life span of the blood substitute and make its properties approach those of real blood.

reaction of oxygen binding to hemoglobin at the lungs can be summarized as

$$Hb + O_2 \rightleftharpoons HbO_2$$

The amount of oxygen bound to hemoglobin depends primarily on the P_{O_2} of the surrounding plasma (Fig. 17-20b,c ■). Dissolved O_2 in plasma diffuses into red blood cells, where it binds to hemoglobin. This removes dissolved O_2 from the plasma, so more oxygen diffuses in from the alveoli. The transfer of oxygen from air to plasma to red blood cells and onto hemoglobin occurs so rapidly that blood leaving the alveoli normally picks up as much oxygen as the P_{O_2} of the plasma and number of red blood cells permit.

The amount of oxygen bound to hemoglobin at any given P_{O_2} is expressed as a percentage:

(Actual amount of O_2 bound/maximum that could be bound) × 100 = percent saturation of hemoglobin

The **percent saturation of hemoglobin** refers to the percentage of available binding sites that are bound to oxygen. If all binding sites of all hemoglobin molecules are occupied by oxygen molecules, the blood is 100% oxygenated, or *saturated*, with oxygen. If half the available binding sites are carrying oxygen, the hemoglobin is 50% saturated, and so on.

The relationship between the plasma P_{O_2} and oxygen-hemoglobin binding can be explained with the following analogy. The hemoglobin molecules carrying oxygen are like students moving books from an old library to a new one. Each student (a hemoglobin molecule) can carry a maximum of four books (100% saturation). The librarian in charge controls how many books (O_2 molecules) each student will carry, just as P_{O_2} determines the amount of oxygen that binds to hemoglobin.

At the same time, the total number of books being carried depends on the number of available students, just as the amount of oxygen delivered to the tissues depends on the number of available hemoglobin molecules. For example, if there are 100 students and the librarian gives each of them four books, then 400 books will be carried to the new library.

If the librarian has only three books for each (decreased P_{O_2}), then only 300 books will go to the new library, even though each student can carry four (decreased percent saturation of hemoglobin).

If the librarian is handing out four books per student but only 50 students show up (fewer hemoglobin molecules), then only 200 books will get to the new library, even though the students will be carrying the maximum number of books they can carry.

Once arterial blood reaches the tissues, the exchange process that took place in the lungs reverses. As dissolved oxygen diffuses out of the plasma into the cells, the resultant drop in plasma P_{O_2} disturbs the equilibrium of the oxygen-hemoglobin binding reaction by removing oxygen from the left side of the equation. The

...continued from page 510

Edna has been admitted to the hospital for tests related to her COPD. One of the tests is a hematocrit, an indicator of the number of red blood cells in her blood. The results of this test show that Edna has higher-than-normal numbers of red blood cells.

Question 4: *Why does Edna have an increased hematocrit? (Hint: Because of Edna's COPD, her arterial P_{O_2} is low.)*

reaction shifts to the left according to the law of mass action [∞ p. 86], and the hemoglobin molecules release their oxygen stores.

Like oxygen loading at the lungs, this process takes place very rapidly and goes to equilibrium. The P_{O_2} of the cell determines how much oxygen unloads from hemoglobin. As cells increase their metabolic activity, their P_{O_2} decreases and hemoglobin releases more oxygen to them.

P_{O_2} Determines Hemoglobin Binding of Oxygen

The physical relationship between P_{O_2} and oxygen binding to hemoglobin can be studied *in vitro* in the laboratory. Researchers expose samples of hemoglobin to varying P_{O_2} levels and quantitatively determine the amount of oxygen that binds. **Oxyhemoglobin (HbO_2) dissociation curves,** such as the one shown in Figure 17-21 ■, are the result of these *in vitro* binding studies.

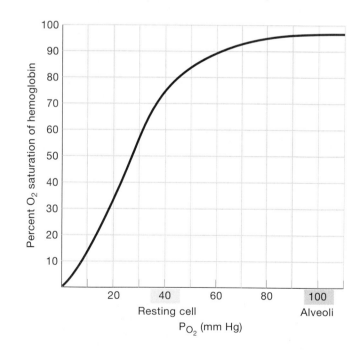

Graph question:
(a) When the P_{O_2} is 20 mm Hg, what is the percent O_2 saturation of hemoglobin?
(b) At what P_{O_2} is hemoglobin 50% saturated with O_2?

■ **Figure 17-21** **Oxygen-hemoglobin dissociation curve**

The shape of the HbO_2 dissociation curve reflects the properties of the hemoglobin molecule and its affinity for oxygen. If you look at the curve, you find that at normal alveolar and arterial P_{O_2} (100 mm Hg), 98% of the hemoglobin is bound to oxygen. In other words, as blood passes through the lungs, hemoglobin picks up nearly the maximum amount of oxygen that it can carry.

Notice that the curve is nearly flat at higher P_{O_2} levels (i.e., slope approaches zero). Above 100 mm Hg, large changes in P_{O_2} cause only minor changes in the percent saturation. Hemoglobin will not be 100% saturated until the P_{O_2} reaches nearly 650 mm Hg, a partial pressure far higher than anything we encounter in everyday life.

The flattening of the dissociation curve at higher P_{O_2} also means that alveolar P_{O_2} can fall a good bit below 100 mm Hg without having a significant effect on hemoglobin saturation. As long as P_{O_2} in the pulmonary capillaries stays above 60 mm Hg, hemoglobin will be more than 90% saturated and maintain near-normal levels of oxygen transport. However, once the P_{O_2} falls below 60 mm Hg, the curve becomes steeper. The steep slope means that a small decrease in P_{O_2} causes a relatively large release of oxygen.

For example, if P_{O_2} falls from 100 mm Hg to 60 mm Hg, the percent saturation of hemoglobin goes from 98% saturation to about 88% saturation, a decrease of 10%. This is equivalent to a saturation change of 2.5% for each 10 mm Hg change.

If P_{O_2} falls further, from 60 to 40 mm Hg, the percent saturation goes from 88% to 75%, a decrease of 6.5% for each 10 mm Hg. Notice that the slope of the curve is steeper in the 40–60 mm Hg range than in the 60–100 mm Hg range. If P_{O_2} decreases from 40 mm Hg to 20 mm Hg, the slope of the curve becomes even steeper. Hemoglobin saturation declines from 75% to 35%, a change of 20% for each 10 mm Hg.

What is the physiological significance of the HbO_2 curve's shape? At a P_{O_2} of 40 mm Hg, the partial pressure of resting cells, hemoglobin is still 75% saturated and has released only one-fourth of the oxygen it is capable of carrying. The 75% of the oxygen that remains bound serves as a reservoir that cells can draw upon if metabolism increases.

When metabolically active tissues use additional oxygen, their P_{O_2} decreases and additional oxygen is released by hemoglobin. At a P_{O_2} of 20 mm Hg, an average value for exercising muscle, hemoglobin saturation falls to about 35%. With this 20 mm Hg decrease in P_{O_2} (40 mm Hg to 20 mm Hg), hemoglobin releases an additional 40% of the oxygen it is capable of carrying. This is another example of the built-in reserve capacity of the body.

✔ Can a person breathing 100% oxygen at sea level achieve 100% saturation of her hemoglobin?

✔ What effect does hyperventilation have on the percent saturation of arterial hemoglobin? (Hint: Fig. 17-16 ■)

Temperature, pH, and Metabolites Affect Oxygen-Hemoglobin Binding

Any factor that changes the conformation of the hemoglobin protein may affect its ability to bind oxygen. In humans, physiological changes in pH, P_{CO_2}, and temperature of the blood all alter the oxygen-binding capability of hemoglobin. Changes in binding affinity are reflected by changes in the shape of the HbO_2 dissociation curve.

Increased temperature and P_{CO_2} or decreased pH decrease the affinity of hemoglobin for oxygen and shift the oxygen-hemoglobin dissociation curve to the right (Fig. 17-22 ■). When these factors change in the opposite direction, binding affinity increases and the HbO_2 curve shifts to the left.

Notice that when the curve shifts in either direction, the changes are much more pronounced in the steep part of the curve. This means that oxygen binding at the lungs is not greatly affected, but oxygen delivery at the tissues will be significantly altered.

Let's examine one specific example, the shift that takes place when pH goes from 7.4 (normal) to 7.2 (more acidic). A pH of 7.2 is lower than normal for the body (normal range 7.38–7.42) but is compatible with life. At a P_{O_2} of 40 mm Hg and pH of 7.2, hemoglobin molecules release 15% more oxygen than they do at a pH of 7.4.

When does the body undergo shifts in pH? Anaerobic exercise is one example of a situation in which body pH decreases. Anaerobic metabolism in exercising muscle fibers produces lactic acid, which in turn releases H^+ into the cytoplasm and extracellular fluid. As H^+ concentrations increase, pH falls, the affinity of hemoglobin for oxygen decreases, and the HbO_2 dissociation curve shifts to the right. More oxygen is released at the tissues as the blood becomes more acidic (pH decreases). A shift in the hemoglobin saturation curve that results from a change in pH is called the **Bohr effect.**

An additional factor that affects oxygen-hemoglobin binding is **2,3-diphosphoglycerate (2,3-DPG;** also called 2,3-bisphosphoglycerate or 2,3-BPG), a compound made from an intermediate of the glycolysis pathway. **Chronic hypoxia** (extended periods of low oxygen) triggers an increase in 2,3-DPG production in red blood cells. Like the other factors discussed, 2,3-DPG lowers the binding affinity of hemoglobin and shifts the HbO_2 dissociation curve to the right (Fig. 17-23 ■). Ascent to high altitude and anemia are two situations that increase 2,3-DPG production.

Changes in hemoglobin structure also will change oxygen binding. For example, fetal hemoglobin has a unique protein chain for two of its subunits. This change enhances the ability of fetal hemoglobin to bind oxygen in the low-oxygen environment of the placenta. The altered binding affinity is reflected by the different shape of the fetal HbO_2 dissociation curve (Fig. 17-24 ■).

Figure 17-25 ■ summarizes all the factors that influence oxygen transport in the blood.

(a) Effect of pH

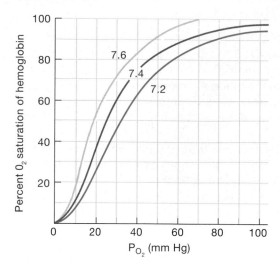

Graph question: (a) At a P_{O_2} of 20 mm Hg, how much more oxygen is released at an exercising muscle cell whose pH is 7.2 than by a cell with a pH of 7.4? (b) What happens to oxygen release when the exercising muscle cell warms up?

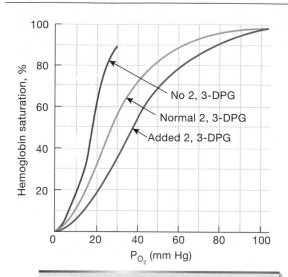

Graph question: Blood stored in blood banks loses its normal content of 2, 3-DPG. Is this good or bad? Explain.

■ **Figure 17-23** **2,3-DPG alters hemoglobin affinity for oxygen**

(b) Effect of temperature

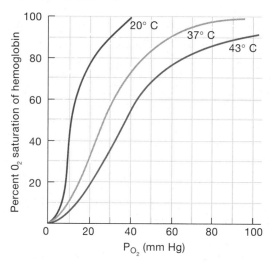

(c) Effect of P_{CO2}

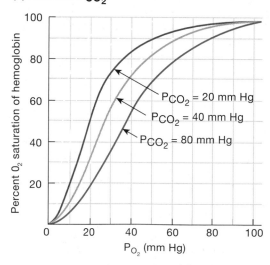

■ **Figure 17-22** **Physical factors alter oxygen binding to hemoglobin**

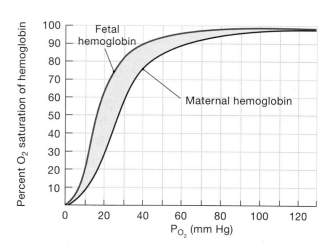

Graph question:
(a) Because of incomplete gas exchange across the thick membranes of the placenta, hemoglobin in fetal blood leaving the placenta is 80% saturated with oxygen. What is the P_{O_2} of that placental blood?
(b) Blood in the vena cava of the fetus has a P_{O_2} around 10 mm Hg. What is the percent O_2 saturation of maternal hemoglobin at the same P_{O_2}?

■ **Figure 17-24** **Differences in the oxygen-binding properties of maternal and fetal hemoglobin**

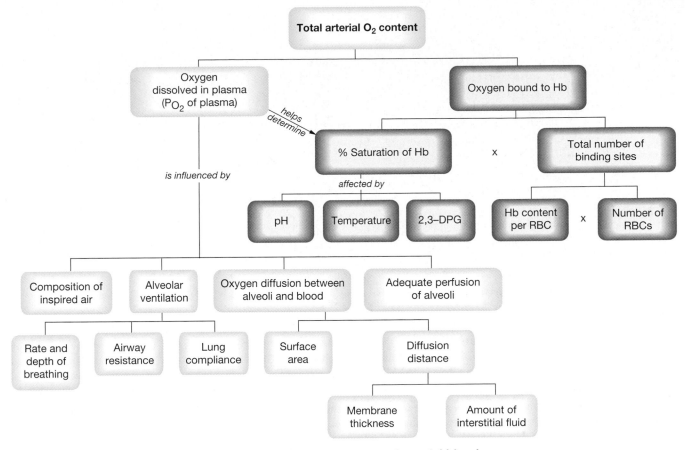

Figure 17-25 **Factors contributing to the total oxygen content of arterial blood**

✔ A muscle that is actively contracting may have an intracellular P_{O_2} of 25 mm Hg. What happens to oxygen binding to hemoglobin at this P_{O_2}? What is the P_{O_2} of the venous blood leaving the active muscle?

Carbon Dioxide Is Transported Three Ways

Gas transport in the blood is a two-way process because the removal of carbon dioxide from the body is as essential as the uptake of oxygen. The hemoglobin molecule, as we will see, is as important in CO_2 removal as it was in the transport of oxygen.

Why is removing CO_2 so important? The reason is that elevated P_{CO_2} (*hypercapnia*) causes the pH disturbance known as *acidosis* (see Chapter 19). Abnormally high P_{CO_2} levels also depress central nervous system function, causing confusion, coma, or even death. Thus, CO_2 is a potentially toxic waste product that must be removed by the lungs.

Carbon dioxide is produced as a by-product of cellular respiration [∞ p. 94]. CO_2 is more soluble in body fluids than oxygen, but the cells produce far more CO_2 than can dissolve in the plasma. Only about 7% of carbon dioxide carried by venous blood is in the form of dissolved CO_2. The remaining 93% diffuses into red blood cells, where 70% is converted to bicarbonate ion,

as explained below, and 23% binds to hemoglobin (Hb·CO_2). Figure 17-26 ■ summarizes these three methods of carbon dioxide transport in the blood.

CO₂ and Bicarbonate About 70% of the CO_2 that enters the blood is transported to the lungs as bicarbonate ions (HCO_3^-) dissolved in the plasma. The conversion of carbon dioxide to bicarbonate serves two purposes:

…continued from page 524

Edna is discharged from the hospital after three days with prescriptions for bronchodilator drugs to keep her airways open. Four days later, however, she becomes dizzy and confused. She falls in her kitchen, but manages to dial 9-1-1. The paramedics who arrive ask her if she has any medical conditions. She replies that she has COPD. The paramedics start Edna on high-flow oxygen and prepare to transport her to the hospital. A paramedic-in-training, along for the ride, asks one of the paramedics about the danger of giving oxygen to Edna. "Thinking has changed on that," the paramedic replies. "Experts think it's more dangerous to withhold oxygen from these patients, especially if they have severe hypercapnia, like this patient."

Question 5: *Without performing blood tests, why does the paramedic suspect that Edna has severe hypoxia and hypercapnia (abnormally high levels of P_{CO_2})?*

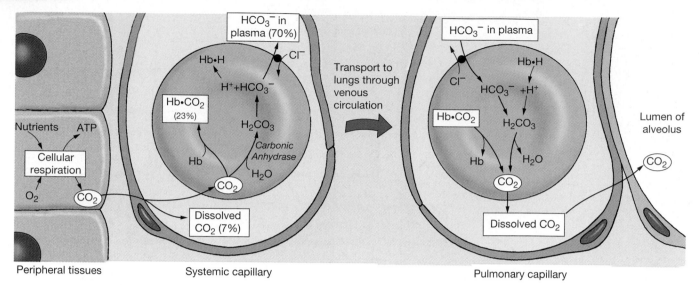

■ **Figure 17-26** **Carbon dioxide transport**

(1) it provides an additional means by which carbon dioxide can be transported from cells to lungs, and (2) the bicarbonate is available to act as a buffer for metabolic acids [∞ p. 28], thereby helping stabilize the body's pH.

But how does CO_2 turn into bicarbonate? The rapid conversion depends on the presence of **carbonic anhydrase**, an enzyme found concentrated in red blood cells. Let's see how this happens.

Dissolved CO_2 in plasma diffuses into red blood cells, where it reacts with water in the presence of carbonic anhydrase to form *carbonic acid*. Carbonic acid then dissociates into a hydrogen ion and a bicarbonate ion. The equation for the reaction is written as follows:

$$CO_2 + H_2O \underset{\text{carbonic acid}}{\overset{\text{carbonic anhydrase}}{\rightleftharpoons}} H_2CO_3 \rightleftharpoons H^+ + HCO_3^-$$

Because of the dissociation of carbonic acid, we sometimes ignore the intermediate step and summarize the reaction as:

$$CO_2 + H_2O \overset{\text{carbonic anhydrase}}{\rightleftharpoons} H^+ + HCO_3^-$$

This reaction is reversible. The rate in either direction depends on the relative concentrations of the substrates and obeys the law of mass action.

When CO_2 moves into red blood cells, carbonic anhydrase converts it into HCO_3^- and H^+ (Fig. 17-26 ■). This reaction will continue until an equilibrium state is reached. (Water is always in excess in the body, so water concentration plays no role in the dynamic equilibrium of this reaction.) To keep the reaction going, the products, H^+ and HCO_3^-, must be removed from the cytoplasm of the red blood cell. If the product concentrations are kept low, the reaction cannot reach equilibrium.

Two separate steps remove free H^+ and HCO_3^-. In the first step, bicarbonate ions produced from CO_2 leave the red blood cell on an antiport protein [∞ p. 125]. This

transport process, known as the **chloride shift,** exchanges one HCO_3^- for one Cl^-. The one-for-one exchange maintains electrical neutrality so that the cell's membrane potential is not affected.

The transfer of HCO_3^- into the plasma makes additional buffer available to moderate pH changes caused by production of metabolic acids. Bicarbonate is the most important extracellular buffer in the body.

Hemoglobin and H^+ The second step removes free H^+ from the red blood cell cytoplasm when hemoglobin acts as a buffer and binds hydrogen ions (Hb·H). Hemoglobin's buffering of H^+ is an important step that prevents large changes in the body's pH.

If blood P_{CO_2} is elevated much above normal, the hemoglobin buffer cannot soak up all H^+ produced from the reaction of CO_2 and water. In those cases, excess H^+ accumulates in the plasma, causing the condition known as **respiratory acidosis**. Further information on the role of the respiratory system in maintaining pH homeostasis is found in Chapter 19.

As H^+ and HCO_3^- disappear from the red blood cell cytoplasm, the reaction continues and more CO_2 converts to carbonic acid. This CO_2 in turn is replaced by CO_2 that diffuses into the red blood cell from the plasma. Then, as plasma P_{CO_2} decreases, additional CO_2 diffuses out of cells.

Hemoglobin and CO_2 Although most carbon dioxide that enters the red blood cells is converted to bicarbonate ions, about 23% of the CO_2 in venous blood binds directly to hemoglobin. When oxygen leaves its binding sites on the hemoglobin molecule, CO_2 binds with the free hemoglobin at exposed amino groups (NH_2), forming **carbaminohemoglobin**. This reaction can be summarized as

$$CO_2 + Hb \rightleftharpoons Hb\cdot CO_2 \text{ (carbaminohemoglobin)}$$

The formation of carbaminohemoglobin is facilitated by the presence of both CO_2 and H^+ produced from CO_2. Both factors decrease hemoglobin's binding affinity for oxygen (see Fig. 17-22 ■).

CO_2 Removal at the Lungs

When venous blood reaches the lungs, the processes that took place in the systemic capillaries reverse (Fig. 17-26 ■). The P_{CO_2} of alveoli is lower than that of venous blood in pulmonary capillaries. Carbon dioxide diffuses out of plasma into the alveoli, and the plasma P_{CO_2} begins to fall.

The decrease in plasma P_{CO_2} allows dissolved CO_2 to diffuse out of red blood cells. As CO_2 levels in the red blood cells decrease, the equilibrium of the CO_2-bicarbonate reaction is disturbed.

Removal of CO_2 causes H^+ to leave the hemoglobin molecules. The chloride shift reverses: Cl^- returns to the plasma in exchange for HCO_3^- that moves back into the red blood cells. The HCO_3^- and H^+ re-form into carbonic acid that is converted into water and CO_2. The CO_2 is then free to diffuse out of the red blood cell and into the alveoli.

In summary, Figure 17-27 ■ shows the combined transport of CO_2 and O_2 in the blood. At the alveoli, O_2 diffuses down its pressure gradient into the plasma and, from there, into the red blood cells. Hemoglobin binds O_2, increasing the amount of oxygen that can be transported to the cells.

At the cells, the process reverses. P_{O_2} in the cells is lower than that in the arterial blood, so O_2 diffuses from the plasma to the cells. The decrease in plasma P_{O_2} causes hemoglobin to release O_2, giving up additional oxygen to enter the cells.

Carbon dioxide from aerobic metabolism simultaneously enters the blood, dissolving in the plasma. From there, it enters red blood cells, where most is converted to HCO_3^- and H^+. The HCO_3^- is returned to the plasma in exchange for a Cl^-, while the H^+ binds to hemoglobin. A small fraction of the CO_2 also binds directly to hemoglobin. At the lungs, the process reverses as CO_2 diffuses out into the alveoli.

In order to understand fully how the respiratory system coordinates delivery of oxygen to the lungs with its transport in the circulation, we will now consider the central nervous system control of ventilation.

✔ How would an obstruction of the airways affect alveolar ventilation, arterial P_{CO_2}, and the body's pH?

▶ REGULATION OF VENTILATION

Breathing is a rhythmic process that occurs without conscious thought. In that respect, it resembles the rhythmic beating of the heart. However, skeletal muscles, unlike autorhythmic cardiac muscles, are not able to contract spontaneously. Instead, skeletal muscle contraction

must be initiated by somatic motor neurons, which in turn are controlled by the central nervous system.

In the respiratory system, contraction of the diaphragm and intercostals is initiated by a group of neurons in the medulla oblongata (Fig. 17-28 ■). These neurons take the form of a network, or **central pattern generator,** that has intrinsic rhythmic activity [∞ p. 392]. It has been postulated that the rhythmic activity arises from a pacemaker neuron, a single neuron with an unstable membrane potential that periodically reaches threshold. However, to date, no pacemaker has been definitively identified, and scientists now believe that the rhythmicity may arise within a network of neurons with unstable membrane potentials.

Direct study of the brain centers controlling ventilation is difficult because of the complexity of the neuronal network and its anatomical location. Consequently, some of our understanding of the control of ventilation has come from observing patients with brain damage. Other information has come from animal experiments in which the neural connections between major parts of the brain stem were severed.

In order to better understand the complex relationships between the parts of the brain that control ventilation, let's review some of these clinical and experimental findings and see what deductions researchers made about the control of rhythmic breathing.

1. *Observation.* If the brain stem is severed below the medulla, all respiratory movement ceases. If the brain stem is cut above the level of the pons, ventilation is normal.

   Hypothesis 1. The control centers for ventilation lie in the medulla or the pons or both.

2. *Observation.* If the medulla is completely separated from the pons and higher brain centers, ventilation becomes gasping and irregular in depth, but the respiratory rhythm remains.

   Hypothesis 2. The primary control center for ventilation lies in the medulla.

   Hypothesis 3. The medulla contains neurons that set the rhythm of ventilation.

   Hypothesis 4. The normal smooth pattern of ventilation depends on communication between neurons in the pons and the medulla.

From these observations and hypotheses, the following model for the control of ventilation has been proposed. Some parts of the model are well supported with experimental evidence. Other aspects are still under investigation. The model states that:

1. Respiratory neurons in the medulla control inspiration and expiration.
2. Neurons in the pons influence the rate and depth of ventilation.

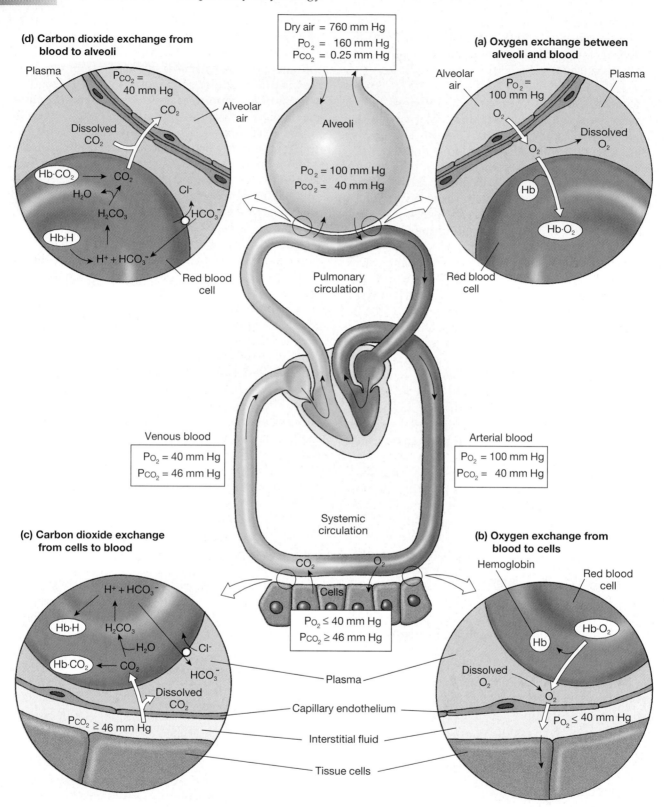

(d) Carbon dioxide exchange from blood to alveoli

Plasma

$P_{CO_2} = 40$ mm Hg

CO_2

Dissolved CO_2

Alveolar air

$Hb \cdot CO_2 \rightarrow CO_2$

H_2O

H_2CO_3

Cl^-

$Hb \cdot H$

HCO_3^-

$H^+ + HCO_3^-$

Red blood cell

Dry air $= 760$ mm Hg
$P_{O_2} = 160$ mm Hg
$P_{CO_2} = 0.25$ mm Hg

Alveoli

$P_{O_2} = 100$ mm Hg
$P_{CO_2} = 40$ mm Hg

Pulmonary circulation

(a) Oxygen exchange between alveoli and blood

Alveolar air

$P_{O_2} = 100$ mm Hg

Plasma

O_2

Dissolved O_2

O_2

Hb

$Hb \cdot O_2$

Red blood cell

Venous blood

$P_{O_2} = 40$ mm Hg
$P_{CO_2} = 46$ mm Hg

Arterial blood

$P_{O_2} = 100$ mm Hg
$P_{CO_2} = 40$ mm Hg

Systemic circulation

(c) Carbon dioxide exchange from cells to blood

$H^+ + HCO_3^-$

$Hb \cdot H$

H_2CO_3

H_2O

$Hb \cdot CO_2 \leftarrow CO_2$

Cl^-

HCO_3^-

Dissolved CO_2

$P_{CO_2} \geq 46$ mm Hg

CO_2

O_2

Cells

$P_{O_2} \leq 40$ mm Hg
$P_{CO_2} \geq 46$ mm Hg

Plasma

Capillary endothelium

Interstitial fluid

Tissue cells

(b) Oxygen exchange from blood to cells

Hemoglobin

Red blood cell

$Hb \cdot O_2$

Hb

O_2

Dissolved O_2

O_2

$P_{O_2} \leq 40$ mm Hg

■ **Figure 17-27 Summary of gas transport**

3. The rhythmic pattern of breathing arises from a network of spontaneously discharging neurons.
4. Ventilation is subject to modulation by various chemical factors and by higher brain centers. (The experimental evidence for this point will be discussed later.)

Neurons in the Medulla Control Breathing

Classic descriptions of the brain's respiratory control of ventilation divided groups of neurons into various control centers. The most recent descriptions have become less specific about assigning function to particular cen-

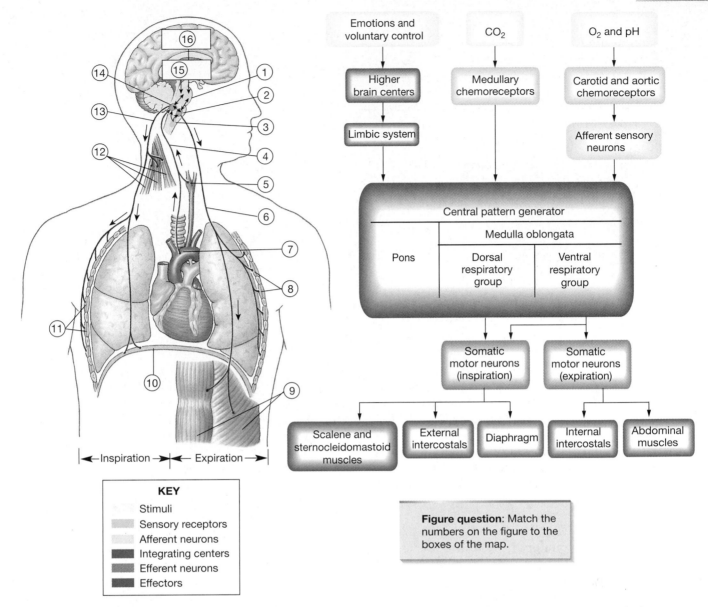

■ **Figure 17-28** **Reflex control of ventilation** Chemoreceptors monitor blood gases and pH. Control centers in the brain stem regulate activity in somatic motor neurons leading to respiratory muscles.

ters and now simply refer to the network of neurons in the brain stem as the central pattern generator.

The central pattern generator is unique in that it functions automatically throughout a person's life, and yet it can be controlled voluntarily, up to a point. Complicated synaptic interactions between neurons in the network create the rhythmic cycles of inspiration and expiration, influenced continuously by sensory input from receptors for CO_2, O_2, and H^+. Ventilation pattern depends in large part on the levels of those three substances in the blood.

Although there is still much to be learned about the central pattern generator, we do know that respiratory neurons are found concentrated in two nuclei in the medulla oblongata. The **dorsal respiratory group** (DRG) contains mostly **inspiratory neurons** (I neurons) that control the external intercostal muscles and the diaphragm (Fig. 17-28 ■). The **ventral respiratory group**

(VRG) contains neurons that control the muscles used for active expiration (E neurons) and for greater-than-normal inspiration (I+ neurons), such as occurs during vigorous exercise.

During quiet respiration, the inspiratory neurons of the dorsal respiratory group gradually increase stimulation of the inspiratory muscles for 2 seconds. This increase is sometimes called ramping because of the shape of the graph of inspiratory neuron activity (Fig. 17-29 ■).

A few inspiratory neurons fire to begin the ramp. The firing of these neurons recruits other inspiratory neurons to fire in an apparent positive feedback loop. As more neurons fire, more skeletal muscle fibers are recruited. The rib cage expands smoothly as the diaphragm contracts.

At the end of 2 seconds, the inspiratory neurons abruptly stop firing and the respiratory muscles relax. Over the next 3 seconds, passive expiration occurs because

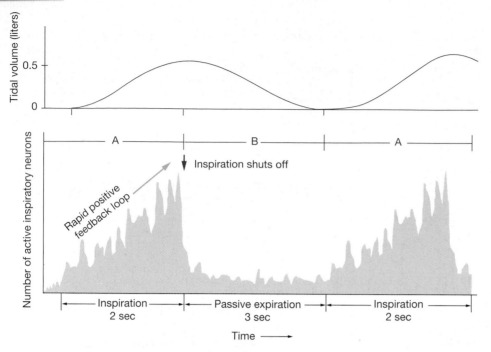

■ **Figure 17-29 Rhythmic breathing** During inspiration, the activity of inspiratory neurons increases steadily, apparently through a positive feedback mechanism. At the end of inspiration, the activity shuts off abruptly and expiration takes place through recoil of elastic lung tissue.

of elastic recoil of the inspiratory muscles and the elastic lung tissue. However, there is some motor neuron activity during passive expiration, suggesting that perhaps muscles in the upper airways contract to slow the flow of air out of the respiratory system.

The expiratory neurons and I+ neurons of the ventral respiratory group remain mostly inactive during quiet respiration. They function primarily during forced breathing, when inspiratory movements are exaggerated, or during active expiration.

In forced breathing, the increased activity of dorsal respiratory group inspiratory neurons in some way activates the I+ neurons of the ventral respiratory group. These I+ neurons in turn stimulate accessory inspiratory muscles such as the sternocleidomastoid. Contraction of the accessory inspiratory muscles enhances expansion of the thorax by raising the sternum and upper ribs.

In active expiration, expiratory (E) neurons from the ventral respiratory group activate the internal intercostal and abdominal muscles. There seems to be reciprocal inhibition between the inspiratory and expiratory neurons. The I neurons inhibit the E neurons during inspiration, and the E neurons inhibit I neurons during active expiration. In addition, the cycling of respiratory activity is subject to continuous modulation as CO_2 in the blood changes.

Carbon Dioxide, Oxygen, and pH Influence Ventilation

Sensory input from two sets of chemoreceptors modifies the rhythmicity of the central pattern generator. Carbon dioxide is the primary stimulus for changes in ventilation. Oxygen and plasma pH play lesser roles.

The chemoreceptors for oxygen and carbon dioxide are strategically associated with the arterial circulation.

If too little oxygen is present in the arterial blood destined for the tissues, the rate and depth of breathing increase. Or, if the rate of carbon dioxide production by the cells exceeds the rate of carbon dioxide removal by the lungs, ventilation is intensified to match carbon dioxide removal to production. These homeostatic reflexes operate constantly, keeping arterial P_{O_2} and P_{CO_2} within a narrow range.

Peripheral chemoreceptors located in the *carotid* and *aortic bodies* sense changes in the oxygen concentration, pH, and P_{CO_2} of the plasma (Fig. 17-28 ■). (These are the same carotid and aortic bodies that contain the baroreceptors involved in reflex control of blood pressure [∞ p. 463].)

Central chemoreceptors monitor cerebrospinal fluid (CSF) composition and respond to changes in the concentration of CO_2 in the cerebrospinal fluid. These receptors lie on the ventral surface of the medulla, close to neurons involved in respiratory control.

The Carotid and Aortic Bodies The carotid and aortic chemoreceptors are sensitive to the P_{O_2}, P_{CO_2}, and pH of arterial blood. When these receptors are activated by decreases in P_{O_2} or pH or an increase in P_{CO_2}, they send action potentials through sensory neurons to the brain stem. Sensory information is integrated within the medullary control centers. The respiratory centers respond by sending signals through somatic motor neurons to the skeletal muscles that control ventilation. The response to decreased P_{O_2} and pH or increased P_{CO_2} is the same: an increase in ventilation.

Under most circumstances, oxygen is not an important factor in modulating ventilation. Arterial P_{O_2} must be less than 60 mm Hg before ventilation is stimulated. This decrease in P_{O_2} is equivalent to ascending to an altitude of 3000 m. (For reference, Denver is located at an altitude of 1609 m.)

Because the peripheral chemoreceptors respond only to dramatic changes in arterial P_{O_2}, they do not play a role in the everyday regulation of ventilation. However, unusual physiological conditions (such as ascending to high altitude) and pathological conditions such as chronic obstructive pulmonary disease (COPD) can reduce arterial P_{O_2} to low levels that activate the peripheral chemoreceptors.

The peripheral chemoreceptors in the carotid and aortic bodies are more responsive to increases in plasma H^+ concentrations and P_{CO_2} than to changes in P_{O_2}. Any condition that reduces plasma pH or increases P_{CO_2} stimulates ventilation by way of peripheral chemoreceptors.

Central Chemoreceptors   The most important chemical controller of ventilation is carbon dioxide, mediated through central chemoreceptors located in the medulla (Fig. 17-31 ■). These receptors set the respiratory pace, providing continuous input into the central pattern generator.

When arterial P_{CO_2} increases, carbon dioxide crosses the blood-brain barrier quite rapidly and activates the central chemoreceptors. These receptors signal the central pattern generator to increase the rate and depth of ventilation, thereby increasing alveolar ventilation and removing carbon dioxide from the blood (Fig. 17-32 ■).

Although we say that the central chemoreceptors respond to carbon dioxide, they actually respond to pH changes in the cerebrospinal fluid. Carbon dioxide that diffuses across the blood-brain barrier into the cerebrospinal fluid is converted to bicarbonate and H^+ ($CO_2 + H_2O \rightleftharpoons H^+ + HCO_3^-$). Experiments indicate that the H^+ is able to trigger the chemoreceptor reflex. However, pH changes in the plasma do not influence the central chemoreceptor directly. Free H^+ in the plasma is unable to cross the blood-brain barrier and therefore has no direct effect on the central chemoreceptors.

Decreases in arterial P_{CO_2} also influence ventilation. If alveolar P_{CO_2} falls, as it might during hyperventilation, plasma P_{CO_2} and P_{CO_2} in the cerebrospinal fluid decrease. As a result, central chemoreceptor activity declines, and the central pattern generator slows the ventilation rate.

When ventilation decreases, carbon dioxide begins to accumulate in alveoli and the plasma. Eventually, the arterial P_{CO_2} rises above the threshold level for the chemoreceptors. At that point, the receptors fire, and the central pattern generator again increases ventilation.

The central chemoreceptors initially respond strongly to an increase in plasma P_{CO_2} by increasing ventilation. However, if P_{CO_2} remains elevated for several days, ventilation decreases. The mechanism for this adaptation is not certain, but may occur because the blood-brain barrier begins to transport bicarbonate ions into the cerebrospinal fluid. The bicarbonate acts as a buffer, removing

The Carotid Oxygen Sensors   The carotid and aortic bodies contain chemoreceptors that adjust ventilation in response to changes in blood levels of oxygen. But what is the mechanism that translates dissolved oxygen concentrations into electrical signals? In the last decade, researchers have come closer to answering that question. Cells in the carotid bodies called **glomus cells** [*glomus*, a ball-shaped mass] have been identified as the chemoreceptors (Fig. 17-30 ■). The glomus cells contain gated potassium channels, called K_{O_2} channels, with an extracellular oxygen sensor. When the sensor is combined with oxygen, the channel stays open and K^+ leaves the cell, hyperpolarizing it. If oxygen levels in the blood decrease, fewer sensors are combined with O_2 and more K_{O_2} channels close. The decrease in K^+ permeability depolarizes the cells, opens voltage-gated Ca^{2+} channels, and allows Ca^{2+} to enter the cells. Just as in the axon terminal, Ca^{2+} entry triggers exocytosis of vesicles, in this case containing dopamine. Dopamine initiates action potentials in sensory neurons leading to the central nervous system, signaling the respiratory control centers to increase ventilation.

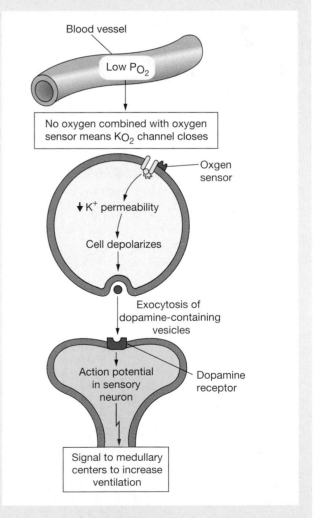

Figure 17-30   Carotid body oxygen sensor

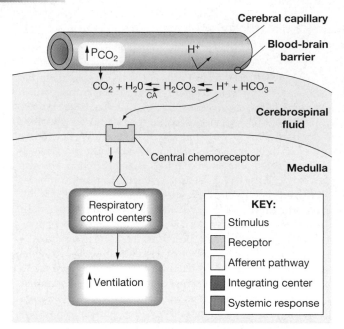

Figure 17-31 **Central chemoreceptor monitors CO₂ in cerebrospinal fluid**

tral chemoreceptors adapt to high P_{CO_2}. In some situations, low P_{O_2} becomes the primary chemical stimulus for ventilation.

For example, patients with severe chronic lung disease such as emphysema have chronic hypercapnia and hypoxia. Their arterial P_{CO_2} may rise to 50–55 mm Hg (normal 35–45), while their P_{O_2} falls to 45–50 mm Hg (normal 75–100). Because these levels are chronic, the central chemoreceptors gradually adapt to the elevated P_{CO_2}.

Most of the chemical stimulus for ventilation then comes from low P_{O_2}, sensed by the carotid and aortic chemoreceptors. If these patients are given too much oxygen, they may stop breathing because their chemical stimulus for ventilation is eliminated.

On occasion, homeostatic compensation for decreased P_{O_2} will adversely affect plasma P_{CO_2} levels. For example, at altitudes above 3000 m, the P_{O_2} of inspired air drops below 60 mm Hg. This decrease in P_{O_2} activates the carotid and aortic bodies and increases ventilation in an attempt to bring arterial P_{O_2} back into a normal range.

This hyperventilation cannot significantly increase arterial P_{O_2} because the inspired air is low in oxygen. However, hyperventilation will cause arterial P_{CO_2} to decrease. The decrease in arterial P_{CO_2}, monitored by the central chemoreceptors, then *depresses* ventilation. Within a few days, acclimatization to the lower arterial P_{CO_2} takes place, so ventilation again increases.

H⁺ from the cerebrospinal fluid and decreasing H⁺ stimulation of the central pattern generator.

Fortunately for people with chronic lung diseases, the response of peripheral chemoreceptors to low arterial oxygen remains intact over time, even though cen-

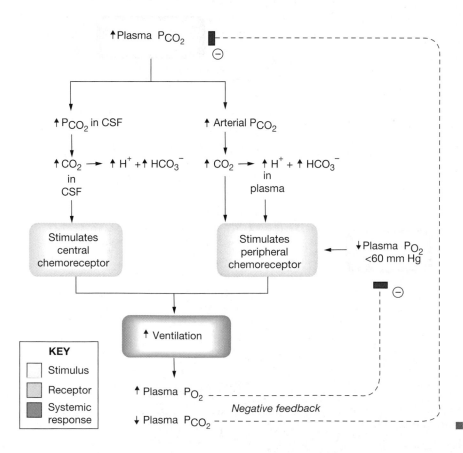

Figure 17-32 **Chemoreceptor reflex**

High Altitude   In 1981 a group of 20 physiologists, physicians, and climbers, supported by 42 Sherpa assistants, formed the American Medical Research Expedition to Mt. Everest. The purpose of the expedition was to study human physiology at extreme altitudes, starting with the base camp at 5400 m and continuing on to the summit at 8848 m. From the work of these scientists and others, we now have a good picture of the physiology of high-altitude acclimatization. The body's immediate homeostatic response to the hypoxia of high altitude is hyperventilation. Hyperventilation enhances alveolar ventilation, but may not help arterial P_{O_2} significantly when atmospheric P_{O_2} is low. However, hyperventilation does lower plasma P_{CO_2} and increase pH, causing a state of alkalosis. This pH change increases HbO_2 binding, seen on the HbO_2 saturation curve as a left shift (Fig. 17-23 ■). Increased HbO_2 binding might seem counterproductive, but it actually allows hemoglobin to pick up more oxygen at the lower P_{O_2} of the lungs. After an acclimatization period of hours to days, 2,3-DPG production by red blood cells increases, shifting the HbO_2 saturation curve back to the right and offsetting the effects of the respiratory alkalosis. Thus, at altitudes up to 6300 m, the HbO_2 saturation curve is essentially normal. The hypoxia of high altitude also triggers the release of erythropoietin from the kidney and liver. This hormone stimulates red blood cell production. Even though the P_{O_2} of the blood remains low, the total oxygen-carrying capacity is thus increased. In some individuals, the increased hematocrit (*polycythemia*) that results from living at high altitudes may be maladaptive. Hematocrits greater than 45% increase the viscosity of the blood enough to impede blood flow to the brain and peripheral tissues. The decrease in blood flow counteracts the increased oxygen-carrying capacity of the blood. Altitude-induced polycythemia is considered to be the cause of the condition known as chronic mountain sickness, with symptoms of headache, fatigue, mental status changes, and poor exercise tolerance. People who exhibit symptoms of elevated hematocrits may be helped by phlebotomy (blood withdrawal). In Leadville, Colorado (3100 m), the 60 people who suffer from chronic mountain sickness are the main donors to the local blood bank!

Mechanoreceptor Reflexes Protect the Lungs from Inhaled Irritants

In addition to the chemoreceptor reflexes that help regulate ventilation, the body has protective reflexes that respond to physical injury or irritation of the respiratory tract and to overinflation of the lungs. The major protective reflex is *bronchoconstriction,* mediated through parasympathetic neurons that innervate bronchiolar smooth muscle. Inhaled particles or noxious gases stimulate **irritant receptors** in the airway mucosa. The irritant receptors send signals through sensory neurons to control centers in the central nervous system that trigger bronchoconstriction. Protective reflex responses also include coughing and sneezing.

The **Hering-Breuer inflation reflex** is designed to prevent overexpansion of the lungs during strenuous exercise. If tidal volume exceeds 1 liter, stretch receptors in the lung signal the brain stem to terminate inspiration. However, because normal tidal volume is only 500 mL, this reflex does not operate during quiet breathing and mild exertion.

Higher Brain Centers Affect Patterns of Ventilation

Conscious and unconscious thought processes also affect respiratory activities. Higher centers in the hypothalamus and cerebrum can alter the activity of the central pattern generator and change ventilation rate and depth. Voluntary control of ventilation falls into this category. But higher brain center control is not a *requirement* for ventilation. Even if the brain stem above the pons is severely damaged, essentially normal respiratory cycles continue.

Respiration can also be affected by stimulation of portions of the limbic system. As a result, emotional and autonomic activities such as fear and excitement may affect the pace and depth of respiration. In some of these situations, the neural pathway goes directly to the somatic motor neurons, bypassing the central pattern generator in the brain stem.

Although we can temporarily alter our respiratory performance, we cannot override the chemoreceptor reflexes. Holding one's breath is a good example. We can hold our breath voluntarily only until elevated P_{CO_2} in the blood and cerebrospinal fluid activates the chemoreceptor reflex, forcing us to inhale.

Small children having temper tantrums sometimes attempt to manipulate parents by threatening to hold their breath until they die. However, the chemoreceptor reflexes make it impossible for the children to carry out that threat. Extremely strong-willed children can continue holding their breath until they turn blue and pass out from hypoxia, but once they are unconscious, normal breathing will automatically resume.

Breathing is intimately linked to cardiovascular function. The integrating centers for both functions are located in the brain stem, and interneurons project between the two centers, allowing signaling back and forth. In Chapter 19, we will examine the integration of cardiovascular, respiratory, and renal function as the three systems work together to maintain fluid and acid-base homeostasis. But first, in Chapter 18, we examine the physiology of the kidneys.

PROBLEM CONCLUSION

The emergency room physician examines Edna and concludes that her dizziness and falling are the result of a middle ear infection and are not related to her COPD. Edna is given antibiotics and sent home.

In this running problem, you learned about chronic obstructive pulmonary disease. Further check your understanding of this running problem by checking your answers against those in the summary table.

	Question	Facts	Integration and Analysis
1	What does narrowing of the airways do to the resistance of the airways to air flow?	The relationship between radius and resistance is the same for air flow as it was for blood flow in the circulatory system: as radius decreases, resistance increases [∞ p. 409].	When resistance increases, the body must use more energy to create air flow or blood flow.
2	Why do people with chronic bronchitis have a higher-than-normal rate of respiratory infections?	Cigarette smoke paralyzes the cilia that sweep debris and mucus out of the airways. Without the action of cilia, mucus and trapped particles pool in the airways.	Because of the lack of the ciliary "housekeeping," bacteria can become established in the lower respiratory tract more easily.
3	Name the muscles that patients with emphysema use to exhale forcefully.	Normal expiration depends on elastic recoil of muscles and elastic tissue in the lungs.	Forceful expiration uses the internal intercostal muscles and abdominal muscles.
4	Why does Edna have an increased hematocrit?	The body is heavily dependent on the oxygen carried by hemoglobin in red blood cells. Because of Edna's COPD, her arterial P_{O_2} is low. The major stimulus for red blood cell synthesis is hypoxia.	Low arterial oxygen levels will trigger the synthesis of additional red blood cells. This provides more binding sites for oxygen transport.
5	Without performing blood tests, why does the paramedic suspect that Edna has severe hypoxia and hypercapnia (abnormally high levels of P_{CO_2})?	Abnormally high P_{CO_2} or low P_{O_2} can depress the function of the central nervous system.	Edna's altered mental state suggests lack of oxygen to the brain or excessively high CO_2 levels.

CHAPTER REVIEW

SUMMARY

1. Aerobic metabolism in living cells consumes oxygen and produces carbon dioxide. (p. 498)

2. Gas exchange requires a large, thin, moist exchange surface, a pump to move air, and a circulatory system to transport gases to the cells. (p. 498)

3. Respiratory system functions include gas exchange, pH regulation, vocalization, and protection from foreign substances. (p. 498)

The Respiratory System

4. **Cellular respiration** refers to the metabolic processes of the cell that consume oxygen. **External respiration** is exchange of gases between the atmosphere and cells. It includes ventilation, gas exchange at the lung and cells, and transport of gases in the blood. **Ventilation** is the movement of air into and out of the lungs. (p. 498)

5. The **respiratory system** consists of anatomical structures involved in ventilation and gas exchange. (p. 498)

6. The **upper respiratory tract** includes the mouth, nasal cavity, **pharynx,** and **larynx.** The **lower respiratory tract** includes the **trachea, bronchi, bronchioles,** and exchange surfaces of the **alveoli.** (p. 499)

7. The thoracic cage is bounded by the ribs, spine, and **diaphragm.** Two sets of **intercostal muscles** connect the ribs. (p. 499)

8. The **lungs** are contained within a double-walled **pleural sac** that contains a small quantity of **pleural fluid.** (p. 499)

9. The two **primary bronchi** enter the lungs. Each primary bronchus divides into progressively smaller bronchi and finally into collapsible **bronchioles.** (p. 499)

10. The alveoli consist mostly of thin-walled **type I alveolar cells** for gas exchange. **Type II alveolar cells** produce surfactant. A network of capillaries surrounds each alveolus. (p. 502)

11. Blood flow through the lungs equals the cardiac output. Resistance to blood flow in the pulmonary circulation is low. Pulmonary arterial pressure averages 25/8 mm Hg. (p. 502)

Gas Laws

12. The total pressure of a mixture is the sum of the pressures of the individual gases in the mixture (**Dalton's law**). **Partial pressure** is the pressure contributed by a single gas such as oxygen. (p. 503)

13. Bulk flow of air occurs down pressure gradients, as does the movement of individual gas. (p. 504)

14. **Boyle's law** states that as volume increases, pressure decreases. The body creates pressure gradients by changing thoracic volume. (p. 504)

15. The amount of a gas that dissolves in a liquid is proportional to the partial pressure of the gas and to the **solubility** of the gas. Carbon dioxide is 20 times more soluble in body fluids than oxygen. (p. 505)

Ventilation

16. The upper respiratory system filters, warms, and humidifies inhaled air. (p. 506)

17. Air flow in the respiratory system is directly proportional to the pressure gradient and inversely related to the resistance of the airways. (p. 507)

18. A single **respiratory cycle** consists of an inspiration and an expiration. (p. 507)

19. During **inspiration, alveolar pressure** decreases and air flows into the lungs. Inspiration requires contraction of the inspiratory muscles and the diaphragm. (p. 507)

20. **Expiration** is usually passive, resulting from elastic recoil of the lungs. (p. 508)

21. **Active expiration** requires contraction of internal intercostal and abdominal muscles. (p. 509)

22. Intrapleural pressures are always subatmospheric because the pleural cavity is a sealed compartment. (p. 510)

23. **Compliance** is the ease with which the chest wall and lungs expand. Loss of compliance increases the work of breathing. **Elastance** is the ability of a stretched lung to resume its normal volume. (p. 511)

24. **Surfactant** decreases surface tension in the fluid lining the alveoli. This prevents smaller alveoli from collapsing into larger ones. (p. 511)

25. The diameter of the bronchioles determines resistance to air flow. (p. 513)

26. Increased CO_2 in expired air dilates bronchioles. Parasympathetic neurons cause **bronchoconstriction** in response to irritant stimuli. There is no significant sympathetic innervation of bronchioles, but epinephrine causes **bronchodilation.** (p. 513)

27. **Tidal volume** is the amount of air taken in during a single normal inspiration. **Vital capacity** is tidal volume plus **expiratory** and **inspiratory reserve volumes.** The air left in the lungs at the end of maximum expiration is the **residual volume.** (p. 515)

28. **Total pulmonary ventilation** = tidal volume $\times$ ventilation rate. **Alveolar ventilation** = ventilation rate $\times$ (tidal volume − dead space volume). (p. 516)

29. Alveolar gas composition changes very little during a normal respiratory cycle. **Hyperventilation** increases alveolar P_{O_2} and decreases alveolar P_{CO_2}. **Hypoventilation** has the opposite effect. (p. 517)

30. Air flow is matched to blood flow around the alveoli by local mechanisms. CO_2 dilates bronchioles, and decreased O_2 constricts pulmonary arterioles. (p. 519)

Gas Exchange in the Lungs

31. Oxygen and carbon dioxide diffuse into and out of the blood. The composition of inspired air and effectiveness of alveolar ventilation affect partial pressure gradients at the alveoli. (p. 520)

32. Changes in alveolar surface area and membrane thickness, and interstitial distance also affect gas exchange. (p. 521)

33. Normal alveolar and arterial P_{O_2} is 100 mm Hg. Normal P_{CO_2} is 40 mm Hg. (p. 521)

Gas Exchange in the Tissues

34. Gas exchange in the tissues depends on pressure gradients between the blood and cells. Venous blood has a P_{O_2} of 40 mm Hg and P_{CO_2} of 46 mm Hg. (p. 522)

Gas Transport in the Blood

35. Oxygen and carbon dioxide first dissolve in plasma. Their solubility is limited, so red blood cells assist with gas transport. (p. 522)

36. More than 98% of blood oxygen is bound to hemoglobin. (p. 522)

37. The P_{O_2} of the plasma determines how much oxygen will bind to hemoglobin. (p. 522)

38. Oxygen-hemoglobin binding is affected by pH, temperature, and **2,3-diphosphoglycerate (2,3-DPG).** (p. 525)

39. Venous blood carries 7% of its carbon dioxide dissolved in plasma, 23% as **carbaminohemoglobin,** and 70% as bicarbonate ion in the plasma. (p. 527)

40. Carbon dioxide is converted to carbonic acid by **carbonic anhydrase** inside red blood cells. Carbonic acid dissociates into H^+ and HCO_3^-. H^+ ions bind to hemoglobin. HCO_3^- leaves the red blood cell in exchange for Cl^- (the **chloride shift**). Plasma HCO_3^- acts as a buffer. (p. 528)

Regulation of Ventilation

41. Respiratory control resides in a **central pattern generator,** a network of neurons in the pons and medulla oblongata. (p. 529)

42. The medullary **dorsal respiratory group** contains inspiratory neurons that control somatic motor neurons to the inspiratory muscles and diaphragm. The **ventral respiratory group** of neurons assists in greater-than-normal inspiration and active expiration. (p. 531)

43. Carbon dioxide is the primary stimulus for changes in ventilation. **Central chemoreceptors** in the medulla respond to changes in P_{CO_2}. (p. 532)

44. **Peripheral chemoreceptors** in the carotid and aortic bodies monitor P_{O_2}, P_{CO_2}, and pH in the blood. Ventilation increases with a decrease in pH, increase in P_{CO_2}, or if P_{O_2} falls below 60 mm Hg. (p. 532)

45. Protective reflexes monitored by peripheral mechanoreceptors prevent injury to the lungs from overinflation or irritants. (p. 535)

46. Conscious and unconscious thought processes can affect respiratory activity. (p. 535)

QUESTIONS

LEVEL ONE   Reviewing Facts and Terms

1. List four functions of the respiratory system.

2. Give two different definitions for the word *respiration.*

3. Which sets of muscles are used for normal quiet inspiration? For normal quiet expiration? For active expiration?

4. What is the function of pleural fluid?

5. Name the anatomical structures that an oxygen molecule passes on its way from the atmosphere to the blood.

6. Diagram the structure of an alveolus and give the function of each part. How are capillaries associated with an alveolus?

7. Trace the path of the pulmonary circulation. About how much blood is found here at any given moment? What is a typical arterial blood pressure for the pulmonary circuit and how does this compare to that of the systemic circulation?

8. List three factors that influence the movement of gas molecules from air into solution. Which of these factors is usually not significant in humans?

9. What happens to inspired air as it is conditioned during its passage through the airways?

10. During inspiration, most of the volume change of the chest is due to movement of the _____.

11. Describe the changes in alveolar and intrapleural pressure during one respiratory cycle.

12. What is the function of surfactants in general? In the respiratory system?

13. Of the three factors that contribute to the resistance of air flow through a tube, which plays the largest role in changing resistance in the human respiratory system?

14. Match the following items with their correct effect on the bronchioles:

    (a) histamine   1. bronchoconstriction
    (b) epinephrine   2. bronchodilation
    (c) acetylcholine   3. no effect
    (d) increased P_{CO_2}

15. Define these four terms and explain how they relate to one another: tidal volume, inspiratory reserve volume, residual volume, expiratory reserve volume. Estimate the value for each volume in yourself.

16. More than _____% of the oxygen in arterial blood is transported bound to hemoglobin. How is the remaining oxygen transported to the cells?

17. Name four factors that influence the amount of oxygen that binds to hemoglobin. Which of these four factors is the most important?

18. Describe the structure of a hemoglobin molecule. What element is essential for hemoglobin synthesis?

19. The centers for control of ventilation are found in the _____ and _____ of the brain. What do the dorsal and ventral respiratory groups of neurons control? What is a central pattern generator?

20. Name the chemoreceptors that influence ventilation and explain how they do so. What chemical is the most important controller of ventilation?

21. Describe the protective reflexes of the respiratory system. What does the Hering-Breuer reflex prevent? How is it initiated?

LEVEL TWO   Reviewing Concepts

22. **Concept map:** Construct a map of gas transport using the following terms. You may add additional terms.

alveoli	hemoglobin saturation
arterial blood	oxyhemoglobin
carbaminohemoglobin	P_{CO_2}
carbonic anhydrase	plasma
chloride shift	P_{O_2}
dissolved CO_2	pressure gradient
dissolved O_2	red blood cell
hemoglobin	venous blood

23. Compare and contrast the following sets of concepts:

    (a) compliance and elastance
    (b) inspiration and expiration
    (c) intrapleural and alveolar pressures
    (d) total pulmonary ventilation and alveolar ventilation
    (e) transport of oxygen and carbon dioxide in arterial blood

24. For each of the following parameters, decide if it will increase, decrease, or not change in the situations given.

    (a) airway resistance with bronchodilation
    (b) intrapleural pressure during inspiration

    (c) air flow with bronchoconstriction
    (d) HbO_2 binding with decreased pH
    (e) bronchiolar diameter with increased P_{CO_2}
    (f) tidal volume with decreased compliance
    (g) alveolar pressure during expiration

25. Define the following terms: pneumothorax, spirometer, hypoxia, COPD, auscultation, hypoventilation, hypercapnia, bronchoconstriction, minute volume.

26. The cartoon coyote is blowing up a balloon as part of an attempt to once more catch a roadrunner. He first breathes in as much air as he can, then blows out all that he can into the balloon.

    (a) The volume of air in the balloon is equal to the _____ _____ of the coyote's lungs. This volume can be measured directly, as above, or by adding what respiratory volumes together?

    (b) In 10 years, when the coyote is still chasing the roadrunner, will he still be able to put as much air into the balloon in one breath? Explain.

LEVEL THREE   Problem Solving

27. Marco tries to hide at the bottom of a swimming hole by breathing in and out through a garden hose that greatly increases his dead space. What happens to the following parameters in his arterial blood, and why?

    (a) P_{CO_2} (b) P_{O_2} (c) bicarbonate ion   (d) pH

28. A hospitalized patient with severe chronic obstructive lung disease has a P_{CO_2} of 55 mm Hg and a P_{O_2} of 50 mm Hg. To elevate his blood oxygen, he is given pure oxygen through a nasal tube. The patient immediately stops breathing. Explain why this might occur.

29. You are a physiologist on the manned space flight to a distant planet. You find intelligent humanoid creatures inhabiting the planet, and they willingly submit to your tests. Some of the data you have collected are described below. The first graph shows the oxygen dissociation curve for the oxygen-carrying pigment in the blood of the humanoid named Bzork. Bzork's normal alveolar P_{O_2} is 85 mm Hg. His normal cell P_{O_2} is 20 mm Hg but it drops to 10 mm Hg with exercise.

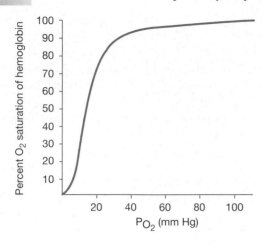

(a) What is the percent saturation for Bzork's oxygen-carrying pigment in blood at the alveoli? At an exercising cell?

(b) What conclusions can you draw about Bzork's oxygen requirement during normal activity and during exercise, based on the graph?

The next experiment on Bzork involves his ventilatory re-

sponse to different conditions. The data from that experiment are graphed below. Line A represents ventilation with an alveolar P_{O_2} of 50 mm Hg. Line B represents ventilation with an alveolar P_{O_2} of 85 mm Hg. Line C represents ventilation with an alveolar P_{O_2} of 85 mm Hg after Bzork has been given beer to drink. Interpret the results of experiments A and C.

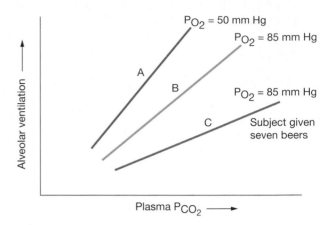

LEVEL FOUR   Quantitative Problems

30. A container of gas with a movable piston has a volume of 500 mL and a pressure of 60 mm Hg. The piston is moved, and the new pressure is measured as 150 mm Hg. What is the new volume of the container?

31. You have a mixture of gases in dry air, with an atmospheric pressure of 760 mm Hg. Calculate the partial pressure of each gas for each of the examples below:

    (a) 21% oxygen, 78% nitrogen, 0.3% carbon dioxide
    (b) 40% oxygen, 13% nitrogen, 45% carbon dioxide, 2% hydrogen
    (c) 10% oxygen, 15% nitrogen, 1% argon, 25% carbon dioxide

32. Li is a tiny woman, with a tidal volume of 400 mL and a respiratory rate of 12 breaths per minute at rest. What is her total pulmonary ventilation? Just before a physiology exam, her ventilation increases to 18 breaths per minute from nervousness. Now what is her total pulmonary ventilation? Assuming her dead space is 120 mL, what is her alveolar ventilation in each case?

33. You collected the following data on your classmate Neelesh.

    Minute volume = 5000 mL/min

    3 breaths/15 seconds

    Vital capacity 4800 mL

    Expiratory reserve volume = 1000 mL

    What are Neelesh's tidal volume and inspiratory reserve volume?

34. You are given the following information on a patient.

    Blood volume = 5.2 liters. Hematocrit = 47%. Hemoglobin concentration = 12 gm/dL whole blood

    Total amount of oxygen carried in this person's blood = 1015 mL oxygen

    Arterial plasma P_{O_2} = 100 mm Hg.

    You know that at a P_{O_2} of 100 mm Hg, plasma contains 0.3 mL oxygen/dL and that hemoglobin is 98% saturated. Each hemoglobin molecule can bind to a maximum of four molecules of oxygen. Using the information above, calculate the maximum oxygen-carrying capacity of hemoglobin (when hemoglobin is 100% saturated). Units will be mL oxygen/gm hemoglobin.

E X P L < MediaLab >

Introduction

The book, *Into Thin Air* by Jon Krakauer, chronicles an ill-fated trek to the top of Mt. Everest. To reach the summit of Mt. Everest, climbers just pass through the "death zone" located about 8000 meters. Besides the wind and subzero temperatures, why do you suppose this is termed the death zone? In the Media-Lab for this chapter, you will learn why climbing Mt. Everest is such a challenge for the human body and why people with emphysema experience the same respiratory challenges at sea level. After reading the description below, visit the MediaLab for Chapter 17 in your Companion Website and select the appropriate keyword.

Web Exploration 1

Estimated time for completion = 10 minutes

What are the physiological challenges of climbing Mt. Everest (29,028 ft)? Recall the last time you flew in a plane. Just before take-off, the flight attendant informs you that if there is a loss of cabin pressure, the oxygen masks will drop from the overhead compartment and you must put your mask on immediately. At what altitude do planes cruise on long flights? About 30,000–33,000 feet, just a little higher that the summit of Mt. Everest.

Select the keyword **EVEREST** from the website to learn more about the atmosphere at that altitude. To help you understand the implications for surviving at this high altitude, consider the following questions. What is the percentage of oxygen in the air on Mt. Everest? What is the barometric pressure at the top of Mt. Everest? What is the driving force for oxygen transfer from the alveoli into the blood and onto hemoglobin? How will the conditions on Mt. Everest affect this driving force? What are the consequences to the body? How does the body try to adapt to these new environmental conditions? What happens to respiratory rate? What is stimulating this change

in respiratory rate? Over a period of days, what happens to the hematocrit? How is this adaptive? How long does it take for the body to acclimatize? What is the usual period of time a climber must spend at each campsite?

Web Exploration 2

Estimated time for completion = 15 minutes

The running problem accompanying this chapter deals with a case of emphysema. You learned that one change in emphysema is a loss of elastin in the lung tissue. But there are other changes in lung anatomy associated with this disease. Select the keyword **EMPHYSEMA 1** from the website. Look at the first picture that appears. What gross anatomic changes do you observe?

Now select **EMPHYSEMA 2** from the website. Select slide #9. This is a histological section from a patient with emphysema. What does this slide show? What is the consequence of this change in alveolar tissue? What determines the amount of oxygen that diffuses across the alveoli? How is this affected by emphysema? Could this explain why Edna has a high hematocrit?

Based on your mental model of ventilation, predict how emphysema will alter Edna's lung volumes and capacities. Begin by looking at a normal spirogram found in your book (Figure 17-14, p. 515). Draw out a new spirogram for Edna.

Now return to the Emphysema 1 website and compare your prediction to the actual values. Which volumes have changed? What accounts for this change? How has the compliance of Edna's lungs changed? Draw a graph comparing the compliance of a normal lung and the compliance of Edna's lung. Select **COMPLIANCE** at the Website to see if your prediction was correct. How does this change in compliance affect a patient's ability to expire? Pink Puffers got this name because they often need to purse their lips to aid expiration. Why would exhaling through pursed lips help them?

18 The Kidneys

■ *"Plasma undergoes modification to urine in the nephron."* — *Arthur Grollman, in* Clinical Physiology:
The Functional Pathology of Disease, *1957.* ■

A section of latex-injected kidney showing blood vessels (red) and
a Bowman's capsule and renal tubule (yellow)

About 100 A.D. Areteaus the Cappadocian wrote,
"Diabetes is a wonderful affection, not very frequent among men, being a melting down of the
flesh and limbs into urine. . . . The patients never stop
making water [urinating], but the flow is incessant, as if
from the opening of aqueducts." Physicians have known
for centuries that *urine,* the fluid waste produced by the
kidneys, reflects the function of the body. To aid them
in their diagnosis of illness, they even carried special
flasks for collection and inspection of patients' urine.

The first step in examining a urine sample was to
determine its color. Was it dark yellow (concentrated),
pale straw (dilute), red (indicating the presence of blood),

PROBLEM
Gout

Michael Moustakakis, 43, had spent the last two days on the sofa from a throbbing, relentless pain in his left big toe. When the pain began, Michael thought he had a mild sprain or perhaps the beginnings of arthritis. But then the pain intensified, and the toe joint became hot and red. Finally Michael hobbled into his doctor's office, feeling a little silly about his problem. But upon hearing his symptoms, the doctor seemed to know instantly what was wrong. "Sounds to me like you have gout," said Dr. Garcia.

...continued on page 545

• •

or black (indicating the presence of hemoglobin metabolites)? One form of malaria was called *blackwater fever* because metabolized hemoglobin from abnormal breakdown of red blood cells turned victims' urine black or dark red.

Physicians also inspected urine samples for clarity, smell, taste, and froth (abnormal presence of proteins). Physicians who did not want to taste the urine themselves would allow their students the "privilege" of tasting it for them. A physician without students might expose insects to the urine and study their reaction.

Probably the most famous example of using urine for diagnosis is the taste test for diabetes mellitus, historically known as the honey-urine disease. Diabetes is an endocrine disorder characterized by the presence of glucose in the urine. The urine of diabetics tasted sweet and attracted insects, making the diagnosis clear.

Although today we have much more sophisticated tests for glucose in the urine, the first step of a *urinalysis* is still to examine the color, clarity, and odor of the urine. In this chapter, you will learn why we can tell so much about the body's function by what is present in the urine.

▶ FUNCTIONS OF THE KIDNEYS

If you asked people on the street, "What is the most important function of the kidney?" they would probably say, "The removal of wastes." Actually, the most important function of the kidney is the homeostatic regulation of the water and ion content of the blood, also called "salt and water balance" or "fluid and electrolyte balance." Waste removal is important, but disturbances in blood volume or ion levels will cause serious medical problems before the accumulation of metabolic wastes reaches toxic levels.

The kidneys maintain normal concentrations of ions and water by balancing intake of those substances with excretion in the urine, obeying the principle of mass balance [∞ p. 7]. We can divide kidney function into six general areas:

1. **Regulation of extracellular fluid volume.** When extracellular fluid volume decreases, blood pressure also decreases [∞ p. 451]. If volume and blood pressure fall too low, the body cannot maintain adequate blood flow to the brain and other essential organs. The kidneys work in an integrated fashion with the cardiovascular system to ensure that blood pressure and tissue perfusion remain within an acceptable range.

2. **Regulation of osmolarity.** The body integrates kidney function with behavioral drives such as thirst to maintain blood osmolarity at a value close to 290 mOsM. We will examine the reflex pathways for regulation of volume and osmolarity in Chapter 19.

3. **Maintenance of ion balance.** The kidneys keep concentrations of key ions within a normal range by balancing dietary intake with urinary loss. Sodium (Na^+) is the major ion involved in the regulation of extracellular fluid volume and osmolarity. Potassium (K^+) and calcium (Ca^{2+}) concentrations are also closely regulated.

   We will discuss renal control of sodium and potassium balance in Chapter 19 but defer the discussion of calcium until Chapter 21, when we look at all aspects of calcium homeostasis.

4. **Homeostatic regulation of pH.** The pH of plasma is normally kept within a narrow range [∞ p. 27]. If extracellular fluid becomes too acidic, the kidneys remove H^+ and conserve bicarbonate ions (HCO_3^-), which act as a buffer [∞ p. 28]. Conversely, when extracellular fluid becomes too alkaline, the kidneys remove HCO_3^- and conserve H^+. The kidneys do not correct pH disturbances as rapidly as the lungs do, but they play a significant role in pH homeostasis, as you will learn in the next chapter.

5. **Excretion of wastes and foreign substances.** The kidneys remove two types of wastes: those that are by-products of metabolism, and foreign substances such as drugs and environmental toxins. Metabolic wastes include creatinine from muscle metabolism [∞ p. 359] and the nitrogenous wastes *urea* [∞ p. 96] and *uric acid*. A metabolite of hemoglobin called *urobilinogen* gives urine its characteristic yellow color.

   Examples of foreign substances that are actively removed by the kidneys include the artificial sweetener *saccharin* and the anion *benzoate,* part of the preservative *potassium benzoate* that you ingest each time you drink a diet soft drink.

6. **Production of hormones.** Although the kidneys are not endocrine glands, they play important roles in three endocrine pathways. Kidney cells synthesize *erythropoietin*, the cytokine/hormone that regulates red blood cell synthesis [∞ p. 479]. They also release *renin*, an enzyme that regulates the production of hormones involved in sodium balance and blood pressure homeostasis. Renal enzymes help convert vitamin D_3 into a hormone that regulates Ca^{2+} balance.

The kidneys, like many organs, are supplied with a tremendous reserve capacity. By most estimates, you must lose nearly three-fourths of your kidney function before homeostasis begins to be affected. Many people function perfectly normally with only one kidney, including the 1-in-1000 people born with only one kidney (the mate fails to develop during gestation) or those people who donate a kidney for transplantation.

✔ Ion regulation is a key feature of kidney function. What happens to the resting membrane potential of a neuron if extracellular K^+ levels decrease? [∞ p. 234]

✔ What happens to the force of cardiac contraction if plasma Ca^{2+} levels decrease substantially? [∞ p. 417]

▶ ANATOMY OF THE URINARY SYSTEM

The **urinary system** [*ouron,* urine] is composed of the kidneys and accessory structures. The study of kidney function is called **renal physiology,** from the Latin word *renes,* meaning "kidneys."

The Urinary System Consists of Kidneys, Ureters, Bladder, and Urethra

Let's begin by following the route a drop of water would take on its way to excretion in the urine. In the first step of urine production, water and solutes move from plasma into hollow tubules (*nephrons)* that make up the bulk of the paired **kidneys** (Fig. 18-1a ■). These tubules modify the composition of the fluid as it passes through.

The modified fluid leaves the kidney and passes into a hollow tube called a **ureter.** There are two ureters, one leading from each kidney to the **urinary bladder.** The bladder expands and fills with urine until, by reflex action, it contracts and expels urine through a single tube, the **urethra.**

The urethra in males exits the body through the shaft of the penis. In females, the urethral opening is found anterior to the openings of the vagina and anus. *Micturition,* or urination, is the process by which urine is **excreted,** or removed from the body.

The kidneys are the site of urine formation. They lie on either side of the spine at the level of the eleventh and twelfth ribs, just above the waist (Fig. 18-1b ■). Although they are below the diaphragm, they are technically outside the abdominal cavity (**retroperitoneal**), sandwiched between the membranous **peritoneum,** which lines the abdomen, and the bones and muscles of the back.

The concave surface of each kidney faces the spine. The renal blood vessels, nerves, lymphatics, and ureters all emerge from this surface. **Renal arteries,** which branch off the abdominal aorta, supply blood to the kidneys. **Renal veins** carry blood from the kidneys back to the inferior vena cava.

At any given time, the kidneys receive 20%–25% of the cardiac output, even though they are only 0.4% of the total

---

Urinary Tract Infections Because of the shorter length of the female urethra and its proximity to bacteria from the large intestine, women are more prone than men to develop bacterial infections of the bladder and kidneys, or **urinary tract infections** (UTIs). The most common cause of UTIs is the bacterium *Escherichia coli,* a normal inhabitant of the human large intestine. *Escherichia coli* is not harmful while restricted to the lumen of the large intestine, but it is pathogenic [*patho-,* disease + *-genic,* causing] if it gets into the urethra. The most common symptoms of a UTI are pain or burning with urination and frequency of urination. A urine sample from a patient with a UTI will often contain many red and white blood cells, neither of which is commonly found in normal urine. UTIs are treated with antibiotics.

---

body weight (4.5–6 ounces each). This high rate of blood flow through the kidneys is critical to renal function.

The Nephron Is the Functional Unit of the Kidney

A cross section through a kidney shows that the interior is arranged in two layers: an outer **cortex** and inner **medulla** (Fig. 18-1c ■). The layers are formed by the organized arrangement of microscopic tubules called **nephrons** (Fig. 18-1i ■). About 80% of our nephrons are almost completely contained within the cortex (*cortical* nephrons) but the others dip down into the medulla (*juxtamedullary* nephrons).

The nephron is the functional unit of the kidney, with about 1 million nephrons in each kidney. (A *functional unit* is the smallest structure that can carry out all the functions of an organ.) Each nephron has both **vascular elements,** the blood vessels of the nephron, and **tubular elements.**

Vascular Elements of the Nephron Blood enters the kidney through the *renal artery,* flowing into smaller arteries and then into arterioles in the cortex (Fig. 18-1d, e ■). At this point, the arrangement of blood vessels becomes unique. Blood flows from an arteriole into a set of capillaries, then into a *second* arteriole and a second set of capillaries. This arrangement creates a *portal system,* one of three in the body [∞ p. 406]. Fluid filters out of the blood in the first set of capillaries, then re-enters the blood at the second set.

Arterial blood arriving at a nephron passes from the **afferent arteriole** into a ball-like network of capillaries known as the **glomerulus** [*glomus,* a ball-shaped mass; plural *glomeruli*] (Fig. 18-1f ■). All glomeruli are located in the cortex of the kidney. The glomerulus is the site where plasmalike fluid filters out of the capillaries and into the lumen of the tubule. This process is similar to the filtration of water and molecules out of systemic capillaries in other tissues, as you will see.

Blood leaving the glomerulus flows into an **efferent arteriole.** It then enters a series of **peritubular capillaries** [*peri-*, around] that surround the tubule (Fig. 18-1g ■). Long capillaries that dip into the medulla are part of the **vasa recta** (Fig. 18-1h ■). Fluid moves out of the tubule and back into the blood as it passes through the peritubular capillaries and vasa recta. Finally, renal capillaries join to form venules and small veins, exiting the kidney through the **renal vein.**

Tubular Elements of the Nephron   The tubular portion of the nephron begins with a hollow ball-like structure called **Bowman's capsule** that surrounds the glomerulus (Fig. 18-1j ■). The capillary endothelium of the glomerulus is fused to the epithelium of Bowman's capsule so that fluid filtering out of the capillaries passes directly into the lumen of the tubule. The combination of glomerulus and Bowman's capsule is called the **renal corpuscle.**

From Bowman's capsule, filtered fluid flows into the **proximal tubule** [*proximal,* close or near], then into the **loop of Henle,** a hairpin-shaped segment that dips down into the medulla and then back up to reenter the cortex. The loop of Henle is divided into two *limbs,* a **descending limb,** in which fluid flows into the medulla, and an **ascending limb,** in which fluid flows back to the cortex.

Fluid leaving the loop of Henle goes into the **distal tubule** [*distal,* distant or far]. Notice in Figure 18-1g ■ how the kidney tubule twists and folds back on itself so that the distal tubule passes between afferent and efferent arterioles. This region is known as the **juxtaglomerular apparatus** [*juxta-,* beside]. The proximity of tubule and arterioles allows paracrine communication between the structures, a key feature of kidney autoregulation.

Because the twisted configuration of the kidney tubule makes it difficult to follow fluid flow, we will unfold the nephron in the remaining figures in this chapter so that fluid flows from left to right across the figure.

The distal tubules of up to eight nephrons drain into a single larger tube called the **collecting duct.** (The distal tubules and collecting duct together form the **distal nephron.**) Collecting ducts pass from the cortex through the medulla and drain into the **renal pelvis.** From the renal pelvis, the filtered and modified fluid, now called **urine,** goes into the ureter on its way to excretion.

▶ OVERVIEW OF KIDNEY FUNCTION

Imagine drinking a 12-ounce soft drink every three minutes around the clock: by the end of 24 hours, you would have consumed the equivalent of 90 2-liter bottles. The thought of putting 180 liters into our intestinal tract is staggering, but that is how much plasma moves into the nephrons in a day! Yet the average volume of urine leaving the kidneys is only 1.5 L/day. More than 99% of the fluid that enters nephrons must find its way back into the blood, or the body would rapidly dehydrate.

The Three Processes of the Nephron Are Filtration, Reabsorption, and Secretion

There are three basic processes that take place in the nephron: filtration, reabsorption, and secretion (Fig. 18-2 ■, p. 548). **Filtration** is the movement of fluid from blood into the lumen of the nephron. Filtration takes place only in the renal corpuscle, where the walls of glomerular capillaries and Bowman's capsule are modified to allow bulk flow of fluid.

Once filtered fluid passes into the lumen of the nephron, it becomes part of the body's external environment, just as substances in the lumen of the intestinal tract are part of the external environment (∞ Fig. 1-2, p. 3). Thus, anything that filters into the nephron is destined for removal in the urine unless it is reabsorbed into the body.

After filtered fluid leaves Bowman's capsule, it is modified by the addition of solute and the removal of solutes and water. **Reabsorption** is the process of moving material from the lumen of the nephron back to the blood as it flows through peritubular capillaries.

Secretion removes selected molecules from the blood and adds them to the fluid in the lumen. Although secretion and filtration both move substances from blood into the lumen, secretion is a more selective process that usually uses membrane proteins to move molecules across the tubule epithelium.

Volume and Osmolarity Change as Fluid Flows Through the Nephron

Now let's follow some fluid through the nephron to learn what happens to it in various segments of the tubule. An average of 180 liters a day filters into Bowman's capsule (Table 18-1, p. 548). This filtrate is almost identical in composition to plasma and is nearly isosmotic, about 300 mOsM [∞ p. 137].

As the 180 liters of filtrate flow through the proximal tubule, about 70% of the volume is reabsorbed, leaving 54 liters. The proximal tubule cells transport solutes out of the lumen, and water follows by osmosis. Fluid leaving

---

...continued from page 543

Gout is a metabolic disease characterized by high blood concentrations of uric acid (*hyperuricemia*). If uric acid concentrations reach a critical level (7.5 mg/dL), crystals of monosodium urate form in peripheral joints, particularly in the feet, ankles, and knees. These crystals trigger an inflammatory reaction and cause periodic attacks of excruciating pain. Uric acid crystals may also form kidney stones in the renal pelvis.

Question 1: *Trace the route followed by kidney stones when they are excreted.*

546

ANATOMY SUMMARY The Urinary System

■ Figure 18-1

(a) The urinary system

Kidney

Ureter

Urinary
bladder

Urethra

**(e) Afferent arterioles and glomeruli
are all found in the cortex**

Afferent arterioles

Glomerulus

Arcuate
artery

Arcuate
vein

Cortical
nephron

Renal
artery

Renal
vein

**(d) Renal arteries take
blood to the cortex**

**(b) The kidneys are located retroperitoneally
at the level of the lower ribs.**

Diaphragm

Inferior
vena cava

Right
kidney

Aorta

Ureter

Peritoneum
(cut)

Rectum
(cut)

Left adrenal
gland

Left
kidney

Renal
artery

Renal
vein

Urinary
bladder

Nephrons

Cortex

Medulla

Renal pelvis

Ureter

Capsule

**(c) In cross section, the kidney is divided into an
outer cortex and an inner medulla. Urine leaving
the nephrons flows into the renal pelvis prior to
passing through the ureter into the bladder.**

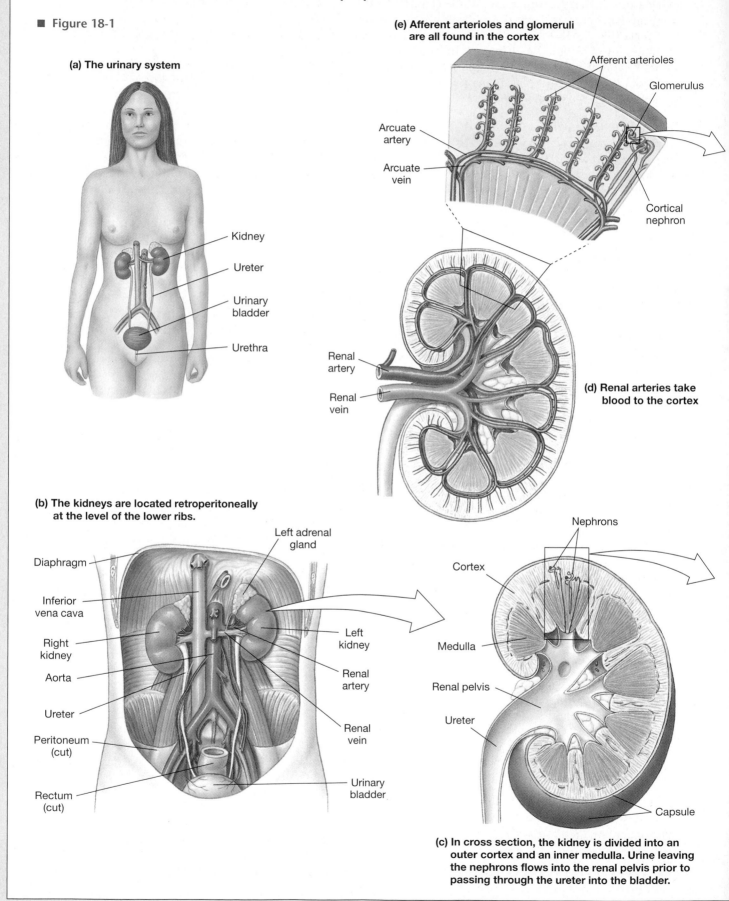

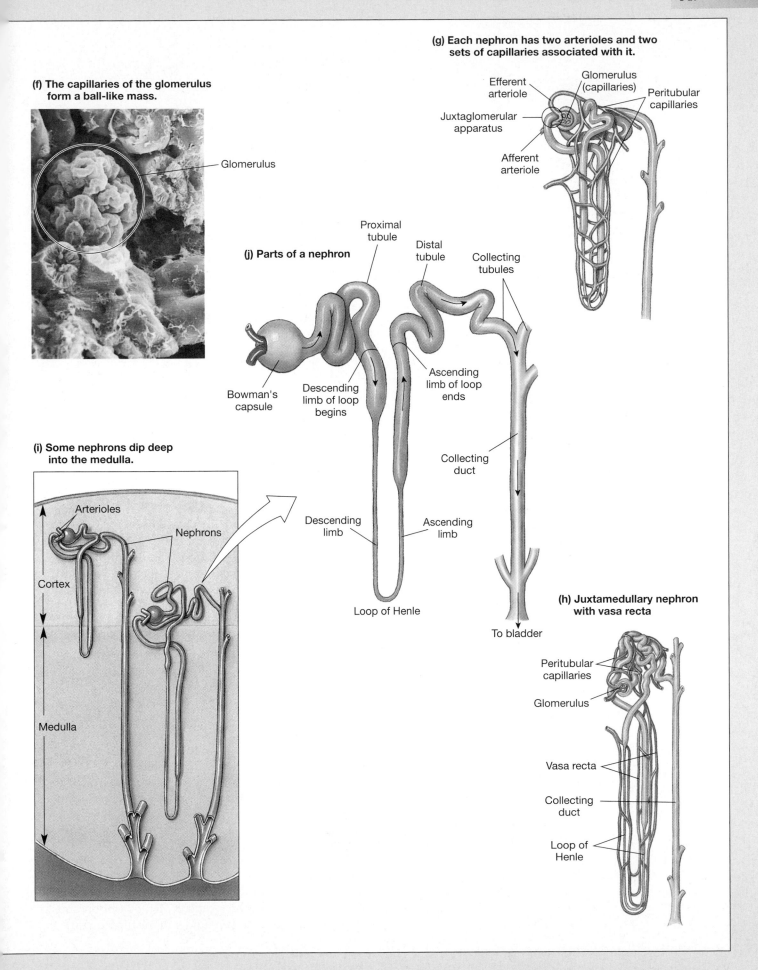

(f) The capillaries of the glomerulus form a ball-like mass.

Glomerulus

(g) Each nephron has two arterioles and two sets of capillaries associated with it.

Efferent arteriole

Glomerulus (capillaries)

Peritubular capillaries

Juxtaglomerular apparatus

Afferent arteriole

(j) Parts of a nephron

Proximal tubule

Distal tubule

Collecting tubules

Bowman's capsule

Descending limb of loop begins

Ascending limb of loop ends

Collecting duct

Descending limb

Ascending limb

Loop of Henle

To bladder

(i) Some nephrons dip deep into the medulla.

Arterioles

Nephrons

Cortex

Medulla

(h) Juxtamedullary nephron with vasa recta

Peritubular capillaries

Glomerulus

Vasa recta

Collecting duct

Loop of Henle

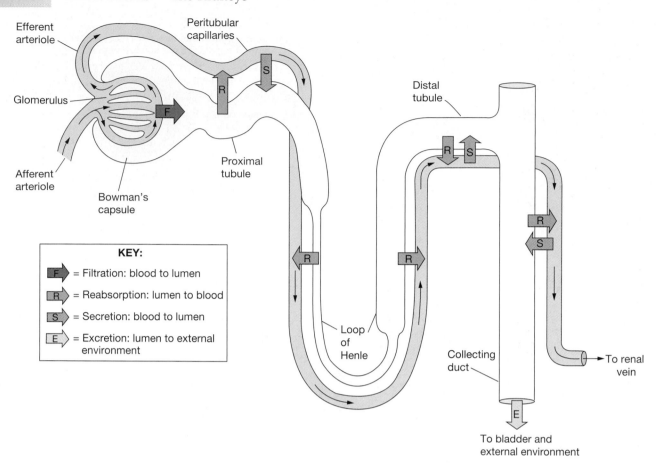

■ **Figure 18-2** **Filtration, reabsorption, secretion, and excretion**

the proximal tubule therefore has the same concentration as filtrate but its composition has changed. Thus, the primary function of the proximal tubule is the *bulk reabsorption of isosmotic fluid.*

Proximal tubule fluid then passes into the loop of Henle, the primary site for creation of dilute fluid. As the filtrate passes through the loop, proportionately more solute is reabsorbed than water. This process leaves hyposmotic fluid in the lumen. By the time filtrate leaves the loop, it averages 100 mOsM and its volume has fallen from 54 L/day to about 18 L/day. Now 90% of the original filtered volume has been reabsorbed.

From the loop of Henle, filtrate passes into the distal tubule and the collecting duct. In these two segments,

TABLE 18-1 Changes in Volume and Osmolarity Along the Nephron

Location in Nephron	Volume of Fluid	Osmolarity of Fluid
Bowman's capsule	180 L/day	300 mOsM
End of proximal tubule	54 L/day	300 mOsM
End of loop of Henle	18 L/day	100 mOsM
End of collecting duct (final urine)	1.5 L/day (average)	50–1200 mOsM

the fine regulation of salt and water balance takes place under the control of several hormones. Reabsorption and, to a lesser extent, secretion in the collecting duct determine the final composition and concentration of the fluid.

Once tubule fluid leaves the collecting duct, it can no longer be modified. In humans, the composition of the urine does not change during storage in the bladder.

By the end of the collecting duct, tubule fluid has a volume of 1.5 L/day and an osmolarity that can range from 50 to 1200 mOsM. Both the volume and osmolarity of urine depend on the body's need to conserve or excrete water and solute.

A word of caution here: it is very easy to confuse *secretion* with *excretion*. Try to remember the origins of the two prefixes. *Se-* means *apart*, as in to separate something from its source. In the nephron, secreted solutes are moved from plasma to tubule lumen.

Ex- means *out*, or *away*, as in out of or away from the body. Excretion refers to the removal of a substance from the body. Besides the kidneys, other organs that carry out excretory processes include the lungs (CO_2) and intestines (undigested food, bilirubin).

Figure 18-2 ■ summarizes filtration, reabsorption, secretion, and excretion. Filtration takes place in the renal corpuscle as fluid moves from the capillaries of the glomerulus into Bowman's capsule. Reabsorption and secretion occur along the remainder of the tubule, transferring material between lumen and peritubular capillaries. The quantity and composition of the substances being reabsorbed or secreted vary in different segments of the nephron. Fluid that remains in the lumen at the end of the nephron is excreted as urine.

The amount of any substance excreted in the urine reflects how it was handled during its passage through the nephron (Fig. 18-3 ■). The amount excreted is equal to the amount filtered into the tubule, minus the amount reabsorbed, plus any additional amount of the substance secreted into the lumen:

Amount excreted = amount filtered − amount reabsorbed + amount secreted

This equation is a useful way to think about renal handling of solutes. In the following sections, we will look in more detail at the important processes of filtration, reabsorption, secretion, and excretion.

✔ Name one way in which filtration and secretion are alike. Name one way in which they differ.

✔ A water molecule enters the renal corpuscle from the blood and ends up in the urine. Name all anatomical structures that the molecule passes through on its trip to the outside world.

✔ What would happen to the body if filtration continued at a normal rate but reabsorption dropped to one-half of normal?

▶ FILTRATION

The filtration of plasma into the kidney tubule is the first step in urine formation. This relatively nonspecific process creates a filtrate whose composition is like that of plasma minus the plasma proteins. Under normal conditions, blood cells are not filtered and they remain in the capillary. Filtered fluid is composed only of water and dissolved solutes.

The Renal Corpuscle Consists of the Glomerulus and Bowman's Capsule

Filtration takes place in the renal corpuscle, which consists of the glomerular capillaries surrounded by Bowman's capsule. Substances leaving the plasma must pass through three *filtration barriers* before reaching the tubule lumen: the glomerular capillary endothelium, the basal lamina (basement membrane), and the epithelium of Bowman's capsule (Fig. 18-4d ■).

The first barrier is the capillary endothelium. **Glomerular capillaries** are *fenestrated capillaries* [∞ p. 458] with large pores that allow most components of plasma to filter out of the blood (Fig. 18-4d ■). The pores are small enough, however, to prevent blood cells from leaving the capillary. Their negatively charged surface proteins also help repel negatively charged plasma proteins.

Glomerular **mesangial cells** lie between and around the glomerular capillaries (Fig. 18-4c ■). Mesangial cells have bundles of actinlike filaments in their cytoplasm that allow them to contract and alter blood flow through the capillaries. In addition, mesangial cells secrete cytokines associated with immune and inflammatory processes. Disruptions of mesangial cell function have been linked to several disease processes in the kidney.

The second filtration barrier is an acellular layer of extracellular matrix called the **basal lamina,** which separates the capillary endothelium from the epithelial lining of Bowman's capsule. The basal lamina consists of negatively charged glycoproteins and a collagen-like material that act like a coarse sieve, excluding most plasma proteins from the fluid that filters through it.

The third filtration barrier is the epithelium of Bowman's capsule. The portion of the capsule epithelium that surrounds each capillary consists of specialized cells called **podocytes** [*podos*, foot]. The podocytes have long, fingerlike cytoplasmic extensions that extend from the main cell body (Fig. 18-4a, b ■). Adjacent "fingers," or **foot processes,** wrap around the capillaries and interlace, leaving narrow filtration slits closed by a membrane.

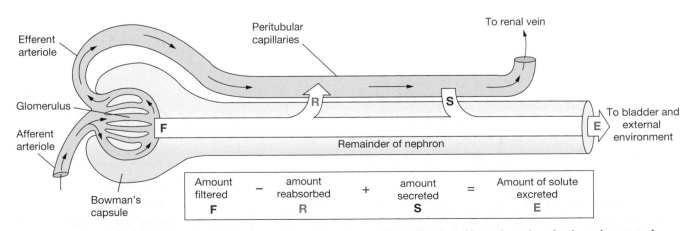

■ **Figure 18-3   The excretion of a substance depends on the amount that was filtered, reabsorbed, and secreted**

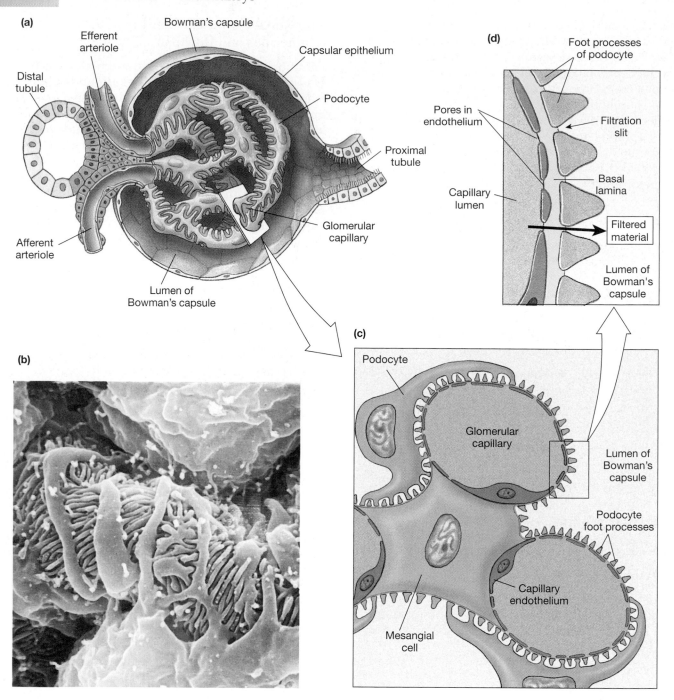

(a) Efferent arteriole / Bowman's capsule / Capsular epithelium / Distal tubule / Podocyte / Proximal tubule / Glomerular capillary / Afferent arteriole / Lumen of Bowman's capsule

(b)

(c) Podocyte / Glomerular capillary / Lumen of Bowman's capsule / Podocyte foot processes / Capillary endothelium / Mesangial cell

(d) Foot processes of podocyte / Pores in endothelium / Filtration slit / Basal lamina / Capillary lumen / Filtered material / Lumen of Bowman's capsule

■ **Figure 18-4   Structure of the renal corpuscle**   (a) The epithelium of Bowman's capsule that surrounds the glomerular capillaries is modified into special cells known as podocytes. (b, c) The foot processes of the podocytes surround each capillary, leaving slits through which filtration takes place. (d) Filtered substances must pass through pores in the capillary endothelium, the acellular basal lamina, and the filtration slits before reaching the lumen of Bowman's capsule.

Podocytes, like mesangial cells, contain contractile fibers. These are connected to the basal lamina by integrins.

When you visualize plasma filtering out of the glomerular capillaries, it is easy to imagine that all the plasma in the capillary moves into Bowman's capsule. However, total filtration of the plasma would leave behind a sludge of blood cells and proteins that could not flow out of the glomerulus. Normally, only about one-fifth of the plasma that flows through the kidneys at any one time filters into the nephrons. The remaining four-fifths of the plasma, along with the plasma proteins and

blood cells, flow into the peritubular capillaries (Fig. 18-5 ■). The percentage of total plasma volume that filters is called the **filtration fraction.**

Filtration Occurs Because of Hydrostatic Pressure in the Capillaries

What drives filtration across the walls of the glomerular capillaries? It is similar in many ways to filtration of fluid out of systemic capillaries [∞ p. 459]. The following forces influence glomerular filtration:

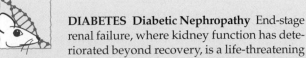

1. The *hydrostatic pressure of blood flowing through glomerular capillaries* forces fluid through the leaky endothelium. Capillary blood pressure averages 55 mm Hg and favors filtration. Although pressure declines along the length of the capillaries, it remains higher than the opposing pressures. Consequently, filtration takes place along nearly the entire length of glomerular capillaries.

2. The *colloid osmotic pressure inside glomerular capillaries* is higher than that of fluid within Bowman's capsule due to the presence of proteins in the plasma. The osmotic pressure gradient averages 30 mm Hg and favors fluid movement back into the capillaries.

3. Bowman's capsule is an enclosed space (unlike the interstitial fluid), so the presence of fluid within the capsule creates *hydrostatic fluid pressure* that opposes fluid movement into Bowman's capsule. Fluid filtering out of the capillaries must displace the fluid already in the lumen. Hydrostatic fluid pressure in the capsule averages 15 mm Hg opposing filtration.

The three pressures—capillary blood pressure, colloid osmotic pressure, and capsule fluid pressure—that influence glomerular filtration are summarized in Figure 18-6 ■. The net driving force is 10 mm Hg, favoring filtration. Although this pressure may not seem very high, when combined with the very leaky nature of the fenestrated capillaries, it results in the filtration of 180 L of fluid per day into the nephrons.

✔ The osmotic pressure of plasma in the efferent arterioles is higher than that in the afferent. Why?

Glomerular Filtration Rate Averages 180 Liters per Day

Filtration efficiency is described by the **glomerular filtration rate (GFR)**, the amount of fluid that filters into Bowman's capsule per unit time. Average GFR is 125 mL/min, or 180 L/day, an incredible rate considering that the total plasma volume is only about 3 liters. This rate means that the kidneys filter the entire plasma volume 60 times a day, or 2.5 times every hour. If most of the filtrate were not reabsorbed during its passage through the nephron, we would run out of plasma in only 24 minutes of filtration!

GFR is influenced by two factors: (1) the net filtration pressure just described and (2) the *filtration coefficient*. Filtration pressure is determined primarily by renal blood flow and pressure. The filtration coefficient has two components: surface area of the glomerular capillaries available for filtration, and permeability of the capillary-Bowman's capsule interface. In this respect, GFR

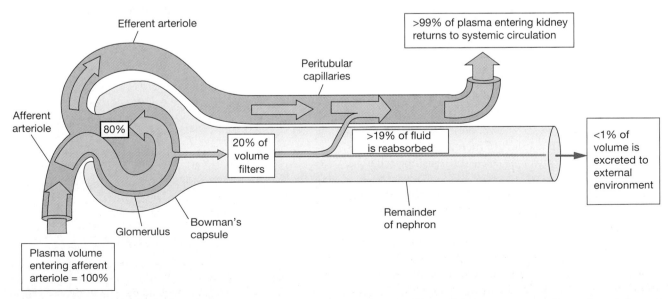

■ **Figure 18-5   The filtration fraction**   Only 20% of the plasma that passes through the glomerulus is filtered. Less than 1% of the filtered fluid is eventually excreted.

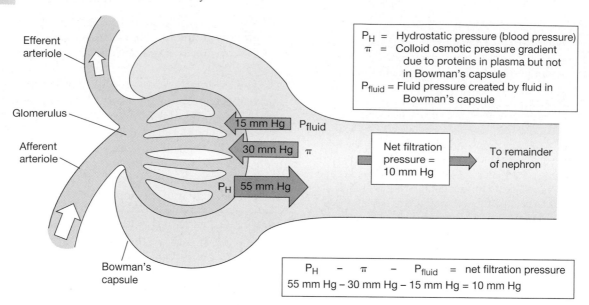

Figure 18-6 **Filtration pressure in the renal corpuscle**

is similar to gas exchange at the alveoli, where the rate of gas exchange depends on partial pressure differences, surface area of the alveoli, and permeability of the alveolar membrane.

Blood Pressure and Renal Blood Flow Influence GFR

Blood pressure provides the hydrostatic pressure that drives glomerular filtration. Therefore it would seem reasonable to assume that if blood pressure rises, GFR goes up, or that if blood pressure falls, GFR goes down. But that is not usually the case. GFR is remarkably constant over a wide range of blood pressures. As long as mean arterial blood pressure remains between 80 and 180 mm Hg, GFR averages 180 L/day (Fig. 18-7 ■).

✔ If average blood pressure is 120/80 mm Hg and mean blood pressure is diastolic pressure + 1/3 the pulse pressure, what is an average value for mean blood pressure?

The control of GFR is accomplished primarily by regulation of blood flow through the renal arterioles (Fig. 18-8 ■). If the resistance of the renal arterioles increases, renal blood flow decreases and blood is diverted to other organs [∞ p. 409]. The effect of increased resistance on GFR, however, depends upon *where* the resistance change takes place.

If resistance increases in the *afferent* arteriole (Fig. 18-8b ■), hydrostatic pressure decreases on the downstream side of the constriction. This translates into a decrease in GFR. If resistance increases in the *efferent* arteriole, blood "dams up" in front of the increased resistance and hydrostatic pressure in the glomerular capillaries increases (Fig. 18-8c ■). GFR goes up. The opposite occurs with decreased resistance in the affer-

ent or efferent arterioles. Most regulation occurs at the afferent arteriole.

GFR Is Subject to Autoregulation

Autoregulation of GFR is a local control process in which the kidney maintains a relatively constant GFR in the face of normal fluctuations in blood pressure. We do not completely understand the process, but several different mechanisms are at work. The **myogenic response** is the intrinsic ability of vascular smooth muscle to respond to pressure changes. **Tubuloglomerular feedback** is a paracrine signaling mechanism through which changes in fluid flow through the distal tubule influence GFR.

Myogenic Response The myogenic response of afferent arterioles is similar to autoregulation in other systemic arterioles. When smooth muscle in the arteriole wall stretches because of increased blood pressure,

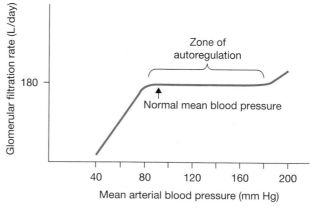

Figure 18-7 **Autoregulation of glomerular filtration rate** Autoregulation of glomerular filtration maintains a nearly constant GFR when mean arterial blood pressure is between 80 and 180 mm Hg.

(a) Renal blood flow and GFR change if resistance in the arterioles changes.

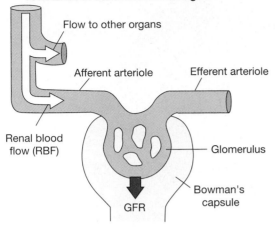

Flow to other organs

Afferent arteriole

Efferent arteriole

Renal blood flow (RBF)

Glomerulus

Bowman's capsule

GFR

(b) Vasoconstriction of the afferent arteriole increases resistance and decreases renal blood flow, capillary blood pressure (P$_H$), and GFR.

Flow diverted to other organs

↑Resistance ↓RBF

↓P$_H$

↓GFR

(c) Increased resistance of efferent arteriole decreases renal blood flow but increases P$_H$ and GFR.

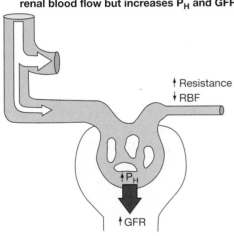

↑Resistance ↓RBF

↑P$_H$

↑GFR

(d)

> **Figure question:** What happens to capillary blood pressure, GFR, and RBF when the afferent arteriole dilates?

↓Resistance ?RBF

? P$_H$

? GFR

■ **Figure 18-8   Resistance changes in renal arterioles alter GFR and renal blood flow**

stretch-sensitive ion channels open, and the muscle cells depolarize, then contract [∞ p. 454]. Vasoconstriction increases resistance to flow, so blood flow through the arteriole diminishes. The decrease in blood flow decreases filtration pressure within the glomerulus.

If blood pressure decreases, you might expect the opposite response to occur. The tonic level of arteriolar contraction disappears, and the vessel becomes maximally dilated. However, vasodilation is not as effective at maintaining GFR as vasoconstriction because the afferent arteriole is normally fairly relaxed.

Consequently, if mean blood pressure drops below 80 mm Hg, GFR will decrease. This decrease is adaptive in the sense that if less plasma is filtered, the potential for fluid loss in the urine is decreased. In other words, a decrease in GFR helps the body conserve blood volume.

Tubuloglomerular Feedback   Tubuloglomerular feedback is a local control pathway in which fluid flow through the distal tubule influences GFR. The configuration of the nephron causes the distal tubule to pass be-

tween the afferent and efferent arterioles (Fig. 18-9 ■). The tubule and arteriolar walls are modified in the regions where they contact each other.

The modified portion of the distal tubule epithelium is called the **macula densa.** The adjacent walls of the arterioles have specialized smooth muscle cells called **juxtaglomerular cells (JG cells;** also known as granular cells). The JG cells secrete renin, an enzyme involved in salt and water balance. The complex of macula densa and JG cells is called the *juxtaglomerular apparatus.*

When fluid flow through the distal tubule increases as a result of increased GFR, the macula densa cells send a paracrine message to the neighboring afferent arteriole (Fig. 18-10 ■). The afferent arteriole constricts, increasing resistance and decreasing GFR.

The exact stimulus that signals increased tubule fluid flow is unclear. Experimental evidence indicates that some aspect of NaCl absorption at the macula densa is involved, because inhibitors of NaCl transport decrease the response. Other experiments show that there are several different paracrine signals, such as nitric

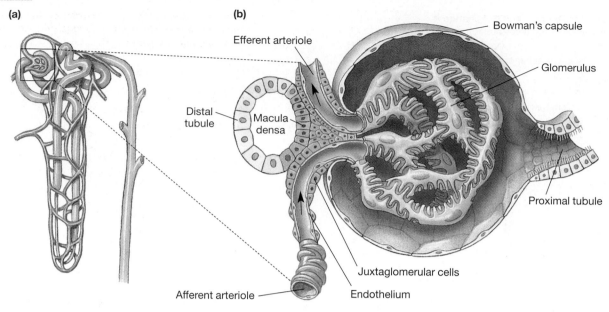

■ **Figure 18-9** **The juxtaglomerular apparatus** Macula densa cells in the distal tubule sense fluid flow through the tubule and release paracrines that affect juxtaglomerular and smooth muscle cells in the afferent arteriole.

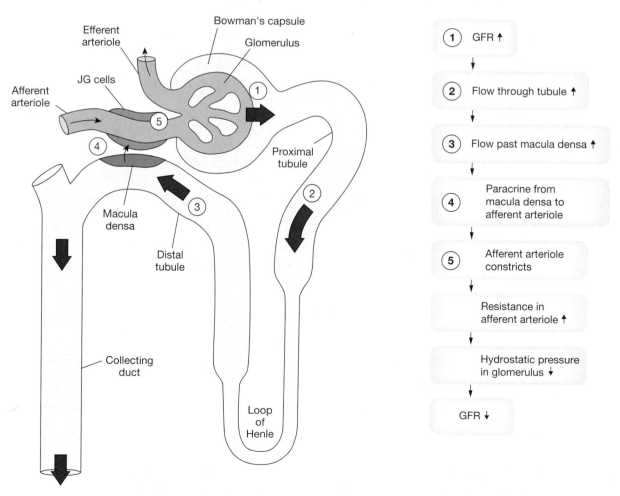

① GFR ↑

② Flow through tubule ↑

③ Flow past macula densa ↑

④ Paracrine from macula densa to afferent arteriole

⑤ Afferent arteriole constricts

Resistance in afferent arteriole ↑

Hydrostatic pressure in glomerulus ↓

GFR ↓

■ **Figure 18-10** **Tubuloglomerular feedback**

oxide and adenosine, that pass from the macula densa to the arteriole.

Hormones and Autonomic Neurons Also Influence GFR

Systemic signals such as hormones and the autonomic nervous system also affect GFR by changing resistance in the arterioles or by altering the filtration coefficient. Integrating centers outside the kidney initiate these signals, often in response to changes in systemic blood pressure or volume.

Both the afferent and efferent arterioles are innervated by sympathetic neurons. Norepinephrine binding to α receptors on vascular smooth muscle causes vasoconstriction. If sympathetic activity is moderate, there is little effect on GFR.

If systemic blood pressure drops sharply, as occurs with hemorrhage or severe dehydration, sympathetically induced vasoconstriction of the arterioles decreases GFR and renal blood flow. This is an adaptive response that helps conserve fluid volume.

A variety of hormones influence arteriolar resistance. Among the most important are angiotensin II, a potent vasoconstrictor, and prostaglandins, which act as vasodilators.

These same hormones may affect the filtration coefficient by acting on podocytes or mesangial cells. Podocytes change the size of filtration slits. If the slits widen, more surface area is available for filtration, and GFR increases. Contraction of mesangial cells apparently changes the glomerular capillary surface area available for filtration. We still have much to learn about these processes, and physiologists are actively investigating them.

✔ Systemic blood pressure remains constant but the afferent arteriole of a nephron constricts. What happens to renal blood flow and GFR in that nephron?

✔ A person with cirrhosis of the liver has lower-than-normal levels of plasma proteins and a higher-than-normal GFR. Explain why a decrease in plasma protein concentration would cause an increase in GFR.

▶ REABSORPTION

Each day, 180 liters of filtered fluid pass from the glomerular capillaries into the nephrons, yet only about 1.5 liters are excreted in the urine. More than 99% of what filters therefore must be reabsorbed as the fluid moves through the nephron. Most reabsorption takes place in the proximal tubule. Finely regulated reabsorption takes place in the distal segments of the nephron and allows the kidneys to return ions and water to the plasma selectively as needed to maintain homeostasis.

One question you might be asking is, "Why bother to filter 180 L/day and then reabsorb 99% of it? Why not simply filter and excrete the 1% that needs to be eliminated?" There are two reasons. First, many foreign substances are filtered into the nephron but not reabsorbed. The high daily filtration rate helps clear such substances from the plasma very rapidly.

Second, filtering ions and water into the tubule simplifies their regulation. If a portion of filtrate that reaches the distal nephron is not needed to maintain homeostasis, it passes into the urine. With a high GFR, this excretion can occur quite rapidly. However, if the ions and water are needed, they are reabsorbed.

Reabsorption May Be Active or Passive

Reabsorption of water and solutes from the tubule lumen to the extracellular fluid depends upon active transport. Filtrate moving out of Bowman's capsule into the proximal tubule has the same solute concentrations as extracellular fluid. Therefore, the tubule cells must use active transport to create concentration gradients. Water osmotically follows solutes as they are reabsorbed.

Most reabsorption involves *transepithelial transport,* in which substances cross both the apical and basolateral membranes of the tubule epithelial cell [∞ p. 131]. (A few substances can pass between the cells because this is a leaky epithelium, but we will ignore that route in this discussion.) The concentration gradient of a molecule determines the mechanism required to transport it across a membrane.

Molecules moving down their concentration (or electrochemical) gradient use open leak channels or facilitated diffusion carriers. Molecules that need to be pushed against their concentration gradient are moved by either primary or secondary active transport. Sodium is directly or indirectly involved in many instances.

Active Transport of Sodium Filtrate entering the proximal tubule is similar in ion composition to plasma, with a higher Na^+ concentration than is found within cells. Subsequently, Na^+ in the filtrate moves into the proximal tubule cell through open leak channels, moving down its electrochemical gradient (Fig. 18-11 ■). Once inside the tubule cell, Na^+ is actively transported into the extracellular fluid by the Na^+-K^+-ATPase on the basolateral membrane. The result is Na^+ reabsorption across the epithelium.

Secondary Active Transport: Symport with Sodium Sodium-linked secondary active transport in the nephron is responsible for the reabsorption of many substances, including glucose, amino acids, ions, and various organic metabolites. Figure 18-12 ■ shows Na^+-dependent glucose movement across the epithelium [∞ Fig. 5-28, p. 132].

The apical membrane contains a Na^+-glucose cotransporter that brings glucose into the cytoplasm against its concentration gradient by harnessing the energy of Na^+ moving down its electrochemical gradient.

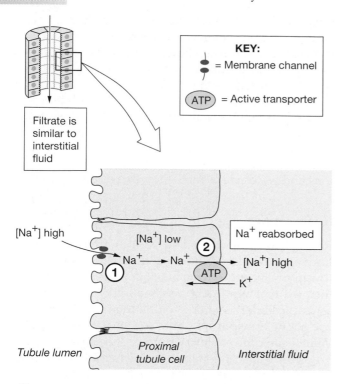

(1) Na⁺ enters cell through open channels, moving down its electrochemical gradient.

(2) Na⁺ is pumped out the basolateral side of cell by the Na⁺-K⁺-ATPase.

■ **Figure 18-11** **Sodium reabsorption in the proximal tubule**

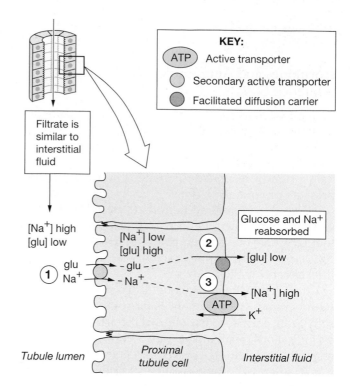

(1) Na⁺ moving down its gradient pulls glucose into the cell against its gradient.

(2) Glucose diffuses out basolateral side of cell.

(3) Na+ is pumped out by Na⁺-K⁺-ATPase.

■ **Figure 18-12** **Sodium-linked glucose reabsorption in the proximal tubule**

On the basolateral side of the cell, Na^+ is pumped out by the Na^+-K^+-ATPase, while glucose diffuses out with the aid of a facilitated diffusion carrier. The same basic pattern holds for the other molecules absorbed by Na^+-dependent transport: an apical symport protein and a basolateral facilitated diffusion carrier.

Passive Reabsorption: Urea Reabsorption

Urea has no active transporters in the proximal tubule, but it will move across the epithelium if there is a urea concentration gradient. Initially, urea concentrations in the filtrate and extracellular fluid are equal. However, the active transport of Na^+ and other solutes creates a urea concentration gradient by the following process.

When Na^+ and other solutes are reabsorbed from the lumen of the nephron, the transfer of osmotically active particles makes the extracellular fluid more concentrated than the fluid remaining in the lumen (Fig. 18-13 ■). In response to the osmotic gradient, water moves by osmosis across the epithelium. Up to this point, no urea molecules have moved out of the lumen because there has been no urea concentration gradient. Now, however, when water leaves the lumen, the *concentration* of urea goes up because the same amount of urea is contained within a smaller volume. Once a concentration gradient for urea exists, urea diffuses out of the lumen into the extracellular fluid.

Transcytosis: Plasma Proteins

Filtration of plasma at the glomerulus normally excludes most plasma proteins, but some smaller protein hormones and enzymes can pass through the filtration barrier. Most filtered proteins are reabsorbed in the proximal tubule. Normally, only trace amounts of protein appear in urine.

Proteins are too large to be reabsorbed by carriers or through channels. Instead they enter epithelial cells by

...*continued from page 545*

Uric acid, the molecule that causes gout, is a normal product of purine metabolism. Increased uric acid production may be associated with cell and tissue breakdown, or it may occur as a result of inherited enzyme defects. Uric acid in the blood filters freely into Bowman's capsule but is almost totally reabsorbed in the proximal tubule. Some uric acid that appears in the urine is secreted into the proximal tubule.

Question 2: *Why would uric acid levels in the blood go up when cell breakdown increases? (Hint: Purines are part of what biomolecules?)*

Question 3: *Based on what you have learned about uric acid, predict two different ways a person may develop hyperuricemia.*

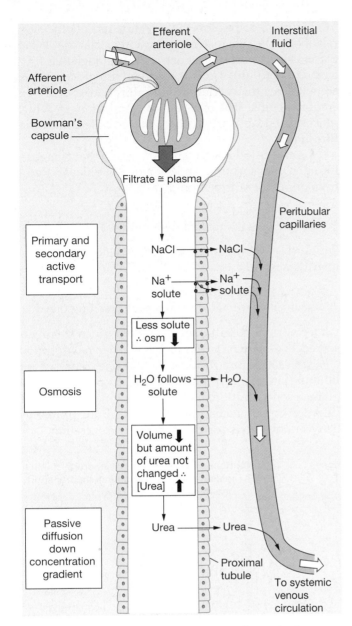

■ **Figure 18-13** **Passive reabsorption of urea in the prox-imal tubule**

endocytosis at the apical membrane. Once in the cell, they may be digested and released as amino acids or delivered intact to the extracellular fluid (transcytosis, [∞ p. 133]).

Saturation of Renal Transport Plays an Important Role in Kidney Function

Most transport in the nephron is mediated by membrane proteins and exhibits saturation, specificity, and competition, the three characteristics of mediated transport [∞ p. 122].

Saturation refers to the maximum rate of transport that occurs when all available carriers are occupied with substrate (i.e., have become saturated with substrate). At substrate concentrations below the saturation point, transport rate is directly related to substrate concentra-

tion (Fig. 18-14 ■). At substrate concentrations equal to or above the saturation point, transport occurs at a maximum rate. The transport rate at saturation is called the **transport maximum,** or **T$_m$.**

Glucose reabsorption in the nephron provides us with an excellent example of the consequences of saturation. At normal plasma glucose concentrations, all glucose that enters the nephron is reabsorbed before it reaches the end of the proximal tubule. The tubule epithelium is well supplied with carriers to capture the glucose as it flows past.

But what happens if blood glucose concentrations become excessive, as they do in the disease diabetes mellitus? In that case, glucose is filtered faster than the carriers can reabsorb it. The carriers become saturated and are unable to reabsorb all the glucose that flows through the tubule. As a result, some glucose escapes reabsorption and is excreted in the urine.

Consider the following analogy. Assume that the carriers are like seats on a train at Disney World. Instead of boarding the stationary train from a stationary platform, passengers step onto a moving sidewalk that rolls them past the train. As the passengers see an open seat, they grab it. But if more people are allowed onto the moving sidewalk than there are seats in the train, some

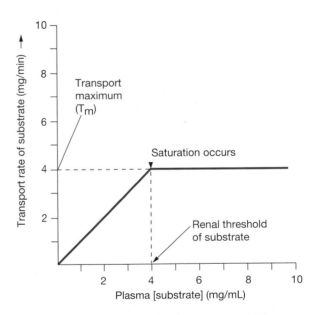

Graph questions: What is the transport rate at the following plasma substrate concentrations: 3 mg/mL, 5 mg/mL, 8 mg/mL? At what plasma substrate concentration is the transport rate 4 mg/min?

■ **Figure 18-14** **Saturation of mediated transport** The transport rate of a substance is proportional to the plasma concentration of the substance, up to the point at which transporters become saturated. Once saturation occurs, transport rate reaches a maximum, known as the transport maximum. The plasma concentration of substrate at which the transport maximum occurs is called the renal threshold.

people will not find seats. And because the sidewalk is moving these people past the train toward an exit, the extra people cannot wait for the next train. Instead, they are transported out the exit.

In the same fashion, glucose molecules entering the kidney tubule as part of the filtrate are like passengers stepping onto the moving sidewalk. In order to be reabsorbed, each glucose molecule must bind to a transporter as the filtrate flows through the proximal tubule. If only a few glucose molecules filter at a time, each one will find a free transporter and be reabsorbed, just as a small number of people on the moving sidewalk all find seats on the train. However, if glucose molecules filter into the tubule faster than the glucose carriers can transport them, some glucose remains in the lumen and is excreted in the urine.

Figure 18-15 ■ is a graphic representation of glucose handling by the kidney. Graph (a) shows that the filtration rate of glucose from plasma into Bowman's capsule is proportional to the plasma concentration of glucose. Because filtration does not exhibit saturation, the graph continues infinitely in a straight line. The concentrations of glucose in plasma and in filtrate are always equal.

Figure 18-15b ■ plots the reabsorption rate of glucose in the proximal tubule against the plasma concentration of glucose. Reabsorption exhibits a maximum transport rate (T_m) when the carriers reach saturation.

Figure 18-15c ■ plots the excretion rate of glucose in relation to the plasma concentration of glucose. When plasma glucose concentrations are low enough that 100% of the filtered glucose is reabsorbed, no glucose is excreted. Once the carriers reach saturation, glucose excretion begins. The plasma concentration at which glucose first appears in the urine is called the **renal threshold.**

Figure 18-15d ■ is a composite graph that compares filtration, reabsorption, and excretion of glucose. Recall from our earlier discussion that

Amount excreted =
amount filtered − amount reabsorbed + amount secreted

For glucose, which is not secreted, the equation can be rewritten as

Amount glucose excreted =
amount glucose filtered − amount glucose reabsorbed

Under normal conditions, all filtered glucose is reabsorbed. In other words, filtration is equal to reabsorption.

Notice in Figure 18-15d ■ that the lines representing filtration and reabsorption are identical up to the plasma glucose concentration that equals the renal threshold. If filtration equals reabsorption, the algebraic difference between the two is zero, and there is no excretion.

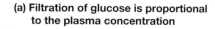

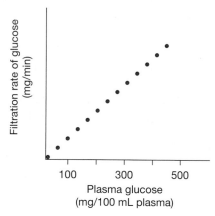

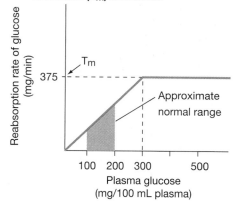

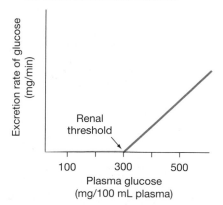

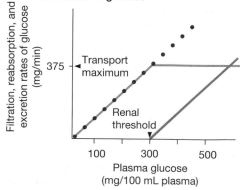

■ **Figure 18-15 Glucose handling by the nephron**

Once the renal threshold is reached, filtration begins to exceed reabsorption. Notice on the graph that the filtration and reabsorption lines diverge once the transport maximum for reabsorption is reached. The difference between the filtration line and the reabsorption line represents the excretion rate:

$$\text{Excretion} = \underset{\text{(increasing)}}{\text{filtration}} - \underset{\text{(constant)}}{\text{reabsorption}}$$

Excretion of glucose in the urine is called **glucosuria,** or **glycosuria** [-*uria*, in the urine], and usually indicates an elevated blood glucose concentration. Rarely, glucose appears in the urine even though the blood glucose concentrations are normal. This situation is due to a genetic disorder in which the tubule does not make enough carriers.

Peritubular Capillary Pressures Favor Reabsorption

The capillary reabsorption we have just discussed refers to the movement of solutes and water from the tubule lumen to the interstitial fluid. But how does that fluid get into the capillary? The answer is the low hydrostatic pressure along the entire length of the peritubular capillaries, which favors reabsorption.

Peritubular capillaries have an average hydrostatic pressure of 10 mm Hg, in contrast to the glomerular capillaries, where hydrostatic pressure averages 55 mm Hg. Colloid osmotic pressure, which favors movement of fluid into the capillaries, is 30 mm Hg. As a result, the pressure gradient in peritubular capillaries is 20 mm Hg, favoring the reabsorption of fluid. Fluid that is reabsorbed passes from the capillaries to the venous circulation and returns to the heart.

▶ SECRETION

Secretion is the transfer of molecules from the extracellular fluid into the lumen of the nephron. Secretion, like reabsorption, depends mostly on membrane transport systems. The secretion of K^+ and H^+ by the nephron is important in the homeostatic regulation of those ions (see Chapter 19). In addition, many organic compounds are secreted in the kidney. These compounds include both metabolites produced in the body and substances brought into the body.

Secretion enables the nephron to enhance excretion of a molecule. If a substance is filtered and not reabsorbed, it will be excreted very effectively. But if it is filtered, not reabsorbed, *and* the nephron removes more of it from the peritubular capillaries by secretion, excretion will be even more effective.

Secretion is an active process because it requires moving substrates against their concentration gradients. Most organic compounds are transported across the tubule epithelium by secondary active transport.

Competition Decreases Penicillin Secretion

An interesting and important example of an organic molecule secreted by the kidney tubule is the antibiotic **penicillin.** Many people today take antibiotics for granted, but until the early decades of the twentieth century, infections were a leading cause of death.

In 1928, Alexander Fleming discovered a substance in the *Penicillium* bread mold that retarded the growth of bacteria. But the antibiotic was difficult to isolate, so it did not become available for clinical use until the late 1930s. During World War II, penicillin made a major difference in the number of deaths and amputations caused by infected wounds. The only means of producing penicillin, however, was to isolate it from bread mold, and supplies were limited.

Demand for the drug was heightened by the fact that kidneys secrete penicillin. Renal secretion is so efficient at clearing foreign molecules from the blood that within three to four hours of administering a dose of penicillin, about 80% has been excreted in the urine. During the war, the drug was in such short supply that it was common procedure to collect the urine from patients being treated with penicillin so that the antibiotic could be isolated and reused.

This was not a satisfactory solution, however, so researchers looked for a way to slow penicillin secretion. They hoped to find a molecule that could compete with penicillin for the organic acid transporter responsible for secretion [∞ p. 122]. That way, when presented with both drugs, the carrier would bind preferentially to the competitor and secrete it, leaving penicillin behind in the blood. A synthetic compound named *probenecid* was the answer. When probenecid is administered concurrently with penicillin, the transporter removes probenecid preferentially, prolonging the activity of penicillin. Once mass-produced synthetic penicillin became available and supply was no longer a problem, the medical use of probenecid declined.

● ●

...continued from page 556

Michael finds it amazing that a metabolic problem could lead to pain in his big toe. "There's a way to treat gout, right?" he asks. Dr. Garcia explains that the treatment includes anti-inflammatory agents, lots of water, and avoidance of alcohol, which can trigger gout attacks. "In addition, I would like to put you on a *uricosuric agent*, like probenecid, that will enhance renal excretion of uric acid," replies Dr. Garcia. "By enhancing excretion, we can reduce uric acid levels in your blood and thus provide relief." Michael agrees to try these measures, and on his way home, stops off at a convenience store for a case of his favorite bottled water.

Question 4: *Uric acid is reabsorbed into the proximal tubule on a membrane transporter. Uricosuric agents are also organic acids. Knowing those two facts, explain how uricosuric agents might enhance excretion of uric acid.*

● ●

▶ EXCRETION

Urine output is the result of all the processes that take place in the kidney. By the time fluid reaches the end of the nephron, it bears little resemblance to the filtrate that started in Bowman's capsule. Glucose, amino acids, and useful metabolites are gone, having been reabsorbed into the blood. Organic wastes are more concentrated. The concentrations of ions and water in the urine are highly variable depending on the state of the body.

Although excretion tells us what the body is eliminating, excretion by itself cannot tell us the details of renal function. Recall that for any substance,

Amount excreted =
amount filtered − amount reabsorbed + amount secreted

Simply looking at the excretion rate of a substance tells us nothing about how the kidney handles that substance. The excretion rate of a substance depends on (1) its filtration rate and (2) whether the substance is reabsorbed or secreted as it passes through the nephron.

These three processes in the kidney are often of interest, however. For example, clinicians use information about a person's glomerular filtration rate as an indicator of overall kidney function. Pharmaceutical companies developing drugs must provide the Food and Drug Administration with complete information on how the human kidney handles each new compound.

But how can investigators dealing with living humans assess filtration, reabsorption, and secretion at the level of the individual nephron? They have no way to do this directly: the kidneys are not easily accessible, and the nephrons are microscopic. Scientists therefore had to develop a technique that would allow them to assess renal function using only analysis of the urine and blood. From their work came the concept of clearance.

Clearance Is a Noninvasive Way to Measure GFR

Clearance of a solute describes how many milliliters of plasma passing through the kidneys have been totally cleared of that solute in a given period of time. Before we jump into the mathematical expression of clearance, let's look at an example that shows how clearance relates to kidney function.

For our example, we will use **inulin,** a polysaccharide isolated from the tuberous roots of dahlia plants. (Careful! Inulin is not the same as *insulin*, the protein hormone that regulates glucose metabolism.) Scientists discovered from experiments with isolated nephrons that inulin injected into the plasma filters freely into the nephron. As it passes through the kidney tubule, inulin is neither reabsorbed nor secreted. Thus, 100% of the inulin that filters is excreted.

How does this relate to clearance? Let's take a look (Fig. 18-16 ■). Assume for the sake of this discussion that 100% of a filtered volume of plasma is reabsorbed. (This is not too far off the actual value, which is more than

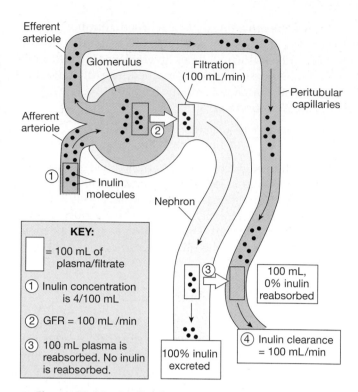

■ **Figure 18-16   Inulin clearance**

99%.) Inulin has been injected so that the plasma concentration is 4 inulin per 100 mL plasma.

If the GFR is 100 mL plasma filtered per minute, we can calculate the filtration rate, or *filtered load,* of inulin using the following equation:

Filtered load of a substance =
plasma concentration of the substance × GFR

Filtered load of inulin =
(4 inulin/100 mL plasma) × 100 mL plasma filtered/min

Filtered load of inulin = 4 inulin filtered/min

As inulin and the filtered plasma pass along the nephron, the plasma is reabsorbed but the inulin remains in the tubule. The reabsorbed plasma has no inulin in it, so it has been totally *cleared* of inulin. The *inulin clearance* therefore is 100 mL of plasma cleared/min. At the same time, the excretion rate of inulin is 4 inulin excreted per minute.

What good is this information? For one thing, we can use it to calculate the glomerular filtration rate. Notice from Figure 18-16 ■ that inulin clearance (100 mL plasma cleared/min) is equal to the GFR (100 mL plasma filtered/min).

For any substance (such as inulin) that is freely filtered but neither reabsorbed nor secreted, its clearance is equal to the GFR.

Now let's show mathematically that inulin clearance is equal to the GFR. We saw above that:

(1) Filtered load of inulin =
plasma concentration of inulin × GFR

But we know that 100% of the inulin that filters is excreted. In other words:

(2) Filtered load of inulin = excretion rate of inulin

We can substitute excretion rate for filtered load in equation (1) by using algebra (if A = B and A = C, then B = C):

(3) Excretion rate of inulin =
     plasma concentration of inulin × GFR

This equation can be rearranged to read:

$$(4) \quad GFR = \frac{\text{excretion rate of inulin}}{\text{plasma concentration of inulin}}$$

It turns out that the right side of the equation is identical to the clearance equation for inulin. The equation for clearance is:

(5) Clearance of a substance =
$$\frac{\text{urine excretion rate of the substance (mg/min)}}{\text{plasma concentration of the substance (mg/mL plasma)}}$$

For inulin:

$$(6) \quad \text{Clearance of inulin} = \frac{\text{excretion rate of inulin}}{\text{plasma concentration of inulin}}$$

By using algebra again:

(7) GFR = clearance of inulin

So why is this important? For one thing, you have just learned how we can measure GFR in a living human by taking only blood and a timed urine sample. Try the example in the concept check box to see if you understand.

Inulin is not practical for routine clinical applications because it does not occur naturally in the body and must be administered by continuous intravenous infusion. As a result, inulin use is restricted to research. Unfortunately, no substance that occurs naturally in the human body is handled by the kidney exactly like inulin.

In clinical settings, physicians use creatinine to estimate GFR. **Creatinine** is the breakdown product of phosphocreatine, an energy-storage compound found primarily in muscles. It is constantly produced by the body and does not need to be administered. Normally, the production and breakdown rates of phosphocreatine are relatively constant, and plasma concentration of creatinine does not vary much.

Although creatinine is always present in the plasma and is easy to measure, it is not the perfect molecule for estimating GFR because a small amount is secreted into the urine. However, the error is small enough that, in most people, *creatinine clearance* is routinely used to estimate GFR.

✔ If plasma creatinine = 1.8 mg/100 mL plasma, urine creatinine = 1.5 mg/mL urine, and urine volume is 1100 mL in 24 hours, what is the creatinine clearance? What is this person's GFR?

Knowing the GFR Helps Us Determine How the Kidney Handles a Solute

Once we know a person's GFR, we can determine how the kidney handles any solute by measuring the solute's plasma concentration and excretion rate. If we assume that the solute is freely filtered at the glomerulus, we know that:

Filtered load of a solute =
plasma concentration of the solute × GFR

If we compare the filtered load of the solute with its excretion rate, then we know how the nephron handled that substance (Table 18-2). For example, if less of the substance appears in the urine than was filtered, then there was net reabsorption (excreted = filtered − reabsorbed). If *more* of the substance appears in the urine than was filtered, there must have been net secretion of the substance into the lumen (excreted = filtered + secreted). If the same amount of the substance is filtered

…continued from page 559

Three weeks later, Michael is back in Dr. Garcia's office. The anti-inflammatory drugs and probenecid had eliminated the pain in his toe, but last night, he had gone to the hospital with a very painful kidney stone. "We'll have to wait until the analysis comes back," says Dr. Garcia, "But I will guess that it is a uric acid stone. Did you drink as much water as I told you to?" Sheepishly, Michael admits that he had good intentions, but could never find the time at work to get out his bottled water. "You have to drink enough water while on this drug to produce 3000 milliliters or more of urine a day. That's over three quarts. Otherwise, you may end up with another uric acid kidney stone." Michael agrees that this time, he will follow instructions to the letter.

Question 5: *Explain why not drinking enough water while taking uricosuric agents may cause uric acid stone formation in the urinary tract.*

TABLE 18-2   Renal Handling of Solutes

For any molecule X that is freely filtered at the glomerulus:	
If filtration rate is greater than excretion rate,	there is net reabsorption of X.
If excretion rate is greater than filtration rate,	there is net secretion of X.
If filtration and excretion rate are the same,	X passes through the nephron without net reabsorption or secretion.
If the clearance of X is less than inulin clearance,	there is net reabsorption of X.
If the clearance of X is equal to inulin clearance,	X is neither reabsorbed nor secreted.
If the clearance of X is greater than inulin clearance,	there is net secretion of X.

and excreted, then the substance is handled like inulin: neither reabsorbed nor secreted.

For example, suppose glucose is present in the plasma at 100 mg glucose/dL plasma, and the GFR is calculated from creatinine clearance to be 125 mL plasma/min:

Filtered load of glucose =
(100 mg glucose/100 mL plasma) $\times$ 125 mL plasma/min

Filtered load of glucose = 125 mg glucose/min

But there is no glucose present in the urine: glucose excretion rate is zero. Because glucose was filtered at a rate of 125 mg/min but excreted at a rate of 0 mg/min, it must have been totally reabsorbed.

Clearance Can Be Used to Determine Renal Handling of a Substance

Clearance values are also used to determine how the nephron handles a solute filtered into it. In this method, researchers calculate creatinine or inulin clearance, then compare the clearance of the solute being investigated with the creatinine or inulin clearance (Table 18-2). If the clearance of the solute is less than the inulin clearance, the solute has been reabsorbed.

On the other hand, if the clearance of the solute is higher than the inulin clearance, some solute has been secreted into the nephron. More plasma was cleared of the solute than was filtered, so the additional material must have been removed from the plasma by secretion.

Figure 18-17 ■ shows clearance of three different molecules: glucose, urea, and penicillin. These examples are similar to the earlier figure on inulin clearance (Fig. 18-16 ■), where inulin clearance was 100 mL plasma/min. All solutes have the same concentration, 4 molecules/100 mL plasma. GFR is 100 mL/min, and the entire 100 mL of plasma volume that filters is reabsorbed.

For any solute, its clearance reflects how the kidney tubule handles it. For example, 100% of the glucose that filters is reabsorbed and glucose clearance is zero (Fig. 18-17a ■). Urea is partially reabsorbed (Fig. 18-17b ■, point 3). Four molecules filter, but only two are reabsorbed. Consequently, urea clearance is 50 mL plasma cleared per minute. Urea and glucose clearance are both less than inulin clearance, which tells you that urea and glucose have been reabsorbed.

Penicillin (Fig. 18-17c ■) is filtered, not reabsorbed, and additional penicillin molecules are secreted from plasma in the peritubular capillaries. In this example, an extra 50 mL of plasma have been cleared of penicillin in addition to the original 100 mL that filtered. The penicillin clearance therefore is 150 mL plasma cleared per minute. Penicillin clearance is greater than inulin clearance, which tells you that there is net secretion of penicillin.

Note that a comparison of clearance values tells you only the *net* handling of the solute. It does not tell you if a molecule is both reabsorbed and secreted. For example, nearly all K^+ filtered is reabsorbed in the proximal tubule

and loop of Henle, then a small amount is secreted back into the lumen by the distal nephron. On the basis of K^+ clearance, it appears that there was only net reabsorption.

Clearance calculations are simple because all you need to know are urine excretion rates and their plasma concentrations, both easily obtained. If you also know either inulin or creatinine clearance, then you can determine the renal handling of any compound.

▶ MICTURITION

Once fluid leaves the collecting ducts, it flows into the renal pelvis then down the ureter to the bladder with the help of rhythmic smooth muscle contractions. The bladder is a hollow organ whose walls contain well-developed layers of smooth muscle. In the bladder, urine is stored until released in the process known as urination, voiding, or more formally, **micturition** [*micturire*, to desire to urinate].

The bladder is a hollow sac that can expand to hold a volume of about 500 mL. The neck of the bladder is continuous with the urethra, a single tube through which urine passes to reach the external environment. The opening between the bladder and urethra is closed by two rings of muscle called **sphincters** (Fig. 18-18a ■).

The *internal sphincter* is a continuation of the bladder wall and consists of smooth muscle. Its normal tone keeps it contracted. The *external sphincter* is a ring of skeletal muscle controlled by somatic motor neurons. Tonic stimulation from the central nervous system maintains contraction of the external sphincter except during urination.

Micturition is a simple spinal reflex that is subject to both conscious and unconscious control from higher brain centers. As the bladder fills with urine and its walls expand, stretch receptors send signals via sensory neurons to the spinal cord (Fig. 18-18b ■). There, the information is integrated and transferred to two different sets of neurons. The stimulus of a full bladder *excites* parasympathetic neurons leading to the smooth muscle in the bladder wall. The smooth muscle contracts, increasing the pressure on the bladder contents. Simultaneously, somatic motor neurons leading to the external sphincter are *inhibited*. Contraction of the bladder occurs in a wave that pushes urine downward toward the urethra. Pressure exerted by urine forces the internal sphincter open while the external sphincter relaxes. Urine passes into the urethra and out of the body, aided by gravity.

This simple micturition reflex occurs primarily in infants who have not yet been toilet trained. A person who has been toilet trained acquires a learned reflex that keeps the micturition reflex inhibited until the person consciously desires to urinate. The learned reflex involves additional sensory fibers in the bladder that signal the degree of fullness. Centers in the brain stem and cerebral cortex receive that information and override the basic micturition reflex by directly inhibiting the parasympathetic fibers and by reinforcing contraction of the exter-

(a) Glucose clearance

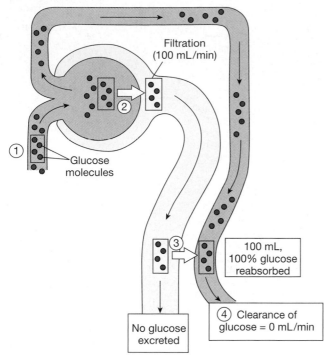

KEY:

☐ = 100 mL of plasma/filtrate

① Plasma concentration is 4/100 mL

② GFR = 100 mL /min

③ 100 mL plasma is reabsorbed.

④ Clearance depends on renal handling of solute

(b) Urea clearance

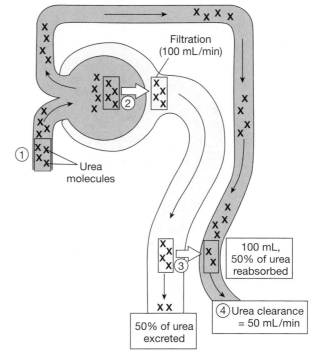

(c) Penicillin clearance

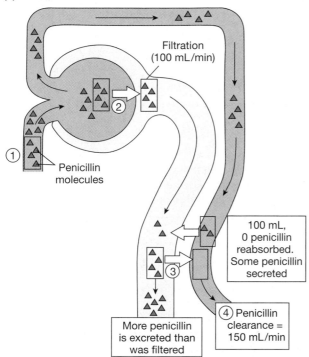

■ **Figure 18-17   The relationship between clearance and excretion**   The figure represents the events taking place in one minute. For simplicity, 100% of the filtered volume was reabsorbed.

nal sphincter. When an appropriate time to urinate arrives, those same centers remove the inhibition and facilitate the reflex by inhibiting the external sphincter.

In addition to conscious control of urination, various subconscious factors can affect the micturition reflex. "Bashful bladder" is a condition in which a person is un-

able to urinate in the presence of other people despite the conscious intent to do so. The sound of running water facilitates micturition and is often used to help patients urinate if the urethra is irritated from insertion of a *catheter*, a tube inserted into the bladder to drain it passively.

(a) Bladder at rest

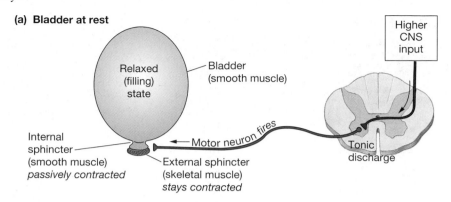

(b) Micturition

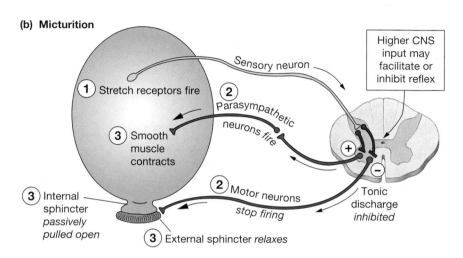

■ **Figure 18-18   The micturition reflex**

PROBLEM CONCLUSION

In this running problem, you learned that gout, which often presents as a debilitating pain in the big toe, is a metabolic problem whose cause and treatment may be linked to kidney function. Further check your understanding of this running problem by comparing your answers against those in the summary table.

	Question	Facts	Integration and Analysis
1	Trace the route followed by kidney stones when they are excreted.	Kidney stones often form in the renal pelvis.	From the renal pelvis, a stone passes down the ureter and into the urinary bladder. From there, it leaves the body through the urethra.
2	Why would uric acid levels in the blood go up when cell breakdown increases?	Purines include adenine and guanine, which are components of DNA, RNA, and ATP.	When a cell dies, the nucleus and other components are broken down into biomolecules. Degradation of DNA, RNA, and ATP would increase purine production, which in turn would increase uric acid.
3	Based on what you have learned about uric acid, predict two different ways a person may develop hyperuricemia.	Hyperuricemia is a condition of elevated uric acid levels in the blood. Uric acid is made from purines. It is filtered, reabsorbed, and secreted by the kidney tubule.	Hyperuricemia results from overproduction of uric acid or from failure of the kidneys to excrete uric acid. Because uric acid normally gets into the urine by secretion, a defect in the renal secretory mechanism would lead to hyperuricemia.

	Question	Facts	Integration and Analysis
4	Uric acid is reabsorbed in the proximal tubule on a membrane transporter. Uricosuric agents are organic acids. Knowing those two facts, explain how uricosuric agents might enhance excretion of uric acid.	Mediated transport exhibits competition, where related molecules compete for one transporter. Usually, one molecule will bind preferentially and therefore inhibit transport of the second molecule.	Uricosuric agents such as probenecid may compete with uric acid for the proximal tubule transporter. This would block uric acid access to the transporter, leaving it in the lumen and increasing its excretion.
5	Explain why not drinking enough water while taking uricosuric agents may cause uric acid stone formation in the urinary tract.	Uric acid crystals form when uric acid concentrations exceed a critical level.	If a person drinks large volumes of water, the excess water will be excreted by the kidneys. Large amounts of water will dilute the urine and thus prevent high concentrations of uric acid.

CHAPTER REVIEW

SUMMARY

Functions of the Kidneys

1. The kidneys regulate extracellular fluid volume and osmolarity, maintain ion balance, regulate pH, excrete wastes and foreign substances, and participate in endocrine pathways. (p. 543)

Anatomy of the Urinary System

2. The **urinary system** is composed of the kidneys, ureters, bladder, and urethra. (p. 544)

3. Each **kidney** has about 1 million **nephrons.** In cross-section, a kidney is arranged into an outer **cortex** and inner **medulla.** (p. 544)

4. Blood flow through a nephron goes from **afferent arteriole** to **glomerulus** to **efferent arteriole** to **peritubular capillaries.** The **vasa recta** capillaries dip into the medulla. (p. 544)

5. Fluid filters from the glomerulus into **Bowman's capsule.** From there, it flows through the **proximal tubule, loop of Henle, distal tubule,** and **collecting duct.** Fluid from collecting ducts drains into the **renal pelvis.** From there, **urine** goes through the **ureter** to the **urinary bladder.** (p. 545)

Overview of Kidney Function

6. **Filtration** is the movement of fluid from plasma into Bowman's capsule. **Reabsorption** is the movement of filtered material from nephron to the blood. **Secretion** is the movement of selected molecules from the blood into the nephron. (p. 545)

7. Average urine volume is 1.5 L/day. Osmolarity varies between 50 and 1200 mOsM. (p. 548)

8. The amount of a solute excreted equals the amount filtered minus the amount reabsorbed plus the amount secreted. (p. 549)

Filtration

9. Filtered solutes must pass through glomerular capillary endothelium, **basal lamina,** and Bowman's capsule epithelium before reaching the lumen of Bowman's capsule. (p. 549)

10. Filtration allows most components of plasma to enter the tubule but excludes blood cells and most plasma proteins. (p. 549)

11. The capsule epithelium has specialized cells called **podocytes** that wrap around the capillaries and create **filtration slits. Mesangial cells** are associated with glomerular capillaries. (p. 549)

12. One-fifth of renal plasma flow filters into the nephrons. The percentage of total plasma volume that filters is called the **filtration fraction.** (p. 550)

13. Hydrostatic pressure in glomerular capillaries averages 55 mm Hg, favoring filtration. Opposing filtration are colloid osmotic pressure of 30 mm Hg and hydrostatic capsule

fluid pressure averaging 15 mm Hg. The net driving force is 10 mm Hg, favoring filtration. (p. 551)

14. The **glomerular filtration rate (GFR)** is the amount of fluid that filters into Bowman's capsule per unit time. Average GFR is 125 mL/min, or 180 L/day. (p. 551)

15. Blood pressure in glomerular capillaries can be altered by changing resistance in renal arterioles. (p. 552)

16. Autoregulation of glomerular filtration is accomplished by a **myogenic response** of vascular smooth muscle to respond to pressure changes and by **tubuloglomerular feedback.** When fluid flow through the distal tubule increases, the **macula densa** cells send a paracrine signal to the afferent arteriole, which constricts. (p. 552)

17. Reflex control of GFR is mediated through systemic signals such as hormones and the autonomic nervous system. (p. 555)

Reabsorption

18. Most reabsorption takes place in the proximal tubule. Finely regulated reabsorption takes place in the later segments of the nephron. (p. 555)

19. Reabsorption of water and solutes depends upon active transport of solutes from the lumen to the extracellular fluid. Most reabsorption involves transepithelial transport. (p. 555)

20. Na^+ in the lumen moves into tubule cells through apical leak channels, then is actively transported to the extracellular fluid by Na^+-K^+-ATPase on the basolateral membrane. (p. 555)

21. Glucose, amino acids, ions, and various organic metabolites are reabsorbed by Na^+-linked secondary active transport. (p. 555)

22. The active transport of Na^+ and other solutes creates concentration gradients for passive reabsorption of urea and other solutes. (p. 556)

23. Most renal transport is mediated by membrane proteins and exhibits saturation, specificity, and competition. The **transport maximum,** or T_m, is the transport rate at saturation. (p. 557)

24. The **renal threshold** is the plasma concentration at which a substance first appears in the urine. (p. 558)

25. Peritubular capillaries reabsorb fluid along their entire length. (p. 559)

Secretion

26. Secretion enhances excretion by removing solutes from the peritubular capillaries. K^+, H^+, and a variety of organic compounds are secreted. (p. 559)

27. Molecules that compete for renal carriers slow the secretion of a molecule. (p. 559)

Excretion

28. The excretion rate of a solute depends on (1) its filtered load and (2) whether it is reabsorbed or secreted as it passes through the nephron. (p. 560)

29. **Clearance** describes how many milliliters of plasma passing through the kidneys have been totally cleared of a solute in a given period of time. (p. 560)

30. **Inulin** clearance is equal to GFR. In clinical settings, **creatinine** is used to measure GFR. (p. 561)

31. Clearance can be used to determine how the nephron handles a solute filtered into it. (p. 561)

Micturition

32. The external sphincter of the bladder is skeletal muscle that is tonically contracted except during urination. (p. 562)

33. Micturition is a simple spinal reflex subject to conscious and unconscious control. Parasympathetic neurons cause contraction of the smooth muscle in the bladder wall. Somatic motor neurons leading to the external sphincter are simultaneously inhibited. (p. 562)

QUESTIONS

LEVEL ONE   Reviewing Facts and Terms

1. List and explain the significance of the five characteristics of urine that can be found by physical examination.

2. List and explain the six major kidney functions.

3. At any given time, what percentage of the cardiac output goes to the kidneys?

4. List the major structures of the urinary system in sequence

from the kidneys to the urine leaving the body. Give a brief function for each structure.

5. Arrange these structures in the order that a drop of water entering the nephron would encounter them:

(a) afferent arteriole
(b) Bowman's capsule
(c) collecting duct
(d) distal tubule
(e) glomerulus
(f) loop of Henle
(g) proximal tubule
(h) renal pelvis

6. Name the three filtration barriers that solutes must cross as they move from plasma to the lumen of Bowman's capsule. What components of blood are usually excluded by these layers?

7. What force(s) promotes glomerular filtration? What force(s) opposes it? What is meant by the term *net driving force?*

8. What does the abbreviation GFR stand for? What is a typical numerical value for GFR in milliliters per minute? In liters per day?

9. Identify the following structures, then explain their significance in renal physiology:

(a) juxtaglomerular apparatus
(b) macula densa
(c) mesangial cell
(d) podocyte
(e) sphincters in the bladder
(f) renal cortex

10. In which segment of the nephron does most reabsorption take place? When a molecule or ion is reabsorbed from the lumen of the nephron, where does it go? If a solute is filtered and not reabsorbed from the tubules, where does it go?

11. Match the molecule with its primary mode(s) of transport across the kidney epithelium.

(a) Na^+
(b) glucose
(c) urea
(d) plasma proteins
(e) water

1. transcytosis
2. primary active transport
3. secondary active transport
4. facilitated diffusion
5. movement through open channels
6. simple diffusion through the phospholipid bilayer

12. List three solutes secreted by the kidney tubule.

13. What solute, normally present in the body, is used to estimate GFR in humans?

14. What is micturition?

LEVEL TWO   Reviewing Concepts

15. Map the following terms. You may add additional terms if you like.

α receptor
afferent arteriole
autoregulation
basal lamina
Bowman's capsule
capillary blood pressure
capsule fluid pressure
colloid osmotic pressure
efferent arteriole
endothelium
epithelium
GFR

glomerulus
JG cells
macula densa
mesangial cell
myogenic autoregulation
norepinephrine
paracrine
plasma proteins
podocyte
resistance
vasoconstriction

16. Define, compare, and contrast the following sets of terms and/or events:

(a) filtration, secretion, and excretion
(b) saturation, transport maximum, and renal threshold
(c) probenecid, creatinine, inulin, and penicillin
(d) clearance, excretion, and glomerular filtration rate

17. What are the advantages of a kidney that filters a large volume of fluid and then reabsorbs 99% of it?

18. If the afferent arteriole of a nephron constricts, what happens to GFR in that nephron? If the efferent arteriole of a nephron constricts, what happens to GFR in that nephron? Assume that no autoregulation takes place.

19. Diagram the micturition reflex. How is this reflex altered by toilet training? How do higher brain centers influence micturition?

LEVEL THREE   Problem Solving

20. You have been asked to study kidney function in a new species of rodent found in the Amazonian jungle. You isolate some nephrons and expose them to inulin. The graph below shows the results of your studies. (a) How is the rodent nephron handling inulin? Is inulin filtered? excreted? net reabsorption? net secretion? (b) On the graph, accurately draw in a line to show the net reabsorption or secretion. (Hint: Excretion = filtration − reabsorption + secretion)

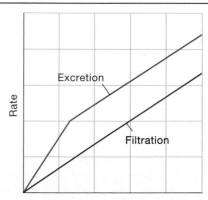

Plasma concentration of X

21. Darlene weighs 50 kg. Assume that her total blood volume is 8% of her body weight, her heart pumps her total blood volume once a minute, and her renal blood flow is 25% of her cardiac output. Calculate the volume of blood that flows through Darlene's kidneys each minute.

22. Dwight was competing for a spot on the Olympic equestrian team. As his horse, Nitro, cleared a jump, the footing gave way, causing the horse to somersault, landing on Dwight and crushing him. The doctors feared kidney damage and ran the following test. Dwight's serum creatinine level was 2 mg/100 mL. His 24-hour urine specimen contained 1 L and had a creatinine concentration of 20 mg/mL. How many milligrams of creatinine are in the specimen?

What is Dwight's creatinine clearance? What is his GFR? Evaluate these results and comment on Dwight's kidney function.

23. You are a physiologist taking part in an archeological expedition to search for Atlantis. One of the deep-sea submersibles has come back with a mermaid, on which you are running a series of experiments. You have determined that the mermaid's GFR is 250 mL/min. The mermaid's kidney reabsorbs glucose with a transport maximum (T_m) of 50 mg/min. What is the mermaid's renal threshold for glucose? When the mermaid's plasma concentration of glucose is 15 mg/mL, what is its glucose clearance?

E   X   P   L   O   R   E   <MediaLab>

Introduction

This chapter introduced you to the kidney. The components of urine are constantly changing and reflect the kidney's functions of regulating ions and water and removing wastes. This MediaLab will help you take a histological tour of the nephron. You may also explore the concept of renal clearance and look at how your body handles the over-the-counter drug, Pepcid AC. After reading the description below, visit the MediaLab for Chapter 18 in your Companion Website and select the appropriate keyword.

Web Exploration 1

Estimated time for completion = 5 minutes

Take a look at Figure 18-2, which shows you a schematic of the nephron and indicates where filtration, reabsorption, secretion and excretion take place. Based on what you learned about cell and tissue structure in Chapter 3, predict the histology of the proximal and distal convoluted tubule. What are the physical characteristics of a transporting epithelium?

Select the keyword **UROLOGY** from the Website. Compare the tissues found on slides 11, 12, and 13. What is the function of the brush border found on the cells of the proximal convoluted tubule (PCT)? Compare the size of the green reabsorption arrows on the PCT and DCT in Figure 18-2. Might the brush border have anything to do with this size difference? How? Reabsorption of the sodium in the PCT is by active transport. If you were able to stain the cells in slide 13 for mitochondria, what do you predict the distribution pattern would be? Why? Compare the metabolic rate of the cells of

the PCT with the cells of the DCT? Do you think there would there be a difference? Why or why not?

Web Exploration 2

Estimated time for completion = 15 minutes

How would you determine what dose of a new drug to give to patients? First you would need to determine what level of the drug you need to maintain in the blood to obtain the desired physiological outcome. Next you would need to determine how rapidly the body removes the drug. Drugs are metabolized and destroyed by the liver but they also are removed from the blood stream by the kidney. Select the keyword **PEPCID** from your Website. Scroll down on the linked page till you reach the pharmacokinetics section. How rapidly is Pepcid eliminated from the body? What is its clearance? Each drug will have a different clearance rate and this will be one determinant of the dose administered.

Select the keyword **CLEARANCE** from your Website and review the concept of clearance as it is presented here. Halfway down the page you'll find a table showing the clearance of 3 drugs: A, B, and C. Use the clearance values to predict if each of the drugs is being filtered, reabsorbed, and/or secreted and if so, to what degree.

In the very last example you are asked to consider the effect plasma protein binding would have on clearance of a drug. Compare drug transport to what you learned about transport of hormones in the blood. Would you expect free hormones to be filtered by the kidney? What evidence do you have that the kidney excretes hormones?

19 Integrative Physiology II: Fluid and Electrolyte Balance

■ "At a 10% loss of body fluid, the patient will show signs of confusion, distress, and hallucinations and at 20%, death will occur."— *Poul Astrup, in* Salt and Water in Culture and Medicine, *1993.* ■

CHAPTER OUTLINE

BACKGROUND BASICS

The American businesswoman in Tokyo finished her workout and stopped at the snack bar of the fitness club to ask for a sports drink. The attendant handed her a bottle labeled "Pocari Sweat®." Although the thought of drinking sweat is not very appealing, the physiological basis for the name is sound.

During exercise, the body secretes sweat, a dilute solution of water and ions, particularly Na^+, K^+, and

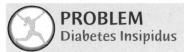

PROBLEM
Diabetes Insipidus

"Urine without flavor." That is the literal meaning of the term *diabetes insipidus*, a disorder in which the mechanism that concentrates urine is inadequate or absent. Christopher Godell, age 1, had been exhibiting two symptoms of this rare disease: intense thirst, particularly for ice water, and copious excretion of dilute urine. His mother took him to the pediatrician because of problems she noted. "Christopher is always thirsty, and always urinating," she explained. "I know this isn't normal."

...continued on page 576

• •

Cl^-. To maintain homeostasis, the body must replace any substance that it has lost to the external environment. Therefore, the replacement fluid after exercise should resemble sweat.

In this chapter, we explore how the body maintains salt and water balance, also known as fluid and electrolyte balance. The homeostatic control mechanisms for fluid and electrolyte balance are aimed at maintaining four parameters: volume, osmolarity, concentrations of individual ions, and pH of body fluids.

▶ FLUID AND ELECTROLYTE HOMEOSTASIS

The human body is in a state of constant flux. Over the course of a day we ingest about two liters of food and drink that contain from 6 to 15 grams of NaCl. In addition, we bring in varying amounts of other electrolytes, including K^+, H^+, Ca^{2+}, HCO_3^- and phosphate ions. The body's task is to maintain *mass balance* [∞ p. 7]: what comes in must be excreted if the body does not need it.

The body has several potential routes for excreting ions and water. The kidneys are the primary route for water loss and for removal of many ions. Under normal conditions, small amounts of these substances are lost in the feces as well. In addition, the lungs help remove H^+ and HCO_3^- by excretion of CO_2.

Although physiological mechanisms that maintain fluid and electrolyte balance are important, behavioral mechanisms also play an essential role. *Thirst* is critical because drinking is the only normal way to replace lost water. *Salt appetite* is a behavior that leads people and animals to seek and ingest salt (sodium chloride, NaCl).

Why are we concerned with homeostasis of these substances? Water and Na^+ are associated with extracellular fluid (ECF) volume and osmolarity. Disturbances in K^+ balance can cause serious problems with cardiac and muscle function by disrupting the membrane potential of excitable cells. Ca^{2+} is involved in a variety of body processes, from exocytosis and muscle contraction

to bone formation and blood clotting. And H^+ and HCO_3^- are the ions whose balance determines body pH.

ECF Osmolarity Affects Cell Volume

Why is maintaining osmolarity so important to the body? The answer lies in the fact that water crosses most cell membranes freely. If the osmolarity of the ECF changes, water moves into or out of cells and changes intracellular volume. If ECF osmolarity decreases due to excess water intake, water moves into the cells and they swell. If ECF osmolarity increases due to salt intake, water moves out of the cells and they shrink.

Changes in cell volume, either shrinking or swelling, can impair cell function. For example, when cells swell, ion channels in the membrane open, disrupting membrane potential and cell signaling. The brain, encased in the rigid skull, is particularly vulnerable to damage from swelling. Some cells can regulate their own volume in response to swelling or shrinking, but this capability is limited. In general, maintenance of ECF osmolarity within a normal range is essential to maintain homeostasis.

Fluid and Electrolyte Balance Requires Integration Among Multiple Systems

The process of fluid and electrolyte balance is truly integrative because it involves the respiratory and cardiovascular systems in addition to renal and behavioral responses. Adjustments made by the lungs and cardiovascular system are primarily under nervous control and can be made quite rapidly. Homeostatic compensation by the kidneys occurs more slowly because the kidneys are primarily under endocrine and neuroendocrine control.

For example, small changes in blood pressure due to fluid gain or loss are quickly corrected by the cardiovascular control center in the brain [∞ p. 463]. If volume changes are persistent or of larger magnitude, the kidneys step in to help maintain homeostasis.

Regulation of Cell Volume   The regulation of cell volume is so important that many cells have independent mechanisms for volume regulation. For example, renal tubule cells in the medulla of the kidney are constantly exposed to high extracellular fluid osmolarity, yet they maintain normal cell volume. They do this by synthesizing internal organic solutes so that their intracellular osmolarity matches that of the extracellular fluid. The organic solutes used to raise intracellular osmolarity include sugar alcohols such as sorbitol, and certain amino acids such as taurine. In addition, cells can regulate their volume by changing their ionic composition. In some instances, changes in cell volume are believed to act signals that initiate certain cellular responses. For example, swelling of liver cells activates protein and glycogen synthesis, and shrinking of liver cells causes breakdown of protein and glycogen.

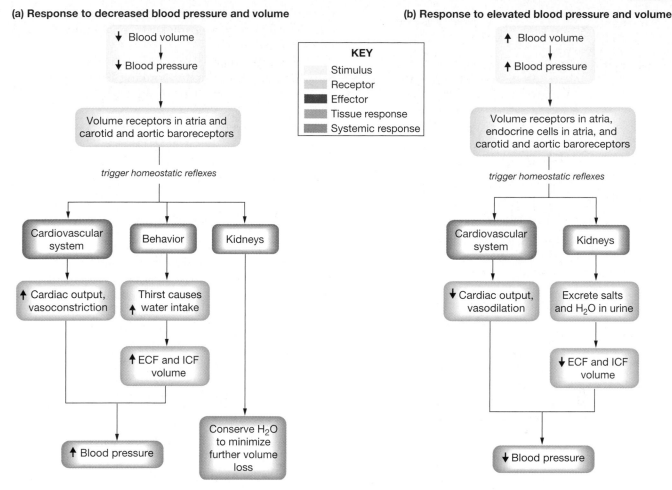

(a) Response to decreased blood pressure and volume

(b) Response to elevated blood pressure and volume

■ Figure 19-1   **Integrated responses to changes in blood volume and blood pressure**

Figure 19-1 ■ summarizes the integrated response of the body to changes in blood volume and blood pressure. Signals from carotid, aortic, and atrial baroreceptors initiate a quick neural response mediated through the cardiovascular control center and a slower response elicited from the kidneys. In addition, low blood pressure stimulates thirst. In both situations, renal function integrates with the cardiovascular system to keep blood pressure within a normal range.

Because of the overlap in their functions, a change made by one system is likely to have consequences that affect the other. Endocrine pathways initiated by the kidneys have direct effects on the cardiovascular system. Hormones released by myocardial cells act on the kidneys. Sympathetic pathways from the cardiovascular control center affect not only cardiac output and vasoconstriction but also glomerular filtration and hormone secretion by the kidneys.

Thus, the maintenance of blood pressure, blood volume, and ECF osmolarity forms a network of interwoven control pathways. This integration of function in multiple systems is one of the more difficult concepts in physiology but, at the same time, one of the most exciting areas of physiological research.

▶ WATER BALANCE AND THE REGULATION OF URINE CONCENTRATION

Water is the most abundant molecule in the body, constituting about 50% of total body weight in a 17- to 39-year-old woman and 60% of total body weight in a man of the same age. A 60-kg (132 lb) woman therefore has about 30 liters of body water. The prototype 70-kg man has about 42 liters of water volume. Two-thirds of his water (about 28 liters) is inside the cells, about 3 liters are in the plasma, and the remaining 11 liters are in the interstitital fluid [∞ p. 134].

Daily Water Intake and Excretion Are Balanced

To maintain constant body water content, we must take in the same amount of water that we excrete: intake must equal output. There are multiple sources for water gain and loss during a day (Fig. 19-2 ■). On the average, an adult ingests a little more than 2 liters of water in food and drink in a day. Normal cellular respiration (glucose + $O_2 \rightleftharpoons CO_2 + H_2O$) adds about 0.3 liters of water, bringing the total daily intake to approximately 2.5 liters.

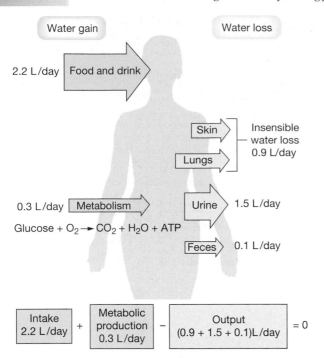

| Intake 2.2 L/day | + | Metabolic production 0.3 L/day | − | Output (0.9 + 1.5 + 0.1)L/day | = 0 |

■ **Figure 19-2   Water balance in the body**

Notice that the only means by which water normally enters the body from the external environment is by absorption through the digestive tract. Unlike some animals, we cannot absorb significant amounts of water directly through our skin. If fluids must be rapidly replaced or an individual is unable to eat and drink, fluid can be added directly to the plasma by means of **intravenous (IV) injection,** a medical procedure.

The major route of water loss from the body is the urine, which has a daily volume of about 1.5 liters. A small volume of water, about 100 mL, is lost in the feces. We also experience **insensible water loss** (called insensible because we are not normally aware of it). Humans are adapted for life in a terrestrial habitat, and our epidermis is modified with an outer layer of keratin to reduce evaporative water loss [∞ p. 68]. Despite this epidermal protection, however, we still lose about 900 mL of pure water per day across the skin surface and by exhaling humidified air. Thus, the 2.5 liters of water that we take in are balanced by 2.5 liters that leave the body. Of all the water lost, only water excreted in the urine can be regulated.

Although urine is normally the major route of water loss, in certain situations other routes of water loss can become significant. Excessive sweating is one example. Another example is water loss through diarrhea, a condition that can pose a major threat to the maintenance of water balance, particularly in infants.

Pathological water loss disrupts homeostasis in two ways. Volume depletion from the extracellular compartment decreases blood volume and blood pressure. If blood pressure cannot be maintained, the tissues do not get adequate oxygen. If the fluid lost is hyposmotic to

the body, as in excessive sweating, the solutes left behind in the body raise osmolarity, potentially disrupting cell function.

Normally, water balance takes place automatically. Salty food makes us thirsty. Drinking 42 ounces of a soft drink means an extra trip to the bathroom. Salt and water balance is a subtle process that we are only peripherally aware of, like breathing and the beating of the heart.

Now that we have seen *why* regulation of osmolarity is important, let's see *how* the body accomplishes that goal.

The Kidneys Conserve Water

Figure 19-3 ■ summarizes the role of the kidneys in water balance. In this figure, the mug with a hollow handle represents the body. The handle represents the kidneys, where body fluid filters into the nephrons and is reabsorbed. Some solutes and water leave in the urine, but the amount that leaves can be regulated, as represented by the little gates.

If additional fluid is added to the mug, normal volume can easily be restored by allowing the extra water to leak out of the handle. If fluid is lost from the mug, however, the only way to replace it is to add water from a source outside the mug. During volume loss, fluid still flows through the handle, but even maximal reabsorption only conserves the water that is in the mug. Reabsorption of water by the kidneys can do nothing to increase the volume. This figure underscores the fact that

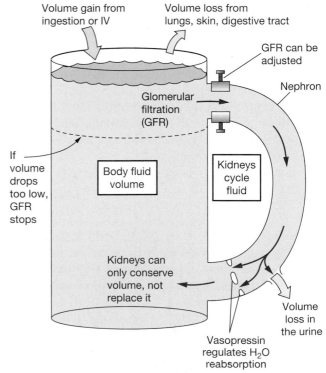

■ **Figure 19-3   Role of the kidneys in water balance**   The kidneys can only conserve fluid. They cannot restore lost fluid volume.

the kidneys cannot replenish lost water. They can only conserve it.

Urine Concentration Is Determined in the Loop of Henle and Collecting Duct

The concentration, or osmolarity, of urine is a measure of water excretion by the kidneys. When the body needs to eliminate excess water, the kidneys put out copious amounts of dilute urine with osmolarities as low as 50 mOsM. Removal of excess water in dilute urine is known as **diuresis** [*diourein,* to pass in urine]. Drugs that cause diuresis are known as *diuretics.* When the kidneys are conserving water, the urine becomes quite concentrated, up to four times as concentrated as the blood (1200 mOsM vs. 300 mOsM).

The kidneys control urine concentration by varying the amounts of water and Na$^+$ reabsorbed in the distal regions of the nephron. (For a review of renal anatomy, see Figure 18-1, p. 546.) To produce dilute urine, the kidney must reabsorb solute without allowing water to follow osmotically. This means that the cell membranes across which solute is transported must not be permeable to water.

On the other hand, to produce concentrated urine the nephron must reabsorb water while leaving solute in the lumen. At one time, scientists believed that water was actively transported on carriers, just as Na$^+$ and other ions are. However, once they developed micropuncture techniques to sample fluid inside kidney tubules, they discovered that water is reabsorbed only by osmosis.

Osmosis will not take place unless a concentration gradient exists. Through an ingenious arrangement of blood vessels and tubules, to be discussed later, the renal medulla maintains a high osmotic concentration in its interstitial fluid. This high *medullary interstitial osmolarity* allows urine to be concentrated.

Let's follow some fluid through the nephron to see where these changes in fluid osmolarity take place (Fig. 19-4 ■). You may recall from the last chapter that reabsorption in the proximal tubule is isosmotic. Fluid entering the loop of Henle has an osmolarity of about 300 mOsM. Fluid leaving the loop of Henle, on the other hand, is hyposmotic, around 100 mOsM. Cells in the ascending limb of the loop are impermeable to water. When they pump Na$^+$, K$^+$, and Cl$^-$ out of the tubule

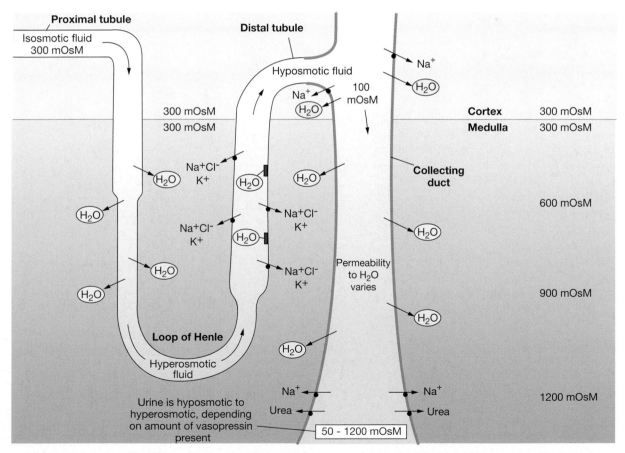

■ **Figure 19-4   Osmolarity changes as fluid flows through the nephron**   Fluid leaving the proximal tubule is isosmotic to ECF. Active transport of solute out of the ascending limb of the loop of Henle creates hyposmotic fluid. The osmolarity of the final urine depends on the presence or absence of vasopressin in the collecting duct. Urea is transported out of the collecting duct to help keep interstitial osmolarity high.

lumen, water cannot follow. This creates a hyposmotic solution in the lumen.

Once the hyposmotic fluid leaves the loop of Henle and passes through the distal segments of the nephron, its concentration will be determined by the permeability of those tubule membranes to water. If the membrane is not permeable, water remains in the tubule and dilute urine will be excreted. A small amount of additional solute can be reabsorbed in the collecting duct. As a result, the most dilute urine we can form is about 50 mOsM.

Conversely, if the urine is to become more concentrated, the tubule epithelium must become permeable to water. When the collecting duct membrane is permeable to water, osmosis draws water out of the lumen and into the interstitial fluid. At maximal water permeability, removal of water from the tubule leaves behind a concentrated urine with an osmolarity of 1200 mOsM.

Vasopressin Regulates Urine Osmolarity

How does the collecting duct alter its permeability to water and determine the final concentration of urine? The process involves adding or removing water pores in the apical membrane under the direction of the posterior pituitary hormone **vasopressin** [∞ p. 198]. Because

vasopressin causes the body to retain water, it is also known as **antidiuretic hormone** or **ADH.**

When vasopressin acts on the cell, water pores are present in the apical membrane, and water moves out of the lumen by osmosis (Fig. 19-5a ■). Osmotic movement of water occurs because the cells and interstitial fluid of the kidney medulla are more concentrated than fluid in the tubule.

In the absence of vasopressin, the collecting duct is impermeable to water (Fig. 19-5b ■). Although a concentration gradient is present across the epithelium, water remains in the tubule, producing dilute urine.

The water permeability of the collecting duct is not an all-or-none phenomenon, as the paragraph above might suggest. Permeability is variable, depending on how much vasopressin is present. The graded effect of vasopressin allows the body to match urine concentration closely to the body's needs.

Vasopressin and Aquaporins   Most membranes in the body are freely permeable to water. What makes the cells of the collecting duct different? The answer lies with its water pores. Water pores are **aquaporins,** a family of membrane channels with at least ten different isoforms that occur in mammalian tissues. The kidney has at least

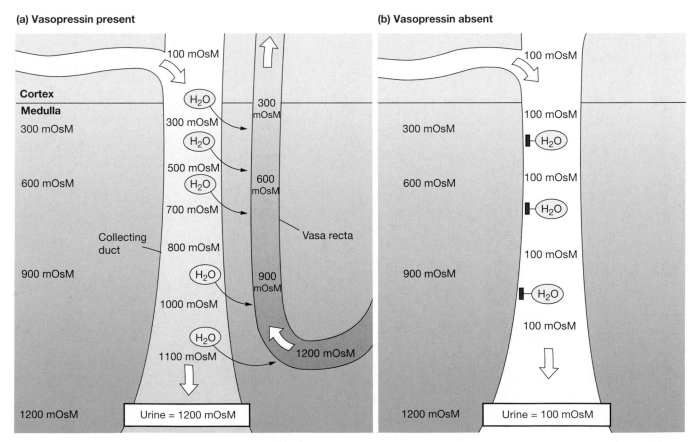

(a) Vasopressin present

(b) Vasopressin absent

■ **Figure 19-5   Water movement in the collecting duct in the presence and absence of vasopressin**   (a) In the presence of maximal vasopressin, the collecting duct is freely permeable to water. Water moves by osmosis into the more concentrated interstitial fluid and is carried away by the vasa recta capillaries to reenter the main circulation. (b) In the absence of vasopressin, the collecting duct is impermeable to water and the urine is dilute.

DIABETES **Diabetes and Osmotic Diuresis** The primary sign of diabetes mellitus is an elevated blood glucose concentration. In untreated diabetics, if blood glucose levels exceed the renal threshold for glucose reabsorption [∞ p. 558], glucose will be excreted in the urine. This may not seem like a big deal, but any additional non-reabsorbable solute that remains in the lumen will force additional water to be excreted, causing *osmotic diuresis.* For example, suppose the nephrons need to excrete 300 milliosmoles of NaCl. If the urine is maximally concentrated at 1200 mOsM, then the NaCl will be excreted in a volume of 0.25 L. If the NaCl is joined by 300 mosmoles of glucose that must be excreted, the volume of urine doubles, to 0.5 L. Osmotic diuresis in untreated diabetics (primarily type 1) will cause *polyuria* (excessive urination) and *polydipsia* [*dipsios,* thirsty] due to dehydration and high plasma osmolarity.

six different types of aquaporins, including aquaporin-2 (AQP2), the water channel regulated by vasopressin.

AQP2 in collecting duct cells may be found in two locations: on the apical membrane facing the tubule lumen and in the membrane of cytoplasmic vesicles. (Two other types of aquaporin water channels are present in the basolateral membrane, but they are not regulated by vasopressin.)

When vasopressin levels and collecting duct water permeability are low, the tubule cell has few water pores in its apical membrane (Fig. 19-6 ■). It stores its AQP2 water pores in cytoplasmic vesicles.

When vasopressin arrives from the posterior pituitary, it binds to its receptor on the basolateral side of the collecting duct cell. Binding activates a G-protein/cAMP second messenger system [∞ p. 161]. Subsequent phosphorylation of intracellular proteins causes the AQP2 vesicles to move to the apical membrane and fuse with it. Exocytosis inserts the AQP2 water pores into the apical membrane. Now the cell is permeable to water. This process, in which parts of the cell membrane are alternately added by exocytosis and withdrawn by endocytosis, is known as **membrane recycling** [∞ Fig. 5-26, p. 130].

✔ Will the apical membrane of a collecting duct cell have more water channels when vasopressin is present or when it is absent?

✔ The collecting duct cells are surrounded by extremely high osmolarity on their basolateral sides, yet do not shrivel up. How can they maintain normal cell volume in the face of such high ECF osmolarity? (Hint: read the box on regulation of cell volume, p. 570.)

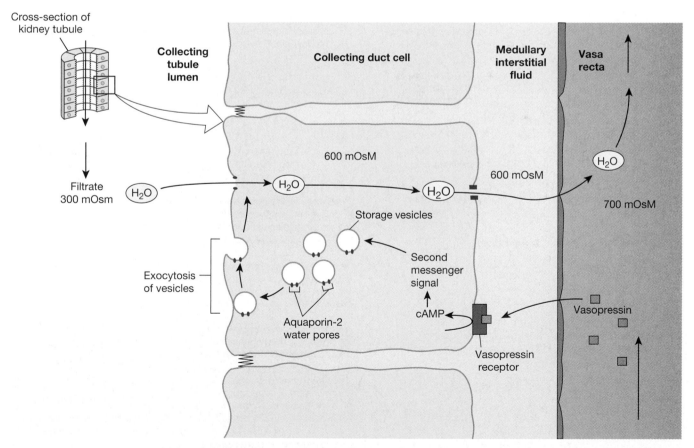

■ **Figure 19-6   The mechanism of action of vasopressin** In the collecting duct, vasopressin binds to a membrane receptor and activates a cAMP second messenger system. In response, the cell inserts AQP2 water pores into its apical membrane. In the absence of vasopressin, the pores are withdrawn and stored in cytosolic vesicles.

Changes in Blood Pressure and Osmolarity Trigger Water Balance Reflexes

What stimuli control vasopressin secretion? There are two: ECF osmolarity and blood pressure (Fig. 19-7 ■). The most potent stimulus for vasopressin release is an increase in blood osmolarity. Osmolarity is monitored by **osmoreceptors,** stretch-sensitive cells that activate sensory neurons when osmolarity increases above threshold.

The primary osmoreceptors for vasopressin release are found in the hypothalamus. When osmolarity is below the threshold value of 280 mOsM, the osmoreceptors do not fire, and vasopressin secretion ceases (Fig. 19-8 ■). If osmolarity rises above 280 mOsM, the osmoreceptors stimulate release of vasopressin.

Water reabsorption in the kidneys conserves water and can decrease osmolarity when coupled to excretion of solute. However, vasopressin cannot restore lost fluid volume. Only the ingestion or infusion of water can replace water that has been lost.

Decreases in blood pressure and blood volume are less powerful stimuli for vasopressin release. The primary receptors for volume are stretch-sensitive receptors in the atria. Blood pressure is monitored by the same baroreceptors in the carotid and aortic bodies that initiate cardiovascular responses [∞ p. 463]. When blood pressure

...continued from page 570

Diabetes insipidus is caused either by a deficiency in vasopressin or by inability of nephrons to respond to vasopressin. Diabetes insipidus due to vasopressin deficiency (central diabetes insipidus) can be caused by head injury, trauma, or various drugs that damage the pituitary or hypothalamus. In the other form of diabetes insipidus, called *nephrogenic diabetes insipidus*, the collecting duct cells fail to increase their water permeability in response to vasopressin.

Question 1:   *Christopher's doctor sends him for a test that measures the amount of vasopressin in his blood. The test shows that Christopher has normal levels of vasopressin. Is your diagnosis central or nephrogenic diabetes insipidus?*

or blood volume is low, these receptors signal the hypothalamus to secrete vasopressin and conserve fluid.

✔ A scientist monitoring the activity of osmoreceptors noticed that infusion of hyperosmotic saline (NaCl) caused increased firing of the osmoreceptors. Infusion of hyperosmotic urea (a penetrating solute) had no effect. If osmoreceptors fire only when cell volume decreases, explain why hyperosmotic urea did not affect them.

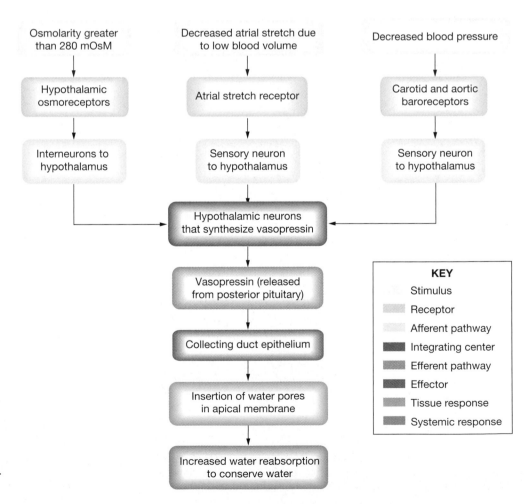

■ **Figure 19-7   Factors affecting vasopressin release**

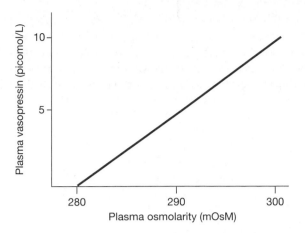

■ **Figure 19-8   The effect of plasma osmolarity on vasopressin secretion**

✔ If vasopressin increases water reabsorption by the nephron, would vasopressin secretion be increased or decreased with dehydration?

✔ If vasopressin secretion is suppressed, will urine be dilute or concentrated?

The Loop of Henle Is a Countercurrent Multiplier

Vasopressin is the signal, but the key to the kidney's ability to produce concentrated urine is the high osmolarity of the medullary interstitium. Without it, there would be no concentration gradient for osmotic movement of water out of the collecting duct.

What creates this high ECF osmolarity? And why isn't the high osmolarity diluted by water being reabsorbed from the collecting duct and descending limb of the loop of Henle (see Fig. 19-4 ■, p. 573)? The answer to these questions can be found in the anatomical arrangement of the loop of Henle and its associated blood vessels, the vasa recta, into a countercurrent exchange system.

Bed-Wetting and Vasopressin   Bed-wetting, or **nocturnal enuresis,** is a problem that affects about 8% of school-age children. For many years, enuresis was thought to be of psychological origin, but now it is felt that most enuretic children have a physiological cause for their bed-wetting. In normal children, vasopressin secretion shows a circadian rhythm, with levels increasing at night. The result is nocturnal production of a smaller volume of more concentrated urine. Many children with enuresis fail to increase vasopressin secretion at night, with the result that their urine output stays elevated to the point that the bladder fills to its maximum capacity and empties spontaneously during sleep. These children can be successfully treated with a nasal spray of desmopressin, a vasopressin derivative, at bedtime.

Countercurrent Systems Exchange Molecules or Heat   **Countercurrent exchange systems** require arterial and venous blood vessels that pass very close to each other, with fluid flow moving in opposite directions (the countercurrent). This anatomical arrangement allows transfer of heat or molecules from one vessel to the other. Because the countercurrent heat exchanger is easier to understand, we first examine how it works, then translate the same principle to the kidney.

The countercurrent heat exchanger in mammals and birds evolved to reduce heat loss in flippers, tails, and other limbs that are poorly insulated and that have a high surface-to-volume ratio. Without such a system, warm blood flowing out from the body core into the limb easily loses heat to the surrounding environment (Fig. 19-9a ■).

With the countercurrent heat exchanger (Fig. 19-9b ■), blood entering the limb in arteries transfers its heat to blood going back into the body in veins. As cooler blood at the tip of the limb flows back into the body, it picks up

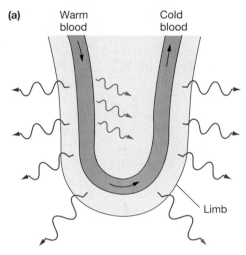

(a)

If blood vessels are not close to each other, heat is dissipated to the external environment.

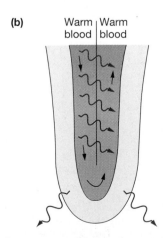

(b)

Countercurrent heat exchanger allows warm blood entering the limb to transfer heat directly to blood flowing back into the body.

■ **Figure 19-9   A countercurrent heat exchanger**

heat from the warm blood flowing out into the limb. This arrangement reduces heat loss to the environment.

The countercurrent system of the kidney, the loop of Henle, works on the same principle, except that it transfers solute instead of heat from the circulating fluid. However, because the kidney forms a closed system, the solutes are not lost to the environment as heat is. Instead, the solutes concentrate in the interstitial fluid. This process is aided by active transport of solute out of the ascending limb, which makes the ECF osmolarity even greater. For this reason, the loop of Henle is known as a *countercurrent multiplier*.

The countercurrent multiplier system in the renal medulla is shown in Figure 19-10 ■. Fluid from the proximal tubule flows into the descending limb of the loop of Henle. The descending limb is permeable to water but does not transport ions. As the loop dips into the medulla, water moves by osmosis into the progressively more concentrated interstitial fluid, leaving solutes behind in the lumen.

The tubule fluid becomes progressively more concentrated as it moves deeper into the medulla. At the tips of the longest loops of Henle, the tubule fluid reaches a concentration of 1200 mOsM. Fluid in shorter loops that do not extend into the most concentrated regions of the medulla does not reach such a high concentration.

When the fluid reverses direction and enters the ascending limb of the loop, the properties of the tubule epithelium change. The tubule becomes impermeable to water while actively transporting Na$^+$, K$^+$ and Cl$^-$ out of the tubule into the interstitial fluid. The loss of solute

from the lumen causes the osmolarity of the tubule fluid to decrease steadily, from 1200 mOsM at the bottom of the loop to 100 mOsM at the cortex. The net result of the countercurrent multiplier in the kidney is to produce hyperosmotic interstitial fluid in the medulla and hyposmotic fluid leaving the loop of Henle.

The Vasa Recta Removes Water   It is easy to see how transport of solute from the ascending limb dilutes the fluid in the loop of Henle and helps concentrate fluid in the medulla. Still, why doesn't the water leaving the descending limb of the loop dilute the concentrated interstitial fluid of the medulla? The answer lies in the close anatomical association of the loop of Henle and those peritubular capillaries known as the **vasa recta.**

These capillaries, like the loop of Henle, dip down into the medulla and then go back up to the cortex, forming hairpin loops. Although textbooks traditionally show a single nephron with a single loop of capillary, each kidney has millions of collecting ducts and loops of Henle packed between millions of vasa recta capillaries. Thus, functionally, blood flow in the limbs of the vasa recta is in the opposite direction from fluid flow in the loop, as shown in Figure 19-10 ■.

Water or solutes that leave the tubule move into the vasa recta if a concentration or osmotic gradient exists. For example, assume that blood entering the medulla in the vasa recta is 300 mOsM, isosmotic with the cortex. As the blood flows into the medulla, it loses water and picks up solutes transported out of the ascending limb of the loop of Henle, carrying them farther into the medulla. By

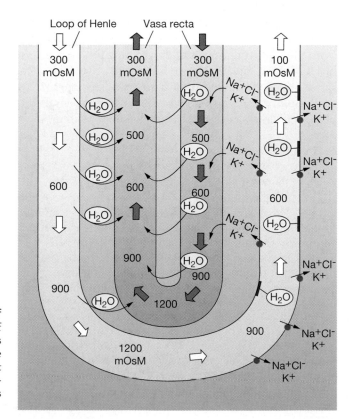

■ **Figure 19-10   Countercurrent exchange in the medulla of the kidney**   Filtrate entering the descending limb of the loop of Henle becomes progressively more concentrated as it loses water. As fluid moves through the ascending limb, Na$^+$, K$^+$, and Cl$^-$ are pumped out. Water cannot follow, creating hyposmotic fluid that enters the distal tubule. Blood in the vasa recta removes water leaving the loop of Henle so that the medullary interstitial fluid remains concentrated.

the time the blood reaches the bottom of the vasa recta loop, it has a high osmotic concentration, similar to that of the surrounding interstitial fluid.

Then, as the blood in the vasa recta flows out of the medulla, its high osmolarity attracts the water that is being lost from the tubule. The movement of water into the vasa recta decreases the osmolarity of the blood while simultaneously preventing the water from diluting the concentrated interstitial fluid.

As blood in the vasa recta leaves the medulla, it removes water reabsorbed from the loop of Henle. Without the vasa recta, water moving out of the descending loops of Henle would eventually dilute the medullary interstitial fluid. The vasa recta thus are an important part of keeping the medullary concentration high.

Urea Increases Osmolarity of the Medullary Interstitium The high concentration in the medullary interstitium is only partly due to NaCl. Nearly half the solute in the medullary interstitial fluid is urea. Where does it come from? For many years scientists thought urea only crossed cell membranes by passive transport. However, in recent years we have learned that there are several different transporters for urea in the collecting duct. One is a facilitated diffusion carrier, and the other is a Na^+-dependent secondary active transporter. The transporters help concentrate urea in the medullary interstitium, where it contributes to the high interstitial osmolarity.

▶ SODIUM BALANCE AND THE REGULATION OF ECF VOLUME

As noted in the introduction to this chapter, we ingest a lot of NaCl—an average of 9 grams in one day. This is about 2 teaspoons of salt or 155 milliosmoles of Na^+ and 155 milliosmoles of Cl^-. Let's see what would happen to our bodies if the kidneys could not get rid of this Na^+.

Our normal plasma Na^+ concentration is 135–145 milliosmoles Na^+ per liter. Sodium distributes freely between plasma and interstitial fluid, so this value also represents our ECF Na^+ concentration. If we add 155 milliosmoles Na^+ to the ECF, how much water would we have to add to keep ECF sodium at 140 mOsM?

$$155 \text{ mosmol/ ? liters} = 140 \text{ mosmol/liter}$$

We would have to add over a liter of water to the ECF to compensate for the addition of the Na^+. Normal ECF volume is about 14 liters, so that increase would represent about an 8% gain! Imagine what that volume increase would do to blood pressure.

On the other hand, suppose we add the NaCl and don't drink water. What would happen to osmolarity? If we assume that normal total body osmolarity is 300 mOsM and volume is 42 L, the addition of 155 milliosmoles of Na^+ and 155 milliosmoles of Cl^- would increase total body osmolarity to 307 mOsM—a substantial increase. In addition, because NaCl is a nonpenetrating

solute, it would stay in the ECF. Higher osmolarity in the ECF would draw water from the cells, shrinking them and disrupting normal cell function.

Fortunately, our homeostatic mechanisms maintain mass balance: anything extra that comes into the body will be excreted. Figure 19-11 ■ shows a generalized homeostatic pathway for sodium balance in response to salt ingestion. Here's how it works.

The addition of NaCl to the body raises osmolarity. This stimulus triggers two responses: vasopressin secretion and thirst. Vasopressin release causes the kidneys to conserve water and concentrate the urine. Thirst prompts us to drink water or other fluids.

Ingested water decreases osmolarity, but the combination of salt and water intake increases ECF volume and blood pressure. The increase in ECF volume and pressure then triggers another series of control pathways, which bring blood pressure and osmolarity back into the normal range by excreting extra salt and water.

The kidneys are the primary means of Na^+ excretion. Normally, only a small amount leaves the body in feces and perspiration. However, in situations such as vomiting, diarrhea, and heavy sweating, we may lose significant amounts of Na^+ and Cl^- through nonrenal routes.

Although we speak of ingesting and losing salt (NaCl), the body monitors only Na^+ for homeostasis. (Chloride usually accompanies Na^+.) Sodium regulation is mediated by aldosterone from the adrenal cortex. Interestingly, the stimuli that set the aldosterone pathway in motion are more closely tied to blood volume, blood pressure, and osmolarity than to Na^+ itself.

Aldosterone Controls Sodium Balance

The regulation of blood Na^+ levels takes place through one of the most complicated endocrine pathways of the body. The reabsorption of Na^+ in the distal tubule and collecting duct of the kidney is regulated by the steroid hormone **aldosterone**: the more aldosterone, the more Na^+ reabsorption. Because one target of aldosterone is increased activity of the Na^+-K^+-ATPase, aldosterone also causes K^+ secretion.

Aldosterone is a steroid hormone synthesized in the adrenal cortex, the outer portion of the glands that sit on top of each kidney [∞ p. 192]. Like typical steroid hormones, aldosterone is secreted into the blood and transported on a protein carrier to its target. The primary site of aldosterone action is the last third of the distal tubule and the portion of the collecting duct that runs through the kidney cortex (the *cortical collecting duct*).

The target cell of aldosterone is the **principal cell**, or **P cell** (Fig. 19-12 ■). Principal cells look much like other polarized transporting epithelial cells, with Na^+-K^+-ATPase pumps on the basolateral membrane, and various channels and transporters on the apical membrane [∞ p. 131]. The apical membranes of principal cells contain leak channels for both Na^+ and K^+.

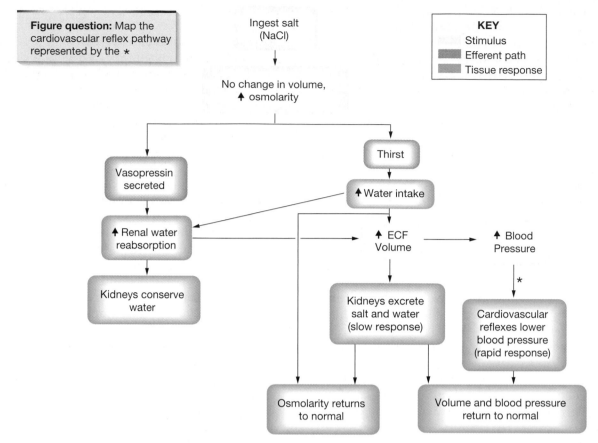

Figure question: Map the cardiovascular reflex pathway represented by the ✶

Ingest salt
(NaCl)

KEY
Stimulus
Efferent path
Tissue response

No change in volume,
↑ osmolarity

Thirst

Vasopressin secreted

↑ Water intake

↑ Renal water reabsorption

↑ ECF Volume

↑ Blood Pressure

Kidneys conserve water

Kidneys excrete salt and water (slow response)

✶

Cardiovascular reflexes lower blood pressure (rapid response)

Osmolarity returns to normal

Volume and blood pressure return to normal

■ Figure 19-11   **Homeostatic responses to eating salt**

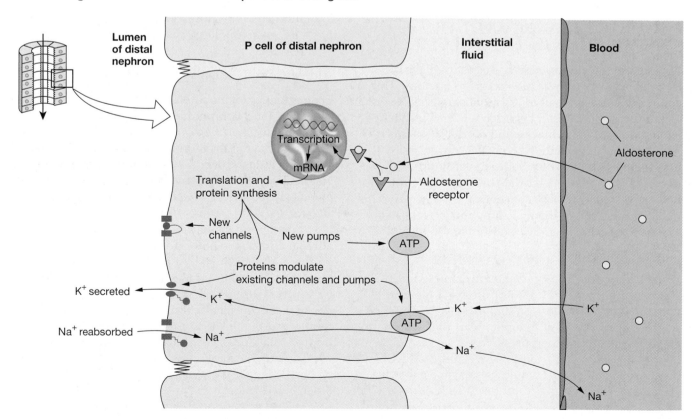

■ Figure 19-12   **The response of principal cells to aldosterone**   Aldosterone combines with a cytoplasmic receptor in the P cell. The hormone-receptor complex moves into the nucleus and initiates transcription, translation, and new protein synthesis. Aldosterone-induced proteins can modify existing pumps and channels.

Aldosterone enters cells by simple diffusion. In target cells, aldosterone combines with a cytoplasmic receptor. In the earliest response phase, apical Na^+ channels increase their open time under the influence of an as-yet-unidentified signal molecule. As intracellular Na^+ levels rise, the Na^+-K^+-ATPase speeds up, transporting cytoplasmic Na^+ into the ECF. The net result is a rapid increase in Na^+ reabsorption that does not require the synthesis of new channel or ATPase proteins. In the slower phase of aldosterone action, newly synthesized channels and pumps are inserted into epithelial cell membranes.

Note that Na^+ and water reabsorption are separately regulated in the distal nephron. Water does not automatically follow Na^+ reabsorption: vasopressin must be present. In contrast, in the proximal tubule, Na^+ reabsorption is automatically followed by water reabsorption because the proximal tubule epithelium is always freely permeable to water.

✔ In Figure 19-12 ■, what force(s) causes Na^+ and K^+ to cross the apical membrane?

Blood Pressure, Osmolarity, and K^+ Influence Aldosterone Secretion

What controls aldosterone secretion? There are three primary stimuli: increased K^+, increased osmolarity, and a trophic hormone called *angiotensin II* (ANG II).

The simplest stimuli for aldosterone release act directly on the adrenal cortex cells: increased plasma K^+ and increased osmolarity (Table 19-1). An increase in K^+ concentration stimulates aldosterone production, which results in K^+ secretion by the nephron. This reflex protects the body from *hyperkalemia* (elevated blood K^+).

The other stimulus that acts directly on the adrenal gland is osmolarity: an increase in ECF osmolarity inhibits aldosterone secretion. Less aldosterone means less Na^+ reabsorption and therefore increased Na^+ excretion, which helps decrease osmolarity. The direct inhibitory effect of high osmolarity on the adrenal cortex is an important part of the body's response to dehydration.

The Renin-Angiotensin-Aldosterone Pathway

The primary signal for aldosterone release is a trophic hormone known as **angiotensin II**. ANG II is one component of the **renin-angiotensin-aldosterone system**

(RAAS), a complex, multistep pathway for maintaining blood pressure.

The RAAS pathway begins when the *juxtaglomerular cells* (JG cells) in the afferent arterioles of nephrons [∞ p. 545] secrete an enzyme called **renin** (Fig. 19-13 ■). Renin converts an inactive plasma protein, **angiotensinogen,** into **angiotensin I (ANG I).** (The suffix *-ogen* indicates an inactive precursor.) When ANG I in the blood encounters an enzyme called **angiotensin converting enzyme (ACE),** ANG I is converted into ANG II.

This conversion was originally thought to take place only in the lungs, but ACE is now known to occur on the endothelium of blood vessels throughout the body. When ANG II in the blood reaches the adrenal gland, it causes synthesis and release of aldosterone. Finally, at the distal nephron, aldosterone initiates a series of intracellular reactions that cause the tubule to reabsorb Na^+.

The stimuli that begin the RAAS pathway are all related directly or indirectly to low blood pressure (Fig. 19-14 ■):

1. The *JG cells* are directly sensitive to pressure. They respond to low pressure in renal arterioles by secreting renin.
2. *Sympathetic neurons,* activated by the cardiovascular control center when blood pressure decreases, terminate on the JG cells and stimulate renin secretion.
3. *Paracrine feedback* from the macula densa in the distal tubule to the JG cells stimulates renin release [∞ p. 553]. When fluid flow through the distal tubule is relatively high, the macula densa cells release nitric oxide (NO), which inhibits renin release. If fluid flow in the distal tubule decreases because of a decrease in blood pressure, glomerular filtration rate (GFR), or both, macula densa cells decrease NO release and the JG cells secrete renin.

Sodium reabsorption does not directly raise low blood pressure, but retention of Na^+ does help stimulate fluid intake and volume expansion (see Fig. 19-11 ■). When blood volume increases, blood pressure also increases.

TABLE 19-1 Factors Affecting Aldosterone Release

Direct, at the adrenal cortex	
Increased extracellular K^+ concentration	Stimulates
Increased osmolarity	Inhibits
Indirect, through the RAAS pathway	
Decreased blood pressure	Stimulates
Decreased flow past the macula densa	Stimulates

Nitric Oxide and Renin Release Nitric oxide (NO) apparently controls renin release by activating a K^+ channel in the membrane of the JG cell. NO diffuses into the JG cell and activates cytosolic guanylyl cyclase [∞ p. 163] to form cGMP. Acting through a signal cascade, cGMP activates a K^+ channel that allows K^+ to leak out of the cell. Loss of K^+ from the cytoplasm hyperpolarizes the cell. When NO is absent, the K^+ channel closes and the cell depolarizes. Voltage-gated Ca^{2+} channels open and Ca^{2+} entry is the signal for exocytosis of renin-containing secretory vesicles. This mechanism is similar to the process by which insulin is released from pancreatic beta cells [∞ Fig. 5-39, p. 146].

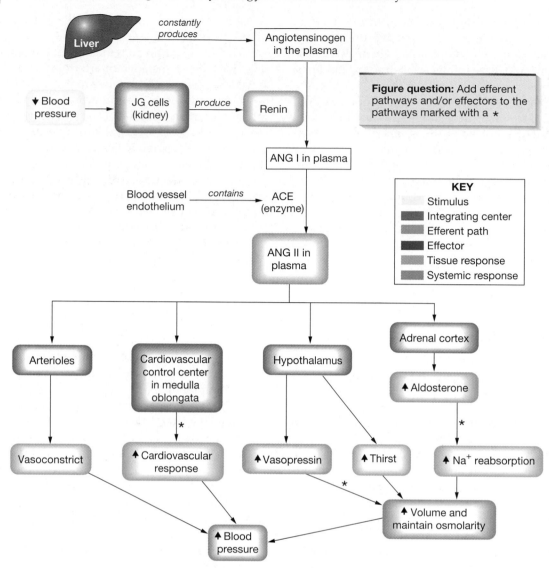

■ **Figure 19-13**   **The renin-angiotensin-aldosterone pathway**

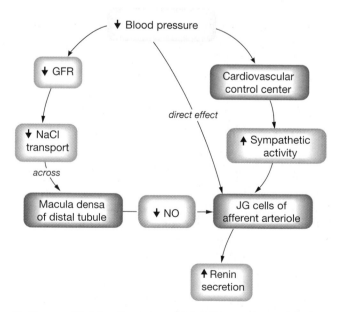

■ **Figure 19-14   Decreased blood pressure stimulates renin secretion**

The RAAS pathway does not end here, however. Angiotensin II is a remarkable hormone with additional effects directed at raising blood pressure. These actions make ANG II an important hormone in its own right, not merely an intermediate step in the aldosterone control pathway.

Angiotensin II Influences Blood Pressure Through Multiple Pathways

Angiotensin II has significant effects on fluid balance and blood pressure beyond stimulating aldosterone secretion. ANG II increases blood pressure both directly and indirectly through four additional pathways (see Fig. 19-13 ■):

1. Activation of ANG II receptors in the brain increases vasopressin secretion. Fluid retention in the kidney under the influence of vasopressin helps conserve blood volume.

2. ANG II stimulates thirst. Fluid ingestion is a behavioral response that expands blood volume and raises blood pressure.

3. ANG II is one of the most potent vasoconstrictors known in humans. Vasoconstriction causes blood pressure to increase without a change in blood volume.

4. ANG II receptors are found in the cardiovascular control center. Their activation increases sympathetic output to the heart and blood vessels, which increases cardiac output and vasoconstriction. These responses both increase blood pressure.

Once these blood pressure-raising effects of ANG II became known, it was not surprising that pharmaceutical companies started looking for drugs to block ANG II. Their research produced a new class of antihypertensive drugs called *ACE (angiotensin converting enzyme) inhibitors.* These drugs block the ACE-mediated conversion of ANG I to ANG II, thereby helping to relax blood vessels and lower blood pressure. Less ANG II also means less aldosterone release, a decrease in Na^+ reabsorption and, ultimately, a decrease in ECF volume. All these responses contribute to lowering blood pressure.

However, the ACE inhibitors have not been without side effects. ACE inactivates a cytokine called *bradykinin.* When ACE is inhibited by drugs, bradykinin levels increase and in some patients, this creates a dry, hacking cough. One solution has been the development of drugs called *sartans* that block the blood pressure–raising effects of ANG II by binding to *AT_1 receptors,* a subtype of ANG II receptor.

✔ A man comes to the doctor with high blood pressure. Tests show that he also has elevated levels of renin in his blood and atherosclerotic plaques that have nearly blocked blood flow through his renal arteries. How does decreased blood flow in his renal arteries cause renin secretion to increase?

✔ Describe the pathways through which elevated renin causes high blood pressure in this man.

Atrial Natriuretic Peptide Promotes Na^+ and Water Excretion

Once it was known that aldosterone and vasopressin increase Na^+ and water reabsorption, scientists speculated that other hormones might cause Na^+ loss (**natriuresis**) [*natrium,* sodium + *ourein,* to urinate] and water loss (diuresis) in the urine. If found, these hormones might be used clinically to lower blood volume and blood pressure in patients with essential hypertension [∞ p. 468]. During years of searching, however, evidence for the other hormones was not forthcoming.

Then, in 1981, a group of Canadian researchers found that injections of homogenized rat atria caused rapid but short-lived excretion of Na^+ and water in the

urine. They hoped that they had found the missing hormone, one whose activity would complement that of aldosterone and vasopressin. As it turned out, they had discovered the first in an entire family of hormones, some produced in the myocardial cells of the heart and others found in the brain.

Atrial natriuretic peptide (ANP; also known as *atriopeptin*) is a peptide hormone produced in specialized myocardial cells in the atria of the heart. It is synthesized as part of a large pro-hormone that is cleaved into several active hormone fragments [∞ p. 191]. Atrial natriuretic peptide is secreted when atrial cells stretch more than is normal, as would occur with increased blood volume.

At the systemic level, ANP enhances Na^+ excretion and urinary water loss, but the exact mechanisms by which it does so still are not clear. Atrial natriuretic peptide increases GFR, apparently by making more surface area available for filtration (Fig. 19-15 ■). In addition, ANP has been reported to decrease NaCl and water reabsorption in the collecting duct. The mechanism by which ANP affects tubular reabsorption is not known.

Atrial natriuretic peptide also has several indirect effects on kidney function that alter Na^+ and water excretion. It inhibits the release of renin, aldosterone, and vasopressin. These actions all reinforce its natriuretic-diuretic effect. At this time, however, scientists are not convinced that ANP is the primary hormone responsible for regulating Na^+ excretion, so the search for other natriuretic factors continues.

Atrial natriuretic peptide and related peptides are also secreted by neurons in the brain. These neurons originate in the hypothalamus and terminate in the cardiovascular control center of the medulla. When secreted as a neurotransmitter or neuromodulator in this location, ANP apparently helps lower blood pressure,

Uroguanylin: A Na^+-Regulating Hormone? Scientists in the 1970s observed that when large amounts of salt are ingested, the body responds by excreting more Na^+ in the urine. On the other hand, if the same salt load is given intravenously, Na^+ excretion barely increases. From this observation, scientists proposed a feedforward mechanism, in which sensors in the intestine or hepatic portal system monitor the amount of Na^+ ingested and increase urinary Na^+ output when Na^+ intake is excessive. Now investigators have identified a family of candidate hormones, the guanylin peptides, named because they act through a guanylyl cyclase-cGMP second messenger system. One of these peptides is uroguanylin, found in the intestinal mucosa, plasma, and urine. Mice treated with uroguanylin increase their urinary output of Na^+, K^+, and water. Humans eating high-salt diets have elevated amounts of uroguanylin and salt in their urine. These observations suggest that uroguanylin may be one of the Na^+-regulating hormones that scientists have been looking for.

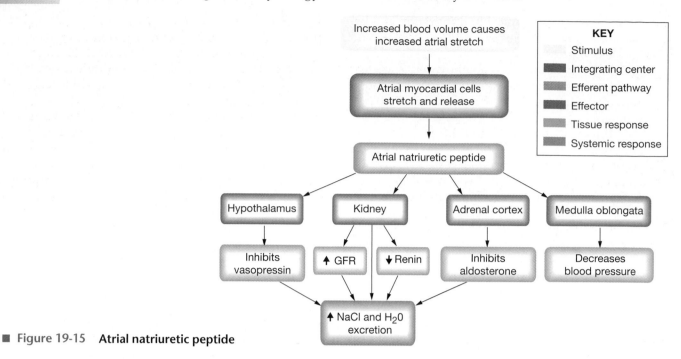

■ **Figure 19-15** **Atrial natriuretic peptide**

the very effect that makes it of interest to drug companies. Unfortunately, ANP has a short half-life in the body, and patients would have to take ANP continuously to maintain its effects. Researchers are hoping to learn more about the cellular mechanism of ANP action so that they can produce a longer-acting agonist that can be used to treat hypertension.

▶ POTASSIUM BALANCE

The regulation of body potassium levels is essential to maintain a state of well-being. Most K^+ is found within the cells: only about 2% is in the ECF. Although only a small fraction of the body's K^+ is in the ECF, potassium homeostasis keeps plasma K^+ concentrations within a narrow range. As you learned in Chapter 8, changes in the concentration of extracellular K^+ affect the resting membrane potential of cells by altering the K^+ concentration gradient between the cytoplasm and the ECF (∞ Fig. 8-18, p. 234).

If plasma (and ECF) K^+ concentrations decrease (**hypokalemia**), the concentration gradient becomes larger, more K^+ leaves the cell, and the resting membrane potential becomes more negative. If plasma (and ECF) K^+ increases (**hyperkalemia**), the concentration gradient decreases and more K^+ remains in the cell, depolarizing it.

Because of the effect of plasma K^+ on excitable tissues such as the heart, clinicians are always concerned about keeping plasma K^+ within the normal range (3.5–5 meq/L). If K^+ falls below 3 meq/L or rises above 6 meq/L, the excitable tissues of muscle and nerve begin to show altered function.

Hypokalemia causes muscle weakness because it is more difficult for hyperpolarized neurons and muscles

to fire action potentials. The danger in this condition lies in the failure of respiratory muscles and the heart. Fortunately, skeletal muscle weakness is usually significant enough to lead the patient to seek treatment before cardiac problems occur. Mild hypokalemia may be corrected by oral intake of K^+ supplements and K^+-rich foods, such as orange juice and bananas.

Hyperkalemia is a more dangerous potassium disturbance because depolarization of excitable tissues makes them more excitable initially. Subsequently, the cells are unable to repolarize fully and actually become *less* excitable. In this state, they have action potentials that are either smaller than normal or nonexistent. Cardiac muscle excitability affected by changes in plasma K^+ and hyperkalemia can lead to life-threatening cardiac arrhythmias.

Under normal conditions, the body very carefully matches K^+ excretion to K^+ ingestion. If plasma K^+ concentration goes up, aldosterone is secreted through the direct effect of hyperkalemia on the adrenal cortex. Aldosterone prompts the kidneys to excrete more K^+ while simultaneously retaining Na^+. (Remember that when plasma K^+ concentrations change, anions such as Cl^- are also added or subtracted to the ECF in a 1:1 ratio, maintaining overall electrical neutrality.)

Disturbances in K^+ balance may result from kidney disease, loss of K^+ in diarrhea, or the use of certain types of diuretics that prevent the kidneys from fully reabsorbing K^+. Inappropriate correction of dehydration can also create K^+ imbalance.

For example, a golfer went out to play a round of golf when the temperature was above 100° F. He was aware of the risk of dehydration, so he took water along to replace his fluid loss. The replacement of sweat loss with pure water kept the golfer's ECF volume normal

●●●●●●●●●●●●●●●●●●●●●●●●●●●●●●●

...continued from page 576

Nephrogenic diabetes insipidus is associated with low water permeability in the collecting duct despite adequate levels of vasopressin. Normally, vasopressin stimulates a cAMP second messenger pathway, which prompts insertion of water pores into apical membranes of the tubule cells. In the absence of vasopressin, the cells remove water pores and store them in the cytoplasm.

Question 2: *Scientists recently discovered that nephrogenic diabetes insipidus can be caused by two different defects. One defect affects the vasopressin V_2 receptor, leading to the inability of vasopressin to bind to the receptor and activate adenylyl cyclase. In the other defect, the V_2 receptors are normal. What could be the problem in the other defect?*

●●●●●●●●●●●●●●●●●●●●●●●●●●●●●●●

but dropped his total blood osmolarity and his K^+ and Na^+ concentrations. He was unable to finish the round of golf because of muscle weakness, and he required medical attention that included ion replacement therapy.

Potassium balance is also closely tied to acid-base balance, as you will learn in the last section of the chapter. Correction of a pH disturbance requires close attention to plasma K^+ levels. Similarly, correction of K^+ imbalance may alter body pH.

▶ BEHAVIORAL MECHANISMS IN SALT AND WATER BALANCE

Although neural, neuroendocrine, and endocrine reflexes play key roles in salt and water homeostasis, behavioral responses are critical in restoring the normal state, especially when ECF volume decreases or osmolarity increases. Drinking water is normally the only way to restore lost water, and eating salt is the only way to raise the body's Na^+ content. Both behaviors are essential for normal salt and water balance. The clinician must recognize the absence of these behaviors in patients who are unconscious or otherwise unable to obey behavioral urges, and must adjust treatment accordingly. The study of the biological basis for behaviors, including drinking and eating, is a field known as *physiological psychology*.

Drinking Replaces Fluid Loss

Thirst is one of the most powerful urges known in humans. In 1952, the Swedish physiologist Bengt Andersson showed that stimulating certain regions of the hypothalamus triggered drinking behavior. This discovery led to the identification of hypothalamic osmoreceptors that initiate drinking when osmolarity rises above 280 mOsM. This is an example of a behavior initiated by an internal stimulus.

Interestingly, although increased osmolarity triggers thirst, the act of drinking is sufficient to relieve thirst. The ingested water does not need to be absorbed for

thirst to be quenched. There are as-yet-unidentified receptors in the mouth and pharynx (*oropharynx receptors*) that respond to cold water by decreasing thirst and decreasing vasopressin release, although plasma osmolarity remains high. This oropharynx reflex is one reason that surgery patients are allowed to suck on ice chips: the ice alleviates their thirst without putting significant amounts of fluid into the digestive system.

A similar reflex exists in camels. When led to water, they will drink just enough to replenish their water deficit. The oropharynx receptors apparently act as a feedforward "metering" system that helps prevent wide swings in osmolarity by matching water intake to water need.

In humans, the reflex response of drinking to relieve increased osmolarity is complicated by cultural rituals. Drinking takes place during social events, not just in response to the stimulus of thirst. As a result, our bodies must be capable of eliminating excess fluid ingested in various social situations.

✔ Incorporate the thirst reflex into Figure 19-7 ■.

Low Na^+ Stimulates Salt Appetite

Thirst is not the only urge associated with fluid balance. Salt appetite is a craving for salty foods that occurs when plasma Na^+ concentrations drop. It can be observed in deer and cattle attracted to salt blocks or naturally occurring salt licks. In humans, salt appetite is linked to aldosterone and angiotensin, hormones that regulate Na^+ balance. The centers for salt appetite are in the hypothalamus close to the center for thirst.

Avoidance Behaviors Help Prevent Dehydration

Other behaviors play a role in fluid balance by preventing or triggering dehydration. Desert animals avoid the heat of the day and become active only at night, when environmental temperatures fall and humidity rises. Humans, especially now that we have air conditioning, are not always so wise.

The midday nap, or *siesta*, is a cultural adaptation in tropical countries that keeps people indoors during the hottest part of the day, thereby helping prevent dehydration and overheating. In the United States, we have abandoned this civilized custom and are active continuously during daylight hours, even when the temperature soars during summer in the South and Southwest. Fortunately, our homeostatic mechanisms usually keep us out of trouble.

▶ INTEGRATED CONTROL OF VOLUME AND OSMOLARITY

The disruption of salt and water balance is corrected using an integrated response from two systems. The cardiovascular system responds to changes in blood volume and the kidneys respond to changes in blood volume or osmolarity. Maintaining homeostasis throughout the day is a continuous process in which the amounts of salt and water in the body shift, according to whether you just drank a soft drink or sweated through an aerobics class.

In that respect, maintaining fluid balance is like driving a car down the highway: small adjustments keep the car in the center of the lane. But just as exciting movies are made from wild car chases, not from sedate driving, the exciting part of fluid homeostasis is the body's response to crisis situations such as severe dehydration or hemorrhage. In this section, we examine challenges to salt and water balance.

Osmolarity and ECF Volume Can Change Independently

Normally, blood volume and osmolarity in the body are homeostatically maintained within a narrow range. However, under some circumstances, fluid loss exceeds fluid gain or vice versa, and the body goes out of balance. Common examples of fluid loss include excessive sweating, vomiting, diarrhea, and **hemorrhage** (excessive blood loss). All of these situations may require medical intervention. In contrast, fluid gain is seldom a medical emergency.

Volume and osmolarity of the ECF can each have three possible states: normal, increased, or decreased. The relation of volume and osmolarity changes can be represented by a matrix as shown in Figure 19-16 ■. The center box represents the normal state and the surrounding boxes represent the most common examples of the possible changes.

In all cases, the appropriate homeostatic compensation for the change attempts to follow the principle of mass balance: whatever fluid and solute were added to the body must be removed, or what was lost must be re-

placed. However, perfect compensation is not always possible. Let's begin at the upper right corner and move right to left.

1. **Increased volume and increased osmolarity.** This situation might occur if you ate salty food and drank liquids at the same time, such as popcorn and a soft drink at the movies. The net result may be ingestion of hypertonic saline that increases ECF volume and osmolarity. The appropriate homeostatic response is excretion of hypertonic urine. For homeostasis to be maintained, the osmolarity and volume of the urinary output must match the salt and water input from the popcorn and soft drink.

2. **Increased volume but no change in osmolarity.** If the proportion of salt and water in the ingested food is equivalent to an isotonic NaCl solution, volume will increase but osmolarity will not change. The appropriate response is excretion of isotonic urine whose volume equals that of the ingested fluid.

3. **Increased volume and decreased osmolarity.** This situation would happen if you drank pure water without ingesting any solute. The goal here would be to excrete a very dilute urine to maximize water loss while conserving salts. However, because our kidneys cannot excrete pure water, there will always be some loss of solute in the urine. In this example, urinary output cannot exactly match input, so compensation is imperfect.

4. **Normal volume with increased osmolarity.** This disturbance might occur if you ate salted popcorn without drinking anything. The ingestion of salt without water increases ECF osmolarity and causes some water to shift from cells to the ECF. The homeostatic response is intense thirst, which prompts ingestion of water to dilute the added solute. The kidneys help by creating a highly concentrated urine of minimal volume, conserving water while removing the excess NaCl. Once water is ingested, the disturbance becomes that described in situation 1 or situation 2 above.

5. **Normal volume with decreased osmolarity.** This scenario might occur when a person who is dehydrated (situation 6: decreased volume, increased osmolarity) replaces the fluid loss with pure water, like the golfer described earlier. The decreased volume is corrected, but the replacement fluid has no solute to replace that which was lost. Consequently, a new imbalance is created.

This situation led to the development of electrolyte-containing sports beverages. If people working out in hot weather replace their sweat loss with pure water, they may restore volume but they run the risk of diluting their plasma K^+ and Na^+ concentrations to dangerously low levels (*hypokalemia* and *hyponatremia*).

Osmolarity

		Decrease	No change	Increase
Volume	Increase	Drinking large amount of water	Ingestion of isotonic saline	Ingestion of hypertonic saline
	No change	Replacement of sweat loss with plain water	Normal volume and osmolarity	Eating salt without drinking water
	Decrease	Incomplete compensation for dehydration	Hemorrhage	Dehydration (e.g., sweat loss or diarrhea)

■ **Figure 19-16** **Disturbances of volume and osmolarity**

6. **Decreased volume and increased osmolarity.** Dehydration is a common example of this fluid and electrolyte disturbance. Dehydration has multiple causes. During prolonged heavy exercise, water loss from the lungs can double, while sweat loss may increase from 0.1 liter to as much as 5 liters! The fluid secreted by sweat glands is hyposmotic, so the fluid left behind in the body becomes hyperosmotic.

   Diarrhea [*diarhein,* to flow through], excessive watery feces, is a pathological condition involving major water and solute loss, this time from the digestive tract. In both sweating and diarrhea, if too much fluid is lost from the circulatory system, the blood volume decreases to the point that the heart can no longer pump blood effectively to the brain. In addition, cell shrinkage caused by increased osmolarity disrupts cell function.

7. **Decreased volume with no change in osmolarity.** This situation occurs with excessive blood loss, or *hemorrhage.* Blood loss represents the loss of isosmotic fluid from the extracellular compartment, similar to scooping a cupful of seawater out of a large bucketful of seawater. If a blood transfusion is not immediately available, the best replacement solution is one that is isosmotic and remains in the ECF, such as isotonic NaCl.

8. **Decreased volume and decreased osmolarity.** This situation might also result from incomplete compensation of dehydration, but it is uncommon, so we will ignore it.

Dehydration Triggers Renal and Cardiovascular Responses

To put together an integrated response to changes in volume and osmolarity, you must first have a clear idea of which pathways become active in response to various stimuli. Table 19-2 is a summary of the many pathways involved in the homeostasis of salt and water balance. For details of individual pathways, refer to the figures cited in Table 19-2.

Severe dehydration is an excellent example of how the body works to maintain blood volume and cell volume in the face of decreased volume and increased osmolarity. It also illustrates the role of neural and endocrine integrating centers. In dehydration, the adrenal cortex receives two opposing signals. One says, "Secrete aldosterone. " The other says, "Do not secrete aldosterone." The rule of thumb that fluid balance integrating centers use when faced with conflicting signals is: *Attempt to restore osmolarity to normal before restoring volume.*

The body has multiple mechanisms to deal with diminished blood volume, but high osmolarity causes cells to shrink and presents a more immediate threat to well-being. Thus, faced with decreased volume and increased osmolarity, the adrenal cortex will not secrete aldos-

terone. (If secreted, aldosterone would cause Na^+ reabsorption that could worsen the already-high osmolarity associated with dehydration.)

In severe dehydration, compensatory mechanisms are aimed at restoring normal blood pressure, ECF volume, and osmolarity by (1) conserving fluid to prevent additional loss, (2) triggering cardiovascular reflexes to increase blood pressure, and (3) stimulating thirst so that normal fluid volume and osmolarity can be restored. Figure 19-17 ■ maps the interwoven nature of these responses. This figure is complex and intimidating at first glance, so we will discuss it step by step.

At the top of the map (in yellow) are the two stimuli caused by dehydration: decreased volume and increased osmolarity. Decreased ECF volume causes decreased blood pressure.

Blood pressure acts as a stimulus for several pathways, mediated through the baroreceptors of the carotid and aortic bodies, stretch receptors in the atria, and the pressure-sensitive JG cells:

1. The carotid and aortic baroreceptors signal the cardiovascular control center (CVCC) to raise blood pressure. Sympathetic output from the cardiovascular control center increases while parasympathetic output decreases.
   a. Heart rate goes up as control of the SA node shifts from predominantly parasympathetic to sympathetic.
   b. The force of ventricular contraction also increases under sympathetic stimulation. Increased ventricular contraction combines with increased heart rate to increase cardiac output.
   c. Simultaneously, sympathetic input causes arteriolar vasoconstriction, increasing peripheral resistance.
   d. Sympathetic vasoconstriction of afferent arterioles in the kidneys decreases GFR, helping conserve fluid.
   e. Increased sympathetic activity at the JG cells of the kidneys increases renin secretion.
2. Decreased peripheral blood pressure directly decreases GFR. A lower GFR conserves ECF by filtering less of it into the nephron.
3. In addition, lower GFR decreases fluid flow past the macula densa, which releases less NO. Removal of paracrine feedback causes the JG cells to release renin.
4. The JG cells respond directly to decreased blood pressure. The combination of decreased blood pressure, increased sympathetic input, and signals from the macula densa stimulates renin release and ensures increased production of angiotensin II.
5. Decreased arterial pressure, decreased blood volume, increased osmolarity and ANG II all stimulate vasopressin and thirst centers of the hypothalamus.

The redundancy in the control pathways ensures that all four main compensatory mechanisms are activated:

TABLE 19-2 Reflexes Triggered by Changes in Volume, Blood Pressure, and Osmolarity

Stimulates	Organ or Tissue Involved	Response	Figure
Decreased blood pressure			
Direct effects	JG cells	Renin secretion	19-14
	Glomerulus	Decreased GFR	18-6, 18-7
Reflexes			
Carotid and aortic baroreceptors	Cardiovascular control center	Increased sympathetic, decreased parasympathetic output	15-22
Carotid and aortic baroreceptors	Hypothalamus	Thirst stimulation	19-15
Carotid and aortic baroreceptors	Hypothalamus	Vasopressin secretion	19-7
Atrial baroreceptors	Hypothalamus	Thirst stimulation	19-17
Atrial baroreceptors	Hypothalamus	Vasopressin secretion	19-17
Increased blood pressure			
Direct effects	Glomerulus	Increased GFR (transient)	
	Atrial cells	ANP secretion	19-15
Reflexes			
Carotid and aortic baroreceptors	Cardiovascular control center	Decreased sympathetic, increased parasympathetic output	15-21
Carotid and aortic baroreceptor	Hypothalamus	Thirst inhibition	
Carotid and aortic baroreceptor	Hypothalamus	Vasopressin inhibition	19-7
Atrial baroreceptors	Hypothalamus	Thirst inhibition	
Atrial baroreceptors	Hypothalamus	Vasopressin inhibition	19-7
Increased osmolarity			
Direct effects	Adrenal cortex	Decreased aldosterone secretion	19-17
Reflexes			
Osmoreceptors	Hypothalamus	Thirst stimulation	19-17
Osmoreceptors	Hypothalamus	Vasopressin secretion	19-17
Decreased osmolarity			
Direct effects	None	None	
Reflexes			
Osmoreceptors	Hypothalamus	Decreased vasopressin secretion	

cardiovascular responses, ANG II, vasopressin, and thirst.

1. **Cardiovascular responses** combine enhanced cardiac output and increased peripheral resistance to raise blood pressure. Note, however, that this increase in blood pressure does *not necessarily* mean that blood pressure returns to normal. If dehydration is severe, compensation may be incomplete, and blood pressure may remain below normal.

2. **Angiotensin II** has a variety of effects aimed at raising blood pressure including stimulation of thirst, vasopressin release, direct vasoconstriction, and reinforcement of the cardiovascular control center output. ANG II also goes to the adrenal cortex and attempts to stimulate aldosterone release. In dehydration, however, Na^+ reabsorption would worsen the already-high osmolarity. Consequently, high osmolarity at the adrenal cortex directly inhibits aldosterone release, blocking the action of ANG II. The

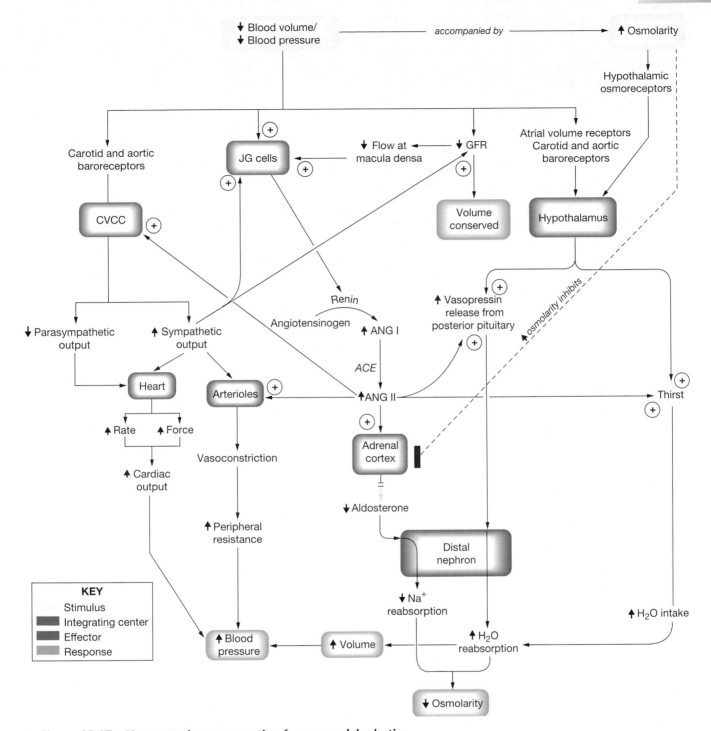

■ **Figure 19-17**   **Homeostatic compensation for severe dehydration**

result of the RAAS pathway in dehydration permits the beneficial blood pressure-enhancing effects of ANG II while avoiding the detrimental effects of Na^+ reabsorption. This is a beautiful example of integrated function.

3. **Vasopressin** increases the water permeability of the collecting ducts, allowing water reabsorption to conserve fluid. Without fluid replacement, however, vasopressin cannot bring volume and osmolarity back to normal.

4. **Oral (or intravenous) intake of water** is the only mechanism for replacing lost fluid volume and for restoring osmolarity to normal.

The net result of all four mechanisms is shown in green at the bottom of Figure 19-17 ■. The net result includes: (1) restoration of blood volume by water conservation and fluid intake, (2) maintenance of blood pressure through increased blood volume, enhanced cardiac output, and vasoconstriction, and (3) restoration of

...continued from page 585

"Christopher is always thirsty," Mrs. Godell told the doctor. "And he constantly urinates. With all the water he drinks, sometimes I think he'll drown!" The doctor explained that people with diabetes insipidus will not "drown." On the contrary, they run the risk of severe dehydration, despite the copious amounts of water they drink.

Question 3: Why do people with diabetes insipidus become dehydrated even though they are drinking a lot?

Question 4: Explain why intense thirst is a major symptom of diabetes insipidus.

normal osmolarity by decreased Na^+ reabsorption and increased water reabsorption and intake.

Using the pathways listed in Table 19-2, see if you can create reflex maps for the other disturbances of volume and osmolarity shown in Figure 19-16 ■.

▶ ACID-BASE BALANCE

Acid-base balance, also called pH homeostasis, is one of the essential functions of the body. The pH of a solution is a measure of its H^+ concentration [∞ p. 27]. A normal arterial plasma sample has an H^+ concentration of 0.00004 meq/L, minute compared with other ions. (For example, the plasma concentration of Na^+ is about 135 meq/L.)

Because H^+ concentration is so low, it is commonly expressed on a logarithmic pH scale of 0–14, where a pH of 7.0 is neutral. If pH falls below 7.0, H^+ concentration has increased above 1×10^{-7} M and the solution is considered acidic. If pH rises above 7.0, the solution has an H^+ concentration lower than 1×10^{-7} M and is considered alkaline (basic).

The normal pH of the body is 7.40, slightly alkaline. A change of 1 pH unit represents a 10-fold change in H^+ concentration. If you need to review pH and the logarithmic scale upon which it is based, please see Appendix A.

Enzymes and the Nervous System Are Particularly Sensitive to Changes in pH

The normal range of pH in the plasma is 7.38–7.42. Extracellular pH usually reflects intracellular pH and vice versa. Because monitoring intracellular conditions is difficult, plasma values are used clinically as an indicator of whole body pH.

Body fluids that are "outside" the body's internal environment, such as those in the lumen of the gastrointestinal tract or kidney tubule, can have a much greater range of pH. Acidic secretions in the stomach may create gastric pH as low as 1. The pH of urine varies between 4.5 and 8.5, depending on the body's need to excrete H^+ or HCO_3^-.

The concentration of H^+ in the body is closely regulated. Intracellular proteins such as enzymes and membrane channels are particularly sensitive to pH because the function of these proteins depends on their three-dimensional shape [∞ p. 32]. Changes in H^+ concentration alter the tertiary structure of proteins by interacting with hydrogen bonds in the molecule, disrupting the protein's three-dimensional structure and activity.

Abnormal pH may significantly affect the activity of the nervous system. If pH is too low, the condition known as **acidosis,** neurons become less excitable and CNS depression results. Patients become confused and disoriented, then slip into a coma. If CNS depression progresses, the respiratory centers cease to function, causing death.

If pH is too high, the condition known as **alkalosis,** neurons become hyperexcitable, firing action potentials at the slightest signal. This condition shows up first as sensory changes, such as numbness or tingling, then as muscle twitches. If alkalosis is severe, muscle twitches turn into sustained contractions (*tetanus*) that paralyze respiratory muscles.

Disturbances of acid-base balance are associated with disturbances in K^+ balance. This is partly due to a renal transporter that moves the two ions in an antiport fashion. In acidosis, the kidneys excrete H^+ and reabsorb K^+ using an H^+-K^+-ATPase. Conversely, when the body goes into a state of alkalosis, the kidneys reabsorb H^+ and excrete K^+. Potassium imbalance usually shows up as disturbances in excitable tissues, especially the heart.

Acids and Bases in the Body Come from Many Sources

In day-to-day functioning, the body is challenged by intake and production of acids more than bases. Hydrogen ions come both from food and from internal metabolism. Maintaining mass balance requires that acid intake and production be balanced by acid excretion. Hydrogen balance in the body is summarized in Figure 19-18 ■.

Acid Input Many metabolic intermediates and foods are organic acids that ionize and contribute H^+ to body fluids. Examples of organic acids include amino acids, fatty acids, intermediates in the citric acid cycle, and lactic acid produced by anaerobic metabolism. Often these organic acids are recognizable by the suffix *-ate,* which denotes the anion (A) half of an acid:

$$HA \rightleftarrows A^- + H^+$$
Example: pyruvic acid $\rightleftarrows$ pyruvate $+ H^+$

Metabolic production of organic acids each day creates a significant amount of H^+ that must be excreted to maintain mass balance.

Under extraordinary circumstances, metabolic organic acid production can increase significantly and create a crisis. For example, severe anaerobic conditions,

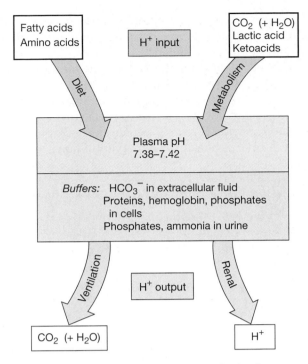

■ Figure 19-18   Hydrogen balance in the body

Base Input   If acid-base physiology focuses on acids, it is for good reasons. First, the diet and metabolism have few significant sources of bases. Some fruits and vegetables contain anions that metabolize to HCO_3^-, but their influence is far outweighed by the contribution of acidic fruits, amino acids, and fatty acids. Second, in acid-base disturbances, those due to excess acid are more common than those due to excess base. For these reasons, the body spends most of its time removing excess acid.

pH Homeostasis Depends on Buffers, the Lungs, and the Kidneys

How does the body cope with minute-to-minute changes in pH? There are three mechanisms: (1) buffers, (2) ventilation, and (3) renal regulation of H^+ and HCO_3^-. Buffers are the first line of defense, always present and waiting to prevent wide swings in pH. Ventilation, the second line of defense, is a rapid, reflexively controlled response that can take care of 75% of most pH disturbances. The final defense of pH lies with the kidneys. They are slower than buffers or the lungs but are very effective at coping with any remaining pH disturbance under normal conditions. These three mechanisms help the body balance acid so effectively that normal body pH varies only slightly. Let's take a closer look at each of them.

Buffer Systems Include Proteins, Phosphate Ions, and HCO_3^-

A *buffer* is a molecule that moderates but does not prevent changes in pH by combining with or releasing H^+. In the absence of buffers, the addition of acid to a solution will cause a sharp change in pH. In the presence of a buffer, the pH change will be moderated or may even be unnoticeable. Because acid production is the major challenge to pH homeostasis, most physiological buffers combine with H^+.

Buffers are found both within the cells and in the plasma. Intracellular buffers include cellular proteins, phosphate ions (HPO_4^{2-}), and hemoglobin. As we have seen, hemoglobin in red blood cells buffers the H^+ produced by the reaction of CO_2 with H_2O (∞ Fig. 17-26, p. 528).

Each H^+ ion buffered by hemoglobin leaves a matching bicarbonate (HCO_3^-) ion inside the red blood cell. The HCO_3^- then leaves the red blood cell by exchanging with a plasma Cl^- ion. This exchange is the *chloride shift* described in Chapter 17 [∞ p. 528].

The large amounts of bicarbonate produced from metabolic CO_2 create the most important extracellular buffer system of the body. Plasma HCO_3^- concentration averages 24 meq/L, which is approximately 600,000 times as concentrated as plasma H^+. Although H^+ and HCO_3^- are created in a 1:1 ratio from CO_2 and H_2O, intracellular buffering of the H^+ by hemoglobin is a major reason why the two ions do not appear in the plasma in

such as circulatory collapse, produce so much lactic acid that normal homeostatic mechanisms cannot keep pace, resulting in a state of *lactic acidosis.* In diabetes mellitus, abnormal fat and amino acid metabolism creates strong acids known as **ketoacids.** These acids cause a state of metabolic acidosis known as *ketoacidosis.*

The biggest source of acid on a daily basis is the production of CO_2 during aerobic respiration. Carbon dioxide is not an acid because it does not contain any hydrogen atoms. However, CO_2 from respiration combines with water to form carbonic acid (H_2CO_3), which dissociates into H^+ and HCO_3^- :

$$CO_2 + H_2O \rightleftarrows H_2CO_3 \rightleftarrows H^+ + HCO_3^-$$

This reaction takes place in all cells and in the plasma, but at a slow rate. However, in certain cells of the body, the reaction goes very rapidly due to the presence of large amounts of *carbonic anhydrase* [∞ p. 528]. This enzyme catalyzes the conversion of CO_2 and H_2O to H_2CO_3.

The production of H^+ from CO_2 and H_2O is the single biggest source of acid input under normal conditions. By some estimates, CO_2 from resting metabolism produces 12,500 meq of H^+ in one day. If this amount of acid were placed in a volume of water equal to the plasma volume, it would create an H^+ concentration of 4167 meq/L, over *100 million* (10^8) times as concentrated as the normal plasma H^+ concentration of 0.00004 meq/L!

These numbers show that CO_2 from aerobic respiration has the potential to seriously affect pH in the body. Fortunately, homeostatic mechanisms normally prevent CO_2 from accumulating in the body.

the same concentration. The HCO_3^- in plasma can then buffer H^+ that comes from nonrespiratory sources such as metabolism.

The relationship between CO_2, HCO_3^-, and H^+ in the plasma is expressed by the following equation:

(1) $CO_2 + H_2O \rightleftarrows \underset{\text{carbonic acid}}{H_2CO_3} \rightleftarrows H^+ + HCO_3^-$

According to the law of mass action, any change in CO_2, H^+, or HCO_3^- will cause the reaction to shift until a new equilibrium is reached. (Water is always in excess in the body and does not contribute to the equilibrium state of the reaction.)

For example, if CO_2 increases (shown in blue), the equation shifts to the right, creating one additional H^+ and one additional HCO_3^- from each CO_2 and water:

(2) $\uparrow CO_2 + H_2O \rightarrow H_2CO_3 \rightarrow \uparrow H^+ + \uparrow HCO_3^-$

Once a new equilibrium state is reached, both H^+ and HCO_3^- levels have increased. The addition of H^+ makes the solution more acidic and lowers its pH. It does not matter that a HCO_3^- buffer molecule has also been produced: HCO_3^- acts as a buffer only when it binds to H^+.

Now suppose H^+ is added to the plasma from some metabolic source such as lactic acid. In this case, plasma HCO_3^- will act as a buffer, combining with some of the additional H^+ until the reaction reaches a new equilibrium state. Before equilibrium

(3) $CO_2 + H_2O \rightleftarrows H_2CO_3 \rightleftarrows \uparrow H^+ + HCO_3^-$
     (additional H^+ from lactic acid)

The increase in H^+ shifts the equation to the left:

(4) $CO_2 + H_2O \leftarrow H_2CO_3 \leftarrow \uparrow H^+ + HCO_3^-$

The net result at equilibrium is that H^+ is still elevated, but not as much as initially. HCO_3^- is decreased because some has been used as a buffer. The buffered H^+ is converted to CO_2 and H_2O, increasing the amounts of both:

(5) $\uparrow CO_2 + \uparrow H_2O \rightleftarrows H_2CO_3 \rightleftarrows \uparrow H^+ + \downarrow HCO_3^-$

The law of mass action is a useful way to think about the relationship between changes in the concentrations of H^+, HCO_3^-, and CO_2 as long as you remember certain qualifications.

First, a change in HCO_3^- concentration according to the reaction may not show up clinically as a HCO_3^- concentration outside the normal range. This is because HCO_3^- is 600,000 times as concentrated in the plasma as H^+ is. If both H^+ and HCO_3^- are added to the plasma, you may observe changes in pH but not in HCO_3^- concentrations because so much HCO_3^- was present initially. Both H^+ and HCO_3^- experience an *absolute* increase, but because so many HCO_3^- were in the plasma to begin with, the *relative increase* in HCO_3^- goes unnoticed.

As an analogy, think of two football teams playing in a packed stadium before 80,000 fans. If 10 more players (H^+) run out onto the field, everyone notices. But if 10 people (HCO_3^-) come into the stands at the same time, no one pays any attention because there were already so many people watching the game that 10 more make no significant difference.

The second qualification for the law of mass action is that when the reaction shifts to the left and increases plasma CO_2, a nearly instantaneous increase in ventilation takes place (in a normal person). If extra CO_2 is ventilated off, arterial P_{CO_2} may remain normal or even fall below normal as a result of hyperventilation.

Ventilation Can Compensate for pH Disturbances

The increase in ventilation just described is a *respiratory compensation* for acidosis. Ventilation and acid-base status are intimately linked, as shown by the now-familiar equation below:

$$CO_2 + H_2O \rightleftarrows H_2CO_3 \rightleftarrows H^+ + HCO_3^-$$

Changes in ventilation can correct disturbances in acid-base balance, but they can also cause them. Because of the dynamic equilibrium between CO_2 and H^+, any change in plasma P_{CO_2} will affect both H^+ and HCO_3^- content of the blood.

For example, if a person hypoventilates and P_{CO_2} increases, the equation shifts to the right. More carbonic acid is formed and H^+ goes up, creating a more acidotic state:

(6) $\uparrow CO_2 + H_2O \rightarrow H_2CO_3 \rightarrow \uparrow H^+ + \uparrow HCO_3^-$

If a person hyperventilates, blowing off CO_2 and decreasing the plasma P_{CO_2}, the equation shifts to the left. H^+ combines with HCO_3^-, thereby raising the pH:

(7) $\downarrow CO_2 + H_2O \leftarrow H_2CO_3 \leftarrow \downarrow H^+ + \downarrow HCO_3^-$

In these two examples, you can see that a change in P_{CO_2} will affect the H^+ concentration and therefore the pH of the plasma.

The body uses ventilation as a method for adjusting pH only if a stimulus associated with pH triggers the reflex response. Two stimuli can do so: H^+ and CO_2.

Ventilation is affected directly by H^+ through carotid and aortic chemoreceptors (Fig. 19-19 ■). These are located in the carotid and aortic bodies along with the oxygen and blood pressure sensors we discussed previously [∞ pp. 463, 532]. An increase in plasma H^+ stimulates the chemoreceptors, which in turn signal the medullary control centers to increase ventilation. Increased alveolar ventilation allows the lungs to excrete more CO_2 and convert H^+ to carbonic acid.

The central chemoreceptors of the medulla oblongata cannot respond directly to changes in plasma pH because H^+ does not cross the blood-brain barrier. However, changes in pH change P_{CO_2}, and CO_2 stimulates the central chemoreceptors [∞ Fig. 17-31, p. 534]. Dual control of ventilation through the central and peripheral chemoreceptors helps the body respond rapidly to changes in either pH or plasma CO_2.

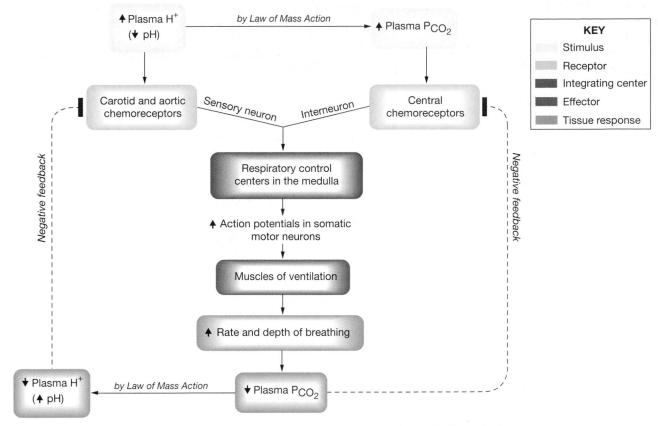

■ **Figure 19-19**   **The reflex pathway for respiratory compensation of metabolic acidosis**

✔ In Figure 19-19 ■, name the muscles of ventilation that might be involved in this reflex.

✔ In equation 6 above, HCO_3^- is shown as increasing in amount. Why doesn't the HCO_3^- buffer the increased H^+ and prevent acidosis from occurring?

Kidneys Excrete or Reabsorb H^+ and HCO_3^-

The kidneys take care of the 25% of compensation that the lungs cannot handle. They alter pH two ways: (1) directly by excreting or reabsorbing H^+ and (2) indirectly by changing the reabsorption or excretion of HCO_3^- buffer.

In acidosis, the kidney excretes H^+ into the tubule lumen using direct and indirect active transport (Fig. 19-20 ■). Ammonia and phosphate ions in the kidney act as buffers, trapping large amounts of H^+ as NH_4^+ and $H_2PO_4^-$ and allowing more H^+ to be excreted. Phosphate ions (HPO_4^{2-}) are present in filtered fluid and combine with H^+ secreted by the tubule:

$$HPO_4^{2-} + H^+ \rightleftarrows H_2PO_4^-$$

Ammonia is made from amino acids, as described in the next section.

Even with these buffers, urine can become quite acidic, down to a pH of about 4.5. While H^+ is being excreted, the kidneys make new HCO_3^- from CO_2 and

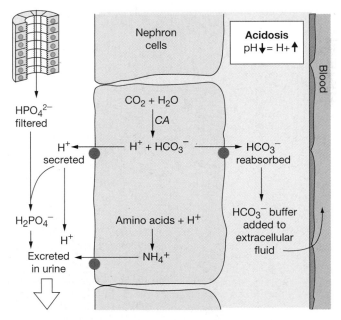

■ **Figure 19-20   Renal compensation for acidosis**   The transporters shown in this figure are generic transporters. The actual transporters are shown in Figures 19-21 and 19-22.

H_2O. The HCO_3^- is reabsorbed into the blood to act as a buffer and increase pH.

In times of alkalosis, the kidney reverses the general process described above, excreting HCO_3^- and reabsorbing

H^+ in an effort to bring pH back into the normal range. Renal compensations are slower than respiratory compensations, and their effect on pH may not be noticed for 24–48 hours. However, once activated, renal compensations effectively handle all but severe acid-base disturbances.

The cellular mechanisms for renal handling of H^+ and HCO_3^- resemble transport processes in other epithelia. However, these mechanisms involve some membrane transporters that you have not encountered before:

1. The **apical Na^+-H^+ antiporter** is an indirect active transporter that brings Na^+ into the epithelial cell in exchange for moving H^+ against its concentration gradient into the lumen.
2. The **basolateral Na^+-HCO_3^- symporter** moves Na^+ and HCO_3^- out of the epithelial cell and into the interstitial fluid. This indirect active transporter couples the energy of HCO_3^- diffusing down its concentration gradient to the uphill movement of Na^+ from the cell to the ECF.
3. The **H^+-ATPase** uses energy from ATP to acidify the urine, pushing H^+ against its concentration gradient into the lumen of the distal nephron.
4. The **H^+-K^+-ATPase** puts an H^+ into the urine in exchange for a reabsorbed K^+. This exchange contributes to the potassium imbalance that sometimes accompanies acid-base disturbances.
5. A **Na^+-NH_4^+ antiport** moves an ammonium ion from the cell to the lumen in exchange for Na^+.

In addition to these transporters, the renal tubule also uses the ubiquitous Na^+-K^+-ATPase and the same HCO_3^--Cl^- antiport protein that is responsible for the chloride shift in red blood cells.

The Proximal Tubule: Hydrogen Ion Excretion and Bicarbonate Reabsorption

The kidneys filter the equivalent of a pound of baking soda bicarbonate each day, and most of it must be reabsorbed to maintain the body's buffer capacity. The proximal tubule reabsorbs most filtered HCO_3^- by indirect methods because there is no membrane transporter to bring HCO_3^- into the tubule cell. In Figure 19-21 ■, the numbers correspond to the steps below. In these figures, you will see how the transporters listed in the previous section function together.

In the proximal tubule, there are two pathways for bicarbonate reabsorption:

1. H^+ is secreted into the lumen in exchange for a filtered Na^+, using the Na^+-H^+ antiport protein.

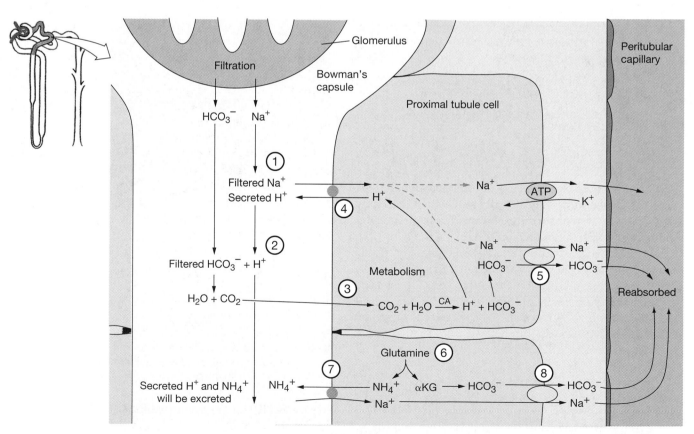

■ **Figure 19-21** **Reabsorption of filtered HCO_3^- in the proximal tubule and excretion of H^+** See text for explanations of the numbers.

2. The secreted H^+ then combines with filtered HCO_3^- to form CO_2 in the lumen.

3. CO_2 diffuses into the proximal tubule cell and combines with water to form H_2CO_3, which dissociates to form H^+ and HCO_3^- in the cytoplasm.

4. The H^+ can be secreted again, replacing the H^+ that combined with the filtered HCO_3^-.

5. The HCO_3^- is transported out of the cell on the basolateral side by the HCO_3^- -Na^+ symporter.

The net result of this process is reabsorption of filtered Na^+ and HCO_3^- and excretion of H^+.

A second way to reabsorb bicarbonate and excrete H^+ comes from metabolism of the amino acid *glutamine* (Fig. 19-21 ■).

6. Glutamine loses its two amino groups, which become ammonia (NH_3). The ammonia buffers H^+ to become ammonium ion.

7. Ammonium is transported into the lumen in exchange for Na^+.

8. The α-ketoglutarate molecule made from deamination of glutamine is metabolized further to HCO_3^-, which is transported into the blood along with Na^+.

The net result of these pathways is reabsorption of sodium bicarbonate—baking soda.

The Distal Nephron: H^+ and HCO_3^- Handling Depend on the Acid-Base State of the Body   The distal nephron plays a significant role in the fine regulation of acid-base balance. Special cells known as **intercalated cells,** or **I cells,** are interspersed among the principal cells in this section of the nephron. They are responsible for acid-base regulation.

Intercalated cells are characterized by high concentrations of carbonic anhydrase in their cytoplasm. This enzyme allows them to convert large amounts of CO_2 into H^+ and HCO_3^-. The H^+ ions are pumped out of the intercalated cell by the H^+-ATPase or by the ATPase that exchanges one H^+ for one K^+. Bicarbonate leaves the cell by means of the HCO_3^--Cl^- antiport exchanger.

Intercalated cells come in two variants whose transporters are found on different faces of the epithelial cell. In times of acidosis, type A intercalated cells secrete H^+ and conserve bicarbonate. During periods of alkalosis, type B intercalated cells secrete HCO_3^- and reabsorb H^+.

Figure 19-22a ■ shows how the type A intercalated cell works in times of acidosis, secreting H^+ and reabsorbing HCO_3^-. The process is similar to H^+ secretion in the proximal tubule except for the specific H^+ transporters. The distal nephron uses a H^+-ATPase and a H^+-K^+-ATPase rather than a Na^+-H^+ antiport protein.

In times of alkalosis, when the H^+ concentration of the body is too low, H^+ is reabsorbed and HCO_3^- buffer is excreted in the urine (Fig. 19-22b ■). Once again, the ions are produced by the dissociation of H_2CO_3 formed from H_2O and CO_2. Hydrogen ions are reabsorbed into the ECF on the basolateral side, while HCO_3^- is secreted into the lumen. The polarity of the two types of cells is reversed, with the transport proteins found on the opposite sides of the cell.

The H^+-K^+-ATPase of the distal nephron helps create parallel disturbances of acid-base and K^+ balance. In acidosis, when plasma H^+ is high, the kidney excretes H^+ and reabsorbs K^+. Thus, acidosis is often accompanied by hyperkalemia. (There are also nonrenal events that contribute to elevated ECF K^+ concentrations in acidosis.)

The reverse is true for alkalosis, when blood H^+ levels are low. The mechanism that allows the distal nephron to retain H^+ simultaneously causes it to excrete K^+, so alkalosis goes hand in hand with hypokalemia.

✔ Why does the K^+-H^+ transporter of the distal nephron require ATP to secrete H^+ but the Na^+-H^+ exchanger of the proximal tubule does not?

✔ In hypokalemia, the intercalated cells of the distal nephron reabsorb K^+ from the tubule lumen. What happens to pH in hypokalemia as a result?

Acid-Base Disturbances May Be Respiratory or Metabolic in Origin

The three compensatory mechanisms (buffers, ventilation, and renal excretion) take care of most variations in plasma pH. But under some circumstances, the production or loss of H^+ or HCO_3^- is so extreme that compensatory mechanisms fail to maintain pH homeostasis. In these states, the pH of the blood moves out of the normal range of 7.38–7.42. If the body fails to keep pH within a range of pH 7.00–7.70, acidosis or alkalosis can be fatal.

Acid-base problems are classified both by the direction of the pH change (acidosis or alkalosis) and by the underlying cause (metabolic or respiratory). As you learned earlier, changes in P_{CO_2} due to hyper- or hypoventilation cause pH to shift. These disturbances are said to be of respiratory origin. If the pH problem arises from acids or bases of non-CO_2 origin, the problem is said to be a metabolic problem.

Note that by the time an acid-base disturbance shows up as a change in plasma pH, the body's buffers are ineffectual. The loss of buffering ability leaves the body with only two options: respiratory compensation and renal compensation.

If the problem is of respiratory origin, only one homeostatic compensation is available—the kidneys. On the other hand, in metabolic acid-base disturbances, both respiratory and renal mechanisms can compensate.

The combination of an initial pH disturbance and compensatory changes is one factor that makes analysis of acid-base disorders in the clinical setting so difficult. In this book, we will concentrate on the simple disturbances with a single cause.

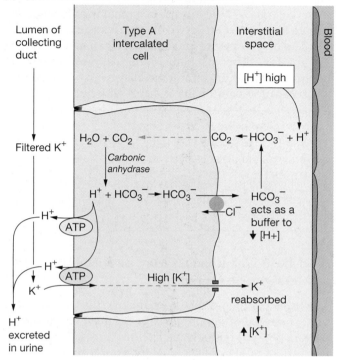

(a) Type A intercalated cell function in acidosis

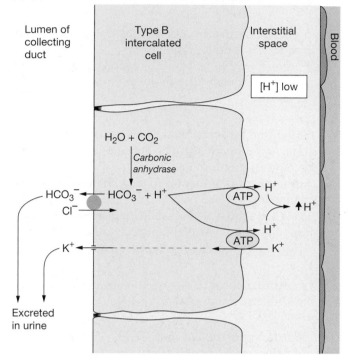

(b) Type B intercalated cell function in alkalosis

■ **Figure 19-22   Role of the intercalated cell in acidosis and alkalosis**   Intercalated cells in the collecting duct secrete or reabsorb H^+ and HCO_3^- according to the needs of the body.

Respiratory Acidosis   A state of respiratory acidosis occurs when alveolar hypoventilation results in CO_2 retention and elevated plasma P_{CO_2}. Some situations in which this occurs include respiratory depression due to drugs or alcohol, increased airway resistance in asthma, impaired gas exchange in fibrosis or severe pneumonia, or muscle weakness in muscular dystrophy and other muscle diseases. The most common cause of respiratory acidosis is *chronic obstructive pulmonary disease* (COPD), such as emphysema, in which inadequate gas exchange is compounded by loss of alveolar exchange area.

No matter what the cause of respiratory acidosis, plasma CO_2 levels increase (shown in blue), leading to elevated H^+ and HCO_3^-:

$$\uparrow CO_2 + H_2O \rightarrow H_2CO_3 \rightarrow \uparrow H^+ + \uparrow HCO_3^-$$

The hallmark of simple respiratory acidosis is decreased pH and elevated bicarbonate levels (Table 19-3). Because the problem is of respiratory origin, the body cannot carry out respiratory compensation. (However, depending on the problem, mechanical ventilation can be used to assist breathing.)

Any compensation must be through renal mechanisms that excrete H^+ and reabsorb HCO_3^-. The excretion of H^+ raises plasma pH. Reabsorption of HCO_3^- provides additional buffer that combines with H^+, lowering free H^+ and increasing the pH.

In chronic obstructive pulmonary disease, renal compensation mechanisms can moderate the pH change, al-

TABLE 19-3   Plasma P_{CO_2}, Ions, and pH in Acid-Base Disturbances

Disturbance	P_{CO_2}	H^+	pH	HCO_3^-
Acidosis				
Respiratory	↑		↑	↓	↑
Metabolic	Normal or ↓*	↑	↓	↓
Alkalosis				
Respiratory	↓	↓	↑	↓
Metabolic	Normal or ↑*	↓	↑	↑

*These values are different from what you would expect from the law of mass action because almost instantaneous respiratory compensation occurs to keep P_{CO_2} from changing significantly.

though they may not be able to return it to normal. If you look at pH and HCO_3^- in patients with compensated respiratory acidosis, you find that those values are both higher than they were when the acidosis first occurred.

Metabolic Acidosis   Metabolic acidosis is a disturbance of mass balance that occurs when the dietary and metabolic input of acid exceeds acid excretion. Metabolic causes of acidosis include lactic acidosis, which is a result of anaerobic metabolism, and ketoacidosis, which occurs when there is excessive breakdown of fats or certain amino acids. The metabolic pathway that produces ketoacids in diabetes mellitus is described in Chapter 21. Ingested substances that cause metabolic acidosis include methanol, aspirin, and ethylene glycol (antifreeze).

Metabolic acidosis can also occur if the body loses HCO_3^-. The most common example is loss of HCO_3^- from the intestines with diarrhea. The exocrine pancreas produces HCO_3^- by a mechanism that is similar to the renal mechanism illustrated in Figure 19-20 ■. The H^+ that is made at the same time is released into the blood. Normally, HCO_3^- is secreted into the small intestine, then reabsorbed, buffering the H^+ made at the same time. However, if HCO_3^- is not reabsorbed due to diarrhea, a state of acidosis results.

Metabolic acidosis is expressed in the equation below. Hydrogen ion concentration increases because of H^+ contributed by the metabolic acids. This increase shifts the equation to the left, increasing CO_2 and using up HCO_3^- buffer:

$$\uparrow CO_2 + H_2O \leftarrow H_2CO_3 \leftarrow \uparrow H^+ + \downarrow HCO_3^-$$

The decrease in HCO_3^- concentration is an important way to distinguish metabolic from respiratory acidosis (Table 19-3).

You would think from looking at the equation above that metabolic acidosis would be accompanied by elevated P_{CO_2}. But unless the subject has a lung disease as well, respiratory compensation takes place almost instantaneously. Both elevated CO_2 and H^+ stimulate ventilation through the pathways described earlier. As a result, P_{CO_2} decreases to normal or even below-normal levels due to hyperventilation.

Uncompensated metabolic acidosis is rarely seen clinically. Indeed, a common sign of metabolic acidosis is hyperventilation, evidence of respiratory compensation occurring in response to the acidosis.

The same renal compensations discussed for respiratory acidosis take place in metabolic acidosis: excretion of H^+ and reabsorption of HCO_3^-. Renal compensations take several days to reach full effectiveness, so they are not usually seen in acute disturbances.

Respiratory Alkalosis   States of alkalosis are much less common than acidotic conditions. Respiratory alkalosis occurs as a result of hyperventilation, when alveolar ventilation goes up without a matching increase in metabolic CO_2 production. Consequently, plasma P_{CO_2} drops and alkalosis results:

$$\downarrow CO_2 + H_2O \leftarrow H_2CO_3 \leftarrow \downarrow H^+ + \downarrow HCO_3^-$$

The decrease in CO_2 pulls the equation to the left, so both H^+ and HCO_3^- concentrations drop. Low HCO_3^- levels in alkalosis indicate a respiratory disorder.

The primary clinical cause of respiratory alkalosis is excessive artificial ventilation. Fortunately, this condition is easily corrected by adjusting the ventilator.

The most common physiological cause of respiratory alkalosis is hysterical hyperventilation due to anxiety. In these situations, the neurological symptoms caused by alkalosis can be partially reversed by having the patient breathe into a paper bag. In doing so, the patient rebreathes exhaled CO_2, which raises arterial P_{CO_2} and corrects the problem.

Because this problem is of respiratory origin, the body can only carry out renal compensation. Filtered bicarbonate is not reabsorbed in the proximal tubule and is therefore excreted. In the distal nephron, bicarbonate is secreted and H^+ is reabsorbed. These compensations decrease the body's HCO_3^- and increase its H^+, which helps correct the alkalosis.

Metabolic Alkalosis   Metabolic alkalosis has two common causes: excessive vomiting of acidic stomach contents and excessive ingestion of bicarbonate-containing antacids. In both cases, the resulting alkalosis reduces H^+ concentration:

$$\downarrow CO_2 + H_2O \rightarrow H_2CO_3 \rightarrow \downarrow H^+ + \uparrow HCO_3^-$$

The decrease in H^+ shifts the reaction to the right. Carbon dioxide (P_{CO_2}) decreases and HCO_3^- levels go up.

Just as in metabolic acidosis, respiratory compensation is very rapid. The increase in pH and drop in P_{CO_2} depress ventilation. Less CO_2 is blown off, raising the P_{CO_2} and creating more H^+ and HCO_3^-.

The ventilatory compensation helps correct the pH problem but elevates HCO_3^- levels even more. However, the ventilatory compensation is limited because hypoventilation causes hypoxia. Once the arterial P_{O_2} drops below 60 mm Hg, hypoventilation ceases.

The renal response to metabolic alkalosis is the same as that for respiratory alkalosis: HCO_3^- is excreted and H^+ is reabsorbed.

This chapter has used fluid and acid-base balance to integrate function between various systems of the body. Changes in body fluid volume, reflected by changes in blood pressure, trigger both cardiovascular and renal homeostatic responses. Disturbances of acid-base balance are met with compensatory responses from both the respiratory and renal systems. Because of the interwoven responsibilities of these three systems, a pathology in one system is likely to cause disturbances in the other two. Recognition of this fact is an important aspect of treatment for many clinical conditions.

PROBLEM CONCLUSION

The doctor told Mrs. Godell that the treatment for Christopher's nephrogenic diabetes insipidus was to ensure that he had adequate fluid intake at all times. He also prescribed a thiazide diuretic. Giving a diuretic to someone with a dehydrating condition seems paradoxical, because traditionally, diuretics have been used to rid the body of excess water. But in patients with nephrogenic diabetes insipidus, thiazide diuretics actually reduce the volume of dilute urine by indirectly increasing reabsorption of NaCl in the proximal tubule.

In this running problem, you learned about a rare disease characterized by the inability to concentrate urine. You also learned that the inability to concentrate urine can be due to defects in either the brain or the kidney. Further check your understanding of this running problem by comparing your answers with those in the summary table.

	Question	Facts	Integration and Analysis
1	Does Christopher have central or nephrogenic diabetes insipidus?	Central diabetes insipidus is caused by a defect in the hypothalamus that leads to decreased levels of vasopressin. Nephrogenic diabetes insipidus is caused by a defect in the kidneys. Christopher has normal blood levels of vasopressin.	Nephrogenic diabetes insipidus is caused by a defect in the kidney, therefore vasopressin levels are normal. Christopher has nephrogenic diabetes insipidus.
2	Nephrogenic diabetes insipidus can be caused by two different defects. One defect affects the vasopressin receptor. What could the other defect be?	Vasopressin initiates a cAMP second messenger pathway and prompts insertion of water pores into the apical membranes of tubule cells.	The second defect could be any part of the process that occurs after the vasopressin receptor is activated. For example, problems with the cytoskeleton might prevent membrane recycling, or the water pores themselves could be defective.
3	Why do people with diabetes insipidus become dehydrated even though they are drinking a lot?	In diabetes insipidus, the collecting duct is unable to reabsorb water.	Inability to reabsorb water in the collecting duct leads to excretion of large volumes of dilute urine. If patients do not continuously drink water, they can lose enough fluid to go into a state of severe dehydration.
4	Explain why intense thirst is a main symptom of diabetes insipidus.	Thirst is triggered when osmolarity rises above 280 mOsM or when blood pressure decreases.	Excretion of large volumes of dilute urine decreases blood volume and pressure and increases osmolarity. Decreased blood pressure stimulates ANG II and thirst. Increased blood osmolarity triggers osmoreceptors for thirst.

CHAPTER REVIEW

SUMMARY

Fluid and Electrolyte Homeostasis

1. The renal, respiratory, and cardiovascular systems control fluid and electrolyte balance. Behaviors such as drinking also play an important role. (p. 570)

2. Adjustments made by the lungs and cardiovascular system are more rapid than compensation by the kidneys. (p. 570)

Water Balance and the Regulation of Urine Concentration

3. Most water intake comes from food and drink. The major route of water loss is 1.5 liters/day in urine. Smaller amounts are lost in feces, by evaporation from the skin, and in exhaled humidified air. (p. 571)

4. Water reabsorption in the kidneys conserves water but cannot restore lost volume. (p. 572)

5. To produce dilute urine, the nephron must reabsorb solute without water. To concentrate urine, the nephron must reabsorb water without solute. (p. 573)

6. Fluid leaving the ascending limb of the loop of Henle is dilute. The final concentration of urine depends on the water permeability of the collecting duct. (p. 573)

7. The hypothalamic hormone **vasopressin** controls collecting duct permeability to water in a graded fashion. When vasopressin is absent, water permeability is nearly zero. (p. 574)

8. Vasopressin causes the cell to insert **aquaporin** water pores in the apical membrane of collecting duct cells. (p. 574)

9. An increase in ECF osmolarity or a decrease in blood pressure will stimulate vasopressin release from the posterior pituitary. Osmolarity is monitored by hypothalamic **osmoreceptors.** Blood pressure and volume are sensed by receptors in the carotid and aortic bodies and atria respectively. (p. 576)

10. The loop of Henle is a **countercurrent multiplier** that creates high osmolarity in the medullary interstitial fluid by actively transporting Na^+, Cl^-, and K^+ out of the tubule. This high medullary osmolarity is necessary for formation of concentrated urine as fluid flows through the collecting duct. (p. 577)

11. The **vasa recta** capillaries carry away water leaving the tubule so that it does not dilute the medullary interstitium. (p. 578)

12. Urea contributes to the high osmolarity in the renal medulla. (p. 579)

Sodium Balance and the Regulation of ECF Volume

13. The total amount of Na^+ in the body is a primary determinant of ECF volume. (p. 579)

14. The steroid hormone **aldosterone** increases Na^+ reabsorption and K^+ excretion. (p. 579)

15. Aldosterone acts on **principal cells** (P cells) of the distal nephron. It enhances activity of the Na^+-K^+-ATPase and increases open time of Na^+ and K^+ leak channels. Aldosterone also stimulates the synthesis of new pumps and channels. (p. 579)

16. Aldosterone secretion can be controlled directly at the adrenal cortex. Increased K^+ stimulates aldosterone secretion but increased ECF osmolarity inhibits it. (p. 581)

17. Aldosterone secretion is also stimulated by **angiotensin II.**

JG cells in the kidney secrete **renin,** which converts **angiotensinogen** in the blood to **angiotensin I. Angiotensin converting enzyme (ACE)** converts ANG I to ANG II. (p. 581)

18. The stimuli that release renin are related directly or indirectly to low blood pressure. (p. 581)

19. ANG II has additional effects that raise blood pressure, including increased vasopressin secretion, stimulation of thirst, vasoconstriction, and activation of the cardiovascular control center. (p. 582)

20. **Atrial natriuretic peptide** (ANP) enhances Na^+ excretion and urinary water loss by increasing GFR, inhibiting tubular reabsorption of NaCl, and inhibiting the release of renin, aldosterone, and vasopressin. (p. 583)

Potassium Balance

21. Potassium homeostasis keeps plasma K^+ concentrations in a narrow range. **Hyperkalemia** and **hypokalemia** cause problems with excitable tissues, especially the heart. (p. 584)

Behavioral Mechanisms in Salt and Water Balance

22. **Thirst** is triggered by hypothalamic osmoreceptors and relived by drinking. (p. 585)

23. **Salt appetite** is triggered by aldosterone and angiotensin. (p. 585)

Integrated Control of Volume and Osmolarity

24. Homeostatic compensations for changes in salt and water balance attempt to follow the law of mass balance. Fluid and solute added to the body must be removed, or fluid and solute lost must be replaced. However, perfect compensation is not always possible. (p. 586)

Acid-Base Balance

25. The body's pH is closely regulated because pH affects intracellular proteins such as enzymes and membrane channels. (p. 590)

26. Acid intake and production are the biggest challenge to body pH. The biggest source of acid is CO_2 from respiration that combines with water to form carbonic acid (H_2CO_3). (p. 590)

27. The body copes with changes in pH by using buffers, ventilation, and renal excretion or reabsorption of H^+ and HCO_3^-. (p. 591)

28. Bicarbonate produced from CO_2 is the most important extracellular buffer system of the body. Bicarbonate buffers organic acids produced by metabolism. (p. 591)

29. Ventilation can correct disturbances in acid-base balance because changes in plasma P_{CO_2} affect both the H^+ and the HCO_3^- content of the blood. An increase in P_{CO_2} stimulates central chemoreceptors. An increase in plasma H^+ stimulates carotid and aortic chemoreceptors. Increased ventilation excretes CO_2 and decreases H^+. (p. 592)

30. In **acidosis,** the kidneys excrete H^+ and reabsorb HCO_3^-. In **alkalosis,** the kidneys excrete HCO_3^- and reabsorb H^+. (p. 593)

31. **Intercalated cells** in the collecting duct are responsible for the fine regulation of acid-base balance. (p. 595)

QUESTIONS

LEVEL ONE   Reviewing Facts and Terms

1. What is an electrolyte? Name five electrolytes whose concentrations must be regulated by the body.

2. List five organs and four hormones important in maintaining fluid and electrolyte balance.

3. Compare the routes by which water enters the body to the ways the body loses water.

4. List the receptors that regulate osmolarity, blood volume, blood pressure, ventilation, and pH. Where are they located, what stimulates them, and what compensatory mechanisms are triggered by these receptors?

5. How do the two limbs of the loop of Henle differ in their permeability? How is this possible?

6. Which ion is a primary determinant of ECF volume? Which ion is the determinant of extracellular pH?

7. What happens to the resting membrane potential of excitable cells when plasma concentrations of K^+ decrease? Which organ is most likely to be affected by changes in K^+ concentration?

8. Appetite for what two substances is important in regulating fluid volume and osmolarity?

9. Write out the words for the following abbreviations: ADH, ANP, ACE, ANG II, JG cell, P cell, I cell.

10. Make a list of all the different membrane transporters in the kidney. For each transporter, tell (a) which section(s) of the nephron contain(s) the transporter; (b) whether the transporter is on the apical or basolateral membrane, or both; (c) whether it participates in reabsorption, excretion, or both.

11. List and briefly explain three reasons why monitoring and regulating the ECF pH are important. What three mechanisms does the body use to cope with changing pH?

12. Which is more likely to accumulate in the body, acids or bases? List some sources of each one.

13. What is a buffer? List three intracellular buffers. Name the primary extracellular buffer.

14. Name two ways that the kidneys alter the plasma pH. Which compounds serve as urinary buffers?

15. Write the equation that shows how CO_2 is related to pH. What enzyme increases the rate of this reaction? Name two different cell types that possess high concentrations of this enzyme.

16. When ventilation increases, what happens to arterial P_{CO_2}? To pH? To plasma H^+ concentration?

LEVEL TWO   Reviewing Concepts

17. **Concept map:** Map the homeostatic reflexes that occur in response to each of the following situations:

    (a) decreased volume, normal osmolarity
    (b) increased volume, increased osmolarity
    (c) normal volume, increased osmolarity

18. Figures 19-19 ■ and 19-20 ■ show the respiratory and renal compensations for acidosis. Draw similar maps for alkalosis.

19. Explain how the loop of Henle and vasa recta work together to create dilute fluid.

20. Diagram the mechanism by which vasopressin alters the composition of the urine.

21. Make a table and specify for each substance listed (a–f): hormone or enzyme? steroid or peptide? cell or tissue that produces it? target cell or tissue? response of the target to the substance?

    (a) ANP   (b) aldosterone   (c) renin   (d) ANG II
    (e) vasopressin   (f) angiotensin-converting enzyme

22. Name the four main compensatory mechanisms for restoring low blood pressure to normal. Why do you think there are so many homeostatic pathways for raising low blood pressure?

23. Compare and contrast the following sets of terms:

    (a) principal and intercalated cells
    (b) renin, angiotensin II, aldosterone, ACE
    (c) respiratory acidosis with metabolic acidosis, including their causes and compensations
    (d) water reabsorption in the proximal tubule, distal tubule, and ascending limb of the loop of Henle
    (e) respiratory alkalosis with metabolic alkalosis, including their causes and compensations

20 Digestion

■ "Give me a good digestion, Lord, and also something to digest."—*Anonymous, A Pilgrim's Grace* ■

CHAPTER OUTLINE

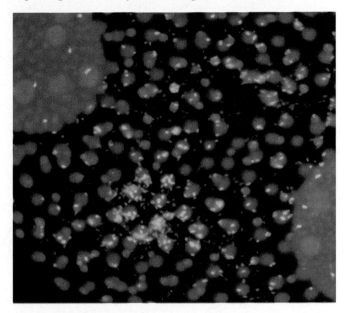

Vibrio cholerae (green), the bacterium that causes the diarrheal disease cholera, occurs in fresh water. Here it is shown attached to the surface of freshwater algae (red).

BACKGROUND BASICS

A shotgun wound to the stomach seems an unlikely beginning to the scientific study of digestive processes. But in 1822 at Fort Mackinac, young Canadian trapper Alexis St. Martin narrowly escaped death when a gun discharged only three feet from him, tearing open his chest and abdomen and leaving a hole in his stomach wall. American army surgeon William Beaumont attended St. Martin and nursed him back to health over the next two years.

LEVEL THREE   Problem Solving

24. Karen has bulimia, in which she induces vomiting to avoid weight gain. When the doctor sees her, her weight is 89 lb and respiration is 6 breaths/min (normal 12). Her blood HCO_3^- is 62 meq/L (normal 24–29). Her arterial blood pH is 7.61 and her P_{CO_2} is 61 mm Hg.

    (a) What is her acid-base condition called?

    (b) Explain why her plasma bicarbonate level is so high.

    (c) Why is she hypoventilating? What effect will this have on the pH and total oxygen content of her blood? Explain your answers.

25. Hannah, a 31-year-old woman, decided to have colonic irrigation, a procedure during which large volumes of distilled water were infused into her rectum. Over the course of the treatment she absorbed 3000 mL of water. About 12 hours later, her roommate found her in convulsions and took her to the emergency room. Her blood pressure was 140/90, her plasma Na^+ concentration was 106 meq/L (normal is 135 meq/L), and her plasma osmolarity was 270 mOsM. In a concept map or flow chart, diagram all the homeostatic responses that her body was using to compensate for the changes in blood pressure and osmolarity.

LEVEL FOUR   Quantitative Problems

26. The Henderson-Hasselbalch equation is a mathematical expression of the relationship between pH, HCO_3^- concentration, and concentration of dissolved CO_2. One variant of the equation uses P_{CO_2} instead of dissolved CO_2:

$$pH = 6.1 + \log [HCO_3^-]/0.03 \times P_{CO_2}$$

    (a) If arterial blood has a P_{CO_2} of 40 mm Hg and its HCO_3^- concentration is 24 mM (normal), what is its pH? (You will need a log table or calculator with a logarithmic function capability.)

    (b) What is the pH of venous blood with the same HCO_3^- but a P_{CO_2} of 46 mm Hg?

E X P L O R E <MediaLab>

Introduction

Ever notice how your urine is dark yellow one day and practically the color of water the next? How much does the volume, color, and composition of your urine reflect your activities on a given day? The MediaLab in this chapter focuses on how your kidneys alter the concentration (and the color) of your urine. You can explore the role of antidiuretic hormone (ADH) in forming concentrated urine and the use of diuretic drugs to decrease your blood volume. You can also learn about what happens to your body when the kidneys fail to perform their function, a condition known as renal failure. After reading the description below, visit the MediaLab for Chapter 19 in your Companion Website and select the appropriate keyword.

Web Exploration 1

Estimated time for completion = 10 minutes

What color is your urine? Is it always the same color? How do color and volume of urine correlate? What was your water intake and activity level on the days when your urine was dark amber? How does this differ from the times when your urine is almost the color of water? In this chapter you learned that urine concentration is regulated by antidiuretic hormone (ADH) (a.k.a. vasopressin or AVP). Select the keyword **ADH** from your Website. How many types of water channels have been cloned and what are the names of these channels? Draw a picture of a nephron and collecting duct and indicate where you would find each type of channel. For water to move out of the nephron/collecting duct, there needs to be a driving force (an osmotic gradient) and a way across the epithelial cell membranes (the barrier). You have just identified "the way across." Now describe what creates the driving force for water movement in each region.

Web Exploration 2

Estimated time for completion = 15 minutes

We think of heart attacks and strokes as life-threatening failures of major physiological systems, yet renal failure is no less traumatic for your body. Stop and consider the numerous tasks your kidney performs on a day-to-day basis. Imagine if you had to take over the kidney's function. What would you have to do?

Make a list of the liquids you have consumed today and estimate their total volume. Then estimate how much water you have lost. What are the major ways your body loses water? Now do the same for salt intake and excretion. Urea is a major component of urine. What part of your diet produces urea? Will you have to alter intake of this food if your kidneys are failing to perform their function?

In renal failure an individual may have to be put on dialysis until a kidney transplant is possible. Select the keyword **RENAL FAILURE** from the Website. What is kidney failure? What are the symptoms associated with renal failure? Based on your understanding of kidney function, offer a physiological explanation for each of these symptoms.

How is serum creatinine used to monitor kidney function? Explain why serum creatinine and not serum sodium is used. Why are renal patients at an increased risk for osteoporosis and anemia? Can you figure out why body itching is a symptom of renal failure? What are the dietary restrictions for a person in renal failure? Imagine you are a health educator in a dialysis unit. Explain to a patient why each of the diet restrictions is critical.

The gaping wound over the stomach failed to heal properly, leaving a *fistula,* or opening, into the lumen. St. Martin was destitute and unable to care for himself, so Beaumont "retained St. Martin in his family for the special purpose of making physiological experiments." In a legal document, St. Martin even agreed to "obey, suffer, and comply with all reasonable and proper experiments of the said William [Beaumont] in relation to . . . the exhibiting . . . of his said stomach and the power and properties . . . and states of the contents thereof."

Beaumont's observations on digestion and the state of St. Martin's stomach under various conditions created a sensation. In 1832, just before Beaumont's observations were published, the nature of gastric juice [*gaster,* stomach] and digestion in the stomach was a subject of much debate. Beaumont's careful observations went far toward solving the mystery.

Like physicians of old who tasted urine when making a diagnosis, Beaumont tasted the mucous lining of the stomach and the gastric juices. He described them both as "saltish," but mucus was not at all acid and gastric fluid was very acid. Beaumont collected copious amounts of gastric fluid through the fistula, and in controlled experiments he confirmed that gastric fluid digested meat, aided by a combination of hydrochloric acid and another active factor that we now know is the enzyme pepsin.

These observations and others about motility and digestion within the stomach became the foundation of what we know about digestive physiology. Although research today is being conducted more at the cellular and molecular level, researchers still create surgical fistulas in experimental animals to observe and sample the contents of the lumen of the digestive tract.

Why is the digestive system of such great interest? Gastrointestinal diseases today account for nearly one-tenth of the money spent on health care. Many of these conditions, such as heartburn, indigestion, gas, and constipation, are troublesome rather than major health risks, but their significance should not be underestimated. Go into any drugstore and look at the number of over-the-counter medications for digestive disorders to get a feel for the impact that gastrointestinal diseases have on our society. In this chapter we examine the digestive system and the remarkable way that it transforms the food we eat into nutrients for the body's use.

FUNCTION AND PROCESSES OF THE DIGESTIVE SYSTEM

The **gastrointestinal tract,** or *GI tract,* is a long tube whose function is to move nutrients, water, and electrolytes from the external to the internal environment. The food we eat is mostly in the form of macromolecules such as proteins and complex carbohydrates, so our digestive systems must secrete powerful enzymes to digest them into smaller molecules, or nutrients, that can

PROBLEM
Peptic Ulcers

Your stomach is one of the most hostile environments on earth. Bathed in acid strong enough to digest nails, its walls protected with a sticky mucus, and constantly churning, the stomach was once thought to be uninhabitable by any life form. Recently, however, scientists have discovered a remarkable organism that actually thrives in the stomach's formidable environment. This hardy bacterium, called *Helicobacter pylori,* has evolved a set of defenses so unique that it escaped detection until 1982. But this bacterium is not a silent passenger in the stomach: it is associated with many cases of peptic ulcers worldwide. Proper drug treatment kills *H. pylori* and allows ulcers in the stomach and duodenum to heal. That is good news to Tonya Berry, whose physician has just informed her that she has a peptic ulcer. "I thought it was stress," she said.

...continued on page 608

be absorbed. At the same time, however, the enzymes must not digest the cells of the GI tract itself. If protective mechanisms against autodigestion fail, we develop raw patches known as *ulcers* on the walls of the gastrointestinal tract.

Another challenge the digestive system faces daily is matching input with output. Various exocrine glands and cells secrete about 7 liters of fluid per day into the lumen of the tract. These secretions contain digestive enzymes, mucus, and water. They are critical to proper digestive function, but they must be reabsorbed or the body will rapidly dehydrate. *Diarrhea,* or watery stools, can become an emergency if fluid loss from the GI tract depletes the extracellular fluid volume to the point that the circulatory system is unable to maintain adequate blood pressure.

These physiological challenges are met by coordinating the four basic processes of the digestive system: digestion, absorption, motility, and secretion (Fig. 20-1 ■). **Digestion** is the chemical and mechanical breakdown of foods into smaller units that can be taken across the intestinal epithelium into the body. **Absorption** is the active or passive transfer of substances from the lumen of the GI tract to the extracellular fluid (ECF). **Motility** [*movere,* move + *tillis,* characterized by] is movement of material in the GI tract as a result of muscle contraction. **Secretion** refers to (1) the transfer of water and ions from the ECF to the lumen, and (2) the release of synthesized material by GI epithelial cells.

For most nutrients, absorption is not regulated, so "what you eat is what you get." In contrast, motility and secretion are continuously regulated to maximize the availability of absorbable material. Motility is regulated because if food moves through the system too rapidly, there is not enough time for everything in the lumen to be digested and absorbed. Secretion is regulated because if digestive enzymes are not secreted in adequate amounts, food in the GI tract cannot be broken down into an absorbable form.

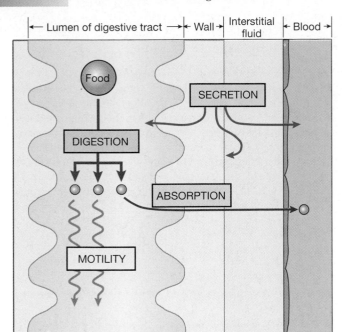

■ Figure 20-1 **Processes of the digestive system**

When digested nutrients have been absorbed and reach the cells, cellular metabolism directs their use or storage (see Chapter 21). Some of the same chemical signal molecules that alter digestive motility and secretion also participate in the control of metabolism, providing an integrating link between the two steps.

Although we tend to think of the digestive system primarily in terms of its digestive function, repelling foreign invaders is another challenge the system faces. The GI tract passes through the interior of the body, but the lumen and its contents are part of the external environment [∞ p. 3]. Indeed, the digestive system provides the largest amount of contact between the internal environment and the outside world, with a total surface area estimated to be the size of a tennis court.

The GI tract faces daily conflict between the need to absorb water and nutrients and the need to keep bacteria, viruses, and other pathogens from entering the body. As a result, the transporting epithelium of the GI tract is combined with an array of physiological defense mechanisms including mucus, digestive enzymes, acid, and the largest collection of lymphoid tissue in the body, the **gut-associated lymphoid tissue** (GALT). By one estimate, 80% of all lymphocytes [∞ p. 476] in the body are found in the small intestine.

▶ ANATOMY OF THE DIGESTIVE SYSTEM

The digestive system begins with the oral cavity (mouth and pharynx), which serves as a receptacle for food (Fig. 20-2 ■). The oral cavity is where the first stages of digestion begin with chewing and secretion of saliva by

three pairs of **salivary glands:** *sublingual* under the tongue, *submandibular* under the mandible, or jawbone, and *parotid* that lie near the hinge of the jaw.

Once swallowed, food moves into the GI tract, a long tube with muscular walls and a lining of transporting epithelium. The tube is closed off by skeletal muscle sphincters at the ends. At intervals along its length, rings of smooth muscle function as sphincters to separate the tract into segments with distinct functions.

Food moves through the GI tract propelled by waves of muscle contraction. Along the way, secretions are added to the food by secretory epithelium, liver, and pancreas, creating a soupy mixture known as **chyme.**

Digestion takes place primarily in the lumen of the tube. The products of digestion are absorbed across the epithelium of the wall and pass into the extracellular space. From there, they move into the blood or lymph for distribution throughout the body. Any waste remaining at the end of the GI tract leaves the body through the opening known as the anus.

✔ What is the difference between digestion and metabolism [∞ p. 88]?

✔ What is the difference between absorption and secretion?

The Digestive System Consists of the GI Tract and Accessory Glandular Organs

A piece of food that enters the mouth and is swallowed passes into the **esophagus,** a narrow tube that travels through the thorax to the abdomen (Fig. 20-2a ■). The esophagus walls are skeletal muscle initially but transition to smooth muscle about two-thirds of the way down the length. Just below the diaphragm, the esophagus ends at the **stomach,** a baglike organ that can hold 2–3 liters of food and fluid when fully (but uncomfortably) expanded.

The stomach is roughly divided into three sections: the upper **fundus,** central **body,** and lower **antrum** (Fig. 20-2b ■). The stomach continues digestion, mixing food with acid and enzymes to create chyme. The opening between the stomach and **small intestine** is guarded by the pyloric valve, or **pylorus.** This thickened band of smooth muscle relaxes to allow only small amounts of chyme into the small intestine at any one time.

In this way, the stomach acts as an intermediary between the behavioral act of eating and the physiological acts of digestion and absorption in the intestine. By regulating the rate at which chyme reaches the intestine, the stomach ensures that the intestine will not be overwhelmed with more than it can digest and absorb.

Most digestion takes place in the small intestine, which is divided into three sections: the **duodenum** (the first 25 cm), **jejunum,** and **ileum** (the last two together are about 260 cm long). Digestion is carried out by intestinal

enzymes, aided by exocrine secretions from two accessory glandular organs, the **pancreas** and the **liver.** Their secretions enter the initial section of the duodenum through ducts. A tonically contracted sphincter (the *sphincter of Oddi*) keeps pancreatic fluid and bile from entering the small intestine except during a meal.

Digestion is essentially completed in the small intestine, and nearly all digested nutrients and secreted fluids are absorbed there, leaving about 1.5 liters of chyme per day to pass into the **large intestine** (Fig. 20-2a ■). In the **colon,** the proximal section of the large intestine, watery chyme is converted into semisolid **feces** [*faeces,* dregs] through the absorption of water and electrolytes.

When feces are propelled into the terminal section known as the **rectum,** distension of the rectal wall triggers the *defecation reflex.* The final opening through which gastrointestinal material must pass is the **anus,** with its external anal sphincter of skeletal muscle, which is under voluntary control.

In a living person, the GI tract from mouth to anus is about 450 cm, or nearly 15 feet, long! Of this length, 395 cm, or about 13 feet, consists of the large and small intestines. Try to imagine 13 feet of rope ranging from 1 to 3 inches in diameter all coiled up inside your abdomen from the belly button down. The tight arrangement of the abdominal organs helps explain why you feel the need to loosen your belt after consuming a large meal!

Measurements of intestinal length made during autopsies are nearly double those given here because after death, the longitudinal muscles of the intestinal tract relax. This accounts for the wide variation in intestinal length that you may encounter in different references.

The GI Tract Wall Has Four Layers

The basic structure of the gastrointestinal wall is similar in the stomach and the intestines, although variations exist between different sections of the GI tract (Fig. 20-2c, d, e, f ■). The gut wall consists of four layers: an inner *mucosa,* a middle layer known as the *submucosa,* and an outer section, the *muscularis externa,* composed of layers of smooth muscle covered with connective tissue (the *serosa*).

The Mucosa   The **mucosa** is the inner lining of the gastrointestinal tract, created from (1) a single layer of epithelial cells, (2) the **lamina propria,** subepithelial connective tissue that holds the epithelium in place, and (3) the **muscularis mucosae,** a thin layer of smooth muscle. Several structural modifications increase the amount of surface area to enhance absorption.

First, the entire wall is crumpled into folds known as *rugae* in the stomach and *plicae* in the small intestine. The intestinal mucosa also projects into the lumen in small fingerlike extensions known as **villi.** Additionally, the surface area of each individual intestinal cell is increased by **microvilli** [∞ p. 49] along the apical membrane (Fig. 20-2g ■). Microvilli create the *brush border,* a feature also seen in the epithelial cells of the nephron.

Additional surface area is added by tubular invaginations of the surface that extend down into the supporting connective tissue. These invaginations are called **gastric glands** in the stomach and **crypts** in the intestine. Some of the deepest invaginations form secretory **submucosal glands** that open into the lumen through ducts.

Epithelial cells that line the lumen are the most variable feature of the GI tract, changing from section to section. The cells include transporting epithelial cells, endocrine and exocrine secretory cells, and stem cells. Transporting epithelial cells move ions and water into the lumen and absorb ions, water, and nutrients. At the *mucosal* (apical) surface, secretory cells release enzymes, mucus, and paracrines into the lumen of the digestive system. On their *serosal* (basolateral) sides, they secrete hormones or paracrines into the interstitial fluid, where they act on neighboring cells.

GI *stem cells* are rapidly dividing undifferentiated cells in the crypts or glands that continuously produce new epithelium. As stem cells divide, the newly formed cells are pushed up toward the luminal surface of the epithelium. The average life span of a GI epithelial cell is only a few days, a good indicator of the rough life such cells lead.

The cell-to-cell junctions that tie GI epithelial cells together vary [∞ p. 55]. In the stomach, the junctions form a tight barrier so that little can pass between the cells. In the small intestine, junctions are not as tight. This intestinal epithelium is considered "leaky" because some water and solutes can be absorbed by going *between* the cells instead of *through* them. We have recently learned that these junctions have plasticity; that is, their "tightness" can be regulated to some extent by the nutritional status of the body.

The subepithelial connective tissue, the *lamina propria,* contains nerve fibers and small blood and lymph vessels into which absorbed nutrients pass (Fig. 20-2c ■). The lamina propria also contains wandering immune cells such as macrophages and lymphocytes, on patrol for invaders that enter through breaks in the epithelium.

In the intestine, collections of lymphoid tissue form small nodules and larger **Peyer's patches** (Fig. 20-2e ■) that create visible bumps in the mucosa. These lymphoid aggregations are a major part of the gut-associated lymphoid tissue (GALT).

The *muscularis mucosae* separates the mucosa from the submucosa. Contraction of this thin smooth muscle layer may somehow alter the effective surface area for absorption by moving the villi back and forth, like the waving tentacles of a sea anemone.

✔ Is the lumen of the digestive tract on the apical or basolateral side of intestinal epithelium?

ANATOMY SUMMARY The Digestive System

■ Figure 20-2

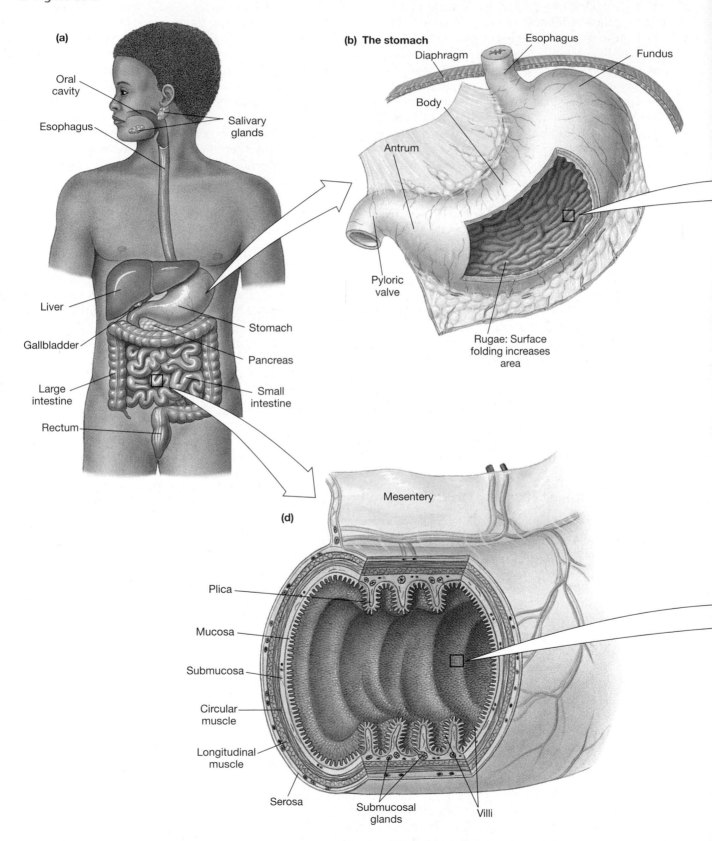

(a)

Oral cavity

Esophagus

Liver

Gallbladder

Large intestine

Rectum

Salivary glands

Stomach

Pancreas

Small intestine

(b) The stomach

Diaphragm

Esophagus

Fundus

Body

Antrum

Pyloric valve

Rugae: Surface folding increases area

(d)

Mesentery

Plica

Mucosa

Submucosa

Circular muscle

Longitudinal muscle

Serosa

Submucosal glands

Villi

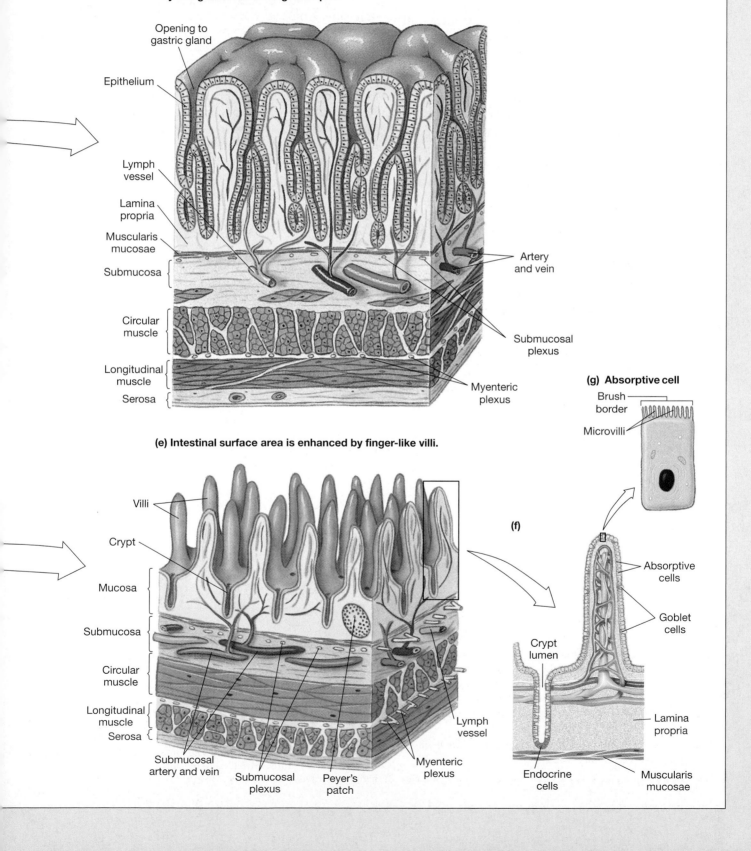

(c) In the stomach, surface area is increased by invaginations called gastric pits.

Opening to gastric gland

Epithelium

Lymph vessel

Lamina propria

Muscularis mucosae

Submucosa

Circular muscle

Longitudinal muscle

Serosa

Artery and vein

Submucosal plexus

Myenteric plexus

(e) Intestinal surface area is enhanced by finger-like villi.

Villi

Crypt

Mucosa

Submucosa

Circular muscle

Longitudinal muscle

Serosa

Submucosal artery and vein

Submucosal plexus

Peyer's patch

Myenteric plexus

Lymph vessel

(f)

(g) Absorptive cell

Brush border

Microvilli

Absorptive cells

Goblet cells

Crypt lumen

Endocrine cells

Lamina propria

Muscularis mucosae

The Submucosa The middle layer of the gut wall is composed of connective tissue with larger blood and lymph vessels (Fig. 20-2c, e ■). The submucosa also contains the **submucosal plexus** [*plexus,* interwoven], one of the two major nerve networks of the **enteric nervous system** [∞ p. 215]. The enteric nervous system is a unique division of the nervous system that helps coordinate digestive function, as we will see.

The Muscularis Externa and Serosa The outer wall of the intestinal tract consists mostly of two layers of smooth muscle: an inner circular layer and an outer longitudinal layer (Fig. 20-2c, d, e ■). Contraction of the circular layer decreases the diameter of the lumen. Contraction of the longitudinal layer shortens the tube.

The stomach has an incomplete third layer of oblique muscle between the circular muscles and the submucosa. The second nerve network of the enteric nervous system, the **myenteric plexus** [*myo-,* muscle + *enteron,* intestine], lies between the two muscle layers of the digestive tract.

A connective tissue membrane called the **serosa** surrounds the entire digestive tract. The serosa is a continuation of the **peritoneal membrane** (*peritoneum*) that lines the abdominal cavity. The peritoneum also forms sheets of **mesentery** that hold the intestines in place so that they do not become tangled as they move.

Now let's take a brief look at the processes of motility, secretion, digestion, and absorption. Gastrointestinal physiology is a rapidly expanding field, and this textbook does not attempt to be all-inclusive. Instead, it focuses on certain broad aspects of digestive physiology, with additional material placed in tables for reference.

✔ Why is the digestive system associated with the largest collection of lymphoid tissue in the body?

▶ MOTILITY

Motility in the gastrointestinal tract serves two purposes: moving food from the mouth to the anus and mechanically mixing food to break it into uniformly small

particles. This mixing maximizes exposure of the particles to digestive enzymes by increasing their surface area. Gastrointestinal motility is determined by the properties of its smooth muscle and modified by chemical input from nerves, hormones, and paracrines.

GI Smooth Muscle Contracts Spontaneously

Most of the gastrointestinal tract is composed of single-unit smooth muscle, with groups of cells electrically connected by gap junctions [∞ p. 57] to create contracting segments. Different regions exhibit different types of contraction. **Tonic contractions** that are sustained for minutes or hours occur in some smooth muscle sphincters as well as in the anterior portion of the stomach. **Phasic contractions** with contraction-relaxation cycles that last only a few seconds occur in the posterior region of the stomach and the small intestine.

Cycles of smooth muscle contraction and relaxation are associated with spontaneous cycles of depolarization and repolarization known as **slow wave potentials,** or simply *slow waves.* Slow wave potentials differ from myocardial pacemaker potentials because they do not reach threshold with each cycle (Fig. 20-3 ■). A slow wave that does not reach threshold will not cause a contraction in the muscle fiber.

When a slow wave potential does reach threshold, voltage-gated Ca^{2+} channels in the muscle fiber open, Ca^{2+} enters, and the cell fires one or more action potentials. The depolarization phase of the action potential, like that of myocardial autorhythmic cells, is due to Ca^{2+} entry into the cell. In addition, Ca^{2+} entry initiates muscle contraction [∞ p. 371].

Contraction of smooth muscle, like that of cardiac muscle, is graded according to the amount of Ca^{2+} that enters the fiber. The higher the amplitude of the slow wave, the more action potentials fire, and the greater the contraction force in the muscle. Similarly, the longer the

...continued from page 603

A peptic ulcer is a raw, inflamed area in the stomach or duodenum that extends from a break in the surface epithelium past the muscularis mucosa. Most peptic ulcers occur in the duodenum (duodenal ulcers). Other peptic ulcers are located within the stomach itself (gastric ulcers). Until 1982, the year *H. pylori* was discovered, peptic ulcers were believed to be caused by high levels of acid and enzyme production in the stomach. However, since the 1980s, scientists have found *H. pylori* in 90% of people with duodenal ulcers and 70% of people with gastric ulcers. To confirm the diagnosis of an *H. pylori*-caused ulcer, Tonya Berry will undergo tests to isolate the bacterium from her stomach.

Question 1: *Why are peptic ulcers found in the duodenum rather than in the jejunum or ileum?*

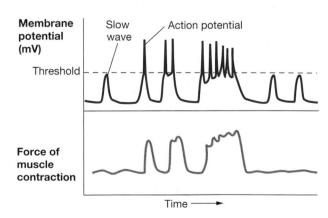

■ **Figure 20-3** **Slow wave potentials in GI smooth muscle** When slow wave potentials exceed threshold, action potentials fire and smooth muscle contracts. The force and duration of muscle contraction is directly related to the amplitude and number of action potentials per unit time.

duration of the slow wave, the longer the duration of contraction. Both amplitude and duration can be modified by neurotransmitters, hormones, or paracrines.

Slow wave frequency varies by region of the digestive tract, ranging from 3 waves/min in the stomach to 12 waves/min in the duodenum. Current research indicates that slow waves originate in modified smooth muscle cells called *interstitial cells of Cajal* (see Emerging Concepts box) and pass into smooth muscle cells through gap junctions. The interstitial cells of Cajal are also associated with enteric neurons and may act as an intermediary between the neurons and smooth muscle.

GI Smooth Muscle Exhibits Different Patterns of Contraction

Muscle contractions in the gastrointestinal tract occur in three general patterns. Between meals, when the tract is largely empty, a series of contractions begins in the stomach and passes slowly from section to section, taking about 90 minutes to reach the large intestine. This pattern, known as the **migrating motor complex,** is a "housekeeping" function that sweeps food remnants and bacteria out of the upper GI tract and into the large intestine.

Muscle contractions during and following a meal fall into one of two other patterns. **Peristaltic contractions** are progressive waves of contraction that move from one section of the GI tract to another, just like the human "waves" that ripple around a football stadium or basketball arena.

In peristalsis, circular muscles contract just behind a mass, or *bolus,* of food. This pushes the bolus forward into a *receiving segment,* where the circular muscles are relaxed (Fig. 20-4a ■). The receiving segment then contracts, continuing the forward movement. Peristaltic

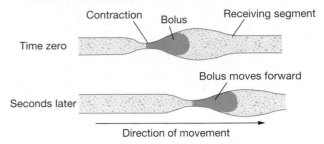

(a) Peristaltic contractions are responsible for forward movement

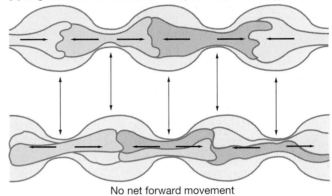

(b) Segmental contractions are responsible for mixing

No net forward movement

■ **Figure 20-4   Contractions in the GI tract**

contractions push a bolus forward at speeds between 2 and 25 cm/sec.

Peristalsis propels material forward through the esophagus, and mixes food as it digests in the stomach. In normal digestion, intestinal peristaltic waves are limited to short distances. Hormones, paracrines, and the autonomic nervous system influence peristalsis in all regions of the GI tract.

In **segmental contractions,** short (1–5 cm) segments of intestine contract and relax (Fig. 20-4b ■). In the contracting segment, circular muscles contract while longitudinal muscles relax. These contractions may occur randomly along the intestine or may occur at regular intervals. Alternating segmental contractions of the type shown churn intestinal contents back and forth, mixing them and keeping them in contact with the absorptive epithelium. When segmental contractions occur sequentially, in an oral-to-aboral direction [*ab-,* away], digesting material is propelled short distances.

Motility disorders are among the more common gastrointestinal problems. They range from esophageal spasms and delayed gastric emptying to constipation and diarrhea. *Irritable bowel syndrome* is a chronic functional disorder characterized by altered bowel habits and abdominal pain.

✔ Why are some sphincters of the digestive system tonically contracted?

Interstitial Cells of Cajal   Current experimental evidence suggests that the rhythm of slow wave potentials arises not within the smooth muscle itself but from a network of associated cells known as the *interstitial cells of Cajal* (named for the Spanish neuroanatomist Santiago Ramón y Cajal). These specialized smooth muscle cells are associated with the smooth muscle layers and intrinsic nerve plexus of the gastrointestinal wall. The interstitial cells of Cajal (ICC) apparently act as the pacemakers for smooth muscle slow waves, with their depolarizations spreading to the muscle fibers through gap junctions. ICCs depend on a tyrosine kinase receptor pathway [∞ p. 160] called *Kit* to maintain normal activity. Mutant mice with weakly functional *Kit* receptors exhibit abnormal gastrointestinal motility. This observation has researchers now trying to establish a link between ICCs and functional bowel disorders such as irritable bowel syndrome and chronic constipation.

▶ SECRETION

In a typical day, 9 liters of fluid pass through the lumen of an adult's gastrointestinal tract—the equivalent of three 3-liter soft drink bottles! Only about 2 liters of that volume enter the system through the mouth. The remaining 7 liters of fluid come from body water secreted along with enzymes and mucus (Fig. 20-5 ■). About half the secreted fluid comes from accessory organs and glands such as the salivary glands, pancreas, and liver. The remaining 3.5 liters are secreted by the epithelial cells of the digestive tract itself.

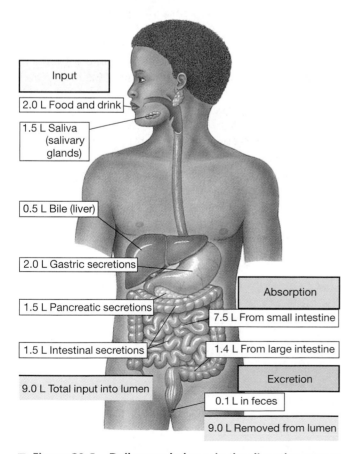

Input
2.0 L Food and drink
1.5 L Saliva (salivary glands)
0.5 L Bile (liver)
2.0 L Gastric secretions
1.5 L Pancreatic secretions
1.5 L Intestinal secretions
9.0 L Total input into lumen

Absorption
7.5 L From small intestine
1.4 L From large intestine

Excretion
0.1 L in feces
9.0 L Removed from lumen

■ **Figure 20-5   Daily mass balance in the digestive system**

The significance of the secreted volume is more apparent if you realize that the secreted fluid equals one-sixth of the entire body water volume, or more than twice the plasma volume. In other words, if the fluid secreted into the lumen is not reabsorbed as it passes along the GI tract, the body will rapidly dehydrate.

Digestive Enzymes Are Secreted into the Mouth, Stomach, and Intestine

Digestive enzymes are secreted either by exocrine glands (salivary glands and the pancreas) or by epithelial cells in the mucosa of the stomach and small intestine. Enzymes are proteins, which means they are synthesized on the rough endoplasmic reticulum, packaged by the Golgi apparatus into secretory vesicles, and stored within the cell until needed. On demand, they are released by exocytosis [∞ p. 130].

Some digestive enzymes are secreted in an inactive **proenzyme** form and are known collectively as *zymogens*. These enzymes must be activated in the GI lumen before they can carry out digestion. Late activation prevents the enzymes from digesting the cells in which they are synthesized. It also allows them to be stockpiled within the cells until needed.

The control pathways for enzyme release vary but include a variety of neural, hormonal, and paracrine signals. Usually, stimulation of parasympathetic neurons in the vagus nerve enhances enzyme secretion.

Specialized Cells Secrete Mucus

Mucus is a viscous secretion composed primarily of glycoproteins that collectively are called **mucins**. The primary functions of mucins are to form a protective coating over the GI mucosa and to lubricate the contents of the gut.

Mucus is made in specialized exocrine cells called **mucous cells** in the stomach and **goblet cells** in the intestine. Goblet cells comprise between 10% and 24% of the intestinal cell population. In the mouth, the minor salivary glands secrete about 70% of the salivary mucus.

The release signals for mucus include parasympathetic innervation, a variety of neuropeptides found in the enteric nervous system, and cytokines from immunocytes. Parasitic infections and inflammatory processes in the gut both cause substantial increases in mucus output as the body attempts to increase its protective barrier.

The Digestive System Secretes Ions and Water

A large portion of the 7 liters of fluid secreted by the digestive system each day is composed of water and ions, particularly Na^+, K^+, Cl^-, HCO_3^-, and H^+. The ions are first secreted into the lumen of the tract, then reabsorbed. Water follows osmotic gradients created by the transfer of solutes from one side of the epithelium to another.

...*continued from page 608*

Protein digestion in the stomach releases large amounts of urea. Urea is a nitrogenous breakdown product of amino acids. *Helicobacter pylori* takes advantage of this sea of urea and uses it for protection against the acidic environment of the stomach. The outer membrane of *H. pylori* is studded with enzymes called ureases. Urease converts urea into carbon dioxide and ammonia, which is a base.

Question 2: *How does the conversion of urea into ammonia protect* H. pylori *from the hostile environment of the stomach?*

Gastrointestinal epithelial cells, like those in the kidney, have distinct apical and basolateral membranes [∞ p. 60], each with proteins for active transport, facilitated diffusion, and ion movement through open channels. Many of these transporters are similar to those of the renal tubule [∞ p. 555].

The basolateral membrane contains the ubiquitous Na^+-K^+-ATPase. Cotransporters include a Na^+-K^+-$2Cl^-$ symporter, Cl^--HCO_3^- antiporter, and an H^+-K^+-ATPase. Ion channels include Na^+, K^+, and Cl^- channels.

▶ DIGESTION AND ABSORPTION

Digestion of macromolecules into absorbable units is accomplished by a combination of mechanical and enzymatic breakdown [∞ p. 79]. Chewing and churning create smaller pieces of food, which expose more surface area to digestive enzymes. *Bile,* a complex chemical mixture secreted by the liver, serves a similar purpose by dispersing lipids (more commonly called *fats* in digestive physiology) as fine droplets with greater surface area.

The pH at which different digestive enzymes function best [∞ p. 84] reflects the location where they are most active. Enzymes that act in the stomach work well at acidic pH. Those that are secreted into the small intestine work best at alkaline pH.

Most nutrient absorption takes place in the small intestine, with additional absorption of water and ions in the large intestine. Like secretion, absorption of nutrients and ions across the GI epithelium uses many of the same transport proteins that the kidney tubule epithelium uses.

Carbohydrates Are Digested to Monosaccharides

About half the calories the average American ingests are in the form of carbohydrates, mainly *starch* and *sucrose* (table sugar). Other dietary carbohydrates include the glucose polymers *glycogen* and *cellulose,* disaccharides such as *lactose* and *maltose,* and the simple sugars *glucose* and *fructose* [∞ Fig. 2-13, p. 30].

The complex carbohydrates we eat include cellulose, glycogen, and starch. We are unable to digest cellulose because we lack the necessary enzymes. As a result, the cellulose in plant matter becomes what is known as fiber, or roughage, in our diet. Glycogen, found mainly in meat, is usually degraded during cooking.

Starch, therefore, is the primary complex carbohydrate that we must digest (Fig. 20-6 ■). **Amylase** is the enzyme that breaks long glucose polymers into smaller glucose chains and into the disaccharide maltose (Table 20-1). Maltose and the other disaccharides must be broken down into their component monosaccharides before they can be absorbed. Enzymes known as **disaccharidases** (maltase, sucrase, and lactase) are responsible for this final step.

Proteins Are Digested into Small Peptides and Amino Acids

Unlike carbohydrates, which are ingested in forms ranging from the simple to the complex, most proteins are ingested as proteins or large polypeptides [∞ Fig. 2-16, p. 33]. Not all proteins are equally digestible, however.

Plant proteins are the least digestible. Egg protein is among the most digestible, with 85%–90% of the protein in a form that can be digested and absorbed. Surprisingly, between 30% and 60% of the protein found in the intestinal lumen comes not from ingested food but from the sloughing of dead cells and from protein secretions such as enzymes and mucus.

The enzymes for protein digestion are classified into two broad groups: endopeptidases (also called *proteases*) and exopeptidases. **Endopeptidases** attack peptide bonds in the interior of the amino acid chain and make smaller peptide fragments from a long chain (Fig. 20-7 ■). These enzymes are secreted as inactive proenzymes from epithelial cells in the stomach, intestine, and pancreas. **Exopeptidases** release single amino acids from peptides by chopping them off the ends, one at a time.

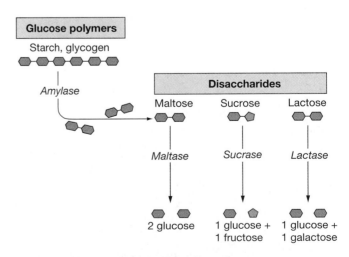

■ **Figure 20-6   Carbohydrate digestion**

TABLE 20-1   Digestive Enzymes

Biomolecule	Digested by	Into	Enzyme Location
Carbohydrate			
Starch	Amylase	Maltose	Saliva, pancreas
Disaccharides			Intestine
Maltose	Maltase	2 Glucose	
Lactose	Lactase	Glucose + galactose	
Sucrose	Sucrase	Glucose + fructose	
Proteins			
Interior peptide bonds	Endopeptidases	Peptides	Stomach, intestine, pancreas
Terminal peptide bonds	Exopeptidases	Amino acids	Stomach, intestine, pancreas
—Works at NH_2-terminal end	Aminopeptidase		
—Works at COOH-terminal end	Carboxypeptidase		
Fats			
Triglycerides	Lipase	Monoglyceride and free fatty acid	Mouth, stomach, pancreas
Phospholipids	Phospholipase		Pancreas

Fat Digestion Is Assisted by Bile

Fats and related molecules in the Western diet include triglycerides, cholesterol, phospholipids, long-chain fatty acids, and the fat-soluble vitamins [∞ Figure 2-14, p. 31]. Nearly 90% of our fat calories come from triglycerides because they are the primary form of lipid in both plants and animals.

Fat digestion is complicated by the fact that most lipids are not particularly water-soluble. As a result, they form large clumps in the aqueous chyme solution of the intestinal tract. Enzymatic fat digestion must therefore be aided by nonenzyme secretions that help make fat globules into smaller particles.

Enzymatic fat digestion is carried out by **lipases.** These enzymes remove two fatty acids from triglycerides, resulting in a monoglyceride and two free fatty acids (Fig. 20-8 ■). Phospholipids are digested by pancreatic *phospholipase.* Free cholesterol does not have to be digested before being absorbed.

Nucleic Acids Are Digested into Bases and Monosaccharides

The nucleic acid polymers DNA and RNA do not make up a significant part of most diets. They are digested by pancreatic and intestinal enzymes, first into their component nucleotides and then into bases and monosaccharides [∞ Figure 2-17, p. 34]. The bases are absorbed by active transport, and the monosaccharides absorbed by facilitated diffusion and secondary active transport like other simple sugars.

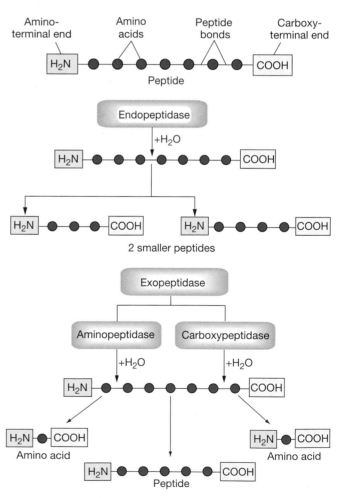

■ Figure 20-7   **Endopeptidases and exopeptidases**   Endopeptidases cleave interior peptide bonds. Exopeptidases cleave terminal peptide bonds, releasing single amino acids.

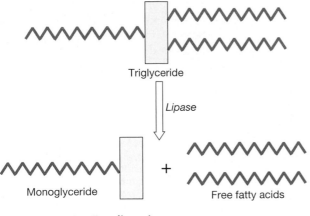

Monoglyceride + Free fatty acids

■ **Figure 20-8** **Fat digestion**

❱ REGULATION OF GI FUNCTION

The digestive system has remarkably complex regulatory mechanisms that we are nowhere close to understanding. Neural control of the GI tract does not strictly follow the sensory neuron-brain-autonomic neuron pattern we might expect. Instead, the enteric nerve plexus acts as a "little brain," allowing some reflexes to begin and end in the gastrointestinal tract without communicating with the rest of the nervous system.

At the same time, the influence of emotions on the GI tract illustrates the link between the enteric nervous system and the "big brain," or *cephalic* brain. The digestive system manifests various emotional responses. They range from travelers' constipation to "butterflies in the stomach" to psychologically induced diarrhea.

Between the two extremes of gastrointestinal control are reflexes that start with stimuli in the gastrointestinal tract, are integrated in the central nervous system, and then are carried out by autonomic neurons synapsing with neurons of the enteric nerve plexus. Complicating these neural pathways are a host of unusual neurotransmitters, hormones, and paracrines.

One particularly difficult part of mastering the complexities of GI regulation is learning the terminology. Some of the hormones and neurotransmitters involved in digestive regulation were first identified in other systems of the body. As a result, their names have nothing to do with their function in the gastrointestinal system.

The Enteric Nervous System Is Known as the Little Brain

The enteric nervous system was first recognized over a century ago, when scientists noted that an increase in intraluminal pressure caused a reflex wave of peristaltic contraction to sweep along sections of isolated intestine removed from the body. In the 1920s, British physiologist Johannis Langley speculated that the nerve networks (*plexuses*) and ganglia of the GI tract were actually a third division of the autonomic nervous system. He named the intestinal nerve network the enteric nervous system (ENS).

The peristaltic reflex continued to be noted in physiology texts over the years, but Langley's hypothesis of a third autonomic division was largely ignored until the early 1970s. Since then, physiologists have come to recognize that the enteric nervous system can receive stimuli, integrate sensory information, and act upon it by changing motility or secretion, despite the absence of an identifiable integrating center like the brain or spinal cord.

In this respect, the enteric nervous system is much like the nervous systems of jellyfish and sea anemones (the phylum Cnidaria). You might have seen sea anemones being fed at an aquarium. As the piece of shrimp or fish drifts down close to the tentacles, they begin to wave, picking up chemical "odors" through the water. Once the food contacts the tentacles, it is directed toward the mouth, passed from one tentacle to another until it disappears into the digestive cavity.

This purposeful reflex is accomplished without a brain, eyes, or a nose. The anemone's nervous system consists of a network composed of sensory receptors, sensory neurons, interneurons, and efferent neurons that control the muscles and secretory cells of the anemone's body. Somehow, the neurons of the network are linked in a way that allows them to integrate information and act upon it.

In the same way that the anemone captures its food, the enteric nervous system receives stimuli and acts upon them. Reflexes that originate within the enteric nervous system and are integrated there without outside input are called **short reflexes** (Fig. 20-9 ■). The primary behaviors controlled by the enteric nervous system are related to motility, secretion, and growth.

The myenteric plexus of the ENS, located between the circular and longitudinal muscle layers, controls motility in the gut. The smaller submucosal plexus contains sensory neurons that receive signals from the lumen. Submucosal neurons send information to myenteric motor neurons to influence motility. They directly control secretion by GI cells.

The Enteric Nervous System Coordinates with the CNS

Although the enteric nervous system can work in isolation, it is also in contact with the cephalic brain. Signals originating in the GI tract can be sent to the central nervous system via sensory neurons. Following CNS integration of the information, several hundred autonomic efferent neurons carry signals back to the millions of neurons in the enteric nervous system.

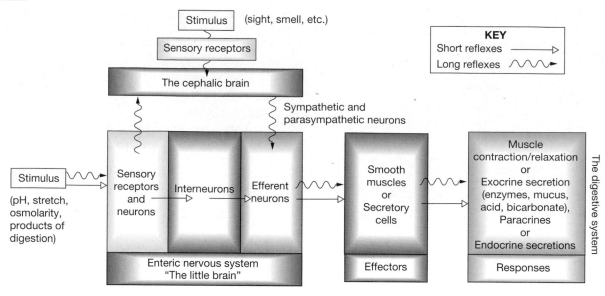

■ Figure 20-9   The enteric nervous system   Short reflexes are completely contained within the digestive system. Long reflexes either originate outside the digestive system or are partially integrated outside the ENS.

Gastrointestinal reflexes can also originate completely outside the enteric nervous system. These reflexes, usually part of the **cephalic phase** of digestion, begin with stimuli such as the sight, smell, sound, or thought of food. An example of a cephalic reflex is the way your mouth waters and your stomach growls when you smell dinner cooking.

Cephalic reflexes act in a feedforward manner, preparing the digestive system to receive the food that the sensory system is anticipating. Cephalic reflexes and other digestive reflexes integrated in the CNS are called **long reflexes** (Fig. 20-9 ■).

The autonomic divisions of the nervous system are active in long reflexes. In general, we say that parasympathetic cholinergic neurons to the GI tract, carried mostly in the vagus nerve, are excitatory and enhance GI functions. Sympathetic adrenergic neurons usually inhibit GI functions.

Although many physiologists follow Langley's classification of the enteric nervous system as a third subdivision of the autonomic nervous system, that classification may change in coming years. Both anatomically and functionally, the ENS is much more like the brain, so much so that it has been nicknamed "the little brain." Some similarities between the ENS and the brain are:

■ **Neurotransmitters and neuromodulators.** The neurons of the enteric nervous system release more than 30 neurotransmitters and neuromodulators, most of which are identical to molecules found in the brain. These neurotransmitters are sometimes called "non-adrenergic, non-cholinergic" to distinguish them from the traditional autonomic neurotransmitters, norepinephrine and acetylcholine. Among the best-known neurotransmitters and neuromodulators in the enteric nervous system are serotonin, vasoactive intestinal peptide (VIP), and nitric oxide (NO).

■ **Support cells.** The support cells of neurons within the enteric nervous system are more similar to the astroglia of the brain than to the Schwann cells of the peripheral nervous system.

■ **Diffusion barrier.** The capillaries that surround ganglia within the enteric nervous system are not very permeable and create a diffusion barrier that is similar to the blood-brain barrier of cerebral blood vessels.

■ **Integrating center.** Reflexes that originate with sensory receptors in the GI tract can be integrated and acted upon without neural signals leaving the enteric nervous system. Thus, the neuron network of the enteric nervous system is its own integrating center, much like the brain and spinal cord.

The enteric nervous system has generated much interest in recent years. It was thought that if we could explain how it integrates simple behaviors, we could use the gut as a simple model system for central nervous system function. But study of enteric nervous system function is difficult because there is no discrete command center for enteric reflexes.

Instead, in an interesting twist, GI physiologists are applying information gleaned from studies of the brain and spinal cord to the function of the enteric nervous system. The complex interactions between the enteric and central nervous systems, the endocrine system, and the immune system promise to provide scientists with questions to investigate for many years to come.

...continued from page 611

Tonya undergoes endoscopy, a procedure in which a fiberoptic scope is inserted into her esophagus and maneuvered into her stomach. Small pincers attached at the end of the scope pluck a piece of her stomach lining, which is sent to the pathology laboratory for study. Tonya is awake but groggy from a tranquilizer during the procedure. That afternoon, she feels fine, but her stomach hurts. She chews a tablet of antacid for relief.

Question 3:   *Some antacids contain sodium bicarbonate ($NaHCO_3$), while many others contain aluminum hydroxide ($Al(OH)_3$). Antacids neutralize hydrochloric acid (HCl) secreted by the stomach by acting as buffers. For both types of antacids, write the chemical equation showing how the antacids buffer hydrochloric acid.*

Digestive Hormones Control GI Function, Metabolism, and Eating Behavior

The hormones of the gastrointestinal tract occupy an interesting place in the history of endocrinology. In 1902, two Canadian physiologists, W. M. Bayliss and E. H. Starling, discovered that acidic chyme entering the small intestine from the stomach caused the release of pancreatic juices even if all nerves to the pancreas were cut. Because the only communication remaining between the intestine and the pancreas was the blood supply that ran between them, they postulated the existence of some blood-borne (*humoral*) factor that was released by the intestine.

When duodenal extracts applied directly to the pancreas stimulated secretion, they knew they were dealing with a chemical produced by the duodenum. They named the substance *secretin*. Starling further proposed that the general name *hormone,* from the Greek word meaning "I excite," be given to all humoral agents that act at a site distant from their release.

A few years later, in 1905, J. S. Edkins postulated the existence of a gastric hormone that stimulated gastric acid secretion. It took more than 30 years for a relatively pure extract of the gastric hormone to be made successfully, and it was 1964 before the hormone named *gastrin* was finally purified.

Why was research on the digestive hormones so slow to develop? A major reason is the way hormones of the GI tract are secreted by isolated endocrine cells scattered among other cells of the mucosal epithelium. In the early years, the only way to obtain these hormones was to make a crude extract of the entire epithelium, a procedure that also liberated digestive enzymes and paracrines made in adjacent cells. Thus, it was very difficult to tell if the physiological effect elicited by the extract came from one hormone, from more than one hormone, or from a paracrine such as histamine.

Today, although researchers have sequenced over 30 peptides from the GI mucosa, only 5 are widely accepted as hormones. A few peptides have well-defined paracrine effects, but most fall into a long list of candidate hormones. In addition, we know of nonpeptide regulatory molecules, such as histamine, a paracrine. Because of the uncertainty associated with the field, we will restrict our focus to the major regulatory molecules.

GI Peptides Include Hormones, Neurocrines, and Cytokines

The gastrointestinal peptides are generally divided into three groups:

1. The *gastrin family* includes the hormones **gastrin** and **cholecystokinin** (CCK), with several variants of each.
2. The *secretin family* includes **secretin,** the neurocrine [∞ p. 155] molecule **vasoactive intestinal peptide** (VIP), and **GIP,** a hormone known originally as *gastric inhibitory peptide* because it inhibited gastric acid secretion in early experiments. More recent studies, however, found that GIP administered in lower physiological doses does not block acid secretion. A new name was proposed, **glucose-dependent insulinotropic peptide.** This name retains the original initials of the hormone (*GIP*) and more accurately describes its actions (see below). Another member of the secretin family is the candidate hormone, **glucagon-like peptide 1** (GLP-1).
3. The third group of peptides includes those that do not fit into the two families above. The primary member of this group is the hormone **motilin.**

The sources, targets, and effects of some GI peptides are summarized in Table 20-2. In the following sections we focus on common themes that govern the release and function of GI hormones and other peptides.

GI Peptides Are Synthesized in Multiple Sites
Gastrin is the primary hormone secreted by the stomach. Secretin, CCK, GIP, motilin, and GLP-1 are secreted by cells in the intestinal mucosa. However, with the introduction of immunohistochemical techniques for identifying specific chemicals within cells, we have discovered that many GI peptides are not restricted to the GI tract.

For example, CCK and VIP both function as neurotransmitters in the brain. **Somatostatin,** a peptide paracrine secreted by cells in the gastric epithelium, is also secreted by cells in the endocrine pancreas and by neurons of the hypothalamus. Hypothalamic somatostatin also goes by the name of *growth hormone-inhibiting hormone* [∞ p. 201]. Multiple release sites add to the difficulty of establishing physiological functions for these peptides.

GI Peptides Affect Motility and Secretion   As a group, the GI peptides excite or inhibit motility and secretion (Fig. 20-10 ■). Motility effects include changes in peristaltic activity, contraction of the gallbladder for bile

TABLE 20-2   The Digestive Hormones

	Gastrin	Cholecystokinin (CCK)	Secretin
Secreted by	G cells in stomach antrum	Endocrine cells of small intestine; neurons of brain and gut	Endocrine cells in small intestine
Target(s)	ECL cells; parietal cells	Gallbladder, pancreas, gastric smooth muscle	Pancreas, stomach
Effects			
Endocrine secretion	None	None	None
Exocrine secretion	Stimulates gastric acid	Stimulates pancreatic enzyme secretion; potentiates bicarbonate secretion	Stimulates bicarbonate secretion and pepsin release; inhibits gastric acid
Motility	None	Stimulates gallbladder contraction for bile release; inhibits gastric emptying	Inhibits gastric emptying
Other	Enhanced mucosal cell growth	Stimulates satiety	None
Stimulus for release	Peptides and amino acids in lumen; gastrin-releasing peptide and ACh in nervous reflexes	Fatty acids and some amino acids	Acid in small intestine
Release inhibited by	pH < 1.5; somatostatin	NA	Somatostatin
Other information	NA	Some effects may be due to CCK as a neuropeptide rather than as a hormone	NA

	Glucose-dependent insulinotropic peptide (GIP)	Motilin	Glucagon-like Peptide 1
Secreted by	Endocrine cells in small intestine	Endocrine cells in small intestine	Endocrine cells in small intestine
Target(s)	Beta cells of endocrine pancreas	Smooth muscle of antrum and duodenum	Endocrine pancreas
Effects			
Endocrine secretion	Stimulates insulin release (feedforward mechanism)	None	Stimulates insulin release; inhibits glucagon release
Exocrine secretion	Inhibits acid secretion	None	Possibly inhibits acid secretion
Motility	None	Stimulates migrating motor complex	Possibly inhibits gastric emptying
Other	None	None	None
Stimulus for release	Glucose, fatty acids, and amino acids in small intestine	Fasting: periodic release every 1.5–2 hours by neural stimulus	Mixed meal that includes carbohydrates or fats in the lumen
Release inhibited by	NA	NA	NA
Other information	Acid inhibition questionable at physiological concentrations	Changes associated with both constipation and diarrhea, but relationship is unclear	Related to but not identical to pancreatic glucagon; may act together with GIP

release, and delay of gastric emptying. Control of gastric emptying is used to regulate the rate at which food enters the small intestine so that digestion and absorption can be maximally effective. Secretory processes under endocrine control include the release of enzymes from the stomach and the release of pancreatic juices.

Some GI peptides interact with other hormones. For example, GIP release is stimulated by glucose entering the duodenum. GIP then acts in a feedforward fashion to enhance insulin release so that the body is prepared to handle the glucose load as soon as it is absorbed. The peptide GLP-1 is also a powerful stimulus

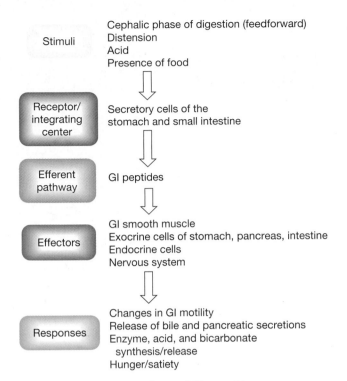

Stimuli

Cephalic phase of digestion (feedforward)
Distension
Acid
Presence of food

Receptor/
integrating
center

Secretory cells of the
stomach and small intestine

Efferent
pathway

GI peptides

Effectors

GI smooth muscle
Exocrine cells of stomach, pancreas, intestine
Endocrine cells
Nervous system

Responses

Changes in GI motility
Release of bile and pancreatic secretions
Enzyme, acid, and bicarbonate
 synthesis/release
Hunger/satiety

■ **Figure 20-10   Functions of GI peptides**

for insulin release while simultaneously inhibiting glucagon release.

The most interesting and difficult pathways to study are those in which peptides act on the brain to change behavior. In the digestive system, the hormone CCK is the main candidate for behavioral effects. In experimental studies, CCK was found to enhance **satiety,** the state of feeling that hunger has been satisfied. From these data, it appears that CCK has the potential to be used as a "natural" appetite suppressant for obesity control.

Eating Triggers Peptide Release   The stimuli for GI peptide secretion arise primarily from the ingestion of food. The biomolecules that stimulate release are those that you might predict by looking at the effects of the hormone.

For example, CCK, which stimulates gallbladder contraction and bile release, is secreted in response to fatty meals. CCK also slows gastric emptying, which allows slower digestion and absorption of fats to go to completion.

As mentioned above, GIP, which causes insulin release, is released in response to glucose in the intestine. Secretin triggers pancreatic bicarbonate secretion in response to acid in the small intestine. In each of these examples, the stimulus for hormone release and the action of the hormone are directly related.

Input from the nervous system may also influence peptide release. For example, increased parasympathetic activity following ingestion of a meal stimulates the release of gastrin in the stomach.

Paracrines in the GI Tract Have Diverse Effects on Digestion

New information about paracrines in the gastrointestinal tract is being published almost daily. Paracrines may be (1) molecules found in the lumen that combine with cell membrane receptors to elicit a response or (2) molecules secreted into the extracellular fluid or lumen by epithelial cells in the wall of the gut.

An interesting example of a paracrine found in the lumen is a molecule formed from inactive procolipase. In the lumen, this proenzyme splits into colipase and a peptide that has been named **enterostatin.** Enterostatin, like CCK, is under a great deal of scrutiny because it seems to act as a satiety signal for fat ingestion.

Paracrines are also secreted into the extracellular fluid on the basolateral side of the GI epithelium. Histamine is a well-studied paracrine that has multiple effects on the stomach, as we will discuss later. Another paracrine is serotonin, also known as 5-hydroxytryptamine (5-HT). This molecule, better known as a CNS neurotransmitter, is released from epithelial cells in response to distension. Serotonin stimulates sensory neurons in the enteric plexus and triggers a peristaltic reflex.

▶ INTEGRATION OF GI FUNCTION

In the remainder of this chapter we will follow some food as it passes through the GI tract. As a preview, look at Figure 20-11 ■, a summary of the main events that occur in each section of the GI tract. Food processing is traditionally divided into three phases: a cephalic phase, a gastric phase, and an intestinal phase.

The Cephalic Phase Begins Digestion

Digestive processes in the body begin before food ever enters our mouths. Simply smelling, seeing, or even *thinking* about food can make our mouths water and our stomachs rumble. These long reflexes that begin in our cephalic brain create a feedforward response that is known as the *cephalic phase* of digestion (Fig. 20-12 ■).

These anticipatory stimuli and the stimulus of food in the oral cavity activate neurons in the medulla oblongata. The medulla in turn sends an efferent signal through autonomic neurons to the salivary glands and through the vagus nerve to the enteric nervous system. In response to these signals, the stomach, intestine, and accessory glandular organs begin secretion and increase motility in anticipation of the food to come.

Chemical and Mechanical Digestion Begins in the Mouth

When food first enters the mouth, it is met by a flood of the secretion we call **saliva,** a dilute solution of water, ions, and proteins such as mucins, lysozyme, and

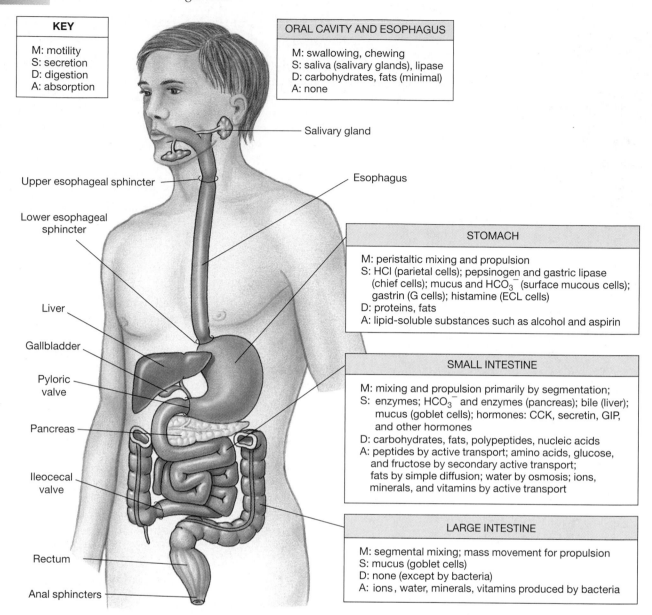

KEY

M: motility
S: secretion
D: digestion
A: absorption

ORAL CAVITY AND ESOPHAGUS

M: swallowing, chewing
S: saliva (salivary glands), lipase
D: carbohydrates, fats (minimal)
A: none

Salivary gland

Upper esophageal sphincter

Esophagus

Lower esophageal sphincter

Liver

Gallbladder

Pyloric valve

Pancreas

Ileocecal valve

Rectum

Anal sphincters

STOMACH

M: peristaltic mixing and propulsion
S: HCl (parietal cells); pepsinogen and gastric lipase
   (chief cells); mucus and HCO_3^- (surface mucous cells);
   gastrin (G cells); histamine (ECL cells)
D: proteins, fats
A: lipid-soluble substances such as alcohol and aspirin

SMALL INTESTINE

M: mixing and propulsion primarily by segmentation;
S: enzymes; HCO_3^- and enzymes (pancreas); bile (liver);
   mucus (goblet cells); hormones: CCK, secretin, GIP,
   and other hormones
D: carbohydrates, fats, polypeptides, nucleic acids
A: peptides by active transport; amino acids, glucose,
   and fructose by secondary active transport;
   fats by simple diffusion; water by osmosis; ions,
   minerals, and vitamins by active transport

LARGE INTESTINE

M: segmental mixing; mass movement for propulsion
S: mucus (goblet cells)
D: none (except by bacteria)
A: ions, water, minerals, vitamins produced by bacteria

■ **Figure 20-11**   **Summary of the processes in the digestive system**

immunoglobulins. Salivary secretion is under autonomic control and can be triggered by multiple stimuli such as the sight, smell, touch, and even thought of food.

Saliva serves multiple purposes. Its water and mucus soften and lubricate food to make it easier to swallow. You can appreciate this function if you've ever tried to swallow a dry soda cracker without chewing it thoroughly. Saliva also dissolves food so that we can taste it.

Saliva begins chemical digestion with the secretion of *salivary amylase* and a small amount of lipase. Amylase breaks starch into maltose after it is activated by chloride ions, which are abundantly available in saliva. If you chew on an unsalted soda cracker for a long time,

you may be able to taste the conversion of flour starch in the cracker into sweeter maltose.

Small nonsalivary glands in the mouth secrete a second digestive enzyme, *lingual lipase.* This enzyme does not contribute significantly to fat digestion in the mouth but remains active in the stomach.

The final function of saliva is protection. *Lysozyme* is an antibacterial enzyme, and *immunoglobulins* are proteins that disable bacteria and viruses. In addition, saliva helps wash down the teeth and keep the tongue free of food particles.

Mechanical digestion of food begins in the oral cavity with chewing. The lips, tongue, and teeth all contribute to the **mastication** of food, creating a softened, moistened mass, or *bolus,* that can be easily swallowed.

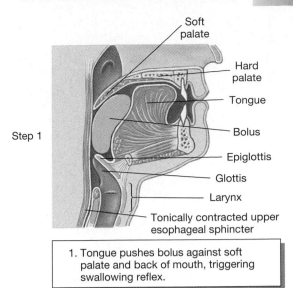

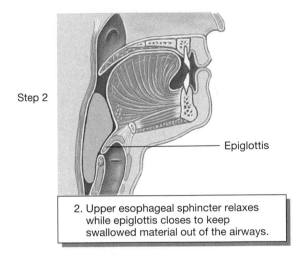

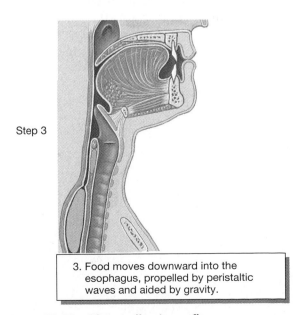

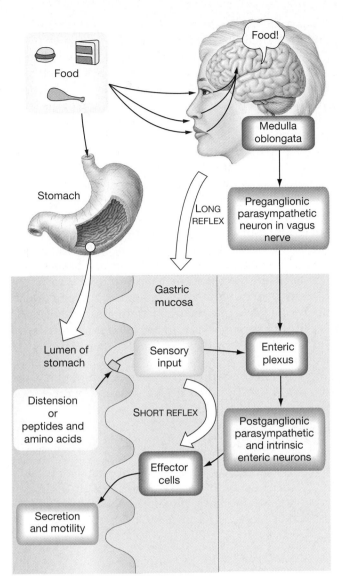

■ **Figure 20-12   Long and short reflexes in the stomach**
The sight, smell, and taste of food initiate long reflexes that prepare
the stomach. Distension of the stomach and peptides or amino
acids in the stomach initiate short reflexes. Both types of reflexes
result in acid and enzyme secretion and increased motility.

1. Tongue pushes bolus against soft
   palate and back of mouth, triggering
   swallowing reflex.

2. Upper esophageal sphincter relaxes
   while epiglottis closes to keep
   swallowed material out of the airways.

3. Food moves downward into the
   esophagus, propelled by peristaltic
   waves and aided by gravity.

Swallowing Moves Food from the Mouth to the Stomach

Swallowing, or **deglutition,** is a reflex action that pushes a bolus of food or liquid into the esophagus (Fig.
20-13 ■). The stimulus for swallowing is pressure created when the bolus is pushed against the soft palate and
back of the mouth with the tongue. Sensory input to a
swallowing center in the medulla oblongata begins the
reflex.

First, the **epiglottis** folds down over the opening of
the larynx to prevent food and liquid from entering the
airways. At the same time, respiration is inhibited and
the upper esophageal sphincter relaxes as the bolus en-

■ **Figure 20-13   The swallowing reflex**

ters the esophagus. Waves of peristaltic contraction then push the bolus toward the stomach, aided by gravity.

The lower end of the esophagus lies just below the diaphragm and is separated from the stomach by the lower esophageal sphincter. This area is not a true sphincter but a region of higher muscle tension that acts as a barrier between the esophagus and the stomach. When food is swallowed, the tension relaxes, allowing the bolus to pass into the stomach.

If the lower esophageal sphincter does not stay contracted, gastric acid and pepsin can irritate the lining of the esophagus, leading to the pain and irritation of heartburn. The walls of the esophagus expand during inspiration, when the intrapleural pressure drops [∞ p. 510]. Expansion creates subatmospheric pressure in the esophageal lumen that can literally suck acidic contents out of the stomach if the sphincter is relaxed. The churning action of the stomach when filled with food can also squirt acid back into the esophagus if the sphincter is not contracted fully.

The Stomach Is the Gatekeeper of the GI Tract

About 3.5 liters of food, drink, and saliva enter the fundus of the stomach each day. The stomach has three general functions:

1. **Storage.** It stores food and regulates its passage into the small intestine, where most digestion and absorption take place.
2. **Digestion.** It chemically and mechanically digests food into the soupy mixture of uniformly small particles called chyme.
3. **Protection.** It protects the body by destroying many of the bacteria and other pathogens that are swallowed with food or trapped in airway mucus. At the same time, the stomach must protect itself from being damaged by its own secretions.

Digestive activity in the stomach begins with the long **vagal reflex** of the cephalic phase, before food even arrives. Then, when food enters the stomach, stimuli within the gastric lumen initiate a series of short reflexes. These short reflexes are the **gastric phase** of digestion.

In gastric phase reflexes, distension of the stomach and the presence of peptides or amino acids in the lumen activate endocrine cells and enteric neurons. Hormones, neurocrines, and paracrines then influence motility and secretion.

Motility in the Stomach

When food arrives, the stomach relaxes and expands to hold the increased volume. The upper half of the stomach remains relatively quiet, holding food until it is needed. The storage function of the stomach is perhaps the least obvious aspect of digestion. However, at times we ingest more than we need from a nutritional standpoint. In these situations,

it falls to the stomach to regulate the rate at which food enters the small intestine.

Without such regulation, the small intestine would not be able to digest and absorb the load presented to it, and significant amounts of unabsorbed chyme would pass into the large intestine. The epithelium of the large intestine is not designed for large-scale nutrient absorption, so most of the chyme would pass out in the feces, resulting in diarrhea. This "dumping syndrome" is one of the less pleasant side effects of surgery that removes portions of either the stomach or small intestine.

While the upper stomach is quietly holding food, the lower stomach is busy with digestion. In the distal half of the stomach, a series of peristaltic waves pushes the food down toward the pyloric valve, mixing it with acid and digestive enzymes. As large food particles are digested to the more uniform texture of chyme, each contractile wave squirts a small amount of chyme through the pyloric valve into the duodenum. Enhanced gastric motility during a meal is primarily under neural control and is stimulated by distension of the stomach.

Digestion in the Stomach

Chemical digestion in the stomach is accomplished by enzymes and hydrochloric acid.

- *Gastric acid* (HCl) kills bacteria and other microorganisms that are swallowed. Acid also *denatures* proteins [∞ p. 83], destroying their tertiary structure by breaking disulfide and hydrogen bonds. Unfolding protein chains makes the peptide bonds between amino acids accessible to enzymes.

Olestra, the No-Calorie Fat Substitute Although nutritionists have determined that the typical Western diet contains too much fat, many consumers are reluctant to give up fat in their diet. As a result, food manufacturers such as Procter and Gamble have been developing low-calorie fat substitutes. In 1996, a fat substitute named Olestra® was approved by the Food and Drug Administration. Olestra is sucrose polyester: a sucrose molecule with 6–8 fatty acids attached to it. This unusual arrangement of naturally occurring substances tastes and feels like dietary fats but cannot be digested by intestinal enzymes or absorbed across the intestinal epithelium. Foods containing Olestra were introduced with considerable fanfare, but, within weeks, unhappy consumers were reporting unpleasant side effects. Intestinal cramping, gas, and diarrhea were significant enough in some people to cause them to stop using Olestra. In addition, nutritionists are concerned that excessive consumption of Olestra will lead to deficiencies of the fat-soluble vitamins (A, D, E, and K) that are normally absorbed along with dietary fats. The Food and Drug Administration requires a warning label on foods with Olestra and is monitoring its effects.

- *Pepsin* is a protease that carries out the initial digestion of proteins. It is particularly effective on collagen and therefore plays an important role in digesting meat.
- *Gastric lipase* is co-secreted with pepsin and begins fat digestion, aided by lingual lipase. About 10% of fat digestion takes place in the stomach.
- The carbohydrate digestion that began in the mouth will continue in the stomach until mixing exposes the amylase to gastric acid. Salivary amylase is inactivated at low pH.

Secretion in the Stomach   The stomach secretes a variety of substances from an assortment of cells. We will discuss the functions and relationships of these secretions in the following section.

- **Parietal cells** deep in the gastric glands secrete hydrochloric acid. Acid secretion in the stomach averages 1–3 liters per day and can create a luminal pH as low as 1. The cytoplasmic pH of the parietal cell is about 7.2, which means the cells are pumping H^+ against a gradient that is 2.5 *million* times as concentrated as the H^+ concentration in the cell.

  The cellular pathway for acid secretion is shown in Figure 20-14 ∎. In an unusual twist, the H^+ comes directly from water by an unknown mechanism. An H^+-K^+-ATPase pumps the H^+ into the lumen. Cl^- moves into the lumen through an open chloride channel. The OH^- from water combines with CO_2 to form HCO_3^-, which is absorbed into the blood.

- Parietal cells also secrete a protein known as **intrinsic factor**. Intrinsic factor complexes with vitamin B_{12} and is essential for B_{12} absorption in the intestine. In the absence of intrinsic factor, vitamin B_{12} deficiency causes the condition known as *pernicious anemia*. In this state, red blood cell synthesis (erythropoiesis), which depends on vitamin B_{12}, is severely diminished. Lack of intrinsic factor cannot be remedied directly, but patients with pernicious anemia can be given vitamin B_{12} shots.

- **Chief cells** in the gastric glands secrete the inactive enzyme **pepsinogen**. Pepsinogen is cleaved to active pepsin in the lumen of the stomach by the action of H^+.

- **D cells,** which are closely associated with the parietal cells, secrete the paracrine **somatostatin.**

- **Enterochromaffin-like** (ECL) **cells** secrete the paracrine histamine.

- **G cells,** found deep within the gastric glands, secrete the hormone **gastrin.** The release of gastrin is stimulated by the presence of amino acids and peptides in the stomach, by distension of the stomach, and by nervous reflexes mediated by **gastrin-releasing peptide.** Coffee, even decaffeinated, is also a good stimulant of gastrin release, one reason that people with excess acid secretion syndromes are advised to avoid coffee. Gastrin release is inhibited by somatostatin and by a luminal pH below 1.5.

- **Mucous cells** in the neck of the gastric glands secrete both mucus and bicarbonate. These substances create a barrier that protects the stomach from digesting itself (autodigestion). The mucus forms a physical barrier, while the bicarbonate creates a chemical buffer barrier underlying the mucus (Fig. 20-15 ∎).

Researchers using micro pH electrodes have shown that the bicarbonate layer just above the cell surface in the stomach has a pH that is close to 7, even when the pH in the lumen is highly acidic at pH 2. Mucus secretion is increased when the stomach is irritated, such as by the ingestion of aspirin (acetylsalicylic acid) or alcohol.

The various secretions of the stomach, their stimuli for release, and their functions are summarized in Figure 20-16 ∎.

Integrated Stomach Processes   Now that you have an overview of digestion and secretion in the stomach, let's see how everything functions together when you eat a meal.

When food comes into the mouth, the feedforward cephalic vagal reflex begins secretion in the stomach. Once food enters the stomach, the gastric phase stimuli continue to promote secretion and increased motility. The coordinated function of the gastric secretory cells is shown in Figure 20-17 ∎.

- Parasympathetic neurons from the vagus nerve stimulate G cells to release gastrin into the blood.

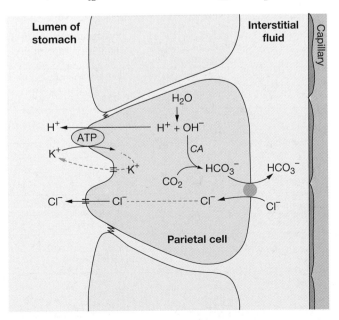

■ **Figure 20-14   Acid secretion by parietal cells**   CA = carbonic anhydrase

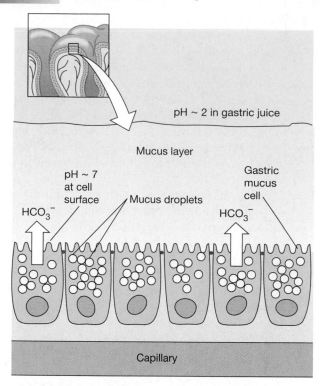

- Histamine is released from ECL cells in response to gastrin and acetylcholine from the enteric nervous system. Histamine diffuses to its target, the parietal cells.
- Histamine stimulates acid secretion by combining with histamine H_2 receptors on parietal cells.
- Acid in the lumen stimulates pepsinogen release from chief cells through a short reflex. In the lumen, acid converts pepsinogen into pepsin, and protein digestion begins.
- Acid also triggers somatostatin release from D cells. Somatostatin acts in a negative feedback fashion to inhibit gastric acid, gastrin, and pepsinogen secretion.

Under normal conditions, the gastric mucosa is protected from enzymes and acid by the mucus-bicarbonate barrier. However, in *Zollinger-Ellison syndrome*, patients secrete excessive levels of gastrin, usually from gastrin-secreting tumors in the pancreas. As a result, hyperacidity in the stomach overwhelms the normal protective mechanisms and causes a *peptic ulcer*. In peptic ulcers, acid and pepsin destroy the mucosa, creating holes that extend into the submucosa and muscularis of the stomach, duodenum, or esophagus.

Excess acid secretion is an uncommon cause of peptic ulcers. By far the most common causes are nonsteroidal anti-inflammatory drugs (NSAIDs) such as aspirin and *Helicobacter pylori*, a bacterium that creates inflammation of the gastric mucosa.

For many years the primary therapy for excess acid secretion or *dyspepsia*, was the ingestion of *antacids*, agents that could neutralize acid in the gastric lumen.

■ **Figure 20-15   The mucus-bicarbonate barrier of the gastric mucosa**   Mucus is a physical barrier. Bicarbonate is a chemical barrier that neutralizes acid.

Amino acids or peptides into the lumen do the same. They also trigger a short reflex for gastrin release.
- Gastrin in turn promotes acid release, both directly and indirectly by stimulating histamine release.

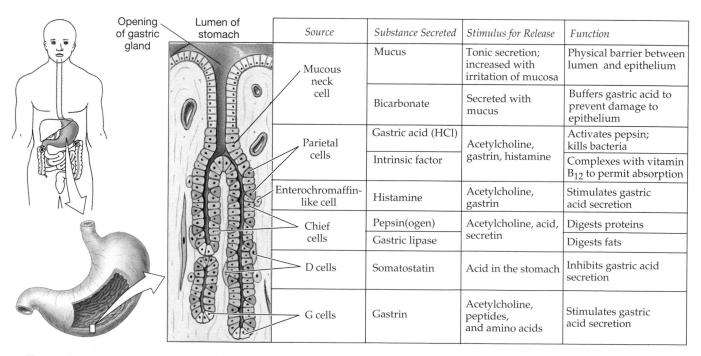

Source	Substance Secreted	Stimulus for Release	Function
Mucous neck cell	Mucus	Tonic secretion; increased with irritation of mucosa	Physical barrier between lumen and epithelium
	Bicarbonate	Secreted with mucus	Buffers gastric acid to prevent damage to epithelium
Parietal cells	Gastric acid (HCl)	Acetylcholine, gastrin, histamine	Activates pepsin; kills bacteria
	Intrinsic factor		Complexes with vitamin B_{12} to permit absorption
Enterochromaffin-like cell	Histamine	Acetylcholine, gastrin	Stimulates gastric acid secretion
Chief cells	Pepsin(ogen)	Acetylcholine, acid, secretin	Digests proteins
	Gastric lipase		Digests fats
D cells	Somatostatin	Acid in the stomach	Inhibits gastric acid secretion
G cells	Gastrin	Acetylcholine, peptides, and amino acids	Stimulates gastric acid secretion

■ Figure 20-16   **Cell types of the gastric mucosa**

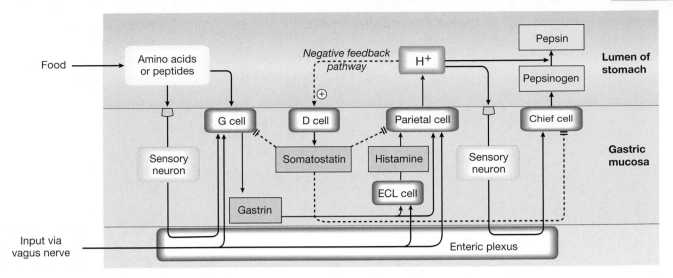

■ Figure 20-17   Integration of secretion in the stomach

But as molecular biologists explored the mechanism for acid secretion by parietal cells, the potential for new therapies became obvious. Today we have two new classes of drugs to fight hyperacidity: H_2 histamine receptor antagonists (cimetidine and ranitidine, for example) and *proton pump inhibitors* (PPIs) that block the H^+-K^+-ATPase (omeprazole and lansoprazole, for example).

The Intestinal Phase and Gastric Function: Feedback Signals

As chyme enters the small intestine, the **intestinal phase** of digestion begins. Before we move on to see what takes place in the small intestine itself, we should look at some reflexes from this phase. These reflexes regulate the delivery rate of chyme from the stomach and are both neural and hormonal (Fig. 20-18 ■).

The presence of acidic chyme in the duodenum releases the hormone secretin. Secretin then inhibits acid production and gastric motility, slowing gastric emptying. In addition, secretin stimulates production of pancreatic HCO_3^- to neutralize the acidic chyme that has entered the intestine.

If a meal contains fats, then the hormone cholecystokinin is secreted into the bloodstream. CCK also slows gastric motility and acid secretion. Fat digestion is slow-er than either protein or carbohydrate digestion, so it is crucial that the stomach only allow small amounts of fat into the intestine at one time.

If the meal contains carbohydrates, then the hormone GIP is released. In experiments, large doses of GIP inhibit gastric acid secretion, but there is some question whether GIP release reaches these levels with a normal meal.

The mixture of acid, enzymes, and digested food in chyme usually forms a hyperosmotic solution. Osmoreceptors in the intestine wall are sensitive to the osmolarity of the entering chyme. When stimulated by high osmolarity, they inhibit gastric emptying in a reflex mediated by some unknown blood-borne substance.

The net result of all three phases of gastric function is the digestion of proteins in the stomach by pepsin, the formation of chyme by the action of pepsin, acid and peristaltic contractions, and the controlled entry of chyme into the small intestine so that further digestion and absorption can take place.

Most Digestion and Absorption Occurs in the Intestine

As chyme from the stomach passes through the small intestine, additional enzymes and fluid are added by the intestinal epithelium, the exocrine portion of the pancreas, and the liver. Most digestion and absorption take place in the duodenum and jejunum. The small intestine and its accessory organs fulfill these roles:

- Bicarbonate secretions, primarily by the pancreas, neutralize acidic chyme to prevent damage to the intestinal mucosa. Intestinal goblet cells secrete mucus for protection and lubrication.
- The pancreas and intestinal epithelium secrete fluid containing enzymes for the digestion of proteins, carbohydrates, and fats. The liver adds *bile,*

...continued from page 615

Tonya's stomach still hurts later that night. Her roommate suggests that she try one of the new "acid blockers" that she has seen advertised on TV. These drugs are known as H_2 or histamine receptor blockers because they bind competitively to H_2-type histamine receptors on parietal cells. Fortunately, Tonya's roommate has just bought a package of these drugs. Two hours after taking the drug, Tonya feels much better.

Question 4:   How does blocking H_2 receptors stop acid production in the stomach?

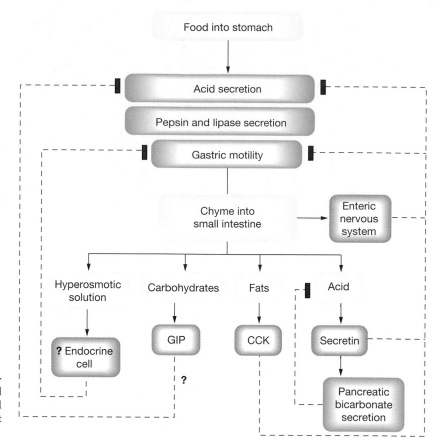

■ **Figure 20-18** **The intestinal phase of gastric function** Once food passes into the small intestine, feedback loops slow gastric motility and acid secretion so that intestinal digestion will not be overwhelmed with too much food at once.

a complex mixture that contains wastes to be excreted and agents to aid fat digestion.

- Motility in the intestine mixes chyme with enzymes and exposes chyme to the mucosa for absorption. Forward movement through the intestine must be slow enough to allow digestion and absorption to go to completion.

- Absorption in the intestine takes place through transporting **absorptive cells,** also called *enterocytes* (Fig. 20-2g ■). Most digested nutrients are absorbed into the blood of the hepatic portal system [∞ p. 406]. This specialized region of the circulation has two capillary beds: one set to pick up absorbed nutrients at the intestine and the other set to deliver the nutrients directly to the liver (Fig. 20-19 ■).

    Digested fats, however, go into the lymphatic system rather than into the blood, because the intestinal capillaries have a basement membrane around them that most lipids are unable to cross.

- As described in the previous section, endocrine secretions from the intestine coordinate gastric emptying. Other endocrine and neural signals stimulate enzyme and bicarbonate secretion or release of bile from the gallbladder. GIP acts in a feedforward fashion to initiate insulin release from the endocrine pancreas.

Motility Once food enters the small intestine, it is slowly propelled forward by a combination of peristaltic and segmental contractions. Parasympathetic innervation, gastrin, and CCK all promote intestinal motility. Sympathetic innervation inhibits it.

Secretion Secretions in the small intestine come from the intestinal epithelium, the exocrine pancreas, and the liver.

Enzyme secretion The pancreas and intestinal epithelial cells secrete enzymes for the intestinal phase. Pancreatic enzymes enter the intestine in a watery fluid and are active in the lumen. Intestinal enzymes, on the other hand, remain bound to the apical membranes of brush border cells, anchored by transmembrane protein "stalks." **Brush border enzymes** are stationary and will not be swept out of the small intestine along with the chyme as it is propelled forward. The brush border enzymes include peptidases, disaccharidases, and a protease called **enteropeptidase** (previously called *enterokinase*).

A variety of enzymes is secreted by exocrine cells in the pancreas, including pancreatic amylase, lipase, and several proteases and exopeptidases. Some of these enzymes are secreted as inactive zymogens that must be activated upon arrival in the intestine. The activation process is a cascade. Brush border enteropeptidase con-

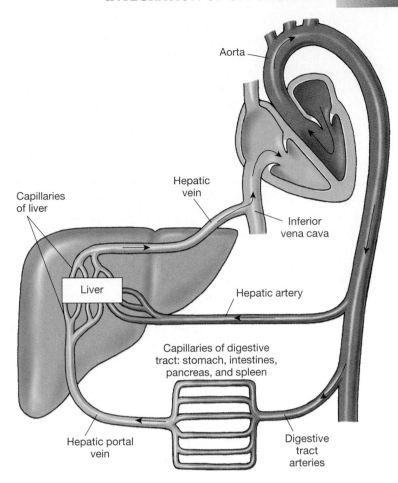

■ **Figure 20-19   The hepatic portal system**

verts inactive **trypsinogen** to active **trypsin.** Trypsin then converts the other inactive pancreatic enzymes to their active forms (Fig. 20-20 ■).

What signals release the enzymes? The brush border enzymes are bound to the membrane and are always in place. The pancreas releases enzymes in response to distension and the presence of food in the intestine. The pancreatic reflex is mediated both by the nervous system and by the hormone CCK.

Bicarbonate secretion   Bicarbonate secretion into the small intestine is essential to neutralize the highly acidic chyme that enters from the stomach. Most of this bicarbonate comes from the pancreas, which secretes a watery solution of $NaHCO_3$. The cells that produce bicarbonate are different than the enzyme-secreting cells, and they respond to different signals. Bicarbonate secretions are released in response to neural stimuli and the hormone secretin.

In the pancreas, as in the renal tubule cells and red blood cells, bicarbonate production requires high levels of the enzyme *carbonic anhydrase* [∞ p. 528]. Bicarbonate produced from CO_2 and water is secreted by a Cl^--HCO_3^- antiport protein into the pancreatic duct for

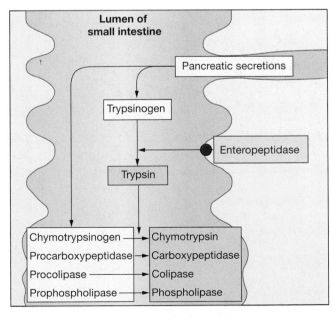

■ **Figure 20-20   Activation of pancreatic secretions** Some pancreatic secretions are inactive when secreted. The brush border enzyme enteropeptidase converts trypsinogen to trypsin. Trypsin in turn activates the other proenzymes and procolipase.

delivery to the small intestine (Fig. 20-21 ■). A basolateral Na^+-H^+ antiport transports H^+ into the extracellular fluid. The H^+ thus reabsorbed helps balance the HCO_3^- put into the blood by parietal cells.

Sodium secretion in these tissues is a passive process, driven by electrochemical gradients. To create these gradients, the epithelial cells transport Cl^- from the extracellular fluid into the lumen. The basolateral membrane contains a Na^+-K^+-$2Cl^-$ symport protein that brings Cl^- into the cell. On the apical membrane, a gated channel known as the **cystic fibrosis transmembrane regulator,** or **CFTR chloride channel,** allows Cl^- to enter the lumen. The Cl^- then re-enters the cell in exchange for HCO_3^-.

The movement of negative charge from the ECF to the lumen attracts positive Na^+, which moves down its electrochemical gradient through leaky junctions *between* the cells. This type of transepithelial movement, which we have not discussed before, is called the *paracellular* pathway. The secretion of Na^+ and HCO_3^- into the lumen creates an osmotic gradient, so water follows by osmosis. The net result is secretion of a watery sodium bicarbonate solution from the pancreas.

In cystic fibrosis, an inherited defect causes the CFTR channel to be defective or absent from the membrane. As a result, secretion of chloride and fluid ceases, but goblet cells continue to secrete mucus. The result is thickened mucus that clogs the small pancreatic ducts and accumulates in the airways of the respiratory system, where the CFTR channel is also found [∞ p. 115].

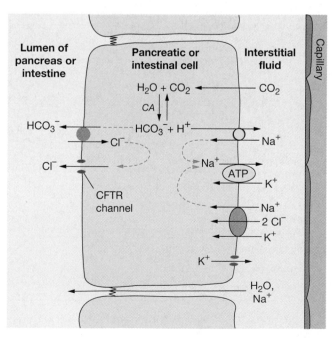

■ **Figure 20-21 Bicarbonate secretion** H^+ and HCO_3^- are formed with the aid of carbonic anhydrase (CA). HCO_3^- is exchanged for Cl^- at the apical membrane while H^+ is exchanged for Na^+ at the basolateral membrane.

Bile secretion **Bile** is a non-enzyme solution secreted from **hepatocytes,** or liver cells (see *Focus on the Liver,* Fig. 22-22 ■). The key components of bile are (1) bile salts for fat digestion, (2) bile pigments, such as bilirubin, that are the waste products of hemoglobin degradation, and (3) cholesterol. **Bile acids** are steroids that combine with amino acids to form **bile salts.** Bile salts are detergents with polar side chains that allow them to interact with both lipids and water, creating water-soluble droplets (Fig. 20-23 ■).

Bile is secreted into hepatic ducts that lead to the **gallbladder,** which stores and concentrates the solution. During a meal, contraction of the gallbladder sends bile into the duodenum through the **common bile duct,** along with a watery solution of bicarbonate and digestive enzymes from the pancreas. Gall bladder contraction is stimulated by CCK.

Carbohydrate Digestion and Absorption

Carbohydrate digestion in the small intestine converts digestible polysaccharides and disaccharides into monosaccharides so that they can be absorbed. Amylase from the pancreas continues to digest starch into maltose. Maltose and disaccharides from food, such as sucrose (table sugar) and lactose (milk sugar), must then be digested by the appropriate brush border disaccharidase (maltase, sucrase, and lactase) to complete the digestion process. The end products of carbohydrate digestion are glucose, galactose, and fructose.

Intestinal glucose and galactose absorption uses the same transporters as the renal proximal tubule: an apical Na^+-glucose symport protein and a basolateral GLUT2-facilitated diffusion transporter (Fig. 20-24 ■). The glucose transporters move galactose as well as glucose. Most fructose absorption, on the other hand, is not Na^+-dependent. Fructose moves by facilitated diffusion on the GLUT5 transporter [∞ p. 122].

If glucose is the major metabolic substrate for aerobic respiration, why don't intestinal epithelial cells use the glucose they absorb for their own metabolism? How can they keep intracellular glucose concentrations high so that facilitated diffusion will move glucose into the extracellular space? From what we can tell, the metabolism of intestinal cells differs from that of most other cells: they do not use glucose as their preferred energy substrate. Current studies suggest that these cells use the amino acid glutamine as their main source of energy, thus allowing absorbed glucose to pass unchanged into the bloodstream.

Protein Digestion and Absorption

Bicarbonate in the intestine neutralizes acidic chyme, which increases luminal pH. Pepsin is inactivated at this higher pH, but pancreatic proteases and at least 17 different additional proteases and peptidases in the brush border continue protein digestion. The primary products of protein digestion are free amino acids, dipeptides, and tripeptides, all of which can be absorbed.

FOCUS ON...The Liver

Figure 20-22

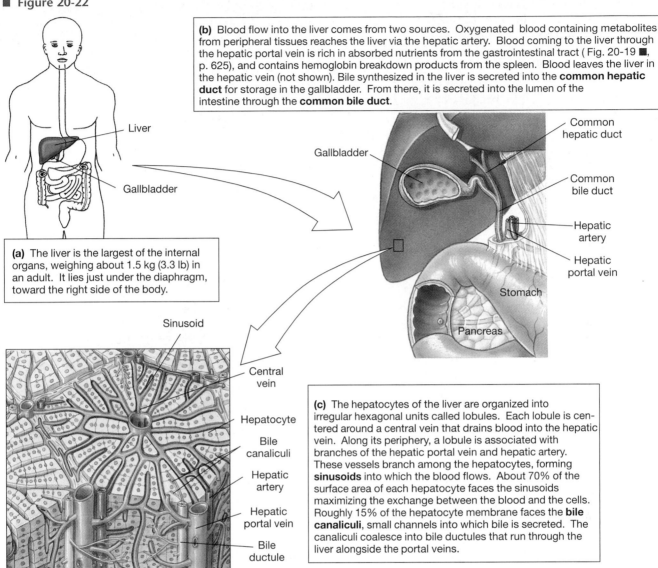

(a) The liver is the largest of the internal organs, weighing about 1.5 kg (3.3 lb) in an adult. It lies just under the diaphragm, toward the right side of the body.

(b) Blood flow into the liver comes from two sources. Oxygenated blood containing metabolites from peripheral tissues reaches the liver via the hepatic artery. Blood coming to the liver through the hepatic portal vein is rich in absorbed nutrients from the gastrointestinal tract (Fig. 20-19 ■, p. 625), and contains hemoglobin breakdown products from the spleen. Blood leaves the liver in the hepatic vein (not shown). Bile synthesized in the liver is secreted into the **common hepatic duct** for storage in the gallbladder. From there, it is secreted into the lumen of the intestine through the **common bile duct**.

(c) The hepatocytes of the liver are organized into irregular hexagonal units called lobules. Each lobule is centered around a central vein that drains blood into the hepatic vein. Along its periphery, a lobule is associated with branches of the hepatic portal vein and hepatic artery. These vessels branch among the hepatocytes, forming **sinusoids** into which the blood flows. About 70% of the surface area of each hepatocyte faces the sinusoids maximizing the exchange between the blood and the cells. Roughly 15% of the hepatocyte membrane faces the **bile canaliculi**, small channels into which bile is secreted. The canaliculi coalesce into bile ductules that run through the liver alongside the portal veins.

(d) Blood entering the liver brings nutrients and foreign substances from the digestive tract, bilirubin from hemoglobin breakdown, and metabolites from peripheral tissues of the body. In turn, the liver excretes some of these in the bile and stores or metabolizes others. Some of the liver's products are wastes to be excreted by the kidney; others are essential nutrients, such as glucose. In addition, the liver synthesizes an assortment of plasma proteins.

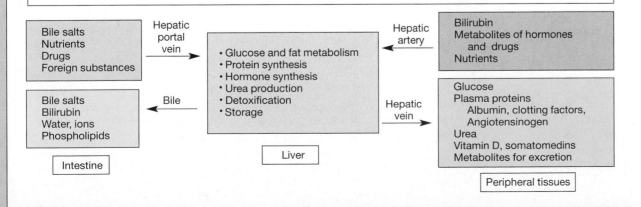

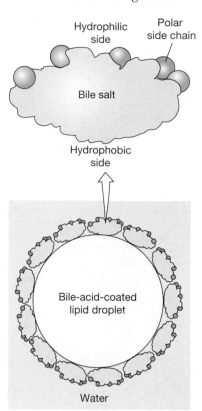

Hydrophilic side   Polar side chain

Bile salt

Hydrophobic side

Bile-acid-coated lipid droplet

Water

■ **Figure 20-23**   **Bile salts**   Bile salts have a hydrophobic side that associates with lipids and a hydrophilic side that associates with water.

Amino acid structure is so variable [∞ Figure 2-15, p. 32] that there are multiple amino acid transport systems in the intestine. Most free amino acids are carried by Na^+-dependent cotransport proteins similar to those in the proximal tubule of the kidney (Fig. 20-25 ■). A few amino acid transporters are H^+-dependent.

Lactose Intolerance   Lactose, or milk sugar, is a disaccharide composed of glucose and galactose. Ingested lactose must be digested before it can be absorbed, a task accomplished by the intestinal brush border enzyme *lactase*. Generally lactase is found only in juvenile mammals, except in some humans of northern European descent. Those people inherit a dominant gene that allows them to produce lactase after childhood. Scientists believe the gene provided a selective advantage to their ancestors, who developed a dairy-oriented culture in which milk and milk products played an important role. In non-Western cultures, where dairy products are not part of the diet after weaning, most adults lack the gene and synthesize less intestinal lactase. Decreased lactase activity is associated with a condition known as *lactose intolerance*. If a person with lactose intolerance drinks milk or eats dairy products, diarrhea may result. In addition, bacteria in the large intestine ferment lactose to gas and organic acids, leading to bloating and flatulence. The simplest remedy is to remove milk from the diet, although milk predigested with lactase is available.

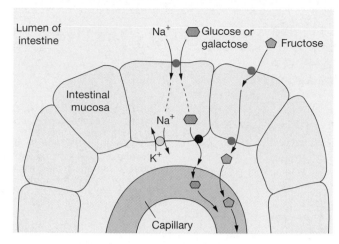

■ **Figure 20-24**   **Carbohydrate absorption**

Di- and tripeptides are carried into the mucosal cell using H^+-dependent cotransport. Once inside the epithelial cell, di- and tripeptides have two possible fates. Most are digested by cytoplasmic peptidases into amino acids, which are then transported across the basolateral membrane and into the circulation.

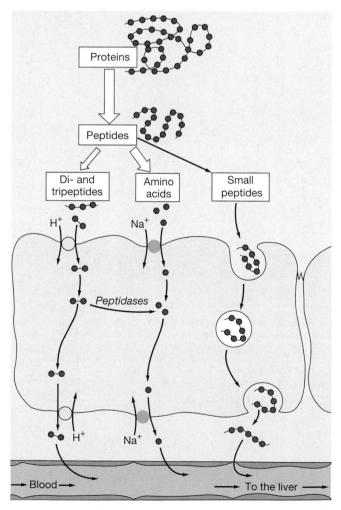

■ **Figure 20-25**   **Peptide absorption**   Amino acids are transported with Na^+. Di- and tripeptides are transported with H^+. Larger peptides are carried intact across the cell by transcytosis.

Undigested di- and tripeptides are transported intact across the basolateral membrane. Interestingly, the transport system that moves these peptides also is responsible for intestinal uptake of certain drugs, including beta-lactam antibiotics, angiotensin-converting enzyme inhibitors, and thrombin inhibitors.

Some larger peptides are absorbed by transcytosis [∞ p. 133] after binding to a receptor on the luminal surface of the epithelium. The discovery that ingested proteins can be absorbed as peptides has implications in medicine. Peptide fragments may act as *antigens,* substances that stimulate antibody formation and result in allergic reactions. Consequently, the intestinal absorption of large peptides may be a significant factor in the development of food allergies and intolerances.

It has been shown that in newborns, large peptide absorption takes place primarily in intestinal crypt cells. At birth, intestinal villi are very small, and the crypts are well exposed to the luminal contents. As the villi grow and the crypts have less access to chyme, the high peptide absorption rates present at birth decline steadily. If parents delay the ingestion of allergy-inducing peptides, the gut has a chance to mature, lessening the likelihood of antibody formation.

One of the most common antigens responsible for food allergies is gluten, a component of wheat. Interestingly, the incidence of childhood gluten allergies has decreased since the 1970s, when the recommendation was made that infants not be fed gluten-based cereals until they are several months of age.

In another medical application, intestinal absorption of intact peptides is being used by pharmaceutical companies to develop undigestible peptide drugs that can be given orally instead of by injection. Probably the best-known example is DDAVP (1-deamino-8-D-arginine vasopressin), the synthetic analog of vasopressin used clinically. If the natural hormone vasopressin is ingested, it is digested rather than absorbed intact. By changing the structure of the hormone slightly, scientists created a synthetic hormone with the same activity that can be absorbed without being digested.

Fat Digestion and Absorption   Fats enter the small intestine from the stomach in the form of a coarse emulsion created by acid lipase digestion and mechanical mixing (Fig. 20-26 ■). An **emulsion** consists of small droplets suspended in a liquid, similar to the emulsion you create when you shake up a bottle of vinegar and oil for salad dressing. In the duodenum, bile salts coat the coarse emulsion to help stabilize the droplets.

The digestion of triglycerides in the emulsion is carried out by pancreatic lipase. Lipase alone is unable to penetrate the bile salt coating and requires the presence of **colipase,** a protein cofactor secreted from the pancreas. Colipase displaces some bile salts, allowing lipase access to fats inside the bile salt coating.

Lipase digests triglycerides trapped in the interior of fat droplets into monoglycerides and free fatty acids. As enzymatic and mechanical digestion proceed, fatty acids, bile salts, monoglycerides, phospholipids, and cholesterol form small cylindrical **micelles** [∞ p. 111].

Fats are absorbed primarily by simple diffusion across the apical membrane of the intestinal cells. At the brush border surface, fatty acids and monoglycerides move out of micelles and into the epithelial cell. Cholesterol is transported on a specific, energy-dependent membrane transporter.

Once in the cytoplasm of the epithelial cell, monoglycerides and fatty acids move to the smooth endoplasmic reticulum, where they are resynthesized into triglycerides. The triglycerides then combine with cholesterol and proteins into large droplets called **chylomicrons.**

Because of their size, chylomicrons have to be packaged into secretory vesicles and leave the cell by exocytosis. Once in the extracellular space, their size also prevents them from crossing the basement membrane that surrounds the capillaries. Instead, chylomicrons are absorbed into the lymph vessels of the villi, the **lacteals.** They pass through the lymphatic system and finally enter the venous blood just before it flows into the heart [∞ p. 461].

Some shorter fatty acids (10 or fewer carbons) are not assembled into chylomicrons. They can therefore cross the capillary basement membrane and go directly into the blood.

Bile salts are not altered during fat digestion. When they reach the terminal section of the small intestine, the ileum, they encounter cells that transport them back into the hepatic portal circulation. From there, they return to the liver, are taken back up into the hepatocytes, and resecreted. This recirculation of bile salts is essential to maintain fat digestion because the body's pool of bile salts will cycle from two to five times for each meal.

Bilirubin and other wastes secreted in bile are not reabsorbed and pass into the large intestine for excretion.

Vitamins and Minerals   In general, the fat-soluble vitamins (A, D, E, and K) are absorbed along with fats. The water-soluble vitamins are absorbed by mediated transport. The major exception is vitamin B_{12}, also known as *cobalamin.* This vitamin is made by bacteria but we obtain most of our dietary supply from seafood, meat, and milk products. The intestinal transporter for B_{12} recognizes the vitamin only when it is complexed with intrinsic factor. This transporter is found only in the ileum.

Mineral absorption usually occurs by active transport. Iron and calcium are two of the few substances whose intestinal absorption is linked to their concentration in the body. For both minerals, a decrease in body concentrations of the mineral leads to enhanced uptake at the intestine. (Calcium balance is described in detail in the next chapter.)

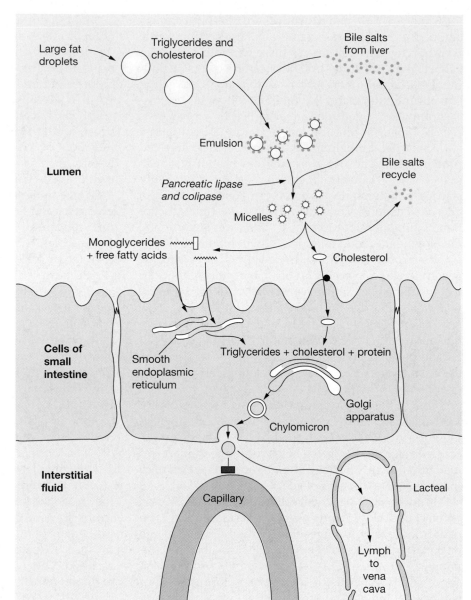

■ **Figure 20-26   Fat digestion and absorption**   Bile salts coat fat droplets to create emulsions and micelles. Absorbed fats combine with cholesterol and proteins in the intestinal cells to form chylomicrons that are released into the lymphatic system.

✔ The recipes for oral rehydration therapy usually include sugar (sucrose) and ions. The sugar not only acts as an energy source but also enhances the intestinal absorption of an ion. Which ion? Explain how sugar enhances this ion's transport.

✔ Does bile digest triglycerides into monoglycerides and free fatty acids?

Absorption of Ions and Water   About 5.5 liters of fluid and food per day enter the small intestine. Another 3.5 liters of pancreatic and intestinal secretions are added, for a total of 9 liters in the lumen. Of this total, most (7.5 liters) is reabsorbed from the small intestine.

The absorption of organic nutrients, which takes place mostly in the duodenum and jejunum, creates an osmotic gradient for water absorption. Na^+ is actively reabsorbed across the intestinal epithelium using transporters similar to those in the kidney [∞ p. 555]. One apical transporter carries Cl^- with Na^+. Another is a Na^+-K^+-2 Cl^- symporter that brings all three types of ions into the cell. K^+ and Cl also pass through the cell junctions in the small intestine.

By the end of the ileum, about 1.5 liters of unabsorbed chyme remain. It passes into the large intestine through the **ileocecal valve.** This is a tonically contracted region of muscularis that narrows the opening between the ileum and the **cecum,** the initial section of the large intestine (Fig. 20-27 ■). The ileocecal valve relaxes each time a peristaltic wave reaches it. It also relaxes at the same time that food leaves the stomach as part of the *gastroileal reflex.*

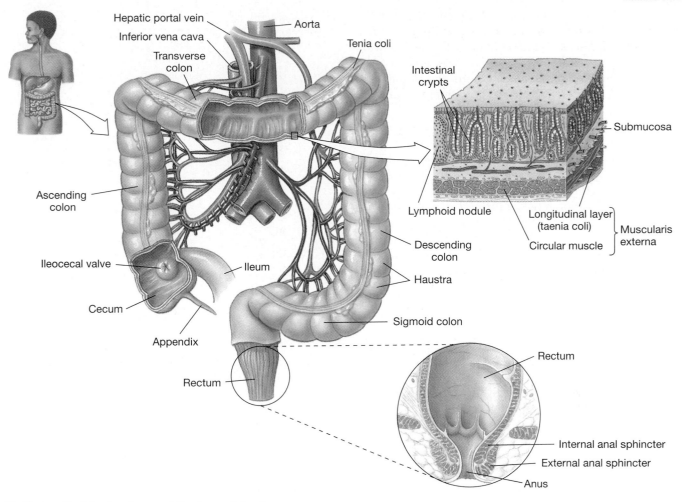

■ **Figure 20-27   Anatomy of the large intestine**

The Large Intestine Concentrates Waste for Excretion

The large intestine has seven regions. Chyme from the ileum enters the cecum, a blind-end pouch. The *appendix* is a small fingerlike projection at the end of the ventral end of the cecum. Material moves from the cecum upward through the **ascending colon,** horizontally across the body through the **transverse colon,** then down through the **descending colon** and **sigmoid colon** [*sigmoeides,* shaped like a sigma, σ]. The **rectum** is the short (about 12 cm) terminal section of the large intestine. It is separated from the external environment by the **anus,** an opening closed by two sphincters, an internal smooth muscle sphincter and external skeletal muscle sphincter.

The wall of the colon differs from that of the small intestine. The muscularis has an inner circular layer but a discontinuous longitudinal muscle layer that is concentrated into three bands called the **tenia coli.** Contractions of the tenia pull the wall into bulging pockets called **haustra.**

The mucosa of the colon has two regions, like that of the small intestine. The luminal surface lacks villi and appears smooth. It is composed of transporting epithelial cells called *colonocytes* and mucus-secreting goblet cells. The crypts contain stem cells that divide to produce new epithelium, goblet cells, endocrine cells, and maturing colonocytes (*crypt cells*).

Motility in the Large Intestine   Chyme that enters the colon continues to be mixed by segmental contractions. Forward movement is minimal during mixing contractions and depends primarily on a unique colonic contraction known as **mass movement.** A wave of contraction decreases the diameter of a segment of colon and sends a substantial bolus of material forward. These contractions occur 3–4 times a day and are associated with eating and distension of the stomach through the *gastrocolic reflex.* Mass movement is responsible for the sudden distension of the rectum that triggers defecation.

The **defecation reflex** removes undigested feces from the body. Defecation resembles urination in that it is a spinal reflex triggered by distension of the organ wall. The movement of fecal material into the normally empty rectum triggers the reflex. The smooth muscle anal sphincter relaxes and peristaltic contractions in the rectum push material toward the anus. At the same time, the external sphincter, which is under voluntary control, is consciously relaxed if the situation is appropriate. Defecation is often aided by conscious abdominal contractions and forced expiratory movements against a closed glottis, the *Valsalva maneuver.*

Defecation, like urination, is subject to emotional influence. Stress may increase intestinal motility and cause psychosomatic diarrhea in some individuals but decrease motility and cause *constipation* in others. When feces are retained in the colon, either through consciously ignoring a defecation reflex or through decreased motility, continued water absorption creates hard, dry feces that are difficult to expel. One treatment for constipation is glycerin suppositories, small bullet-shaped wads that are inserted through the anus into the rectum. Glycerin attracts water and helps soften the feces to promote defecation.

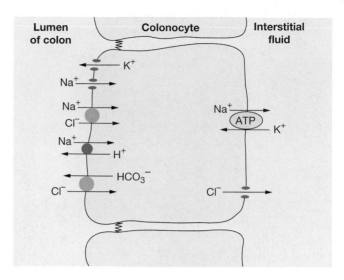

■ **Figure 20-28   NaCl reabsorption by colonocytes**

Digestion and Absorption in the Large Intestine

According to the traditional view of the large intestine, no significant digestion of organic molecules takes place there. However, in recent years this view has been revised. We now know that the numerous bacteria that inhabit the colon break down significant amounts of undigested complex carbohydrates and proteins through fermentation. The end products include lactate and short-chain fatty acids such as butyric acid. The fatty acids are used by colonocytes as their preferred energy substrate.

Colonic bacteria also produce significant amounts of absorbable vitamins, especially vitamin K. Intestinal gases, such as hydrogen sulfide, that escape from the gastrointestinal tract are a less useful product. Some starchy foods, such as beans, are notorious for their tendency to produce intestinal gas (**flatus**).

Secretion and Absorption of Fluid and Electrolytes

The colon is responsible for absorbing most of the water that enters it in the form of chyme. An average of 1.5 L/day enter the large intestine, but only about 0.1 liters is lost daily in feces under normal conditions. The colonocytes on the mucosal surface absorb NaCl using a variety of membrane transporters (Fig. 20-28 ■). At the same time, some K^+ is secreted at the apical membrane. This process is influenced by aldosterone, which promotes Na^+ reabsorption and K^+ secretion, just as it does in the kidney [∞ p. 579].

Crypt cells in the colon (and also the small intestine) secrete Cl into the lumen (Fig. 20-29 ■). Chloride enters the cell by secondary active transport, then exits via the apical CFTR channel. Sodium and water follow by the paracellular pathway. The secreted fluid mixes with mucus secreted by goblet cells to help lubricate the contents of the gut.

Diarrhea Can Cause Dehydration   Diarrhea is a pathological state in which intestinal secretion of fluid is not balanced by absorption, resulting in watery stools. Diarrhea will occur if normal intestinal water absorption mechanisms are disrupted or if there are unabsorbed osmotically active solutes that "hold" water in the lumen. Substances that cause *osmotic diarrhea* include undigested lactose and sorbitol, a sugar alcohol from plants. Sorbitol is used as an "artificial" sweetener in some chewing gums and in foods made for diabetics. Another unabsorbed solute that can cause osmotic diarrhea, intestinal cramping, and gas is Olestra, the "fake fat" made from vegetable oil and sugar [∞ p. 620].

In clinical settings, patients who need to have their bowels cleaned out before surgery or other procedures are often given four liters of an isotonic solution of unabsorbable polyethylene glycol and electrolytes to drink.

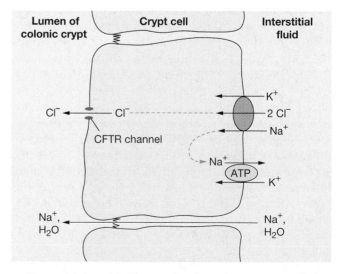

■ **Figure 20-29   NaCl secretion by colonic crypt cells**

The large volume of unabsorbed solution passes into the colon, where it triggers copious diarrhea that removes all solid waste from the GI tract.

Secretory diarrheas occur when bacterial toxins such as cholera toxin from *Vibrio cholerae* and *Escherichia coli* enterotoxin enhance colonic Cl⁻ secretion (Fig. 20-29 ■). When excessive fluid secretion is coupled with increased motility, diarrhea results. Secretory diarrhea in response to intestinal infection can be viewed as adaptive because it helps flush pathogens out of the lumen. However, it also has the potential to cause dehydration if fluid loss is excessive.

The World Health Organization estimates that in developing countries, four million people a year die from diarrhea. In the United States, diarrhea in children causes about 200,000 hospitalizations a year. Oral replacement fluids for treatment of diarrheal salt and water loss can prevent the *morbidity* (sickness) and *mortality* (death) associated with diarrhea. Oral rehydration solutions usually contain glucose as well as Na⁺, K⁺, and Cl⁻ because the inclusion of glucose enhances Na⁺ absorption. If dehydration is severe, intravenous fluid therapy may be necessary.

▶ IMMUNE FUNCTIONS OF THE GI TRACT

As you learned in the beginning of the chapter, the GI tract is the largest immune organ in the body. It is continuously exposed to disease-causing organisms, and it must prevent them from entering the body through delicate absorptive tissues. The first lines of defense are the enzymes and immunoglobulins of saliva and the high acid environment of the stomach. If pathogens or toxic material make it into the small intestine, sensory receptors and the immune cells of the GALT respond. Two common responses are diarrhea, just described, and vomiting.

M Cells Sample the Contents of the Gut

The immune system of the intestinal mucosa consists of immune cells scattered throughout the mucosa, clusters of immune cells in Peyer's patches (see Fig. 20-2e ■), and specialized epithelial cells called **M cells** that overlie the Peyer's patches. The M cells provide information about the contents of the lumen to the immune cells of the GALT.

M cells have fewer and more widely spaced microvilli than the typical intestinal cell. The apical surface also contains clathrin-coated pits [∞ p. 129] with membrane receptors. When *antigens*, molecules that activate the immune response, bind to the luminal receptors, the M cell transports them to its basolateral membrane using transcytosis [∞ p. 133]. Macrophages and lymphocytes [∞ p. 476] are waiting in the extracellular compartment for the M cell to present them with antigens.

If the antigen is a substance that threatens the body, the immune system swings into action. It secretes cytokines to attract additional immune cells that can attack the invader. It also secretes cytokines that trigger an inflammatory response. A third response to cytokines is increased Cl⁻, fluid, and mucus secretion to flush the invader from the GI tract.

In *inflammatory bowel diseases* (ulcerative colitis and Crohn's disease), the immune response is triggered inappropriately by the normal contents of the gut. One experimental therapy for these diseases that seems to be successful is to block the action of cytokines released by the gut-associated lymphoid tissues.

One question that has puzzled scientists for years was how certain pathogenic bacteria cross the barrier created by the intestinal epithelium. With the discovery of M cells, we may have the answer. It appears that some bacteria, such as *Salmonella* and *Shigella*, have evolved surface molecules that bind to the M cell receptors. The M cells then obligingly transport the bacteria across the epithelial barrier and deposit them inside the body, where the immune system immediately reacts. Both bacteria cause diarrhea. *Salmonella* also causes fever and vomiting.

Vomiting Is a Protective Reflex

Vomiting, or *emesis,* is the forceful expulsion of gastric and duodenal contents from the mouth. It is a protective reflex that is designed to remove toxic material from the GI tract before it can be absorbed. However, excessive or prolonged vomiting with loss of gastric acid can cause metabolic alkalosis [∞ p. 597].

The vomiting reflex is coordinated through a vomiting center in the medulla. It begins with stimulation of sensory receptors and is often, but not always, accompanied by nausea. A variety of stimuli from all over the body can trigger vomiting. They include chemicals in the blood, such as cytokines and certain drugs, pain, and disturbed equilibrium, such as occurs in a moving car or rocking boat. Tickling the back of the pharynx can induce vomiting.

Efferent signals from the vomiting center initiate a wave of reverse peristalsis that begins in the small intestine and moves upward. The motility wave is aided by abdominal contraction that increases intra-abdominal pressure. The stomach relaxes so that the increased pressure forces gastric and intestinal contents back into the esophagus and out of the mouth.

During vomiting, respiration is inhibited. The epiglottis and soft palate close off the trachea and nasopharynx to prevent the *vomitus* from being inhaled (*aspirated*). Should acid or small food particles get into the airways, they could damage the respiratory system and cause *aspiration pneumonia.*

PROBLEM CONCLUSION

Tonya's lab results are in. Tests performed on her stomach tissue confirm that she has an *H. pylori* infection that is probably the cause of her duodenal ulcer. Tonya's physician gives her a double antibiotic to take for seven days and changes her H_2 blocker to a proton pump inhibitor.

In this running problem, you learned about peptic ulcers and an unusual bacterium that is associated with most duodenal ulcers. You also learned that antacids and histamine-receptor blockers reduce the level of acid in the stomach.

Further check your understanding of this running problem by comparing your answers with those in the summary table.

	Question	Facts	Integration and Analysis
1	Why are peptic ulcers found in the duodenum rather than in the jejunum or ileum?	Peptic ulcers are caused by gastric acid and stomach enzymes. Chyme moving out of the stomach passes into the duodenum.	If protective mechanisms in the intestine have failed, the duodenum is the first part of the small intestine to be exposed to gastric acid and enzymes.
2	How does the conversion of urea into ammonia protect *H. pylori* from the hostile environment of the stomach?	A base is an anion that combines with H^+. Ammonia combines with H^+ to create ammonium ions, NH_4^+. This reduces acidity (increases pH).	*H. pylori* has enzymes that produce ammonia, a base. Ammonia combines with H^+ to reduce acidity. In this way, the bacteria protect themselves from being killed by the acidic environment of the stomach.
3	Write the chemical equation showing how the antacids sodium bicarbonate ($NaHCO_3$) and aluminum hydroxide ($Al(OH)_3$) buffer hydrochloric acid (HCl).	Buffers are molecules that combine with H^+. Bicarbonate (HCO_3^-) and OH^- are buffers.	$NaHCO_3 + HCl \rightarrow NaCl + H_2CO_3$ $Al(OH)_3 + 3\ HCl \rightarrow AlCl_3 + 3\ H_2O$
4	How does blocking H_2 receptors stop acid production in the stomach?	Histamine is a paracrine that stimulates acid secretion by the stomach.	If histamine receptors are blocked by H_2 blockers, histamine cannot combine with them and stimulate acid secretion.

CHAPTER REVIEW

SUMMARY

Function and Processes of the Digestive System

1. The gastrointestinal (GI) tract moves nutrients, water, and electrolytes from the external to the internal environment. (p. 603)

2. The four processes of the digestive system are digestion, absorption, motility, and secretion. **Digestion** is chemical and mechanical breakdown of foods into absorbable units. **Absorption** is transfer of substances from the lumen of the GI tract to the ECF. **Motility** is movement of material through the GI tract. **Secretion** refers to the transfer of fluid and electrolytes from ECF to lumen or the release of substances from cells. (p. 603)

3. The GI tract contains the largest collection of lymphoid tissue in the body, the **gut-associated lymphoid tissue (GALT).** (p. 604)

Anatomy of the Digestive System

4. **Chyme** is a soupy substance created as ingested food is broken down by mechanical and chemical digestion. (p. 604)

5. Food entering the digestive system passes through mouth, pharynx, **esophagus, stomach (fundus, body, antrum),** **small intestine (duodenum, jejunum, ileum), large intestine, rectum, and anus.** (p. 604)

6. The salivary glands, **pancreas,** and **liver** add exocrine secretions with enzymes and mucus to the lumen. (p. 605)

7. The wall of the GI tract consists of four layers: an inner

mucosa, a middle submucosa, a smooth muscle layer, and an outer **serosa.** (p. 605)

8. The **mucosa** consists of epithelium, the **lamina propria,** lymphoid tissue, and the **muscularis mucosa.** Small **villi** increase the surface area, as do **microvilli** on the cells. (p. 605)

9. The submucosa contains blood and lymph vessels and the **submucosal plexus** of the **enteric nervous system.** (p. 608)

10. The outer wall consists of two layers of circular and longitudinal muscle. The **myenteric plexus** lies between the muscle layers. (p. 608)

Motility

11. Motility moves food from mouth to anus and mechanically mixes food. (p. 608)

12. GI smooth muscle cells are electrically connected by gap junctions. Some segments are **tonically contracted** but others exhibit **phasic contractions.** (p. 608)

13. Intestinal muscle exhibits spontaneous **slow wave potentials.** When a slow wave reaches threshold, it fires action potentials and contracts. (p. 608)

14. Slow waves originate in the **interstitial cells of Cajal.** (p. 608)

15. Between meals, the **migrating motor complex** moves remnants from the upper GI tract to the lower regions. (p. 609)

16. **Peristaltic contractions** are progressive waves of contraction. Peristalsis is mediated by the enteric nervous system and altered by hormones, paracrines, and neurocrines. (p. 609)

17. **Segmental contractions** are primarily mixing contractions. (p. 609)

Secretion

18. About 2 liters of fluid per day enter the GI tract through the mouth. Another 7 liters are secreted by the body. Nearly all of this volume is reabsorbed. (p. 610)

19. Digestive enzymes are secreted by exocrine glands or by gastric and intestinal epithelium. Some enzymes are secreted as zymogens. (p. 610)

20. Mucus from **mucous cells** and **goblet cells** forms a protective coating and lubricates contents of the gut. (p. 610)

21. The digestive system secretes and absorbs water and ions. (p. 610)

Digestion and Absorption

22. Most nutrient absorption takes place in the small intestine. The large intestine absorbs water and ions. (p. 611)

23. **Amylase** digests starch to maltose. **Disaccharidases** digest disaccharides to monosaccharides. (p. 611)

24. **Proteases** break proteins into smaller peptides. **Exopeptidases** remove amino acids from peptides. (p. 611)

25. Fat digestion requires **lipase** and **colipase.** (p. 612)

Regulation of GI Function

26. The enteric nervous system is called the little brain because it can integrate information without input from the CNS. (p. 612)

27. **Short reflexes** originate within the enteric nervous system and are integrated there. **Long reflexes** are integrated in the CNS. Long reflexes may originate within or outside the enteric nervous system. (p. 613)

28. Generally, parasympathetic innervation is excitatory for GI function and sympathetic innervation is inhibitory. (p. 614)

29. Digestive hormones are divided into the gastrin family (**gastrin, cholecystokinin**), secretin family (**secretin, vasoactive intestinal peptide, glucose-dependent insulinotropic peptide,** and **glucagon-like peptide 1**), and hormones that do not fit into either of those two families (**motilin**). (p. 615)

30. GI peptides excite or inhibit motility and secretion. Most stimuli for GI peptide secretion arise from the ingestion of food. (p. 615)

31. GI paracrines, such as histamine, interact with hormones and neural reflexes. (p. 617)

Integration of GI Function

32. In the **cephalic phase** of digestion, the sight, smell, or taste of food initiates GI reflexes. (p. 617)

33. Mechanical digestion begins with chewing, or **mastication. Saliva** moistens and lubricates food. Salivary amylase digests carbohydrates. (p. 617)

34. Swallowing, or **deglutition,** is a reflex integrated by a medullary center. (p. 619)

35. The stomach digests proteins and fats and regulates movement of chyme into the intestine. (p. 620)

36. The stomach secretes mucus and bicarbonate from **mucous cells,** gastric acid from **parietal cells,** pepsinogen from **chief cells,** somatostatin from **D cells,** histamine from **ECL cells,** and gastrin from **G cells.** (p. 620)

37. Acid in the intestine, CCK, and secretin delay gastric emptying. (p. 621)

38. Intestinal enzymes are part of the **brush border.** Some pancreatic enzymes need to be activated after secretion. (p. 624)

39. The pancreas also secretes a watery $NaHCO_3$ solution that neutralizes gastric acid. (p. 625)

40. Epithelial cells secrete Cl^- using the **cystic fibrosis transmembrane regulator,** or **CFTR chloride channel.** (p. 626)

41. Bile from the liver contains bile salts, bilirubin, and cholesterol. (p. 626)

42. Glucose absorption uses the Na^+-glucose symporter and GLUT2 transporter. Fructose uses the GLUT5 transporter. (p. 626)

43. Amino acids are absorbed by Na^+-dependent cotransport. Di- and tripeptides are absorbed using H^+-dependent cotransport. Some larger peptides are absorbed intact by transcytosis. (p. 628)

44. Fat digestion requires bile, which emulsifies fats. As enzymatic and mechanical digestion proceed, fat droplets form **micelles.** (p. 629)

45. Fat absorption occurs primarily by simple diffusion. (p. 629)

46. Monoglycerides and fatty acids reassemble into triglycerides in the intestinal cell, then combine with cholesterol and proteins into **chylomicrons.** Chylomicrons are absorbed into the lymph. (p. 629)

47. Fat-soluble vitamins are absorbed along with fats. Water-soluble vitamins are absorbed by mediated transport. Vitamin B_{12} absorption requires **intrinsic factor** secreted by the stomach. Mineral absorption usually occurs by active transport. (p. 629)

48. Most volume reabsorption takes place in the small intestine. The large intestine concentrates the 1.5 liters of chyme that enter it daily. (p. 630)

49. Undigested material in the colon moves forward by **mass movement.** The **defecation reflex,** a spinal reflex subject to higher control, is triggered by sudden distension of the rectum. (p. 631)

50. Colonic bacteria use fermentation to digest organic material. (p. 632)

51. Cells of the colon can reabsorb or secrete fluid. Excessive fluid secretion or decreased reabsorption causes diarrhea. (p. 632)

Immune Functions of the GI Tract

52. Protective mechanisms of the GI tract include acid, mucus, vomiting, and diarrhea. (p. 633)

53. **M cells** sample gut contents and present antigens to cells of the GALT. (p. 633)

54. **Vomiting** is a protective reflex integrated in the medulla. (p. 633)

QUESTIONS

LEVEL ONE   Reviewing Concepts

1. Define the four basic processes of the digestive system and give an example of each.

2. For most nutrients, the process of _____ is not regulated, whereas _____ and _____ are continuously regulated. Why do you think these differences exist? Defend your answer.

3. Match each of the following descriptions with the appropriate term(s):

   (a) chyme is produced here
   (b) organ where most digestion occurs
   (c) initial section of small intestine
   (d) organ that adds exocrine secretions to the duodenum via a duct
   (e) sphincter between stomach and intestine
   (f) enzymes produced here
   (g) distension of the walls of this organ triggers the defecation reflex

   1. colon
   2. stomach
   3. small intestine
   4. duodenum
   5. ileum
   6. jejunum
   7. pancreas
   8. pylorus
   9. rectum
   10. liver

4. List the four layers found in the GI tract walls. What type of tissue predominates in each?

5. Describe the functional types of epithelium found lining the stomach and intestines.

6. What are Peyer's patches?

7. What purposes does motility serve in the gastrointestinal tract? What types of tissue contribute to gut motility? What types of contraction do the tissues undergo?

8. What is a zymogen? What is a proenzyme? List two specific examples.

9. Match each cell with the product(s) it secretes. Items may be used more than once.

   (a) parietal cells
   (b) goblet cells
   (c) brush border cells
   (d) pancreatic cells
   (e) D cells
   (f) ECL cells
   (g) chief cells
   (h) G cells

   1. enzymes
   2. histamine
   3. mucus
   4. pepsinogen
   5. gastrin
   6. somatostatin
   7. HCO_3^-
   8. HCl
   9. intrinsic factor

10. How does each of these factors affect digestion? Briefly explain how and where the factor exerts its effects.

    (a) emulsification   (b) neural activity
    (c) pH   (d) size of food particles

11. Most digested nutrients are absorbed into the _____ of the _____ system, delivering nutrients to the_____ (organ). However, digested fats go into the _____ system because intestinal capillaries have a _____ around them that most lipids are unable to cross.

12. What is the enteric nervous system and what is its function?

13. What are short reflexes? What types of responses do they regulate? What is meant by the term *long reflex*?

14. What role do paracrines play in digestion? Give specific examples.

LEVEL TWO   Reviewing Concepts

15. **Map exercise:** List the three major groups of biomolecules across the top of a large piece of paper. Down the left side of the paper write: mouth, stomach, small intestine, absorption. For each biomolecule, fill in how it is digested in each of the GI tract regions listed and how it is absorbed.

16. Define, compare, and contrast the following terms:

    (a) mastication, deglutition
    (b) microvilli, villi
    (c) peristalsis, segmentation, migrating motor complex, reflux esophagitis, mass movements, defecation, vomiting, diarrhea
    (d) chyme, feces
    (e) short reflexes, long reflexes

    (f) submucosal plexus, myenteric plexus, enteric nervous system, vagus nerve
    (g) cephalic, gastric, and intestinal phases of digestion

17. Diagram the mechanisms by which Na^+, K^+, and Cl^- are transported out of the intestine. Diagram the mechanisms by which H^+ and HCO_3^- are secreted into the lumen.

18. Compare the enteric nervous system with the cephalic brain. Give some specific examples concerning neurotransmitters, neuromodulators, and supporting cells.

19. List and briefly describe the actions of the members of each of the three groups of GI hormones.

20. Explain how H_2 receptor antagonists and proton pump inhibitors decrease gastric acid secretion.

LEVEL THREE   Problem Solving

21. Erica's baby, Justin, has had a severe bout of diarrhea and is now dehydrated. Is his blood more likely to be acidotic or alkalotic? Why?

22. Mary Littlefeather arrives in her physician's office complaining of severe, steady pain in the upper right quadrant of her abdomen. The pain began shortly after she ate a meal of fried chicken, french fries, and peas. Lab tests and an ultrasound reveal the presence of gallstones in the common bile duct that drains the liver, gallbladder, and pancreas into the small intestine.

    (a) Why was Mary's pain precipitated by the meal she ate?
    (b) Which of the following will be affected by the gallstones: micelle formation in the intestine, carbohydrate digestion in the intestine, protein absorption in the intestine. Explain your reasoning.

E X P L O R E <MediaLab>

Introduction

The digestive tract is a long tube whose different regions are modified to perform specific functions. As food traverses this path, it is digested and absorbed. In this MediaLab you will be introduced to the bacterium *Helicobacter pylori,* the major cause of gastric (stomach) ulcers. You will follow the scientific experiments that led up to this discovery and learn how these results were received by the scientific and medical communities. You'll also learn about Olestra, a new food additive that provides the flavor of fat but none of the calories. You will learn why the body can't absorb this fat and what consequences this will cause as the undigested Olestra travels through the colon. To gain a better appreciation for the volume of liquid that is secreted and absorbed by the digestive tract, you will study the disease cholera. One of the symptoms of cholera is massive diarrhea, induced by the cholera toxin. After reading the description below, visit the MediaLab for Chapter 20 in your Companion Website and select the appropriate keyword.

Web Exploration 1

Estimated time for completion = 10 minutes
Sales of Pepcid are $600 million annually. Tagamet generates $1.2 billion and Zantac leads the group with $3 billion in sales each year. How is it that histamine-2 (H_2) receptor antagonists have become the most lucrative group of drugs ever sold? As you read in the running problem in this chapter, most peptic and duodenal ulcers are caused by the infectious agent *Helicobacter pylori.* The discovery of *Helicobacter pylori* is an amazing story and is an excellent example of the process of basic medical science.

Select *Helicobacter pylori* from the Website to explore this topic in more detail. What findings prompted Warren and Marshall to investigate *H. pylori* as a possible causative agent in ulcers? With *H. pylori* as the prime suspect, what was War-

ren and Marshall's next step? Why was it unsuccessful for so long? What bit of serendipity helped advance their research efforts? What was the ultimate test that Marshall performed to prove that *H. pylori* causes gastritis?

Based on results from epidemiological studies, which parts of the world have the highest infection rate for this bacterium? When you get to the immune system in Chapter 22, you will see that our immune system is usually quite thorough in removing and destroying foreign organisms from our body. However, it appears that *H. pylori* evades total removal by the immune system. What do scientists propose is the mechanism by which *H. pylori* evades destruction?

Based on the information in the graph entitled "Expected costs and days of treatment for ulcer therapy," what is the most cost-efficient manner to treat this pathology? How should the information on this topic impact future sales of H_2-antagonists? Why is a vagotomy a surgical alternative for treatment of peptic ulcers?

Web Exploration 2

Estimated time for completion = 10 minutes
Cholera is a devastating disease that can rapidly wipe out entire villages or towns in developing countries. Cholera is an excellent example of how a foreign organism gains access to the body, disrupts function at a cellular level, and creates dysfunction at the systemic level that causes loss of homeostasis and eventual death. How does cholera kill? Can cholera be prevented?

Select the keyword **CHOLERA** from your Website and review the linked material to help you answer the following questions. Is cholera a virus or bacterium? How is cholera transmitted between humans? How does cholera enter the body? Why isn't cholera destroyed by the acid environment of the stomach?

Energy Balance, Metabolism, and Growth

■ "Probably no single abnormality has contributed more to our knowledge of the intermediary metabolism of animals than has the disease known as diabetes."—*Helen R. Downes*, The Chemistry of Living Cells, 1955. ■

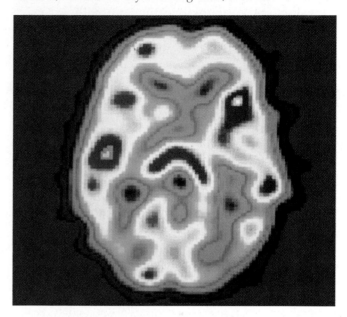

PET scans of the brain use glucose metabolism to indicate regions of neural activity (red and yellow).

BACKGROUND BASICS

The magazine covers at the grocery store check-out stands tell visitors to this country a lot about Americans. Headlines screaming "Lose 10 pounds in a week without dieting" or "CCK: the hormone that makes you thin" vie for attention with glossy full-color photos of luscious desserts dripping with chocolate, whipped cream, fat, and calories. As one magazine article put it, we are a nation obsessed with staying trim and with eating, mutually exclusive occupations.

You learned in the last chapter that "what you eat is what you get" in terms of digestion and absorption. Most of the food that we put into our mouths ends up in our bloodstream. The body's chemical reactions, known as metabolism, determine what happens to those nutrients. Are they destined to burn off as heat? Become muscle? Or turn into extra pounds that make it difficult to zip up blue jeans? In this chapter we examine energy balance in the body and the pathways that determine the fate of those ingested nutrients.

ENERGY BALANCE AND METABOLISM

▶ ENERGY BALANCE

Recall that we introduced energy and metabolism in Chapter 4, Cell Processes. **Metabolism** is the sum of all chemical reactions in the body. These reactions extract energy from nutrients, use energy for work, and store excess energy so that it can be used later. Some metabolic reactions release energy. Others require energy (*exergonic* and *endergonic reactions;* ∞ p. 77).

Metabolic pathways that synthesize small molecules into larger ones are called **anabolic** pathways [*ana-,* completion + *metabole,* change]. Pathways that break large molecules into smaller components are considered **catabolic** [*cata-,* down or back]. The classification of a pathway is its end result, not what happens in one or two steps.

For example, in the first step of *glycolysis* [∞ p. 90], a catabolic pathway, glucose adds a phosphate to become a larger molecule, glucose 6-phosphate. But what happens by the end of glycolysis? One 6-carbon glucose has been converted to two 3-carbon pyruvate molecules, making the net result of glycolysis catabolic.

PROBLEM
Hyperparathyroidism

Broken bones, kidney stones, abdominal groans, and psychic moans. Medical students memorize this saying when they learn about a condition called *hyperparathyroidism,* a disease in which parathyroid glands [∞ p. 188] work overtime and produce excess parathyroid hormone (PTH). Dr. Adiaha Spinks suddenly recalls this saying as she examines Professor Bob Magruder, who has arrived at the office in pain from a kidney stone lodged in his ureter. When questioned about his symptoms, Dr. Bob also mentions pain in his shin bones, muscle weakness, stomach upset, and a vague feeling of depression. "I thought it was all just the stress of getting my textbook published," he says. But to Dr. Spinks, his combination of symptoms sounds suspiciously like Dr. Bob might be suffering from hyperparathyroidism.

...continued on page 646

Energy Input Equals Energy Output

The *first law of thermodynamics* states that energy in the universe is constant. By extension, this statement means that all energy that goes into a biological system such as the human body can be accounted for:

(1) Energy intake = energy output

Energy intake for humans consists of the nutrients we eat, digest, and absorb. However, because the digestive system itself does not regulate energy intake, we must depend on behavioral mechanisms such as hunger and satiety to tell us when and how much to eat. Psychological and social aspects of eating, such as parents who say "Clean your plate," complicate the physiological control of food intake.

Energy output by the body either does work (which includes storing energy) or is returned to the environment as heat. Therefore, we can rewrite equation 1 to read

(2) Energy intake = heat + work

In the human body, at least half the energy released in chemical reactions is lost as heat. Much of this heat production is unregulated "waste" heat. But because humans are homeothermic animals, we also regulate heat production and heat loss to maintain a relatively constant body temperature. We will discuss thermoregulation later in this chapter.

Biological work takes one of three forms [∞ p. 75]:

1. *Transport work,* which is used to move molecules from one side of a membrane to another
2. *Mechanical work,* which uses intracellular fibers and filaments to create movement
3. *Chemical work,* which is used for growth, maintenance, and storage of information and energy

Transport processes bring material into and out of the body and transfer them between compartments within the body. Movement includes both external work, such as movement created by skeletal muscle contraction, and internal work, such as the cytoplasmic movement of vesicles.

Chemical work in the body can be subdivided into synthesis and storage (Fig. 21-1 ■). Storage includes both *short-term energy storage* in high-energy phosphate compounds such as ATP, and *long-term energy storage* in the chemical bonds of glycogen and fat. Cells have a limited capacity for glycogen storage. Therefore, most excess energy ends up as fat droplets in adipose tissue [∞ p. 65].

Age, gender, and genetics greatly affect the pattern of fat deposition in our bodies. Despite the uncontrollable nature of these factors, many Americans try to achieve the "ideal" figure as portrayed in the media. For many, this means losing weight and fat. Shrinking the size of adipose tissue and moving energy out of fat stores requires that energy output exceed energy intake.

Many components of energy output, including synthesis, transport, and heat production, are not under conscious control. Therefore movement such as exercise is the only way people can voluntarily increase energy output. People can control their energy intake, however, by watching what they eat.

Although energy balance is a very simple concept, it is a difficult one for people to accept. Behavior modifications such as eating less and exercising more are among the most frequent instructions health care professionals give to patients. They are also the most difficult instructions to follow, and patient compliance is usually low.

Body Temperature Is a Balance Between Heat Production, Gain, and Loss

The development of *obesity* (excess body fat) may be linked to the efficiency with which the body converts food energy into cell and tissue components. According to one theory, people who are more efficient in transferring energy from food to fat are the ones who put on weight. In contrast, people who are less metabolically efficient can eat the same number of calories and not gain weight because more food energy is released as heat. Much of what we know about the regulation of energy balance comes from studies on body temperature regulation. Body temperature results from a balance between heat production, gain, and loss.

Humans Regulate Body Temperature Within a Narrow Range   Temperature regulation in humans is linked to metabolic heat production (*thermogenesis*). Humans are **homeothermic** animals who regulate their internal body temperatures within a relatively narrow range. Average body temperature is 37° C (98.6° F) with a normal range of 35.5°–37.7° C (96°–99.9° F).

These average values are subject to considerable variation, both from one individual to the next and throughout the day in a single individual. The site at which temperature is measured also makes a difference, because the temperature at the core of the body may be much

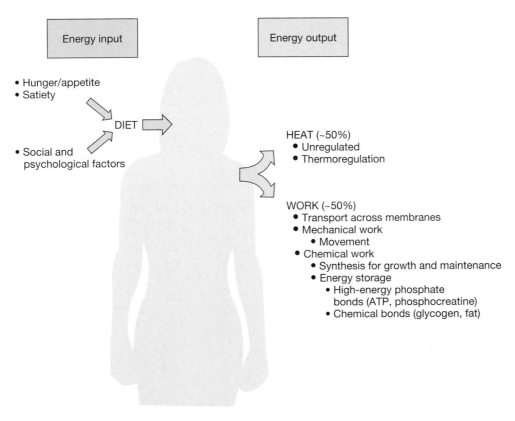

■ Figure 21-1   **Energy balance**

higher than that at the surface of the skin. Oral temperatures are about 0.5° C less than rectal temperatures.

Several factors affect body temperature in a given individual. Body temperature increases with exercise or after a meal because of *diet-induced thermogenesis*. Temperature also cycles throughout the day: the lowest (basal) body temperature occurs in the early morning and the highest in the early evening. Women of reproductive age exhibit a monthly cycle of temperature. Basal body temperatures are about 0.5° higher in the second half of the menstrual cycle, after ovulation, than before ovulation.

Heat Gain and Loss Are Balanced Temperature balance in the body, like energy balance and mass balance, depends on a dynamic equilibrium between heat input and heat output (Fig. 21-2 ■). *Internal heat production* comes from normal metabolism and from heat released during muscle contraction. *External heat input* comes from the environment through *radiation* or *conduction*.

All objects with a temperature above absolute zero give off radiant energy (radiation) with infrared or visible wavelengths. This energy can be absorbed by other objects and constitutes **radiant heat gain.** You absorb radiant energy each time you go out and sit in the sun or sit in front of a fire on a chilly night. **Conductive heat gain** is the transfer of heat between objects that are in contact with each other, such as the skin and a heating pad or the skin and warm water.

We lose heat from the body in four ways: conduction, radiation, convection, and evaporation. **Conductive heat loss** is the loss of body heat to a cooler object, such as cool water or a cold stone bench. **Radiant heat loss** from the human body is estimated to account for nearly half the heat lost from a person at rest in a normal room. **Thermography** is an imaging technique in diagnostic medicine that measures radiant heat loss. For example, some cancerous tumors can be visually identified because they have higher metabolic activity and give off more heat than surrounding tissues.

Radiant and conductive heat loss is enhanced by **convective heat loss,** heat that is carried away by warm air rising from the surface of the body. Convective air currents are created wherever a temperature difference in the air exists: hot air rises and is replaced by cooler air. Convection helps move warmed air away from the skin's surface. Clothing, which traps air and prevents convective air currents, therefore helps retain heat close to the body.

The fourth type of heat loss in the body is **evaporative heat loss,** which takes place as water evaporates at the skin's surface or in the respiratory tract. The conversion of water from a liquid to a gaseous (vapor) state requires the input of substantial amounts of energy. Thus, when water on the body evaporates, it acts as a "heat sink."

You can demonstrate the effect of evaporative cooling by wetting one arm but not the other. The wet arm

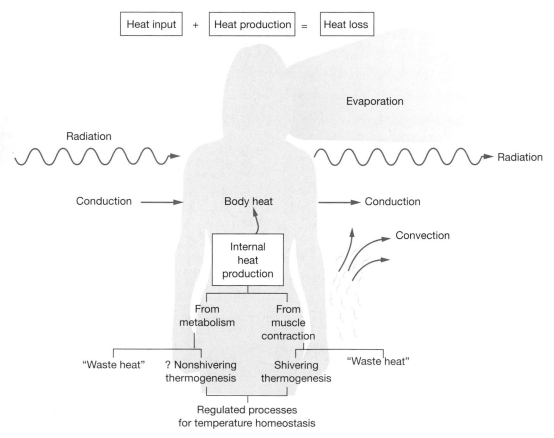

■ **Figure 21-2 Heat balance**

will feel much cooler as it dries because heat is drawn from the skin to vaporize the water. Similarly, the half-liter of water vapor that leaves the body through lungs and skin each day takes with it a significant amount of body heat. Evaporative heat loss is affected by the humidity of surrounding air, with less evaporation taking place at higher humidities.

Conductive, convective, and evaporative heat loss from the body are all enhanced by bulk flow of air across the body, such as air moved across the skin by a fan or breeze. The effect of wind on body temperature regulation is described in the winter as the "wind chill factor," a combination of absolute environmental temperature and the effect of convective heat loss.

Body Temperature Is Homeostatically Regulated

The human body is usually warmer than its environment and therefore loses heat. However, the energy generated by normal metabolism is sufficient to maintain body temperature when the environmental temperature stays between 27.8° and 30° C (82° and 86° F). This range is known as the **thermoneutral zone.**

In temperatures above the thermoneutral zone, the body gains heat because heat production exceeds heat loss. Below the thermoneutral zone, heat loss exceeds heat production. In both cases, the body must use homeostatic compensation to maintain internal temperature.

A human without clothing can thermoregulate at environmental air temperatures between 10° and 55° C (50° and 131° F). Because we are seldom exposed to the higher end of that temperature range, the main physiological challenge in thermal regulation comes from cold environments.

Humans have been described by some physiologists as tropical animals because we are genetically adapted for life in warm climates. But we have retained a certain amount of genetic flexibility. The physiological mechanisms by which we thermoregulate show some ability to adapt to changing conditions.

Body temperature regulation is under control of centers in the hypothalamus. Sensory **thermoreceptors** are located peripherally in the skin and centrally close to the hypothalamus. These sensors monitor skin temperature and body core temperature and send that information to the hypothalamus.

The hypothalamic "thermostat" compares the input signals with the desired temperature setpoint and coordinates an appropriate physiological response to raise or lower the core temperature (Fig. 21-3 ■). Heat loss is promoted by dilation of blood vessels in the skin and by sweating. Heat is generated by shivering and possibly by nonshivering thermogenesis.

Alterations in Cutaneous Blood Flow Conserve or Release Heat   Heat loss across the surface of the skin is regulated by controlling blood flow in **cutaneous blood vessels** near the skin's surface. These blood vessels can pick up heat by convection and transfer it to the body core, or they can lose heat to the surrounding air.

Blood flow through cutaneous blood vessels varies from close to zero when heat needs to be conserved to nearly one-third of cardiac output when heat needs to be released to the environment. Local control influences cutaneous blood flow to a certain degree, possibly through vasodilators produced by the vascular endothelium. However, neural regulation is the primary determining factor.

Most arterioles in the body are under tonic sympathetic control [∞ p. 455]. If core body temperature drops, the hypothalamus selectively activates sympathetic neurons innervating cutaneous blood vessels. The vessels constrict, increasing their resistance to blood flow and diverting blood to lower-resistance vessels in the interior of the body. This response keeps warmer core blood away from the cooler skin surface to reduce heat loss.

In warm temperatures, the opposite happens: cutaneous arterioles dilate. Interestingly, this process is **active vasodilation** rather than passive relaxation from decreased sympathetic input.

Active vasodilation is mediated through special sympathetic neurons that secrete acetylcholine. This **sympathetic cholinergic vasodilator system** selectively dilates cutaneous blood vessels to enhance heat loss at the skin's surface. Sympathetic neurons controlling sweat glands also trigger production of **bradykinin,** a paracrine vasodilator substance that may contribute to the response.

✔ What neurotransmitter and neurotransmitter receptor mediate cutaneous vasoconstriction?

✔ What observations might have prompted researchers who discovered the sympathetic cholinergic vasodilator system to classify the neurons as sympathetic rather than parasympathetic? [Hint: ∞ p. 329]

In some portions of the body, especially hands, feet, and face, cutaneous blood flow is regulated by opening and closing **arteriovenous anastomoses** [*anastomosis*, an opening]. These short blood vessels provide a direct link between arterioles and venules in the skin. When the anastomoses are open, blood flow bypasses the cutaneous capillaries that lie in the superficial layers of the epidermis. When the anastomoses are closed, circulation is diverted to the skin's surface, where body heat can be released into the environment.

Sweat Contributes to Heat Loss   Surface heat loss is enhanced by the evaporation of sweat. By some estimates the human integument has between 2 and 3 million sweat glands. The highest concentrations are found on the forehead, scalp, *axillae* (armpits), palms of hands, and soles of feet.

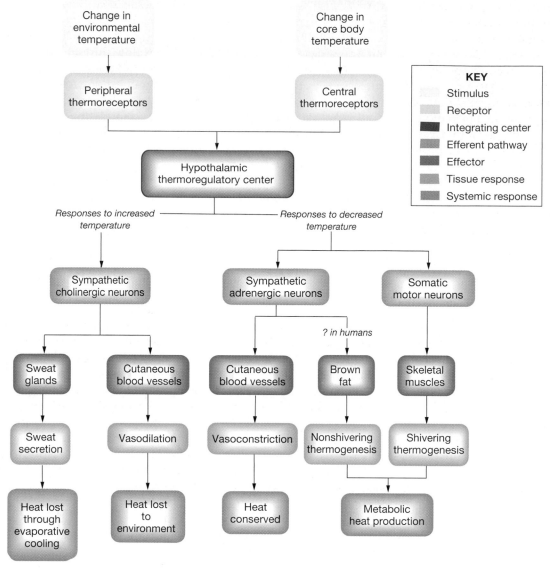

■ **Figure 21-3   Thermoregulatory reflexes**

Sweat glands are made of transporting epithelium. Deep in the gland, it secretes an isotonic solution similar to interstitial fluid into the lumen. As the fluid travels through the duct to the skin, NaCl is reabsorbed, resulting in hypotonic sweat. A typical value for sweat production is 1.5 L/hr. With acclimatization to hot weather, some people sweat at rates of 4–6 L/hr. However, this high rate can only be maintained for short periods unless they are drinking to replace lost fluid volume. Sweat production is regulated by cholinergic sympathetic neurons.

Cooling by evaporative heat loss depends on the vaporization of water from sweat on the skin's surface. Because water evaporates rapidly in dry environments but slowly or not at all in humid ones, the body's ability to withstand high temperature is directly related to the relative humidity of the air. Evaporative cooling slows in humid environments because sweat does not evaporate from the skin's surface as rapidly as it does in dry envi-

ronments. The combination of heat and humidity is reported by meteorologists as the *heat index* or *humidex*. Air moving across a sweaty skin surface enhances vaporization even with high humidity, one reason that fans are so useful in hot weather.

The Body Produces Heat Through Movement and Metabolism   Heat production by the body falls into two broad categories. The first is unregulated heat production from voluntary muscle contraction and normal metabolic pathways. The second is regulated heat production used to maintain temperature homeostasis in low ambient temperatures. Regulated heat production is further divided into two subcategories: shivering thermogenesis and nonshivering thermogenesis.

In **shivering thermogenesis,** the body uses **shivering** (rhythmic tremors caused by skeletal muscle contraction) to generate heat. Signals from the hypothalamic control center initiate these skeletal muscle tremors.

Shivering generates five to six times as much heat as resting muscle. Shivering can be suppressed partially by voluntary control.

Nonshivering thermogenesis is metabolic heat production by pathways other than shivering. In laboratory animals such as the rat, cold exposure significantly increases heat production in brown fat. The mechanism for brown fat heat production is *mitochondrial uncoupling*. In this process, energy flowing through the electron transport system [∞ p. 94] is released as heat rather than being trapped in ATP. Mitochondrial uncoupling in response to cold is promoted by thyroid hormones and by increased sympathetic activity.

The importance of nonshivering thermogenesis in adult humans is unsettled. Humans are born with significant amounts of brown fat, found primarily in the *interscapular* area between the shoulder blades. In newborns, nonshivering thermogenesis in this brown fat contributes significantly to raising and maintaining body temperature. However, as the child ages, white fat gradually replaces most brown fat. The brown fat that remains is very difficult to study because the experimental methodology for doing so requires that fat be extracted from the body.

Another complicating factor in studying the possible association between brown fat and human temperature regulation is the fact that we usually do not stay out in the cold for extended periods. Even people working in the polar regions alternate between indoor and outdoor work. This means that human cold acclimatization is episodic rather than constant. On the other hand, most animal experiments expose their subjects to constant cold conditions.

The body's responses to high and low temperature are summarized in Figure 21-4 ■. In cold environments, the body tries to reduce heat loss while increasing heat production. In hot temperatures, the opposite takes place.

Notice that voluntary behavioral responses play a significant role in temperature regulation. We slow our activity level during hot weather, thereby decreasing muscle heat production. In cold weather, we put on extra clothing, hug our hands in our armpits, or curl up in a ball to slow heat loss.

The Body's Thermostat Can Be Reset   Variations in body temperature regulation can be both physiological and pathological. Examples of physiological variation include the circadian rhythm of body temperature mentioned earlier, menstrual cycle variations in temperature, postmenopausal hot flashes, and fever. All these processes seem to arise from a common mechanism: resetting of the hypothalamic thermostat.

Hot flashes appear to be transient decreases in the thermostat's setpoint caused by the absence of estrogen. When the setpoint lowers, a room temperature that had previously been comfortable suddenly feels too hot. This discomfort triggers the usual thermoregulatory responses to heat, including sweating and cutaneous vasodilation, which leads to flushing of the skin.

Fever for many years was thought to be a pathological response to infection, but it is now considered part

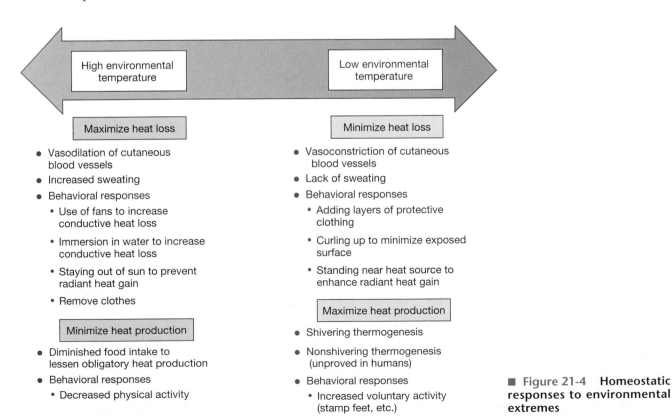

■ **Figure 21-4** **Homeostatic responses to environmental extremes**

of the body's normal immune response. Toxins from bacteria and other pathogens trigger the release of chemicals known as **pyrogens** from various immunocytes [*pyr*, fire]. Pyrogens are fever-producing cytokines that also have many other effects.

Experimentally, some interleukins (IL-1, IL-6), some interferons, and tumor necrosis factor have all been shown to induce fever. They do so by acting on the hypothalamic thermostat, resetting it to a higher set-point. Normal room temperature feels too cold and the patient begins to shiver, creating additional heat. Pyrogens may also increase nonshivering thermogenesis, causing body temperature to rise.

The adaptive significance of fever is still unclear, but it seems to enhance the activity of white blood cells involved in the immune response. For this reason, some people question whether patients with a fever should be given aspirin and other fever-reducing drugs simply for the sake of comfort. High fever can be dangerous, however, as a fever of 41° C (106° F) for more than a brief period causes brain damage.

Pathological conditions in which body temperature strays outside the normal range include different states of hyperthermia and hypothermia. Heat exhaustion and heat stroke are the most common forms of hyperthermia. **Heat exhaustion** is marked by severe dehydration and body core temperatures of 37.5°–39° C (99.5°–102.2° F). Patients may have muscle cramps, nausea, and headache. They are usually pale and sweating profusely. Heat exhaustion often occurs in people who are physically active in hot, humid climates to which they are not acclimatized. It also occurs in the elderly who have lost the ability to thermoregulate.

Heat stroke is a more severe form of hyperthermia, with higher body core temperatures. The skin is usually flushed and dry. Rapid cooling is important, as enzymes and other proteins begin to denature at temperatures above 41° C (106° F). Mortality in heat stroke is nearly 50%.

Malignant hyperthermia, in which body temperature becomes abnormally elevated, is a genetically linked condition. A defective Ca^{2+} channel in skeletal muscle releases too much Ca^{2+} into the cytoplasm. As cell transporters work to move the Ca^{2+} back into mitochondria and the sarcoplasmic reticulum, the heat released from ATP hydrolysis substantially raises the body temperature. Some investigators have suggested that a mild version of this process plays a role in nonshivering thermogenesis in mammals.

Hypothermia, a condition in which body temperature falls abnormally low, is also a dangerous condition. As body core temperature drops, enzymatic reactions slow down and the person loses consciousness. When metabolism slows, oxygen consumption also decreases.

Drowning victims in cold water can sometimes be revived without brain damage if they have gone into a state of hypothermia. This observation led to the development of induced hypothermia for certain surgical procedures such as heart surgery. The patient is cooled to 21°–24° C (70°–75° F) so that tissue oxygen demand can be met by artificial oxygenation of the blood as it passes through a bypass pump. After surgery is complete, the patient is gradually rewarmed.

✔ Why do people with water beds have to heat them to sleep comfortably?

✔ Will a person who is exercising outside overheat faster if the weather is dry or if it is humid?

Energy Balance Is Reflected by an Individual's Oxygen Consumption

To compile an energy balance sheet for the human body, we must be able to estimate both the energy content of food (energy intake) and the energy expenditure from heat loss and various types of work.

Oxygen Consumption Can Be Equated to Metabolic Rate

The most direct way to measure the energy content of food is by **direct calorimetry.** In this procedure, food is burned in an instrument called a *bomb calorimeter,* and the released heat is trapped and measured. The energy content of food is usually measured in **kilocalories** (kcal). One kilocalorie is the amount of heat needed to raise the temperature of 1 liter of water by 1° C. A kilocalorie is the same as a Calorie (with a capital C).

Although direct calorimetry is a quick way of measuring total energy content, the metabolic energy content of food is actually slightly less because not all foods can be fully digested and absorbed. The metabolic energy content of proteins and carbohydrates is 4 kcal/g. Fats contain more than twice as much energy at 9 kcal/g.

The caloric content of food can be calculated by multiplying the number of grams of each component by its metabolic energy content (kcal/g): 2 g fat × 9 kcal/g = 18 kcal. In the United States you can find examples of

---

...continued from page 640

Hyperparathyroidism, or excessive production of PTH, causes breakdown of bone and the release of calcium phosphate into the blood. High levels of Ca^{2+} in the blood can affect the function of excitable tissues such as muscles and neurons. Surprisingly, however, most people with hyperparathyroidism have no symptoms. The condition is usually discovered during blood work performed for a routine health evaluation.

Question 1:   *What role does Ca^{2+} play in the normal function of muscles and neurons? (Hint: See Chapters 8 and 12.)*

Question 2:   *What is the technical term for "elevated levels of calcium in the blood"? (Use your knowledge of word roots to construct this term.)*

the energy content for various foods on the Nutrition Facts label of food packages.

Estimating an individual's energy expenditure, or **metabolic rate,** is more complex than figuring the caloric content of food. The heat released by the body can be measured by enclosing a person in a sealed compartment. Energy intake in food equals work plus heat, so the difference between the heat produced and the person's caloric intake would be the energy used for chemical, mechanical, and transport work. Practically speaking, however, total body heat release is not a very easy way to measure energy use.

Probably the most common method for estimating energy expenditure is to measure a person's **oxygen consumption,** the amount of oxygen consumed per kilocalorie of food metabolized. Studies have shown that oxygen consumption for different foods is relatively constant at a rate of 1 liter oxygen consumed for each 4.5–5 kcal of energy. Thus, the rate of oxygen consumption is generally equated to metabolic rate. Measurement of oxygen consumption is one form of **indirect calorimetry.**

Another method of estimating metabolic energy use is to measure carbon dioxide production, either alone or in combination with oxygen consumption. Aerobic metabolism both consumes O_2 and produces CO_2. However, the ratio of CO_2 produced to O_2 consumed varies with the composition of the diet.

This ratio of CO_2:O_2, known as the **respiratory quotient,** or **RQ,** varies from a high of 1.0 with a pure carbohydrate diet to 0.8 for pure protein and 0.7 for pure fat. The average American diet has an RQ of about 0.82. Whether measured by oxygen consumption or by carbon dioxide production, metabolic rate can be highly variable from one person to another or from day to day in a single individual.

Many Factors Influence Metabolic Rate   What factors affect metabolic rate in humans? They include age, gender, amount of lean muscle mass, activity level, diet, hormones, and genetics. An individual's lowest metabolic rate is considered the **basal metabolic rate,** or BMR. In reality, metabolic rate would be lowest when an individual is asleep. But because measuring the metabolic rate of a sleeping person is difficult, BMR is measured after a 12-hour fast in a person who is awake but resting.

- **Age and gender.** Adult males have an average BMR of 1.0 kcal per hour per kilogram body weight. Adult females have a lower rate than males: 0.9 kcal/hr/kg. The difference arises because women have a higher percentage of adipose tissue and less lean muscle mass.

    Metabolic rates in both sexes decline with age. Some of this decline is due to decreases in lean muscle mass.
- **Amount of lean muscle mass.** Muscle has a higher oxygen consumption rate at rest than fat (recall that most of the volume of an adipose tissue cell is occupied by metabolically inactive lipid droplets). This is one reason that advice on how to lose weight often includes weight training in addition to aerobic exercise. By adding to lean muscle mass, people raise their basal metabolic rate and burn off more calories at rest.
- **Physical activity level.** Physical activity and muscle contraction increase metabolic rate over the basal rate. You are probably familiar with tables that show the caloric expenditure per hour of various activities. Sitting and lying down consume relatively little energy. Rowing in a race and bicycling are among the energetically most expensive activities.
- **Effect of diet.** We now know that resting (nonactive) metabolic rate increases after a meal, a phenomenon termed **diet-induced thermogenesis.** In other words, there is an energetic cost to the digestion and assimilation of food. Diet-induced thermogenesis is related to the type and amount of food ingested. Fats cause relatively little diet-induced thermogenesis. Proteins increase heat production the most. This phenomenon may support the claim of some nutritionists that eating one calorie of fat is different than eating one calorie of protein, although they contain the same amount of energy when measured by direct calorimetry.
- **Effect of hormones.** Basal metabolic rate is increased by thyroid hormones and by catecholamines (epinephrine and norepinephrine). These same factors mediate cold-induced nonshivering thermogenesis in laboratory animals, so scientists are now looking for a common mechanism. If we can understand how energy is diverted from fat storage into heat production, we may be able to develop new thermogenic drugs for the treatment of obesity.

In the end, the two factors that a person can control to balance energy expenditure and avoid weight gain are energy intake (eating) and activity. If a person's activity includes strength training that increases lean muscle mass, resting metabolic rate goes up. The addition of lean muscle mass to the body creates additional energy use, which in turn decreases the number of calories that go into storage.

Energy Is Stored in Fat and Glycogen

A person's daily energy requirement, expressed as caloric intake, varies with the needs and activity of the body. For example, large male athletes in an intensive training program may need more than 10,000 kcal per day. On the other hand, a 70-kg man engaged in normal activities may only require 2000 kcal/day. Let's assume that our average man's energy requirement could be met by ingesting only glucose. Glucose has an energy content of 4 kcal/g, therefore to get 2000 kcal he would have to consume 500 g, or 1.1 pounds, of glucose in a day. Our bodies cannot absorb crystalline glucose, however, so those 500 g of glucose would have to be dissolved in water. If the glucose was made as an isosmotic 5% solution, the 500 g would have to be dissolved in 10 liters of water: a substantial volume to drink during a day!

Fortunately, we do not usually ingest glucose as our primary fuel. Proteins, complex carbohydrates, and fats also provide energy. Glycogen, a glucose polymer [∞ p. 29], is a more compact form of energy than an equal number of individual glucose molecules. Glycogen also requires less water for hydration.

Our cells therefore convert glucose to glycogen for storage. Normally we keep about 100 g of glycogen in the liver and 200 g in skeletal muscles. But even 300 g of glycogen can provide only enough energy for 10 to 15 hours of the body's energy needs. The brain alone requires 150 g of glucose per day.

Consequently, the body keeps most of its energy reserves in compact, high-energy fat molecules. One gram of fat has 9 kcal, more than twice the energy content of an equal amount of carbohydrates or proteins. This feature makes adipose tissue a very efficient way for the body to keep large amounts of energy stored in minimal space. Metabolically, however, the energy in fat is harder to access, and the metabolism of fats is slower than that of carbohydrates.

▶ METABOLISM

In the human body, we divide metabolism into two states. The period of time following a meal, when the products of digestion are being absorbed, used, and stored, is called the **fed, or absorptive, state.** This is an *anabolic* state in which the energy of nutrient biomolecules is transferred to high-energy compounds or is stored in the chemical bonds of other molecules.

Once nutrients from a recent meal are no longer in the bloodstream and available for use by the tissues, the body enters a **fasted, or postabsorptive, state.** As the pool of available nutrients in the blood decreases, the body taps into its stored reserves. The postabsorptive state is a *catabolic* state, in which cells degrade large molecules into smaller molecules. The energy released by breaking chemical bonds is used to do work.

Energy from Ingested Nutrients May Be Used Immediately or Stored

The biomolecules we eat are destined to meet one of three fates:

1. **Energy.** They are metabolized immediately so that energy in their chemical bonds is trapped in ATP, phosphocreatine, and other high-energy compounds.

This energy can then be used to do mechanical or transport work.
2. **Synthesis.** They are used for synthesis of basic components needed for growth and maintenance of cells and tissues.
3. **Storage.** If the food ingested exceeds the body's requirements for energy and synthesis, excess nutrients go into storage as glycogen and fat. Storage makes energy available for times of fasting.

The specific fate of an absorbed nutrient depends on whether the biomolecule is a carbohydrate, protein, or fat. Figure 21-5 ■ is a schematic diagram that follows these biomolecules from the diet into the three *nutrient pools* of the body: the fatty acid pool, the glucose pool, and the amino acid pool. Nutrient pools are nutrients that are available for immediate use. They are located primarily in the plasma.

Fats are absorbed primarily as fatty acids and glycerol. Fatty acids form the primary pool of fats in the blood. They can be used as an energy source by many tissues but are also easily stored as fat (triglycerides) in adipose tissue.

Carbohydrates are absorbed mostly as glucose. Plasma glucose concentration is the most closely regulated of the three nutrient pools because glucose is the only fuel that the brain can metabolize, except in times of starvation.

Notice in Figure 21-5 ■ that if the glucose pool drops below a certain level, the only tissue that has access to glucose is the brain. This conservation measure ensures that the body's control center has adequate amounts of energy. Just as the circulatory system gives top priority to blood supply to the brain, metabolism also gives top priority to the brain.

If the glucose pool is within the normal range, then most tissues use glucose for their energy source. If the glucose pool increases even further, then glucose goes into storage as glycogen. The synthesis of glycogen from glucose is known as **glycogenesis.** Glycogen stores are limited, however, and additional excess glucose is converted to fat (**lipogenesis**).

If plasma glucose concentrations decrease, the body will convert glycogen back to glucose through the process called **glycogenolysis.** The balance between oxidative metabolism, glycogen synthesis, glycogen breakdown, and fat synthesis allows the body to maintain plasma glucose concentrations within a narrow range.

If for some reason plasma glucose concentrations exceed a critical level, as occurs in diabetes mellitus, excess glucose is excreted in the urine. Glucose excretion occurs only when the renal threshold for glucose reabsorption is exceeded [∞ p. 558].

The amino acid pool of the body is used primarily for protein synthesis. However, if glucose intake is low, amino acids can be converted into glucose through the pathways known as **gluconeogenesis.** This word literally means

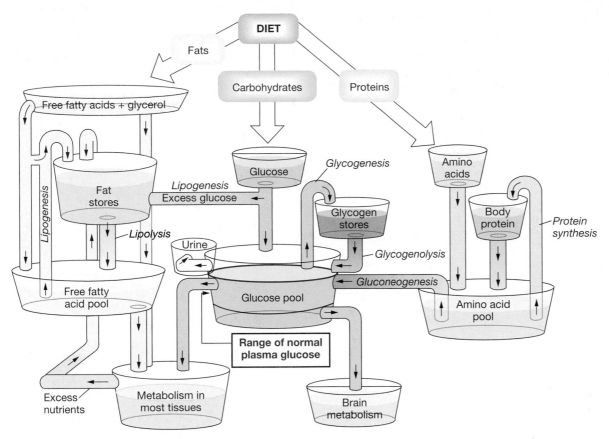

■ **Figure 21-5**   **Summary of metabolism**   Adapted from L. L. Langley, *Homeostasis* (New York: Reinhold, 1965).

"the birth of new glucose" and refers to the synthesis of glucose from a noncarbohydrate precursor.

Amino acids are the main source for glucose through the gluconeogenesis pathways, but glycerol from triglycerides can also be used. Both gluconeogenesis and glycogenolysis are important backup sources for glucose during periods of fasting.

Hormones Control Metabolic Pathways by Changing Enzyme Activity

The most important biochemical pathways for energy production are illustrated in Figure 21-6 ■. This figure does not include all of the metabolic intermediates in each pathway [∞ p. 91]. Instead, it emphasizes the points at which different pathways intersect, because these intersections are often key points at which metabolism is controlled.

One significant feature of metabolic regulation is the use of different enzymes to catalyze forward and reverse reactions. This dual control, sometimes called *push-pull control,* allows close regulation of a reaction. The flow of nutrients through metabolic pathways is mediated in just this way.

Figure 21-7 ■ shows how hormones regulate the flow of nutrients through metabolic pathways by altering enzyme activity. In the first example (Fig. 21-7a ■), enzyme 1 catalyzes the reaction of A to B. Enzyme 2 catalyzes the

reverse reaction. If the activity of the two enzymes is roughly equal, as soon as A is converted into B, B is converted back into A. Turnover of the two substrates is rapid, but there is no net production of product B.

Figure 21-7b ■ shows an example in which the activity of one set of enzymes is enhanced while the enzymes catalyzing the reverse reaction are inhibited. This type of dual control is exerted by the pancreatic hormone *insulin,* which stimulates enzymes for glycogen synthesis and inhibits enzymes for glycogen breakdown. The net result is glycogen synthesis from glucose.

The reverse pattern is shown in Figure 21-7c ■, where *glucagon,* another pancreatic hormone, stimulates the enzymes of glycogenolysis while inhibiting the enzymes for glycogenesis. The net effect of glucagon is glucose synthesis from glycogen.

Anabolic Metabolism Dominates in the Fed State

As we've noted, the *fed state* or *absorptive state* is anabolic: absorbed nutrients are being used for energy, synthesis, and storage. Table 21-1 summarizes the fates of nutrients in the fed state.

Carbohydrates Provide Energy   Glucose absorbed after a meal goes into the circulation of the hepatic portal system and is taken directly to the liver, where about

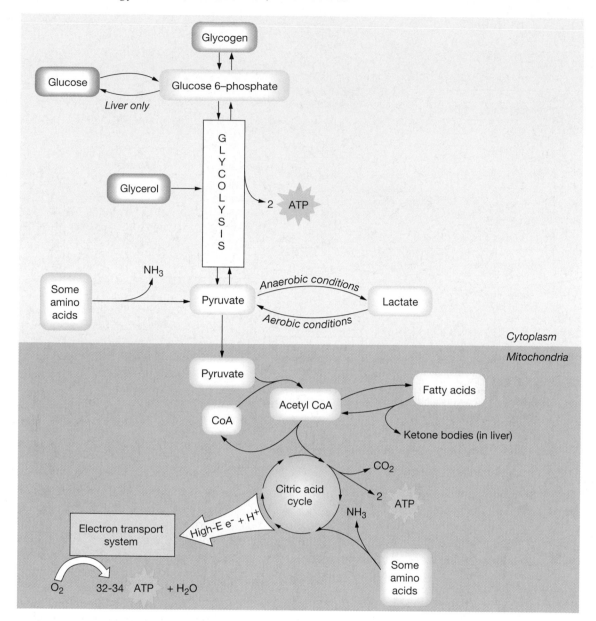

■ Figure 21-6   **Summary of biochemical pathways for energy production**

30% of all ingested glucose is metabolized. The remaining 70% continues in the bloodstream for distribution to the brain, muscles, and other organs and tissues of the body.

Glucose moves from interstitial fluid into cells via facilitated diffusion GLUT transporters [∞ p. 122]. Most glucose absorbed from a meal goes immediately into glycolysis and the citric acid cycle to make ATP. Some glucose is used by the liver for lipoprotein synthesis. Glucose that is not required for energy and synthesis is stored either as glycogen or as fat. The body's ability to store glycogen is limited, so most excess glucose is converted to triglycerides and stored in adipose tissue.

New Proteins Are Made from Amino Acids Most amino acids absorbed from a meal go to the tissues for protein synthesis. Like glucose, amino acids are taken first to the liver by the hepatic portal system. The liver

uses them to synthesize lipoproteins and plasma proteins, such as albumin, clotting factors, and angiotensinogen.

Amino acids not taken up by the liver enter systemic venous blood and are distributed throughout the body. They are used by cells to create structural or functional proteins such as cytoskeletal elements, enzymes, and hormones. Amino acids are also incorporated into non-protein molecules such as amine hormones and neuro-transmitters.

Amino acids can be used for energy if glucose intake is low. On the other hand, if more protein is ingested than is needed for synthesis and energy needs, excess amino acids are converted to fat. Some bodybuilders spend large amounts of money on special amino acid supplements that are advertised to build bigger muscles. What the athletes may not realize is that these amino acids do not automatically go into protein syn-

(a)

| | Without regulation of enzymatic activity, the pathway will simply cycle back and forth. |

(b)

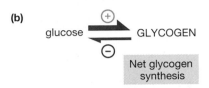

| Fed state metabolism under the influence of insulin |

(c)

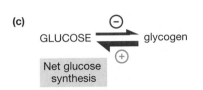

| Fasted state metabolism under the influence of glucagon |

■ **Figure 21-7** **Dual (push-pull) control of metabolism** Push-pull control of metabolism requires different enzymes to catalyze the forward and reverse reactions.

thesis. When amino acid intake exceeds the body's need for protein synthesis, excess amino acids are burned to create energy or are stored as fat.

Fats Store Energy Most ingested fats are assembled into lipoprotein complexes called *chylomicrons* in the intestinal epithelium [∞ p. 629]. Chylomicrons reach the venous circulation via the lymphatics, bypassing the liver (Fig. 21-8 ■). As they circulate through the blood, **lipoprotein lipase** bound to the capillary endothelium of muscles and adipose tissue converts triglycerides in

---

TABLE 21-1 **Fate of Nutrients in the Fed State**

Carbohydrates, absorbed primarily as glucose

1. Used immediately for energy through aerobic pathways*
2. Used for lipoprotein synthesis in the liver
3. Stored as glycogen in liver and muscle
4. Excess converted to fat and stored in adipose tissue (glucose → pyruvate → acetyl CoA → fatty acids)

Proteins, absorbed primarily as amino acids

1. Most amino acids go to tissues for protein synthesis*
2. If needed for energy, amino acids convert in liver to intermediates for aerobic metabolism
3. Excess converted to fat and stored in adipose tissue (amino acids → acetyl CoA → fatty acids)

Fats, absorbed primarily as triglycerides

1. Stored as fats primarily in liver and adipose tissue*

*Primary fate.

---

the chylomicron to free fatty acids and monoglycerides. These molecules then diffuse into the cells, where they are reassembled into triglycerides for storage.

Chylomicron remnants that remain in the circulation are composed primarily of lipoproteins and cholesterol. The combination of lipid with proteins makes cholesterol more soluble in plasma, but the complexes are unable to diffuse through cell membranes as the free fatty acids can. Instead, lipoprotein complexes must be brought into cells using receptor-mediated endocytosis [∞ p. 129]. Chylomicron remnants, for example, are taken up and metabolized by the liver.

Once in liver cells, cholesterol from the remnants joins the liver's pool of lipids. If cholesterol is in excess, some may be excreted in the bile. The remainder is added to newly made cholesterol and fatty acids, then packaged again into lipoprotein complexes for secretion back into the blood.

There are several lipoproteins that form these complexes, ranging from *very low-density lipoprotein (VLDL)* to *high-density lipoprotein (HDL)*. The complexes contain varying ratios of protein, triglyceride, phospholipids, and cholesterol. The more protein the complex contains, the heavier its weight.

Most lipoprotein in the blood is **low-density lipoprotein, or LDL.** LDL-cholesterol is the form in which most cells get cholesterol through receptor-mediated endocytosis. LDL-cholesterol is sometimes known as the "bad cholesterol" because elevated concentrations of plasma LDL are associated with the development of atherosclerosis.

The second most common lipoprotein is **high-density lipoprotein, or HDL.** HDL-cholesterol is sometimes called the "good cholesterol" because HDL takes cholesterol into the liver cells, where it is metabolized or excreted. A simpler way to remember the difference between the two types is to let *L* stand for *lethal* and *H* stand for *healthy*. The ratio of HDL-cholesterol to LDL-cholesterol in the plasma is as important as the absolute concentration. The ratio is used by clinicians to predict a person's risk of developing atherosclerosis.

Apolipoproteins, Cholesterol Metabolism, and Heart Disease The protein components of the lipoproteins are called **apoproteins**, or **apolipoproteins** [*apo-*, derived from]. LDL complexes contain apoprotein B (apoB), a molecule that combines with specific receptors to allow cholesterol and triglycerides to be endocytosed into most cells of the body. HDL complexes contain apoE, which activates receptors in the liver. Several inherited forms of hypercholesterolemia (elevated plasma cholesterol levels) have been linked to defective forms of either apoE or apoB. These abnormal apoproteins may help explain the accelerated development of atherosclerosis in people with hypercholesterolemia.

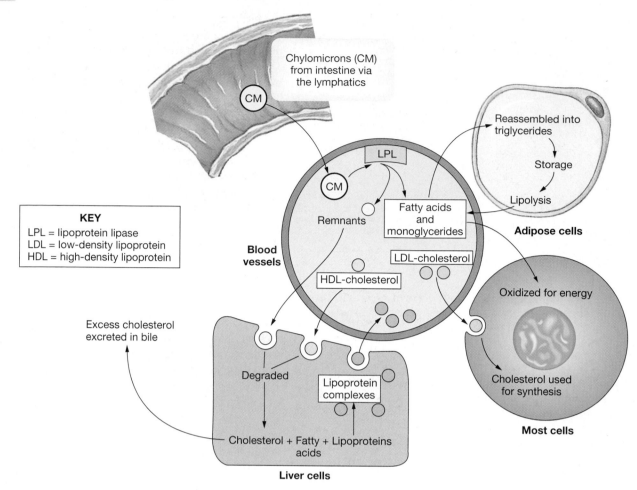

■ Figure 21-8 **Transport and fate of dietary fats**

Catabolic Metabolism Dominates in the Fasted State

Once all nutrients from a meal have been digested, absorbed, and distributed to various cells, plasma concentrations of glucose begin to fall. This is the signal for the body to shift from fed (absorptive) state to fasted (postabsorptive) state metabolism. The goal of the fasted state is to maintain plasma glucose concentrations within an acceptable range so that the brain and neurons have adequate fuel.

The liver is the primary source of glucose production during the fasted state. Liver glycogen can provide enough glucose through glycogenolysis to meet four to five hours of the body's energy needs (Fig. 21-9 ■).

The glycogen stores of skeletal muscle can contribute additional plasma glucose in the fasted state but not through direct conversion of glycogen to glucose. Muscle cells, like most other cells, lack the enzyme that makes glucose from glucose 6-phosphate, so they metabolize glucose 6-phosphate to pyruvate (aerobic conditions) or lactate (anaerobic conditions). Pyruvate and lactate are then transported to the liver, which uses them to make glucose.

Additional glucose or ATP can be made from amino acids, particularly those in muscle proteins. Enzymes remove amino groups from the amino acids (deamination,

∞ p. 96) and convert the amino groups to urea, which is excreted. Some deaminated amino acids become citric acid cycle intermediates and make ATP. This alternative ATP source thus spares plasma glucose for use by the brain. Other amino acids are processed to pyruvate,

§ **Ketogenic Diets** Some people intentionally restrict or eliminate their intake of carbohydrates in an effort to lose weight. Diets that are very low in carbohydrate and high in fat and protein shift metabolism to β-oxidation of fats and ketone body production. These diets are known therefore as *ketogenic diets*. People who go on these diets are often pleased by the initial rapid weight loss, but this is due to glycogen breakdown and water loss, not fat use. Sticking with ketogenic diets decreases caloric intake and eventually results in loss of body fat, but any diet that restricts calories will do the same. Ketogenic diets have risks associated with them, including dehydration, electrolyte loss, inadequate intake of calcium and vitamins, gout, and kidney problems. The only recommended use of ketogenic diets is for children under 10 who have epilepsy that is not responding fully to drug therapy. For reasons that we do not understand, maintaining a state of ketosis in these children decreases the incidence of seizures.

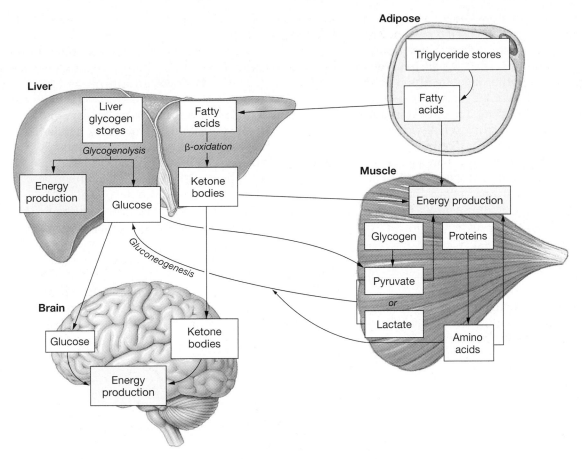

■ Figure 21-9   **Fasted-state metabolism**

which goes to the liver and is made into glucose, as previously described.

In the fasted state, adipose tissue breaks down its stores of triglycerides into fatty acids and glycerol. Glycerol goes to the liver and can be converted to glucose. Fatty acids are released into the blood and are used as a source of energy by many tissues.

The long carbon chains of fatty acids are chopped into two-carbon units through the process of β-oxidation [∞ p. 97]. These two-carbon units then feed into the citric acid cycle through acetyl CoA. If there is a large demand for fatty acid breakdown, β-oxidation in the liver results in the formation of molecules known as **ketone bodies.** Ketone bodies are transported from the liver to

other cells that convert them to acetyl CoA. Acetyl CoA then enters the citric acid cycle.

Ketone bodies are the only fuel besides glucose that the brain can use, so in cases of prolonged starvation, ketones become a significant source of energy. Unfortunately, the ketones *acetoacetic acid* and *β-hydroxybutyric acid* are moderately strong acids, so excessive ketone production leads to a state of acidosis known as **ketoacidosis.** People in ketoacidosis have a fruity odor to their breath due to acetone, a volatile ketone whose odor you might recognize from nail polish remover.

With this summary of metabolic pathways as background, we now turn to the endocrine and neural regulation of metabolism.

HOMEOSTATIC CONTROL OF METABOLISM

The endocrine system has primary responsibility for metabolic regulation, although the nervous system does have some influence, particularly in terms of governing food intake. Hour-to-hour regulation depends on the ratio of insulin to glucagon, two hormones secreted by endocrine cells of the pancreas (see Fig. 21-7 ■). Both hormones have short half-lives and must be continuously secreted if they are to have a sustained effect.

▶ PANCREATIC HORMONES

The endocrine cells of the pancreas are less than 2% of the organ's total mass, with most pancreatic tissue devoted to production and exocrine secretion of digestive enzymes and bicarbonate [∞ p. 625]. In 1869, the German anatomist Paul Langerhans described small clusters of cells scattered throughout the body of the pancreas

(Fig. 21-10 ■). These clusters, now known as the **islets of Langerhans,** contain four distinct cell types, each associated with secretion of a different peptide hormone.

Nearly three-quarters of the islet cells are **beta cells,** which produce insulin. Another 20% are **alpha cells,** which secrete glucagon. Most of the remaining cells are somatostatin-secreting **D cells,** but a few cells produce a mystery peptide known as *pancreatic polypeptide.* These rare cells are called either **PP cells** or **F cells.**

Like all endocrine glands, the islets are closely associated with capillaries into which the hormones are released. Both sympathetic and parasympathetic neurons terminate on the islets, providing a means by which the nervous system can influence metabolism.

The Ratio of Insulin to Glucagon Is a Key to Metabolic Regulation

Insulin and glucagon act in antagonistic fashion to keep plasma glucose concentrations within an acceptable range. Both hormones are present in the blood most of the time.

It is the ratio of the two hormones that determines which hormone dominates.

In the fed state, when the body is absorbing nutrients, insulin dominates and the body undergoes net anabolism (Fig. 21-11a ■). Ingested glucose is used for energy production, with excess glucose stored as glycogen or fat. Proteins go primarily to protein synthesis.

In the fasted state, the goal of metabolic regulation is to prevent low plasma glucose concentrations (**hypoglycemia**). When glucagon predominates (Fig. 21-11b ■), the liver uses glycogen and nonglucose intermediates to synthesize glucose for release into the blood.

Figure 21-12 ■ shows plasma glucose, glucagon, and insulin concentrations over the course of a day. In a normal person, fasting plasma glucose is maintained around 90 mg/dL plasma. After absorption of nutrients from a meal, plasma glucose rises. The increase in glucose stimulates insulin release, which in turn promotes glucose transfer into cells. Plasma glucose concentrations thus fall back toward the fasting level shortly after each meal.

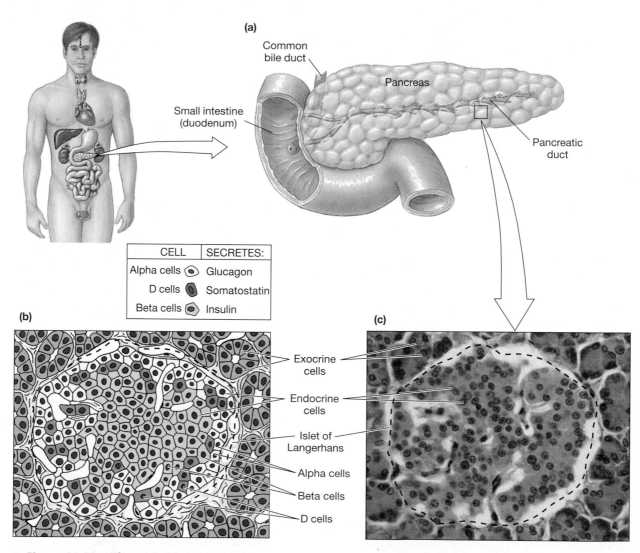

CELL	SECRETES:
Alpha cells	Glucagon
D cells	Somatostatin
Beta cells	Insulin

■ **Figure 21-10** **The endocrine pancreas**

(a) Fed state: insulin dominates

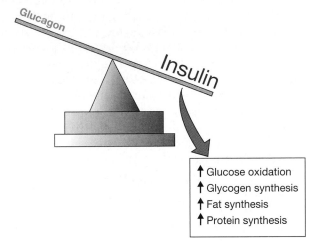

↑ Glucose oxidation
↑ Glycogen synthesis
↑ Fat synthesis
↑ Protein synthesis

(b) Fasted state: glucagon dominates

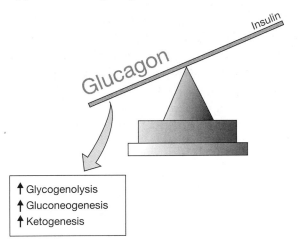

↑ Glycogenolysis
↑ Gluconeogenesis
↑ Ketogenesis

■ **Figure 21-11   Metabolism is controlled by insulin and glucagon**

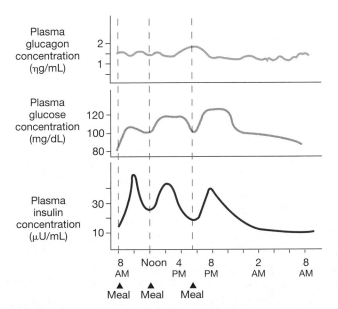

■ **Figure 21-12   Glucose, glucagon, and insulin levels in a 24-hour period**

During an overnight fast, plasma glucose concentrations fall to a low but steady value, and insulin secretion also decreases. Glucagon secretion remains relatively steady during the 24-hour period, supporting the theory that it is the ratio of insulin to glucagon that determines the direction of metabolism.

Insulin Is the Dominant Hormone of the Fed State

Insulin is a typical peptide hormone, produced as an inactive prohormone and activated prior to secretion [∞ Fig. 7-4c, p. 192]. Like other peptide hormones, insulin combines with a membrane receptor on its target cells. The insulin receptor of target tissues has tyrosine kinase activity that initiates complex intracellular cascades [∞ p. 160]. The details of these cascades are still not well understood.

What influences insulin secretion? We know of at least six different stimuli.

1. **Increased glucose concentrations.** The primary stimulus for insulin release is plasma glucose concentrations greater than 100 mg/dL. Glucose absorbed from the small intestine reaches the beta cells, where it is taken up by a GLUT-2 transporter. With more glucose available as substrate, ATP production increases and ATP-gated K^+ channels close. The cell depolarizes, voltage-gated Ca^{2+} channels open, and Ca^{2+} entry through them initiates exocytosis of insulin [∞ Fig. 5-39, p. 146].

2. **Increased amino acid concentrations.** Increased plasma amino acid concentrations following a meal also trigger insulin secretion.

3. **Feedforward GIP secretion.** The intestinal hormone known as *glucose-dependent insulinotropic peptide,* or GIP [∞ p. 616], releases insulin in a feedforward fashion. When you eat a meal containing carbohydrate or protein, glucose or amino acids in the lumen of the small intestine release GIP from endocrine cells. GIP is carried through the circulation to beta cells, arriving there before the first absorbed glucose. The anticipatory release of insulin in response to GIP prevents a sudden surge in plasma glucose concentrations when the meal is absorbed.

4. **Parasympathetic activity.** The feedforward release of insulin by GIP is reinforced by increased parasympathetic activity that accompanies digestion of a meal. Cholinergic neurons terminate on the beta cells and enhance insulin secretion.

5. **Sympathetic activity.** Insulin secretion is inhibited by the sympathetic division of the nervous system.

6. **GLP-1 secretion.** Another hormone from the small intestine, *glucagon-like peptide 1,* or GLP-1, is a powerful stimulus for insulin release. GLP-1 may not play a significant role in normal metabolism, however, because it is released from cells in the ileum,

the distal section of the small intestine. Usually glucose absorption is complete before intestinal contents reach that region and there is no stimulus for GLP-1 release. However, researchers are actively investigating the possibility that pharmacological GLP-1 can be used to enhance insulin release in one form of diabetes.

Insulin Promotes Anabolism

The primary targets of insulin are the liver, adipose tissue, and skeletal muscles (Table 21-2). The usual target cell response to insulin is increased glucose metabolism. In some target tissues, insulin also regulates the GLUT transporters. Other tissues, including the brain and transporting epithelia of the kidney and intestine, do not require insulin for glucose uptake and metabolism.

How does insulin lower plasma glucose? It exerts its action in the following three ways:

1. **Insulin increases glucose transport into most, but not all, insulin-sensitive cells.** Adipose tissue and resting skeletal muscle require insulin for glucose uptake. Without insulin, their GLUT-4 transporters are withdrawn from the membrane and stored in cytoplasmic vesicles (Fig. 21-13a ■). When the insulin receptor is activated, the resulting signal transduction cascade causes vesicles to move to the cell membrane and insert the transporters by exocytosis (Fig. 21-13b ■). The cells then take up glucose from the interstitial fluid by facilitated diffusion.

   Insulin affects liver cells (*hepatocytes*) in a different way. Glucose transport into liver is not directly regulated by insulin. Liver cells have GLUT-2 transporters that are always present in the cell membrane. Instead, in the fed state (Fig. 21-14a ■), insulin activates *hexokinase,* the enzyme that phosphorylates glu-

cose to glucose 6-phosphate. This reaction keeps free intracellular glucose concentrations low relative to the extracellular fluid, so glucose continuously diffuses into the hepatocyte on the GLUT-2 transporter.

In the fasted state, when insulin is low, hepatocytes need transporters to move glucose out of the cell. In this situation (Fig. 21-14b ■), hepatocytes are converting glycogen stores and amino acids to glucose. The newly formed glucose then diffuses *out* across the hepatocyte membrane and into the blood. If the absence of insulin prompted hepatocytes to withdraw GLUT transporters, this process could not take place.

2. **Insulin enhances cellular utilization and storage of glucose** (Fig. 21-15 ■). Insulin activates enzymes for glucose utilization (*glycolysis*) and glycogen and fat synthesis (*glycogenesis* and *lipogenesis*). Insulin simultaneously inhibits enzymes for glycogen breakdown (*glycogenolysis*), glucose synthesis (*gluconeogenesis*), and fat breakdown (*lipolysis*) to ensure that metabolism moves in the anabolic direction. If more glucose has been ingested than is needed for energy and synthesis, the excess is made into glycogen or fatty acids.

3. **Insulin enhances cellular utilization of amino acids.** Insulin activates enzymes for protein synthesis and inhibits enzymes that promote protein breakdown. If a meal includes protein, amino acids are used for protein synthesis by both the liver and muscle. Excess amino acids are converted to fatty acids, then made into triglycerides by the liver. The triglycerides are secreted into the blood as lipoprotein complexes. Excess triglyceride is stored in adipose tissue.

Thus, insulin is an anabolic hormone because it promotes glycogen, protein, and fat synthesis. When insulin is absent or deficient, cells go into catabolic metabolism.

TABLE 21-2   Insulin

Cell of origin	Beta cells of pancreas
Chemical nature	51–amino acid peptide
Biosynthesis	Typical peptide
Transport in the circulation	Dissolved in plasma
Half-life	5 minutes
Factors affecting release	Plasma [glucose] > 100 mg/dL; GIP (feedforward reflex); ↑ blood amino acids; parasympathetic stimulates, sympathetic inhibits
Target cells or tissues	Liver, muscle, and adipose tissue primarily; brain, kidney, and intestine not insulin-dependent
Target receptor	Membrane receptor with tyrosine kinase activity
Whole body or tissue action	↓ Plasma [glucose] by ↑ transport into cells or ↑ metabolic use of glucose
Action at cellular level	↑ Glycogen synthesis; ↑ aerobic metabolism of glucose; ↑ protein and triglyceride synthesis
Action at molecular level (including second messenger)	Inserts GLUT transporters into membranes of muscle and adipose cells; alters enzyme activity. Complex signal transduction pathway.
Feedback regulation	↓ Plasma [glucose] shuts off insulin release
Other information	Growth hormone and cortisol are antagonistic

(a) In the absence of insulin, glucose cannot enter the cell

(b) Insulin allows glucose to enter

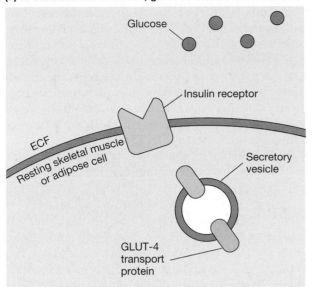

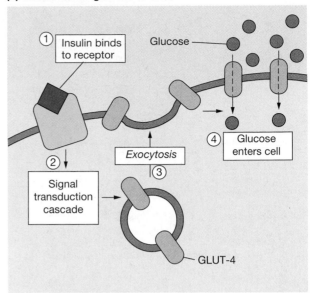

■ **Figure 21-13   Insulin alters glucose transport in adipose tissue and resting skeletal muscle**   (a) Without insulin, GLUT-4 transporters are in vesicles in the cytoplasm. (b) Insulin receptor activation causes the GLUT-4 transporters to be inserted into the cell membrane by exocytosis.

Diabetes Mellitus Is a Family of Metabolic Diseases

Like most hormones secreted by discrete endocrine glands, insulin was first known through the diseases that result from its abnormal secretion or activity. For insulin, the hallmark pathology is a family of diseases known as **diabetes mellitus.** Diabetes is characterized by abnormally elevated plasma glucose concentrations, caused by inadequate insulin secretion, abnormal target cell responsiveness [∞ p. 205], or both. The chronic hyper-

glycemia and associated metabolic abnormalities in turn cause the many complications of the disease, including damage to blood vessels, eyes, kidneys, and the nervous system.

Diabetes has been known to affect humans since ancient times, and written accounts of the disease highlight the calamitous consequences of insulin deficiency. Arataeus the Cappadocian (81–138 A.D.) wrote of the "wonderful" nature of this disease that consisted of the "melting down of the flesh . . . into urine," accompanied by terrible thirst that could not be quenched. The copious

(a) Hepatocyte-fed state

(b) Hepatocyte-fasted state

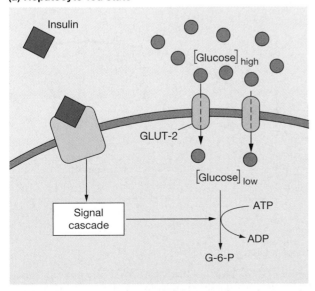

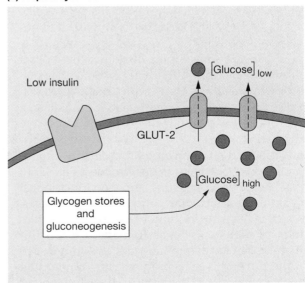

■ **Figure 21-14   Insulin action on hepatocytes**   Insulin does not directly regulate glucose transport in hepatocytes. Instead, it affects the glucose concentration gradient.

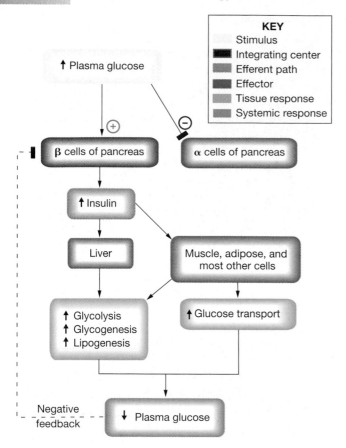

KEY
Stimulus
Integrating center
Efferent path
Effector
Tissue response
Systemic response

↑ Plasma glucose

(+) β cells of pancreas   (−) α cells of pancreas

↑ Insulin

Liver

Muscle, adipose, and most other cells

↑ Glycolysis
↑ Glycogenesis
↑ Lipogenesis

↑ Glucose transport

Negative feedback

↓ Plasma glucose

■ **Figure 21-15** **Fed-state metabolism**

production of glucose-laden urine gave the disease its name. *Diabetes* refers to the flow of fluid through a siphon, and *mellitus* comes from the word for honey. In the Middle Ages, diabetes was known as "the pissing evil."

The severe type of diabetes described by Arataeus is **insulin-dependent diabetes mellitus, or type 1 diabetes.** It is a condition of insulin deficiency due to beta cell destruction. Type 1 diabetes is most commonly an *autoimmune* disease, in which the body fails to recognize the beta cells as "self" and destroys them with antibodies and white blood cells.

Type 1 diabetes is a complex disorder whose onset in genetically susceptible individuals is sometimes preceded by a viral infection. Many type 1 diabetics develop their disease in childhood, giving rise to the old name of *juvenile-onset diabetes.* About 10% of all diabetics have type 1 diabetes.

Because individuals with type 1 diabetes are insulin-deficient, the only treatment is insulin injections. Until the arrival of genetic engineering, most pharmaceutical insulin came from pork, cow, and sheep pancreases. However, once the gene for human insulin was cloned, biotechnology companies began to manufacture artificial human insulin for therapeutic use. In addition, scientists are developing techniques for implanting encapsulated beta cells in the body, in the hope that individuals with type 1 diabetes will no longer need to rely on regular insulin injections.

The other variant of diabetes mellitus is **type 2 diabetes mellitus,** formerly called *non-insulin-dependent diabetes mellitus,* or NIDDM. Type 2 diabetes is also known as *insulin-resistant diabetes* because, in most patients, insulin levels in the blood are normal or even elevated until late in the disease process. Many type 2 diabetics eventually become insulin-deficient and require insulin injections, which is why the old name of NIDDM was dropped in 1997. Type 2 diabetes is actually a whole family of diseases with a variety of causes.

Type 2 diabetics comprise 90% of the diabetic population. The disease is more common in females and people over the age of 40. About 80% of Type 2 diabetics are obese.

A common hallmark of type 2 diabetes is a delayed response to an ingested glucose load. This response can be demonstrated by a procedure known as a *glucose tolerance test* (Fig. 21-16 ■). First, the subject's fasting plasma glucose concentration is determined (time 0). Then the person consumes either a special glucose drink or a typical meal. Plasma glucose is measured periodically for two or more hours.

Normal subjects have a fasting plasma glucose of 109 mg/dL or less. They show a slight increase in plasma glucose concentration in the immediate **postprandial period** [*prandium,* a meal], but the level rapidly returns to normal following insulin secretion.

In diabetic patients, however, fasting plasma glucose concentrations are usually above normal, and they rise even higher as glucose is absorbed. The cells of the body are slow to remove glucose from the blood, so plasma glucose stays elevated for two or more hours. This slow response suggests that insulin, even if present in the blood, is unable to carry out its normal function.

An abnormal glucose tolerance test simply shows that the body's response to an ingested glucose load is not normal. The test cannot distinguish between prob-

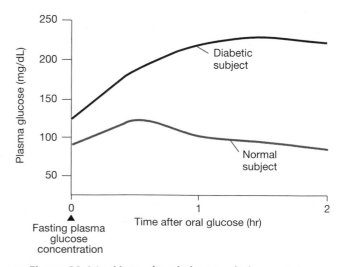

■ **Figure 21-16** **Normal and abnormal glucose tolerance tests** At time 0, the subject consumes a glucose drink or meal. In diabetic subjects, plasma glucose concentrations are above 200 mg/dL after two hours.

lems with insulin synthesis, problems with insulin release, or the responsiveness of target tissues to insulin. Most obese patients with type 2 diabetes are believed to have a defect in their target cell signal transduction pathway at some point past the insulin receptor itself.

Type 1 Diabetics Are Ketosis-Prone

The events that follow ingestion of carbohydrate in an insulin-dependent diabetic create a picture of what happens to metabolism in the absence of insulin (Fig. 21-17 ■). Digestion and absorption of glucose proceed normally because glucose transport by the intestine is insulin-independent. When the absorbed glucose load reaches the liver, transport into hepatocytes is limited because metabolic pathways that use glucose have not been stimulated: glucose diffuses into the cells, but as the intracellular glucose concentration increases, diffusion slows.

Glucose that is normally absorbed by the liver now remains in the blood, increasing plasma glucose concentrations. This condition of elevated plasma glucose is called **hyperglycemia.** Tissues that are not insulin-dependent, such as the brain, carry on glucose metabolism as usual. However, resting skeletal muscles and adipose tissue, normally the largest consumers of glucose, are unable to take up glucose from the blood.

Without insulin to stimulate transport and glucose metabolism, the cells go into fasted-state metabolism. Muscles break down their proteins and adipose tissue uses its fat stores to provide a source of energy. This catabolism leads to the symptom of "melting down of the flesh" described by Arataeus. Brain centers that control food intake do not sense the hyperglycemia, leading to excessive eating, or **polyphagia** (see p. 663).

Meanwhile, without insulin the liver is unable to recognize that plasma glucose concentrations are high. Consequently, it initiates glycogenolysis and gluconeogenesis to produce more glucose. When the liver dumps this glucose into the blood, the hyperglycemia worsens. In addition, because the liver cannot use glucose for its energy needs, it turns to β-oxidation of fatty acids.

These derangements of fat and glucose metabolism create some of the most serious symptoms of uncontrolled insulin deficiency. If the hyperglycemia of diabetes exceeds the renal threshold for glucose, glucose reabsorption in the proximal tubule of the kidney becomes saturated. This means that some filtered glucose is not reabsorbed, and it is excreted in the urine (*glucosuria*) [∞ p. 558].

Additional solute in the tubule lumen causes less water to be reabsorbed from the collecting duct because the glucose "holds" the water in the lumen with it. Reduced water reabsorption creates large volumes of urine (*polyuria*) and, if unchecked, will cause dehydration. The loss of water in the urine due to unreabsorbed solutes is known as **osmotic diuresis.**

Dehydration resulting from osmotic diuresis leads to decreased circulating blood volume. As blood pressure falls, homeostatic mechanisms for maintaining blood pressure are triggered [∞ p. 463]. These include secretion of ADH, thirst that causes constant drinking (*polydipsia*), and cardiovascular compensations.

If cardiovascular compensation fails, blood pressure falls to the point that perfusion of peripheral tissues becomes inadequate. When the tissues no longer receive enough oxygen to support aerobic glycolysis, metabolism shifts to anaerobic glycolysis, which creates lactic acid. The lactic acid leaves the cells and enters the blood, contributing to a state of metabolic acidosis.

The primary cause of *metabolic acidosis* in uncontrolled insulin-deficient diabetes is the production of acidic *ketone bodies* by the liver. Because the liver is limited in its ability to send fatty acids through the citric acid cycle, it produces ketones that it dumps into the circulation. The ketones can be absorbed by other tissues, such as muscle and brain, and converted back into acetyl CoA for use in the citric acid cycle.

Unfortunately, high ketone concentrations (**ketosis**) also cause metabolic acidosis because many ketones are strong acids. The condition of *diabetic ketoacidosis* causes patients to exhibit symptoms of metabolic acidosis: increased ventilation, acidification of the urine, and hyperkalemia [∞ p. 234]. If untreated, the combination of acidosis and hypoxia from circulatory collapse can cause coma and even death. The treatment for a patient in diabetic ketoacidosis is insulin replacement, accompanied by fluid and electrolyte therapy to replenish lost volume and ions.

Type 2 Diabetics Often Have Elevated Insulin Levels

In type 2 diabetes, the acute symptoms are not as nearly as severe because insulin is present, often in higher than normal amounts. The cells, although resistant to insulin's action, are able to carry out a certain amount of glucose metabolism. The liver, for example, does not have to turn to ketone production, so ketosis is rare in type 2 diabetes.

Nevertheless, metabolism overall is not normal, and patients with this condition develop a variety of diabetes-related problems because of abnormal glucose and fat metabolism. Complications of type 2 diabetes include atherosclerosis, neurological changes, and problems with the eyes and kidneys. Type 2 diabetes mellitus is estimated to affect 14 million Americans and is one of the major health care problems in this country.

Because many people with type 2 diabetes are asymptomatic when diagnosed, they can be very difficult to treat. People who come in for their yearly check-up feeling fine, only to be told that they have diabetes, can be very resistant to making dramatic lifestyle changes when they do not feel sick. Unfortunately, by the time diabetic symptoms appear, damage to tissues and organs is well under way. Patient compliance at that point can slow the progress of the disease but cannot re-

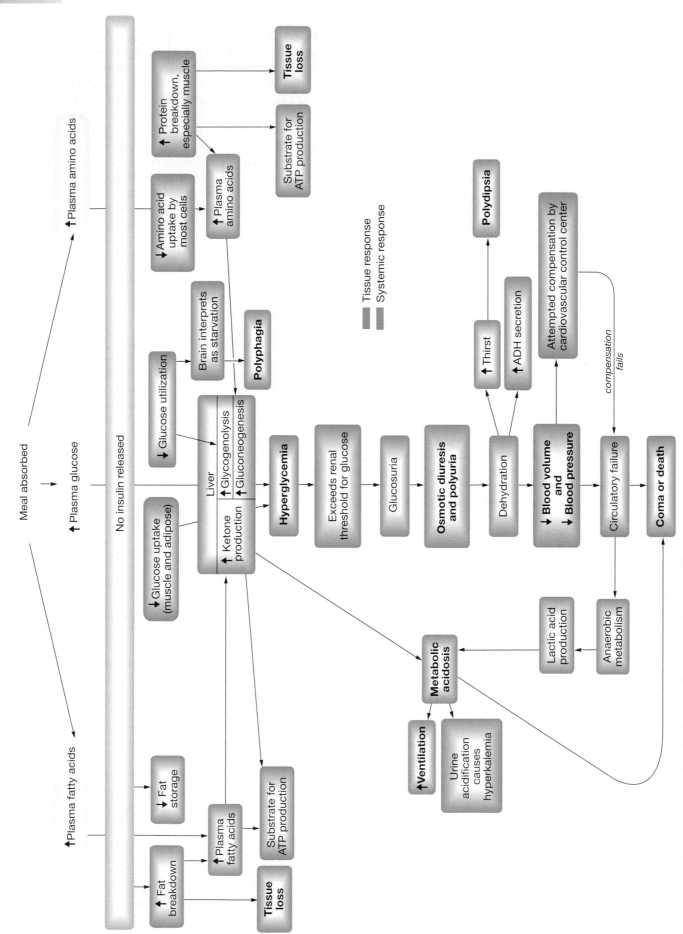

■ Figure 21-17 Acute pathophysiology of type 1 diabetes mellitus

DIABETES Diabetes and Amylin In 1987 researchers discovered that insulin is not the only hormone secreted by pancreatic beta cells. A second peptide named *amylin* is cosecreted with insulin. The functions of amylin are still being investigated, but at this time it appears that the hormone helps regulate glucose homeostasis following a meal. Amylin slows gastric emptying and gastric acid secretion, which delays digestion and absorption of carbohydrates. The combined actions of amylin and GIP thus set up a self-regulating cycle. Glucose in the intestine causes GIP release. GIP goes to the beta cells and initiates insulin and amylin secretion. Amylin then goes back to the GI tract to slow the rate at which food enters the intestine. In rodents, amylin also decreases food intake. Studies are underway with an amylin agonist called *pramlintide* to see if this drug can be used to treat diabetes or obesity.

...*continued from page 646*

High blood Ca^{2+} concentrations lead to high Ca^{2+} concentrations in the kidney filtrate. Calcium-based kidney stones occur when calcium phosphate or calcium oxalate crystals form and aggregate with organic material in the lumen of the kidney tubule. Once Dr. Bob's kidney stone passes into the urine, Dr. Spinks sends it for a chemical analysis.

Question 3: *Only the free Ca^{2+} in the blood filters into Bowman's capsule at the nephron. A significant portion of plasma Ca^{2+} cannot be filtered. Use what you have learned about filtration at the glomerulus to speculate on why some Ca^{2+} cannot filter [⬭ p. 549].*

uptake by muscle. Additional drugs to treat type 2 diabetes are being tested in clinical trials.

✔ Why must insulin be administered as a shot and not as an oral pill?

✔ Patients who are admitted to the hospital with acute diabetic ketoacidosis and dehydration are given insulin and fluids that contain K^+ and other ions. The patient's acidosis is usually accompanied by hyperkalemia, so why is K^+ included in the rehydration fluids? (Hint: Dehydrated patients may have a high *concentration* of K^+, but a low total amount. If fluid is given without K^+, what happens to the concentration of K^+ as the volume returns to normal?)

verse the pathological changes. The goal of treatment is correction of hyperglycemia to prevent complications.

The first therapy recommended for most type 2 diabetics is to lose weight and exercise. For some patients, losing weight will reverse their insulin resistance. Exercise decreases hyperglycemia because exercising skeletal muscle does not require insulin for glucose uptake.

Drug therapy for type 2 diabetes takes several forms. Some drugs act by stimulating insulin secretion. The *sulfonylureas* act by closing the K_{ATP} channel of beta cells. One infrequent but potentially dangerous side effect of these drugs is *hypoglycemia,* where plasma glucose falls so low that brain function is affected.

Some of the other oral drugs prevent hyperglycemia by acting on the liver and other cells. One class of drugs reduces hepatic glucose production. The drugs known as *α-glucosidase inhibitors* block intestinal enzymes that digest complex carbohydrates and therefore decrease glucose absorption. Another class of drugs promotes glucose

Glucagon Promotes Glucose Synthesis

Glucagon, secreted by the pancreatic alpha cells, is generally antagonistic to insulin in its effects on metabolism (Table 21-3). When plasma glucose concentrations decline after a meal, insulin secretion slows and the effects of glucagon on tissue metabolism take on greater significance. It appears that it is the ratio of insulin to

TABLE 21-3   Glucagon

Cell of origin	Alpha cells of pancreas
Chemical nature	29–amino acid peptide
Biosynthesis	Typical peptide
Transport in the circulation	Dissolved in plasma
Half-life	3–4 minutes
Factors affecting release	Stimulated by plasma [glucose] < 200 mg/dL, with maximum secretion below 50 mg/dL; ↑ blood amino acids.
Target cells or tissues	Liver primarily
Target receptor	Membrane receptor
Whole body or tissue action	↑ Plasma [glucose] by glycogenolysis and gluconeogenesis; ↑ lipolysis leads to ketogenesis in liver
Action at molecular level (including second messenger)	Acts via cAMP; alters existing enzymes and stimulates synthesis of new enzymes
Feedback regulation	↑ Plasma [glucose] shuts off glucagon secretion
Other information	Glucagon is structurally similar to the digestive hormones secretin, VIP, GIP, and GLP-1

glucagon that determines the direction of metabolism rather than an absolute amount of either hormone.

The primary stimulus for glucagon release is plasma glucose concentration. When plasma glucose concentrations fall below 100 mg/dL, glucagon secretion rises dramatically. At glucose concentrations above 100 mg/dL, when insulin is being secreted, glucagon secretion is inhibited and remains at a low but relatively constant level (see Fig. 21-12 ■). The strong relationship between insulin secretion and glucagon inhibition has led to speculation that the alpha cells are being regulated by some factor linked to insulin rather than by plasma glucose concentrations directly.

The liver is the primary target tissue of glucagon (Fig. 21-18 ■). Glucagon stimulates glycogenolysis, converting glycogen to glucose. In addition, glucagon stimulates gluconeogenesis. These pathways combine to increase glucose output by the liver. It is estimated that during an overnight fast, 75% of the glucose produced by the liver comes from glycogen stores and the remaining 25% from gluconeogenesis.

Glucagon secretion is also stimulated by an increase in plasma amino acid concentration. This signal helps prevent hypoglycemia after ingestion of a pure protein meal. Let's see how that might occur.

If a meal contains protein but no carbohydrate, amino acids in the food cause insulin secretion although no glucose has been absorbed. Cells increase their glucose uptake in response to insulin, and plasma glucose concentrations fall. Unless something counteracts this process, the brain's fuel supply is threatened by hypoglycemia.

Co-secretion of glucagon in this situation prevents hypoglycemia by stimulating hepatic glucose output. Thus, although only amino acids were ingested, both glucose and amino acids are made available to peripheral tissues.

Another instance in which glucagon is secreted is in type 2 diabetes. Although type 2 diabetics are hyperglycemic, they often have elevated glucagon levels. This seems contradictory until you realize that the pancreatic alpha cells, like muscle and adipose cells, require insulin for glucose uptake. Thus, in diabetes, the alpha cells do not take up glucose, which prompts them to secrete glucagon. Glucagon then contributes to the hyperglycemia of diabetes by promoting glycogenolysis and gluconeogenesis.

▶ NEURALLY MEDIATED ASPECTS OF METABOLISM

The nervous system regulates metabolism several ways. Autonomic neurons are closely involved with both digestive function and metabolism. During and following a meal, parasympathetic activity to beta cells is increased, stimulating insulin secretion.

In times of stress, sympathetic input to the endocrine pancreas is emphasized and reinforced by catecholamine release from the adrenal medulla (Table 21-4). Epinephrine and norepinephrine inhibit insulin secretion and switch metabolism to gluconeogenesis to provide extra fuel for the nervous system and skeletal muscles.

In addition, the nervous system links metabolism with external stimuli, such as environmental tempera-

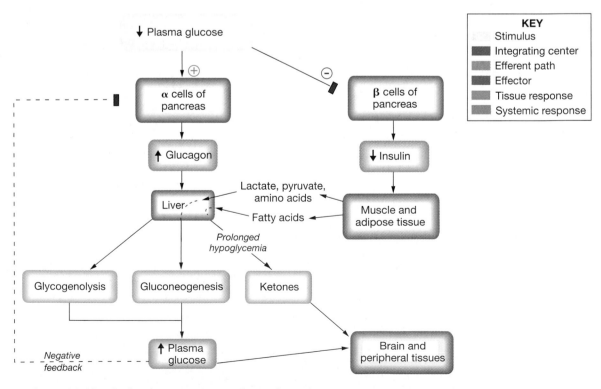

■ **Figure 21-18** **Endocrine response to hypoglycemia**

TABLE 21-4   Catecholamines (Epinephrine and Norepinephrine)

Origin	Adrenal medulla (90% epinephrine and 10% norepinephrine)
Chemical nature	Amines made from tyrosine
Biosynthesis	Typical peptide
Transport in the circulation	Some bound to sulfate
Half-life	2 minutes
Factors affecting release	Primarily fight-or-flight reaction through CNS and autonomic nervous system; hypoglycemia
Target cells or tissues	Mainly neurons, pancreatic endocrine cells, heart and blood vessels
Target receptor	α and β adrenergic membrane receptors
Whole body or tissue action	$\uparrow$ Plasma [glucose]; activate fight-or-flight and stress reactions; $\uparrow$ glucagon and $\downarrow$ insulin secretion
Action at molecular level (including second messenger)	α receptors work via $\uparrow$ intracellular Ca^{2+} levels; β receptors via cAMP
Onset and duration of action	Rapid and brief

ture, and internal stimuli, such as emotions or stress. Hypothalamic neurohormones regulate secretion of hormones such as cortisol, growth hormone, and thyroid hormones, which have extensive metabolic effects, as we will see later in this chapter. Sympathetic neurons are directly responsible for some aspects of temperature regulation, such as sweating and cutaneous vasodilation as discussed earlier in this chapter.

The Brain Controls Food Intake

One of the most interesting areas in which the central nervous system interacts with metabolism is the control of food intake. Digestion and metabolism are not selective processes. If food is eaten, it will be digested, absorbed, and assimilated, even if it is in excess and ends up stored as fat. So the act of eating is the main point at which the body exerts control of its energy input.

The hypothalamus contains two centers that regulate food intake: a **feeding center** that is tonically active and a **satiety center** that stops food intake by inhibiting the feeding center. If the feeding center is destroyed, animals will cease to eat. If the satiety center is destroyed, animals will overeat and become obese.

• •

...continued from page 661

Dr. Bob's blood work reveals that his Ca^{2+} level is 12.3 mg/dL plasma (normal 8.5–10.5 mg/dL). These results support the suspected diagnosis of hyperparathyroidism. "Do you take vitamin D or use a lot of antacids?" Dr. Spinks asks. "Those could raise your blood calcium." Dr. Bob denies the use of either substance. "Well, we need to do one more test before we can say conclusively that you have hyperparathyroidism," Dr. Spinks says.

Question 4: *What one test could definitively prove that Dr. Bob has hyperparathyroidism?*

• •

Efferent messages from the feeding and satiety centers cause changes in eating behavior and create sensations of hunger and fullness. But hunger and satiety are also affected by psychological, environmental, and social factors. As a result, we still do not fully understand what controls food intake.

The model that is emerging from studies using transgenic and knockout mice is very complex. It involves a number of brain neuropeptides and "brain-gut" peptides secreted from both those locations. The model also includes three hormones: insulin, the intestinal hormone cholecystokinin (CCK), and **leptin** [*leptos,* thin], a protein hormone synthesized in adipocytes.

The two classic theories for regulation of food intake are the glucostatic theory and the lipostatic theory. The **glucostatic theory** states that glucose utilization by hypothalamic centers regulates food intake. When blood glucose concentrations decrease, the satiety center is suppressed and the feeding center is dominant. When glucose utilization is up, the satiety center inhibits the feeding center.

This theory explains the *polyphagia* (excessive eating) associated with untreated type 1 diabetes mellitus. Although most neurons in the brain are not affected by insulin, glucose utilization in the satiety center is insulin sensitive. Therefore, in the absence of insulin, the satiety center, like most other cells of the body, is unable to use plasma glucose. It perceives the absence of intracellular glucose as starvation and allows the feeding center to increase food intake.

The **lipostatic theory** of energy balance proposes that a signal from the body's fat stores to the brain modulates eating behavior so that the body maintains a particular weight. If fat stores are increased, eating decreases. In times of starvation, eating increases. Obesity results from disruption of this pathway.

The discovery of leptin provided evidence for the signal between adipose tissue and the brain. Leptin is synthesized in adipocytes under control of the *obese (ob)*

TABLE 21-5 Some Peptides That Modulate Food Intake

Name	Source	Effect on Food Intake
CCK	GI hormone	Decrease
Galanin	Hypothalamus	Increase
Melanin-concentrating hormone (MCH)	Hypothalamus	Increase
Corticotropin-releasing hormone (CRH)	Hypothalamus	Decrease
α-Melanocyte-stimulating hormone (α-MSH)	Hypothalamus	Decrease
CART (cocaine- and amphetamine-regulated transcript)	Hypothalamus	Decrease

gene. Mice who lack the *ob* gene and leptin protein become obese, as do mice with defective leptin receptors. But that is only part of the story.

The newest chemical in the energy balance picture is **neuropeptide Y (NPY)**, a brain neurotransmitter that seems to be the actual stimulus for food intake. In normal-weight animals, leptin inhibits NPY in a negative feedback pathway (Fig. 21-19 ■). However, the complete pathway is much more complex. Other neuropeptides and hormones (Table 21-5) also influence NPY and the

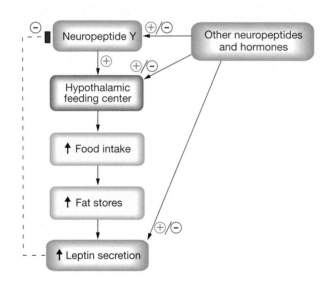

■ **Figure 21-19 Roles of leptin and neuropeptide Y in food intake**

hypothalamic feeding and satiety centers, creating a cascade that researchers are still sorting out.

Appetite and eating are influenced by sensory input through the nervous system in addition to chemical signals. The simple acts of swallowing and chewing food help create a sensation of fullness. The sight, smell, and taste of food can either stimulate or suppress appetite.

One interesting study tried to determine if chocolate craving could be attributed to psychological factors or physiological stimuli. Subjects were given either dark chocolate, white chocolate with none of the pharmacological agents of cocoa, cocoa capsules, or placebo capsules. The results showed that white chocolate was the best substitute for the real thing and that aroma plays a significant role in satisfying chocolate craving.

Psychological factors, such as eating or not eating when stressed, are known to play a significant role in regulating food intake. The eating disorder **anorexia nervosa** probably has both a psychological component and a physiological one. The concept of **appetite** is closely linked to the psychology of eating, and it may explain why dieters who crave smooth, cold ice cream cannot be satisfied by a crunchy carrot stick. A considerable amount of money is directed to research on eating behaviors.

✔ Studies show that most obese humans have elevated leptin levels in their blood. Develop some theories that could explain why leptin in these people is not decreasing food intake.

LONG-TERM ENDOCRINE CONTROL OF METABOLISM

You have learned how insulin and glucagon, with their short half-lives, are responsible for the minute-to-minute control of plasma glucose concentration. In the remainder of the chapter we examine three other hormones that play a role in long-term regulation of metabolism. They are cortisol from the adrenal cortex, thyroid hormones from the thyroid gland, and growth hormone from the anterior pituitary.

In a normal person, the hormones that influence metabolism are difficult to study because their effects are subtle and their interactions with one another are complex. As a result, much of what we know about these

hormones comes from studying pathological conditions in which a hormone is either over- or undersecreted [∞ p. 203]. Yet, despite all the research devoted to this topic, we still do not completely understand how these hormones work on the cellular and molecular levels.

▶ ADRENAL GLUCOCORTICOIDS

The adrenal glands were introduced in connection with the autonomic division of the nervous system [Chapter 11, ∞ p. 332] and the hormone aldosterone [Chapter 19, ∞ p. 579]. The adrenal medulla secretes catecholamines, mostly epinephrine, to mediate rapid metabolic responses in "fight-or-flight" situations. The adrenal hormone cortisol modulates stress in a slower fashion.

The Adrenal Cortex Secretes Steroid Hormones

The adrenal cortex forms the outer three-quarters of the adrenal gland (Fig. 21-20 ■). The outer section of the cortex (*zona glomerulosa*) secretes aldosterone. The inner section (*zona reticularis*) secretes mostly androgens, the sex hormones that are dominant in men. The middle section (*zona fasciculata*) secretes mostly glucocorticoids, named for their ability to increase plasma glucose concentrations. Cortisol is the main glucocorticoid secreted by the adrenal cortex (Table 21-6).

Cortisol is a typical steroid hormone synthesized from cholesterol [∞ Fig. 7-6, p. 192]. Its control pathway follows the 3-part hypothalamic-pituitary-peripheral gland pattern to which you were introduced in Chapter 7 [∞ p. 204]. Cortisol release is regulated by **adrenocorticotropic hormone (ACTH)** from the anterior pituitary (Fig. 21-21 ■). ACTH secretion in turn is controlled by the brain, which acts through hypothalamic **corticotropin-releasing hormone (CRH)**. Cortisol acts as a negative feedback signal, inhibiting ACTH and CRH secretion.

Cortisol Is Essential for Life

Cortisol is continuously secreted with a strong diurnal rhythm (Fig. 21-22 ■). Secretion normally peaks in the morning and diminishes during the night. Cortisol secretion also increases with stress.

Cortisol is essential for life. Animals whose adrenal glands have been removed will die if exposed to any significant environmental stress. Overall, the net effects of cortisol are catabolic.

1. **Cortisol promotes gluconeogenesis** in the liver. The glucose produced is both released into the blood and stored as glycogen.
2. **Cortisol causes the breakdown of skeletal muscle proteins** to provide substrate for gluconeogenesis.
3. **Cortisol enhances fat breakdown** (lipolysis) so that fatty acids are available to peripheral tissues for energy use.
4. **Cortisol suppresses the immune system** through multiple pathways.
5. **In large doses, cortisol is catabolic on bone tissue,** causing breakdown of the calcified bone matrix. People who take therapeutic cortisol for extended periods therefore have a higher than normal incidence of broken bones.

The most important metabolic effect of cortisol is its protective effect against hypoglycemia. If plasma glucose falls below a certain concentration, the adrenal cortex begins to secrete cortisol even without stimulation from ACTH. In the absence of cortisol, glucagon alone is unable to respond adequately to a hypoglycemic chal-

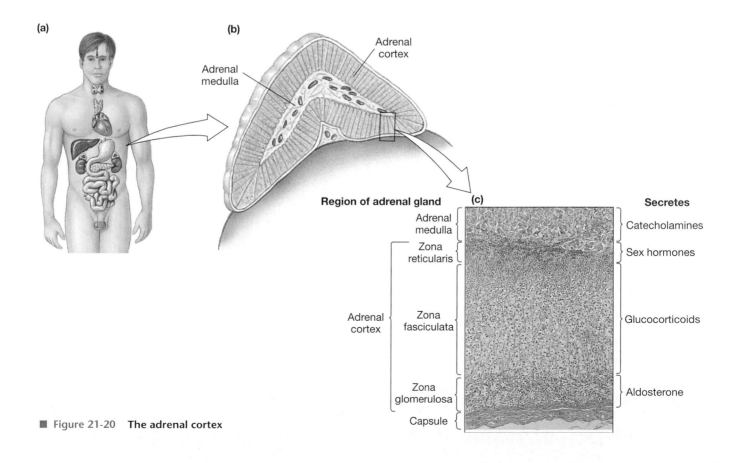

■ Figure 21-20 **The adrenal cortex**

TABLE 21-6   Cortisol

Origin	Adrenal cortex
Chemical nature	Steroid
Biosynthesis	From cholesterol; made on demand, not stored
Transport in the circulation	On corticosteroid-binding globulin (made in liver)
Half-life	60–90 minutes
Factors affecting release	Circadian rhythm of tonic secretion; stress enhances
Control axis	Corticotropin-releasing hormone (CRH) (hypothalamus) → adrenocorticotropic hormone (ACTH) (anterior pituitary) → cortisol (adrenal cortex)
Target cells or tissues	Most tissues except brain and heart
Target receptor	Intracellular
Whole body or tissue action	↑ Plasma [glucose]; ↓ immunocyte activity; permissive for glucagon and catecholamines
Action at cellular level	↑ Gluconeogenesis and glycogenolysis; ↑ protein catabolism; blocks cytokine production by immunocytes
Action at molecular level	Initiates transcription, translation, and new protein synthesis
Feedback regulation	Negative feedback to anterior pituitary and hypothalamus

lenge. Cortisol must be present to ensure that glucagon and catecholamines are effective. Because cortisol is required for full activity of those hormones, it is said to have a *permissive effect* on them [∞ p. 203].

✔ The illegal use of anabolic steroids by bodybuilders and athletes periodically receives much attention. Do these illegal steroids include cortisol? Explain.

Cortisol Is a Useful Therapeutic Drug

The immunosuppressant effects of cortisol make it a useful drug for treating a variety of disorders. Cortisol and its synthetic agonists are used to suppress allergic reactions, such as bee stings, poison ivy, and pollen allergies. They also suppress inflammation and prevent rejection of transplanted organs. However, glucocorticoids also have potentially serious side effects because of their

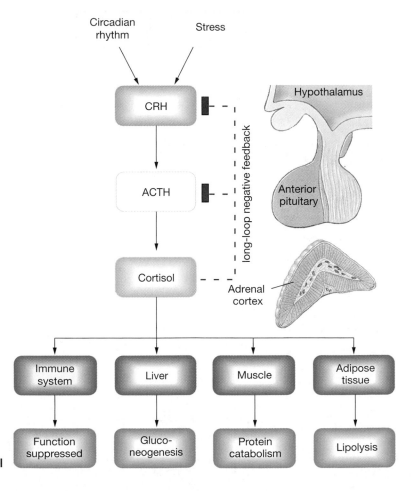

■ Figure 21-21   **The control pathway for cortisol**

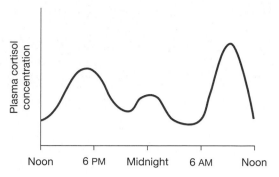

Plasma cortisol concentration

Noon   6 PM   Midnight   6 AM   Noon

■ **Figure 21-22   Circadian rhythm of cortisol secretion**

metabolic actions. Once nonsteroidal anti-inflammatory drugs (NSAIDs) such as ibuprofen were developed, the use of glucocorticoids for treating minor inflammatory problems was discontinued.

Cortisol therapy with high doses for more than a week has the potential to cause hypercortisolism, also known as *Cushing's syndrome,* after Dr. Harvey Cushing who first described the condition in 1932. Most signs of hypercortisolism can be predicted from the normal actions of the hormone.

Excess gluconeogenesis causes hyperglycemia, mimicking diabetes. Muscle protein breakdown and lipolysis cause tissue wasting. Paradoxically, excess cortisol deposits extra fat in the trunk and face. The classic appearance of patients with hypercortisolism is thin arms and legs, obesity in the trunk, and a "moon face," with plump cheeks (Fig. 21-23 ■).

Cushing's syndrome has three common causes:

1. An adrenal tumor autonomously secretes cortisol. The tumor is not under control of pituitary ACTH. This condition is *primary hypersecretion* [∞ p. 205].
2. A pituitary tumor autonomously secretes ACTH, which prompts the adrenal gland to oversecrete cortisol (*secondary hypercortisolism*). The tumor is not subject to the normal negative feedback pathway shown in Figure 21-21 ■. This condition is also called *Cushing's disease* because it was the actual disease

described by Dr. Cushing. (Hypercortisolism from any cause is called Cushing's *syndrome.*)

3. *Iatrogenic* (physician-caused) hypercortisolism occurs secondary to cortisol therapy for some other condition.

Hyposecretion pathologies are far less common than Cushing's syndrome. *Addison's disease* is hyposecretion of all adrenal steroid hormones, usually following autoimmune destruction of the adrenal cortex. Hereditary defects in enzymes for adrenal steroid production cause several related syndromes. These are often marked by excess androgen secretion because substrate that cannot be made into cortisol or aldosterone is converted to androgens. In newborn girls, excess androgens cause masculinization of the external genitalia, or *adrenogenital syndrome.*

✔ For each of the three causes of hypercortisolism, tell whether the ACTH level will be normal, greater than normal, or less than normal.

✔ Would someone with Addison's disease have normal, low, or high levels of ACTH in the blood?

▶ **THYROID HORMONES**

The thyroid gland is a butterfly-shaped gland that lies across the trachea at the base of the throat, just below the larynx (Fig. 21-24 ■). It is one of the larger endocrine glands, weighing between 15 and 20 g. The thyroid gland has two distinct endocrine cell types: the C ("clear") cells, which secrete a calcium-regulating hormone called calcitonin, and follicle cells, which secrete thyroid hormones. C cells are also known as *parafollicular cells.*

Thyroid Hormones Contain Iodine

The thyroid **follicles** (also called *acini*) are spherical structures whose walls are a single layer of epithelial cells. **Colloid,** consisting of the glycoprotein **thyroglobulin**

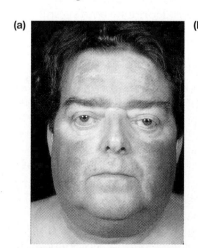

(a)

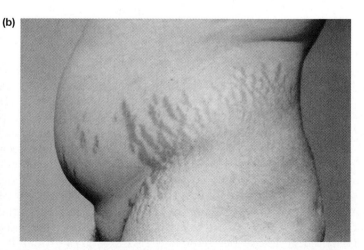

(b)

■ **Figure 21-23   Cushing's syndrome**

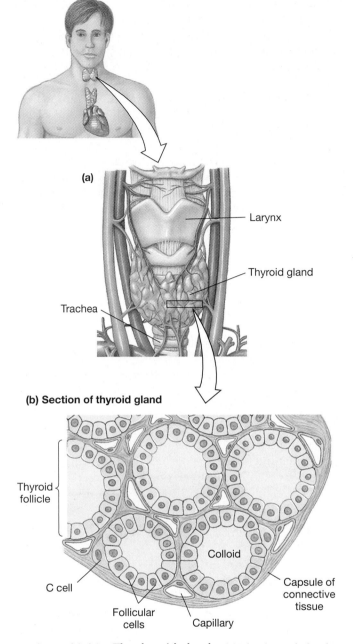

(a)

Larynx

Thyroid gland

Trachea

(b) Section of thyroid gland

Thyroid follicle

C cell

Colloid

Capsule of connective tissue

Follicular cells

Capillary

■ **Figure 21-24   The thyroid gland**   (a) The thyroid gland is a butterfly-shaped gland, located just below the larynx. (b) The gland consists of C cells, which secrete calcitonin, and follicles filled with glycoprotein colloid.

Tyrosine

Thyroxine (T$_4$)

(2 tyrosine + 4 I)

Tyrosine

Triiodothyronine (T$_3$)

(2 tyrosine + 3 I)

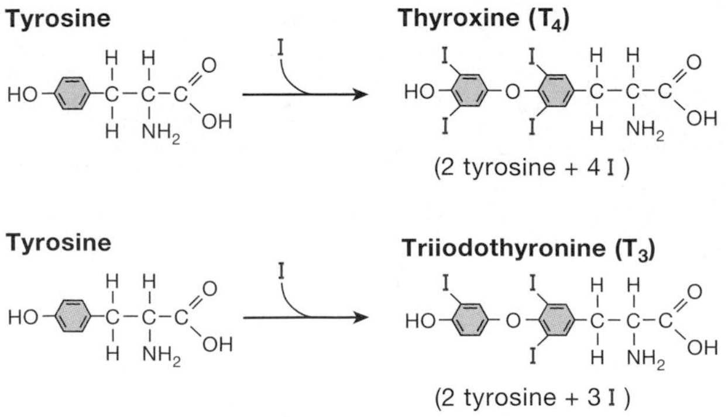

■ **Figure 21-25   Thyroid hormones**

concentrate iodide and transport it into the colloid. There, enzymes add iodide to tyrosine residues of thyroglobulin. The final products are **triiodothyronine,** called **T$_3$,** and **tetraiodothyronine,** also known as **thyroxine,** or **T$_4$.** At this point, the hormones are still attached to thyroglobulin.

Now thyroglobulin in the colloid moves back into the follicular cells by endocytosis. Enzymes then free T$_3$ and T$_4$ from the protein. Both hormones are lipophilic, so they diffuse out of the follicular cells and into the plasma. The colloid of the follicles holds a 2–3 month supply of thyroid hormones at any one time.

Because T$_3$ and T$_4$ are lipophilic molecules, they have limited solubility in plasma. As a result, thyroid hormones bind to plasma proteins such as **thyroxine-binding globulin (TBG).**

Most plasma hormone is in the form of T$_4$. For years, it was thought that T$_4$ (thyroxine) was the active hormone. However, we now know that T$_3$ is three to five times more active biologically, and it is the active hormone in target cells.

About 85% of active T$_3$ is made in target cells by enzymatic removal of an iodine from T$_4$. Tissue activation of the hormone adds another layer of control because regulation of the tissue enzyme **deiodinase** allows individual target tissues to control their exposure to active thyroid hormone. Thyroid receptors, with multiple isoforms, are in the nucleus of target cells. Hormone binding initiates transcription, translation, and synthesis of new proteins.

Thyroid Hormones Affect Quality of Life

The thyroid hormones are not essential for life, but they do affect the quality of life. Thyroid hormones are thermogenic and increase oxygen consumption in most tissues. The exact mechanism is unclear, but is at least partly related to changes in ion transport across the cell and mitochondrial membranes. In addition, thyroid hormones interact with other hormones to modulate protein, carbohydrate, and fat metabolism. The overall effect in adults is catabolic, which provides substrates for oxidative metabolism.

In children, thyroid hormones are necessary for full expression of growth hormone and for normal growth

and enzymes for thyroid hormone synthesis, fills the hollow center of each follicle. Colloid is manufactured in follicular cells.

Thyroid hormones are amines derived from the amino acid tyrosine (Fig. 21-25 ■). They are unusual because they contain iodine (Table 21-7). Currently, thyroid hormones are the only known use for iodine in the body, although a few other tissues are known to concentrate the mineral.

Figure 21-26 ■ summarizes thyroid hormone synthesis. The digestive tract absorbs iodine from the diet in its ionic form, **iodide, I$^-$.** The thyroid follicles actively

TABLE 21-7   Thyroid Hormones

Cell of origin	Thyroid follicle cells
Chemical nature	Iodinated amine
Biosynthesis	From iodine and tyrosine; formed and stored on parent protein thyroglobulin in colloid of follicle
Transport in the circulation	Bound to thyroxine-binding globulin and albumins
Half-life	6–7 days for thyroxine (T_4); about 1 day for triiodothyronine (T_3)
Factors affecting release	Tonic release
Control axis	Thyrotropin-releasing hormone (TRH) (hypothalamus) $\rightarrow$ thyroid-stimulating hormone (TSH) (anterior pituitary) $\rightarrow$ T_3 and T_4 (thyroid) $\rightarrow$ T_4 deiodinates in tissues to form more T_3
Target cells or tissues	Most cells of the body
Target receptor	Nuclear receptor
Whole body or tissue action	$\uparrow$ Oxygen consumption (thermogenesis); protein catabolism in adults but anabolism in children; normal development of nervous system
Action at cellular level	Increases activity of metabolic enzymes and Na^+-K^+-ATPase
Action at molecular level	Production of new enzymes
Feedback regulation	T_3 has negative feedback effect on anterior pituitary and hypothalamus

and development, especially in the nervous system. In the first few years after birth, T_3 and T_4 are needed for proper myelination of neurons and for synapse formation. Cytological studies suggest that thyroid hormones regulate microtubule assembly, which is an essential part of neuronal growth. Because of the importance of thyroid hormones in children, the United States and Canada test all babies at birth for thyroid deficiency.

The actions of thyroid hormones are most memorable in people who suffer from either hypersecretion or hyposecretion. Physiological effects that are subtle in normal people often become exaggerated in patients with endocrine pathologies.

Hyperthyroidism   Hypersecretion of thyroid hormones causes changes in metabolism, the nervous system, and the heart.

1. Hyperthyroidism increases oxygen consumption and metabolic heat production. Because of the internal heat generated, these patients have warm, sweaty skin and may complain of being intolerant of heat.
2. Excess thyroid hormone increases protein catabolism and may cause muscle weakness. Patients often report weight loss.
3. The effects of excess thyroid hormone on the nervous system include hyperexcitable reflexes and psychological disturbances ranging from irritability and insomnia to psychosis. The mechanism for psychological disturbances is unclear, but morphological changes in the hippocampus and effects on β-adrenergic receptors have been suggested.
4. Thyroid hormones are known to influence β-adrenergic receptors in the heart, and these effects are exaggerated with hypersecretion. A common sign of hyperthyroidism is rapid heartbeat and increased force of contraction due to up-regulation of $β_1$ receptors on the myocardium [∞ p. 437].

■ **Figure 21-26   Thyroid hormone synthesis**   Follicle cells synthesize thyroglobulin and actively take up iodide from the blood. In the colloid, iodide is added to tyrosines producing T_3 and T_4. The follicle cells then take up thyroglobulin by endocytosis. Intracellular enzymes separate the thyroid hormones from the protein so that they can diffuse into the blood.

Hypothyroidism   Hyposecretion of thyroid hormones affects the same systems altered by hyperthyroidism.

1. Decreased thyroid hormone secretion slows the metabolic rate and oxygen consumption. Patients become cold-intolerant because they are generating less internal heat.

2. Hyposecretion of thyroid hormone decreases protein synthesis. In adults, this causes brittle nails, thinning hair, and dry, thin skin. Hypothyroidism also causes accumulation of *mucopolysaccharides* under the skin. These molecules attract water and cause the puffy appearance of *myxedema* (Fig. 21-27 ■). Hypothyroid children have slow bone and tissue growth and are shorter than normal for their age.

3. Nervous system changes in adults include slowed reflexes, slow speech and thought processes, and feelings of fatigue. Deficient thyroid hormone secretion in infancy causes **cretinism,** a condition marked by decreased mental capacity.

4. The primary cardiovascular change in hypothyroidism is slow heart rate, or *bradycardia.*

TSH Controls the Thyroid Gland

Problems with thyroid hormone secretion can arise in the thyroid gland itself or along the control pathway. The normal control path and feedback for thyroid hormones are illustrated in Figure 21-28 ■. Secretion by the thyroid gland is under the control of the anterior pituitary hormone **thyrotropin,** also known as **thyroid-stimulating hormone,** or **TSH.** TSH secretion in turn is influenced by **thyrotropin-releasing hormone,** or **TRH,** from the hypothalamus. The thyroid hormones normally act as a negative feedback signal to prevent oversecretion.

The trophic action of TSH on the thyroid gland causes enlargement, or **hypertrophy,** of the follicular cells. In pathological conditions with elevated TSH levels, the thyroid gland can enlarge to weights of hundreds of grams, forming a huge mass called a **goiter.** A large goiter can almost encircle the neck and will extend downward behind the sternum (Fig. 21-29 ■).

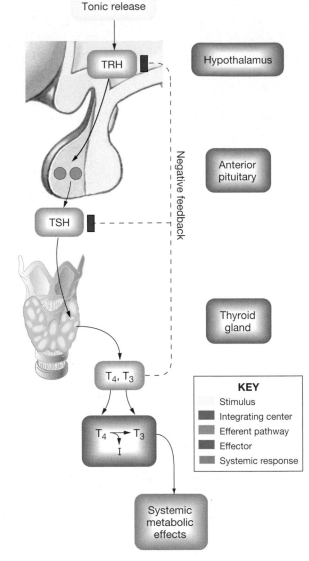

■ **Figure 21-28   Thyroid hormone pathway**

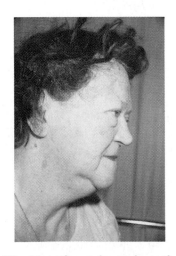

■ Figure 21-27   **Myxedema due to hypothyroidism**

■ Figure 21-29   **Goiter due to excessive TSH stimulation**

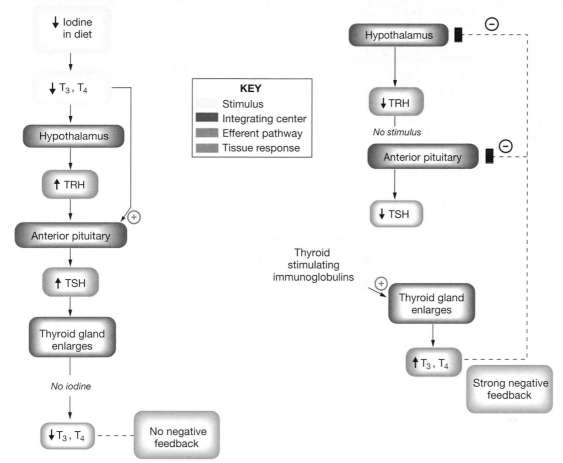

(a) Hypothyroidism due to low iodine

↓ Iodine in diet

↓ T₃ , T₄

Hypothalamus

↑ TRH

Anterior pituitary (+)

↑ TSH

Thyroid gland enlarges

No iodine

↓ T₃ , T₄ ---- No negative feedback

KEY
Stimulus
Integrating center
Efferent pathway
Tissue response

(b) Hyperthyroidism due to Graves' disease

Hypothalamus ⊖

↓ TRH

No stimulus

Anterior pituitary ⊖

↓ TSH

Thyroid stimulating immunoglobulins

(+) Thyroid gland enlarges

↑ T₃ , T₄

Strong negative feedback

■ **Figure 21-30   Two causes of goiter**

Goiters are the result of excess TSH stimulation of the thyroid gland, but knowing that someone has a goiter does not tell you what their pathology is. Let's see how both hypothyroidism and hyperthyroidism can be associated with goiter.

For our first example, let's use primary hypothyroidism [∞ p. 205] caused by a lack of iodine in the diet. Without iodide, the thyroid gland cannot make thyroid hormones (Fig. 21-30a ■). Low levels of T₃ and T₄ in the blood remove negative feedback on the hypothalamus and anterior pituitary. TSH secretion rises dramatically, and TSH stimulation enlarges the thyroid gland (goiter). But despite hypertrophy, the gland cannot obtain iodine to make hormone, so the patient remains hypothyroid. These patients will exhibit the signs of hypothyroidism previously described.

An enlarged thyroid gland may also be indicative of hypersecretion. In Graves' disease, the body produces antibodies called **thyroid-stimulating immunoglobulins,** or **TSI** (Fig. 21-30b ■). These antibodies combine with TSH receptors on the thyroid gland and activate them, enlarging the gland (goiter) and stimulating thyroid hormone production.

In Graves' disease, negative feedback shuts down the body's TRH and TSH secretion but does nothing to block the TSH-like activity of TSI on the thyroid gland. The result is goiter, hypersecretion of thyroid hormone, and symptoms of hormone excess. Graves' disease is often accompanied by **exophthalmus** (Fig. 21-31 ■), a

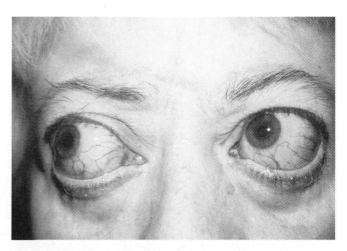

■ **Figure 21-31   Exophthalmus**

bug-eyed appearance caused by immune-mediated enlargement of muscles and tissue in the eye socket.

Therapy for thyroid disorders depends on the cause of the problem. Hypothyroidism is treated with oral thyroxine (T_4). Hyperthyroidism can be treated by surgical removal of all or part of the gland, by destruction of thyroid cells with radioactive iodine (^{131}I), or by drugs that block hormone synthesis (thiourea drugs) or peripheral conversion of T_4 to T_3 (propylthiouracil).

✔ A patient who had her thyroid gland removed due to cancer was given pills containing only T_4. The more active form of the hormone is T_3. Why did she not show signs of being hypothyroid?

✔ Why would excessive production of thyroid hormone, which uncouples mitochondrial ATP production and proton transport, cause a person to become intolerant of heat?

ENDOCRINE CONTROL OF GROWTH

Growth in human beings is a continuous process that begins before birth. Growth rates in children are not steady. The first two years of life and the adolescent years are marked by spurts of rapid growth and development. Normal growth is a complex process that depends on a number of factors, including:

1. **Growth hormone and other hormones.** Without adequate amounts of growth hormone, children simply will not grow. Thyroid hormones, insulin, and the sex hormones at puberty also play both direct and permissive roles. A deficiency in any one of these hormones will lead to abnormal growth and development.
2. **An adequate diet** that includes protein, sufficient energy (caloric intake), vitamins, and minerals. Many amino acids can be manufactured in the body from other precursors, but the essential amino acids [∞ p. 31] needed for protein synthesis must come from dietary sources. Among the minerals, calcium in particular is needed for proper bone formation.
3. **Absence of stress.** Cortisol from the adrenal cortex is released in times of stress and has significant catabolic, antigrowth effects. Children who are subjected to stressful environments may exhibit a condition known as *failure to thrive* that is marked by abnormally slow growth.

4. **Genetics.** Each human's potential adult size is genetically determined at conception.

▶ GROWTH HORMONE

Growth hormone, also known as **somatotropin,** is secreted by the anterior pituitary. It is unique among the anterior pituitary hormones because it acts directly on some target tissues but is also a trophic hormone (Table 21-8). The trophic action of growth hormone stimulates secretion of **insulin-like growth factors** (IGFs; formerly called *somatomedins*) from the liver and other tissues (Fig. 21-32 ■). IGFs act with growth hormone to stimulate bone and soft tissue growth.

Growth Hormone Is Anabolic

The stimuli for growth hormone release are complex and not well understood, but they include circulating nutrients, stress, and other hormones interacting with a daily rhythm of secretion. These stimuli are integrated in the hypothalamus, which releases two trophic factors: **growth hormone–releasing hormone (GHRH)** and *growth hormone–inhibiting hormone,* better known as **somatostatin.** These factors together regulate growth hormone secretion.

TABLE 21-8   Growth Hormone

Cell of origin	Anterior pituitary
Chemical nature	191–amino acid peptide; several closely related forms
Biosynthesis	Typical peptide
Transport in the circulation	Half is dissolved in plasma, half is bound to a binding protein whose structure is identical to that of the GH receptor
Half-life	18 minutes
Factors affecting release	Circadian rhythm of tonic secretion; influenced by circulating nutrients, stress, and other hormones in a complex fashion
Target cells or tissues	Trophic on liver for insulin-like growth factor production; also acts directly on many cells
Target receptor	Membrane receptor with tyrosine kinase activity
Whole body or tissue action (with IGFs)	Bone and cartilage growth; soft tissue growth; ↑ plasma [glucose]
Action at cellular level	Protein synthesis
Action at molecular level	Receptor linked to kinases that phosphorylate proteins to initiate transcription

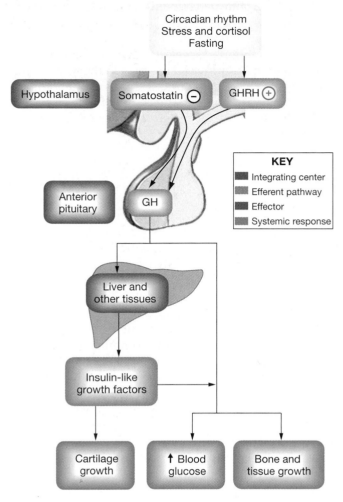

■ Figure 21-32   **Growth hormone pathway**

Growth hormone (GH) is released throughout life, although its biggest role is in children. Peak GH secretion occurs during the teenage years. On a daily basis, GH is released in response to pulses of GHRH from the hypothalamus. In adults, the largest pulse of GH release occurs in the first two hours of sleep. It is speculated that GHRH has sleep-inducing properties, but the role of GH itself in sleep cycles is unclear.

Nearly half the GH in blood is bound to a plasma **growth hormone binding protein.** The binding protein protects plasma GH from being filtered into the urine. This extends its half-life from 7 to 18 minutes. In addition, the bound portion acts as a growth hormone reservoir in the blood. Researchers have hypothesized that genetic determination of binding protein concentration plays a role in determining adult height.

Metabolically, growth hormone and IGFs are anabolic for proteins, directing energy and amino acids into protein synthesis, an essential part of tissue growth. Growth hormone also acts with the insulin-like growth factors to stimulate bone. IGFs are responsible for cartilage growth. GH increases fat breakdown and hepatic glucose output, raising plasma fatty acid and glucose concentrations.

...continued from page 663

The results of Dr. Bob's last test confirm that he has hyperparathyroidism. He goes on a low-calcium diet, avoiding milk, cheese, and other dairy products, but several months later, he returns to the emergency room with another painful calcium phosphate kidney stone. Dr. Spinks sends him to an endocrine specialist, who recommends surgical removal of the overactive parathyroid glands. "We can't tell which of the parathyroid glands is most active," the specialist says, "and we'd like to leave you with some parathyroid hormone of your own. So I will take out all four glands, but we'll reimplant two of them in the muscle of your forearm. In many patients, the implanted glands secrete just enough PTH to maintain calcium homeostasis. And if they secrete too much PTH, it is much easier to take them out of your arm than do major surgery on your neck again."

Question 5:   *Why can't Dr. Bob simply take replacement PTH by mouth? (Hint: PTH is a peptide hormone.)*

Growth Hormone Is Essential for Normal Growth in Children

The pathologies that reflect the actions of growth hormone are most obvious in children. Severe growth hormone deficiency in childhood leads to **dwarfism.** Dwarfism can result from a problem with growth hormone synthesis or defective tissue hormone receptors. Unfortunately, bovine or porcine growth hormone is not effective as replacement therapy, as only primate growth hormone is active in humans. Prior to 1985, when genetically engineered human growth hormone became available, donated human pituitaries harvested at autopsy were the only source for growth hormone. Fortunately, severe growth hormone deficiency is a relatively rare condition.

Oversecretion of growth hormone in children leads to **giantism.** Once bone growth stops in late adolescence, growth hormone cannot further increase height, but GH and IGFs do act on cartilage and soft tissues. Adults with excessive secretion of growth hormone develop a condition known as **acromegaly** characterized by coarsening of facial features and growth of hands and feet (Fig. 21-33 ■).

✔ Why don't adults with hypersecretion of growth hormone grow taller?

Genetically Engineered Human Growth Hormone Created Ethical Dilemmas

When genetically engineered human growth hormone became available in the mid-1980s, the medical profession was faced with a dilemma. Obviously the hormone should be used to treat children who would otherwise be dwarfs. But what about those with only partial GH deficiency or genetically short children with normal GH secretion? What about children whose parents want

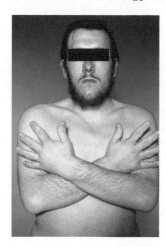

■ **Figure 21-33   Acromegaly**

them to be taller for athletics? These questions are complicated by the difficulty of accurately determining exactly which children have partial growth hormone deficiency.

In 1997 the American Academy of Pediatrics issued a policy statement with recommendations and cautions about the use of human growth hormone in short children. This statement points out that long-term risks associated with GH treatment are unknown and that GH therapy itself has the potential to create psychological problems in children if the results are less than optimum.

◗ TISSUE AND BONE GROWTH

Growth can be divided into two general areas: soft tissue growth and bone growth. In children, bone growth is usually assessed by measuring height, and tissue growth by measuring weight. Multiple hormones have direct or permissive effects on growth. In addition, we are just beginning to understand how paracrine growth factors interact with classic hormones to influence the development and differentiation of tissues.

Tissue Growth Requires Hormones and Paracrines

Soft tissue growth requires adequate amounts of growth hormone, thyroid hormone, and insulin. Growth hormone and insulin-like growth factors are required for tissue protein synthesis and cell division. Under the influence of these hormones, cells undergo both **hypertrophy** (increased cell size) and **hyperplasia** (increased cell number).

Thyroid hormones play a permissive role in growth and contribute directly to the development of the nervous system. At the target tissue level, thyroid hormone interacts synergistically with growth hormone for protein synthesis and development of the nervous system.

Children with untreated hypothyroidism (cretinism) will not grow to a normal height despite having normal secretion of growth hormone.

Insulin supports tissue growth by stimulating protein synthesis and providing energy in the form of glucose. Because insulin is permissive for growth hormone, children who are insulin-deficient will fail to grow normally although they may have normal growth and thyroid hormone concentrations.

Bone Growth Requires Adequate Amounts of Calcium in the Diet

Bone growth, like soft tissue development, requires the proper hormones and adequate amounts of protein and calcium. Bone has calcified extracellular matrix formed when calcium phosphate crystals precipitate and attach to a collagenous lattice support. The most common form of calcium phosphate is **hydroxyapatite**, $Ca_{10}(PO_4)_6(OH)_2$.

Although the large amount of inorganic matrix in bone makes some people think of it as nonliving, bone is a dynamic tissue, constantly being formed and broken down, or **resorbed.** Spaces in the collagen and calcium matrix are occupied by living cells. They are well supplied with oxygen and nutrients by blood vessels that run through adjacent channels. Bone generally forms one of two patterns: dense, or **compact,** bone and spongy, or **trabecular,** bone, which has many open spaces between struts of calcified lattice (Fig. 21-34 ■).

Bone growth occurs when matrix is deposited faster than it is resorbed. Special bone-forming cells called **osteoblasts** produce collagen, enzymes, and other proteins that make up the organic portion of bone matrix. Bone diameter increases when matrix deposition occurs on the outer surface of the bone.

Linear growth of long bones occurs at special regions called the **epiphyseal plates,** found between the ends (*epiphyses*) and shaft (*diaphysis*) of the bone (Fig. 21-35 ■). On the epiphyseal side of the plate are continuously dividing columns of **chondrocytes,** cartilage-producing cells.

As the collagen layer thickens, the older collagen calcifies and older chondrocytes degenerate, leaving spaces that osteoblasts invade. The osteoblasts then lay down bone matrix on top of the cartilage base. As new bone is added at the ends, the shaft lengthens. Long bone growth continues as long as the epiphyseal plate is active. When osteoblasts complete their work, they revert to a less active form known as **osteocytes.**

Growth of long bone is under the influence of growth hormone and the insulin-like growth factors. In the absence of these hormones, normal bone growth will not occur. Long bone growth is also influenced by sex steroid hormones. The growth spurt of adolescent boys is attributed to increased androgen production. For girls, the picture is less clear-cut. In experiments, estrogens both stimulate and inhibit linear growth.

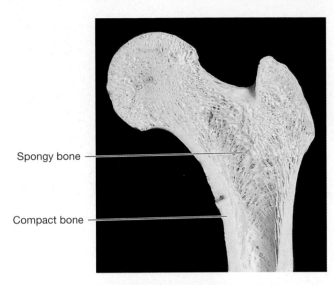

■ **Figure 21-34   Compact and spongy bone**

In all adolescents, the sex hormones eventually inactivate the epiphyseal plate so that long bone growth stops. Because the epiphyseal plates of various bones close in a constant, ordered sequence, X-rays that show which plates are open and which have closed can be used to calculate a child's "bone age."

Linear bone growth ceases in adults, but bones are dynamic tissues that are constantly being remodeled throughout life under the control of hormones that regulate calcium metabolism in the body.

▶ CALCIUM BALANCE

Although 99% of all calcium in the body, nearly 2.5 pounds, is found in the bones, the small fraction of nonbone calcium is the most critical to physiological functioning of the body (Table 21-9). A tremendous concentration gradient exists between the extracellular Ca^{2+} (2.5 mmol/L) and the free cytoplasmic Ca^{2+} (0.001 mmol/L). If cell membrane channels for Ca^{2+} open, Ca^{2+} moves into the cell, creating a Ca^{2+} signal or Ca^{2+} "spark."

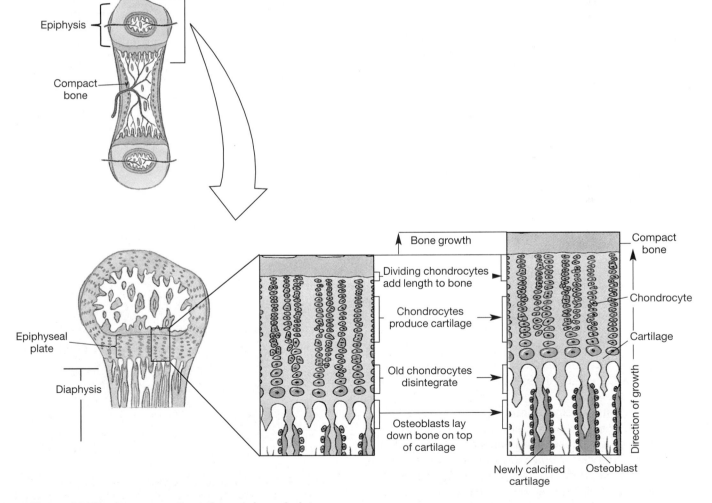

■ **Figure 21-35   Bone growth at the epiphyseal plate**

TABLE 21-9 **Functions of Calcium in the Body**

Compartment	Function	Percentage of All Calcium in Body
Extracellular	Calcified matrix of bone	99%
Extracellular fluid	"Cement" for tight junctions	
	Myocardial and smooth muscle contraction	
	Release of neurotransmitters at synapses	
	Excitability of neurons due to effect on Na^+ permeability	
	Cofactor for coagulation cascade	0.1%
Intracellular	Signal in second messenger pathways	
	Muscle contraction	0.9%

These Ca^{2+} signals initiate exocytosis of synaptic and secretory vesicles, contraction in muscle fibers, and altered activity of enzymes and transporters. Removal of Ca^{2+} from the cytosol requires active transport.

In the extracellular compartment, Ca^{2+} is part of the intercellular cement that holds cells together at tight junctions. It is also a cofactor in several steps of the coagulation cascade [∞ p. 490]. However, the most significant effect of altered plasma Ca^{2+} concentrations is related to the excitability of neurons.

If plasma Ca^{2+} falls too low (**hypocalcemia**), neuronal permeability to Na^+ increases, neurons depolarize, and the nervous system becomes hyperexcitable. In its most extreme form, hypocalcemia causes sustained contraction (tetany) of the respiratory muscles, resulting in asphyxiation. **Hypercalcemia** has the opposite effect, causing depressed neuromuscular activity.

Calcium Concentrations in the Blood Are Closely Regulated

Because calcium is critical to so many physiological functions, the body's plasma Ca^{2+} concentration is very closely regulated. Calcium is absorbed by active transport in the small intestine. In the plasma, about half the Ca^{2+} is carried bound to plasma proteins. Protein-bound Ca^{2+} is too large to be filtered at the kidneys but unbound Ca^{2+} is freely filtered at the glomerulus. It is reabsorbed along the length of the nephron, but hormonally regulated reabsorption takes place only in the distal nephron.

Bone is the largest reservoir of Ca^{2+} in the body, and it forms a pool that can be tapped if plasma Ca^{2+} concentrations fall. A small fraction of bone Ca^{2+} is readily exchangeable and in equilibrium with the Ca^{2+} of the interstitial fluid.

Osteoclasts are large, mobile, multinucleate cells derived from hematopoietic stem cells [∞ p. 477]. They are responsible for dissolving bone, a process known as *resorption*. Osteoclasts attach around their periphery to a section of calcified matrix, much like a suction cup (Fig. 21-36 ■). The central region of the cell next to the bone secretes acid (with the aid of H^+-ATPase) and protease

enzymes that work at low pH. The combination of acid and enzymes dissolves the calcified matrix and its collagen support, freeing Ca^{2+}, which can enter the blood. The bulk of bone is more slowly exchanged with the extracellular fluid.

Three Hormones Control Calcium Balance

Three hormones regulate the movement of Ca^{2+} between bone, kidney, and intestine: parathyroid hormone, vitamin D_3, and calcitonin (Fig. 21-37 ■). Changes in plasma Ca^{2+} concentrations are monitored by the four **parathyroid glands,** which are found on the dorsal surface of the thyroid gland (Fig. 21-38 ■). Their significance was first recognized in the 1890s by physiologists studying the role of the thyroid gland.

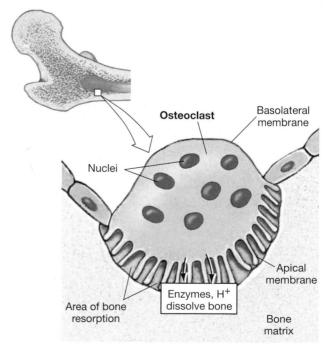

■ **Figure 21-36** **Osteoclasts and bone resorption** Osteoclasts secrete acid and enzymes at their apical membrane, dissolving the adjacent bone.

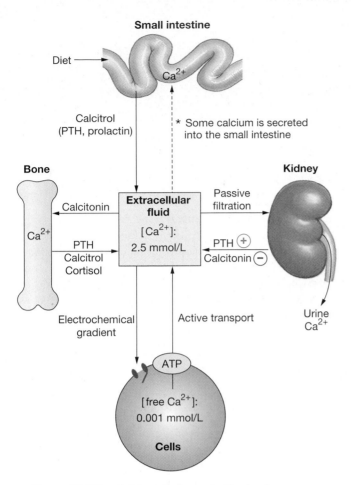

Small intestine

Diet

Ca^{2+}

Calcitrol
(PTH, prolactin)

* Some calcium is secreted
into the small intestine

Bone

Calcitonin

Ca^{2+}

PTH
Calcitrol
Cortisol

Electrochemical
gradient

**Extracellular
fluid**
$[Ca^{2+}]$:
2.5 mmol/L

Passive
filtration

Kidney

PTH $\oplus$
Calcitonin $\ominus$

Active transport

Urine
Ca^{2+}

ATP

[free Ca^{2+}]:
0.001 mmol/L

Cells

■ **Figure 21-37   Calcium balance in the body**

The scientists noted that dogs and cats died a few days after total removal of the thyroid, but rabbits died only if the little parathyroid "glandules" alongside the thyroid were removed. If the parathyroid glands were

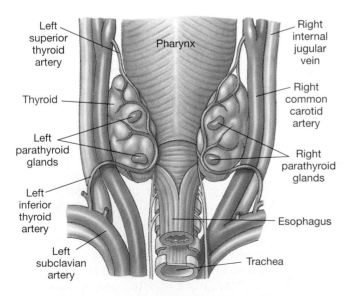

Left
superior
thyroid
artery

Pharynx

Right
internal
jugular
vein

Thyroid

Right
common
carotid
artery

Left
parathyroid
glands

Right
parathyroid
glands

Left
inferior
thyroid
artery

Esophagus

Left
subclavian
artery

Trachea

■ **Figure 21-38   Posterior view of the thyroid gland show-ing the four associated parathyroid glands**

left behind when the thyroid was surgically removed from dogs and cats, the animals lived. The scientists concluded that the parathyroids contained a substance that was essential for life, although the thyroid gland did not.

Parathyroid Hormone   We now know that parathyroid glands secrete **parathyroid hormone** (parathormone, or **PTH**), a peptide whose main effect is to increase plasma Ca^{2+} concentrations (Table 21-10). The stimulus for PTH release is a decrease in plasma Ca^{2+}, sensed by an extracellular Ca^{2+}-sensing membrane receptor. Parathyroid hormone raises Ca^{2+} concentrations in three ways:

1. **PTH mobilizes calcium from bone.** Increased bone resorption by osteoclasts takes about 12 hours to become measurable. Curiously, although osteoclasts are responsible for dissolving the calcified matrix and would be logical targets for PTH, they do not have PTH receptors. PTH effects are mediated by a collection of paracrines including *osteoprotegerin ligand* (OPGL) and *osteoclast differentiation factor* (also called TRANCE or RANKL). These paracrine factors are receiving intense scrutiny as potential pharmacological agents.

2. **PTH enhances renal reabsorption of calcium.** Calcium reabsorption takes place in the distal nephron and is accomplished by enhancing movement through an apical Ca^{2+} channel (ECaC) and basolateral Na^+-Ca^{2+} antiport and Ca^{2+}-ATPase. PTH simultaneously enhances renal excretion of phosphate by reducing its reabsorption.

   The opposing effects of PTH on calcium and phosphate are needed to keep their combined concentrations below a critical level. If the concentrations exceed that level, calcium phosphate crystals form and precipitate out of solution. High concentrations of calcium phosphate in the urine are one cause of kidney stones.

3. **PTH indirectly increases intestinal absorption of calcium** by its influence on vitamin D_3, a process described below. By acting on bone, kidney, and intestine, PTH raises plasma Ca^{2+} concentrations. The increase in plasma Ca^{2+} acts as negative feedback and shuts off PTH secretion.

Calcitrol   Intestinal absorption of calcium is enhanced by the action of a hormone known as **1,25-dihydroxycholecalciferol** ($1,25(OH)_2D_3$), also known as **calcitrol**. Calcitrol is made from vitamin D_3 obtained through diet or made in the body (Table 21-11). The body makes its own vitamin D_3 by the action of sunlight on a precursor made from acetyl CoA (Fig. 21-39 ■). Vitamin D_3 is then modified in two steps, first in the liver, then in the kidneys, to make calcitrol.

TABLE 21-10   Parathyroid Hormone (PTH)

Cell of origin	Parathyroid glands
Chemical nature	84–amino acid peptide
Biosynthesis	Continuous production, little stored
Transport in the circulation	Dissolved in plasma
Half-life	Less than 20 minutes
Factors affecting release	↓ Plasma Ca^{2+}
Target cells or tissues	Kidney, bone, intestine
Target receptor	Membrane receptor
Whole body or tissue action	↑ Plasma Ca^{2+}
Action at cellular level	↑ Vitamin D synthesis; ↑ renal reabsorption of Ca^{2+}; ↑ bone resorption
Action at molecular level (including second messenger)	Via cAMP; rapidly alters transport of Ca^{2+} but also initiates protein synthesis in osteoclasts
Onset of action	2–3 hours for bone, with increased osteoclast activity requiring 12 hours; 1–2 days for intestinal absorption; within minutes for kidney transport
Feedback regulation	Negative feedback by ↑ plasma Ca^{2+}
Other information	Osteoclasts have no PTH receptors, so are affected by PTH-induced paracrines. PTH is essential for life; absence causes hypocalcemic tetany.

TABLE 21-11   Vitamin D (Calcitrol, 1,25-Dihydroxycholecalciferol)

Cell of origin	Complex biosynthesis; see below
Chemical nature	Steroid
Biosynthesis	Vitamin D_3 formed by sunlight on precursor molecules or ingested in food; converted in 2 steps (liver and kidney) to $1,25 (OH)_2D_3$
Transport in the circulation	Bound to plasma proteins
Stimulus for synthesis	↓ Ca^{2+}; indirectly via PTH; prolactin also stimulates synthesis
Target cells or tissues	Intestine, bone, and kidney
Target receptor	Nuclear
Whole body or tissue action	↑ Plasma Ca^{2+}
Action at molecular level	Stimulates production of calbindin, a Ca^{2+}-binding protein; associated with intestinal transport by unknown mechanism
Feedback regulation	Plasma Ca^{2+} shuts off PTH secretion

Calcitrol reinforces the plasma Ca^{2+}-increasing effect of parathyroid hormone by enhancing Ca^{2+} uptake across the small intestine. In addition, it facilitates renal reabsorption of Ca^{2+} and helps mobilize Ca^{2+} out of bone into the extracellular fluid.

The production of calcitrol is regulated at the kidney by the action of PTH. Decreased plasma Ca^{2+} concentrations increase PTH secretion, which in turn stimulates calcitrol synthesis. Intestinal and renal absorption of Ca^{2+} raises blood calcium concentrations, turning off PTH and subsequently decreasing calcitrol synthesis. Prolactin, the hormone responsible for milk production in breast-feeding women, also stimulates calcitrol synthesis. This action ensures maximal absorption of Ca^{2+} from the diet at a time when metabolic demands for calcium are high.

Calcitonin   The third hormone involved with calcium metabolism is **calcitonin,** a peptide produced by the C cells (parafollicular cells) of the thyroid gland (Table 21-12). Its actions are opposite to those of parathyroid hormone. Calcitonin is released when plasma Ca^{2+} goes up. Experimentally in animals, it decreases bone resorption and increases renal calcium excretion.

From what we can tell, calcitonin plays only a minor role in daily calcium balance in adult humans. Patients whose thyroid glands have been removed show no disturbance in calcium balance. People with thyroid tumors that secrete large amounts of calcitonin also show no ill effects.

Medically, calcitonin has been used to treat patients with Paget's disease, a genetically linked condition in which osteoclasts are overactive and bone is weakened by resorption. Calcitonin in these patients stabilizes the abnormal bone loss. This action of calcitonin has led scientists to speculate that the hormone in humans is most important during childhood growth, when net bone deposition is needed, and during pregnancy and lactation (breast-feeding), when the mother's body must supply calcium for both herself and the child.

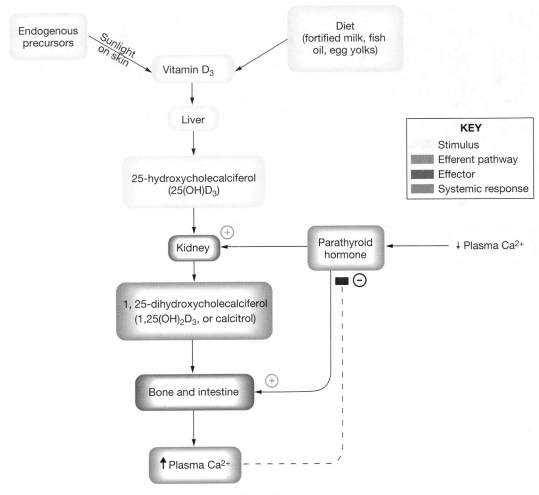

Osteoporosis Is a Disease of Bone Loss

One of the best-known pathologies of bone function is **osteoporosis,** a metabolic disorder characterized by bone resorption exceeding bone deposition. The result is fragile, weakened bones that are more easily fractured (Fig.

21-40 ■). Osteoporosis is most common in women after menopause, when estrogen concentrations drop. However, older men also develop osteoporosis. Bone loss and small fractures and compression in the spinal column lead to the stooped, hunchback appearance that is characteristic of advanced osteoporosis in the elderly.

TABLE 21-12   Calcitonin

Cell of origin	C ("clear") cells of thyroid gland (parafollicular cells)
Chemical nature	32–amino acid peptide
Biosynthesis	Typical peptide
Transport in the circulation	Dissolved in plasma
Half-life	< 10 minutes
Factors affecting release	↑ Plasma $[Ca^{2+}]$
Target cells or tissues	Bone and kidney
Target receptor	Membrane receptor
Whole body or tissue action	Prevents bone resorption; enhances kidney excretion
Action at molecular level (including second messenger)	Receptor coupled to G proteins; signal transduction pathways appear to vary during cell cycle
Other information	Experimentally decreases plasma $[Ca^{2+}]$ but has little apparent physiological effect in adults; no obvious symptoms with either hypo- or hypersecretion; possible effect on skeletal development; possible protection of bone calcium stores during pregnancy and lactation

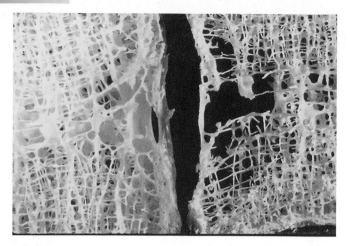

■ **Figure 21-40   Normal bone and bone loss in osteo-porosis**

Osteoporosis is a complex disease with genetic and environmental components. Risk factors include small, thin body type, postmenopausal age without estrogen replacement therapy, Caucasian or Asian ancestry, smoking, and low dietary Ca^{2+} intake.

Estrogen replacement therapy decreases the risk of osteoporosis but increases the risk of endometrial and possibly other cancers. Current research shows that estrogen protects bone mass by inhibiting paracrines that promote bone loss. Drugs called *bisphosphonates* are also effective in decreasing bone mass loss.

To prevent osteoporosis in later years, younger women need to maintain adequate dietary calcium intake and participate in weight-bearing exercise that increases bone density. Loss of bone mass actually begins by age 30, long before most women start to think they are at risk.

PROBLEM CONCLUSION

Dr. Bob had the surgery, and the implanted glands produced an adequate amount of PTH. He must have his plasma Ca^{2+} levels checked regularly for the rest of his life.

In this running problem, you learned about hyperparathyroidism and its effect on the body. You also learned one way this condition can be treated. Further check your understanding of this running problem by comparing your answers to those in the summary table.

	Question	Facts	Integration and Analysis
1	What role does calcium play in the normal function of muscles and neurons?	Calcium triggers neurotransmitter release [∞ p. 235] and uncovers the myosin-binding sites on actin filaments of muscle [∞ p. 354].	Muscle weakness is the opposite effect of what you would predict from knowing the role of calcium in muscles and neurons. However, calcium also affects the sodium permeability of neurons, and it is this effect that leads to muscle weakness and the central nervous system effects.
2	What is the technical term for "elevated levels of calcium in the blood"?	Prefix for "elevated levels": *hyper-* Suffix for "in the blood": *-emia*	Hypercalcemia is the technical term for elevated levels of calcium in the blood.
3	A significant portion of plasma Ca^{2+} cannot be filtered into Bowman's capsule. Speculate on why some plasma Ca^{2+} cannot filter.	Filtration at the glomerulus is a selective process that excludes blood cells and most plasma proteins [∞ p. 549].	A significant amount of Ca^{2+} is bound to plasma proteins and therefore does not filter at the glomerulus.
4	What one test could definitively prove that Dr. Bob has hyperparathyroidism?	Hyperparathyroidism is a condition in which excessive amounts of PTH are secreted.	A test for the amount of PTH in the blood would confirm the diagnosis of hyperparathyroidism.
5	Why can't Dr. Bob take replacement PTH by mouth?	PTH is a peptide hormone.	Ingested peptides will be digested by proteolytic enzymes. Thus PTH taken orally will not be absorbed intact into the body and will not be effective.

CHAPTER REVIEW

SUMMARY

Energy Balance and Metabolism

Energy Balance

1. **Metabolism** is the chemical reactions that extract, use, or store energy. (p. 640)

2. **Anabolic** pathways synthesize small molecules into larger ones. **Catabolic** pathways break large molecules into smaller ones. (p. 640)

3. In energy balance, energy intake equals energy output. (p. 640)

4. Body temperature homeostasis is controlled by the hypothalamus. (p. 642)

5. Heat loss takes place by radiation, conduction, convection, and evaporation. Heat loss is promoted by cutaneous vasodilation and sweating. (p. 642)

6. Heat is generated by **shivering** and **nonshivering thermogenesis.** (p. 644)

7. The energy content of food is measured by **direct calorimetry** and reported in **kilocalories** (kcal). (p. 646)

8. Measuring **oxygen consumption** is the most common method of estimating energy expenditure by an organism. (p. 647)

9. The **respiratory quotient (RQ)** is the ratio of CO_2 produced to O_2 consumed. RQ varies with diet. (p. 647)

10. **Basal metabolic rate,** or BMR, is an individual's lowest metabolic rate. (p. 647)

11. **Diet-induced thermogenesis** is an increase in heat production after eating. (p. 647)

12. Glycogen and fat are the two primary forms of energy storage in the human body. (p. 647)

Metabolism

13. Metabolism is divided into the **fed (absorptive) state** and **fasted (postabsorptive) state.** The fed state is anabolic. The fasted state is catabolic. (p. 648)

14. Plasma glucose concentration is closely regulated because glucose is normally the only fuel that the brain can metabolize. (p. 648)

15. **Glycogenesis** is synthesis of glycogen from glucose. **Glycogenolysis** is glycogen breakdown. (p. 648)

16. **Gluconeogenesis** is synthesis of glucose from noncarbohydrate precursors, especially amino acids. (p. 648)

17. Ingested fats circulate as chylomicrons. **Lipoprotein lipase** removes triglycerides, leaving chylomicron remnants to be taken up and metabolized by the liver. (p. 651)

18. The liver secretes lipoprotein complexes such as **low-density lipoprotein** and **high-density lipoprotein.** (p. 651)

19. The goal of fasted state metabolism is to maintain adequate plasma glucose concentrations. (p. 652)

20. In the fasted state, the liver produces glucose from glycogen and amino acids. Beta oxidation of fatty acids forms acidic **ketone bodies.** (p. 652)

Homeostatic Control of Metabolism

Pancreatic Hormones

21. Hour-to-hour metabolic regulation depends on the ratio of insulin to glucagon. Insulin dominates the fed state and decreases plasma glucose. Glucagon dominates the fasted state and increases plasma glucose. (p. 653)

22. The **islets of Langerhans** secrete insulin from **beta cells,** glucagon from **alpha cells,** and somatostatin from **D cells.** (p. 654)

23. Increased plasma glucose, amino acids, GIP, and parasympathetic input stimulate insulin release. Sympathetic signals inhibit insulin secretion. (p. 655)

24. Major insulin target tissues are liver, adipose, and skeletal muscles. Some tissues are insulin-independent. (p. 656)

25. Insulin increases glucose transport into muscle and adipose tissue, increases glucose utilization, and increases storage of glucose and fat. (p. 656)

26. **Diabetes mellitus** is a collection of diseases marked by abnormal secretion or activity of insulin that causes **hyperglycemia.** In **type 1 diabetes,** pancreatic beta cells are destroyed by antibodies. In **type 2 diabetes,** target tissues fail to respond normally to insulin. (p. 657)

27. Type 1 diabetes is marked by catabolism of muscle and adipose tissue, glucosuria, polyuria, and metabolic ketoacidosis. Type 2 diabetics have less severe symptoms. In both types, complications include atherosclerosis, neurological changes, and problems with the eyes and kidneys. (p. 658)

28. Glucagon stimulates glycogenolysis and gluconeogenesis. (p. 661)

Neurally Mediated Aspects of Metabolism

29. The hypothalamus contains a tonically active **feeding center** and a **satiety center** that inhibits the feeding center. (p. 662)

30. Blood glucose concentrations (the **glucostatic theory**) and body fat content (the **lipostatic theory**) influence food intake. Hypothalamic centers respond to a variety of peptides including **leptin** and **neuropeptide Y.** (p. 663)

Long-Term Endocrine Control of Metabolism

Adrenal Glucocorticoids

31. **Cortisol,** thyroid hormones, and growth hormone play a role in long-term regulation of metabolism. (p. 664)

32. Cortisol is catabolic. It promotes gluconeogenesis, breakdown of skeletal muscle proteins and adipose tissue, and suppression of the immune system. (p. 665)

Thyroid Hormones

33. The thyroid **follicle** has a hollow center filled with **colloid** containing **thyroglobulin** and enzymes. (p. 667)

34. Thyroid hormones are made from tyrosine and iodine. Tetraiodothyronine (**thyroxine, or T$_4$**) is converted in target tissues to the more active hormone **triiodothyronine** (T$_3$). (p. 668)

35. Thyroid hormones are not essential for life, but they influence metabolic rate and protein, carbohydrate, and fat metabolism. (p. 668)

36. Thyroid hormone secretion is controlled by **thyrotropin (thyroid-stimulating hormone, or TSH)** and **thyrotropin-releasing hormone, or TRH.** (p. 670)

Endocrine Control of Growth

Growth Hormone

37. Normal growth requires growth hormone, thyroid hormones, insulin, and sex hormones at puberty. Growth also requires adequate diet and absence of stress. (p. 672)

38. **Growth hormone** is secreted by the anterior pituitary and stimulates secretion of **insulin-like growth factors** from the liver and other tissues. These hormones interact to stimulate bone and soft tissue growth. (p. 672)

39. Secretion of growth hormone is controlled by **growth hormone–releasing hormone** (GHRH) and **growth hormone–inhibiting hormone (somatostatin).** (p. 672)

Tissue and Bone Growth

40. Bone is composed of hydroxyapatite crystals attached to a collagenous support. Bone is a dynamic tissue with living cells. (p. 674)

41. **Osteoblasts** synthesize bone. Long bone growth occurs at **epiphyseal plates,** where **chondrocytes** produce cartilage. (p. 674)

Calcium Balance

42. Calcium acts as an intracellular signal for second messenger pathways, exocytosis, and muscle contraction. (p. 675)

43. Decreased plasma Ca^{2+} stimulates **parathyroid hormone** (PTH) secretion by **parathyroid glands.** (p. 677)

44. PTH mobilizes Ca^{2+} from bone, enhances renal Ca^{2+} absorption, and increases intestinal Ca^{2+} absorption through its effect on **calcitrol.** (p. 677)

45. **Calcitonin** from the thyroid gland plays only a minor role in daily calcium balance in humans. (p. 678)

QUESTIONS

LEVEL ONE   Reviewing Facts and Terms

1. Define metabolic, anabolic, and catabolic pathways.

2. List and briefly explain the three forms of biological work.

3. Define a kilocalorie. What is measured with these units? What is direct calorimetry?

4. What is the respiratory quotient (RQ)? What is a typical RQ value for an American diet?

5. Define basal metabolic rate (BMR). Under what conditions is it measured? Why does the average BMR differ in adult males and females? List at least four factors other than gender that may affect BMR in humans.

6. What are the three general fates of biomolecules in the body?

7. What are the main differences between metabolism in the absorptive and postabsorptive states?

8. What is a *nutrient pool*? What are the three primary nutrient pools of the body?

9. What is the primary goal of fasted-state metabolism?

10. Excess energy in the body is stored in what forms?

11. What are the three possible fates for ingested proteins? For ingested fats?

12. Name the hormones that influence glucose metabolism and explain what effect each hormone has on blood glucose concentrations.

13. What noncarbohydrate molecules can be made into glucose? What are the pathways called through which these molecules are converted to glucose?

14. Under what circumstances are ketone bodies formed?

From what biomolecule are ketone bodies formed? How are they used by the body, and why is their formation potentially dangerous?

15. Name two stimuli that increase insulin secretion and one stimulus that inhibits insulin secretion.

16. What are the two types of diabetes mellitus? How do their cause and basic symptoms differ?

17. What factors release glucagon? What organ is the primary target of glucagon? What effect(s) do(es) glucagon produce?

18. List four conditions that are necessary for people to achieve their full growth. Include five specific hormones known to exert an effect on growth.

19. Name the thyroid hormones. Which one has the highest activity? How and where is most of it produced?

20. Define the following terms and explain their physiological significance:

(a) lipoprotein lipase
(b) osteoporosis
(c) hydroxyapatite
(d) neuropeptide Y
(e) trabecular bone
(f) leptin
(g) epiphyseal plates

LEVEL TWO   Reviewing Concepts

21. **Concept map:** Draw a concept map that compares the fed state and the fasted state. For each state, compare metabolism in skeletal muscles, the brain, adipose tissue, and the liver. Indicate which hormones are active in each stage and at what points they exert their influence.

22. Examine the graphs of insulin and glucagon secretion in Figure 21-12 ■. Why have some researchers concluded it is the ratio of these two hormones that determines whether glucose will be stored or removed from storage?

23. Define, compare, and contrast or relate the terms within each set below:

(a) glucose, glycogenolysis, glycogenesis, gluconeogenesis, glucagon, glycolysis
(b) shivering, nonshivering, and diet-induced thermogenesis
(c) lipoproteins, chylomicrons, cholesterol, HDL, LDL, VLDL
(d) cortisol, glucocorticoids, ACTH, CRH
(e) thyroid, C cells, follicle, colloid
(f) thyroglobulin, tyrosine, iodide, TBG, deiodinase, TSH, TRH
(g) somatotropin, IGF, GHRH, somatostatin, growth hormone binding protein
(h) giantism, acromegaly, dwarfism

(i) hyperplasia, hypertrophy
(j) osteoblast, osteoclast, chondrocyte, osteocyte
(k) vitamin D, calcitrol, 1,25-dihydroxycholecalciferol, calcitonin, estrogen, PTH
(l) conductive heat loss, radiant heat loss, convective heat loss, evaporative heat loss

24. Explain the physiological processes that lead to the following symptoms in a type 1 diabetic:

(a) hyperglycemia
(b) glucosuria
(c) excretion of copious quantities of urine (polyuria)
(d) ketosis
(e) dehydration
(f) severe thirst

25. Both insulin and glucagon are released following ingestion of a protein meal that raises plasma amino acid levels. Why is the secretion of both hormones necessary?

26. Explain the current theory of the control of food intake. Use the following terms in your explanation: hypothalamus, feeding and satiety centers, appetite, leptin, NPY, neuropeptides, CCK.

27. Compare human thermoregulation in hot environments to thermoregulation in cold environments.

LEVEL THREE   Problem Solving

28. Scott is a bodybuilder who consumes large amounts of amino acid supplements in the belief that this will increase his muscle mass. He believes that the amino acids he consumes are stored in his body until he needs them. Is Scott correct? Explain.

29. One diagnostic test to determine the cause of hypercorti-

solism is a dexamethasone suppression test. Dexamethasone blocks secretion of ACTH by the pituitary. If a patient with hypercortisolism is given dexamethasone and cortisol secretion is not affected, what does this tell you about the site of the pathology? What does it tell you if dexamethasone decreases cortisol secretion?

LEVEL FOUR Quantitative Problems

30. One way to estimate obesity is to calculate a person's **body mass index** (BMI). A body mass index greater than 25 for men and 30 for women is considered a sign of obesity. To calculate your BMI, divide your body weight in kilograms by the square of your height in meters: kg/m^2. To convert weight from pounds, use 1 kg/2.2 lb. To convert height from inches, use 1 m/39.24 in.

E X P L O R E <MediaLab>

Introduction

This chapter examined the endocrine control of growth and metabolism. It is a complex field but has far-reaching implications for homeostasis, growth, and development. The hormones that you studied can have synergistic and antagonistic actions on many physiological functions. This MediaLab will challenge your ability to integrate these concepts. After reading the description below, visit the MediaLab for Chapter 21 in your Companion Website and select the appropriate keyword.

Web Exploration 1

Estimated time for completion = 5 minutes
Billions of research dollars are spent to learn about what makes us hungry and why we eat. The control of body weight is a national obsession, as evidenced by countless number of diet books and drugs available with the promise of 'getting the body you have always dreamed of.' What influences our food intake, and how is it regulated? There are countless numbers of influences (culture, psychosocial and genetic), but what internal factors make you desire to eat or stop eating? Can you think of the glands or organs that are involved in informing us that we should eat or not? To what factors do they respond? Select the keyword **SATIETY CENTERS** from your Website and scroll down to the "Physiological Control Mechanisms" section to visualize the signals that interact to control our food intake. How might an elevated serum lipid level effect the desire to eat? How does low serum glucose influence hunger? Where are these signals processed? Using the same site scroll down to "Hypothalamic Centers." What experiments support the idea that the ventromedial and lateral nuclei of the hypothalamus are involved in feeding behavior?

Web Exploration 2

Estimated time for completion = 15 minutes
Normally, growth hormone is responsible for the increase in height during a child's development and progressively decreases as we enter adulthood. Select the keyword **GROWTH HORMONE** from your Website to see how the pattern of growth hormone secretion changes as we age. What are the major target organs for the effects of growth hormone? If growth hormone deficiency in childhood results in dwarfism, can you predict what an excess of growth hormone would do during this developmental period? Specifically, can you list the physical characteristics that you might expect? What if a normally developed adult began to produce excess growth hormone? List those physical characteristic along with the effects of growth hormone oversecretion in a developing child. Are there similarities? Differences? Think about bone development to help you address these questions. After you have made your predictions, visit the **CLINICAL FEATURES** website to learn about the two conditions of giantism and acromegaly.

22 The Immune System

■ "Although, at first sight, the immune system may appear to be autonomous, it is connected by innumerable structural and functional bridges with the nervous system and the endocrine system, so as to constitute a multisystem."
—*Branislav D. Janković*, Neuroimmunomodulation: The State of the Art, *1994* ■

SEM of alveolar macrophage attacking *E. coli* bacteria (turquoise). A red blood cell is at the upper right.

Since the 1980s, it seems that human civilization has become afflicted with plagues of infectious diseases. Some of them, such as tuberculosis, rabies, and malaria, are old enemies that we thought we had under control. Others, such as those caused by the human immunodeficiency virus (HIV), the Hanta virus, and the Ebola virus, seem to be cropping up out of nowhere. The conflict between humans and the invaders that cause disease is literally a fight for survival on both parts. The viruses and parasites (such as the protozoan that causes malaria) need a host such as the human body to reproduce. But if they kill off all the hosts, they too will die out. We as hosts must fight off the invaders so that we can continue to survive as a species.

The ability of the body to protect itself from viruses, bacteria, and other disease-causing entities is known as **immunity**, from the Latin word *immunis*, meaning *exempt*. The human immune system consists of lymphoid tissues of the body, the immune cells, and chemicals, both intracellular and secreted, that coordinate and execute the immune functions.

▶ OVERVIEW OF IMMUNE SYSTEM FUNCTION

The key features of the immune system are specificity and memory. *Specificity* and *memory* together enable the body to distinguish "self" from "non-self" and mount a targeted response to specific invaders when they are encountered a second time.

The immune system serves three major functions:

1. **It protects the body from disease-causing invaders** known as **pathogens**. Microscopic invaders, or **microbes**, include bacteria, viruses, fungi, and one-celled protozoans. Larger pathogens include multicellular parasites such as hookworms and tapeworms.

   In addition, virtually any molecule or cell not of the body has the potential to elicit an immune response. Pollens, chemicals, and foreign bodies such as splinters are examples of substances to which the body may react. Substances that trigger an immune response from the body and that can react with products of that response are known as **antigens**.

2. **The immune system removes dead or damaged tissue and cells** such as tissue damaged by injury (cuts, disease) or old red blood cells. Scavenger cells of the immune system patrol the extracellular compartment, gobbling up and digesting dead or dying cells.

3. **The immune system tries to recognize and remove abnormal cells** created when normal cell growth and development go wrong. For example, the diseases we call *cancer* result from abnormal cells that multiply repeatedly, crowding out normal cells and disrupting function. Scientists believe that cancer cells form on a regular basis, but are usually detected by the immune system and destroyed before they get out of control.

Sometimes the body's mechanisms for distinguishing self from non-self fails, and the immune system attacks normal cells. *Autoimmune diseases* are the result. Type 1 diabetes mellitus, in which antibodies destroy pancreatic beta cells, is an example of an autoimmune disease in humans.

▶ PATHOGENS OF THE HUMAN BODY

The foreign invaders that challenge the human body include viruses, bacteria, one-celled protozoans, fungi, and multicellular parasites. In the United States, our most

PROBLEM
Treatment for AIDS

In 1996, researchers developed a strategy that might offer long-term suppression of the HIV virus, the virus that causes acquired immunodeficiency syndrome (AIDS). In that year, scientists confirmed that an AIDS "cocktail" consisting of several drugs, including a *protease inhibitor*, could be effective against the HIV virus that causes AIDS. But questions remain. How long can the new drug combination keep HIV at bay? Will the new drugs work for everyone with the infection? George, 28, pushed these questions to the back of his mind. After three years of relatively good health following his positive test for HIV infection, his T lymphocyte cell count had begun to slide. It was only a matter of time before George became gravely ill.

…continued on page 696

prevalent infectious diseases are viral and bacterial infections. Worldwide, parasites are an additional significant public health concern. For example, malaria, a protozoan parasite whose life cycle alternates between human and mosquito hosts, is estimated to infect as many as 100 million people in the world.

Many parasitic organisms, such as the protozoan that causes malaria, are injected into the body by biting insects. Others enter the body through the digestive tract, brought in by contaminated food and water. Some, such as the fungi that cause valley fever and histoplasmosis, are inhaled. A few, such as the blood fluke *Schistosoma*, actually burrow through the host's skin until they reach a capillary. Once in the body, parasites may enter host cells in an effort to hide from the immune response, or they may remain in the extracellular fluid (ECF).

Bacteria and Viruses Require Different Defense Mechanisms

The most common infections in the United States are caused by bacteria and viruses. In general, these microbes differ from each other in several ways (Table 22-1):

1. **Structure.** Bacteria are cells, with a cell membrane that is usually surrounded by a cell wall. Some bacteria also produce an additional protective outer layer known as a *capsule*. Viruses are not cells. They consist of a nucleic acid (DNA or RNA) core enclosed in one or two coats (envelopes) of up to 3000 protective proteins (Fig. 22-1a ■).

2. **Living conditions.** Bacteria can survive outside a host if they have the required nutrients, temperature, pH, and so on. Viruses are parasitic and they must have a host cell to survive.

3. **Reproduction.** Bacteria can reproduce without a host. Viruses must use the intracellular machinery of the host cell to reproduce.

TABLE 22-1 Bacteria and Viruses

	Bacteria	*Viruses*
Structure	Cells; usually surrounded by cell wall	Not cells; nucleic acid core with protein envelope(s)
Living conditions	Can survive outside a host	Parasitic; must have a host cell to survive
Reproduction	Can reproduce without a host	Must use the genetic material and organelles of host cell to reproduce
Susceptibility to antibiotics	Can be killed or inhibited by antibiotics	Cannot be killed with antibiotics; must be destroyed by host's immune system

4. **Susceptibility to antibiotics.** Most bacteria can be killed by the drugs we call antibiotics. These drugs act directly on bacteria and destroy them or inhibit their growth. Viruses cannot be killed by antibiotics. To kill a virus, the body's immune system must find and destroy the infected host cell. The immune system must also attack viral offspring released after a virus reproduces in the host cell.

Viruses Must Reproduce Inside Host Cells

The life cycle of a virus begins when the virus invades the host cell (Fig. 22-1b ■). To cross the human host cell membrane, the virus binds to membrane receptors, triggering

(a) Influenza, an RNA virus

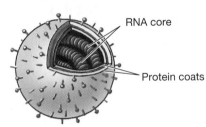

RNA core

Protein coats

(b) General steps of viral reproduction

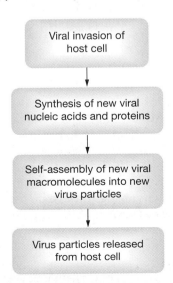

Viral invasion of host cell

↓

Synthesis of new viral nucleic acids and proteins

↓

Self-assembly of new viral macromolecules into new virus particles

↓

Virus particles released from host cell

■ **Figure 22-1   Viruses**

endocytosis of the entire virus particle. In an alternative scenario, the virus envelope fuses with the host cell membrane, allowing the core of the virus to enter the cytoplasm.

Once inside, the virus's nucleic acid takes over the cell. Viruses use their own nucleic acid and the host cell's resources to make new viral nucleic acid and viral proteins. These components assemble into additional virus particles that are released from the host cell to infect other cells.

Viruses can be released from host cells in one of two ways. (1) The virus causes the host cell to lyse (a kind of explosion), releasing virus particles to the ECF, or (2) virus particles surround themselves with host cell membrane and bud from the host cell surface.

Viruses do other kinds of damage to host cells. In taking over the host cell, they may totally disrupt the host cell's own metabolism so that the host cell dies. Some viruses (*Herpes simplex* and *H. zoster*, which cause cold sores and chicken pox, respectively) become resident in the cell and reproduce only sporadically.

Other viruses incorporate their genomes (DNA) into the host cell DNA. Viruses with this characteristic include *human immunodeficiency virus* (HIV), the virus that causes AIDS, and **oncogenic viruses** that cause cancer.

▶ THE IMMUNE RESPONSE

The defense system of the body has two components. Physical and chemical barriers, such as skin, mucus, and stomach acid, first try to keep pathogens out of the

Retroviruses and AIDS   The RNA viruses known as **retroviruses** reproduce in an interesting manner. Because these viruses do not have their own DNA to act as a template for mRNA synthesis, they produce an enzyme called **reverse transcriptase.** This enzyme allows the viral RNA to make the complementary viral DNA. This is the reverse of the usual process—normally, DNA codes for the synthesis of RNA. Once the virus-controlled cell has made viral DNA, it produces viral mRNA. The viral mRNA then interrupts the host cell's protein synthesis, using host ribosomes to make viral proteins. The disease we call AIDS (**acquired immunodeficiency syndrome**) is attributed to viral infection by the HIV retrovirus.

body's internal environment. If that line of defense fails, then the internal *immune response* takes over.

Whatever the nature of the foreign material that comes into the body, the internal immune response includes the same basic steps: (1) *detection* and *identification* of the foreign substance, (2) *communication* with other immune cells to rally an organized response, (3) *recruitment* of assistance and *coordination* of the response among all participants, and (4) *destruction* or *suppression* of the invader.

Not all invaders can be destroyed. In some cases, the best compromise the body can reach is to control the damage and keep the invader from spreading. Examples of pathogens that are suppressed rather than destroyed include the bacterium that causes tuberculosis; the malaria parasite, which hides inside liver cells; and the *Herpes* viruses that cause occasional outbreaks of cold sores or genital lesions.

The immune system is distinguished by its extensive use of chemical signaling. Detection, identification, communication, recruitment, coordination, and the attack on the invader are all chemically mediated processes. *Antibodies*, which are proteins produced by certain immune cells, play a key role in the identification and targeting of pathogens. Many steps in the immune process involve the binding of a ligand to a receptor [∞ p. 114]. Other steps depend on chemical messengers that travel between cells. These messengers are **cytokines**, molecules released by one cell that affect the growth or activity of another cell [∞ p. 156].

The human immune response is generally divided into two categories: innate immunity and acquired immunity. **Innate immunity** takes place within minutes to hours and forms the nonspecific responses of the body to invasion. *Inflammation*, indicated on the skin by a red, warm, swollen area, is a hallmark reaction of innate immunity.

Acquired immunity, also called *adaptive immunity*, depends on antibodies that are directed at specific invaders. Acquired immunity is sometimes known as *humoral immunity* because antibodies are circulated in the blood. One characteristic of acquired immunity is that a response following first exposure to a pathogen may take days. With repeated exposure, the immune system remembers prior exposure and reacts more rapidly.

As we will see later in this chapter, acquired immunity overlaps innate immunity. Although we have described them as if they are separate entities, in reality they are two interconnected parts of a single process. The innate response is the more rapid response. It is reinforced by the more specific acquired response, which amplifies the efficacy of the innate response. Communication and coordination between the different pathways of immunity is vital for maximum protective effect.

Because of the huge numbers of molecules involved in human immunity, and because of the complex interactions between the different components of the immune system, the field of immunology is continuously expanding. In this chapter, you will learn the basics so that you can add the details with more advanced study.

▶ ANATOMY OF THE IMMUNE SYSTEM

The immune system is probably the least anatomically identifiable system of the body because most of it is integrated into the tissues of other organs such as the skin and gastrointestinal tract. Yet the mass of all the immune cells in the body equals the mass of the brain! The immune system has two anatomical components: the lymphoid tissues of the body and the cells that are responsible for the immune response.

Lymphoid Tissues Are Distributed Throughout the Body

The two primary lymphoid tissues are the *thymus gland* (see *Focus on the Thymus Gland*, p. 701) and the *bone marrow*, both sites where cells involved in the immune response form and mature (Fig. 22-2a ■). Secondary lymphoid tissues are divided into **encapsulated** tissues and unencapsulated **diffuse lymphoid tissues**.

Both the **spleen** (Figs. 22-2 ■ and 22-3 ■) and the **lymph nodes** (Fig. 22-2b ■) have outer walls formed from fibrous collagenous capsules. The spleen traps and removes aging red blood cells from the circulation with the aid of phagocytic cells. It also contains immune cells positioned so that they encounter foreign invaders in the circulation.

The lymph nodes are associated with the vessels of the lymphatic circulation. Recall from Chapter 15 that there is net flow of fluid out of capillaries and into the interstitial space [∞ p. 459]. This fluid is picked up by the lymph capillaries. As the fluid called lymph moves through the lymph vessels, it passes through lymph nodes.

In the lymph nodes, clusters of *macrophages* [∞ p. 476] intercept pathogens that have entered the extracellular compartment through breaks in the skin or mucous membranes. By trapping bacteria, viruses, and other microbes, the lymph nodes help prevent those pathogens from spreading throughout the body. You have probably noticed that if you have a sinus infection or a sore throat, the lymph nodes in your neck become swollen. If you have an infected cut on your hand, the lymph nodes in your armpit may become sore. These sore, swollen nodes result from active immune cells that have collected within them to fight the infection.

Unencapsulated lymphoid tissues are aggregations of immune cells that appear throughout other organ systems of the body (see Fig. 22-2 ■). They include the **tonsils** at the posterior nasopharynx, the **gut-associated lymphoid tissue** (GALT) that lies just under the epithelium of the esophagus and intestines, and clusters of lymphoid tissue associated with the skin and respiratory, urinary, and reproductive tracts.

In each case, these tissues contain immune cells that are in position to intercept pathogens before they get into

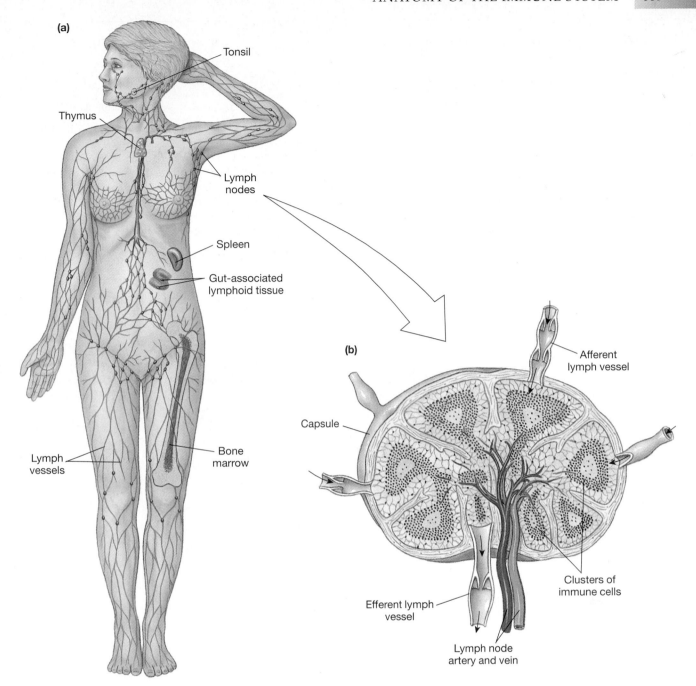

■ Figure 22-2   Anatomy of the immune system

the general circulation. Because of the large surface area of the digestive tract epithelium, some authorities consider the GALT to be the largest immune organ in the body. Anatomically, the cells of the immune system can be found in highest concentrations wherever they are most likely to encounter antigens that penetrate the epithelia.

Leukocytes Are the Primary Cells of the Immune System

The primary cells of the immune system are the white blood cells, or **leukocytes**, which were briefly introduced in Chapter 16 [∞ p. 476]. Most leukocytes are much larg-

er than red blood cells. They are also not nearly as numerous. A microliter (1 μL) of whole blood contains about 5 million red blood cells but only about 7000 leukocytes.

Although most leukocytes circulate in the blood, they usually leave the capillaries and carry out their job *extravascularly* (outside the vessels). Some types of leukocytes can live for several months out in the tissues, but others may live for only hours or days.

Leukocytes are divided into six basic groups: (1) *eosinophils*, (2) *basophils* in the blood and the related *mast cells* in the tissues, (3) *neutrophils*, (4) *monocytes* and their derivative *macrophages*, (5) *lymphocytes* and their derivatives, including *plasma cells* and *cytotoxic T cells*, and

FOCUS ON... The Spleen

■ Figure 22-3

The spleen is the largest lymphoid organ in the body, located in the upper-left quadrant of the abdomen close to the stomach.

The outer surface of the spleen is a connective tissue capsule that extends into the interior to create an open framework that supports the blood vessels and lymphoid tissue. Regions of *white pulp* resemble the interior of lymph nodes and are composed mainly of lymphocytes.

Darker regions of *red pulp* are closely associated with extensive blood vessels and open venous sinuses.

The red pulp contains many macrophages that act as a filter by trapping and destroying foreign material circulating in the blood. In addition, the macrophages ingest old, damaged, and abnormal red blood cells, breaking down their hemoglobin molecules into amino acids, iron, and bilirubin that is transported to the liver for excretion [∞ Fig. 16-8, p. 485].

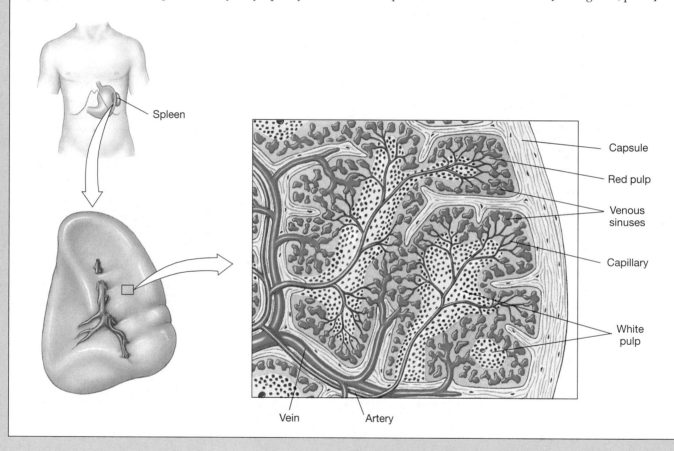

(6) *dendritic cells*. Dendritic cells are not usually found in the blood and therefore were not included with leukocytes in Chapter 16.

Leukocytes can be distinguished from one another in stained tissue samples by the shape and size of the nucleus, the staining characteristics of the cytoplasm and cytoplasmic inclusions, and the regularity of the cell border.

Immune Cell Names Reflect Cell Function or Appearance The terminology associated with immune cells can be very confusing. Some cell types have several variants, and others have been given multiple names (Fig. 22-4 ■). In addition, immune cells can be grouped functionally or morphologically.

One morphological group is the **granulocytes**, white blood cells whose cytoplasm contains prominent granules. The names of the different cell types come from the staining properties of the granules. Neutrophil granules do not stain darkly with standard blood stains and are therefore "neutral." Basophil granules stain dark blue with basic (alkaline) dye, and eosinophil granules stain dark pink with the acidic dye *eosin*. In all three types of granulocytes, the contents of the granules are released into the extracellular space in response to some type of signal, a process known as *degranulation*.

Functional groups of leukocytes include the phagocytes and immunocytes. The term **immunocyte** is an older name given to lymphocytes and their derivative

	Basophils and Mast Cells	Neutrophils	Eosinophils	Monocytes and Macrophages	Lymphocytes and Plasma Cells	Dendritic Cells
	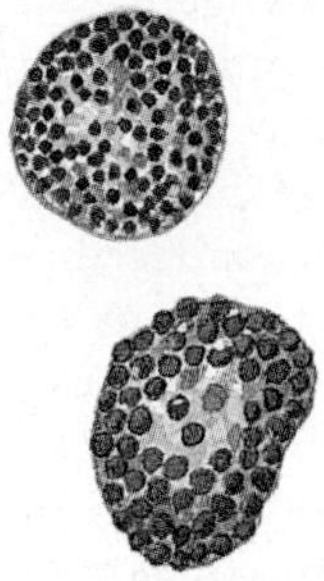	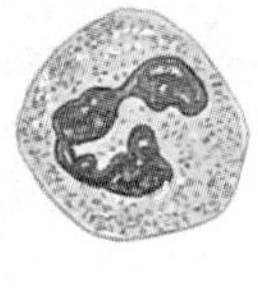	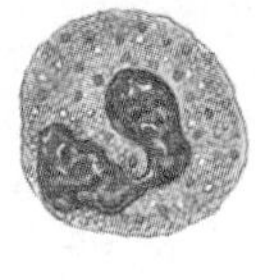	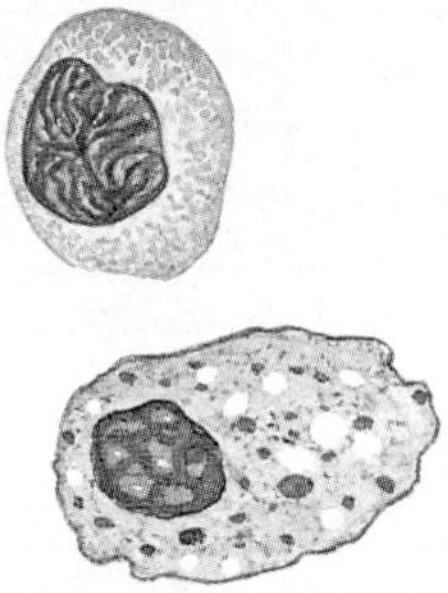	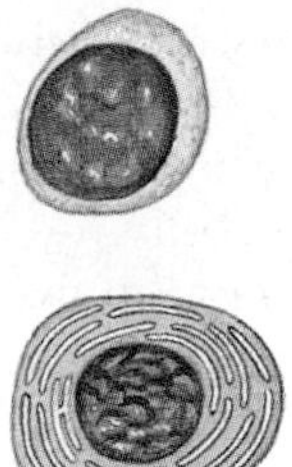	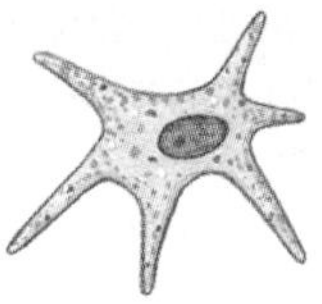
% of WBCs in blood	Rare	50–70%	1–3%	1–6%	20–35%	NA
Subtypes and nicknames		Called "polys" or "segs" Immature forms called "bands" or "stabs"		Called the mononuclear phagocyte system	B lymphocytes, Memory cells Plasma cells T lymphocytes Cytotoxic T cells Helper T cells Natural killer cells	Also called Langerhans cells, veiled cells
Primary function(s)	Release chemicals that mediate inflammation and allergic responses	Ingest and destroy invaders	Destroy invaders, particularly parasites	Ingest and destroy invaders Antigen presentation	Specific responses to invaders, including antibody production	Recognize pathogens and activate other immune cells by antigen presentation
Classifications	*Phagocytes*					
	Granulocytes					
			Cytotoxic cells		*Cytotoxic cells (some types)*	
					Antigen-presenting cells	

■ **Figure 22-4 Cells of the immune system**

cells. Although all leukocytes participate in various aspects of the immune response, the immunocytes are the cells primarily responsible for the acquired response.

Phagocytes are white blood cells that engulf and ingest their targets by phagocytosis. This group incudes the neutrophils, macrophages, monocytes (which are macrophage precursors), and eosinophils. **Cytotoxic cells** kill the cells they are attacking. This group includes eosinophils and some types of lymphocytes. **Antigen-presenting cells** display fragments of foreign proteins on their cell surface. This group includes lymphocytes, dendritic cells, macrophages, and monocytes.

The terminology associated with macrophages has changed along with the history of histology and immunology. For many years, tissue macrophages were known as the **reticuloendothelial system** and were not associated with white blood cells. To confuse matters, the cells were named when they were first described in different tissues, before they were all identified as macrophages. Thus, histiocytes in skin, Kupffer cells in the liver, osteoclasts in bone, microglia in brain, and reticuloendothelial cells in spleen are all names for specialized macrophages. The new name for the reticuloendothelial system is the **mononuclear phagocyte system**, a term that refers both to macrophages in the tissues and to their parent monocytes circulating in the blood.

Let's take a closer look at the six basic types of leukocytes. Figure 22-4 ■, introduced above, makes a good reference.

Eosinophils Fight Parasites and Contribute to Allergic Reactions

Eosinophils, easily recognized by the bright pink-staining granules in their cytoplasm, are associated with allergic reactions and parasitic diseases. They are known to attach to large parasites such as the blood fluke *Schistosoma* and release substances that damage or kill them. Because eosinophils kill pathogens, they are called *cytotoxic* [*cyto-*, cell] cells.

In allergic reactions such as those to pollen, eosinophils congregate near the focus of antigen entry. Although eosinophils have been shown to ingest these particles *in vitro*, there is no evidence that they do so *in vivo.* The granules of eosinophils contain enzymes that generate toxic oxidative substances. When the granules are released, the oxidants attack the invader.

Normally, few eosinophils are found in the peripheral circulation, where they account for only 1%–3% of all leukocytes. The life span of an eosinophil in the blood is estimated to be only 6–12 hours. Most functioning eosinophils are found in the digestive tract, lungs, and connective tissue of the skin.

Basophils Release Histamine and Other Chemicals
Basophils are rare in the circulation but are easily recognized in a stained blood smear by the large, dark blue granules in their cytoplasm. They are very similar to the **mast cells** of tissues. The granules of mast cells and basophils contain histamine, **heparin** (an *anticoagulant* that inhibits blood clotting), and other chemicals involved in the immune response.

Mast cells, like eosinophils, are found concentrated in the connective tissue of skin, lungs, and the gastrointestinal tract. In these locations, they are ideally situated to intercept pathogens that are inhaled or ingested or that enter through breaks in the epidermis.

Neutrophils Eat Bacteria and Release Cytokines
Neutrophils are the most abundant white blood cell (50%–70% of the total) and the most easily identified, with a segmented nucleus of three to five lobes connected by thin strands of nuclear material. Because of the segmented nucleus, neutrophils are also called **polymorphonuclear leukocytes** ("polys") or "segs."

Neutrophils, like other blood cells, are formed in the bone marrow. Occasionally, an immature neutrophil makes its way into the circulation, where it can be identified by its horseshoe-shaped nucleus. These immature neutrophils go by the nicknames of "bands" and "stabs."

Neutrophils are phagocytic cells that typically ingest and kill five to twenty bacteria during their programmed life span of one to two days. Most neutrophils spend their lives in the blood, but they can leave the circulation if attracted to an extravascular site of damage or infection. In addition to ingesting bacteria and foreign particles, neutrophils release a variety of cytokines, including fever-causing pyrogens [∞ p. 646] and chemical mediators of the inflammatory response.

Monocytes Are Tissue Scavengers
Until the late 1930s, scientists did not realize that circulating **monocytes** are the precursor cells of tissue **macrophages**. Monocytes are not very common in the blood (1%–6% of all white blood cells). By some estimates, they spend only eight hours there in transit from the bone marrow to their permanent positions in the tissues.

Once out of the blood, rounded monocytes enlarge and differentiate into macrophages. Some tissue macrophages spend their lives patrolling the tissues, creeping along by amoeboid motion. Others find a location and remain fixed in place.

In either case, macrophages are the primary scavengers for the tissues. They are larger and more effective than neutrophils, ingesting up to 100 bacteria during their life span. Macrophages also remove larger particles such as old red blood cells and dead neutrophils.

Macrophages play an important role in the development of acquired immune responses. When they ingest molecular or cellular antigens and digest them, fragments of antigen are processed and inserted into the macrophage membrane as part of surface protein complexes (Fig. 22-5 ■). Immune cells that process antigen in this fashion are called **antigen-presenting cells** (APCs). Dendritic cells and some types of lymphocytes also act as antigen-presenting cells.

Lymphocytes Secrete Antibodies and Remember Pathogens
The **lymphocytes** and their derivatives are the key cells that mediate the acquired immune response of the body. Only 5% of all lymphocytes are found in the circulation, where they constitute 20%–35% of all white blood cells. Most lymphocytes are found in the lymphoid tissues such as lymph nodes, the spleen, bone marrow, and the mucosa of the gastrointestinal tract. These are all locations where lymphocytes are likely to encounter invaders.

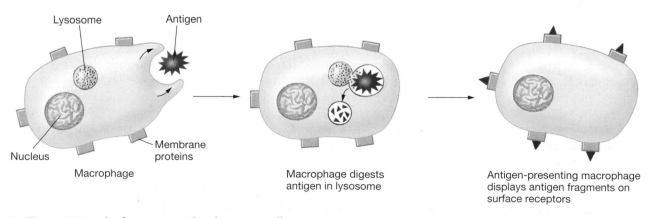

Lysosome   Antigen

Nucleus

Membrane proteins

Macrophage

Macrophage digests antigen in lysosome

Antigen-presenting macrophage displays antigen fragments on surface receptors

■ **Figure 22-5   Antigen-presenting immune cells**

By one estimate, the adult body contains a trillion (10^{12}) lymphocytes at any one time. Although most of these cells look alike under the microscope, they have major differences in their function and specificity, as you will learn later in the chapter.

Dendritic Cells Activate Lymphocytes **Dendritic cells** are antigen-presenting cells characterized by long, thin processes that resemble the dendrites of the neuron. Dendritic cells are found in the skin (where they are called Langerhans cells) and in various organs. When dendritic cells recognize and capture antigens, particularly viruses, they migrate to the lymphoid tissues. There they present antigen fragments to lymphocytes to activate them. Dendritic cells in the lymph system are called *veiled cells*.

▶ INNATE IMMUNITY

The body's first line of defense is to exclude pathogens by physical and chemical barriers. If those barriers fail to keep invaders out, the defenses of innate immunity provide the second line of defense. The innate immune response consists of patrolling and stationary leukocytes that attack and destroy invaders. These cells respond in the same fashion to any material they identify as foreign, which is why the innate response is considered nonspecific.

Physical and Chemical Barriers Are the Body's First Line of Defense

Physical barriers of the body include the skin [∞ p. 68], the protective mucous linings of the gastrointestinal and genitourinary tracts, and the ciliated epithelium of the respiratory tract. The digestive and respiratory systems are most vulnerable to microbial invasion because these regions have extensive areas of thin epithelium in direct contact with the external environment.

In women, the reproductive tract is also vulnerable, but to a lesser degree. The opening to the uterus is normally sealed by a mucus plug that keeps bacteria from ascending from the vagina into the uterine cavity. An infamous example of how this physical barrier can fail was the Dalkon shield, an intrauterine contraceptive device (IUD) with a multi-ply string that extended from the IUD through the mucus plug to the vagina. The string acted as a wick for bacteria, allowing them to bypass the mucus plug that would normally exclude them. Because of bacterial invasions, a number of women with IUDs suffered from pelvic infections and inflammatory reactions that blocked their Fallopian tubes and left them unable to conceive.

In the respiratory system, inhaled particulate matter is trapped by mucus lining the upper respiratory system. The mucus is then transported upward on the mucus escalator to be expelled or swallowed [∞ p. 506]. Swallowed pathogens may be disabled by the acidity of the stomach.

In addition, respiratory tract secretions and tears contain **lysozyme**, an enzyme with antibacterial activity. Lysozyme attacks components of unencapsulated bacterial cell walls and breaks them down. It cannot digest the capsules of encapsulated bacteria, however.

Pathogens that enter the body past the physical barriers of skin and mucus are dealt with by immune cells and the innate immune response. When patrolling and stationary phagocytes in the tissues detect invaders, their response is twofold: destroy or suppress the invader by ingesting it, and attract additional immune cells by secreting cytokines. Molecules that attract immune cells are called **chemotaxins**. They include bacterial toxins, products of tissue injury (fibrin and collagen fragments), and cytokines released by leukocytes.

Phagocytes Recognize Foreign Material and Ingest It

The primary phagocytic cells of the immune system are the tissue macrophages and the neutrophils. If the invader is in the tissue, phagocytes leave the circulation (*extravasation*) by squeezing through pores in the capillary endothelium, a process known as **diapedesis** [*diapedesis*, leaping through]. Once the phagocytes reach the pathogen, they identify it by chemical cues, then ingest it.

Phagocytosis, the process in which cells engulf and digest particulate matter, is similar in macrophages and neutrophils [∞ p. 128]. It is a receptor-mediated event to ensure that only unwanted particles are ingested.

In the simplest reactions, surface molecules on the pathogen act as ligands that bind directly to receptors on the phagocyte membrane (Fig. 22-6a ■). In a sequence similar to a zipper closing, the ligands and receptors combine sequentially, so that the phagocyte surrounds the unwanted foreign particle. The process is aided by actin filaments that push arms of the phagocytic cell around the invader.

Phagocytes recognize many different types of foreign particles, both organic and inorganic, by mechanisms that we still do not fully understand. Phagocytes ingest unencapsulated bacteria, cell fragments, carbon, and asbestos particles, among other materials. In the laboratory, scientists use the ingestion rate of tiny polystyrene beads as an indicator of phagocytic activity.

Not all foreign material can be instantly recognized by phagocytes, however. For example, certain bacteria have evolved to hide their surface markers behind a polysaccharide capsule (Fig. 22-6b ■). If pathogens lack markers, they must be "tagged" so that the phagocytes will recognize them as something to be ingested. This tagging occurs when the pathogen becomes coated with proteins, including antibodies, that occur in the blood of the host.

The tagging proteins are known collectively as **opsonins**, a term derived from the Greek word *opsonin*, meaning "to buy provisions." In the body, opsonins convert unrecognizable particles into "food" for phagocytes.

(a) Pathogen binds directly to phagocyte receptors

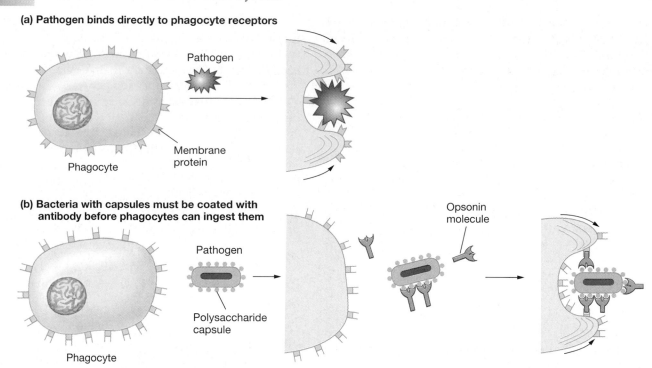

(b) Bacteria with capsules must be coated with antibody before phagocytes can ingest them

■ Figure 22-6   **Phagocytosis**

Opsonins act as a bridge between pathogens and phagocyte by binding to receptors on the phagocyte.

The ingested particle ends up in a cytoplasmic vesicle called a **phagosome** (Fig. 22-7 ■). Phagosomes fuse with intracellular lysosomes and peroxisomes [∞ p. 53]. Enzymes within these organelles produce **hypochlorous acid** (HOCl), the active ingredient in household bleach, and **peroxides**. These powerful oxidizing agents destroy ingested pathogens when the phagosomes and vesicles fuse.

If an area of infection attracts a large number of phagocytes, the material known as **pus** may form. This thick whitish to greenish substance is a collection of living and dead neutrophils and macrophages, along with tissue fluid, cell debris, and other remnants of the immune process.

DIABETES: PPARgamma, Macrophages, and Fat Peroxisomes are organelles responsible for the oxidation of fatty acids, and cells that carry out lots of β-oxidation have lots of peroxisomes. How do cells regulate peroxisomes? This question was a mystery until 1990, when scientists found a steroid-binding receptor in the cell nucleus. This receptor, named the *peroxisome proliferator-activated receptor* or PPAR, binds to lipids and lipid-derived molecules and turns on a variety of genes. One subtype of PPAR, called PPARgamma, has become the subject of intense research. It has been linked to adipocyte differentiation, type 2 diabetes, and "foam cells," which are fat-filled macrophages in atherosclerotic plaques. PPARgamma may be the elusive link between obesity, type 2 diabetes, and atherosclerosis that has evaded scientists for so long.

Chemical Mediators Create the Inflammatory Response

Inflammation, indicated on the skin by a red, warm, swollen area, is a hallmark reaction of the innate immune pathway. The inflammatory response is created when activated tissue macrophages release cytokines. These chemicals attract other immune cells, increase capillary permeability, and cause fever. Immune cells attracted to the site in turn release their own cytokines. Some representative examples of chemicals involved in the innate immune response are described below. They are also listed in Table 22-2, another convenient "scorecard" for following the players in these complex processes.

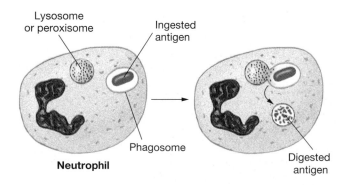

■ Figure 22-7   **Digestion in phagocytes** Phagosomes containing ingested material fuse with lysosomes or peroxisomes and enzymes digest the material within the phagosome.

TABLE 22-2 Chemicals of the Immune Response

Functional classes

Acute phase proteins: Liver proteins released during the acute phase that act as opsonins; also enhance the inflammatory response

Chemotaxins: Molecules that attract phagocytes to a site of infection

Cytokines: Molecules released by one cell that affect the growth or activity of another cell

Opsonins: Proteins that coat pathogens so that phagocytes will recognize and ingest them: antibodies, acute phase proteins, and complement proteins

Pyrogens: Fever-producing substances

Specific chemicals and their functions

Antibodies (immunoglobulins, gamma globulins): Fight specific invaders

Bradykinin: Stimulates pain receptors; vasodilator

Complement: Plasma and cell membrane proteins that act as opsonins, cytolytic agents, and mediators of inflammation

Granzymes: Cytotoxic enzymes that trigger cells into committing suicide

Heparin: An anticoagulant

Histamine: Vasodilator and bronchoconstrictor released by mast cells

Hypochlorous acid (HOCl): Powerful oxidant found with peroxide in peroxisomes of phagocytic cells

Interferons: Proteins that inhibit viral reproduction and modulate the immune response

Interleukin-1: Mediates the inflammatory response; secreted primarily by activated macrophages

Kinins: Inactive plasma proteins that activate in a cascade to form bradykinin

Lysozyme: An extracellular enzyme that attacks bacteria

Major histocompatibility complex (MHC): A family of membrane protein complexes used for cell recognition

Membrane attack complex: A membrane pore protein made in the complement cascade; allows ions and water to enter bacteria and lyse them

Perforin: A membrane pore protein made by natural killer and cytolytic T cells and inserted into the membrane of target cells; large enough to allow granzymes to enter the cell

Peroxide: Powerful oxidant found in lysosomes of phagocytic cells

T-cell receptors: Membrane receptors on T lymphocytes that recognize and bind antigen presented by MHC receptors

Acute Phase Proteins Are Released Early in the Immune Response
In the time immediately following an injury or pathogen invasion (the *acute phase*), the body responds by increasing the concentration of various plasma proteins. Some of these proteins, produced mostly by the liver, are given the general name of **acute phase proteins**. They include molecules that act as opsonins, antiprotease molecules to help prevent tissue damage, and other proteins whose function is less clear.

Normally, the levels of these proteins fall back to normal as the immune response proceeds, but in *chronic inflammatory diseases* such as rheumatoid arthritis, elevated levels of acute phase proteins may persist.

Histamine Initiates Inflammation
Histamine is an amine derived from the amino acid histidine [∞ p. 32]. Histamine is found primarily in the granules of mast cells and basophils, and it is the active molecule that helps initiate the inflammatory response when mast cells degranulate. Histamine's actions bring more leukocytes to the injury site to kill bacteria and remove cellular debris.

Histamine opens pores in capillaries, allowing plasma proteins to escape into the interstitial space. This process causes local edema, or swelling. In addition, histamine dilates blood vessels, increasing blood flow to the area. The result of histamine release is a hot, red, swollen area around a wound or infection site.

Mast cell degranulation is triggered by different cytokines in the immune response. Because mast cells are concentrated under mucous membranes that line the airways and digestive tract, the inhalation or ingestion of certain antigens can trigger histamine release. The resultant edema in the nasal passages leads to one annoying symptom of seasonal pollen allergies, the stuffy nose. Fortunately, pharmacologists have developed a variety of drugs called *antihistamines*. These drugs antagonize histamine by blocking mast cell degranulation or by blocking the action of histamine at its receptor.

Scientists used to believe that histamine was the primary chemical responsible for the bronchoconstriction of asthma and for the severe systemic allergic reaction called *anaphylactic shock*. However, research has shown that mast cells release other powerful cytokines in addition to histamine, including leukotrienes, platelet-activating factor, and prostaglandins. These cytokines work with histamine to create the bronchoconstriction and hypotension that are characteristic of anaphylactic shock.

Interleukins Have Widespread Systemic Effects
Interleukins are cytokines that were initially thought to mediate communication between the different types of leukocytes in the body. However, as scientists learn more about them, they have discovered that many different tissues in the body respond to interleukins.

Interleukin-1 (IL-1) is secreted by activated macrophages and other immune cells. Its main role is to mediate the inflammatory response, but it also has widespread systemic effects on immune function and metabolism. Interleukin-1 modulates the immune response by:

1. Altering blood vessel endothelium to allow easier passage of white cells and proteins during the inflammatory response.

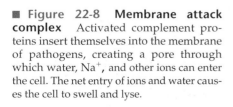

...continued from page 686
HIV is a retrovirus that uses an enzyme called reverse transcriptase to reproduce. Two drugs in the new AIDS "cocktail," AZT and ddC, are antiretroviral drugs that work by inhibiting reverse transcriptase. The other drug in the cocktail is a protease inhibitor. Protease is an enzyme needed to cleave key proteins during one step of HIV replication.

Question 1:   *Why can't retroviruses reproduce without reverse transcriptase?*

2. Stimulating production of acute phase proteins by the liver.
3. Inducing fever by acting on the hypothalamic thermostat [∞ p. 646]. IL-1 is a known pyrogen of the body.
4. Stimulating cytokine and endocrine secretion by a variety of other cells.

Bradykinin Stimulates Pain Receptors   *Kinins* are a group of inactive plasma proteins that participate in a cascade similar to the coagulation cascade [∞ p. 490]. The end product of the kinin cascade is **bradykinin**, a molecule with the same vasodilator effects as histamine. Bradykinin also stimulates pain receptors, creating the tenderness associated with inflammation.

Complement Proteins Are Opsonins and Chemotaxins   **Complement** is a collective term for a group of more than 25 plasma proteins and cell membrane proteins. The complement pathway is a cascade of events similar to the blood coagulation cascade. Various intermediates of the complement cascade act as opsonins, chemical attractants for leukocytes, and agents to cause degranulation of mast cells.

The complement cascade terminates with the formation of **membrane attack complex**, a group of lipid-soluble proteins that insert themselves into the cell membranes of pathogens and form pores (Fig. 22-8 ■). These pores allow Na^+ to enter the cell, and water follows osmotically. As a result, the cell swells and lyses.

✔ Why does the action of histamines on capillary permeability result in swelling?

● ACQUIRED IMMUNITY

Acquired immune responses are those responses in which the body recognizes a specific foreign substance and selectively reacts to it. Acquired immunity is mediated primarily by lymphocytes, which have membrane receptors that allow them to react to only one type of invader. The process of acquired immunity overlaps with the process of innate immunity. Cytokines released by the inflammatory response attract lymphocytes to the site of an immune reaction. The lymphocytes release additional cytokines that enhance the inflammatory response.

Acquired immunity can be subdivided into active immunity and passive immunity. **Active immunity** occurs when the body is exposed to a pathogen and creates its own antibodies. Active immunity can occur naturally, when the pathogen invades the body, or artificially, as when we are given immunizations that contain dead or disabled pathogens. **Passive immunity** occurs when we acquire antibodies made by another animal. The transfer of antibodies from mother to fetus across the placenta is one example. Immunoglobulin shots are another.

Lymphocytes Are the Primary Cells Involved in the Acquired Immune Response

If a specific type of lymphocyte fights each pathogen that enters the body, millions of different types of lymphocytes must be ready to combat millions of different foreign pathogens. But where do all those lymphocytes come from?

On a microscopic level, lymphocytes look alike. At the molecular level, however, the cells can be distinguished by their membrane receptors, which make them specific for a particular ligand (Fig. 22-9 ■). Each unique type of lymphocyte, along with the other cells exactly like it, forms a **clone** [*klon*, a twig].

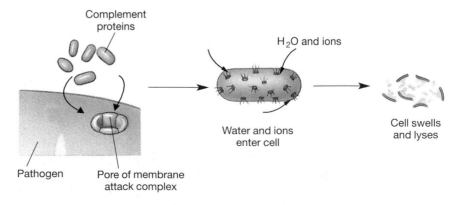

■ **Figure 22-8 Membrane attack complex**   Activated complement proteins insert themselves into the membrane of pathogens, creating a pore through which water, Na^+, and other ions can enter the cell. The net entry of ions and water causes the cell to swell and lyse.

Complement proteins

H_2O and ions

Water and ions enter cell

Cell swells and lyses

Pathogen

Pore of membrane attack complex

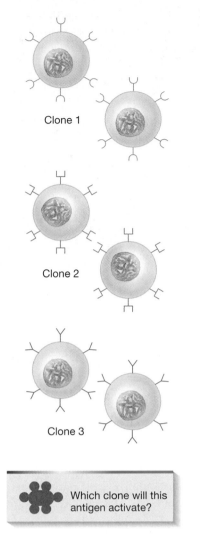

Clone 1

Clone 2

Clone 3

Which clone will this antigen activate?

■ **Figure 22-9** **Lymphocyte clones** A group of lymphocytes that are specific to one antigen is known as a clone. This figure shows three pairs of lymphocytes. Each pair is from a different clone.

All lymphocytes go through the same general life cycle. At birth, each clone of lymphocytes is represented by only a few cells, called **naive lymphocytes** [*naif*, natural]. The small number of cells in each naive clone is not enough to fight off foreign invaders, so the first exposure to an antigen activates the appropriate clone and causes it to divide (Fig. 22-10 ■). This process, called **clonal expansion**, creates additional cells in the clone.

The newly formed lymphocytes in the clone differentiate into **effector cells**, which create the immune response, and into **memory cells**. Effector cells are short-lived and die within a few days. However, memory cells have longer lives and continue to reproduce themselves. With second and subsequent exposures to the antigen, the memory cells activate, creating a more rapid and stronger secondary response to the antigen.

All lymphocytes secrete cytokines that act on immune cells, on non-immune cells of the body, and, sometimes, on the pathogens themselves. The primary lymphocyte cytokines are the interleukins, discussed earlier, and the interferons.

Interferons were named for their ability to interfere with viral reproduction in the body. **Alpha-interferon** (interferon-α) and **beta-interferon** (interferon-β) target uninfected cells and activate pathways to prevent viral replication. **Gamma-interferon** (interferon-γ) activates macrophages and other immune cells in addition to stimulating uninfected cells.

There are three main types of lymphocytes: B lymphocytes, T lymphocytes, and natural killer cells. B lymphocytes secrete antibodies. T lymphocytes and natural killer cells attack and destroy infected cells. Let's take a closer look at each of these cell types.

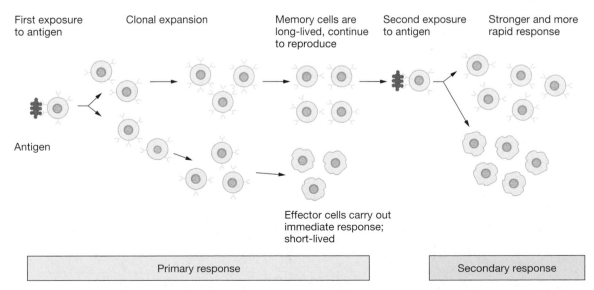

First exposure to antigen   Clonal expansion   Memory cells are long-lived, continue to reproduce   Second exposure to antigen   Stronger and more rapid response

Antigen

Effector cells carry out immediate response; short-lived

Primary response   Secondary response

■ **Figure 22-10** **Clonal expansion and memory** Upon first exposure to an antigen, naive lymphocytes reproduce. They then differentiate into short-lived effector cells that carry out the primary immune response and into long-lived memory cells. When memory cells are reexposed to the appropriate antigen, the clone expands more rapidly to create additional effector and memory cells.

Some B Lymphocytes Secrete Antibodies

The first class of lymphocytes, **B lymphocytes** (also called **B cells**) develops in the bone marrow, a convenient way to remember their name. The primary function of B lymphocytes is the secretion of **antibodies**, or immunoglobulins. A special type of B lymphocyte secretes these proteins that bind to pathogens and target them for destruction. The term *antibody* describes what the molecules do: they work against the foreign body. The equivalent term *immunoglobulin* describes what the molecules are: globular proteins that participate in the humoral immune response.

Mature B lymphocytes insert antibody molecules into their cell membranes so that antibodies become surface receptors marking the members of each clone (see Fig. 22-9 ■). When a clone of B cells activates in response to antigen exposure, some differentiate into plasma cells.

Plasma cells are the effector cells for the B lymphocytes. They do not have antibody proteins bound in their membranes. Instead, they synthesize and secrete additional antibody molecules at incredible rates, estimated to be as high as 2000 molecules of antibody per second! Plasma cell antibodies form the soluble antibodies of the plasma.

After each invader has been successfully repulsed, a few *memory cells* of the clone remain behind, so that the next exposure to the same antigen is greeted by a more rapid response. This phenomenon can be demonstrated by measuring the concentration of antibodies in the circulation.

Figure 22-11 ■ shows the antibody response after the first and second exposure to an antigen. Notice that the primary response is slower and lower in magnitude because the body has not encountered the antigen previously. The *secondary response* is enhanced by lymphocytes that carry a molecular memory of the first exposure to the antigen.

The existence of the enhanced secondary response is used in immunizations. Patients are given a pathogen that has been altered so that it no longer harms the host cell but can be recognized as foreign by immune cells. The altered pathogen causes the creation of memory cells keyed to that particular pathogen. If the immunized person is later infected by the natural pathogen, a stronger and more rapid immune response results.

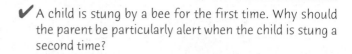

✔ A child is stung by a bee for the first time. Why should the parent be particularly alert when the child is stung a second time?

Antibodies Are Proteins Secreted by Plasma Cells

Historically, antibodies were among the first aspects of the immune system to be discovered, and traditionally they were named by what they do—agglutinins, precipitins, hemolysins, and so on.

When plasma proteins are separated by a biochemical technique called *electrophoresis*, the antibodies move together in a group given the general name of **gamma globulins**. Today, antibodies are divided into five general classes of immunoglobulins (Ig): IgG, IgA, IgE, IgM, and IgD (pronounced *eye-gee-[letter]*).

IgG makes up 75% of the plasma antibody in adults because these immunoglobulins are produced in secondary immune responses. Maternal IgGs cross the placental membrane, so these antibodies give infants immunity in the first few months of life.

IgA antibodies are found in the external secretions of the body such as saliva, tears, intestinal and bronchial mucus, and breast milk. In these locations, they disable pathogens before they reach the internal environment.

IgE is associated with allergic responses. When mast cell receptors combine with IgE and antigen, the mast cells degranulate and release chemical mediators such as histamine.

IgM is associated with primary immune responses and with the antibodies that react to blood group antigens. **IgD** appears on the surface of B lymphocytes along with IgM, but its physiological role is unclear.

✔ Because antibodies are proteins, they are too large to cross cell membranes on transport proteins or through channels. How then do IgA and other antibodies become part of external secretions such as saliva, tears, and mucus?

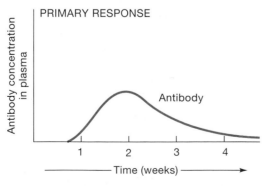

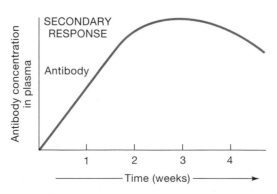

■ **Figure 22-11   Memory in the immune system**   Antibody production in response to the first exposure to an antigen is both slower and weaker than the secondary response upon reexposure.

Antibodies Are Y-Shaped Proteins   The basic anti-body molecule has four polypeptide chains linked into a Y-shape (Figure 22-12 ■). Each side of the Y is identical, with one light chain attached to one heavy chain. The two arms, or **Fab regions**, contain antigen-binding sites that confer the antibody's specificity. The stem of the Y-shaped antibody molecule is known as the **Fc region**. A *hinge region* between the arms and the stem allows flexible positioning of the arms.

In any given antibody molecule, all light chains are identical and all heavy chains are identical. However, the chains vary widely among different antibodies. In addition, two classes of immunoglobulins are polymers formed from multiple antibody molecules. IgM has five Y-shaped antibody molecules. IgA immunoglobulins have from one to four antibody molecules.

Soluble Antibodies Act as Opsonins   Most antibodies are found in the plasma, where they make up about 20% of the plasma proteins in the healthy individual. These antibodies are most effective against extracellular pathogens, such as bacteria, some parasites, antigenic macromolecules, and viruses before they have invaded their host cells.

The primary function of soluble antibodies is to bind antigens to make them more visible and "tasty" to macrophages. They do this in the following ways:

1. **Acting as opsonins.** Soluble antibodies coat antigens to facilitate recognition and phagocytosis by immune cells (Fig. 22-13 ■, ②). Antibodies are not toxic by themselves, but they label antigens so that other immune cells will attack them. All antibody-bound substances are eventually ingested by phagocytes.
2. **Making antigens clump.** Antibodies cause antigens to clump for easier phagocytosis (Fig. 22-13 ■, ③).
3. **Binding to and activating other immune cells.** Antigen-bound antibodies use the Fc end of the antibody to bind to other immune cells. Through this mechanism, antibody-bound antigen can be recognized by a single Fc receptor, rather than requiring millions of different receptors to recognize millions of antigens.

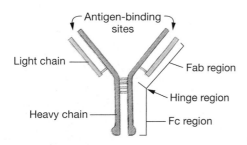

■ **Figure 22-12   Antibody structure**   An antibody molecule is composed of two identical light chains and two identical heavy chains, linked at intervals by disulfide bonds. The arms contain binding sites for antigens. Flexible hinge regions allow movement of the arms.

This type of Fc binding occurs, for example, on cytotoxic and phagocytic cells (Fig. 22-13 ■, ③ and ④). Binding activates the cells.

Soluble antibodies also enhance inflammation. These functions are triggered by the Fc end of the molecule:

4. **Activating complement.** Antigen-bound antibodies activate complement, proteins that play important roles in innate immunity (Fig. 22-13 ■, ⑥).
5. **Activating mast cells.** Antigen-bound antibodies activate mast cells, both directly and via complement, causing them to degranulate and release chemicals that mediate the inflammatory response (Fig. 22-13 ■, ⑤).

Antigen-Binding to Antibodies on B Cells Activates the B Cells   The surface of each B lymphocyte is covered with as many as 100,000 antibody molecules whose Fc ends are inserted into the lymphocyte membrane. This leaves the arms (Fab regions) free to bind to antigens in the extracellular fluid. Antigen binding activates the B lymphocyte, causing it to differentiate into either a memory cell or an antibody-secreting plasma cell (Fig. 22-13 ■, ①).

B cell antibodies bind to viruses or to toxins produced by bacteria, preventing these substances from affecting host cells. In addition to disabling the pathogen, some B cells then become memory cells to fight off a subsequent invasion.

For example, *Corynebacterium diphtheriae* is a bacterium that causes diphtheria, an upper respiratory infection. In this disease, the bacterial toxin kills host cells, leaving ulcers with a characteristic grayish membrane. Immunity to the disease occurs when the host develops antibodies that disable the toxin.

Researchers developed a vaccine for diphtheria by creating an inactivated toxic preparation that did not affect living cells. When given to a person, the vaccine triggers antibody production without causing any symptoms of the disease. As a result, diphtheria has been almost eliminated in countries with good immunization programs.

T Lymphocytes Must Make Direct Contact with Their Target Cells

The second class of lymphocyte, **T lymphocytes** (**T cells**), develops in the thymus gland from immature precursor cells that migrate there from the bone marrow. T lymphocytes are responsible for **cell-mediated immunity**. In this process, the T lymphocyte finds cells that are presenting antigen fragments on their surface. These antigen-presenting cells are signaling the T lymphocytes that they are infected and need to be eliminated. The T lymphocyte then binds to the antigen and either kills the antigen-presenting cell (*cytotoxic T cells*) or secretes cytokines that enhance the immune response (*helper T cells*).

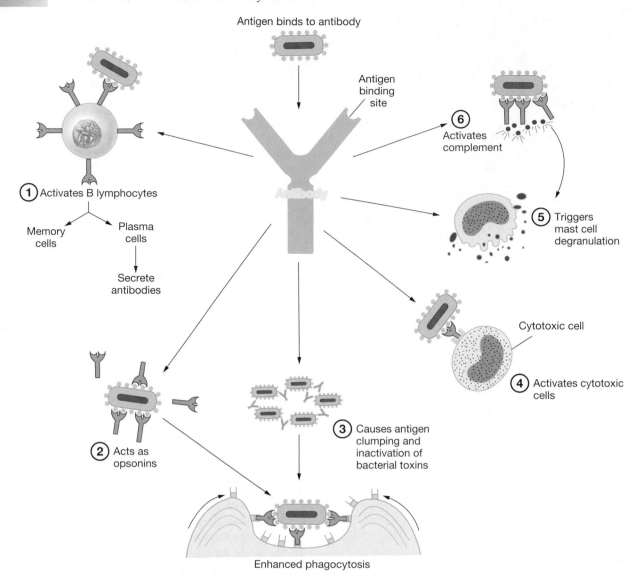

Antigen binds to antibody

Antigen binding site

① Activates B lymphocytes

Memory cells

Plasma cells

Secrete antibodies

② Acts as opsonins

③ Causes antigen clumping and inactivation of bacterial toxins

④ Activates cytotoxic cells

Cytotoxic cell

⑤ Triggers mast cell degranulation

⑥ Activates complement

Enhanced phagocytosis

■ **Figure 22-13** **Functions of antibodies**

T lymphocytes bind to their target cells by using membrane **T-cell receptors** (Fig. 22-14 ■). T-cell receptors are not antibodies like B-cell receptors, although the peptide chains are closely related. When antigen binds to T-cell receptors, the T cell is activated.

T cells cannot bind to free-floating antigens as B cells do. Instead, they only bind antigen that is presented as part of a membrane protein complex called the *major histocompatibility complex*. Cells that present antigen on these complexes are signaling T lymphocytes that they are infected and need to be eliminated.

Major Histocompatibility Complexes Incorporate Antigen Fragments The **major histocompatibility complexes**, or **MHC**, are a family of membrane protein complexes encoded by a specific set of genes. These proteins were named when they were discovered to play a role in foreign tissue rejection following organ or tissue

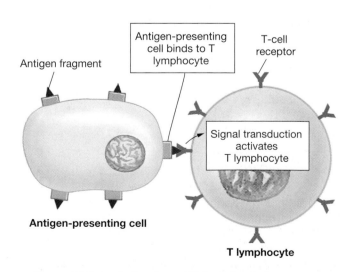

Antigen fragment

Antigen-presenting cell binds to T lymphocyte

T-cell receptor

Signal transduction activates T lymphocyte

Antigen-presenting cell

T lymphocyte

■ **Figure 22-14** **Activation of T lymphocytes**

FOCUS ON... The Thymus Gland

■ Figure 22-15

The thymus gland is a two-lobed organ located in the thorax just above the heart. During embryonic development, undifferentiated lymphocytes are released from the bone marrow and migrate to the thymus, where they differentiate into T lymphocytes and continue to divide. During the maturation process, T lymphocytes synthesize their T-cell receptors and insert them into the membrane. Those cells that would be self-reactive are eliminated, while those that do not react with self tissues multiply to form clones.

The epithelial cells of the thymus secrete peptide factors that influence the development of T lymphocytes, including interleukins, thymosins, thymopoietin, and thymulin. Although these peptides are sometimes classified as hormones, there is little evidence that they are released into the circulation to act on tissues outside the thymus gland.

The thymus gland reaches its greatest size during adolescence. Then it shrinks and is largely replaced by adipose tissue as a person ages. New T lymphocyte production in the thymus is low in adults, but the number of T lymphocytes in the blood does not decrease, suggesting that mature T lymphocytes either have long life spans or can reproduce themselves.

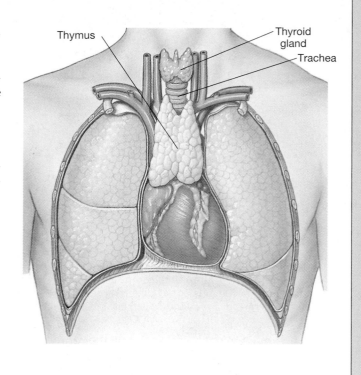

transplants. We know now that MHC is present in all nucleated cells of the body, where it is responsible for the extracellular presentation of processed antigen fragments.

All MHC proteins are related, but they vary from person to person because of the huge number of alleles that a person can inherit. There are so many alleles that it is unlikely that any two people other than identical twins will inherit exactly the same set. Major histocompatibility complexes are one reason that tissues cannot be transplanted from one person to another without first establishing compatibility. More recently, it has been discovered that the inheritance of certain MHC alleles is linked to some disorders in which the body fails to recognize body components as "self."

How do antigen fragments get to the outside of antigen-presenting cells? Free antigen in the ECF cannot simply bind to unoccupied MHC receptors on the cell surface. A phagocyte must ingest the antigen and digest it. Fragments of digested antigen then combine with an MHC complex and are inserted into the cell membrane by exocytosis (see Figs. 22-5 and 22-14 ■). A similar process ocurrs when normal cells are invaded by viruses. In both instances, the antigen-containing extracellular markers tell lymphocytes to destroy the infected cell.

There are two types of MHC molecules. **MHC class I molecules** are found on all nucleated cells. When virus-

es and bacteria invade the cell, they are digested into peptide fragments inside vesicles called *proteasomes*. The fragments are loaded onto MHC-1 "platforms" and inserted into the cell membrane. T lymphocytes that bind to the MHC class I-antigen complexes then kill the cell to prevent the invading pathogen from reproducing.

MHC class II molecules are found primarily on antigen-presenting immune cells (macrophages, monocytes, B lymphocytes, and dendritic cells). When the antigen-presenting cell engulfs and digests an antigen, the fragments are returned to the cell membrane combined with MHC class II proteins. The MHC II-antigen complex then binds to and activates other immune cells.

T Lymphocytes Differentiate in the Thymus T-cell precursors differentiate in the thymus into two major groups: cytotoxic T cells,* and helper T cells (Fig. 22-16 ■). The existence of a third group, suppressor T cells, has been proposed, but at this time it is unclear whether these cells represent a distinct line.

Cytotoxic T (T_c) cells attack and destroy cells that display MHC class I-antigen complexes. Although this

---

*Cytotoxic T cells are sometimes called "killer T cells," but because that name is very similar to "natural killer cells," it should probably be avoided.

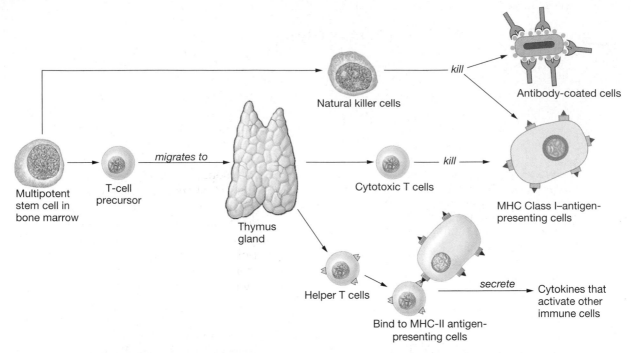

■ **Figure 22-16   T lymphocytes and NK cells**   T lymphocytes and natural killer cells develop from the same precursor. Cytotoxic T cells and natural killer cells destroy their target cells. Helper T cells recognize antigen presented on MHC II molecules and secrete cytokines to activate other immune cells.

may be an extreme response, it prevents the reproduction of intracellular invaders such as viruses, some parasites, and some bacteria.

How do cytotoxic T cells kill their targets? They do so in two ways. First they can release a cytotoxic pore-forming molecule called **perforin** along with **granzymes**, enzymes related to the digestive enzymes trypsin and chymotrypsin. When granzymes enter the target cell through perforin channels, they trigger the cell into committing suicide (**apoptosis**). Second, cytotoxic T cells can induce apoptosis by activating **Fas**, a "death receptor" protein on the target cell membrane.

Helper T (T_H) cells do not directly attack pathogens and infected cells, but they play an essential role in the immune response by secreting cytokines that influence other cells. The cytokines secreted by helper T cells include gamma-interferon, which activates macrophages; interleukins that activate antibody production and cytotoxic T lymphocytes; colony-stimulating factors [∞ p. 479] that enhance leukocyte production in the bone marrow; and interleukins that promote growth of mast cells and eosinophils. The HIV virus that causes AIDS preferentially infects and destroys helper T cells, leaving the host unable to respond to pathogens that in other situations could be easily suppressed.

Natural Killer Lymphocytes Fight Viruses

A third class of lymphocyte, the **natural killer cells**, or **NK cells**, was once considered a subset of T lymphocytes because both cell types develop from the same precursor. However, the NK cells are now considered a distinct cell line. Sometimes these cells are called *large granular lymphocytes* to distinguish them from the smaller B and T lymphocytes.

Natural killer cells attack and kill certain tumor cells and virus-infected cells. For example, in Chédiak-Higashi syndrome, characterized by a lack of natural killer cells, patients have an increased incidence of lymphoma, a cancer of the lymphoid tissue.

NK cells act more rapidly than other lymphocytes, with a response that can be measured within hours of a primary viral infection. NK cells secrete antiviral cytokines such as gamma-interferon (IFN-γ) and apparently also lyse virus-infected cells before the virus can reproduce.

 Cell Suicide, or Apoptosis   Apoptosis is a programmed form of cell suicide that the body uses to eliminate cells that it no longer needs. Cells that are undergoing apoptosis do not swell like dying cells that have been attacked by membrane attack complex. Instead, they shrink and fragment into small pieces that are ingested by neighboring phagocytes. Apoptosis is not limited to cells that are attacked by the immune system. Keratinocytes in the skin [∞ p. 68] are programmed to die about three weeks after they migrate toward the surface layers of epidermis. T lymphocytes that would attack self tissues normally die from apoptosis before being released from the thymus gland. Similarly, genetically altered cells often commit suicide before reproducing. One apparent cause of cancer is failure of these abnormal cells to kill themselves.

...continued from page 696

Early in the AIDS epidemic, researchers studied the effectiveness of a drug made with CD4. CD4 is a naturally occurring protein found on the surface of helper T cells, the primary target of the HIV virus. When the HIV virus binds to CD4, the virus is taken into the cell and begins replicating. Researchers found that adding high amounts of CD4 to test tubes containing HIV and helper T cells reduced the rate of HIV replication. But in human subjects, the results of clinical trials using the CD4 drug were disappointing. The drug did not slow the rate at which HIV infected helper T cells.

Question 2: *What is the rationale for flooding the body with CD4 protein to prevent HIV replication?*

Like cytotoxic T cells, natural killer cells kill their targets by cell lysis and apoptosis. However, natural killer cells do not have to bind to MHC-antigen complexes. They can also bind to the Fc stem of antibodies coating target cells. This nonspecific response of natural killer cells is called **antibody-dependent cell–mediated cytotoxicity** and is similar to the process by which macrophages recognize and interact with antibody-coated pathogens.

▶ IMMUNE RESPONSE PATHWAYS

How does the body respond to different kinds of immune challenges? Although the details depend on the particular challenge, the basic pattern is the same. The innate response starts first, reinforced by the more specific acquired response. The two pathways are interconnected, so cooperation and communication is essential. In the following sections, we examine the body's responses in four situations: an extracellular bacterial infection, a viral infection, an allergic response to pollen, and the transfusion of incompatible blood.

Inflammation Is the Typical Response to Bacterial Invasion

What happens when bacteria invade? If the passive barricades of the skin and mucous membranes fail, bacteria reach the extracellular fluid. There they usually cause an inflammatory response that represents the combined effects of many cells working to get rid of the invader.

Inflammation is characterized by a red, swollen area that is tender or painful. In addition to the nonspecific inflammatory response, lymphocytes attracted to the area produce antibodies keyed to the specific type of bacterium. The entry of bacteria sets off a series of reactions (Fig. 22-17 ■):

1. Components of the bacterial cell wall activate the complement system. Some complement proteins are chemical signals (*chemotaxins*) that attract leukocytes from the circulation to help fight the infection.

2. The complement cascade ends with formation of membrane attack complex molecules that insert themselves into the bacterial wall. The subsequent entry of Na^+ and water lyses the bacteria, aided by the enzyme lysozyme. This is a purely chemical process that does not involve immune cells.

3. If the bacteria are not encapsulated, macrophages can begin to phagocytose the bacteria immediately.

4. If the bacteria are encapsulated, opsonins must coat the capsule before they can be ingested by phagocytes. Opsonins enhance the process for all bacteria, even those that are not encapsulated. The molecules that act as opsonins include complement, acute phase proteins, and antibodies.

5. Complement causes degranulation of mast cells and basophils. Cytokines secreted by the mast cells act as additional chemotaxins, attracting more immune cells. Vasoactive chemicals such as histamine dilate blood vessels and increase capillary permeability. The enhanced blood supply to the site creates the red, warm appearance of inflammation. Plasma proteins that escape into the interstitial space pull water with them, leading to tissue edema.

6. Some elements of the acquired immune response are called into play with bacterial infections. If antibodies to the bacteria are already in the body, they will help the innate response by acting as opsonins and neutralizing bacterial toxins.

7. Memory B cells attracted to the infection site will be activated if they encounter an antigen they recognize. But this process is much slower than the innate responses described above, and it may be as long as several days before plasma cells begin secreting antibodies against the invader.

   If the infection is a novel one, phagocytes that ingest the bacteria present bacterial antigen to naive B cells, triggering clonal expansion, antibody production, and the formation of memory cells. This process is even slower, usually requiring at least four days.

8. If the initial wound damaged blood vessels underlying the skin, platelets and the proteins of the coagulation cascade will also be recruited to minimize the damage [∞ p. 490]. Once the bacteria are removed by the immune response, the injured site is repaired under the control of growth factors and other cytokines.

Intracellular Defense Mechanisms Are Needed to Fight Viral Infections

What happens when viruses invade the body? First, they go through an extracellular phase of immune response similar to that described for bacteria. In the early stages of a viral infection, innate immune responses and antibodies can help control the invasion of the virus.

Once the viruses enter the host's cells, humoral immunity in the form of antibodies is no longer effective.

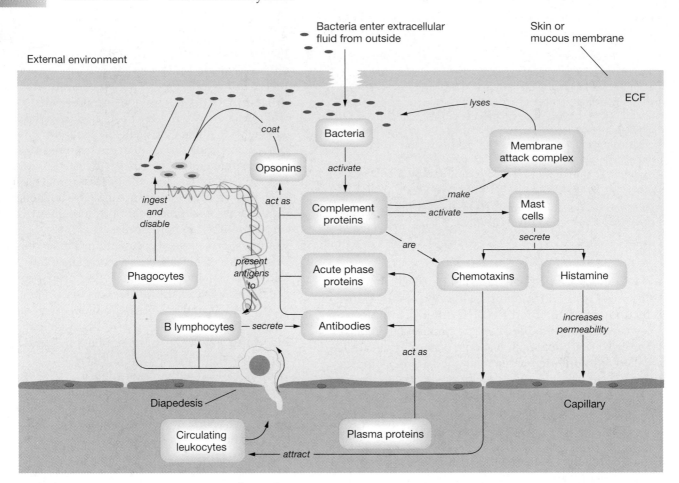

■ **Figure 22-17**   **Immune responses to bacteria**

Cytotoxic T lymphocytes are the main defense against intracellular viruses. They look for infected host cells, then destroy them.

For years, cell-mediated immunity of the T lymphocytes and humoral immunity controlled by B lymphocytes were considered independent processes. We now know that the two types of immunity are linked. Figure 22-18 ■ shows the steps of a viral infection and shows how the two types of lymphocytes coordinate to destroy viruses and virus-infected cells. In this figure, we have assumed prior exposure to the virus and pre-existing antibodies in the circulation. Antibodies can play an important defensive role in the early extracellular stages of a viral infection.

1. Antibodies act as opsonins, coating the viral particles to make them better targets for macrophages.
2. Antibody-bound viruses are prevented from entering their target cells. However, once the virus is inside the target host cell, antibodies are no longer useful.
3. Macrophages that ingest viruses insert fragments of viral antigen into MHC class II molecules on their membranes. Macrophages also secrete a variety of cytokines. Some of these cytokines initiate the in-

flammatory response. They produce alpha-interferon that causes host cells to make antiviral proteins. These keep invading viruses from replicating. Other macrophage cytokines stimulate NK cells and helper T cells.

4. Helper T cells bind to viral antigen on macrophage MHC class II molecules. Activated helper T cells are the primary mediators for additional antibody production by B lymphocytes and for activation of cytotoxic T cells.
5. Cytotoxic T cells and NK cells use viral antigen-MHC class I complexes to recognize the infected host cells. When cytotoxic cells bind to the infected host cell, they secrete the contents of their granules onto the cell surface. Perforin molecules insert pores into the host cell membrane so that granzymes can enter the cell, inducing it to commit suicide and undergo apoptosis. Destruction of infected host cells is a key step in halting the reproduction of invading viruses.

Antibodies Against a Virus May Not Work in Subsequent Infections   Although Figure 22-18 ■ showed antibodies from a previous infection protecting the body, there is no guarantee that an antibody produced during

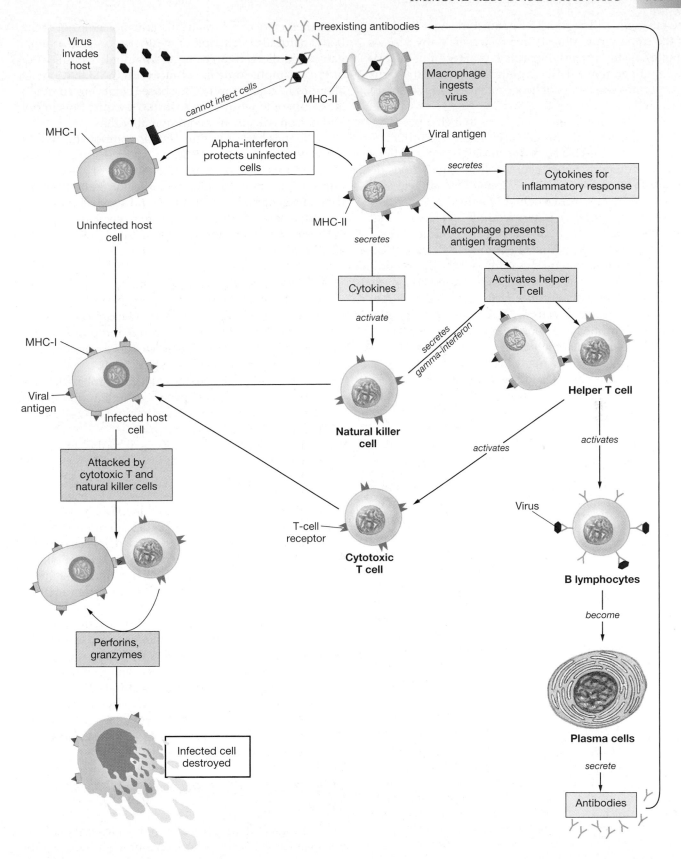

■ **Figure 22-18**   **Immune responses to viruses**   This figure assumes prior exposure to the virus and preexisting antibodies.

one infection will be effective against the next invasion by the same virus. Why is this the case? Many viruses mutate constantly, and the protein coat forming the primary antigen may change significantly when the virus incorporates sequences of host protein as it leaves an infected cell.

The influenza virus is one virus that changes yearly. Consequently, the annual vaccines against influenza must be developed based on the virologists' predictions of what mutations have occurred. If the predictions do not match the mutations, the vaccine for that year will not keep someone from catching the flu.

The rapid mutation of viruses is also one reason that researchers have not yet developed an effective vaccine against HIV, the virus that leads to many cases of **acquired immunodeficiency syndrome** (AIDS). In addition, there is evidence that antibodies coating HIV virus particles can actually *enhance* virus uptake into target cells. HIV virus infects cells of the immune system, particularly T lymphocytes, monocytes, and macrophages. When HIV wipes out the helper T cells, cell-mediated immunity against the virus is lost. The general loss of immune response in AIDS leaves these patients susceptible to a variety of viral, bacterial, fungal, and parasitic infections.

Allergic Responses Are Inflammatory Responses Triggered by Specific Antigens

An **allergy** is an inflammatory immune response to a nonpathogenic antigen. Left alone, the antigen, called an **allergen**, is not harmful to the body. But if an individual is sensitive to the allergen, the body creates an inflammatory response designed to get rid of the allergen. Allergic inflammatory responses have symptoms ranging from mild tissue damage to fatal reactions.

The immune response in allergies is called **sensitivity** or **hypersensitivity** to the allergen. **Immediate hypersensitivity** reactions are mediated by antibodies and occur within minutes of exposure to antigen. **Delayed hypersensitivity reactions** are mediated by T lymphocytes and may take several days to develop.

●●●●●●●●●●●●●●●●●●●●●●●●●●●●●●●●●●●●●

…continued from page 703

George remembered the day that his HIV test came back positive. In an HIV test, a sample of blood is tested for the presence of antibodies to HIV. A positive test indicates that HIV virus particles are circulating in the body and that the immune system has responded by making antibodies against the virus. After his initial shock, George asked his physician this astute question: "If I'm making antibodies against HIV, why can't these antibodies kill the virus?"

Question 3:   *Answer George's question. Hint: Recall that HIV eventually destroys helper T cells. Look at Figure 22-18 ■ to see what consequences this would have.*

●●●●●●●●●●●●●●●●●●●●●●●●●●●●●●●●●●●●●

Allergens can be practically any molecule not of the body: naturally occurring or synthetic, organic or inorganic. The allergen can be ingested, inhaled, or injected, or it can simply come in contact with the skin. Contact allergies to copper and other base metals are common among people who wear costume jewelry. Poison ivy and poison oak are also common allergens.

Allergies have a strong genetic component, so that if parents have a ragweed allergy, chances are good that their children will too. The development of allergies requires exposure to the allergen, a factor that will be affected by geographical, cultural, and social conditions.

What happens during an allergic reaction to pollen? Let's follow the development of an allergy with Figure 22-19 ■.

1. The first step, the sensitization phase ①, is equivalent to the primary immune response discussed previously (see Fig. 22-10 ■ and the right side of Fig. 22-18 ■). The allergen is ingested and processed by an antigen-presenting cell such as a macrophage, which in turn activates a helper T cell.

   The helper T cell activates B lymphocytes, resulting in plasma cell production of antibodies to the allergen. Memory T and memory B cells store the record of the allergen exposure.
2. Upon reexposure ②, equivalent to the secondary immune response, the body reacts more strongly and more rapidly.

The type of allergic reaction depends on the antibody or immune cell involved. Allergen binding to IgE is the most common response to inhaled, ingested, or injected allergens. When allergens bind to IgE antibodies on mast cells, the cells degranulate, releasing histamine and other cytokines. The result is an inflammatory reaction.

The severity of the reaction varies, ranging from localized reactions near the site of allergen entry to systemic reactions such as total body rashes. The most severe IgE-mediated allergic reaction is called **anaphylaxis** [*ana-*, without + *phylax*, guard]. In an anaphylactic reaction, massive release of histamine and other cytokines causes widespread vasodilation, circulatory collapse, and severe bronchoconstriction. Unless treated promptly, anaphylaxis can result in death.

MHC Proteins Allow Recognition of Foreign Tissue

As surgical techniques have been developed that allow transplantation of organs between human beings or from animals to humans, physicians have had to wrestle with the problem of donor tissue rejection by the recipient's immune system. The ubiquitous major histocompatibility complex proteins (MHC) are the primary tissue antigens that determine whether tissue transferred between

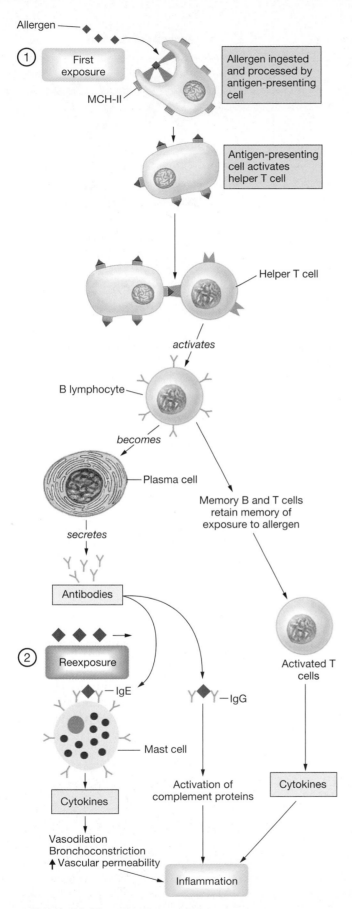

Figure 22-19   Allergic responses

Allergen

① First exposure

MCH-II

Allergen ingested and processed by antigen-presenting cell

Antigen-presenting cell activates helper T cell

Helper T cell

activates

B lymphocyte

becomes

Plasma cell

Memory B and T cells retain memory of exposure to allergen

secretes

Antibodies

② Reexposure

IgE

IgG

Mast cell

Activated T cells

Cytokines

Activation of complement proteins

Cytokines

Vasodilation
Bronchoconstriction
↑Vascular permeability

Inflammation

donor and recipient will be recognized as foreign by the recipient's immune system.

The MHC proteins are also known as **human leukocyte antigens (HLA)** and are classified in an international HLA system. A tissue graft or transplanted organ is more likely to be successful if the donor and recipient share HLA antigens. Incompatible matches trigger antibody production in much the same way as described for allergens.

One of the most widely used examples of donated tissue is blood transfusion. Human red blood cells contain antigenic membrane proteins and glycoproteins that are encoded by a number of different genes but lack the MHC proteins found on nucleated cells. In the absence of the usual cell marker for recognizing foreign tissue, the ABO blood group antigens and the Rh* antigens become the two surface proteins that are most likely to cause a rejection reaction after a blood transfusion.

The ABO blood groups consist of four blood types created by combinations of two different antigens found on red blood cells. The antigens are membrane glycoproteins designated A and B (Fig. 22-20a ■). Each person inherits two alleles for ABO antigen, one from each parent. The three possible alleles are designated A, B, and O (neither A nor B will be expressed). Alleles A and B are dominant to allele O, leading to four possible physical expressions of antigen: A, B, AB, and O (Table 22-3).

Problems with blood transfusions arise because plasma normally contains antibodies to the ABO group antigens. These antibodies are thought to be stimulated early in life by common substances such as bacteria and food components. They can be measured in the blood of infants as early as age 3–6 months.

People express antibodies to the red blood cell antigen(s) that they do not possess. Thus, people with blood type A have anti-B antibodies in their plasma, people with no antigens on their red blood cells (blood type O) have both anti-A and anti-B antibodies, and people with both antigens on their red blood cells (blood type AB) have no antibodies.

How does the body respond to a transfusion of incompatible blood? If a person with blood type O is mistakenly given a transfusion of type A blood, for example, an immune reaction takes place. The anti-A antibodies of the type O recipient bind to the transfused type A red blood cells, causing them to clump (*agglutinate*). This reaction is easily seen in a blood sample and forms the basis for the blood-typing tests often performed in student laboratories.

Antibody binding also activates the complement system, resulting in production of membrane attack complex that causes the transfused cells to swell and leak hemoglobin. Free hemoglobin in the plasma can result in acute renal failure as the kidneys try to filter the large molecules from the blood.

---

*Rh stands for Rhesus (monkeys). This group is also called the D antigen.

(a)

Blood type	Antigen on red blood cell	Antibodies in plasma
O	No A or B antigens	Anti-A and anti-B
A	A antigens	Anti-B
B	B antigens	Anti-A
AB	A and B antigens	None to A or B

(b)   A mixture of type O and type A blood

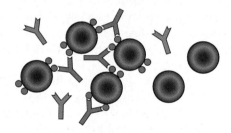

■ **Figure 22-20   ABO blood groups**   (a) The antigens and antibodies characteristic of each blood type are shown. (b) When red blood cells with group A or group B antigens on their membranes are mixed with plasma containing antibodies to their specific blood group antigens, the antibodies cause the blood cells to clump, or agglutinate.

✔ A person with AB blood type is transfused with type O blood. What happens and why?

✔ A person with O blood type is transfused with type A blood. What happens? Why?

Recognition of Self Is an Important Function of the Immune System

One amazing property of immunity is the ability of the body to distinguish its own cells from foreign cells. The lack of immune response by lymphocytes to cells of the

TABLE 22-3   ABO Blood Groups

Blood Group	Possible Genotypes	Red Blood Cell Antigens	Serum Antibody	Frequency in U.S. Population (%)			
				White	Black	Asian	American Indian
O	OO		anti-A, anti-B	45	49	40	79
A	AA, AO	A	anti-B	40	27	28	16
B	BB, BO	B	anti-A	11	20	27	4
AB	AB	A and B	no antibodies to A or B	4	4	5	<1

body is known as **self-tolerance** and appears to be due to the elimination of self-reactive lymphocytes. During development of T lymphocytes in the thymus, T lymphocytes develop that can combine with MHC-self-antigen complexes. But by a mechanism that is not understood, these self-reactive T lymphocytes are eliminated, dying by apoptosis. If self-reactive lymphocytes arise after the thymus has become inactive, they are usually either destroyed or rendered ineffective.

When self-tolerance fails, the body makes antibodies against its own components through T cell-activated B lymphocytes. The body's attack on its own cells leads to **autoimmune diseases** [*auto-*, self]. The antibodies produced in autoimmune diseases are specific against a particular antigen and are usually restricted to a particular organ or tissue type. Table 22-4 lists some common autoimmune diseases of humans.

Why does self-tolerance suddenly fail? We still do not know. We have learned, however, that autoimmune diseases often begin in association with an infection. One potential trigger for autoimmune diseases is foreign antigens that are similar to human antigens. When the body makes antibodies to the foreign antigen, those antibodies have enough cross-reactivity with human tissues to do some damage.

One example of an autoimmune disease is type 1 diabetes mellitus, in which the body makes *islet cell antibodies* that destroy pancreatic beta cells but leave the other endocrine cells untouched. Another autoimmune condition, discussed in the last chapter, is the hyperthryoidism of Graves' disease. The body makes *thyroid-stimulating immunoglobulins* that mimic thyroid-stimulating hormone and cause the thyroid gland to oversecrete hormone.

TABLE 22-4 Some Common Autoimmune Diseases in Humans

Disease	Antibodies Produced Against
Graves' disease (hyperthyroidism)	Thyroid-stimulating hormone receptor on the thyroid cells
Insulin-dependent diabetes mellitus	Pancreatic beta cell antigens
Multiple sclerosis	Myelin of CNS neurons
Myasthenia gravis	Acetylcholine receptor of motor endplate
Rheumatoid arthritis	Collagen?
Systemic lupus erythematosus	Intracellular nucleic acid protein complexes (antinuclear antibodies)
Guillian-Barré syndrome (acute inflammatory demyelinating polyneuropathy)	Myelin of peripheral nerves

Immune Surveillance Allows the Body to Remove Abnormal Cells

What is the role of the human immune system in protecting the body against cancer? This questions holds considerable interest for scientists. One theory, known as **immune surveillance**, proposes that cancerous cells develop on a regular basis but are detected and destroyed before they can spread. Although some evidence supports that theory, it does not explain why so many cancerous tumors develop every year, or why people with depressed immune function do not develop cancerous tumors in all types of tissues.

We do know that many types of tumors lack surface antigens that allow them to be recognized as abnormal by the immune system. One active area of cancer research is looking at ways to activate the immune system to fight cancerous tumor cells.

▶ NEURO-ENDOCRINE-IMMUNE INTERACTIONS

One of the most fascinating and rapidly evolving areas in physiology involves the link between the mind and the body. Although serious scientists scoffed at the topic for many years, the interaction of emotions and somatic illnesses has been described for centuries. The Old Testament says, "A merry heart doeth good like a medicine; but a broken spirit drieth the bones" (Proverbs 17:22). Most societies have tales of people who lost the will to live and subsequently died without obvious illness, or of people given up for dead who made remarkable recoveries.

Today, the medical field is beginning to pay attention to the reality of the mind-body connection. Psychosomatic illnesses and the placebo effect have been accepted for many years. Why should we not investigate the possibility that cancer patients can enhance their immune function by visualizing their immune cells gobbling up the abnormal cancer cells?

The study of brain-immune interactions is now a recognized field known as **neuroimmunomodulation**. At this time, the field is still in its infancy, and the results

...continued from page 706

George began taking the new AIDS drug cocktail. On the day he filled the prescription, his T-cell count was 338. A normal T-cell count is 1000. Within four weeks, his T-cell count was 489. Within two months, it was inching toward 900. The amount of virus in George's blood (called the "viral load") had also decreased significantly. George was thrilled. "I don't know if this will last forever," he told his doctor. "But I'm going to enjoy this reprieve while I can."

Question 4: *What correlation can be made between T-cell count and the viral load of HIV in George's blood?*

of many studies raise more questions than they answer. For example, in one experiment, mice were repeatedly injected with a chemical that suppressed the activity of their lymphocytes. During each injection, they were also exposed to the odor of camphor, a chemical that does not affect the immune system. After a period of conditioning, the mice were exposed only to the camphor odor. When the researchers checked the mice's lymphocyte function, they found that the odor of camphor had suppressed the activity of the immune cells, just as the chemical suppressant had done previously. The pathways through which this conditioning occurred are still largely unexplained.

What do we know at this point about the relationship between the immune, nervous, and endocrine systems?

1. **The three systems share common signal molecules and common receptors for those molecules.** Chemicals that we once thought were exclusive to a single cell or tissue are being discovered all over the body. Immune cells secrete hormones, and leukocyte cytokines are produced by non-immune cells. For example, lymphocytes secrete thyrotropin (TSH), ACTH, growth hormone, and prolactin, as well as hypothalamic corticotropin-releasing hormone (CRH).

   Receptors for hormones and cytokines are being discovered in multiple tissues. Neurons in the brain have receptors for cytokines produced by immune cells. Natural killer cells have opiate receptors and beta adrenergic receptors. Everywhere we turn, the nervous, endocrine, and immune systems seem to share chemical signal molecules and their receptors.

2. **Hormones and neuropeptides can alter the function of immune cells.** It has been known for years that increased cortisol levels due to stress were associated with decreased antibody production, depressed lymphocyte proliferation, and diminished activity of natural killer cells. Substance P, a neuropeptide, has been shown to induce degranulation of mast cells in the mucosa of the intestines and the respiratory tract. And sympathetic innervation of the bone marrow increases antibody synthesis and production of cytotoxic T cells.

3. **Cytokines from the immune system can affect neuroendocrine function.** Stressors such as bacterial and viral infections or tumors can induce stress responses in the central nervous system through cytokines released by immune cells. Interleukin-1 is probably the best-studied cytokine for this response, but the induction of cortisol release by lymphocyte ACTH is receiving considerable attention.

   Previously, it was believed that cortisol secretion depended on neural signals translated through the hypothalamic CRH-anterior pituitary ACTH pathway. Now it appears that pathogenic stressors can activate the cortisol pathway by causing immune cells to secrete ACTH.

The interaction of the three systems is summarized in the model shown in Figure 22-21 ■. The nervous, endocrine, and immune systems are closely linked through bidirectional communication carried out using cytokines, hormones, and neuropeptides. The brain is linked to the immune system by autonomic neurons, CNS neuropeptides, and leukocyte cytokines. The nervous system controls endocrine glands by secretion of hypothalamic releasing hormones, but immune cells secrete some of the same trophic hormones. Hormones from endocrine glands feed back to influence both the nervous and immune systems.

The result is a complex network of chemical signals subject to modulation by outside factors. These outside factors include physical and emotional stimuli integrated through the brain, pathogenic stressors integrated through the immune system, and a variety of miscellaneous factors including magnetic fields, chemical factors from brown adipose tissue, and melatonin from the pineal gland. It will probably take years for us to decipher these complex pathways.

Stress Alters Immune System Function

One area of interest is the link between inability to cope with stress and the development of illnesses. The modern study of stress is attributed to Hans Selye, beginning in 1936. He defined **stress** as nonspecific stimuli that disturb homeostasis and elicit an invariable stress response that he termed the *General Adaptation Syndrome*.

Selye's stress response consisted of stimulation of the adrenal glands, followed by suppression of the immune system due to high levels of circulating glucocorticoids [∞ p. 666]. For many years, Selye's experiment was the benchmark for defining stress. However, since the 1970s, the definition of stress and the stress response has broadened.

Stressors, the events or items that create stress, are highly variable and difficult to define in experimental

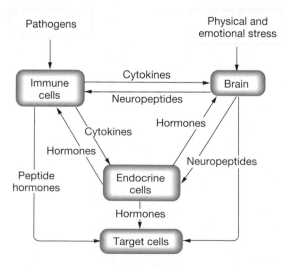

■ **Figure 22-21** **Model for interaction between nervous, endocrine, and immune systems**

settings. Acute stress is different from chronic stress. A person's reaction to stress is affected by loneliness and by whether the person feels in control of the stressful situation. Many stressors are sensed and interpreted by the brain, leading to modulation of the stressor by experience and expectations. A stressor to one person may not affect another.

Most physical and emotional stressors are integrated through the central nervous system. The two classic stress responses are (1) the rapid, neurally mediated **fight-or-flight reaction,** and (2) the elevation of adrenal cortisol levels with associated suppression of the immune response. Fight-or-flight is a rapid reaction to acute stress. The cortisol response is a better indicator of chronic or repetitive stress.

One difficulty in studying the body's response to stress is the complexity of integrating information from three levels of response (Fig. 22-22 ■). Scientists have gathered a good deal of information at each of three levels: cellular and molecular factors related to stress, systemic responses to stress, and clinical studies describing the relationships between stress and illness. In some cases, we have linked two different levels together, although in other cases, the experimental evidence is scanty or even contradictory. What we still do not understand is the big picture, the way that all three levels of response overlap.

Experimental progress is slow because many traditional study animals, such as rodents, are not good models for studying stress in humans. At the same time, controlled experiments that create stress in humans are difficult to design and execute for both practical and ethical reasons. Much of our research is observational.

For example, a recent study followed natural killer cell activity over an 18-month period in two groups of people: spouse caregivers of Alzheimer's disease patients, a group that reports high psychosocial stress, and

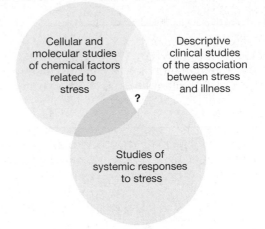

■ Figure 22-22   **Neuroimmunomodulation**   The field of neuroimmunomodulation is still in its infancy, and we do not understand how the nervous, immune, and endocrine systems interact to affect health and well-being.

a matched group of controls. For a subset of the caregivers, high perceived stress at the beginning of the study was associated with low NK cell activity at the end. However, there are many variables that must be taken into account before we can conclude that stress was the causative factor, including age, gender, exercise, alcohol use, and prescription drug use.

In spite of the difficulties, scientists are slowly beginning to piece together the intricate tapestry of mind-body interactions. When we can explain why left-handed people have a higher incidence of autoimmune and allergic diseases than do right-handed people, why students tend to get sick just before exam time, and why relaxation therapy enhances the immune response, we will be well on the way to understanding the complex connections between the brain and the immune system.

PROBLEM CONCLUSION

In this running problem, you learned about a treatment for AIDS that consists of two antiretroviral drugs and a protease inhibitor. The combination of these drugs seems to stem the tide of HIV replication more effectively than any of the drugs used alone.

Further check your understanding of this running problem by comparing your answers to those in the summary table.

	Question	Facts	Integration and Analysis
1	Why can't retroviruses reproduce without reverse transcriptase?	Retroviruses have no DNA to use as a template to make viral mRNA. Reverse transcriptase is an enzyme that allows retroviruses to make complementary DNA from their own viral RNA.	Without reverse transcriptase, the retroviruses cannot make viral DNA. Without viral DNA, the virus cannot make viral mRNA or viral proteins. Therefore, without reverse transcriptase, a retrovirus cannot replicate.

	Question	Facts	Integration and Analysis
2	What is the rationale for flooding the body with CD4 protein in order to prevent HIV replication?	CD4 is the receptor on helper T cells to which HIV binds. Binding causes the virus to be taken into the T cell.	Introducing large amounts of CD4 protein into the body might "trick" HIV into binding to these free proteins instead of to CD4 on helper T cells. This would reduce the number of virus particles that could invade the T cells.
3	George asks, "If I'm making antibodies against HIV, why can't these antibodies kill the virus"? Answer George's question.	Viruses can only be affected by antibodies when they are outside the host cells. HIV bound to antibody is taken up into helper T cells.	HIV enters helper T cells and, over time, will destroy them.
4	What correlation can be made between T-cell count and the viral load blood?	George's T-cell count increases while the viral load decreases.	Apparently, the drug cocktail is effective in decreasing the rate of T-cell infection by HIV. This has two results. As fewer T cells are killed by the virus, the T-cell count rises. As the drugs suppress viral replication, the viral load decreases.

CHAPTER REVIEW

SUMMARY

Overview of Immune System Function

1. **Immunity** is the ability of the body to protect itself from disease-causing entities. The human immune system consists of lymphoid tissues, immune cells, and cytokines. (p. 686)

2. **Antigens** are substances that trigger an immune response and react with products of that response. (p. 686)

Pathogens of the Human Body

3. Bacteria are cells that can survive and reproduce outside of a living host. Antibiotic drugs kill bacteria. (p. 686)

4. Viruses are DNA or RNA cores enclosed in protective proteins. They invade host cells and reproduce using the host's intracellular machinery. The immune system must find and destroy infected host cells to kill a virus. (p. 687)

The Immune Response

5. The immune response includes: (1) *detection* and *identification*, (2) *communication* with other immune cells, (3) *recruitment* of assistance and *coordination* of the response, and (4) *destruction* or *suppression* of the invader. (p. 688)

6. **Innate immunity** is the rapid, nonspecific immune response. **Acquired immunity** is slower and directed at specific invaders. The two types of immunity overlap. (p. 688)

Anatomy of the Immune System

7. Immune cells form and mature in the **thymus gland** and bone marrow. The **spleen** and **lymph nodes** are **encapsulated** lymphoid tissues. **Unencapsulated lymphoid tissues** include **tonsils** and the **gut-associated lymphoid tissue (GALT)**. (p. 688)

8. **Eosinophils** are **cytotoxic cells** that kill parasites. **Basophils** and **mast cells** release histamine, **heparin**, and other cytokines. (p. 691)

9. **Neutrophils** are phagocytic cells that release pyrogens and cytokines that mediate inflammation. (p. 692)

10. **Monocytes** are the precursors of tissue **macrophages**. **Antigen-presenting cells** display antigen fragments in proteins on their membrane. (p. 692)

11. **Lymphocytes** and **plasma cells** mediate the acquired immune response. (p. 692)

12. **Dendritic cells** capture antigens, particularly viruses, and present antigen fragments to lymphocytes. (p. 693)

Innate Immunity

13. Physical and chemical barriers, such as skin and mucus, are the first line of defense. (p. 693)

14. **Chemotaxins** attract immune cells to pathogens. (p. 693)

15. Phagocytosis requires that surface molecules on the pathogen bind to membrane receptors on the phagocyte. **Opsonins** are tagging proteins that coat foreign material. (p. 693)

16. **Inflammation** results from cytokines released by activated tissue macrophages. (p. 694)

17. Inflammatory cytokines include **acute phase proteins**, histamine, **interleukins**, **bradykinin**, and **complement proteins**. (p. 695)

18. The complement cascade forms **membrane attack complex**, pore-forming molecules that lyse pathogens. (p. 696)

Acquired Immunity

19. Lymphocytes mediate the acquired immune response. Lymphocytes of one type form a **clone**. (p. 696)

20. First exposure to an antigen activates a **naive lymphocyte** clone and causes it to divide (**clonal expansion**). Newly formed lymphocytes differentiate into **effector cells** or **memory cells**. Additional exposure to the antigen creates faster and stronger secondary response. (p. 697)

21. Lymphocyte cytokines include **interleukins** and **interferons**, which interfere with viral replication. **Gamma-interferon** activates macrophages and other immune cells. (p. 697)

22. **B lymphocytes** secrete antibodies. **T lymphocytes** and **natural killer cells** attack and destroy infected cells. (p. 698)

23. **Antibodies** bind to pathogens. Mature B lymphocytes are covered with antibodies that act as surface receptors. Activated B lymphocytes differentiate into **plasma cells**, which secrete soluble antibodies. (p. 698)

24. The five classes of immunoglobulins (Ig) are IgG, IgA, IgE, IgM, and IgD, collectively known as **gamma globulins**. (p. 698)

25. The antibody molecule is Y-shaped and has four polypeptide chains. The arms, or **Fab regions**, contain antigen-binding sites. The stem is known as the **Fc region** and can bind to immune cells. Some immunoglobulins are polymers. (p. 699)

26. Soluble antibodies act as opsonins, serve as a bridge between antigens and leukocytes, activate complement proteins and mast cells, and disable viruses and toxins. (p. 699)

27. B cell antibodies bind antigens to activate B lymphocytes and initiate the production of additional antibodies. (p. 699)

28. **T lymphocytes** are responsible for **cell-mediated immunity**, in which the lymphocytes bind to target cells using **T-cell receptors**. (p. 699)

29. T lymphocytes react to antigen bound to **major histocompatibility complex** proteins on the target cell. **MHC class I molecules** cause T lymphocytes to kill the cell displaying MHC I-antigen complexes. **MHC class II molecules** with antigen activate leukocytes to secrete cytokines. (p. 700)

30. **Cytotoxic T cells** release **perforin**, a pore-forming molecule that allows **granzymes** to enter the target cell and trigger it into committing suicide (**apoptosis**). (p. 701)

31. **Helper T (T$_H$) cells** secrete cytokines that influence other cells. (p. 702)

32. **Natural killer cells** kill certain tumor cells and virus-infected cells. They bind to the Fc stem of antibodies. This is called **antibody-dependent cell-mediated cytotoxicity**. (p. 702)

Immune Response Pathways

33. Bacteria usually cause a nonspecific inflammatory response. In addition, lymphocytes produce antibodies keyed to the specific type of bacterium. (p. 703)

34. Innate immune responses and antibodies help control the early stages of a viral infection. Once the viruses enter the host's cells, cytotoxic T cells must destroy infected host cells. (p. 703)

35. An allergy is an inflammatory immune response to a nonpathogenic **allergen**. The body gets rid of the allergen by using an inflammatory response. (p. 706)

36. **Immediate hypersensitivity reactions** to allergens are mediated by antibodies and occur within minutes of exposure to antigen. **Delayed hypersensitivity reactions** are mediated by T lymphocytes and may take several days to develop. (p. 706)

37. The major histocompatibility complex (MHC) proteins found on all nucleated cells determine tissue compatibility. Lymphocytes will attack any MHC complexes that they do not recognize as self proteins. (p. 706)

38. Human red blood cells lack MHC proteins but contain

other antigenic membrane proteins and glycoproteins such as the ABO and Rh antigens. (p. 707)

39. **Self-tolerance** is the ability of the body to distinguish its own cells and not create an immune response. When self-tolerance fails, the body makes antibodies against its own components, creating an **autoimmune disease**. (p. 708)

40. The theory of **immune surveillance** proposes that cancerous cells develop on a regular basis but are detected and destroyed by the immune system before they can spread. (p. 709)

Neuro-Endocrine-Immune Interactions

41. The nervous, endocrine, and immune systems are linked together by signal molecules and receptors. The study of these interactions is a field known as **neuroimmunomodulation**. (p. 709)

42. The link between inability to cope with stress and the development of illnesses is believed to result from neuroimmunomodulation. (p. 710)

QUESTIONS

LEVEL ONE   Reviewing Facts and Terms

1. Define immunity. What is meant by memory and specificity in the immune system?

2. Name the anatomical components of the immune system.

3. List and briefly summarize the three main functions of the immune system.

4. Name two ways viruses are released from the host cell. Give three ways in which viruses damage host cells or disrupt their function.

5. If a pathogen has invaded the body, what four steps occur if the body is able to destroy the pathogen? What compromise may be made if the pathogen cannot be destroyed?

6. Define the following terms and explain their significance:
   (a) anaphylaxis
   (b) agglutinate
   (c) extravascular
   (d) degranulation
   (e) acute phase protein
   (f) clonal expansion
   (g) immune surveillance

7. How are histiocytes, Kupffer cells, osteoclasts, and microglia related?

8. What is the mononuclear phagocytic system and what role does it play in the immune system?

9. Match the cell type with its description:
   (a) lymphocyte
   (b) neutrophil
   (c) monocyte
   (d) dendritic cell
   (e) eosinophil
   (f) basophil

   1. most abundant leukocyte; phagocytic; lives 1–2 days
   2. cytotoxic; associated with allergic reactions and parasitic infestations
   3. precursor of macrophages; relatively rare in blood smears
   4. related to mast cells; releases cytokines such as histamine and heparin
   5. most of these cells reside in the lymphoid tissues
   6. these cells capture antigens, then migrate to the lymphoid tissues

10. Give examples of physical and chemical barriers to infection.

11. Name the three main types of lymphocytes and their subtypes. Explain the functions and interactions of each group.

12. Define self-tolerance. Which cells are responsible? Relate self-tolerance to autoimmune disease.

13. What is meant by the term *neuroimmunomodulation*?

14. Explain the terms *stress*, *stressor*, and *general adaptation syndrome*.

LEVEL TWO   Reviewing Concepts

15. **Concept map:** Map the following terms. You may add additional terms.

acquired immunity	cytotoxic cell
antibody	dendritic cell
antigen	eosinophil
antigen-presenting cell	Fab region
B lymphocyte	Fc region
bacteria	helper T cell
basophil	innate immunity
chemotaxin	memory cell
cytokine	MHC class I
cytotoxic T cell	MHC class II

monocyte	phagosome
natural killer cell	plasma cell
neutrophil	T lymphocyte
opsonin	viruses
phagocyte	

16. Why do lymph nodes often swell and become tender or even painful when you are sick?

17. Summarize the effects of histamine, interleukin-1, acute phase proteins, bradykinin, complement, and gamma interferon. Are these chemicals antagonistic or synergistic to one another?

18. Compare and contrast the following sets of terms:

    (a) pathogen, microbe, pyrogen, antigen, antibody, antibiotic

    (b) infection, inflammation, allergy, autoimmune disease

    (c) virus, bacteria, allergen, retrovirus

    (d) chemotaxin, opsonin, cytokine, interleukin, bradykinin, interferon

    (e) diapedesis, phagocytosis

    (f) acquired immunity, innate immunity, humoral immunity, cell-mediated immunity

    (g) immediate and delayed hypersensitivity

    (h) membrane attack complex, granzymes, perforin

19. Diagram and label the parts of an antibody molecule. What is the significance of each part?

20. Diagram the steps in the process of inflammation.

21. Diagram the steps in the process of fighting viral infections.

22. Diagram the steps in the process of an allergic reaction.

LEVEL THREE   Problem Solving

23. One ABO blood type is called the "universal donor" and one is called the "universal recipient." Which types are these and why are they given those names?

24. Maxie Moshpit claims that Snidley Superstar is the father of her child. Maxie has blood type O. Snidley has blood type B. Baby Moshpit has blood type O. Can you rule out Snidley as the father of the baby with only this information? Explain.

25. Every semester around exam time more students visit the Student Health Center with colds and viral infections. The physicians attribute the increased incidence of illness to stress. Can you explain the biological basis for that relationship? What are some other possible explanations?

26. Barbara is troubled with rheumatoid arthritis, characterized by painful, swollen joints from inflamed connective tissue. She notices that it flares up when she is tired and stressed. This condition is regarded as an autoimmune disease. Outline the physiological reactions that would lead to this condition.

27. The differential white cell count [∞ Fig. 16-2, p. 478] is an important diagnostic tool in medicine. If a patient's "diff" shows an increase in the number of immature neutrophils ("bands") in the circulation, do you think the cause is more likely to be a bacterial or a viral infection? If the count shows an increase in esosinophils (*eosinophilia*), is the cause more likely to be a viral or a parasitic infection? Explain your reasoning.

E X P L <MediaLab>

Introduction

In this chapter you learned how the body protects itself from pathogens and removes damaged tissue or abnormal cells using the cells and chemicals of the immune system. An immune response requires detection of the foreign substance, communication with other immune cells, and coordination of the response. This MediaLab will help you gain an appreciation of these mechanisms and how they can be applied to everyday events. After reading the description below, visit the MediaLab for Chapter 22 in your Companion Website and select the appropriate keyword.

Web Exploration 1

Estimated time for completion = 10 minutes

Our immune system functions to protect us from pathogens such as bacteria, viruses, and toxins. What is the immune system's initial response when our body is invaded by a pathogen? How is this initial response mediated?

The immune response can be illustrated by something as simple (and painful) as a bee sting. We have heard of or may know of someone who is particularly sensitive to bee stings. It is possible that the first few times they were stung by a bee, their reaction was no greater than that of an average person. However, with several bee stings over a period of time, their reaction increased in intensity. Can you postulate what might mediate this delayed hypersensitivity to repeated exposure of the antigen from the bee sting?

When someone's sensitivity to an allergen such as bee venom increases, these people are at risk for developing **ANAPHYLAC-TIC SHOCK.** Select this keyword from your Website and learn more about anaphylaxis. What are the primary symptoms of this condition? What other commonly encountered allergens can cause anaphylaxis? To visualize other allergic responses, visit the **SKIN REACTIONS** link on your Companion Website.

Web Exploration 2

Estimated time for completion = 15 minutes

It is estimated that 1 out of 5 Americans has some kind of auto-immune disease. Autoimmune diseases are a major problem in this country, and 86 billion dollars are spent every year in search of treatments. The mechanisms that lead to an immune response against self-tissue is a topic of intense investigation and appear to vary depending upon the disease in question. What we do know is that loss of self-tolerance may be organ-specific, localized, or systemic.

To explore autoimmune diseases, select **IMMUNE TOLER-ANCE** on your Website and scroll down to "Loss of Immune Tolerance." This site will help you learn more about what might be happening.

One devastating disease thought to be autoimmune in origin is multiple sclerosis (MS). MS is caused by the destruction of myelin tissue in the brain. Given what we have learned in Chapter 9 on the central nervous system, can you predict what symptoms a person with MS is likely to experience? If MS is autoimmune, what can you think of that could destroy myelin? To help you explore this topic further, diagram or propose a mechanism of how the immune system might mediate this process. Now select **MS** from your Website to see if what you have diagrammed correlates with the mechanism proposed by scientists.

23 Integrative Physiology III: Exercise

■ "One fascinating aspect of the physiology of the human being at work is that it provides basic information about the nature and the range of the functional capacity of different organ systems."
—*Per-Olof Åstrand and Kaare Rodahl*, Textbook of Work Physiology, 1977. ■

CHAPTER OUTLINE

BACKGROUND BASICS

One of the most common challenges to the body's homeostasis is not a disease but an everyday activity, exercise. The American College of Sports Medicine defines **exercise** as any muscular activity that generates force and disrupts homeostasis. Running is an example of a dynamic endurance exercise. Weight lifting and strength training are examples of dynamic resistive exercise.

In this chapter we examine dynamic exercise as a challenge to homeostasis that is met with an integrated response from multiple systems of the body. In many ways, exercise is the ideal teaching example of physiological integration. Everyone is familiar with it. It is an activity that does not require special environmental conditions, unlike high-altitude activities or deep-sea diving. Exercise is a normal physiological state rather than a pathological one, although it can be affected by disease and, if excessive, can result in injury.

Note that this chapter does not intend to be a complete description of exercise physiology. For further information, readers should consult an exercise physiology textbook.

◗ METABOLISM AND EXERCISE

Why does exercise disrupt homeostasis? The reason is that exercising muscle demands a steady supply of ATP, produced through metabolism. Metabolism in turn demands oxygen and fuel, such as glucose. Let's start by examining muscle metabolism.

Exercise begins with skeletal muscle contraction, an active process that requires ATP for energy. But where does the ATP for muscle contraction come from? A small amount is in the muscle fiber cytoplasm when contraction begins (Fig. 23-1 ■). As this ATP is transformed into ADP, another phosphate compound, **phosphocreatine** (PCr), transfers energy from its high-energy phosphate bond to ADP. The transfer replenishes the muscle's supply of ATP [∞ p. 35].

Together, the combination of muscle ATP and phosphocreatine is adequate to support only about 15 seconds of intense exercise. Consequently, the muscle fiber must manufacture additional ATP from energy stored in the chemical bonds of nutrients. Some of these mole-

PROBLEM
Heat Stroke

"Dangerous heat continues for fifth straight day," reads the morning headline. Had Colleen Warren, 19, taken time to scan the story, she might have avoided the life-threatening health emergency she faced a few hours later. But Colleen overslept and was rushing to get to field hockey practice. The weather during this first week of the fall semester was unbearably hot and humid, just as it has been all summer. Today, the temperature was predicted to reach 105°F, with high humidity. In her haste to leave, Colleen forgot her water bottle. "Oh well," she thought as she hopped on her bike to peddle the two miles to the practice field.

...continued on page 720

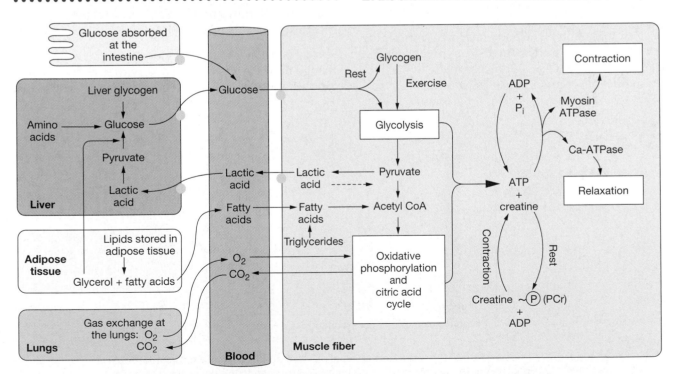

■ **Figure 23-1   Energy metabolism in skeletal muscle**   Glucose for ATP production comes from the diet or from glycogen stores in muscle and liver. Oxygen for aerobic metabolism enters the body at the lungs. Fatty acids from adipose tissue and in muscle can be used for ATP synthesis only if oxygen is available. When oxygen is not available, glucose is metabolized to lactic acid. The liver is able to convert lactic acid back to glucose. Some energy in the muscle is stored in the form of phosphocreatine (PCr).

cules are contained within the muscle fiber itself. Others must be mobilized from the liver and adipose tissue, then transported to muscles through the circulation.

The primary substrates for energy production are carbohydrates and fats. Recall from Chapter 4 that the most efficient production of ATP occurs through aerobic pathways such as the glycolysis-citric acid cycle pathway [∞ p. 90]. If the cell has adequate amounts of oxygen for oxidative phosphorylation, then both glucose and fatty acids can be metabolized to provide ATP (Fig. 23-1 ■).

If the oxygen requirement of a muscle fiber exceeds its oxygen supply, energy production from fatty acids decreases dramatically and glucose metabolism shifts to anaerobic pathways. Anaerobic metabolism is also known as **glycolytic metabolism**, because in low oxygen conditions, anaerobic glycolysis is the primary pathway for ATP production. When a cell lacks oxygen for oxidative phosphorylation, the final product of glycolysis, pyruvate, is converted to lactic acid [∞ p. 92]. In general, exercise that depends on anaerobic metabolism cannot be sustained for an extended period.

Anaerobic metabolism has the advantage of speed, producing ATP 2.5 times as rapidly as aerobic pathways do (Fig. 23-2 ■). But this advantage comes with two distinct disadvantages: (1) anaerobic metabolism provides only 2 ATP per glucose, compared with the average of 30–32 ATP per glucose for oxidative metabolism, and (2) it contributes to a state of metabolic acidosis by producing lactic acid. (However, the CO_2 generated during metabolism is a more significant source of acid.)

Where does glucose for aerobic and anaerobic ATP production come from? The body has three sources: the plasma glucose pool, intracellular stores of glycogen in muscles and liver, and "new" glucose made in the liver from nonglucose precursors such as amino acids (*gluco-*

neogenesis; [∞ p. 98]). Muscle and liver glycogen stores provide enough energy substrate for about 2000 kcal (equivalent to about 20 miles of running in the average person), more than adequate for the exercise that most of us do. However, glucose alone cannot provide sufficient ATP for endurance athletes such as marathon runners. To meet their energy demands, they rely on the energy stored in fats, estimated at 70,000 kcal per person.

In reality, aerobic exercise of any duration uses both fatty acids and glucose as substrates for ATP production. About 30 minutes after aerobic exercise begins, the concentration of free fatty acids in the blood increases significantly, showing that fats are being mobilized from adipose tissue. However, the breakdown of fatty acids through the process of β-oxidation [∞ p. 97] is slower than glucose metabolism through glycolysis, so muscle fibers use a combination of fatty acids and glucose to meet their energy needs.

At lower exercise intensity, most of the energy for ATP production comes from fats (Fig. 23-3 ■). As the intensity of exercise increases and ATP is consumed more rapidly, the muscle fibers begin to use a larger proportion of glucose. When exercise reaches about 70% of maximum, carbohydrates become the primary source of energy.

Aerobic training increases both fat and glycogen stores within the muscle fibers themselves. Training also enhances the ability of exercising muscle to use fatty acids as a source of ATP by increasing the activity of enzymes for β-oxidation.

Hormones Regulate Metabolism During Exercise

Several hormones that affect glucose and fat metabolism change their pattern of secretion with exercise. Plasma concentrations of glucagon, cortisol, the catecholamines

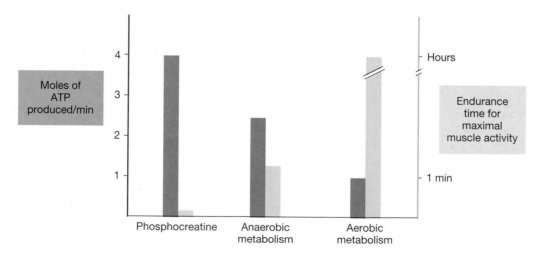

■ **Figure 23-2** **Speed of ATP production compared with ability to sustain maximal muscle activity** Phosphocreatine in muscle fibers has the highest rate of energy production. However, a muscle fiber contains only enough phosphocreatine to sustain about 10 seconds of maximal exercise. Anaerobic metabolism can product ATP at a rate 2.5 times as fast as aerobic metabolism, but can sustain maximal muscle activity for only about a minute. Aerobic metabolism of glucose and fatty acids is the slowest way for the muscle fiber to produce ATP, but aerobic exercise can be sustained for hours.

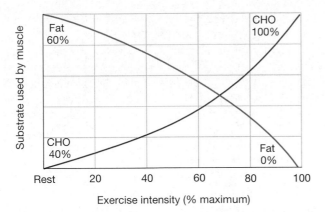

■ **Figure 23-3** **Use of carbohydrates and fats with increasing exercise** At low intensity exercise, muscles rely more on fats than glucose. Above exercise intensities exceeding 70% of maximum, glucose becomes the main energy source. Data from G. A. Brooks and J. Mercier, "Balance of carbohydrate and lipid utilization during exercise: The 'crossover' concept." *Journal of Applied Physiology* 76:2253–2261 (1994).

(epinephrine and norepinephrine), and growth hormone all increase during exercise. Cortisol and the catecholamines, along with growth hormone, promote the conversion of triglycerides to glycerol and fatty acids. Glucagon, catecholamines, and cortisol also mobilize liver glycogen and raise plasma glucose levels. A hormonal environment that favors the conversion of glycogen into glucose is desirable, because glucose is a major energy substrate for exercising muscle.

Curiously, although plasma glucose concentrations rise with exercise, the secretion of insulin decreases. This response is contrary to what you would predict, because normally an increase in plasma glucose stimulates insulin release. During exercise, however, insulin secretion is suppressed, probably by sympathetic input onto the beta cells of the pancreas.

What could be the advantage of lower insulin levels during exercise? For one thing, less insulin means that cells other than muscle fibers reduce their glucose uptake, sparing blood glucose for the muscles to use. Actively contracting muscle cells, on the other hand, are not affected by low levels of insulin because they do not require insulin for glucose uptake. Experiments have shown that electrical stimulation of a muscle fiber in a glucose solution increases its glucose uptake in proportion to its contractile activity.

Oxygen Consumption Is Related to Exercise Intensity

The activities we call exercise range widely in intensity and duration, from the rapid burst of energy exerted by a sprinter or power lifter to the sustained effort of a marathoner. Physiologists traditionally quantify the intensity of a period of exercise by measuring oxygen consumption ($\dot{V}_{O_2}$). **Oxygen consumption** refers to the fact that oxygen is used up, or consumed, during oxidative phosphorylation, when it combines with hydrogen in the mitochondria to form water [∞ p. 94].

Oxygen consumption is a measure of cellular respiration and is usually measured in liters of oxygen consumed per minute. A person's *maximal rate of oxygen consumption* ($\dot{V}_{O_2max}$) is an indicator of a person's ability to perform endurance exercise. The greater the $\dot{V}_{O_2max}$, the greater the person's predicted ability to do work.

A metabolic hallmark of exercise is an increase in oxygen consumption that persists even after the activity ceases (Fig. 23-4 ■). When exercise commences, oxygen consumption increases rapidly, but it is not instantaneously matched by the oxygen supply to the muscles. During the lag time, ATP is provided by muscle ATP reserves, phosphocreatine, and aerobic metabolism supported by oxygen stored on muscle myoglobin and blood hemoglobin [∞ pp. 361, 483].

The use of these muscle stores creates an *oxygen deficit* because their replacement requires aerobic metabolism and oxygen uptake. Once exercise stops, oxygen consumption is slow to resume its resting level. The **excess postexercise oxygen consumption** (EPOC; formerly called the oxygen debt) represents oxygen being used to metabolize lactate, restore ATP and phosphocreatine levels, and replenish the oxygen bound to myoglobin. Other factors that play a role in elevating postexercise oxygen consumption include increased body temperature and circulating catecholamines.

Several Factors Limit Exercise

What are the factors that limit exercise capacity? To some extent, the answer depends on the type of exercise. Resistive exercise such as strength training is likely to be limited by the availability of oxygen, so the muscles rely heavily on anaerobic energy production. The picture is more complex with aerobic or endurance exercise. Is the limiting factor for aerobic exercise the ability of the exercising muscle

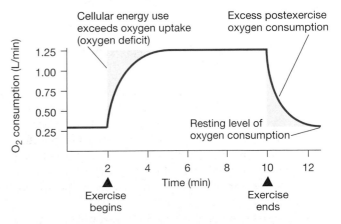

■ **Figure 23-4** **Changes in oxygen consumption during and after exercise** The cells' use of energy begins as soon as exercise begins, but oxygen supply as measured by oxygen consumption lags behind, creating an oxygen deficit. The oxygen deficit is reflected by elevated oxygen consumption that persists after exercise stops.

...continued from page 717

By the time Colleen reached the practice field, she was already sweating and her face was flushed. At 9 A.M., the air temperature was 80°F, and the humidity was 54%. Colleen took a quick drink of water from the large water container and ran out to the field.

Question 1: *"Humidity" is the percentage of water present in air. Why is the thermoregulatory mechanism of sweating less efficient in humid environments?*

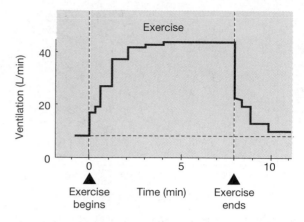

■ **Figure 23-5** **Changes in ventilation with submaximal exercise** Ventilation rate jumps as soon as exercise begins, despite the fact that neither arterial P_{CO_2} nor P_{O_2} has changed. Modified from P. Dejours, *Handbook of Physiology* (Washington, D.C.: American Physiological Society, 1964).

to use oxygen efficiently? The ability of the cardiovascular system to deliver oxygen to the tissues? Or the ability of the respiratory system to provide oxygen to the blood?

One possible limiting factor in exercise is the ability of muscle fibers to use oxygen. If muscle mitochondria are limited in number, or if they have insufficient oxidative enzyme content, the muscle fibers are unable to produce ATP rapidly, even with adequate supply of oxygen and substrates. Although data suggest that muscle metabolism is not the limiting factor for maximum exercise capacity, it has been shown to influence submaximal exercise capacity—a finding that explains the increase in size and number of muscle mitochondria with endurance training.

The question of whether the pulmonary system or the cardiovascular system limits maximal exercise was resolved when research showed that ventilation is only 65% of its maximum when cardiac output has reached 90% of its maximum. From that information, exercise physiologists concluded that the ability of the cardiovascular system to deliver oxygen and nutrients to the muscle at a rate that supports aerobic metabolism is a major factor in determining maximum oxygen consumption. Let's now examine the reflexes that integrate breathing and cardiovascular function during exercise.

▶ VENTILATORY RESPONSES TO EXERCISE

Think about what happens to your breathing when you exercise. Exercise is associated with both increased rate and increased depth of breathing, resulting in enhanced alveolar ventilation [∞ p. 516]. **Exercise hyperventilation**, or **hyperpnea**, results from a combination of feedforward signals from central command neurons in the motor cortex and sensory feedback from peripheral receptors.

When exercise begins, mechanoreceptors and proprioreceptors [∞ p. 387] in muscles and joints send information about movement to the motor cortex. Descending pathways from the motor cortex to the respiratory control center of the medulla oblongata then immediately increase ventilation (Fig. 23-5 ■).

As muscle contraction continues, sensory information from a variety of sources feeds back to the control center to ensure that ventilation and oxygen use by the tissue remain closely matched. The sensory receptors involved in the secondary response probably include central and carotid body chemoreceptors that monitor P_{CO_2}, pH, and P_{O_2} [∞ p. 532], proprioceptors in the joints, and possibly receptors located within the exercising muscle itself. Pulmonary stretch receptors [∞ p. 536] were once thought to play a role, but recipients of heart-lung transplants display a normal ventilatory response to exercise although the neural connections between lung and brain are absent.

Exercise hyperventilation maintains nearly normal arterial P_{O_2} and P_{CO_2} by steadily increasing alveolar ventilation in proportion to the level of exercise. The compensation is so effective that when arterial P_{O_2}, P_{CO_2}, and pH are monitored during mild to moderate exercise, they show no significant change (Fig. 23-6 ■). This observation eliminates the traditional explanation for increased ventilation during exercise. Low arterial P_{O_2}, elevated arterial P_{CO_2}, and decreased plasma pH are the usual stimuli for the carotid and central chemoreceptors. The fact that these stimuli are not present in mild to moderate exercise means that the chemoreceptors or medullary respiratory control center, or both, must be responding to other exercise-induced signals.

Several factors have been postulated as the stimulus for exercise hyperventilation, including sympathetic input to the carotid body and changes in plasma K^+ concentration. During even mild exercise, plasma K^+ levels increase as repeated action potentials in the muscle fibers allow K^+ to move out of cells and into the extracellular fluid. Carotid chemoreceptors are known to respond to increased K^+ concentration by increasing ventilation. However, because K^+ concentration changes slowly, this mechanism does not explain the sharp initial rise in ventilation at the onset of activity.

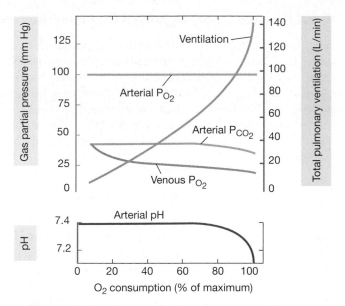

■ **Figure 23-6 Changes in blood gas, partial pressures, and arterial pH with exercise** Pulmonary ventilation increases steadily as exercise increases, but arterial P_{O_2} remains constant, indicating that ventilation is matched to blood flow through the lungs. Venous P_{O_2} drops as exercise increases, showing that the cells are removing more oxygen from hemoglobin as oxygen consumption increases. Arterial P_{CO_2} does not change until oxygen consumption reaches 80% of maximum. At that level, ventilation increases sharply. As the person begins to hyperventilate, arterial (and alveolar) P_{CO_2} declines. Blood pH also remains fairly constant except at near-maximal exercise levels.

It appears likely that the initial change in ventilation is caused by sensory input from muscle mechanoreceptors combined with parallel descending pathways from the motor cortex. Once exercise is underway, sensory input keeps ventilation matched to metabolic needs.

✔ If venous P_{O_2} decreases as exercise intensity increases, what do you know about the P_{O_2} of the muscle cells as exercise intensity increases?

•••

...continued from page 720

By 10 A.M., the temperature had risen to 93° F, with 50% humidity. Colleen felt dizzy and nauseated, but she pressed on. She had made the team by the skin of her teeth, and felt she had to prove herself to her teammates. She took only a short drink of water at the break and hustled back to the field. At 10:07 A.M. Colleen collapsed on the field. One of Colleen's teammates, with training in first aid, felt Colleen's skin. It was hot and dry. A call went in to emergency medical services.

Question 2: *Individuals with a heat emergency called* heat exhaustion *have cool, moist skin. Hot, dry skin indicates a more serious emergency called* heat stroke. *Why is heat exhaustion less serious than heat stroke? (Hint: How does skin temperature relate to the body's ability to regulate body temperature?)*

•••

▶ CARDIOVASCULAR RESPONSES TO EXERCISE

When exercise begins, mechanosensory input from working limbs combines with descending pathways from the motor cortex to activate the cardiovascular control center in the medulla oblongata. The center responds with sympathetic discharge that increases cardiac output and causes vasoconstriction in many peripheral arterioles.

Cardiac Output Increases During Exercise

During strenuous exercise, cardiac output rises dramatically. In untrained individuals, cardiac output goes up fourfold, from 5 L/min to 20 L/min. In trained athletes, it may go up six to eight times, reaching as much as 35 L/min. As you learned in Chapter 14, cardiac output is influenced by several factors, including (1) heart rate, (2) force of contraction, and (3) venous return [∞ p. 437]:

Cardiac output (CO) = heart rate × stroke volume

Or, if the factors that influence heart rate and stroke volume are considered,

CO = (SA node rate + influence of autonomic nervous system) × (venous return + force of contraction)

Let's examine which of those factors has the greatest effect on cardiac output during exercise in a healthy heart. During exercise, venous return is enhanced by skeletal muscle contraction and deep inspiratory movements [∞ p. 436]. It is therefore tempting to postulate that, owing to Starling's law of the heart, the cardiac muscle fibers simply stretch in response to increased venous return, thereby increasing contractility.

However, overfilling the ventricles is potentially dangerous, because overstretching may damage the fibers. One way the body offsets increased venous return is by speeding up heart rate. If the interval between contractions is less, the heart will not have as much time to fill and therefore is not likely to be damaged by excessive stretch.

The initial change in heart rate at the onset of exercise is due to decreased parasympathetic activity at the sinoatrial (SA) node [∞ p. 434]. As cholinergic inhibition lessens, heart rate rises from its resting rate to around 100 beats per minute, the intrinsic pacemaker rate of the SA node. At that point, sympathetic output from the cardiovascular control center escalates.

Sympathetic stimulation has two effects on the heart. First, it increases contractility so that the heart squeezes out more blood per stroke (increased stroke volume). Second, sympathetic innervation increases heart rate so that the heart does not have much time to relax, protecting it from overfilling. The combination of faster heart rate and greater stroke volume increases cardiac output during exercise.

Peripheral Blood Flow Redistributes to Muscle During Exercise

At rest, skeletal muscles receive less than a fourth of the cardiac output, or about 1.2 L/min. During exercise, because of local and reflex reactions, a significant shift in peripheral blood flow takes place. With strenuous exercise by highly trained athletes, the combination of increased cardiac output and vasodilation can increase blood flow through exercising muscle to more than 22 L/min! The relative distribution of blood flow to the tissues also shifts. About 88% of the cardiac output is diverted to the exercising muscle, up from 21% at rest (Fig. 23-7 ■).

The redistribution of blood flow during exercise results from a combination of vasodilation in skeletal muscle arterioles and vasoconstriction in other tissues. At the onset of exercise, sympathetic signals from the cardiovascular control center cause vasoconstriction in peripheral tissues. As muscles become active, changes in the microenvironment of the muscle tissue take place. Tissue O_2 concentrations decrease, while temperature, CO_2, and acid in the interstitial fluid around the muscle fibers increase. All of these factors act as paracrines to cause local vasodilation that overrides the sympathetic signal for vasoconstriction. The net result is shunting of blood flow from inactive tissues to the exercising muscles, where it is needed.

Blood Pressure Rises Slightly During Exercise

What happens to blood pressure during exercise? You learned in Chapter 15 that peripheral blood pressure is determined by a combination of cardiac output and peripheral resistance [∞ p. 463]. Cardiac output increases during exercise and therefore contributes to increased blood pressure. The changes resulting from peripheral resistance are harder to predict, however, because some peripheral arterioles are constricting while others are dilating:

$$\text{Mean arterial blood pressure} = \text{cardiac output} \times \text{peripheral resistance}$$

Skeletal muscle vasodilation decreases peripheral resistance to blood flow. At the same time, sympathetically induced vasoconstriction in nonexercising tissues offsets the vasodilation, but only partially. Consequently, total peripheral resistance to blood flow falls dramatically as exercise commences, reaching a minimum at about 75% of $\dot{V}_{O_2\text{max}}$ (Fig. 23-8a ■).

If no other compensation occurred, this decrease in peripheral resistance would dramatically lower arterial blood pressure. However, increased cardiac output cancels out decreased peripheral resistance. When blood pressure is monitored during exercise, mean arterial blood pressure actually increases slightly as exercise intensity increases (Fig. 23-8b ■). The fact that it increases at all, however, suggests that the normal baroreceptor

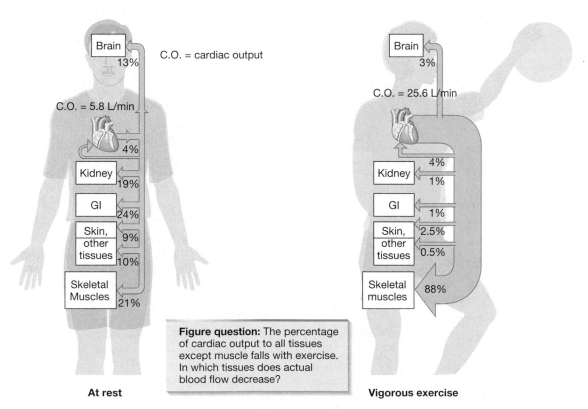

Figure question: The percentage of cardiac output to all tissues except muscle falls with exercise. In which tissues does actual blood flow decrease?

At rest   **Vigorous exercise**

■ Figure 23-7   **Distribution of cardiac output at rest and during exercise**   The shunting of blood from nonexercising tissues to muscle is accomplished by sympathetically induced vasoconstriction in nonexercising tissues combined with vasodilation in exercising muscle.

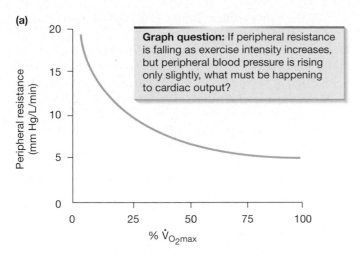

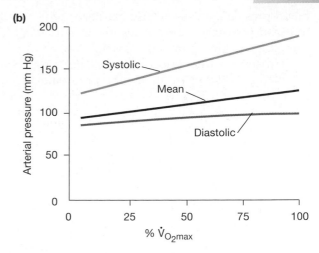

Graph question: If peripheral resistance is falling as exercise intensity increases, but peripheral blood pressure is rising only slightly, what must be happening to cardiac output?

■ **Figure 23-8   Peripheral resistance and arterial blood pressure during exercise**   (a) Peripheral resistance falls as the level of exercise increases, primarily because of vasodilation in exercising muscle. (b) Mean arterial blood pressure rises slightly during exercise, despite the drop in resistance. This rise is due to increased cardiac output that results from an increase in heart rate and stroke volume.

reflexes that control blood pressure are not functioning during exercise.

✔ In Figure 23-8b ■, why does the line for mean pressure lie closer to the diastolic pressure line instead of being evenly centered between systolic and diastolic pressures? (Hint: what is the equation for calculating mean blood pressure?)

The Baroreceptor Reflex Adjusts to Exercise

Normally, homeostasis of blood pressure is regulated through peripheral baroreceptors in the carotid and aortic bodies: an increase in blood pressure initiates responses that return blood pressure to normal. But during exercise, blood pressure increases without activating homeostatic compensation. Why is the normal baroreceptor reflex absent during exercise?

There are several theories. According to one, signals from the motor cortex during exercise reset arterial baroreceptor threshold to a higher pressure. Blood pressure can then increase slightly during exercise without triggering the homeostatic counterregulatory responses.

Another theory suggests that signals in baroreceptor afferent neurons are blocked in the spinal cord by presynaptic inhibition [∞ p. 243] at some point before the afferent neurons synapse with central nervous system neurons. This central inhibition inactivates the baroreceptor reflex during exercise.

• •

…continued from page 721

When the paramedics arrived, they took Colleen's blood pressure. It was extremely low. Her body temeprature was 103° F. The paramedics quickly transported Colleen to the hospital.

Question 3:   *Why was Colleen's blood pressure so low?*

• •

A third theory is based on the postulated existence of muscle chemoreceptors that are sensitive to metabolites, probably H^+, produced during strenuous exercise. When stimulated, these chemoreceptors signal the CNS that tissue blood flow is not adequate to remove muscle metabolites or keep the muscle in aerobic metabolism. The chemoreceptor input is reinforced by sensory input from mechanoreceptors in the working limbs. The CNS response to this sensory input is to override the baroreceptor reflex and raise blood pressure to enhance muscle perfusion. The same hypothetical muscle chemoreceptors may play a role in ventilatory responses to exercise.

❱ FEEDFORWARD RESPONSES TO EXERCISE

Interestingly, there is a significant *feedforward* element [∞ p. 175] in the physiological responses to exercise. It is easy to explain physiological changes that occur with exercise as reactions to the disruption of homeostasis. However, many of these changes occur in the absence of the normal stimuli or before the stimuli are present. For example, ventilation rates jump as soon as exercise begins, even though experiments have shown that arterial P_{O_2} and P_{CO_2} do not change (see Figs. 23-5 and 23-6 ■).

How does the feedforward response work? As exercise begins, proprioceptors in the muscles and joints send information to the motor cortex of the brain. Descending signals from the motor cortex go not only to the exercising muscles but also along parallel pathways to the cardiovascular and respiratory control centers and to the limbic system of the brain.

Output from the limbic system and cardiovascular control center triggers generalized sympathetic discharge. As a result, an immediate slight increase in blood pressure marks the beginning of exercise. Sympathetic discharge causes widespread vasoconstriction, increasing blood pressure. Once exercise has begun, this increase in blood pressure compensates for decreases in blood pressure resulting from muscle vasodilation.

However, at the onset of exercise, muscle arterioles are not yet dilated by metabolic paracrines. Consequently, the anticipatory sympathetic vasoconstriction has the potential to elevate blood pressure to potentially lethal levels. To prevent this dangerous elevation of blood pressure, a unique system of special sympathetic neurons *dilate* (instead of constrict) selected arterioles in the exercising muscles. Vasodilation decreases peripheral resistance in the muscle, and the blood pressure does not rise excessively. The sympathetic fibers to the muscle that control this response are called the **cholinergic vasodilator system**. These sympathetic neurons are notable because they use acetylcholine (not norepinephrine) as their neurotransmitter.

The anticipatory dilation of muscle arterioles by the cholinergic vasodilator system does not improve muscle oxygen and glucose supply. The arterioles involved in the cholinergic vasodilator reflex do not supply the capillaries that provide nutrition to the muscle fibers. Increased oxygen and nutrient supply to the muscle fibers must await the onset of exercise because the muscle arterioles leading to those capillaries are controlled by the more common adrenergic sympathetic neurons.

As exercise proceeds, reactive compensations become superimposed on the feedforward changes. For example, when exercise reaches 50% of aerobic capacity, muscle chemoreceptors sense the buildup of lactic acid and metabolites and send this information along to central command centers in the brain. The command centers then maintain changes in ventilation and circulation that were initiated in a feedforward manner. Thus, the integration of systems in exercise involves both common reflex pathways and some unique centrally mediated reflex pathways.

▶ TEMPERATURE REGULATION

As exercise continues, heat released through metabolism creates an additional challenge to homeostasis. Most of the energy released during metabolism is not converted into ATP but is released as heat. (Energy conversion from organic substrates to ATP is only 20%–25% efficient.) With continued exercise, heat production exceeds heat loss, and core body temperature rises. In endurance events, body temperature can reach 40°–42° C (104°–108° F), which we would normally call a fever.

This rise in body temperature with exercise triggers two thermoregulatory mechanisms: sweating and increased cutaneous blood flow [∞ p. 643]. Both mechanisms help regulate body temperature but have the potential to disrupt homeostasis in other ways.

Sweating lowers body temperature through evaporative cooling. At the same time, however, loss of fluid from the extracellular compartment has the potential to cause dehydration and significantly decrease circulating blood volume. Because sweat is a hypotonic fluid, the extra water loss increases body osmolarity. The combination of decreased ECF volume and increased osmolarity during extended exercise sets in motion the complex homeostatic pathway for dehydration described in Chapter 19, including thirst and renal conservation of water [∞ Fig. 19-17, p. 589].

The other thermoregulatory mechanism, increased blood flow to the skin, allows loss of body heat to the environment through convective heat loss [∞ p. 642]. However, increased sympathetic output during exercise tends to vasoconstrict cutaneous blood vessels and oppose the thermoregulatory response. The primary control of vasodilation in the skin during exercise appears to come from the sympathetic cholinergic vasodilator system. Activation of these neurons as body core temperature rises permits dilation of cutaneous blood vessels without altering sympathetic vasoconstriction in other tissues of the body.

Although vasodilation of cutaneous vessels is essential for thermoregulation, it disrupts homeostasis by decreasing peripheral resistance and diverting blood flow from the muscles. In the face of these contradictory demands, the body initially gives preference to thermoregulation. However, if central venous pressure falls below a critical minimum, the body abandons thermoregulation in the interest of maintaining blood flow to the brain.

The degree to which the body can adjust to both demands depends on the type of exercise being performed and its intensity and duration. Strenuous exercise in hot, humid environments can severely impair normal thermoregulatory mechanisms and cause *heat stroke*, a potentially fatal condition. Unless prompt measures are taken to cool the body, core temperatures can go as high as 43° C (109° F).

It is possible for the body to adapt to repeated exercise in hot environments, however, through **acclimatization**. In this process, physiological mechanisms shift to

…continued from page 723

When Colleen arrived in the emergency room, she was quickly diagnosed with heat stroke, a life-threatening condition. She was immersed in a tub of cool water and given intravenous fluids. Heat stroke can occur in athletes who overexert in excessively hot weather. Heat stroke is also common in elderly individuals, whose thermoregulatory mechanisms are less efficient than those in younger people.

Question 4: *What steps could Colleen have taken to avoid heat stroke? (Start at 9 A.M., when Colleen set out for the practice field.)*

fit a change in environmental conditions. As the body adjusts to exercise in the heat, sweating begins sooner and is more copious, doubling or tripling in volume, enhancing evaporative cooling. With acclimatization, sweat also becomes more dilute, as salt is reabsorbed from the sweat glands under the influence of increased aldosterone. Salt loss in an unacclimatized person exercising in the heat may reach 30 g NaCl per day, but that value decreases to as little as 3 g after a month of acclimatization.

▶ EXERCISE AND HEALTH

Physical activity has many positive effects on the human body. Our lifestyles have changed dramatically since humans were hunter-gatherers, but our bodies still seem to work best with a certain level of physical activity. Several common pathological conditions—including high blood pressure, strokes, and diabetes mellitus—can be improved by physical activity. Even so, developing regular exercise habits is one lifestyle change that many people find difficult to make. In this section, we look at the effect that exercise has on several common conditions.

Exercise Lowers the Risk of Cardiovascular Disease

As early as the 1950s, scientists showed that men who are physically active have a lower rate of heart attacks than do men who lead a sedentary lifestyle. These studies started many investigations into the exact relationship between cardiovascular disease and exercise. Subsequently, it has been demonstrated that exercise has positive benefits, including lowering blood pressure, decreasing plasma triglyceride levels, and raising plasma high-density lipoprotein (HDL) levels. High blood pressure (hypertension) is a major risk factor for strokes, while elevated triglycerides and LDL-cholesterol levels are associated with development of atherosclerosis and increased risk of heart attack.

Overall, exercise reduces the risk of death or illness from a variety of cardiovascular diseases. Even mild exercise, such as walking, has significant health benefits and could help the estimated 40 million American adults with sedentary lifestyles that put them at risk for developing cardiovascular disease, complications of obesity, or diabetes mellitus.

Type 2 Diabetes Mellitus May Improve with Exercise

Regular exercise is now widely accepted as effective in preventing and alleviating type 2 diabetes mellitus. With chronic exercise, skeletal muscle fibers up-regulate both the number of glucose transporters and the number of insulin receptors on their membrane. The addition of glucose transporters decreases the muscle's dependence on insulin for glucose uptake.

Increasing the number of insulin receptors makes the fibers more sensitive to insulin. A smaller amount of insulin then can achieve a response that previously required more insulin. Because the cells are responding to lower insulin levels, the endocrine pancreas secretes less insulin. This lessens the stress on the pancreas, resulting in a lower incidence of type 2 diabetes mellitus.

Figure 23-9 ■ shows the results of an experiment in which men with impaired glucose tolerance were started on an exercise regimen. Their response to glucose ingestion was compared with that of normal subjects. In each experiment, the individual ingested 100 g of glucose, and researchers followed changes in his plasma glucose levels for 120 minutes (this procedure is known as a *glucose tolerance test*). Simultaneous measurements were made of plasma insulin.

After only seven days of exercise, both the glucose tolerance test and insulin secretion in exercising diabetic subjects had shifted to a pattern that was more like that of the normal subjects. Because of the beneficial effect of exercise on glucose transport and metabolism, patients with type 2 diabetes are encouraged to begin a regular exercise program.

Stress and the Immune System May Be Influenced by Exercise

Another area that is receiving much attention is the interaction of exercise with the immune system. Epidemiological evidence looking at large populations of people associates exercise with a reduced incidence of disease and with increased longevity. There is also widespread public belief that exercise boosts immunity, prevents cancer, and helps HIV-infected patients combat AIDS.

But so far, the results of rigidly controlled research studies have not supported those claims. Indeed, there is good evidence suggesting that strenuous exercise is a form of stress and will suppress the immune response. Some immune suppression may be due to corticosteroid release. Other studies suggest that it may be due to release of interferon-γ during strenuous exercise.

Researchers have proposed that the relationship between exercise and decreased immunity forms a J-shaped curve (Fig. 23-10 ■). People who exercise moderately have better immune systems than those who are sedentary, but people who exercise strenuously may see a decrease in immune function because of the stress of the exercise.

Another area of exercise physiology filled with interesting though contradictory results is the effect of exercise on stress, depression, and other psychological parameters. The literature shows an inverse relationship between exercise and depression: people who exercise

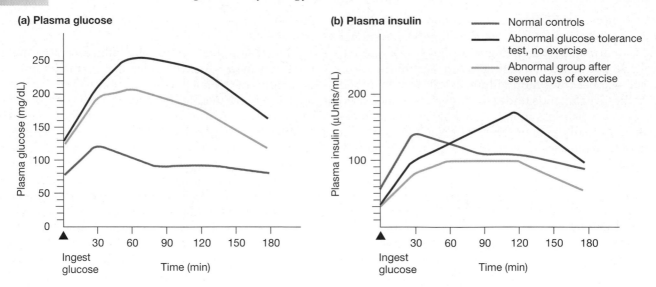

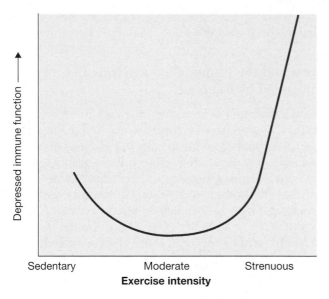

■ **Figure 23-9** **The effect of exercise on glucose tolerance and insulin secretion** In this experiment, a group of men with abnormal glucose tolerance tests (red line, part a) exercised for seven days and were tested again. After seven days of exercise, their glucose tolerance test (green line, part a) had shifted in the direction of normal controls, as had their pattern of insulin secretion (green line, part b). Exercise is known to enhance glucose uptake and increase muscle sensitivity to insulin. Data from B. R. Seals et al., "Glucose Tolerance in Older and Younger Athletes and Sedentary Men," *Journal of Applied Physiology* 56(6):1521–1525 (1984); and M. A. Rogers et al., "Improvement in Glucose Tolerance After One Week of Exercise in Patients with Mild NIDDM," *Diabetes Care* 11:613–618 (1988).

■ **Figure 23-10** **Immune function and exercise** The J-shaped curve of this graph shows one theory of the relationship between immune function and exercise. Moderate exercise enhances immunity, but strenuous exercise is a form of stress that depresses immunity.

regularly are significantly less likely to be clinically depressed than are people who do not exercise regularly. Although the association exists, assigning cause and effect to the two parameters is difficult. Are the exercisers less depressed because they exercise? Or do the depressed subjects exercise less because they are depressed?

Many published studies appear to show that regular exercise is effective in reducing depression. But a careful analysis of experimental design suggests that the conclusions of those studies may be overstated. The subjects in many of the experiments were being treated concurrently with drugs or psychotherapy, and it is difficult to attribute improvement in their condition solely to exercise.

In addition, participation in exercise studies gives subjects a period of social interaction, another factor that might play a role in the reduction of stress and depression. The assertion that exercise reduces depression would be better supported by a demonstration of the effects of exercise on biological parameters such as disturbances of monoamine neurotransmitters [∞ p. 236]. To date, however, little work of that type has been done.

PROBLEM CONCLUSION

Colleen was kept overnight in the hospital for observation. She was unable to practice with the team for the remainder of the season because victims of heat stroke are more sensitive to high temperatures for some time after the episode.

In this running problem, you learned about a dangerous heat-related emergency called heat stroke. Further check your understanding of this running problem by comparing your answers to those in the summary table.

	Question	Facts	Integration and Analysis
1	Why is the thermoregulatory mechanism of sweating less efficient in humid environments?	"Humidity" is the percentage of water present in air. Thermoregulation in warm environments includes sweating and evaporative cooling.	Evaporation is slower in humid air, therefore, evaporative cooling is less effective in high humidity.
2	Individuals with a heat emergency called heat exhaustion have cool, moist skin. Hot, dry skin indicates a more serious emergency called heat stroke. Why is heat exhaustion less serious than heat stroke?	Sweating helps the body regulate temperature through evaporative cooling. As evaporation occurs, the skin surface cools.	Cool, moist skin indicates that the sweating mechanism is still functioning. Hot, dry skin indicates that the sweating mechanism has failed. Subjects with hot, dry skin are likely to have higher internal temperatures.
3	Why was Colleen's blood pressure so low?	Blood volume decreases when the body loses large amounts of water through sweating.	Colleen had been sweating but not replacing the fluid she lost. This caused her blood volume to decrease, with a corresponding decrease in blood pressure.
4	What steps could Colleen have taken to avoid heat stroke? (Start at 9 A.M., when Colleen set out for the practice field.)	Heat stroke is caused by dehydration as a result of excessive sweating. This results in a drop in blood pressure. Peripheral blood vessels constrict in an effort to maintain pressure and blood flow to the brain. Constricted blood vessels in the skin cannot release excess heat. In addition, the sweating response is inhibited to prevent further fluid loss.	Colleen could have avoided heat stroke by taking the following measures: (1) consuming large amounts of fluid (water plus electrolytes) before and during practice; (2) avoiding excessive exercise such as bicycling to practice; (3) stopping exercising at the first signs of heat emergency (dizziness and nausea); (4) seeking shade and drinking large amounts of fluid to replenish lost fluids; (5) using cool wet towels or ice to bring body temperature lower.

CHAPTER REVIEW

SUMMARY

1. **Exercise** is any muscular activity that generates force and disrupts homeostasis. (p. 717)

Metabolism and Exercise

2. Exercising muscle requires a steady supply of ATP. A small amount of ATP is stored in the muscle fiber and additional ATP can be rapidly formed from phosphocreatine. (p. 717)

3. Carbohydrates and fats are the primary substrates for energy production. Glucose can be metabolized through both oxidative and anaerobic pathways, but fatty acid metabolism requires adequate amounts of oxygen. (p. 718)

4. Anaerobic **glycolytic metabolism** converts glucose to lactic acid. Glycolytic metabolism is 2.5 times more rapid than aerobic pathways but is not as efficient at ATP production. (p. 718)

5. Glucagon, cortisol, catecholamines, and growth hormone influence glucose and fatty acid metabolism during exercise. These hormones favor the conversion of glycogen to glucose. (p. 718)

6. Although plasma glucose concentrations rise with exercise, the secretion of insulin decreases. This response reduces glucose uptake by most cells, making more glucose available for exercising muscle. (p. 719)

7. The intensity of exercise is indicated by oxygen consumption ($\dot{V}_{O_2}$). A person's maximal rate of oxygen consumption ($\dot{V}_{O_2max}$) is an indicator of that person's ability to perform endurance exercise. (p. 719)

8. Oxygen consumption increases rapidly at the onset of exercise. **Excess postexercise oxygen consumption** is due to ongoing metabolism, increased body temperature, and circulating catecholamines. (p. 719)

9. Muscle mitochondria increase in size and number with endurance training. (p. 720)

10. In maximal exertion, the ability of the cardiovascular system to deliver oxygen and nutrients appears to be the primary factor limiting exertion. (p. 720)

Ventilatory Responses to Exercise

11. Exercise hyperventilation results from feedforward signals from the motor cortex and sensory feedback from peripheral sensory receptors. (p. 720)

12. Arterial P_{O_2}, P_{CO_2}, and pH show no significant change during mild to moderate exercise. Exercise hyperventilation must be due to some other factor such as increased plasma K^+ levels. (p. 720)

Cardiovascular Responses to Exercise

13. Cardiac output increases with exercise because of increased venous return and sympathetic stimulation of heart rate and contractility. (p. 721)

14. Blood flow through exercising muscle increases dramatically when skeletal muscle arterioles dilate. Arterioles in other tissues constrict. (p. 722)

15. Decreased tissue O_2 and glucose or increased muscle temperature, CO_2, and acid act as paracrines and cause local vasodilation. (p. 722)

16. Mean arterial blood pressure increases slightly as exercise intensity increases. The normal baroreceptor reflexes that control blood pressure do not function during exercise. (p. 722)

Feedforward Responses to Exercise

17. Many physiological changes at the beginning of exercise occur without the normal stimuli. For example, a **cholin-** **ergic vasodilator system** dilates selected muscle arterioles (p. 723)

Temperature Regulation

18. Heat released during exercise is dissipated by sweating and increased cutaneous blood flow. (p. 724)

Exercise and Health

19. Physical activity has many positive effects and can help prevent or decrease the risk of developing high blood pressure, strokes, and type 2 diabetes mellitus. (p. 725)

QUESTIONS

LEVEL ONE   Reviewing Facts and Terms

1. Exercise is defined as a _____ activity that generates _____ and disrupts _____.

2. The most efficient ATP production is through *aerobic/anaerobic* pathways. When these pathways are being used, then *glucose/fatty acids/both/neither* can be metabolized to provide ATP.

3. If skeletal muscle's requirement for _____ is not met, metabolism shifts to _____ that results in the production of lactic acid.

4. List three sources of glucose that can be metabolized to ATP, either directly or indirectly.

5. List four hormones that promote the conversion of triglyc-erides into fatty acids. What effects do these hormones have on plasma glucose levels?

6. What is meant by the term *oxygen deficit*? How is oxygen deficit related to excessive postexercise oxygen consumption?

7. What organ system is the limiting factor for maximal exertion?

   What role is K^+ hypothesized to play in exercise?

8. In endurance events, body temperature can reach 40°–42° C. What is normal body temperature? What two thermoregulatory mechanisms are triggered by this change in temperature during exercise?

LEVEL TWO   Reviewing Concepts

9. **Concept map:** Map the metabolic, cardiovascular, and respiratory changes that occur with exercise. Include the signals to and from the nervous system and show what specific areas signal and coordinate the exercise response.

10. What causes insulin secretion to decrease during exercise and why is this decrease adaptive?

11. State two advantages and two disadvantages of anaerobic glycolysis.

12. Compare and contrast the following terms, especially as they relate to exercise:

    (a) ATP, ADP, PCr
    (b) myoglobin, hemoglobin

13. Match the response to the brain area that controls it. Brain areas may be used once, more than once, or not at all. Some responses may have more than one answer.

    (a) pons
    (b) medulla oblongata
    (c) midbrain
    (d) motor cortex
    (e) hypothalamus
    (f) cerebellum
    (g) brain not involved; uses local control

    1. changes in cardiac output
    2. vasoconstriction
    3. exercise hyperventilation
    4. increased stroke volume
    5. increased heart rate
    6. coordination of skeletal muscle movement

14. Specify whether each parameter listed below stays the same, increases, or decreases when a person becomes better conditioned for athletic activities:

    (a) heart rate during exercise
    (b) resting heart rate
    (c) cardiac output during exercise
    (d) resting cardiac output
    (e) breathing rate during exercise
    (f) blood flow to muscles during exercise
    (g) blood pressure during exercise
    (h) total peripheral resistance during exercise

15. Why doesn't increased venous return during exercise overstretch the heart muscles?

16. Diagram the three theories that explain why the normal baroreceptor reflex is absent during exercise.

17. What is the cholinergic vasodilator system? What is the primary function of this system?

18. List and briefly discuss the benefits of a lifestyle that includes regular exercise.

19. Explain how exercise decreases blood glucose in type 2 diabetes mellitus.

LEVEL THREE   Problem Solving

20. You have decided to manufacture a new sports drink that will help athletes, from football players to gymnasts. List at least five different ingredients you would put in your drink and tell why each one is important for the athlete.

21. You are a well-conditioned athlete. At rest, your heart rate is 60 beats per minute and your stroke volume is 70 mL/beat. What is your cardiac output? At one point during exercise, your heart rate goes up to 120 beats/min. Does your cardiac output increase proportionately? Explain.

E X P L < MediaLab >

Introduction

In this chapter you learned about exercise and the physiological challenges that it presents. Exercise integrates information from all preceding 22 chapters. It creates metabolic challenges and requires muscular, cardiovascular, and ventilatory responses. The concepts of neural control, feedback, and tissue-to-tissue communications are crucial for understanding exercise. The Web Explorations in this MediaLab will help expand your knowledge about how all the body's systems integrate when the body is performing work. After reading the description below, visit the MediaLab for Chapter 23 in your Companion Website and select the appropriate keyword.

Web Exploration 1

Estimated time for completion = 15 minutes

It doesn't matter whether you are a marathon runner or just a couch potato. If you work out long enough and hard enough, you will eventually tire and be out of breath. What limits our ability to endure a long workout? Are you really out of breath? To focus your thoughts, think about what is required to keep muscles in movement for a long period of time. Now, make a list of the events that need to occur in order to sustain repeated muscle contraction. Consider both large and small-scale events and be sure to include molecule(s) used for energy and how they are synthesized in your list.

Once you have constructed your list, consider the following information. A steady supply of oxygen is certainly required in order for muscle to metabolize glucose, but what limits this supply of oxygen? List the two candidate organs you have in mind. What is happening to these two organs during exercise? Ventilation increases and so does heart rate, but which of these functions will ultimately restrict you? To explore this topic further, select **VENTILATION CAPACITY** from your Website then scroll down to "Ventilation Issues in Endurance Performance." This site contains information to help you learn more about our lung function during exercise.

Web Exploration 2

Estimated time for completion = 5 minutes

What is happening to our muscles when we feel soreness during or after exercise? You have probably heard that lactic acid buildup is responsible, but what does that mean in terms of physiology? Select the keyword **MUSCLE CONTRACTION** from your Website to watch a movie of a muscle contraction. Now, think of what might happen inside the muscle cell when the demand for movement exceeds the availability of oxygen. Chapter 12 will help with this question.

After you've considered this point, answer these questions. How does regular exercise delay the buildup of lactic acid? Select **LACTIC ACID** from your Website and scroll to "Measuring the Lactate Threshold" to see a diagram of lactate production in relation to heart rate and oxygen consumption. What is the fate of the lactate that is released into the blood stream? List the changes that occur in a muscle after regular exercise. Can you describe how these changes might affect muscle cell performance and lactic acid production? What other mechanisms alleviate the effects of lactic acid production within the muscle cell?

24 Reproduction and Development

■ "Birth, and copulation, and death. That's all the facts when you come to brass tacks."—*T. S. Eliot*, Sweeney Agonistes ■

CHAPTER OUTLINE

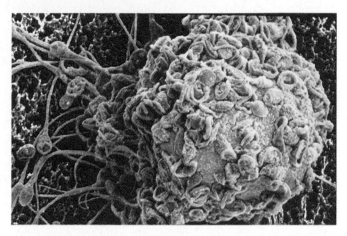

SEM of oocyte at fertilization showing zona pellucida (blue), sperm (pink), and partially cleared corona radiata cells (green).

BACKGROUND BASICS

Imagine growing up as a girl, then at the age of 12 or so, finding that your voice is deepening and your genitals are developing into those of a man. This scenario actually happens to a small number of men with a condition known as *pseudohermaphroditism* [*pseudes*, false + *hermaphrodites*, the dual-sex offspring of Hermes and Aphrodite]. These men have the internal sex organs of a male but inherit a gene that causes a deficiency in one of the male hormones. Consequently, they are born with external genitalia that appear feminine, and they are raised as girls. At puberty, when they begin to secrete male hormones, they develop some, but not all, of the characteristics of men. As you can imagine, a conflict arises: change gender, or remain female? Most choose to continue life as men.

Reproduction is one area of physiology in which we humans like to think of ourselves as significantly advanced over other animals. We mate for pleasure as well as procreation, and the female human is always sexually receptive, not just during her fertile periods. But just how different are we?

Like other terrestrial animals, we have internal fertilization that allows motile flagellated sperm to remain in an aqueous environment. Because internal fertilization requires cooperation of the two parties, we have mating and courtship rituals as do other animals. Fetal development is also internal, protecting the growing embryo from dehydration by cushioning it with a layer of fluid.

Humans are *sexually dimorphic*; that is, the males and females are physically distinct. This distinction is sometimes blurred by dress and hairstyle, but these are cultural acquisitions. Although everyone agrees that male and female humans are physically dimorphic, we are still debating whether we are behaviorally and psychologically dimorphic as well.

Sex hormones play a significant role in the behavior of other mammals, acting on adults as well as influencing the brain of the developing embryo. The degree to which the results of nonprimate and primate animal studies can be extrapolated to humans is controversial. Human fetuses are exposed to sex hormones while in the uterus, but it is unclear how much gender imprinting occurs. Does the preference of little girls for dolls and of little boys for toy guns have a biological basis or a cultural basis? We have no answer yet, but it appears that at least part of our brain structure is influenced by sex hormones before we ever leave the womb.

In this chapter we deal with human reproduction and development. In talking about reproduction, we are always faced with the age-old question of the chicken or the egg: which came first, and where do we start? In this book, we will begin with the fertilized egg, or **zygote**, and examine sex determination and sexual differentiation.

▶ SEX DETERMINATION

The male and female sex organs consist of the gonads, the internal accessory ducts and glands, and the external reproductive structures known collectively as the **genitalia**. **Gonads** [*gonos*, seed] are the organs that produce **gametes** [*gamein*, to marry], the reproductive cells (eggs and sperm) that unite to form a new individual. The male gonads are the **testes** (singular *testis*), which produce **sperm** (spermatozoa). The female gonads are the **ovaries**, which produce eggs, or **ova** (singular *ovum*). The undifferentiated gonadal cells destined to produce eggs and sperm are called **germ cells**.

The secret of sexual development is locked within the human genome [∞ Appendix B]. Each nucleated cell of the body except egg and sperm contains 46 chromosomes (the *diploid number*): 22 matched pairs of **autosomes** and one pair of **sex chromosomes**, each of which is designated either X or Y. The 22 pairs of autosomal chromosomes direct development of the human body form and of variable characteristics such as hair color and blood type.

The two sex chromosomes contain genes that direct development of internal and external sex organs. The X chromosome is larger than the Y chromosome and includes many genes that are missing from the Y chromosome. Figure 24-1 ■ shows a set of human chromosomes laid out as pairs of *homologous* (corresponding) autosomes, with the sex chromosomes at the lower right.

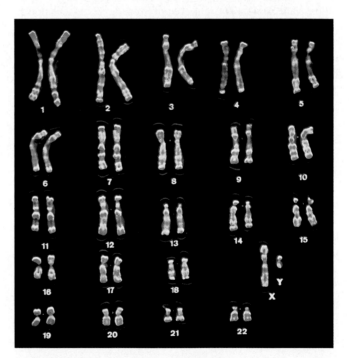

■ **Figure 24-1 Human chromosomes** Humans have 23 pairs of chromosomes, which in this figure have been arranged in homologous pairs. The presence of an X and a Y chromosome means that these chromosomes came from a male.

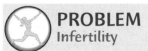

PROBLEM
Infertility

Peggy and Larry have just about everything to make them happy: successful careers, a loving marriage, a comfortable home. But one thing is missing. After five years of marriage, they have been unable to have a child. Today, Peggy and Larry have their first appointment with Dr. Charles Coddington, an infertility specialist. "Finding the cause of your infertility is going to require some painstaking detective work," he explains. Dr. Coddington will begin his workup of Peggy and Larry by asking detailed questions about their reproductive histories. Based on the answers to these questions, he will then order tests to pinpoint the problem.

…continued on page 741

§ **X-Linked Inherited Disorders**   Normally, a person inherits two copies of the gene for a given trait, one copy from the mother and one from the father. But many traits that are found on the X chromosome have no matching gene on the much smaller Y chromosome (X-linked genes). Females, who inherit two X chromosomes, always get two copies of X-linked genes, so the expression of X-linked traits follows the usual pattern of gene dominance and recession. Males, on the other hand, receive only one copy of the gene on the X chromosome, so they *always* exhibit the characteristics of the X-linked gene. If the maternally inherited gene is defective, male offspring will exhibit the mutation. Among the X-linked diseases that have been identified are Duchenne muscular dystrophy [∞ p. 369], hemophilia [∞ p. 493], and color-blindness.

Eggs and sperm each have a half set of 23 chromosomes (the *haploid number*) because of meiotic divisions during their creation. When egg and sperm unite, the resulting zygote contains a unique set of 46 chromosomes. Thus, humans inherit one chromosome of each pair from the mother and the other from the father.

The Sex Chromosomes Determine Genetic Sex

The sex chromosomes that a person inherits determine the genetic sex of the individual. Normal females are XX, and normal males are XY (Fig. 24-2 ■). Females inherit one X chromosome from each parent. Males inherit a Y chromosome from their father and an X from their mother. The Y chromosome is essential for development of the male reproductive organs.

If sex chromosomes are abnormally distributed during fertilization, the presence or absence of a Y chromosome determines whether the embryo will become male. A zygote that inherits only a Y chromosome (YO) will

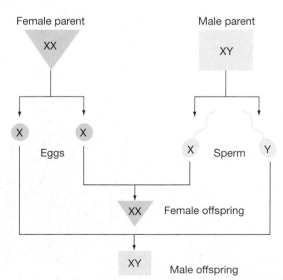

■ **Figure 24-2   Inheritance of X and Y chromosomes**   Each egg produced by a female (XX) has an X chromosome. Each sperm produced by a male (XY) can have either an X or a Y chromosome.

die because the larger X chromosome contains unique genes that are essential for development. A zygote that gets an extra sex chromosome will develop according to whether or not a Y chromosome is present. For instance, an XXY zygote will become male.

In the absence of a Y chromosome, an embryo will develop into a female. Therefore, a zygote that gets only one X chromosome (XO; Turner's syndrome) will develop into a female. Two X chromosomes are needed for normal female reproductive function, however.

Once the ovaries develop, one X chromosome in each cell of the body inactivates and condenses into a clump of nuclear chromatin known as a *Barr body*. The selection of which X chromosomes become inactive is random, so some cells have an active maternal X chromosome and others have an active paternal X chromosome. Because inactivation occurs early in development, before cell division is complete, all cells of a given tissue will usually have the same active X chromosome, either maternal or paternal.

Sexual Differentiation Occurs in the Second Month of Development

The gender of an early embryo is difficult to determine because reproductive structures do not begin to differentiate until the seventh week of development. Before differentiation, the embryonic tissues are considered *bipotential* because they cannot be morphologically identified as male or female (Fig. 24-3 ■). The bipotential external genitalia consist of a *genital tubercle, urethral folds, urethral groove*, and *labioscrotal swellings*. These structures differentiate into the male and female external genitalia as development progresses (Table 24-1).

In the early embryo, bipotential gonadal tissue is associated with two pairs of accessory ducts: **Wolffian ducts** from the embryonic kidney and **Müllerian ducts**. As development proceeds along either male or female lines, one pair of ducts develops while the other degenerates.

What is it that directs some single-cell zygotes to become males and others to become females? Gender determination depends on the presence or absence of the **SRY gene**, found on the Y chromosome. This gene produces a *testis-determining factor* that binds to DNA and activates additional genes. The products of these genes together direct the development of testes from the bipotential gonad (Fig. 24-4 ■).

The developing testes then secrete hormones to continue development of the male reproductive tract. **Müllerian inhibiting substance** (**MIS**; also called **anti-Müllerian hormone**), a glycoprotein from testicular *Sertoli cells*, causes the Müllerian ducts to regress. **Testosterone**, a male sex hormone from testicular *Leydig cells*, and its derivative **dihydrotestosterone (DHT)** convert the Wolffian ducts into the *male accessory structures*: epididymis, vas deferens, and seminal vesicle (see Table 24-1).

Testosterone and DHT also cause the external genitalia to take on the male pattern. Later in fetal development,

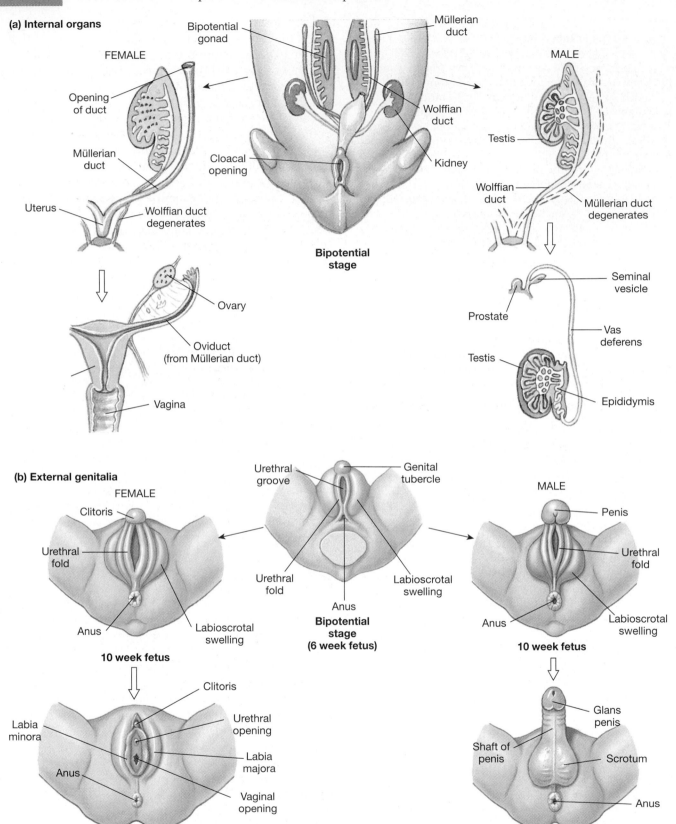

(a) Internal organs

Bipotential gonad

Müllerian duct

FEMALE

Opening of duct

Müllerian duct

Uterus

Wolffian duct degenerates

Cloacal opening

Wolffian duct

Kidney

Bipotential stage

MALE

Testis

Wolffian duct

Müllerian duct degenerates

Ovary

Oviduct (from Müllerian duct)

Vagina

Prostate

Testis

Seminal vesicle

Vas deferens

Epididymis

(b) External genitalia

Urethral groove

Genital tubercle

FEMALE

Clitoris

Urethral fold

Anus

Labioscrotal swelling

Urethral fold

Anus

Bipotential stage (6 week fetus)

Labioscrotal swelling

MALE

Penis

Urethral fold

Anus

Labioscrotal swelling

10 week fetus

10 week fetus

Clitoris

Labia minora

Urethral opening

Labia majora

Anus

Vaginal opening

At birth

Glans penis

Shaft of penis

Scrotum

Anus

At birth

■ **Figure 24-3   Sexual development in the human embryo**   (a) At six weeks of fetal development, the internal and external genitalia have the potential to develop into male or female reproductive structures. (b) By 10 weeks, testis-determining factor in a male embryos has directed development of the bipotential gonads into testes. The testes produce testosterone and Müllerian inhibiting substance, which causes the Müllerian ducts to disappear. In the female, absence of Müllerian inhibiting substance allows the Müllerian duct to become the Fallopian tube, uterus, and upper part of the vagina. The Wolffian ducts disappear. (c) Testosterone and its metabolite DHT cause development of the external genitalia and male accessory ducts and glands. By birth, the testes have descended from the abdominal cavity into the scrotum. In the absence of male hormones, the external genitalia are feminized.

TABLE 24-1   Sexual Differentiation

Male	Bipotential Structure	Female
Glans penis	← Genital tubercle →	Clitoris
Shaft of penis	← Urethral folds and groove →	Labia minora, opening of vagina and urethra
Shaft of penis and scrotum	← Labioscrotal swellings →	Labia majora
Regresses	← Gonad (cortex) →	Forms ovary
Forms testis	← Gonad (medulla) →	Regresses
Becomes epididymis, vas deferens, and seminal vesicle (testosterone present)	← Wolffian duct →	Regresses (testosterone absent)
Regresses (Müllerian inhibiting substance present)	← Müllerian duct →	Becomes Fallopian tube, uterus, cervix, and upper 1/3 of vagina (Müllerian inhibiting substance absent)

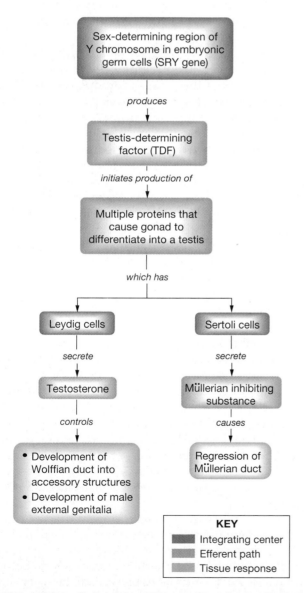

■ **Figure 24-4   Role of the SRY gene in male development**

testosterone controls migration of the testes from the abdomen into the *scrotum*, or scrotal sac.

In female embryos, who have no SRY gene, the primitive ducts follow a different developmental course. In the absence of MIS, Müllerian ducts develop into the upper portion of the *vagina*, the *uterus*, and the *Fallopian tubes* (after the anatomist Fallopius who first described them; also known as *oviducts*). Without testosterone and DHT, the Wolffian ducts degenerate, and external genitalia take on the female pattern. Thus, absence of the SRY gene, Müllerian inhibiting substance, testosterone, and DHT creates a female.

Exposure to testosterone during embryonic development is known to masculinize certain aspects of brain function, such as the brain's responsiveness to certain hormones. One controversial aspect of the masculinizing effects of testosterone is its influence on human sexual behavior and gender identity. It is well documented that in many nonhuman mammals, adult sexual behavior

Pseudohermaphrodites and DHT   The importance of dihydrotestosterone in male development came to the attention of scientists through studies of the male pseudohermaphrodites described in the opening of this chapter. These men inherit a defective gene for the enzyme *5-alpha-reductase*, which catalyzes the conversion of testosterone to dihydrotestosterone (DHT). The defect, like that of sickle cell anemia, results from the substitution of a single amino acid in the peptide sequence. Because of deficient DHT, the male external genitalia and prostate gland fail to develop fully during fetal development, despite normal testosterone secretion by the testes. At birth, the infants are usually considered female and are raised as such. However, at puberty, the testes again begin to secrete testosterone, causing masculinization of the external genitalia, pubic hair growth (although scanty facial and body hair), and deepening voice. By studying the defect in 5-alpha-reductase in these individuals, we have been able to separate the effects of testosterone from those of DHT.

depends on the absence or presence of testosterone during critical periods of brain development. However, a similar cause/effect relationship has never been proved in humans. In human behavior, it is very difficult to separate biological influences from environmental factors, and it will probably be years before this question is resolved.

✔ Why was King Henry VIII of England wrong to blame his wives when they were unable to produce a male heir to the throne?

✔ What sex will a zygote become if it inherits only one X chromosome (XO)?

✔ If the testes are removed from an early male embryo, why does it develop a uterus and Fallopian tubes rather than the normal male accessory structures? Will the embryo have male or female external genitalia?

▶ BASIC PATTERNS OF REPRODUCTION

The male testis and female ovary share some similarities, as might be expected of organs with the same origin. Both produce hormones and gametes, although the gametes themselves and the timing of gamete production are quite different. Let's examine some of the similarities and differences between male and female reproduction.

The timing of gamete production, or **gametogenesis**, is very different in males and females. Women are born with all the eggs, or **oocytes**, they will ever have. During the reproductive years, the eggs mature. They are released from the ovaries roughly once a month for about 40 years. Then female reproductive cycles cease (**menopause**).

Men, on the other hand, manufacture sperm continuously from the time they reach reproductive maturity. Sperm and testosterone production diminish with age, but do not cease the way women's reproductive cycles do.

The gametes themselves are also very different. Eggs are some of the largest cells in the body [∞ Fig. 3-2, p. 44]. They are nonmotile and must be moved through the reproductive tract on currents created by smooth muscle contraction or the beating of cilia. Sperm, on the other hand, are quite small. They are the only flagellated cells of the body and are highly motile so that they can swim up the female reproductive tract in their search for an egg to fertilize.

Gametogenesis Begins *in Utero* and Resumes During Puberty

Figure 24-5 ■ compares the male and female patterns of gametogenesis. In both genders, germ cells of the embryonic gonads first undergo a series of mitotic divisions to increase their numbers. After that, the germ cells are ready to begin **meiosis**, the cell division process through which gametes are formed. Through meiotic divisions, the normal number of 46 chromosomes (written 46,XX for females or 46,XY for males) is reduced by half. All ova are 23,X. Sperm can be either 23,X or 23,Y.

When meiosis begins, the cell's DNA replicates to create the equivalent of 92 chromosomes. However, cell and chromosomal division do not take place. Each of the duplicated chromosomes is in the form of two identical **sister chromatids** linked together at a region known as the **centromere**. The resultant cells, called *primary gametes*, still have 46 chromosomes, but each chromosome is duplicated and contains twice the normal amount of DNA.

In the **first meiotic division**, the chromosomes sort themselves into two *secondary gametes*, each of which has 23 chromosomes. Each chromosome is still composed of two sister chromatids. In the **second meiotic division**, the sister chromatids separate and go to different cells. The result of that division is two cells with 23 single chromatids.

The process of gametogenesis as described above is identical in males and females, but the timing of mitosis and meiosis is very different. The testes of a newborn boy contain immature spermatogonia that are capable of replicating through mitosis. However, the gonads become inactive after birth and do not resume mitotic activity until **puberty**, the period in the early teen years when the gonads mature.

Some of the germ cells, known as **spermatogonia** (singular *spermatogonium*), continually duplicate themselves throughout the male's reproductive life. Other spermatogonia are destined to develop into sperm. They undergo meiosis and become **primary spermatocytes**.

In the first meiotic division, one primary spermatocyte divides into two **secondary spermatocytes**. In the second meiotic division, each secondary spermatocyte divides into two **spermatids**. One spermatid has 23 single chromosomes, the haploid number characteristic of a gamete. The spermatids then mature into sperm. Thus, one primary spermatocyte creates four sperm.

In the embryonic ovary, germ cells are called **oögonia** (singular *oögonium*). Oögonia complete mitotic replication and the first stage of meiosis by the fifth month of embryonic development (Fig. 24-5 ■). At this point, the embryonic ovary contains half a million primary gametes, or **primary oocytes**. Germ cell mitosis ceases before birth, and no further oocytes can be formed.

The oocyte's first meiotic division takes place following puberty. Each primary oocyte divides into two cells, a large **egg (secondary oocyte)** and a tiny **first polar body**. Despite the size difference, each cell contains 23 duplicated chromosomes. The first polar body disintegrates.

Meanwhile the egg begins the second meiotic division. After its sister chromatids separate, meiosis pauses again. The final step of meiosis, in which sister chromatids go to separate cells, does not take place unless the egg is fertilized.

The ovary releases the mature egg during a process known as **ovulation**. If the egg is not fertilized, meiosis never goes to completion, and the egg disintegrates or passes out of the body. If fertilization by sperm occurs, the final step of meiosis takes place. Half the sister chromatids remain in the fertilized egg (zygote), while the

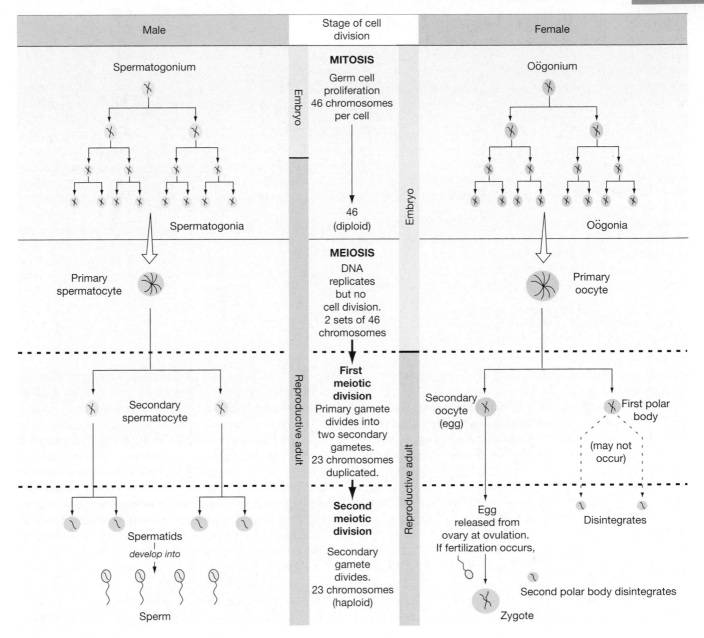

■ **Figure 24-5   Gametogenesis**   Mitosis is completed *in utero* in the female but continues after puberty in the male.

other half are released in a **second polar body**. The second polar body, like the first, degenerates. In the female, each primary oocyte thus gives rise to only one egg.

Gametogenesis in both males and females is under the control of hormones from the brain and from endocrine cells in the gonads. Some of these hormones are identical in males and females, and others are different.

✔ At what stage of development is the gamete in a newborn male? In a newborn female?

✔ Compare the amount of DNA in the first polar body with the amount of DNA in the second polar body.

✔ How many gametes are formed from one primary oocyte? From one primary spermatocyte?

The Brain Directs Reproduction

The pathways that regulate reproduction begin with secretion of peptide, protein, and glycoprotein hormones by the hypothalamus and anterior pituitary. These trophic hormones control gonadal secretion of the steroid sex hormones, including **androgens**, **estrogens**, and **progesterone**. These hormones are closely related and arise from the same steroid precursors (Fig. 24-6 ■). Both sexes produce androgens and estrogens, but androgens predominate in males, and estrogens are dominant in females.

In men, most testosterone is secreted by the testes, but about 5% comes from the adrenal cortex. Testosterone is converted in peripheral tissues to its more potent derivative dihydrotestosterone (DHT). Some of the

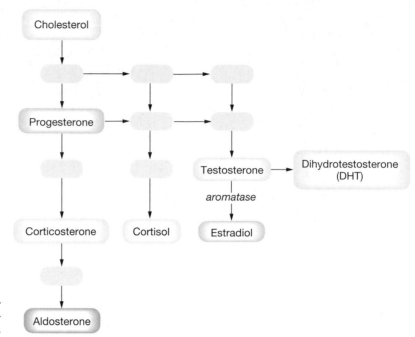

■ **Figure 24-6   Synthesis pathway for steroid hormones**   The blank boxes represent intermediate compounds whose names have been omitted for simplicity.

physiological effects attributed to testosterone are actually the result of DHT activity.

The testes also contain the enzyme **aromatase**, which converts testosterone to **estradiol**, the main estrogen in humans. A small amount of additional estrogen is made in peripheral tissues. Although males synthesize estrogens, their feminizing effects are usually not obvious.

In women, the ovary produces estrogens (estradiol and *estrone*) and progesterone, the other female hormone. The ovary and the adrenal cortex produce small amounts of androgens.

Control Pathways for Sex Steroids Are Similar in Males and Females   The hormonal control of reproduction in the two sexes follows the basic hypothalamus-anterior pituitary-peripheral gland pattern (Fig. 24-7 ■). **Gonadotropin releasing hormone** (GnRH*) from the hypothalamus controls secretion of two anterior pituitary **gonadotropins: follicle stimulating hormone (FSH)** and **luteinizing hormone (LH)**. FSH and LH in turn act trophically on the gonads.

FSH, along with steroid sex hormones, is required to initiate and maintain gametogenesis. LH acts primarily on endocrine cells, stimulating production of the steroid sex hormones.

Although primary control of gonadal function arises in the brain, the gonads also influence their own function. Both ovary and testis secrete peptide hormones that act directly on the pituitary. **Inhibins** are hormones that inhibit FSH secretion. A family of related peptides, called **activins**, stimulate secretion of FSH. Activins also promote spermatogenesis, oocyte maturation, and embryonic ner-

vous system development. These gonadal peptides are produced in nongonadal tissues as well, and their other functions in the body are still being investigated.

Feedback Pathways Include Both Long-Loop and Short-Loop Feedback   The feedback pathways for trophic hormones from the hypothalamus and anterior pituitary follow the general pattern of long-loop and short-loop feedback described in Chapter 7 [∞ p. 200]. Gonadal steroids suppress secretion of GnRH, FSH, and LH in a long-loop response. The pituitary gonadotropins inhibit GnRH release by the short-loop path (Fig. 24-7 ■).

When circulating levels of gonadal steroids are low, the pituitary secretes FSH and LH (Table 24-2). Once steroid secretion reaches a certain level, negative feedback inhibits gonadotropin release. Androgens always maintain negative feedback on gonadotropin release: as androgen levels go up, FSH and LH secretion diminishes.

On the other hand, only lower concentrations of estrogen have a negative feedback effect. If estrogen secretion increases above a threshold level for at least 36 hours, the feedback changes to positive feedback, and

> **Inhibins and Activins**   The family of hormones known as *inhibins* and *activins* are part of a large family of related molecules known as the *transforming growth factor-β* family. Müllerian inhibiting substance is also a member of this family. Interestingly, although inhibin and activin have opposite effects on FSH secretion, their structures are similar. Inhibin is composed of two peptide subunits, α and β. Activin is composed of two β subunits. These two hormones are an interesting example of how a limited number of genes can create a variety of signal molecules.

*GnRH is sometimes called *luteinizing hormone releasing hormone* (LHRH) because it was first thought to have its primary effect on LH.

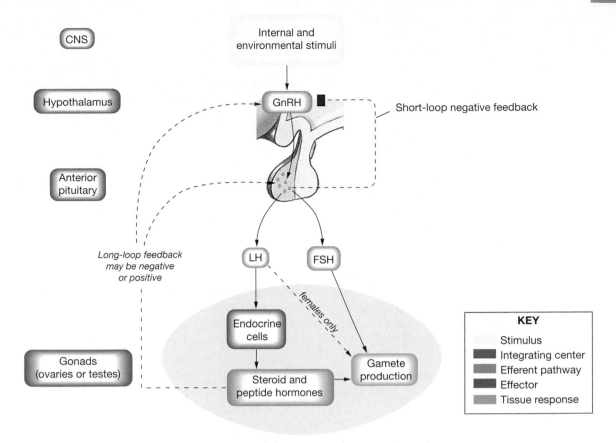

■ Figure 24-7   Hormonal control of reproduction

gonadotropin release (particularly LH) is *stimulated*. The paradoxical effects of estrogen on gonadotropin release play a significant role in the female reproductive cycle, as you will learn later in this chapter.

We still do not fully understand the mechanism of the change from negative to positive feedback with estrogen. Some evidence suggests that high levels of estrogen increase the number of GnRH receptors, making the anterior pituitary more sensitive to GnRH (up-regulation of receptors; [∞ p. 165]). Other evidence points to a direct effect of estrogen on hypothalamic release of GnRH.

The Hypothalamus Releases Gonadotropin in Small Pulses
Tonic GnRH release from the hypothalamus occurs in small pulses every 1–3 hours in both males and females. The region of the hypothalamus that contains the GnRH neuron cell bodies has been called a **pulse generator** because it coordinates the periodic pulsatile secretion of GnRH from a group of neurons.

Scientists were puzzled for a time about why tonic GnRH release occurred in pulses rather than in a steady fashion, but several studies have shown the significance of the pulses. Children who suffer from a deficiency of GnRH will not mature sexually in the absence of gonadotropin stimulation of the gonads. If treated with steady infusions of GnRH through drug-delivery pumps, these children still fail to mature sexually.

But if the pumps are adjusted to deliver GnRH in pulses similar to those that occur naturally, the children will go through puberty. Apparently, steady high levels of GnRH cause down-regulation of the GnRH receptors on gonadotropin cells, making the pituitary unable to respond to GnRH.

This down-regulation is the basis for the therapeutic use of GnRH in treating certain disorders. For example, patients with prostate and breast cancers stimulated by androgens or estrogens may be given GnRH agonists to slow the growth of the cancer cells. It seems para-

TABLE 24-2   Effects of Sex Steroids on Gonadotropin Release

Steroid Hormone	Effect	Gonadotropin Level
Low estrogen or androgen	Absence of negative feedback	Increases
Moderate estrogen or androgen	Negative feedback	Decreases
High androgen	Negative feedback	Decreases
Sustained high estrogen	Positive feedback	Increases

doxical to give these patients a drug that stimulates se-cretion of androgens and estrogens, but after a brief in-crease in FSH and LH, the pituitary becomes insensitive to GnRH. FSH and LH secretion then decreases, and go-nadal output of steroid hormones also falls. In essence, the GnRH agonist creates chemical castration that re-verses when the drug is no longer administered.

Some of the least-understood influences on repro-ductive hormones and gametogenesis are environmental effects. In men, factors that influence gametogenesis are difficult to monitor short of requesting periodic sperm counts. Disruption of the normal reproductive cycle in women is easier to study because physiological uterine bleeding in the menstrual cycle is easily monitored.

Some factors that affect reproductive function in women are stress, nutritional status, and changes in the day-night cycle such as those which occur with travel across time zones or with shift work. The hormone **mela-tonin** from the pineal gland mediates reproduction in seasonally breeding animals such as birds and deer, so researchers are investigating whether melatonin plays a role in seasonal and daily rhythms in humans.

Environmental estrogens are also receiving a lot of at-tention. These are naturally occurring compounds, such as the *phytoestrogens* of plants, or synthetic compounds that have been released into the environment. Some of these compounds bind to estrogen receptors and mimic estrogen's effects. Others are anti-estrogens that block estrogen receptors or interfere with second messenger pathways or protein synthesis.

Now that you have learned the basic patterns of hor-mone secretion and gamete development, let's look in detail at the male and female reproductive systems.

◗ MALE REPRODUCTION

The male reproductive system consists of the testes, the accessory glands and ducts, and the external genitalia. The external genitalia consist of the **penis** and the **scrotum**, a saclike structure that contains the testes (Fig. 24-8 ■). The **urethra** serves as a common passageway for sperm and urine, although not simultaneously. It runs through the ventral aspect of the shaft of the penis, sur-rounded by a spongy column of tissue known as the

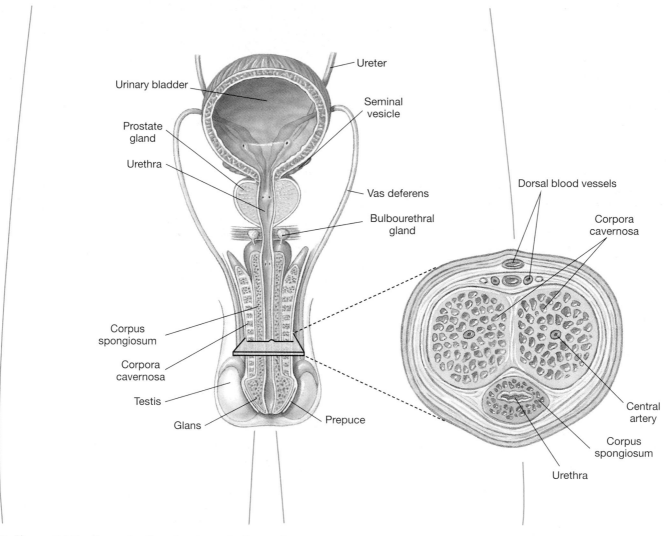

■ **Figure 24-8   Reproductive structures in the male**

corpus spongiosum [*corpus*, body; plural *corpora*]. The corpus spongiosum along with two columns of tissue called the **corpora cavernosa** constitute the erectile tissue of the penis.

The tip of the penis is enlarged into a region called the **glans** that at birth is covered by a layer of skin called the **foreskin**, or **prepuce**. In some cultures, the foreskin is removed surgically in a procedure called **circumcision**. In the United States, this practice goes through cycles of popularity. Proponents of the procedure claim that it is necessary for good hygiene and they cite evidence suggesting that the incidence of penile cancer, sexually transmitted diseases, and urinary tract infections is lower in circumcised men. Opponents claim that it is cruel to subject newborn boys to an unnecessary and possibly painful procedure.

The scrotum is an external sac into which the testes migrate during fetal development. This location outside the abdominal cavity is necessary because normal sperm development requires a temperature that is 2°–3° F lower than body core temperature. Men who have borderline or low sperm counts are advised to switch from jockey-style underwear, which keeps the scrotum close to the body, to boxer shorts, which allow the testes to stay cooler.

The failure of one or both testes to descend is known as **cryptorchidism** [*crypto*, hidden + *orchis*, testicle] and occurs in 1%–3% of all male births. If left alone, about 80% of cryptorchid testes spontaneously descend later. Those that remain in the abdomen through puberty become sterile and unable to produce sperm.

Although cryptorchid testes lose their gametogenic potential, they can produce androgens, indicating that hormone production is not as temperature-sensitive as sperm production. Because undescended testes are prone to become cancerous, authorities recommend that they be moved to the scrotum with testosterone treatment or surgery if necessary.

The male accessory glands and ducts include the **prostate gland**, the **seminal vesicles**, and the **bulbourethral (Cowper's) glands**. The bulbourethral glands and seminal vesicles empty their secretions into the urethra through ducts. The individual glands of the prostate open directly into the urethral lumen.

The prostate gland is the best known of the three accessory glands because of its medical significance. Cancer of the prostate is the second most common form of cancer in men (next to lung cancer), and *benign prostatic hypertrophy* (enlargement) creates problems for many men after age 50. Because the prostate gland completely encircles the urethra, its enlargement causes difficult urination by narrowing the passageway.

The Testes Produce Sperm and Testosterone

The human testes are paired ovoid structures about 5 cm by 2.5 cm (Fig. 24-9 ■). The word *testis* means "witness" in Latin, and its application to the male gonad comes from the fact that in ancient Rome, men taking an oath placed one hand on their genitals.

The testes have a tough outer fibrous capsule that encloses masses of coiled **seminiferous tubules** clustered into 250–300 compartments. Between the tubules is *interstitial tissue* consisting primarily of blood vessels and testosterone-producing *Leydig (interstitial) cells*. The seminiferous tubules constitute nearly 80% of the testicular mass in an adult. Each individual tubule is 0.3–1 meter long and, if stretched out and laid end to end, the entire mass would extend for about the length of two and a half football fields.

The seminiferous tubules leave the testis and join the **epididymis** [*epi-*, upon + *didymos*, twin], a single duct that forms a tightly coiled cord on the surface of the testicular capsule (Fig. 24-9b ■). The epididymis becomes the **vas deferens** [*vas*, vessel + *deferre*, to carry away from], also known as the **ductus deferens**. This duct passes into the abdomen, where it eventually empties into the urethra, the passageway from the urinary bladder to the external environment (see Fig. 24-8 ■).

The Seminiferous Tubules Contain Developing Sperm A seminiferous tubule is composed of two types of cells: spermatocytes in various stages of becoming sperm, and **Sertoli cells** (Fig. 24-9d, e ■). The developing spermatocytes stack in columns from the outer edge of the tubule to the inner lumen. Between each column of spermatocytes is a single Sertoli cell that extends from the outer edge of the tubule to the lumen.

Surrounding the outside of the tubule is a basement membrane (*basal lamina*). It acts as a barrier, preventing certain large molecules in the interstitial fluid from en-

...continued from page 732

Infertility can be caused by problems in either the man or woman. Sometimes, however, both partners have problems that contribute to their infertility. In general, male infertility is caused by low sperm counts, abnormalities in sperm morphology, or abnormalities in the reproductive structures that carry sperm. Female infertility may be caused by problems in hormonal pathways that govern maturation and release of eggs or by abnormalities of the reproductive structures (cervix, uterus, ovaries, oviducts). Because tests of male fertility are simple to perform, Dr. Coddington first analyzes Larry's sperm. In this test, trained technicians examine a fresh sperm sample under a microscope. They note the shape and motility of the sperm and estimate the concentration of sperm in the sample.

Question 1: *Name the male reproductive structures that carry sperm from the testes to the external environment.*

Question 2: *A new technique for the treatment of male infertility involves retrieval of sperm from the epididymis. The retrieved sperm can then be used to fertilize an egg, which is then implanted in the uterus. What causes of male infertility might make this treatment necessary?*

ANATOMY SUMMARY Male Reproduction

■ Figure 24-9

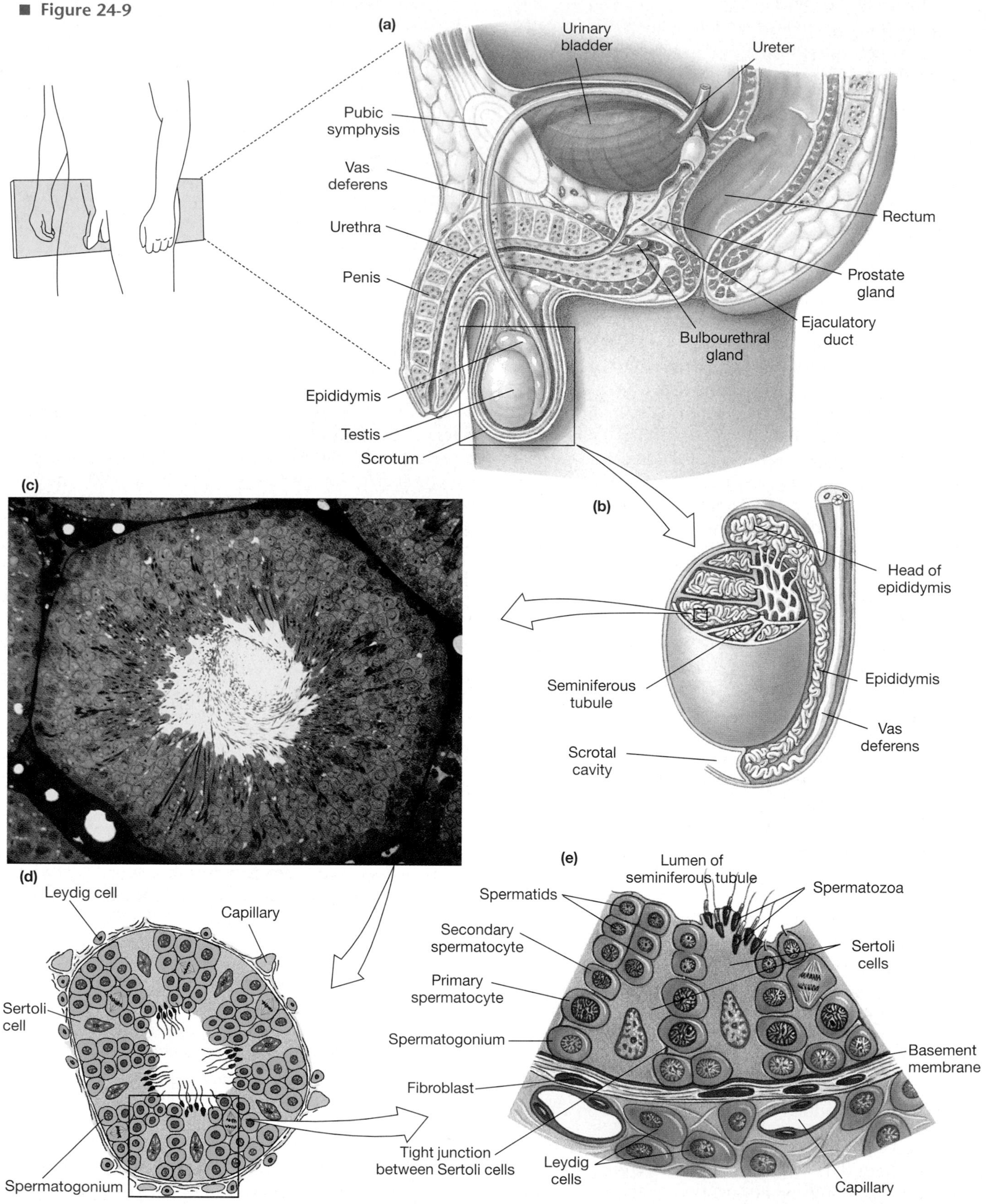

(a)

Urinary bladder

Ureter

Pubic symphysis

Vas deferens

Urethra

Penis

Epididymis

Testis

Scrotum

Rectum

Prostate gland

Ejaculatory duct

Bulbourethral gland

(b)

Head of epididymis

Seminiferous tubule

Epididymis

Vas deferens

Scrotal cavity

(c)

(d)

Leydig cell

Capillary

Sertoli cell

Spermatogonium

(e)

Lumen of seminiferous tubule

Spermatids

Spermatozoa

Secondary spermatocyte

Primary spermatocyte

Spermatogonium

Fibroblast

Tight junction between Sertoli cells

Sertoli cells

Basement membrane

Leydig cells

Capillary

tering the tubule, yet allowing testosterone to pass easily (Fig. 24-9c ■). Adjacent Sertoli cells in a tubule are linked to each other by tight junctions that form an additional barrier between the lumen of the tubule and the interstitial fluid outside the tubule.

The tight junctions are sometimes called the **blood-testis barrier** because functionally they behave much like the impermeable capillaries of the blood-brain barrier, preventing movement of material between two compartments. However, the testes have three functional compartments: (1) the interstitial fluid surrounding the tubules, (2) a basal compartment in the tubules between the basement membrane and basal ends of the Sertoli cells, and (3) an inner compartment that includes the lumen of the tubule. Because of the barriers between these three compartments, fluid in the tubule lumen has a composition different from that of interstitial fluid, with low concentrations of glucose and high concentrations of K^+ and steroid hormones.

Sperm Production Is Continuous   *Spermatogonia*, the germ cells that undergo meiotic division to become sperm, are found clustered near the basal ends of the Sertoli cells, just under the basement membrane of the seminiferous tubules. In this basal compartment, they undergo mitotic replication to produce additional germ cells. Some of the spermatogonia remain near the outer edge of the tubule to produce future spermatogonia. Others enter meiosis and become primary spermatocytes.

As the spermatocytes differentiate into sperm, they move inward toward the tubule lumen, continuously surrounded by the Sertoli cells. The tight junctions of the blood-testis barrier break and reform around the migrating cells, ensuring that the barrier remains intact. By the time the spermatocytes reach the luminal end of the Sertoli cell, they have divided twice and become spermatids.

The spermatids remain embedded in the apical membrane of the Sertoli cells while they complete the transformation into sperm. In this process, they lose most of their cytoplasm and develop a flagellated tail (Fig. 24-10 ■). The chromatin of the nucleus condenses into a dense structure, while a lysosome-like vesicle called an **acrosome** flattens out to cover most of the surface of the nucleus. The acrosome contains enzymes that are essential for fertilization.

Mitochondria to produce energy for sperm movement concentrate in the midpiece of the sperm body, along with microtubules that extend into the tail [∞ p. 50]. The result is a small, motile gamete that bears little resemblance to the parent spermatid. Sperm are released into the lumen of the tubule, along with secreted fluid. From there, they are free to move out of the testis.

The entire process takes about 64 days from spermatogonium division until sperm are released into the tubule. However, at any given time, different regions of the tubule contain spermatocytes in different stages of development. The staggering of developmental stages allows sperm production to remain nearly constant at a

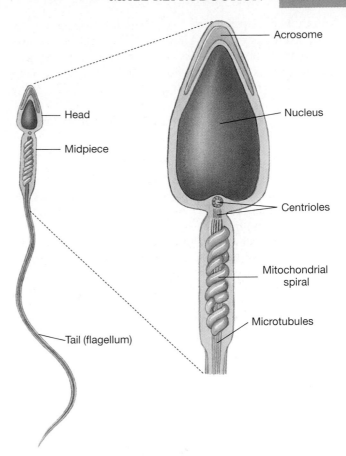

■ **Figure 24-10   Sperm structure**

rate of *200 million* sperm per day. That may sound like an extraordinarily high number, but it is about the number of sperm released in a single ejaculation.

Sperm released from the Sertoli cells are not yet mature or capable of swimming. They are pushed through the lumen of the seminiferous tubule by other sperm and by bulk flow of fluid secreted by the Sertoli cells. Sperm entering the epididymis complete their maturation during the 12 or so days of their transit time, aided by protein secretions from epididymal cells.

The Sertoli Cells Secrete Proteins   The function of Sertoli cells is to regulate sperm development. Another name for Sertoli cells is *sustentacular cells* because they provide sustenance or nourishment for the developing spermatogonia. The Sertoli cells manufacture and secrete proteins that range from the hormones inhibin and activin to growth factors, enzymes, and androgen-binding protein.

Androgen-binding protein is secreted into the seminiferous tubule lumen where it binds to testosterone secreted by Leydig cells. Protein binding makes testosterone less lipophilic and concentrates it within the luminal fluid.

Leydig Cells Secrete Testosterone   The **Leydig cells**, located in the interstitial tissue outside the seminiferous tubules, secrete testosterone. They are first active in the fetus, when testosterone is needed to direct development of male characteristics. After birth, the cells become inactive until puberty. At that time, they resume their production of testosterone. The Leydig cells also convert some testosterone to estradiol.

Spermatogenesis Requires Gonadotropins and Testosterone

The hormonal control of spermatogenesis follows the general pattern described previously. Gonadotropin releasing hormone (GnRH) from the hypothalamus controls the release of LH and FSH from the anterior pituitary. Those hormones in turn stimulate the testes to produce testosterone. The gonadotropins were named originally from their effect on the female ovary, but the same names have been retained in the male.

GnRH release is pulsatile, peaking every 1.5 hours, and LH release follows the same pattern. FSH levels are not as obviously related to GnRH secretion because FSH secretion is also influenced by inhibin and activin.

The target of FSH is the Sertoli cells, because male germ cells do not have FSH receptors (Fig. 24-11 ■). FSH stimulates paracrine factors needed for mitotic replication of spermatogonia and for spermatogenesis. In addition, FSH stimulates production of androgen-binding protein and inhibin.

The primary target of LH is the Leydig cells, which produce testosterone. In turn, testosterone feeds back to inhibit LH release by the anterior pituitary. Testosterone is essential for spermatogenesis, but its actions appear to be mediated by the Sertoli cells. Spermatocytes have no androgen receptors and cannot respond directly to testosterone, but they do have receptors for androgen-binding protein. Sertoli cells, which produce androgen-binding protein, have androgen receptors.

Spermatogenesis is a very difficult process to study *in vivo* or *in vitro*, and the available animal models may not accurately reflect the situation in the human testis. Therefore, it may be some time before we can say with certainty how testosterone and FSH regulate spermatogenesis.

✔ Because GnRH agonists cause down-regulation of GnRH receptors, what would be the advantage and disadvantage of using these drugs as a male contraceptive?

Male Accessory Glands Contribute Secretions to Semen

The male reproductive tract has several accessory glands whose primary function is to secrete various fluid mixtures. When sperm leave the vas deferens during ejaculation, they are joined by these secretions, resulting in a sperm-fluid mixture known as **semen**. About 99% of the volume of semen is fluid added from the accessory glands, the bulbourethral glands, seminal vesicles, and prostate.

The contributions of the accessory glands to the composition of semen are shown in Table 24-3. Semen provides a liquid medium for sperm delivery, including

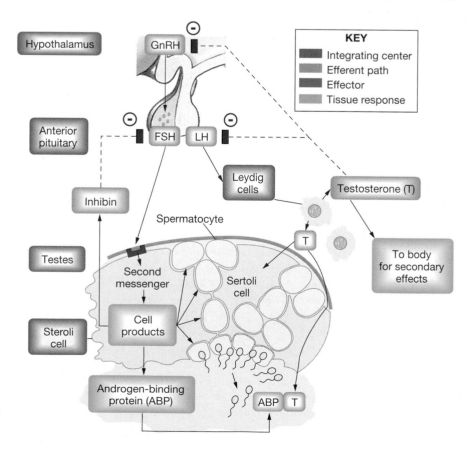

■ **Figure 24-11** **Hormonal control of spermatogenesis**

TABLE 24-3 Composition of Semen

Component	Function	Source
Sperm	Gamete	Seminiferous tubules
Mucus	Lubricant	Bulbourethral glands
Water	Provides liquid medium	All accessory glands
Buffers	Neutralize acidic environment of the vagina	Prostate, bulbourethral glands
Nutrients	Nourish sperm	
Fructose		Seminal vesicles
Citric acid		Prostate
Vitamin C		Seminal vesicles
Carnitine		Epididymis
Enzymes	Clot semen in vagina, then liquefy the clot	Seminal vesicles and prostate
Zinc	Unknown; possible association with fertility	Unknown
Prostaglandins	Smooth muscle contraction; may aid sperm transport	Seminal vesicles

mucus for lubrication, buffers to neutralize the usually acidic environment of the vagina, and nutrients for sperm metabolism. The seminal vesicles contribute *prostaglandins* [∞ p. 30], lipid-related molecules that appear to influence sperm motility and transport within both the male and female reproductive tracts. Prostaglandins were originally believed to come from the prostate gland, and the name was well established by the time their true source was discovered.

In addition to providing a medium for sperm, the accessory gland secretions help protect the reproductive tract from pathogens that might ascend the urethra from the external environment. They physically flush out the urethra and supply immunoglobulins, lysozyme, and other compounds with antibacterial action. One interesting component of semen is zinc. Its role in reproduction is unclear, but concentrations of zinc below a certain level are associated with male infertility.

Androgens Influence Secondary Sex Characteristics

The androgens have a number of effects on the body in addition to gametogenesis. These effects are divided into primary and the secondary sex characteristics. **Primary sex characteristics** are the internal sexual organs and external

DHT and the Prostate Fetal development of the prostate gland, like that of the external genitalia, is under the control of dihydrotestosterone (DHT). The discovery of the role of DHT in prostate growth led to the development of finasteride, a 5-alpha-reductase inhibitor that blocks production of DHT. This drug provides the first nonsurgical treatment for benign prostatic hypertrophy. Studies are currently under way to determine if lowering (but not eliminating) DHT will decrease the incidence of cancer of the prostate gland.

genitalia that distinguish males from females. As you have already learned, androgens are responsible for their differentiation during embryonic development and for their growth during the maturation period known as puberty.

The **secondary sex characteristics** are other features of the body, such as body shape, that distinguish males from females. The male body shape is sometimes described as an inverted triangle, with broad shoulders and narrow waist and hips. The female body is more pear-shaped, with broad hips and narrow shoulders. Androgens are responsible for typically male traits such as beard and body hair growth, muscular development, thickening of the vocal chords with subsequent lowering of the voice, and behavioral effects such as the sex drive (**libido**).

Androgens are anabolic hormones that promote protein synthesis, which gives them their street name of *anabolic steroids*. The illicit use of these drugs by athletes has been widespread despite possible adverse side effects such as liver tumors and infertility. One of the more interesting side effects researchers have noted is the apparent addictiveness of anabolic steroids.

Withdrawal from taking the drugs may be associated with behavioral changes that include excessive aggression ('roid rage), depression, or psychosis. These psychiatric disturbances suggest that human brain function can be modulated by sex steroids, just as the brain function of other animals can. Fortunately, many side effects of anabolic steroids are reversible once use is stopped.

✔ Explain why the use of exogenous anabolic steroids by males might shrink the testes and make the man temporarily infertile.

▶ FEMALE REPRODUCTION

Female reproduction is a more complicated topic than male reproduction due to the cyclic nature of gamete production in the ovary.

The Female Reproductive Tract Includes Ovaries and Uterus

The female external genitalia are known collectively as the **vulva**, or **pudendum** [*vulva*, womb; *pudere*, to be ashamed]. They are shown in Figure 24-12a ■, the view seen by a health care worker who is about to do a pelvic exam or Pap smear [∞ p. 43].

Starting at the periphery are the **labia majora** [*labium*, lip], folds of skin that arise from the same embryonic tissue as the scrotum. Within the labia majora are the **labia minora**, derived from embryonic tissues that in the male give rise to the shaft of the penis (see Fig. 24-3 ■). The **clitoris** is a small bud of erectile, sensory tissue at the anterior end of the vulva, enclosed by the labia minora and an additional fold of tissue equivalent to the foreskin of the penis.

In females, the urethra opens to the outside between the clitoris and the **vagina** [*vagina*, sheath], a cavity that acts as the receptacle for the penis during copulation. At birth, the external opening of the vagina is partially closed by a thin ring of tissue called the **hymen**, or *maidenhead*. Contrary to what some young women believe, the hymen is external to the vagina, not found within it, so the normal use of tampons during menstruation will not rupture the hymen. However, it can be stretched by normal activities such as horseback riding and therefore is not an accurate indicator of a woman's virginity.

Now let's follow the path of sperm deposited in the vagina during intercourse. To continue into the female reproductive tract, sperm first must pass through the narrow opening of the **cervix**, the neck of the uterus that protrudes slightly into the upper end of the vagina (Fig. 24-12b, c ■). The cervical canal is lined with mucous glands whose secretions create a barrier between the vagina and uterus. Sperm that make it through the cervical canal find themselves in the lumen of the uterus, or *womb*, a hollow, muscular organ slightly smaller than a woman's clenched fist.

The uterus is the structure in which fertilized eggs implant and develop during pregnancy. It is composed of three tissue layers: a thin outer connective tissue covering, a thick middle layer of smooth muscle known as the **myometrium** (Fig. 24-12f ■), and an inner layer known as the **endometrium** [*metra*, womb]. The endometrial lining is an epithelium with glands that dip into a connective tissue layer below. The thickness and character of the endometrium vary during the menstrual cycle. Cells of the epithelial lining alternately proliferate and slough off, accompanied by a small amount of bleeding in the process known as **menstruation**.

Sperm swimming upward through the uterus leave its cavity through openings into the two **Fallopian tubes**. The Fallopian tubes are 20–25 cm long and about the diameter of a drinking straw. Their walls have two layers of smooth muscle, longitudinal and circular, similar to the walls of the intestine. A ciliated epithelium lines the inside of the tubes (Fig. 24-12c ■). Fluid movement created by the cilia and aided by muscular contractions transports eggs toward the uterus. Pathological conditions in which ciliary function is absent are associated with infertility or with pregnancies in which the embryo implants in the Fallopian tube rather than the uterus.

The dilated open end of the Fallopian tube divides into fingerlike projections called **fimbriae** that form a fringe around the open edge [*fimbriae*, fringe]. The fimbriae are held close to the adjacent ovary by connective tissue (Fig. 24-12c ■). The close association of Fallopian tube and ovary ensures that eggs released from the surface of the ovary will be swept into the tube rather than floating off into the abdominal cavity.

The Ovary Produces Eggs and Hormones

The ovary is an elliptical structure, about 2–4 cm long (Fig. 24-12d ■). It has an outer connective tissue layer and an inner connective tissue framework known as the **stroma** [*stroma*, mattress]. Most of the ovary is taken up by a thick *cortex* filled with ovarian follicles in various stages of development or decline. The small central *medulla* contains nerves and blood vessels.

The ovary, like the testis, produces both gametes and hormones. As discussed earlier, about 7 million oögonia in the embryonic ovary develop into half a million *primary oocytes*. Each oocyte is contained in a **primary follicle** with an outer layer of granulosa cells (Fig. 24-12e ■). As the primary follicle develops, it gains an additional outer layer of cells known as the **theca** [*theke*, case or cover]. The thecal cells are separated from granulosa cells by a basal lamina.

A Menstrual Cycle Lasts About One Month

Female humans produce gametes in monthly cycles (average 28 days; normal range 24–35 days). These cycles are commonly called **menstrual cycles** because they are marked by a 3–7 day period of bloody uterine discharge known as the **menses** [*menses*, months], or menstruation. The menstrual cycle is described according to changes that occur in follicles of the ovary (**ovarian cycle**) and in the endometrial lining of the uterus (**uterine cycle**). Figure 24–13 ■, p. 750, is a summary figure showing hormonal and morphological changes that take place during a typical menstrual cycle.

The ovarian cycle is divided into three phases:

1. **Follicular phase.** The first part of the ovarian cycle, known as the **follicular phase,** is a period of follicular growth in the ovary. This phase is the most variable in length and lasts from 10 days to 3 weeks.
2. **Ovulation.** Once one or more follicles has ripened, the ovary releases the egg(s) during ovulation.
3. **Luteal phase.** The phase of the ovarian cycle following ovulation is known as the postovulatory, or **luteal phase.** The second name comes from the

transformation of the follicular remnants into a **corpus luteum** [*corpus*, body + *luteus*, yellow], named from its yellow pigment and lipid deposits. The corpus luteum secretes hormones to continue the preparations for pregnancy. If a pregnancy does not occur, the corpus luteum ceases to function after about two weeks, and the ovarian cycle begins again.

The endometrial lining of the uterus goes through its own cycle regulated by ovarian hormones.

1. **Menses.** The beginning of the follicular phase in the ovary corresponds to menstrual bleeding from the uterus.
2. **Proliferative phase.** The later part of the ovary's follicular phase corresponds to the **proliferative phase** in the uterus, during which the endometrium adds a new layer of cells in anticipation of pregnancy.
3. **Secretory phase.** After ovulation, hormones from the corpus luteum convert the thickened endometrium into a secretory structure. Thus, the luteal phase of the ovary corresponds to the **secretory phase** of the uterus. If no pregnancy occurs, the superficial layers of the endometrium are lost during menstruation as the uterine cycle begins again.

Hormonal Control of the Menstrual Cycle Is Complex

The ovarian and uterine cycles are under the primary control of:

- Gonadotropin releasing hormone (GnRH) from the hypothalamus
- FSH and LH from the anterior pituitary
- Estrogen, progesterone, and inhibin from the ovary

During the follicular phase, the dominant steroid hormone is estrogen. In the luteal phase, progesterone is dominant, although estrogen is still present. Now let's go through an ovarian cycle in detail.

...continued from page 741

The results of Larry's sperm analysis are normal. Dr. Coddington is therefore able to rule out sperm abnormalities as a cause of Peggy's and Larry's infertility. Peggy is instructed to track her body temperature daily and record the results on a chart. This temperature tracking is designed to determine if she is ovulating. Following ovulation, body temperature rises slightly and remains elevated through the remainder of the menstrual cycle.

Question 3: *For what causes of female infertility is temperature tracking useful? For what causes is it not useful?*

Early Follicular Phase The first day of menstruation is day 1 of a cycle. This point was chosen to start the cycle because the bleeding of menstruation is an easily monitored physical sign. Just before the beginning of each cycle, gonadotropin secretion from the anterior pituitary increases. Under the influence of FSH, several follicles in the ovaries begin to mature (Table 24-4, p. 751).

As the follicles grow, their granulosa cells under the influence of FSH and their thecal cells under the influence of LH start to produce steroid hormones. Thecal cells synthesize androgens that diffuse into the neighboring granulosa cells, where aromatase converts them to estrogens (Fig. 24-14a ■, p. 752).

Gradually increasing estrogen levels in the circulation have several effects. Estrogen exerts a negative feedback effect on pituitary FSH and LH secretion, thus preventing the development of additional follicles in the same cycle. At the same time, estrogen stimulates estrogen production by the granulosa cells. This positive feedback loop allows the follicles to continue estrogen production even though FSH and LH decrease.

As the follicles enlarge, granulosa cells secrete fluid that collects in a cavity known as the **antrum**. The antral fluid contains hormones and enzymes needed for ovulation. At each stage of development, some follicles undergo *atresia*, hormonally regulated cell death. Only a few follicles reach the final stage, and usually only one dominant follicle continues to ovulation.

In the uterus, menstruation ends during the early follicular phase. Under the influence of estrogens from the ovary, the endometrium begins to grow, or proliferate. This period is characterized by an increase in cell number and by enhanced blood supply to bring nutrients and oxygen to the thickening endometrium. Estrogen causes the mucous glands of the cervix to produce clear, watery mucus.

Late Follicular Phase As the follicular phase nears its end, ovarian estrogen secretion peaks. By this point of the cycle, only one follicle is still developing. As the follicular phase ends, the granulosa cells of the dominant follicle begin to secrete inhibin and progesterone in addition to estrogen (Fig. 24-14b ■). Estrogen, which had exerted a negative feedback effect on GnRH and gonadotropins earlier in the follicular phase, changes to positive feedback.

Immediately before ovulation, the persistently high levels of estrogen, aided by rising levels of progesterone, enhance pituitary responsiveness to GnRH. As a result, LH secretion increases dramatically, a phenomenon known as the *LH surge*. FSH surges also, but to a lesser degree, presumably because it is being suppressed by inhibin and estrogen.

The LH surge is an essential part of ovulation. Without it, the final steps of oocyte maturation cannot take place. Meiosis, which had paused in the embryo, resumes in the developing follicle with the first meiotic division (see Fig. 24-5 ■). The division converts a primary oocyte into a polar body, which is extruded, and an egg,

ANATOMY SUMMARY   Female Reproduction

■ **Figure 24-12**

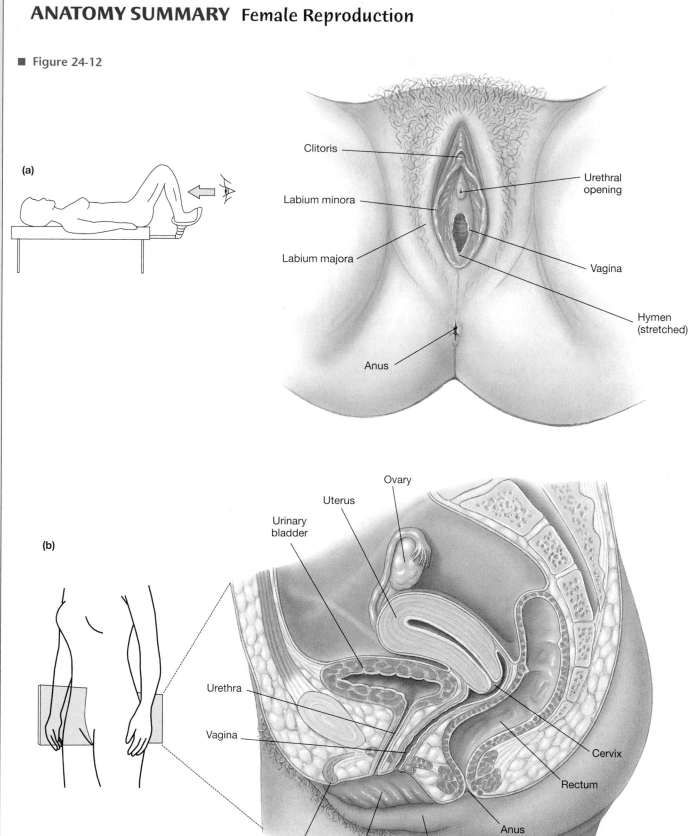

(a)

Clitoris

Urethral opening

Labium minora

Labium majora

Vagina

Hymen (stretched)

Anus

(b)

Ovary

Uterus

Urinary bladder

Urethra

Vagina

Clitoris

Labium minora

Cervix

Rectum

Anus

Labium majora

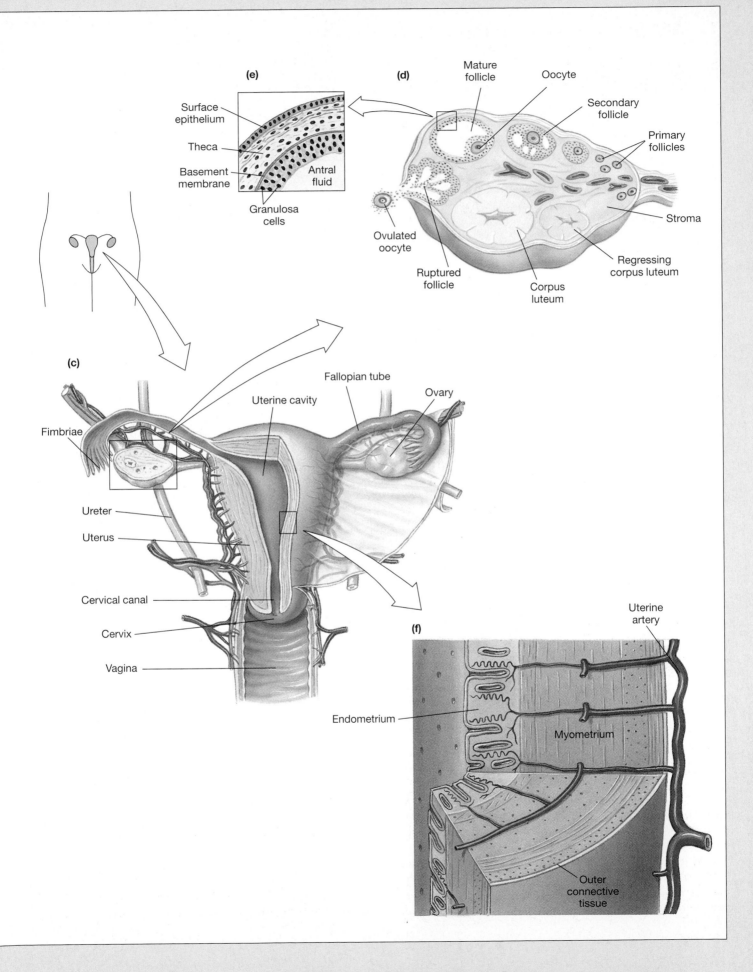

(e)
Surface epithelium
Theca
Basement membrane
Granulosa cells
Antral fluid

(d)
Mature follicle
Oocyte
Secondary follicle
Primary follicles
Stroma
Regressing corpus luteum
Corpus luteum
Ruptured follicle
Ovulated oocyte

(c)
Fimbriae
Ureter
Uterus
Cervical canal
Cervix
Vagina
Uterine cavity
Fallopian tube
Ovary

(f)
Uterine artery
Endometrium
Myometrium
Outer connective tissue

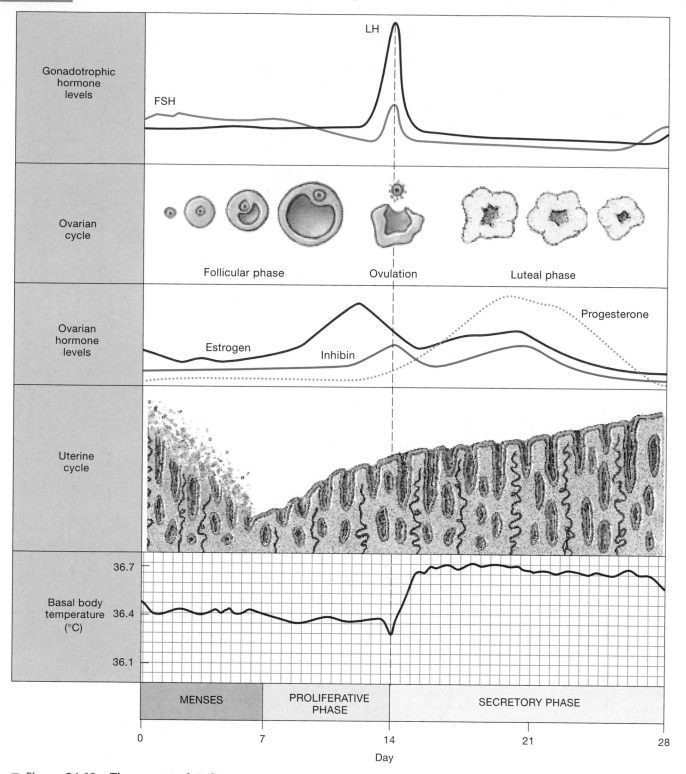

■ **Figure 24-13** **The menstrual cycle** This 28-day menstrual cycle is divided into phases on events in the uterus and ovary.

or secondary oocyte. While this is taking place, antral fluid collects and the follicle grows to its greatest size, preparing to release the egg.

High levels of estrogen in the late follicular phase prepare the uterus for a possible pregnancy. The endometrium grows to a final thickness of 3–4 mm. Just before ovulation the cervical glands produce copious

amounts of thin, stringy mucus to facilitate sperm entry. The stage is set for ovulation.

Ovulation About 16–24 hours after LH peaks, ovulation occurs. The mature follicle secretes *collagenase*, dissolving collagen in the connective tissue that holds the follicular cells together. The breakdown products of col-

TABLE 24-4   Development of Ovarian Follicles

The contents of the follicle are given starting at the center with the ovum and moving outward. The zona pellucida (see Fig. 24-16 ■) is a glycoprotein coat that protects the ovum.

Stage	Ovum	Zona Pellucida	Granulosa Cells	Antrum	Basal Lamina	Theca
Primary follicle (before FSH stimulation)	Primary oocyte	Minimal	Single layer	None	Separates granulosa and theca	Single cell layer plus blood vessels
Secondary follicle (early follicular phase)	Primary oocyte	Increased in width	2–6 cell layer	None	Present	Single cell layer
Tertiary follicle (late follicular phase)	Primary oocyte, then secondary oocyte with division arrested	Present	3–4 cell layer	Develops within granulosa layer and fills with fluid; swells to 15–20 mm diameter	Present	*Inner layer:* secretory and small blood vessels. *Outer layer:* connective tissue, smooth muscle cells, large blood vessels
Corpus luteum (luteal phase)	None	None	Converted to luteal cells	Fills with migrating cells	Disappears	Converted to luteal cells
Corpus albicans (post-luteal phase)	None	None	Cells degenerate	None	None	Cells degenerate

lagen create an inflammatory reaction, attracting leukocytes that secrete prostaglandins into the follicle.

Our understanding of events from this point is unclear, but it is possible that the prostaglandins cause smooth muscle cells in the outer theca to contract, rupturing the follicle wall at its weakest point. The antral fluid spurts out along with the egg, which is surrounded by two to three layers of granulosa cells. The egg is swept into the Fallopian tube and carried away to be fertilized or die.

In addition to promoting rupture of the follicle, the LH surge causes follicular thecal cells to migrate into the antral space, mingling with the former granulosa cells and filling the cavity. Both cell types then transform into *luteal cells* of the corpus luteum. This process, known as *luteinization*, involves biochemical and morphological changes. Newly formed luteal cells accumulate lipid droplets and glycogen granules in their cytoplasm and begin to secrete progesterone. Estrogen synthesis diminishes.

Early to Mid-Luteal Phase   After ovulation, the corpus luteum produces steadily increasing amounts of progesterone and estrogen. Progesterone is the domi-

nant hormone of the luteal phase. Estrogen levels increase but never reach the peak seen before ovulation.

The combination of estrogen and progesterone exerts negative feedback on the hypothalamus and anterior pituitary. Gonadotropin secretion, further suppressed by luteal inhibin production, remains shut down throughout most of the luteal phase (Fig. 24-14c ■).

Under the influence of progesterone, the endometrium of the uterus continues its preparation for pregnancy. The endometrial glands coil, and additional blood vessels grow into the connective tissue layer. Endometrial cells deposit lipids and glycogen in their cytoplasm. These deposits will provide nourishment for a developing embryo while the placenta, the fetal-maternal connection, is developing.

Progesterone also causes cervical mucus to thicken. Thicker mucus creates a plug that blocks the cervical opening, preventing bacteria as well as sperm from entering the uterus.

One interesting effect of progesterone is its thermogenic ability. During the luteal phase of an ovulatory cycle, a woman's basal body temperature, taken immediately upon awakening and before getting out of bed,

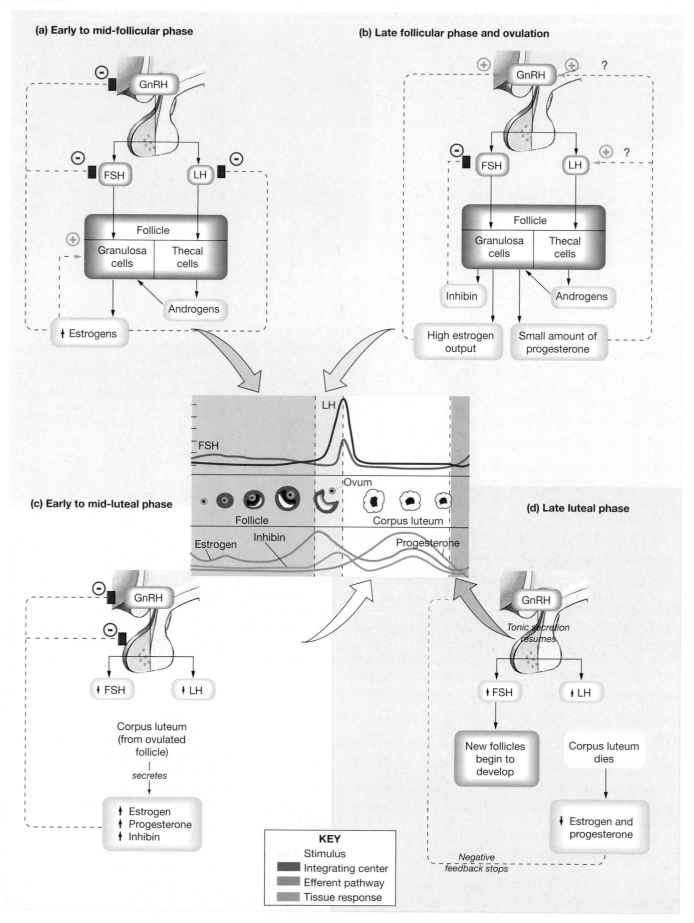

■ Figure 24-14   **Hormonal control of the menstrual cycle**

jumps 0.3°–0.5° F and remains elevated until menstruation. Because this change in the setpoint occurs after ovulation, it cannot be used effectively to predict ovulation. However, it is a simple way to assess if a woman is having ovulatory or *anovulatory* (not ovulating) cycles.

Late Luteal Phase and Menstruation   The corpus luteum has an intrinsic life span of approximately 12 days. If pregnancy does not occur, the corpus luteum spontaneously degenerates into an inactive structure called a **corpus albicans** [*albus*, white]. As the luteal cells shut down, production of progesterone and estrogen drops (Fig. 24-14d ■). This removes the negative feedback signal to the pituitary and hypothalamus, and secretion of FSH and LH increases.

Maintenance of the secretory endometrium depends on the presence of progesterone. When the corpus luteum degenerates and hormone production falls, blood vessels in the surface layer of the endometrium contract. Without oxygen and nutrients, the surface cells die. About two days after the corpus luteum ceases to function, or 14 days after ovulation, the endometrium begins to slough its surface layer, and menstruation begins.

The menstrual discharge from the uterus totals about 40 mL of blood and 35 mL of serous fluid and cellular debris. There are few clots of blood in the menstrual flow because of the presence of *plasmin*, a substance that breaks up clots. Menstruation continues for 3–7 days, well into the follicular phase of the next ovulatory cycle.

Estrogens and Androgens Influence Female Secondary Sex Characteristics

Estrogens control the development of primary sex characteristics in females, just as androgens control them in males. They also control the most prominent secondary sex traits: the female pattern of fat distribution (hips and upper thighs) and breast development. Other female secondary sex characteristics are actually governed by androgens produced in the adrenal cortex. Pubic and *axillary* (armpit) hair growth and libido (sex drive) are under the control of adrenal androgens.

• •

…continued from page 747

Results of the temperature tracking reveal after several months that Peggy is ovulating regularly. Dr. Coddington therefore believes that her ovaries are functioning normally. Other possible causes for this couple's infertility include abnormalities in Peggy's cervix, Fallopian tubes, or uterus. Dr. Coddington next decides to order a postcoital test. In this test, the couple is instructed to have intercourse 12 hours before the physician visit. The cervical mucus is then analyzed. This test will also analyze the interaction between sperm and mucus.

Question 4:   *What abnormalities in the cervix, Fallopian tubes, and uterus could cause infertility?*

• •

✔ What side effects would you predict in women athletes who take anabolic steroids to build muscles?

✔ Aromatase converts androgens to estrogen. What would happen to the ovarian cycle of a woman who is given an aromatase inhibitor?

◗ PROCREATION

Reproduction throughout the animal kingdom is marked by species-specific behaviors designed to ensure that egg and sperm meet. For aquatic animals that release gametes into the water, coordinated timing is everything. Interaction between males and females of these species may be limited to chemical communication by pheromones.

On the other hand, internal fertilization in terrestrial vertebrates requires interactive behaviors and special adaptations of the genitalia. The female must have an internal receptacle for sperm (the vagina) and the male must possess an organ (the penis) capable of placing sperm into the receptacle. The penis of human males is *flaccid* (soft and limp) in its resting state, not capable of penetrating the narrow opening of the female's vagina. The processes through which the penis stiffens and enlarges (**erection**), then releases sperm from the ducts of the reproductive tract (**ejaculation**), make up the main components of the male sex act. Without these events, fertilization cannot take place.

The Human Sexual Response Has Four Phases

The human sex act, also known as sexual intercourse, copulation, or **coitus** [*coitio*, a coming together] is highly variable in some ways and highly stereotyped in others.

Human sexual response in both sexes is divided into four phases: (1) excitement, (2) plateau, (3) orgasm, and (4) resolution. In the excitement phase, various erotic stimuli prepare the genitalia for the act of copulation. For the male, excitement involves erection of the penis. For the female, it includes erection of the clitoris and lubrication of the vagina by secretions from the vaginal walls. In both sexes, erection is a state of vasocongestion in which arterial blood flow into spongy erectile tissue exceeds venous outflow.

Erotic stimuli include sexually arousing tactile stimuli as well as psychological stimuli. The latter vary widely among individuals and among cultures, so what is erotic in one culture may be considered disgusting in another. Regions of the body that possess receptors for sexually arousing tactile stimuli are called **erogenous zones**. They include the genitalia as well as the lips, tongue, nipples, ear lobes, and other regions.

In the plateau phase, the changes that started during excitement intensify. They then peak in an **orgasm** (climax). In both sexes, orgasm is a series of muscular contractions accompanied by intense pleasurable sensations, increased blood pressure, heart rate, and respiration rate. In the female, the walls of the vagina and the uterus contract. In the male, the contractions usually result in the ejaculation of semen from the penis. Pregnancy cannot occur without ejaculation to place sperm in the female reproductive tract. However, female orgasm is not required for pregnancy.

The final phase of the sexual response is resolution, a period during which the physiological parameters that changed in the first three phases slowly return to normal.

The Male Sex Act Is Composed of Erection and Ejaculation

A key element to successful copulation is the ability of the male to achieve and sustain an erection. Sexual excitement from either tactile or psychological stimuli triggers the erection reflex, a spinal reflex that is subject to control from higher centers in the brain. The urination and defecation reflexes are similar types of reflexes [∞ pp. 562, 631].

In its simplest form, the erection reflex begins with tactile stimuli sensed by mechanoreceptors in the glans penis or other erogenous zones (Fig. 24-15 ■). Sensory neurons signal the spinal integration center, which inhibits vasoconstrictive sympathetic input on penile arterioles. Simultaneously, increased parasympathetic input mediated primarily by nitric oxide actively vasodilates the penile arterioles.

As a result of vasodilation, arterial blood flows into the open spaces of the erectile tissue, passively compressing the veins and trapping blood. The erectile tissue becomes engorged, stiffening and lengthening the penis within 5–10 seconds.

The climax of the male sexual act coincides with emission and ejaculation. **Emission** is the movement of

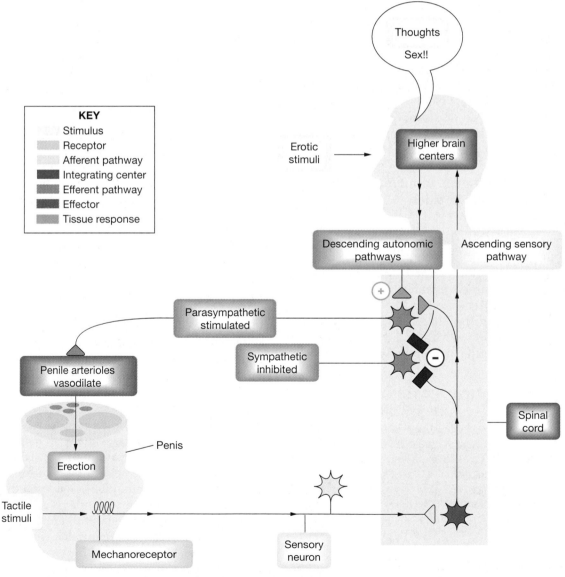

■ **Figure 24-15**   **The erection reflex**   Erection can take place as a simple spinal reflex, without input from higher brain centers. It can also be stimulated (and inhibited) by descending pathways from the cerebral cortex.

Viagra Erectile dysfunction is a matter of global concern, because inability to achieve and sustain an erection disrupts the sex act for both men and women. In 1998 the U.S. Food and Drug Administration approved sildenafil (Viagra®) for the treatment of erectile dysfunction. Erections occur when neurotransmitters released from pelvic nerves increase nitric oxide in vascular smooth muscle. Nitric oxide activates guanylyl cyclase, which increases cGMP and causes vasodilation. Sildenafil blocks the enzyme that degrades cGMP, thereby prolonging the effects of nitric oxide. The maximum effect occurs about one hour after taking the pill. Clinical trials showed that sildenafil corrected erectile dysfunction in 82% of the men who took it. However, it is not without side effects. Men who are taking nitrate-containing drugs should not take sildenafil because drug interactions could create very low blood pressure. The U.S. Federal Aviation Administration issued an order that pilots should not take sildenafil within six hours of flying because 3% of men report impaired color vision, a blue or greenish haze. This occurs because sildenafil also inhibits an enzyme in the retina. And finally, there is no evidence that sildenafil corrects orgasmic dysfunction in women. Although women do have clitoral erections, the female sexual response is more complicated. At least one drug company reportedly is testing a drug that affects brain receptors linked to sexual arousal.

sperm out of the vas deferens and into the urethra, where they are joined by secretions from the accessory glands to make semen. The average semen volume is 3 mL (range 2–6 mL), of which less than 10% is sperm.

During ejaculation, semen in the urethra is expelled to the exterior by a series of rapid muscular contractions accompanied by sensations of intense pleasure, the orgasm. A sphincter at the base of the bladder contracts to prevent sperm from entering the bladder and urine from joining the semen.

Both erection and ejaculation can occur in the absence of mechanical stimulation. Sexually arousing thoughts, sights, sounds, emotions, and dreams can all initiate sexual arousal and even lead to orgasm in both men and women. In addition, nonsexual penile erection accompanies rapid eye movement (REM) sleep.

The inability to obtain or sustain an erection is known as *erectile dysfunction* or **impotence**. Impotence may result from a variety of physiological and psychological causes. For example, fear of interruption may result in loss of arousal. Alcohol also inhibits sexual performance, as noted by Shakespeare in *Macbeth* (III, iii, 26–30). When MacDuff asks, "What three things does drink especially provoke?" the porter answers, "Marry, sir, nose-painting, sleep, and urine. Lechery, sir, it provokes and unprovokes: it provokes the desire, but it takes away the performance."

Because of the influence of psychological factors on the erection reflex, clinicians often use the unconscious, nonsexual erection of REM sleep to determine whether impotence is physiological or psychological. In its simplest form, the REM erection test consists of gluing a band of postage stamps around the flaccid penis before the subject goes to sleep. If the perforations between stamps have torn by the time the subject awakens, he knows that he had normal erections during REM sleep.

Contraceptives Attempt to Prevent Pregnancy

One disadvantage of sexual intercourse for pleasure rather than reproduction is the possibility of an unplanned pregnancy. On average, 80% of young women who have intercourse without using any form of birth control will get pregnant within a year. But many women get pregnant after a single unprotected encounter. Couples who hope to avoid unwanted pregnancies generally use some form of birth control, or **contraception**.

Contraceptive practices fall into several broad groups. **Abstinence**, the avoidance of intercourse, is the surest method to avoid pregnancy (and sexually transmitted diseases). Interventional methods include (1) barrier methods that prevent union of eggs and sperm, (2) methods that prevent implantation of the fertilized egg, and (3) hormonal treatment that decreases or stops gamete production.

Barrier methods of contraception are among the earliest recorded means of birth control. Once people made the association between pregnancy and semen, they concocted a variety of substances to block or kill the sperm. An ancient Egyptian papyrus with the earliest known references to birth control describes the use of vaginal plugs made of leaves, feathers, figs, and alum held together with crocodile and elephant dung. Sea sponges soaked in vinegar and disks of oiled silk have also been used at one time or another. Later, women used douches of garlic, turpentine, and rose petals to rinse the vagina after intercourse. As you can imagine, many of these items also caused vaginal or uterine infections.

The modern version of the female barrier is the **diaphragm**, introduced into the United States in 1916. These rubber domes and a smaller version called a *cervical cap* are usually filled with a spermicidal cream, then inserted into the top of the vagina so that they cover the cervix. One advantage to the diaphragm is that it is nonhormonal. In theory, diaphragms are highly effective (97%–99%) if used properly and regularly. However, in real life they are not always used because they must be inserted close to the time of intercourse, and about 20% of women who depend on diaphragms for contraception are pregnant within the first year.

The male barrier contraceptive is the **condom**, a closed sheath that fits closely over the penis to catch ejaculated semen. Males have used condoms made from animal bladders and intestines for centuries. Condoms lost popularity when oral contraceptives came into widespread use in the 1960s and 1970s, but in recent years they have regained favor because they combine pregnancy protection with protection from many sexually transmitted diseases. However, latex condoms may cause allergic reactions and there is evidence that the HIV virus can pass through pores in some contemporary condoms.

Sterilization is the most effective contraceptive method for sexually active people, but it is a surgical procedure and is not easily reversed. Female sterilization is called **tubal ligation** and consists of tying off and cutting the Fallopian tubes. A woman with a tubal ligation will still ovulate, but the eggs remain in the abdomen. The male form of sterilization is the **vasectomy**, in which the vas deferens is tied and clipped. Sperm are still made in the seminiferous tubules, but since they cannot leave the reproductive tract, they are reabsorbed.

Some contraceptive methods do not prevent fertilization but do keep a fertilized egg from establishing itself in the endometrium. They include **intrauterine devices (IUDs)** as well as chemicals that change the properties of the endometrium. IUDs are plastic devices that are inserted into the uterine cavity, where they create an inflammatory reaction that prevents implantation. They have low failure rates (0.5% per 100 women per year) but a high incidence of side effects ranging from pain and bleeding to infertility from pelvic inflammatory disease and blockage of the Fallopian tubes.

Techniques for decreasing gamete production depend on altering the hormonal milieu of the body. These methods were tried historically when women would eat or drink various plant concoctions for contraception. Some actually worked because of estrogen-like compounds contained in the plants. Modern pharmacology has improved on this method, and now women can choose between oral contraceptive pills, injections lasting three months, or capsules that are inserted under the skin and release hormones.

The oral contraceptives, also known as birth control pills, were first made available in 1960. They rely on various combinations of estrogen and progesterone that inhibit gonadotropin secretion from the pituitary. Without adequate FSH and LH, ovulation is suppressed. In addition, progesterones in the contraceptive pills thicken the cervical mucus and help prevent sperm penetration. These hormonal methods of contraception are highly effective when taken correctly but also carry some risks, including an increased incidence of blood clots and strokes (especially in women who smoke).

Two new forms of hormonal contraception for women may become available soon: the GnRH agonists discussed earlier and mifepristone, also known as RU 486. The GnRH agonists appear to effectively and reversibly block ovulation, as do the estrogen-progesterone combination pills. Mifepristone, a controversial drug because of its use in inducing abortions in Europe, is an antiprogesterone drug that blocks progesterone receptors. In low doses, it blocks ovulation or alters the endometrium so that fertilized eggs do not implant.

Most attempts to devise a male hormonal contraceptive have failed so far because of undesirable side effects. To inhibit sperm production, the contraceptive must block testosterone secretion or action. In doing so, the contraceptive is also likely to decrease the male libido or even cause impotence. Both side effects are unacceptable to the men who would be most interested in using the contraceptive. The search for a male contraceptive pill is concentrating on something that will affect only sperm production in the seminiferous tubules and not testosterone production by the Leydig cells.

The newest methods of contraception being tested are based on development of antibodies to various components of the male and female reproductive systems. These contraceptives would be long-lasting and reversible, and they would be administered as vaccines. One contraceptive vaccine is currently in clinical trials in Sweden.

Infertility Is the Inability to Conceive

While some couples are trying to prevent pregnancy, other couples are spending thousands of dollars trying to get pregnant. Infertility is the inability of a couple to conceive a child after a year of unprotected intercourse. For years, infertile couples had no choice but adoption if they wanted to have a child, but incredible strides have been made in this field since the 1970s. As a result, many infertile couples today are able to have children.

Infertility can arise from a problem in the male, the female, or both. Male infertility usually results from a low sperm count or an abnormally high number of defective sperm. Female infertility can be mechanical (blocked Fallopian tubes or other structural problems), or hormonal, leading to decreased or absent ovulation. One problem involving both partners is production of antibodies to sperm by the female. In addition, not all pregnancies go to a successful conclusion. By some estimates, as many as a third of all pregnancies spontaneously terminate, many within the first weeks, before the woman is even aware that she is pregnant.

Some of the most dramatic advances have been made in the field of *in vitro* fertilization, in which a woman's ovaries are hormonally manipulated to ovulate multiple eggs at one time. The eggs are collected surgically and fertilized outside the body. The developing embryos are then placed in the woman's uterus,

…continued from page 753

Analysis of the cervical mucus in Peggy's and Larry's post-coital test shows that sperm are present but not moving. Dr. Coddington explains that it is likely that Peggy's cervical mucus contains antibodies that destroy Larry's sperm.

Question 5: *Speculate on how this kind of infertility problem might be treated.*

which has been primed for pregnancy by hormonal therapy. Because of the expense and complicated nature of the procedure, multiple embryos are usually placed in the uterus at one time, which may result in multiple births. *In vitro* fertilization has allowed some infertile couples to have children, although the success rate is less than 25%.

▶ PREGNANCY AND PARTURITION

Now let's return to a recently ovulated egg and some sperm deposited in the vagina and follow them through fertilization, pregnancy, and **parturition**, the birth process.

Fertilization Requires Capacitation

Once an egg is released from the ruptured follicle, it is swept into the Fallopian tube by the beating cilia. Meanwhile, sperm deposited in the vagina must go through their final maturation step, capacitation. **Capacitation** of sperm confers the ability to swim rapidly and fertilize an egg. The process is not well understood, but apparently it involves the removal of a glycoprotein *decapacitation factor* from the outer membrane of the sperm head.

Normally, capacitation takes place in the female reproductive tract. This location presents a problem for sperm being used for *in vitro* fertilization. These sperm must be artificially capacitated by physiological salines supplemented with human serum before being exposed to the harvested eggs. Much of what we know about human fertilization has come from infertility research aimed at improving the success rate of *in vitro* fertilization.

Fertilization of an egg by a sperm is the result of a chance encounter, possibly aided by chemical attractants produced by the egg. An egg can be fertilized for about 12–24 hours after ovulation. Sperm remain viable for about two days.

Fertilization normally takes place in the distal part of the Fallopian tube. Of the millions of sperm in a single ejaculation, only about 100 or so reach this point (Fig. 24-16 ■). To fertilize the egg, a sperm must first penetrate an outer layer of loosely connected granulosa cells (the *corona radiata*), then the protective glycoprotein coat of the **zona pellucida**. To get past these barriers, capacitated sperm release powerful enzymes from the acrosome in the sperm head, a process known as the **acrosomal reaction**.

The enzymes dissolve cell junctions and the glycoprotein coat, allowing the sperm to wiggle their way toward the egg. The first sperm to reach the egg quickly finds sperm-binding receptors on the membrane and fuses its membrane to the oocyte membrane (Fig. 24-17 ■). The fused section of membrane opens, and the sperm nucleus sinks into the cytoplasm of the egg.

Fusion of the egg and sperm membranes signals the egg to resume meiosis and complete its second division. The final meiotic division creates a second polar body that is discarded. At this point, the 23 chromosomes of the sperm join the 23 chromosomes of the egg, creating a zygote nucleus with a full set of genetic material.

What prevents more than one sperm from fertilizing an egg? It is another process initiated by egg-sperm fusion, the **cortical reaction**. This chemical reaction prevents **polyspermy**, fertilization by more than one sperm. Membrane-bound **cortical granules** in the peripheral cytoplasm of the egg release their contents into the space

(a)

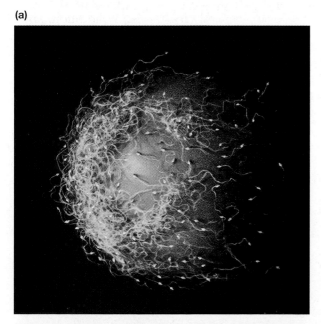

(b)

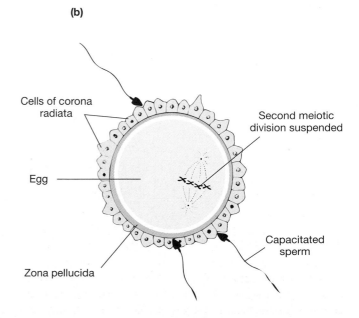

Cells of corona radiata

Egg

Zona pellucida

Second meiotic division suspended

Capacitated sperm

■ **Figure 24-16   Fertilization**   (a) This photograph shows the tremendous difference in the size of human sperm and egg. (b) Capacitated sperm release enzymes from their acrosomes in order to penetrate the cells and glycoprotein zona pellucida surrounding the ovum.

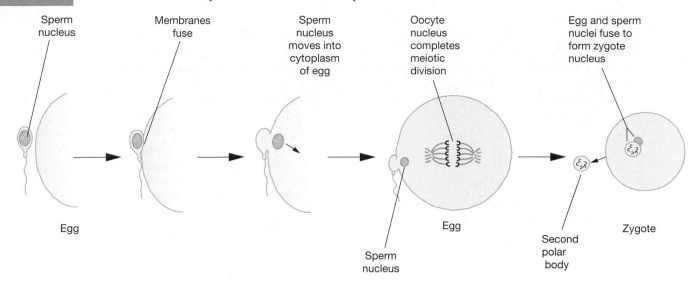

■ **Figure 24-17   Fusion of sperm and egg to form a zygote**

just outside the oocyte membrane. These chemicals rapidly alter the membrane and surrounding zona pellucida so that additional sperm cannot penetrate or bind.

Once the egg is fertilized and becomes a zygote, it begins mitotic division as it slowly makes its way to the uterus, where it will settle for the remainder of the **gestation** period [*gestare*, to carry in the womb].

The Developing Zygote Implants in the Endometrium

The dividing zygote takes about four days to move from the distal end of the Fallopian tube into the uterine cavity (Fig. 24-18 ■). Under the influence of progesterone, the smooth muscle of the tube relaxes, and transport proceeds slowly. By the time the developing embryo reach-

es the uterus, it consists of a hollow ball of about 100 cells called a **blastocyst**.

Some of the outer layer of blastocyst cells will become the **chorion**, an *extraembryonic membrane* that encloses the entire embryo and forms the placenta (Fig. 24-19a ■). The inner cell mass of the blastocyst will develop into the embryo itself and other extraembryonic membranes. These membranes include the **amnion**, which secretes *amniotic fluid* in which the developing embryo floats; the **allantois**, which becomes part of the umbilical cord that links the embryo to the mother; and the **yolk sac**, which degenerates early in human development.

Implantation of the blastocyst into the uterine wall normally takes place about 7 days after fertilization. The outer cells of the blastocyst secrete enzymes that allow them to invade the endometrium, like a parasite bur-

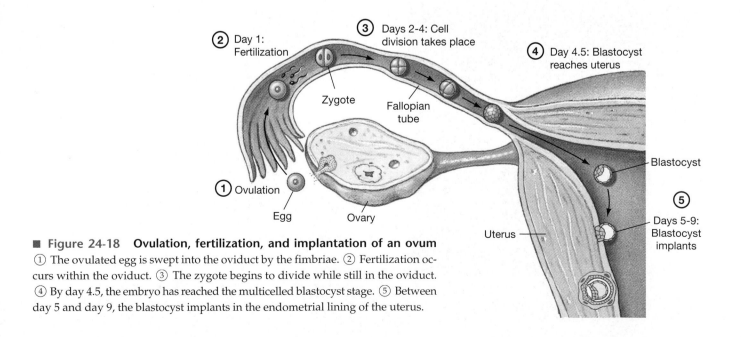

■ **Figure 24-18   Ovulation, fertilization, and implantation of an ovum**
① The ovulated egg is swept into the oviduct by the fimbriae. ② Fertilization occurs within the oviduct. ③ The zygote begins to divide while still in the oviduct. ④ By day 4.5, the embryo has reached the multicelled blastocyst stage. ⑤ Between day 5 and day 9, the blastocyst implants in the endometrial lining of the uterus.

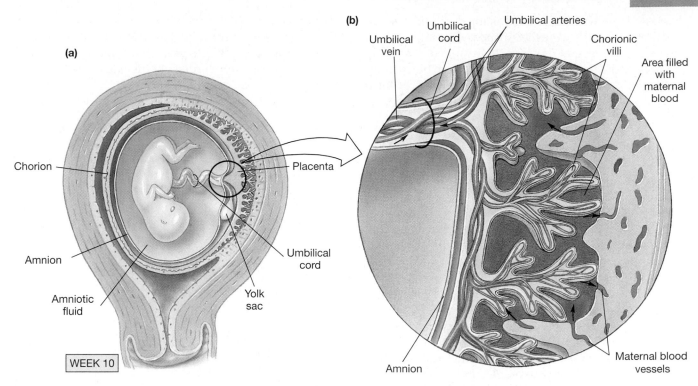

(a)

Chorion

Amnion

Amniotic fluid

Placenta

Umbilical cord

Yolk sac

WEEK 10

(b)

Umbilical vein

Umbilical cord

Umbilical arteries

Chorionic villi

Area filled with maternal blood

Amnion

Maternal blood vessels

■ **Figure 24-19   The placenta**   (a) The developing embryo floats in amniotic fluid contained within the amniotic and chorionic membranes. The embryo obtains oxygen and nutrients from the mother through the placenta and umbilical cord. The umbilical vein carries well-oxygenated blood to the embryo. The umbilical artery returns blood from the embryo to the placenta. (b) The maternal blood bathes the fingerlike chorionic villi that contain the embryonic blood vessels. Some material is exchanged by diffusion, but other material must be transported across the membranes.

rowing into its host. As they do so, the cells of the endometrium grow out around the blastocyst until it is completely engulfed in endometrial cells.

As the embryo grows, cells that will become the placenta form fingerlike **chorionic villi** that penetrate into the vascularized endometrium. Enzymes from the villi break down the walls of maternal blood vessels until the villi are surrounded by pools of maternal blood (Fig. 24-19b ■). The blood of the embryo and that of the mother do not mix, but nutrients, gases, and wastes are exchanged across the membranes of the villi. Many of these substances move by simple diffusion, but some, such as maternal antibodies, are transported across the membrane.

The placenta continues to grow during pregnancy until, by delivery, it is about 20 cm in diameter (the size of a small dinner plate). The placenta receives as much as 10% of the total maternal cardiac output. The tremendous blood flow to the placenta is one reason that the sudden abnormal separation of the placenta from the uterine wall is a medical emergency.

The Placenta Secretes Hormones During Pregnancy

As the embryo implants in the uterine wall and the placenta begins to form, the corpus luteum is coming to the end of its preprogrammed 12-day life span. Unless the embryo sends a hormonal signal, the corpus luteum will disintegrate, progesterone and estrogen levels will drop,

and the embryo will be washed from the body along with the surface layers of endometrium during menstruation. The hormones that prevent this from happening are secreted by the placenta. They include human chorionic gonadotropin, chorionic somatomammotropin, estrogen, and progesterone.

Human Chorionic Gonadotropin   The corpus luteum remains active during early pregnancy because of the hormone **human chorionic gonadotropin** (hCG), secreted by the developing placenta. This peptide is structurally related to LH, and it binds to LH receptors. Under the influence of hCG, the corpus luteum keeps producing progesterone to keep the endometrium intact.

By the seventh week of development, however, the placenta has taken over progesterone production and the corpus luteum is no longer needed. At that point, it finally degenerates. Human chorionic gonadotropin production by the placenta peaks at three months of development, then diminishes.

A second function of hCG is stimulation of testosterone production by the developing testes in male fetuses. As you learned in the opening sections of this chapter, fetal testosterone and its metabolite DHT are essential for expression of male characteristics and for descent of the testes into the scrotum before birth.

Human chorionic gonadotropin is the chemical detected by pregnancy tests. Because hCG can induce ovulation in rabbits, years ago the urine from a woman who suspected she was pregnant was injected into a rabbit. The rabbit's ovaries were then inspected for signs of ovulation. It took several days for the results of this test to get back to the patient. Today, with modern biochemical techniques, women can perform their own pregnancy tests in a few minutes in the privacy of their home.

Human Chorionic Somatomammotropin (hCS)

Another peptide hormone produced by the placenta is human **chorionic somatomammotropin** (hCS), also known as **human placental lactogen** (hPL). This hormone, structurally related to growth hormone and prolactin, was initially believed to be necessary for breast development during pregnancy and for milk production (**lactation**). Although hCS probably does contribute to lactation, women who do not make hCS during pregnancy because of a genetic defect still have adequate breast development and milk production.

The other postulated role for hCS is regulation of the mother's metabolism. During pregnancy, some women develop elevated blood glucose levels (a condition usually associated with diabetes mellitus). This extra glucose is available for fetal needs, moving by facilitated diffusion across the membranes of the placenta into the fetal circulation. After delivery, glucose metabolism in most of these women returns to normal. It seems likely, however, that hCS is not the sole regulator of metabolism during pregnancy, and there are still many unanswered questions about its functions.

Estrogen and Progesterone

Estrogen and progesterone are produced continuously during pregnancy, first by the corpus luteum under the influence of hCG and then by the placenta. With high circulating levels of the steroid hormones, feedback suppression of the pituitary continues throughout pregnancy, preventing another set of follicles from beginning development.

Estrogen during pregnancy contributes to the development of the milk-secreting ducts of the breasts. Progesterone is essential for maintaining the endometrium and in addition helps suppress uterine contractions. The placenta makes a variety of other hormones including inhibin, prolactin, and prorenin, but at present, the function of most of them is unclear.

Pregnancy Ends with Labor and Delivery

Parturition, the birth process, normally occurs in the 38th to 40th week of gestation. What triggers this process? For many years, researchers developed animal models of the signals that initiate parturition, only to discover in recent years that many of those theories do not apply to humans. Parturition begins with **labor**, the rhythmic contractions of the uterus designed to push the fetus out into the world. Signals that initiate these contractions could begin with either the mother or the fetus, or they could be a combination of signals from both.

In many nonhuman mammals, a drop in estrogen and progesterone levels marks the beginning of parturition. A drop in progesterone levels is logical, as progesterone inhibits uterine contractions. However, in humans, these hormones do not decrease until labor is well under way. One possible explanation may be that there is inactivation of progesterone receptors. In this fashion, although total progesterone concentration remains elevated through the early part of labor, the *effective* concentration of progesterone decreases.

Another possibility for the induction of labor is that the fetus somehow signals that it has completed development. The ACTH-cortisol axis plays this role in sheep and may be a contributing factor in humans. Human fetuses that develop without most of the brain (*anencephaly*) lack a pituitary gland and functional adrenal cortex, and they also have gestational periods that are highly variable.

Although we do not know what initiates parturition, we do understand the sequence of events. Once the contractions of labor begin, a positive feedback loop consisting of mechanical and hormonal factors is set into motion. The fetus is normally oriented head down (Fig. 24-20a ■). At the beginning of labor, in late pregnancy, it

---

DIABETES: Gestational Diabetes   Carbohydrate and fat metabolism change during pregnancy under the influence of maternal and placental hormones. In about 3% of all pregnancies, the mother develops elevated fasting blood glucose levels, similar to those observed in diabetes mellitus. This condition is called *gestational diabetes*. We still do not know what causes it, but we do know that mothers with gestational diabetes and their babies are at higher risk of developing type 2 diabetes later in life. The babies also have a higher rate of complications around birth. The elevated glucose in the mother's blood crosses the placenta and the babies secrete more insulin in response. As a result, they tend to weigh more at birth. They are also at risk of developing hypoglycemia around birth due to their elevated insulin levels.

…continued from page 756

Assisted reproductive technologies (ART) are one treatment option currently available to infertile couples. All ART techniques involve either artificially stimulating the ovaries to produce eggs or using an egg from an egg donor. The eggs are harvested surgically and often fertilized *in vitro*. The zygote may be immediately placed in the Fallopian tube or may be allowed to develop into an early embryo before being returned to the uterus. A different technique used to overcome infertility is intrauterine insemination, a procedure in which sperm that have been washed to remove antigenic material are introduced into the uterus through a tube inserted through the cervix.

Question 6:   *Based on the results of their infertility workup, which intervention—ART or intrauterine insemination—should be recommended for Peggy and Larry? Why?*

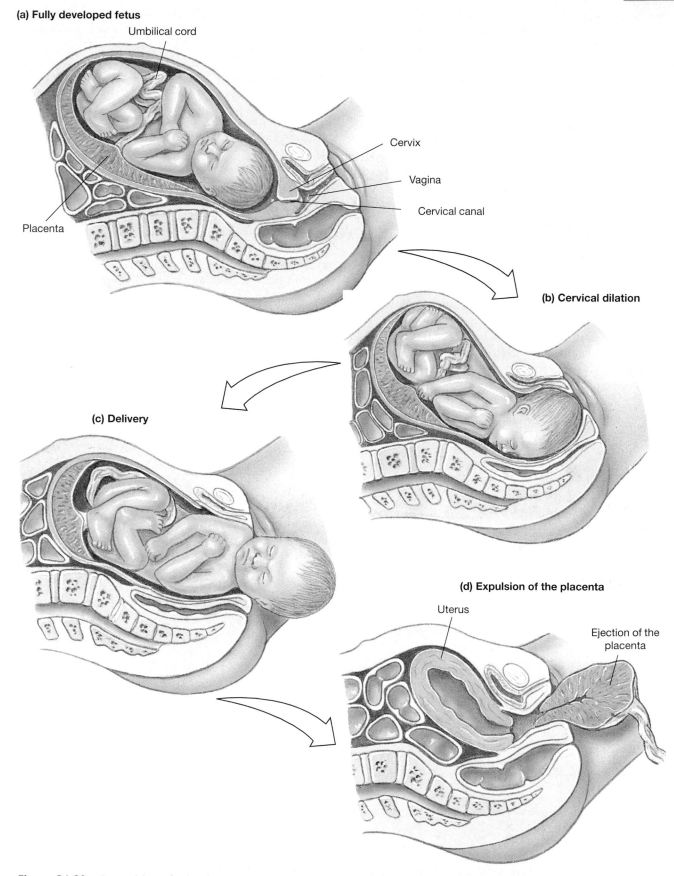

(a) Fully developed fetus

Umbilical cord

Cervix

Vagina

Cervical canal

Placenta

(b) Cervical dilation

(c) Delivery

(d) Expulsion of the placenta

Uterus

Ejection of the placenta

■ **Figure 24-20** **Parturition: the birth process** (a) As labor begins, the fetus is normally head down in the uterus. (b) The rhythmic uterine contractions of labor push the head of the fetus against the softened cervix, stretching and dilating it. (c) Once the cervix is fully dilated and stretched, the uterine contractions push the fetus out of the uterus through the vagina. (d) Shortly after the fetus is delivered, the placenta detaches from the uterine wall and is expelled.

repositions itself lower in the abdomen ("the baby has dropped") and begins to push on the softened cervix (Fig. 24-20b ■). The peptide hormone **relaxin**, secreted by the ovaries and placenta, apparently helps soften the cervix and also loosens ligaments holding the pelvic bones together.

Cervical stretch triggers uterine contractions that move in a wave from the top of the uterus down, pushing the fetus farther into the pelvis. The cervix stretches and opens (dilates) even more and starts a positive feedback cycle of escalating contractions (Fig. 24-21 ■). In addition, neurons from the stretching cervix signal the hypothalamus-posterior pituitary to secrete oxytocin [∞ p. 198]. This peptide hormone causes uterine muscle contraction and enhances the contractions of labor.

Could oxytocin be the signal that starts parturition? As the pregnancy nears full term, the number of uterine oxytocin receptors increases. However, studies have shown that secretion of the hormone does not increase until labor begins. Synthetic oxytocin is often used to induce labor in pregnant women, but it is not always effective. Apparently, the start of labor requires other conditions in addition to adequate amounts of oxytocin.

Prostaglandins, another chemical factor involved in labor and delivery, are produced in the uterus in response to oxytocin secretion. Prostaglandins are very effective at causing uterine muscle contractions at any time. They are the primary cause of menstrual cramps and have been used to induce abortion in early pregnancy. During labor and delivery, prostaglandins reinforce the uterine contractions induced by oxytocin.

As the contractions of labor intensify, the fetus moves down though the birth canal and out into the world, still attached to the placenta (Fig. 24-20c ■). The placenta then detaches from the uterine wall and is expelled a short time later (Fig. 24-20d ■). The contractions

of the uterus clamp the maternal blood vessels and help prevent excessive bleeding, although typically about 240 mL of blood is lost.

The Mammary Glands Secrete Milk During Lactation

When an infant is born, it loses its source of maternal nourishment through the placenta and must rely on an external source of food instead. Primates, who normally have only one or two offspring at a time, have two functional mammary glands. A mammary gland is composed of about 20 milk-secreting lobules, each made of branched hollow ducts of secretory epithelium surrounded by *myoepithelial* contractile cells (Fig. 24-22 ■). Interestingly, the mammary gland epithelium is closely related to the secretory epithelium of sweat glands, so milk and sweat secretion share some common features.

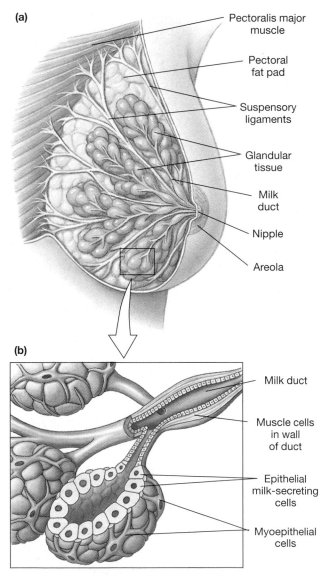

■ **Figure 24-22   Mammary glands**   The epithelial cells of the mammary glands secrete milk into the lumen of the gland.

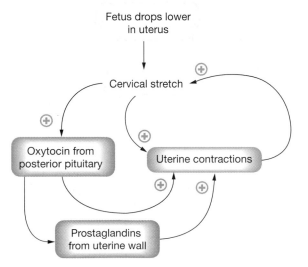

■ **Figure 24-21   The positive feedback loop of parturition** The initiation of this loop is dependent on a uterus that is ready to respond to these stimuli. The positive feedback loop stops when the fetus is delivered and the cervix is no longer being stretched.

The breasts first begin to develop under the influence of estrogen during puberty. The milk ducts grow and branch, while fat is deposited behind the glandular tissue. During pregnancy, the glands develop further under the direction of estrogen, aided by growth hormone and cortisol. The final development step also requires progesterone, which converts the duct epithelium into a secretory structure. This process is similar to progesterone's effect on the uterus, in which progesterone makes the endometrium into a secretory tissue during the luteal phase.

Although estrogen and progesterone stimulate mammary development, they inhibit the actual secretion of milk. Milk production is under the control of the hormone prolactin from the anterior pituitary [∞ p. 200]. Prolactin is an unusual pituitary hormone in that it is primarily controlled by an inhibiting hormone from the hypothalamus. There is good evidence that the **prolactin inhibiting hormone** (PIH) is actually dopamine, an amine hormone related to epinephrine and norepinephrine.

During the later stages of pregnancy, PIH secretion falls, and prolactin levels reach 10 or more times those found in nonpregnant women. Prior to delivery, when estrogen and progesterone are also high, the mammary glands produce only small amounts of a thin, low-fat secretion called **colostrum**. After delivery, when estrogen and progesterone decrease, the glands produce greater amounts of milk that contains 4% fat and substantial amounts of calcium. Proteins in colostrum and milk include maternal immunoglobulins, secreted into the duct and absorbed intact by the infant's intestinal epithelium [∞ p. 629]. This transfers some of the mother's immunity to the infant during early weeks of life.

Suckling, the mechanical stimulus of the infant's nursing at the breast, inhibits release of PIH from the hypothalamus (Fig. 24-23 ■). In the absence of this inhibition, the pituitary increases spontaneous prolactin secretion, resulting in milk production. Pregnancy is not a requirement for lactation, and some women who have adopted babies have been successful in breast-feeding.

The ejection of milk from the glands, known as the **let-down reflex**, requires the presence of oxytocin from the posterior pituitary. Oxytocin initiates smooth muscle contraction in the uterus and breasts. In the *postpartum* (after delivery) uterus, oxytocin-induced contractions help return the uterus to a prepregnancy size.

In the lactating breast, oxytocin causes contraction of myoepithelial cells surrounding the mammary glands. This contraction creates high pressure that sends the milk literally squirting into the infant's mouth. Although prolactin release requires the mechanical stimulus of suckling, oxytocin release can be stimulated by various cerebral stimuli, including the thought of the child. Many nursing mothers experience inappropriate milk release triggered by hearing someone else's child cry.

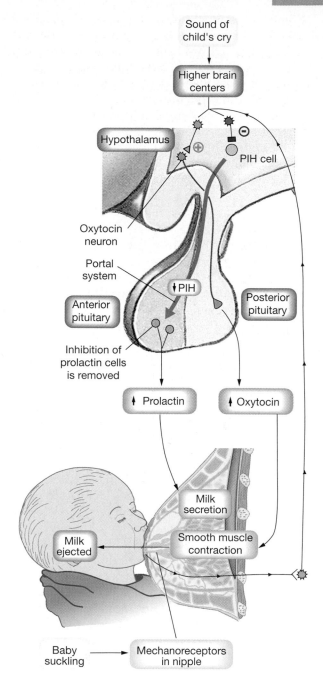

■ **Figure 24-23   The hormonal control of milk secretion and release**

Prolactin Has Other Physiological Roles

Although we discussed prolactin in the context of nursing mothers, all nonnursing women and men have tonic prolactin secretion that shows a diurnal cycle, peaking during sleep. Prolactin is related to growth hormone, and plays a role in other reproductive and nonreproductive processes. For example, prolactin is synthesized in the uterine endometrium during normal menstrual cycles. Male knockout mice who lack prolactin or a prolactin receptor have decreased fertility.

Some of the most interesting research has established a role for prolactin in neuroimmunomodulation [∞ p. 709]. Prolactin and growth hormone both appear to be necessary for normal differentiation of T lymphocytes in the thymus gland, an observation supported by impaired immune function in animals with *hypoprolactinemia*. In contrast, several autoimmune diseases, including multiple sclerosis, systemic lupus erythematosus, and autoimmune thyroiditis, have been linked to elevated levels of prolactin.

◗ GROWTH AND AGING

The reproductive years begin with the events surrounding puberty and end with decreasing gonadal hormone production.

Puberty Marks the Beginning of the Reproductive Years

Puberty is the period when a person makes the transition from being nonreproductive to being reproductive. In girls, the onset of puberty is marked by budding breasts and the first menstrual period, or **menarche**, a time of ritual significance in many cultures. In the United States, the average age at menarche is 12 years (normal range is considered 8 to 13 years).

In boys, the signs of puberty are more subtle. They include growth and maturation of the external genitalia, development of secondary sex characteristics such as pubic and facial hair, change in body shape, and growth in height. The age range for male puberty is 9 to 14 years.

Puberty requires the maturation of the hypothalamic-pituitary control pathway. Before puberty, the child has low levels of both steroid sex hormones and gonadotropins. Because low sex hormone levels normally enhance gonadotropin release, the combination of low steroids and low gonadotropins indicates that the pituitary is not yet sensitive to steroid levels in the blood.

What is the signal that initiates sexual development? The basis for the onset of puberty, like many other areas in reproductive physiology, is still uncertain. One theory says that it is the genetically programmed maturation of hypothalamic neurons. We know that puberty has a genetic base because inherited patterns of maturation are common. If a woman did not start her menstrual periods until she was 16, it is likely that her daughters will also have late menarche.

More recent theories have included the adipose tissue hormone *leptin* [∞ p. 663] as a contributing factor to the onset of puberty. Undernourished women with low leptin levels often stop having menstrual periods (*amenorrhea*) and knockout mice without leptin are infertile. The improvement in nutrition over the last century presumably increased prepubertal leptin secretion, which could interact with other factors to initiate puberty.

Menopause and Andropause Are a Consequence of Aging

Several centuries ago in America, most people died of acute illnesses while still reproductively active. Now modern medicine has overcome most acute illnesses, and we are living well past the time that most of us are likely to have children. Women's reproductive cycles stop completely at the time known as **menopause**.

In men, testosterone production decreases with age. The existence of **andropause** in men is still controversial because the physical and psychological symptoms of aging men are not clearly linked to the decline in testosterone. Many men remain reproductively active as they age, and it is not uncommon for men in their fifties or sixties to have children with younger women. Postmenopausal women also remain sexually active, although not reproductively active, and some report a more fulfilling sex life once the fear of unwanted pregnancy has been removed.

The physiology of menopause has been well studied. After about 40 years of menstrual cycles, a woman's periods become irregular and finally cease. The failure of reproductive cycles is not due to the pituitary but to the ovaries, which can no longer respond to gonadotropins. In the absence of negative feedback, gonadotropin levels increase dramatically in an effort to stimulate the ovaries into maturing more follicles.

The absence of estrogen in postmenopausal women leads to symptoms of varying severity. These may include hot flashes [∞ p. 645], atrophy of genitalia and breasts, and osteoporosis due to loss of calcium from the bones [∞ p. 679]. Hormone replacement therapy for women in menopause consists of estrogen or a combination of estrogen and progesterone. This treatment continues to be controversial because of a possibility of increased risk of breast and uterine cancer, although studies that contradict one another are published annually.

The newest drug therapy for menopause is the *selective estrogen receptor modulators*, or SERMs [∞ p. 165]. These drugs bind with different affinities to the two estrogen receptor subtypes. This allows the therapy to mimic the beneficial effects of estrogen on bone while avoiding the potentially detrimental effects on breast and uterus.

PROBLEM CONCLUSION

In this running problem, you learned how the cause of infertility is diagnosed in a typical couple. You also learned how interventions are used to overcome some causes of infertility.

Further check your understanding of the running problem by checking your answers against those in the summary table.

	Question	Facts	Integration and Analysis
1	Name the male reproductive structures that carry sperm from the testes to the external environment.	The male reproductive structures include the testes, accessory glandular organs, and a series of ducts.	Sperm leaving the seminiferous tubules pass into the epididymis, vas deferens, and finally the urethra.
2	For what causes of male infertility might retrival of sperm from the epididymis be necessary?	The epididymis is the first duct the sperm enter upon leaving the seminiferous tubules.	If the infertility problem is due to blockage or congenital defects in the vas deferens or urethra, removal of sperm from the epididymis might be useful. If the problem is due to low sperm count or abnormal sperm morphology, this technique would probably not be useful.
3	For what causes of female infertility is temperature tracking useful? For what causes is it not useful?	Basal body temperature rises slightly following ovulation.	Temperature tracking is a useful way to tell if someone is ovulating but it cannot be used to predict structural problems in the female reproductive tract.
4	What abnormalities in the cervix, Fallopian tubes, and uterus could cause infertility?	The cervix, Fallopian tubes, and uterus are hollow structures through which sperm must pass.	Any blockage of these organs due to disease or congenital defects would prevent normal movement of sperm and cause infertility. Hormonal problems might cause the endometrium to develop incompletely, preventing implantation of the embryo.
5	Speculate on how infertility due to cervical mucus antibodies to sperm might be treated.	Antibodies in the cervical mucus react with antigenic material in the semen or on the sperm, causing the sperm to become immobile.	If the antigenic material can be removed from the semen, this might help the problem of sperm immobilization. A semen sample can be washed to remove nonsperm components. If the antigens are part of the sperm, this method will not work.
6	Should ART or intrauterine insemination be recommended for Peggy and Larry? Why?	ART is used when ovulation is abnormal. Intrauterine insemination is used when ovulation is normal.	Intrauterine insemination can be used to overcome cervical factors, as this technique bypasses the cervix. Peggy can ovulate; therefore, intrauterine insemination should be recommended for Peggy and Larry.

CHAPTER REVIEW

SUMMARY

Sex Determination

1. The sex organs consist of gonads, accessory ducts and glands, and **genitalia**. (p. 732)

2. The male **testes** produce **sperm**. The female **ovaries** produce eggs, or **ova**. Embryonic cells that will produce gametes are called **germ cells**. (p. 732)

3. Humans have 46 chromosomes. (p. 732)

4. The genetic sex of an individual depends on the **sex chromosomes**: females are XX and males are XY. In the absence of a Y chromosome, an embryo will develop into a female. (p. 733)

5. The **SRY gene** on the Y chromosomes produces a **testis determining factor**. (p. 733)

6. Testicular **Sertoli cells** secrete **Müllerian inhibiting substance** that causes the **Müllerian ducts** to regress. **Leydig cells** secrete **testosterone** that converts **Wolffian ducts** to male accessory structures. (p. 733)

7. Absence of testosterone and Müllerian inhibiting substance causes Müllerian ducts to develop into **Fallopian tubes (oviducts), uterus**, and **vagina**. In females, the Wolffian ducts regress. (p. 735)

Basic Patterns of Reproduction

8. **Gametogenesis** begins with mitotic divisions of **spermatogonia** and **oogonia**. The first step of meiosis creates **primary spermatocytes** and **primary oocytes**. The first meiotic division creates two identical **secondary spermatocytes** or a large secondary oocyte (egg) and a tiny **first polar body**. (p. 736)

9. The second meiotic division in males creates haploid **spermatids** that mature into sperm. In females, the second meiotic division of the oocyte does not take place unless the egg is fertilized. (p. 736)

10. In both sexes, **gonadotropin releasing hormone** (GnRH) controls the secretion of **follicle stimulating hormone** (FSH) and **luteinizing hormone** (LH) from the anterior pituitary. FSH and steroid sex hormones regulate gametogenesis in gonadal gamete-producing cells. LH stimulates production of steroid sex hormones. (p. 738)

11. The steroid sex hormones include **androgens, estrogens**, and **progesterone**. **Aromatase** converts androgens to estrogens. Peptide hormones **inhibin** and **activin** inhibit and stimulate secretion of FSH, respectively. (p. 738)

12. Gonadal steroids generally suppress secretion of GnRH, FSH, and LH in a long-loop response. However, if estrogen secretion increases above a threshold level for at least 36 hours, its feedback changes to positive and stimulates gonadotropin release. (p. 738)

13. Tonic GnRH release occurs in small pulses every 1–3 hours from a region of the hypothalamus called a **pulse generator**. (p. 739)

Male Reproduction

14. The **corpus spongiosum** and **corpora cavernosa** comprise the erectile tissue of the penis. The **glans** is covered by the **foreskin**. The urethra runs though the **penis**. (p. 741)

15. The testes migrate into the scrotum during fetal development. Failure of one or both testes to descend is known as **cryptorchidism**. (p. 741)

16. The testes consist of **seminiferous tubules** and interstitial tissue with blood vessels and Leydig cells. The seminiferous tubules join the **epididymis**, which becomes the **vas deferens**. The vas deferens empties into the urethra. (p. 741)

17. A seminiferous tubule has spermatocytes and Sertoli cells, which form a **blood-testis barrier**. (p. 743)

18. Spermatogonia in the tubule enter meiosis, becoming primary spermatocytes, spermatids, and finally sperm in about 64 days. (p. 743)

19. Sertoli cells regulate sperm development. They also produce inhibin, activin, growth factors, enzymes, and **androgen-binding protein**. (p. 743)

20. The Leydig cells produce testosterone. About 5% of testosterone comes from the adrenal cortex. (p. 743)

21. FSH stimulates Sertoli cell production of androgen-binding protein, inhibin, and paracrines. The Leydig cells produce testosterone under the direction of LH. (p. 744)

22. The **prostate gland, seminal vesicles**, and **bulbourethral glands** secrete the fluid component of **semen**. (p. 744)

23. **Primary sex characteristics** are internal sexual organs and external genitalia. **Secondary sex characteristics** are other features of the body, such as body shape. (p. 745)

Female Reproduction

24. The female external genitalia are the **labia majora, labia minora**, and **clitoris**. The urethra opening is between the clitoris and the vagina. (p. 746)

25. The uterine tissue layers are outer connective tissue, **myometrium**, and inner **endometrium**. (p. 746)

26. The **Fallopian tubes** are lined with ciliated epithelium. The bulk of an ovary consists of ovarian follicles. (p. 746)

27. Eggs are produced in monthly **menstrual cycles**. (p. 746)

28. In the ovarian cycle, the **follicular phase** is a period of follicular growth. **Ovulation** is the release of an egg from its follicle. In the **luteal phase**, the ruptured follicle becomes a **corpus luteum**. (p. 746)

29. The **menses** begin the uterine cycle. This is followed by a **proliferative phase**, with endometrial thickening. Following ovulation, the endometrium goes into a **secretory phase**. (p. 747)

30. Follicular **granulosa cells** secrete estrogen. As the follicular phase ends, a surge in LH is necessary for oocyte maturation. (p. 747)

31. The corpus luteum secretes progesterone and some estrogen, which exert negative feedback on the hypothalamus-anterior pituitary. (p. 751)

32. Estrogens and androgen control primary and secondary sex characteristics in females. (p. 753)

Procreation

33. The human sex act is divided into four phases; (1) excitement, (2) plateau, (3) orgasm, and (4) resolution. (p. 753)

34. The male **erection** reflex is a spinal reflex that can be influenced by higher brain centers. Parasympathetic input mediated by nitric oxide actively vasodilates the penile arterioles. (p. 754)

35. **Emission** is the movement of sperm out of the vas deferens and into the urethra. **Ejaculation** is the expulsion of semen to the exterior. (p. 754)

36. Contraceptive methods include **abstinence, barrier methods**, prevention of implantation of the fertilized egg, and hormonal treatment that decreases or stops gamete production. (p. 755)

37. Infertility can arise from a problem in the male, the female, or both. *In vitro* fertilization has allowed a number of infertile couples to have children, although the success rates are not high. (p. 756)

Pregnancy and Parturition

38. Sperm must go through **capacitation** before they can fertilize an egg. (p. 757)

39. Fertilization normally takes place in the Fallopian tube. Capacitated sperm release acrosomal enzymes (the **acrosomal reaction**) to dissolve cell junctions and the **zona pellucida**. The first sperm to reach the egg fertilizes it. (p. 757)

40. Fusion of egg and sperm membranes initiates a **cortical reaction** that prevents **polyspermy**. (p. 757)

41. The developing embryo is a hollow **blastocyst** when it reaches the uterus. Once the embryo implants, it develops **chorionic, amniotic**, and **allantoic membranes**. (p. 758)

42. The **chorionic villi** of the placenta are surrounded by pools of maternal blood where nutrients, gases, and wastes are exchanged. (p. 759)

43. The corpus luteum remains active during early pregnancy due to **human chorionic gonadotropin** from the placenta. (p. 759)

44. The placenta also secretes estrogen, progesterone, and **human chorionic somatomammotropin**. This hormone plays a role in maternal metabolism. (p. 760)

45. Estrogen during pregnancy contributes to the development of the milk-secreting ducts of the breasts. Progesterone is essential for maintaining the endometrium and, along with **relaxin**, helps suppress uterine contractions. (p. 760)

46. **Parturition** normally occurs in the 38th–40th week of gestation. It begins with **labor** and ends with delivery of the fetus and placenta. A positive feedback loop of oxytocin secretion causes uterine muscle contraction. (p. 760)

47. Following delivery, the mammary glands produce milk under the influence of prolactin. Milk is released during nursing by oxytocin. (p. 762)

48. Prolactin plays a role in immune function in both sexes. (p. 763)

Growth and Aging

49. **Puberty** is the period when a person makes the transition from being nonreproductive to being reproductive. (p. 764)

50. The cessation of reproductive cycles in women is known as the **menopause**. The existence of a male **andropause** is controversial. (p. 764)

QUESTIONS

LEVEL ONE   Reviewing Facts and Terms

1. Match each item below with all the terms that it applies to:

   (a) X or Y
   (b) inactivated X chromosome
   (c) XX
   (d) XY
   (e) XX or XY
   (f) autosomes

   1. chromosomes other than sex chromosomes
   2. fertilized egg
   3. sperm or ova
   4. sex chromosomes
   5. germ cells
   6. male chromosomes
   7. female chromosomes
   8. Barr body

2. The Y chromosome contains a region for male sex determination that is known as the _____ gene.

3. List the functions of the gonads. How do the products of gonadal function differ in males and females?

4. Trace the anatomical routes to the external environment followed by a newly formed sperm and by an ovulated egg. Name all structures that the gametes pass on their way.

5. Define the following terms and give their significance to reproductive physiology:

   (a) aromatase
   (b) blood-testis barrier
   (c) androgen binding protein
   (d) first polar body
   (e) acrosome

6. Decide if the statements below are true or false, and defend your answer.

   (a) All testosterone is produced in the testes.
   (b) Each sex hormone is produced only by members of one gender.
   (c) Anabolic steroid use appears to be addictive, and withdrawal symptoms include psychological disturbances.
   (d) High levels of estrogen in the late follicular phase help to prepare the uterus for menstruation.
   (e) Progesterone is the dominant hormone of the luteal phase of the female cycle.

7. What is semen? What are its main components and where are they produced?

8. List and give a specific example of the various methods of contraception. Which is/are most effective? Least effective?

LEVEL TWO   Reviewing Concepts

9. **Concept maps:** Map the following groups of terms. You may add additional terms.

List 1	List 2
Wolffian ducts	myometrium
Müllerian ducts	endometrium
SRY	granulosa cells
Müllerian inhibiting substance	corpus luteum
testosterone	follicle
DHT	antrum
Leydig cells	thecal cells
Sertoli cells	
spermatogonia	
spermatocytes	
sperm	
spermatids	

10. Diagram the hormonal control of gametogenesis in males.

11. Diagram the menstrual cycle in the female, including the ovarian cycle and the uterine cycle. Include all hormones.

12. Why are X-linked traits exhibited more frequently by males than females?

13. Define and relate the terms in each group below.

   (a) gamete, zygote, germ cell, embryo
   (b) coitus, erection, ejaculation, orgasm, emission, erogenous zones
   (c) capacitation, decapacitation factor, zona pellucida, acrosomal reaction, cortical reaction, cortical granules
   (d) puberty, menarche, menopause, andropause

14. Compare the actions of the hormones below in males and females.

   (a) FSH
   (b) inhibin
   (c) activin
   (d) GnRH
   (e) LH
   (f) DHT
   (g) estrogen
   (h) testosterone
   (i) progesterone

15. Compare and contrast the events of the four phases of sexual intercourse in each gender.

16. Discuss the roles of the following hormones in pregnancy, labor and delivery, and mammary gland development and lactation.

   (a) human chorionic gonadotropin
   (b) luteinizing hormone
   (c) human placental lactogen
   (d) estrogen
   (e) progesterone
   (f) relaxin
   (g) prolactin

LEVEL THREE   Problem Solving

17. Down syndrome is a chromosomal defect known as "trisomy" (three copies instead of two) of the twenty-first chromosome. The extra chromosome usually comes from the mother. Speculate what causes trisomy, using what you have learned about the events surrounding fertilization.

18. Sometimes eggs fail to leave the ovary at ovulation, even though they appear to have gone through all of the correct stages of development. This condition results in benign ovarian cysts, and the unruptured follicles can be palpated as bumps on the surface of the ovary. If the cysts persist, symptoms of this condition often mimic pregnancy, with missed menstrual periods and tender breasts. Can you explain or diagram how these symptoms occur?

E X P L O R E <MediaLab>

Introduction

In this chapter you learned how the human race perpetuates itself through reproduction. Feedback mechanisms are crucial to coordinate and dictate development of zygotes into males and females. The discussion delved into the numerous signals that regulate growth and development during gestation and puberty. This MediaLab and its Web Explorations will help you investigate the physiology of reproduction and development. After reading the descriptions below, visit the MediaLab for Chapter 24 in your Companion Website and select the appropriate keywords.

Web Exploration 1

Estimated time for completion = 10 minutes

Sex determination requires the gonad to develop from an indifferent state into a testis or an ovary. Testis formation is currently believed to be a more active process than ovary formation, and the gene *Sry,* plays a key role in this process. Which chromosome encodes *Sry* and is this significant? What would happen to a genotypic male (XY) if *Sry* were deleted? Can you predict the phenotype if a genotypic female (XX) abnormally expressed *Sry?* The *Sry* protein (or testis determining factor) is a "transcription factor" (i.e., it binds to DNA and controls the expression of target genes). Select the keyword **SEX DETERMINATION** from your Website to find out more about *Sry.*

Once the testis has formed, it produces two hormones that are essential for male development. What are these hormones and what are their actions? Can you predict what might happen if there was a decrease in the production of these hormones or defects in the receptors they act on?

Web Exploration 2

Estimated time for completion = 10 minutes

The onset of puberty is normally controlled by activation of the hypothalamic GnRH pulse generator. The timing of this activation can vary between boys and girls, between different races, and between different children, so adolescents go through puberty at different ages. In some children, puberty may occur abnormally early (precocious) or not at all (hypogonadism). This may reflect more significant problems, rather than simple variations in timing. Consider each level within the reproductive axis (hypothalamus, pituitary, gonad) and try to list abnormalities that might cause early puberty or apparent pubertal failure. To see the normal milestones of pubertal development, select the keyword **PUBERTY** from your Website.

APPENDIX A

Physics and Math

Richard D. Hill and Daniel Biller, University of Texas

❱ INTRODUCTION

This appendix discusses selected aspects of **biophysics,** the study of physics as it applies to biological systems. Because living systems are in a continual exchange of force and energy, it is necessary to define these important concepts. According to the seventeenth-century scientist Sir Isaac Newton, a body at rest tends to stay at rest, and a body in motion tends to continue moving in a straight line unless the body is acted upon by some force (Newton's First Law). Newton further defined **force** as an influence, measurable in both intensity and direction, that operates on a body in such a manner as to produce an alteration of its state of rest or motion. Put another way, force gives **energy** to a quantity, or mass, thereby enabling it to do work. In general, a driving force multiplied by a quantity yields energy, or work. Some relevant examples of this principle include:

Mechanical force × distance = mechanical energy or mechanical work

Gas pressure × volume of gas = mechanical energy or mechanical work

Osmotic pressure × molar value = osmotic energy or osmotic work

Electrical potential × charge = electrical energy or electrical work

Temperature × entropy = heat energy or mechanical work

Chemical potential × concentration = chemical energy or chemical work

Energy exists in two general forms: kinetic energy and potential energy. **Kinetic energy** [*kinein,* to move] is the energy possessed by a mass in motion. **Potential energy** is energy possessed by a mass because of its position. Kinetic energy (KE) is equal to one-half the mass (m) of a body in motion multiplied by the square of the velocity (v) of the body:

$$KE = 1\backslash 2\ mv^2$$

Potential energy (PE) is equal to the mass (m) of a body multiplied by acceleration due to gravity (g) times the height (h) of the body above the earth's surface:

$$PE = mgh \qquad \text{where } g = 10 \text{ m/s}^2$$

Both kinetic and potential energy are measured in joules.

❱ BASIC UNITS OF MEASUREMENT

For physical concepts to be useful in scientific endeavors, they must be measurable and should be expressed in standard units of measurement. Some fundamental units of measure include the following:

Length (l): Length is measured in meters (m).
Time (t): Time is measured in seconds (s).
Mass (m): Mass is measured in kilograms (kg), and is defined as the weight of a body in a gravitational field.
Temperature (T): Temperature is measured in degrees Kelvin (°K),
    where °K = degrees Celsius (°C) +273.15
    and °C = (degrees Fahrenheit −32)/1.8
Electric current (I): Electric current is measured in amperes (A).
Amount of substance (n): The amount of a substance is measured in moles (mol).

Using these fundamental units of measure, we can now establish standard units for physical concepts (Table A-1). Although these are the standard units for these concepts at this time, they are not the only units ever used to describe them. For instance, force can also be measured in dynes, energy can be measured in calories, pressure can be

TABLE A-1 Standards Units for Physical Concepts

Measured Concept	Standard (SI*) Unit	Mathematical Derivation/ Definition
Force	Newton (N)	$1 \text{ N} = 1 \text{ kg m/s}^2$
Energy/Work/Heat	Joule (J)	$1 \text{ J} = 1 \text{ N} \bullet \text{m}$
Power	Watt (W)	$1 \text{ W} = 1 \text{ J/s}$
Electrical charge	Coulomb (C)	$1 \text{ C} = 1 \text{ A} \bullet \text{s}$
Potential	Volt (V)	$1 \text{ V} = 1 \text{ J/C}$
Resistance	Ohm (Ω)	$1 \text{ Ω} = 1 \text{ V/A}$
Capacitance	Farad (F)	$1 \text{F} = 1 \text{ C/V}$
Pressure	Pascal (Pa)	$1 \text{ Pa} = 1 \text{ N/m}^2$

*SI = Système International d'Unites

measured in torr or mm Hg, and power can be measured in horsepower. However, all of these units can be converted into a standard unit counterpart, and vice versa.

The remainder of this appendix will discuss some biologically relevant applications of physical concepts. This discussion will include topics such as bioelectrical principles, osmotic principles, and behaviors of gases and liquids relevant to living organisms.

▶ BIOELECTRICAL PRINCIPLES

Living systems are composed of different molecules, many of which exist in a charged state. Cells are filled with charged particles such as proteins and organic acids, and ions are in continual flux across the cell membrane. Therefore, electrical forces are important to life.

When molecules gain or lose electrons, they develop positive or negative charges. A basic principle of electricity is that opposite charges attract and like charges repel. A force must act on a charged particle (a mass) to bring about changes in its position. Therefore, there must be a force acting on charged particles to cause attraction or repulsion, and this electrical force can be measured. Electrical force increases as the strength (number) of charges increases, and it decreases as the distance between the charges increases. This observation has been called **Coulomb's law**, and can be written:

$$F = \frac{q_1 q_2}{\epsilon d^2}$$

where q_1 and q_2 are the electrical charges (coulombs), d is the distance between the charges (meters), ϵ is the dielectric constant, and F is the force of attraction or repulsion, depending on the type of charge on the particles.

When opposite charges are separated, a force acts over a distance to draw them together. As the charges move together, work is being done by the charged particles and energy is being released (Fig. A-1 ■). Conversely, to separate the united charges, energy must be added and work done. If charges are separated and kept apart, they have the potential to do work. This electrical potential is called **voltage.** Voltage is measured in **volts (V).**

If electrical charges are separated and there is a potential difference between them, then the force between the charges will allow electrons to flow. Electron flow is called an electric **current.** The amount of current that flows depends on the nature of the material between the charges. If that material hinders the electron flow, then it is said to offer **resistance (R),** measured in ohms. Current is inversely proportional to resistance, such that current decreases as resistance increases. If a material offers a high resistance, then that material is called an **insulator.** If resistance is low, and current flows relatively freely, then the material is called a **conductor.** Current, voltage, and resistance are related by **Ohm's law,** which states:

$$V = IR$$

where V = potential difference in volts
I = current in amperes
R = resistance in ohms

If you separate two opposite charges, there will be an electric force between them.

If you increase the number of charges that are separated, the force increases.

If you increase the distance between the charges, the force decreases.

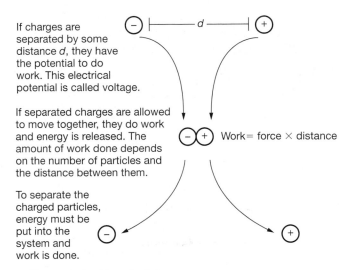

If charges are separated by some distance d, they have the potential to do work. This electrical potential is called voltage.

If separated charges are allowed to move together, they do work and energy is released. The amount of work done depends on the number of particles and the distance between them.

Work = force × distance

To separate the charged particles, energy must be put into the system and work is done.

■ Figure A-1   **Electrical force**

In biological systems, pure water is not a good conductor, but water containing dissolved NaCl is a fairly good conductor because ions provide charges to carry the current. In biological membranes, the lipids have few or no charged groups, so they offer high resistance to current flow across them. Thus, different cells can have different electrical properties depending on their membrane lipid composition and the permeability of their membranes to ions.

▶ OSMOTIC PRINCIPLES

Freezing point, vapor pressure, boiling point, and osmotic pressure are properties of solutions collectively called **colligative properties.** These properties depend on the number of solute particles present in a solution. **Osmotic pressure** is the force that drives the diffusion of water across a membrane. Because there are no solutes in pure water, it has no osmotic pressure. However, if one adds a solute like NaCl, the greater the concentration (c) of a solute dissolved in water, the greater the osmotic pressure. The osmotic pressure (π) varies directly with the concentration of solute (number of particles (n) per volume (V)):

$$\pi = (n/V)RT$$
$$\pi = cRT$$

where R is the universal gas constant (8.314 joules/°K • mol) and T is the absolute temperature (in °Kelvin).

Osmotic pressure can be measured by determining the mechanical pressure that must be applied to a solution so that osmosis ceases.

Water balance in the body is under the control of osmotic pressure gradients (concentration gradients). Most cellular membranes allow water to pass freely because water is a small, uncharged molecule. Therefore, by making cellular membranes selectively permeable to solutes, the body can control the movement of water by controlling the movement of solutes.

▶ THE RESTING MEMBRANE POTENTIAL

We have now discussed two very important biophysical concepts: electrical force and osmotic force. It is the interaction between these two forces that sets up the membrane potential across a cell membrane. Each living cell membrane effectively acts as a charge separator so that there is a potential difference (**membrane potential**, or $\mathbf{V_m}$) across it. The selective permeability of the cell membrane sets up osmotic gradients that also contribute to the membrane potential by affecting the flow of certain ions either into or out of the cell. Voltage across a nerve cell membrane is about -70 mV; across a cardiac contractile cell, the membrane potential is about -90 mV.

Two important physical equations describe the relationship between osmotic and electrical forces and membrane potentials. These equations are the Nernst equation, described in Chapter 5 of the text, and the Goldman equation. The Nernst equation describes the maximum membrane potential that a single ion could produce if the membrane were permeable to only that one ion. In reality, the membrane potential is a result of the many ion concentration gradients and permeabilities. The Goldman equation takes this fact into consideration.

The Goldman Equation

The Goldman equation is used to calculate the membrane potential that results from the contribution of all ions that can cross the membrane. The equation takes membrane permeability into account because an ion can contribute to the membrane potential only if the membrane is permeable to it. An ion's contribution to the membrane potential is proportional to its ability to cross the membrane. The Goldman equation for cells that are permeable to Na^+, K^+, and Cl^- is

$$V_m = \frac{RT}{F} \ln \frac{P_K[K^+]_{out} + P_{Na}[Na^+]_{out} + P_{Cl}[Cl^-]_{in}}{P_K[K^+]_{in} + P_{Na}[Na^+]_{in} + P_{Cl}[Cl^-]_{out}}$$

where

V_m = voltage across the membrane
P = relative permeability of the membrane to an ion
R = universal constant (8.314 joules/°K • mol)
T = temperature in °K

$[I]$ = ion concentration (in or out of cell)
F = Faraday constant (96,000 coulombs/mol)

With the Goldman equation, you can calculate membrane potentials based on given ion concentrations and membrane permeabilities. You can also predict the changes expected in the membrane potential if these ion concentrations or the membrane permeability change.

▶ RELEVANT BEHAVIORS OF GASES AND LIQUIDS

The respiratory and circulatory systems of the human body have developed according to the physical laws that govern the behaviors of gases and liquids. This section will discuss some of the important laws that govern these behaviors and how our body systems utilize these laws.

Gases   The **ideal gas law** states:

$$PV = nRT$$

where

P = pressure of gases in the system
V = volume of the system
n = number of moles in gas
T = temperature
R = universal constant (8.314 J/°K • mol)

If n and T are kept constant for all pressures and volumes in a system, then any two pressures and volumes in that system are related to Boyle's Law,

$$P_1V_1 = P_2V_2$$

where P represents pressure and V represents volume.

This principle is relevant to the human lungs because the concentration of gas in the lungs is relatively equal to that in the atmosphere. In addition, body temperature is maintained at a constant temperature by homeostatic mechanisms. Therefore, if the volume of the lungs is changed, then the pressure in the lungs will change inversely. For example, an increase in pressure causes a decrease in volume, and vice versa.

Liquids   **Fluid pressure** (or hydrostatic pressure) is the pressure exerted by a fluid on a real or hypothetical body. In other words, the pressure exists whether or not there is a body submerged in the fluid. Fluid exerts a pressure (P) on an object submerged in it at a certain depth from the surface (h). **Pascal's law** allows us to find the fluid pressure at a specified depth for any given fluid. It states:

$$P = \rho g h$$

where

P = fluid pressure (measured in pascals, Pa)
ρ = density of the fluid
g = acceleration due to gravity (10 m/s^2)
h = depth below the surface of the fluid

Fluid pressure is unrelated to the shape of the container in which the fluid is situated.

▶ REVIEW OF LOGARITHMS

Understanding logarithms ("logs") is important in biology because of the definition of pH:

$$pH = -\log_{10}[H^+]$$

This equation is read as "pH is equal to the negative log to the base 10 of the hydrogen ion concentration." But what is a logarithm?

A logarithm is the exponent to which you would have to raise the base (10) to get the number in which you are interested. For example, to get the number 100, you would have to square the base (10):

$$10^2 = 100$$

The base 10 was raised to the second power; therefore, the log of 100 is 2:

$$\log 100 = 2$$

Some other simple examples include:

$10^1 = 10$ The log of 10 is 1.
$10^0 = 1$ The log of 1 is 0.
$10^{-1} = 0.1$ The log of 0.1 is −1.

What about numbers that fall between the powers of 10? If log of 10 is 1 and log of 100 is 2, the log of 70 would be between 1 and 2. The actual value can be looked up on a log table or ascertained with most calculators.

To calculate pH, you need to know another rule of logs that says:

$$-\log x = \log (1/x)$$

and a rule of exponents that says:

$$1/10^x = 10^{-x}$$

Suppose you have a solution whose hydrogen ion concentration [H$^+$] is 10^{-7} mEq/L. What is the pH of this solution?

$$pH = -\log [H^+]$$
$$pH = -\log (10^{-7})$$

Using the rule of logs, this can be rewritten as

$$pH = \log (1/10^{-7})$$

Using the rule of exponents, this can be rewritten as

$$pH = \log 10^7$$

The log of 10^7 is 7, so the solution has a pH of 7.

Natural logarithms (ln) are logs in the base e. The mathematical constant e is approximately equal to 2.7183.

Genetics

Richard D. Hill, University of Texas

▶ WHAT IS DNA?

Deoxyribonucleic acid (DNA) is the macromolecule that stores the information necessary to build structural and functional cellular components. It also provides the basis for inheritance when DNA is passed from parent to offspring. The union of these concepts about DNA allows us to devise a working definition of a gene. A **gene** (1) is a segment of DNA that codes for the synthesis of a protein, and (2) acts as a unit of inheritance that can be transmitted from generation to generation. The external appearance (**phenotype**) of an organism is determined to a large extent by the genes it inherits (**genotype**). Thus, one can begin to see how variation at the DNA level can cause variation at the level of the entire organism. These concepts form the basis of **genetics** and evolutionary theory.

▶ NUCLEOTIDES AND BASE-PAIRING

DNA belongs to a group of macromolecules called **nucleic acids. Ribonucleic acid (RNA)** is also a nucleic acid, but it has different functions in the cell that are not discussed here (see Chapter 4, p. 99). Nucleic acids are polymers made from monomers [*mono-*, one] called **nucleotides.** Each nucleotide consists of a *nucleoside* (a pentose, or 5-carbon, sugar covalently bound to a nitrogenous base) and a phosphoric acid with at least one phosphate group (Fig. B-1a ■). Nitrogenous bases in nucleic acids are classified as either **purines** or **pyrimidines.** The purine bases are **guanine (G)** and **adenine (A);** the pyrimidine bases are **cytosine (C), thymine (T),** found in DNA only, and **uracil (U),** found in RNA only. To remember which DNA bases are pyrimidines, look at the first syllable. The word "pyrimidine" and names of the DNA pyrimidine bases all have a "y" in the first syllable.

When nucleotides link together to form polymers such as DNA and RNA, the phosphate group of one nucleotide bonds covalently to the sugar group of the adjacent nucleotide (Fig. B-1b, c ■). The end of the polymer that has an unbound sugar is called the 3′ ("three prime") end. The end of the polymer with the unbound phosphate is called the 5′ end.

▶ DNA STRUCTURE

In humans, many millions of nucleotides are joined together to form DNA. Eukaryotic DNA is commonly in the form of a double-stranded double helix (Fig. B-2 ■) that looks like a twisted ladder or twisted zipper. The sugar-phosphate sides, or backbone, are the same for every DNA molecule, but the sequence of the nucleotides is unique for each individual organism.

The backbone of the double helix is formed by covalent **phosphodiester bonds** that link a deoxyribose sugar from one nucleotide to the phosphate group on the adjacent nucleotide. The "rungs" of the double helix are created when the nitrogenous bases on one DNA strand form hydrogen bonds with nitrogenous bases on the adjoining DNA strand. This phenomenon is called **base-pairing.** The base-pairing rules are as follows:

1. Purines pair only with pyrimidines.
2. Guanine (G) forms three hydrogen bonds with cytosine (C) in both DNA and RNA.
3. Adenine (A) forms with two hydrogen bonds with thymine (T) in DNA or with uracil (U) in RNA.

The number of hydrogen bonds is directly related to the amount of energy necessary to break the base pair. Thus, more energy is required to break G:::C bonds (each ":" represents a hydrogen bond) than A::T bonds. This fact can be useful experimentally to make general conjectures about the similarity of two DNA samples.

The two strands of DNA are bound in **antiparallel** orientation, so that the 3′ end of one strand is bound to the 5′ end of the second strand (see Fig. B-1c ■). This organization has important implications for DNA replication.

DNA Replication Is Semi-Conservative

In order to be transmitted from one generation to the next, DNA must be replicated. Furthermore, the process of replication must be accurate and fast enough for a living system. The base-pairing rules for nitrogenous bases provide a means for making an appropriate replication system.

In DNA replication, special proteins unzip the DNA double helix and build new DNA by pairing new nu-

(a)

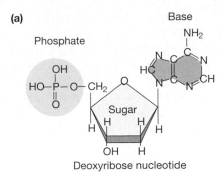

Phosphate

Base

Sugar

Deoxyribose nucleotide

(b)

Phosphate

Base

Sugar

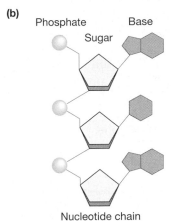

Nucleotide chain

(c)

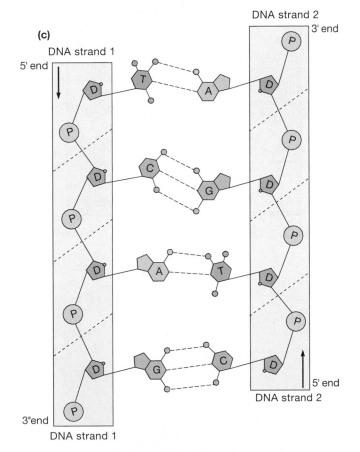

DNA strand 2

3' end

DNA strand 1

5' end

DNA strand 2

5' end

3"end

DNA strand 1

P = phosphate	D = deoxyribose
G = guanine	C = cytosine
A = adeine	T = thymine

■ **Figure B-1 Nucleotides and DNA**   (a) A nucleotide is composed of a pentose sugar, a nitrogenous base, and a phosphate group. (b) Nucleotides link sugar to phosphate to form nucleic acids. The end of the nucleic acid with an unbound sugar is designated the 3' end; the end with the unbound phosphate is the 5' end. (c) Two nucleotide strands link by hydrogen binding between complementary base pairs to form the double helix of DNA. Adenine always pairs to thymine, and guanine pairs to cytosine.

cleotide molecules to the two existing DNA strands. The result of this replication is two double-stranded DNA molecules, such that each DNA molecule contains one DNA strand from the template and one newly synthesized DNA strand. This form of replication is called **semi-conservative replication.**

Replication of DNA is bidirectional. A portion of DNA that is "unzipped" and has enzymes performing replication is called a **replication fork** (Fig. B-2 ■). Replication begins at many points (**replicons**), and it continues along both parent strands simultaneously until all the replication forks join.

Nucleotides bond together to form new strands of DNA with the help of an enzyme called **DNA polymerase.** DNA polymerase can only add nucleotides to the 3' end of a growing strand of DNA. For this reason, DNA is said to replicate in a 5' to 3' direction.

The antiparallel orientation of the DNA strands and the directionality of DNA polymerase force replication into two different modes: **leading strand replication** and **lagging strand replication.** The DNA polymerase can replicate continuously along only one parent strand of DNA: the parent strand in the 3' to 5' orientation. The DNA replicated continuously is called the **leading strand.**

The DNA replication along the other parent strand is discontinuous because of the strand's 5' to 3' orientation. DNA replication on this strand occurs in short fragments called **Okazaki fragments** that are synthesized in the direction away from the replication fork. Another

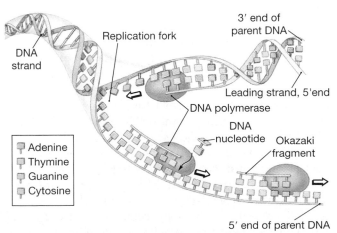

Replication fork

3' end of parent DNA

DNA strand

Leading strand, 5'end

DNA polymerase

DNA nucleotide   Okazaki fragment

▯	Adenine
▯	Thymine
▯	Guanine
▯	Cytosine

5' end of parent DNA

■ **Figure B-2 DNA replication**

enzyme known as **DNA ligase** later connects these fragments into a continuous strand. The DNA replicated in this way is called the **lagging strand.** Because the 5′ ends of the lagging strand of DNA cannot be replicated by DNA polymerase, a specialized enzyme called **telomerase** has arisen to replicate the 5′ ends.

Much of the accuracy of DNA replication comes from base pairing, but on occasion, mistakes in replication happen. However, several quality control mechanisms are in place to keep the error rate at 1 error/10^9 to 10^{12} base pairs. **Genome** (the entire amount of DNA in an organism) sizes in eukaryotes range from 10^9 to 10^{11} base pairs per genome, so this error rate is low enough to prevent many lethal mutations, yet still allows genetic variation to arise.

▶ FUNCTIONS OF DNA

One of the key functions of DNA is its ability to code for the synthesis of proteins that participate in structural or functional aspects of a cell. The processes by which information coded in DNA is converted into proteins are known as *transcription* and *translation.* Both are discussed in Chapter 4. The other key function of DNA is its ability to form genes that act as units of inheritance when transmitted across generations.

Before we begin our discussion of DNA as a unit of inheritance, a few terms need to be introduced. A **chromosome** is one complete molecule of DNA, and each chromosome contains many genes. An **allele** is a form of a gene; interactions between the cell products of alleles determine how that gene will be expressed in the phenotype of an individual. **Somatic** [*soma*, body] **cells** are those cells that comprise the majority of the body (e.g., a skin cell, a liver cell); they are not directly involved with passing on genetic information to future generations. Each somatic cell in a human contains two alleles of each gene, one allele inherited from each parent. For this reason, human somatic cells are called **diploid** ("two chromosome sets"), meaning that they have two complete sets of all their chromosomes. In contrast, **germ cells** pass genetic information directly to the next generation. In human males, the germ cells are the **spermatozoa** (sperm), and in human females, the germ cells are the **oocytes** (eggs). Human germ cells are called **haploid** ("half of the chromosome sets") because each germ cell only contains one chromosome set, which is equal to half of the number of chromosome sets in somatic cells. When a human male germ cell joins with a human female germ cell, the result is a fertilized egg (zygote) containing the diploid number of chromosomes. If this zygote eventually develops into a healthy adult, that adult will have diploid somatic cells and haploid germ cells.

Cells alternate between periods of cell growth and cell division. There are two types of cell division: mitosis and meiosis. **Mitosis** is cell division by somatic cells that results in two daughter cells, each with a diploid set of chromosomes. **Meiosis,** in contrast, is cell division that

results in four daughter cells, each with a haploid set of chromosomes. The daughter cells develop into germ cells.

The period of cell growth is called **interphase.** It is divided into three stages: G_1, a period of cell growth, protein synthesis, and organelle production; **S,** the period during which DNA is replicated in preparation for cell division; and G_2, a period of protein synthesis and final preparations for cell division (Fig. B-3 ■). During interphase, the DNA in the nucleus is not visible under the light microscope without dyes because it is uncoiled and diffuse. However, as a cell prepares for division, it must condense all its DNA to form more manageable packages. Each eukaryotic DNA molecule has millions of base pairs, which, if laid end-to-end, could stretch out to about 6 cm. If this DNA did not coil tightly and condense for cell division, imagine how difficult moving it around during cell division would be.

There is a hierarchy of DNA packaging in the cell (Fig. B-4 ■). Each chromosome begins with a linear molecule of DNA about 2 nm in diameter. Then proteins

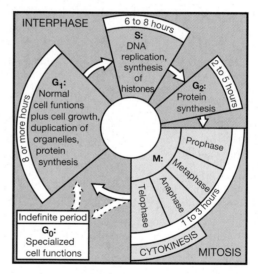

■ **Figure B-3** The cell cycle

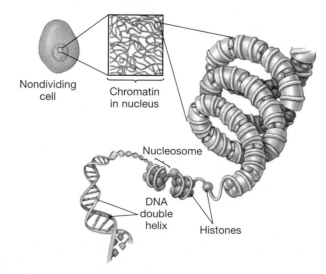

■ **Figure B-4** Levels of organization of DNA

called **histones** associate with the DNA to form a fiber about 10 nm in diameter that looks like "beads on a string." The "beads" are **nucleosomes** made up of histones wrapped in DNA. The beaded string can twist into a **solenoid** [*solen*, pipe] coil about 30 nm in diameter, consisting of about 6 nucleosomes per turn. The 30-nm solenoid structure can form looped domains, and these looped domains can attach to **nonhistone** (structural) **proteins** to form a 250–300-nm fiber called **chromatid fiber.** The chromatid fiber then coils to form the **chromosome fiber** (about 700 nm in diameter) that is visible during cell division. In this state of packaging, the cell is ready for division.

Mitosis Creates Two Identical Daughter Cells

As stated earlier, mitosis is the cellular division of a somatic cell that results in two diploid daughter cells. The steps of mitosis are **prophase, metaphase, anaphase,** and **telophase** (Fig. B-5 ■). The entire somatic cell cycle can be remembered by the acronym, IPMAT, in which the "I" stands for interphase and the other letters stand for the steps of mitosis that follow.

Prophase   During prophase, chromatin becomes condensed and microscopically visible as duplicate chromosomes. The duplicated chromosomes form **sister chromatids,** which are joined to each other at the **centromere.** The centriole pair [∞ p. 49] duplicates and the centriole pairs move to opposing ends of the cell. The **mitotic spindle,** composed of microtubules, assembles between them. The nuclear membrane begins to break down and disappears by the end of prophase.

Metaphase   In metaphase, mitotic spindle fibers extending from the centrioles attach to the centromere of each chromosome. The forty-six chromosomes, each con-sisting of a pair of sister chromatids, line up at the "equa-tor" of the cell.

Anaphase   During anaphase, the spindle fibers pull the sister chromatids apart, so that an identical copy of each chromosome moves toward each pole of the cell. By the end of anaphase, an identical set of forty-six chromosomes is present at each pole. At this point, the cell has a total of ninety-two chromosomes.

Telophase   The actual division of the parent cell into two daughter cells takes place during telophase. In **cytokinesis,** the cytoplasm divides when an actin contractile ring tightens at the midline of the cell. The result is two separate daughter cells, each with a full diploid set of chromosomes. The spindle fibers disintegrate, nuclear envelopes are formed around the chromosomes in each cell, and the chromatin returns to its loosely coiled state.

Mutations Change the Sequence of DNA

Over the course of a lifetime, there are countless opportunities for mistakes to arise in the replication of DNA. A change in a DNA sequence, such as the addition, substitution, or deletion of a base, is a **point mutation.** If a mutation is not corrected, it may cause a change in the gene product. These changes may be relatively minor, or they may result in dysfunctional gene products that could kill the cell or the organism. Only rarely does a mutation result in a beneficial change in a gene product. Fortunately, our cells contain enzymes that can detect and repair damage to DNA.

Some mutations are caused by **mutagens,** which are factors that increase the rate of mutation. Various chemicals, ionizing radiation such as X-rays and atomic radiation, ultraviolet light, and other factors can behave as mutagens. Mutagens either alter the base code of DNA

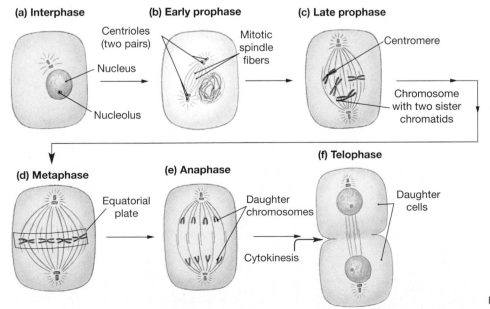

■ **Figure B-5 Mitosis**

or interfere with repair enzymes, thereby promoting mutation.

Mutations that occur in body cells are called **somatic mutations.** Somatic mutations are perpetuated in the somatic cells of an individual, but they are not passed on to subsequent generations. However, **germ-line mutations** can also occur. Because these mutations arise in the germ cells of an individual, they are passed on to future generations.

Oncogenes and Cancer

Proto-oncogenes are normal genes in the genome of an organism that primarily code for protein products that regulate cell growth, cell division, and cell adhesion. Mutations in these proto-oncogenes give rise to **oncogenes,** genes that induce uncontrolled cell proliferation and the condition known as **cancer.** The mutations in proto-oncogenes that give rise to cancer-causing oncogenes are often the result of viral activity.

Anatomical Positions of the Body

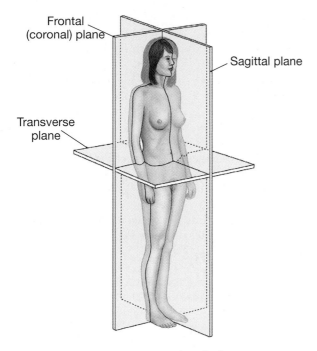

■ Figure C-1   Sectional planes

ANTERIOR (situated in front of): in humans, toward the front of the body (see VENTRAL)

POSTERIOR (situated behind): in humans, toward the back of the body (see DORSAL)

MEDIAL (middle, as in *median strip*): located nearer to the midline of the body (the line that divides the body into mirror-image halves)

LATERAL (side, as in a *football lateral*): located toward the sides of the body

DISTAL (distant): farther away from the point of reference or from the center of the body

PROXIMAL (closer, as in *proximity*): closer to the center of the body

SUPERIOR (higher): located toward the head or the upper part of the body

INFERIOR (lower): located away from the head or from the upper part of the body

PRONE: lying on the stomach, face downward

SUPINE: lying on the back, face up

DORSAL: refers to the back of the body

VENTRAL: refers to the front of the body

IPSILATERAL: on the same side as

CONTRALATERAL: on the opposite side from

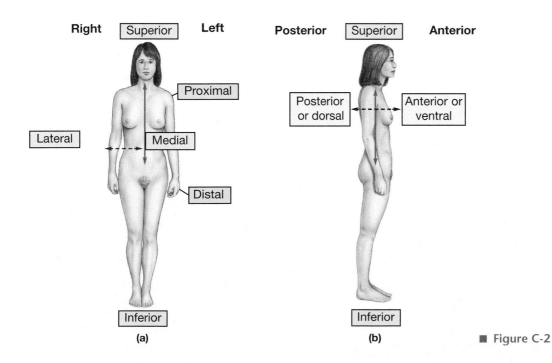

■ Figure C-2

APPENDIX D

Answers to Concept Checks, Graph/Figure Questions, and End-of-Chapter Level 4 Questions

CHAPTER 1 Introduction to Physiology

p. 5

Fig. 1-3: Examples: Humans drive buses; humans eat animals.

p. 8

✓ The remaining 1 mg of salt stays in the body.

p. 13 Quantitative Problems

13. (a) Dependent variable is body length. Independent variable is time. (b) This experiment as described has no control. A control group would have been baby fish of the same brood fed a normal vitamin D diet. (c) The graph of this data should have time (days) on the x-axis and length (mm) on the y-axis. Because the data show a rate of growth, a scatter plot with the points connected by a line is the most appropriate way to graph this data.
14. (a) The independent variable was the varying salinity of the solutions; the dependent variable was the percent change in volume. (b) The control was the volume before soaking. (c) Because the data are discrete points, the most appropriate graph would be a bar graph showing sample salinity on the x-axis and percent change in volume on the y-axis.
15. Total body water volume is 28.05 L.

CHAPTER 2 Atoms, Ions, and Molecules

p. 18

✓ Zinc has 30 protons.
✓ Sodium has 11 electrons.
✓ One atom of calcium has 20 electrons.
✓ The atomic mass of nitrogen is 14.0067 amu.
✓ The element illustrated is lithium.

p. 19

✓ The atomic masses are: (a) 1 (b) 2 (c) 3 amu.

p. 21

✓ The molecular weight of water is 18 amu.

p. 22

✓ Hydrogen cannot form a double bond because it has one electron and requires only one more electron to fill its outer shell. A double bond is formed by the sharing of four electrons.
✓ The outer shell of carbon can hold 8 electrons. The carbon atom contains 4 electrons in that shell, and it obtains the other 4 electrons needed to fill the shell through covalent bonds.
✓ $C_6H_8O_7$
✓ The molecular weight of the citric acid molecule is 204.

p. 23

✓ Fig. 2-7: (a) $C_6H_{13}NO_2$ (b) Leucine has carboxyl and amino groups. Glucose has hydroxyl groups.

p. 24

✓ Table sugar dissolves easily because it is polar; vegetable oil is nonpolar and does not dissolve.

p. 25

✓ The hydrogen ion is called a proton because it has lost its electron, leaving behind one proton.
✓ Dishwashing detergents are surfactants that disrupt the hydrogen bonds of water. This disrupts the surface tension of the water and prevents beading.

p. 26

✓ Na^+ and Cl^- ions form hydrogen bonds with the polar water molecules. This disrupts the ionic bonds that held the NaCl crystal together.

p. 27

✓ A mole of potassium chloride weighs 74.6 grams.
✓ A 1 M solution is equal to a 1000 mM solution, so a 0.1 M solution is equal to a 100 mM solution.
✓ If you add 5 g of glucose to 100 mL of water, the glucose will add volume to the solution and you will no longer have 100 mL of solution. This means you will no longer have a 5% solution. To correctly make the solution, add water to the 5 g of glucose until you have a total solution volume of 100 mL.

p. 28

✓ Acidity is related to increased H^+ concentration. As H^+ concentration increases, pH decreases. Therefore, increasing acidity is associated with decreasing pH.
✓ Urine, stomach acid, and saliva are all contained in the lumens of hollow organs, where they are not part of the body's internal environment (see Fig. 1-2 on p. 3.).

p. 33

✓ Our bodies cannot make the essential amino acids. We must therefore obtain them through our diet. We can synthesize the other amino acids from nutrients.
✓ Proteins are composed of many different amino acids, which can be linked in an almost infinite number of combinations. These combinations allow proteins to take on different forms, such as the globular and fibrous proteins.

p. 41 Quantitative Problems

31. (a) $C_6H_{12}O_6$; m.w. 180; polar (b) CO_2; m.w. 44; polar (c) H_2O; m.w. 18; polar (d) $C_3H_7O_2N$; m.w. 89; polar (e) $C_5H_{10}O_5$; m.w. 150; polar
32. 0.9% solution = 0.9 g/100 mL. Weigh out 9 g NaCl, place into a 1 liter flask, and dissolve in enough water to yield 1 L of solution.

33. (a) 1 M NaCl contains 6.02×10^{23} molecules of NaCl. (b) 1 M solution contains 1000 millimoles. (c) This equals one equivalent. (d) 58.5 g/liter = 5.85% solution

34. 5% glucose = 5 g/100 mL = 0.056 moles or 56 millimoles. 500 mL of this solution would have 138 millimoles.

CHAPTER 3   Cells and Tissues

p. 50

✓ Without a flagellum, a sperm could not control its movement and would be carried along with the fluid flow.

p. 53

✓ Large numbers of mitochondria suggest that the cell has a high energy requirement because mitochondria are the site of greatest energy production in the cell.

✓ Large amounts of smooth endoplasmic reticulum suggest that the tissue synthesizes large amounts of lipids, fatty acids, or steroids.

✓ The rough endoplasmic reticulum is the site of protein synthesis, so pancreatic cells would be more likely to have greater amounts.

p. 62

✓ No, skin would have many layers of cells in order to protect the internal environment. A simple squamous epithelium that is one cell thick with flattened cells would not be a protective epithelium.

✓ The cell is an endocrine cell because it secretes its product into the extracellular space for distribution in the blood.

p. 65

✓ The plasma, or liquid portion of blood, could be considered the matrix.

✓ Cartilage lacks a blood supply, so oxygen and nutrients must reach the cells by diffusion, a slow process.

CHAPTER 4   Cellular Metabolism

p. 76

✓ Potential energy in the body is stored in the chemical bonds of lipids and glycogen.

✓ Kinetic energy is energy in motion: something is happening. Potential energy is stored energy: something is waiting to happen.

p. 79

✓ The reactants are baking soda and vinegar; the product is carbon dioxide.

✓ The foaming-up indicates that energy is being released, so this is an exergonic reaction. The relatively large amount of energy released indicates that the reaction is not readily reversible.

p. 82

✓ The substrates are lactose (lactase), peptides (peptidase), lipids (lipase), and sucrose (sucrase).

✓ In the induced-fit model, both reactants (substrates) and products can combine with the active site. In the lock-and-key model, the enzyme fits with the reactants but not the products.

✓ By having isozymes, one reaction could be catalyzed under a variety of conditions.

p. 84

Fig. 4-12: (a) The enzyme is more active at 30°. (b) When pH goes from 8 to 7.4, the activity increases.

✓ Modulators may affect the rate of enzymatic reactions by binding to the active site (competitive inhibitors) or by binding to a site on the enzyme protein away from the active site (allosteric and covalent modulators).

p. 85

Fig. 4-15: At enzyme concentration A, the rate is 1 mg product formed/sec. At concentration C, the rate is 2.5 mg product formed/sec.

p. 86

Fig. 4-16: When substrate concentration is 200 mg/mL, the rate is 4 mg product formed/sec.

✓ As the amount of enzyme present increases, the rate of the reaction increases.

p. 90

✓ Cells regulate substrate movement through metabolic pathways by (1) controlling the amount of enzyme, (2) producing allosteric and covalent modulators, (3) using two different enzymes to catalyze reversible reactions, (4) isolating enzymes within intracellular organelles, and (5) altering the ADP:ATP ratio in the cell.

p. 93

✓ (a) Exergonic reactions are from glucose to glucose 6-phosphate and from fructose to fructose 6-phosphate. (b) Endergonic reactions are from 1,3-bisphosphoglycerate to 3-phosphoglycerate and from phosphoenol pyruvate to pyruvate. (c) The reactions in (a) are catalyzed by kinases. (d) The reaction from 2-phosphoglycerate to phosphoenol pyruvate is catalyzed by a dehydrogenase. (e) The reaction in (d) is a dehydration reaction.

p. 96

✓ Anaerobic metabolism of glucose can proceed in the absence of oxygen; aerobic metabolism requires oxygen. Anaerobic metabolism produces a maximum yield of 2 ATP/glucose; aerobic metabolism has a maximum yield of 36–38 ATP.

p. 97

✓ The digestion of proteins to polypeptides is a hydrolysis reaction because large molecules are split into smaller molecules by adding water.

p. 98

✓ A hexokinase adds a phosphate group; a phosphatase removes a phosphate group.

p. 103

✓ The polypeptide will have 149 amino acids.

p. 107 Quantitative Problems

31. Exothermic

CHAPTER 5   Membrane Dynamics

p. 113

✓ Two types of lipids found in cell membranes are phospholipid and cholesterol.

✓ Integral proteins span the thickness of the membrane and are tightly bound to it. Associated proteins are loosely bound to polar regions of the membrane components.

p. 116

✓ Membrane proteins serve as structural elements, receptors, enzymes, and transporters.

p. 121

✓ Compartment A will remain yellow.

✓ The skin's thick extracellular matrix is generally impermeable to oxygen. Also, oxygen needs a moist surface to effectively diffuse across a membrane.

p. 122

Fig. 5-16: The line is zero from the origin until the point at which glucose is added. It then increases by the same amount that galactose decreases and becomes constant.

p. 124

✓ Facilitated diffusion of glucose reverses.

p. 131

✓ In phagocytosis, the cell pushes its membrane out to engulf the particle in a large vesicle. In endocytosis, the membrane surface indents and the vesicle is much smaller.

✓ In receptor-mediated endocytosis, receptors cluster in clathrin-coated pits, which facilitate vesicle formation. Potocytosis uses receptors in caveolae to bring molecules into the cell.

p. 133

✓ If ouabain is applied to the apical side, nothing will happen because there are no Na^+-K^+-ATPase molecules on that side. If ouabain is applied to the basolateral side, it will stop the pump. Glucose transport will continue for a time until the Na^+ gradient between the cell and the lumen disappears due to Na^+ entry.

✓ The GLUT2 transporter is illustrated.

✓ Vesicular transport of vesicles from one side of the cell to the other depends on microtubules. If they are not functioning, transcytosis will stop.

p. 135

✓ The baby lost 0.91 kg or 0.91 liter of water.

p. 137

✓ Solution B is hyperosmotic to A.

p. 138

Fig. 5-32: (a) isosmotic and isotonic (b) hyperosmotic and isotonic (c) hyperosmotic and hypotonic

p. 139

✓ Both 5% dextrose and 0.9% NaCl are isosmotic, so neither would affect body osmolarity. Blood loss is loss of fluid from the extracellular compartment, so the best replacement solution is one that remains in the ECF. NaCl is nonpenetrating, so the NaCl solution meets that requirement. Glucose will slowly enter the cells, taking water with it.

✓ If one liter has been lost, you should replace one liter.

p. 144

✓ Ouabain inhibits the Na^+-K^+-ATPase that maintains the resting membrane potential. Over time, Na^+ would leak into the cell, and the resting membrane potential will become more positive.

p. 151 Quantitative Problems

34. Plasma osmolarity = 296 mOsM

35. (a) ICF volume = 29.5 L; interstitial fluid volume = 9.8
    (b) Total solute = 12.432 osmoles; ECF solute = 3.7 osmoles; ICF solute = 8.732 osmoles; plasma solute = 0.799 osmoles.

36. Half-normal saline is approximately 150 mOsM.

37. (a) increases (b) decreases (c) increase (d) decrease

CHAPTER 6 Communication, Integration, and Homeostasis

p. 156

✓ Electrical signals pass through gap junctions. The remainder are chemical signals. Some of them, such as neurotransmitters, create electrical signals in their target cells.

✓ Cytokines, hormones, and neurohormones travel through the blood.

✓ The signal to pounce could not have been a paracrine because the eyes are too far away from the legs and because the response was too rapid for it to have taken place by diffusion.

p. 158

✓ (a) hyperpolarize (b) hyperpolarize (c) depolarize

p. 159

✓ Steroid hormones diffuse across the membranes of their target cells and do not need to rely on signal transduction to get their signal inside.

p. 163

✓ The cell must use active transport to move Ca^{2+} from cytosol to extracellular fluid against its steep concentration gradient.

p. 164

✓ The first messenger is the chemical signal that starts the events leading to a response. Second messengers are intracellular signals that are activated by first messengers that cannot enter the cell.

✓ (a) signal ligand → receptor → second messenger → cell response
    (b) amplifier enzyme → second messenger → protein kinase → phosphorylated protein → cell response

p. 165

✓ Insulin can have different effects in different cells by using different second messenger systems or by having different receptors.

p. 169

Fig. 6-17: (a) 60 beats/min (b) 280 beats/min

p. 177

✓ Sense organs are receptors; sensory neurons are afferent pathways; brain and spinal cord are integrating centers; efferent neurons, hormones, and neurohormones are efferent pathways.

CHAPTER 7 Introduction to the Endocrine System

p. 190

✓ Steroid-producing cells would have extensive smooth endoplasmic reticulum, and protein-producing cells would have lots of rough endoplasmic reticulum.

p. 193

✓ The shorter half-life suggests that aldosterone is not bound to plasma proteins as much as other steroid hormones.

p. 195

Fig. 7-8: The conversion of tyrosine to dopamine adds a hydroxyl (—OH) group to the 6-carbon ring and changes the carboxyl (—COOH) group to a hydrogen. Norepinephrine is made from dopamine by changing one hydrogen to a hydroxyl group. Epinephrine is made from norepinephrine by changing one hydrogen attached to the nitrogen to a methyl (—CH_3) group.

p. 197

✓ The parathyroid cell would have ion channels that are sensitive to plasma Ca^{2+} and that are regulated such that a depolarization occurs when Ca^{2+} is low. The depolarization would trigger secretion of PTH.

✓ Two pathways that match insulin release are 3 and 6. There is insufficient information about the hormone secreted from the digestive tract to match it to a pathway.

p. 200

✓ The pathway that fits the hypothalamic trophic hormone-anterior pituitary hormone-endocrine target is pathway 5.

✓ The pathway that fits the hypothalamic trophic hormone-prolactin-breast pattern is pathway 4.

p. 201

✓ The target of these hypothalamic hormones is endocrine cells in the anterior pituitary.

✓ A decrease in blood glucose is the negative feedback signal.

p. 206

✓ Table 7-2: Secondary problem at pituitary: CRH up, ACTH low, cortisol low.
Primary problem: CRH up, ACTH up, cortisol low.
Secondary problem at hypothalamus: CRH low, ACTH low, cortisol low.

p. 212 Quantitative Problems

30. Half-life is 6 hours.

CHAPTER 8   The Nervous System

p. 220

✓ (a) depolarize   (b) depolarize

✓ If Na^+ leaks into the cell, it will depolarize.

p. 223

Fig. 8-7: The graded potential will be stronger at B.

✓ The trigger zone is the initial segment of the axon.

p. 226

✓ If you could disable the inactivation gates, Na^+ would continue to flood in and the membrane potential would stay depolarized.

p. 230

✓ If you artificially depolarize a cell in the middle of the axon, the signal will travel in both directions, toward the cell body and toward the terminal. Normally signals do not go backward because the channels are in their refractory period.

p. 235

✓ Rough endoplasmic reticulum synthesizes proteins; Golgi packages into vesicles.

✓ Mitochondria travel to axon terminals via axonal transport on microtubules.

✓ Some event between the arrival of the action potential at the presynaptic terminal and the depolarization of the postsynaptic cell is dependent on extracellular calcium. We now know that it is the exocytosis of neurotransmitter.

p. 243

✓ Yes, because the net effect would be a 17 mV depolarization, just above the threshold of -55 mV.

✓ The point of recording is spatially removed from the stimulus. Also the effect takes time due to the relative slowness of ion movement.

✓ Axon terminals convert (transduce) the electrical signal of the action potential into a chemical signal (neurotransmitter release).

p. 250 Quantitative Problems

35. (a) $(12 \times 2$ mV $= +24) + (3 \times -3$ mV$= -9) =$ signal strength of $+15$. Potential $-70 + 15 = -55$. Threshold is -50, so no action potential. (Potential must be equal to or more positive than threshold.) (b) $(11 \times 2 = +22) + (3 \times -3 = -9) = +13$. Potential $-70 + 13 = -57$. Threshold is -60, so action potential will fire. (c) $(14 \times 2 = +28) + (-9) = +19$. Potential $-70 + 19 = -51$. Threshold is -50, so no action potential.

CHAPTER 9   The Central Nervous System

p. 255

✓ When H^+ concentration increases, pH decreases.

✓ Blood will collect in the space between the membranes, pushing on the soft brain tissue under the skull. This is called a subdural hematoma.

✓ Cerebrospinal fluid is more like interstitial fluid because it has little protein and no blood cells.

p. 260

✓ Roots are sections of spinal nerves just before they enter the spinal cord. Horns are extensions of gray matter in the spinal cord. Long projections of white matter (axons) that extend up and down the spinal cord are called tracts. Clusters of tracts are called columns.

✓ Dorsal roots carry sensory information.

p. 262

✓ Activities could include moving eyes, jaw or tongue or walking a straight line.

p. 266

✓ Neurons cross from one side of the body to the other at the pyramids and along the spinal cord.

✓ Spinal cord → medulla → pons, cerebellum → midbrain → diencephalon → cerebrum

CHAPTER 10   Sensory Physiology

p. 283

✓ Myelinated axons have a faster conduction velocity than unmyelinated axons.

✓ Vision = photoreceptor; hearing = mechanoreceptor; taste = chemoreceptor; smell = chemoreceptor; equilibrium = mechanoreceptor; touch-pressure = mechanoreceptor; temperature = thermoreceptor; pain = nociceptor; proprioreception = mechanoreceptor.

p. 287

✓ Neuron B is probably causing K^+ channels or Cl^- channels to open in neurons A and C.

p. 288

✓ Sensory neurons signal intensity of a stimulus by the rate at which they fire action potentials.

✓ Nociceptors, which indicate pain, are telling the body that it is being harmed. If possible, the body's response should be something that stops the pain. Therefore it is important that the pain continue as long as the stimulus is present.

p. 292

✓ The adaptive advantage of a spinal reflex is speed of reaction.

p. 295

✓ Responses to temperature and pain usually do not require rapid responses as balance and touch do.

✓ There are many examples. One would be taste receptors.

✓ The knobby terminals of olfactory cells function as dendrites.

✓ Olfactory neurons are bipolar neurons.

p. 298

✓ Umami is associated with ingestion of protein foods, a biomolecule necessary for body function.

p. 300

✓ A kilohertz is 1000 hertz or waves per second.

p. 306

✓ The location of somatosensory information is coded according to the side of the brain where the sensory pathway terminates. The location of smell and sound is coded by the time difference in the arrival of the stimulus in each hemisphere.

✓ A cochlear implant would not help people with either nerve deafness or conductive hearing loss. It can only help people with sensorineural hearing loss.

p. 308
- ✓ When fluid builds up in the middle ear, the eardrum is unable to move freely and therefore cannot transmit sound through the bones of the middle ear as well.
- ✓ When dancers spot, the endolymph in the ampulla moves with each head rotation, but then stops as the dancer holds his/her head still. This results in less inertia than if the head were continuously turning.

p. 312
- ✓ The aqueous humor acts as a support for the cornea and lens. It also provides a route for bringing nutrients to and removing wastes from the epithelial layer of the cornea, which has no blood supply.
- ✓ The aqueous humor is secreted into the anterior chamber of the eye and usually drains into a small duct that joins the venous circulation. If the drainage is impeded, fluid brings up in the eye, causing the condition known as glaucoma. If untreated, the pressure can impair function of the optic nerves and cause blindness.

p. 314
- ✓ Constriction is mediated by parasympathetic pathways, so an anticholinergic (anti-ACh) drug would cause pupillary dilation.
- ✓ In myopia, the lens bends light too sharply for the length of the eyeball, and the focal distance falls in front of the retina. Myopia is corrected with a concave lens that scatters the light and increases the focal distance.
  Fig. 10-31: (1) Convex lenses focus a beam of light; concave lenses scatter a beam of light passing through them. (2) In hyperopia, the focal point lies behind the retina, so the convex lens focuses it at a shorter distance onto the retina. In myopia, the focal point lies in front of the retina, so scattering the light beam with a concave lens allows the eye's lens to focus the light on the retina.

p. 316
- ✓ Our dark vision is black and white because only rods (black and white vision), not cones (color vision), are sensitive enough to be stimulated by such low levels of photons.

p. 317
  Fig. 10-35: Red cones absorb light over the broadest spectrum, blue cones over the narrowest. At 500 nm, blue and green cones absorb light equally.

p. 319
- ✓ In both areas, the finest discrimination occurs in the region with the smallest visual or receptive fields.

CHAPTER 11   Efferent Peripheral Nervous System: The Autonomic and Somatic Motor Divisions

p. 328
- ✓ Automobile and sports accidents cause many spinal cord injuries.

p. 339
- ✓ The motor end plate has nonspecific monovalent cation (Na^+ and K^+) channels that open when ACh binds to them. The axon contains voltage-gated channels, with separate channels for Na^+ and K^+.

p. 340
- ✓ The preganglionic neurons of sympathetic pathways release acetylcholine onto nicotinic cholinergic receptors. Thus, nicotine will excite the postganglionic neurons of sympathetic pathways, such as those of the pathway that increases heart rate.

- ✓ The monkeys become paralyzed because the nicotinic ACh receptors on skeletal muscles are inactivated, preventing muscle contraction.
- ✓ Anticholinesterase drugs block or slow the breakdown of ACh at the motor end plate. This allows ACh to remain active at the motor end plate for a longer duration and helps offset the decrease in active receptors.

CHAPTER 12   Muscles

p. 347
- ✓ Biceps and triceps in the upper arm; legs: hamstring (flexor) and quadriceps (extensor)

p. 352
- ✓ Ends of the A bands are darkest because they are where the thick and thin filaments overlap.
- ✓ T-tubules allow action potentials to travel from the surface of the muscle fiber to its interior.
- ✓ The banding pattern of organized filaments in the sarcomere forms striations in the muscle.

p. 356
- ✓ The crossbridges do not all unlink at one time, so while some myosin heads are free and swiveling, others are still tightly bound.

p. 359
- ✓ The part of contraction that requires ATP binding is the release of myosin heads from actin. Subsequent related steps are crossbridge formation and the powerstroke. Relaxation occurs when tropomyosin again blocks the myosin binding site. The actual movement of tropomyosin does not require ATP although it does require that Ca^{2+} be pumped back into the sarcoplasmic reticulum using a Ca^{2+}-ATPase.

p. 363
- ✓ Summation in muscle fibers means that the force increases. Temporal summation in neurons means that the membrane potential difference decreases.

p. 364
- ✓ A marathoner probably has more slow twitch muscle fibers, whereas a sprinter probably has more fast-twitch muscle fibers.

p. 365
- ✓ An increase in firing of the somatic motor neuron causes the muscle fiber to increase its force of contraction due to summation.
- ✓ Force in a muscle can be increased by recruiting additional motor units.

p. 369
- ✓ If the muscle insertion is farther from the joint, the leverage is better and a contraction will create more force for a given increase in distance.

p. 373
- ✓ In multi-unit smooth muscle, force is increased by recruiting additional muscle fibers. In single-unit smooth muscle, which is electrically coupled, force is increased by increasing the amount of Ca^{2+} that enters the cell.
- ✓ If there is no calcium to enter the smooth muscle, contraction stops.
- ✓ Tetrodotoxin does not affect action potentials in smooth muscle; therefore, the depolarization of the smooth muscle action potential must not be due to Na^+ entry.

p. 375
- ✓ Pacemaker potentials are always above threshold and create regular rhythms of contraction. Slow wave potentials

are variable in magnitude and are below threshold some of the time.

p. 382 Quantitative Problems

33. (a) The biceps would need to exert 7.5 kg of force to hold the arm stationary, a 125% increase over a biceps inserted at 5 cm. (b) With a 7 kg weight at 20 cm, the biceps needs to exert 28 kg of force. This is greater than if the weight is placed in the hand

CHAPTER 13 Integrative Physiology I: Control of Body Movement

p. 386

✓ Sensor (sensory receptor) → afferent pathway (sensory neuron) → integrating center (central nervous system) → efferent pathway (autonomic or somatic motor neuron) → effector (muscles, glands, some adipose tissue)

✓ Hyperpolarization means that the membrane potential of a neuron becomes more negative and moves farther away from threshold.

p. 392

✓ When you pick up a weight, alpha and gamma neurons, spindle afferents, and Golgi tendon organ afferents are all active.

✓ A stretch reflex is initiated by stretch and causes a reflex contraction. An extensor reflex is a postural reflex initiated by withdrawal from a painful stimulus; the extensor muscles contract but the corresponding flexors are inhibited.

CHAPTER 14 Cardiovascular Physiology

p. 409

✓ The bottom tube has the greatest flow because it has the larger pressure gradient (50 mm Hg versus 40 mm Hg).

p. 410

✓ Tube 3 has the greatest flow because it has the larger radius (less resistance) and the shorter length (less resistance). Tube 4, with the greatest resistance due to length and narrow radius, has the least flow.

✓ If the canals are identical in size and therefore in cross-sectional area, the canal with the faster flow will have the higher rate of flow.

p. 415

✓ Electrical signals within cardiac muscle pass from cell to cell via gap junctions. The connective tissue does not have gap junctions and is not an excitable tissue, so any depolarizations reaching the connective tissue will die out.

✓ Superior vena cava → right atrium → tricuspid (right AV) valve → right ventricle → pulmonary (right semilunar) valve → pulmonary trunk → right or left pulmonary artery → pulmonary arteriole → pulmonary capillary → pulmonary venule → pulmonary vein → left atrium → mitral (bicuspid, left AV) valve → left ventricle → aortic (left semilunar) valve → aorta

p. 417

✓ From this experiment, it is possible to conclude that myocardial cells require extracellular calcium for contraction, but skeletal muscle cells do not.

p. 419

✓ If all calcium channels in the muscle are blocked, there will be no contraction. If only some are blocked, the force of contraction will be less.

p. 421

✓ The rising phase of the action potential of a myocardial contractile cell is due to Na^+ influx through voltage-gated Na^+ channels, just as in the axon. If those channels are blocked with tetrodotoxin, the cell will not depolarize and will not contract.

p. 423

✓ The Ca^{2+} channels in the autorhythmic cells are not the same as the Ca^{2+} channels in the contractile cells. Those in the autorhythmic cells open rapidly when the membrane potential reaches about -50 mV and close when it reaches about $+20$ mV. The Ca^{2+} channels in the contractile cells are slower and do not open until the membrane has depolarized fully.

✓ If tetrodotoxin is applied to a myocardial autorhythmic cell, nothing will happen as there are no voltage-gated Na^+ channels in those cells.

✓ The vagus nerve contains many parasympathetic pathways to the internal organs. When the nerve was cut, parasympathetic innervation to the heart was eliminated and the rate sped up to the rate set by the autorhythmic cells of the SA node.

p. 426

✓ The AV node provides a pathway for action potentials from the SA node to pass into the ventricles. It also slows down the conduction of those action potentials, allowing completion of atrial contraction before ventricular contraction begins.

✓ The fastest pacemaker sets the heart rate. Therefore, the ectopic pacemaker will speed the heart up to 120 beats/min.

p. 427

Fig. 14-21: The heart rate is 80 beats per minute.

p. 429

✓ In (b), notice that there is no regular association between the P waves and the QRS complexes (the P-R segment varies in length). Notice also that not every P wave has an associated QRS complex. Both sets of waves appear at regular intervals, but the atrial rate (P waves) is faster than the ventricular rate (QRS complexes).
In (c), there are identifiable R waves but no P waves. In (d), there are no recognizable waves at all, showing that the depolarizations are not coordinated.

p. 430

Fig. 14-24: (e) From this ECG, it appears that only one of every three P waves is conducted to the ventricles. This is a condition known as partial heart block.

p. 434

✓ After 10 beats, the pulmonary circulation will have gained 10 mL of blood and the systemic circulation will have lost 10 mL.

p. 436

Fig. 14-29: Maximum stroke volume is 160 mL, first achieved when end-diastolic volume is about 330 mL.
Fig. 14-30: At point A, the heart under the influence of norepinephrine has a larger stroke volume and therefore is creating more force.

p. 437

Fig. 14-31: Heart rate is the only parameter controlled by ACh. Heart rate and contractility are both controlled by norepinephrine. The SA node has muscarinic receptors. The SA node and contractile myocardium have β_1 receptors.

p. 441 Quantitative Problems

24. SV/EDV = 0.25. If SV = 40 mL, EDV = 160 mL. SV = EDV − ESV, so 40 mL = 160 mL − ESV, therefore ESV = 120 mL. CO= HR × SV = 100 bpm × 40 mL/ beat = 4000 mL/min or 4 L/ min.

25. (a) 162.2 cm H_2O (b) 108.1 cm H_2O

26. 5200 mL/min or 5.2 L/min

27. 85 mL

28. (a) 1 min (b) 12 sec

CHAPTER 15   Blood Flow and the Control of Blood Pressure

p. 449

✓ Veins from the brain to the heart do not require valves because the flow of blood toward the heart is aided by gravity.

✓ The pressure wave at the carotid artery would arrive slightly ahead of the pressure wave to the left wrist because the distance from the heart to the carotid artery is shorter.

p. 450

✓ Pulse pressure = systolic pressure minus diastolic pressure. (112 − 68) = 44 mm Hg.
   Mean arterial pressure = diastolic pressure + 1/3 pulse pressure. 68 + 1/3 (44) = 82.7 mm Hg

p. 455

✓ Extracellular K^+ acts as a paracrine that dilates arterioles, thereby increasing blood flow into the tissue.

p. 456

✓ Stimulus: Sight, sound, and smell of the *T. rex*. Receptors: Eyes, ears, and nose. Integrating center: cerebral cortex, with descending pathways through limbic system. Divergent pathways go to the medullary cardiovascular control center, which increases sympathetic output to arterioles. A second descending spinal pathway goes to the adrenal medulla, which releases epinephrine. Epinephrine on β_2 receptors of liver, heart, and skeletal muscle arterioles causes vasodilation of those arterioles. Norepinephrine from sympathetic innervation onto α receptors in most other arterioles causes vasoconstriction.

✓ (a) The tissue with the highest blood flow per unit weight is the kidney. (b) The tissue with the least total blood flow is the heart.

p. 457

   Fig. 15-13: Blood flow through the lungs is 5 L/min.

p. 461

✓ Net filtration out of the capillary will increase due to increased hydrostatic pressure.

p. 462

✓ Low protein diets result in a low concentration of plasma proteins, the solute responsible for the osmotic pressure difference between the plasma and the interstitial fluid. With lower osmotic pressure inside the capillaries, reabsorption is reduced while filtration remains constant. This results in a greater net outward movement of fluid and leads to edema.

p. 463

   Fig. 15-20: The SA node has muscarinic cholinergic receptors for ACh and β_1 receptors for catecholamines. The ventricles have β_1 receptors for catecholamines. Sympathetic neurons terminating on arterioles and veins release norepinephrine onto α receptors.

p. 465

✓ Action potentials are created when a cell depolarizes to threshold. Depolarization results from net influx of positive ions or efflux of negative ions. The most likely ion is Na^+, moving into the receptor cell.

   Fig. 15-22: Arterioles and veins: norepinephrine, α receptors. Ventricles: norepinephrine, β_1 receptors. SA node: norepinephrine, β_1 receptors, and ACh, muscarinic receptors.

p. 473 Quantitative Problems

30. Flow $\propto \Delta P \times$ radius^4. If diameter goes from 2 to 4, flow increases 16-fold.

31. Answers will vary. For a 50-kg individual with a resting pulse of 70 bpm, her weight in blood will be pumped in approximately 10 minutes.

32. Mean arterial pressure (MAP) = 1/3 (115 − 73) + 73 or 87 mm Hg. Pulse pressure = 115 − 73 = 42 mm Hg.

33. His cardiac output is 6250 mL/min or 6.25 L/min.

CHAPTER 16   Blood

p. 477

✓ Liver degeneration reduces the total plasma protein concentration, which reduces the colloid osmotic pressure in the capillaries. This increases net capillary filtration and edema results.

p. 486

✓ Some other factor that is essential for red blood cell synthesis must be lacking. These would include iron for hemoglobin production, folic acid, and vitamin B_{12}.

✓ Low atmospheric P_{O_2} at high altitude → low arterial P_{O_2} → sensed by kidney → erythropoietin synthesized and released → acts on bone marrow to increase production of red blood cells

✓ A dehydrated person will have a lower plasma volume, but red blood cell volume will remain constant. This results in an elevated hematocrit.

p. 491

✓ In Figure 16-11, the intrinsic pathway starts at collagen and the extrinsic pathway starts at tissue factor.

p. 495 Quantitative Problems

17. Total blood volume in a 200 lb man is about 6.4 L; his plasma volume is about 3.1 L. A 130 lb woman has total blood volume of 4.1 L and plasma volume of about 2.4 L.

CHAPTER 17   Respiratory Physiology

p. 503

✓ Increased hydraulic pressure results in greater net filtration out of the capillaries. Accumulation of fluid in the interstitial space of the lungs is the condition known as pulmonary edema.

p. 506

✓ The other factor that affects how much gas will dissolve is the solubility of the gas.

✓ 720 mm Hg × 0.78 N_2 = 561.6 mm Hg

p. 507

✓ If cilia cannot move mucus up the mucus escalator and out of the lungs, the collected mucus will trigger a cough reflex in an attempt to clear it out.

p. 509

✓ Scarlett will be more successful if she exhales deeply, as this will decrease her thoracic volume and will pull her lower rib cage inward.

✓ Coughing clears the airways of mucus that has captured bacteria and viruses. Inability to cough decreases the ability to expel the potentially harmful material.

p. 511

✓ A hiccup results from a quick contraction of the diaphragm that causes a rapid decrease in both intrapleural and intrapulmonary pressures.

✓ The knife wound would cause the left lung to collapse if it punctured the pleural membrane. The loss of adhesion between lung and chest wall would release the inward pressure exerted on the chest wall, so the rib cage would expand outward. The right side would be unaffected as the right lung is contained within its own pleural sac.

p. 513

✓ Scar tissue will reduce lung compliance.

✓ If surfactant is not present, the first breath will probably not expand the alveoli fully because of the high cohesiveness of the fluid lining the alveoli.

p. 514

✓ Resistance to air flow is inversely proportional to the radius of the tube through which air is flowing. If the bronchiolar lumen becomes more narrow, the resistance will increase.

✓ In a steam room, the water vapor increases the viscosity of the air and therefore increases the resistance of the air to flow.

✓ The neurotransmitter is acetylcholine and the receptor is muscarinic.

p. 516

✓ Inspiratory reserve volume will decrease in a lung with lower compliance.

✓ Residual volume will increase in a patient who cannot fully exhale.

✓ Residual volume increases in aging subjects if total lung capacity does not change.

p. 518

✓ An increase in tidal volume will increase alveolar P_{O_2}.

p. 519

✓ If blood flow into one small section of lung decreases, the P_{O_2} in those alveoli will increase because oxygen is not leaving the alveoli and entering the blood. The P_{CO_2} in those alveoli will decrease because new CO_2 is not entering the alveoli from the blood. Bronchioles constrict when P_{CO_2} decreases, shunting air to areas of the lung with better blood flow. The compensation will not bring ventilation in this section of lung back to normal because the original pathology still exists and local control is insufficient to correct the disparity.

p. 521

✓ Left ventricular failure or mitral valve dysfunction will cause blood to pool in the lungs when the left heart is unable to pump as well as the right heart. The increase in blood volume will result in an increase in blood pressure.

✓ When alveolar ventilation increases, arterial P_{O_2} will increase because more fresh air is entering the alveoli. Arterial P_{CO_2} will decrease because the fresh air in the alveoli has a lower P_{CO_2}, creating a greater pressure gradient for CO_2 to leave the blood. Venous P_{O_2} and P_{CO_2} will not change because these are determined by the oxygen consumption and CO_2 production in the cells.

p. 524

Fig. 17-21: (a) When P_{O_2} is 20 mm Hg, hemoglobin is about 34% saturated with oxygen. (b) Hemoglobin is 50% saturated with oxygen at a P_{O_2} of 23 mm Hg.

p. 525

✓ Yes. Hemoglobin reaches 100% saturation at 650 mm Hg. At sea level, 100% oxygen equals P_{O_2} of 760 mm Hg.

✓ Hyperventilation causes a minimal increase in percent saturation of arterial hemoglobin.

p. 526

Fig. 17-22: (a) When pH falls from 7.4 to 7.2, hemoglobin saturation decreases by 13%, from 37% saturation to 24%. (b) When an exercising muscle cell warms up, it releases more oxygen at any given P_{O_2}.

Fig. 17-23: Loss of 2,3-DPG is bad because then hemoglobin binds more tightly to oxygen at P_{O_2} values found in cells.

Fig. 17-24: (a) The P_{O_2} of placental blood is about 30 mm Hg. (b) At a P_{O_2} of 10 mm Hg, maternal blood is only about 5% saturated with oxygen.

p. 527

✓ As the P_{O_2} of the exercising muscle falls, more oxygen is released by hemoglobin. The P_{O_2} of the venous blood leaving the muscle will be 25 mm Hg, the same as the P_{O_2} of the muscle cell.

p. 529

✓ An obstruction of the airways would decrease alveolar ventilation and cause CO_2 to be retained in the body. This would increase the H^+ concentration and the pH would fall.

p. 540 Quantitative Problems

30. $P_1V_1 = P_2V_2$. New volume = 200 mL

31. (a) 21% O_2 = 160 mm Hg, 78% nitrogen = 593 mm Hg, 0.3% CO_2 = 3 mm Hg. (b) 40% O_2 = 304 mm Hg, 13% nitrogen = 99 mm Hg, 45% CO_2 = 342 mm Hg, 2% H2 = 15 mm Hg (c) 10% O_2 = 76 mm Hg, 15% nitrogen = 114 mm Hg, 1% argon = 8 mm Hg, 25% CO_2 = 190 mm HG

32. 400 mL/breath × 12 breaths/min = total pulmonary ventilation of 4800 mL/min. Before an exam, ventilation is 7200 mL/min. Alveolar ventilation is (400 − 120) × rate, or 3360 and 5040 mL/min.

33. Tidal volume = 471 mL/breath. IRV = 3383 mL.

34. The maximum O_2 carrying capacity is 1.65 mL O_2/gm Hb.

CHAPTER 18   The Kidneys

p. 544

✓ If extracellular K^+ decreases, more K^+ will leave the cell and the membrane potential will hyperpolarize (become more negative, increase).

✓ If plasma Ca^{2+} decreases, the force of contraction will decrease.

p. 549

✓ Filtration and secretion both represent movement of substances from the extracellular fluid into the lumen of the nephron. Filtration is always passive; secretion may be passive or active.

✓ Glomerulus → Bowman's capsule → proximal tubule → loop of Henle → distal tubule → collecting duct → renal pelvis → ureter → urinary bladder → urethra.

✓ If reabsorption decreases to half of normal, the body would run out of plasma in under an hour.

p. 551

✓ Osmotic pressure is higher in efferent arterioles because some of the fluid volume has been filtered, leaving the same amount of protein in a smaller volume.

p. 552

✓ Average mean blood pressure is 93 mm Hg.

p. 553

Fig. 18-8: Capillary blood pressure, GFR, and RBF all increase.

p. 555

✓ If the afferent arteriole constricts, the resistance in that arteriole increases, flow through that arteriole is diverted to lower resistance arterioles, and GFR in that nephron will decrease.

✓ The primary driving force for GFR is blood pressure, opposed by fluid pressure in Bowman's capsule and osmotic pressure due to plasma proteins. With fewer plasma proteins, the plasma has lower colloid osmotic pressure. With less osmotic pressure opposing GFR, GFR will increase.

p. 557

Fig. 18-14: The transport rate at 3 mg/mL is 3 mg/min; at 5 and 8 mg/mL, it is 4 mg/min. The transport rate is 4 mg/min at any plasma concentration equal to or higher than 4 mg/mL.

p. 561

✓ GFR is equal to creatinine clearance, which in this case is about 92 L/day.

p. 568 Quantitative Problems

21. Renal blood flow (25%) is 1 L/min.

22. 20,000 mg creatinine in the specimen. Clearance = 1000 L plasma/day. Normally creatinine clearance = GFR. However, this value of 1000 L/day is not at all realistic for GFR (normal average is 180 L/day). The test should be repeated to check for laboratory error.

23. For any solute that filters: plasma concentration × GFR = filtration rate. At the transport maximum, filtration rate = reabsorption rate or T_m. By substitution: plasma concentration × GFR = T_m. The renal threshold represents the plasma concentration at which the transporters working at their maximum (Tm). By substitution: renal threshold × GFR = T_m. Mermaid's GFR is 250 mL/min and T_m is 50 mg/min. At 15 mg glucose/mL plasma, plasma concentration × GFR = 3750 mg/min glucose filtered. 3750 mg/min glucose filtered minus 50 mg/min reabsorbed = 3700 mg/min excreted. How many mL of plasma contain 3700 mg? 1 mL plasma/15 mg = ? mL plasma/3700 mg. Clearance = 246.7 mL plasma/min.

CHAPTER 19 Integrative Physiology II: Fluid and Electrolyte Balance

p. 575

✓ Apical membranes on collecting duct cells will have more water pores when ADH is present.

✓ The collecting duct cells concentrate organic solutes which counteract the hypertonic effect of their environment.

p. 576

✓ Hyperosmotic NaCl is hypertonic and causes the osmoreceptors to shrink, but hyperosmotic urea is hypotonic and causes them to swell. Thus, they will not fire.

p. 577

✓ In dehydration, the kidneys try to conserve water. Vasopressin makes the nephron permeable to water and enhances absorption, therefore vasopressin levels would increase with dehydration.

✓ If vasopressin secretion is suppressed, the urine will be dilute.

p. 580

Fig. 19-11: See Fig. 15-21, p. 464.

p. 581

✓ In Fig. 19-12, Na^+ and K^+ are moving down their electrochemical gradients.

p. 582

Fig. 19-13: See Fig. 15-22, p. 465, for the cardiovascular pathway and Fig. 19-12 for aldosterone.

p. 583

✓ Atherosclerotic plaques block blood flow, which decreases GFR and decreases pressure in the afferent arteriole. These are both stimuli for renin release.

✓ Renin secretion begins a cascade that produces angiotensin II. Angiotensin II is a powerful vasoconstrictor, acts on the medullary cardiovascular control center to increase blood pressure, increases vasopressin and aldosterone secretion and thirst, resulting in increased fluid volume in the body. These responses may all contribute to increased blood pressure.

p. 585

✓ On the left side of Fig. 19-7, interneurons lead from osmoreceptors to the hypothalamic thirst centers as well.

p. 593

✓ The muscles of ventilation that are involved include the diaphragm, the external intercostals, and the scalenes.

✓ The bicarbonate is increasing as the reaction shifts to the right and achieves a new equilibrium. At equilibrium, both H^+ and bicarbonate are increased. The bicarbonate cannot act as a buffer when the system is at equilibrium.

p. 595

✓ Both K^+ and H^+ are being moved against their concentration gradients, which requires ATP. In the proximal tubule, the Na^+ is moving down its concentration gradient, providing the energy to push H^+ against its gradient.

✓ When intercalated cells reabsorb K^+, they secrete H^+; therefore pH will increase.

p. 601 Quantitative Problems

26. (a) pH = 6.1 + log {24/(.03 × 40)} = 7.40 (b) 7.34.

CHAPTER 20 Digestion

p. 604

✓ Digestion is the process by which nutrients are broken down into a form that is absorbable by the intestine; it takes place in the lumen of the digestive tract, which is technically outside the body. Metabolism is the sum of all chemical reactions that take place inside the body.

✓ Absorption moves material from the lumen to the extracellular fluid. Secretion moves substances from the cells or extracellular fluid into the lumen.

p. 605

✓ The lumen of the digestive tract is on the apical side of the intestinal epithelium.

p. 608

✓ The digestive system has a large and vulnerable surface area facing the external environment; therefore, it needs the immune cells of the lymphoid tissue to combat invaders.

p. 609
✓ The sphincters are tonically contracted to close off the lumen of the digestive tract from the outside world and to keep material from passing freely between sections of the tract.

p. 630
✓ Glucose from table sugar enhances the intestinal absorption of Na^+ because the transporter for glucose depends on the movement of Na^+ from the lumen into the intestinal cell.
✓ Bile does not digest triglycerides. It simply emulsifies them into smaller globules so that lipase can digest them.

CHAPTER 21 Endocrine Control of Metabolism and Growth

p. 643
✓ Norepinephrine binds to β_1 receptors to elicit vasoconstriction.
✓ Observations prompting researchers to classify the neurons as sympathetic might have been anatomical clues, such as where along the spinal cord the neurons originate.

p. 646
✓ Water has a high specific heat and will draw heat away from the body through conductive heat transfer. If this loss exceeds the person's heat production, they will feel cold.
✓ A person exercising in a humid environment loses the benefit of evaporative cooling and is likely to overheat faster.

p. 661
✓ Insulin is a protein and will be digested if administered as a pill.
✓ Although dehydrated patients may have elevated K^+ concentrations in their blood, their total amount of K^+ is below normal. If normal fluid volume is replaced with no K^+ added, the result will be a below-normal K^+ concentration. Thus, these patients are given fluid with K^+ even though they have an initial hyperkalemia.

p. 664
✓ A target cell might not respond to leptin because it has no leptin receptors (or they are defective), or because there is a problem with the signal transduction/second messenger pathway.

p. 666
✓ Anabolic steroids, used to build muscle, do not include cortisol because cortisol is catabolic on muscle proteins.

p. 667
✓ (1) ACTH less than normal (2) ACTH greater than normal (3) ACTH less than normal
✓ Someone with Addison's disease will have elevated levels of ACTH due to reduced corticosteroids and negative feedback.

p. 672
✓ The T_4 given to the patient is converted to T_3 in the peripheral tissues. T_3 is the more active form of the hormone.
✓ When mitochondria are uncoupled, the energy that is normally captured in the high-energy bond of ATP is released as heat instead. This will raise the body temperature of the person and cause heat intolerance.

p. 673
✓ Adults who hypersecrete growth hormone do not grow taller because the epiphyseal plates of the long bones have closed.

p. 684
30. Answers will vary. A 64-inch tall woman weighing 50 kg will have a BMI of approximately 19.

CHAPTER 22 The Immune System

p. 696
✓ When capillary permeability increases, proteins escape from the plasma to the interstitial fluid. This decreases the force opposing capillary filtration, and causes additional fluid to accumulate in the interstitial space (swelling or edema).

p. 687
Fig. 22-9: The antigen will activate clone 1.

p. 698
✓ The first exposure causes a slower and weaker immune response (see Fig. 22-11). The second exposure will cause a stronger response, which may be severe and cause anaphylaxis.
✓ Protein antibodies can be moved across cells by transcytosis or released from cells by exocytosis.

p. 708
✓ Type O blood has neither A nor B antigens, but even if it did, type AB blood does not produce anti-A or anti-B antibodies. Therefore no immune reaction will occur.
✓ If type A blood is given to a type O recipient, the recipient will synthesize anti-A antibodies and agglutination will occur.

CHAPTER 23 Integrative Physiology III: Exercise

p. 721
✓ If venous P_{O_2} decreases, the cell's P_{O_2} is also decreasing.

p. 722
Fig. 23-7: To calculate blood flow to an organ, multiply the cardiac output (L/min) times the percentage of flow to that organ. When rest and exercise values are compared, actual blood flow decreases only in the kidneys, GI tract, and other tissues.

p. 723
✓ The line for mean blood pressure lies closer to the line for diastolic pressure because the heart spends more time in diastole than systole.
Fig. 23-8: Mean arterial pressure is determined by cardiac output and resistance. If resistance is falling but MAP is increasing, cardiac output must be increasing.

p. 730 Quantitative Problems
21. 60 beats/ minute $\times$ 70 mL/beat = 4200 mL/min cardiac output. If heart rate doubles to 120 beats/min, CO goes to 8400 mL/min, or doubles also.

CHAPTER 24 Reproduction and Development

p. 736
✓ The male donates the chromosome that determines the sex of the zygote.
✓ In the absence of a Y chromosome, the XO fetus will be a female. She will be infertile, however, as two X chromosomes are required for functioning ovaries.
✓ Developing testes secrete testosterone, which (along with its derivative DHT) prompts the development of the male

accessory structures. The external genitalia of the embryo will be female.

p. 737

Fig. 24-5: (1) The gamete in a newborn male is at the spermatogonia stage; it is at the primary oocyte stage in females. (2) The first polar body has twice as much DNA as the second polar body. (3) Each primary oocyte forms one egg; each primary spermatocyte forms four sperm.

p. 744

✓ GnRH agonists cause reduced secretion of GnRH and therefore of FSH and LH. In the absence of the gonadotropins, the testes stop producing sperm. However, they also stop producing testosterone. This causes reduced sex drive, which is usually an undesirable side effect.

p. 745

✓ Exogenous anabolic steroids are usually androgens, which have a negative feedback effect on the anterior pituitary gonadotropins. In the absence of FSH and LH, the testes will shrink and stop producing sperm.

p. 753

✓ Women who take anabolic steroids may experience growth of facial and body hair, deepening of the voice, increased libido, and irregular menstrual cycles.

✓ A woman given an aromatase inhibitor would have decreased estrogen production during the follicular phase of the menstrual cycle.

Glossary/Index

A band Band of striated muscle sarcomere whose length equals that of the thick filament, 351f, 352

abdominal muscles, 509

abducens nerve (VI) Cranial nerve that controls eye movement, 263t

abnormal hemoglobin (HbS), 485

abnormal tissue responsiveness Endocrine pathologies in which the problem lies in the target tissue receptor or pathways, 205

ABO blood group Membrane glycoproteins found on red blood cells that are one determinant of blood type, 707, 708t, 708f

absolute refractory period A period of time immediately following an action potential during which a second action potential cannot be triggered, no matter how large the stimulus, 227, 228f, 421

absorption Transfer of substances from the lumen of the kidney or gastrointestinal tract to the extracellular space, 459, 460f, 548, 603, 604f, 611–12, 618f, 628–30

absorptive cell Small intestinal epithelial cell whose primary function is absorption. *Synonym:* enterocyte, 624

absorptive state (fed state) The time following a meal when the products of digestion are being absorbed, used, and stored, 647–50, 651t, 655–56

abstinence, 755

accessory nerve (XI) Cranial nerve that controls muscles of oral cavity, neck, and shoulders, 263t

acclimation Physiological adaptation to an environmental change under laboratory conditions, 173, 644

acclimatization Physiological adaptation to an environmental change, 173, 724–25

accommodation The process by which the eye adjusts the shape of the lens to keep objects in focus, 313–14

ACE. *See* angiotensin converting enzyme.

acetoacetic acid Metabolic acid formed from oxidation of fatty acids, 653

acetyl CoA. *See* acetyl coenzyme A.

acetyl coenzyme A (acetyl CoA) metabolic intermediate that links glycolysis and β-oxidation to the citric acid cycle, 88, 93, 94f, 97–98

acetylcholine (ACh) Neurotransmitter used by neurons of the central and peripheral nervous system, 158, 236, 237t, 239, 268, 334, 336, 369, 395, 422

acetylcholinesterase Enzyme that breaks down acetylcholine in the synapse, 239, 334, 339–40

acetylsalicylic acid. *See* aspirin.

ACh. *See* acetylcholine.

acid A molecule that ionizes and contributes an H⁺ to a solution, 27–28, 417, 590–91

acid lipase Fat-digesting enzyme of the mouth and stomach, 612

acid phosphatase, 81t

acid-base balance The homeostatic regulation of body pH, 27–28, 590–97, 596–97

acid-base disturbance, 595–97

    renal compensation in, 592–93, 595–97

    respiratory compensation in, 592

acidosis Extracellular pH less than 7.38, 527, 590, 592–93, 596f

acini. *See* follicle, thyroid.

acquired immune deficiency syndrome. *See* AIDS.

acquired immunity Immune responses directed at specific invaders and mediated by antibodies, 688, 696–703

acromegaly Abnormal growth of cartilage and soft tissues due to excess growth hormone secretion in an adult, 673, 674f

acrosomal reaction Release of enzymes from the sperm head when it contacts an egg, 757

acrosome Lysosome-like vesicle of sperm that contains powerful enzymes essential for fertilization, 743

ACTH. *See* adrenocorticotrophic hormone.

actin A globular protein (G-actin) that polymerizes to form thin filaments (F-actin), 350–52, 355f

action potential Rapid and uniform electrical signal conducted down an axon to the axon terminal or along the membrane of a muscle fiber, 65, 215t, 224–32

    axons, 230f, 232f, 227–28

    coding for stimulus intensity, 227–28, 229f

    compared to graded potential, 222t

    conduction of, 231f, 232–33

    distance traveled without losing strength, 224–26

    falling phase of, 225

    frequency of, 227–28

    ion movements, 228f

    in myocardial autorhythmic cells, 421–23

    in myocardial contractile cells, 420f, 424, 436

    refractory period of, 227, 228f, 421f

    rising phase of, 225

    in skeletal muscle, 421f

    in smooth muscle, 375

activating complement, 699

activating mast cells, 699

activation energy Energy needed to initiate a chemical reaction, 77, 79–80

activation gate Sodium channel gate that opens to initiate an action potential, 225

active expiration Expiration that is assisted by contraction of abdominal and internal intercostal muscles, 508–09

active hyperemia An increase in blood flow that accompanies an increase in metabolism, 454–455f

active immunity, 696

active site Region of an enzyme or transport protein to which the substrate binds. *Synonym:* binding site, 80

active transport Movement across a membrane that requires the input of energy from ATP, 118, 121, 124–25

active vasodilation Vasodilation that occurs in response to increased neural activity mediated through ACh, 643

activin Peptide hormone from the gonads that stimulates FSH secretion, 738

acuity Keenness of vision, 315

acupuncture, 295

acute motor axonal polyneuropathy (AMAN), 244

acute phase protein Liver proteins that act as opsonins and enhance the inflammatory response, 695

adaptation of receptors Process in which sensory receptors decrease their response to a stimulus over time, 288

addictive behavior, 327–28, 336, 340

Addison's disease, 667

addition reaction Chemical reaction that adds a functional group to a molecule, 88

adenine Nucleotide base found in ATP, DNA, RNA, and cAMP, 35, 99

adenohypophysis. *See* anterior pituitary.

adenoma, 204

adenosine Nucleoside composed of adenine and ribose, 238, 454

adenosine diphosphate (ADP) Precursor of ATP, composed of adenine, ribose, and two phosphate groups, 35, 57, 354, 489t

adenosine triphosphate (ATP) An energy-storing compound composed of adenine, ribose, and three phosphate groups, 29, 35, 77–78, 88, 96–97, 131, 238, 354, 355f, 717–18

    ATP/ADP ratio, 89–90

    energy for active transport, 124–25

    energy release from, 77–78

    in muscle contraction, 353–54, 359–60, 436

    production of, 90–97

    transfer of energy between reactions, 88

adenylyl cyclase Membrane-bound enzyme that converts ATP to cyclic AMP, 166t

adequate stimulus The form of energy to which a particular receptor is most responsive, 285

ADH. *See* antidiuretic hormone.

adherens junctions Bands that link action microfilaments in adjacent cells together with the help of cadherins, 56

adhesive junction Desmosomes or adherens junctions that hold cells of a tissue together, 55–56

adipocyte(s) Fat cells, 64f, 65, 612, 663–64

adipose tissue, 65, 641, 652f

ADP. *See* adenosine diphosphate.

adrenal cortex Outer portion of adrenal gland that produces steroid hormones, 192, 194f, 584f, 665–66

adrenal gland Endocrine and neuroendocrine gland that sits on top of the kidney, 332, 544

adrenal glucocorticoids, 664–66

adrenal medulla Modified sympathetic ganglion, the inner portion of the adrenal gland that produces catecholamines, 271, 332, 333f, 418, 647, 663t

adrenaline. *See* epinephrine.

adrenergic Adjective pertaining to epinephrine (adrenaline) or norepinephrine, 236, 237t, 418, 514

adrenergic neuron Neuron that secretes norepinephrine, 236–37, 334

adrenergic receptor Receptor that binds to norepinephrine or epinephrine, 164, 238–39, 334–35t, 418

Page numbers followed by an f indicates figures; t indicates tables.

791

plaque Deposition of lipid in arterial walls, accompanied by smooth muscle proliferation, scar tissue formation, and calcification, 55, 467f

plasma The fluid portion of the blood, 117–18, 475–77

plasma cell Type of lymphocyte that secretes antibodies, 475–76, 689, 691f, 698–99

plasma membrane The cell membrane that serves as both a gateway and a barrier for substances moving into and out of the cell, 45, 47f–48f, 110, 117f, 726f

    composition of, 117f, 475–76

    functions of, 114–16, 596t

    movement across, 119f

    structure of, 117f, 475–76

plasma protein, 462, 475

plasma volume, 462

plasmalemma, 45

plasmin Enzyme that breaks down fibrin. *Synonym:* fibrinolysin, 488, 491, 753

plasminogen, 492t

plasticity The ability of the central nervous system to change circuit connections and function in response to sensory input and past experience, 245, 253, 272

plateau phase (1) flattening of the myocardial contractile cell action potential due to Ca^{2+} entry, or (2) intermediate phase of the human sexual response, 423t

    of myocardial action potential, 420f, 422f, 423

    of human sexual response, 753–54

platelet(s) Cell fragments that participate in coagulation. *Synonym:* thrombocyte, 475–81, 487f

    coagulation and, 475–76, 486–93

    production of, 477–81

platelet-activating factor (PAF), 489

platelet adhesion Platelets stick to exposed collagen in wall of damaged blood vessel, 487–88

platelet aggregation Activated platelets stick to each other, 488

platelet count, 478f

platelet plug, 487, 488f, 489f

platelet-activating factor (PAF), 489

platelet-derived growth factor (PDGF), 489t

pleasure, 753–54

pleura The membranes that line the chest cavity and cover the outer surface of the lungs, 499, 500f, 510f

pleural fluid, 499, 502f

pleural sac, 499, 502f

plicae Large folds of the intestinal wall, 605, 606f–07f

pluripotent hematopietic stem cell, 477

pneumonia Bacterial or viral lung infection, 595, 694

pneumothorax Air in the intrapleural space, 510–11

P_{O_2}, 524–25, 532

podocyte(s) Specialized epithelial cells in Bowman's capsule that surround each capillary and form filtration slits, 545, 546f–47f, 549, 550f

Poiseuille, Jean Leonard Marie, 409

Poiseuille's law, 513

polar body, first and second Unused chromosomes that are discarded from the egg as it undergoes meiosis, 736, 746, 749f

polar molecule(s) Molecules that develop regions of partial positive and negative charge when one or more atoms in the molecule have a strong attraction for electrons, 23–24, 736

polarity, of cell Cells restrict certain membrane proteins to particular regions, thereby creating cells with different functions in different areas, 113, 118, 131, 736

polio, 509

polyaclyamide polymer gel, 81

polycythemia Elevated hematocrit, 535

polycythemia vera, 485

polydipsia Excessive drinking, 661

polymer, 29

polymorphonuclear leukocyte, 692

polypeptide A chain of 10–100 amino acids, 32, 237–38

polypeptide neurotransmitter, 236, 237t, 239, 268, 334, 336, 369, 395, 422

polyphagia Excessive eating, 659, 663

polyploid cell A cell with multiple nuclei and therefore greater than the diploid amount of DNA, 487

polyribosome, 45

polysaccharides Complex carbohydrates composed of glucose polymers; used for energy storage and structure, 29, 30f, 611

polyspermy Fertilization of an egg by more than one sperm, 757

polysynaptic reflex Any nervous reflex that has three or more neurons in the pathway, 385–86

polyunsaturated fatty acid A fatty acid with more than one double bond, 29

polyuria Excessive urination, 337

pons Region of the brain stem that contains centers for respiration and serves as a relay station, 257f, 262

popranolol, 237t

population, 2

population coding The number of sensory receptors activated encodes the intensity of a stimulus, 287

pores, 54, 115

porphyrin ring, 484

portal system A specialized region of the circulation consisting of two capillary beds directly connected by a set of blood vessels, 407

positive feedback loop A feedback loop in which the response reinforces the stimulus, triggering a vicious cycle of ever-increasing response, 173, 174f, 762f

positron, 21

positron emission tomography (PET scan), 21, 267f, 274

postcoital test, 753

postabsorptive state A catabolic state, in which the body taps into its stored reserves and the cells degrade large molecules into smaller molecules. *Synonym:* fasted state, 647, 653f, 655f

posterior pituitary gland An extension of the brain that secretes neurosecretory hormones made in the hypothalamus, 198, 199f, 200f–01f, 264f

postganglionic neuron Autonomic neuron that has its cell body in the ganglion and sends its axon to the target tissue, 329, 334

postprandial period The time immediately following a meal, 655

postsynaptic cell The target cell at a synapse, 218, 234, 235f, 240f–42f, 243, 244f,

postsynaptic integration Multiple signals in a postsynaptic cell combine to create a single integrated signal, 242

postsynaptic inhibition, 241

postsynaptic modulation A modulatory neuron, usually inhibitory, synapses on the dendrites or cell body of a post synaptic cell, 241–42

postsynaptic response, 239–44, 245f

post-translational processing Enzymatic changes made to proteins as they pass through the endoplasmic reticulum and Golgi apparatus, 191

postural reflex Reflexes that help us maintain body position, 394–95

postovulatory phase, See luteal phase.

potassium (K^+), 142–43, 422, 435f

potassium benzoate, 543

potassium equilibrium potential (E_K), 143f

potassium homeostasis, 584–85

potassium permanganate, 119f

potassium permeability, 422–23

potential energy Stored energy that has the ability to do work, 75, 76f

potocytosis Cellular uptake of small molecules and ions in vesicles known as caveolae, 129–30

power stroke Movement of the myosin head that is the basis for muscle contraction, 353, 360

PPARgamma, 694

PP cell Pancreatic endocrine cell that secretes pancreatic polypeptide, 654

P-R interval From the beginning of the P wave to the beginning of the QRS complex, 427, 428f, 430

P-R segment From the end of the P wave to the beginning of the QRS complex, 427, 428f, 430

precapillary sphincter Bands of smooth muscle that can alter blood flow through capillary beds, 456, 457f

prefrontal cortex, 267f

preganglionic neuron Autonomic neuron that originates in the central nervous system and terminates in an autonomic ganglion, 329

preload The degree of myocardial stretch created by venous return, 435

preprohormone Inactive molecule composed of one or more copies of a peptide hormone, a signal sequence, and other peptide sequences that may or may not have biological activity, 191, 192f

prepuce, 741

presbyopia Loss of the accommodation reflex with aging, 314

pressure, 171f, 407

pressure gradient, 407

pressure-volume curve, cardiac cycle, 431–34

presynaptic cell The cell releasing neurotransmitter into a chemical synapse, 218

presynaptic facilitation Modulation of the presynaptic neuron that enhances neurotransmitter release, 243

presynaptic inhibition, 243, 723

presynaptic modulation A modulatory neuron terminates on or close to an axon terminal of the presynaptic neuron, 243–44

presynaptic terminal. See axon terminal.

primary active transport The energy for transport comes from the high-energy phosphate bond of ATP, 125–26

primary bronchi The first two airways created by branching of the trachea, 499, 500f

primary endocrine pathology A pathology that arises in the last endocrine gland in the pathway, 205

primary follicle An undeveloped oocyte and its outer layer of granulosa cells, 736, 746, 749f

primary hypersecretion, 667

primary immune response The immune response that occurs with first exposure to a pathogen, 698

primary motor cortex Regions of the frontal lobe that coordinate skeletal muscle movements, 265

primary oocyte Oocyte that has duplicated its DNA but not undergone a meiotic division, 736

primary reponse, 697f, 698

primary sensory neuron The sensory neuron that takes information from the sensory receptor into the spinal cord, 265, 283

primary sex characteristics The internal sexual organs and external genitalia that distinguish each sex, 745

primary somatic sensory cortex, 369

primary spermatocyte Spermatocyte that has duplicated its DNA but not undergone a meiotic division, 746

primary structure, of protein The sequence of amino acids in the peptide chain, 32

primary tissue types, 55

principal cell. See P cell.

probenecid, 559

process Long, thin extensions of nerve cells, 15

process, in physiological systems, 4

procreation The act of creating a new being, 753–56

product, 76

proenzyme An inactive enzyme, 82

progesterone Female sex hormone produced by the corpus luteum, 737, 756, 758

prohormone Inactive protein containing one or more copies of a hormone, 191, 192f

prokaryotic endosymbiont theory, 51

prolactin A peptide hormone from the anterior pituitary that controls milk production in the breast, 763

prolactin inhibiting hormone (PIH) Hypothalamic hormone that inhibits prolactin secretion by the anterior pituitary, 200, 201f, 763–64

proliferative phase Phase of the menstrual cycle when the endometrium grows and thickens, 747, 750f

promoter Section of DNA near the starting end of a gene that must be activated to begin transcription, 101

Photo Credits

About the Author

Author and Consultant   Dee Silverthorn   **Illustrators**
Bill Ober/Claiire Garrison

Owner's Manual and Preface

All   Dee Silverthorn

Chapter 1

p. 1 GJLP/CNRI/Phototake NYC

Chapter 2

p. 15 Paul Axelsen, M.D./Phototake NYC

Chapter 3

p. 42 Todd Derksen   **3-1a & c** Todd Derksen   **3-1b** David
M. Phillips/Visuals Unlimted   **3-6** Dr. Don Fawcett/Photo
Researchers, Inc. **3-7** Fawcett/Hirokawa/Heuser/Science
Source/Photo Researchers, Inc.   **3-8** Robert W. Riess
3-9 CNRI/Science Source/Photo Researchers, Inc. **3-10**
Robert W. Riess   **3-11** Robert W. Riess   **3-12** Dr. Don
Fawcett/Photo Researchers, Inc. **3-13 left** Don W. Fawcett,
M.D., Harvard Medical School **3-13 middle** Biophoto
Associates/Photo Researchers, Inc. **3-14a & b** Todd Derksen
3-14c Photo Researchers, Inc.   **3-15** Todd Derksen   **3-18**
Custom Medical Stock Photo, Inc. **3-19** Todd Derksen   **3-21**
Ward's Natural Science Establishment, Inc. **3-22** John D.
Cunningham/Visuals Unlimited   **3-23** Frederic H. Martini
3-25 Cabisco/Visuals Unlimited

Chapter 4

p. 73 Photo Researchers, Inc.

Chapter 5

p. 109 Richard Feldmann/Phototake NYC

Chapter 7

p. 186 CNRI/Phototake NYC   **7-1** Dee Silverthorn

Chapter 8

p. 214 David Scott/Phototake NYC   **8-16a** Todd Derksen
8-16b David M. Phillips/Visuals Unlimited

Chapter 9

p. 252 McGill University/CNRI/Phototake NYC   **9-13**
Mazziotta et al./Science Photo Library/ Photo Researchers, Inc.

Chapter 10

p. 281 Prof. P. Motta, Department of Anatomy, University
La Sapienza, Rome/Science Photo Library/ Photo
Researchers, Inc. **10-14b** Todd Derksen   **10-27c** Custom
Medical Stock Photo, Inc.

Chapter 11

p. 326 Dr. Flora Love

Chapter 12

p. 345 University of Texas at Austin   **12-1a & c** Todd
Derksen **12-1b** Phototake NYC   **12-7** J. J. Head/Carolina
Biological Supply Company/Phototake NYC   **12-13 left**
D. Comack, ed. *Ham's Histology 9th ed.*   Philadelphia: J. B.
Lippincott, 1987. By permission. **12-13 right** Frederic H.
Martini   **12-24** Biophoto Associates/ Photo Researchers, Inc.

Chapter 14

p. 403 Carole L. Moncman, Ph.D.

Chapter 15

p. 443 Simon Fraser/Newcastle General Hospital/SPF/
Photo Researchers, Inc.

Chapter 16

p. 474 Phototake NYC   **16-6a** Todd Derksen   **16-9** Photo
Researchers, Inc. **16-10a & b** Todd Derksen

Chapter 17

p. 497 Custom Medical Stock Photo, Inc. **17-7b** Frederic H.
Martini

Chapter 18

p. 542 Daniel Casellas, Ph.D.   **18-1f** Todd Derksen
18-4b Todd Derksen

Chapter 20

p. 602 Mark L. Tamplin, Anne L. Gauzens, and Rita
R. Colwell

Chapter 21

p. 639 McGill University/CNRI/Phototake NYC   **21-10c**
Ward's Natural Science Establishment, Inc. **21-20c** Ward's
Natural Science Establishment, Inc. **21-23a** Biophoto
Associates/Science Source/Photo Researchers, Inc. **21-23b**
Biophoto Associates/Photo Researchers, Inc. **21-27** Martin
Rotker/Phototake NYC   **21-29** John Paul Kay/Peter
Arnold, Inc. **21-31** Ralph Eagle/Science Source/Photo
Researchers, Inc. **21-33** John Radcliffe Hospital/Science
Photo Library/Photo Researchers, Inc.   **21-34** Ralph T.
Hutchings   **21-40** Dr. Michael Klein/Peter Arnold, Inc.

Chapter 22

p. 686 Dr. Dennis Kunkel/Phototake NYC

Chapter 24

p. 731 David Scharf/Peter Arnold, Inc.   **24-1** CNRI/Science
Photo Library/Photo Researchers, Inc.   **24-9c** David M.
Phillips/Visuals Unlimited   **24-16a** Francis
Leroy/Biocosmos/Science Photo Library/Custom Medical
Stock Photo, Inc.

Measurements and Conversions

PREFIXES

deci-	(d)	1/10	0.1	1×10^{-1}
centi-	(c)	1/100	0.01	1×10^{-2}
milli-	(m)	1/1000	0.001	1×10^{-3}
micro-	(μ)	1/1,000,000	0.000001	1×10^{-6}
nano-	(n)	1/1,000,000,000	0.000000001	1×10^{-9}
pico-	(p)	1/1,000,000,000,000	0.000000000001	1×10^{-12}
kilo-	(k)		1000	1×10^{3}

METRIC SYSTEM

1 meter (m) = 100 centimeters (cm) = 1000 millimeters (mm)
1 centimeter (cm) = 10 millimeters (mm) = 0.01 meters (m)
1 millimeter (mm) = 1000 micrometers (μm; also called micron, μ)
1 angstrom (Å) = 1/10,000 micrometer = 1×10^{-7} millimeters

1 liter (L) = 1000 milliliters (mL)
1 deciliter (dL) = 100 milliliters (mL) = 0.1 liters (L)
1 cubic centimeter (cc) = 1 milliliter (mL)
1 milliliter (mL) = 1000 microliters (μL)

1 kilogram (kg) = 1000 grams (g)
1 gram (g) = 1000 milligrams (mg)
1 milligram = 1000 micrograms (μg)

CONVERSIONS

1 yard (yd) = 0.92 meters	1 meter = 1.09 yards
1 inch (in) = 2.54 centimeters	1 centimeter = 0.39 inches
1 liquid quart (qt) = 946 milliliters	1 liter = 1.05 liquid quarts

1 fluid ounce (oz) = 8 fluid drams = 29.57 milliliters (mL)

1 pound (lb) = 453.6 grams	1 kilogram = 2.2 pounds

TEMPERATURE

Freezing = 0 degrees Celsius (°C) = 32 degrees Fahrenheit (°F) = 273 degrees Kelvin (°K)

To convert degrees Celsius (°C) to degrees Fahrenheit (°F):   (°C x 9/5) + 32
To convert degrees Fahrenheit (°F) to degrees Celsius (°C):   (°F - 32) x 5/9

NORMAL VALUES OF BLOOD COMPONENTS

Substance or parameter	Normal range	Measured in
Calcium (Ca^{2+})	4.3-5.3 meq/L	Serum
Chloride (Cl^{-})	100-108 meq/L	Serum
Potassium (K^{+})	3.5-5.0 meq/L	Serum
Sodium (Na^{+})	135-145 meq/L	Serum
pH	7.35-7.45	Whole blood
P_{O_2}	75-100 mm Hg	Arterial blood
P_{CO_2}	35-45 mm Hg	Arterial blood
Osmolality	280-296 mosmol/kg water	Serum
Glucose, fasting	70-110 mg/dL	Plasma
Creatinine	0.6-1.5 mg/dL	Serum
Protein, total	6.0-8.0 g/dL	Serum

Modified from W. F. Ganong, *Review of Medical Physiology,* Appleton & Lange, Norwalk, 1995.